AF183159

THE JOURNAL OF

Fourier Analysis and Applications

VOLUME 25 · NUMBER 5 · 2019

Contents

THE JOURNAL OF

Fourier Analysis and Applications

VOLUME 25 · NUMBER 5 · 2019

EDITOR-IN-CHIEF
Hans G. Feichtinger
FOUNDING EDITOR-IN-CHIEF
John J. Benedetto
EXECUTIVE EDITORS
John J. Benedetto
Luis A. Caffarelli
Ronald R. Coifman
Ingrid Daubechies
Charles Fefferman
Karlheinz Gröchenig
Yves Meyer
Robert S. Strichartz

Birkhäuser

Disclaimer

While the advice and information in this journal are believed to be true and accurate at the date of its publication, neither the authors, the editors nor the publisher can accept any legal responsibility for any errors or omissions that may be made. The publisher makes no warranty, express or implied, with respect to the material contained herein. Springer publishes advertisements in this journal in reliance upon the responsibility of the advertiser to comply with all legal requirements relating to the marketing and sale of products or services advertised. Springer and the editors are not responsible for claims made in the advertisements published in the journal. The appearance of advertisements in Springer publications does not constitute endorsement, implied or intended, of the product advertised or the claims made for it by the advertiser.

Special regulations for photocopies in the USA: Photocopies may be made for personal or in-house use beyond the limitations stipulated under Section 107 or 108 of U.S. Copyright Law, provided a fee is paid. All fees should be paid to the *Copyright Clearance Center, Inc.*, 222 Rosewood Drive, Danvers, MA 01923, USA (WEB: http://www.copyright.com; FAX: 978-646-8600; TEL: 978-750-8400), stating the journal's ISSN, the volume, and the first and last page numbers of each article copied. The copyright owner's consent does not include copying for general distribution, promotion, new works or resale. In these cases, specific written permission must first be obtained from the publisher.

The *Canada Institute for Scientific and Technical Information (CISTI)* provides a comprehensive, world-wide document delivery service for all Springer journals. For more information, or to place an order for a copyright-cleared Springer document, please contact Client Assistant, Document Delivery, CISTI, Ottawa K1A 0S2, Canada. TEL: 613-993-9251; FAX: 613-952-8243; E-MAIL: cisti.docdel@nrc.ca.

You may also contact *INFOTRIEVE* (WEB: http://www.infotrieve.com/; E-MAIL: orders@infotrieve.com; FAX: 310-208-5971; TEL: 800-422-4633), *UnCover:* http://www.ingenta.com; *UMI:* http://www.umi.com/; *ISI:* http://www.isinet.com

Responsible for Advertisements
M.J. Mrvica Associates, Inc., 2 West Taunton Avenue, Berlin, NJ 08009, USA. Telephone: (856) 768-9360; Fax: (856) 753-0064; Email: mjmrvica@mrvica.com.

Office of Publication
Birkhäuser Boston, 233 Spring Street, New York, NY 10013, USA. Telephone: 212-460-1500; email: journalsny@springer-sbm.com, website: springeronline.com, Production Editor: Bill Payne.

Aim and Scope

The Journal of Fourier Analysis and Applications will publish results in Fourier analysis, as well as applicable mathematics having a significant Fourier analytic component. Appropriate manuscripts at the highest research level will be accepted for publication. Because of the extensive, intricate, and fundamental relationships between Fourier analysis and so many other subjects, select and readable surveys will also be published. These surveys will include historical articles, research tutorials, and expositions of specific topics.

The Journal of Fourier Analysis and Applications will provide a perspective and means for centralizing and disseminating new information from the vantage point of Fourier analysis. The breadth of Fourier analysis and diversity of its applicability require that each paper should contain a clear and motivated introduction, which is accessible to all of our readers.

Applications include the following topics:

Antenna theory
Crystallography
Fast algorithms
Gabor theory and applications
Image processing
Number theory
Optics
Partial differential equations
Prediction theory
Radar applications
Sampling theory
Spectral estimation
Speech processing
Stochastic processes
Time-frequency analysis
Time series
Tomography
Turbulence
Uncertainty principles
Wavelet theory and applications

Editorial Statement

The Journal of Fourier Analysis and Applications is a journal of the mathematical sciences devoted to Fourier analysis and its applications. The subject of Fourier analysis has had a major impact on the development of mathematics, on the understanding of many engineering and scientific phenomena, and on the solution of some of the most important problems in mathematics and the sciences. The usefulness and influence of Fourier analysis continues to grow, both in the number of topics it affects, and in the significance of the ideas and techniques it brings to bear on these topics. Fourier analysis is a deep subject with its own identity; and it is a subject whose relationship with other disciplines yields a multifaceted personality.

Historically, Fourier series were developed in the analysis of some of the classical partial differential equations (PDE) of mathematical physics; and these series were used to solve such equations. In order to understand Fourier series and what sorts of solutions they could represent, some of the basic notions of analysis were defined, e.g., the concept of function. Since the coefficients of Fourier series are integrals, it is no surprise that Riemann integrals were conceived to deal with uniqueness properties of trigonometric series; and Cantor's set theory was developed because of such uniqueness questions.

A basic problem in Fourier analysis is to show how complicated phenomena, such as sound waves, can be described in terms of elementary harmonics. There are two aspects of the problem: first, to find, or even define properly, the harmonics or spectrum of a given phenomenon, e.g., the spectroscopy problem in optics; second, to determine which phenomena can be constructed from given classes of harmonics, as done, for example, by the mechanical synthesizers in tidal analysis.

Fourier analysis is not only the natural setting for important parts of PDE, but for many other problems in engineering, mathematics, and the sciences. For example, Wiener's Tauberian theorem in Fourier analysis not only characterizes the behavior of the prime numbers, but also provides the proper notion of spectrum for phenomena such as white light; this latter process leads to the Fourier analysis associated with correlation functions in filtering and prediction problems, and these problems, in turn, deal naturally with Hardy spaces in the theory of complex variables.

Nowadays, some of the theory of PDE has given way to the study of Fourier integral operators. Problems in antenna theory are studied in terms of unimodular trigonometric polynomials. Applications of Fourier analysis abound in signal processing, whether with the Fast Fourier Transform (FFT), or filter design, or the adaptive modeling inherent in time-frequency and time-scale methods such as wavelet theory. The coherent states of mathematical physics are translated and modulated Fourier transforms, and these are used, in conjunction with the uncertainty principle, for dealing with signal reconstruction in communications theory.

This is a remarkable period since many engineers, mathematicians, and scientists are really talking (and listening) seriously to one another in areas such as speech and image processing, turbulence, PDE, numerical analysis, etc. Fourier analysis seems to be an endless well of inspiration in certain of these topics, and is making new inroads in others, especially with recent theoretical and algorithmic developments in wavelet theory.

We envisage a journal which draws upon the aforementioned Fourier technologies, and which responds creatively to significant engineering, mathematics, and scientific problems.

John J. Benedetto
Founding Editor-in-Chief

Journal of Fourier Analysis and Applications (2019) 25:2197–2267
https://doi.org/10.1007/s00041-018-09652-y

A Complete Real-Variable Theory of Hardy Spaces on Spaces of Homogeneous Type

Ziyi He[1] · Yongsheng Han[2] · Ji Li[3] · Liguang Liu[4] · Dachun Yang[1] · Wen Yuan[1]

Received: 21 April 2018 / Published online: 20 November 2018
© Springer Science+Business Media, LLC, part of Springer Nature 2018

Abstract

Let (X, d, μ) be a space of homogeneous type, with the upper dimension ω, in the sense of Coifman and Weiss. Assume that η is the smoothness index of the wavelets on X constructed by Auscher and Hytönen. In this article, when $p \in (\omega/(\omega+\eta), 1]$, for the atomic Hardy spaces $H^p_{\mathrm{cw}}(X)$ introduced by Coifman and Weiss, the authors establish their various real-variable characterizations, respectively, in terms of the grand maximal functions, the radial maximal functions, the non-tangential maximal functions, the various Littlewood–Paley functions and wavelet functions. This completely answers the question of Coifman and Weiss by showing that no additional (geometrical) condition is necessary to guarantee the radial maximal function characterization of $H^1_{\mathrm{cw}}(X)$ and even of $H^p_{\mathrm{cw}}(X)$ with p as above. As applications, the authors obtain the finite atomic characterizations of $H^p_{\mathrm{cw}}(X)$, which further induce some criteria for the boundedness of sublinear operators on $H^p_{\mathrm{cw}}(X)$. Compared with the known results, the novelty of this article is that μ is not assumed to satisfy the reverse doubling condition and d is only a quasi-metric, moreover, the range $p \in (\omega/(\omega+\eta), 1]$ is natural and optimal.

Keywords Space of homogeneous type · Hardy space · Maximal function · Atom · Littlewood–Paley function · Wavelet

Mathematics Subject Classification Primary 42B30; Secondary 42B25 · 42B20 · 30L99

Communicated by Loukas Grafakos.

This project is supported by the National Natural Science Foundation of China (Grant Nos. 11771446, 11571039, 11726621, 11761131002, 11671185 and 11871100). Ji Li is supported by ARC DP 160100153.

Extended author information available on the last page of the article

1 Introduction

The real-variable theory of Hardy spaces plays a fundamental role in harmonic analysis. The classical Hardy space on the n-dimensional Euclidean space $\mathbb{R}^n$ was initially developed by Stein and Weiss [50] and later by Fefferman and Stein [11]. Hardy spaces $H^p(\mathbb{R}^n)$ have proved a suitable substitute of Lebesgue spaces $L^p(\mathbb{R}^n)$ with $p \in (0, 1]$ in the study of the boundedness of operators. Indeed, any element in the Hardy space can be decomposed into a sum of some basic elements (which are called *atoms*); see Coifman [5] for $n = 1$ and Latter [36] for general $n \in \mathbb{N}$. Characterizations of Hardy spaces via Littlewood–Paley functions were due to Uchiyama [51]. For more study on classical Hardy spaces on $\mathbb{R}^n$, we refer the reader to the well-known monographs [16–18,41,49]. Modern developments regarding the real-variable theory of Hardy spaces are so deep and vast that we can only list a few literatures here, for example, the theory of Hardy spaces associated with operators (see [2,3,10,30]), Hardy spaces with variable exponents (see [44]), the real-variable theory of Musielak–Orlicz Hardy spaces (see [35,53]), and also Hardy spaces for ball quasi-Banach spaces (see [48]).

In this article, we focus on the real-variable theory of Hardy spaces on spaces of homogeneous type. It is known that the space of homogeneous type introduced by Coifman and Weiss [6,7] provides a natural setting for the study of both function spaces and the boundedness of operators. A *quasi-metric space* (X, d) is a non-empty set X equipped with a *quasi-metric* d, that is, a non-negative function defined on $X \times X$, satisfying that, for any $x,\ y,\ z \in X$,

(i) $d(x, y) = 0$ if and only if $x = y$;
(ii) $d(x, y) = d(y, x)$;
(iii) there exists a constant $A_0 \in [1, \infty)$ such that $d(x, z) \le A_0[d(x, y) + d(y, z)]$.

The *ball* B on X centered at $x_0 \in X$ with radius $r \in (0, \infty)$ is defined by setting

$$B := B(x_0, r) := \{x \in X : d(x, x_0) < r\}.$$

For any ball B and $\tau \in (0, \infty)$, denote by τB the ball with the same center as that of B but of radius τ times that of B. Given a quasi-metric space (X, d) and a non-negative measure μ, we call (X, d, μ) a *space of homogeneous type* if μ satisfies the *doubling condition*: there exists a positive constant $C_{(\mu)} \in [1, \infty)$ such that, for any ball $B \subset X$,

$$\mu(2B) \le C_{(\mu)}\mu(B).$$

The above doubling condition is equivalent to that, for any ball B and $\lambda \in [1, \infty)$,

$$\mu(\lambda B) \le C_{(\mu)}\lambda^{\omega}\mu(B), \tag{1.1}$$

where $\omega := \log_2 C_{(\mu)}$ is called the *upper dimension* of X. If $A_0 = 1$, we call (X, d, μ) a *doubling metric measure space*.

According to [7, pp. 587–588], we *always make* the following assumptions throughout this article. For any point $x \in X$, assume that the balls $\{B(x, r)\}_{r\in(0,\infty)}$ form a *basis* of open neighborhoods of x; assume that μ is Borel regular, which means that open sets are measurable and every set $A \subset X$ is contained in a Borel set E satisfying that $\mu(A) = \mu(E)$; we also assume that $\mu(B(x,r)) \in (0,\infty)$ for any $x \in X$ and $r \in (0,\infty)$. For the presentation concision, we always assume that (X, d, μ) is non-atomic [namely, $\mu(\{x\}) = 0$ for any $x \in X$] and $\operatorname{diam}(X) := \sup\{d(x,y) : x, y \in X\} = \infty$. It is known that $\operatorname{diam}(X) = \infty$ implies that $\mu(X) = \infty$ (see, for example, [45, Lemma 5.1] or [1, Lemma 8.1]).

Let us recall the notion of the atomic Hardy space on spaces of homogeneous type introduced by Coifman and Weiss [7]. For any $\alpha \in (0,\infty)$, the *Lipschitz space* $\mathcal{L}_\alpha(X)$ is defined to be the collection of all measurable functions f such that

$$\|f\|_{\mathcal{L}_\alpha(X)} := \sup_{x \neq y} \frac{|f(x) - f(y)|}{[\mu(B(x, d(x,y)))]^\alpha} < \infty.$$

Denote by $(\mathcal{L}_\alpha(X))'$ the *dual space* of $\mathcal{L}_\alpha(X)$ equipped with the weak-$*$ topology.

Definition 1.1 Let $p \in (0,1]$ and $q \in (p,\infty] \cap [1,\infty]$. A function a is called a *(p,q)-atom* if

(i) $\operatorname{supp} a := \{x \in X : a(x) \neq 0\} \subset B(x_0, r)$ for some $x_0 \in X$ and $r \in (0,\infty)$;
(ii) $[\int_X |a(x)|^q \, d\mu(x)]^{\frac{1}{q}} \le [\mu(B(x_0, r))]^{\frac{1}{q} - \frac{1}{p}}$;
(iii) $\int_X a(x) \, d\mu(x) = 0$.

The *atomic Hardy space* $H^{p,q}_{\mathrm{cw}}(X)$ is defined as the subspace of $(\mathcal{L}_{1/p-1}(X))'$ when $p \in (0,1)$ or of $L^1(X)$ when $p = 1$, which consists of all the elements f admitting an atomic decomposition

$$f = \sum_{j=0}^{\infty} \lambda_j a_j, \tag{1.2}$$

where $\{a_j\}_{j=0}^\infty$ are (p,q)-atoms, $\{\lambda_j\}_{j=0}^\infty \subset \mathbb{C}$ satisfies $\sum_{j=0}^\infty |\lambda_j|^p < \infty$ and the series in (1.2) converges in $(\mathcal{L}_{1/p-1}(X))'$ when $p \in (0,1)$ or in $L^1(X)$ when $p = 1$. Define

$$\|f\|_{H^{p,q}_{\mathrm{cw}}(X)} := \inf \left\{ \left(\sum_{j=0}^\infty |\lambda_j|^p \right)^{\frac{1}{p}} \right\},$$

where the infimum is taken over all the decompositions of f as in (1.2).

It was proved in [7] that the atomic Hardy space $H^{p,q}_{\mathrm{cw}}(X)$ is independent of the choice of q and hence we sometimes write $H^p_{\mathrm{cw}}(X)$ for short. It was also proved in [7] that the dual space of $H^p_{\mathrm{cw}}(X)$ is the Lipschitz space $\mathcal{L}_{1/p-1}(X)$ when $p \in (0,1)$, and the space $\mathrm{BMO}(X)$ of bounded mean oscillation when $p = 1$.

It is well known that the most basic result in the real-variable theory of Hardy spaces is their characterizations in terms of maximal functions. Coifman and Weiss [7, pp. 641–642] observed that a proof of the duality result between $H^1(\mathbb{R}^n)$ and $\mathrm{BMO}(\mathbb{R}^n)$ from Carleson [4] can be extended to the general setting of spaces of homogeneous type provided a certain additional geometrical assumption is added, from which one can then obtain a radial maximal function characterization of $H^1_{\rm cw}(X)$. Coifman and Weiss [7, p. 642] then asked that *to what extent their geometrical condition is necessary for the validity of the radial maximal function characterization of* $H^1_{\rm cw}(X)$. Since then, lots of efforts are made to build various real-variable characterizations of the atomic Hardy spaces on spaces of homogeneous type with few geometrical assumptions. In this article, we completely answer the aforementioned question of Coifman and Weiss by showing that no any additional (geometrical) condition is necessary to guarantee the radial maximal function characterization of $H^1_{\rm cw}(X)$ and even of $H^p_{\rm cw}(X)$ with $p \le 1$ but near to 1.

Recall that a triple (X, d, μ) is said to be *Ahlfors-n regular* if $\mu(B(x, r)) \sim r^n$ for any $x \in X$ and $r \in (0, \operatorname{diam} X)$ with positive equivalence constants independent of x and r. When (X, d, μ) is Ahlfors-n regular, upon assuming the quasi-metric d satisfying that there exists $\theta \in (0, 1)$ such that, for any $x,\ x',\ y \in X$,

$$|d(x, y) - d(x', y)| \lesssim [d(x, x')]^\theta [d(x, y) + d(x', y)]^{1-\theta}, \tag{1.3}$$

Macías and Segovia [43] characterized Hardy spaces via the grand maximal functions, and Li [37] obtained another grand maximal function characterization via test functions introduced in [28]. Also, Duong and Yan [9] characterized Hardy spaces via the Lusin area function associated with certain semigroup.

Recall that an RD-*space* (X, d, μ) is a doubling metric measure space with the measure μ further satisfying the *reverse doubling condition*, that is, there exist a positive constant $\widetilde{C} \in (0, 1]$ and $\kappa \in (0, \omega]$ such that, for any ball $B(x, r)$ with $x \in X$, $r \in (0, \operatorname{diam} X/2)$ and $\lambda \in [1, \operatorname{diam} X/[2r])$,

$$\widetilde{C}\lambda^\kappa \mu(B(x, r)) \le \mu(B(x, \lambda r)).$$

Indeed, any path connected doubling metric measure space is an RD-space (see [27, 57]). Characterizations of Hardy spaces on RD-spaces via various Littlewood–Paley functions were established in [26,27]. Also, characterizations of Hardy spaces on RD-spaces via various maximal functions can be found in [20,21,56]. It should be mentioned that local Hardy spaces can be used to characterize more general scale of function spaces like Besov and Triebel–Lizorkin spaces on RD-spaces (see [57]). For a systematic study of Besov and Triebel–Lizorkin spaces on RD-spaces, we refer the reader to [27]. More on analysis over Ahlfors-n regular metric measure spaces or RD-spaces can be found in [8,19,22,32–34,54,57,58].

The main motivation of studying the real-variable theory of function spaces and the boundedness of operators on spaces of homogeneous type comes from the celebrated work of Auscher and Hytönen [1], in which they constructed an orthonormal wavelet basis $\{\psi^k_\alpha : k \in \mathbb{Z},\ \alpha \in \mathcal{G}_k\}$ of $L^2(X)$ with Hölder continuity exponent $\eta \in (0, 1)$ and exponential decay by using the system of random dyadic cubes. The first creative attempt of using the idea of [1] to investigate the real-variable theory of Hardy spaces on spaces of homogeneous type was due to Han et al. [23] (see also Han et al. [24]). Indeed, in [23], Hardy spaces via wavelets on spaces of homogeneous type were introduced and then these spaces proved to have atomic decompositions. The method used in

[23] is based on a new Calderón reproducing formula on spaces of homogeneous type (see [23, Proposition 2.5]). But there exists an *error* in the proof of [23, Proposition 2.5], namely, since the regularity exponent of the approximations of the identity in [23, p. 3438] is θ [indeed, θ is from the regularity of the quasi-metric d in (1.3)], it follows that the regularity exponent in [23, (2.6)] should be $\min\{\theta, \eta\}$ and hence the correct range of p in [23, Proposition 2.5] (indeed, all results of [23]) seems to be $(\omega/[\omega+\min\{\theta, \eta\}], 1]$ which is not optimal. Moreover, the criteria of the boundedness of Calderón–Zygmund operators on the dual of Hardy spaces were established in [23]. Also, Fu and Yang [14] obtained an unconditional basis of $H^1_{\mathrm{cw}}(X)$ and several equivalent characterizations of $H^1_{\mathrm{cw}}(X)$ in terms of wavelets.

Another motivation of this article comes from the Calderón reproducing formulae established in [29]. Indeed, the work of [29] was partly motivated by the wavelet theory of Auscher and Hytönen in [1] and a corresponding wavelet reproducing formula (which can converge in the distribution space) in [29]. The already existing works (see [20,26,27,56,57]) regarding Hardy spaces on RD-spaces show the feasibility of establishing various real-variable characterizations of the atomic Hardy spaces on spaces of homogeneous type via the Calderón reproducing formulae. It should be mentioned that a characterization of the atomic Hardy spaces via the Littlewood–Paley functions was established in [25] via the aforementioned wavelet reproducing formula; see also [25] for some corresponding conclusions of product Hardy spaces on spaces of homogeneous type.

In this article, motivated by [23,29], for the atomic Hardy spaces $H^p_{\mathrm{cw}}(X)$ with any $p \in (\omega/[\omega + \eta], 1]$, we establish their various real-variable characterizations, respectively, in terms of the grand maximal functions, the radial maximal functions, the non-tangential maximal functions, the various Littlewood–Paley functions and wavelets. Observe that these characterizations are true for $H^p_{\mathrm{cw}}(X)$ with $p \in (\omega/[\omega + \eta], 1]$ and X being any space of homogeneous type *without any additional (geometrical) conditions*, which completely answers the aforementioned question asked by Coifman and Weiss [7, p. 642]. As an application, we obtain the finite atomic characterizations of Hardy spaces, which further induce some criteria for the boundedness of sublinear operators on Hardy spaces. Compared with the known results, the novelty of this article is that μ is not assumed to satisfy the reverse doubling condition and d is only a quasi-metric. Moreover, the range of $p \in (\omega/(\omega + \eta), 1]$ for the various maximal function characterizations and the Littlewood–Paley function characterizations of the atomic Hardy spaces $H^p_{\mathrm{cw}}(X)$ is natural and optimal. The key tool used through this article is those Calderón reproducing formulae from [29].

In addition, we point out that, when X is a doubling metric measure space, the finite atomic characterizations of Hardy spaces are also useful in establishing the bilinear decomposition of the product space $H^1_{\mathrm{cw}}(X) \times \mathrm{BMO}(X)$ and $H^p_{\mathrm{cw}}(X) \times \mathcal{L}_{1/p-1}(X)$, with $p \in (\omega/[\omega + \eta], 1)$ in [13–15,40], and also in the study of the endpoint boundedness of commutators generated by Calderón-Zygmund operators and $\mathrm{BMO}(X)$ functions in [38,39].

The organization of this article is as follows.

In Sect. 2, we recall the notions of the space of test functions and the space of distributions introduced in [26], as well as the random dyadic cubes in [1] and the approximation of the identity with exponential decay introduced in [29]. Then we restate the Calderón reproducing formulae established in [29].

Section 3 concerns Hardy spaces defined via the grand maximal functions, the radial maximal functions and the non-tangential maximal functions. We show that these Hardy spaces are all equivalent to the Lebesgue space $L^p(X)$ when $p \in (1, \infty]$ (see Sect. 3.1), and they are all mutually equivalent when $p \in (\omega/(\omega+\eta), 1]$ (see Sect. 3.2), all in the sense of equivalent (quasi-)norms. The proof for the latter borrows some ideas from [56] and uses the Calderón reproducing formulae built in [29]. Moreover, we prove that the Hardy space $H^{*,p}(X)$ defined via the grand maximal function is independent of the choices of the distribution space $(\mathcal{G}_0^\eta(\beta, \gamma))'$ whenever $\beta,\ \gamma \in (\omega[1/p - 1], \eta)$; see Proposition 3.8 below.

Section 4 is devoted to the atomic characterizations of $H^{*,p}(X)$. Notice that, if a distribution has an atomic decomposition, then it belongs to $H^{*,p}(X)$ obviously by the definition of atoms; see Sect. 4.1. All we remain to do is to establish the converse relationship. In Sect. 4.2, by modifying the definition of the grand maximal function f^* to $f^\star$ so that the level set $\Omega_\lambda := \{x \in X : f^\star(x) > \lambda\}$ with $\lambda \in (0, \infty)$ is open, we then apply the partition of unity to the open set Ω_λ and obtain a Calderón–Zygmund decomposition of $f \in H^{*,p}(X)$. This is further used in Sect. 4.3 to construct an atomic decomposition of f. In Sect. 4.4, we compare the atomic Hardy spaces $H^{p,q}_{\mathrm{at}}(X)$ with $H^{p,q}_{\mathrm{cw}}(X)$ and prove that they are exactly the same space in the sense of equivalent (quasi-)norms.

Section 5 deals with the Littlewood–Paley theory of Hardy spaces. In Sect. 5.1, we show that the Hardy space $H^p(X)$, defined via the Lusin area function, is independent of the choices of exp-ATIs. In Sect. 5.2, we use the homogeneous continuous Calderón reproducing formula and the molecular characterizations of the atomic Hardy spaces (see [39]) to establish the atomic decompositions of elements in $H^p(X)$, and then we connect $H^p(X)$ with $H^{*,p}(X)$. In Sect. 5.3, we characterize Hardy spaces $H^p(X)$ via the Lusin area function with aperture, the Littlewood–Paley g-function and the Littlewood–Paley g_λ^*-function.

In Sect. 6, we consider the Hardy space $H^p_{\mathrm{w}}(X)$ defined via wavelets, which was introduced in [23]. We improve the result of [25, Theorem 4.3] and prove that $H^p_{\mathrm{w}}(X)$ coincides with $H^p(X)$ in the sense of equivalent (quasi-)norms.

In Sect. 7, as an application, we obtain criteria of the boundedness of the sublinear operators from Hardy spaces to quasi-Banach spaces. To this end, we first establish the finite atomic characterizations, namely, we show that, if $q \in (p, \infty) \cap [1, \infty)$, then $\|\cdot\|_{H^{p,q}_{\mathrm{fin}}(X)}$ and $\|\cdot\|_{H^{p,q}_{\mathrm{at}}(X)}$ are equivalent (quasi)-norms on a dense subspace $H^{p,q}_{\mathrm{fin}}(X)$ of $H^{p,q}_{\mathrm{at}}(X)$; the above equivalence also holds true on a dense subspace $H^{p,\infty}_{\mathrm{fin}}(X) \cap \mathrm{UC}(X)$ of $H^{p,\infty}_{\mathrm{at}}(X)$, where $\mathrm{UC}(X)$ denotes the space of all uniformly continuous functions on X.

At the end of this section, we make some conventions on notation. We *always assume* that ω is as in (1.1) and η the smoothness index of wavelets (see [1, Theorem 7.1] or Definition 2.4 below). We assume that δ is a very small positive number, for example, $\delta \le (2A_0)^{-10}$ in order to construct the dyadic cube system and the wavelet system on X (see [31, Theorem 2.2] or Lemma 2.3 below). For any $x,\ y \in X$ and $r \in (0, \infty)$, let

$$V_r(x) := \mu(B(x,r)) \quad \text{and} \quad V(x,y) := \mu(B(x,d(x,y))),$$

where $B(x,r) := \{y \in X : d(x,y) < r\}$. We always let $\mathbb{N} := \{1,2,\ldots\}$ and $\mathbb{Z}_+ := \mathbb{N} \cup \{0\}$. For any $p \in [1,\infty]$, we use p' to denote its *conjugate index*, namely, $1/p + 1/p' = 1$. The symbol C denotes a positive constant which is independent of the main parameters, but it may vary from line to line. We also use $C_{(\alpha,\beta,\ldots)}$ to denote a positive constant depending on the indicated parameters α, β, ... The symbol $A \lesssim B$ means that there exists a positive constant C such that $A \le CB$. The symbol $A \sim B$ is used as an abbreviation of $A \lesssim B \lesssim A$. We also use $A \lesssim_{\alpha,\beta,\ldots} B$ to indicate that here the implicit positive constant depends on α, β, ... and, similarly, $A \sim_{\alpha,\beta,\ldots} B$. We also use the following convention: If $f \le Cg$ and $g = h$ or $g \le h$, we then write $f \lesssim g \sim h$ or $f \lesssim g \lesssim h$, *rather than* $f \lesssim g = h$ or $f \lesssim g \le h$. For any $s,\ t \in \mathbb{R}$, denote the *minimum* of s and t by $s \wedge t$. For any finite set $\mathcal{J}$, we use $\#\mathcal{J}$ to denote its *cardinality*. Also, for any set E of X, we use $\mathbf{1}_E$ to denote its characteristic function and $E^{\complement}$ the set $X \setminus E$.

2 Calderón Reproducing Formulae

This section is devoted to recalling Calderón reproducing formulae obtained in [29]. To this end, we first recall the notions of both the space of test functions and the distribution space.

Definition 2.1 Let $x_1 \in X$, $r \in (0,\infty)$, $\beta \in (0,1]$ and $\gamma \in (0,\infty)$. A function f defined on X is called a *test function of type* (x_1,r,β,γ), denoted by $f \in \mathcal{G}(x_1,r,\beta,\gamma)$, if there exists a positive constant C such that

(i) (the *size condition*) for any $x \in X$,

$$|f(x)| \le C \frac{1}{V_r(x_1) + V(x_1,x)} \left[\frac{r}{r + d(x_1,x)}\right]^{\gamma};$$

(ii) (the *regularity condition*) for any $x,\ y \in X$ satisfying $d(x,y) \le (2A_0)^{-1}[r + d(x_1,x)]$,

$$|f(x) - f(y)| \le C \left[\frac{d(x,y)}{r + d(x_1,x)}\right]^{\beta} \frac{1}{V_r(x_1) + V(x_1,x)} \left[\frac{r}{r + d(x_1,x)}\right]^{\gamma}.$$

For any $f \in \mathcal{G}(x_1,r,\beta,\gamma)$, define the norm

$$\|f\|_{\mathcal{G}(x_1,r,\beta,\gamma)} := \inf\{C \in (0,\infty) : C \text{ satisfies (i) and (ii)}\}.$$

Define

$$\mathring{\mathcal{G}}(x_1,r,\beta,\gamma) := \left\{f \in \mathcal{G}(x_1,r,\beta,\gamma) : \int_X f(x)\,d\mu(x) = 0\right\}$$

equipped with the norm $\|\cdot\|_{\mathring{\mathcal{G}}(x_1,r,\beta,\gamma)} := \|\cdot\|_{\mathcal{G}(x_1,r,\beta,\gamma)}$.

Observe that the above version of $\mathcal{G}(x_1, r, \beta, \gamma)$ was originally introduced by Han et al. [27] (see also [26]).

Fix $x_0 \in X$. For any $x \in X$ and $r \in (0, \infty)$, we know that $\mathcal{G}(x, r, \beta, \gamma) = \mathcal{G}(x_0, 1, \beta, \gamma)$ with equivalent norms, but the positive equivalence constants depend on x and r. Obviously, $\mathcal{G}(x_0, 1, \beta, \gamma)$ is a Banach space. In what follows, we simply write $\mathcal{G}(\beta, \gamma) := \mathcal{G}(x_0, 1, \beta, \gamma)$ and $\mathring{\mathcal{G}}(\beta, \gamma) := \mathring{\mathcal{G}}(x_0, 1, \beta, \gamma)$.

Fix $\epsilon \in (0, 1]$ and $\beta, \gamma \in (0, \epsilon)$. Let $\mathcal{G}_0^\epsilon(\beta, \gamma)$ [resp., $\mathring{\mathcal{G}}_0^\epsilon(\beta, \gamma)$] be the completion of the set $\mathcal{G}(\epsilon, \epsilon)$ [resp., $\mathring{\mathcal{G}}(\epsilon, \epsilon)$] in $\mathcal{G}(\beta, \gamma)$, that is, if $f \in \mathcal{G}_0^\epsilon(\beta, \gamma)$ [resp., $f \in \mathring{\mathcal{G}}_0^\epsilon(\beta, \gamma)$], then there exists $\{\phi_j\}_{j=1}^\infty \subset \mathcal{G}(\epsilon, \epsilon)$ [resp., $\{\phi_j\}_{j=1}^\infty \subset \mathring{\mathcal{G}}(\epsilon, \epsilon)$] such that $\|\phi_j - f\|_{\mathcal{G}(\beta,\gamma)} \to 0$ as $j \to \infty$. If $f \in \mathcal{G}_0^\epsilon(\beta, \gamma)$ [resp., $f \in \mathring{\mathcal{G}}_0^\epsilon(\beta, \gamma)$], we then let

$$\|f\|_{\mathcal{G}_0^\epsilon(\beta,\gamma)} := \|f\|_{\mathcal{G}(\beta,\gamma)} \quad [\text{resp., } \|f\|_{\mathring{\mathcal{G}}_0^\epsilon(\beta,\gamma)} := \|f\|_{\mathcal{G}(\beta,\gamma)}].$$

The *dual space* $(\mathcal{G}_0^\epsilon(\beta, \gamma))'$ [resp., $(\mathring{\mathcal{G}}_0^\epsilon(\beta, \gamma))'$] is defined to be the set of all continuous linear functionals on $\mathcal{G}_0^\epsilon(\beta, \gamma)$ [resp., $\mathring{\mathcal{G}}_0^\epsilon(\beta, \gamma)$] and equipped with the weak-$*$ topology. The spaces $(\mathcal{G}_0^\epsilon(\beta, \gamma))'$ and $(\mathring{\mathcal{G}}_0^\epsilon(\beta, \gamma))'$ are called the *spaces of distributions*.

Let $L^1_{\mathrm{loc}}(X)$ be the space of all locally integrable functions on X. Denote by $\mathcal{M}$ the *Hardy–Littlewood maximal operator*, that is, for any $f \in L^1_{\mathrm{loc}}(X)$ and $x \in X$,

$$\mathcal{M}(f)(x) := \sup_{B \ni x} \frac{1}{\mu(B)} \int_B |f(y)|\, d\mu(y),$$

where the supremum is taken over all balls B of X that contain x. For any $p \in (0, \infty]$, the *Lebesgue space* $L^p(X)$ is defined to be the set of all μ-measurable functions f such that

$$\|f\|_{L^p(X)} := \left[\int_X |f(x)|^p\, d\mu(x)\right]^{\frac{1}{p}} < \infty$$

with the usual modification made when $p = \infty$; the *weak Lebesgue space* $L^{p,\infty}(X)$ is defined to be the set of all μ-measurable functions f such that

$$\|f\|_{L^{p,\infty}(X)} := \sup_{\lambda \in (0,\infty)} \lambda[\mu(\{x \in X : |f(x)| > \lambda\})]^{\frac{1}{p}} < \infty.$$

It is known (see [7]) that $\mathcal{M}$ is bounded on $L^p(X)$ when $p \in (1, \infty]$ and bounded from $L^1(X)$ to $L^{1,\infty}(X)$. Then we state some estimates from [27, Lemma 2.1], which are proved by using (1.1).

Lemma 2.2 *Let* $\beta, \gamma \in (0, \infty)$.

(i) *For any* $x, y \in X$ *and* $r \in (0, \infty)$, $V(x, y) \sim V(y, x)$ *and*

$$\begin{aligned} V_r(x) + V_r(y) + V(x, y) &\sim V_r(x) + V(x, y) \\ &\sim V_r(y) + V(x, y) \sim \mu(B(x, r + d(x, y))), \end{aligned}$$

where the positive equivalence constants are independent of x, y and r.

(ii) *There exists a positive constant C such that, for any $x_1 \in X$ and $r \in (0, \infty)$,*

$$\int_X \frac{1}{V_r(x_1) + V(x_1, x)} \left[\frac{r}{r + d(x_1, x)}\right]^\gamma d\mu(x) \le C.$$

(iii) *There exists a positive constant C such that, for any $x \in X$ and $R \in (0, \infty)$,*

$$\int_{d(x,y)\le R} \frac{1}{V(x, y)} \left[\frac{d(x, y)}{R}\right]^\beta d\mu(y) \le C \quad \text{and}$$
$$\int_{d(x,y)\ge R} \frac{1}{V(x, y)} \left[\frac{R}{d(x, y)}\right]^\beta d\mu(y) \le C.$$

(iv) *There exists a positive constant C such that, for any $x_1 \in X$ and R, $r \in (0, \infty)$,*

$$\int_{d(x,x_1)\ge R} \frac{1}{V_r(x_1) + V(x_1, x)} \left[\frac{r}{r + d(x_1, x)}\right]^\gamma d\mu(x) \le C\left(\frac{r}{r + R}\right)^\gamma.$$

(v) *There exists a positive constant C such that, for any $r \in (0, \infty)$, $f \in L^1_{\mathrm{loc}}(X)$ and $x \in X$,*

$$\int_X \frac{1}{V_r(x) + V(x, y)} \left[\frac{r}{r + d(x, y)}\right]^\gamma |f(y)|\, d\mu(y) \le C\mathcal{M}(f)(x).$$

Next we recall the system of dyadic cubes established in [31, Theorem 2.2] (see also [1]), which is restated in the following version.

Lemma 2.3 *Fix constants $0 < c_0 \le C_0 < \infty$ and $\delta \in (0, 1)$ such that $12A_0^3C_0\delta \le c_0$. Assume that a set of points, $\{z_\alpha^k : k \in \mathbb{Z}, \alpha \in \mathcal{A}_k\} \subset X$ with $\mathcal{A}_k$ for any $k \in \mathbb{Z}$ being a countable set of indices, has the following properties: for any $k \in \mathbb{Z}$,*

(i) *$d(z_\alpha^k, z_\beta^k) \ge c_0\delta^k$ if $\alpha \ne \beta$;*

(ii) *$\min_{\alpha\in\mathcal{A}_k} d(x, z_\alpha^k) \le C_0\delta^k$ for any $x \in X$.*

Then there exists a family of sets, $\{Q_\alpha^k : k \in \mathbb{Z}, \alpha \in \mathcal{A}_k\}$, satisfying

(iii) *for any $k \in \mathbb{Z}$, $\bigcup_{\alpha\in\mathcal{A}_k} Q_\alpha^k = X$ and $\{Q_\alpha^k : \alpha \in \mathcal{A}_k\}$ is disjoint;*

(iv) *if k, $l \in \mathbb{Z}$ and $l \ge k$, then either $Q_\beta^l \subset Q_\alpha^k$ or $Q_\beta^l \cap Q_\alpha^k = \emptyset$;*

(v) *for any $k \in \mathbb{Z}$ and $\alpha \in \mathcal{A}_k$, $B(z_\alpha^k, c_\natural\delta^k) \subset Q_\alpha^k \subset B(z_\alpha^k, C^\natural\delta^k)$, where $c_\natural := (3A_0^2)^{-1}c_0$, $C^\natural := 2A_0C_0$ and z_α^k is called "the center" of Q_α^k.*

Throughout this article, we keep the notation used in Lemma 2.3. Moreover, for any $k \in \mathbb{Z}$, let

$$\mathcal{X}^k := \{z_\alpha^k\}_{\alpha\in\mathcal{A}_k}, \qquad \mathcal{G}_k := \mathcal{A}_{k+1} \setminus \mathcal{A}_k \qquad \text{and} \qquad \mathcal{Y}^k := \{z_\alpha^{k+1}\}_{\alpha\in\mathcal{G}_k} =: \{y_\alpha^k\}_{\alpha\in\mathcal{G}_k}.$$

Next we recall the notion of approximations of the identity with exponential decay introduced in [29].

Definition 2.4 A sequence $\{Q_k\}_{k\in\mathbb{Z}}$ of bounded linear integral operators on $L^2(X)$ is called an *approximation of the identity with exponential decay* (for short, exp-ATI) if there exist constants $C,\ \nu \in (0,\infty)$, $a \in (0,1]$ and $\eta \in (0,1)$ such that, for any $k \in \mathbb{Z}$, the kernel of operator Q_k, which is still denoted by Q_k, satisfying

(i) (the *identity condition*) $\sum_{k=-\infty}^{\infty} Q_k = I$ in $L^2(X)$, where I is the identity operator on $L^2(X)$;

(ii) (the *size condition*) for any $x,\ y \in X$,

$$|Q_k(x,y)| \le C\frac{1}{\sqrt{V_{\delta^k}(x)\,V_{\delta^k}(y)}}\exp\left\{-\nu\left[\frac{d(x,y)}{\delta^k}\right]^a\right\}$$
$$\times\exp\left\{-\nu\left[\frac{\max\{d(x,\mathcal{Y}^k),\ d(y,\mathcal{Y}^k)\}}{\delta^k}\right]^a\right\}; \tag{2.1}$$

(iii) (the *regularity condition*) for any $x,\ x',\ y \in X$ with $d(x,x') \le \delta^k$,

$$|Q_k(x,y) - Q_k(x',y)| + |Q_k(y,x) - Q_k(y,x')|$$
$$\le C\left[\frac{d(x,x')}{\delta^k}\right]^\eta \frac{1}{\sqrt{V_{\delta^k}(x)\,V_{\delta^k}(y)}}\exp\left\{-\nu\left[\frac{d(x,y)}{\delta^k}\right]^a\right\}$$
$$\times\exp\left\{-\nu\left[\frac{\max\{d(x,\mathcal{Y}^k),\ d(y,\mathcal{Y}^k)\}}{\delta^k}\right]^a\right\}; \tag{2.2}$$

(iv) (the *second difference regularity condition*) for any $x,\ x',\ y,\ y' \in X$ with $d(x,x') \le \delta^k$ and $d(y,y') \le \delta^k$, then

$$|[Q_k(x,y) - Q_k(x',y)] - [Q_k(x,y') - Q_k(x',y')]|$$
$$\le C\left[\frac{d(x,x')}{\delta^k}\right]^\eta\left[\frac{d(y,y')}{\delta^k}\right]^\eta \frac{1}{\sqrt{V_{\delta^k}(x)\,V_{\delta^k}(y)}}\exp\left\{-\nu\left[\frac{d(x,y)}{\delta^k}\right]^a\right\}$$
$$\times\exp\left\{-\nu\left[\frac{\max\{d(x,\mathcal{Y}^k),\ d(y,\mathcal{Y}^k)\}}{\delta^k}\right]^a\right\}; \tag{2.3}$$

(v) (the *cancelation condition*) for any $x,\ y \in X$,

$$\int_X Q_k(x,y')\,d\mu(y') = 0 = \int_X Q_k(x',y)\,d\mu(x').$$

Remark 2.5 By [29, Remark 2.8], we know that the factor $\frac{1}{\sqrt{V_{\delta^k}(x)V_{\delta^k}(y)}}$ in (2.1), (2.2) and (2.3) can be replaced by $\frac{1}{V_{\delta^k}(x)}$ or $\frac{1}{V_{\delta^k}(y)}$, and $\max\{d(x,\mathcal{Y}^k), d(y,\mathcal{Y}^k)\}$ by $d(x,\mathcal{Y}^k)$ or by $d(y,\mathcal{Y}^k)$, with $\exp\{-\nu[\frac{d(x,y)}{\delta^k}]^a\}$ replaced by $\exp\{-\nu'[\frac{d(x,y)}{\delta^k}]^a\}$, where $\nu' \in (0,\nu)$ only depends on a and A_0. Moreover, the condition in Definition 2.4(iii) [resp., (iv)] can be replaced by $d(x,x') \le (2A_0)^{-1}[\delta^k + d(x,y)]$ (resp.,

$d(x, x') \le (2A_0)^{-2}[\delta^k + d(x, y)]$ and $d(y, y') \le (2A_0)^{-2}[\delta^k + d(x, y)]$). For their proofs, see [29, Proposition 2.9].

With the above exp-ATI, we have the following homogeneous continuous Calderón reproducing formula established in [29].

Theorem 2.6 *Let* $\{Q_k\}_{k\in\mathbb{Z}}$ *be an* exp-ATI *and* $\beta,\ \gamma \in (0, \eta)$. *Then there exists a sequence* $\{\widetilde{Q}_k\}_{k\in\mathbb{Z}}$ *of bounded linear operators on* $L^2(X)$ *such that, for any* $f \in (\mathring{\mathcal{G}}_0^\eta(\beta, \gamma))'$,

$$f = \sum_{k=-\infty}^{\infty} \widetilde{Q}_k Q_k f,$$

where the series converges in $(\mathring{\mathcal{G}}_0^\eta(\beta, \gamma))'$. *Moreover, there exists a positive constant* C *such that, for any* $k \in \mathbb{Z}$, *the kernel of* $\widetilde{Q}_k$ *satisfies the following conditions:*

(i) *for any* $x,\ y \in X$,

$$\left|\widetilde{Q}_k(x, y)\right| \le C \frac{1}{V_{\delta^k}(x) + V(x, y)} \left[\frac{\delta^k}{\delta^k + d(x, y)}\right]^\gamma;$$

(ii) *for any* $x,\ x',\ y \in X$ *with* $d(x, x') \le (2A_0)^{-1}[\delta^k + d(x, y)]$,

$$\left|\widetilde{Q}_k(x, y) - \widetilde{Q}_k(x', y)\right| \le C \left[\frac{d(x, x')}{\delta^k + d(x, y)}\right]^\beta \frac{1}{V_{\delta^k}(x) + V(x, y)} \left[\frac{\delta^k}{\delta^k + d(x, y)}\right]^\gamma;$$

(iii) *for any* $x \in X$,

$$\int_X \widetilde{Q}_k(x, y)\, d\mu(y) = 0 = \int_X \widetilde{Q}_k(y, x)\, d\mu(y).$$

Next, we recall the homogeneous discrete Calderón reproducing formulae established in [29]. To this end, let $j_0 \in \mathbb{N}$ be a sufficiently large integer such that $\delta^{j_0} \le (2A_0)^{-4}C^\natural$, where $C^\natural$ is as in Lemma 2.3. Based on Lemma 2.3, for any $k \in \mathbb{Z}$ and $\alpha \in \mathcal{A}_k$, we let

$$\mathcal{N}(k, \alpha) := \{\tau \in \mathcal{A}_{k+j_0} :\ Q_\tau^{k+j_0} \subset Q_\alpha^k\}$$

and $N(k, \alpha)$ be the cardinality of the set $\mathcal{N}(k, \alpha)$. For any $k \in \mathbb{Z}$ and $\alpha \in \mathcal{A}_k$, we rearrange the set $\{Q_\tau^{k+j_0} :\ \tau \in \mathcal{N}(k, \alpha)\}$ as $\{Q_\alpha^{k,m}\}_{m=1}^{N(k,\alpha)}$, whose centers are denoted, respectively, by $\{z_\alpha^{k,m}\}_{m=1}^{N(k,\alpha)}$.

Theorem 2.7 *Let* $\{Q_k\}_{k\in\mathbb{Z}}$ *be an* exp-ATI *and* $\beta,\ \gamma \in (0, \eta)$. *For any* $k \in \mathbb{Z}$, $\alpha \in \mathcal{A}_k$ *and* $m \in \{1, \ldots, N(k, \alpha)\}$, *suppose that* $y_\alpha^{k,m}$ *is an arbitrary point in* $Q_\alpha^{k,m}$. *Then, for any* $i \in \{1, 2\}$, *there exists a sequence* $\{\widetilde{Q}_k^{(i)}\}_{k=-\infty}^\infty$ *of bounded linear operators on* $L^2(X)$ *such that, for any* $f \in (\mathring{\mathcal{G}}_0^\eta(\beta, \gamma))'$,

$$f(\cdot) = \sum_{k=-\infty}^{\infty} \sum_{\alpha \in \mathcal{A}_k} \sum_{m=1}^{N(k,\alpha)} \widetilde{Q}_k^{(1)}\left(\cdot, y_\alpha^{k,m}\right) \int_{Q_\alpha^{k,m}} Q_k f(y)\, d\mu(y)$$
$$= \sum_{k=-\infty}^{\infty} \sum_{\alpha \in \mathcal{A}_k} \sum_{m=1}^{N(k,\alpha)} \mu\left(Q_\alpha^{k,m}\right) \widetilde{Q}_k^{(2)}\left(\cdot, y_\alpha^{k,m}\right) Q_k f\left(y_\alpha^{k,m}\right),$$

where the equalities converge in $(\mathring{\mathcal{G}}_0^\eta(\beta, \gamma))'$. *Moreover, for any* $k \in \mathbb{Z}$, *the kernels of* $\widetilde{Q}_k^{(1)}$ *and* $\widetilde{Q}_k^{(2)}$ *satisfy (i), (ii) and (iii) of Theorem* 2.6.

To recall the inhomogeneous discrete Calderón reproducing formulae established in [29], we introduce the following 1-exp-ATI and exp-IATI.

Definition 2.8 A sequence $\{P_k\}_{k=-\infty}^{\infty}$ of bounded linear operators on $L^2(X)$ is called an *approximation of the identity with exponential decay and integration* 1 (for short, 1-exp-ATI) if $\{P_k\}_{k=-\infty}^{\infty}$ has the following properties:

(i) for any $k \in \mathbb{Z}$, P_k satisfies (ii), (iii) and (iv) of Definition 2.4 but without the exponential decay factor

$$\exp\left\{-\nu\left[\frac{\max\{d(x, \mathcal{Y}^k), d(y, \mathcal{Y}^k)\}}{\delta^k}\right]^a\right\};$$

(ii) for any $k \in \mathbb{Z}$ and $x \in X$, $\int_X P_k(x, y)\, d\mu(y) = 1 = \int_X P_k(y, x)\, d\mu(y)$;

(iii) for any $k \in \mathbb{Z}$, letting $Q_k := P_k - P_{k-1}$, then $\{Q_k\}_{k \in \mathbb{Z}}$ is an exp-ATI.

Remark 2.9 The existence of the 1-exp-ATI is guaranteed by [1, Lemma 10.1]. Moreover, by the proofs of [29, Proposition 2.9] and [27, Proposition 2.7(iv)], we know that, for any $f \in L^2(X)$, $\lim_{k\to\infty} P_k f = f$ in $L^2(X)$.

Definition 2.10 A sequence $\{Q_k\}_{k=0}^{\infty}$ of bounded linear operators on $L^2(X)$ is called an *inhomogeneous approximation of the identity with exponential decay* (for short, exp-IATI) if there exists a 1-exp-ATI $\{P_k\}_{k=-\infty}^{\infty}$ such that $Q_0 = P_0$ and $Q_k = P_k - P_{k-1}$ for any $k \in \mathbb{N}$.

Next we recall the following inhomogeneous discrete Calderón reproducing formula established in [29].

Theorem 2.11 *Let* $\{Q_k\}_{k=0}^{\infty}$ *be an* exp-IATI *and* $\beta,\ \gamma \in (0, \eta)$. *Then there exists a sequence* $\{\widetilde{Q}_k\}_{k=0}^{\infty}$ *of bounded linear operators on* $L^2(X)$ *such that, for any* $f \in (\mathcal{G}_0^\eta(\beta, \gamma))'$,

$$f(\cdot) = \sum_{k=0}^{N} \sum_{\alpha \in \mathcal{A}_k} \sum_{m=1}^{N(0,\alpha)} \int_{Q_\alpha^{k,m}} \widetilde{Q}_k(\cdot, y)\, d\mu(y) Q_{\alpha,1}^{k,m}(f)$$
$$+ \sum_{k=N+1}^{\infty} \sum_{\alpha \in \mathcal{A}_k} \sum_{m=1}^{N(k,\alpha)} \mu\left(Q_\alpha^{k,m}\right) \widetilde{Q}_k\left(\cdot, y_\alpha^{k,m}\right) Q_k f\left(y_\alpha^{k,m}\right),$$

where the equality converges in $(\mathcal{G}_0^\eta(\beta,\gamma))'$, *every* $y_\alpha^{k,m}$ *is an arbitrary point in* $Q_\alpha^{k,m}$ *and, for any* $k \in \{0,\dots,N\}$,

$$Q_{\alpha,1}^{k,m}(f) := \frac{1}{\mu(Q_\alpha^{k,m})}\int_{Q_\alpha^{k,m}} Q_k f(u)\,d\mu(u).$$

Moreover, for any $k \in \mathbb{Z}_+$, $\widetilde{Q}_k$ *satisfies (i) and (ii) of Theorem* 2.6 *and, for any* $x \in X$,

$$\int_X \widetilde{Q}_k(x,y)\,d\mu(y) = \int_X \widetilde{Q}_k(y,x)\,d\mu(y) = \begin{cases} 1 & \text{if } k \in \{0,\dots,N\}, \\ 0 & \text{if } k \in \{N+1, N+2, \dots\}, \end{cases}$$

where $N \in \mathbb{N}$ *is some fixed constant independent of* f *and* $y_\alpha^{k,m}$.

3 Hardy Spaces via Various Maximal Functions

Let $\beta,\ \gamma \in (0,\eta)$ and $f \in (\mathcal{G}_0^\eta(\beta,\gamma))'$. Let $\{P_k\}_{k\in\mathbb{Z}}$ be a 1-exp-ATI as in Definition 2.8. Define the *radial maximal function* $\mathcal{M}^+(f)$ of f by setting

$$\mathcal{M}^+(f)(x) := \sup_{k\in\mathbb{Z}} |P_k f(x)|, \quad \forall\, x \in X.$$

Define the *non-tangential maximal function* $\mathcal{M}_\theta(f)$ *of* f *with aperture* $\theta \in (0,\infty)$ by setting

$$\mathcal{M}_\theta(f)(x) := \sup_{k\in\mathbb{Z}}\ \sup_{y\in B(x,\theta\delta^k)} |P_k f(y)|, \quad \forall\, x \in X.$$

Also, define the *grand maximal function* f^* of f by setting

$$f^*(x) := \sup\Big\{|\langle f,\varphi\rangle| :\ \varphi \in \mathcal{G}_0^\eta(\beta,\gamma) \text{ and } \|\varphi\|_{\mathcal{G}(x,r_0,\beta,\gamma)} \le 1 \text{ for some } r_0 \in (0,\infty)\Big\}, \quad \forall\, x \in X.$$

Correspondingly, for any $p \in (0,\infty]$, the *Hardy spaces* $H^{+,p}(X)$, $H_\theta^p(X)$ with $\theta \in (0,\infty)$ and $H^{*,p}(X)$ are defined, respectively, by setting

$$H^{+,p}(X) := \Big\{f \in \big(\mathcal{G}_0^\eta(\beta,\gamma)\big)' :\ \|f\|_{H^{+,p}(X)} := \|\mathcal{M}^+(f)\|_{L^p(X)} < \infty\Big\},$$
$$H_\theta^p(X) := \Big\{f \in \big(\mathcal{G}_0^\eta(\beta,\gamma)\big)' :\ \|f\|_{H_\theta^p(X)} := \|\mathcal{M}_\theta(f)\|_{L^p(X)} < \infty\Big\}$$

and

$$H^{*,p}(X) := \Big\{f \in \big(\mathcal{G}_0^\eta(\beta,\gamma)\big)' :\ \|f\|_{H^{*,p}(X)} := \|f^*\|_{L^p(X)} < \infty\Big\}.$$

Based on [20, Remark 2.9(ii)], we easily observe that, for any $f \in (\mathcal{G}_0^\eta(\beta,\gamma))'$ and $x \in X$,

$$\mathcal{M}^+(f)(x) \le \mathcal{M}_\theta(f)(x) \le Cf^*(x), \tag{3.1}$$

where C is a positive constant only depending on θ.

The aim of this section is to prove that the Hardy spaces $H^{+,p}(X)$, $H_\theta^p(X)$ and $H^{*,p}(X)$ are mutually equivalent when $p \in (\omega/(\omega+\eta), \infty]$ in the sense of equivalent (quasi-)norms (see Sect. 3.2); in particular, they all are equivalent to the Lebesgue space $L^p(X)$ when $p \in (1,\infty]$ in the sense of equivalent norms (see Sect. 3.1). Moreover, we prove that $H^{*,p}(X)$ is independent of the choices of the distribution space $(\mathcal{G}_0^\eta(\beta,\gamma))'$ whenever $\beta, \gamma \in (\omega(1/p-1), \eta)$; see Proposition 3.8 below.

3.1 Equivalence to the Lebesgue Space $L^p(X)$ When $p \in (1,\infty]$

In this section, we show that the Hardy spaces $H^{+,p}(X)$, $H_\theta^p(X)$ and $H^{*,p}(X)$ are all equivalent to the Lebesgue space $L^p(X)$, when $p \in (1,\infty]$, in the sense of both representing the same distributions and equivalent norms. First we give some basic properties of $H^{*,p}(X)$.

Proposition 3.1 *Let $p \in (0,\infty]$. Then $H^{*,p}(X)$ is a (quasi-)Banach space, which is continuously embedded into $(\mathcal{G}_0^\eta(\beta,\gamma))'$, where $\beta, \gamma \in (0,\eta)$.*

Proof Let $f \in H^{*,p}(X)$ and $\varphi \in \mathcal{G}_0^\eta(\beta,\gamma)$ with $\|\varphi\|_{\mathcal{G}(\beta,\gamma)} \le 1$. For any $x \in B(x_0,1)$, by Definition 2.1, we easily know that $\|\varphi\|_{\mathcal{G}(x,1,\beta,\gamma)} \lesssim 1$ with the implicit positive constant independent of x and hence $|\langle f,\varphi\rangle| \lesssim f^*(x)$. Therefore, for any $\varphi \in \mathcal{G}_0^\eta(\beta,\gamma)$ with $\beta, \gamma \in (0,\eta)$, we have

$$|\langle f,\varphi\rangle|^p \lesssim \frac{1}{V_1(x_0)} \int_{B(x_0,1)} [f^*(x)]^p \, d\mu(x) \lesssim \|f^*\|_{L^p(X)}^p \sim \|f\|_{H^{*,p}(X)}^p.$$

This implies that $H^{*,p}(X)$ is continuously embedded into $(\mathcal{G}_0^\eta(\beta,\gamma))'$.

To see that $H^{*,p}(X)$ is a (quasi-)Banach space, we only prove its completeness. Indeed, suppose that $\{f_k\}_{k=1}^\infty$ in $H^{*,p}(X)$ is a Cauchy sequence, which is also a Cauchy sequence in $(\mathcal{G}_0^\eta(\beta,\gamma))'$ with $\beta, \gamma \in (0,\eta)$. By the completeness of $(\mathcal{G}_0^\eta(\beta,\gamma))'$, the sequence $\{f_k\}_{k=1}^\infty$ converges to some element $f \in (\mathcal{G}_0^\eta(\beta,\gamma))'$ as $k \to \infty$. If $\varphi \in \mathcal{G}_0^\eta(\beta,\gamma)$ satisfies $\|\varphi\|_{\mathcal{G}(x,r_0,\beta,\gamma)} \le 1$ for some $x \in X$ and $r_0 \in (0,\infty)$, then $|\langle f_{k+l} - f_k, \varphi\rangle| \le (f_{k+l} - f_k)^*(x)$ for any $k, l \in \mathbb{N}$. Letting $l \to \infty$, we obtain

$$|\langle f - f_k, \varphi\rangle| \le \liminf_{l\to\infty} (f_{k+l} - f_k)^*(x),$$

which further implies that, for any $x \in X$,

$$(f - f_k)^*(x) \le \liminf_{l\to\infty} (f_{k+l} - f_k)^*(x).$$

By the Fatou lemma, we conclude that

$$\|(f - f_k)^*\|_{L^p(X)} \le \liminf_{l\to\infty} \|(f_{k+l} - f_k)^*\|_{L^p(X)} \to 0$$

as $k \to \infty$, which, together with the sublinearity of $\|\cdot\|_{H^{*,p}(X)}$, further implies that $f \in H^{*,p}(X)$ and $\lim_{k\to\infty} \|f - f_k\|_{H^{*,p}(X)} = 0$. Therefore, $H^{*,p}(X)$ is complete. This finishes the proof of Proposition 3.1. □

To show the equivalence of $H^{+,p}(X)$, $H^p_\theta(X)$ and $H^{*,p}(X)$ to the Lebesgue space $L^p(X)$ when $p \in (1,\infty]$ in the sense of both representing the same distributions and equivalent norms, we need the following technical lemma.

Lemma 3.2 *Let $\{P_k\}_{k\in\mathbb{Z}}$ be a* 1-exp-ATI *as in Definition* 2.8. *Assume that $\beta,\ \gamma \in (0,\eta)$. Then the following statements hold true:*

(i) *there exists a positive constant C such that, for any $k \in \mathbb{Z}$ and $\varphi \in \mathcal{G}(\beta,\gamma)$, $\|P_k\varphi\|_{\mathcal{G}(\beta,\gamma)} \le C\|\varphi\|_{\mathcal{G}(\beta,\gamma)}$;*
(ii) *for any $f \in \mathcal{G}(\beta,\gamma)$ and $\beta' \in (0,\beta)$, $\lim_{k\to\infty} P_k f = f$ in $\mathcal{G}(\beta',\gamma)$;*
(iii) *if $f \in \mathcal{G}^\eta_0(\beta,\gamma)$ [resp., $f \in (\mathcal{G}^\eta_0(\beta,\gamma))'$], then $\lim_{k\to\infty} P_k f = f$ in $\mathcal{G}^\eta_0(\beta,\gamma)$ [resp., $(\mathcal{G}^\eta_0(\beta,\gamma))'$].*

Proof The proof of (i) can be obtained by the method used in the proof of [29, Lemma 4.14]. The proof of (ii) is given in [20, Lemma 3.6], whose proof does not rely on the reverse doubling condition of μ and the metric d. We obtain (iii) directly by (i), (ii) and a standard duality argument. This finishes the proof of Lemma 3.2. □

We have the following proposition.

Proposition 3.3 *Let $p \in [1,\infty]$, $\beta,\ \gamma \in (0,\eta)$ and $\{P_k\}_{k\in\mathbb{Z}}$ be a* 1-exp-ATI. *If $f \in (\mathcal{G}^\eta_0(\beta,\gamma))'$ belongs to $H^{+,p}(X)$, then there exists $\widetilde{f} \in L^p(X)$ such that, for any $\varphi \in \mathcal{G}^\eta_0(\beta,\gamma)$,*

$$\langle f,\varphi\rangle = \int_X \widetilde{f}(x)\varphi(x)\,d\mu(x) \tag{3.2}$$

and $\|\widetilde{f}\|_{L^p(X)} \le \|\mathcal{M}^+(f)\|_{L^p(X)}$; moreover, if $p \in [1,\infty)$, then, for almost every $x \in X$, $|\widetilde{f}(x)| \le \mathcal{M}^+(f)(x)$.

Proof Let $f \in (\mathcal{G}^\eta_0(\beta,\gamma))'$ and $\mathcal{M}^+(f) := \sup_{k\in\mathbb{Z}} |P_k f| \in L^p(X)$, where $\{P_k\}_{k\in\mathbb{Z}}$ is a 1-exp-ATI as in Definition 2.8. Then $\{P_k f\}_{k\in\mathbb{Z}}$ is uniformly bounded in $L^p(X)$. If $p \in (1,\infty]$, then $p' \in [1,\infty)$ and $L^{p'}(X)$ is separable. Thus, by the Banach–Alaoglu theorem (see, for example, [46, Theorem 3.17]), we find a function $\widetilde{f} \in L^p(X)$ and a sequence $\{k_j\}_{j=1}^\infty \subset \mathbb{Z}$ such that $k_j \to \infty$ and $P_{k_j} f \to \widetilde{f}$ as $j \to \infty$ in the weak-$*$ topology of $L^p(X)$. By this and the Hölder inequality, for any $g \in L^{p'}(X)$, we have

$$\left|\int_X \widetilde{f}(x)g(x)\,d\mu(x)\right| = \lim_{j\to\infty}\left|\int_X P_{k_j} f(x)g(x)\,d\mu(x)\right| \le \|\mathcal{M}^+(f)\|_{L^p(X)}\|g\|_{L^{p'}(X)},$$

which further implies that $\|\widetilde{f}\|_{L^p(X)} \le \|\mathcal{M}^+(f)\|_{L^p(X)}$.

If $p = 1$, notice that $\|\sup_{k\in\mathbb{Z}} |P_k f|\|_{L^1(X)} = \|\mathcal{M}^+(f)\|_{L^1(X)} < \infty$. Then, by the proof of [52, Theorem III.C.12], $\{P_k f\}_{k\in\mathbb{Z}}$ is relatively compact in $L^1(X)$. Therefore, by the Eberlin–Šmulian theorem (see [52, II.C]), we know that $\{P_k f\}_{k\in\mathbb{Z}}$ is weakly sequentially compact, that is, there exist a function $\widetilde{f} \in L^1(X)$ and a subsequence $\{P_{k_j} f\}_{j=1}^{\infty}$ such that $\{P_{k_j} f\}_{j=1}^{\infty}$ converges to $\widetilde{f}$ weakly in $L^1(X)$. As the arguments for the case $p \in (1, \infty]$, we still have $\|\widetilde{f}\|_{L^1(X)} \leq \|\mathcal{M}^+(f)\|_{L^1(X)}$.

Moreover, for any $\varphi \in \mathcal{G}_0^\eta(\beta, \gamma)$, by the fact $\mathcal{G}_0^\eta(\beta, \gamma) \subset L^p(X)$ for any $p \in [1, \infty]$ and Lemma 3.2(iii), we conclude that

$$\langle f, \varphi\rangle = \lim_{k\to\infty} \langle P_{k_j} f, \varphi\rangle = \lim_{j\to\infty} \int_X P_{k_j} f(x)\varphi(x)\, d\mu(x) = \int_X \widetilde{f}(x)\varphi(x)\, d\mu(x). \tag{3.3}$$

Let $p \in [1, \infty)$. For any $j \in \mathbb{N}$ and $x \in X$, we have $P_{k_j}(x, \cdot) \in \mathcal{G}(\eta, \eta)$ (see the proof of [29, Proposition 2.10]), which, together with (3.3), implies that

$$P_{k_j} f(x) = \langle f, P_{k_j}(x, \cdot)\rangle = \int_X P_{k_j}(x, y)\widetilde{f}(y)\, d\mu(y) = P_{k_j} \widetilde{f}(x).$$

From this and [27, Proposition 2.7(iv)], we deduce that $\{P_{k_j} f\}_{j\in\mathbb{N}}$ converges to $\widetilde{f}$ in the sense of $\|\cdot\|_{L^p(X)}$. Then, by the Riesz theorem, we find a subsequence of $\{P_{k_j} f\}_{j\in\mathbb{N}}$, still denoted by $\{P_{k_j} f\}_{j\in\mathbb{N}}$, such that $P_{k_j} f(x) \to \widetilde{f}(x)$ as $k_j \to \infty$ for almost every $x \in X$. Therefore, $|\widetilde{f}(x)| \leq \mathcal{M}^+(f)(x)$ for almost every $x \in X$. This finishes the proof of Proposition 3.3. □

Finally, we show the following main result of this section.

Theorem 3.4 *Let $p \in (1, \infty]$ and $\beta, \gamma \in (0, \eta)$. Then the following hold true:*

(i) *if $f \in (\mathcal{G}_0^\eta(\beta, \gamma))'$ belongs to $H^{+,p}(X)$, then there exists $\widetilde{f} \in L^p(X)$ such that* (3.2) *holds true and $\|\widetilde{f}\|_{L^p(X)} \leq \|f\|_{H^{+,p}(X)}$;*

(ii) *any $f \in L^p(X)$ induces a distribution on $\mathcal{G}_0^\eta(\beta, \gamma)$ as in* (3.2), *still denoted by f, such that $f \in H^{*,p}(X)$ and $\|f\|_{H^{*,p}(X)} \leq C\|f\|_{L^p(X)}$, where C is a positive constant independent of f.*

Consequently, for any fixed $\theta \in (0, \infty)$, $H^{+,p}(X) = H_\theta^p(X) = H^{*,p}(X) = L^p(X)$ in the sense of both representing the same distributions and equivalent norms.

Proof We obtain (i) directly by Proposition 3.3. Now we prove (ii). Suppose that $p \in (1, \infty]$ and $f \in L^p(X)$. Clearly, f induces a distribution on $\mathcal{G}_0^\eta(\beta, \gamma)$ as in (3.2). By [20, Proposition 3.9], we find that, for almost every $x \in X$, $f^*(x) \lesssim \mathcal{M}(f)(x)$, with the implicit positive constant independent of f and x. Therefore, from the boundedness of $\mathcal{M}$ on $L^p(X)$, we deduce that $\|f^*\|_{L^p(X)} \lesssim \|\mathcal{M}(f)\|_{L^p(X)} \lesssim \|f\|_{L^p(X)}$. This finishes the proof of (ii).

By (i), (ii) and (3.1), we obtain $H^{+,p}(X) = H_\theta^p(X) = H^{*,p}(X) = L^p(X)$, which completes the proof of Theorem 3.4. □

3.2 Equivalence of Hardy Spaces Defined via Various Maximal Functions

The main aim of this section concerns the equivalence of Hardy spaces defined via various maximal functions for the case $p \in (\omega/(\omega + \eta), 1]$. Indeed, our goal is to show the following equivalence theorem.

Theorem 3.5 *Assume that* $p \in (\omega/(\omega+\eta), 1]$ *and* $\theta \in (0, \infty)$. *Then, for any* $f \in (\mathcal{G}_0^\eta(\beta, \gamma))'$ *with* $\beta,\ \gamma \in (\omega(1/p-1), \eta)$,

$$\|f\|_{H^{+,p}(X)} \sim \|f\|_{H_\theta^p(X)} \sim \|f\|_{H^{*,p}(X)},$$

with positive equivalence constants independent of f. *In other words,* $H^{+,p}(X) = H_\theta^p(X) = H^{*,p}(X)$ *with equivalent (quasi-)norms.*

To prove Theorem 3.5, we borrow some ideas from [56]. To this end, we need the following two technical lemmas.

Lemma 3.6 *Assume that* $\phi \in \mathcal{G}_0^\eta(\beta, \gamma)$ *with* $\beta,\ \gamma \in (0, \eta)$. *Let* $\sigma := \int_X \phi(x)\,d\mu(x)$. *If* $\psi \in \mathcal{G}(\eta, \eta)$ *with* $\int_X \psi(x)\,d\mu(x) = 1$, *then* $\phi - \sigma\psi \in \mathring{\mathcal{G}}_0^\eta(\beta, \gamma)$.

Proof Since $\phi \in \mathcal{G}_0^\eta(\beta, \gamma)$ with $\beta,\ \gamma \in (0, \eta)$, it follows that there exists $\{\phi_n\}_{n=1}^\infty \subset \mathcal{G}(\eta, \eta)$ such that $\lim_{n\to\infty} \|\phi - \phi_n\|_{\mathcal{G}(\beta,\gamma)} = 0$. Letting $\sigma_n := \int_X \phi_n(x)\,d\mu(x)$ for any $n \in \mathbb{N}$, by Definition 2.1 and Lemma 2.2(ii), we conclude that $\lim_{n\to\infty} |\sigma - \sigma_n| = 0$, where $\sigma := \int_X \phi(x)\,d\mu(x)$. Let $\varphi_n := \phi_n - \sigma_n\psi$ for any $n \in \mathbb{N}$. Then $\varphi_n \in \mathring{\mathcal{G}}(\eta, \eta)$ and

$$\|\phi - \sigma\psi - \varphi_n\|_{\mathcal{G}(\beta,\gamma)} \le \|\phi - \phi_n\|_{\mathcal{G}(\beta,\gamma)} + |\sigma - \sigma_n|\|\psi\|_{\mathcal{G}(\beta,\gamma)} \to 0 \quad \text{as } n \to \infty.$$

Thus, $\phi - \sigma\psi \in \mathring{\mathcal{G}}_0^\eta(\beta, \gamma)$. This finishes the proof of Lemma 3.6. □

The next lemma comes from [27, Lemma 5.3], whose proof remains true for a quasi-metric d and also does not rely on the reverse doubling condition of μ.

Lemma 3.7 *Let all the notation be as in Theorem 2.7. Let* $k,\ k' \in \mathbb{Z}$, $\{a_\alpha^{k,m}\}_{k\in\mathbb{Z},\ \alpha\in\mathcal{A}_k,\ m\in\{1,\ldots,N(k,\alpha)\}} \subset \mathbb{C}$, $\gamma \in (0, \eta)$ *and* $r \in (\omega/(\omega+\gamma), 1]$. *Then there exists a positive constant* C, *independent of* $k,\ k',\ y_\alpha^{k,m} \in Q_\alpha^{k,m}$ *and* $a_\alpha^{k,m}$ *with* $k \in \mathbb{Z}$, $\alpha \in \mathcal{A}_k$ *and* $m \in \{1, \ldots, N(k,\alpha)\}$, *such that, for any* $x \in X$,

$$\begin{aligned}
&\sum_{\alpha\in\mathcal{A}_k} \sum_{m=1}^{N(k,\alpha)} \mu\left(Q_\alpha^{k,m}\right) \frac{1}{V_{\delta^{k\wedge k'}}(x) + V(x, y_\alpha^{k,m})} \left[\frac{\delta^{k\wedge k'}}{\delta^{k\wedge k'} + d(x, y_\alpha^{k,m})}\right]^\gamma \left|a_\alpha^{k,m}\right| \\
&\quad \le C\delta^{[k-(k\wedge k')]\omega(1-\frac{1}{r})} \left[\mathcal{M}\left(\sum_{\alpha\in\mathcal{A}_k} \sum_{m=1}^{N(k,\alpha)} \left|a_\alpha^{k,m}\right|^r \mathbf{1}_{Q_\alpha^{k,m}}\right)(x)\right]^{\frac{1}{r}}.
\end{aligned}$$

Now we show Theorem 3.5 by using the above two technical lemmas. In what follows, the *symbol* $\epsilon \to 0^+$ means that $\epsilon \in (0, \infty)$ and $\epsilon \to 0$.

Proof of Theorem 3.5 Let $f \in (\mathcal{G}_0^\eta(\beta, \gamma))'$ with $\beta,\ \gamma \in (\omega(1/p-1), \eta)$. Fix $\theta \in (0, \infty)$. By (3.1), we have

$$\|\mathcal{M}^+(f)\|_{L^p(X)} \le \|\mathcal{M}_\theta(f)\|_{L^p(X)} \lesssim \|f^*\|_{L^p(X)}.$$

Thus, the proof of Theorem 3.5 is reduced to showing

$$\|f^*\|_{L^p(X)} \lesssim \|\mathcal{M}^+(f)\|_{L^p(X)}. \tag{3.4}$$

To obtain (3.4), it suffices to prove that, for some $r \in (0, p)$ and any $x \in X$,

$$f^*(x) \lesssim \mathcal{M}^+(f)(x) + \left\{\mathcal{M}\left(\left[\mathcal{M}^+(f)\right]^r\right)(x)\right\}^{\frac{1}{r}}. \tag{3.5}$$

If (3.5) holds true, then, by the boundedness of $\mathcal{M}$ on $L^{p/r}(X)$, we conclude that

$$\|f^*\|_{L^p(X)} \lesssim \|\mathcal{M}^+(f)\|_{L^p(X)} + \left\|\mathcal{M}\left(\left[\mathcal{M}^+(f)\right]^r\right)\right\|_{L^{p/r}(X)}^{\frac{1}{r}} \sim \|\mathcal{M}^+(f)\|_{L^p(X)},$$

which proves (3.4).

We now fix $x \in X$ and show (3.5). Let $\{P_k\}_{k\in\mathbb{Z}}$ be a 1-exp-ATI. For any $k \in \mathbb{Z}$, define $Q_k := P_k - P_{k-1}$. Then $\{Q_k\}_{k\in\mathbb{Z}}$ is an exp-ATI. Assume for the moment that, for any $\varphi \in \mathring{\mathcal{G}}_0^\eta(\beta,\gamma)$ with $\|\varphi\|_{\mathcal{G}(x,\delta^l,\beta,\gamma)} \le 1$ for some $l \in \mathbb{Z}$,

$$|\langle f, \varphi\rangle| \lesssim \left\{\mathcal{M}\left(\left[\mathcal{M}^+(f)\right]^r\right)(x)\right\}^{\frac{1}{r}}. \tag{3.6}$$

We now use (3.6) to show (3.5). For any $\phi \in \mathcal{G}_0^\eta(\beta,\gamma)$ with $\|\phi\|_{\mathcal{G}(x,r_0,\beta,\gamma)} \le 1$ for some $r_0 \in (0,\infty)$, choose $l \in \mathbb{Z}$ such that $\delta^{l+1} \le r_0 < \delta^l$. Clearly, $\|\phi\|_{\mathcal{G}(x,\delta^l,\beta,\gamma)} \lesssim 1$. Let $\sigma := \int_X \phi(y)\,d\mu(y)$ and $\varphi := \phi - \sigma P_l(x,\cdot)$. Notice that $\int_X P_l(x,y)\,d\mu(y) = 1$ and $P_l(x,\cdot) \in \mathcal{G}(\eta,\eta)$ (see the proof of [29, Proposition 2.10]). From Lemma 3.6, it follows that $\varphi \in \mathring{\mathcal{G}}_0^\eta(\beta,\gamma)$. Moreover, $\|\varphi\|_{\mathcal{G}(x,\delta^l,\beta,\gamma)} \lesssim \|\phi\|_{\mathcal{G}(x,\delta^l,\beta,\gamma)} + |\sigma|\|P_l(x,\cdot)\|_{\mathcal{G}(x,\delta^l,\beta,\gamma)} \lesssim 1$. By (3.6), we know that

$$\begin{aligned}
|\langle f, \phi\rangle| &\le |\langle f, \varphi\rangle| + |\sigma||\langle f, P_l(x,\cdot)\rangle| \\
&\lesssim \left\{\mathcal{M}\left(\left[\mathcal{M}^+(f)\right]^r\right)(x)\right\}^{\frac{1}{r}} + |P_l f(x)| \\
&\lesssim \left\{\mathcal{M}\left(\left[\mathcal{M}^+(f)\right]^r\right)(x)\right\}^{\frac{1}{r}} + \mathcal{M}^+(f)(x),
\end{aligned}$$

which is exactly (3.5).

It remains to prove (3.6). For any $\epsilon \in (0,\infty)$, choose $y_\alpha^{k,m} \in Q_\alpha^{k,m}$ such that

$$\left|Q_k f\left(y_\alpha^{k,m}\right)\right| \le \inf_{z\in Q_\alpha^{k,m}} |Q_k f(z)| + \epsilon \le 2 \inf_{z\in Q_\alpha^{k,m}} \mathcal{M}^+(f)(z) + \epsilon.$$

Let $g := f|_{\mathring{\mathcal{G}}_0^\eta(\beta,\gamma)}$ be the restriction of f on $\mathring{\mathcal{G}}_0^\eta(\beta,\gamma)$. Obviously, $g \in (\mathring{\mathcal{G}}_0^\eta(\beta,\gamma))'$ and $\|g\|_{(\mathring{\mathcal{G}}_0^\eta(\beta,\gamma))'} \le \|f\|_{(\mathcal{G}_0^\eta(\beta,\gamma))'}$. By Theorem 2.7, we conclude that

$$\langle f, \varphi\rangle = \langle g, \varphi\rangle = \sum_{k=-\infty}^{\infty} \sum_{\alpha\in\mathcal{A}_k} \sum_{m=1}^{N(k,\alpha)} \mu\left(Q_\alpha^{k,m}\right) \widetilde{Q}_k^*\varphi\left(y_\alpha^{k,m}\right) Q_k g\left(y_\alpha^{k,m}\right)$$

$$= \sum_{k=-\infty}^{\infty} \sum_{\alpha\in\mathcal{A}_k} \sum_{m=1}^{N(k,\alpha)} \mu\left(Q_\alpha^{k,m}\right) \widetilde{Q}_k^* \varphi\left(y_\alpha^{k,m}\right) Q_k f\left(y_\alpha^{k,m}\right),$$

where $\widetilde{Q}_k^*$ denotes the *dual operator* of $\widetilde{Q}_k$. By the proof of [27, (3.2)], which remains true for a quasi-metric d and does not rely on the reverse doubling condition of μ, we find that, for any fixed $\beta' \in (0, \beta \wedge \gamma)$ and any $k \in \mathbb{Z}$,

$$\left|\widetilde{Q}_k^* \varphi\left(y_\alpha^{k,m}\right)\right| \lesssim \delta^{|k-l|\beta'} \frac{1}{V_{\delta^{k\wedge l}}(x) + V(x, y_\alpha^{k,m})} \left[\frac{\delta^{k\wedge l}}{\delta^{k\wedge l} + d(x, y_\alpha^{k,m})}\right]^\gamma. \tag{3.7}$$

Choose $\beta' \in (0, \beta \wedge \gamma)$ such that $\omega/(\omega + \beta') < p$. From this and Lemma 3.7, we deduce that, for any fixed $r \in (\omega/(\omega + \beta'), p)$,

$$\begin{aligned}
|\langle f, \varphi\rangle| &\lesssim \sum_{k=-\infty}^{\infty} \delta^{|k-l|\beta'} \sum_{\alpha\in\mathcal{A}_k} \sum_{m=1}^{N(k,\alpha)} \mu\left(Q_\alpha^{k,m}\right) \\
&\quad \times \frac{\inf_{z\in Q_\alpha^{k,m}} \mathcal{M}^+(f)(z) + \epsilon}{V_{\delta^{k\wedge l}}(x) + V(x, y_\alpha^{k,m})} \left[\frac{\delta^{k\wedge l}}{\delta^{k\wedge l} + d(x, y_\alpha^{k,m})}\right]^\gamma \\
&\lesssim \sum_{k=-\infty}^{\infty} \delta^{|k-l|\beta'} \delta^{[k-(k\wedge l)]\omega\left(1-\frac{1}{r}\right)} \\
&\quad \times \left\{\mathcal{M}\left(\sum_{\alpha\in\mathcal{A}_k} \sum_{m=1}^{N(k,\alpha)} \left[\inf_{z\in Q_\alpha^{k,m}} \mathcal{M}^+(f)(z) + \epsilon\right]^r \mathbf{1}_{Q_\alpha^{k,m}}\right)(x)\right\}^{\frac{1}{r}} \\
&\lesssim \sum_{k=-\infty}^{\infty} \delta^{|k-l|\beta'} \delta^{[k-(k\wedge l)]\omega\left(1-\frac{1}{r}\right)} \left\{\mathcal{M}\left(\left[\mathcal{M}^+(f)+\epsilon\right]^r\right)(x)\right\}^{\frac{1}{r}} \\
&\lesssim \left\{\mathcal{M}\left(\left[\mathcal{M}^+(f)\right]^r\right)(x) + \epsilon^r\right\}^{\frac{1}{r}} \to \left\{\mathcal{M}\left(\left[\mathcal{M}^+(f)\right]^r\right)(x)\right\}^{\frac{1}{r}} \quad \text{as } \epsilon \to 0^+.
\end{aligned} \tag{3.8}$$

This proves (3.6) and hence finishes the proof of Theorem 3.5. □

To conclude this section, we show that the Hardy space $H^{*,p}(X)$ is independent of the choices of $(\mathcal{G}_0^\eta(\beta, \gamma))'$ whenever $\beta,\ \gamma \in (\omega(1/p - 1), \eta)$.

Proposition 3.8 *Let* $p \in (\omega/(\omega + \eta), 1]$ *and* $\beta_1,\ \beta_2,\ \gamma_1,\ \gamma_2 \in (\omega(1/p - 1), \eta)$. *If* $f \in (\mathcal{G}_0^\eta(\beta_1, \gamma_1))'$ *and* $f \in H^{*,p}(X)$, *then* $f \in (\mathcal{G}_0^\eta(\beta_2, \gamma_2))'$.

Proof Let $f \in (\mathcal{G}_0^\eta(\beta_1, \gamma_1))'$ with $\|f\|_{H^{*,p}(X)} < \infty$. We first prove that there exists $\theta \in (0, \infty)$ such that, for any $\varphi \in \mathcal{G}(\eta, \eta)$ with $\|\varphi\|_{\mathcal{G}(\beta_2,\gamma_2)} \le 1$,

$$|\langle f, \varphi\rangle| \lesssim \|\mathcal{M}_\theta(f)\|_{L^p(X)}. \tag{3.9}$$

Notice that $\varphi \in \mathcal{G}(\eta,\eta) \subset \mathcal{G}_0^\eta(\beta_1,\gamma_1)$ and $f \in (\mathcal{G}_0^\eta(\beta_1,\gamma_1))'$. With all the notation involved as in Theorem 2.11, we have

$$\begin{aligned}\langle f,\varphi\rangle &= \sum_{k=0}^{N}\sum_{\alpha\in\mathcal{A}_0}\sum_{m=1}^{N(k,\alpha)}\int_{Q_\alpha^{k,m}}\widetilde{Q}_k^*\varphi(y)\,d\mu(y)Q_{\alpha,1}^{k,m}(f)\\ &\quad+\sum_{k=N+1}^{\infty}\sum_{\alpha\in\mathcal{A}_k}\sum_{m=1}^{N(k,\alpha)}\mu\left(Q_\alpha^{k,m}\right)\widetilde{Q}_k^*\varphi\left(y_\alpha^{k,m}\right)Q_kf\left(y_\alpha^{k,m}\right)=:\mathrm{Z}_1+\mathrm{Z}_2.\end{aligned}$$

Choose $\theta := 2A_0C^\natural$ with $C^\natural$ as in Lemma 2.3(v). By the definition of $Q_\alpha^{k,m}$ and Lemma 2.3(v), we have $Q_\alpha^{k,m} \subset B(z_\alpha^{k,m}, C^\natural\delta^{k+j_0}) \subset B(z, 2A_0C^\natural\delta^k) = B(z,\theta\delta^k)$ for any $z \in Q_\alpha^{k,m}$.

Fix $x \in B(x_0,1)$. Then $\|\varphi\|_{\mathcal{G}(x,1,\beta_2,\gamma_2)} \sim \|\varphi\|_{\mathcal{G}(x_0,1,\beta_2,\gamma_2)} \lesssim 1$. If $k \in \{0,\ldots,N\}$, then we have $\|\varphi\|_{\mathcal{G}(x,\delta^k,\beta_2,\gamma_2)} \sim \|\varphi\|_{\mathcal{G}(x,1,\beta_2,\gamma_2)} \lesssim 1$, where the implicit positive constants are independent of x but can depend on N. Let $\beta_- := \min\{\beta_1,\gamma_1,\beta_2,\gamma_2\}$. By [27, (3.2)], we conclude that, for any $y \in Q_\alpha^{k,m}$,

$$\left|\widetilde{Q}_k^*\varphi(y)\right| \lesssim \frac{1}{V_1(x)+V(x,y)}\left[\frac{1}{1+d(x,y)}\right]^{\beta_-} \sim \frac{1}{V_1(x)+V(x,y_\alpha^{k,m})}\left[\frac{1}{1+d(x,y_\alpha^{k,m})}\right]^{\beta_-}.$$

Moreover, for any $k \in \{0,\ldots,N\}$ and $z \in Q_\alpha^{k,m}$, we have

$$\left|Q_{\alpha,1}^{k,m}(f)\right| \le \frac{1}{\mu(Q_\alpha^{k,m})}\int_{Q_\alpha^{k,m}}[|P_kf(y)|+|P_{k-1}f(y)|]\,d\mu(y) \le 2\mathcal{M}_\theta(f)(z).$$

Thus, we obtain

$$\begin{aligned}|\mathrm{Z}_1| &\lesssim \sum_{k=0}^{N}\sum_{\alpha\in\mathcal{A}_k}\sum_{m=1}^{N(k,\alpha)}\mu\left(Q_\alpha^{k,m}\right)\frac{1}{V_1(x)+V(x,y_\alpha^{k,m})}\\ &\quad\times\left[\frac{1}{1+d(x,y_\alpha^{k,m})}\right]^{\beta_-}\inf_{z\in Q_\alpha^{k,m}}\mathcal{M}_\theta(f)(z).\end{aligned} \tag{3.10}$$

If $k \in \{N+1,N+2,\ldots\}$, then $|Q_kf(y_\alpha^{k,m})| \le 2\inf_{z\in Q_\alpha^{k,m}}\mathcal{M}_\theta(f)(z)$. Again, by $\|\varphi\|_{\mathcal{G}(x,1,\beta_2,\gamma_2)} \lesssim 1$ and [27, (3.2)], we find that, for any fixed $\beta' \in (0,\beta_-)$,

$$\left|\widetilde{Q}_k^*\varphi\left(y_\alpha^{k,m}\right)\right| \lesssim \delta^{k\beta'}\frac{1}{V_1(x)+V(x,y_\alpha^{k,m})}\left[\frac{1}{1+d(x,y_\alpha^{k,m})}\right]^{\beta_-},$$

because now $k \in \mathbb{Z}_+$ and we do not need the cancelation of φ. Therefore, we have

$$|\mathrm{Z}_2| \lesssim \sum_{k=N+1}^{\infty} \delta^{k\beta'} \sum_{\alpha\in\mathcal{A}_k} \sum_{m=1}^{N(k,\alpha)} \mu\left(Q_\alpha^{k,m}\right)$$
$$\times \frac{1}{V_1(x)+V(x,y_\alpha^{k,m})}\left[\frac{1}{1+d(x,y_\alpha^{k,m})}\right]^{\beta_-} \inf_{z\in Q_\alpha^{k,m}} \mathcal{M}_\theta(f)(z). \tag{3.11}$$

Following the estimation of (3.8), from (3.10) and (3.11), we deduce that, for some $r \in (\omega/(\omega+\eta), p)$,

$$|\langle f, \varphi\rangle| \lesssim \left\{\mathcal{M}\left([\mathcal{M}_\theta(f)]^r\right)(x)\right\}^{\frac{1}{r}}.$$

Notice that the above inequality holds true for any $x \in B(x_0, 1)$. Then, by the boundedness of $\mathcal{M}$ on $L^{p/r}(X)$, we further conclude that

$$|\langle f, \varphi\rangle|^p \lesssim \frac{1}{V_1(x_0)} \int_X \left\{\mathcal{M}\left([\mathcal{M}_\theta(f)]^r\right)(x)\right\}^{\frac{p}{r}} d\mu(x) \lesssim \|\mathcal{M}_\theta(f)\|_{L^p(X)}^p,$$

which is exactly (3.9).

Combining (3.9) and (3.1), we find that, for any $\varphi \in \mathcal{G}(\eta, \eta)$,

$$|\langle f, \varphi\rangle| \lesssim \|\mathcal{M}_\theta(f)\|_{L^p(X)} \|\varphi\|_{\mathcal{G}(\beta_2,\gamma_2)} \lesssim \|f\|_{H^{*,p}(X)} \|\varphi\|_{\mathcal{G}(\beta_2,\gamma_2)}. \tag{3.12}$$

Now let $g \in \mathcal{G}_0^\eta(\beta_2, \gamma_2)$. By the definition of $\mathcal{G}_0^\eta(\beta_2, \gamma_2)$, we know that there exist $\{\varphi_j\}_{j=1}^\infty \subset \mathcal{G}(\eta, \eta)$ such that $\|g - \varphi_j\|_{\mathcal{G}(\beta_2,\gamma_2)} \to 0$ as $j \to \infty$, which implies that $\{\varphi_j\}_{j=1}^\infty$ is a Cauchy sequence in $\mathcal{G}(\beta_2, \gamma_2)$. By (3.12), we find that, for any $j,\ k \in \mathbb{N}$,

$$|\langle f, \varphi_j - \varphi_k\rangle| \lesssim \|f\|_{H^{*,p}(X)} \|\varphi_j - \varphi_k\|_{\mathcal{G}(\beta_2,\gamma_2)}.$$

Therefore, $\lim_{j\to\infty}\langle f, \varphi_j\rangle$ exists and the limit is independent of the choice of $\{\varphi_j\}_{j=1}^\infty$. Thus, it is reasonable to define $\langle f, g\rangle := \lim_{j\to\infty}\langle f, \varphi_j\rangle$. Moreover, by (3.12), we conclude that

$$\begin{aligned}|\langle f, g\rangle| &= \lim_{j\to\infty} |\langle f, \varphi_j\rangle| \lesssim \|f\|_{H^{*,p}(X)} \liminf_{j\to\infty} \|\varphi_j\|_{\mathcal{G}(\beta_2,\gamma_2)} \\ &\sim \|f\|_{H^{*,p}(X)} \|g\|_{\mathcal{G}_0^\eta(\beta_2,\gamma_2)}.\end{aligned}$$

This implies $f \in (\mathcal{G}_0^\eta(\beta_2, \gamma_2))'$ and $\|f\|_{(\mathcal{G}_0^\eta(\beta_2,\gamma_2))'} \lesssim \|f\|_{H^{*,p}(X)}$, which completes the proof of Proposition 3.8. □

4 Grand Maximal Function Characterizations of Atomic Hardy Spaces

In this section, we establish the atomic characterizations of $H^{*,p}(X)$ with $p \in (\omega/(\omega+\eta), 1]$.

Definition 4.1 Let $p \in (\omega/(\omega+\eta), 1]$, $q \in (p, \infty] \cap [1, \infty]$ and $\beta, \gamma \in (\omega(1/p-1), \eta)$. The *atomic Hardy space* $H^{p,q}_{\mathrm{at}}(X)$ is defined to be the set of all $f \in (\mathcal{G}^{\eta}_0(\beta, \gamma))'$ such that $f = \sum_{j=1}^{\infty} \lambda_j a_j$, where $\{a_j\}_{j=1}^{\infty}$ is a sequence of (p, q)-atoms and $\{\lambda_j\}_{j=1}^{\infty} \subset \mathbb{C}$ satisfies $\sum_{j=1}^{\infty} |\lambda_j|^p < \infty$. Moreover, let

$$\|f\|_{H^{p,q}_{\mathrm{at}}(X)} := \inf \left\{ \left(\sum_{j=1}^{\infty} |\lambda_j|^p \right)^{\frac{1}{p}} \right\},$$

where the infimum is taken over all the decompositions of f as above.

Observe that the difference between $H^{p,q}_{\mathrm{cw}}(X)$ and $H^{p,q}_{\mathrm{at}}(X)$ mainly lies on the choices of distribution spaces. When (X, d, μ) is a doubling metric measure space, it was proved in [40, Theorem 4.4] that $H^{p,q}_{\mathrm{cw}}(X)$ and $H^{p,q}_{\mathrm{at}}(X)$ coincide with equivalent (quasi-)norms. Since now d is a quasi-metric, for the completeness of this article, we include a proof of their equivalence in Sect. 4.4 below.

The main aim in this section is to prove the following conclusion.

Theorem 4.2 *Let $p \in (\omega/(\omega+\eta), 1]$, $q \in (p, \infty] \cap [1, \infty]$ and $\beta, \gamma \in (\omega(1/p-1), \eta)$. As subspaces of $(\mathcal{G}^{\eta}_0(\beta, \gamma))'$, $H^{*,p}(X) = H^{p,q}_{\mathrm{at}}(X)$ with equivalent (quasi-) norms.*

We divide the proof of Theorem 4.2 into three sections. In Sect. 4.1, we prove that $H^{p,q}_{\mathrm{at}}(X) \subset H^{*,p}(X)$ directly by the definition of $H^{p,q}_{\mathrm{at}}(X)$. The next two sections mainly deal with the proof of $H^{*,p}(X) \subset H^{p,q}_{\mathrm{at}}(X)$. In Sect. 4.2, we obtain a Calderón–Zygmund decomposition for any $f \in H^{*,p}(X)$. Then, in Sect. 4.3, we show that any $f \in H^{*,p}(X)$ has a (p, ∞)-atomic decomposition. In Sect. 4.4, we reveal the equivalent relationship between $H^{p,q}_{\mathrm{at}}(X)$ and $H^{p,q}_{\mathrm{cw}}(X)$.

4.1 Proof of $H^{p,q}_{\mathrm{at}}(X) \subset H^{*,p}(X)$

In this section, we prove $H^{p,q}_{\mathrm{at}}(X) \subset H^{*,p}(X)$, as subspaces of $(\mathcal{G}^{\eta}_0(\beta, \gamma))'$ with $\beta, \gamma \in (\omega(1/p-1), \eta)$. To do this, we need the following technical lemma.

Lemma 4.3 *Let $p \in (\omega/(\omega+\eta), 1]$ and $q \in (p, \infty] \cap [1, \infty]$. Then there exists a positive constant C such that, for any (p, q)-atom a supported on $B := B(x_B, r_B)$, with $x_B \in X$ and $r_B \in (0, \infty)$, and any $x \in X$,*

$$a^*(x) \le C\mathcal{M}(a)(x)\mathbf{1}_{B(x_B, 2A_0 r_B)}(x) + C\left[\frac{r_B}{d(x_B, x)}\right]^{\beta} \frac{[\mu(B)]^{1-\frac{1}{p}}}{V(x_B, x)} \mathbf{1}_{[B(x_B, 2A_0 r_B)]^{\complement}}(x) \tag{4.1}$$

and

$$\|a^*\|_{L^p(X)} \le C, \tag{4.2}$$

where the atom a is viewed as a distribution on $\mathcal{G}^{\eta}_0(\beta, \gamma)$ with $\beta, \gamma \in (\omega(1/p-1), \eta)$.

Proof First, we show (4.1). Let $\varphi \in \mathcal{G}_0^\eta(\beta,\gamma)$ be such that $\|\varphi\|_{\mathcal{G}(x,r,\beta,\gamma)} \le 1$ for some $r \in (0,\infty)$, where $\beta,\ \gamma \in (\omega(1/p-1),\eta)$. When $x \in B(x_B, 2A_0 r_B)$, by Lemma 2.2(v), we find that

$$|\langle a,\varphi\rangle| = \left|\int_X a(y)\varphi(y)\,d\mu(y)\right| \le \int_X |a(y)|\frac{1}{V_r(x)+V(x,y)}\left[\frac{r}{r+d(x,y)}\right]^\gamma d\mu(y)$$
$$\lesssim \mathcal{M}(a)(x),$$

which consequently implies that $a^*(x) \lesssim \mathcal{M}(a)(x)$.

Let $x \notin B(x_B, 2A_0 r_B)$. Then, for any $y \in B$, we have $d(x,x_B) \ge 2A_0 r_B > 2A_0 d(x_B,y)$. Therefore, by the definition of (p,q)-atoms and Definition 2.1(ii), we conclude that

$$\begin{aligned}|\langle a,\varphi\rangle| &= \left|\int_B a(y)\varphi(y)\,d\mu(y)\right| \le \int_B |a(y)||\varphi(y)-\varphi(x_B)|\,d\mu(y)\\ &\le \int_B |a(y)|\left[\frac{d(x_B,y)}{r+d(x,x_B)}\right]^\beta \frac{1}{V_r(x)+V(x,x_B)}\left[\frac{r}{r+d(x,x_B)}\right]^\gamma d\mu(y)\\ &\le \left[\frac{r_B}{d(x_B,x)}\right]^\beta \frac{1}{V(x,x_B)}\|a\|_{L^1(X)} \lesssim \left[\frac{r_B}{d(x_B,x)}\right]^\beta \frac{[\mu(B)]^{1-\frac{1}{p}}}{V(x_B,x)}.\end{aligned}$$

Taking the supremum over all such $\varphi \in \mathcal{G}_0^\eta(\beta,\gamma)$ satisfying $\|\varphi\|_{\mathcal{G}(x,r,\beta,\gamma)} \le 1$ for some $r \in (0,\infty)$, we obtain (4.1).

Now, we use (4.1) to show (4.2). When $q \in (1,\infty]$, from the Hölder inequality and the boundedness of $\mathcal{M}$ on $L^q(X)$, we deduce that

$$\begin{aligned}\int_{B(x_B,2A_0r_B)} [\mathcal{M}(a)(x)]^p\,d\mu(x) &\le [\mu(B(x_B,2A_0r_B))]^{1-p/q}\|\mathcal{M}(a)\|_{L^q(X)}^p\\ &\lesssim [\mu(B)]^{1-p/q}\|a\|_{L^q(X)}^p \lesssim 1.\end{aligned}$$

If $q=1$, then, by $p \in (\omega/(\omega+\eta),1)$ and the boundedness of $\mathcal{M}$ from $L^1(X)$ to $L^{1,\infty}(X)$, we conclude that

$$\begin{aligned}&\int_{B(x_B,2A_0r_B)} [\mathcal{M}(a)(x)]^p\,d\mu(x)\\ &\quad= \int_0^\infty \mu(\{x \in B(x_B,2A_0r_B):\ \mathcal{M}(a)(x)>\lambda\})\,d\lambda^p\\ &\quad\lesssim \int_0^\infty \min\left\{\mu(B),\frac{\|a\|_{L^1(X)}}{\lambda}\right\}d\lambda^p\\ &\quad\lesssim \int_0^{\|a\|_{L^1(X)}/\mu(B)} \mu(B)\,d\lambda^p + \int_{\|a\|_{L^1(X)}/\mu(B)}^\infty \|a\|_{L^1(X)}\lambda^{-1}\,d\lambda^p\\ &\quad\lesssim \|a\|_{L^1(X)}^p[\mu(B)]^{1-p} \lesssim 1.\end{aligned}$$

By the fact $\beta > \omega(1/p - 1)$ and the doubling condition (1.1), we have

$$\int_{d(x,x_B)\geq 2A_0 r_B} \left[\frac{r_B}{d(x_B,x)}\right]^{\beta p} \left[\frac{1}{\mu(B)}\right]^{1-p} \left[\frac{1}{V(x_B,x)}\right]^{p} d\mu(x)$$
$$\lesssim \sum_{k=1}^{\infty} 2^{-k\beta p} 2^{k\omega(1-p)} \int_{2^k A_0 r_B \leq d(x,x_B) < 2^{k+1} A_0 r_B} \frac{1}{V(x_B,x)}\, d\mu(x) \lesssim 1.$$

Combining the last three formulae with (4.1), we obtain (4.2), which then completes the proof of Lemma 4.3. □

Proof of $H^{p,q}_{\mathrm{at}}(X) \subset H^{*,p}(X)$ Assume that $f \in (\mathcal{G}^{\eta}_0(\beta,\gamma))'$ is non-zero and it belongs to $H^{p,q}_{\mathrm{at}}(X)$ with $\beta,\ \gamma \in (\omega(1/p-1),\eta)$. Then $f = \sum_{j=1}^{\infty} \lambda_j a_j$, where $\{a_j\}_{j=1}^{\infty}$ are (p,q)-atoms and $\{\lambda_j\}_{j=1}^{\infty} \subset \mathbb{C}$ satisfy $\sum_{j=1}^{\infty} |\lambda_j|^p \sim \|f\|^p_{H^{p,q}_{\mathrm{at}}(X)}$. By the definition of the grand maximal function, we conclude that, for any $x \in X$,

$$f^*(x) \leq \sum_{j=1}^{\infty} |\lambda_j| a_j^*(x).$$

From this and (4.2), we deduce that

$$\|f^*\|^p_{L^p(X)} \lesssim \sum_{j=1}^{\infty} |\lambda_j|^p \left\|a_j^*\right\|_{L^p(X)} \lesssim \sum_{j=1}^{\infty} |\lambda_j|^p \sim \|f\|^p_{H^{p,q}_{\mathrm{at}}(X)}.$$

This finishes the proof of $H^{p,q}_{\mathrm{at}}(X) \subset H^{*,p}(X)$. □

4.2 Calderón–Zygmund Decomposition of Distributions from $H^{*,p}(X)$

In this section, we obtain a Calderón–Zygmund decomposition of any $f \in H^{*,p}(X)$. First we establish a partition of unity for an open set Ω with $\mu(\Omega) < \infty$.

Proposition 4.4 *Suppose $\Omega \subset X$ is a proper open set with $\mu(\Omega) \in (0,\infty)$ and $A \in [1,\infty)$. For any $x \in \Omega$, let*

$$r(x) := \frac{d(x,\Omega^{\complement})}{2AA_0} \in (0,\infty).$$

Then there exist $L_0 \in \mathbb{N}$ and a sequence $\{x_k\}_{k\in I} \subset \Omega$, where I is a countable index set, such that

(i) *$\{B(x_k, r_k/(5A_0^3))\}_{k\in I}$ is disjoint. Here and hereafter, $r_k := r(x_k)$ for any $k \in I$;*
(ii) *$\bigcup_{k\in I} B(x_k,r_k) = \Omega$ and $B(x_k, Ar_k) \subset \Omega$;*
(iii) *for any $x \in \Omega$, $Ar_k \leq d(x,\Omega^{\complement}) \leq 3AA_0^2 r_k$ whenever $x \in B(x_k,r_k)$ and $k \in I$;*
(iv) *for any $k \in I$, there exists $y_k \notin \Omega$ such that $d(x_k,y_k) < 3AA_0 r_k$;*

(v) *for any given $k \in I$, the number of balls $B(x_j, Ar_j)$ that intersect $B(x_k, Ar_k)$ is at most L_0;*

(vi) *if, in addition, Ω is bounded, then, for any $\sigma \in (0, \infty)$, the set $\{k \in I : r_k > \sigma\}$ is finite.*

Proof We show this proposition by borrowing some ideas from [49, pp. 15–16]. Let $\epsilon := (5A_0^3)^{-1}$ and $\{B(x, \epsilon r(x))\}_{x\in\Omega}$ be a covering of Ω. Now we pick the maximal disjoint subcollection of $\{B(x, \epsilon r(x))\}_{x\in\Omega}$, denoted by $\{B_k\}_{k\in I}$, which is at most countable, because of (1.1) and $\mu(\Omega) \in (0, \infty)$. For any $k \in I$, denote the center of B_k by x_k and $r(x_k)$ by r_k. Then we obtain (i).

Properties (iii) and (iv) can be shown by the definition of r_k, the details being omitted. Now we show (ii). Obviously, $B(x_k, Ar_k) \subset \Omega$ for any $k \in I$. It suffices to prove that $\Omega \subset \bigcup_{k\in I} B(x_k, r_k)$. For any $x \in \Omega$, since $\{B_k\}_{k\in I}$ is maximal, it then follows that there exists $k \in I$ such that $B(x_k, \epsilon r_k) \cap B(x, \epsilon r(x)) \neq \emptyset$. We claim that $r_k \geq r(x)/(4A_0^2)$. If not, then $r_k < r(x)/(4A_0^2)$. Suppose that $x_0 \in B(x_k, \epsilon r_k) \cap B(x, \epsilon r(x))$. Then, for any $y \in B(x_k, 3AA_0 r_k)$, we have

$$\begin{aligned} d(y, x) &\leq A_0[d(y, x_0) + d(x_0, x)] \leq A_0^2[d(y, x_k) + d(x_k, x_0)] + A_0 d(x_0, x) \\ &\leq 6AA_0^3 r_k + A_0\epsilon r(x) \\ &\leq \frac{3}{2}AA_0 r(x) + \frac{1}{5}AA_0 r(x) = \frac{17}{10}AA_0 r(x) \end{aligned}$$

and hence $B(x_k, 3AA_0 r_k) \subset B(x, \frac{17}{10}AA_0 r(x)) \subset \Omega$, which contradicts to (iv). This proves the claim. Further, by the fact that $r(x) \leq 4A_0^2 r_k$, we have

$$d(x, x_k) \leq A_0[d(x, x_0) + d(x_0, x_k)] < A_0\epsilon r(x) + A_0\epsilon r_k \leq 5A_0^3\epsilon r_k = r_k,$$

that is, $x \in B(x_k, r_k)$. This finishes the proof of (ii).

Now we prove (v). Fix $k \in I$. Suppose that $B(x_j, Ar_j) \cap B(x_k, Ar_k) \neq \emptyset$. We claim that $r_j \leq 8A_0^2 r_k$. If not, then $r_j > 8A_0^2 r_k$. Choose $y_0 \in B(x_j, Ar_j) \cap B(x_k, Ar_k)$. For any $y \in B(x_k, 3AA_0 r_k)$, we have

$$\begin{aligned} d(y, x_j) &\leq A_0[d(y, y_0) + d(y_0, x_j)] \leq A_0^2[d(y, x_k) + d(x_k, y_0)] + A_0 d(y_0, x_j) \\ &\leq 3AA_0^3 r_k + AA_0^2 r_k + AA_0 r_j \leq \frac{3}{2}AA_0 r_j, \end{aligned}$$

which further implies that $y \in B(x_j, \frac{3}{2}AA_0 r_j)$. Therefore, $B(x_k, 3AA_0 r_k) \subset B(x_j, \frac{3}{2}AA_0 r_j) \subset \Omega$, which contradicts to (iv), Thus, we have $r_j \leq 8A_0^2 r_k$. By symmetry, we also have $r_k \leq 8A_0^2 r_j$. Let

$$\mathcal{J} := \{j \in I : \; B(x_j, Ar_j) \cap B(x_k, Ar_k) \neq \emptyset\}.$$

Then, for any $j \in \mathcal{J}$, $d(x_j, x_k) < AA_0(r_j + r_k) \leq 9AA_0^3 r_k$, which further implies that

$$B\left(x_j, (5A_0^3)^{-1} r_j\right) \subset B\left(x_k, A_0\left[d(x_j, x_k) + (5A_0^3)^{-1} r_j\right]\right) \subset B(x_k, 11AA_0^4 r_k).$$

Then, from the fact $d(x_j, x_k) \lesssim \min\{r_j, r_k\}$ and (1.1), we deduce that

$$\mu\left(B\left(x_j, (5A_0^3)^{-1}r_j\right)\right) \sim \mu(B(x_j, r_j)) \sim \mu(B(x_k, r_k)) \sim \mu(B(x_k, 11AA_0^4 r_k))$$

with the positive equivalence constants depending on A. Thus, we obtain (v) by (i).

Finally we prove (vi). Since Ω is bounded, it follows that there exist $x_0 \in X$ and $R \in (0, \infty)$ such that $\Omega \subset B(x_0, R)$. If (vi) fails, then there exists $\sigma_0 \in (0, \infty)$ such that $\mathcal{K} := \{k \in I : r_k > \sigma_0 R\}$ is infinite. Then, for any $k \in \mathcal{K}$,

$$\mu(B(x_k, r_k/(5A_0^3))) \sim \mu(B(x_k, \sigma_0 R)) \gtrsim \mu(B(x_0, R)) \gtrsim \mu(\Omega) > 0.$$

By this and (i), we have $\mu(\Omega) \geq \sum_{k\in\mathcal{K}} \mu(B(x_k, r_k/(5A_0^3))) = \infty$. That is a contradiction. This proves (vi) and hence finishes the proof of Proposition 4.4. □

Proposition 4.5 *Let $\Omega \subset X$ be an open set and $\mu(\Omega) < \infty$. Suppose that sequences $\{x_k\}_{k\in I}$ and $\{r_k\}_{k\in I}$ are as in Proposition 4.4 with $A := 16A_0^4$. Then there exist non-negative functions $\{\phi_k\}_{k\in I}$ such that*

(i) *for any $k \in I$, $0 \leq \phi_k \leq 1$ and $\operatorname{supp} \phi_k \subset B(x_k, 2A_0 r_k)$;*
(ii) $\sum_{k\in I} \phi_k = \mathbf{1}_\Omega$;
(iii) *for any $k \in I$, $\phi_k \geq L_0^{-1}$ in $B(x_k, r_k)$, where L_0 is as in Proposition 4.4;*
(iv) *there exists a positive constant C such that, for any $k \in I$, $\|\phi_k\|_{\mathcal{G}(x_k, r_k, \eta, \eta)} \leq C V_{r_k}(x_k)$.*

Proof By [1, Corollary 4.2], for any $k \in I$, we find a function ψ_k such that $\mathbf{1}_{B(x_k,r_k)} \leq \psi_k \leq \mathbf{1}_{B(x_k,2A_0 r_k)}$ and $\|\psi_k\|_{\dot{C}^\eta(X)} \lesssim r_k^{-\eta}$. Here and hereafter, for any $s \in (0, \eta]$ and a measurable function f, define

$$\|f\|_{\dot{C}^s(X)} := \sup_{x\neq y} \frac{|f(x) - f(y)|}{[d(x, y)]^\beta}.$$

Since $A \geq 2A_0$, from (ii) and (v) of Proposition 4.4, it follows that, for any $x \in \Omega$, $1 \leq \sum_{k\in I} \psi_k(x) \leq L_0$. For any $k \in I$ and $x \in X$, let

$$\phi_k(x) := \begin{cases} \psi_k(x)\left[\sum_{j\in I} \psi_j(x)\right]^{-1} & \text{when } x \in \Omega, \\ 0, & \text{when } x \notin \Omega. \end{cases}$$

Then, for any $k \in I$, we have $0 \leq \phi_k \leq 1$, $\operatorname{supp} \phi_k \subset B(x_k, 2A_0 r_k)$ and $\sum_{k\in I} \phi_k(x) = 1$ when $x \in \Omega$. Moreover, for any $k \in I$, we have $\phi_k \geq L_0^{-1}$ in $B(x_k, r_k)$. Thus, we prove (i), (ii) and (iii).

It remains to prove (iv). Fix $k \in I$. For any $y \in X$, we have

$$|\phi_k(y)| \leq \mathbf{1}_{B(x_k,2A_0 r_k)}(y) \lesssim V_{r_k}(x_k) \frac{1}{V_{r_k}(x_k) + V(x_k, y)} \left[\frac{r_k}{r_k + d(x_k, y)}\right]^\eta.$$

Now we prove that ϕ_k satisfies the regularity condition. Suppose that $d(y, y') \le (2A_0)^{-1}[r_k + d(x_k, y)]$. If $|\phi_k(y) - \phi_k(y')| \neq 0$, then $d(x_k, y) < (3A_0)^2 r_k$. If not, then $d(x_k, y) \ge (3A_0)^2 r_k$, so that $\phi_k(y) = 0$ and

$$d(y', x_k) \ge A_0^{-1} d(x_k, y) - d(y, y') \ge (2A_0)^{-1} d(x_k, y) - (2A_0)^{-1} r_k > 2A_0 r_k$$

and hence $\phi_k(y') = 0$, which contradicts to $|\phi_k(y) - \phi_k(y')| \neq 0$. Notice that $\psi_k(y')|\psi_j(y) - \psi_j(y')| \neq 0$ implies that $y' \in B(x_k, 2A_0 r_k)$ and also y or y' belongs to $B(x_j, 2A_0 r_j)$, which further implies that $B(x_k, Ar_k) \cap B(x_j, Ar_j) \neq \emptyset$. Then, by the proof of Proposition 4.4(v), the number of j satisfying $\psi_k(y')|\psi_j(y) - \psi_j(y')| \neq 0$ is at most L_0 and $r_j \sim r_k$. Therefore,

$$\begin{aligned}
|\phi_k(y) - \phi_k(y')| &= \left| \frac{\psi_k(y)}{\sum_{j \in I} \psi_j(y)} - \frac{\psi_k(y')}{\sum_{j \in I} \psi_j(y')} \right| \\
&\le \frac{|\psi_k(y) - \psi_k(y')|}{\sum_{j \in I} \psi_j(y)} + \frac{\psi_k(y') \sum_{j \in I} |\psi_j(y) - \psi_j(y')|}{[\sum_{j \in I} \psi_j(y)][\sum_{j \in I} \psi_j(y')]} \\
&\lesssim \left[\frac{d(y, y')}{r_k}\right]^\eta + \sum_{\{j \in I:\ B(x_k, Ar_k) \cap B(x_j, Ar_j) \neq \emptyset\}} \left[\frac{d(y, y')}{r_j}\right]^\eta \\
&\lesssim \left[\frac{d(y, y')}{r_k}\right]^\eta \\
&\sim V_{r_k}(x_k) \left[\frac{d(y, y')}{r_k + d(x_k, y)}\right]^\eta \frac{1}{V_{r_k}(x_k) + V(x_k, y)} \left[\frac{r_k}{r_k + d(x_k, y)}\right]^\eta.
\end{aligned}$$

Then we obtain the desired regularity condition of ϕ_k. This finishes the proof of (iv) and hence of Proposition 4.5. □

Assume that $f \in (\mathcal{G}_0^\eta(\beta, \gamma))'$ belongs to $f \in H^{*,p}(X)$, where $p \in (\omega/(\omega + \eta), 1]$ and $\beta,\ \gamma \in (\omega(1/p - 1), \eta)$. To obtain the Calderón–Zygmund decomposition of f, we apply Propositions 4.4 and 4.5 to the level set $\{x \in X :\ f^*(x) > \lambda\}$ with $\lambda \in (0, \infty)$. The encountering problem is that such a level set may not be open even in the case that d is a metric. To solve this problem in the case that d is a metric, a variant of the notion of the space of test functions is adopted in [20, Definition 2.5] so that to ensure that the level set is open (see [20, Remark 2.9]). Here, we borrow some ideas from [20].

By the proof of [42, Theorem 2], we know that there exist $\theta \in (0, 1)$ and a metric d' such that $d' \sim d^\theta$. For any $x \in X$ and $r \in (0, \infty)$, define the *d'-ball* $B'(x, r) := \{y \in X :\ d'(x, y) < r\}$. Then (X, d', μ) is a doubling metric measure space. Moreover, for any $x,\ y \in X$ and $r \in (0, \infty)$, we have

$$\mu(B(y, r + d(x, y))) \sim \mu\left(B'\left(y, [r + d(x, y)]^\theta\right)\right) \sim \mu\left(B'\left(y, r^\theta + d'(x, y)\right)\right),$$

where the positive equivalence constants are independent of x and r. Using the metric d', we introduce a variant of the space of test functions as follows.

Definition 4.6 For any $x \in X$, $\rho \in (0,\infty)$ and $\beta',\ \gamma' \in (0,\infty)$, define $G(x,\rho,\beta',\gamma')$ to be the set of all functions f satisfying that there exists a positive constant C such that

(i) (the *size condition*) for any $y \in X$,

$$|f(y)| \le C \frac{1}{\mu(B'(y,\rho+d'(x,y)))}\left[\frac{\rho}{\rho+d'(x,y)}\right]^{\gamma'};$$

(ii) (the *regularity condition*) for any $y,\ y' \in X$ satisfying $d(y,y') \le [\rho + d'(x,y)]/2$, then

$$|f(y)-f(y')| \le C\left[\frac{d'(y,y')}{\rho+d'(y,y')}\right]^{\beta'}\frac{1}{\mu(B'(y,\rho+d'(x,y)))}\left[\frac{\rho}{\rho+d'(x,y)}\right]^{\gamma'}.$$

Also, define

$$\|f\|_{G(x,\rho,\beta',\gamma')} := \inf\{C \in (0,\infty) : \text{ (i) and (ii) hold true}\}.$$

By the previous argument, we find that $\mathcal{G}(x,r,\beta,\gamma) = G(x,r^\theta,\beta/\theta,\gamma/\theta)$ with equivalent norms, where the positive equivalence constants are independent of x and r. For any $\beta,\ \gamma \in (0,\eta)$ and $f \in (\mathcal{G}_0^\eta(\beta,\gamma))'$, define the *modified grand maximal function* of f by setting, for any $x \in X$,

$$f^\star(x) := \sup\left\{|\langle f,\varphi\rangle| : \varphi \in \mathcal{G}_0^\eta(\beta,\gamma) \text{ with } \|\varphi\|_{G(x,r^\theta,\beta/\theta,\gamma/\theta)} \le 1 \text{ for some } r \in (0,\infty)\right\}.$$

Then $f^\star \sim f^*$ pointwisely on X. For any $\lambda \in (0,\infty)$ and $j \in \mathbb{Z}$, define

$$\Omega_\lambda := \{x \in X : f^\star(x) > \lambda\} \quad \text{and} \quad \Omega^j := \Omega_{2^j}.$$

By the argument used in [20, Remark 2.9(ii)], we find that Ω_λ is *open* under the topology induced by d', so is it under the topology induced by d.

Now suppose that $p \in (\omega/(\omega+\eta), 1]$, $\beta,\gamma \in (\omega(1/p-1),\eta)$ and $f \in H^{*,p}(X)$. Then $f^\star \in L^p(X)$ and every Ω^j with $j \in \mathbb{Z}$ has finite measure. Consequently, there exist $\{x_k^j\}_{k\in I_j} \subset X$ with I_j being a countable index set, $\{r_k^j\}_{k\in I_j} \subset (0,\infty)$, $L_0 \in \mathbb{N}$ and a sequence $\{\phi_k^j\}_{k\in I_j}$ of non-negative functions satisfying all the conclusions of Propositions 4.4 and 4.5. For any $j \in \mathbb{Z}$ and $k \in I_j$, define Φ_k^j by setting, for any $\varphi \in \mathcal{G}_0^\eta(\beta,\gamma)$ and $x \in X$,

$$\Phi_k^j(\varphi)(x) := \phi_k^j(x)\left[\int_X \phi_k^j(z)\,d\mu(z)\right]^{-1}\int_X [\varphi(x)-\varphi(z)]\phi_k^j(z)\,d\mu(z).$$

It can be seen that Φ_k^j is bounded on $\mathcal{G}_0^\eta(\beta,\gamma)$ with operator norm depending on j and k; see [20, Lemma 4.9]. Thus, it makes sense to define a distribution b_k^j on $\mathcal{G}_0^\eta(\beta,\gamma)$ by setting, for any $\varphi \in \mathcal{G}_0^\eta(\beta,\gamma)$,

$$\left\langle b_k^j, \varphi\right\rangle := \left\langle f, \Phi_k^j(\varphi)\right\rangle. \tag{4.3}$$

To estimate $(b_k^j)^*$, we have the following result. For its proof, see, for example, [37, Lemma 3.7].

Proposition 4.7 *For any $j \in \mathbb{Z}$ and $k \in I_j$, b_k^j is defined as in* (4.3). *Then there exists a positive constant C such that, for any $j \in \mathbb{Z}$, $k \in I_j$ and $x \in X$,*

$$\left(b_k^j\right)^*(x) \le C2^j \frac{\mu(B(x_k^j, r_k^j))}{\mu(B(x_k^j, r_k^j)) + V(x_k^j, x)} \left[\frac{r_k^j}{r_k^j + d(x_k^j, x)}\right]^\beta \mathbf{1}_{[B(x_k^j, 16A_0^4 r_k^j)]^\complement}(x)$$
$$+ Cf^*(x)\mathbf{1}_{B(x_k^j, 16A_0^4 r_k^j)}(x).$$

The next lemma is exactly [20, Lemma 4.10]. The proof remains true if d is a quasi-metric and μ does not satisfy the reverse doubling condition.

Lemma 4.8 *Let $\beta \in (0, \infty)$, $p \in (\omega/(\omega+\beta), \infty)$, $L_0 \in \mathbb{N}$ and I be a countable index set. Then there exists a positive constant C such that, for any sequences $\{x_k\}_{k\in I} \subset X$ and $\{r_k\}_{k\in I} \subset (0, \infty)$ satisfying $\sum_{k\in I} \mathbf{1}_{B(x_k, r_k)} \le L_0$,*

$$\int_X \left\{\sum_{k\in I} \frac{V_{r_k}(x_k)}{V_{r_k}(x_k) + V(x_k, x)} \left[\frac{r_k}{r_k + d(x_k, x)}\right]^\beta\right\}^p d\mu(x) \le C\mu\left(\bigcup_{k\in I} B(x_k, r_k)\right).$$

Then, by Proposition 4.7 and Lemma 4.8, we have the following result.

Proposition 4.9 *Let $p \in (\omega/(\omega + \eta), 1]$. For any $j \in \mathbb{Z}$ and $k \in I_j$, let b_k^j be as in* (4.3). *Then there exists a positive constant C such that, for any $j \in \mathbb{Z}$,*

$$\int_X \sum_{k\in I_j} \left[\left(b_k^j\right)^*(x)\right]^p d\mu(x) \le C \left\|f^*\mathbf{1}_{\Omega^j}\right\|_{L^p(X)}^p; \tag{4.4}$$

moreover, there exists $b^j \in H^{*,p}(X)$ such that $b^j = \sum_{k\in I_j} b_k^j$ in $H^{*,p}(X)$ and, for any $x \in X$,

$$(b^j)^*(x) \le C2^j \sum_{k\in I_j} \frac{\mu(B(x_k^j, r_k^j))}{\mu(B(x_k^j, r_k^j)) + V(x_k^j, x)} \left[\frac{r_k^j}{r_k^j + d(x_k^j, x)}\right]^\beta + Cf^*(x)\mathbf{1}_{\Omega^j}(x); \tag{4.5}$$

if $g^j := f - b^j$ for any $j \in \mathbb{Z}$, then, for any $x \in X$,

$$(g^j)^*(x) \le C2^j \sum_{k\in I_j} \frac{\mu(B(x_k^j, r_k^j))}{\mu(B(x_k^j, r_k^j)) + V(x_k^j, x)} \left[\frac{r_k^j}{r_k^j + d(x_k^j, x)}\right]^\beta + Cf^*(x)\mathbf{1}_{(\Omega^j)^\complement}(x). \tag{4.6}$$

Proof Fix $j \in \mathbb{Z}$. We first prove (4.4). Indeed, by Proposition 4.7, we find that

$$\begin{aligned}&\int_X \sum_{k\in I_j} \left[\left(b_k^j\right)^*(x)\right]^p d\mu(x)\\ &\lesssim 2^{jp}\int_X \sum_{k\in I_j}\left\{\frac{\mu(B(x_k^j,r_k^j))}{\mu(B(x_k^j,r_k^j))+V(x_k^j,x)}\left[\frac{r_k^j}{r_k^j+d(x_k^j,x)}\right]^\beta\right\}^p d\mu(x)\\ &\quad+\int_{\bigcup_{k\in I_j} B(x_k^j,16A_0^4r_k^j)} [f^*(x)]^p\,d\mu(x).\end{aligned}$$

By Proposition 4.4(ii), we have $\Omega^j = \bigcup_{k\in I_j} B(x_k^j, 16A_0^4 r_k^j)$. Applying this and Lemma 4.8, the first term in the right-hand side of the above formula is bounded by a harmless positive constant multiple of $2^{jp}\mu(\Omega^j)$. Combining this with $f^* \sim f^\star$ implies that

$$\int_X \sum_{k\in I_j} \left[\left(b_k^j\right)^*(x)\right]^p d\mu(x) \lesssim 2^{jp}\mu\left(\Omega^j\right)+\int_{\Omega^j}[f^*(x)]^p\,d\mu(x) \lesssim \|f^*\mathbf{1}_{\Omega^j}\|_{L^p(X)}^p,$$

which proves (4.4).

Next we prove (4.5). By (4.4), the dominated convergence theorem and the completeness of $H^{*,p}(X)$ (see Proposition 3.1), we know that there exists $b^j \in H^{*,p}(X)$ such that $b^j = \sum_{k\in I_j} b_k^j$ in $H^{*,p}(X)$. Moreover, from Proposition 4.7 and $\Omega^j = \bigcup_{k\in I_j} B(x_k^j, 16A_0^4 r_k^j)$, we deduce that, for any $x \in X$,

$$\begin{aligned}(b^j)^*(x) &\le \sum_{k\in I_j}\left(b_k^j\right)^*(x)\\ &\lesssim 2^j\sum_{k\in I_j}\frac{\mu(B(x_k^j,r_k^j))}{\mu(B(x_k^j,r_k^j))+V(x_k^j,x)}\left[\frac{r_k^j}{r_k^j+d(x_k^j,x)}\right]^\beta + f^*(x)\mathbf{1}_{\Omega^j}(x).\end{aligned}$$

This finishes the proof of (4.5).

It remains to prove (4.6). If $x \in (\Omega^j)^\complement$, then, by (4.5), we conclude that

$$\begin{aligned}(g^j)^*(x) &\le f^*(x)+(b^j)^*(x)\\ &\lesssim 2^j\sum_{k\in I_j}\frac{\mu(B(x_k^j,r_k^j))}{\mu(B(x_k^j,r_k^j))+V(x_k^j,x)}\left[\frac{r_k^j}{r_k^j+d(x_k^j,x)}\right]^\beta + f^*(x),\end{aligned}$$

as desired.

Now we consider the case $x \in \Omega^j$. According to Proposition 4.4(v), for any $n \in I_j$, we choose a point $y_n^j \notin \Omega^j$ satisfying $32A_0^5 r_n^j \le d(x_n^j, y_n^j) < 48A_0^5 r_n^j$. Since $x \in \Omega^j$, it follows that there exists $k_0 \in I_j$ such that $x \in B(x_{k_0}^j, r_{k_0}^j)$. Let $\mathcal{J}$ be the set of all $n \in I_j$ such that $B(x_n^j, 16A_0^4 r_n^j) \cap B(x_{k_0}^j, 16A_0^4 r_{k_0}^j) \neq \emptyset$. Then, by the proof of Proposition 4.4(v), $\#\mathcal{J} \le L_0$ and $r_n^j \sim r_{k_0}^j$ whenever $n \in \mathcal{J}$.

Suppose that $\varphi \in \mathcal{G}_0^\eta(\beta,\gamma)$ with $\|\varphi\|_{\mathcal{G}(x,r,\beta,\gamma)} \le 1$ for some $r \in (0,\infty)$. We then estimate $\langle g^j, \varphi\rangle$ by considering the cases $r \le r_{k_0}^j$ and $r > r_{k_0}^j$, respectively.

Case 1) $r \le r_{k_0}^j$. In this case, we write

$$\begin{aligned}\langle g^j, \varphi\rangle &= \langle f, \varphi\rangle - \sum_{n\in I_j}\langle b_n^j, \varphi\rangle = \langle f, \varphi\rangle - \sum_{n\in\mathcal{J}}\langle b_n^j, \varphi\rangle - \sum_{n\notin\mathcal{J}}\langle b_n^j, \varphi\rangle \\ &= \langle f, \widetilde{\varphi}\rangle - \sum_{n\in\mathcal{J}}\langle f, \widetilde{\varphi}_n\rangle - \sum_{n\notin\mathcal{J}}\langle b_n^j, \varphi\rangle,\end{aligned}$$

where $\widetilde{\varphi} := (1 - \sum_{n\in\mathcal{J}} \phi_n^j)\varphi$ and, for any $n \in \mathcal{J}$,

$$\widetilde{\varphi}_n := \phi_n^j \left[\int_X \phi_n^j(z)\,d\mu(z)\right]^{-1} \int_X \varphi(z)\phi_n^j(z)\,d\mu(z).$$

We first consider the term $\sum_{n\notin\mathcal{J}}\langle b_n^j, \varphi\rangle$. Indeed, from $x \in B(x_{k_0}^j, r_{k_0}^j)$, it follows that $x \notin B(x_n^j, 16A_0^4 x_n^j)$ when $n \notin \mathcal{J}$. Applying Proposition 4.7 implies that

$$\left|\left\langle b_n^j, \varphi\right\rangle\right| \le \left|\left(b_n^j\right)^*(x)\right| \lesssim 2^j \frac{\mu(B(x_n^j, r_n^j))}{\mu(B(x_n^j, r_n^j)) + V(x_n^j, x)}\left[\frac{r_n^j}{r_n^j + d(x_n^j, x)}\right]^\beta,$$

and hence

$$\sum_{n\notin\mathcal{J}}\left|\left\langle b_n^j, \varphi\right\rangle\right| \lesssim 2^j \sum_{n\notin\mathcal{J}} \frac{\mu(B(x_n^j, r_n^j))}{\mu(B(x_n^j, r_n^j)) + V(x_n^j, x)}\left[\frac{r_n^j}{r_n^j + d(x_n^j, x)}\right]^\beta,$$

as desired.

Next we consider the term $\sum_{n\in\mathcal{J}}\langle f, \widetilde{\varphi}_n\rangle$. Notice that $\|\widetilde{\varphi}_n\|_{\mathcal{G}(x_n^j, r_n^j,\beta,\gamma)} \lesssim 1$. By $d(x_n^j, y_n^j) \sim r_n^j$, we then have $\|\widetilde{\varphi}_n\|_{\mathcal{G}(y_n^j, r_n^j,\beta,\gamma)} \lesssim 1$. Therefore,

$$|\langle f, \widetilde{\varphi}_n\rangle| \lesssim f^*\left(y_n^j\right) \sim f^\star\left(y_n^j\right) \lesssim 2^j \sim 2^j \frac{\mu(B(x_n^j, r_n^j))}{\mu(B(x_n^j, r_n^j)) + V(x_n^j, x)}\left[\frac{r_n^j}{r_n^j + d(x_n^j, x)}\right]^\beta,$$

where, in the last step, we used the facts that $x \in B(x_{k_0}^j, r_{k_0}^j)$ and $d(x_n^j, x_{k_0}^j) \lesssim r_n^j + r_{k_0}^j \sim r_n^j$ whenever $n \in \mathcal{J}$. Then, summing all $n \in \mathcal{J}$, we obtain the desired estimate.

Finally, we consider the term $\langle f, \widetilde{\varphi}\rangle$. Since $\varphi \in \mathcal{G}_0^\eta(\beta,\gamma)$, it is easy to see that $\widetilde{\varphi} \in \mathcal{G}_0^\eta(\beta,\gamma)$. Once we have proved that

$$\|\widetilde{\varphi}\|_{\mathcal{G}(y_{k_0}^j, r_{k_0}^j,\beta,\gamma)} \lesssim 1, \tag{4.7}$$

then

$$|\langle f,\widetilde{\varphi}\rangle|\lesssim f^*\left(y_{k_0}^j\right)\sim f^\star\left(y_{k_0}^j\right)\lesssim 2^j\sim 2^j\frac{\mu(B(x_{k_0}^j,r_{k_0}^j))}{\mu(B(x_{k_0}^j,r_{k_0}^j))+V(x_{k_0}^j,x)}\left[\frac{r_{k_0}^j}{r_{k_0}^j+d(x_{k_0}^j,x)}\right]^\beta,$$

as desired.

To prove (4.7), we first consider the size condition. For any $z\in B(x_{k_0}^j,16A_0^4r_{k_0}^j)$, by Proposition 4.5, we have $\sum_{n\in\mathcal{J}}\phi_n^j(z)=\sum_{n\in I_j}\phi_n^j(z)=1$ and hence $\widetilde{\varphi}(z)=0$. When $d(z,x_{k_0}^j)\ge 16A_0^4r_{k_0}^j$, by the fact $d(x_{k_0}^j,z)\ge 2A_0d(x,x_{k_0}^j)$, we have

$$\begin{aligned}r_{k_0}^j+d\left(z,y_{k_0}^j\right)&\le r_{k_0}^j+A_0\left[d\left(z,x_{k_0}^j\right)+d\left(x_{k_0}^j,y_{k_0}^j\right)\right]\le(2A_0)^6\left[r_{k_0}^j+d\left(z,x_{k_0}^j\right)\right]\\&\le(2A_0)^7d\left(z,x_{k_0}^j\right)\le(2A_0)^8d(x,z)\le(2A_0)^8[r+d(x,z)]\end{aligned}\tag{4.8}$$

and hence $\mu(B(y_{k_0}^j,r_{k_0}^j))+V(y_{k_0}^j,z)\lesssim V_r(x)+V(x,z)$, which, together with the size condition of φ and the fact that $r\le r_{k_0}^j$, further implies that

$$\begin{aligned}|\widetilde{\varphi}(z)|\le|\varphi(z)|&\le\frac{1}{V_r(x)+V(x,z)}\left[\frac{r}{r+d(x,z)}\right]^\gamma\\&\lesssim\frac{1}{\mu(B(y_{k_0}^j,r_{k_0}^j))+V(y_{k_0}^j,z)}\left[\frac{r_{k_0}^j}{r_{k_0}^j+d(y_{k_0}^j,z)}\right]^\gamma.\end{aligned}$$

This finishes the proof of the size condition.

Now we consider the regularity of $\widetilde{\varphi}$. Suppose that $z,\ z'\in X$ with $d(z,z')\le(2A_0)^{-1}[r_{k_0}^j+d(z,y_{k_0}^j)]$. Due to the size condition, we only need to consider the case $d(z,z')\le(2A_0)^{-9}[r_{k_0}^j+d(z,y_{k_0}^j)]$. If $\widetilde{\varphi}(z)-\widetilde{\varphi}(z')\ne 0$, then either $d(z,x_{k_0}^j)\ge 16A_0^4r_{k_0}^j$ or $d(z',x_{k_0}^j)\ge 16A_0^4r_{k_0}^j$, which always implies that $d(z,x_{k_0}^j)\ge 8A_0^3r_{k_0}^j$.

Indeed, if $d(z,x_{k_0}^j)<8A_0^3r_{k_0}^j$, then $d(z,y_{k_0}^j)\le A_0[d(z,x_{k_0}^j)+d(x_{k_0}^j,y_{k_0}^j)]<(2A_0)^6r_{k_0}^j$ and hence $d(z,z')\le(2A_0)^{-2}r_{k_0}^j$, which further implies that $d(z',x_{k_0}^j)\le A_0[d(z',z)+d(z,x_{k_0}^j)]<16A_0^4r_{k_0}^j$ and it is a contraction.

Notice that $d(z,x_{k_0}^j)\ge 8A_0^3r_{k_0}^j$, which, together with an argument as in the estimation of (4.8), implies $r_{k_0}^j+d(z,y_{k_0}^j)\le(2A_0)^8[r+d(z,x)]$, so that $d(z,z')\le(2A_0)^{-1}[r+d(z,x)]$. By the definition of $\widetilde{\varphi}$, we find that

$$\left|\widetilde{\varphi}(z)-\widetilde{\varphi}(z')\right|\le\left[1-\sum_{n\in\mathcal{J}}\phi_n^j(z)\right]|\varphi(z)-\varphi(z')|+|\varphi(z')|\sum_{n\in\mathcal{J}}\left|\phi_n^j(z)-\phi_n^j(z')\right|.$$

Using the regularity condition of φ and the fact $d(z,z')\le(2A_0)^{-1}[r+d(z,x)]$, we obtain

$$\begin{aligned}
&\left[1-\sum_{n\in\mathcal{J}}\phi_n^j(z)\right]|\varphi(z)-\varphi(z')|\\
&\quad\lesssim\left[\frac{d(z,z')}{r+d(z,x)}\right]^\beta\frac{1}{V_r(x)+V(x,z)}\left[\frac{r}{r+d(x,z)}\right]^\gamma\\
&\quad\lesssim\left[\frac{d(z,z')}{r_{k_0}^j+d(z,y_{k_0}^j)}\right]^\beta\frac{1}{\mu(B(y_{k_0}^j,r_{k_0}^j))+V(y_{k_0}^j,z)}\left[\frac{r_{k_0}^j}{r_{k_0}^j+d(y_{k_0}^j,z)}\right]^\gamma,
\end{aligned}$$

where, in the last step, we used $r_{k_0}^j+d(z,y_{k_0}^j)\lesssim r+d(z,x)$, $r\le r_{k_0}^j$, $x\in B(x_{k_0}^j,r_{k_0}^j)$ and $d(y_{k_0}^j,x_{k_0}^j)\sim r_{k_0}^j$.

We now estimate $|\varphi(z')|\sum_{n\in\mathcal{J}}|\phi_n^j(z)-\phi_n^j(z')|$. If $\varphi(z')|\phi_n^j(z)-\phi_n^j(z')|\neq 0$, then $z'\notin B(x_{k_0}^j,16A_0^4r_{k_0}^j)$ and z or z' belongs to $B(x_n^j,2A_0r_n^j)$. When $n\in\mathcal{J}$, we have $r_n^j\sim r_{k_0}^j\sim r_{k_0}^j+d(y_{k_0}^j,z)$. Also, $r_{k_0}^j+d(z,y_{k_0}^j)\lesssim r+d(z,x)\sim r+d(z',x)$. By these, $\#\mathcal{J}\le L_0$ and $r\le r_{k_0}^j$, we conclude that

$$\begin{aligned}
&|\varphi(z')|\sum_{n\in\mathcal{J}}\left|\phi_n^j(z)-\phi_n^j(z')\right|\\
&\quad\lesssim\frac{1}{V_r(x)+V(x,z')}\left[\frac{r}{r+d(z,x)}\right]^\gamma\sum_{n\in\mathcal{J}}\left[\frac{d(z,z')}{r_n^j}\right]^\eta\\
&\quad\lesssim\left[\frac{d(z,z')}{r_{k_0}^j+d(y_{k_0}^j,z)}\right]^\beta\frac{1}{\mu(B(y_{k_0}^j,r_{k_0}^j))+V(y_{k_0}^j,z)}\left[\frac{r_{k_0}^j}{r_{k_0}^j+d(y_{k_0}^j,z)}\right]^\gamma.
\end{aligned}$$

This finishes the proof of the regularity condition and hence of (4.7). Thus, we complete the proof of Case 1).

Case 2) $r>r_{k_0}^j$. In this case, we write

$$\left|\left\langle g^j,\varphi\right\rangle\right|\le|\langle f,\varphi\rangle|+\sum_{n\in\mathcal{J}}\left|\left\langle b_n^j,\varphi\right\rangle\right|+\sum_{n\notin\mathcal{J}}\left|\left\langle b_n^j,\varphi\right\rangle\right|.$$

The estimation of $\sum_{n\notin\mathcal{J}}|\langle b_n^j,\varphi\rangle|$ has already been given in Case 1).

From $x\in B(x_{k_0}^j,r_{k_0}^j)$ and $d(y_{k_0}^j,x_{k_0}^j)\sim r_{k_0}^j\lesssim r$, it follows that $\|\varphi\|_{\mathcal{G}(y_{k_0}^j,r,\beta,\gamma)}\lesssim 1$ and hence

$$|\langle f,\varphi\rangle|\lesssim f^*\left(y_{k_0}^j\right)\lesssim 2^j\sim 2^j\frac{\mu(B(x_{k_0}^j,r_{k_0}^j))}{\mu(B(x_{k_0}^j,r_{k_0}^j))+V(x_{k_0}^j,x)}\left[\frac{r_{k_0}^j}{r_{k_0}^j+d(x_{k_0}^j,x)}\right]^\beta.$$

If $n\in\mathcal{J}$, then $r_n^j\sim r_{k_0}^j$ and hence $d(y_n^j,x_{k_0}^j)\lesssim r_{k_0}^j$. This, together with the fact $r_{k_0}^j<r$ and $x\in B(x_{k_0}^j,r_{k_0}^j)$, implies that $\|\varphi\|_{\mathcal{G}(y_n^j,r,\beta,\gamma)}\lesssim 1$. Thus, by Proposition 4.7, we have

$$\sum_{n\in\mathcal{J}}\left|\left\langle b_n^j,\varphi\right\rangle\right| \lesssim \sum_{n\in\mathcal{J}}\left(b_n^j\right)^*\left(y_n^j\right)$$
$$\lesssim 2^j\sum_{n\in\mathcal{J}}\frac{\mu(B(x_n^j,r_n^j))}{\mu(B(x_n^j,r_n^j))+V(x_n^j,x)}\left[\frac{r_n^j}{r_n^j+d(x_n^j,x)}\right]^\beta.$$

Then we obtain the desired estimate for $\langle g^j,\varphi\rangle$ in the case $r > r_{k_0}^j$.

Combining the two cases above, we find that, for any $x\in\Omega^j$,

$$(g^j)^*(x)\lesssim 2^j\sum_{k\in I_j}\frac{\mu(B(x_k^j,r_k^j))}{\mu(B(x_k^j,r_k^j))+V(x_k^j,x)}\left[\frac{r_k^j}{r_k^j+d(x_k^j,x)}\right]^\beta.$$

Thus, (4.6) holds true. This finishes the proof of Proposition 4.9. □

4.3 Atomic Characterizations of $H^{*,p}(X)$

In this section, we prove $H^{*,p}(X)\subset H_{\mathrm{at}}^{p,q}(X)$ and complete the proof of Theorem 4.2. First, we obtain dense subspaces of $H^{*,p}(X)$ as follows.

Lemma 4.10 ([20, Proposition 4.12]) *Let $p\in(\omega/(\omega+\eta),1]$, $\beta,\ \gamma\in(\omega(1/p-1),\eta)$ and $q\in[1,\infty)$. If regard $H^{*,p}(X)$ as a subspace of $(\mathcal{G}_0^\eta(\beta,\gamma))'$, then $L^q(X)\cap H^{*,p}(X)$ is dense in $H^{*,p}(X)$.*

In the next two lemmas, we suppose that $f\in L^2(X)\cap H^{*,p}(X)$. Based on Proposition 3.3 and (3.1), we may follow [20, Remark 4.14] and assume that there exists a positive constant C such that, for any $x\in X$, $|f(x)|\le Cf^*(x)$. With all the notation as in the previous section, for any $j\in\mathbb{Z}$ and $k\in I_j$, define

$$m_k^j := \frac{1}{\|\phi_k^j\|_{L^1(X)}}\int_X f(\xi)\phi_k^j(\xi)\,d\mu(\xi) \quad\text{and}\quad b_k^j := \left(f-m_k^j\right)\phi_k^j. \tag{4.9}$$

Then we have the following technical lemma.

Lemma 4.11 ([20, Proposition 4.13]) *For any $j\in\mathbb{Z}$ and $k\in I_j$, let m_k^j and b_k^j be as in* (4.9). *Then*

(i) *there exists a positive constant C, independent of j and $k\in I_j$, such that $|m_k^j|\le C2^j$;*
(ii) *b_k^j induces the same distribution as defined in* (4.3);
(iii) *$\sum_{k\in I_j} b_k^j$ converges to some function b^j in $L^2(X)$, which induces a distribution that coincides with b^j as in Proposition* 4.9;
(iv) *let $g^j := f-b^j$. Then $g^j = f\mathbf{1}_{(\Omega^j)^\complement}+\sum_{k\in I_j} m_k^j\phi_k^j$. Moreover, there exists a positive constant C, independent of j, such that, for any $x\in X$, $|g^j(x)|\le C2^j$.*

For any $j \in \mathbb{Z}$, $k \in I_j$ and $l \in I_{j+1}$, define

$$L_{k,l}^{j+1} := \frac{1}{\|\phi_l^{j+1}\|_{L^1(X)}} \int_X \left[f(\xi) - m_l^{j+1} \right] \phi_k^j(\xi) \phi_l^{j+1}(\xi)\, d\mu(\xi). \tag{4.10}$$

Then $L_{k,l}^{j+1}$ has the following properties.

Lemma 4.12 *For any $j \in \mathbb{Z}$, $k \in I_j$ and $l \in I_{j+1}$, let $L_{k,l}^{j+1}$ be as in (4.10). Then*

(i) *there exists a positive constant C, independent of j, k and l, such that*

$$\sup_{x\in X} \left| L_{k,l}^{j+1} \phi_l^{j+1}(x) \right| \le C2^j;$$

(ii) $\sum_{k\in I_j} \sum_{l\in I_{j+1}} L_{k,l}^{j+1} \phi_l^{j+1} = 0$ *both in $(\mathcal{G}_0^\eta(\beta,\gamma))'$ and everywhere.*

Proof We first show (i). Indeed, for any $j \in \mathbb{Z}$, $k \in I_j$, $l \in I_{j+1}$ and $x \in X$,

$$\left| L_{k,l}^{j+1} \phi_l^{j+1}(x) \right| \le \left| m_l^{j+1} \right| \phi_l^{j+1}(x) + \phi_l^{j+1}(x) \left| \int_X f(\xi) \frac{\phi_k^j(\xi)\phi_l^{j+1}(\xi)}{\|\phi_l^{j+1}\|_{L^1(X)}}\, d\mu(\xi) \right| =: \mathrm{Y}_1 + \mathrm{Y}_2.$$

By Lemma 4.11(i) and the definition of ϕ_l^{j+1}, it is easy to obtain $\mathrm{Y}_1 \lesssim 2^j$.

Now we consider Y_2. If $\phi_k^j \phi_l^{j+1}$ is a non-zero function, then $B(x_k^j, 2A_0 r_k^j) \cap B(x_l^{j+1}, 2A_0 r_l^{j+1}) \neq \emptyset$, which further implies that $r_l^{j+1} \le 3A_0 r_k^j$. Otherwise, if $r_l^{j+1} > 3A_0 r_k^j$, then, for any $y \in B(x_k^j, 48A_0^5 r_k^j)$,

$$\begin{aligned} d\left(y, x_l^{j+1}\right) &\le A_0 \left[d\left(y, x_k^j\right) + d\left(x_k^j, x_l^{j+1}\right) \right] < 48A_0^6 r_k^j + A_0^2 \left(2A_0 r_k^j + 2A_0 r_l^{j+1} \right) \\ &< 16A_0^5 r_l^{j+1} + \frac{2}{3} A_0^2 r_l^{j+1} + 2A_0^3 r_l^{j+1} < 20A_0^5 r_l^{j+1}, \end{aligned}$$

which implies that $B(x_k^j, 48A_0^5 r_k^j) \subset B(x_l^{j+1}, 20A_0^5 r_l^{j+1}) \subset \Omega^{j+1} \subset \Omega^j$ and hence contradicts to Proposition 4.4(v).

Define $\varphi := \phi_k^j \phi_l^{j+1} / \|\phi_l^{j+1}\|_{L^1(X)}$. According to Proposition 4.4(iv) with $A := 16A_0^4$, we can choose $y_l^{j+1} \in (\Omega^{j+1})^\complement$ such that $d(y_l^{j+1}, x_l^{j+1}) \le 48A_0^5 r_l^{j+1}$. We now show $\varphi \in \mathcal{G}(y_l^{j+1}, r_l^{j+1}, \eta, \eta)$ and $\|\varphi\|_{\mathcal{G}(y_l^{j+1}, r_l^{j+1}, \eta, \eta)} \lesssim 1$. Notice that supp $\varphi \subset B(x_l^{j+1}, 2A_0 r_l^{j+1})$. Moreover, by this and the choice of y_l^{j+1}, we conclude that, for any $x \in B(x_l^{j+1}, 2A_0 r_l^{j+1})$,

$$\begin{aligned} |\varphi(x)| \lesssim |\phi_l^{j+1}(x)| &\lesssim \frac{1}{\mu(B(x_l^{j+1}, r_l^{j+1})) + V(x_l^{j+1}, x)} \left[\frac{r_l^{j+1}}{r_l^{j+1} + d(x_l^{j+1}, x)} \right]^\eta \\ &\sim \frac{1}{\mu(B(y_l^{j+1}, r_l^{j+1})) + V(y_l^{j+1}, x)} \left[\frac{r_l^{j+1}}{r_l^{j+1} + d(y_l^{j+1}, x)} \right]^\eta. \end{aligned}$$

This shows the size condition of φ.

To consider the regularity condition of φ, we suppose that $x,\ x' \in X$ satisfying $d(x,x') \le (2A_0)^{-1}[r_l^{j+1} + d(y_l^{j+1},x)]$. Due to the size condition, we may assume $d(x,x') \le (2A_0)^{-3}[r_l^{j+1} + d(y_l^{j+1},x)]$. We claim that $\varphi(x) - \varphi(x') \ne 0$ implies that $d(x, x_l^{j+1}) \le 96A_0^6 r_l^{j+1}$.

Indeed, if $d(x,x_l^{j+1}) > 96A_0^6 r_l^{j+1}$, then $\varphi(x) = 0$. By $d(x_l^{j+1}, y_l^{j+1}) \le 48A_0^5 r_l^{j+1}$, we find that $d(x, y_l^{j+1}) > 48A_0^5 r_l^{j+1}$ and hence $d(x,x') \le (2A_0)^{-2} d(x,y_l^{j+1}) \le (2A_0)^{-1} d(x,x_l^{j+1})$. Consequently, $d(x', x_l^{j+1}) \ge A_0^{-1} d(x, x_l^{j+1}) - d(x,x') > 48A_0^5 r_l^{j+1}$ and $\varphi(x') = 0$. This contradicts to $\varphi(x) - \varphi(x') \ne 0$.

By the above claim, $r_l^{j+1} \le 3A_0 r_k^j$ and $d(y_l^{j+1}, x_j^{j+1}) \sim r_l^{j+1}$, we know that

$$\begin{aligned}
&|\varphi(x) - \varphi(x')| \\
&\lesssim \frac{1}{\mu(B(x_l^{j+1}, r_l^{j+1}))}\left[\phi_k^j(x)\left|\phi_l^{j+1}(x) - \phi_l^{j+1}(x')\right| + \left|\phi_k^j(x) - \phi_k^j(x')\right|\phi_l^{j+1}(x')\right] \\
&\lesssim \frac{1}{\mu(B(x_l^{j+1}, r_l^{j+1}))}\left\{\left[\frac{d(x,x')}{r_l^{j+1}}\right]^\eta + \left[\frac{d(x,x')}{r_k^j}\right]^\eta\right\} \\
&\sim \left[\frac{d(x,x')}{r_l^{j+1} + d(y_l^{j+1},x)}\right]^\eta \frac{1}{\mu(B(y_l^{j+1}, r_l^{j+1})) + V(y_l^{j+1},x)}\left[\frac{r_l^{j+1}}{r_l^{j+1} + d(y_l^{j+1},x)}\right]^\eta.
\end{aligned}$$

Thus, we obtain $\varphi \in \mathcal{G}(y_l^{j+1}, r_l^{j+1}, \eta, \eta)$ and $\|\varphi\|_{\mathcal{G}(y_l^{j+1}, r_l^{j+1}, \eta, \eta)} \lesssim 1$, which further implies that $\|\varphi\|_{\mathcal{G}(y_l^{j+1}, r_l^{j+1}, \beta, \gamma)} \lesssim 1$ and hence

$$\mathrm{Y}_2 = |\langle f, \varphi\rangle| \lesssim f^*\left(y_l^{j+1}\right) \lesssim 2^j.$$

This finishes the proof of (i).

Next we prove (ii). If $L_{k,l}^{j+1} \ne 0$, then the proof in (i) implies $B(x_k^j, 2A_0 r_k^j) \cap B(x_l^{j+1}, 2A_0 r_l^{j+1}) \ne \emptyset$ and $r_l^{j+1} \le 3A_0 r_k^j$. Further, for any $y \in B(x_l^{j+1}, 2A_0 r_l^{j+1})$, we have

$$\begin{aligned}
d\left(y, x_k^j\right) &\le A_0\left[d\left(y, x_l^{j+1}\right) + d\left(x_k^j, x_l^{j+1}\right)\right] < 2A_0^2 r_l^{j+1} + A_0^2\left(2A_0 r_k^j + 2A_0 r_l^{j+1}\right) \\
&< 6A_0^3 r_k^j + 2A_0^3 r_k^j + 6A_0^4 r_k^j \le 14A_0^4 r_k^j < 16A_0^4 r_k^j,
\end{aligned}$$

which implies that $B(x_l^{j+1}, 2A_0 r_l^{j+1}) \subset B(x_k^j, 16A_0^4 r_k^j) \subset \Omega^j$ by Proposition 4.4(v). Thus, for any $k \in I_j$ and $x \in X$, we find that

$$\sum_{l \in I_{j+1}} \left|L_{k,l}^{j+1}\phi_l^{j+1}\right| \lesssim 2^j \mathbf{1}_{B(x_k^j, 16A_0^4 r_k^j)}(x) \tag{4.11}$$

and hence

$$\sum_{k\in I_j}\sum_{l\in I_{j+1}}\left|L_{k,l}^{j+1}\phi_l^{j+1}(x)\right|\lesssim 2^j\sum_{k\in I_j}\mathbf{1}_{B(x_k^j,16A_0^4r_k^j)}(x)\lesssim 2^j\mathbf{1}_{\Omega^j}(x).$$

Consequently,

$$\begin{aligned}\sum_{k\in I_j}\sum_{l\in I_{j+1}}L_{k,l}^{j+1}\phi_l^{j+1}&=\sum_{l\in I_{j+1}}\left(\sum_{k\in I_j}L_{k,l}^{j+1}\right)\phi_l^{j+1}\\&=\sum_{l\in I_{j+1}}\frac{\phi_l^{j+1}}{\|\phi_l^{j+1}\|_{L^1(X)}}\int_X\left[f(\xi)-m_l^{j+1}\right]\phi_l^{j+1}(\xi)\sum_{k\in I_j}\phi_k^j(\xi)\,d\mu(\xi)\\&=\sum_{l\in I_{j+1}}\frac{\phi_l^{j+1}}{\|\phi_l^{j+1}\|_{L^1(X)}}\int_X\left[f(\xi)-m_l^{j+1}\right]\phi_l^{j+1}(\xi)\,d\mu(\xi)\\&=\sum_{l\in I_{j+1}}\frac{\phi_l^{j+1}}{\|\phi_l^{j+1}\|_{L^1(X)}}\int_X b_l^{j+1}(\xi)\,d\mu(\xi)=0.\end{aligned}$$

By the fact that $\sum_{k\in I_j}\sum_{l\in I_{j+1}}\int_X|L_{k,l}^{j+1}\phi_l^{j+1}(\xi)|\,d\mu(\xi)\lesssim 2^j\mu(\Omega^j)<\infty$ and the dominated convergence theorem, we find that $\sum_{k\in I_j}\sum_{l\in I_{j+1}}L_{k,l}^{j+1}\phi_l^{j+1}=0$ in $L^1(X)$ and hence in $(\mathcal{G}_0^\eta(\beta,\gamma))'$. This finishes the proof of Lemma 4.12. □

Now we show the other side of Theorem 4.2.

Proof of $H^{*,p}(X)\subset H_{\mathrm{at}}^{p,q}(X)$ By Lemma 4.10, we first suppose $f\in L^2(X)\cap H^{*,p}(X)$. We may also assume $|f(x)|\lesssim f^*(x)$ for any $x\in X$. We use the same notation as in Lemmas 4.11 and 4.12. For any $j\in\mathbb{N}$, let $h^j:=g^{j+1}-g^j=b^j-b^{j+1}$. Then $f-\sum_{j=-m}^m h^j=b^{m+1}+g^{-m}$. For any $m\in\mathbb{Z}$, by Lemma 4.11, we conclude that $\|g^{-m}\|_{L^\infty(X)}\lesssim 2^{-m}$. Moreover, by (4.5), we find that $\|(b^{m+1})^*\|_{L^p(X)}\lesssim\|f^*\mathbf{1}_{(\Omega^{m+1})^\complement}\|_{L^p(X)}\to 0$ as $m\to\infty$. Thus, $f=\sum_{j=-\infty}^\infty h^j$ in $(\mathcal{G}_0^\eta(\beta,\gamma))'$. Besides, by the definition of b_k^m, we know that supp $b^{m+1}\subset\Omega^{m+1}$, which then implies that $\sum_{j=-\infty}^\infty h^j$ converges almost everywhere. Notice that, by Lemma 4.12(ii), for any $j\in\mathbb{Z}$, we have

$$\begin{aligned}h^j=b^j-b^{j+1}&=\sum_{k\in I_j}b_k^j-\sum_{l\in I_{j+1}}b_l^{j+1}+\sum_{k\in I_j}\sum_{l\in I_{j+1}}L_{k,l}^{j+1}\phi_l^{j+1}\\&=\sum_{k\in I_j}\left[b_k^j-\sum_{l\in I_{j+1}}\left(b_l^{j+1}\phi_k^j-L_{k,l}^{j+1}\phi_l^{j+1}\right)\right]=:\sum_{k\in I_j}h_k^j,\end{aligned}\tag{4.12}$$

which converges in $(\mathcal{G}_0^\eta(\beta,\gamma))'$ and almost everywhere. Moreover, for any $j\in\mathbb{Z}$ and $k\in\mathbb{N}$,

$$
\begin{aligned}
h_k^j &= b_k^j - \sum_{l\in I_{j+1}} \left(b_l^{j+1}\phi_k^j - L_{k,l}^{j+1}\phi_l^{j+1} \right) \\
&= \left(f - m_k^j\right)\phi_k^j - \sum_{l\in I_{j+1}} \left[\left(f - m_l^{j+1}\right)\phi_k^j - L_{k,l}^{j+1} \right]\phi_l^{j+1} \\
&= f\phi_k^j \mathbf{1}_{(\Omega^{j+1})^\complement} - m_k^j\phi_k^j + \phi_k^j \sum_{l\in I_{j+1}} m_l^{j+1}\phi_l^{j+1} + \sum_{l\in I_{j+1}} L_{k,l}^{j+1}\phi_l^{j+1}.
\end{aligned}
$$

The fourth term is supported on $B_k^j := B(x_k^j, 16A_0^4 r_k^j)$, which is deduced from (4.11). Thus, supp $h_k^j \subset B_k^j$. Moreover, by Lemmas 4.11(i) and 4.12(i), we conclude that there exists a positive constant C, independent of j and k, such that $\|h_k^j\|_{L^\infty(X)} \le C2^j$. Now, let

$$
\lambda_k^j := C2^j \left[\mu\left(B_k^j\right)\right]^{\frac{1}{p}} \quad \text{and} \quad a_k^j := \left(\lambda_k^j\right)^{-1} h_k^j. \tag{4.13}
$$

Then a_k^j is a (p,∞)-atom supported on B_k^j and $f = \sum_{j=-\infty}^{\infty}\sum_{k\in I_j}\lambda_k^j a_k^j$ in $(\mathcal{G}_0^\eta(\beta,\gamma))'$. Moreover, we have

$$
\begin{aligned}
\sum_{j=-\infty}^{\infty}\sum_{k\in I_j}\left|\lambda_k^j\right|^p &\lesssim \sum_{j=-\infty}^{\infty} 2^{-jp}\sum_{k\in I_j}\mu\left(B_k^j\right) \\
&\lesssim \sum_{j=-\infty}^{\infty} 2^{-jp}\mu\left(\Omega^j\right) \sim \|f^\star\|_{L^p(X)}^p \sim \|f^*\|_{L^p(X)}^p,
\end{aligned}
$$

which further implies that $\|f\|_{H_{\mathrm{at}}^{p,\infty}(X)} \lesssim \|f\|_{H^{*,p}(X)}$.

When $f \in H^{*,p}(X)$, using Lemma 4.10 and a standard density argument and following the proof in [43, pp. 301–302], we obtain the atomic decomposition of f, the details being omitted. This finishes the proof of $H^{*,p}(X) \subset H_{\mathrm{at}}^{p,q}(X)$ and hence of Theorem 4.2. □

Remark 4.13 By the argument used in the proof of $H^{*,p}(X) \subset H_{\mathrm{at}}^{p,q}(X)$, we find that, if $f \in L^q(X) \cap H^{*,p}(X)$ with $q \in [1,\infty]$, then $f = \sum_{j=1}^{\infty}\sum_{k\in I_j} h_k^j$ in $(\mathcal{G}_0^\eta(\beta,\gamma))'$ and almost everywhere, where, for any $j \in \mathbb{Z}$ and $k \in I_j$, h_k^j is as in (4.12).

4.4 Relationship Between $H_{\mathrm{at}}^{p,q}(X)$ and $H_{\mathrm{cw}}^{p,q}(X)$

In this section, we consider the relationship between $H_{\mathrm{at}}^{p,q}(X)$ and $H_{\mathrm{cw}}^{p,q}(X)$. To see this, we need the following two technical lemmas.

Lemma 4.14 ([7, p. 592]) *Let $p \in (0,1)$, $q \in (p,\infty] \cap [1,\infty]$ and a be a (p,q)-atom. Then, for any $\varphi \in \mathcal{L}_{1/p-1}(X)$, $|\langle a,\varphi\rangle| \le \|\varphi\|_{\mathcal{L}_{1/p-1}(X)}$.*

Lemma 4.15 *Let $\beta \in (0,\eta]$ and $\gamma \in (0,\infty)$. If $\varphi \in \mathcal{G}(\beta,\gamma)$, then $\varphi \in \mathcal{L}_{\beta/\omega}(X)$ and there exists a positive constant C, independent of φ, such that $\|\varphi\|_{\mathcal{L}_{\beta/\omega}(X)} \le C\|\varphi\|_{\mathcal{G}(\beta,\gamma)}$.*

Proof Suppose that $\|\varphi\|_{\mathcal{G}(\beta,\gamma)} \le 1$. If $d(x,y) \le (2A_0)^{-1}[1+d(x_0,x)]$, then, by the regularity condition of φ and (1.1), we have

$$\begin{aligned}|\varphi(x)-\varphi(y)| &\le \left[\frac{d(x,y)}{1+d(x_0,x)}\right]^{\beta}\frac{1}{V_1(x_0)+V(x_0,x)}\left[\frac{1}{1+d(x_0,x)}\right]^{\gamma}\\ &\lesssim \left[\frac{\mu(B(x,d(x,y)))}{\mu(B(x,1+d(x_0,x)))}\right]^{\beta/\omega} \lesssim [V(x,y)]^{\beta/\omega}.\end{aligned}$$

If $d(x,y) > (2A_0)^{-1}[1+d(x_0,x)]$, then, from the size condition of φ, we deduce that

$$\begin{aligned}|\varphi(x)-\varphi(y)| &\lesssim 1 \sim [\mu(B(x_0,1))]^{\beta/\omega} \lesssim [\mu(B(x_0,1+d(x_0,x)))]^{\beta/\omega}\\ &\sim [\mu(B(x,1+d(x_0,x)))]^{\beta/\omega} \lesssim [V(x,y)]^{\beta/\omega}.\end{aligned}$$

Thus, for any $x,\ y\in X$, we always have $|\varphi(x)-\varphi(y)| \lesssim \|\varphi\|_{\mathcal{G}(\beta,\gamma)}[V(x,y)]^{\beta/\omega}$. This implies $\varphi\in\mathcal{L}_{\beta/\omega}(X)$ and $\|\varphi\|_{\mathcal{L}_{\beta/\omega}(X)} \lesssim \|\varphi\|_{\mathcal{G}(\beta,\gamma)}$, which completes the proof of Lemma 4.15. □

Now we establish the relationship between two kinds of atomic Hardy spaces.

Theorem 4.16 *Let $p\in(\omega/(\omega+\eta),1]$, $q\in(p,\infty]\cap[1,\infty]$ and $\beta,\ \gamma\in(\omega(1/p-1),\eta)$. If regard $H^{p,q}_{\rm at}(X)$ as a subspace of $(\mathcal{G}^{\eta}_0(\beta,\gamma))'$, then $H^{p,q}_{\rm cw}(X)=H^{p,q}_{\rm at}(X)$ with equal (quasi-)norms.*

Proof We only consider the case $p\in(\omega/(\omega+\eta),1)$. The proof of $p=1$ is similar and the details are omitted.

We first prove $H^{p,q}_{\rm cw}(X)\subset H^{p,q}_{\rm at}(X)$. By Lemma 4.15, we have $\mathcal{G}^{\eta}_0(\beta,\gamma)\subset \mathcal{G}(\omega(1/p-1),\gamma)\subset\mathcal{L}_{1/p-1}(X)$ and hence $(\mathcal{L}_{1/p-1}(X))'\subset(\mathcal{G}^{\eta}_0(\beta,\gamma))'$. For any $f\in H^{p,q}_{\rm cw}(X)$, by Definition 1.1, we know that there exist (p,q)-atoms $\{a_j\}_{j=1}^{\infty}$ and $\{\lambda_j\}_{j=1}^{\infty}\subset\mathbb{C}$ with $\sum_{j=1}^{\infty}|\lambda_j|^p<\infty$ such that $f=\sum_{j=1}^{\infty}\lambda_j a_j$ in $(\mathcal{L}_{1/p-1}(X))'$ and hence in $(\mathcal{G}^{\eta}_0(\beta,\gamma))'$. Let $g:=f|_{\mathcal{G}^{\eta}_0(\beta,\gamma)}$. Then, for any $\varphi\in\mathcal{G}^{\eta}_0(\beta,\gamma)\subset\mathcal{L}_{1/p-1}(X)$, we have

$$\langle g,\varphi\rangle=\langle f,\varphi\rangle=\sum_{j=1}^{\infty}\lambda_j\langle a_j,\varphi\rangle.$$

Thus, $g=\sum_{j=1}^{\infty}\lambda_j a_j$ in $(\mathcal{G}^{\eta}_0(\beta,\gamma))'$ and $\|g\|_{H^{p,q}_{\rm at}(X)}\le(\sum_{j=1}^{\infty}|\lambda_j|^p)^{\frac{1}{p}}$. If we take the infimum over all the atomic decompositions of f as above, we obtain $\|g\|_{H^{p,q}_{\rm at}(X)}\le \|f\|_{H^{p,q}_{\rm cw}(X)}$. Thus, $H^{p,q}_{\rm cw}(X)\subset H^{p,q}_{\rm at}(X)$.

To show $H^{p,q}_{\rm cw}(X)\supset H^{p,q}_{\rm at}(X)$, following the proof of [7, p. 593, Theorem B], we conclude that the dual space of $H^{p,q}_{\rm at}(X)$ is $\mathcal{L}_{1/p-1}(X)$ in the following sense: every bounded linear functional on $H^{p,q}_{\rm at}(X)$ is a mapping of the form

$$f\mapsto\sum_{j=1}^{\infty}\lambda_j\int_X a_j(x)g(x)\,d\mu(x),$$

where $g \in \mathcal{L}_{1/p-1}(X)$ and f has an atomic decomposition

$$f = \sum_{j=1}^{\infty} \lambda_j a_j \tag{4.14}$$

in $(\mathcal{G}_0^{\eta}(\beta, \gamma))'$ with (p, q)-atoms $\{a_j\}_{j=1}^{\infty}$ and $\{\lambda_j\}_{j=1}^{\infty} \subset \mathbb{C}$ satisfying $\sum_{j=1}^{\infty} |\lambda_j|^p < \infty$. Therefore, it is reasonable to define the pair $\langle f, g\rangle$ as follows:

$$\langle f, g\rangle := \sum_{j=1}^{\infty} \lambda_j \int_X a_j(x) g(x)\, d\mu(x).$$

In this way, we find that (4.14) also converges in $(\mathcal{L}_{1/p-1}(X))'$, and hence $f \in H_{\mathrm{cw}}^{p,q}(X)$ and $\|f\|_{H_{\mathrm{cw}}^{p,q}(X)} \le (\sum_{j=1}^{\infty} |\lambda_j|^p)^{\frac{1}{p}}$. Taking the infimum over all the atomic decompositions of f as above, we obtain $\|f\|_{H_{\mathrm{cw}}^{p,q}(X)} \le \|f\|_{H_{\mathrm{at}}^{p,q}(X)}$. Thus, $H_{\mathrm{at}}^{p,q}(X) \subset H_{\mathrm{cw}}^{p,q}(X)$, which completes the proof of Theorem 4.16. □

5 Littlewood–Paley Function Characterizations of Atomic Hardy Spaces

In this section, we consider the Littlewood–Paley function characterizations of Hardy spaces. Differently from Sects. 3 and 4, we use $(\mathring{\mathcal{G}}_0^{\eta}(\beta, \gamma))'$ as underlying spaces to introduce Hardy spaces. Let $p \in (\omega/(\omega + \eta), 1]$, $\beta, \gamma \in (\omega(1/p - 1), \eta)$, $f \in (\mathring{\mathcal{G}}_0^{\eta}(\beta, \gamma))'$ and $\{Q_k\}_{k\in\mathbb{Z}}$ be an exp-ATI. For any $\theta \in (0, \infty)$, define the *Lusin area function of* f*, with aperture* θ*,* $\mathcal{S}_\theta(f)$, by setting, for any $x \in X$,

$$\mathcal{S}_\theta(f)(x) := \left[\sum_{k=-\infty}^{\infty} \int_{B(x,\theta\delta^k)} |Q_k f(y)|^2 \frac{d\mu(y)}{V_{\theta\delta^k}(x)}\right]^{\frac{1}{2}}. \tag{5.1}$$

In particular, when $\theta = 1$, we write $\mathcal{S}_\theta$ simply as $\mathcal{S}$. Define the *Hardy space* $H^p(X)$ via the Lusin area function by setting

$$H^p(X) := \left\{f \in \left(\mathring{\mathcal{G}}_0^{\eta}(\beta, \gamma)\right)' : \|f\|_{H^p(X)} := \|\mathcal{S}(f)\|_{L^p(X)} < \infty\right\}.$$

In Sect. 5.1, we show that $H^p(X)$ is independent of the choices of exp-ATIs. In Sect. 5.2, we connect $H^p(X)$ with $H^{*,p}(X)$ by considering the molecular and the atomic characterizations of elements in $H^p(X)$. Sect. 5.3 deals with equivalent characterizations of $H^p(X)$ via the *Littlewood–Paley g-function*

$$g(f)(x) := \left[\sum_{k=-\infty}^{\infty} |Q_k f(x)|^2\right]^{\frac{1}{2}} \tag{5.2}$$

and the *Littlewood–Paley g_λ^*-function*

$$g_\lambda^*(f)(x) := \left\{ \sum_{k=-\infty}^{\infty} \int_X |Q_k f(y)|^2 \left[\frac{\delta^k}{\delta^k + d(x,y)} \right]^\lambda \frac{d\mu(y)}{V_{\delta^k}(x) + V_{\delta^k}(y)} \right\}^{\frac{1}{2}}, \quad (5.3)$$

where $f \in (\mathring{\mathcal{G}}_0^\eta(\beta,\gamma))'$ with $\beta,\ \gamma \in (\omega(1/p-1),\eta)$, $x \in X$ and $\lambda \in (0,\infty)$.

5.1 Independence of exp-ATIs

In this section, we show that $H^p(X)$ is independent of the choices of exp-ATIs. If $\mathcal{E} := \{E_k\}_{k\in\mathbb{Z}}$ and $Q := \{Q_k\}_{k\in\mathbb{Z}}$ are two exp-ATIs, then we denote by $\mathcal{S}_\mathcal{E}$ and $\mathcal{S}_Q$ the Lusin area functions via $\mathcal{E}$ and Q, respectively.

Theorem 5.1 *Let $\mathcal{E} := \{E_k\}_{k\in\mathbb{Z}}$ and $Q := \{Q_k\}_{k\in\mathbb{Z}}$ be two* exp-ATIs. *Suppose that $p \in (\omega/(\omega+\eta), 1]$ and $\beta,\ \gamma \in (\omega(1/p-1),\eta)$. Then there exists a positive constant C such that, for any $f \in (\mathring{\mathcal{G}}_0^\eta(\beta,\gamma))'$,*

$$C^{-1}\|\mathcal{S}_Q(f)\|_{L^p(X)} \le \|\mathcal{S}_\mathcal{E}(f)\|_{L^p(X)} \le C\|\mathcal{S}_Q(f)\|_{L^p(X)}.$$

To show Theorem 5.1, the Fefferman–Stein vector-valued maximal inequality is necessary.

Lemma 5.2 ([22, Theorem 1.2] or [47, Theorem 1.3]) *Suppose that $p \in (1,\infty)$ and $u \in (1,\infty]$. Then there exists a positive constant C such that, for any sequence $\{f_j\}_{j=1}^\infty$ of measurable functions,*

$$\left\| \left\{ \sum_{j=1}^\infty [\mathcal{M}(f_j)]^u \right\}^{\frac{1}{u}} \right\|_{L^p(X)} \le C \left\| \left(\sum_{j=1}^\infty |f_j|^u \right)^{\frac{1}{u}} \right\|_{L^p(X)}$$

with the usual modification made when $u = \infty$.

Proof of Theorem 5.1 By symmetry, we only need to prove

$$\|\mathcal{S}_\mathcal{E}(f)\|_{L^p(X)} \lesssim \|\mathcal{S}_Q(f)\|_{L^p(X)}.$$

For any $k \in \mathbb{Z}$, $f \in (\mathring{\mathcal{G}}_0^\eta(\beta,\gamma))'$ with $\beta,\ \gamma$ as in Theorem 5.1, and $z \in X$, define

$$m_k(f)(z) := \left[\frac{1}{V_{\delta^k}(z)} \int_{B(z,\delta^k)} |Q_k f(u)|^2\, d\mu(u) \right]^{\frac{1}{2}}.$$

Now suppose that $l \in \mathbb{Z}$, $x \in X$ and $y \in B(x, \delta^l)$. By Theorem 2.7, we conclude that

$$E_l f(y) = \sum_{k=-\infty}^{\infty} \sum_{\alpha \in \mathcal{A}_k} \sum_{m=1}^{N(k,\alpha)} E_l \widetilde{Q}_k\left(y, y_\alpha^{k,m}\right) \int_{Q_\alpha^{k,m}} Q_k f(u)\, d\mu(u),$$

where all the notation is as in Theorem 2.7 and $\{\widetilde{Q}_k\}_{k=-\infty}^{\infty}$ satisfies the conditions of Theorem 2.7. Notice that, if $z \in Q_\alpha^{k,m}$, then $Q_\alpha^{k,m} \subset B(z, \delta^k)$ and $\mu(Q_\alpha^{k,m}) \sim V_{\delta^k}(z)$. Therefore, we have

$$\left|\frac{1}{\mu(Q_\alpha^{k,m})} \int_{Q_\alpha^{k,m}} Q_k f(u)\, d\mu(u)\right| \lesssim \left[\frac{1}{V_{\delta^k}(z)} \int_{B(z,\delta^k)} |Q_k f(u)|^2\, d\mu(y)\right]^{\frac{1}{2}} \sim m_k(f)(z),$$

which further implies that

$$\left|\frac{1}{\mu(Q_\alpha^{k,m})} \int_{Q_\alpha^{k,m}} Q_k f(u)\, d\mu(u)\right| \lesssim \inf_{z \in Q_\alpha^{k,m}} m_k(f)(z).$$

Moreover, by the proof of (3.7), we find that, for any fixed $\beta' \in (0, \beta)$,

$$\begin{aligned}\left|E_l \widetilde{Q}_k\left(y, y_\alpha^{k,m}\right)\right| &\lesssim \delta^{|k-l|\beta'} \frac{1}{V_{\delta^{k\wedge l}}(y) + V(y, y_\alpha^{k,m})} \left[\frac{\delta^{k\wedge l}}{\delta^{k\wedge l} + d(y, y_\alpha^{k,m})}\right]^\gamma \\ &\sim \delta^{|k-l|\beta'} \frac{1}{V_{\delta^{k\wedge l}}(x) + V(x, y_\alpha^{k,m})} \left[\frac{\delta^{k\wedge l}}{\delta^{k\wedge l} + d(x, y_\alpha^{k,m})}\right]^\gamma,\end{aligned}$$

where only the regularity condition of $\widetilde{Q}_k$ on the first variable is used. Therefore, by Lemma 3.7, for any fixed $r \in (\omega/(\omega + \gamma), 1]$, we have

$$\begin{aligned}|E_l f(y)| &\lesssim \sum_{k=-\infty}^{\infty} \delta^{|k-l|\beta'} \sum_{\alpha \in \mathcal{A}_k} \sum_{m=1}^{N(k,\alpha)} \mu\left(Q_\alpha^{k,m}\right) \frac{1}{V_{\delta^{k\wedge l}}(x) + V(x, y_\alpha^{k,m})} \\ &\quad \times \left[\frac{\delta^{k\wedge l}}{\delta^{k\wedge l} + d(x, y_\alpha^{k,m})}\right]^\gamma \inf_{z \in Q_\alpha^{k,m}} m_k(f)(z) \\ &\lesssim \sum_{k=-\infty}^{\infty} \delta^{|k-l|\beta'} \delta^{[k-(k\wedge l)]\omega(1-\frac{1}{r})} \\ &\quad \times \left\{\mathcal{M}\left(\sum_{\alpha \in \mathcal{A}_k} \sum_{m=1}^{N(k,\alpha)} \inf_{z \in Q_\alpha^{k,m}} [m_k(f)(z)]^r \mathbf{1}_{Q_\alpha^{k,m}}\right)(x)\right\}^{\frac{1}{r}}.\end{aligned}$$

Choose β' and r such that $r \in (\omega/(\omega+\beta'), p)$. Then, by the Hölder inequality, we conclude that

$$
\begin{aligned}
\left[\mathcal{S}_{\mathcal{E}}(f)(x)\right]^2 &= \sum_{l=-\infty}^{\infty} \int_{B(x,\delta^l)} |E_l f(y)|^2 \frac{dy}{V_{\delta^l}(x)} \\
&\lesssim \sum_{l=-\infty}^{\infty} \left[\sum_{k=-\infty}^{\infty} \delta^{|k-l|\beta'} \delta^{[k-(k\wedge l)]\omega(1-\frac{1}{r})} \right. \\
&\quad \left. \times \left\{ \mathcal{M}\left(\sum_{\alpha\in\mathcal{A}_k} \sum_{m=1}^{N(k,\alpha)} \inf_{z\in Q_\alpha^{k,m}} [m_k(f)(z)]^r \mathbf{1}_{Q_\alpha^{k,m}} \right)(x) \right\}^{\frac{1}{r}} \right]^2 \\
&\lesssim \sum_{l=-\infty}^{\infty} \sum_{k=-\infty}^{\infty} \delta^{|k-l|\beta'} \delta^{[k-(k\wedge l)]\omega(1-\frac{1}{r})} \\
&\quad \times \left\{ \mathcal{M}\left(\sum_{\alpha\in\mathcal{A}_k} \sum_{m=1}^{N(k,\alpha)} \inf_{z\in Q_\alpha^{k,m}} [m_k(f)(z)]^r \mathbf{1}_{Q_\alpha^{k,m}} \right)(x) \right\}^{\frac{2}{r}} \\
&\lesssim \sum_{k=-\infty}^{\infty} \left\{ \mathcal{M}\left(\sum_{\alpha\in\mathcal{A}_k} \sum_{m=1}^{N(k,\alpha)} \inf_{z\in Q_\alpha^{k,m}} [m_k(f)(z)]^r \mathbf{1}_{Q_\alpha^{k,m}} \right)(x) \right\}^{\frac{2}{r}} \\
&\lesssim \sum_{k=-\infty}^{\infty} \left\{ \mathcal{M}\left([m_k(f)]^r\right)(x) \right\}^{\frac{2}{r}}.
\end{aligned}
$$

Therefore, from Lemma 5.2, we deduce that

$$
\begin{aligned}
\|\mathcal{S}_{\mathcal{E}}(f)\|_{L^p(X)} &\lesssim \left\| \left(\sum_{k=-\infty}^{\infty} \left\{ \mathcal{M}\left([m_k(f)]^r\right) \right\}^{\frac{2}{r}} \right)^{\frac{r}{2}} \right\|_{L^{p/r}(X)}^{\frac{1}{r}} \\
&\lesssim \left\| \left\{ \sum_{k=-\infty}^{\infty} [m_k(f)]^2 \right\}^{\frac{1}{2}} \right\|_{L^p(X)} \sim \|\mathcal{S}_Q(f)\|_{L^p(X)}.
\end{aligned}
$$

This finishes the proof of Theorem 5.1. □

5.2 Atomic Characterizations of $H^p(X)$

The main aim of this section is to obtain the atomic characterizations of $H^p(X)$ when $p \in (\omega/(\omega+\eta), 1]$.

For any $p \in (\omega/(\omega+\eta), 1]$, $q \in (p, \infty] \cap [1, \infty]$ and $\beta, \gamma \in (\omega(1/p-1), \eta)$, we define the *homogeneous atomic Hardy space* $\mathring{H}_{\mathrm{at}}^{p,q}(X)$ in the same way as $H_{\mathrm{at}}^{p,q}(X)$, but with the distribution space $(\mathcal{G}_0^\eta(\beta,\gamma))'$ replaced by $(\mathring{\mathcal{G}}_0^\eta(\beta,\gamma))'$. Then the following relationship between $H^{p,q}_{\rm at}(X)$ and $\mathring{H}^{p,q}_{\rm at}(X)$ can be found in [20, Theorem 5.4].

Proposition 5.3 *Suppose $p \in (\omega/(\omega+\eta),1]$, $\beta,\ \gamma \in (\omega(1/p-1),\eta)$ and $q \in (p,\infty]\cap[1,\infty]$. Then $\mathring{H}^{p,q}_{\rm at}(X) = H^{p,q}_{\rm at}(X)$ with equivalent (quasi)-norms. More precisely, if $f \in H^{p,q}_{\rm at}(X)$, then the restriction of f on $\mathring{\mathcal{G}}_0^\eta(\beta,\gamma)$ belongs to $\mathring{H}^{p,q}_{\rm at}(X)$; Conversely, if $f \in \mathring{H}^{p,q}_{\rm at}(X)$, then there exists a unique $\widetilde{f} \in H^{p,q}_{\rm at}(X)$ such that $\widetilde{f} = f$ in $(\mathring{\mathcal{G}}_0^\eta(\beta,\gamma))'$.*

Due to the fact that the kernels $\widetilde{Q}_k$ in the homogeneous continuous Calderón formula in Theorem 2.6 has no compact support, we can only use Theorem 2.6 to decompose an element of $H^p(X)$ into a linear combination of the following *molecules*.

Definition 5.4 Suppose that $p \in (0,1]$, $q \in (p,\infty]\cap[1,\infty]$ and $\vec{\epsilon} := \{\epsilon_m\}_{m=1}^\infty \subset [0,\infty)$ satisfying

$$\sum_{m=1}^\infty m[\epsilon_m]^p < \infty. \tag{5.4}$$

A function $M \in L^q(X)$ is called a $(p,q,\vec{\epsilon})$-*molecule* centered at a ball $B := B(x_0,r_0)$ for some $x_0 \in X$ and $r \in (0,\infty)$ if M has the following properties:

(i) $\|M\mathbf{1}_B\|_{L^q(X)} \le [\mu(B)]^{\frac1q-\frac1p}$;

(ii) for any $m\in\mathbb{N}$, $\|M\mathbf{1}_{B(x_0,\delta^{-m}r_0)\setminus B(x_0,\delta^{-m+1}r_0)}\|_{L^q(X)}\le\epsilon_m[\mu(B(x_0,\delta^{-m}r_0))]^{\frac1q-\frac1p}$;

(iii) $\int_X M(x)\,d\mu(x) = 0$.

By (i) and (ii) of Definition 5.4, the Hölder inequality, (5.4) and the fact $p \in (0,1]$, we find that, if M satisfies (i) and (ii) of Definition 5.4, then $M \in L^1(X)$ and hence Definition 5.4(iii) makes sense.

After carefully checking the proof of [39, Theorem 3.4], we obtain the following molecular characterization of the atomic Hardy space $H^{p,q}_{\rm cw}(X)$ of Coifman and Weiss [7], the details being omitted.

Proposition 5.5 *Suppose that $p \in (0,1]$, $q \in (p,\infty]\cap[1,\infty]$ and $\vec{\epsilon} := \{\epsilon_l\}_{l=1}^\infty$ satisfies (5.4). Then $f \in (\mathcal{G}_0^\eta(\beta,\gamma))'$ belonging to $H^{p,q}_{\rm cw}(X)$ if and only if there exist $(p,q,\vec{\epsilon})$-molecules $\{M_j\}_{j=1}^\infty$ and $\{\lambda_j\}_{j=1}^\infty \subset \mathbb{C}$, with $\sum_{j=1}^\infty |\lambda_j|^p < \infty$, such that*

$$f = \sum_{j=1}^\infty \lambda_j M_j \tag{5.5}$$

converges in $(\mathcal{L}_{1/p-1}(X))'$ when $p \in (0,1)$ or in $L^1(X)$ when $p = 1$. Moreover, there exists a positive constant C, independent of f, such that, for any $f \in H^{p,q}_{\rm cw}(X)$,

$$C^{-1}\|f\|_{H^{p,q}_{\rm cw}(X)} \le \inf\left\{\left(\sum_{j=1}^\infty |\lambda_j|^p\right)^{\frac1p}\right\} \le C\|f\|_{H^{p,q}_{\rm cw}(X)},$$

where the infimum is taken over all the molecular decompositions of f as in (5.5).

Let $p \in (\omega/(\omega+\eta), 1]$ and $q \in (p, \infty] \cap [1, \infty]$. By Proposition 5.3, $\mathring{H}^{p,q}_{\rm at}(X) = H^{p,q}_{\rm cw}(X)$ and the already known fact that $H^{p,q}_{\rm cw}(X)$ is independent of the choice of $q \in (p, \infty] \cap [1, \infty]$, we know that $\mathring{H}^{p,q}_{\rm at}(X) = \mathring{H}^{p,2}_{\rm at}(X)$. With this observation, we show $\mathring{H}^{p,q}_{\rm at}(X) \subset H^p(X)$ as follows.

Proposition 5.6 *Let $p \in (\omega/(\omega+\eta), 1]$, $\beta,\ \gamma \in (\omega(1/p-1), \eta)$, $q \in (p, \infty] \cap [1, \infty]$ and $\{Q_k\}_{k\in\mathbb{Z}}$ be an* exp-ATI. *Let $\theta \in (0, \infty)$ and $\mathcal{S}_\theta$ be as in* (5.1). *Then there exists a positive constant C, independent of θ, such that, for any distribution $f \in (\mathring{\mathcal{G}}^{\eta}_0(\beta, \gamma))'$ belonging to $\mathring{H}^{p,2}_{\rm at}(X)$,*

$$\|\mathcal{S}_\theta(f)\|_{L^p(X)} \le C \max\left\{\theta^{-\omega/2}, \theta^{\omega/p}\right\} \|f\|_{\mathring{H}^{p,2}_{\rm at}(X)}. \tag{5.6}$$

In particular, $\mathring{H}^{p,q}_{\rm at}(X) = \mathring{H}^{p,2}_{\rm at}(X) \subset H^p(X)$.

Proof Let $\beta,\ \gamma \in (\omega(1/p-1), \eta)$. It suffices to show (5.6) for the case $\theta \in [1, \infty)$, because both (5.6) with $\theta = 1$ and $\mathcal{S}_\theta(f) \lesssim \theta^{-\omega/2}\mathcal{S}(f)$ for any $f \in (\mathring{\mathcal{G}}^{\eta}_0(\beta, \gamma))'$ whenever $\theta \in (0, 1)$ imply that (5.6) also holds true for any $\theta \in (0, 1)$.

We start with the proof of the fact that the Littlewood–Paley g-function as in (5.2) is bounded on $L^2(X)$. Indeed, for any $h \in L^2(X)$, we write

$$\|g(h)\|^2_{L^2(X)} = \sum_{k=-\infty}^{\infty} \int_X |Q_k h(z)|^2\, d\mu(z) = \sum_{k=-\infty}^{\infty} \left\langle Q_k^* Q_k h, h\right\rangle.$$

By Theorem 2.6 and the proof of [27, (3.2)], we find that, for any fixed $\beta' \in (0, \beta\wedge\gamma)$, any $k_1,\ k_2 \in \mathbb{Z}$ and $x,\ y \in X$, we have

$$\left|Q_{k_1} Q^*_{k_2}(x, y)\right| \lesssim \delta^{|k_1-k_2|\beta'} \frac{1}{V_{\delta^{k_1\wedge k_2}}(x) + V(x, y)} \left[\frac{\delta^{k_1\wedge k_2}}{\delta^{k_1\wedge k_2} + V(x, y)}\right]^{\gamma}. \tag{5.7}$$

Notice that, in (5.7), only the regularity of Q_k with respect to the second variable is used. Thus, by Lemma 2.2(v) and the boundedness of $\mathcal{M}$ on $L^2(X)$, we conclude that, for any $k_1,\ k_2 \in \mathbb{Z}$,

$$\left\|\left(Q^*_{k_1} Q_{k_1}\right)\left(Q^*_{k_2} Q_{k_2}\right)\right\|_{L^2(X)\to L^2(X)} \lesssim \left\|Q_{k_1} Q^*_{k_2}\right\|_{L^2(X)\to L^2(X)} \lesssim \delta^{|k_1-k_2|\beta'}.$$

Therefore, by the fact that $Q^*_k Q_k$ is self-adjoint and the Cotlar–Stein lemma (see [49, pp. 279–280] and [29, Lemma 4.5]), we obtain the boundedness of $\sum_{k=-\infty}^{\infty} Q^*_k Q_k$ on $L^2(X)$ and hence the boundedness of g on $L^2(X)$.

Suppose that a is a $(p, 2)$-atom supported on a ball $B := B(x_0, r_0)$ with $x_0 \in X$ and $r_0 \in (0, \infty)$. By the Fubini theorem and the boundedness of g on $L^2(X)$, we find that

$$\|\mathcal{S}_\theta(a)\|_{L^2(X)} \lesssim \left\|\left\{\sum_{k\in\mathbb{Z}} |Q_k a|^2\right\}^{1/2}\right\|_{L^2(X)} \sim \|g(a)\|_{L^2(X)} \lesssim \|a\|_{L^2(X)} \lesssim [\mu(B)]^{\frac{1}{2}-\frac{1}{p}},$$

which further implies that

$$\int_{B(x_0,4A_0^2\theta r_0)} [S_\theta(a)(x)]^p \, d\mu(x)$$
$$\le \|S_\theta(a)\|_{L^2(X)}^p \left[\mu\left(B\left(x_0, 4A_0^2\theta r_0\right)\right)\right]^{1-\frac{p}{2}} \lesssim \theta^{\omega(1-\frac{p}{2})}. \tag{5.8}$$

Let $x \notin B(x_0, 4A_0^2\theta r_0)$ and $y \in B(x, \theta\delta^k)$. Since now $\theta \in [1, \infty)$, for any $u \in B = B(x_0, r_0)$, we have $d(u, x_0) < (4A_0^2\theta)^{-1}d(x_0, x) < (2A_0)^{-1}[\delta^k + d(x_0, y)]$ and hence

$$|Q_k a(y)| = \left|\int_X Q_k(y, u)a(u)\, d\mu(u)\right| \le \int_B |Q_k(y, u) - Q_k(y, x_0)||a(u)|\, d\mu(u)$$
$$\lesssim \int_B \left[\frac{d(x_0, u)}{\delta^k + d(x_0, y)}\right]^\eta \frac{1}{V_{\delta^k}(x_0) + V(x_0, y)} \left[\frac{\delta^k}{\delta^k + d(x_0, y)}\right]^\gamma |a(u)|\, d\mu(u)$$
$$\lesssim [\mu(B)]^{1-\frac{1}{p}} \left[\frac{r_0}{\delta^k + d(x_0, y)}\right]^\eta \frac{1}{V_{\delta^k}(x_0) + V(x_0, y)} \left[\frac{\delta^k}{\delta^k + d(x_0, y)}\right]^\gamma.$$

On another hand, if $\delta^k < (4A_0^2\theta)^{-1}d(x_0, x)$, then $d(x_0, y) \ge (4A_0)^{-1}d(x_0, x)$ and hence

$$|Q_k a(y)| \lesssim [\mu(B)]^{1-\frac{1}{p}} \left[\frac{r_0}{d(x_0, x)}\right]^\eta \frac{1}{V(x_0, x)} \left[\frac{\delta^k}{d(x_0, x)}\right]^\gamma,$$

which further implies that

$$\sum_{\delta^k<(4A_0^2\theta)^{-1}d(x_0,x)} \int_{B(y,\theta\delta^k)} |Q_k a(y)|^2 \frac{d\mu(y)}{V_{\theta\delta^k}(x)}$$
$$\lesssim [\mu(B)]^{2-\frac{2}{p}} \left[\frac{r_0}{d(x_0, x)}\right]^{2\eta} \left[\frac{1}{V(x_0, x)}\right]^2 \sum_{\delta^k<(4A_0^2\theta)^{-1}d(x_0,x)} \left[\frac{\delta^k}{d(x_0, x)}\right]^{2\gamma}$$
$$\lesssim [\mu(B)]^{2-\frac{2}{p}} \left[\frac{r_0}{d(x_0, x)}\right]^{2\eta} \left[\frac{1}{V(x_0, x)}\right]^2.$$

If $\delta^k \ge (4A_0^2\theta)^{-1}d(x_0, x)$, then $V(x_0, x) \lesssim \mu(B(x_0, \theta\delta^k)) \lesssim \theta^\omega V_{\delta^k}(x_0)$ and

$$|Q_k a(y)| \lesssim \theta^\omega [\mu(B)]^{1-\frac{1}{p}} \left(\frac{r_0}{\delta^k}\right)^\eta \frac{1}{V(x_0, x)},$$

which further implies that

$$
\begin{aligned}
&\sum_{\delta^k \ge (4A_0^2\theta)^{-1} d(x_0,x)} \int_{B(y,\theta\delta^k)} |Q_k a(y)|^2 \frac{d\mu(y)}{V_{\theta\delta^k}(x)}\\
&\lesssim \theta^{2\omega}[\mu(B)]^{2-\frac{2}{p}} \left[\frac{1}{V(x_0,x)}\right]^2 \sum_{\delta^k \ge (4A_0^2\theta)^{-1} d(x_0,x)} \left(\frac{r_0}{\delta^k}\right)^{2\eta}\\
&\sim \theta^{2\omega+2\eta}[\mu(B)]^{2-\frac{2}{p}} \left[\frac{r_0}{d(x_0,x)}\right]^{2\eta} \left[\frac{1}{V(x_0,x)}\right]^2.
\end{aligned}
$$

Therefore, when $x \notin B(x_0, 4A_0^2\theta r_0)$, we have

$$
\mathcal{S}_\theta(a)(x) \lesssim \theta^{\omega+\eta}[\mu(B)]^{1-\frac{1}{p}} \left[\frac{r_0}{d(x_0,x)}\right]^{\eta} \frac{1}{V(x_0,x)}.
$$

Consequently, using $p \in (\eta/(\omega+\eta), 1]$, $B = B(x_0, r_0)$ and (1.1), we obtain

$$
\begin{aligned}
&\int_{[B(x_0,4A_0^2\theta r_0)]^\complement} [\mathcal{S}_\theta(a)(x)]^p \, d\mu(x)\\
&\lesssim \theta^{(\omega+\eta)p}[\mu(B)]^{p-1} \int_{[B(x_0,4A_0^2\theta r_0)]^\complement} \left[\frac{r_0}{d(x_0,x)}\right]^{p\eta} \left[\frac{1}{V(x_0,x)}\right]^p d\mu(x)\\
&\lesssim \theta^{p\omega}[\mu(B)]^{p-1} \sum_{j=2}^{\infty} 2^{-jp\eta}\\
&\quad \times \int_{(2A_0)^j\theta r_0 \le d(x_0,x) < (2A_0)^{j+1}\theta r_0} \left[\frac{1}{\mu(B(x_0,(2A_0)^j\theta r_0))}\right]^p d\mu(x)\\
&\lesssim \theta^{\omega} \sum_{j=2}^{\infty} 2^{-j[p\eta-(1-p)\omega]} \lesssim \theta^{\omega}.
\end{aligned} \tag{5.9}
$$

Combining (5.8) and (5.9) implies that, when $\theta \in [1,\infty)$,

$$
\|\mathcal{S}_\theta(a)\|_{L^p(X)} \lesssim \theta^{\omega/p}. \tag{5.10}
$$

Let $f \in \mathring{H}^{p,2}_{\rm at}(X)$. By the definition of $\mathring{H}^{p,2}_{\rm at}(X)$, we know that, for any $\epsilon \in (0,\infty)$, there exist $(p,2)$-atoms $\{a_j\}_{j=1}^\infty$ and $\{\lambda_j\}_{j=1}^\infty \subset \mathbb{C}$ such that $f = \sum_{j=1}^\infty \lambda_j a_j$ in $(\mathring{\mathcal{G}}_0^\eta(\beta,\gamma))'$ and $\sum_{j=1}^\infty |\lambda_j|^p \le \|f\|^p_{\mathring{H}^{p,2}_{\rm at}(X)} + \epsilon$. By (5.10) and the fact $\mathcal{S}_\theta(f) \le \sum_{j=1}^\infty |\lambda_j| \mathcal{S}_\theta(a_j)$, we conclude that

$$
\begin{aligned}
\|\mathcal{S}_\theta(f)\|^p_{L^p(X)} &\le \sum_{j=1}^\infty |\lambda_j|^p \|\mathcal{S}_\theta(a_j)\|^p_{L^p(X)} \lesssim \theta^\omega \sum_{j=1}^\infty |\lambda_j|^p\\
&\lesssim \theta^\omega [\|f\|^p_{\mathring{H}^{p,2}_{\rm at}(X)} + \epsilon] \to \theta^\omega \|f\|^p_{\mathring{H}^{p,2}_{\rm at}(X)}
\end{aligned}
$$

as $\epsilon \to 0^+$. This finishes the proof of (5.6) and hence of Proposition 5.6. □

Next, we use Proposition 5.5 to show the following converse of Proposition 5.6.

Proposition 5.7 *Let $p \in (\omega/(\omega+\eta), 1]$, $\beta, \gamma \in (\omega(1/p-1), \eta)$ and $f \in (\mathring{\mathcal{G}}_0^\eta(\beta,\gamma))'$ belong to $H^p(X)$. Then there exist a sequence $\{a_j\}_{j=1}^\infty$ of $(p,2)$-atoms and $\{\lambda_j\}_{j=1}^\infty \subset \mathbb{C}$ such that $f = \sum_{j=1}^\infty \lambda_j a_j$ in $(\mathring{\mathcal{G}}_0^\eta(\beta,\gamma))'$ and $\sum_{j=1}^\infty |\lambda_j|^p \le C\|f\|_{H^p(X)}^p$, where C is a positive constant independent of f. Consequently, $H^p(X) \subset \mathring{H}_{\mathrm{at}}^{p,2}(X)$.*

Proof Assume that $f \in (\mathring{\mathcal{G}}_0^\eta(\beta,\gamma))'$ belongs to $H^p(X)$. In this proof, to avoid the confusion of notation between the exp-ATI $\{Q_k\}_{k\in\mathbb{Z}}$ and $Q \in \mathcal{D}$, we use $\{E_k\}_{k\in\mathbb{Z}}$ to denote an exp-ATI and then define $\mathcal{S}(f)$ as in (5.1) but with Q_k therein replaced by E_k. Denote by $\mathcal{D}$ the set of all dyadic cubes. For any $k \in \mathbb{Z}$, we define $\Omega_k := \{x \in X : \mathcal{S}(f)(x) > 2^k\}$ and

$$\mathcal{D}_k := \left\{Q \in \mathcal{D} : \mu(Q \cap \Omega_k) > \frac{1}{2}\mu(Q) \text{ and } \mu(Q \cap \Omega_{k+1}) \le \frac{1}{2}\mu(Q)\right\}.$$

It is easy to see that, for any $Q \in \mathcal{D}$, there exists a unique $k \in \mathbb{Z}$ such that $Q \in \mathcal{D}_k$. A dyadic cube $Q \in \mathcal{D}_k$ is called a *maximal cube in $\mathcal{D}_k$* if $Q' \in \mathcal{D}$ and $Q' \supset Q$, then $Q' \notin \mathcal{D}_k$. Denote the set of all maximal cubes in $\mathcal{D}_k$ at level $j \in \mathbb{Z}$ by $\{Q_{\tau,k}^j\}_{\tau\in I_{j,k}}$, where $I_{j,k} \subset \mathcal{A}_j$ may be empty. The center of $Q_{\tau,k}^j$ is denoted by $z_{\tau,k}^j$. Then $\mathcal{D} = \bigcup_{j,\,k\in\mathbb{Z}} \bigcup_{\tau\in I_{j,k}} \{Q \in \mathcal{D}_k : Q \subset Q_{\tau,k}^j\}$.

From now on, we adopt the notation $E_Q := E_l$ and $\widetilde{E}_Q := \widetilde{E}_l$ whenever $Q = Q_\alpha^{l+1}$ for some $l \in \mathbb{Z}$ and $\alpha \in \mathcal{A}_{l+1}$. Then, by Theorem 2.6, we find that

$$\begin{aligned} f(\cdot) &= \sum_{l=-\infty}^{\infty} \widetilde{E}_l E_l f(\cdot) \\ &= \sum_{l=-\infty}^{\infty} \sum_{\alpha\in\mathcal{A}_{l+1}} \int_{Q_\alpha^{l+1}} \widetilde{E}_l(\cdot,y) E_l f(y)\,d\mu(y) \\ &= \sum_{Q\in\mathcal{D}} \int_Q \widetilde{E}_Q(\cdot,y) E_Q f(y)\,d\mu(y) \\ &= \sum_{k=-\infty}^{\infty} \sum_{j=-\infty}^{\infty} \sum_{\tau\in I_{j,k}} \sum_{Q\in\mathcal{D}_k,\ Q\subset Q_{\tau,k}^j} \int_Q \widetilde{E}_Q(\cdot,y) E_Q f(y)\,d\mu(y) \\ &=: \sum_{k=-\infty}^{\infty} \sum_{j=-\infty}^{\infty} \sum_{\tau\in I_{j,k}} \lambda_{\tau,k}^j b_{\tau,k}^j(\cdot), \end{aligned} \tag{5.11}$$

where all the equalities converge in $(\mathring{\mathcal{G}}_0^\eta(\beta,\gamma))'$,

$$\lambda_{\tau,k}^j := \left[\mu\left(Q_{\tau,k}^j\right)\right]^{\frac{1}{p}-\frac{1}{2}} \left[\sum_{Q\in\mathcal{D}_k,\ Q\subset Q_{\tau,k}^j} \int_Q |E_Q f(y)|^2\,d\mu(y)\right]^{\frac{1}{2}}$$

and

$$b^j_{\tau,k}(\cdot) := \frac{1}{\lambda^j_{\tau,k}} \sum_{Q\in\mathcal{D}_k,\ Q\subset Q^j_{\tau,k}} \int_Q \widetilde{E}_Q(\cdot, y)E_Q f(y)\,d\mu(y). \tag{5.12}$$

For any $Q \in \mathcal{D}_k$ and $Q \subset Q^j_{\tau,k}$, assume that $Q = Q^{l+1}_\alpha$ for some $l \in \mathbb{Z}$ and $\alpha \in \mathcal{A}_{l+1}$. Since δ is assumed to satisfy $\delta < (2A_0)^{-10}$, it then follows that $2A_0C^\natural\delta < 1$ so that $Q = Q^{l+1}_\alpha \subset B(y, \delta^l)$ for any $y \in Q$. By this and the fact that $\mu(Q\cap\Omega_{k+1}) \le \frac{1}{2}\mu(Q)$, we obtain

$$\begin{aligned}\mu(B(y,\delta^l)\cap[Q^j_{\tau,k}\setminus\Omega_{k+1}]) &\ge \mu(B(y,\delta^l)\cap[Q\setminus\Omega_{k+1}])\\ &= \mu(Q\setminus\Omega_{k+1}) \ge \frac{1}{2}\mu(Q) \sim V_{\delta^l}(y).\end{aligned}$$

Thus, we have

$$\begin{aligned}&\sum_{Q\in\mathcal{D}_k,\ Q\subset Q^j_{\tau,k}} \int_Q |E_Q f(y)|^2\,d\mu(y)\\ &\lesssim \sum_{l=j-1}^{\infty} \sum_{\alpha\in\mathcal{A}_{l+1},\ \mathcal{D}_k\ni Q^{l+1}_\alpha\subset Q^j_{\tau,k}} \int_{Q^{l+1}_\alpha} \frac{\mu(B(y,\delta^l)\cap(Q^j_{\tau,k}\setminus\Omega_{k+1}))}{V_{\delta^l}(y)} |E_l f(y)|^2\,d\mu(y)\\ &\lesssim \sum_{l=j-1}^{\infty} \int_{Q^j_{\tau,k}} \frac{\mu(B(y,\delta^l)\cap(Q^j_{\tau,k}\setminus\Omega_{k+1}))}{V_{\delta^l}(y)} |E_l f(y)|^2\,d\mu(y)\\ &\sim \int_X \sum_{l=j-1}^{\infty} \int_{B(y,\delta^l)\cap(Q^j_{\tau,k}\setminus\Omega_{k+1})} |E_l f(y)|^2 \frac{d\mu(x)}{V_{\delta^l}(y)}\,d\mu(y)\\ &\lesssim \int_{Q^j_{\tau,k}\setminus\Omega_{k+1}} [\mathcal{S}(f)(x)]^2\,d\mu(x) \lesssim 2^{2k}\mu\left(Q^j_{\tau,k}\right).\end{aligned}$$

From this and the fact $\mu(Q^j_{\tau,k}) < 2\mu(Q^j_{\tau,k}\cap\Omega_k)$, it follows that

$$\begin{aligned}\sum_{k=-\infty}^{\infty}\sum_{j=-\infty}^{\infty}\sum_{\tau\in I_{j,k}} \left(\lambda^j_{\tau,k}\right)^p &\lesssim \sum_{k=-\infty}^{\infty} 2^{kp} \sum_{j=-\infty}^{\infty}\sum_{\tau\in I_{j,k}} \mu\left(Q^j_{\tau,k}\right)\\ &\lesssim \sum_{k=-\infty}^{\infty} 2^{kp} \sum_{j=-\infty}^{\infty}\sum_{\tau\in I_{j,k}} \mu\left(Q^j_{\tau,k}\cap\Omega_k\right)\\ &\lesssim \sum_{k=-\infty}^{\infty} 2^{kp}\mu(\Omega_k) \sim \|\mathcal{S}(f)\|^p_{L^p(X)}.\end{aligned} \tag{5.13}$$

Choose $\gamma' \in (\omega(1/p-1), \gamma)$ and let $\vec{\epsilon} := \{\delta^{m[\gamma'-\omega(1/p-1)]}\}_{m\in\mathbb{N}}$. Assume for the moment that every $b^j_{\tau,k}$ as in (5.12) is a $(p, 2, \vec{\epsilon})$-molecule centered at a ball

$B^j_{\tau,k} := B(z^j_{\tau,k}, 4A_0^2\delta^{j-1})$, whose proof is given in Lemma 5.8 below. Further, applying Proposition 5.5, we conclude that $\|b^j_{\tau,k}\|_{H^{p,2}_{\rm cw}(X)} \lesssim 1$. Thus, $b^j_{\tau,k}$ can be written as a linear combination of $(p,2)$-atoms, which converges in $(\mathcal{L}_{1/p-1}(X))'$ when $p \in (\omega/(\omega+\eta), 1)$ or in $L^1(X)$ when $p = 1$, and hence converges in $(\mathring{\mathcal{G}}^\eta_0(\beta,\gamma))'$ because $\mathring{\mathcal{G}}^\eta_0(\beta,\gamma) \subset \mathcal{L}_{1/p-1}(X)$ (see Lemma 4.15). Invoking this, (5.11) and (5.13), we find that $f \in \mathring{H}^{p,2}_{\rm at}(X)$ and $\|f\|_{\mathring{H}^{p,2}_{\rm at}(X)} \lesssim \|\mathcal{S}(f)\|_{L^p(X)}$. This finishes the proof of Proposition 5.7 modulo the proof of Lemma 5.8. □

Lemma 5.8 *Let all the notation be as in the proof of Proposition* 5.7. *Then every* $b^j_{\tau,k}$ *as in* (5.12) *is a harmless positive constant multiple of a* $(p, 2, \vec{\epsilon})$*-molecule centered at the ball* $B^j_{\tau,k} := B(z^j_{\tau,k}, 4A_0^2\delta^{j-1})$, *where* $\vec{\epsilon} := \{\delta^{m[\gamma'-\omega(1/p-1)]}\}_{m\in\mathbb{N}}$ *and* $\gamma' \in (\omega(1/p-1), \gamma)$.

Proof Let $b^j_{\tau,k}$ be as in (5.12). For any $h \in L^2(X)$ with $\|h\|_{L^2(X)} \le 1$, by the Fubini theorem and the Hölder inequality, we conclude that

$$\begin{aligned}
&\left|\int_X b^j_{\tau,k}(x)h(x)\,d\mu(x)\right| \\
&\quad\le \frac{1}{\lambda^j_{\tau,k}} \sum_{Q\in\mathcal{D}_k,\ Q\subset Q^j_{\tau,k}} \int_Q |E_Q f(y)| \left|\int_X \widetilde{E}_Q(x,y)h(x)\,d\mu(x)\right| d\mu(y) \\
&\quad\le \frac{1}{\lambda^j_{\tau,k}} \left[\sum_{Q\in\mathcal{D}_k,\ Q\subset Q^j_{\tau,k}} \int_Q |E_Q f(y)|^2\,d\mu(y)\right]^{\frac12} \\
&\qquad\times \left[\sum_{Q\in\mathcal{D}_k,\ Q\subset Q^j_{\tau,k}} \int_X \left|\widetilde{E}^*_Q h(y)\right|^2 d\mu(y)\right]^{\frac12} \\
&\quad\le \left[\mu\left(Q^j_{\tau,k}\right)\right]^{\frac12-\frac1p} \|\widetilde{g}(h)\|_{L^2(X)},
\end{aligned}$$

where $\widetilde{g}(h) := [\sum_{l=-\infty}^{\infty} |\widetilde{E}^*_l h|^2]^{1/2}$. Noticing that the kernel of $\widetilde{E}^*_l$ has the regularity with respect to the second variable, we follow the argument used in the beginning of the proof of Proposition 5.6 to deduce that $\widetilde{g}$ is bounded on $L^2(X)$. Thus, we have

$$\left|\int_X b^j_{\tau,k}(x)h(x)\,d\mu(x)\right| \lesssim \left[\mu\left(Q^j_{\tau,k}\right)\right]^{\frac12-\frac1p} \|h\|_{L^2(X)} \lesssim \left[\mu\left(B^j_{\tau,k}\right)\right]^{\frac12-\frac1p}.$$

Taking supremum over all $h \in L^2(X)$ with $\|h\|_{L^2(X)} \le 1$, we further find that

$$\left\|b^j_{\tau,k}\right\|_{L^2(X)} \lesssim \left[\mu\left(B^j_{\tau,k}\right)\right]^{\frac12-\frac1p}.$$

Let $\gamma' \in (\omega(1/p-1), \gamma)$. Fix $m \in \mathbb{N}$ and let $R_m := (\delta^{-m} B^j_{\tau,k}) \setminus (\delta^{-m+1} B^j_{\tau,k})$. Then, for any $x \in R_m$, by the Hölder inequality and the size condition of $\{\widetilde{E}_l\}_{l\in\mathbb{Z}}$, we conclude that

$$
\begin{aligned}
\left|b^j_{\tau,k}(x)\right| &\le \frac{1}{\lambda^j_{\tau,k}} \sum_{Q\in\mathcal{D}_k,\ Q\subset Q^j_{\tau,k}} \int_Q \left|\widetilde{E}_Q(x,y)E_Q f(y)\right| d\mu(y)\\
&\lesssim \frac{1}{\lambda^j_{\tau,k}} \sum_{l=j-1}^{\infty} \sum_{\alpha\in\mathcal{A}_{l+1},\ \mathcal{D}_k\ni Q^{l+1}_\alpha\subset Q^j_{\tau,k}} \int_{Q^{l+1}_\alpha} \frac{1}{V_{\delta^l}(x)+V(x,y)}\\
&\quad\times \left[\frac{\delta^l}{\delta^l+d(x,y)}\right]^{\gamma} |E_l f(y)|\, d\mu(y)\\
&\lesssim \frac{1}{\lambda^j_{\tau,k}} \left\{ \sum_{l=j-1}^{\infty} \sum_{\alpha\in\mathcal{A}_{l+1},\ Q^{l+1}_\alpha\subset Q^j_{\tau,k}} \int_{Q^{l+1}_\alpha} \frac{1}{V_{\delta^l}(x)+V(x,y)}\right.\\
&\quad\left.\times \left[\frac{\delta^l}{\delta^l+d(x,y)}\right]^{2\gamma'} d\mu(y)\right\}^{\frac12}\\
&\quad\times \left\{ \sum_{l=j-1}^{\infty} \sum_{\substack{\alpha\in\mathcal{A}_{l+1}\\ \mathcal{D}_k\ni Q^{l+1}_\alpha\subset Q^j_{\tau,k}}} \int_{Q^{l+1}_\alpha} \frac{1}{V_{\delta^l}(x)+V(x,y)}\right.\\
&\quad\left.\times \left[\frac{\delta^l}{\delta^l+d(x,y)}\right]^{2(\gamma-\gamma')} |E_l f(y)|^2\, d\mu(y)\right\}^{\frac12}\\
&=: \frac{1}{\lambda^j_{\tau,k}} \Upsilon(x) \mathrm{Z}(x).
\end{aligned}
$$

Notice that, for any $x \in R_m$, we have $4A_0^2\delta^{j-m-1} \le d(x, z^j_{\tau,k}) < 4A_0^2\delta^{j-m-2}$ and, for any $y \in Q^{l+1}_\alpha \subset Q^j_{\tau,k}$, we have $\delta^l + d(x,y) \sim d(x,y) \sim \delta^{-m+j}$ and hence

$$
\begin{aligned}
\Upsilon(x) &\lesssim \left[\sum_{l=j-1}^{\infty} \sum_{\alpha\in\mathcal{A}_{l+1},\ Q^{l+1}_\alpha\subset Q^j_{\tau,k}} \int_{Q^{l+1}_\alpha} \frac{1}{\mu(B(y,\delta^{-m+j}))} \left(\frac{\delta^l}{\delta^{-m+j}}\right)^{2\gamma'} d\mu(y)\right]^{\frac12}\\
&\lesssim \left[\sum_{l=j-1}^{\infty} \left(\frac{\delta^l}{\delta^{-m+j}}\right)^{2\gamma'} \int_{Q^j_{\tau,k}} \frac{1}{\mu(B(z^j_{\tau,k},\delta^{-m+j}))} d\mu(y)\right]^{\frac12}\\
&\lesssim \delta^{m\gamma'} \left[\frac{\mu(B^j_{\tau,k})}{\mu(\delta^{-m} B^j_{\tau,k})}\right]^{\frac12}.
\end{aligned}
$$

Thus, for any $x \in R_m$, we have

$$\left|b^j_{\tau,k}(x)\right| \lesssim \frac{1}{\lambda^j_{\tau,k}} \delta^{m\gamma'} \left[\frac{\mu(B^j_{\tau,k})}{\mu(\delta^{-m} B^j_{\tau,k})}\right]^{\frac{1}{2}} \mathrm{Z}(x),$$

which, together with the Fubini theorem and Lemma 2.2(ii), implies that

$$\begin{aligned}
\left\|b^j_{\tau,k}\mathbf{1}_{R_m}\right\|_{L^2(X)} &\lesssim \frac{1}{\lambda^j_{\tau,k}} \delta^{m\gamma'} \left[\frac{\mu(B^j_{\tau,k})}{\mu(\delta^{-m} B^j_{\tau,k})}\right]^{\frac{1}{2}} \left\{\int_{R_m} [\mathrm{Z}(x)]^2 \, d\mu(x)\right\}^{\frac{1}{2}} \\
&\lesssim \frac{1}{\lambda^j_{\tau,k}} \delta^{m\gamma'} \left[\frac{\mu(B^j_{\tau,k})}{\mu(\delta^{-m} B^j_{\tau,k})}\right]^{\frac{1}{2}} \left\{\sum_{Q\in\mathcal{D}_k,\ Q\subset Q^j_{\tau,k}} \int_Q \left|E_Q f(y)\right|^2 \, d\mu(y)\right\}^{\frac{1}{2}} \\
&\lesssim \delta^{m\gamma'} \left[\frac{\mu(B^j_{\tau,k})}{\mu(\delta^{-m} B^j_{\tau,k})}\right]^{\frac{1}{2}} \left[\mu\left(B^j_{\tau,k}\right)\right]^{\frac{1}{2}-\frac{1}{p}} \\
&\lesssim \delta^{m[\gamma'-\omega(\frac{1}{p}-1)]} \left[\mu\left(\delta^{-m} B^j_{\tau,k}\right)\right]^{\frac{1}{2}-\frac{1}{p}}.
\end{aligned}$$

The cancelation of $b^j_{\tau,k}$ follows directly from that of $\widetilde{E}_l$, the details being omitted.

Letting $\epsilon_m := \delta^{m[\gamma'-\omega(\frac{1}{p}-1)]}$ for any $m \in \mathbb{N}$, we find that $\{\epsilon_m\}_{m=1}^\infty$ satisfies (5.4) and $b^j_{\tau,k}$ is a harmless positive constant multiple of a $(p, 2, \vec{\epsilon})$-molecule. This finishes the proof of Lemma 5.8. □

Combining Propositions 5.6 and 5.7, we immediately obtain the following main result of this section, the details being omitted.

Theorem 5.9 *Suppose that* $p \in (\omega/(\omega+\eta), 1]$, $\beta,\ \gamma \in (\omega(1/p-1), \eta)$ *and* $q \in (p, \infty] \cap [1, \infty]$. *As subspaces of* $(\mathring{\mathcal{G}}^\eta_0(\beta,\gamma))'$, *it holds true that* $\mathring{H}^{p,q}_{at}(X) = H^p(X)$ *with equivalent (quasi-)norms.*

5.3 Hardy Spaces via Various Littlewood–Paley Functions

In this section, we characterize Hardy spaces $H^p(X)$ via the Lusin area functions with apertures, the Littlewood–Paley g-functions and the Littlewood–Paley g^*_λ-functions, respectively. We first consider the Littlewood–Paley g-function characterizations.

Theorem 5.10 *Let* $p \in (\omega/(\omega+\eta), 1]$ *and* $\beta,\ \gamma \in (\omega(1/p-1), \eta)$. *Then there exists a constant* $C \in [1, \infty)$ *such that, for any* $f \in (\mathring{\mathcal{G}}^\eta_0(\beta,\gamma))'$,

$$C^{-1}\|f\|_{H^p(X)} \le \|g(f)\|_{L^p(X)} \le C\|f\|_{H^p(X)}. \tag{5.14}$$

Proof Let $f \in (\mathring{\mathcal{G}}_0^\eta(\beta,\gamma))'$ with $\beta,\ \gamma \in (\omega(1/p-1),\eta)$. With $\{Q_k\}_{k\in\mathbb{Z}}$ being an exp-ATI, we define $\mathcal{S}(f)$ and $g(f)$, respectively, as in (5.1) and (5.3). If $f \in H^p(X) = \mathring{H}^{p,2}_{\rm at}(X)$, then, following the proof of (5.6), we also obtain

$$\|g(f)\|_{L^p(X)} \lesssim \|f\|_{\mathring{H}^{p,2}_{\rm at}(X)} \sim \|f\|_{H^p(X)}.$$

To finish the proof of (5.14), it remains to prove $\|f\|_{H^p(X)} \lesssim \|g(f)\|_{L^p(X)}$. Indeed, for any $x \in X$, we have

$$\begin{aligned}
\mathcal{S}(f)(x) &= \left[\sum_{k\in\mathbb{Z}}\sum_{\alpha\in\mathcal{A}_k}\sum_{m=1}^{N(k,\alpha)}\int_{B(x,\delta^k)}|Q_kf(y)|^2\mathbf{1}_{Q_\alpha^{k,m}}(x)\frac{d\mu(y)}{V_{\delta^k}(x)}\right]^{\frac12}\\
&\le \left\{\sum_{k\in\mathbb{Z}}\sum_{\alpha\in\mathcal{A}_k}\sum_{m=1}^{N(k,\alpha)}\left[\sup_{z\in B(z_\alpha^{k,m},\delta^{k-1})}|Q_kf(z)|^2\right]\mathbf{1}_{Q_\alpha^{k,m}}(x)\right\}^{\frac12}, \qquad (5.15)
\end{aligned}$$

where $Q_\alpha^{k,m}$ is as in Sect. 2 and $z_\alpha^{k,m}$ the center of $Q_\alpha^{k,m}$. With all the notation as in Theorem 2.7, we know that, for any $z \in B(z_\alpha^{k,m},\delta^{k-1})$,

$$Q_kf(z) = \sum_{k'\in\mathbb{Z}}\sum_{\alpha'\in\mathcal{A}_{k'}}\sum_{m'=1}^{N(k',\alpha')}\mu\left(Q_{\alpha'}^{k',m'}\right)Q_k\widetilde{Q}_{k'}\left(z,y_{\alpha'}^{k',m'}\right)Q_{k'}f\left(y_{\alpha'}^{k',m'}\right),$$

where $y_{\alpha'}^{k',m'}$ is an arbitrary point in $Q_{\alpha'}^{k',m'}$. Fix $\beta' \in (0,\beta\wedge\gamma)$. Then, similarly to the proof of (3.7) (see also [27, (3.2)]), we conclude that, for any $z \in B(z_\alpha^{k,m},\delta^{k-1})$

$$\left|Q_k\widetilde{Q}_{k'}\left(z,y_{\alpha'}^{k',m'}\right)\right| \lesssim \delta^{|k-k'|\beta'}\frac{1}{V_{\delta^{k\wedge k'}}(z)+V(z,y_{\alpha'}^{k',m'})}\left[\frac{1}{\delta^{k\wedge k'}+d(z,y_{\alpha'}^{k',m'})}\right]^\gamma. \qquad (5.16)$$

The variable z in (5.16) can be replaced by any $x \in Q_\alpha^{k,m}$, because $\max\{d(z,x), d(z,z_\alpha^{k,m})\} \lesssim \delta^k \lesssim \delta^{k\wedge k'}$. Further, from Lemma 3.7, we deduce that, for any fixed $r \in (\omega/(\omega+\eta),1]$, any $k' \in \mathbb{Z}$ and $z \in B(z_\alpha^{k,m},\delta^{k-1})$,

$$\begin{aligned}
&\left|\sum_{\alpha'\in\mathcal{A}_{k'}}\sum_{m'=1}^{N(k',\alpha')}\mu\left(Q_{\alpha'}^{k',m'}\right)Q_k\widetilde{Q}_{k'}\left(z,y_{\alpha'}^{k',m'}\right)Q_{k'}f\left(y_{\alpha'}^{k',m'}\right)\right|\\
&\quad\lesssim \delta^{(k\wedge k'-k)\omega(\frac1r-1)}\left[\mathcal{M}\left(\sum_{\alpha'\in\mathcal{A}_{k'}}\left|Q_{k'}f\left(y_{\alpha'}^{k',m'}\right)\right|^r\mathbf{1}_{Q_{\alpha'}^{k',m'}}\right)(x)\right]^{\frac1r}
\end{aligned}$$

and hence

$$|Q_k f(z)| \lesssim \sum_{k'\in\mathbb{Z}} \delta^{|k-k'|\beta'} \delta^{(k\wedge k'-k)\omega(\frac{1}{r}-1)} \times \left[\mathcal{M}\left(\sum_{\alpha'\in\mathcal{A}_{k'}} \sum_{m'=1}^{N(k',\alpha')} \left|Q_{k'} f\left(y_{\alpha'}^{k',m'}\right)\right|^r \mathbf{1}_{Q_{\alpha'}^{k',m'}}\right)(x)\right]^{\frac{1}{r}}. \quad (5.17)$$

Combining (5.15) and (5.17), choosing r and β' such that $r \in (\omega/(\omega+\beta'), p)$ and applying the Hölder inequality, we further conclude that, for any $x \in X$,

$$\begin{aligned}
[\mathcal{S}(f)(x)]^2 &\lesssim \sum_{k\in\mathbb{Z}} \sum_{\alpha\in\mathcal{A}_k} \sum_{m=1}^{N(k,\alpha)} \left\{ \sum_{k'\in\mathbb{Z}} \delta^{|k-k'|\beta'} \delta^{(k\wedge k'-k)\omega(\frac{1}{r}-1)} \right. \\
&\quad \left. \times \left[\mathcal{M}\left(\sum_{\alpha'\in\mathcal{A}_{k'}} \sum_{m'=1}^{N(k',\alpha')} \left|Q_{k'} f\left(y_{\alpha'}^{k',m'}\right)\right|^r \mathbf{1}_{Q_{\alpha'}^{k',m'}}\right)(x)\right]^{\frac{1}{r}} \mathbf{1}_{Q_\alpha^{k,m}}(x) \right\}^2 \\
&\lesssim \sum_{k\in\mathbb{Z}} \sum_{\alpha\in\mathcal{A}_k} \sum_{m=1}^{N(k,\alpha)} \sum_{k'\in\mathbb{Z}} \delta^{|k-k'|\beta'} \delta^{(k\wedge k'-k)\omega(\frac{1}{r}-1)} \\
&\quad \times \left[\mathcal{M}\left(\sum_{\alpha'\in\mathcal{A}_{k'}} \sum_{m'=1}^{N(k',\alpha')} \left|Q_{k'} f\left(y_{\alpha'}^{k',m'}\right)\right|^r \mathbf{1}_{Q_{\alpha'}^{k',m'}}\right)(x)\right]^{\frac{2}{r}} \mathbf{1}_{Q_\alpha^{k,m}}(x) \\
&\lesssim \sum_{k\in\mathbb{Z}} \sum_{k'\in\mathbb{Z}} \delta^{|k-k'|[\beta'-\omega(\frac{1}{r}-1)]} \left[\mathcal{M}\left(\sum_{\alpha'\in\mathcal{A}_{k'}} \sum_{m'=1}^{N(k',\alpha')} \left|Q_{k'} f\left(y_{\alpha'}^{k',m'}\right)\right|^r \mathbf{1}_{Q_{\alpha'}^{k',m'}}\right)(x)\right]^{\frac{2}{r}} \\
&\lesssim \sum_{k'\in\mathbb{Z}} \left[\mathcal{M}\left(\sum_{\alpha'\in\mathcal{A}_{k'}} \sum_{m'=1}^{N(k',\alpha')} \left|Q_{k'} f\left(y_{\alpha'}^{k',m'}\right)\right|^r \mathbf{1}_{Q_{\alpha'}^{k',m'}}\right)(x)\right]^{\frac{2}{r}}.
\end{aligned}$$

From this and Lemma 5.2, we deduce that

$$\begin{aligned}
\|f\|_{H^p(X)} &= \|[\mathcal{S}(f)]^r\|_{L^{p/r}(X)}^{\frac{1}{r}} \\
&\lesssim \left\| \left\{ \sum_{k'\in\mathbb{Z}} \left[\mathcal{M}\left(\sum_{\alpha'\in\mathcal{A}_{k'}} \sum_{m'=1}^{N(k',\alpha')} \left|Q_{k'} f\left(y_{\alpha'}^{k',m'}\right)\right|^r \mathbf{1}_{Q_{\alpha'}^{k',m'}}\right)\right]^{\frac{2}{r}} \right\}^{\frac{r}{2}} \right\|_{L^{p/r}(X)}^{\frac{1}{r}}
\end{aligned}$$

$$\lesssim \left\| \left\{ \sum_{k'\in\mathbb{Z}} \left[\sum_{\alpha'\in\mathcal{A}_{k'}} \sum_{m'=1}^{N(k',\alpha')} \left| Q_{k'} f\left(y_{\alpha'}^{k',m'}\right)\right|^r \mathbf{1}_{Q_{\alpha'}^{k',m'}} \right]^{\frac{2}{r}} \right\}^{\frac{r}{2}} \right\|_{L^{p/r}(X)}^{\frac{1}{r}}$$

$$\sim \left\| \left[\sum_{k'\in\mathbb{Z}} \sum_{\alpha'\in\mathcal{A}_{k'}} \sum_{m'=1}^{N(k',\alpha')} \left| Q_{k'} f\left(y_{\alpha'}^{k',m'}\right)\right|^2 \mathbf{1}_{Q_{\alpha'}^{k',m'}} \right]^{\frac{1}{2}} \right\|_{L^p(X)}.$$

By this and the arbitrariness of $y_{\alpha'}^{k',m'}$, we finally conclude that

$$\|f\|_{H^p(X)} \lesssim \left\| \left[\sum_{k'\in\mathbb{Z}} \sum_{\alpha'\in\mathcal{A}_{k'}} \sum_{m'=1}^{N(k',\alpha')} \inf_{z\in Q_{\alpha'}^{k',m'}} \left| Q_{k'} f(z)\right|^2 \mathbf{1}_{Q_{\alpha'}^{k',m'}} \right]^{\frac{1}{2}} \right\|_{L^p(X)} \lesssim \|g(f)\|_{L^p(X)}.$$

This finishes the proof of $\|f\|_{H^p(X)} \lesssim \|g(f)\|_{L^p(X)}$ and hence of Theorem 5.10. □

To consider the g_λ^*-function characterization of $H^p(X)$, we need a new kind of Littlewood–Paley functions. For any $\beta,\ \gamma \in (0,\eta)$, $\theta \in (0,\infty)$ and $f \in (\mathring{\mathcal{G}}_0^\eta(\beta,\gamma))'$, define the *Littlewood–Paley auxiliary function* $\mathcal{S}_\theta^{(1)}(f)$ *of* f *with aperture* θ by setting, for any $x \in X$,

$$\mathcal{S}_\theta^{(1)}(f)(x) := \left[\sum_{k=-\infty}^{\infty} \int_{B(y,\delta^k)} |Q_k f(y)|^2 \frac{d\mu(y)}{V_{\delta^k}(y)} \right]^{1/2}. \tag{5.18}$$

It is obvious that, for any $f \in (\mathring{\mathcal{G}}_0^\eta(\beta,\gamma))'$, $\mathcal{S}_1^{(1)}(f) \sim \mathcal{S}(f)$ and, for any $\theta \in [1,\infty)$, by the Fubini theorem, we have

$$\left\|\mathcal{S}_\theta^{(1)}(f)\right\|_{L^2(X)} \lesssim \theta^{\omega/2}\|\mathcal{S}(f)\|_{L^2(X)} \sim \theta^{\omega/2}\left\|\mathcal{S}_1^{(1)}(f)\right\|_{L^2(X)},$$

with the implicit positive constants independent of f. For the case $p \in (0,2)$, we have the following lemma.

Lemma 5.11 *Let* $\beta,\ \gamma \in (0,\eta)$ *and* $p \in (0,2)$. *Then there exists a positive constant* C *such that, for any* $\theta \in [1,\infty)$ *and* $f \in (\mathring{\mathcal{G}}_0^\eta(\beta,\gamma))'$,

$$\left\|\mathcal{S}_\theta^{(1)}(f)\right\|_{L^p(X)} \le C\theta^{\omega/p}\left\|\mathcal{S}_1^{(1)}(f)\right\|_{L^p(X)}. \tag{5.19}$$

Proof For any $\theta \in [1,\infty)$, any non-negative function g and any $x \in X$, define

$$\widetilde{\mathcal{M}}(g)(x) := \sup_{k\in\mathbb{Z}} \sup_{d(x,y)<\theta\delta^k} \frac{1}{V_{\delta^k}(y)} \int_{B(y,\delta^k)} g(z)\,d\mu(z).$$

Notice that, for any $k \in \mathbb{Z}$, $y \in B(x, \theta\delta^k)$ and $z \in B(y, \delta^k)$,

$$d(x, z) \le A_0[d(x, y) + d(y, z)] < 2A_0\theta\delta^k,$$

which further implies $B(y, \delta^k) \subset B(x, 2A_0\theta\delta^k)$. By this and (1.1), we find that, for any $k \in \mathbb{Z}$, $x \in X$ and $y \in B(x, \theta\delta^k)$,

$$\begin{aligned}\frac{1}{V_{\delta^k}(y)} \int_{B(y,\delta^k)} g(z)\, d\mu(z) &\le \frac{V_{2A_0\theta\delta^k}(x)}{V_{\delta^k}(y)} \frac{1}{V_{2A_0\theta\delta^k}(x)} \int_{B(x,2A_0\theta\delta^k)} g(z)\, d\mu(z)\\ &\lesssim \frac{V_{2A_0\theta\delta^k}(y)}{V_{\delta^k}(y)} \mathcal{M}(g)(x) \lesssim \theta^\omega \mathcal{M}(g)(x),\end{aligned}$$

which, together with the boundedness of $\mathcal{M}$ from $L^1(X)$ to $L^{1,\infty}(X)$, further implies that, for any $r \in (0, \infty)$,

$$\mu\left(\left\{x \in X : \widetilde{\mathcal{M}}(g)(x) > r\right\}\right) \lesssim \frac{\theta^\omega}{r} \|g\|_{L^1(X)}. \tag{5.20}$$

For any $t \in (0, \infty)$ and $f \in (\mathring{\mathcal{G}}_0^\eta(\beta, \gamma))'$ with $\beta,\ \gamma \in (0, \eta)$, define $E_t := \{x \in X : \mathcal{S}_1^{(1)}(f)(x) > t\}$ and $\widetilde{E}_t := \{x \in X : \widetilde{\mathcal{M}}(\mathbf{1}_{E_t})(x) > 1/2\}$. We claim that, for any $t \in (0, \infty)$,

$$\int_{\widetilde{E}_t^\complement} \left[\mathcal{S}_\theta^{(1)}(f)(x)\right]^2 d\mu(x) \lesssim \theta^\omega \int_{E_t^\complement} \left[\mathcal{S}_1^{(1)}(f)(x)\right]^2 d\mu(x). \tag{5.21}$$

Assuming this for the moment, we continue the proof of (5.19). Indeed, by the Chebyshev inequality, (5.20) and (5.21), we find that, for any $t \in (0, \infty)$,

$$\begin{aligned}\mu\left(\left\{x \in X : \mathcal{S}_\theta^{(1)}(f)(x) > t\right\}\right) &\le \mu\left(\widetilde{E}_t\right) + \mu\left(\left\{x \in \widetilde{E}_t^\complement : \mathcal{S}_\theta^{(1)}(f)(x) > t\right\}\right)\\ &\lesssim \theta^\omega \mu(E_t) + t^{-2} \int_{\widetilde{E}_t^\complement} \left[\mathcal{S}_\theta^{(1)}(f)(x)\right]^2 d\mu(x)\\ &\lesssim \theta^\omega \mu(E_t) + t^{-2}\theta^\omega \int_{E_t^\complement} \left[\mathcal{S}_1^{(1)}(f)(x)\right]^2 d\mu(x)\\ &\lesssim \theta^\omega \left[\mu(E_t) + t^{-2} \int_0^t s\mu(E_s)\, ds\right].\end{aligned}$$

By this, the Tonelli theorem and the fact $p \in (0, 2)$, we conclude that

$$\begin{aligned}\left\|\mathcal{S}_\theta^{(1)}(f)\right\|_{L^p(X)}^p &= p \int_0^\infty t^{p-1} \mu\left(\left\{x \in X : \mathcal{S}_\theta^{(1)}(f)(x) > t\right\}\right) dt\\ &\lesssim \theta^\omega \left[\int_0^\infty t^{p-1} \mu(E_t)\, dt + \int_0^\infty t^{p-3} \int_0^t s\mu(E_s)\, ds\, dt\right]\\ &\sim \theta^\omega \left\|\mathcal{S}_1^{(1)}(f)\right\|_{L^p(X)}^p + \theta^\omega \int_0^\infty s\mu(E_s) \int_s^\infty t^{p-3}\, dt\, ds\end{aligned}$$

$$\sim \theta^{\omega} \left\| \mathcal{S}_1^{(1)}(f) \right\|_{L^p(X)}^p + \theta^{\omega} \int_0^{\infty} s^{p-1} \mu(E_s)\, ds \sim \theta^{\omega} \left\| \mathcal{S}_1^{(1)}(f) \right\|_{L^p(X)}^p .$$

This finishes the proof of (5.19) under the assumption (5.21).

It remains to show (5.21). Fix $t \in (0, \infty)$ and, for any $y \in X$, let $\rho(y) := \inf_{x \in \widetilde{E}_t^{\complement}} d(x, y)$. Then, for any $k \in \mathbb{Z}$ and $x,\ y \in X$, $x \in \widetilde{E}_\lambda^{\complement} \cap B(y, \theta\delta^k)$ implies that $\rho(y) < \theta\delta^k$. By this and the Tonelli theorem, we find that

$$\begin{aligned} \int_{\widetilde{E}_t^{\complement}} \left[\mathcal{S}_\theta^{(1)}(f)(x)\right]^2 d\mu(x) &= \int_{\widetilde{E}_t^{\complement}} \sum_{k=-\infty}^{\infty} \int_{B(x,\theta\delta^k)} |Q_k f(y)|^2 \frac{d\mu(y)}{V_{\delta^k}(y)}\, d\mu(x) \\ &= \sum_{k=-\infty}^{\infty} \int_{\rho(y)<\theta\delta^k} |Q_k f(y)|^2 \mu\left(\widetilde{E}_t^{\complement} \cap B(y, \theta\delta^k)\right) \frac{d\mu(y)}{V_{\delta^k}(y)} \\ &\lesssim \theta^{\omega} \sum_{k=-\infty}^{\infty} \int_{\rho(y)<\theta\delta^k} |Q_k f(y)|^2 \mu(B(y, \delta^k))\, \frac{d\mu(y)}{V_{\delta^k}(y)}. \end{aligned} \tag{5.22}$$

When $\rho(y) < \theta\delta^k$, we have $\widetilde{E}_t^{\complement} \cap B(y, \theta\delta^k) \neq \emptyset$. Choose $y_0 \in \widetilde{E}_t^{\complement} \cap B(y, \theta\delta^k)$. Then we have

$$\begin{aligned} \mu(E_t \cap B(y, \delta^k)) &= \int_{B(y,\delta^k)} \mathbf{1}_{E_t}(z)\, d\mu(z) \le \mu(B(y, \delta^k))\widetilde{\mathcal{M}}\left(\mathbf{1}_{E_t}\right)(y_0) \\ &\le \frac{1}{2}\mu(B(y, \delta^k)). \end{aligned}$$

Thus, $\mu(E_t^{\complement} \cap B(y, \delta^k)) \ge \frac{1}{2}\mu(B(y, \delta^k))$. By this, (5.22) and the Tonelli theorem, we further conclude that

$$\begin{aligned} \int_{\widetilde{E}_t^{\complement}} \left[\mathcal{S}_\theta^{(1)}(f)(x)\right]^2 d\mu(x) &\lesssim \theta^{\omega} \sum_{k=-\infty}^{\infty} \int_{\rho(y)<\theta\delta^k} |Q_k f(y)|^2 \mu(B(y, \delta^k))\, \frac{d\mu(y)}{V_{\delta^k}(y)} \\ &\lesssim \theta^{\omega} \sum_{k=-\infty}^{\infty} \int_X |Q_k f(y)|^2 \mu\left(E_t^{\complement} \cap B(y, \delta^k)\right) \frac{d\mu(y)}{V_{\delta^k}(y)} \\ &\sim \sum_{k=-\infty}^{\infty} \int_{E_t^{\complement}} \int_{B(x,\delta^k)} |Q_k f(y)|^2 \frac{d\mu(y)}{V_{\delta^k}(y)}\, d\mu(x) \\ &\sim \theta^{\omega} \int_{E_t^{\complement}} \left[\mathcal{S}_1^{(1)}(f)(x)\right]^2 d\mu(x). \end{aligned}$$

This finishes the proof of (5.21) and hence of Lemma 5.11. □

Using Lemma 5.11, we now establish the Littlewood–Paley g_λ^*-function characterization of $H^p(X)$ when $p \in (\omega/(\omega + \eta), 1]$.

Theorem 5.12 *Let* $p \in (\omega/(\omega+\eta), 1]$, $\beta, \gamma \in (\omega(1/p-1), \eta)$ *and* $\lambda \in (2\omega/p, \infty)$. *Then there exists a constant* $C \in [1, \infty)$ *such that, for any* $f \in (\mathring{\mathcal{G}}_0^\eta(\beta, \gamma))'$,

$$C^{-1}\|f\|_{H^p(X)} \le \|g_\lambda^*(f)\|_{L^p(X)} \le C\|f\|_{H^p(X)}. \tag{5.23}$$

Proof Fix $p \in (\omega/(\omega+\eta), 1]$ and $\lambda \in (2\omega/p, \infty)$. For any $f \in (\mathring{\mathcal{G}}_0^\eta(\beta, \gamma))'$ with $\beta, \gamma \in (\omega(1/p-1), \eta)$, by the definition of $g_\lambda^*(f)$ [see (5.3)], we find that

$$[\mathcal{S}(f)]^2 \lesssim [g_\lambda^*(f)]^2 \lesssim \sum_{j=0}^\infty 2^{(1-j)\lambda}\left[\mathcal{S}_{2^j}^{(1)}(f)\right]^2. \tag{5.24}$$

By (5.24), we easily obtain the first inequality of (5.23). For the second one, by (5.24), the fact $p < 2$, (5.19) and $\lambda > 2\omega/p$, we conclude that, for any $f \in (\mathring{\mathcal{G}}_0^\eta(\beta, \gamma))'$ with $\beta, \gamma \in (\omega(1/p-1), \eta)$,

$$\begin{aligned}
\|g_\lambda^*(f)\|_{L^p(X)}^p &= \left\|[g_\lambda^*(f)]^2\right\|_{L^{p/2}(X)}^{p/2} \lesssim \sum_{j=0}^\infty 2^{-j\lambda p/2}\left\|\left[\mathcal{S}_{2^j}^{(1)}(f)\right]^2\right\|_{L^{p/2}(X)}^{p/2} \\
&\sim \sum_{j=0}^\infty 2^{-j\lambda p/2}\left\|\mathcal{S}_{2^j}^{(1)}(f)\right\|_{L^p(X)}^p \lesssim \left\|\mathcal{S}_1^{(1)}(f)\right\|_{L^p(X)}^p \sum_{j=0}^\infty 2^{-j\lambda p/2}2^{j\omega} \\
&\sim \|\mathcal{S}(f)\|_{L^p(X)}^p.
\end{aligned}$$

This finishes the proof of the second inequality of (5.23) and hence of Theorem 5.12. □

Remark 5.13 If X is a homogeneous group, Folland and Stein [12, Corollary 7.4] showed that, for any given $p \in (0, 2]$ and any $f \in \mathscr{S}'(X)$, $\|g_\lambda^*(f)\|_{L^p(X)} \le C\|\mathcal{S}(f)\|_{L^p(X)}$ whenever $\lambda \in (2\omega/p, \infty)$ with the positive constant C independent of f by observing that λ in (5.3) equals to 2λ with λ as in the Littlewood–Paley g_λ^*-function in [12], where $\mathscr{S}'(X)$ denotes the space of tempered distributions on X. Observe that Theorem 5.12 in this case coincides with Folland and Stein [12, Corollary 7.4], whose range of λ is the known best possible. Moreover, Lemma 5.11 and its proof have their own interest in dealing with Littlewood–Paley g_λ^*-functions on spaces of homogeneous type. For example, using this method, one can improve the range of λ in [26, Proposition 3.4(ii)] to $(2n/p, \infty)$, which then coincides with Theorem 5.12 and whose range of λ is then the known best possible.

6 Wavelet Characterizations of Hardy Spaces

In this section, we characterize the Hardy space $H^p(X)$ via the wavelet orthogonal system $\{\psi_\alpha^k : k \in \mathbb{Z}, \alpha \in \mathcal{G}_k\}$ introduced in [1, Theorem 7.1]. The sequence $\{D_k\}_{k\in\mathbb{Z}}$ of operators on $L^2(X)$ associated with integral kernels

$$D_k(x, y) := \sum_{\alpha \in \mathcal{G}_k} \psi_\alpha^k(x) \psi_\alpha^k(y), \qquad \forall\, x,\ y \in X \tag{6.1}$$

turns out to be an exp-ATI; see [25,29]. Thus, all the conclusions in Sect. 5 hold true for $\{D_k\}_{k\in\mathbb{Z}}$.

For any $f \in (\mathring{\mathcal{G}}_0^\eta(\beta, \gamma))'$ with $\beta,\ \gamma \in (0, \eta)$, define the *wavelet Littlewood–Paley function* $S(f)$ by setting, for any $x \in X$,

$$S(f)(x) := \left\{ \sum_{k\in\mathbb{Z}} \sum_{\alpha \in \mathcal{G}_k} \left[\mu\left(Q_\alpha^{k+1}\right)\right]^{-1} \left|\left\langle \psi_\alpha^k, f \right\rangle\right|^2 \mathbf{1}_{Q_\alpha^{k+1}}(x) \right\}^{\frac{1}{2}}.$$

For any $p \in (0, \infty)$, define the corresponding *wavelet Hardy space* $H_\mathrm{w}^p(X)$ by

$$H_\mathrm{w}^p(X) := \left\{ f \in \left(\mathring{\mathcal{G}}_0^\eta(\beta, \gamma)\right)' :\ \|f\|_{H_\mathrm{w}^p(X)} := \|S(f)\|_{L^p(X)} < \infty \right\}.$$

For any $p \in (\omega/(\omega + \eta), \infty)$, the $L^p(X)$-norm equivalence between the wavelet Littlewood–Paley function $S(f)$ and the Littlewood–Paley g-function $g(f)$ was proved in [25, Theorem 4.3] whenever f is a distribution. The proof of [25, Theorem 4.3] seems *problematic* because the authors therein used an unknown fact that, when $f \in (\mathring{\mathcal{G}}(\beta, \gamma))'$ and $n \in \mathbb{N}$,

$$\sum_{|k|\le n} \sum_{\alpha \in \mathcal{G}_k} \left\langle f, \psi_\alpha^k \right\rangle \psi_\alpha^k \in L^2(X). \tag{6.2}$$

Although (6.2) may not be true for distributions, it is obviously true when $f \in L^2(X)$. Indeed, the argument used in the proof of [25, Theorem 4.3] proves the following result.

Theorem 6.1 *Suppose $p \in (\omega/(\omega + \eta), \infty)$ and $\beta,\ \gamma \in (0, \eta)$. Then there exists a positive constant C such that, for any $f \in (\mathring{\mathcal{G}}_0^\eta(\beta, \gamma))'$,*

$$\|\mathcal{G}(f)\|_{L^p(X)} \le C\|S(f)\|_{L^p(X)} \tag{6.3}$$

and, if $f \in L^2(X)$, then

$$C^{-1}\|S(f)\|_{L^p(X)} \le \|\mathcal{G}(f)\|_{L^p(X)} \le C\|S(f)\|_{L^p(X)}. \tag{6.4}$$

Here and hereafter, $\mathcal{G}(f)$ is defined as in (5.2), but with Q_k therein replaced by D_k in (6.1).

To show that (6.4) holds true for all distributions, we need the following basic property of $H_\mathrm{w}^p(X)$.

Proposition 6.2 *Let $p \in (\omega/(\omega + \eta), 1]$ and $\beta,\ \gamma \in (\omega(1/p - 1), \eta)$. Then $H_\mathrm{w}^p(X)$ is a (quasi-) Banach space that can be continuously embedded into $(\mathring{\mathcal{G}}_0^\eta(\beta, \gamma))'$.*

Proof Assume that $f \in (\mathring{\mathcal{G}}_0^\eta(\beta,\gamma))'$ belongs to $H_{\rm w}^p(X)$. By (6.3), Theorems 5.10 and 5.9, we have $\|f\|_{\mathring{H}_{\rm at}^{p,2}(X)} \lesssim \|f\|_{H_{\rm w}^p(X)}$. Consequently, for any $\epsilon \in (0,\infty)$, there exist $(p,2)$-atoms $\{a_j\}_{j=1}^\infty$ and $\{\lambda_j\}_{j=1}^\infty \subset \mathbb{C}$ satisfying $(\sum_{j=1}^\infty |\lambda_j|^p)^{\frac{1}{p}} \le \|f\|_{\mathring{H}_{\rm at}^{p,2}(X)} + \epsilon$ such that $f = \sum_{j=1}^\infty \lambda_j a_j$ in $(\mathring{\mathcal{G}}_0^\eta(\beta,\gamma))'$. Combining this with Lemmas 4.14 and 4.15, we find that, for any $\varphi \in \mathring{\mathcal{G}}_0^\eta(\beta,\gamma)$,

$$
\begin{aligned}
|\langle f,\varphi\rangle| &\le \sum_{j=1}^\infty |\lambda_j||\langle a_j,\varphi\rangle| \lesssim \sum_{j=1}^\infty |\lambda_j|\|\varphi\|_{\mathcal{L}_{1/p-1}(X)} \lesssim \|\varphi\|_{\mathring{\mathcal{G}}_0^\eta(\beta,\gamma)} \left[\sum_{j=1}^\infty |\lambda_j|^p\right]^{1/p}\\
&\lesssim \|\varphi\|_{\mathring{\mathcal{G}}_0^\eta(\beta,\gamma)}[\|f\|_{H_{\rm w}^p(X)} + \epsilon].
\end{aligned}
$$

Letting $\epsilon \to 0^+$, we obtain $\|f\|_{(\mathring{\mathcal{G}}_0^\eta(\beta,\gamma))'} \lesssim \|f\|_{H_{\rm w}^p(X)}$. Thus, $H_{\rm w}^p(X)$ can be continuously embedded into $(\mathring{\mathcal{G}}_0^\eta(\beta,\gamma))'$.

To prove that $H_{\rm w}^p(X)$ is a (quasi-)Banach space, we only prove its completeness. Let $\{f_n\}_{n=1}^\infty$ be a Cauchy sequence in $H_{\rm w}^p(X)$. Then $\{f_n\}_{n=1}^\infty$ is also a Cauchy sequence in $(\mathring{\mathcal{G}}_0^\eta(\beta,\gamma))'$, so it converges to some element f in $(\mathring{\mathcal{G}}_0^\eta(\beta,\gamma))'$. For any $n \in \mathbb{N}$ and $x \in X$, applying the Fatou lemma twice, we conclude that

$$
\begin{aligned}
S(f-f_n)(x) &= S\left(\lim_{m\to\infty}[f_m - f_n]\right)(x)\\
&= \left[\sum_{k\in\mathbb{Z}}\sum_{\alpha\in\mathcal{G}_k} \left\{\mu(Q_\alpha^{k+1})\right\}^{-1} \left|\left\langle \psi_\alpha^k, \lim_{m\to\infty}[f_m - f_n]\right\rangle\right|^2 \mathbf{1}_{Q_\alpha^{k+1}}(x)\right]^{\frac{1}{2}}\\
&= \left[\sum_{k\in\mathbb{Z}}\sum_{\alpha\in\mathcal{G}_k} \lim_{m\to\infty} \left\{\mu(Q_\alpha^{k+1})\right\}^{-1} \left|\left\langle \psi_\alpha^k, f_m - f_n\right\rangle\right|^2 \mathbf{1}_{Q_\alpha^{k+1}}(x)\right]^{\frac{1}{2}}\\
&\le \liminf_{m\to\infty} \left[\sum_{k\in\mathbb{Z}}\sum_{\alpha\in\mathcal{G}_k} \left\{\mu(Q_\alpha^{k+1})\right\}^{-1} \left|\left\langle \psi_\alpha^k, f_m - f_n\right\rangle\right|^2 \mathbf{1}_{Q_\alpha^{k+1}}(x)\right]^{\frac{1}{2}}\\
&= \liminf_{m\to\infty} S(f_m - f_n)(x)
\end{aligned}
$$

and hence

$$
\begin{aligned}
\|f - f_n\|_{H_{\rm w}^p(X)}^p &= \int_X [S(f-f_n)(x)]^p\, d\mu(x)\\
&\le \int_X \liminf_{m\to\infty} [S(f_m - f_n)(x)]^p\, d\mu(x)\\
&\le \liminf_{m\to\infty} \int_X [S(f_m - f_n)(x)]^p\, d\mu(x) = \liminf_{m\to\infty} \|f_m - f_n\|_{H_{\rm w}^p(X)}^p.
\end{aligned}
$$

Letting $n \to \infty$, we find that $f \in H^p_{\mathrm{w}}(X)$ and $\lim_{n\to\infty} \|f - f_n\|_{H^p_{\mathrm{w}}(X)} = 0$. Therefore, $H^p_{\mathrm{w}}(X)$ is complete. This finishes the proof of Proposition 6.2. □

Applying Theorem 6.1 and Proposition 6.2, we establish the following wavelet characterizations of Hardy spaces.

Theorem 6.3 *Suppose* $p \in (\omega/(\omega+\eta), 1]$ *and* $\beta, \gamma \in (\omega(1/p-1), \eta)$. *As subspaces of* $(\mathring{\mathcal{G}}^\eta_0(\beta,\gamma))'$, $H^p(X) = H^p_{\mathrm{w}}(X)$ *with equivalent (quasi-)norms.*

Proof Due to (6.3), Theorems 5.10 and 5.9, we obtain $H^p_{\mathrm{w}}(X) \subset H^p(X)$ and

$$\|\cdot\|_{H^p(X)} \lesssim \|\cdot\|_{H^p_{\mathrm{w}}(X)}.$$

It remains to show $H^p(X) \subset H^p_w(X)$. To this end, by Theorem 5.9, we conclude that $L^2(X) \cap H^p(X)$ is dense in $H^p(X)$. Thus, for any $f \in H^p(X)$, there exist $\{f_n\}_{n=1}^\infty \subset L^2(X) \cap H^p(X)$ such that $\lim_{n\to\infty} \|f - f_n\|_{H^p(X)} = 0$. Obviously, $\{f_n\}_{n=1}^\infty$ is a Cauchy sequence of $H^p(X)$. Noticing that $\{f_n\}_{n=1}^\infty \subset L^2(X)$, we use (6.4) and Theorem 5.10 to conclude that

$$\begin{aligned}\|f_m - f_n\|_{H^p_{\mathrm{w}}(X)} &= \|S(f_m - f_n)\|_{L^p(X)} \sim \|\mathcal{G}(f_m - f_n)\|_{L^p(X)}\\ &\sim \|f_m - f_n\|_{H^p(X)} \to 0\end{aligned}$$

as $m, n \to \infty$, so that $\{f_n\}_{n=1}^\infty$ is also a Cauchy sequence of $H^p_{\mathrm{w}}(X)$. By Proposition 6.2, there exists $\widetilde{f} \in H^p_{\mathrm{w}}(X)$ such that $f_n \to \widetilde{f}$ as $n \to \infty$ in $H^p_{\mathrm{w}}(X)$, also in $(\mathring{\mathcal{G}}^\eta_0(\beta,\gamma))'$. Meanwhile, $f_n \to f$ as $n \to \infty$ in $H^p(X)$, also in $(\mathring{\mathcal{G}}^\eta_0(\beta,\gamma))'$. Therefore, $\widetilde{f} = f$ in $(\mathring{\mathcal{G}}^\eta_0(\beta,\gamma))'$ and $f \in H^p_{\mathrm{w}}(X)$. Moreover,

$$\begin{aligned}\|f\|^p_{H^p_{\mathrm{w}}(X)} &\le \|f - f_n\|^p_{H^p_{\mathrm{w}}(X)} + \|f_n\|^p_{H^p_{\mathrm{w}}(X)} \sim \|f - f_n\|^p_{H^p_{\mathrm{w}}(X)} + \|f_n\|^p_{H^p(X)}\\ &\lesssim \|f\|^p_{H^p(X)}\end{aligned}$$

when n is sufficiently large. Thus, we obtain $H^p(X) \subset H^p_{\mathrm{w}}(X)$ and $\|\cdot\|_{H^p_{\mathrm{w}}(X)} \lesssim \|\cdot\|_{H^p(X)}$. This finishes the proof of Theorem 6.3. □

7 Criteria of the Boundedness of Sublinear Operators

Let $p \in (\omega/(\omega+\eta), 1]$. By the argument used in Sects. 3–6, we conclude that the Hardy spaces $H^{+,p}(X)$, $H^p_\theta(X)$ with $\theta \in (0,\infty)$, $H^{*,p}(X)$, $H^{p,q}_{\mathrm{at}}(X)$, $H^{p,q}_{\mathrm{cw}}(X)$, $\mathring{H}^{p,q}_{\mathrm{at}}(X)$ with $q \in (p,\infty] \cap [1,\infty]$ and $H^p_{\mathrm{w}}(X)$ are essentially the same space in the sense of equivalent (quasi-)norms. From now on, we simply use $H^p(X)$ to denote anyone of them if there is no confusion. In this section, we establish some criteria of the boundedness of sublinear operators on Hardy spaces via first establishing finite atomic characterizations of $H^p(X)$.

7.1 Finite Atomic Characterizations of Hardy Spaces

For any $p \in (\omega/(\omega+\eta), 1]$ and $q \in (p, \infty] \cap [1, \infty]$, we say $f \in H^{p,q}_{\text{fin}}(X)$ if there exist $N \in \mathbb{N}$, a sequence $\{a_j\}_{j=1}^N$ of (p,q)-atoms and $\{\lambda_j\}_{j=1}^N \subset \mathbb{C}$ such that

$$f = \sum_{j=1}^N \lambda_j a_j.$$

Also, define

$$\|f\|_{H^{p,q}_{\text{fin}}(X)} := \inf \left\{ \left(\sum_{j=1}^N |\lambda_j|^p \right)^{\frac{1}{p}} \right\},$$

where the infimum is taken over all the decompositions of f as above. It is easy to see that $H^{p,q}_{\text{fin}}(X)$ is a dense subset of $H^{p,q}_{\text{at}}(X)$ and $\|\cdot\|_{H^{p,q}_{\text{at}}(X)} \le \|\cdot\|_{H^{p,q}_{\text{fin}}(X)}$. Denote by the symbol $\mathrm{UC}(X)$ the *space of all uniformly continuous functions on* X, that is, a function $f \in \mathrm{UC}(X)$ if and only if, for any fixed $\epsilon \in (0,\infty)$, there exists $\sigma \in (0, \infty)$ such that $|f(x) - f(y)| < \epsilon$ whenever $d(x, y) < \sigma$. The next theorem characterizes $H^{p,q}_{\text{at}}(X)$ via $H^{p,q}_{\text{fin}}(X)$.

Theorem 7.1 *Suppose $p \in (\omega/(\omega+\eta), 1]$. Then the following statements hold true:*

(i) *if $q \in (p, \infty) \cap [1, \infty)$, then $\|\cdot\|_{H^{p,q}_{\text{fin}}(X)}$ and $\|\cdot\|_{H^{p,q}_{\text{at}}(X)}$ are equivalent (quasi)-norms on $H^{p,q}_{\text{fin}}(X)$;*

(ii) *$\|\cdot\|_{H^{p,\infty}_{\text{fin}}(X)}$ and $\|\cdot\|_{H^{p,\infty}_{\text{at}}(X)}$ are equivalent (quasi)-norms on $H^{p,q}_{\text{fin}}(X) \cap \mathrm{UC}(X)$;*

(iii) *$H^{p,\infty}_{\text{fin}}(X) \cap \mathrm{UC}(X)$ is a dense subspace of $H^{p,\infty}_{\text{at}}(X)$.*

Proof First, we prove (i). It suffices to show that $\|f\|_{H^{p,q}_{\text{fin}}(X)} \lesssim \|f\|_{H^{p,q}_{\text{at}}}$ for any $f \in H^{p,q}_{\text{fin}}(X)$ with $q \in (p, \infty) \cap [1, \infty)$. We may as well assume that $\|f\|_{H^{*,p}(X)} = 1$. Let all the notation be as in the proof that $H^{*,p}(X) \subset H^{p,q}_{\text{at}}(X)$ of Theorem 4.2. Then

$$f = \sum_{j\in\mathbb{Z}} \sum_{k \in I_j} \lambda_k^j a_k^j = \sum_{j\in\mathbb{Z}} \sum_{k \in I_j} h_k^j = \sum_{j\in\mathbb{Z}} h_j$$

both in $(\mathcal{G}_0^\eta(\beta, \gamma))'$ and almost everywhere. Here and hereafter, for any $j \in \mathbb{Z}$ and $k \in I_j$, the quantities h_j, h_k^j, λ_k^j and a_k^j are as in (4.12) and (4.13). Since $f \in H^{p,q}_{\text{fin}}(X)$, it follows that there exist $x_1 \in X$ and $R \in (0, \infty)$ such that $\operatorname{supp} f \subset B(x_1, R)$. We claim that there exists a positive constant $\tilde{c}$ such that, for any $x \notin B(x_1, 16A_0^4 R)$,

$$f^\star(x) \le \tilde{c}[\mu(B(x_1, R))]^{-\frac{1}{p}}. \tag{7.1}$$

We admit (7.1) temporarily and use it to prove (i) and (ii). Let j' be the maximal integer such that $2^{j} \le \tilde{c}[\mu(B(x_1, R))]^{-\frac{1}{p}}$ and define

$$h := \sum_{j\le j'}\sum_{k\in I_j}\lambda_k^j a_k^j \quad\text{and}\quad \ell := \sum_{j>j'}\sum_{k\in I_j}\lambda_k^j a_k^j. \tag{7.2}$$

In what follows, for the sake of convenience, we elide the fact whether or not I_j is finite and simply write the summation $\sum_{k\in I_j}$ in (7.2) as $\sum_{k=1}^{\infty}$. If $j > j'$, then $\Omega^j = \{x \in X : f^\star(x) > 2^j\} \subset B(x_1, 16A_0^4R)$, which implies that supp $\ell \subset B(x_1, 16A_0^4R)$ because supp $a_k^j \subset \Omega^j$. From $f = h+\ell$, it then follows that supp $h \subset B(x_1, 16A_0^4R)$. Noticing that

$$\|h\|_{L^\infty(X)} \le \sum_{j\le j'}\left\|h^j\right\|_{L^\infty(X)} \lesssim \sum_{j\le j'} 2^j \sim [\mu(B(x_1,R))]^{-\frac{1}{p}}$$

and $\int_X h(x)\,d\mu(x) = 0$, we conclude that h is a harmless constant multiple of a (p,∞)-atom.

Next we deal with ℓ. For any $N := (N_1, N_2) \in \mathbb{N}^2$, define

$$\ell_N := \sum_{j=j'+1}^{N_1}\sum_{k=1}^{N_2}\lambda_k^j a_k^j = \sum_{j=j'+1}^{N_1}\sum_{k=1}^{N_2} h_k^j.$$

Then ℓ_N is a finite linear combination of (p,∞)-atoms and $\sum_{j=j'+1}^{N_1}\sum_{k=1}^{N_2}|\lambda_k^j|^p \lesssim 1$. Notice that supp $(\ell - \ell_N) \subset B(x_1, 16A_0^4R)$ and $\int_X[\ell(x) - \ell_N(x)]\,d\mu(x) = 0$. It suffices to show that $\|\ell - \ell_N\|_{L^q(X)} \to 0$ can be sufficiently small when N_1 and N_2 are big enough. Noticing that $\ell = \sum_{j=N_1+1}^{\infty} h^j + \sum_{j=j'+1}^{N_1}\sum_{k=1}^{\infty} h_k^j$, we have

$$\|\ell - \ell_N\|_{L^q(X)} \le \left\|\sum_{j=N_1+1}^{\infty} h^j\right\|_{L^q(X)} + \sum_{j=j'+1}^{N_1}\left\|\sum_{k=N_2+1}^{\infty} h_k^j\right\|_{L^q(X)}.$$

For any $j \in \mathbb{Z}$ and $k \in \mathbb{N}$, we recall that supp $h_k^j \subset B_k^j \subset \Omega^j$ and $\|h^j\|_{L^\infty(X)} \lesssim 2^j$. By $f = \sum_{j=-\infty}^{\infty} h^j$ and supp $(\sum_{j=N_1+1}^{\infty} h^j) \subset \Omega^{N_1}$, we conclude that, for any $z \in \Omega^{N_1}$,

$$\left|\sum_{j=N_1+1}^{\infty} h^j(z)\right| = \left|f(z) - \sum_{j\le N_1} h^j(z)\right| \le |f(z)| + \sum_{j\le N_1}\left|h^j(z)\right| \lesssim |f(z)| + 2^{N_1}.$$

Notice that, by [20, Proposition 3.9], there exists a constant $\widetilde{C} > 1$ such that $f^\star \le \widetilde{C}\mathcal{M}(f)$. With $f_1 := f\mathbf{1}_{\{x\in X:\,|f(x)|>2^{N_1-1}/\widetilde{C}\}}$ and $f_2 := f - f_1$, we have

$$\begin{aligned}2^{N_1q}\mu\left(\Omega^{N_1}\right) &\le 2^{N_1q}\mu\left(\left\{x \in X : \widetilde{C}\mathcal{M}(f)(x) > 2^{N_1}\right\}\right)\\ &\le 2^{N_1q}\mu\left(\left\{x \in X : \widetilde{C}\mathcal{M}(f_1)(x) > 2^{N_1-1}\right\}\right) \lesssim \|f_1\|_{L^q(X)}^q \to 0\end{aligned}$$

as $N_1 \to \infty$, because $\mathcal{M}$ is bounded from $L^q(X)$ to $L^{q,\infty}(X)$ and $f \in H^{p,q}_{\mathrm{fin}}(X) \subset L^q(X)$. Therefore,

$$\left\| \sum_{j=N_1+1}^{\infty} h^j \right\|_{L^q(X)}^q \lesssim \int_{\Omega^{N_1}} \left[|f(z)|^q + 2^{N_1 q} \right] d\mu(z)$$
$$\lesssim \| f \mathbf{1}_{\Omega^{N_1}} \|_{L^q(X)}^q + 2^{N_1 q} \mu\left(\Omega^{N_1}\right) \to 0$$

as $N_1 \to \infty$. Then, for any $\epsilon \in (0,\infty)$, we choose $N_1 \in \mathbb{N}$ such that $\| \sum_{j=N_1+1}^{\infty} h^j \|_{L^q(X)} < \epsilon/2$.

If we fix $N_1 \in \mathbb{N}$ and $N_1 \geq j > j'$, then the fact $\sum_{k=1}^{\infty} |h_k^j| \lesssim 2^j \mathbf{1}_{\Omega^j} \in L^q(X)$ implies that

$$\lim_{N_2 \to 0} \left\| \sum_{k=N_2+1}^{\infty} h_k^j \right\|_{L^q(X)} = 0.$$

So, we further choose $N_2 \in \mathbb{N}$ such that $\sum_{j=j'+1}^{N_1} \| \sum_{k=N_2+1}^{\infty} h_k^j \|_{L^q(X)} < \epsilon/2$. In this way, we have $\|\ell - \ell_N\|_{L^q(X)} < \epsilon$ for large N. Then there exist a positive constant $C_\flat$, independent of N and ϵ, and a (p,q)-atom $a_{(N)}$ such that $\ell - \ell_N = C_\flat \epsilon a_{(N)}$. Therefore, we obtain $\|f\|_{H^{p,q}_{\mathrm{fin}}(X)} \lesssim 1 \sim \|f\|_{H^{p,q}_{\mathrm{at}}(X)}$ and complete the proof of (i) under the assumption (7.1).

To obtain (ii), we only need to prove that $\|f\|_{H^{p,\infty}_{\mathrm{fin}}(X)} \lesssim \|f\|_{H^{p,\infty}_{\mathrm{at}}}$ whenever $f \in H^{p,\infty}_{\mathrm{fin}}(X) \cap \mathrm{UC}(X)$. We may also assume that $\|f\|_{H^{*,p}(X)} = 1$. Notice that $f \in L^\infty(X)$ and $\|f^\star\|_{L^\infty(X)} \lesssim \|\mathcal{M}(f)\|_{L^\infty(X)} \leq c_0 \|f\|_{L^\infty(X)}$, where c_0 is a positive constant independent of f. Let $j'' > j'$ be the largest integer such that $2^j \leq c_0 \|f\|_{L^\infty(X)}$. We write $f = h + \ell$ with h as in (7.2) but now $\ell = \sum_{j' < j \leq j''} \sum_{k=1}^{\infty} h_k^j$. As in the proof of (i), we know that h is a harmless positive constant multiple of some (p,∞)-atom.

Now we consider ℓ. Notice that $f \in \mathrm{UC}(X)$. Then, for any $\epsilon \in (0,\infty)$, there exists $\sigma \in (0,\infty)$ such that $|f(x) - f(y)| \leq \epsilon$ whenever $d(x,y) \leq \sigma$. Split $\ell = \ell_1^\sigma + \ell_2^\sigma$ with

$$\ell_1^\sigma := \sum_{(j,k) \in G_1} h_k^j = \sum_{(j,k) \in G_1} \lambda_k^j a_k^j \quad \text{and} \quad \ell_2^\sigma := \sum_{(j,k) \in G_2} h_k^j,$$

where

$$G_1 := \{(j,k) : \ 12 A_0^3 r_k^j \geq \sigma, \ j' < j \leq j''\} \quad \text{and}$$
$$G_2 := \{(j,k) : \ 12 A_0^3 r_k^j < \sigma, \ j' < j \leq j''\}.$$

Notice that, for any $j' < j \leq j''$, Ω^j is bounded. Thus, by Proposition 4.4(vi), we find that G_1 is a finite set, which further implies that ℓ_1^σ is a finite linear combination of (p,∞)-atoms and

$$\sum_{(j,k)\in G_1} \left|\lambda_k^j\right|^p \lesssim 1.$$

To consider ℓ_2^σ, it is obvious that supp $\ell_2^\sigma \subset B(x_1, 16A_0^4 R)$ and that $\int_X \ell_2^\sigma(x)\,d\mu(x) = 0$, so it remains to estimate $\|\ell_2^\sigma\|_{L^\infty(X)}$. For any $(j,k) \in G_2$, applying the definition of h_k^j in (4.12) implies that

$$\left|h_k^j\right| \le \left|b_k^j\right| + \sum_{l\in I_{j+1}} \left|b_l^{j+1}\phi_k^j\right| + \sum_{l\in I_{j+1}} \left|L_{k,l}^{j+1}\phi_l^{j+1}\right|.$$

By the definition of b_k^j, we have supp $b_k^j \subset B(x_k^j, 2A_0 r_k^j)$. Moreover, for any $x \in B(x_k^j, 2A_0 r_k^j)$,

$$\begin{aligned}\left|b_k^j(x)\right| &\le \left|f(x) - \frac{1}{\|\phi_k^j\|_{L^1(X)}}\int_{B(x_k^j,2A_0r_k^j)} f(\xi)\phi_k^j(\xi)\,d\mu(\xi)\right| \\ &\le \left|f(x) - f\left(x_k^j\right)\right| + \frac{1}{\|\phi_k^j\|_{L^1(X)}}\int_{B(x_k^j,2A_0r_k^j)} \left|f(\xi) - f\left(x_k^j\right)\right|\phi_k^j(\xi)\,d\mu(\xi) \lesssim \epsilon.\end{aligned} \tag{7.3}$$

If $b_l^{j+1}\phi_k^j \ne 0$, then $B(x_k^j, 2A_0 r_k^j) \cap B(x_l^{j+1}, 2A_0 r_l^{j+1}) \ne \emptyset$, which further implies that $r_l^{j+1} \le 6A_0^2 r_k^j$. Thus, for any $x \in B(x_l^{j+1}, 2A_0 r_l^{j+1})$, we have $d(x, x_l^{j+1}) < 12A_0^3 r_k^j$ and hence an argument similar to the estimation of (7.3) gives

$$\begin{aligned}\left|b_l^{j+1}(x)\right| &= \left|f(x) - \frac{1}{\|\phi_l^{j+1}\|_{L^1(X)}}\int_{B(x_l^{j+1},2A_0r_l^{j+1})} f(\xi)\phi_l^{j+1}(\xi)\,d\mu(\xi)\right|\phi_l^{j+1}(x) \\ &\lesssim \epsilon\phi_l^{j+1}(x),\end{aligned}$$

so that

$$\sum_{l\in I_{j+1}} \left|b_l^{j+1}(x)\phi_k^j(x)\right| \lesssim \epsilon\phi_k^j(x) \sum_{l\in I_{j+1}} \phi_l^{j+1}(x) \sim \epsilon\phi_k^j(x) \lesssim \epsilon.$$

Using the definition of $L_{k,l}^{j+1}$ and arguing similarly to the estimation of (7.3), we conclude that, for any $x \in X$,

$$\sum_{l\in I_{j+1}} \left|L_{k,l}^{j+1}\phi_k^j(x)\right| \lesssim \epsilon,$$

where $L_{k,l}^{j+1}$ is as in (4.10). Summarizing all gives $\|h_k^j\|_{L^\infty(X)} \lesssim \epsilon$. Recalling that supp $h_k^j \subset B_k^j$ and $\sum_{k=1}^\infty \mathbf{1}_{B_k^j} \le L_0$, we obtain $\|\ell_2^\sigma\|_{L^\infty(X)} \lesssim \epsilon$. Therefore, there exist a positive constant $\widetilde{C}_\flat$, independent of σ and ϵ, and a (p,∞)-atom $a_{(\sigma)}$ such that $\ell_2^\sigma = \widetilde{C}_\flat \epsilon a_{(\sigma)}$. This proves that $\|f\|_{H^{p,\infty}_{\mathrm{fin}}(X)} \lesssim 1$ and hence finishes the proof of (ii) under the assumption (7.1).

Now we prove (7.1). Let $x \notin B(x_1, 16A_0^4R)$. Suppose that $\varphi \in \mathcal{G}_0^\eta(\beta,\gamma)$ with $\|\varphi\|_{\mathcal{G}(x,r,\beta,\gamma)} \lesssim 1$ for some $r \in (0,\infty)$. First we consider the case $r \geq 4A_0^2 d(x,x_1)/3$. For any $y \in B(x, d(x,x_1))$, we have $\|\varphi\|_{\mathcal{G}(y,r,\beta,\gamma)} \lesssim 1$, which implies that $|\langle f,\varphi\rangle| \lesssim f^*(y)$ and hence

$$|\langle f,\varphi\rangle| \lesssim \left\{\frac{1}{\mu(B(x,d(x,x_1)))}\int_{B(x,d(x,x_1))}[f^*(y)]^p\,d\mu(y)\right\}^{\frac{1}{p}} \lesssim [\mu(B(x_1,R))]^{-\frac{1}{p}}. \tag{7.4}$$

Next we consider the case $r < 4A_0^2 d(x_1,x)/3$. Choose a function ξ satisfying $\mathbf{1}_{B(x_1,(2A_0)^{-4}d(x_1,x))} \leq \xi \leq \mathbf{1}_{B(x_1,(2A_0)^{-3}d(x_1,x))}$ and $\|\xi\|_{\dot{C}^\eta(X)} \lesssim [d(x_1,x)]^{-\eta}$. Since $\operatorname{supp} f \subset B(x_1,R)$, it follows that $f\xi = f$. Let $\widetilde{\varphi} := \varphi\xi$. For any $y \in B(x, d(x,x_1))$, assuming for the moment that

$$\|\widetilde{\varphi}\|_{\mathcal{G}(y,r,\beta,\gamma)} \lesssim 1, \tag{7.5}$$

we obtain

$$|\langle f,\varphi\rangle| = \left|\int_X f(z)\varphi(z)\,d\mu(z)\right| = \left|\int_X f(z)\xi(z)\varphi(z)\,d\mu(z)\right| = |\langle f,\widetilde{\varphi}\rangle| \lesssim f^*(y),$$

which implies that (7.4) remains true in this case. Therefore, by the arbitrariness of φ and the fact that $f^* \sim f^\star$, we obtain (7.1).

Now we fix $y \in B(x_1, d(x_1,x))$ and prove (7.5). First we consider the size condition. Indeed, if $\widetilde{\varphi}(z) \neq 0$, then $d(z,x_1) < (2A_0)^{-3}d(x_1,x)$ and hence $d(z,y) < (16A_0^2/7)d(x,z)$, which implies that

$$\begin{aligned}|\widetilde{\varphi}(z)| \leq |\varphi(z)| &\leq \frac{1}{V_r(x)+V(x,z)}\left[\frac{r}{r+d(x,z)}\right]^\gamma\\ &\sim \frac{1}{V_r(y)+V(y,z)}\left[\frac{r}{r+d(y,z)}\right]^\gamma.\end{aligned}$$

To consider the regularity condition of $\widetilde{\varphi}$, we may assume that $d(z,z') \leq (2A_0)^{-10}[r + d(y,z)]$ due to the size condition. For the case $d(z,x_1) > (2A_0)^{-1}d(x_1,x)$, we have $\widetilde{\varphi}(z) = 0$ and, by $y \in B(x_1, d(x_1,x))$ and $r < 4A_0^2 d(x_1,x)/3$, we further obtain

$$\begin{aligned}d(z,z') &\leq (2A_0)^{-10}[r+d(y,z)] \leq (2A_0)^{-10}[r + A_0 d(y,x_1) + A_0 d(x_1,z)]\\ &\leq (2A_0)^{-10}[4A_0^2 d(x_1,x) + A_0 d(x_1,z)] \leq (2A_0)^{-2}d(x_1,z),\end{aligned}$$

which further implies that $d(z',x_1) \geq \frac{1}{A_0}d(x_1,z) - d(z,z') \geq (2A_0)^{-2}d(x_1,x)$ and hence $\widetilde{\varphi}(z') = 0$. So we only need to consider the case $d(z,x_1) \leq (2A_0)^{-1}d(x_1,x)$. Then we have $(2A_0)^{-1}d(x_1,x) \leq d(z,x) \leq 2A_0 d(x_1,x)$ and

$$d(y,z) \le A_0^2[d(y,x_1)+d(x_1,x)+d(x,z)] \le 2A_0^2 d(x_1,x) + A_0^2 d(x,z) \le (2A_0)^3 d(x,z),$$

which implies that $d(z,z') \le (2A_0)^{-1}[r+d(x,z)]$ and $r+d(y,z) \lesssim \min\{r+d(x,z), r+d(x,z'), d(x_1,x)\}$. Therefore, by the regularity of φ and the definition of ξ, we conclude that

$$\begin{aligned}
\left|\widetilde{\varphi}(z)-\widetilde{\varphi}(z')\right| &\le \xi(z)|\varphi(z)-\varphi(z')| + |\varphi(z')||\xi(z)-\xi(z')| \\
&\lesssim \left[\frac{d(z,z')}{r+d(x,z)}\right]^\beta \frac{1}{V_r(x)+V(x,z)}\left[\frac{r}{r+d(x,z)}\right]^\gamma \\
&\quad + \frac{1}{V_r(x)+V(x,z')}\left[\frac{r}{r+d(x,z')}\right]^\gamma \left[\frac{d(z,z')}{d(x_1,x)}\right]^\beta \\
&\lesssim \left[\frac{d(z,z')}{r+d(y,z)}\right]^\beta \frac{1}{V_r(y)+V(y,z)}\left[\frac{r}{r+d(y,z)}\right]^\gamma .
\end{aligned}$$

This proves (7.5) and hence finishes the proofs of (i) and (ii).

Now we prove (iii). According to [23, pp. 3347–3348] (see also [27, Theorem 2.6]), there exists a sequence $\{S_k\}_{k\in\mathbb{Z}}$ of bounded operators on $L^2(X)$ with their kernels satisfying the following conditions:

(i) $S_k(x,y)=0$ if $d(x,y)\ge C_\sharp\delta^k$ and, for any $x,\ y\in X$,

$$|S_k(x,y)| \lesssim \frac{1}{V_{\delta^k}(x)+V_{\delta^k}(y)},$$

where $C_\sharp$ is a fixed positive constant greater than 1;

(ii) for any $x,\ x',\ y\in X$ with $d(x,x')\le C_\sharp\delta^k$,

$$|S_k(x,y)-S_k(x',y)|+|S_k(y,x)-S_k(y,x')| \lesssim \left[\frac{d(x,x')}{\delta^k}\right]^\theta \frac{1}{V_{\delta^k}(x)+V_{\delta^k}(y)},$$

where θ is as in [23, Theorem 2.4];

(iii) for any $x\in X$, $\int_X S_k(x,y)\,d\mu(y) = 1 = \int_X S_k(y,x)\,d\mu(y)$.

For any $g\in\bigcup_{p\in[1,\infty]} L^p(X)$ and $x\in X$, define

$$S_k g(x) := \int_X S_k(x,y)g(y)\,d\mu(y).$$

Then, for any (p,∞)-atom a supported on $B(z,r)$ with $z\in X$ and $r\in(0,\infty)$, we observe that $S_k a$ has the following properties:

(a) $\|S_k a\|_{L^\infty(X)} \lesssim \|a\|_{L^\infty(X)}$ and $\lim_{k\to\infty}\|S_k a - a\|_{L^2(X)} = 0$;
(b) when k is sufficiently large, supp $S_k(a)\subset B(z,2A_0 r)$;
(c) $\int_X S_k a(x)\,d\mu(x) = 0$;
(d) $S_k a\in \mathrm{UC}(X)$.

Consequently, $S_k a$ is a harmless constant multiple of a (p, ∞)-atom and hence of a $(p, 2)$-atom. Thus, $\|S_k a - a\|_{H^{p,\infty}_{\rm at}(X)} \sim \|S_k a - a\|_{H^{p,2}_{\rm at}(X)} \to 0$ as $k \to \infty$. For any $f \in H^{p,\infty}_{\rm at}(X)$, there exists a sequence $\{f_n\}_{n\in\mathbb{N}} \subset H^{p,\infty}_{\rm fin}(X)$ such that $\lim_{n\to\infty} \|f_n - f\|_{H^{p,q}_{\rm at}(X)} = 0$. Then, for any $n \in \mathbb{N}$, by the above (a)–(d), we find that $S_k(f_n) \in H^{p,\infty}_{\rm fin}(X) \cap {\rm UC}(X)$ and that $\lim_{k\to\infty} \|S_k f_n - f_n\|_{H^{p,\infty}_{\rm at}(X)} = 0$. This proves that $\|S_k f_n - f\|_{H^{p,\infty}_{\rm at}(X)} \to 0$ as $n,\ k \to \infty$, which completes the proof of (iii) and hence of Theorem 7.1. □

7.2 Criteria of the Boundedness of Sublinear Operators on Hardy Spaces

In this section, applying the finite atomic characterizations of Hardy spaces, we obtain two criteria on the boundedness of sublinear operators on Hardy spaces.

Recall that a complete vector space $\mathcal{B}$ is called a *quasi-Banach space* if its quasi-norm $\|\cdot\|_{\mathcal{B}}$ satisfies the following condition:

(i) for any $f \in \mathcal{B}$, $\|f\|_{\mathcal{B}} = 0$ if and only if f is the zero element in $\mathcal{B}$;
(ii) for any $\lambda \in \mathbb{C}$ and $f \in \mathcal{B}$, $\|\lambda f\|_{\mathcal{B}} = |\lambda| \|f\|_{\mathcal{B}}$;
(iii) there exists $C \in [1, \infty)$ such that, for any $f,\ g \in \mathcal{B}$, $\|f + g\|_{\mathcal{B}} \le C(\|f\|_{\mathcal{B}} + \|g\|_{\mathcal{B}})$.

Next we recall the definition of r-quasi-Banach spaces (see, for example, [20,35,53–55]).

Definition 7.2 Suppose that $r \in (0, 1]$ and $\mathcal{B}_r$ is a quasi-Banach space with its quasi-norm $\|\cdot\|_{\mathcal{B}_r}$. The space $\mathcal{B}_r$ is called an *r-quasi-Banach space* if there exists $\kappa \in [1, \infty)$ such that, for any $m \in \mathbb{N}$ and $\{f_j\}_{j=1}^m \subset \mathcal{B}_r$,

$$\left\| \sum_{j=1}^m f_j \right\|_{\mathcal{B}_r}^r \le \kappa \sum_{j=1}^m \|f_j\|_{\mathcal{B}_r}^r .$$

Obviously, when $p \in (0, 1]$, $L^p(X)$ and $H^{*,p}(X)$ are p-quasi-Banach-spaces. Let $\mathcal{Y}$ be a linear space and $\mathcal{B}_r$ an r-quasi-Banach space with $r \in (0, 1]$. An operator $T : \mathcal{Y} \to \mathcal{B}_r$ is said to be *$\mathcal{B}_r$-sublinear* if there exists a positive constant $\kappa \in [1, \infty)$ such that

(i) for any $f,\ g \in \mathcal{Y}$, $\|T(f) - T(g)\|_{\mathcal{B}_r} \le \kappa \|T(f - g)\|_{\mathcal{B}_r}$;
(ii) for any $m \in \mathbb{N}$, $\{f_j\}_{j=1}^m \subset \mathcal{Y}$ and $\{\lambda_j\}_{j=1}^m \subset \mathbb{C}$,

$$\left\| T\left(\sum_{j=1}^m \lambda_j f_j \right) \right\|_{\mathcal{B}_r}^r \le \kappa \sum_{j=1}^m |\lambda_j|^r \|T(f_j)\|_{\mathcal{B}_r}^r$$

(see, for example, [35, Definition 2.5], [53, Definition 1.6.7], [55, Remark 1.1(3)], [54, Definition 1.6] and [20, Definition 5.8]).

The next theorem gives us some criteria for $\mathcal{B}_r$-sublinear operators that can be extended to bounded $\mathcal{B}_r$-sublinear operators from Hardy spaces to $\mathcal{B}_r$. It can be proved by following the proof of [20, Theorem 5.9] with some slight modifications, the details being omitted.

Theorem 7.3 *Let* $p \in (\omega/(\omega+\eta), 1]$ *and* $r \in [p, 1]$. *Suppose that* $\mathcal{B}_r$ *is an* r*-quasi-Banach space and either of the following holds true:*

(i) $q \in (p, \infty) \cap [1, \infty)$ *and* $T : H_{\mathrm{fin}}^{p,q}(X) \to \mathcal{B}_r$ *is a* $\mathcal{B}_r$*-sublinear operator with*

$$\sup\{\|T(a)\|_{\mathcal{B}_r} : a \text{ is any } (p, q)\text{-atom}\} < \infty;$$

(ii) $T : H_{\mathrm{fin}}^{p,\infty}(X) \cap \mathrm{UC}(X) \to \mathcal{B}_r$ *is a* $\mathcal{B}_r$*-sublinear operator with*

$$\sup\{\|T(a)\|_{\mathcal{B}_r} : a \text{ is any } (p, \infty)\text{-atom}\} < \infty.$$

Then T *can be uniquely extended to a bounded* $\mathcal{B}_r$*-sublinear operator from* $H_{\mathrm{at}}^{p,q}(X)$ *to* $\mathcal{B}_r$.

References

1. Auscher, P., Hytönen, T.: Orthonormal bases of regular wavelets in spaces of homogeneous type. Appl. Comput. Harmon. Anal. **34**, 266–296 (2013)
2. Bui, T.A., Duong, X.T.: Hardy spaces associated to the discrete Laplacians on graphs and boundedness of singular integrals. Trans. Am. Math. Soc. **366**, 3451–3485 (2014)
3. Bui, T.A., Duong, X.T., Ly, F.K.: Maximal function characterizations for new local Hardy type spaces on spaces of homogeneous type. Trans. Am. Math. Soc. **370**, 7229–7292 (2018)
4. Carleson, L.: Two remarks on H^1 and BMO. Adv. Math. **22**, 269–277 (1976)
5. Coifman, R.R.: A real variable characterization of H^p. Stud. Math. **51**, 269–274 (1974)
6. Coifman, R.R., Weiss, G.: Analyse Harmonique Non-commutative sur Certains Espaces Homogènes, (French) Étude de certaines intégrales singulières. Lecture Notes in Mathematics, vol. 242. Springer, New York (1971)
7. Coifman, R.R., Weiss, G.: Extensions of Hardy spaces and their use in analysis. Bull. Am. Math. Soc. **83**, 569–645 (1977)
8. Deng, D., Han, Y.: Harmonic Analysis on Spaces of Homogeneous Type. With a Preface by Yves Meyer. Lecture Notes in Mathematics, vol. 1966. Springer, Berlin (2009)
9. Duong, X.Y., Yan, L.: Hardy spaces of spaces of homogeneous type. Proc. Am. Math. Soc. **131**, 3181–3189 (2003)
10. Duong, X.T., Yan, L.: Duality of Hardy and BMO spaces associated with operators with heat kernel bounds. J. Am. Math. Soc. **18**, 943–973 (2005)
11. Fefferman, C., Stein, E.M.: H^p spaces of several variables. Acta Math. **129**, 137–193 (1972)
12. Folland, G.B., Stein, E.M.: Hardy Spaces on Homogeneous Groups. Mathematical Notes, vol. 28. University of Tokyo Press, Tokyo (1982)
13. Fu, X., Chang, D.-C., Yang, D.: Recent progress in bilinear decompositions. Appl. Anal. Optim. **1**, 153–210 (2017)
14. Fu, X., Yang, D.: Wavelet characterizations of the atomic Hardy space H^1 on spaces of homogeneous type. Appl. Comput. Harmon. Anal. **44**, 1–37 (2018)
15. Fu, X., Yang, D., Liang, Y.: Products of functions in BMO($\mathcal{X}$) and $H^1_{\mathrm{at}}(\mathcal{X})$ via wavelets over spaces of homogeneous type. J. Fourier Anal. Appl. **23**, 919–990 (2017)
16. García-Cuerva, J., Rubio de Francia, J.L.: Weighted Norm Inequalities and Related Topic. North-Holland Mathematics Studies, vol. 116. North-Holland Publishing Co., Amsterdam (1985)
17. Grafakos, L.: Classical Fourier Analysis. Graduate Texts in Mathematics, vol. 249, 3rd edn. Springer, New York (2014)
18. Grafakos, L.: Modern Fourier Analysis. Graduate Texts in Mathematics, vol. 250, 3rd edn. Springer, New York (2014)

19. Grafakos, L., Liu, L., Maldonado, D., Yang, D.: Multilinear analysis on metric spaces. Diss. Math. (Rozprawy Mat.) **497**, 1–121 (2014)
20. Grafakos, L., Liu, L., Yang, D.: Maximal function characterizations of Hardy spaces on RD-spaces and their applications. Sci. China Ser. A **51**, 2253–2284 (2008)
21. Grafakos, L., Liu, L., Yang, D.: Radial maximal function characterizations for Hardy spaces on RD-spaces. Bull. Soc. Math. Fr. **137**, 225–251 (2009)
22. Grafakos, L., Liu, L., Yang, D.: Vector-valued singular integrals and maximal functions on spaces of homogeneous type. Math. Scand. **104**, 296–310 (2009)
23. Han, Ya., Han, Yo, Li, J.: Criterion of the boundedness of singular integrals on spaces of homogeneous type. J. Funct. Anal. **271**, 3423–3464 (2016)
24. Han, Ya., Han, Yo, Li, J.: Geometry and Hardy spaces on spaces of homogeneous type in the sense of Coifman and Weiss. Sci. China Math. **60**, 2199–2218 (2017)
25. Han, Y., Li, J., Ward, L.D.: Hardy space theory on spaces of homogeneous type via orthonormal wavelet bases. Appl. Comput. Harmon. Anal. **45**, 120–169 (2018)
26. Han, Y., Müller, D., Yang, D.: Littlewood-Paley characterizations for Hardy spaces on spaces of homogeneous type. Math. Nachr. **279**, 1505–1537 (2006)
27. Han, Y., Müller, D., Yang, D.: A theory of Besov and Triebel–Lizorkin spaces on metric measure spaces modeled on Carnot–Carathéodory spaces, Abstr. Appl. Anal., Art. ID 893409, 250 pp (2008)
28. Han, Y.S., Sawyer, E.T.: Littlewood–Paley theory on spaces of homogeneous type and the classical function spaces. Mem. Am. Math. Soc. **110**, 530 (1994). vi+126 pp
29. He, Z., Liu, L., Yang, D., Yuan, W.: New Calderón reproducing formulae with exponential decay on spaces of homogeneous type, Sci. China Math. (2018), https://doi.org/10.1007/s11425-018-9346-4.
30. Hofmann, S., Mayboroda, S.: Hardy and BMO spaces associated to divergence form elliptic operators. Math. Ann. **344**, 37–116 (2009)
31. Hytönen, T., Kairema, A.: Systems of dyadic cubes in a doubling metric space. Colloq. Math. **126**, 1–33 (2012)
32. Hu, G., Yang, D., Zhou, Y.: Boundedness of singular integrals in Hardy spaces on spaces of homogeneous type. Taiwan. J. Math. **13**, 91–135 (2009)
33. Koskela, P., Yang, D., Zhou, Y.: A characterization of Hajłasz–Sobolev and Triebel–Lizorkin spaces via grand Littlewood–Paley functions. J. Funct. Anal. **258**, 2637–2661 (2010)
34. Koskela, P., Yang, D., Zhou, Y.: Pointwise characterizations of Besov and Triebel–Lizorkin spaces and quasiconformal mappings. Adv. Math. **226**, 3579–3621 (2011)
35. Ky, L.D.: New Hardy spaces of Musielak–Orlicz type and boundedness of sublinear operators. Integral Eq. Oper. Theory **78**, 115–150 (2014)
36. Latter, R.H.: A characterization of $H^p(\mathbf{R}^n)$ in terms of atoms. Stud. Math. **62**, 93–101 (1978)
37. Li, W.: A maximal function characterization of Hardy spaces on spaces of homogeneous type. Approx. Theory Appl. (N.S.) **14**, 12–27 (1998)
38. Liu, L., Chang, D.-C., Fu, X., Yang, D.: Endpoint boundedness of commutators on spaces of homogeneous type. Appl. Anal. **96**, 2408–2433 (2017)
39. Liu, L., Chang, D.-C., Fu, X., Yang, D.: Endpoint estimates of linear commutators on Hardy spaces over spaces of homogeneous type. Math. Methods Appl. Sci. **41**, 5951–5984 (2018)
40. Liu, L., Yang, D., Yuan, W.: Bilinear decompositions for products of Hardy and Lipschitz spaces on spaces of homogeneous type. Diss. Math. (Rozprawy Mat.) **533**, 1–93 (2018)
41. Lu, S.Z.: Four Lectures on Real H^p Spaces. World Scientific Publishing Co. Inc., River Edge (1995)
42. Macías, R.A., Segovia, C.: Lipschitz functions on spaces of homogeneous type. Adv. Math. **33**, 257–270 (1979)
43. Macías, R.A., Segovia, C.: A decomposition into atoms of distributions on spaces of homogeneous type. Adv. Math. **33**, 271–309 (1979)
44. Nakai, E., Sawano, Y.: Hardy spaces with variable exponents and generalized Campanato spaces. J. Funct. Anal. **262**, 3665–3748 (2012)
45. Nakai, E., Yabuta, K.: Pointwise multipliers for functions of weighted bounded mean oscillation on spaces of homogeneous type. Math. Japon. **46**, 15–28 (1997)
46. Rudin, W.: Functional Analysis. International Series in Pure and Applied Mathematics, 2nd edn. McGraw-Hill Inc., New York (1991)
47. Sawano, Y.: Sharp estimates of the modified Hardy–Littlewood maximal operator on the nonhomogeneous space via covering lemmas. Hokkaido Math. J. **34**, 435–458 (2005)

48. Sawano, Y., Ho, P.-K., Yang, D., Yang, S.: Hardy spaces for ball quasi-Banach function spaces. Diss. Math. (Rozprawy Mat.) **525**, 1–102 (2017)
49. Stein, E.M.: Harmonic Analysis: Real-Variable Methods, Orthogonality, and Oscillatory Integrals. Princeton Mathematical Series, vol. 43. Princeton University Press, Princeton, NJ (1993)
50. Stein, E.M., Weiss, G.: On the theory of harmonic functions of several variables. I. The theory of H^p-spaces. Acta Math. **103**, 25–62 (1960)
51. Uchiyama, A.: Characterization of $H^p(\mathbf{R}^n)$ in terms of generalized Littlewood–Paley g-functions. Stud. Math. **81**, 135–158 (1985)
52. Wojtaszczyk, P.: Banach Spaces for Analysts. Cambridge Studies in Advanced Mathematics, vol. 25. Cambridge University Press, Cambridge (1991)
53. Yang, D., Liang, Y., Ky, L.D.: Real-Variable Theory of Musielak–Orlicz Hardy Spaces. Lecture Notes in Mathematics, vol. 2182. Springer, Cham (2017)
54. Yang, D., Zhou, Y.: Boundedness of sublinear operators in Hardy spaces on RD-spaces via atoms. J. Math. Anal. Appl. **339**, 622–635 (2008)
55. Yang, D., Zhou, Y.: A boundedness criterion via atoms for linear operators in Hardy spaces. Constr. Approx. **29**, 207–218 (2009)
56. Yang, D., Zhou, Y.: Radial maximal function characterizations of Hardy spaces on RD-spaces and their applications. Math. Ann. **346**, 307–333 (2010)
57. Yang, D., Zhou, Y.: New properties of Besov and Triebel–Lizorkin spaces on RD-spaces. Manuscr. Math. **134**, 59–90 (2011)
58. Zhuo, C., Sawano, Y., Yang, D.: Hardy spaces with variable exponents on RD-spaces and applications. Diss. Math. (Rozprawy Mat.) **520**, 1–74 (2016)

Affiliations

Ziyi He[1] · Yongsheng Han[2] · Ji Li[3] · Liguang Liu[4] · Dachun Yang[1] · Wen Yuan[1]

✉ Dachun Yang
dcyang@bnu.edu.cn

Ziyi He
ziyihe@mail.bnu.edu.cn

Yongsheng Han
hanyong@auburn.edu

Ji Li
ji.li@mq.edu.au

Liguang Liu
liuliguang@ruc.edu.cn

Wen Yuan
wenyuan@bnu.edu.cn

1 Laboratory of Mathematics and Complex Systems (Ministry of Education of China), School of Mathematical Sciences, Beijing Normal University, Beijing 100875, People's Republic of China

2 Department of Mathematics, Auburn University, Auburn, AL 36849-5310, USA

3 Department of Mathematics, Macquarie University, Sydney, NSW 2109, Australia

4 School of Mathematics, Renmin University of China, Beijing 100872, People's Republic of China

Journal of Fourier Analysis and Applications (2019) 25:2268–2323
https://doi.org/10.1007/s00041-018-09656-8

The Solution to the Frame Quantum Detection Problem

Sara Botelho-Andrade[1] · Peter G. Casazza[1] · Desai Cheng[1] · Tin T. Tran[1]

Received: 7 December 2017 / Published online: 10 December 2018
© Springer Science+Business Media, LLC, part of Springer Nature 2018

Abstract

We will give a complete solution to the frame quantum detection problem. We will solve both cases of the problem: the quantum injectivity problem and quantum state estimation problem. We will answer the problem in both the real and complex cases and in both the finite dimensional and infinite dimensional cases. Finite Dimensional Case:

(1) We give two complete classifications of the sets of vectors which solve the injectivity problem - for both the real and complex cases. We also give methods for constructing them.
(2) We show that the frames which solve the injectivity problem are open and dense in the family of all frames.
(3) We show that the Parseval frames which give injectivity are dense in the Parseval frames.
(4) We classify all frames for which the state estimation problem is solvable, and when it is not solvable, we give the best approximation to a solution.

Infinite Dimensional Case:

(1) We give a classification of all frames which solve the injectivity problem and give methods for constructing solutions.
(2) We show that the frames solving the injectivity problem are neither open nor dense in all frames.
(3) We give necessary and sufficient conditions for a frame to solve the state estimation problem for all measurements in ℓ_1 and show that there is no injective frame for which the state estimation problem is solvable for all measurements in ℓ_2.
(4) When the state estimation problem does not have an exact solution, we give the best approximation to a solution.

Keywords Quantum detection · Quantum injectivity · Quantum state estimation · Hilbert space frame theory · Positive operator valued measures · Separated sets

---

Communicated by Hans G. Feichtinger.

The authors were supported by NSF DMS 1609760, NSF ATD 1321779, and ARO W911NF-16-1-0008.

Extended author information available on the last page of the article

Mathematics Subject Classification 42C15 · 46L10 · 47A05

1 Introduction and Preliminaries

In this paper we will give a complete answer to the frame quantum detection problem including the injectivity problem and state estimation problem. We will answer the problem in both the real and complex cases and in both the finite dimensional and infinite dimensional cases.

Important Notation Throughout the paper we will let $\{e_i\}_{i=1}^n$ be the canonical orthonormal basis of $\mathbb{R}^n$ or $\mathbb{C}^n$ and $\{e_i\}_{i=1}^\infty$ will denote the canonical orthonormal basis of real or complex ℓ_2. Also, ι will be used to denote the complex unit.

For a vector x_k in $\mathbb{R}^n$ or $\mathbb{C}^n$, we denote its coordinates as

$$x_k = (x_{k1}, x_{k2}, \ldots, x_{kn}).$$

Similarly, for x_k belonging to ℓ_2, we write

$$x_k = (x_{k1}, x_{k2}, \ldots, x_{ki}, \ldots).$$

To explain exactly what we will solve, we need to introduce the basics of quantum detection. Let $L^\infty(\mathbb{H})$ be the space of bounded linear operators on a finite or infinite dimensional (real or complex) Hilbert space $\mathbb{H}$. Let $\{e_i\}_{i\in I}$ be an orthonormal basis for $\mathbb{H}$. For an operator $T \in L_0(\mathbb{H})$, the finite rank operators on $\mathbb{H}$, the *trace* of T is given by: $\text{tr}(T) = \sum_{i\in I}\langle Te_i, e_i\rangle$, which is finite and independent of the orthonormal basis. The trace induces a scalar product by $\langle T, S\rangle_{HS} = \text{tr}(TS^*)$. The closure of $L_0(\mathbb{H})$ with respect to this scalar product, denoted $L^2(\mathbb{H})$ is the space of the Hilbert-Schmidt operators on $\mathbb{H}$. For any $T \in L^\infty(\mathbb{H})$ we denote by $|T| = \sqrt{TT^*}$, the positive square root of TT^*. We say that T is a *trace class operator* if $\text{tr}(|T|) < \infty$. The set of all trace class operators is denoted by $L^1(\mathbb{H})$ and forms a Banach space under the *trace norm* $\|T\|_1 = \text{tr}(|T|)$.

Let

$$Sym(\mathbb{H}) = \{T : T \in L^\infty(\mathbb{H}),\ T = T^*\},$$

denote the real Banach space of self-adjoint operators on $\mathbb{H}$ and let

$$Sym^+(\mathbb{H}) = \{T = T^* \geq 0\},$$

denote the real cone of positive self-adjoint operators on $\mathbb{H}$. The main objects to analyze these operators are the *positive operator-valued measures*.

1.1 Positive Operator-Valued Measures

In quantum mechanics, the definition of a von Neumann measurement can be generalized using positive operator-valued measures (POVMs) [9,11,14]. The advantage of this is that it allows one to distinguish more accurately among elements of a set of non-orthogonal quantum states.

Let X denote a set of outcomes (e.g. this could be a finite or infinite subset of $\mathbb{Z}^d$ or $\mathbb{R}^d$). Let β denote a sigma algebra of subsets of X.

Definition 1.1 A **positive operator-valued measure (POVM)** is a function $\Pi : \beta \to Sym^+(\mathbb{H})$ satisfying:

(1) $\Pi(\emptyset) = 0$ (the zero operator).
(2) For every at most countable disjoint family $\{U_i\}_{i\in I} \subset \beta$, $x, y \in \mathbb{H}$ we have

$$\langle \Pi\left(\cup_{i\in I} U_i\right) x, y\rangle = \sum_{i\in I} \langle \Pi(U_i)x, y\rangle.$$

(3) $\Pi(X) = I$ (the identity operator).

1.2 Quantum Systems

A **quantum system** is defined as a von Neumann algebra $\mathcal{A}$ of operators acting on $\mathbb{H}$. The set of **states** on $\mathbb{H}$ is

$$\mathcal{S}(\mathbb{H}) = \{T \in L^1(\mathbb{H}),\ T = T^* \geq 0,\ \mathrm{tr}(T) = 1\},$$

and it represents the reservoir of **quantum states** for any quantum system.

The set of **quantum states** $\mathcal{S}(\mathcal{A})$ associated to a quantum system $\mathcal{A}$ is obtained by identifying states that differ by a null state with respect to $\mathcal{A}$. Thus, the set of quantum states are in one-to-one correspondence with the linear functionals on $\mathcal{A}$ of the form:

$$\rho : \mathcal{A} \to \mathbb{C},\ \text{ for some } S \in \mathcal{S}(\mathbb{H}),\ \rho(T) = \mathrm{tr}(TS),\quad \text{for every } T \in \mathcal{A}.$$

A quantum state $\rho \in \mathcal{S}(\mathcal{A})$ is called a **pure state** if it is an extreme point in the convex *weak** compact set of quantum states $\mathcal{S}(\mathcal{A})$. We say a POVM Π is *associated to a von Neumann algebra* $\mathcal{A}$ if $\Pi : \beta \to \mathcal{A} \cap Sym^+(\mathbb{H})$.

Given a quantum state ρ, the **quantum measurement** performed by the POVM Π is the map $p : \beta \to \mathbb{R}$ defined by $p(U) = \rho(\Pi(U)) = \mathrm{tr}(\Pi(U)T)$, where $T \in \mathcal{S}(\mathbb{H})$ is in the equivalence class associated to ρ.

1.3 The Quantum Detection Problem

Let $L(\beta, \mathbb{R})$ denote the set of real-valued bounded functions defined on β. Given a POVM Π associated to a von Neumann algebra $\mathcal{A}$, the **quantum detection problem** is formulated as follows.

Quantum Detection Problem Is there a unique quantum state $\rho \in \mathcal{S}(\mathcal{A})$ compatible with the set of quantum measurements performed by the POVM Π?

Specifically, the quantum detection problem asks two questions:

(1) *Injectivity, or state separability* Is the following map injective

$$\mathbb{M} : \mathcal{S}(\mathcal{A}) \to L(\beta, \mathbb{R}), \quad \mathbb{M}(\rho)(U) = \rho(\Pi(U))?$$

(2) *Range analysis, or state estimation* Assume $\mathbb{M}$ is injective. Then, given a map $p \in L(\beta, \mathbb{R})$, determine if p is in the range of $\mathbb{M}$, hence is of the form $p = \mathbb{M}(\rho)$ for some unique $\rho \in \mathcal{S}(\mathcal{A})$. If not, find a quantum state ρ that best approximates p in some sense (e.g. robustness to noise).

We point out that in the context of quantum detection in quantum mechanics, a significant amount of work has been put into computing the probability of detection error [10,13,14,17,19]. We will not address this question here.

1.4 Frame POVMs

In this section we introduce the Hilbert space frame version of the quantum detection problem. For a background on frame POVMs we recommend [1,11,12,16]. For a background on Hilbert space frame theory we recommend [4,6,8].

Definition 1.2 A family of vectors $\{x_k\}_{k \in I}$ is a **frame** for a real or complex, finite or infinite dimensional Hilbert space $\mathbb{H}$ if there are constants $0 < A \leq B < \infty$ satisfying:

$$A\|x\|^2 \leq \sum_{k \in I} |\langle x, x_k \rangle|^2 \leq B\|x\|^2, \text{ for all } x \in \mathbb{H}.$$

We have

(1) A, B are the **lower and upper frame bounds** of the frame.
(2) If $A = B$ this is a **tight frame**. If $A = B = 1$ this is a **Parseval frame**.
(3) If we only assume we have $0 < B < \infty$, this is called a **B-Bessel sequence**. Note that $\|x_k\|^2 \leq B$, for all $k \in I$.

We define the **analysis operator** of the frame as $T : \mathbb{H} \to \ell_2(I)$ by

$$T(x) = (\langle x, x_1 \rangle, \langle x, x_2 \rangle, \ldots) = \sum_{k \in I} \langle x, x_k \rangle e_k.$$

The **synthesis operator** T^* is given by:

$$T^* \left(\{a_k\}_{k \in I}\right) = \sum_{k \in I} a_k x_k.$$

The **frame operator** is $S = T^*T$. This is a positive, self-adjoint invertible operator on $\mathbb{H}$ satisfying:

$$S(x) = \sum_{k \in I} \langle x, x_k \rangle x_k.$$

It is known that for any frame $\{x_k\}_{k\in I}$, $\{S^{-1/2}x_k\}_{k\in I}$ is a Parseval frame. It is also known that a frame is Parseval if and only if its frame operator is the identity operator.

Definition 1.3 A frame $\{x_k\}_{k\in I}$ is said to be **bounded** if there is a constant $C > 0$ such that

$$\|x_k\| \geq C, \text{ for all } k \in I.$$

If $\{x_k\}_{k\in I}$ is a Parseval frame for a Hilbert space $\mathbb{H}$, it naturally induces a POVM Π on $X = I$ with $\beta = 2^I$ (the power set of I):

$$\Pi(U) = \sum_{k\in U} x_k x_k^*, \text{ where } x_k^* : \mathbb{H} \to \mathbb{C},\ x_k^*(x) = \langle x, x_k\rangle,$$

with strong convergence for any $U \subset I$.

Given a state $T \in \mathcal{S}(\mathbb{H})$ (i.e. a unit-trace, trace class, positive, self-adjoint operator on $\mathbb{H}$), the frame induced quantum measurement is given by the function

$$p : \beta \to \mathbb{R},\quad p(U) = \sum_{k\in U} \operatorname{tr}(T x_k x_k^*) = \sum_{k\in U} \langle T x_k, x_k\rangle.$$

For the von Neumann algebra $\mathcal{A} = L^\infty(\mathbb{H})$, the quantum states coincide with the convex set of states $\mathcal{S}(\mathbb{H})$. In this case, the injectivity problem and the state estimation problem ask:

Injectivity Problem Is there a Parseval frame $\chi = \{x_k\}_{k\in I}$ so that the map $\mathbb{M} : \mathcal{S}(\mathbb{H}) \to L(\beta, \mathbb{R})$ defined by $\mathbb{M}(T)(U) = \sum_{k\in U} \langle T x_k, x_k\rangle$ for $U \subset I$ is injective?

State Estimation Problem Given an injective Parseval frame $\{x_k\}_{k\in I}$ and a function $p : \beta \to \mathbb{R}$, is there any $T \in \mathcal{S}(\mathbb{H})$ so that $\mathbb{M}(T) = p$? If not, find a quantum state T that best approximates p.

1.5 Generalizing Quantum Detection

We will work on a much more general quantum detection problem. In particular, we will work with

(1) Self-adjoint operators which may not be positive.
(2) Operators which are not trace one but are Hilbert Schmidt.
(3) Frames which are not Parseval.

We will see that solving the problem in this more general form will also solve the original problem.

First, we need a definition.

Definition 1.4 A family of vectors $\mathcal{X} = \{x_k\}_{k\in I}$ in a Hilbert space $\mathbb{H}$ is said to be **injective** if whenever a Hilbert Schmidt self-adjoint operator T satisfies

$$\langle Tx_k, x_k \rangle = 0, \text{ for all } k \in I,$$

then $T = 0$.

Now we will show that we do not need to find Parseval frames for the quantum detection problem. If we have a frame giving injectivity, then its canonical Parseval frame is injective.

Proposition 1.5 *Let $\{x_k\}_{k\in I}$ be a frame for $\mathbb{H}$ which gives injectivity. If F is a bounded invertible operator on $\mathbb{H}$, then $\{Fx_k\}_{k\in I}$ also gives injectivity.*

Proof Let T be a Hilbert Schmidt self-adjoint operator such that

$$\langle TFx_k, Fx_k \rangle = 0, \text{ for all } k.$$

Then $\langle F^*TFx_k, x_k \rangle = 0$, for all k. Note that F^*TF is also a Hilbert Schmidt self-adjoint operator. Therefore, $F^*TF = 0$ and hence $T = 0$. □

Corollary 1.6 *Let $\{x_k\}_{k\in I}$ be a frame with frame operator S. If $\{x_k\}_{k\in I}$ gives injectivity, then the canonical Parseval frame $\{S^{-1/2}x_k\}_{k\in I}$ also gives injectivity.*

2 The Solution for the Finite Dimensional Case

In this section we will solve the finite dimensional injectivity problem and the state estimation problem for both the real and complex cases. These problems were originally solved by Scott [18] (See also [2]) where the solutions are called **informationally complete quantum measurements**. We will have to redo this here since we need much more information about the solutions and need proofs in a format that will easily generalize to infinite dimensions.

2.1 Solution to the Injectivity Problem

First, we will see that we do not need to work with positive operators via the following theorem.

Theorem 2.1 *Given a family of vectors $\mathcal{X} = \{x_k\}_{k=1}^m$ in $\mathbb{H}^n$, the following are equivalent:*

(1) *Whenever T, S are positive and self-adjoint, and*

$$\langle Tx_k, x_k \rangle = \langle Sx_k, x_k \rangle, \textit{ for all } k,$$

then $T = S$.

(2) *Whenever T, S are self-adjoint, and*

$$\langle Tx_k, x_k \rangle = \langle Sx_k, x_k \rangle, \textit{ for all } k,$$

then $T = S$.

(3) *$\mathcal{X}$ is injective.*

Proof (1) $\Rightarrow$ (2): Let T, S be self-adjoint operators such that

$$\langle Tx_k, x_k\rangle = \langle Sx_k, x_k\rangle, \text{ for all } k.$$

Set

$$m_1 := \inf_{\|x\|=1} \langle Tx, x\rangle, \quad m_2 := \inf_{\|x\|=1} \langle Sx, x\rangle$$

then $m_1, m_2 \in \mathbb{R}$. Set $m = \min\{m_1, m_2\}$.

Now let $P = T - mI$, $Q = S - mI$. Then for any $x \in \mathbb{H}$, $\|x\| = 1$, we have

$$\langle Px, x\rangle = \langle (T - mI)x, x\rangle = \langle Tx, x\rangle - m \geq 0.$$

Hence, P is positive. Similarly, Q is positive.

We have

$$\begin{aligned}\langle Px_k, x_k\rangle &= \langle (T - mI)x_k, x_k\rangle \\ &= \langle Tx_k, x_k\rangle - m\|x_k\|^2 \\ &= \langle Sx_k, x_k\rangle - m\|x_k\|^2 \\ &= \langle Qx_k, x_k\rangle.\end{aligned}$$

By (1) we get $P = Q$ and therefore $T = S$.

(2) $\Rightarrow$ (3): Let T be any self-adjoint operator such that

$$\langle Tx_k, x_k\rangle = 0, \text{ for all } k.$$

Then

$$\langle Tx_k, x_k\rangle = \langle Sx_k, x_k\rangle, \text{ for all } k,$$

where $S = 0$. It follows that $T = 0$.

(3) $\Rightarrow$ (1): Let any positive self-adjoint operators T, S satisfy

$$\langle Tx_k, x_k\rangle = \langle Sx_k, x_k\rangle, \text{ for all } k.$$

Then

$$\langle (T - S)x_k, x_k\rangle = 0, \text{ for all } k.$$

Since $T - S$ is a self-adjoint operator, $T = S$ by (3). □

Remark 2.2 If we further require that the operators are trace one, then to prove injectivity, we only need to show that if T is trace zero and $\langle Tx_k, x_k\rangle = 0$ for all $k = 1, 2, \ldots$, then $T = 0$. Since if T, S are trace one and

$$\langle Tx_k, x_k\rangle = \langle Sx_k, x_k\rangle, \text{ for all } k.$$

then

$$\langle (T - S)x_k, x_k\rangle = 0, \text{ for all } k \text{ and } \mathrm{tr}(T - S) = 0.$$

2.1.1 The Real Case

We start with a proposition which shows where our classification of the quantum detection problem comes from.

Proposition 2.3 *Given a self-adjoint operator* $T = (a_{ij})_{i,j=1}^n$ *on* $\mathbb{R}^n$ *and a vector* $x = (x_1, x_2, \ldots, x_n) \in \mathbb{R}^n$, *we have*

$$\langle Tx, x\rangle = \sum_{i=1}^n \sum_{j=1}^n a_{ij}x_i x_j = \sum_{i=1}^n a_{ii}x_i^2 + 2\sum_{i=1}^n \sum_{j=i+1}^n a_{ij}x_i x_j.$$

Proof First we compute:

$$Tx = \left(\sum_{j=1}^n a_{1j}x_j, \sum_{j=1}^n a_{2j}x_j, \ldots, \sum_{j=1}^n a_{nj}x_j \right).$$

So,

$$\langle Tx, x\rangle = \sum_{j=1}^n a_{1j}x_1 x_j + \sum_{j=1}^n a_{2j}x_2 x_j + \cdots + \sum_{j=1}^n a_{nj}x_n x_j.$$

Using the fact that T is self-adjoint:

$$\begin{aligned}
\langle Tx, x\rangle &= \sum_{i=1}^n \sum_{j=1}^n a_{ij}x_i x_j \\
&= \sum_{i=1}^n a_{ii}x_i^2 + \sum_{1\le i<j\le n} a_{ij}x_i x_j + \sum_{1\le j<i\le n} a_{ij}x_i x_j \\
&= \sum_{i=1}^n a_{ii}x_i^2 + \sum_{1\le i<j\le n} a_{ij}x_i x_j + \sum_{1\le i<j\le n} a_{ji}x_j x_i \\
&= \sum_{i=1}^n a_{ii}x_i^2 + 2\sum_{1\le i<j\le n} a_{ij}x_i x_j.
\end{aligned}$$

□

This proposition leads us to the following definition:

Definition 2.4 To a vector $x = (x_1, x_2, \ldots, x_n) \in \mathbb{R}^n$ we associate a vector $\tilde{x}$ in $\mathbb{R}^{\frac{n(n+1)}{2}}$ by:

$$\tilde{x} = (x_1x_1, x_1x_2, \ldots, x_1x_n; x_2x_2, x_2x_3, \ldots, x_2x_n; \ldots; x_{n-1}x_{n-1}, x_{n-1}x_n; x_nx_n).$$

To a self-adjoint operator $T = (a_{ij})_{i,j=1}^n$ on $\mathbb{R}^n$, we associate a vector $\tilde{T}$ in $\mathbb{R}^{\frac{n(n+1)}{2}}$ by:

$$\tilde{T} = (a_{11}, 2a_{12}, \ldots, 2a_{1n}; a_{22}, 2a_{23}, \ldots, 2a_{2n}; \ldots; a_{(n-1)(n-1)}, 2a_{(n-1)n}; a_{nn}).$$

Proposition 2.3 now becomes:

Corollary 2.5 *Given a self-adjoint operator $T = (a_{ij})_{i,j=1}^n$ on $\mathbb{R}^n$ and a vector $x = (x_1, x_2, \ldots, x_n) \in \mathbb{R}^n$, we have*

$$\langle Tx, x\rangle = \langle \tilde{T}, \tilde{x}\rangle.$$

We are now able to give a classification of the frames χ which give injectivity for the quantum detection problem.

Theorem 2.6 *Let $\chi = \{x_k\}_{k=1}^m$ be a frame for $\mathbb{R}^n$. The following are equivalent:*

(1) *χ gives injectivity.*
(2) *We have that $\{\tilde{x}_k\}_{k=1}^m$ spans $\mathcal{K} := \mathbb{R}^{\frac{n(n+1)}{2}}$.*

Proof (1) $\Rightarrow$ (2): Let a vector

$$a = (a_{11}, a_{12}, \ldots, a_{1n}; a_{22}, a_{23}, \ldots, a_{2n}; \ldots; a_{(n-1)(n-1)}, a_{(n-1)n}; a_{nn}) \in \mathcal{K}$$

be such that $\langle a, \tilde{x}_k\rangle = 0$ for all k.

Define an operator $T = (b_{ij})_{i,j=1}^n$ on $\mathbb{R}^n$, where $b_{ii} = a_{ii}$ for $i = 1, 2, \ldots, n$ and $b_{ij} = b_{ji} = \frac{1}{2}a_{ij}$ for $i < j$. Then T is a self-adjoint operator.

For any $x = (x_1, x_2, \ldots, x_n) \in \mathbb{R}^n$ we have

$$\begin{aligned}
\langle Tx, x\rangle &= \sum_{i=1}^n \sum_{j=1}^n b_{ij}x_ix_j \\
&= \sum_{i=1}^n b_{ii}x_i^2 + 2\sum_{1\le i<j\le n} b_{ij}x_ix_j \\
&= \sum_{i=1}^n a_{ii}x_i^2 + \sum_{1\le i<j\le n} a_{ij}x_ix_j \\
&= \langle a, \tilde{x}\rangle.
\end{aligned}$$

Therefore, $\langle Tx_k, x_k\rangle = \langle a, \tilde{x}_k\rangle = 0$ for all k. This implies $T = 0$ and hence $a = 0$.
(2) $\Rightarrow$ (1): Let $T = (a_{ij})_{i,j=1}^n$ be a self-adjoint operator such that

$$\langle Tx_k, x_k\rangle = 0, \text{ for all } k.$$

Then by Corollary 2.5,

$$\langle \tilde{T}, \tilde{x}_k\rangle = \langle Tx_k, x_k\rangle = 0, \text{ for all } k.$$

Since $\{\tilde{x}_k\}_{k=1}^m$ spans $\mathcal{K}$, we have that $\tilde{T} = 0$ and so $T = 0$. □

The theorem gives a lower limit on the number of vectors needed to achieve injectivity.

Corollary 2.7 *If a frame $\mathcal{X} = \{x_k\}_{k=1}^m$ gives injectivity in $\mathbb{R}^n$, then $m \geq \dfrac{n(n+1)}{2}$.*

As a consequence (See the related [7]):

Corollary 2.8 *Given a frame $\{x_k\}_{k=1}^m$ for $\mathbb{R}^n$, the following are equivalent:*

(1) *The family $\{x_k x_k^*\}_{k=1}^m$ spans the class of self-adjoint operators.*
(2) *The family of vectors $\{\tilde{x}_k\}_{k=1}^m$ spans $\mathbb{R}^{\frac{n(n+1)}{2}}$.*

Proof This is immediate since for every $x \in \mathbb{R}^n$ and self-adjoint operator T, we have

$$\langle T, xx^*\rangle = \text{tr}(Txx^*) = \langle Tx, x\rangle.$$

□

Remark 2.9 For any of the frames $\{x_k\}_{k=1}^m$ giving injectivity, if S is the frame operator, then $\{S^{-1/2}x_k\}_{k=1}^m$ is a Parseval frame giving injectivity by Corollary 1.6.

Normally in the frame quantum detection problem, there is the added assumption that the trace of the operators is one. We will now see that with this assumption, we can eliminate one measurement. We start with a simple example.

Example 2.10 Let $\mathcal{X} = \{(1,0), (1,1)\}$ in $\mathbb{R}^2$. Then $\mathcal{X}$ gives injectivity in $\mathbb{R}^2$ for all self-adjoint operators of trace one.

Indeed, let

$$T = \begin{bmatrix} a & b \\ b & c \end{bmatrix}$$

be a self-adjoint matrix of trace zero such that

$$\langle T(1,0), (1,0)\rangle = \langle T(1,1), (1,1)\rangle = 0.$$

Then

$$a = \langle T(1,0), (1,0)\rangle = 0.$$

Since $a + c = 0$ and

$$\langle T(1,1), (1,1)\rangle = \langle (a+b, b+c), (1,1)\rangle = a + 2b + c,$$

then $b = c = 0$. So $T = 0$.

For the classification of all frames which give injectivity with this added assumption, we will need:

Definition 2.11 Let $x = (x_1, x_2, \dots, x_n) \in \mathbb{R}^n$. Define

$$\tilde{x} = (x_1x_2, \dots, x_1x_n; x_2^2 - x_1^2, x_2x_3, \dots, x_2x_n; \dots; x_{n-1}^2 - x_1^2, x_{n-1}x_n; x_n^2 - x_1^2).$$

Now we can prove the trace one version of our classification.

Theorem 2.12 *Let $\mathcal{X} = \{x_k\}_{k=1}^m$ be a frame for $\mathbb{R}^n$. The following are equivalent:*

(1) *$\mathcal{X}$ gives injectivity for all self-adjoint operators of trace one.*
(2) *We have that $\{\tilde{x}_k\}_{k=1}^m$ spans $\mathcal{K} := \mathbb{R}^{\frac{n(n+1)}{2}-1}$.*

Proof Note that we are trying to show that when two positive, self-adjoint operators T, S of trace one satisfy

$$\langle Tx_k, x_k\rangle = \langle Sx_k, x_k\rangle, \text{ for all } k = 1, 2, \dots, m,$$

then $T = S$. This is clearly equivalent to showing that if T is a self-adjoint operator of trace zero and $\langle Tx_k, x_k\rangle = 0$ for all $k = 1, 2, \dots,$ then $T = 0$.

(1) $\Rightarrow$ (2): Let a vector

$$a = (a_{12}, \dots, a_{1n}; a_{22}, \dots, a_{2n}; \dots; a_{(n-1)(n-1)}, a_{(n-1)n}; a_{nn}) \in \mathcal{K}$$

be such that $\langle a, \tilde{x}_k\rangle = 0$ for all k.

Define an operator $T = (b_{ij})_{i,j=1}^n$, where $b_{11} = -\sum_{i=2}^n a_{ii}$, $b_{ii} = a_{ii}$ for $i = 2, 3, \dots, n$ and $b_{ij} = b_{ji} = \dfrac{1}{2}a_{ij}$ for $i < j$. Then T is self-adjoint and $\text{tr}(T) = 0$.

For any $x = (x_1, x_2, \dots, x_n) \in \mathbb{R}^n$ we have

$$\begin{aligned}\langle Tx, x\rangle &= \sum_{i=1}^n \sum_{j=1}^n b_{ij}x_ix_j \\ &= \sum_{i=1}^n b_{ii}x_i^2 + 2\sum_{1\le i<j\le n} b_{ij}x_ix_j\end{aligned}$$

$$
\begin{aligned}
&= \left(-\sum_{i=2}^{n} a_{ii}\right) x_1^2 + \sum_{i=2}^{n} a_{ii} x_i^2 + \sum_{1 \le i < j \le n} a_{ij} x_i x_j \\
&= \langle a, \tilde{x} \rangle.
\end{aligned}
$$

Therefore, $\langle Tx_k, x_k \rangle = \langle a, \tilde{x}_k \rangle = 0$ for all k. This implies $T = 0$ and hence $a = 0$.

(2) $\Rightarrow$ (1): Let $T = (a_{ij})_{i,j=1}^n$ be a self-adjoint operator with $\text{tr}(T) = 0$ and such that $\langle Tx_k, x_k \rangle = 0$ for all k. Then $a_{11} = -\sum_{i=2}^n a_{ii}$.

Define

$$
\tilde{T} = (2a_{12}, 2a_{13}, \ldots, 2a_{1n}; a_{22}, 2a_{23}, \ldots, 2a_{2n}; \ldots; a_{(n-1)(n-1)}, 2a_{(n-1)n}; a_{nn}).
$$

Then $\tilde{T} \in \mathcal{K}$ and

$$
\langle \tilde{T}, \tilde{x}_k \rangle = \langle Tx_k, x_k \rangle = 0, \text{ for all } k.
$$

Since $\{\tilde{x}_k\}_{k=1}^m$ spans $\mathcal{K}$, then $\tilde{T} = 0$. Hence $T = 0$. □

2.1.2 The Complex Case

We need to adjust some of the definitions for the real case so they will apply to the complex case.

Definition 2.13 Given $x = (x_1, x_2, \ldots, x_n) \in \mathbb{C}^n$, define

$$
\begin{aligned}
\tilde{x} = (&|x_1|^2, \text{Re}(\bar{x}_1 x_2), \text{Im}(\bar{x}_1 x_2), \ldots, \text{Re}(\bar{x}_1 x_n), \text{Im}(\bar{x}_1 x_n); \\
&|x_2|^2, \text{Re}(\bar{x}_2 x_3), \text{Im}(\bar{x}_2 x_3), \ldots, \text{Re}(\bar{x}_2 x_n), \text{Im}(\bar{x}_2 x_n); \ldots; \\
&|x|_{n-1}^2, \text{Re}(\bar{x}_{n-1} x_n), \text{Im}(\bar{x}_{n-1} x_n); |x_n|^2) \in \mathbb{R}^{n^2}.
\end{aligned}
$$

Now we can give our classification theorem for injectivity in the quantum detection problem for the complex case.

Theorem 2.14 *Let $\mathcal{X} = \{x_k\}_{k=1}^m$ be a frame for $\mathbb{C}^n$. The following are equivalent:*

(1) *$\mathcal{X}$ gives injectivity.*
(2) *We have that $\{\tilde{x}_k\}_{k=1}^m$ spans $\mathbb{R}^{n^2}$.*

Proof (1) $\Rightarrow$ (2): Let a be any vector

$$
\begin{aligned}
a = (&a_{11}, u_{12}, v_{12}, \ldots, u_{1n}, v_{1n}; a_{22}, u_{23}, v_{23}, \ldots, u_{2n}, v_{2n}; \ldots; \\
&a_{(n-1)(n-1)}, u_{(n-1)n}, v_{(n-1)n}; a_{nn}) \in \mathbb{R}^{n^2}
\end{aligned}
$$

such that $\langle a, \tilde{x}_k \rangle = 0$ for all k.

Define an operator $T = (b_{ij})_{i,j=1}^n$ with $b_{ii} = a_{ii}$ for $i = 1, 2, \ldots, n$ and $b_{ij} = \bar{b}_{ji} = \frac{1}{2}(u_{ij} - \iota v_{ij})$ for $i < j$. Then T is a self-adjoint operator.

For any $x = (x_1, x_2, \ldots, x_n) \in \mathbb{C}^n$ we have

$$\begin{aligned}
\langle Tx, x\rangle &= \sum_{i=1}^{n}\sum_{j=1}^{n} b_{ij}\bar{x}_i x_j \\
&= \sum_{i=1}^{n} b_{ii}|x_i|^2 + \sum_{1\le i<j\le n} b_{ij}\bar{x}_i x_j + \sum_{1\le j<i\le n} b_{ij}\bar{x}_i x_j \\
&= \sum_{i=1}^{n} b_{ii}|x_i|^2 + \sum_{1\le i<j\le n} b_{ij}\bar{x}_i x_j + \sum_{1\le j<i\le n} \bar{b}_{ji}\bar{x}_i x_j \\
&= \sum_{i=1}^{n} b_{ii}|x_i|^2 + \sum_{1\le i<j\le n} b_{ij}\bar{x}_i x_j + \sum_{1\le i<j\le n} \bar{b}_{ij}\bar{x}_j x_i \\
&= \sum_{i=1}^{n} b_{ii}|x_i|^2 + 2\sum_{1\le i<j\le n} \mathrm{Re}(b_{ij}\bar{x}_i x_j) \\
&= \sum_{i=1}^{n} b_{ii}|x_i|^2 + 2\sum_{1\le i<j\le n} (\mathrm{Re}(b_{ij})\mathrm{Re}(\bar{x}_i x_j) - \mathrm{Im}(b_{ij})\mathrm{Im}(\bar{x}_i x_j)) \\
&= \sum_{i=1}^{n} a_{ii}|x_i|^2 + \sum_{1\le i<j\le n} \big(u_{ij}\mathrm{Re}(\bar{x}_i x_j) + v_{ij}\mathrm{Im}(\bar{x}_i x_j)\big) \\
&= \langle a, \tilde{x}\rangle.
\end{aligned}$$

Therefore, $\langle Tx_k, x_k\rangle = \langle a, \tilde{x}_k\rangle = 0$ for all k. This implies $T = 0$ and hence $a = 0$.

(2) $\Rightarrow$ (1): Let $T = (a_{ij})_{i,j=1}^n$ be a self-adjoint operator such that

$$\langle Tx_k, x_k\rangle = 0, \text{ for all } k.$$

Define

$$\begin{aligned}
\tilde{T} = (&a_{11}, 2\mathrm{Re}(a_{12}), -2\mathrm{Im}(a_{12}), \ldots, 2\mathrm{Re}(a_{1n}), -2\mathrm{Im}(a_{1n}); \\
&a_{22}, 2\mathrm{Re}(a_{23}), -2\mathrm{Im}(a_{23}), \ldots, 2\mathrm{Re}(a_{2n}), -2\mathrm{Im}(a_{2n}); \ldots; \\
&a_{(n-1)(n-1)}, 2\mathrm{Re}(a_{(n-1)n}), -2\mathrm{Im}(a_{(n-1)n}); a_{nn}) \in \mathbb{R}^{n^2}.
\end{aligned}$$

Then we have

$$\langle \tilde{T}, \tilde{x}_k\rangle = \langle Tx_k, x_k\rangle = 0, \text{ for all } k.$$

Since $\{\tilde{x}_k\}_{k=1}^m$ spans $\mathcal{K}$ we have that $\tilde{T} = 0$ and so $T = 0$; i.e. $\{x_k\}_{k=1}^m$ gives injectivity. □

Corollary 2.15 *If a frame* $\mathcal{X} = \{x_k\}_{k=1}^m$ *gives injectivity in* $\mathbb{C}^n$, *then* $m \geq n^2$.

Similar to the real case, we have a classification for injectivity for positive self-adjoint operators of trace one in a complex Hilbert space. This requires another definition to fit this case.

Definition 2.16 Given $x = (x_1, x_2, \ldots, x_n) \in \mathbb{C}^n$, define

$$\begin{aligned}\tilde{x} = (&\mathrm{Re}(\bar{x}_1 x_2), \mathrm{Im}(\bar{x}_1 x_2), \ldots, \mathrm{Re}(\bar{x}_1 x_n), \mathrm{Im}(\bar{x}_1 x_n);\\ &|x_2|^2 - |x_1|^2, \mathrm{Re}(\bar{x}_2 x_3), \mathrm{Im}(\bar{x}_2 x_3), \ldots, \mathrm{Re}(\bar{x}_2 x_n), \mathrm{Im}(\bar{x}_2 x_n); \ldots;\\ &|x_{n-1}|^2 - |x_1|^2, \mathrm{Re}(\bar{x}_{n-1} x_n), \mathrm{Im}(\bar{x}_{n-1} x_n); |x_n|^2 - |x_1|^2) \in \mathbb{R}^{n^2-1}.\end{aligned}$$

Now we classify the frames which give injectivity in the complex case for operators of trace one.

Theorem 2.17 *Let* $\mathcal{X} = \{x_k\}_{k=1}^m$ *be a frame for* $\mathbb{C}^n$. *The following are equivalent:*

(1) $\mathcal{X}$ *gives injectivity for all self-adjoint operators of trace one.*
(2) *We have that* $\{\tilde{x}_k\}_{k=1}^m$ *spans* $\mathbb{R}^{n^2-1}$.

Proof (1) $\Rightarrow$ (2): Let a vector

$$\begin{aligned}a = (&u_{12}, v_{12}, \ldots, u_{1n}, v_{1n}; a_{22}, u_{23}, v_{23}, \ldots, u_{2n}, v_{2n}; \ldots;\\ &a_{(n-1)(n-1)}, u_{(n-1)n}, v_{(n-1)n}; a_{nn}) \in \mathbb{R}^{n^2-1}\end{aligned}$$

be such that $\langle a, \tilde{x}_k \rangle = 0$ for all k.

Define an operator $T = (b_{ij})_{i,j=1}^n$ with $b_{11} = -\sum_{i=2}^n a_{ii}$, $b_{ii} = a_{ii}$ for $i = 2, \ldots, n$ and $b_{ij} = \bar{b}_{ji} = \frac{1}{2}(u_{ij} - \iota v_{ij})$ for $i < j$. Then T is a self-adjoint operator and $\mathrm{tr}(T) = 0$.

For any $x = (x_1, x_2, \ldots, x_n) \in \mathbb{C}^n$ we have

$$\begin{aligned}\langle Tx, x \rangle &= \sum_{i=1}^n b_{ii} |x_i|^2 + 2 \sum_{1 \leq i < j \leq n} (\mathrm{Re}(b_{ij}) \mathrm{Re}(\bar{x}_i x_j) - \mathrm{Im}(b_{ij}) \mathrm{Im}(\bar{x}_i x_j))\\ &= \left(-\sum_{i=2}^n a_{ii} \right) |x_1|^2 + \sum_{i=2}^n a_{ii} |x_i|^2 + \sum_{1 \leq i < j \leq n} \big(u_{ij} \mathrm{Re}(\bar{x}_i x_j) + v_{ij} \mathrm{Im}(\bar{x}_i x_j) \big)\\ &= \langle a, \tilde{x} \rangle.\end{aligned}$$

Therefore, $\langle T x_k, x_k \rangle = \langle a, \tilde{x}_k \rangle = 0$ for all k. This implies $T = 0$ and hence $a = 0$.

(2) $\Rightarrow$ (1): Let $T = (a_{ij})_{i,j=1}^n$ be a self-adjoint operator such that $\mathrm{tr}(T) = 0$ and $\langle T x_k, x_k \rangle = 0$ for all k. Then $a_{11} = -\sum_{i=2}^n a_{ii}$.

Define

$$\begin{aligned}\tilde{T} = (&2\mathrm{Re}(a_{12}), -2\mathrm{Im}(a_{12}), \ldots, 2\mathrm{Re}(a_{1n}), -2\mathrm{Im}(a_{1n});\\ &a_{22}, 2\mathrm{Re}(a_{23}), -2\mathrm{Im}(a_{23}), \ldots, 2\mathrm{Re}(a_{2n}), -2\mathrm{Im}(a_{2n}); \ldots;\\ &a_{(n-1)(n-1)}, 2\mathrm{Re}(a_{(n-1)n}), -2\mathrm{Im}(a_{(n-1)n}); a_{nn}) \in \mathbb{R}^{n^2-1}.\end{aligned}$$

Then we have that

$$\langle \tilde{T}, \tilde{x}_k \rangle = \langle Tx_k, x_k \rangle = 0, \text{ for all } k.$$

Since $\{\tilde{x}_k\}_{k=1}^m$ spans $\mathbb{R}^{n^2-1}$ then $\tilde{T} = 0$. Hence $T = 0$. □

Now we will give another classification of injectivity for the quantum detection problem. This classification has the disadvantage that the requirements are quite complex and difficult to verify in practice. The advantage here is in the other direction. That is, if a frame gives injectivity in the quantum detection problem, then it must satisfy these complex requirements.

Theorem 2.18 *Let $\mathcal{X} = \{x_k\}_{k=1}^m$ be a frame for a real or complex Hilbert space $\mathbb{H}^n$. The following are equivalent:*

(1) *$\mathcal{X}$ gives injectivity.*
(2) *For every orthonormal basis $\mathcal{E} = \{e_j\}_{j=1}^n$ for $\mathbb{H}^n$ we have:*

$$H(\mathcal{E}) =: \text{span}\{(|\langle x_k, e_1\rangle|^2, |\langle x_k, e_2\rangle|^2, \ldots, |\langle x_k, e_n\rangle|^2) : k = 1, 2, \ldots, m\} = \mathbb{R}^n.$$

Proof (1) $\Rightarrow$ (2): We prove the contrapositive. Suppose that (2) fails. Then there is an orthonormal basis $\mathcal{E} = \{e_j\}_{j=1}^n$ so that $H(\mathcal{E}) \neq \mathbb{R}^n$. Hence there is a non-zero vector $\lambda = (\lambda_1, \lambda_2, \ldots, \lambda_n) \in \mathbb{R}^n$ such that $\lambda \perp H(\mathcal{E})$.

Define an operator on $\mathbb{H}^n$ by

$$Te_j = \lambda_j e_j, \; j = 1, 2, \ldots, n.$$

Then T is a non-zero self-adjoint operator and satisfies $\langle Tx_k, x_k\rangle = 0$, for all $k = 1, 2, \ldots, m$, which is a contradiction.

(2) $\Rightarrow$ (1): Let T be a self-adjoint operator such that $\langle Tx_k, x_k\rangle = 0$, for all k. Let $\mathcal{E} = \{e_j\}_{j=1}^n$ be an eigenbasis for T with respective eigenvalues $\{\lambda_j\}_{j=1}^n$. Then for every $k = 1, 2, \ldots, m$ we have

$$\langle Tx_k, x_k\rangle = \sum_{j=1}^n \lambda_j |\langle x_k, e_j\rangle|^2 = 0.$$

That is,

$$(\lambda_1, \lambda_2, \ldots, \lambda_n) \perp H(\mathcal{E}) = \mathbb{R}^n \text{ by assumption (2).}$$

Therefore, $\lambda_j = 0$ for all $j = 1, 2, \ldots, n$ and so $T = 0$. □

Finally in this subsection, we notice that if a family of vectors gives injectivity in a Hilbert space $\mathbb{H}^n$, then it is a frame for $\mathbb{H}^n$.

Proposition 2.19 *Let $\{x_k\}_{k=1}^m$ be a family of vectors in $\mathbb{H}^n$ which is injective. Then* $\text{span}\{x_k\}_{k=1}^m = \mathbb{H}^n$.

Proof Suppose by contradiction that $W := \text{span}\{x_k\}_{k=1}^m \neq \mathbb{H}^n$. Let P be the orthogonal projection onto $W^\perp$. Then $\langle Px_k, x_k\rangle = 0$ for all k, but $P \neq 0$, a contradiction. □

2.2 Constructing the Solutions to the Injectivity Problem

In this subsection, we will construct large classes of frames which give injectivity for the quantum detection problem in both the real and complex cases.

Theorem 2.20 *Let $\{x_k\}_{k=1}^n$ be a linearly independent set in $\mathbb{R}^n$ such that the first coordinates of these vectors are non-zero. Now choose $(n-1)$ linearly independent vectors $\{x_k\}_{k=n+1}^{2n-1}$ in $\mathbb{R}^n$ such that each vector is zero in the first coordinate and is non-zero in the second coordinate. Continuing this procedure we get a frame $\{x_k\}_{k=1}^{\frac{n(n+1)}{2}}$ which gives injectivity.*

Proof We will show that $\{\tilde{x}_k\}_{k=1}^{\frac{n(n+1)}{2}}$ is a basis for $\mathbb{R}^{\frac{n(n+1)}{2}}$.

Indeed, suppose that $\sum_{k=1}^{\frac{n(n+1)}{2}} \alpha_k \tilde{x}_k = 0$ for some scalars $\{\alpha_k\}$. Since after n, all tilde vectors are zero in the first coordinate, then we get

$$\sum_{k=1}^{n} \alpha_k x_{k1} x_k = 0.$$

Since $\{x_k\}_{k=1}^n$ are linearly independent, $\alpha_k x_{k1} = 0$ for all k and since $x_{k1} \neq 0, \alpha_k = 0$ for $k = 1, 2, \ldots, n$.

Now do this argument for the next $(n-1)$ vectors and continue we get $\alpha_k = 0$ for all $k = 1, 2, \ldots, \frac{n(n+1)}{2}$. □

A simple example satisfying the construction is the following.

Example 2.21 The frame

$$\{e_i\}_{i=1}^n \cup \{e_i + e_j : i < j\}_{i,j=1}^n$$

gives injectivity.

For the complex case, we have the following construction. The proof is as in the real case.

Theorem 2.22 *Let $\{x_k\}_{k=1}^{2n-1}$ be a basis for $\mathbb{R}^{2n-1}$, where*

$$x_k = (u_{k1}, u_{k2}, v_{k2}, \ldots, u_{kn}, v_{kn})$$

and $u_{k1} \neq 0$, $k = 1, \ldots, 2n-1$.

Define $(2n-1)$ vectors $\{z_k\}_{k=1}^{2n-1}$ in $\mathbb{C}^n$ by

$$z_k = (u_{k1}, u_{k2} + \iota v_{k2}, \ldots, u_{kn} + \iota v_{kn}).$$

Now let $\{x_k\}_{k=2n}^{4n-4}$ *be a basis for* $\mathbb{R}^{2n-3}$, *where*

$$x_k = (u_{k2}, u_{k3}, v_{k3}, \ldots, u_{kn}, v_{kn})$$

and $u_{k2} \neq 0$, $k = 2n, \ldots, 4n-4$.

Define $(2n-3)$ *vectors* $\{z_k\}_{k=2n}^{4n-4}$ *in* $\mathbb{C}^n$ *by*

$$z_k = (0, u_{k2}, u_{k3} + \iota v_{k3}, \ldots, u_{kn} + \iota v_{kn}).$$

Continuing this procedure we get n^2 *vectors* $\{z_k\}_{k=1}^{n^2}$ *in* $\mathbb{C}^n$ *and they give injectivity.*

As we have seen, we can get Parseval frames giving injectivity by taking $\{S^{-1/2}x_k\}_{k=1}^m$, where $\{x_k\}_{k=1}^m$ gives injectivity and has frame operator S. But the above construction can be adjusted to directly construct Parseval frames giving injectivity.

Theorem 2.23 *Let* $\{\lambda_{ij}\}_{i=1,}^{n}{}_{j=i}^{n}$ *be non-negative numbers satisfying:*

(1) $\lambda_{ij} = 0$ *if and only if* $j < i$.
(2) *For each* $j = 1, 2, \ldots, n$ *we have* $\sum_{i=1}^n \lambda_{ij} = 1$.

Let $\mathcal{E} = \{e_j\}_{j=1}^n$ *be the canonical basis of* $\mathbb{R}^n$. *Let* $\{x_k\}_{k=1}^{\frac{n(n+1)}{2}}$ *be vectors in* $\mathbb{R}^n$ *which satisfy:*

(1) $\{x_k\}_{k=1}^n$ *is a linearly independent set with* $x_{k1} \neq 0$ *for all* $k = 1, \ldots, n$ *and it has frame operator* S_1 *with eigenvectors* $\mathcal{E}$ *and respective eigenvalues* $\{\lambda_{1j}\}_{j=1}^n$ *(See* [5].*)*
(2) $\{x_k\}_{k=n+1}^{2n-1}$ *is a linearly independent set with* $x_{k1} = 0$, *for all* k, $x_{k2} \neq 0$ *for all* k, *and it has frame operator* S_2 *with eigenvectors* $\mathcal{E}$ *and respective eigenvalues* $\{\lambda_{2j}\}_{j=1}^n$.
(3) *continue.*

Then the vectors $\{x_k\}_{k=1}^{\frac{n(n+1)}{2}}$ *form a Parseval frame for* $\mathbb{R}^n$ *which is injective.*

Proof This is injective by Theorem 2.20. To see that it is Parseval, observe that the frame operator of this frame is $\sum_{i=1}^n S_i$. Now, let $y \in \mathbb{R}^n$ and compute:

$$\begin{aligned}\sum_{i=1}^n S_i y &= \sum_{i=1}^n S_i \left(\sum_{j=1}^n \langle y, e_j \rangle e_j \right) \\ &= \sum_{i=1}^n \sum_{j=1}^n \langle y, e_j \rangle S_i e_j \\ &= \sum_{i=1}^n \sum_{j=1}^n \langle y, e_j \rangle \lambda_{ij} e_j\end{aligned}$$

$$= \sum_{j=1}^{n} \langle y, e_j \rangle e_j \sum_{i=1}^{n} \lambda_{ij}$$
$$= \sum_{j=1}^{n} \langle y, e_j \rangle e_j = y.$$

□

Similarly, we have the following theorem for the complex case.

Theorem 2.24 *Fix* $\{\lambda_{ij}\}_{i=1,j=i}^{n\ n}$ *be non-negative numbers satisfying:*

(1) $\lambda_{ij} = 0$ *if and only if* $j < i$.
(2) *For each* $j = 1, 2, \ldots, n$ *we have* $\sum_{i=1}^{n} \lambda_{ij} = 1$.

Let $\mathcal{E} = \{e_j\}_{j=1}^{n}$ *be the canonical basis of* $\mathbb{C}^n$. *Let* $\{z_k\}_{k=1}^{n^2}$ *be vectors in* $\mathbb{C}^n$ *which satisfy:*

(1) *For each* $k = 1, \ldots, 2n-1$, z_k *has the form*

$$z_k = (u_{k1}, u_{k2} + \iota v_{k2}, \ldots, u_{kn} + \iota v_{kn}),$$

where $u_{k1} \neq 0$ *and the set* $\{(u_{k1}, u_{k2}, v_{k2}, \ldots, u_{kn}, v_{kn})\}_{k=1}^{2n-1}$ *is linearly independent in* $\mathbb{R}^{2n-1}$. *Moreover* $\{z_k\}_{k=1}^{2n-1}$ *has frame operator* S_1 *with eigenvectors* $\mathcal{E}$ *and respective eigenvalues* $\{\lambda_{1j}\}_{j=1}^{n}$.

(2) *For each* $k = 2n, \ldots, 4n-4$, z_k *has the form*

$$z_k = (0, u_{k2}, u_{k3} + \iota v_{k3}, \ldots, u_{kn} + \iota v_{kn}),$$

where $u_{k2} \neq 0$ *and the set* $\{(u_{k2}, u_{k3}, v_{k3}, \ldots, u_{kn}, v_{kn})\}_{k=2n}^{4n-4}$ *is linearly independent in* $\mathbb{R}^{2n-3}$. *Moreover* $\{z_k\}_{k=2n}^{4n-4}$ *has frame operator* S_2 *with eigenvectors* $\mathcal{E}$ *and respective eigenvalues* $\{\lambda_{2j}\}_{j=1}^{n}$.

(3) *continue.*

Then the vectors $\{z_k\}_{k=1}^{n^2}$ *form a Parseval frame for* $\mathbb{C}^n$ *which is injective.*

Remark 2.25 We can easily vary the above construction to find frames which give injectivity and have any previously prescribed eigenvalues for their frame operators.

We recall:

Definition 2.26 Two orthonormal bases $\{x_k\}_{k=1}^{n}$ and $\{y_k\}_{k=1}^{n}$ are **mutually unbiased** if

$$|\langle x_k, y_j \rangle| = \frac{1}{\sqrt{n}}, \text{ for all } i, j = 1, 2, \ldots, n.$$

A family of orthonormal bases is **mutually unbiased** if each pair is mutually unbiased.

It is known that the maximal number of mutually unbiased bases in $\mathbb{H}^n$ is n+1 and this is rarely achieved. It holds if $n = p^m$ for a prime p. It is observed in [18] and [1] that a maximal family of mutually unbiased bases will give injectivity in the quantum detection problem.

2.3 The Solutions are Open and Dense

In this section we will show that the family of m-element frames which solve the quantum detection injectivity problem is open and dense in the family of all m-element frames. For this, we need to measure the distance between m-element frames. There is a standard metric measuring the distance between frames.

Definition 2.27 Given frames $\mathcal{X} = \{x_k\}_{k=1}^m$ and $\mathcal{Y} = \{y_k\}_{k=1}^m$ for a real or complex Hilbert space $\mathbb{H}^n$, the **distance** between them is

$$d(\mathcal{X}, \mathcal{Y})^2 = \sum_{k=1}^{m} \|x_k - y_k\|^2.$$

Theorem 2.28 *The set of all m-element frames on $\mathbb{H}^n$ that give injectivity in the frame quantum detection problem is dense in the space of all m-element frames on $\mathbb{H}^n$.*

Proof We will prove the real case. The complex case is similar.

Let a frame $\{x_k\}_{k=1}^{\frac{n(n+1)}{2}} \subset \mathbb{R}^n$ give injectivity. By Theorem 2.6, this is equivalent to the determinant of the matrix whose rows are $\tilde{x}_k, k = 1, 2, \ldots, \frac{n(n+1)}{2}$ being non-zero.

The determinant of this matrix is a polynomial of $\frac{n^2(n+1)}{2}$ variables x_{ki} for $1 \le k \le \frac{n(n+1)}{2}$ and $1 \le i \le n$. Since the complement of the zero set of this polynomial is dense in $\mathbb{R}^{\frac{n^2(n+1)}{2}}$, the set of all $\frac{n(n+1)}{2}$-element frames which give injectivity is dense in the space of all $\frac{n(n+1)}{2}$-element frames on $\mathbb{R}^n$.

Now let any m-element frame $\{x_k\}_{k=1}^m$ in $\mathbb{R}^n$ with $m \ge \frac{n(n+1)}{2}$ and $\delta > 0$. Then there exists a subframe containing $\frac{n(n+1)}{2}$ vectors. We can assume that this subframe is $\{x_k\}_{k=1}^{\frac{n(n+1)}{2}}$. By denseness above, there is an injective frame $\{y_k\}_{k=1}^{\frac{n(n+1)}{2}}$ such that

$$\sum_{k=1}^{n(n+1)/2} \|x_k - y_k\|^2 < \delta.$$

Now define a new frame $\{\phi_k\}_{k=1}^m$, where $\phi_k = y_k$ for $k = 1, \ldots, \frac{n(n+1)}{2}$ and $\phi_k = x_k$ for $k > \frac{n(n+1)}{2}$. Then the frame $\{\phi_k\}_{k=1}^m$ is injective and

$$\sum_{k=1}^{m} \|x_k - \phi_k\|^2 < \delta.$$

The conclusion of the theorem then follows. □

Remark 2.29 In the real case it is known that the complement of the zero set of a nontrivial polynomial of n variables is dense in $\mathbb{R}^n$. In the complex case, we see that given a polynomial $P(z_1, \ldots, z_n)$ on $\mathbb{C}^n$, we may write P as

$$P'(x_1, y_1, \ldots, x_n, y_n) + \iota P''(x_1, y_1, \ldots, x_n, y_n)$$

where $z_j = x_j + \iota y_j$. Hence P' and P'' are polynomials on $\mathbb{R}^{2n}$. P has a zero if and only if P' and P'' have a common zero. We see that the complement of the intersection of the zero sets of P' and P'' is dense in $\mathbb{R}^{2n}$ and hence is dense in $\mathbb{C}^n$ after natural identification of $\mathbb{R}^{2n}$ with $\mathbb{C}^n$.

Theorem 2.30 *The family of all m-element frames on $\mathbb{H}^n$ that give injectivity in the frame quantum detection problem is open in the space of all m-element frames on $\mathbb{H}^n$.*

Proof As above we will prove the real case and the complex case follows similarly.

Denote by $\mathcal{F}$ the space of all $\frac{n(n+1)}{2}$-element frames for $\mathbb{R}^n$. Consider the map:

$$f : \mathcal{F} \longrightarrow \mathbb{R}$$
$$\mathcal{X} = \{x_k\}_{k=1}^{\frac{n(n+1)}{2}} \longmapsto f(\mathcal{X}) = \det\left\{\tilde{x}_1, \tilde{x}_2, \ldots, \tilde{x}_{\frac{n(n+2)}{2}}\right\}.$$

Then f is a continuous function. Since $f^{-1}(0)$ is a closed set, by Theorem 2.6, the set of all $\frac{n(n+1)}{2}$-element frames is open in $\mathcal{F}$.

Now let $\mathcal{X} = \{x_k\}_{k=1}^m$ in $\mathbb{R}^n$, $(m \geq \frac{n(n+1)}{2})$ be an m-element frame which gives injectivity. Then there is a subframe $\mathcal{Y}$ containing $\frac{n(n+1)}{2}$ vectors, which is also injective. Therefore, there exists $\epsilon > 0$ such that every $\frac{n(n+1)}{2}$-element frame in the ball $B(\mathcal{Y}, \epsilon)$ is injective. This implies that every m-element frame in the ball $B(\mathcal{X}, \epsilon)$ is also injective. The proof is now complete. □

To show that the Parseval frames giving injectivity in the quantum detection problem are dense in the Parseval frames, we will first prove a very general problem about frames.

Theorem 2.31 *Let $\mathcal{P}$ be a property of Hilbert space frames and assume:*

(1) *The set of all m-element frames in $\mathbb{H}^n$ having property $\mathcal{P}$ is dense in the set of all m-element frames.*
(2) *If a frame $\{x_k\}_{k=1}^m$ with frame operator S has property $\mathcal{P}$, then $\{S^{-1/2}x_k\}_{k=1}^m$ has property $\mathcal{P}$.*

Then the set of all m-element Parseval frames with property $\mathcal{P}$ is dense in the set of all m-element Parseval frames.

Proof Fix $\epsilon > 0$ and let $\delta > 0$ so that

$$2m\delta^2 + 8(m\delta)^2 m(1+\delta)^2 < \epsilon, \; 2m\delta < 1.$$

Let $\{x_k\}_{k=1}^m$ be any Parseval frame for $\mathbb{H}^n$. By denseness, we can choose a frame $\{y_k\}_{k=1}^m$ having property $\mathcal{P}$ and satisfying $\|x_k - y_k\| \leq \delta$, for all $k = 1, 2, \ldots, m$. Since $\|x_k\| \leq 1$, we have that $\|y_k\| \leq 1 + \delta$. Let S_1 be the frame operator of $\{y_k\}_{k=1}^m$. Then,

$$
\begin{aligned}
\langle S_1 x, x\rangle^{1/2} &= \left(\sum_{k=1}^{m} |\langle x, y_k\rangle|^2\right)^{1/2} \\
&\le \left(\sum_{k=1}^{m} |\langle x, x_k\rangle|^2\right)^{1/2} + \left(\sum_{k=1}^{m} |\langle x, x_k - y_k\rangle|^2\right)^{1/2} \\
&\le \|x\| + \|x\| \left(\sum_{k=1}^{m} \|x_k - y_k\|^2\right)^{1/2} \\
&\le \|x\|(1 + m\delta).
\end{aligned}
$$

Therefore

$$\sum_{k=1}^{m} |\langle x, y_k\rangle|^2 \le \|x\|^2(1 + m\delta)^2.$$

Similarly,

$$\sum_{k=1}^{m} |\langle x, y_k\rangle|^2 \ge \|x\|^2(1 - m\delta)^2.$$

I.e. $(1 - m\delta)^2 I \le S_1 \le (1 + m\delta)^2 I$. Hence, $(1 - m\delta)I \le S_1^{1/2} \le (1 + m\delta)I$ and so $(1 + m\delta)^{-1} I \le S_1^{-1/2} \le (1 - m\delta)^{-1} I$. Finally,

$$I - (1 - m\delta)^{-1} I \le I - S_1^{-1/2} \le I - (1 + m\delta)^{-1} I,$$

and so

$$-2m\delta I \le \frac{-m\delta}{1 - m\delta} I \le I - S_1^{-1/2} \le \frac{m\delta}{1 + m\delta} I \le 2m\delta I.$$

Now, $\{S_1^{-1/2} y_k\}_{k=1}^m$ is a Parseval frame with property $\mathcal{P}$ and

$$
\begin{aligned}
\sum_{k=1}^{m} \|x_k - S_1^{-1/2} y_k\|^2 &\le 2\sum_{k=1}^{m} \|x_k - y_k\|^2 + 2\sum_{k=1}^{m} \|(I - S_1^{-1/2}) y_k\|^2 \\
&\le 2m\delta^2 + 2\sum_{k=1}^{m} (2m\delta)^2 \|y_k\|^2 \\
&\le 2m\delta^2 + 8(m\delta)^2 m(1 + \delta)^2 < \epsilon.
\end{aligned}
$$

□

Corollary 2.32 *The set of all m-element Parseval frames which give injectivity is dense in the set of all m-element Parseval frames.*

2.4 Solution to the State Estimation Problem

In this section we will give a classification of injective Parseval frames for which the state estimation problem is solvable.

Recall that for an injective Parseval frame $\{x_k\}_{k\in I}$ and $\beta = 2^I$, the map $\mathbb{M}$ which maps a quantum state $T \in \mathcal{S}(\mathbb{H})$ to a function $p \in L(\beta, \mathbb{R})$ is injective. Given a function $p \in L(\beta, \mathbb{R})$, if $p = \mathbb{M}(T)$ for some $T \in \mathcal{S}(\mathbb{H})$, then for any $U \in \beta$, we must have

$$p(U) = \mathbb{M}(T)(U) = \sum_{k\in U}\langle Tx_k, x_k\rangle = \sum_{k\in U} p(\{k\}).$$

Thus, p must be additive and is determined by its value at the singleton sets $\{k\}$ for all $k \in I$. Therefore, for the state estimation problem in the finite case, we will ask:

The State Estimation Problem Given an injective Parseval frame $\{x_k\}_{k=1}^m$ on $\mathbb{H}^n$ and a measurement vector $a = (a_1, a_2, \ldots, a_m) \in \mathbb{R}^m$, can we find a positive self-adjoint trace one operator T so that

$$\langle Tx_k, x_k\rangle = a_k, \text{ for all } k?$$

Remark 2.33 We will not require the operator T of the problem to be positive and trace one. This will be considered as a special case of the problem. Hence, we will say that the state estimation problem is solvable if there exists a self-adjoint operator T so that

$$\langle Tx_k, x_k\rangle = a_k, \text{ for all } k.$$

We will give a complete classification of injective Parseval frames for which the state estimation problem is solvable. Recall that for a vector $x \in \mathbb{R}^n$, the vector $\tilde{x}$ is defined as in the Definition 2.4.

Theorem 2.34 *Let $\mathcal{X} = \{x_k\}_{k=1}^m$ be an injective Parseval frame for $\mathbb{R}^n$, and $a = (a_1, a_2, \ldots, a_m) \in \mathbb{R}^m$, the following are equivalent:*

(1) *The state estimation problem is solvable.*
(2) $\operatorname{rank}(A) = \operatorname{rank}(B)$, *where A is a matrix whose k-row is $\tilde{x}_k$, and $B = [A, a]$.*

Proof Note that a self-adjoint operator T is determined by the values $\langle Te_i, e_j\rangle$ for all $i \le j$. Then the state estimation problem is solvable if and only if there exists a self-adjoint operator T so that

$$\begin{aligned} a_k &= \langle T x_k, x_k \rangle \\ &= \left\langle T \left(\sum_{i=1}^{n} \langle x_k, e_i \rangle e_i \right), \sum_{j=1}^{n} \langle x_k, e_j \rangle e_j \right\rangle \\ &= \sum_{i=1}^{n} \sum_{j=1}^{n} \langle x_k, e_i \rangle \langle x_k, e_j \rangle \langle T e_i, e_j \rangle \end{aligned}$$

for all k. This is equivalent to the linear system with unknowns $\langle T e_i, e_j \rangle$:

$$\sum_{i=1}^{n} x_{ki}^2 \langle T e_i, e_i \rangle + 2 \sum_{i<j} x_{ki} x_{kj} \langle T e_i, e_j \rangle = a_k, \ k = 1, 2, \ldots, m$$

having a solution, and hence is equivalent to $\operatorname{rank}(A) = \operatorname{rank}(B)$. □

In the case where the number of frame vectors equals $\frac{n(n+1)}{2}$, we have the following corollary.

Corollary 2.35 *Let* $\mathcal{X} = \{x_k\}_{k=1}^{\frac{n(n+1)}{2}} \subset \mathbb{R}^n$ *be an injective Parseval frame. Then the state estimation problem has a unique solution for all choices of vectors* $a = (a_1, a_2, \ldots, a_{\frac{n(n+1)}{2}})$.

Proof By Theorem 2.6, $\mathcal{X}$ is injective is equivalent to $\{\tilde{x}_k\}_{k=1}^{\frac{n(n+1)}{2}}$ is linearly independent. Hence

$$\operatorname{rank} A = \operatorname{rank} B = \frac{n(n+1)}{2}.$$

The conclusion then follows by Theorem 2.34. □

For the completeness of the state estimation problem, we will state the classification in the case that the operator T is required to be positive, self-adjoint operator of trace one. First, we need to recall *Sylvester's Criterion* [15].

Theorem 2.36 *A self-adjoint matrix* T *is positive if and only if all of its principal minors are nonnegative.*

Now we have the following classification:

Theorem 2.37 *Let* $\mathcal{X} = \{x_k\}_{k=1}^{m}$ *be an injective Parseval frame for* $\mathbb{R}^n$*, and* $a = (a_1, a_2, \ldots, a_m) \in \mathbb{R}^m$*, the following are equivalent:*

(1) *The state estimation problem is solvable for a positive, self-adjoint operator of trace one.*
(2) *The linear system*

$$\sum_{i=1}^{n} x_{ki}^2 \langle T e_i, e_i \rangle + 2 \sum_{i<j} x_{ki} x_{kj} \langle T e_i, e_j \rangle = a_k, \ k = 1, 2, \ldots, m$$

has a solution $\{\langle Te_i, e_j\rangle : i \le j\}$, *which determines a self-adjoint matrix* T *such that all of its principal minors are nonnegative, and* $\sum_{i=1}^{n}\langle Te_i, e_i\rangle = 1$.

Remark 2.38 All of the theorems above still hold for the complex case with the corresponding $\tilde{x}_k$, defined as in Definition 2.13. We state one of them here, the other are similar to the real case.

Theorem 2.39 *Let* $\mathcal{X} = \{x_k\}_{k=1}^m$ *be an injective Parseval frame for* $\mathbb{C}^n$, *and* $a = (a_1, a_2, \ldots, a_m) \in \mathbb{R}^m$, *the following are equivalent:*

(1) *The state estimation problem is solvable.*
(2) $\text{rank}(A) = \text{rank}(B)$, *where* A *is a matrix whose* k*-row is* $\tilde{x}_k$, *and* $B = [A, a]$.

Proof In the complex case, a self-adjoint operator T is determined by the values of the real part and imaginary part of $\langle Te_j, e_i\rangle$ for all $i \le j$. Then the state estimation problem is solvable if and only if there exists a self-adjoint operator T so that

$$\begin{aligned} a_k &= \langle Tx_k, x_k\rangle \\ &= \left\langle T\left(\sum_{i=1}^{n}\langle x_k, e_i\rangle e_i\right), \sum_{j=1}^{n}\langle x_k, e_j\rangle e_j\right\rangle \\ &= \sum_{i=1}^{n}\sum_{j=1}^{n}\langle x_k, e_i\rangle\overline{\langle x_k, e_j\rangle}\langle Te_i, e_j\rangle \\ &= \sum_{i=1}^{n}|x_{ki}|^2\langle Te_i, e_i\rangle + 2\sum_{i<j}[\text{Re}(\bar{x}_{ki}x_{kj})\text{Re}\langle Te_j, e_i\rangle - \text{Im}(\bar{x}_{ki}x_{kj})\text{Im}\langle Te_j, e_i\rangle] \end{aligned}$$

for all k.

This is equivalent to the following linear system:

$$\sum_{i=1}^{n}|x_{ki}|^2\langle Te_i, e_i\rangle + 2\sum_{i<j}[\text{Re}(\bar{x}_{ki}x_{kj})\text{Re}\langle Te_j, e_i\rangle - \text{Im}(\bar{x}_{ki}x_{kj})\text{Im}\langle Te_j, e_i\rangle] = a_k,$$

$k = 1, 2, \ldots, m$ with unknowns $\text{Re}\langle Te_j, e_i\rangle, \text{Im}\langle Te_j, e_i\rangle, i \le j$ having a solution, and hence is equivalent to $\text{rank}(A) = \text{rank}(B)$. □

Remark 2.40 If a frame $\{x_k\}_{k=1}^m$ has $m > \frac{n(n+1)}{2}$ in the real case, or $m > n^2$ in the complex case, because of redundancy, it is unlikely the state estimation is solvable. This is because if two of the $x_k's$ are equal while the corresponding $a_k's$ are not, then the problem is not solvable. More generally, if some of the x_k are linearly dependent then at least one of the corresponding a_k is uniquely determined. However, in this case there is a natural way to find the best estimate for the problem. We consider the real case. Note that there always exists a subset $I \subset \{1, 2, \ldots, m\}$ of size $\frac{n(n+1)}{2}$, and a self-adjoint operator T so that $\langle Tx_k, x_k\rangle = a_k$, for all $k \in I$. Therefore, if the state estimation problem is not solvable, it is natural to find such T so that the the distance to the measurement vector a:

$$\sum_{k=1}^{m} |\langle Tx_k, x_k\rangle - a_k|^2$$

is minimum.

To do this, let $\mathcal{S}$ be the set of all bases of $\mathbb{R}^{\frac{n(n+1)}{2}}$ that are subsets of $\{\tilde{x}_k\}_{k=1}^m$. This set is obviously finite. Since each element $\{\tilde{x}_k\}_{k\in I}$ in $\mathcal{S}$ determines a unique self-adjoint operator T satisfying $\langle Tx_k, x_k\rangle = a_k$, for all $k \in I$, we can find the quantum state T that gives the best approximation to the measurement vector a by choosing the set which minimizes the distance above.

3 The Solution for the Infinite Dimensional Case

In infinite dimensions we will work with both the trace class operators and the Hilbert Schmidt operators. I.e. Operators $T = (a_{ij})_{i,j=1}^{\infty}$ with $\sum_{i,j=1}^{\infty} |a_{ij}|^2 < \infty$. This class contains the trace class operators. As in the finite case, we will solve the following frame injectitivity problem:

Injectivity Problem For what frames $\{x_k\}_{k=1}^{\infty}$ in real or complex infinite dimensional Hilbert space $\mathbb{H}$ do we have the property: Whenever T, S are Hilbert Schmidt positive self-adjoint operators on $\mathbb{H}$ and $\langle Tx_k, x_k\rangle = \langle Sx_k, x_k\rangle$, for all $k = 1, 2, \ldots$, then $T = S$.

Remark 3.1 We will not require our operators to be trace class and trace one. These requirements will be considered as a special case of our problem.

3.1 The Solution to the Injectivity Problem

In this subsection we will solve the injectivity problem for infinite dimensional Hilbert spaces. Similar to the finite case, we first show that we only need to work with self adjoint operators. Note that the proof "(1) implies (2)" of Theorem 2.1 is not true for the infinite case. So we will give another proof here. The other implications are as in the finite case.

Theorem 3.2 *Given a family of vectors $\mathcal{X} = \{x_k\}_{k=1}^{\infty}$ in a real or complex Hilbert space $\mathbb{H}$, the following are equivalent:*

(1) *Whenever T, S are Hilbert Schmidt, positive and self-adjoint, and*

$$\langle Tx_k, x_k\rangle = \langle Sx_k, x_k\rangle, \text{ for all } k,$$

then $T = S$.

(2) *Whenever T, S are Hilbert Schmidt self-adjoint, and*

$$\langle Tx_k, x_k\rangle = \langle Sx_k, x_k\rangle, \text{ for all } k,$$

then $T = S$.

(3) *$\mathcal{X}$ is injective.*

Proof We will show that (1) implies (2). Let T, S be Hilbert Schmidt self-adjoint operators such that

$$\langle Tx_k, x_k\rangle = \langle Sx_k, x_k\rangle, \text{ for all } k.$$

Set $R = T - S$. Then R is also a Hilbert Schmidt self-adjoint operator. Let $\{e_j\}_{j=1}^{\infty}$ be an orthonormal basis for $\mathbb{H}$ and let $\{u_j\}_{j=1}^{\infty}$ be an eigenbasis for R with respective eigenvalues $\{\lambda_j\}_{j=1}^{\infty}$. Define operators U and D on $\mathbb{H}$ by $Ue_j = u_j$ and $De_j = \lambda_j e_j$, for $j = 1, 2, \ldots$. Then U is a unitary operator, D is Hilbert Schmidt self-adjoint operator, and

$$R = UDU^*.$$

Now let $r_j = |\lambda_j|, s_j = |\lambda_j| - \lambda_j, j = 1, 2, \ldots$ be non-negative numbers. Then $\lambda_j = r_j - s_j$. Let D_1, D_2 be operators defined by

$$D_1 e_j = r_j e_j, \ \ D_2 e_j = s_j e_j \text{ for } j = 1, 2, \ldots.$$

Note that since R is Hilbert Schmidt, $\sum_{j=1}^{\infty} \lambda_j^2$ converges. Hence D_1, D_2 are Hilbert-Schmidt positive self-adjoint and we have

$$R = UDU^* = U(D_1 - D_2)U^* = UD_1U^* - UD_2U^*.$$

Moreover, $UD_1U^*, \ UD_2U^*$ are Hilbert Schmidt positive self-adjoint operators. Since

$$0 = \langle Rx_k, x_k\rangle = \langle UD_1U^*x_k, x_k\rangle - \langle UD_2U^*x_k, x_k\rangle,$$

we have that $UD_1U^* = UD_2U^*$. Thus, $R = 0$ and hence $T = S$. □

If our operators are trace class, then we will have the following theorem. The proof of Theorem 3.2 is still valid here by noticing that $\sum_{j=1}^{\infty} |\lambda_j| < \infty$ for the trace class operator R.

Theorem 3.3 *Given a family of vectors $\mathcal{X} = \{x_k\}_{k=1}^{\infty}$ in a infinite dimensional Hilbert space $\mathbb{H}$, the following are equivalent:*

(1) *Whenever T, S are trace class positive and self-adjoint, and*

$$\langle Tx_k, x_k\rangle = \langle Sx_k, x_k\rangle, \text{ for all } k,$$

then $T = S$.

(2) *Whenever T, S are trace class self-adjoint, and*

$$\langle Tx_k, x_k\rangle = \langle Sx_k, x_k\rangle, \text{ for all } k,$$

then $T = S$.

(3) *Whenever T is trace class self-adjoint, and*

$$\langle Tx_k, x_k\rangle = 0, \textit{ for all } k,$$

then $T = 0$.

Similar to the finite case, we will first give the following classification of injectivity for Hilbert Schmidt operators.

Theorem 3.4 *Let $\mathcal{X} = \{x_k\}_{k=1}^{\infty}$ be a frame for an infinite dimensional real or complex Hilbert space $\mathbb{H}$. The following are equivalent:*

(1) *$\mathcal{X}$ is injective.*
(2) *For every orthonormal basis $\mathcal{E} = \{e_j\}_{j=1}^{\infty}$ for $\mathbb{H}$ we have:*

$$H(\mathcal{E}) =: \overline{\text{span}}\{(|\langle x_k, e_1\rangle|^2, |\langle x_k, e_2\rangle|^2, \ldots) : k = 1, 2, \ldots\} = \ell_2.$$

Proof (1) $\Rightarrow$ (2): We prove the result by way of contradiction. Suppose that (2) is false. Then there is an orthonormal basis $\mathcal{E} = \{e_j\}_{j=1}^{\infty}$ so that $H(\mathcal{E}) \neq \ell_2$. Hence there is a non-zero vector $\lambda = (\lambda_1, \lambda_2, \ldots) \in \ell_2$ such that $\lambda \perp H(\mathcal{E})$.

Define an operator on $\mathbb{H}$ by

$$Te_j = \lambda_j e_j, \text{ for all } j = 1, 2, \ldots.$$

Then T is a non-zero Hilbert Schmidt operator. We also have: $\langle Tx_k, x_k\rangle = 0$, for all $k = 1, 2, \ldots$. This is a contradiction.

(2) $\Rightarrow$ (1): Let T be a Hilbert Schmidt self-adjoint operator such that

$$\langle Tx_k, x_k\rangle = 0, \text{ for all } k.$$

Since T is Hilbert Schmidt and hence compact, there is an eigenbasis $\mathcal{E} = \{e_j\}_{j=1}^{\infty}$ for T with respective eigenvalues $\{\lambda_j\}_{j=1}^{\infty}$. Then for every $k = 1, 2, \ldots$, we have

$$\langle Tx_k, x_k\rangle = \sum_{j=1}^{\infty} \lambda_j |\langle x_k, e_j\rangle|^2 = 0.$$

Since T is Hilbert Schmidt then

$$\sum_{j=1}^{\infty} |\lambda_j|^2 = \sum_{j=1}^{\infty} \|Te_j\|^2 < \infty.$$

That is, $(\lambda_1, \lambda_2, \ldots) \in \ell_2$. Since

$$(\lambda_1, \lambda_2, \ldots) \perp H(\mathcal{E}) = \ell_2 \text{ by assumption (2).}$$

Therefore, $\lambda_j = 0$ for all $j = 1, 2, \ldots$, and so $T = 0$. □

If we consider operators which are trace class, then we have the following classification for the infinite dimensions.

Theorem 3.5 *Let $\mathcal{X} = \{x_k\}_{k=1}^{\infty}$ be a frame for an infinite dimensional real or complex Hilbert space $\mathbb{H}$. The following are equivalent:*

(1) *$\mathcal{X}$ satisfies the following property: The only trace class self-adjoint operator T such that*

$$\langle Tx_k, x_k\rangle = 0, \ \textit{for all } k,$$

is $T = 0$.

(2) *For every $\lambda = (\lambda_1, \lambda_2, \ldots) \in \ell_1$ and for every orthonormal basis $\{e_j\}_{j=1}^{\infty}$ for $\mathbb{H}$, if $\sum_{j=1}^{\infty} \lambda_j |\langle x_k, e_j\rangle|^2 = 0$ for all k then $\lambda = 0$.*

Proof (1) $\Rightarrow$ (2): We prove the result by way of contradiction. Suppose that (2) is false. Then there is an $\lambda = (\lambda_1, \lambda_2, \ldots) \in \ell_1$ and an orthonormal basis $\{e_j\}_{j=1}^{\infty}$ so that $\sum_{j=1}^{\infty} \lambda_j |\langle x_k, e_j\rangle|^2 = 0$ for all k but $\lambda \neq 0$.

Define an operator on $\mathbb{H}$ by

$$Te_j = \lambda_j e_j, \ \text{for all } j = 1, 2, \ldots.$$

Then T is a non-zero self-adjoint operator. Moreover,

$$|T|e_j = \sqrt{TT^*}e_j = |\lambda_j|e_j, \ \text{for all } j.$$

Therefore,

$$\sum_{j=1}^{\infty} \langle |T|e_j, e_j\rangle = \sum_{j=1}^{\infty} |\lambda_j| < \infty.$$

Thus, T is a non-zero trace class self-adjoint operator. Moreover, we have that $\langle Tx_k, x_k\rangle = \sum_{j=1}^{\infty} \lambda_j |\langle x_k, e_j\rangle|^2 = 0$, for all $k = 1, 2, \ldots$. This is a contradiction.

(2) $\Rightarrow$ (1): Let T be a trace class self-adjoint operator such that

$$\langle Tx_k, x_k\rangle = 0, \ \text{for all } k.$$

Since T is trace class and hence compact, there is an eigenbasis $\{e_j\}_{j=1}^{\infty}$ for T with respective eigenvalues $\{\lambda_j\}_{j=1}^{\infty}$. Then for every $k = 1, 2, \ldots$, we have

$$\sum_{j=1}^{\infty} \lambda_j |\langle x_k, e_j\rangle|^2 = \langle Tx_k, x_k\rangle = 0, \ \text{for all } k.$$

Since T is trace class then

$$\sum_{j=1}^{\infty} |\lambda_j| = \sum_{j=1}^{\infty} |\langle Te_j, e_j\rangle| < \infty.$$

That is, $\lambda = (\lambda_1, \lambda_2, \ldots) \in \ell_1$. By assumption (2) we get $\lambda = 0$ and hence $T = 0$. □

Finally, by normalizing the trace, we can give a classification for the Injectivity problem if we require further that our operators are trace one. First, we need to justify Theorem 3.2 so that we can use it for this case.

Theorem 3.6 *Given a family of vectors* $\mathcal{X} = \{x_k\}_{k=1}^{\infty}$ *in the real or complex Hilbert space* $\mathbb{H}$, *the following are equivalent:*

(1) *Whenever* T, S *are trace class positive and self-adjoint of trace one, and*

$$\langle Tx_k, x_k\rangle = \langle Sx_k, x_k\rangle, \textit{ for all } k,$$

then $T = S$.

(2) *Whenever* T, S *are trace class self-adjoint of trace one, and*

$$\langle Tx_k, x_k\rangle = \langle Sx_k, x_k\rangle, \textit{ for all } k,$$

then $T = S$.

(3) *Whenever* T *is trace class self-adjoint of trace zero, and*

$$\langle Tx_k, x_k\rangle = 0, \textit{ for all } k,$$

then $T = 0$.

Proof (1) $\Rightarrow$ (2): Let T, S be trace class self-adjoint operators of trace one such that

$$\langle Tx_k, x_k\rangle = \langle Sx_k, x_k\rangle, \text{ for all } k.$$

Set $R = T - S$ then R is a trace class self-adjoint operator of trace zero. Let $\{e_j\}_{j=1}^{\infty}$ be an orthonormal basis for $\mathbb{H}$ and let $\{u_j\}_{j=1}^{\infty}$ be an eigenbasis for R with respective eigenvalues $\{\lambda_j\}_{j=1}^{\infty}$. Then $\sum_{j=1}^{\infty} \lambda_j = 0$. Define operators U and D on $\mathbb{H}$ by $Ue_j = u_j$ and $De_j = \lambda_j e_j$, for $j = 1, 2, \ldots$. Then U is an unitary operator and D is a trace class self-adjoint operator of trace zero, and

$$R = UDU^*.$$

Now define non-negative numbers

$$r_1 = \frac{1 + |\lambda_1|}{A}, s_1 = \frac{1 + |\lambda_1| - \lambda_1}{A}, r_j = \frac{|\lambda_j|}{A}, s_j = \frac{|\lambda_j| - \lambda_j}{A}, j = 2, 3, \ldots,$$

where

$$A = 1 + \sum_{j=1}^{\infty} |\lambda_j| = 1 + \sum_{j=1}^{\infty} |\lambda_j| - \sum_{j=1}^{\infty} \lambda_j > 0$$

then $\lambda_j = r_j - s_j$ for all j. Let D_1, D_2 be operators defined by

$$D_1 e_j = r_j e_j,\ \ D_2 e_j = s_j e_j \text{ for } j = 1, 2, \ldots.$$

Then D_1, D_2 are trace class positive self-adjoint of trace one and we have

$$R = UDU^* = U(D_1 - D_2)U^* = UD_1U^* - UD_2U^*.$$

Moreover, UD_1U^*, UD_2U^* are trace class positive self-adjoint operators of trace one. Since

$$0 = \langle Rx_k, x_k \rangle = \langle UD_1U^* x_k, x_k \rangle - \langle UD_2U^* x_k, x_k \rangle,$$

then $UD_1U^* = UD_2U^*$. Thus, $R = 0$ and hence $T = S$.

(2) $\Rightarrow$ (3): Let T be any trace class operator of trace zero such that

$$\langle Tx_k, k_k \rangle = 0 \text{ for all } k.$$

Define an operator S on $\mathbb{H}$ by

$$Se_1 = e_1, Se_j = 0,\ \text{ for } j = 2, 3, \ldots.$$

Then S and $T + S$ are trace class self-adjoint operators of trace one.

Since $\langle (T + S)x_k, x_k \rangle = \langle Sx_k, x_k \rangle$ for all k, $T + S = S$ and hence $T = 0$.

(3) $\Rightarrow$ (1): Let T, S are trace class positive self-adjoint operators of trace one such that

$$\langle Tx_k, x_k \rangle = \langle Sx_k, x_k \rangle \text{ for all } k.$$

Then $\langle (T - S)x_k, x_k \rangle = 0$ for all k. Since $T - S$ is a trace class seft-adjoint operator of trace zero, $T = S$ by (3). □

Now we are ready to give a classification for the Injectivity problem for operators of trace one. First, we need a definition.

Definition 3.7 We define a subspace of the real space ℓ_1 as follows:

$$W := \left\{ (\lambda_1, \lambda_2, \ldots) \in \ell_1 : \sum_{j=1}^{\infty} \lambda_j = 0 \right\}.$$

Theorem 3.8 *Let $\mathcal{X} = \{x_k\}_{k=1}^{\infty}$ be a frame for an infinite dimensional real or complex Hilbert space $\mathbb{H}$. The following are equivalent:*

(1) *If T is a trace class self-adjoint operator of trace zero such that*

$$\langle Tx_k, x_k\rangle = 0, \text{ for all } k,$$

then $T = 0$.

(2) *For every $\lambda = (\lambda_1, \lambda_2, \ldots) \in W$ and for every orthonormal basis $\{e_j\}_{j=1}^{\infty}$ for $\mathbb{H}$, if $\sum_{j=1}^{\infty} \lambda_j |\langle x_k, e_j\rangle|^2 = 0$ for all k then $\lambda = 0$.*

Proof (1) $\Rightarrow$ (2): We prove the contrapositive. Suppose that (2) is false. Then there is an $\lambda = (\lambda_1, \lambda_2, \ldots) \in W$ and an orthonormal basis $\{e_j\}_{j=1}^{\infty}$ so that $\sum_{j=1}^{\infty} \lambda_j |\langle x_k, e_j\rangle|^2 = 0$ for all k but $\lambda \neq 0$.

Define an operator on $\mathbb{H}$ by

$$Te_j = \lambda_j e_j, \text{ for all } j = 1, 2, \ldots.$$

Then T is a non-zero trace class self-adjoint operator of trace zero. Moreover, we have that $\langle Tx_k, x_k\rangle = \sum_{j=1}^{\infty} \lambda_j |\langle x_k, e_j\rangle|^2 = 0$, for all $k = 1, 2, \ldots$. This is a contradiction.

(2) $\Rightarrow$ (1): Let T be a trace class self-adjoint operator of trace zero such that

$$\langle Tx_k, x_k\rangle = 0, \text{ for all } k.$$

Let $\{e_j\}_{j=1}^{\infty}$ be an eigenbasis for T with respective eigenvalues $\{\lambda_j\}_{j=1}^{\infty}$. Then for every $k = 1, 2, \ldots$, we have

$$\sum_{j=1}^{\infty} \lambda_j |\langle x_k, e_j\rangle|^2 = \langle Tx_k, x_k\rangle = 0, \text{ for all } k.$$

Since T is trace class,

$$\sum_{j=1}^{\infty} |\lambda_j| = \sum_{j=1}^{\infty} |\langle Te_j, e_j\rangle| < \infty.$$

Moreover, $\sum_{j=1}^{\infty} \lambda_j = 0$. Thus, $\lambda = (\lambda_1, \lambda_2, \ldots) \in W$. By assumption (2) we get $\lambda = 0$ and hence $T = 0$. □

From now on, we will also work in the direct sum of infinitely many copies of ℓ_2.

Definition 3.9 Denote by $\tilde{\mathbb{H}}$ the direct sum of the real Hilbert spaces ℓ_2:

$$\tilde{\mathbb{H}} = \left(\sum_{i=1}^{\infty} \oplus \ell_2\right)_{\ell_2}.$$

To avoid confusion with earlier notation, a vector in this direct sum will be written in the form:

$$\vec{x} = (\vec{x}_1, \vec{x}_2, \dots, \vec{x}_n, \dots),$$

and we have

$$\langle \vec{x}, \vec{y} \rangle = \sum_{i=1}^{\infty} \langle \vec{x}_i, \vec{y}_i \rangle.$$

We also need the following lemma for both the real and complex cases.

Lemma 3.10 *Let $A = (a_{ij})_{i,j=1}^{\infty}$ be a real or complex infinite matrix such that $\sum_{i,j=1}^{\infty} |a_{ij}|^2 < \infty$. Then the operator T_A defined in ℓ_2 by*

$$T_A(x_1, x_2, \dots) = (y_1, y_2, \dots),$$

where

$$y_i = \sum_{j=1}^{\infty} a_{ij} x_j, \, i = 1, 2, \dots,$$

is a bounded operator. Moreover, T_A is self-adjoint if and only if $a_{ji} = \bar{a}_{ij}$ for all i, j.

Proof Let $x = \{x_i\}_{i=1}^{\infty} \in \ell_2$. For each $i = 1, 2 \dots$, we have

$$|y_i|^2 \le \left(\sum_{j=1}^{\infty} |a_{ij} x_j| \right)^2 \le \left(\sum_{j=1}^{\infty} |a_{ij}|^2 \right) \left(\sum_{j=1}^{\infty} |x_j|^2 \right) = \left(\sum_{j=1}^{\infty} |a_{ij}|^2 \right) \|x\|^2.$$

Hence,

$$\|T_A x\|^2 = \sum_{i=1}^{\infty} |y_i|^2 \le \left(\sum_{i=1}^{\infty} \sum_{j=1}^{\infty} |a_{ij}|^2 \right) \|x\|^2.$$

This shows that T_A is a bounded operator on ℓ_2.

Suppose that T is self-adjoint. Then

$$a_{ji} = \langle T_A e_i, e_j \rangle = \langle e_i, T_A e_j \rangle = \overline{\langle T_A e_j, e_i \rangle} = \bar{a}_{ij},$$

for all i, j.

Conversely, if $a_{ji} = \bar{a}_{ij}$ for all i, j, then

$$\langle T_A^* e_i, e_j \rangle = \langle e_i, T_A e_j \rangle = \overline{\langle T_A e_j, e_i \rangle} = \bar{a}_{ij} = a_{ji} = \langle T_A e_i, e_j \rangle,$$

for all i, j. Hence $T_A^* = T_A$. □

3.1.1 The Real Case

Now we will solve the infinite dimensional injectivity problem in the real case. To avoid confusion between coordinates of a vector in ℓ_2 and vectors in $\tilde{\mathbb{H}}$ we define:

Definition 3.11 For $x = \{x_i\}_{i=1}^{\infty} \in \ell_2$, we define

$$\tilde{x} = (\vec{x}_1, \vec{x}_2, \ldots, \vec{x}_n, \ldots) \in \tilde{\mathbb{H}},$$

where

$$\vec{x}_1 = (x_1x_1, x_1x_2, \ldots);\ \vec{x}_2 = (x_2x_2, x_2x_3, \ldots); \ldots; \vec{x}_n = (x_nx_n, x_nx_{n+1}, \ldots); \ldots$$

We first observe that these vectors are actually in $\tilde{\mathbb{H}}$.

Lemma 3.12 *If $x = \{x_i\}_{i=1}^{\infty} \in \ell_2$, then $\tilde{x} \in \tilde{\mathbb{H}}$.*

Proof We have that

$$\sum_{j=i}^{\infty} |x_i x_j|^2 = |x_i|^2 \sum_{j=i}^{\infty} |x_j|^2 \le |x_i|^2 \|x\|^2,$$

for $i = 1, 2, \ldots$. Hence $\vec{x}_i \in \ell_2$ for all i.

Moreover, since

$$\sum_{i=1}^{\infty} \|\vec{x}_i\|^2 \le \sum_{i=1}^{\infty} |x_i|^2 \|x\|^2 = \|x\|^4,$$

then $\tilde{x} \in \tilde{\mathbb{H}}$. □

Now we are ready for the classification of the solutions to the injectivity problem in the infinite dimensional case.

Theorem 3.13 *Let $\mathcal{X} = \{x_k\}_{k=1}^{\infty}$ be a frame in the real Hilbert space ℓ_2. The following are equivalent:*

(1) *$\mathcal{X}$ is injective.*
(2) $\overline{\text{span}}\{\tilde{x}_k\}_{k=1}^{\infty} = \tilde{\mathbb{H}}$.

Proof (1) $\Rightarrow$ (2): Let any $a = (\vec{a}_1, \vec{a}_2, \ldots) \in \tilde{\mathbb{H}}$ be such that $a \perp \overline{\text{span}}\{\tilde{x}_k\}_{k=1}^{\infty}$. Then $\langle a, \tilde{x}_k \rangle = 0$ for all k.

We denote

$$\vec{a}_1 = (a_{11}, a_{12}, \ldots);\ \vec{a}_2 = (a_{22}, a_{23}, \ldots); \ldots, \vec{a}_n = (a_{nn}, a_{n(n+1)}, \ldots); \ldots.$$

Define an infinite matrix $B = (b_{ij})_{i,j=1}^{\infty}$, where $b_{ii} = a_{ii}$ for all i and $b_{ij} = b_{ji} = \frac{1}{2} a_{ij}$ for all $i < j$.

Then by Lemma 3.10, the operator T_B defined by B is a Hilbert Schmidt self-adjoint operator.

For any $x = \{x_i\}_{i=1}^{\infty} \in \ell_2$, we have

$$\begin{aligned}
\langle T_B x, x\rangle &= \sum_{i=1}^{\infty}\sum_{j=1}^{\infty} b_{ij}x_i x_j \\
&= \sum_{i=1}^{\infty} b_{ii}x_i^2 + 2\sum_{i<j} b_{ij}x_i x_j \\
&= \sum_{i=1}^{\infty} a_{ii}x_i^2 + \sum_{i<j} a_{ij}x_i x_j \\
&= \sum_{i=1}^{\infty} \langle \vec{a}_i, \vec{x}_i\rangle \\
&= \langle a, \tilde{x}\rangle.
\end{aligned}$$

Hence, $\langle T_B x_k, x_k\rangle = \langle a, \tilde{x}_k\rangle = 0$ for all k. This implies $T_B = 0$ by (1) and therefore $a = 0$.

(2) $\Rightarrow$ (1): Let T be a Hilbert Schmidt self-adjoint operator on ℓ_2 such that $\langle Tx_k, x_k\rangle = 0$ for all k, and recall that $\{e_i\}_{i=1}^{\infty}$ is the canonical orthonormal basis for ℓ_2.

Denote

$$a_{ij} = \langle Te_j, e_i\rangle, i, j = 1, 2\ldots,$$

and

$$\tilde{T} = (\vec{a}_1, \vec{a}_2, \ldots, \vec{a}_n, \ldots),$$

where

$$\begin{aligned}
&\vec{a}_1 = (a_{11}, 2a_{12}, 2a_{13}, \ldots); \quad \vec{a}_2 = (a_{22}, 2a_{23}, 2a_{24}, \ldots); \ldots; \\
&\vec{a}_n = (a_{nn}, 2a_{n(n+1)}, 2a_{n(n+2)}, \ldots); \ldots.
\end{aligned}$$

Since T is a Hilbert Schmidt operator, $\tilde{T} \in \tilde{\mathbb{H}}$. Moreover, we have

$$\langle \tilde{T}, \tilde{x}_k\rangle = \langle Tx_k, x_k\rangle = 0, \text{ for all } k.$$

Since $\overline{\text{span}}\{\tilde{x}_k\}_{k=1}^{\infty} = \tilde{\mathbb{H}}$, we get $\tilde{T} = 0$. So $T = 0$. □

Remark 3.14 We have that (2) $\Rightarrow$ (1) in the theorem holds for trace class operators. But in general (1) $\Rightarrow$ (2) fails since the operators we construct may not be trace class.

3.1.2 The Complex Case

For the complex case of the injectivity problem, we need a new variation of the tilde vectors.

Definition 3.15 For $x = \{x_i\}_{i=1}^{\infty} \in \ell_2$, we define

$$\tilde{x} = (\vec{x}_1, \vec{x}_2, \ldots, \vec{x}_n, \ldots),$$

where

$$\begin{aligned}
\vec{x}_1 &= (|x_1|^2, \mathrm{Re}(\bar{x}_1 x_2), \mathrm{Im}(\bar{x}_1 x_2), \mathrm{Re}(\bar{x}_1 x_3), \mathrm{Im}(\bar{x}_1 x_3), \ldots);\\
\vec{x}_2 &= (|x_2|^2, \mathrm{Re}(\bar{x}_2 x_3), \mathrm{Im}(\bar{x}_2 x_3), \mathrm{Re}(\bar{x}_2 x_4), \mathrm{Im}(\bar{x}_2 x_4), \ldots); \ldots;\\
\vec{x}_n &= (|x_n|^2, \mathrm{Re}(\bar{x}_n x_{n+1}), \mathrm{Im}(\bar{x}_n x_{n+1}), \mathrm{Re}(\bar{x}_n x_{n+2}), \mathrm{Im}(\bar{x}_n x_{n+2}), \ldots); \ldots.
\end{aligned}$$

We first need to verify that our vectors are in $\tilde{\mathbb{H}}$.

Lemma 3.16 *If $x = \{x_i\}_{i=1}^{\infty} \in \ell_2$, then $\tilde{x} \in \tilde{\mathbb{H}}$.*

Proof For each $i = 1, 2, \ldots,$ we have

$$\begin{aligned}
\|\vec{x}_i\|^2 &= |x_i|^4 + \sum_{j=i+1}^{\infty} |\mathrm{Re}(\bar{x}_i x_j)|^2 + \sum_{j=i+1}^{\infty} |\mathrm{Im}(\bar{x}_i x_j)|^2\\
&= |x_i|^4 + \sum_{j=i+1}^{\infty} |\bar{x}_i x_j|^2\\
&= |x_i|^2 \left(|x_i|^2 + \sum_{j=i+1}^{\infty} |x_j|^2 \right)\\
&\le |x_i|^2 \|x\|^2.
\end{aligned}$$

It follows that:

$$\sum_{i=1}^{\infty} \|\vec{x}_i\|^2 \le \sum_{i=1}^{\infty} |x_i|^2 \|x\|^2 = \|x\|^4.$$

This implies $\tilde{x} \in \tilde{\mathbb{H}}$. □

Now we give the classification theorem for injectivity in the infinite dimensional case.

Theorem 3.17 *Let $\mathcal{X} = \{x_k\}_{k=1}^{\infty}$ be a frame in the complex Hilbert space ℓ_2. The following are equivalent:*

(1) *$\mathcal{X}$ gives injectivity.*
(2) $\overline{\mathrm{span}}\{\tilde{x}_k\}_{k=1}^{\infty} = \tilde{\mathbb{H}}$.

Proof (1) $\Rightarrow$ (2): Let any $a = (\vec{a}_1, \vec{a}_2, \ldots) \in \tilde{\mathbb{H}}$ be such that $a \perp \overline{\text{span}}\{\tilde{x}_k\}_{k=1}^{\infty}$. Then $\langle a, \tilde{x}_k\rangle = 0$ for all k.

Denote

$$\vec{a}_1 = (a_{11}, u_{12}, v_{12}, u_{13}, v_{13}, \ldots);\ \vec{a}_2 = (a_{22}, u_{23}, v_{23}, u_{24}, v_{24}, \ldots); \ldots;$$
$$\vec{a}_n = (a_{nn}, u_{n(n+1)}, v_{n(n+1)}, u_{n(n+2)}, v_{n(n+2)}, \ldots); \ldots.$$

Define an infinite matrix $B = (b_{ij})_{i,j=1}^{\infty}$, where $b_{ii} = a_{ii}$ for all i and $b_{ij} = \bar{b}_{ji} = \frac{1}{2}(u_{ij} - \iota v_{ij})$ for all $i < j$.

We have

$$\sum_{i,j=1}^{\infty} |b_{ij}|^2 = \sum_{i=1}^{\infty} |a_{ii}|^2 + 2\sum_{i<j} |b_{ij}|^2 = \sum_{i=1}^{\infty} |a_{ii}|^2 + \frac{1}{2}\sum_{i<j}\left(|u_{ij}|^2 + |v_{ij}|^2\right) < \infty.$$

Then by Lemma 3.10, the operator T_B defined by B is Hilbert Schmidt and self-adjoint.

For any $x = \{x_i\}_{i=1}^{\infty} \in \ell_2$, we have

$$\begin{aligned}
\langle T_B x, x\rangle &= \sum_{i=1}^{\infty}\sum_{j=1}^{\infty} b_{ij}\bar{x}_i x_j \\
&= \sum_{i=1}^{\infty} b_{ii}|x_i|^2 + 2\sum_{i<j} \text{Re}(b_{ij}\bar{x}_i x_j) \\
&= \sum_{i=1}^{\infty} b_{ii}|x_i|^2 + 2\sum_{i<j} (\text{Re}(b_{ij})\text{Re}(\bar{x}_i x_j) - \text{Im}(b_{ij})\text{Im}(\bar{x}_i x_j)) \\
&= \sum_{i=1}^{\infty} a_{ii}|x_i|^2 + \sum_{i<j} \left(u_{ij}\text{Re}(\bar{x}_i x_j) + v_{ij}\text{Im}(\bar{x}_i x_j)\right) \\
&= \sum_{i=1}^{\infty} \langle \vec{a}_i, \vec{x}_i\rangle \\
&= \langle a, \tilde{x}\rangle.
\end{aligned}$$

Hence, $\langle T_B x_k, x_k\rangle = \langle a, \tilde{x}_k\rangle = 0$ for all k. This implies $T_B = 0$ by (1) and therefore $a = 0$.

(2) $\Rightarrow$ (1): Let T be a Hilbert Schmidt self-adjoint operator such that $\langle T x_k, x_k\rangle = 0$ for all k.

Denote

$$a_{ij} = \langle T e_j, e_i\rangle, i, j = 1, 2, \ldots,$$

and

$$\tilde{T} = (\vec{a}_1, \vec{a}_2, \ldots, \vec{a}_n, \ldots),$$

where

$$\vec{a}_1 = (a_{11}, 2\mathrm{Re}(a_{12}), -2\mathrm{Im}(a_{12}), 2\mathrm{Re}(a_{13}), -2\mathrm{Im}(a_{13}), \ldots);$$
$$\vec{a}_2 = (a_{22}, 2\mathrm{Re}(a_{23}), -2\mathrm{Im}(a_{23}), 2\mathrm{Re}(a_{24}), -2\mathrm{Im}(a_{24}), \ldots); \ldots;$$
$$\vec{a}_n = (a_{nn}, 2\mathrm{Re}(a_{n(n+1)}), -2\mathrm{Im}(a_{n(n+1)}), 2\mathrm{Re}(a_{n(n+2)}), -2\mathrm{Im}(a_{n(n+2)}), \ldots); \ldots$$

Since T is Hilbert Schmidt, $\tilde{T} \in \tilde{\mathbb{H}}$.

For any $x = \sum_{j=1}^{\infty} x_j e_j$ we have

$$\begin{aligned}
\langle Tx, x\rangle &= \sum_{i=1}^{\infty}\sum_{j=1}^{\infty} \bar{x}_i x_j \langle Te_j, e_i\rangle \\
&= \sum_{i=1}^{\infty}\sum_{j=1}^{\infty} \bar{x}_i x_j a_{ij} \\
&= \sum_{i=1}^{\infty} a_{ii}|x_i|^2 + 2\sum_{i<j} \mathrm{Re}(a_{ij}\bar{x}_i x_j) \\
&= \sum_{i=1}^{\infty} a_{ii}|x_i|^2 + 2\sum_{i<j} (\mathrm{Re}(a_{ij})\mathrm{Re}(\bar{x}_i x_j) - \mathrm{Im}(a_{ij})\mathrm{Im}(\bar{x}_i x_j)) \\
&= \langle \tilde{T}, \tilde{x}\rangle.
\end{aligned}$$

Hence

$$\langle \tilde{T}, \tilde{x}_k\rangle = \langle Tx_k, x_k\rangle = 0, \text{ for all } k.$$

Since $\overline{\mathrm{span}}\{\tilde{x}_k\}_{k=1}^{\infty} = \tilde{\mathbb{H}}$, $\tilde{T} = 0$. So $T = 0$. This completes the proof. □

As a consequence we have:

Corollary 3.18 *For a frame $\{x_k\}_{k=1}^{\infty}$ in ℓ_2 the following are equivalent:*

(1) *The family $\{x_k x_k^*\}_{k=1}^{\infty}$ spans the family of real self-adjoint Hilbert Schmidt operators on ℓ_2.*
(2) *The family $\{\tilde{x}_k\}_{k=1}^{\infty}$ spans $\tilde{\mathbb{H}}$.*

Remark 3.19 As we have seen in the proof of Theorem 3.13 for the real case and Theorem 3.17 for the complex case, for a vector $a \in \tilde{\mathbb{H}}$, there is a Hilbert Schmidt self-adjoint operator T so that

$$\langle Tx, x\rangle = \langle a, \tilde{x}\rangle, \text{ for all } x \in \ell_2.$$

Conversely, for a Hilbert Schmidt self-adjoint operator T, there is a vector $\tilde{T} \in \tilde{\mathbb{H}}$ satisfying

$$\langle \tilde{T}, \tilde{x} \rangle = \langle Tx, x \rangle, \quad \text{for all } x \in \ell_2.$$

Is is easy to see that the canonical orthonormal basis is not injective. Actually, the family $\{\tilde{e}_i\}_{i=1}^{\infty}$ forms an orthonormal set in $\tilde{\mathbb{H}}$. We will see in general that any frame in ℓ_2 so that the corresponding tilde vectors form a frame sequence in $\tilde{\mathbb{H}}$ cannot be injective.

Theorem 3.20 *For any Bessel sequence $\{x_k\}_{k=1}^{\infty}$ for the real or complex space ℓ_2, the family $\{\tilde{x}_k\}_{k=1}^{\infty}$ is a Bessel sequence in $\tilde{\mathbb{H}}$. However, $\{\tilde{x}_k\}_{k=1}^{\infty}$ is not a frame for $\tilde{\mathbb{H}}$.*

Proof We may assume that $\|x_k\| \le 1$ for all k. Let B be the Bessel bound of $\{x_k\}_{k=1}^{\infty}$.

Given any finite real scalar sequence $\{a_k\}_{k=1}^{\infty}$ we will compute the real case and the complex case separately.

The real case Using Definition 3.11 for the tilde vector $\tilde{x}_k$, we have

$$\begin{aligned}
\left\| \sum_{k=1}^{\infty} a_k \tilde{x}_k \right\|^2 &= \sum_{i=1}^{\infty} \sum_{j=i}^{\infty} \left(\sum_{k=1}^{\infty} a_k x_{ki} x_{kj} \right)^2 \\
&\le \sum_{i=1}^{\infty} \sum_{j=1}^{\infty} \left(\sum_{k=1}^{\infty} a_k x_{ki} x_{kj} \right)^2 \\
&= \sum_{i=1}^{\infty} \left\| \sum_{k=1}^{\infty} a_k x_{ki} x_k \right\|^2 .
\end{aligned}$$

Using the fact that $\{x_k\}_{k=1}^{\infty}$ is Bessel with bound B, we get

$$\begin{aligned}
\left\| \sum_{k=1}^{\infty} a_k \tilde{x}_k \right\|^2 &\le B \sum_{i=1}^{\infty} \sum_{k=1}^{\infty} (a_k x_{ki})^2 \\
&= B \sum_{k=1}^{\infty} a_k^2 \sum_{i=1}^{\infty} x_{ki}^2 \\
&= B \sum_{k=1}^{\infty} a_k^2 \|x_k\|^2 \\
&\le B \sum_{k=1}^{\infty} a_k^2.
\end{aligned}$$

The complex case Now we need to use Definition 3.15 for the tilde vectors $\tilde{x}_k$. We have that

$$\begin{aligned}
\left\|\sum_{k=1}^{\infty} a_k \tilde{x}_k\right\|^2 &= \sum_{i=1}^{\infty}\left(\sum_{k=1}^{\infty} a_k |x_{ki}|^2\right)^2 + \sum_{i=1}^{\infty}\sum_{j=i+1}^{\infty}\left(\sum_{k=1}^{\infty} a_k \mathrm{Re}(\bar{x}_{ki} x_{kj})\right)^2 \\
&\quad + \sum_{i=1}^{\infty}\sum_{j=i+1}^{\infty}\left(\sum_{k=1}^{\infty} a_k \mathrm{Im}(\bar{x}_{ki} x_{kj})\right)^2 \\
&= \sum_{i=1}^{\infty}\left(\sum_{k=1}^{\infty} a_k |x_{ki}|^2\right)^2 + \sum_{i=1}^{\infty}\sum_{j=i+1}^{\infty}\left(\mathrm{Re}\left(\sum_{k=1}^{\infty} a_k \bar{x}_{ki} x_{kj}\right)\right)^2 \\
&\quad + \sum_{i=1}^{\infty}\sum_{j=i+1}^{\infty}\left(\mathrm{Im}\left(\sum_{k=1}^{\infty} a_k \bar{x}_{ki} x_{kj}\right)\right)^2 \\
&\le 2\sum_{i=1}^{\infty}\left(\sum_{k=1}^{\infty} a_k |x_{ki}|^2\right)^2 + 2\sum_{i=1}^{\infty}\sum_{j=i+1}^{\infty}\left|\sum_{k=1}^{\infty} a_k \bar{x}_{ki} x_{kj}\right|^2 \\
&\le 2\sum_{i=1}^{\infty}\left(\sum_{k=1}^{\infty} a_k |x_{ki}|^2\right)^2 + 2\sum_{i=1}^{\infty}\sum_{j=1, j\neq i}^{\infty}\left|\sum_{k=1}^{\infty} a_k \bar{x}_{ki} x_{kj}\right|^2 \\
&= 2\sum_{i=1}^{\infty}\left\|\sum_{k=1}^{\infty} a_k \bar{x}_{ki} x_k\right\|^2 \le 2B\sum_{i=1}^{\infty}\sum_{k=1}^{\infty} a_k^2 |x_{ki}|^2 \\
&= 2B\sum_{k=1}^{\infty} a_k^2 \sum_{i=1}^{\infty} |x_{ki}|^2 = 2B\sum_{k=1}^{\infty} a_k^2 \|x_k\|^2 \le 2B\sum_{k=1}^{\infty} a_k^2.
\end{aligned}$$

Hence, $\{\tilde{x}_k\}_{k=1}^{\infty}$ is a Bessel sequence for the both cases.

Now we will show that $\{\tilde{x}_k\}_{k=1}^{\infty}$ fails to have a lower frame bound. We will prove the real case, the complex case is similar. By our assumption, we have that

$$\sum_{i=1}^{\infty} |x_{ki}|^2 < \infty, \text{ for all } k.$$

Also,

$$\sum_{k=1}^{\infty} |x_{ki}|^2 = \sum_{k=1}^{\infty} |\langle e_i, x_k\rangle|^2 < \infty, \text{ for all } i.$$

Fix $\epsilon > 0$ and choose n so that

$$\sum_{k=n}^{\infty} |x_{k1}|^2 < \epsilon.$$

Now choose m so that

$$\sum_{k=1}^{n-1} |x_{km}|^2 < \epsilon.$$

Let

$$\tilde{e}_{1m} := (e_m; \vec{0}; \vec{0}, \ldots) \in \tilde{\mathbb{H}}.$$

Then we have

$$\begin{aligned}
\sum_{k=1}^{\infty} |\langle \tilde{e}_{1m}, \tilde{x}_k \rangle|^2 &= \sum_{k=1}^{\infty} |x_{k1}|^2 |x_{km}|^2 \\
&= \sum_{k=1}^{n-1} |x_{k1}|^2 |x_{km}|^2 + \sum_{k=n}^{\infty} |x_{k1}|^2 |x_{km}|^2 \\
&\leq \sum_{k=1}^{n-1} |x_{km}|^2 + \sum_{k=n}^{\infty} |x_{k1}|^2 \\
&< 2\epsilon.
\end{aligned}$$

It follows that $\{\tilde{x}_k\}_{k=1}^{\infty}$ does not have a lower frame bound. □

Corollary 3.21 *Let $\{x_k\}_{k=1}^{\infty}$ be a frame for ℓ_2. If $\{\tilde{x}_k\}_{k=1}^{\infty}$ is a frame sequence, then $\{x_k\}_{k=1}^{\infty}$ cannot be injective.*

Proof We will prove the real case, the complex case is similar.

Suppose by way of contradiction that $\{x_k\}_{k=1}^{\infty}$ is injective. Then by Theorem 3.13, $\{\tilde{x}_k\}_{k=1}^{\infty}$ spans $\tilde{\mathbb{H}}$. Thus, it is a frame for $\tilde{\mathbb{H}}$, which contradicts Theorem 3.20. □

3.2 Constructing the Solutions to the Injectivity Problem

For the construction of solutions to the injectivity problem, we will follow the outline for the finite dimensional case. But this construction is much more complicated because of problems with convergence, problems with keeping the upper frame bound finite, and the fact that we cannot show spanning in ℓ_2 by just checking linear independence. Also, we proved in the finite dimensional case that the $\tilde{x}_i$ span by showing they are independent and have enough vectors to span $\tilde{\mathbb{H}}$. This does not work in the infinite dimensional case. Note that the following construction works for trace class operators and for Hilbert Schmidt operators.

Theorem 3.22 *Let $\{e_i\}_{i=1}^{\infty}$ be the canonial basis for the real Hilbert space ℓ_2 and let $a_i \neq 0$ for $i = 1, 2, \ldots$ be such that $\sum_{i=1}^{\infty} a_i^2 < \infty$. Define*

$$x_k = a_k(e_1 + e_{k+1}), \ \textit{for}\ k = 1, 2, \ldots.$$

Let L be the right shift operator on ℓ_2. Then the family

$$\{e_i\}_{i=1}^{\infty} \cup \{\frac{1}{2^i} L^i x_k\}_{i=0,k=1}^{\infty,\ \infty}$$

is a frame for ℓ_2 which gives injectivity.

Proof First we need to see that our family of vectors forms a frame for ℓ_2. Since our family contains an orthonormal basis for ℓ_2, we automatically have a lower frame bound. So we need to check that our family is Bessel, and since $\{e_i\}_{i=1}^{\infty}$ is already Bessel, we only need to check that $\{\frac{1}{2^i} L^i x_k\}_{i=0,k=1}^{\infty,\ \infty}$ is Bessel.

For any $x \in \ell_2$, we have

$$\begin{aligned}\sum_{i=0}^{\infty}\sum_{k=1}^{\infty}\left|\left\langle x, \frac{1}{2^i} L^i x_k\right\rangle\right|^2 &\le \sum_{i=0}^{\infty}\sum_{k=1}^{\infty}\frac{1}{4^i}\|x\|^2\|L^i x_k\|^2 \\ &\le \sum_{i=0}^{\infty}\sum_{k=1}^{\infty}\frac{1}{4^i}\|x\|^2 4a_k^2 \\ &= \left(\sum_{i=0}^{\infty}\frac{1}{4^{i-1}}\sum_{k=1}^{\infty}a_k^2\right)\|x\|^2.\end{aligned}$$

So our family is a Bessel sequence.

To see our frame is injective, let T be a Hilbert Schmidt self-adjoint operator such that

$$\langle Te_k, e_k\rangle = 0 \text{ and } \langle T(L^i x_k), L^i x_k\rangle = 0, \text{ for } i = 0, 1 \ldots;\ k = 1, 2, \ldots.$$

Note that

$$L^i x_k = a_k(e_{1+i} + e_{1+i+k}) \text{ for all } i, k.$$

Hence

$$\begin{aligned}\langle T(L^i x_k), L^i x_k\rangle &= a_k^2\langle Te_{1+i}, e_{1+i}\rangle + 2a_k^2\langle Te_{1+i}, e_{1+i+k}\rangle + a_k^2\langle Te_{1+i+k}, e_{1+i+k}\rangle \\ &= 2a_k^2\langle Te_{1+i}, e_{1+i+k}\rangle,\end{aligned}$$

for all i, k.

This implies $\langle Te_j, e_k\rangle = 0$ for all $j, k = 1, 2, \ldots$, and hence $T = 0$. □

The complex version of this construction looks like:

Theorem 3.23 *Let $\{e_i\}_{i=1}^{\infty}$ be the canonical orthonormal basis for complex ℓ_2, and let $\{a_i\}_{i=1}^{\infty}, \{b_i\}_{i=1}^{\infty} \in \ell_2$, $|a_i|, |b_i| \neq 0$ for all i. Then the following frame gives injectivity:*

$$\{e_i\}_{i=1}^{\infty} \cup \left\{\frac{1}{2^i} L^i(a_k(e_1 + e_{k+1}))\right\}_{i=0,k=1}^{\infty} \cup \left\{\frac{1}{2^i} L^i(b_k(e_1 + \iota e_{k+1}))\right\}_{i=0,k=1}^{\infty}.$$

The above frames are unbounded. The following theorem shows that we can easily adjust unbounded injective frames to produce bounded injective frames.

Theorem 3.24 *Every injective frame $\mathcal{X} = \{e_i\}_{i=1}^{\infty} \cup \{x_k\}_{k=1}^{\infty}$ of finitely supported vectors, induces a bounded injective frame.*

Proof Recall that for each k, we denote

$$x_k = (x_{k1}, x_{k2}, \ldots, x_{ki}, \ldots).$$

Choose integer $n_1 < n_2 < \cdots$ so that

$$\max\{i : x_{ki} \neq 0\} < n_k.$$

For $k = 1, 2, \ldots$ let

$$y_{2k} = x_k + e_{n_k} \text{ and } y_{2k+1} = x_k - e_{n_k}, \text{ for } k = 1, 2, \ldots.$$

It is clear that $\mathcal{Y} = \{e_i\}_{i=1}^{\infty} \cup \{y_k\}_{k=1}^{\infty}$ is still a frame and $\|y_k\| \geq 1$ for all $k = 1, 2, \ldots$. Since $\tilde{y}_{2k} + \tilde{y}_{2k+1} = 2\tilde{x}_k + 2\tilde{e}_{n_k}$, and $\{\tilde{e}_{n_k}\}_{k=1}^{\infty}$ are vectors in our set, it follows that $\{\tilde{x}_k\}_{k=1}^{\infty}$ is in our set of vectors and so $\mathcal{Y}$ is injective. □

3.3 The Solutions are Neither Open Nor Dense

In this section we will show that the solutions to the injectivity problem in infinite dimensions are neither open nor dense in the class of frames.

First we need a definition:

Definition 3.25 Given frames $\mathcal{X} = \{x_k\}_{k=1}^{\infty}$ and $\mathcal{Y} = \{y_k\}_{k=1}^{\infty}$ for ℓ_2, we define the **distance between them** by

$$d^2(\mathcal{X}, \mathcal{Y}) = \sum_{k=1}^{\infty} \|x_k - y_k\|^2.$$

Note that this distance may be infinity.

The following theorem shows that the frames which give injectivity are not open in the family of frames for ℓ_2.

Theorem 3.26 *Let $\mathcal{X} = \{e_i\}_{i=1}^{\infty} \cup \{\frac{1}{2^i} L^i x_k\}_{i=0,k=1}^{\infty}$ be the injective frame for the real space ℓ_2 as in Theorem 3.22. Then for any $\epsilon > 0$, there is a frame $\mathcal{Y}$ such that $d(\mathcal{X}, \mathcal{Y}) < \epsilon$, and $\mathcal{Y}$ is not injective.*

Proof Let any $\epsilon > 0$. Since the series $\sum_{i=0}^{\infty} \sum_{k=1}^{\infty} \|\frac{1}{2^i} L^i x_k\|^2$ converges, for any ϵ, there exists n_0 such that

$$\sum_{i=n_0+1}^{\infty} \sum_{k=1}^{\infty} \left\| \frac{1}{2^i} L^i x_k \right\|^2 < \epsilon^2.$$

Let $y_{ik} = \frac{1}{2^i} L^i x_k$ for $i = 0, 1, \ldots, n_0$ and $k = 1, 2, \ldots$, and $y_{ik} = 0$ otherwise. It is clear that

$$\mathcal{Y} = \{e_i\}_{i=1}^{\infty} \cup \{y_{ik}\}_{i=0,k=1}^{\infty}$$

cannot give injectivity by Theorem 3.13 while

$$d^2(\mathcal{X}, \mathcal{Y}) = \sum_{i=n_0+1}^{\infty} \sum_{k=1}^{\infty} \left\| \frac{1}{2^i} L^i x_k \right\|^2 < \epsilon^2.$$

This completes the proof. □

Remark 3.27 There is a perturbation theory for frames which looks like it should apply here. The problem is that although our vectors form a frame for ℓ_2, their tilde vectors do not form a frame to $\tilde{\mathbb{H}}$ and so the theory does not apply.

To show the solutions are not dense, we need the definition of a Riesz sequence in ℓ_2.

Definition 3.28 A family of vectors $\{x_i\}_{i=1}^{\infty}$ in the real or complex Hilbert space ℓ_2 is called a **Riesz sequence** if there are constants $0 < A \leq B < \infty$ so that for all $\{a_i\}_{i=1}^{\infty} \subset \ell_2$ we have:

$$A \sum_{i=1}^{\infty} |a_i|^2 \leq \left\| \sum_{i=1}^{\infty} a_i x_i \right\|^2 \leq B \sum_{i=1}^{\infty} |a_i|^2.$$

The constants A, B are called the **lower and upper Riesz bounds**. If the vectors span ℓ_2, this is called a **Riesz basis**.

Remark 3.29 It is known [6,8] that a Riesz basis is a frame and the Riesz bounds are the frame bounds. Also, $\{x_i\}_{i=1}^{\infty}$ is a Riesz sequence if and only if the operator $T : \ell_2 \to \ell_2$ given by $Te_i = x_i$ is a bounded, linear, invertible operator (on its range).

Also, we need a perturbation result from frame theory.

Proposition 3.30 *Assume $\chi = \{x_i\}_{i=1}^{\infty}$ are vectors in the real or complex space ℓ_2 satisfying:*

$$\sum_{i=1}^{\infty} \|e_i - x_i\|^2 < \epsilon^2.$$

Then χ is a Riesz sequence in ℓ_2 with lower Riesz bound $(1 - \epsilon)^2$.

Proof We compute for scalars $\{a_i\}_{i=1}^{\infty}$,

$$\begin{aligned}\left\|\sum_{i=1}^{\infty} a_i x_i\right\| &\geq \left\|\sum_{i=1}^{\infty} a_i e_i\right\| - \left\|\sum_{i=1}^{\infty} a_i (e_i - x_i)\right\| \\ &\geq \left(\sum_{i=1}^{\infty} |a_i|^2\right)^{1/2} - \sum_{i=1}^{\infty} |a_i| \|e_i - x_i\| \\ &\geq \left(\sum_{i=1}^{\infty} |a_i|^2\right)^{1/2} - \left(\sum_{i=1}^{\infty} |a_i|^2\right)^{1/2} \left(\sum_{i=1}^{\infty} \|e_i - x_i\|^2\right)^{1/2} \\ &\geq \left(\sum_{i=1}^{\infty} |a_i|^2\right)^{1/2} (1 - \epsilon)\end{aligned}$$

The upper Riesz bound is done similarly. □

We also need a theorem from [3].

Theorem 3.31 *Let Y, Z be subspaces of a Banach space X. If $T : Y \to Z$ is a surjective linear operator with $\|I - T\| < 1$, then $codim_X Y = codim_X Z$.*

The next theorem shows that the solution set of the infinite dimensional injectivity problem is not dense in the class of all frames for ℓ_2.

Theorem 3.32 *Let $\{e_k\}_{k=1}^{\infty}$ be the canonical basis for the real space ℓ_2 and $\mathcal{X} = \{x_k\}_{k=1}^{\infty} \subset \ell_2$ be such that*

$$\sum_{k=1}^{\infty} \|e_k - x_k\|^2 \leq \frac{1}{8},$$

Then $\mathcal{X}$ is not injective.

Proof Will will show that $\mathcal{X}$ does not satisfy Theorem 3.13. Note that $codim_{\tilde{\mathbb{H}}} \{\tilde{e}_k\}_{k=1}^{\infty}$ is infinite. Also, $\{\tilde{e}_k\}_{k=1}^{\infty}$ is an orthonormal sequence in $\tilde{\mathbb{H}}$. We have that

$$\sum_{k=1}^{\infty} \|e_k - x_k\|^2 = \sum_{k=1}^{\infty} \left((1 - x_{kk})^2 + \sum_{i \neq k} x_{ki}^2\right) \leq \frac{1}{8}.$$

In particular, $\|x_k\|^2 \leq 2$. Let

$$X = \overline{\text{span}}\{\tilde{e}_k\}_{k=1}^{\infty}, \text{ and } Y = \overline{\text{span}}\{\tilde{x}_k\}_{k=1}^{\infty}.$$

For each $k = 1, 2, \dots$ we have

$$\begin{aligned}\|\tilde{e}_k - \tilde{x}_k\|^2 &= (1 - x_{kk}^2)^2 + \sum_{j \ge k+1} (x_{kk}x_{kj})^2 + \sum_{i \ne k}\sum_{j \ge i} (x_{ki}x_{kj})^2 \\ &\le (1 - x_{kk})^2(2\|x_k\|^2 + 2) + \|x_k\|^2 \sum_{j \ge k+1} x_{kj}^2 + \|x_k\|^2 \sum_{i \ne k} x_{ki}^2 \\ &\le 6\left((1 - x_{kk})^2 + \sum_{i \ne k} x_{ki}^2\right).\end{aligned}$$

Hence,

$$\sum_{k=1}^{\infty} \|\tilde{e}_k - \tilde{x}_k\|^2 \le 6 \sum_{k=1}^{\infty} \left((1 - x_{kk})^2 + \sum_{i \ne k} x_{ki}^2\right) \le \frac{3}{4}.$$

It follows that $\{\tilde{x}_k\}_{k=1}^{\infty}$ is a Riesz sequence.

Now we define $T : X \to Y$ by: for $x = \sum_{k=1}^{\infty} \langle x, \tilde{e}_k \rangle \tilde{e}_k \in X$,

$$Tx = \sum_{k=1}^{\infty} \langle x, \tilde{e}_k \rangle \tilde{x}_k.$$

Since T is mapping a Riesz sequence to a Riesz sequence, it follows that T is bounded and surjective. Now,

$$\begin{aligned}\|(I - T)x\| &= \left\| \sum_{k=1}^{\infty} \langle x, \tilde{e}_k \rangle (\tilde{e}_k - \tilde{x}_k) \right\| \\ &\le \sum_{k=1}^{\infty} |\langle x, \tilde{e}_k \rangle| \|\tilde{e}_k - \tilde{x}_k\| \\ &\le \left(\sum_{k=1}^{\infty} |\langle x, \tilde{e}_k \rangle|^2 \right)^{1/2} \left(\sum_{k=1}^{\infty} \|\tilde{e}_k - \tilde{x}_k\|^2 \right)^{1/2} \\ &\le \frac{\sqrt{3}}{2} \|x\|.\end{aligned}$$

Hence, $\|I - T\| < 1$ and by Theorem 3.31, $codim_{\tilde{\mathbb{H}}} Y = codim_{\tilde{\mathbb{H}}} X = \infty$. □

3.4 The Solution to the State Estimation Problem

For the infinite dimensional case, the state estimation problem asks:

State Estimation Problem Given an injective Parseval frame $\{x_k\}_{k=1}^{\infty}$ for ℓ_2, and a sequence of real numbers $a = \{a_k\}_{k=1}^{\infty}$, does there exist a Hilbert Schmidt self-adjoint operator T so that

$$\langle Tx_k, x_k\rangle = a_k, \text{ for all } k?$$

Remark 3.33 This problem is rarely solvable.

(1) If $x_k x_k^* = x_l x_l^*$, but $a_k \neq a_l$ for some k, l, then the problem has no solution.
(2) Recall a set of vectors $\{x_i\}_{i=1}^{\infty}$ is ω **-independent** if $\sum_{i=1}^{\infty} c_i x_i = 0$ implies $c_i = 0$ for all $i = 1, 2, \ldots$. If $\{x_k x_k^*\}_{k=1}^{\infty}$ is not ω-independent and $\sum_{k=1}^{\infty} c_k x_k x_k^* = 0$ but not all c_k are zero, then for $\langle Tx_k, x_k\rangle = a_k$ we need

$$\sum_{k=1}^{\infty} c_k a_k = \left\langle T, \sum_{k=1}^{\infty} c_k x_k x_k^* \right\rangle = 0.$$

For the solution of the state estimation problem we will need the notion of a separated sequence in ℓ_2.

Definition 3.34 A family of vectors $\{x_i\}_{i=1}^{\infty}$ in ℓ_2 is **separated** if for every $j \in \mathbb{N}$,

$$x_j \notin \overline{\text{span}}\{x_i\}_{i\neq j}.$$

It is δ-separated if the projection P_j onto $\overline{\text{span}}\{x_i\}_{i\neq j}$ satisfies

$$\|(I - P_j)x_j\| \geq \delta.$$

Remark 3.35 A Riesz basic sequence $\{x_k\}_{k\in I}$ is δ-separated since it is clear from the definition that

$$dist(x_k, span\{x_i\}_{i\neq k}) \geq \delta.$$

In general, a Bessel sequence which is δ-separated may not be a Riesz sequence. To see this let

$$\mathbb{H} = \left(\sum_{n=1}^{\infty} \oplus \mathbb{H}_n\right)_{\ell_2},$$

where $\mathbb{H}^n$ is an n-dimensional Hilbert space with orthonormal basis $\{e_{in}\}_{i=1}^n$. Let P be the orthogonal projection onto the one dimensional subspace spanned by $\sum_{i=1}^n e_{in}$. Then $\{(I - P)e_{in}\}_{i=1,n=1}^{n-1,\ \infty}$ as a family of vectors in $\mathbb{H}$ is δ-separated, 1-Bessel, but not a Riesz sequence. (Careful: We have thrown away the vectors $(I - P)e_{nn}$ above.)

Note also that a δ-separated sequence may not be Bessel.

Example 3.36 Let $x_i = e_1 + e_{i+1}$, $i = 1, 2, \ldots$. Then $\{x_i\}_{i=1}^{\infty}$ is not a Bessel sequence. However, it is δ-separated.

Indeed, let P_j be the projection onto $\overline{\text{span}}\{x_i\}_{i\neq j}$. Then

$$\begin{aligned}\|x_j - P_j x_j\|^2 &= \|e_1 + e_j - P_j(e_1 + e_j)\|^2 = \|e_j + e_1 - P_j e_1\|^2 \\ &= 1 + \|e_1 - P_j e_1\|^2 \geq 1,\end{aligned}$$

for all j. So $\{x_i\}_{i=1}^{\infty}$ is δ-separated, where $\delta = 1$.

The next proposition presents a fundamental property of separated sequences.

Proposition 3.37 *If a family of vectors $\{x_i\}_{i=1}^{\infty}$ is separated, then there are vectors $\{y_i\}_{i=1}^{\infty}$ satisfying:*

$$\langle y_i, x_j \rangle = \delta_{ij}, \text{ for all i,j.}$$

If it is δ-separated then, $\sup_i \|y_i\| < \infty$.

Proof Fix j and let P_j be the orthogonal projection onto $\overline{\text{span}}\{x_i\}_{i\neq j}$. Note that $P_j x_j \neq x_j$ and so $(I - P_j)x_j \neq 0$.

Clearly,

$$\langle (I - P_j)x_j, x_i \rangle = 0 \text{ for } i \neq j.$$

So let

$$y_j = \frac{(I - P_j)x_j}{\|(I - P_j)x_j\|^2},$$

we get the desired sequence.

For the δ-separated case, we have that $\|(I - P_j)x_j\| \geq \delta$ and the result follows. □

For the next result, we will need:

Proposition 3.38 *Let $\{x_i\}_{i=1}^{\infty}$ be a bounded sequence in a Hilbert space $\mathbb{H}$. The following are equivalent:*

(1) *For some $\delta > 0$, $\{x_i\}_{i=1}^{\infty}$ is δ-separated.*
(2) *$\{x_i\}_{i=1}^{\infty}$ is separated and $\{x_i\}_{i=n}^{\infty}$ is δ_1-separated for some $\delta_1 > 0$, for some $n \geq 1$.*

Proof We just need to show that (2) ⇒ (1). So assume $\{x_i\}_{i=1}^{\infty}$ is separated and $\{x_i\}_{i=n}^{\infty}$ is δ_1-separated. Let P_j be the projection onto $\overline{\text{span}}\{x_i\}_{i\neq j}$, for $j = 1, 2, \ldots$, and let Q_j be the projection onto $\overline{\text{span}}\{x_i\}_{n\leq i\neq j}$, for $j = n, n+1, \ldots$. So

$$\|(I - Q_j)x_j\| \geq \delta_1, \text{ for all } j \geq n.$$

We need to show that there exists a $\delta > 0$ so that

$$\|(I - P_j)x_j\| \geq \delta, \text{ for all } j \geq 1.$$

We will do this in steps.

Step 1 There exists a $\delta_2 > 0$ so that

$$\|(I - P_j)x_j\| \geq \delta_2, \text{ for all } j \geq n.$$

We will do this by way of contradiction. So assume there are natural numbers $n \leq n_1 < n_2 < \cdots$ satisfying:

$$\|x_{n_j} - P_{n_j}(x_{n_j})\| < \frac{1}{j}.$$

It follows that there are vectors $y_j \in \text{span}\{x_i\}_{i=1}^{n-1}$ and $z_j \in \text{span}\{x_i\}_{n \leq i \neq n_j < \infty}$ so that $\|x_{n_j} - (y_j + z_j)\| < \frac{1}{j}$.

Claim 1 There are an $\epsilon > 0$ and $n_0 \in \mathbb{N}$ so that $\|y_j\| \geq \epsilon$, for all $j \geq n_0$.

We prove the claim by way of contradiction. If the claim fails, there are integers $j_1 < j_2 < \cdots$ so that $\|y_{j_k}\| < \frac{1}{k}$ for all $k = 1, 2, \ldots$. It follows that

$$\|x_{n_{j_k}} - z_{j_k}\| \leq \|x_{n_{j_k}} - (z_{j_k} + y_{j_k})\| + \|y_{j_k}\| < \frac{2}{k}, \text{ for all } k,$$

which contradicts the fact that $\{x_i\}_{i=n}^{\infty}$ is δ-separated.

Claim 2 There is a constant $K > 0$ so that $\|y_j\| \leq K$, for all $j \geq n_0$.

Define

$$\gamma = \inf\{\|u - v\| : u \in \text{span}\{x_i\}_{i=1}^{n-1}, v \in \overline{\text{span}}\{x_i\}_{i=n}^{\infty}, \|u\| = 1\}.$$

We will show that $\gamma > 0$. Indeed, if $\gamma = 0$ then there are sequences $\{u_j\}_{j=1}^{\infty} \subset \text{span}\{x_i\}_{i=1}^{n-1}$, $\|u_j\| = 1$, for all j, and $\{v_j\}_{j=1}^{\infty} \subset \overline{\text{span}}\{x_i\}_{i=n}^{\infty}$ so that

$$\|u_j - v_j\| \to 0 \text{ as } j \to \infty.$$

By switching to a subsequence if necessary, we may assume $u_j \to u \in \text{span}\{x_i\}_{i=1}^{n-1}$ and $u \neq 0$. Since

$$\|v_j - u\| \leq \|v_j - u_j\| + \|u_j - u\|,$$

we conclude that $v_j \to u \in \overline{\text{span}}\{x_i\}_{i=n}^{\infty}$. Thus,

$$u \in \text{span}\{x_i\}_{i=1}^{n-1} \cap \overline{\text{span}}\{x_i\}_{i=n}^{\infty}.$$

Since $u \in \text{span}\{x_i\}_{i=1}^{n-1}$, $u \neq 0$, we can write $u = \sum_{i=1}^{n-1} \alpha_i x_i$ for some scalars $\alpha_i's$ not all zero. Without loss of generality, we can assume $\alpha_1 \neq 0$. Then

$$x_1 = \frac{1}{\alpha_1}\left(u - \sum_{i=2}^{n-1} \alpha_i x_i\right) \in \overline{\text{span}}\{x_i\}_{i=2}^{\infty},$$

which contradicts the fact that $\{x_i\}_{i=1}^{\infty}$ is separated. So, $\gamma > 0$.

Now we have

$$\|\frac{y_j + z_j}{\|y_j\|}\| \geq \gamma, \text{ for all } j \geq n_0,$$

and $\sup_{j\geq 1}\|x_j\|$ is finte. Therefore, there is some $K > 0$ such that

$$\|y_j\| \leq \frac{1}{\gamma}\|y_j + z_j\| \leq \frac{1}{\gamma}(\|y_j + z_j - x_{n_j}\| + \|x_{n_j}\|) \leq K, \text{ for all } j \geq n_0.$$

The Claim 2 is proven.

Now since $\epsilon \leq \|y_j\| \leq K$ for all $j \geq n_0$, it has a convergent subsequence $y_{j_k} \to y \in \text{span}\{x_i\}_{i=1}^{n-1}$, and $y \neq 0$.

From the fact that

$$\|x_{n_{j_k}} - z_{j_k} - y\| \leq \|x_{n_{j_k}} - z_{j_k} - y_{j_k}\| + \|y_{j_k} - y\| \leq \frac{1}{j_k} + \|y_{j_k} - y\|,$$

we conclude $x_{n_{j_k}} - z_{j_k} \to y \in \overline{\text{span}}\{x_i\}_{i=n}^{\infty}$ as $k \to \infty$. Thus,

$$y \in \text{span}\{x_i\}_{i=1}^{n-1} \cap \overline{\text{span}}\{x_i\}_{i=n}^{\infty}$$

By the same argument as in the proof of Claim 2, this leads to a contradiction with the fact that $\{x_i\}_{i=1}^{\infty}$ is separated.

Step 2 There exists a $\delta > 0$ so that

$$\|(I - P_j)x_j\| \geq \delta, \text{ for all } j \geq 1.$$

Since $\{x_i\}_{i=1}^{\infty}$ is separated, for each $i = 1, 2, \ldots, n-1$, there exists $\epsilon_i > 0$ so that $\|(I - P_i)x_i\| \geq \epsilon_i$. Combined with Step 1, we have that $\{x_i\}_{i=1}^{\infty}$ is δ-separated, where $\delta = \min_{i=1,\ldots,n-1}\{\epsilon_i, \delta_2\}$. The proof of the Proposition is completed. □

Now we give a complete classification of when the state estimation problem is solvable for all measurement vectors in ℓ_1. Note that we have done it in complete generality and not assumed that $\{x_k\}_{k=1}^{\infty}$ is injective.

Theorem 3.39 *Let $\mathcal{X} = \{x_k\}_{k=1}^{\infty}$ be a frame for the real or complex space ℓ_2. The following are equivalent:*

(1) *For every real vector $a = (a_1, a_2, \ldots) \in \ell_1$, there is a Hilbert Schmidt self-adjoint operator T so that*

$$\langle Tx_k, x_k\rangle = a_k, \text{ for all } k = 1, 2, \ldots.$$

(2) *The sequence* $\{\tilde{x}_k\}_{k=1}^{\infty}$ *is* δ*-separated.*

Proof (1) $\Rightarrow$ (2): By (1), for each $k = 1, 2, \ldots$, there is a Hibert Schmidt self-adjoint operator R_k, and hence a vector $\tilde{R}_k \in \tilde{\mathbb{H}}$ so that

$$\langle \tilde{R}_k, \tilde{x}_l \rangle = \langle R_k x_l, x_l \rangle = \begin{cases} 1 & \text{if } k = l \\ 0 & \text{if } k \neq l. \end{cases}$$

It follows that $\tilde{x}_l \notin \overline{\text{span}}\{\tilde{x}_k\}_{k \neq l}$ and hence $\{\tilde{x}_k\}_{k=1}^{\infty}$ is separated. We now proceed by way of contradiction. Suppose that $\{\tilde{x}_k\}_{k=1}^{\infty}$ is not δ-separated. Then $\{\tilde{x}_k\}_{k=n}^{\infty}$ is not δ_n-separated for all n. Then for $n = 1$, there is $k_1 \geq 1$ such that

$$\|\tilde{x}_{k_1} - P_{k_1}(\tilde{x}_{k_1})\| < \frac{1}{2}.$$

Since $P_{k_1}(\tilde{x}_{k_1}) \in \overline{\text{span}}\{\tilde{x}_k\}_{k=1, k \neq k_1}^{\infty}$, there are some scalars $\alpha_k, k \in I$, where I is a finite subset of $\{k : k \geq 1, k \neq k_1\}$ such that

$$\left\| P_{k_1}(\tilde{x}_{k_1}) - \sum_{k \in I} \alpha_k \tilde{x}_k \right\| < \frac{1}{2}.$$

Let $y_1 = \sum_{k \in I} \alpha_k \tilde{x}_k$. Then

$$\|\tilde{x}_{k_1} - y_1\| < 1.$$

Now let $n_2 > \max\{k_1, k\}_{k \in I}$. Since $\{\tilde{x}_k\}_{k=n_2}^{\infty}$ is not δ_{n_2}-separated, similar to the above, there are numbers $n_2 \leq k_2 < n_3$ and a vector

$$y_2 \in \text{span}\{\tilde{x}_k : n_2 \leq k \neq k_2 < n_3\}$$

such that

$$\|\tilde{x}_{k_2} - y_2\| < \frac{1}{2^3}.$$

Continuing this procedure we can choose $1 = n_1 \leq k_1 < n_2 \leq k_2 < n_3 < \cdots$ and vectors

$$y_m \in \text{span}\{\tilde{x}_k : n_m \leq k \neq k_m < n_{m+1}\},$$

such that

$$\|\tilde{x}_{k_m} - y_m\| < \frac{1}{m^3},$$

for all m. Now let $a = \{a_k\}_{k=1}^{\infty} \in \ell_1$, where

$$a_k = \begin{cases} \frac{1}{m^2} & \text{if } k = k_m \\ 0 & \text{otherwise} . \end{cases}$$

Then by assumption, there exists a Hilbert Schmidt self-adjoint operator T and a vector $\tilde{T} \in \tilde{\mathbb{H}}$ so that $\langle \tilde{T}, \tilde{x}_k \rangle = \langle T x_k, x_k \rangle = a_k$ for all k. But then

$$\frac{1}{m^2} = \langle \tilde{T}, \tilde{x}_{k_m} \rangle = \langle \tilde{T}, \tilde{x}_{k_m} - y_m \rangle \leq \|\tilde{T}\| \|\tilde{x}_{k_m} - y_m\| \leq \|\tilde{T}\| \frac{1}{m^3},$$

which implies $\|\tilde{T}\| \geq m$ for all m, a contradiction.

(2) $\Rightarrow$ (1): Since $\{\tilde{x}_k\}_{k=1}^{\infty}$ is δ-separated, by Proposition 3.37, there are vectors $\{\tilde{T}_k\}_{k=1}^{\infty}$ in $\tilde{\mathbb{H}}$ satisfying

$$\langle \tilde{T}_k, \tilde{x}_l \rangle = \begin{cases} 1 & \text{if } k = l \\ 0 & \text{if } k \neq l \end{cases}$$

for all $k, l \geq 1$, and $\sup\limits_{k \geq 1} \|\tilde{T}_k\| < \infty$. Now, fix $a = (a_1, a_2, \ldots) \in \ell_1$ and let

$$\tilde{T} = \sum_{k=1}^{\infty} a_k \tilde{T}_k.$$

This series converges since $a \in \ell_1$ and $\sup\limits_{k \geq 1} \|\tilde{T}_k\| < \infty$. Now, let T be the Hilbert Schmidt self-adjoint operator that corresponds with $\tilde{T}$. Then we have

$$\langle T x_k, x_k \rangle = \langle \tilde{T}, \tilde{x}_k \rangle = a_k, \text{ for all } k = 1, 2, \ldots .$$

This completes the proof. □

Now we show that there is no injective frame for which the state estimation problem is solvable for all measurements taken from ℓ_2. Note that for a Hilbert Schmidt self-adjoint operator T on the Hilbert space ℓ_2, the corresponding vector $\tilde{T}$ is defined as in the proof of Theorem 3.13 for the real case and Theorem 3.17 for the complex case.

Theorem 3.40 *There is no injective frame $\mathcal{X} = \{x_k\}_{k=1}^{\infty}$ in the real or complex space ℓ_2 so that for every $a = \{a_k\}_{k=1}^{\infty} \in \ell_2$, there is a Hilbert Schmidt operator T so that*

$$\langle T x_k, x_k \rangle = a_k, \text{ for all } k = 1, 2, \ldots .$$

Proof We will proceed by way of contradiction. The proof is divided into steps.

Suppose that there is an injective frame $\mathcal{X} = \{x_k\}_{k=1}^{\infty}$ for which the state estimation problem is solvable for all choices $\{a_k\}_{k=1}^{\infty} \in \ell_2$.

Step I There are vectors $\tilde{R}_k \in \tilde{\mathbb{H}}$, $k = 1, 2, \ldots$ so that $\langle \tilde{R}_k, \tilde{x}_l \rangle = \delta_{kl}$.

This is immediate because by assumption, for each $k = 1, 2, \ldots$, there is a Hilbert Schmidt self-adjoint operator R_k so that

$$\langle \tilde{R}_k, \tilde{x}_l \rangle = \langle R_k x_l, x_l \rangle = \begin{cases} 1 & \text{if } k = l \\ 0 & \text{if } k \neq l. \end{cases}$$

Denote $E_n = \text{span}\{\tilde{x}_k\}_{k=1}^n$ and let P_n be the projection onto E_n.

Step II If there is a real vector $\{a_k\}_{k=1}^\infty \in \ell_2$ satisfying $\sup_n \| \sum_{k=1}^n a_k \tilde{R}_k \| = \infty$, then there is a real vector $\{b_k\}_{k=1}^\infty \in \ell_2$ and $n_1 < n_2 < \cdots$ so that

$$\left\| P_{n_j} \left(\sum_{k=1}^{n_j} b_k \tilde{R}_k \right) \right\| \geq j.$$

Indeed, since $\sup_n \| \sum_{k=1}^n a_k \tilde{R}_k \| = \infty$, we can choose a sequence $m_1 < m_2 < \cdots$ so that

$$\left\| \sum_{k=1}^{m_j} a_k \tilde{R}_k \right\| \geq 2j.$$

For any $j > 1$, we have

$$\begin{aligned} \left\| \sum_{k=1}^{m_1} a_k \tilde{R}_k - \sum_{k=m_1+1}^{m_j} a_k \tilde{R}_k \right\| &\geq \left\| \sum_{k=1}^{m_j} a_k \tilde{R}_k \right\| - 2 \left\| \sum_{k=1}^{m_1} a_k \tilde{R}_k \right\| \\ &\geq 2j - 2 \left\| \sum_{k=1}^{m_1} a_k \tilde{R}_k \right\|. \end{aligned}$$

Combining this with the fact that $E_1 \subset E_2 \subset \ldots$ and $\cup_{n=1}^\infty E_n$ is dense in $\tilde{\mathbb{H}}$, we can choose j large enough so that

$$\left\| P_{m_j} \left(\sum_{k=1}^{m_1} a_k \tilde{R}_k \right) \right\| \geq \frac{1}{2} \left\| \sum_{k=1}^{m_1} a_k \tilde{R}_k \right\|,$$

and

$$\left\| \sum_{k=1}^{m_1} a_k \tilde{R}_k - \sum_{k=m_1+1}^{m_j} a_k \tilde{R}_k \right\| \geq 4.$$

Since

$$\left\| P_{m_j}\left(\sum_{k=1}^{m_1} a_k \tilde{R}_k\right) + P_{m_j}\left(\sum_{k=m_1+1}^{m_j} a_k \tilde{R}_k\right)\right\|^2$$
$$+ \left\| P_{m_j}\left(\sum_{k=1}^{m_1} a_k \tilde{R}_k\right) - P_{m_j}\left(\sum_{k=m_1+1}^{m_j} a_i \tilde{R}_k\right)\right\|^2$$
$$= 2\left(\left\| P_{m_j}\left(\sum_{k=1}^{m_1} a_k \tilde{R}_k\right) \|^2 + \| P_{m_j}\left(\sum_{k=m_1+1}^{m_j} a_k \tilde{R}_k\right)\right\|^2\right)$$
$$\geq 2\left\| P_{m_j}\left(\sum_{k=1}^{m_1} a_k \tilde{R}_k\right)\right\|^2,$$

we can choose $b_i = a_i$ for $i = 1, \ldots, m_1$ and $b_i \in \{a_i, -a_i\}$ for $i = m_1 + 1, \ldots, m_j$ so that

$$\left\| P_{m_j}\left(\sum_{k=1}^{m_j} b_k \tilde{R}_k\right)\right\| \geq \left\| P_{m_j}\left(\sum_{k=1}^{m_1} b_k \tilde{R}_k\right)\right\| \geq \frac{1}{2}\left\|\sum_{k=1}^{m_1} b_k \tilde{R}_k\right\| \geq 1 \text{ and } \left\|\sum_{k=1}^{m_j} b_k \tilde{R}_k\right\| \geq 4.$$

Setting $n_1 = m_j$,

$$\left\| P_{n_1}\left(\sum_{k=1}^{n_1} b_k \tilde{R}_k\right)\right\| \geq 1 \text{ and } \left\|\sum_{k=1}^{n_1} b_k \tilde{R}_k\right\| \geq 4.$$

Now for m_j above, by the same argument, there is $m_l > m_j$ and $b_i \in \{a_i, -a_i\}$ for $i = m_j + 1, \ldots, m_l$ so that

$$\left\| P_{m_l}\left(\sum_{k=1}^{m_l} b_k \tilde{R}_k\right)\right\| \geq \left\| P_{m_l}\left(\sum_{k=1}^{m_j} b_k \tilde{R}_k\right)\right\| \geq \frac{1}{2}\left\|\sum_{k=1}^{m_j} b_k \tilde{R}_k\right\| \geq 2$$

and

$$\left\|\sum_{k=1}^{m_l} b_k \tilde{R}_k\right\| \geq 6.$$

Set $n_2 = m_l$ we get

$$\left\| P_{n_2}\left(\sum_{k=1}^{n_2} b_k \tilde{R}_k\right)\right\| \geq 2.$$

Continuing this process inductively, the result follows.

Step III For all vectors $\{a_k\}_{k=1}^{\infty} \in \ell_2$, $\sup_n \| \sum_{k=1}^{n} a_k \tilde{R}_k \|$ is finite.

Suppose by contradiction that there is a vector $\{a_k\}_{k=1}^{\infty} \in \ell_2$ so that $\sup_n \| \sum_{k=1}^{n} a_k \tilde{R}_k \| = \infty$. Let $\{b_k\}_{k=1}^{\infty}$ be the vector in Step II, then there exists a vector $\tilde{T} \in \tilde{\mathbb{H}}$ so that $\langle \tilde{T}, \tilde{x}_k \rangle = b_k$, for all $k = 1, 2, \ldots$. It follows that

$$P_{n_j} \tilde{T} = P_{n_j} \left(\sum_{k=1}^{n_j} b_k \tilde{R}_k \right)$$

for all $j = 1, 2, \ldots$. Hence,

$$\infty = \sup_j \left\| P_{n_j} \left(\sum_{k=1}^{n_j} b_k \tilde{R}_k \right) \right\| = \sup_j \| P_{n_j} \tilde{T} \| \le \| \tilde{T} \|,$$

which is a contradiction.

Step IV $\{\tilde{R}_k\}_{k=1}^{\infty}$ is a Bessel sequence in $\tilde{\mathbb{H}}$.

For each $n \in \mathbb{N}$, define an operator

$$T_n : \ell_2 \longrightarrow \tilde{\mathbb{H}}$$
$$x = (a_1, a_2, \ldots) \longmapsto T_n(x) = \sum_{k=1}^{n} a_k \tilde{R}_k$$

Then T_n is a bounded linear operator for all n.

By Step III, $\sup_n \| \sum_{k=1}^{n} a_k \tilde{R}_k \|$ is finite for all $x = \{a_k\}_{k=1}^{\infty}$. By the Uniform Boundedness Principle, $\sup_n \|T_n\| \le B$, for some $B > 0$. For any $n, m \in \mathbb{N}, m > n$, we have

$$\left\| \sum_{k=n+1}^{m} a_k \tilde{R}_k \right\|^2 = \left\| T_m \left(\sum_{k=n+1}^{m} a_k e_k \right) \right\|^2 \le B^2 \sum_{k=n+1}^{m} a_k^2.$$

It follows that $\sum_{k=1}^{\infty} a_k \tilde{R}_k$ converges, and hence $\{\tilde{R}_k\}_{k=1}^{\infty}$ is Bessel.

Step V We arrive at a contradiction.

We have shown that under our assumption, $\{\tilde{R}_k\}_{k=1}^{\infty}$ is B^2-Bessel for some B. Now choose any $a = \{a_k\}_{k=1}^{\infty} \in \ell_2$. We have that

$$\left\| \sum_{k=1}^{\infty} a_k \tilde{R}_k \right\|^2 \le B^2 \sum_{k=1}^{\infty} a_k^2.$$

By Theorem 3.20, $\sum_{k=1}^{\infty} a_k \tilde{x}_k$ converges. Now, we have

$$\begin{aligned}\left\|\sum_{k=1}^{\infty} a_k \tilde{x}_k\right\| &= \sup_{\|x\|\le 1} \left|\left\langle x, \sum_{k=1}^{\infty} a_k \tilde{x}_k \right\rangle\right| \\ &\ge \frac{1}{B\|a\|} \left|\left\langle \sum_{k=1}^{\infty} a_k \tilde{R}_k, \sum_{l=1}^{\infty} a_l \tilde{x}_l \right\rangle\right| \\ &= \frac{1}{B\|a\|} \left| \sum_{k,l=1}^{\infty} a_k a_l \langle \tilde{R}_k, \tilde{x}_l \rangle \right| \\ &= \frac{1}{B}\|a\|.\end{aligned}$$

It follows that $\{\tilde{x}_k\}_{k=1}^{\infty}$ has a positive lower Riesz bound and since this family is injective, it is a Riesz basis. Hence by Theorem 3.20, it is a frame sequence. But then by Corollary 3.21, $\{x_k\}_{k=1}^{\infty}$ cannot be injective, a contradiction. The proof of our theorem is now complete. □

Remark 3.41 As in the finite dimensional case, it is often the case that the state estimation problem is not solvable. But again there is a natural way to get a good estimation to the solution. Given a frame $\{x_k\}_{k=1}^{\infty}$ and $\{a_k\}_{k=1}^{\infty} \in \ell_2$, choose m so that $\sum_{k=m+1}^{\infty} a_k^2 \le \epsilon$. Then apply the argument in the finite case to get the best solution for $\{a_k\}_{k=1}^{m}$.

References

1. Benedetto, J.J., Kebo, A.: The role of frame force in quantum detection. J. Fourier Anal. Appl. **14**, 443–474 (2008)
2. Bodmann, B., Haas, J.: A short history of frames and quantum designs. arXiv:1709.01958
3. Casazza, P.G., Kalton, N.J.: Generalizing the Paley–Wiener perturbation theory for Banach spaces. Proc. AMS **127**(2), 519–527 (1999)
4. Casazza, P.G., Kutyniok, G.: Finite Frames: Theory and Applications. Birkhauser, Boston (2012)
5. Casazza, P.G., Leon, M.: Existence and construction of finite frames with a given frame operator. Int. J. Pure Appl. Math. **63**(2), 149–158 (2010)
6. Casazza, P.G., Lynch, R.: A brief introduction to Hilbert space frame theory and its applications. In: K. Okoudjou (Ed.) Proceedings of Symposia in Applied Mathematics—Finite Frame Theory; AMS Short Course 2015, vol. 73, pp. 1–51 (2016). arXiv:1509.07347
7. Casazza, P.G., Pinkham, E., Toumanen, B.: Riesz outer product Hilbert space frames: quantitative bounds, topological properties and full geometric characterization. J. Math. Anal. Appl. **441**(1), 475–498 (2016). https://doi.org/10.1016/j.jmaa.2016.04.001. arXiv:1410.7755
8. Christensen, O.: Frames and Riesz Bases. Birkhauser, Boston (2016)
9. Eldar, Y.C.: Von Neumann measurement is optimal for detecting linearly independent 0 quantum states. Phys. Rev. A (3) **68**(5) (2003)
10. Eldar, Y.C., Bolcskei, H.: Geometrically uniform frames. IEEE Trans. Inf. Theory **49**(4), 993–1006 (2003)
11. Eldar, Y.C., Je, G.D.F.: Optimal tight frames and quantum measurement. IEEE Trans. Inf. Theory **48**(3), 599–610 (2002)
12. Han, D., Larson, D.R., Liu, B., Lin, R.: Operator-valued measures, dilatons, and the theory of frames. Memoirs of AMS **229**(1075), (2013)

13. Hauladen, C.W., Wooters, W.K.: A "pretty good" measurement for distinguishing quantum states. J. Mod. Opt. **41**(12), 2385–2390 (1994)
14. Helstrom, C.W.: Quantum detection and estimation theory. J. Stat. Phys. **1**, 231–252 (1969)
15. Horn, R.A., Johnson, C.R.: Matrix Analysis. Cambridge University Press, Cambridge (1985)
16. Moran, B., Howard, S., Cochran, D.: Positive-operator-valued measures: a general setting for frames. Excursions Harmon. Anal. Book Ser. **2**, 49–64 (2012)
17. Peres, A., Terno, D.R.: Optimal distinction between non-orthogonal quantum states. J. Phys. A **31**(34), 7105–7111 (1998)
18. Scott, A.J.: Tight informationally complete quantum measurements. J. Math. Phys. **39**(42), 13507–13530 (2006)
19. Yuen, H.P., Kennedy, R.S., Lax, M.: Optimum testing of multiple hypotheses in quantum detection theory. IEEE Trans. Inf. Theory **IT–21**, 125–134 (1975)

Affiliations

Sara Botelho-Andrade[1] · Peter G. Casazza[1] · Desai Cheng[1] · Tin T. Tran[1]

✉ Peter G. Casazza
Casazzap@missouri.edu

Sara Botelho-Andrade
sandrade102087@gmail.com

Desai Cheng
chengdesai@yahoo.com

Tin T. Tran
tinmizzou@gmail.com

1 Department of Mathematics, University of Missouri, Columbia, MO 65211-4100, USA

Journal of Fourier Analysis and Applications (2019) 25:2324–2355
https://doi.org/10.1007/s00041-018-09657-7

Bounds for Anisotropic Carleson Operators

Joris Roos[1,2]

Received: 30 October 2017 / Published online: 10 December 2018
© Springer Science+Business Media, LLC, part of Springer Nature 2018

Abstract
We prove weak (2, 2) bounds for maximally modulated anisotropically homogeneous smooth multipliers on $\mathbb{R}^n$. These can be understood as generalizing the classical one-dimensional Carleson operator. For the proof we extend the time-frequency method by Lacey and Thiele to the anisotropic setting. We also discuss a related open problem concerning Carleson operators along monomial curves.

Keywords Carleson operators · Anisotropic dilations · Time-frequency analysis · Singular integrals

Mathematics Subject Classification 42B20 · 42B25 · 44A12

1 Introduction

Let us consider $\mathbb{R}^n$ equipped with the dilations given by

$$\delta_\lambda(x) = \delta_\lambda^\alpha(x) = (\lambda^{\alpha_1} x_1, \ldots, \lambda^{\alpha_n} x_n), \tag{1.1}$$

where $\alpha = (\alpha_1, \ldots, \alpha_n)$ with $\alpha_i \geq 1$ for $i = 1, \ldots, n$. We write $|\alpha| = \sum_{i=1}^n \alpha_i$ and fix

$$\rho_\alpha(x) = \max\left\{ |x_i|^{\frac{1}{\alpha_i}} \,:\, i = 1, \ldots, n \right\}.$$

For an integer $\nu \geq 0$, we say that a function m on $\mathbb{R}^n$ is in the class $\mathscr{M}_\alpha^\nu$ if

Communicated by Loukas Grafakos.

✉ Joris Roos
jroos@math.wisc.edu

1 Mathematical Institute, University of Bonn, Endenicher Allee 60, 53115 Bonn, Germany

2 Present Address: Department of Mathematics, University of Wisconsin-Madison, 480 Lincoln Dr, Madison, WI 53706, USA

(a) m is bounded and contained in $C^\nu(\mathbb{R}^n \setminus \{0\})$, and
(b) $m(\delta_\lambda(\xi)) = m(\xi)$ for all $\xi \neq 0$ and $\lambda > 0$.

Let us denote

$$\|m\|_{\mathcal{M}_\alpha^\nu} = \sup_{|\beta|\leq\nu} \sup_{\rho_\alpha(\xi)=1} |\partial^\beta m(\xi)|.$$

Define the Carleson operator associated with the multiplier m as

$$\mathscr{C}_m f(x) = \sup_{N\in\mathbb{R}^n} \left| \int_{\mathbb{R}^n} \widehat{f}(\xi) e^{ix\xi} m(\xi - N) d\xi \right|.$$

Replacing α by $c\alpha$ for some scalar c does not modify the classes $\mathcal{M}_\alpha^\nu$. Thus we make the assumption $\alpha_1 = 1$ for normalization. For technical reasons we also assume that the α_i are positive integers.

Then we can state our main result as follows.

Theorem 1.1 *There exists a positive integer ν_0 depending only on α and a constant $C \in (0, \infty)$ depending only on α such that for all $m \in \mathcal{M}_\alpha^{\nu_0}$ we have*

$$\|\mathscr{C}_m f\|_{2,\infty} \leq C \|m\|_{\mathcal{M}_\alpha^{\nu_0}} \|f\|_2. \tag{1.2}$$

The proof of this theorem is based on the time-frequency techniques of Lacey and Thiele [9]. In the one-dimensional case $n = 1, \alpha = 1$ we recover the weak $(2, 2)$ bound for Carleson's operator, which immediately implies Carleson's theorem on almost everywhere convergence of Fourier series [1] (up to a standard transference argument, see [7]). In the isotropic case $\alpha = (1, \ldots, 1)$ the theorem follows from a result of Sjölin [13]. Pramanik and Terwilleger [12] study weak $(2, 2)$ bounds for the isotropic case in $\mathbb{R}^n$ using the method of Lacey and Thiele. This is extended to strong (p, p) for $1 < p < \infty$ by Grafakos, Tao and Terwilleger in [5]. We speculate that Theorem 1.1 could also be extended to strong (p, p) for $1 < p < \infty$ using the methods from [5]. However we don't pursue this here to keep the exposition simple.

Our present proof shows that (1.2) holds for all $\nu_0 \geq 3|\alpha|+2$. It would be interesting to know what the minimum regularity requirement is for (1.2) to hold. The same question is also open in the isotropic case. It seems plausible that (1.2) should at least hold for all $\nu_0 \geq |\alpha| + 1$, because curiously, the only place in the current proof that requires more than that is a tail estimate in the single tree lemma (see (7.5)). All the main terms can be bounded using only $\nu_0 \geq |\alpha| + 1$.

The method of Lacey and Thiele involves several ingredients. The first step is a reduction to a discrete dyadic model operator that involves summation over certain regions in phase space which are called tiles. This is detailed in Sect. 3. In this step we encounter a complication which is caused by the absence of rotation invariance in the anisotropic case. We resolve this using an anisotropic cone decomposition (see Lemma 3.3). The next step is a certain procedure of combinatorial nature the purpose of which is to organize the tiles into certain collections (which are called trees) each of which is associated with a component of the operator that behaves more like a classical singular integral operator (see Lemma 4.4). The combinatorial part of the argument requires only little modification compared to the original procedure in [9] (see Sects. 4–6). The core component and most difficult part of the proof is the single tree estimate (see Sect. 7). Due to an extra dependence on the linearizing function, the estimate is more technical than the corresponding estimate in [9]. It is complicated further by the presence of anisotropic dilations.

In Sect. 2 we discuss a related open problem on Carleson operators along monomial curves, which served as a main motivation for this work. We demonstrate how Theorem 1.1 can be applied to a certain family of rougher multipliers that can be seen as a toy model for the Carleson operators along monomial curves. We also discuss the particular case of the parabolic Carleson operator (2.4), which exhibits some additional symmetries and is related to Lie's quadratic Carleson operator [10]. Some partial progress on Carleson operators along monomial curves was obtained in [6]. Also see the related work [11] on Carleson operators along paraboloids in dimensions $n \geq 3$.

2 Carleson Operators Along Monomial Curves

In this section we set $\alpha = (1, d)$ for a positive integer $d \geq 2$. Let us consider the multiplier of the Hilbert transform along the curve (t, t^d) in the form

$$m_d(\xi, \eta) = p.v. \int_{\mathbb{R}} e^{i\xi t - i\eta t^d} \frac{dt}{t}, \quad (\xi, \eta) \in \mathbb{R}^2. \tag{2.1}$$

It is currently an open problem to decide whether $\mathscr{C}_{m_d}$ satisfies any L^p bounds. The multiplier m_d satisfies the anisotropic dilation symmetry

$$m_d(\lambda\xi, \lambda^d \eta) = m_d(\xi, \eta)$$

for $\lambda > 0$ and $(\xi, \eta) \neq 0$. However, Theorem 1.1 does not apply because m_d is too rough to be in the class $\mathscr{M}_\alpha^\nu$ for any positive integer ν.

Next we discuss a family of toy model operators. In the following discussion we focus on the intersection of the quadrant $\{\xi \geq 0, \eta \geq 0\}$ with the region $\eta^{\frac{1}{d}} \leq 2\xi$. The other quadrants can be treated similarly (though depending on the parity of d the phase might not have a critical point in each quadrant; this is an inconsequential subtlety that we will ignore). Our restriction to the region $\eta^{\frac{1}{d}} \leq 2\xi$ is natural because stationary phase considerations show that m_d is smooth away from the axis $\eta = 0$. For $\xi > 0$ and $\eta > 0$ we define

$$m_{d,1}(\xi, \eta) = \left(\eta^{\frac{1}{d}} \xi^{-1}\right)^{\frac{d'}{2}} e^{i\left(\eta^{\frac{1}{d}} \xi^{-1}\right)^{-d'}} \psi(\eta^{\frac{1}{d}} \xi^{-1}), \tag{2.2}$$

where $\frac{1}{d} + \frac{1}{d'} = 1$ and ψ is a smooth cutoff function supported in $[-2, 2]$ and equal to one on a slightly smaller interval. We extend $m_{d,1}$ continuously by setting it equal to zero on the remainder of $\mathbb{R}^2$.

From a standard computation using the stationary phase principle (see [14, Ch. VIII.1, Prop. 3]) we can see that, up to a negligible constant rescaling, this term constitutes the main contribution to the oscillatory integral in (2.1). The remainder term from stationary phase is smoother in the variables ξ and η and is therefore simpler to handle. We will ignore it for the purpose of this discussion.

From the definition we see directly that $m_{d,1}$ (and therefore also m_d) is only Hölder continuous of class $C^{\frac{1}{2(d-1)}}$ along the axis $\eta = 0$ while it is infinitely differentiable away from that axis.

Let us from here on denote

$$\zeta = \zeta(\xi, \eta) = \eta^{\frac{1}{d}} \xi^{-1}$$

for $\xi > 0, \eta > 0$. Since we do not know how to handle $m_{d,1}$ we introduce a family of modified, less oscillatory multipliers which is defined on the quadrant $\{\xi > 0, \eta > 0\}$ by

$$m_{d,\delta}(\xi, \eta) = \zeta^{\frac{d'}{2}} e^{i\zeta^{-d'\delta}} \psi(\zeta), \tag{2.3}$$

where $0 \le \delta \le 1$ is a parameter (and we again extend $m_{d,\delta}$ to the rest of $\mathbb{R}^2$ by zero). The multiplier $m_{d,\delta}$ still fails to be in $\mathscr{M}_\alpha^\nu$ for every positive integer ν and $\delta > 0$. However, we can nevertheless apply Theorem 1.1 to bound $\mathscr{C}_{m_{d,\delta}}$ for small enough δ. For this purpose we assume that ψ takes the form

$$\psi = \sum_{j \le 0} \varphi_j,$$

where $\varphi_j(x) = \varphi(2^{-j}x)$ and φ is a smooth bump function supported in $[1/2, 2]$ and satisfying $\sum_{j\in\mathbb{Z}} \varphi_j(x) = 1$ for all $x \neq 0$. Then we have the following consequence of Theorem 1.1.

Corollary 2.1 *There exists $\delta_0 > 0$ such that for all $0 \le \delta < \delta_0$ we have*

$$\|\mathscr{C}_{m_{d,\delta}} f\|_{2,\infty} \lesssim \|f\|_2.$$

Proof Let us write

$$\begin{aligned} m_{d,\delta,j}(\xi, \eta) &= \zeta^{\frac{d'}{2}} e^{i\zeta^{-d'\delta}} \varphi_j(\zeta) \quad (\text{for } \xi > 0, \eta > 0), \\ T_j f(x, y) &= \int_{\mathbb{R}^2} \widehat{f}(\xi, \eta) m_{d,\delta,j}(\xi, \eta) e^{ix\xi + iy\eta} d(\xi, \eta). \end{aligned}$$

By the change of variables $\eta \mapsto 2^{jd}\eta$, we see that

$$\begin{aligned} T_j f(x, y) &= 2^{j\frac{d'}{2}} 2^{jd} \int_{\mathbb{R}^2} \widehat{f}(\xi, 2^{jd}\eta) \widetilde{m}_{d,\delta,j}(\xi, \eta) e^{ix\xi + i2^{jd}y\eta} d(\xi, \eta) \\ &= 2^{j\frac{d'}{2}} \mathrm{D}_{2^{jd}} \widetilde{T}_j \mathrm{D}_{2^{-jd}} f(x, y), \end{aligned}$$

where $\mathrm{D}_\lambda f(x, y) = f(x, \lambda y)$ and

$$\widetilde{m}_{d,\delta,j}(\xi, \eta) = \zeta^{\frac{d'}{2}} e^{i2^{-jd'\delta}\zeta^{-d'\delta}} \varphi(\zeta) \quad (\text{for } \xi > 0, \eta > 0),$$
$$\widetilde{T}_j f(x, y) = \int_{\mathbb{R}^2} \widehat{f}(\xi, \eta)\widetilde{m}_{d,\delta,j}(\xi, \eta) e^{ix\xi + iy\eta} d(\xi, \eta).$$

We have

$$\|\widetilde{m}_{d,\delta,j}\|_{\mathscr{M}_\alpha^\nu} \lesssim 2^{-jd'\delta\nu}$$

for every integer $\nu \geq 0$ (where the implied constant depends on ν, d, δ and φ). Using Theorem 1.1 we therefore obtain

$$\|\mathscr{C}_{m_{d,\delta}}\|_{L^2 \to L^{2,\infty}} \lesssim \sum_{j \leq 0} 2^{j\frac{d'}{2}} \|\mathscr{C}_{\widetilde{m}_{d,\delta,j}}\|_{L^2 \to L^{2,\infty}} \lesssim \sum_{j \leq 0} 2^{jd'(\frac{1}{2} - \delta\nu_0)}.$$

Thus, setting $\delta_0 = \frac{1}{2}\nu_0^{-1}$ yields the claim. □

An improvement for the bound on ν_0 in Theorem 1.1 will give an improvement for δ_0. However, even if we could show Theorem 1.1 for all $\nu_0 > |\alpha|/2$, we would still have $\delta_0 < \frac{1}{|\alpha|} < 1$. Therefore additional insight is likely to be required to bound the operator $\mathscr{C}_{m_{d,1}}$ (and thus $\mathscr{C}_{m_d}$).

For the case of the parabola, $d = 2$, there are some additional obstructions in bounding $\mathscr{C}_{m_2}$. Let us write the parabolic Carleson operator as

$$\mathscr{C}^{\text{par}} f(x, y) = \sup_{N \in \mathbb{R}^2} \left| p.v. \int_{\mathbb{R}} f(x - t, y - t^2) e^{iN_1 t + iN_2 t^2} \frac{dt}{t} \right|. \tag{2.4}$$

Apart from the linear modulation symmetries given by

$$\mathscr{C}^{\text{par}} f = \mathscr{C}^{\text{par}} \mathrm{M}_N f$$

for $N \in \mathbb{R}^2$, there are additional modulation symmetries. For a polynomial in two variables, $P = P(x, y)$, we write the corresponding polynomial modulation as

$$\mathrm{M}_P f(x, y) = e^{iP(x,y)} f(x, y).$$

Then we have that

$$\begin{aligned} \mathscr{C}^{\text{par}} \mathrm{M}_{Nx^2} f &= \mathscr{C}^{\text{par}} f, \\ \mathscr{C}^{\text{par}} \mathrm{M}_{Nx(y+x^2)} f &= \mathscr{C}^{\text{par}} f, \\ \mathscr{C}^{\text{par}} \mathrm{M}_{N(y+x^2)^2} f &= \mathscr{C}^{\text{par}} f \end{aligned} \tag{2.5}$$

hold for all $N \in \mathbb{R}$. The quadratic modulation symmetry (2.5) suggests a connection to Lie's quadratic Carleson operator [10]. Indeed, even a certain partial L^2 bound for $\mathscr{C}^{\text{par}}$

would immediately imply an L^2 bound for the quadratic Carleson operator (see [6]). These are all the polynomial modulation symmetries of the operator $\mathscr{C}^{\mathrm{par}}$ (up to linear combination). To see that, introduce the change of variables $\tau(x, y) = (x, y + x^2)$ and observe that

$$\mathscr{C}^{\mathrm{par}} f = \mathscr{C}^{\mathrm{sh}}(f \circ \tau^{-1}) \circ \tau,$$

where

$$\mathscr{C}^{\mathrm{sh}} f(x, y) = \sup_{N \in \mathbb{R}^2} \left| p.v. \int_{\mathbb{R}} f(x - t, y - 2xt) e^{iN_1 t + iN_2 t^2} \frac{dt}{t} \right|.$$

It is easy to check that we have

$$\mathscr{C}^{\mathrm{sh}} \mathrm{M}_P f = \mathscr{C}^{\mathrm{sh}} f$$

for a polynomial P if and only if P is of degree at most two. This shows that the list of polynomial modulation symmetries that we gave for $\mathscr{C}^{\mathrm{par}}$ is complete. Also since τ is measure preserving, L^p bounds for $\mathscr{C}^{\mathrm{par}}$ and $\mathscr{C}^{\mathrm{sh}}$ are equivalent.

3 Reduction to a Model Operator

Before we begin we need to introduce some more notation and definitions. Denote

$$\mathrm{dist}_\alpha(A, B) = \inf_{x \in A, y \in B} \rho_\alpha(x - y)$$

and $\mathrm{dist}_\alpha(A, x) = \mathrm{dist}_\alpha(A, \{x\})$. For $a, b \in \mathbb{R}^n$ we write

$$[a, b] = \prod_{i=1}^{n} [a_i, b_i]$$

and similarly (a, b), $[a, b)$. We will refer to all such sets as *rectangles*. For a rectangle $I \subset \mathbb{R}^n$ we define $c(I)$ to be its center. By an *anisotropic cube* we mean a rectangle $[a, b]$ such that $b_i - a_i = \lambda^{\alpha_i}$ holds for all $i = 1, \ldots, n$ and some $\lambda > 0$. We define the collection of *anisotropic dyadic cubes* by

$$\mathcal{D}^\alpha = \left\{ [\delta_{2^k}(\ell), \delta_{2^k}(\ell + 1)) \,:\, \ell \in \mathbb{Z}^n,\ k \in \mathbb{Z} \right\}.$$

Every two anisotropic dyadic cubes have the property that they are either disjoint or contained in one another. Moreover, for every $I \in \mathcal{D}^\alpha$ there exists a unique dyadic cube $I^+ \in \mathcal{D}^\alpha$ such that $|I^+| = 2^{|\alpha|}|I|$ and $I \subset I^+$. We call I^+ the *parent* of I and say that I is a *child* of I^+.

Definition 3.1 A *tile* P is a rectangle in $\mathbb{R}^n \times \mathbb{R}^n$ of the form

$$P = I_P \times \omega_P,$$

where $I_P, \omega_P \in \mathcal{D}^\alpha$ and $|I_P| \cdot |\omega_P| = 1$.

The set of tiles is denoted by $\overline{\mathcal{P}}$. Given a tile P we denote its *scale* by $k(P) = |I_P|^{1/|\alpha|}$. For $r \in \{0, 1\}^n$ and a tile P with $\omega_P = [\delta_{2^{-k(P)}}(\ell), \delta_{2^{-k(P)}}(\ell + 1)]$ we define the *semi-tile* $P(r)$ by

$$P(r) = I_P \times \omega_{P(r)}, \text{ where } \omega_{P(r)} = \Big[\delta_{2^{-k(P)}}\Big(\ell + \frac{1}{2}r\Big), \delta_{2^{-k(P)}}\Big(\ell + \frac{1}{2}(r+1)\Big)\Big].$$

The model operator is built up using a large family of wave packets adapted to tiles. It is convenient to generate this family by letting the symmetry group of our operator act on a single bump function. For this purpose, let ϕ be a Schwartz function on $\mathbb{R}^n$ such that $0 \le \widehat{\phi} \le 1$ with $\widehat{\phi}$ being supported in $[-\frac{b_0}{2}, \frac{b_0}{2}]^n$ and equal to 1 on $[-\frac{b_1}{2}, \frac{b_1}{2}]^n$, where $0 < b_1 < b_0 \ll 1$ are some fixed, small numbers whose ratio is not too large (it becomes clear what precisely is required in Sect. 7). For example, we may set $b_0 = \frac{1}{10}$, $b_1 = \frac{9}{100}$. We denote translation, modulation and dilation of a function f by

$$\begin{aligned}
\mathrm{T}_y f(x) &= f(x - y), \qquad (y \in \mathbb{R}^n)\\
\mathrm{M}_\xi f(x) &= e^{ix\xi} f(x), \qquad (\xi \in \mathbb{R}^n)\\
\mathrm{D}^p_\lambda f(x) &= \lambda^{-\frac{|\alpha|}{p}} f\left(\delta_{\lambda^{-1}}(x)\right), \qquad (\lambda, p > 0),
\end{aligned}$$

where $|\alpha| = \sum_{i=1}^n \alpha_i$.

Given a tile P and $N \in \mathbb{R}^n$ we define the wave packets ϕ_P, ψ_P^N on $\mathbb{R}^n$ by

$$\phi_P(x) = \mathrm{M}_{c(\omega_{P(0)})}\mathrm{T}_{c(I_P)}\mathrm{D}^2_{2^{k(P)}}\phi(x) \tag{3.1}$$

$$\widehat{\psi_P^N}(\xi) = \mathrm{T}_N m(\xi) \cdot \widehat{\phi_P}(\xi) \tag{3.2}$$

We think of ϕ_P as being essentially time-frequency supported in the semi-tile $P(0)$. More precisely, we have that $\widehat{\phi_P}$ is compactly supported in (a small cube centrally contained in) $\omega_{P(0)}$ and $|\phi_P|$ decays rapidly outside of I_P.

For $N \in \mathbb{R}^n$ and $r \neq 0$ we introduce the dyadic model sum operator

$$A_N^{r,m} f(x) = \sum_{P \in \overline{\mathcal{P}}} \langle f, \phi_P\rangle \psi_P^N(x) \mathbf{1}_{\omega_{P(r)}}(N). \tag{3.3}$$

This reduces to the model sum of Lacey and Thiele [9] in the case $n = \alpha = 1$ and to that of Pramanik and Terwilleger [12] in the isotropic case $\alpha = (1, \ldots, 1)$.

Theorem 3.2 *For every large enough integer ν_0 there exists $C > 0$ depending only on ν_0, α and the choice of ϕ such that for all multipliers $m \in \mathscr{M}_\alpha^{\nu_0}$ we have*

$$\| \sup_{N \in \mathbb{R}^n} |A_N^{r,m} f| \|_{2,\infty} \le C \|m\|_{\mathscr{M}_\alpha^{\nu_0}} \|f\|_2. \tag{3.4}$$

The proof of the theorem is contained in Sects. 4–7. We conclude this section by showing that Theorem 3.2 implies Theorem 1.1. For this purpose we employ the averaging procedure of Lacey and Thiele [9] combined with an anisotropic cone decomposition of the multiplier m. By an *anisotropic cone* we mean a subset $\Theta \subsetneq \mathbb{R}^n$ of the form

$$\Theta = \{\delta_t(\xi) \, : \, t > 0, \, \xi \in Q\}$$

for some cube $Q \subset \mathbb{R}^n$. Let us denote $\mathcal{B}_s = \{x \, : \, \rho_\alpha(x) \le s\}$. Let

$$\mathbf{A}^{r,m} f(x) = \lim_{R\to\infty} \frac{1}{R^{2|\alpha|}} \int_{\mathcal{B}_R} \int_{\mathcal{B}_R} \int_0^1 \mathrm{M}_{-\eta}\mathrm{T}_{-y}\mathrm{D}^2_{2^{-s}} A^{r,m}_{2^{-s}\eta} \mathrm{D}^2_{2^s}\mathrm{T}_y\mathrm{M}_\eta f(x) ds dy d\eta. \tag{3.5}$$

Lemma 3.3 *For every $r \in \{0,1\}^n$ and every test function f, the function $\mathbf{A}^{r,m} f(x)$ is well-defined and also a test function. We have*

$$\widehat{\mathbf{A}^{r,m} f}(\xi) = \theta_r(\xi) m(\xi) \widehat{f}(\xi)$$

for some smooth function θ_r that is independent of m. Moreover, there exists a constant $\varepsilon_0 > 0$ and an anisotropic cone Θ_r such that

$$\theta_r(\xi) > \varepsilon_0 \quad \textit{for all } \xi \in \Theta_r.$$

and

$$(-\infty, \varepsilon_0]^n \subset \bigcup_{r\in\{0,1\}^n\setminus\{0\}} \Theta_r. \tag{3.6}$$

Proof By expanding definitions we see that

$$\left(\mathrm{M}_{-\eta}\mathrm{T}_{-y}\mathrm{D}^2_{2^{-s}} A^{r,m}_{2^{-s}\eta} \mathrm{D}^2_{2^s}\mathrm{T}_y\mathrm{M}_\eta f\right)^\wedge(\xi)$$

is equal to a universal constant times

$$\begin{aligned} m(\xi) \sum_{P\in\overline{\mathcal{P}}} &\left\langle \widehat{f}, \mathrm{T}_{-\eta+\delta_{2^s}(c(\omega_{P(0)}))}\mathrm{M}_{y-\delta_{2^{-s}}(c(I_P))}\mathrm{D}^2_{2^{s-k(P)}}\widehat{\phi}\right\rangle \\ &\times \mathrm{T}_{-\eta+\delta_{2^s}(c(\omega_{P(0)}))}\mathrm{M}_{y-\delta_{2^{-s}}(c(I_P))}\mathrm{D}^2_{2^{s-k(P)}}\widehat{\phi}(\xi)\mathbf{1}_{\omega_{P(r)}}(\delta_{2^{-s}}(\eta)), \end{aligned}$$

where we have used that $m(\delta_{2^{-s}}(\xi)) = m(\xi)$. The previous display equals

$$\begin{aligned} m(\xi) \sum_{k\in\mathbb{Z}} \sum_{\ell\in\mathbb{Z}^n} \sum_{u\in\mathbb{Z}^n} & 2^{-|\alpha|(s-k)} \int_{\mathbb{R}^n} \widehat{f}(\zeta) e^{i(y-\delta_{2^{-s+k}}(u+\frac{1}{2}))(\xi-\zeta)} \\ &\times \overline{\widehat{\phi}}\Big(\delta_{2^{-s+k}}(\zeta+\eta) - \Big(\ell + \frac{1}{4}\Big)\Big) d\zeta \\ &\times \widehat{\phi}\Big(\delta_{2^{-s+k}}(\xi+\eta) - \Big(\ell + \frac{1}{4}\Big)\Big) \mathbf{1}_{\omega_{P(r)}}(\delta_{2^{-s}}(\eta)). \end{aligned}$$

Applying the Poisson summation formula to the summation in u and using the Fourier support information of the function ϕ we see that the previous display equals (up to a universal constant)

$$m(\xi)\widehat{f}(\xi)\sum_{k\in\mathbb{Z}}\sum_{\ell\in\mathbb{Z}^n}|\widehat{\phi}|^2\Big(\delta_{2^{-s+k}}(\xi+\eta)-\Big(\ell+\frac{1}{4}\Big)\Big)\mathbf{1}_{\omega_{P(r)}}(\delta_{2^{-s}}(\eta)).$$

Observe that the expression no longer depends on the variable y. It remains to compute the function $\theta_r(\xi)=c\cdot\lim_{R\to\infty}I_R(\xi)$, where c is a universal constant and

$$I_R(\xi)=\frac{1}{R^{|\alpha|}}\int_{\mathcal{B}_R}\int_0^1\sum_{k\in\mathbb{Z}}\sum_{\ell\in\mathbb{Z}^n}|\widehat{\phi}|^2\Big(\delta_{2^{-s+k}}(\xi+\eta)-\Big(\ell+\frac{1}{4}\Big)\Big)\mathbf{1}_{\omega_{P(r)}}(\delta_{2^{-s}}(\eta))dsd\eta.$$

Note the formula

$$\int_0^1\sum_{k\in\mathbb{Z}}F(2^{k-s})ds=\frac{1}{\log 2}\int_0^\infty F(t)\frac{dt}{t},$$

which follows from a change of variables $2^{k-s}\to t$. Using this we have

$$I_R(\xi)=\frac{c}{R^{|\alpha|}}\int_{\mathcal{B}_R}\int_0^\infty\sum_{\ell\in\mathbb{Z}^n}|\widehat{\phi}|^2\Big(\delta_t(\xi+\eta)-\Big(\ell+\frac{1}{4}\Big)\Big)\mathbf{1}_{Q_r}(\delta_t(\eta)-\ell)\frac{dt}{t}d\eta,$$

where $Q_r=\left[\frac{1}{2}r,\frac{1}{2}(r+1)\right)=\prod_{i=1}^n\left[\frac{1}{2}r_i,\frac{1}{2}(r_i+1)\right)$ and $c=(\log 2)^{-1}$ (c may change from line to line in this proof). To simplify our expression further we perform the change of variables

$$\delta_t(\xi+\eta)-\ell\to\zeta$$

in the integration in η. This yields

$$I_R(\xi)=c\int_{\mathbb{R}^n}\int_0^\infty\chi(\zeta)\mathbf{1}_{Q_r}(\zeta-\delta_t(\xi))\left(\sum_{\ell\in\mathbb{Z}^n}\frac{\mathbf{1}_{\rho_\alpha(\zeta+\ell-\delta_t(\xi))\le tR}}{(tR)^{|\alpha|}}\right)\frac{dt}{t}d\zeta \tag{3.7}$$

where we have set

$$\chi(\zeta)=|\widehat{\phi}|^2\Big(\zeta-\frac{1}{4}\Big).$$

Observe that the integrand in (3.7) is supported in a compact subset of $\mathbb{R}^n\times(0,\infty)$ (which depends on ξ). By counting the ℓ for which the summand is non-zero we see that for every fixed $\zeta,\xi\in\mathbb{R}^n$ and $t>0$ the sum

$$\sum_{\ell\in\mathbb{Z}^n}\frac{\mathbf{1}_{\rho_\alpha(\zeta+\ell-\delta_t(\xi))\le tR}}{(tR)^{|\alpha|}}$$

converges to a universal constant as $R \to \infty$. Thus, from Lebesgue's dominated convergence theorem we conclude that

$$\theta_r(\xi) = c \int_{\mathbb{R}^n} \int_0^\infty \chi(\zeta) \mathbf{1}_{Q_r}(\zeta - \delta_t(\xi)) \frac{dt}{t} d\zeta. \tag{3.8}$$

Evidently we have $\theta_r(\delta_t(\xi)) = \theta_r(\xi)$ for every $t > 0$ and $\xi \in \mathbb{R}^n$. From our choice of ϕ we get that χ is supported on $Q^{(0)}$ and equal to one on $Q^{(1)}$, where

$$Q^{(j)} = \Big[\frac{1}{4} - \frac{b_j}{2}, \frac{1}{4} + \frac{b_j}{2}\Big]$$

for $j = 0, 1$. Let us set

$$\Theta_r = \{\delta_t(\xi) \,:\, \xi \in Q^{(1)} - Q_r\}.$$

Then we can read off (3.8) that θ_r is greater than some positive constant on $\Theta_r^{(1)}$. Note that

$$Q^{(j)} - Q_r = \Big[-\frac{1}{2}r - \Big(\frac{1}{4} + \frac{b_j}{2}\Big), -\frac{1}{2}r + \Big(\frac{1}{4} + \frac{b_j}{2}\Big)\Big].$$

Looking at the anisotropic cone generated by each of the regions $Q^{(1)} - Q_r$ we see that (3.6) is satisfied for sufficiently small ε_0. □

In the isotropic case $\alpha = (1, \ldots, 1)$ we can assume without loss of generality that the multiplier m is supported in some arbitrarily chosen cone. Due to the lack of rotation invariance this assumption becomes invalid in the anisotropic setting.

Proof of Theorem 1.1 Let $m \in \mathscr{M}_\alpha^{\nu_0}$. Without loss of generality we may assume that m is supported in the "quadrant" $(-\infty, 0]^n$. By (3.6) we can choose smooth functions $(\varrho_r)_r$ such that ϱ_r is supported in Θ_r and

$$\sum_{r \in \{0,1\}^n \setminus \{0\}} \varrho_r(\xi) = 1$$

for $\xi \in (-\infty, 0]^n$. By the triangle inequality and Lemma 3.3, we have

$$\|\mathscr{C}_m f\|_{2,\infty} \le \sum_{r \in \{0,1\}^n \setminus \{0\}} \| \sup_{N \in \mathbb{R}^n} |\mathbf{A}^{r, \theta_r^{-1} \varrho_r m} \mathrm{M}_N f| \|_{2,\infty}.$$

Here θ_r^{-1} refers to the function $\xi \mapsto (\theta_r(\xi))^{-1}$, which is bounded on Θ_r. By (3.5) and Minkowski's integral inequality, the previous is no greater than

$$\sum_{r \in \{0,1\}^n \setminus \{0\}} \limsup_{R \to \infty} \frac{1}{R^{2|\alpha|}} \int_{\mathcal{B}_R} \int_{\mathcal{B}_R} \int_0^1 \| \sup_{N \in \mathbb{R}^n} |A_N^{r, \theta_r^{-1} \varrho_r m} \mathrm{D}_{2^s}^2 \mathrm{T}_y \mathrm{M}_\eta f| \|_{2,\infty} ds\, dy\, d\eta,$$

which by Theorem 3.2 is bounded by

$$C \sum_{r\in\{0,1\}^n\setminus\{0\}} \|\theta_r^{-1}\varrho_r m\|_{\mathscr{M}_\alpha^{\nu_0}} \|f\|_2 \lesssim \|m\|_{\mathscr{M}_\alpha^{\nu_0}} \|f\|_2.$$

□

4 Boundedness of the Model Operator

In this section we describe the proof of Theorem 3.2. We follow [9]. First, we perform some preliminary reductions. Given a measurable function $N : \mathbb{R}^n \to \mathbb{R}^n$ we define

$$Tf(x) = A^r_{N(x)} f(x).$$

Note that the estimate (3.4) is equivalent to showing

$$\|Tf\|_{2,\infty} \le C\|m\|_{\mathscr{M}_\alpha^{\nu_0}} \|f\|_2$$

with C not depending on the choice of the measurable function N. By duality, it is equivalent to show

$$|\langle Tf, \mathbf{1}_E\rangle| \lesssim \|m\|_{\mathscr{M}_\alpha^{\nu_0}} |E|^{\frac{1}{2}} \|f\|_2,$$

where E is an arbitrary measurable set. By scaling, we may assume without loss of generality that $\|f\|_2 = 1$ and $|E| \le 1$. Thus, by the triangle inequality, it suffices to show that

$$\sum_{P\in\mathcal{P}} |\langle f, \phi_P\rangle \langle \mathbf{1}_{E\cap N^{-1}(\omega_{P(r)})}, \psi_P^{N(\cdot)}\rangle| \lesssim \|m\|_{\mathscr{M}_\alpha^{\nu_0}}, \tag{4.1}$$

for all finite sets of tiles $\mathcal{P} \subset \overline{\mathcal{P}}$, with the implied constant being independent of $f, E, N, \mathcal{P}$. Throughout this and the following sections we fix $r \in \{0, 1\}^n \setminus \{0\}$. Before we continue we need to introduce certain collections of tiles called trees. There is a partial order on tiles defined by

$$P \le P' \quad \text{if} \quad I_P \subset I_{P'} \quad \text{and} \quad c(\omega_{P'}) \in \omega_P.$$

Observe that two tiles are comparable with respect to $\le$ if and only if they have a non-empty intersection.

Definition 4.1 A finite collection $\mathbf{T} \subset \overline{\mathcal{P}}$ of tiles is called a *tree* if there exists $P \in \mathbf{T}$ such that $P' \le P$ for every $P' \in \mathbf{T}$. In that case, P is uniquely determined and referred to as the *top* of the tree $\mathbf{T}$. We denote the top of a tree $\mathbf{T}$ by $P_\mathbf{T} = I_\mathbf{T} \times \omega_\mathbf{T}$ and write $k_\mathbf{T} = |I_\mathbf{T}|^{1/|\alpha|}$.

A tree $\mathbf{T}$ is called a 1-*tree* if $c(\omega_\mathbf{T}) \notin \omega_{P(r)}$ for all $P \in \mathbf{T}$ and it is called a 2-*tree* if $c(\omega_\mathbf{T}) \in \omega_{P(r)}$ for all $P \in \mathbf{T}$. These names are due to historical reasons (see [9]).

The notion of a tree was first introduced in this context by Fefferman [4]. For a tile $P \in \overline{\mathcal{P}}$ we write

$$E_P = E \cap N^{-1}(\omega_P) \text{ and } E_{P(r)} = E \cap N^{-1}(\omega_{P(r)}).$$

The *mass* of a single tile P is defined as

$$\mathcal{M}(P) = \sup_{P' \geq P} \int_{E_{P'}} w_{P'}^{\nu_1}(x)dx, \tag{4.2}$$

where ν_1 is a fixed large positive number depending only on $|\alpha|$ that is to be determined later and

$$w_P^{\nu}(x) = \mathrm{T}_{c(I_P)}\mathrm{D}^1_{2^{k(P)}} w^{\nu}(x),$$

where the weight w^{ν} takes the form

$$w^{\nu}(x) = (1 + \rho_\alpha(x))^{-\nu}.$$

For convenience we also write $w_P = w_P^{\nu_1}$. For a collection of tiles $\mathcal{P} \subset \overline{\mathcal{P}}$ we define their mass as

$$\mathcal{M}(\mathcal{P}) = \sup_{P \in \mathcal{P}} \mathcal{M}(P). \tag{4.3}$$

The *energy* of a collection of tiles $\mathcal{P}$ is defined as

$$\mathcal{E}(\mathcal{P}) = \sup_{\mathbf{T} \subset \mathcal{P} \text{ 2-tree}} \left(\frac{1}{|I_T|} \sum_{P \in \mathbf{T}} |\langle f, \phi_P \rangle|^2 \right)^{1/2}. \tag{4.4}$$

These quantities and the following lemmas originate in [9].

Lemma 4.2 (Mass lemma) *If $\nu_1 > |\alpha|+1$, then there exists $C > 0$ depending only on α such that for every finite set of tiles $\mathcal{P} \subset \overline{\mathcal{P}}$ there is a decomposition $\mathcal{P} = \mathcal{P}_{\text{light}} \cup \mathcal{P}_{\text{heavy}}$ such that*

$$\mathcal{M}(\mathcal{P}_{\text{light}}) \leq 2^{-2} \mathcal{M}(\mathcal{P}) \tag{4.5}$$

and $\mathcal{P}_{\text{heavy}}$ is a union of a set $\mathcal{T}$ of trees such that

$$\sum_{\mathbf{T} \in \mathcal{T}} |I_{\mathbf{T}}| \leq \frac{C}{\mathcal{M}(\mathcal{P})}. \tag{4.6}$$

Lemma 4.3 (Energy lemma) *There exists $C > 0$ depending only on α such that for every finite set of tiles $\mathcal{P} \subset \overline{\mathcal{P}}$ there is a decomposition $\mathcal{P} = \mathcal{P}_{\text{low}} \cup \mathcal{P}_{\text{high}}$ such that*

$$\mathcal{E}(\mathcal{P}_{\text{low}}) \le 2^{-1}\,\mathcal{E}(\mathcal{P}) \tag{4.7}$$

and $\mathcal{P}_{\text{high}}$ is a union of a set $\mathcal{T}$ of trees such that

$$\sum_{\mathbf{T}\in\mathcal{T}} |I_{\mathbf{T}}| \le \frac{C}{\mathcal{E}(\mathcal{P})^2}. \tag{4.8}$$

Lemma 4.4 (Tree estimate) *There exists $C > 0$ depending only on α such that if $m \in \mathscr{M}_\alpha^{\nu_0}$, then the following inequality holds for every tree $\mathbf{T}$:*

$$\sum_{P\in\mathbf{T}} |\langle f, \phi_P\rangle\langle \psi_P^{N(\cdot)}, \mathbf{1}_{E_{P(r)}}\rangle| \le C\|m\|_{\mathscr{M}_\alpha^{\nu_0}} |I_{\mathbf{T}}|\mathcal{E}(\mathbf{T})\mathcal{M}(\mathbf{T}) \tag{4.9}$$

The proofs of these lemmas are contained in Sects. 5, 6, and 7, respectively. By iterated application of these lemmas we obtain a proof of (4.1). This argument is literally the same as in [9], but we include it here for convenience of the reader. Let $\mathcal{P}$ be a finite collection of tiles. We will decompose $\mathcal{P}$ into disjoint sets $(\mathcal{P}_\ell)_{\ell\in\mathcal{N}}$ (where $\mathcal{N}$ is some finite set of integers) such that each $\mathcal{P}_\ell$ satisfies

$$\mathcal{M}(\mathcal{P}_\ell) \le 2^{2\ell} \quad \text{and} \quad \mathcal{E}(\mathcal{P}_\ell) \le 2^\ell \tag{4.10}$$

and is equal to the union of a set of trees $\mathcal{T}_\ell$ such that

$$\sum_{\mathbf{T}\in\mathcal{T}_\ell} |I_{\mathbf{T}}| \le C2^{-2\ell}. \tag{4.11}$$

This is achieved by the following procedure:

(1) Initialize $\mathcal{P}^{\text{stock}} := \mathcal{P}$ and choose an initial ℓ that is large enough such that

$$\mathcal{M}(\mathcal{P}^{\text{stock}}) \le 2^{2\ell} \quad \text{and} \quad \mathcal{E}(\mathcal{P}^{\text{stock}}) \le 2^\ell. \tag{4.12}$$

(2) If $\mathcal{M}(\mathcal{P}^{\text{stock}}) > 2^{2(\ell-1)}$, then apply Lemma 4.2 to decompose $\mathcal{P}^{\text{stock}}$ into $\mathcal{P}_{\text{light}}$ and $\mathcal{P}_{\text{heavy}}$. We add[1] $\mathcal{P}_{\text{heavy}}$ to $\mathcal{P}_\ell$ and update $\mathcal{P}^{\text{stock}} := \mathcal{P}_{\text{light}}$ (thus, we now have $\mathcal{M}(\mathcal{P}^{\text{stock}}) \le 2^{2(\ell-1)}$).
(3) If $\mathcal{E}(\mathcal{P}^{\text{stock}}) > 2^{\ell-1}$, then apply Lemma 4.3 to decompose $\mathcal{P}^{\text{stock}}$ into $\mathcal{P}_{\text{low}}$ and $\mathcal{P}_{\text{high}}$. We add $\mathcal{P}_{\text{high}}$ to $\mathcal{P}_\ell$ and update $\mathcal{P}^{\text{stock}} := \mathcal{P}_{\text{low}}$ (thus, we now have $\mathcal{E}(\mathcal{P}^{\text{stock}}) \le 2^{\ell-1}$).
(4) If $\mathcal{P}^{\text{stock}}$ is not empty, then replace ℓ by $\ell - 1$ and go to Step (2).

[1] We can think of all the $\mathcal{P}_\ell$ as being initialized by the empty set.

Then we can finish the proof of (4.1) by using (4.10), (4.11), (4.9) and keeping in mind that we always have $\mathcal{M}(\mathcal{P}) \leq \|w^{\nu_1}\|_1$:

$$\sum_{P\in\mathcal{P}} |\langle f, \phi_P\rangle\langle \mathbf{1}_{E\cap N^{-1}(\omega_{P(r)})}, \psi_P^{N(\cdot)}\rangle| = \sum_{\ell\in\mathcal{N}}\sum_{\mathbf{T}\in\mathcal{T}_\ell}\sum_{P\in\mathbf{T}} |\langle f, \phi_P\rangle\langle \mathbf{1}_{E\cap N^{-1}(\omega_{P(r)})}, \psi_P^{N(\cdot)}\rangle|$$
$$\lesssim \|m\|_{\mathscr{M}_\alpha^{\nu_0}} \sum_{\ell\in\mathcal{N}} 2^\ell \min(1, 2^{2\ell}) \sum_{\mathbf{T}\in\mathcal{T}_\ell} |I_{\mathbf{T}}| \lesssim \|m\|_{\mathscr{M}_\alpha^{\nu_0}} \sum_{\ell\in\mathbb{Z}} 2^{-\ell}\min(1, 2^{2\ell}) \lesssim \|m\|_{\mathscr{M}_\alpha^{\nu_0}}.$$

To conclude this section we collect several standard auxiliary estimates for m, K, ϕ_P, ψ_P^N which are used during the remainder of the proof. First, from the definition of $\mathscr{M}_\alpha^\nu$ we have the symbol estimate

$$|\partial_i^\nu m(\xi)| \leq \|m\|_{\mathscr{M}_\alpha^\nu} \rho_\alpha(\xi)^{-\nu\alpha_i} \tag{4.13}$$

for every integer $0 \leq \nu \leq \nu_0$ and $i = 1, \ldots, n$. If we let K denote the corresponding kernel (that is, $\widehat{K} = m$), we have

$$|K(x)| \lesssim \|m\|_{\mathscr{M}_\alpha^{\lfloor\frac{|\alpha|}{2}\rfloor+1}} \rho_\alpha(x)^{-|\alpha|} \tag{4.14}$$

for $x \neq 0$. This is a consequence of the anisotropic Hörmander–Mikhlin theorem (see [3]). For every integer $\nu \geq 0$ and $N \notin \omega_{P(0)}$ we have

$$|\psi_P^N(x)| \lesssim \|m\|_{\mathscr{M}_\alpha^\nu} |I_P|^{1/2} w_P^\nu(x), \tag{4.15}$$

where the implicit constant depends only on ν, α and the choice of ϕ. We defer the proof of this estimate to Sect. 8.

The next estimates concern the interaction of two wave packets associated with distinct tiles. Let $P, P' \in \overline{\mathcal{P}}$ be tiles. The idea is that if P, P' are disjoint (or equivalently, incomparable with respect to $\leq$) then their associated wave packets are almost orthogonal, i.e. $\langle \phi_P, \phi_{P'}\rangle$ is negligibly small. Indeed, if ω_P and $\omega_{P'}$ are disjoint, then we even have $\langle \phi_P, \phi_{P'}\rangle = 0$. However, as an artifact of the Heisenberg uncertainty principle, in the case that only I_P and $I_{P'}$ are disjoint, we need to deal with tails. The precise estimate we need is as follows. Assume that $|I_P| \geq |I_{P'}|$. Then for every integer $\nu \geq 0$ we have that

$$|\langle \phi_P, \phi_{P'}\rangle| \lesssim |I_P|^{-\frac{1}{2}} |I_{P'}|^{\frac{1}{2}} (1 + 2^{-k(P)}\rho_\alpha(c(I_P) - c(I_{P'})))^{-\nu}, \tag{4.16}$$

where the implicit constant depends only on ν and ϕ. See [15, Lemma 2.1] for the version of this estimate for one-dimensional wave packets. Similarly, we have

$$|\langle \psi_P^N, \psi_{P'}^N\rangle| \lesssim \|m\|_{\mathscr{M}_\alpha^\nu}^2 |I_P|^{-\frac{1}{2}} |I_{P'}|^{\frac{1}{2}} (1 + 2^{-k(P)}\rho_\alpha(c(I_P) - c(I_{P'})))^{-\nu}, \tag{4.17}$$

for every integer $\nu \geq 0$ provided that $N \notin \omega_{P(0)} \cup \omega_{P'(0)}$. We prove (4.16) and (4.17) in Sect. 8.

5 Proof of the Mass Lemma

In this section we prove Lemma 4.2. The proof is in essence the same as in [9, Prop. 3.1]. Let $\mathcal{P}$ be a finite set of tiles and set $\mu = \mathcal{M}(\mathcal{P})$. We define the set of heavy tiles by

$$\mathcal{P}_{\text{heavy}} = \left\{P \in \mathcal{P} : \mathcal{M}(P) > \frac{\mu}{4}\right\}$$

and accordingly $\mathcal{P}_{\text{light}} = \mathcal{P} \backslash \mathcal{P}_{\text{heavy}}$. Then (4.5) is automatically satisfied. It remains to show (4.6). By the definition of mass (4.2) we know that for every $P \in \mathcal{P}_{\text{heavy}}$ there exists a $P' = P'(P) \in \overline{\mathcal{P}}$ with $P' \geq P$ such that

$$\int_{E_{P'}} w_{P'}(x)dx > \frac{\mu}{4} \tag{5.1}$$

Note that P' need not be in $\mathcal{P}$. Let $\mathcal{P}'$ be the maximal elements in

$$\{P'(P) \, : \, P \in \mathcal{P}_{\text{heavy}}\}$$

with respect to the partial order $\leq$ of tiles. Then $\mathcal{P}_{\text{heavy}}$ is a union of trees with tops in $\mathcal{P}'$. Therefore it suffices to show

$$\sum_{P \in \mathcal{P}'} |I_P| \leq \frac{C}{\mu} \tag{5.2}$$

First we rewrite (5.1) as

$$\sum_{j=0}^{\infty} \int_{E_P \cap (\delta_{2^j}(I_P) \backslash \delta_{2^{j-1}}(I_P))} w_P(x)dx > C\mu \sum_{j=0}^{\infty} 2^{-j}. \tag{5.3}$$

where we adopt the temporary convention that $\delta_{2^{-1}}(I_P) = \emptyset$ and for $j \geq 0$,

$$\delta_{2^j}(I_P) = \prod_{i=1}^{n} \left[c(I_P)_i - 2^{(k(P)+j)\alpha_i - 1}, c(I_P)_i + 2^{(k(P)+j)\alpha_i - 1}\right).$$

Thus, for every $P \in \mathcal{P}'$ there exists a $j \geq 0$ such that

$$\int_{E_P \cap (\delta_{2^j}(I_P) \backslash \delta_{2^{j-1}}(I_P))} \frac{dx}{\left(1 + 2^{-k(P)} \rho_\alpha(x - c(I_P))\right)^{\nu_1}} > C|I_P|\mu 2^{-j}. \tag{5.4}$$

Note that for $x \in \delta_{2^j}(I_P) \backslash \delta_{2^{j-1}}(I_P)$ we have

$$1 + 2^{-k(P)} \rho_\alpha(x - c(I_P)) \geq C2^j.$$

Using this we obtain from (5.4),

$$|I_P| < C\mu^{-1}|E_P \cap \delta_{2^j}(I_P)|2^{-(\nu_1-1)j}. \tag{5.5}$$

Summarizing, we have shown that for every $P \in \mathcal{P}'$ there exists $j \geq 0$ such that (5.5) holds. This leads us to define for every $j \geq 0$, a set of tiles $\mathcal{P}_j$ by

$$\mathcal{P}_j = \{P \in \mathcal{P}' : |I_P| < C\mu^{-1}|E_P \cap \delta_{2^j}(I_P)|2^{-j(\nu_1-1)}\}.$$

The estimate (5.2) will follow by summing over j if we can show that

$$\sum_{P\in\mathcal{P}_j} |I_P| \leq C2^{-j}\mu^{-1} \tag{5.6}$$

for all $j \geq 0$. To show (5.6) we use a covering argument reminiscent of Vitali's covering lemma. Fix $j \geq 0$. For every tile $P = I_P \times \omega_P$ we have an enlarged tile $\delta_{2^j}(I_P)\times\omega_P$ (this is not a tile anymore). We inductively choose $P_i \in \mathcal{P}_j$ such that $|I_{P_i}|$ is maximal among the $P \in \mathcal{P}_j \backslash \{P_0, \dots, P_{i-1}\}$ and the enlarged tile of P_i is disjoint from the enlarged tiles of $P_0, \dots, P_{i-1}$. Since $\mathcal{P}_j$ is finite, this process terminates after finitely many steps, so that we have selected a subset $\mathcal{P}'_j = \{P_0, P_1, \dots\} \subset \mathcal{P}_j$ of tiles whose enlarged tiles are pairwise disjoint. By construction, for every $P \in \mathcal{P}_j$ there exists a unique $P' \in \mathcal{P}'_j$ such that $|I_P| \leq |I_{P'}|$ and the enlarged tiles of P and P' intersect. We call P *associated* with P'.

Now the claim is that if two tiles $P, Q \in \mathcal{P}_j$ are associated with the same $P' \in \mathcal{P}'_j$, then I_P and I_Q are disjoint. To see this note that ω_P intersects $\omega_{P'}$ by definition. Thus, since $|I_P| \leq |I_{P'}|$, we have $\omega_{P'} \subset \omega_P$. The same holds for Q. Therefore we have $\omega_{P'} \subset \omega_P \cap \omega_Q$. But $P, Q \in \mathcal{P}_j \subset \mathcal{P}'$ are disjoint tiles, so we must have $I_P \cap I_Q = \emptyset$. Moreover, all tiles P associated with P' satisfy $I_P \subset \delta_{2^{j+2}}(I_{P'})$. Therefore we get

$$\begin{aligned}
\sum_{P\in\mathcal{P}_j} |I_P| &= \sum_{P'\in\mathcal{P}'_j} \sum_{\substack{P\in\mathcal{P}_j \\ \text{assoc. with } P'}} |I_P| = \sum_{P'\in\mathcal{P}'_j} \Bigg| \bigcup_{\substack{P\in\mathcal{P}_j \\ \text{assoc. with } P'}} I_P \Bigg| \\
&\leq \sum_{P'\in\mathcal{P}'_j} 2^{(j+2)|\alpha|}|I_{P'}| \leq C\mu^{-1}2^{-j(\nu_1-|\alpha|-1)} \\
&\quad \times \sum_{P'\in\mathcal{P}'_j} |E \cap N^{-1}(\omega_{P'}) \cap \delta_{2^j}(I_{P'})| \\
&\leq C2^{-j}\mu^{-1},
\end{aligned}$$

using that $\nu_1 > |\alpha| + 1$. The penultimate inequality is a consequence of (5.5) and the last inequality follows, because the sets $N^{-1}(\omega_{P'}) \cap \delta_{2^j}(I_{P'})$ are disjoint and $|E| \leq 1$.

6 Proof of the Energy Lemma

In this section we prove Lemma 4.3. We adapt the argument of Lacey and Thiele [9, Prop. 3.2]. The tree selection algorithm of Lacey and Thiele relies on the natural ordering of real numbers. In our situation this can be replaced by any functional on $\mathbb{R}^n$ that separates $\omega_{P(0)}$ from $\omega_{P(r)}$ for every tile $P \in \overline{\mathcal{P}}$ (this was already observed in [12]). Let i_0 be such that $r_{i_0} = 1$ (exists because $r \neq 0$). Let us introduce the projection to the i_0th coordinate: $\pi_0 : \mathbb{R}^n \to \mathbb{R}, x \mapsto x_{i_0}$. Then we have that

$$\pi_0(\xi) < \pi_0(\eta) \tag{6.1}$$

holds for every $\xi \in \omega_{P(0)}, \eta \in \omega_{P(r)}, P \in \overline{\mathcal{P}}$.

Let $\varepsilon = \mathcal{E}(\mathcal{P})$. For a 2-tree $\mathbf{T}_2$ we define

$$\Delta(\mathbf{T}_2) = \left(\frac{1}{|I_{\mathbf{T}_2}|} \sum_{P \in \mathbf{T}_2} |\langle f, \phi_P \rangle|^2 \right)^{1/2}.$$

We will now describe an algorithm to choose the desired collection of trees $\mathcal{T}$ and also an auxiliary collection of 2-trees $\mathcal{T}_2$:

(1) Initialize $\mathcal{T} := \mathcal{T}_2 := \emptyset$ and $\mathcal{P}^{\text{stock}} := \mathcal{P}$.
(2) Choose a 2-tree $\mathbf{T}_2 \subset \mathcal{P}^{\text{stock}}$ such that

   (a) $\Delta(\mathbf{T}_2) \geq \varepsilon/2$, and
   (b) $\pi_0(c(\omega_{\mathbf{T}_2}))$ is minimal among all the 2-trees in $\mathcal{P}^{\text{stock}}$ satisfying (a).

   If no such $\mathbf{T}_2$ exists, then terminate.
(3) Let $\mathbf{T}$ be the maximal tree in $\mathcal{P}^{\text{stock}}$ with top $P_{\mathbf{T}_2}$ (with respect to set inclusion).
(4) Add $\mathbf{T}$ to $\mathcal{T}$ and $\mathbf{T}_2$ to $\mathcal{T}_2$. Also, remove all the elements of $\mathbf{T}$ from $\mathcal{P}^{\text{stock}}$. Then continue again with Step (2).

Since $\mathcal{P}$ is finite it is clear that the algorithm terminates after finitely many steps. Also note for every $\mathbf{T} \in \mathcal{T}$ there exists a unique $\mathbf{T}_2 \in \mathcal{T}_2$ with $\mathbf{T}_2 \subset \mathbf{T}$, and vice versa. After the algorithm terminates we set $\mathcal{P}_{\text{low}} = \mathcal{P}^{\text{stock}}$ and $\mathcal{P}_{\text{high}}$ to be the union of the trees in $\mathcal{T}$. Then, (4.7) is automatically satisfied and it only remains to show

$$\sum_{\mathbf{T}_2 \in \mathcal{T}_2} |I_{\mathbf{T}_2}| \lesssim \varepsilon^{-2}. \tag{6.2}$$

Before we do that we establish a geometric property of the selected trees that will be crucial in the following.

Lemma 6.1 *Let $\mathbf{T}_2 \neq \mathbf{T}_2' \in \mathcal{T}_2$ and $P \in \mathbf{T}_2$, $P' \in \mathbf{T}_2'$. If $\omega_P \subset \omega_{1P'}$, then $I_{P'} \cap I_{\mathbf{T}_2} = \emptyset$.*

Proof Note that $c(\omega_{\mathbf{T}_2}) \in \omega_P \subset \omega_{P'(0)}$ while $c(\omega_{\mathbf{T}_2'}) \in \omega_{P'(r)}$. By (6.1) and condition (b) in Step (2) we therefore conclude that $\mathbf{T}_2$ was chosen before $\mathbf{T}_2'$ during the above algorithm. Let $\mathbf{T}$ be the tree in $\mathcal{T}$ such that $\mathbf{T}_2 \subset \mathbf{T}$. Thus, if $I_{P'}$ was not disjoint from $I_{\mathbf{T}_2} = I_{\mathbf{T}}$, then it would be contained in $I_{\mathbf{T}}$ and therefore $P' \leq P_{\mathbf{T}}$ which means it would have been included into $\mathbf{T}$ during Step (3). That is a contradiction. □

The sum in (6.2) equals

$$\sum_{T_2 \in \mathcal{T}_2} \Delta(\mathbf{T}_2)^{-2} \sum_{P \in \mathbf{T}_2} |\langle f, \phi_P \rangle| \leq 4\varepsilon^{-2} \sum_{P \in \bigcup \mathcal{T}_2} |\langle f, \phi_P \rangle|^2,$$

where $\bigcup \mathcal{T}_2 = \bigcup_{\mathbf{T}_2 \in \mathcal{T}_2} \mathbf{T}_2$. Let us write

$$\sum_{P \in \bigcup \mathcal{T}_2} |\langle f, \phi_P \rangle|^2 = \left\langle \sum_{P \in \bigcup \mathcal{T}_2} \langle f, \phi_P \rangle \phi_P, f \right\rangle \tag{6.3}$$

and use the Cauchy–Schwarz inequality to estimate this by

$$\left\| \sum_{P \in \bigcup \mathcal{T}_2} \langle f, \phi_P \rangle \phi_P \right\|_2, \tag{6.4}$$

where we used that $\|f\|_2 = 1$. So far we have shown that

$$\varepsilon^2 \sum_{\mathbf{T}_2 \in \mathcal{T}_2} |I_{\mathbf{T}_2}| \lesssim \left\| \sum_{P \in \bigcup \mathcal{T}_2} \langle f, \phi_P \rangle \phi_P \right\|_2. \tag{6.5}$$

Thus if we can show that

$$\left\| \sum_{P \in \bigcup \mathcal{T}_2} \langle f, \phi_P \rangle \phi_P \right\|_2^2 \lesssim \varepsilon^2 \sum_{\mathbf{T}_2 \in \mathcal{T}_2} |I_{\mathbf{T}_2}|, \tag{6.6}$$

then (6.2) follows. Expanding the L^2 norm in (6.6) we get that the left hand side is bounded by

$$\sum_{\substack{P, P' \in \bigcup \mathcal{T}_2, \\ \omega_P = \omega_{P'}}} |\langle f, \phi_P \rangle \langle f, \phi_{P'} \rangle \langle \phi_P, \phi_{P'} \rangle| + 2 \sum_{\substack{P, P' \in \bigcup \mathcal{T}_2, \\ \omega_P \subset \omega_{P'(0)}}} |\langle f, \phi_P \rangle \langle f, \phi_{P'} \rangle \langle \phi_P, \phi_{P'} \rangle|. \tag{6.7}$$

Here we have used that $\langle \phi_P, \phi_P' \rangle = 0$ if $\omega_{P(0)} \cap \omega_{P'(0)} = \emptyset$ and therefore either $\omega_P = \omega_{P'}$, $\omega_P \subset \omega_{P'(0)}$, or $\omega_{P'} \subset \omega_{P(0)}$ (the last two cases are symmetric). We treat both sums in this term separately. Estimating the smaller one of $|\langle f, \phi_P \rangle|$ and $|\langle f, \phi_{P'} \rangle$ by the larger one, we obtain that the first sum in (6.7) is

$$\lesssim \sum_{P \in \bigcup \mathcal{T}_2} |\langle f, \phi_P \rangle|^2 \sum_{\substack{P' \in \bigcup \mathcal{T}_2, \\ \omega_P = \omega_{P'}}} |\langle \phi_P, \phi_{P'} \rangle|.$$

Using (4.16) we estimate this by

$$\sum_{P\in\bigcup\mathcal{T}_2}|\langle f,\phi_P\rangle|^2\sum_{\substack{P'\in\bigcup\mathcal{T}_2,\\ \omega_P=\omega_{P'}}}(1+2^{-k(P)}\rho_\alpha(c(I_P)-c(I_{P'})))^{-\nu}. \tag{6.8}$$

Notice that $I_P\cap I_{P'}=\emptyset$ for $P\neq P'$ in the inner sum. This implies

$$\sum_{\substack{P'\in\bigcup\mathcal{T}_2,\\ \omega_P=\omega_{P'}}}(1+2^{-k(P)}\rho_\alpha(c(I_P)-c(I_{P'})))^{-\nu}\lesssim\int_{\mathbb{R}^n}(1+\rho_\alpha(x))^{-\nu}dx\lesssim 1,$$

provided that $\nu>|\alpha|$. Therefore (6.8) is

$$\lesssim\sum_{\mathbf{T}_2\in\mathcal{T}_2}\sum_{P\in\mathbf{T}_2}|\langle f,\phi_P\rangle|^2\leq\varepsilon^2\sum_{\mathbf{T}_2\in\mathcal{T}_2}|I_{\mathbf{T}_2}|, \tag{6.9}$$

as desired. It remains to estimate the second sum in (6.7). To that end it suffices to show that

$$\sum_{P\in\mathbf{T}_2}\sum_{P'\in\mathcal{S}_P}|\langle f,\phi_P\rangle\langle f,\phi_{P'}\rangle\langle\phi_P,\phi_{P'}\rangle|\lesssim\varepsilon^2|I_{\mathbf{T}_2}|, \tag{6.10}$$

for every $\mathbf{T}_2\in\mathcal{T}_2$, where

$$\mathcal{S}_P=\left\{P'\in\bigcup\mathcal{T}_2\ :\ \omega_P\subset\omega_{P'(0)}\right\}.$$

Here we follow the argument given in [8]. Observe that if $P\in\mathbf{T}_2$, then $\mathcal{S}_P\cap\mathbf{T}_2=\emptyset$. Interpreting the singleton $\{P\}$ as a 2-tree we obtain

$$|\langle f,\phi_P\rangle|\leq\varepsilon|I_P|^{1/2} \tag{6.11}$$

for all $P\in\mathcal{P}$. Combining this with (4.16) we can estimate the left hand side of (6.10) by

$$\varepsilon^2\sum_{P\in\mathbf{T}_2}\sum_{P'\in\mathcal{S}_P}|I_{P'}|(1+2^{-k(P)}\rho_\alpha(c(I_P)-c(I_{P'})))^{-\nu}. \tag{6.12}$$

Indeed, Lemma 6.1 implies that $I_{\mathbf{T}_2}\cap I_{P'}=\emptyset$ for every $P\in\mathbf{T}_2$, $P'\in\mathcal{S}_P$. Moreover, it also implies that for $P'\neq P''\in\mathcal{S}_P$ we have $I_{P'}\cap I_{P''}=\emptyset$. These facts facilitate the following estimate:

$$\begin{aligned}&\sum_{P\in\mathbf{T}_2}\sum_{P'\in\mathcal{S}_P}|I_{P'}|(1+2^{-k(P)}\rho_\alpha(c(I_P)-c(I_{P'})))^{-\nu}\\&\lesssim\sum_{P\in\mathbf{T}_2}\sum_{P'\in\mathcal{S}_P}\int_{I_{P'}}(1+2^{-k(P)}\rho_\alpha(c(I_P)-x))^{-\nu}dx\\&\lesssim\sum_{P\in\mathbf{T}_2}\int_{(I_{\mathbf{T}_2})^c}(1+2^{-k(P)}\rho_\alpha(c(I_P)-x))^{-\nu}.\end{aligned}$$

Since $\mathbf{T}_2$ is a tree, the last quantity can be estimated by

$$\sum_{k\le k_{\mathbf{T}_2}} \sum_{u\in Q_k\cap(\mathbb{Z}^n+\frac{1}{2})} \int_{(I_{\mathbf{T}_2})^c} (1+\rho_\alpha(u-\delta_{2^{-k}}(x)))^{-\nu}dx,$$

where $Q_k \in \mathcal{D}^\alpha$ is an anisotropic dyadic rectangle of scale $k_{\mathbf{T}_2}-k$ that is given by a rescaling of $I_{\mathbf{T}_2}$. The previous display is no greater than a constant times

$$\sum_{k\le k_{\mathbf{T}_2}} 2^{k|\alpha|}\left(\sum_{u\in Q_k\cap(\mathbb{Z}^n+\frac{1}{2})} (1+\mathrm{dist}_\alpha((Q_k)^c,u)^{-|\alpha|-\gamma}\right)\left(\int_{\mathbb{R}^n}(1+\rho_\alpha(x))^{-(\nu-|\alpha|-\gamma)}dx\right), \tag{6.13}$$

where $\nu > 2|\alpha|$ and γ is a fixed and sufficiently small positive constant. The integral over x in the previous display is bounded by a constant depending on $\nu-|\alpha|-\gamma > |\alpha|$. To estimate the sum over u we note that for every u in the indicated range there exists a lattice point $v \in \partial Q_k \cap \mathbb{Z}^n$ such that $\mathrm{dist}_\alpha((Q_k)^c,u) \ge \frac{1}{2}\rho_\alpha(v-u)$. Thus we may bound the sum over u by

$$\sum_{v\in\partial Q_k\cap\mathbb{Z}^n} \sum_{u\in\mathbb{Z}^n+\frac{1}{2}} (1+\rho_\alpha(v-u))^{-|\alpha|-\gamma} \lesssim |\partial Q_k\cap\mathbb{Z}^n| \lesssim 2^{(k_{\mathbf{T}_2}-k)|\alpha|_\infty}.$$

Thus, (6.13) is bounded by a constant times

$$2^{k_{\mathbf{T}_2}|\alpha|_\infty}\sum_{k\le k_{\mathbf{T}_2}} 2^{k(|\alpha|-|\alpha|_\infty)} \lesssim 2^{k_{\mathbf{T}_2}|\alpha|} = |I_{\mathbf{T}_2}|.$$

This proves (6.10).

7 Proof of the Tree Estimate

In this section we prove Lemma 4.4. This is the core of the proof. For a rectangle $I=\prod_{i=1}^n I_i \in \mathcal{D}^\alpha$ we denote by $\widetilde{I}$ the enlarged rectangle defined by

$$\widetilde{I} = \prod_{i=1}^n (2^{\alpha_i+1}-1)I_i.$$

Here λI_i is the interval of length $\lambda|I_i|$ with the same center as I_i. Let $\mathcal{J}$ be the partition of $\mathbb{R}^n$ that is given by the collection of maximal anisotropic dyadic rectangles $J \in \mathcal{D}^\alpha$ such that $\widetilde{J}$ does not contain any I_P with $P \in \mathbf{T}$ (maximal with respect to inclusion). Set $\varepsilon = \mathcal{E}(\mathbf{T})$ and $\mu = \mathcal{M}(\mathbf{T})$. Choose phase factors $(\epsilon_P)_P$ of modulus 1 such that

$$\begin{aligned}\sum_{P\in\mathbf{T}}|\langle f,\phi_P\rangle\langle\psi_P^{N(\cdot)},\mathbf{1}_{E_{P(r)}}\rangle|\\&=\int_{\mathbb{R}^n}\sum_{P\in\mathbf{T}}\epsilon_P\langle f,\phi_P\rangle\psi_P^{N(x)}(x)\mathbf{1}_{E_{P(r)}}(x)dx\\&\leq\left\|\sum_{P\in\mathbf{T}}\epsilon_P\langle f,\phi_P\rangle\psi_P^{N}\mathbf{1}_{E_{P(r)}}\right\|_1\leq\mathcal{K}_1+\mathcal{K}_2,\end{aligned}$$

where

$$\mathcal{K}_1=\sum_{J\in\mathcal{J}}\sum_{P\in\mathbf{T},|I_P|\leq|J^+|}\|\langle f,\phi_P\rangle\psi_P^{N(\cdot)}\mathbf{1}_{E_{P(r)}}\|_{L^1(J)},$$

$$\mathcal{K}_2=\sum_{J\in\mathcal{J}}\left\|\sum_{P\in\mathbf{T},|I_P|>|J^+|}\epsilon_P\langle f,\phi_P\rangle\psi_P^{N(\cdot)}\mathbf{1}_{E_{P(r)}}\right\|_{L^1(J)}.$$

We first estimate $\mathcal{K}_1$. This is the easy part, since in the sum defining $\mathcal{K}_1$ we have that I_P is disjoint from $\widetilde{J}$. Again, interpreting the singleton $\{P\}$ as a 2-tree we see that (6.11) holds for all $P\in\mathbf{T}$. This gives

$$\mathcal{K}_1\leq\varepsilon\sum_{J\in\mathcal{J}}\sum_{\substack{P\in\mathbf{T}\\|I_P|\leq|J^+|}}2^{|\alpha|k(P)/2}\|\psi_P^{N(\cdot)}\mathbf{1}_{E_{P(r)}}\|_{L^1(J)}.$$

Using (4.15) the previous display is seen to be no larger than a constant times

$$\begin{aligned}&\|m\|_{\mathcal{M}_\alpha^{\nu_0}}\varepsilon\sum_{J\in\mathcal{J}}\sum_{\substack{P\in\mathbf{T}\\|I_P|\leq|J^+|}}\int_{J\cap E_{P(r)}}w^{\nu_0}(2^{-k(P)}(x-c(I_P)))dx\\&\leq\|m\|_{\mathcal{M}_\alpha^{\nu_0}}\varepsilon\mu\sum_{J\in\mathcal{J}}\sum_{\substack{P\in\mathbf{T}\\|I_P|\leq|J^+|}}2^{|\alpha|k(P)}\sup_{x\in J}w^{2|\alpha|+\frac{2}{3}}(2^{-k(P)}(x-c(I_P))),\end{aligned}\tag{7.1}$$

where we have set $\nu_1=|\alpha|+\frac{4}{3}$. Since I_P is disjoint from $\widetilde{J}$ we can estimate (7.1) as

$$\lesssim\|m\|_{\mathcal{M}_\alpha^{\nu_0}}\varepsilon\mu\sum_{J\in\mathcal{J}}\sum_{\substack{k\in\mathbb{Z},\\2^{k|\alpha|}\leq|J^+|}}2^{|\alpha|k}\sum_{\substack{P\in\mathbf{T},\\k(P)=k}}w^{2|\alpha|+\frac{2}{3}}(2^{-k}\mathrm{dist}_\alpha(J,I_P)).\tag{7.2}$$

Before we proceed, we claim that for every $\nu>|\alpha|$, $k\in\mathbb{Z}$ and fixed $J\in\mathcal{J}$ with $2^{k|\alpha|}\leq|J^+|$ we have

$$\sum_{\substack{P\in\mathbf{T},\\k(P)=k}}w^{\nu}(2^{-k}\mathrm{dist}_\alpha(J,I_P))\lesssim1,\tag{7.3}$$

where the implicit constant blows up as ν approaches $|\alpha|$. To verify the claim, let us assume for simpler notation that J is centered at the origin. Then by disjointness of I_P and $\widetilde{J}$ we have

$$\operatorname{dist}_\alpha(J, I_P) \gtrsim \operatorname{dist}_\alpha(0, I_P) \gtrsim 2^k \rho(m),$$

where $m = (m_1, \ldots, m_n) \in \mathbb{Z}^n$ is such that $I_P = \prod_{i=1}^n [2^{k\alpha_i} m_i, 2^{k\alpha_i}(m_i + 1))$. Thus the sum in (7.3) is

$$\lesssim \sum_{m \in \mathbb{Z}^n} (1 + \rho(m))^{-\nu},$$

which implies the claim.

Estimate (7.2) by

$$\lesssim \|m\|_{\mathscr{M}_\alpha^{\nu_0}} \varepsilon\mu \sum_{J \in \mathcal{J}} w^{|\alpha|+\frac{1}{3}}(2^{-k_{\mathbf{T}}} \operatorname{dist}_\alpha(J, I_T)) \sum_{\substack{k \in \mathbb{Z}, \\ 2^{k|\alpha|} \le |J^+|}} 2^{|\alpha|k}, \tag{7.4}$$

Here $k_{\mathbf{T}}$ is the scale of $I_{\mathbf{T}}$ and we have used (7.3) and

$$2^{-k} \operatorname{dist}_\alpha(J, I_P) \ge 2^{-k_{\mathbf{T}}} \operatorname{dist}_\alpha(J, I_{\mathbf{T}}).$$

Summing the geometric series, (7.4) is

$$\lesssim \|m\|_{\mathscr{M}_\alpha^{\nu_0}} \varepsilon\mu \sum_{J \in \mathcal{J}} w^{|\alpha|+\frac{1}{3}}(2^{-k_{\mathbf{T}}} \operatorname{dist}_\alpha(J, I_P))|J|.$$

The sum in that expression can be estimated as follows:

$$\sum_{J \in \mathcal{J}} w^{|\alpha|+\frac{1}{3}}(2^{-k_{\mathbf{T}}} \operatorname{dist}_\alpha(J, I_P))|J| \lesssim \sum_{J \in \mathcal{J}} \int_J (1 + 2^{-k_{\mathbf{T}}} \rho(x - c(I_{\mathbf{T}})))^{-(|\alpha|+\frac{1}{3})} dx.$$

By disjointness of the J we can bound this by

$$\int_{\mathbb{R}^n} (1 + 2^{-k_{\mathbf{T}}} \rho(x - c(I_{\mathbf{T}})))^{-(|\alpha|+\frac{1}{3})} dx = |I_{\mathbf{T}}| \int_{\mathbb{R}^n} (1 + \rho(x))^{-(|\alpha|+\frac{1}{3})} dx \lesssim |I_{\mathbf{T}}|.$$

To summarize, we showed that

$$\mathcal{K}_1 \lesssim \|m\|_{\mathscr{M}_\alpha^{\nu_0}} \varepsilon\mu |I_{\mathbf{T}}|, \tag{7.5}$$

using that $\nu_0 \ge 3|\alpha| + 2$.

Let us proceed to estimating $\mathcal{K}_2$. This is more difficult. We may assume that the sum runs only over those J for which there is a $P \in \mathbf{T}$ such that $|I_P| > |J^+|$. Then $|I_{\mathbf{T}}| > |J^+|$ and $J \subset \widetilde{I}_{\mathbf{T}}$. From now on let such a J be fixed. Define

$$G_J = J \cap \bigcup_{P \in \mathbf{T}, |I_P| > |J^+|} E_{P(r)} \tag{7.6}$$

Before proceeding we prove the following.

Lemma 7.1 *There exists a constant $C > 0$ independent of J such that*

$$|G_J| \leq C\mu|J| \tag{7.7}$$

Proof By definition of J, there exists $P_0 \in \mathbf{T}$ such that I_{P_0} is contained in $\widetilde{J^+}$. We claim that there exists a tile $P_0 < P' < P_{\mathbf{T}}$ such that $|I_{P'}| = |J^{++}|$. Indeed, note $|I_{P_0}| \leq |J^{++}|$. If there is equality, we simply take $P' = P_0$. Otherwise we take $I_{P'} \in \mathcal{D}^\alpha$ to be the unique dyadic ancestor of I_{P_0} such that $|I_{P'}| = |J^{++}|$ and choose $\omega_{P'}$ accordingly such that it contains $c(\omega_{\mathbf{T}})$. Now we have

$$|\omega_P| = |I_P|^{-1} \leq |J^{++}|^{-1} = |I_{P'}|^{-1} = |\omega_{P'}|$$

for every tile $P \in \mathbf{T}$ with $|I_P| > |J^+|$. This implies $\omega_P \subset \omega_{P'}$ and thus

$$G_J \subset J \cap E_{P'}.$$

As a consequence,

$$|G_J| \leq \int_{E_{P'}} \mathbf{1}_J(x)dx \lesssim |I_{P'}| \int_{E_{P'}} w_{P'}(x)dx \lesssim \mu|J|.$$

□

Let us define

$$F_J = \sum_{\substack{P \in \mathbf{T}, \\ |I_P| > |J^+|}} \epsilon_P \langle f, \phi_P \rangle \psi_P^{N(\cdot)} \mathbf{1}_{E_{P(r)}}. \tag{7.8}$$

Since every tree can be written as the union of a 1-tree and a 2-tree, we may treat each of these cases separately.

7.1 The Case of 1-Trees

Assume that $\mathbf{T}$ is a 1-tree. This is the easier case. The reason is that for every $P, P' \in \mathbf{T}$, $\omega_P \neq \omega_{P'}$ we have that $\omega_{P(r)}$ and $\omega_{P'(r)}$ are disjoint and thus we have good orthogonality of the summands in (7.8). Using (6.11) and (4.15) we see that

$$|F_J(x)| \lesssim \|m\|_{\mathscr{M}_\alpha^{\nu_0}} \varepsilon \sum_{\substack{P \in \mathbf{T}, \\ |I_P| > |J^+|}} (1 + 2^{-k(P)} \rho_\alpha(x - c(I_P)))^{-\nu_0} \mathbf{1}_{E_{P(r)}}(x).$$

Using disjointness of the $E_{P(r)}$ this can be estimated by

$$\|m\|_{\mathscr{M}_\alpha^{\nu_0}} \varepsilon \cdot \sup_{k \in \mathbb{Z}} \sum_{m \in \mathbb{Z}^n + \frac{1}{2}} (1 + 2^{-k} \rho_\alpha(x - \delta_{2^k}(m)))^{-\nu_0}.$$

By an index shift we see that

$$\sum_{m \in \mathbb{Z}^n + \frac{1}{2}} (1 + 2^{-k} \rho_\alpha(x - \delta_{2^k}(m)))^{-\nu_0} = \sum_{m \in \mathbb{Z}^n + \frac{1}{2}} (1 + \rho_\alpha(m + \gamma))^{-\nu_0},$$

where $\gamma \in [0, 1]^n$ depends on k and x. The last sum is $\lesssim 1$ independently of γ. Thus we proved the pointwise estimate

$$|F_J(x)| \lesssim \|m\|_{\mathscr{M}_\alpha^{\nu_0}} \varepsilon. \tag{7.9}$$

Combining this with the support estimate (7.7) we obtain

$$\|F_J\|_{L^1(J)} \lesssim \|m\|_{\mathscr{M}_\alpha^{\nu_0}} \varepsilon \mu |J|. \tag{7.10}$$

Summing over the pairwise disjoint $J \subset \widetilde{I_\mathbf{T}}$ we obtain

$$\mathcal{K}_2 \lesssim \|m\|_{\mathscr{M}_\alpha^{\nu_0}} \varepsilon \mu |I_\mathbf{T}|$$

as desired. Note that we only needed $\nu_0 > |\alpha|$ to obtain this estimate.

7.2 The Case of 2-Trees

Here we assume that $\mathbf{T}$ is a 2-tree. The additional x-dependence present in the wave packets $\psi_P^{N(x)}$ makes this part more difficult than the congruent argument in [9]. This problem arises already in the isotropic case [12]. The goal is again to obtain a pointwise estimate for F_J. In the following we fix $x \in J$ such that $F_J(x) \neq 0$. Observe that the $\omega_{P(r)}$, $P \in \mathbf{T}$ are nested. Let us denote the smallest (resp. largest) $\omega_{P(r)}$ (resp. ω_P) such that $x \in N^{-1}(\omega_{P(r)}) \cap E$ by ω_- (resp. ω_+). Let $k_+ \in \mathbb{Z}$ be such that $|\omega_+| = 2^{k_+|\alpha|}$ and $k_- \in \mathbb{Z}$ such that $|\omega_-| = 2^{-k_-|\alpha|-n}$ (note from the definition that $\omega_- \notin \mathcal{D}^\alpha$ if $\alpha \neq (1, \ldots, 1)$). Then the nestedness property implies

$$F_J(x) = \sum_{\substack{P \in \mathbf{T}, \\ k_+ \leq k(P) \leq k_-}} \epsilon_P \langle f, \phi_P \rangle \psi_P^{N(x)}(x)$$

Define

$$h_x = \mathrm{M}_{c(\omega_+)}\mathrm{D}^1_{2^{k_+}}\phi_+ - M_{c(\omega_-)}D^1_{2^{k_-}}\phi_-,$$

where $\phi_+(x) = b_1^{-n}\phi(b_1^{-1}x)$ and ϕ_- is a Schwartz function satisfying $0 \le \widehat{\phi_-} \le 1$ such that $\widehat{\phi_-}$ is supported on $[-\frac{b_2}{2}, \frac{b_2}{2}]$ and equals to one on $[-\frac{b_3}{2}, \frac{b_3}{2}]$, where $b_{j+2} = \frac{1}{2} + b_j$ for $j = 0, 1$. From the definition we see that $\widehat{h_x}$ is supported on $b_0 b_1^{-1}\omega_+ \cap (2b_3\omega_-)^c$ and equal to one on $\omega_+ \cap (2b_2\omega_-)^c$. In particular, $\widehat{h_x}(\xi)$ equals to one if $\xi \in \operatorname{supp}\phi_P$ and $k_+ \le k(P) \le k_-$ and vanishes if $k(P)$ is outside this range. For technical reasons that become clear further below we need the support of $\widehat{h_x}$ to keep a certain distance to ω_-. We obtain

$$F_J(x) = \sum_{P\in\mathbf{T}} \epsilon_P\langle f, \phi_P\rangle(\psi_P^{N(x)} * h_x)(x).$$

Fix $\xi_0 \in \omega_{\mathbf{T}}$. We decompose

$$F_J(x) = \sum_{P\in\mathbf{T}} \epsilon_P\langle f, \phi_P\rangle(\psi_P^{\xi_0} * h_x)(x) + \sum_{P\in\mathbf{T}} \epsilon_P\langle f, \phi_P\rangle(\psi_P^{N(x)} - \psi_P^{\xi_0}) * h_x)(x) \tag{7.11}$$

$$= G * \mathrm{M}_{\xi_0}K * h_x(x) + G * (\mathrm{M}_{N(x)}K - \mathrm{M}_{\xi_0}K) * h_x(x) \tag{7.12}$$

where

$$G = \sum_{P\in\mathbf{T}} \epsilon_P\langle f, \phi_P\rangle\phi_P. \tag{7.13}$$

Before proceeding with the proof we record the following simple variant of a standard fact about maximal functions (see [2]).

Lemma 7.2 *Let $\lambda > 0$ and w be an integrable function on $\mathbb{R}^n$ which is constant on $\{\rho(y) \le \lambda\}$ and radial and decreasing with respect to ρ, i.e.*

$$w(x) \le w(y)$$

if $\rho_\alpha(x) \ge \rho_\alpha(y)$, with equality if $\rho_\alpha(x) = \rho_\alpha(y)$. Let $x \in \mathbb{R}^n$ and $J \subset \mathbb{R}^n$ be such that $J \subset \{y : \rho_\alpha(x - y) \le \lambda\}$. Then we have

$$|F * w|(x) \le \|w\|_1 \sup_{J\subset I} \frac{1}{|I|}\int_I |F(y)|dy,$$

where the supremum is taken over all anisotropic cubes $I \subset \mathbb{R}^n$.

Proof First we assume that w is a step function. That is,

$$w(y) = \sum_{j=1}^{\infty} c_j \mathbf{1}_{\rho_\alpha(y)\le r_j}$$

with $\lambda \le r_1 < r_2 < \cdots$. Then we have

$$F * w(x) = \sum_j r_j^{|\alpha|} c_j \frac{1}{r_j^{|\alpha|}} \int_{\rho_\alpha(x-y)\le r_j} |F(y)|dy \le \|w\|_1 \sup_{J\subset I} \frac{1}{|I|} \int_I |F(y)|dy.$$

The general case follows by approximation of w by step functions and an application of Lebesgue's dominated convergence theorem. □

Since

$$|h_x(y)| \lesssim 2^{-k_+|\alpha|}|\phi_+|(\delta_{2^{-k_+}}(y)) + 2^{-k_-|\alpha|}|\phi_-|(\delta_{2^{-k_-}}(y)) \tag{7.14}$$

and $x \in J$, $|J| \le 2^{k_+|\alpha|} \le 2^{k_-|\alpha|}$ we have from Lemma 7.2 that

$$|G * \mathrm{M}_{\xi_0} K * h_x(x)| \lesssim \sup_{J\subset I} \frac{1}{|I|} \int_I |G * \mathrm{M}_{\xi_0} K(y)|dy. \tag{7.15}$$

Let us assume for the moment that we also have the estimate

$$|G * (\mathrm{M}_{N(x)} K - \mathrm{M}_{\xi_0} K) * h_x(x)| \lesssim \|m\|_{\mathscr{M}_\alpha^{\nu_0}} \sup_{J\subset I} \frac{1}{|I|} \int_I |G(y)|dy. \tag{7.16}$$

We will first show how to finish the proof from here. At the end of the section we will then show that (7.16) indeed holds.

From (7.11), (7.15), (7.16) and Lemma 7.1 we see that

$$\begin{aligned}&\sum_{\substack{J\in\mathcal{J}, \\ J\subset \tilde{I}_{\mathrm{T}}}} \|F_J\|_{L^1(J)} \\ &\lesssim \mu \sum_{\substack{J\in\mathcal{J}, \\ J\subset \tilde{I}_{\mathrm{T}}}} |J| \left(\sup_{J\subset I} \frac{1}{|I|} \int_I |G * \mathrm{M}_{\xi_0} K(y)|dy + \|m\|_{\mathscr{M}_\alpha^{\nu_0}} \sup_{J\subset I} \frac{1}{|I|} \int_I |G(y)|dy \right)\end{aligned}$$

By disjointness of the $J \in \mathcal{J}$ this is no greater than

$$\mu \left(\left\| \mathcal{M}(G * \mathrm{M}_{\xi_0} K) \right\|_{L^1(\tilde{I}_{\mathrm{T}})} + \|m\|_{\mathscr{M}_\alpha^{\nu_0}} \|\mathcal{M}(G)\|_{L^1(\tilde{I}_{\mathrm{T}})} \right), \tag{7.17}$$

where $\mathcal{M}$ denotes the maximal function defined by

$$\mathcal{M}F(y) = \sup_{y\in I} \frac{1}{|I|} \int_I |F|,$$

where the supremum runs over all anisotropic cubes $I \subset \mathbb{R}^n$. Clearly, $\mathcal{M}$ is a bounded operator $L^2(\mathbb{R}^n) \to L^2(\mathbb{R}^n)$, because it is bounded pointwise by a composition of one-dimensional Hardy–Littlewood maximal functions applied in each component.

Applying the Cauchy-Schwarz inequality and the L^2 boundedness of $\mathcal{M}$ we see that (7.17) is

$$\lesssim \mu |I_T|^{\frac{1}{2}} \left(\left\| G * \mathrm{M}_{\xi_0} K \right\|_2 + \|m\|_{\mathscr{M}_\alpha^{\nu_0}} \|G\|_2 \right).$$

By repeating the arguments that lead to the proof of (6.6), using (4.16) or (4.17), respectively, we obtain that

$$\left\| G * \mathrm{M}_{\xi_0} K \right\|_2 + \|m\|_{\mathscr{M}_\alpha^{\nu_0}} \|G\|_2 \lesssim \|m\|_{\mathscr{M}_\alpha^{\nu_0}} \varepsilon |I_{\mathbf{T}}|^{\frac{1}{2}}.$$

This concludes the proof. It remains to prove (7.16). Let us write

$$R(y) = (\mathrm{M}_{N(x)} K - \mathrm{M}_{\xi_0} K) * h_x(y).$$

We will give two different estimates for R. The first one is only effective if $\rho_\alpha(y)$ is large and the second one if $\rho_\alpha(y)$ is small. Let us start with the first estimate. By Fourier inversion, we can write $R(y)$ (up to a constant) as

$$\int_{\mathbb{R}^n} (m(\xi - N(x)) - m(\xi - \xi_0)) \widehat{h_x}(\xi) e^{i\xi y} d\xi. \tag{7.18}$$

Fix y and let i be such that $\rho_\alpha(y) = |y_i|^{1/\alpha_i}$. Then we integrate by parts in the ith component to see that (7.18) is bounded by

$$\lesssim \rho_\alpha(y)^{-\nu'\alpha_i} \int_{\mathbb{R}^n} \left| \partial_{\xi_i}^{\nu'} \left[(m(\xi - N(x)) - m(\xi - \xi_0)) \widehat{h_x}(\xi) \right] \right| d\xi \tag{7.19}$$

for integer $\nu' \geq 0$, where we have used that $\rho_\alpha(y) \geq 2^{k_-}$ to estimate $|\delta_{2^{-k_-}}(y)| \geq 2^{-k_-} \rho_\alpha(y)$.

Let $\ell \leq \nu_0$ be a non-negative integer. Using (4.13) we obtain

$$\left| \partial_{\xi_i}^{\ell} \left[m(\xi - N(x)) - m(\xi - \xi_0) \right] \right| \leq \|m\|_{\mathscr{M}_\alpha^{\ell}} (\rho_\alpha(\xi - N(x))^{-\ell\alpha_i} + \rho_\alpha(\xi - \xi_0)^{-\ell\alpha_i}). \tag{7.20}$$

Recall that ξ_0 and $N(x)$ are contained in ω_- and the integrand of (7.19) is supported on $b_0 b_1^{-1} \omega_+ \cap (2b_3\omega_-)^c$. Also, there exist $\omega_1, \ldots, \omega_M \in \mathcal{D}^\alpha$ such that

$$\omega_- \subsetneq \omega_1 \subsetneq \cdots \subsetneq \omega_M = \omega_+$$

and $|\omega_j| = 2^{-k_j|\alpha|}$ with $k_1 = k_-$ and $k_{j+1} = k_j - 1$. If $\xi \in (2b_3\omega_-)^c$ we have

$$\min(\rho_\alpha(\xi - N(x)), \rho_\alpha(\xi - \xi_0)) \gtrsim 2^{-k_-}. \tag{7.21}$$

On the other hand, if $\xi \in (b_0 b_1^{-1} \omega_j) \cap \omega_{j-1}^c$ for $j = 2, \ldots, M$, then

$$\min(\rho_\alpha(\xi - N(x)), \rho_\alpha(\xi - \xi_0)) \gtrsim 2^{-k_j}. \tag{7.22}$$

Combining (7.20) and (7.21), (7.22) we get

$$|\partial_{\xi_i}^{\ell}\Big[m(\xi - N(x)) - m(\xi - \xi_0)\Big]| \lesssim \|m\|_{\mathscr{M}_\alpha^\ell} \sum_{j=1}^{M} 2^{k_j \ell \alpha_i} \mathbf{1}_{b_0 b_1^{-1}\omega_j}(\xi). \tag{7.23}$$

We also have

$$|\partial_{\xi_i}^{\ell} \widehat{h_x}(\xi)| \lesssim 2^{k_+ \ell \alpha_i} \mathbf{1}_{b_0 b_1^{-1}\omega_+}(\xi) + 2^{k_- \ell \alpha_i} \mathbf{1}_{2b_3\omega_-}(\xi). \tag{7.24}$$

Thus we see from (7.23) and (7.24) that for all $i = 1, \dots, n$ and $0 \le \ell \le \nu'$ we obtain

$$\int_{\mathbb{R}^n} \Big|\partial_{\xi_i}^{\ell}\Big[m(\xi - N(x)) - m(\xi - \xi_0)\Big]\partial_{\xi_i}^{\nu'-\ell}\widehat{h_x}(\xi)\Big| d\xi \lesssim \|m\|_{\mathscr{M}_\alpha^{\nu'}} 2^{k_-(\nu'\alpha_i - |\alpha|)},$$

provided that $\nu'\alpha_i \ge |\alpha|$. Setting $\nu' = \lceil \frac{\nu_0}{\alpha_i} \rceil$, we have shown that

$$|R(y)| \lesssim \|m\|_{\mathscr{M}_\alpha^{\nu_0}} 2^{-k_-|\alpha|}(2^{-k_-}\rho_\alpha(y))^{-\nu_0}. \tag{7.25}$$

It remains to find a good estimate for $R(y)$ when $\rho_\alpha(y)$ is small. Let us estimate

$$|R(y)| \le R_+(y) + R_-(y),$$

where

$$R_\pm = |(\mathrm{M}_{N(x)}K - \mathrm{M}_{\xi_0}K) * \mathrm{D}^1_{2^{k_\pm}}\phi_\pm|.$$

The first claim is that if $\rho(y) \le 2^{k_\pm + 1}$, then

$$R_\pm(y) \lesssim \|m\|_{\mathscr{M}_\alpha^{\nu_0}} 2^{-k_\pm|\alpha|}. \tag{7.26}$$

(Here and throughout the proof of this claim $\pm$ always stands for a fixed choice of sign, either $+$ or $-$.) To see this, we first estimate $R_\pm(y)$ by

$$2^{-k_\pm|\alpha|}\int_{\mathbb{R}^n} \Big|(e^{i(N(x)-\xi_0)z} - 1)K(z)\phi_\pm(2^{-k_\pm}(y - z))\Big| dz \lesssim 2^{-k_\pm|\alpha|}(\mathbf{I} + \mathbf{II}),$$

where

$$\mathbf{I} = \int_{\rho_\alpha(z) \le 2^{k_\pm + 2}} \Big|(e^{i(N(x)-\xi_0)z} - 1)K(z)\phi_\pm(2^{-k_\pm}(y - z))\Big| dz, \quad \text{and}$$

$$\mathbf{II} = \sum_{j=2}^{\infty} \int_{2^{k_\pm + j} \le \rho_\alpha(z) \le 2^{k_\pm + j + 1}} \Big|K(z)\phi_\pm(2^{-k_\pm}(y - z))\Big| dz.$$

We first estimate **I**. Changing variables $z \mapsto \delta_{2^{k_\pm+2}}(z)$ we see that

$$\mathbf{I} \lesssim \int_{\rho(z)\le 1} \left| (e^{i\delta_{2^{k_\pm+2}}(N(x)-\xi_0)z} - 1)K(z) \right| dz.$$

Using (4.14), the previous display is

$$\lesssim \|m\|_{\mathcal{M}_\alpha^{\nu_0}} \left|\delta_{2^{k_\pm+2}}(N(x)-\xi_0)\right| \int_{\rho(z)\le 1} |z|\rho_\alpha(z)^{-|\alpha|} dz.$$

Using $\rho_\alpha(\delta_{2^{k_\pm+2}}(N(x)-\xi_0)) = 2^{k_\pm+2}\rho_\alpha(N(x)-\xi_0) \lesssim 1$ we can bound this further as

$$\lesssim \|m\|_{\mathcal{M}_\alpha^{\nu_0}} \int_{\rho(z)\le 1} \rho_\alpha(z)^{1-|\alpha|} dz \lesssim \|m\|_{\mathcal{M}_\alpha^{\nu_0}}.$$

This proves that $\mathbf{I} \lesssim \|m\|_{\mathcal{M}_\alpha^{\nu_0}}$. It remains to treat **II**. Here we make use of the fact that we have $\rho_\alpha(y-z) \ge 2^{k_\pm+j-1}$ in the integrand of **II**, because of our assumption $\rho(y) \le 2^{k_\pm}$. Using the decay of $\phi_\pm$ we obtain

$$\mathbf{II} \lesssim \|m\|_{\mathcal{M}_\alpha^{\nu_0}} \sum_{j=2}^{\infty} 2^{-j} \int_{2^{k_\pm+j} \le \rho_\alpha(z) \le 2^{k_\pm+j+1}} \rho_\alpha(z)^{-|\alpha|} dz \lesssim \|m\|_{\mathcal{M}_\alpha^{\nu_0}}.$$

Thus we have proved (7.26). The only further ingredient which we need in order to verify (7.16) is a good estimate for $R_+(y)$ when $2^{k_+ + 1} \le \rho_\alpha(y) \le 2^{k_-}$. In order to do this we need to do a slightly more careful decomposition. Let us write

$$Q_\ell = [\delta_{2^{k_+}}(\ell), \delta_{2^{k_+}}(\ell+1)) = \prod_{i=1}^{n} [2^{k_+\alpha_i}\ell_i, 2^{k_+\alpha_i}(\ell_i+1))$$

for $\ell \in \mathbb{Z}^n$. Assume that $y \in Q_\ell$ with $1 \le |\ell|_\infty < 2^{k_- - k_+}$. We have

$$R_+(y) \le 2^{-k_+|\alpha|} \sum_{s\in\mathbb{Z}^n} \int_{Q_s} |(e^{i(N(x)-\xi_0)z} - 1)K(z)\phi_+(2^{-k_+}(y-z))| dz. \tag{7.27}$$

Moreover, the same estimates that were used to prove (7.26) yield

$$\int_{Q_s} \left| (e^{i(N(x)-\xi_0)z} - 1)K(z)\phi_+(2^{-k_+}(y-z)) \right| dz$$
$$\lesssim \|m\|_{\mathcal{M}_\alpha^{\nu_0}} 2^{k_+ - k_-} (1+\rho_\alpha(s-\ell))^{-|\alpha|-1} (1+\rho_\alpha(s))^{1-|\alpha|}$$

Plugging this inequality into (7.27) we obtain

$$\begin{aligned} R_+(y) &\lesssim \|m\|_{\mathscr{M}_\alpha^{\nu_0}} 2^{-k_-} 2^{k_+(1-|\alpha|)} \sum_{s\in\mathbb{Z}^n} (1+\rho_\alpha(s-\ell))^{-|\alpha|-1}(1+\rho_\alpha(s))^{1-|\alpha|} \\ &\lesssim \|m\|_{\mathscr{M}_\alpha^{\nu_0}} 2^{-k_-}(2^{k_+}\rho_\alpha(\ell))^{1-|\alpha|}, \end{aligned}$$

where the last inequality requires ν to be large enough. Therefore,

$$R_+(y) \lesssim \|m\|_{\mathscr{M}_\alpha^{\nu_0}} 2^{-k_-}\rho_\alpha(y)^{1-|\alpha|}. \tag{7.28}$$

Finally, summarizing (7.25), (7.26) and (7.28) we have shown that

$$|R(y)| \lesssim \|m\|_{\mathscr{M}_\alpha^{\nu_0}}(w_0(y)+w_+(y)+w_-(y)+w_1(y)),$$

where

$$\begin{aligned} w_0(y) &= 2^{-k_-|\alpha|}(2^{-k_-}\rho_\alpha(y))^{-|\alpha|-1}\mathbf{1}_{\rho_\alpha(y)\geq 2^{k_-}}, \\ w_\pm(y) &= 2^{-k_\pm|\alpha|}\mathbf{1}_{\rho_\alpha(y)\leq 2^{k_\pm+1}}, \\ w_1(y) &= 2^{-k_-}\rho_\alpha(y)^{1-|\alpha|}\mathbf{1}_{2^{k_++1}\leq\rho_\alpha(y)\leq 2^{k_-}}. \end{aligned}$$

Each of these functions is integrable with an $L^1(\mathbb{R}^n)$ norm not depending on k_-, k_+, radial and decreasing with respect to ρ in the sense of Lemma 7.2 and constant on $\{\rho(y)\leq 2^{k_+}\}$ or $\{\rho(y)\leq 2^{k_-}\}$. Thus, applying Lemma 7.2 to each of these functions yields (7.16). Note that to prove (7.16) we only required that $\nu_0 \geq |\alpha|+1$.

8 Proofs of Auxiliary Estimates

In this section we prove (4.15), (4.16) and (4.17).
Proof of (4.15). Expanding definitions and using Fourier inversion we see that, up to a universal constant, $|\psi_P^N(x)|$ is equal to

$$2^{k(P)|\alpha|/2}\left|\int_{\mathbb{R}^n} e^{i\xi(x-c(I_P))} m(\xi-N)\widehat{\phi}(\delta_{2^{k(P)}}(\xi-c(\omega_{P(0)})))d\xi\right|.$$

Via a change of variables $\delta_{2^{k(P)}}(\xi-c(\omega_{P(0)}))\to\zeta$ and using that $m(\xi)=m(\delta_{2^{k(P)}}(\xi))$ this becomes

$$2^{-k(P)|\alpha|/2}\left|\int_{\mathbb{R}^n} e^{i\zeta\delta_{2^{-k(P)}}(x-c(I_P))} m(\zeta+\delta_{2^{k(P)}}(c(\omega_{P(0)})-N))\widehat{\phi}(\zeta)d\zeta\right|. \tag{8.1}$$

Let us fix x and P and take i to be such that $\rho_\alpha(x-c(I_P)) = |x_i-c(I_P)_i|^{1/\alpha_i}$. From repeated integration by parts we see that (8.1) is bounded by

$$2^{-k(P)|\alpha|/2}(2^{-k(P)}\rho_\alpha(x-c(I_P)))^{-\nu'\alpha_i}\int_{\mathbb{R}^n}\left|\partial_{\zeta_i}^{\nu'}(m(\zeta+\delta_{2^{k(P)}}(c(\omega_{P(0)})-N))\widehat{\phi}(\zeta))\right|d\zeta,$$

for every integer $\nu' \geq 0$. We set $\nu' = \lceil \nu/\alpha_i \rceil \leq \nu$. Since $N \notin \omega_{P(0)}$ we have $|\zeta + \delta_{2^{k(P)}}(c(\omega_{P(0)}) - N)| \gtrsim 1$. Therefore,

$$\int_{\mathbb{R}^n} \left| \partial_{\zeta_i}^{\nu'} (m(\zeta + \delta_{2^{k(P)}}(c(\omega_{P(0)}) - N))\widehat{\phi}(\zeta)) \right| d\zeta \lesssim \|m\|_{\mathscr{M}_\alpha^{\nu'}} \leq \|m\|_{\mathscr{M}_\alpha^{\nu}}.$$

This concludes the proof of (4.15) in the case that $\rho_\alpha(x - c(I_P)) \geq 1$. In the case $\rho_\alpha(x - c(I_P)) \leq 1$ we simply use the triangle inequality on (8.1). □

Proof of (4.16) *and* (4.17). If $c(I_P) = c(I_{P'})$, the estimates are trivial. Thus we may assume $c(I_P) \neq c(I_{P'})$. We have

$$|\langle \phi_P, \phi_{P'} \rangle| \leq |I_P|^{-\frac{1}{2}} |I_{P'}|^{-\frac{1}{2}} \int_{\mathbb{R}^n} |\phi(\delta_{2^{k(P)}}(x - c(I_P)))\phi(\delta_{2^{k_{P'}}}(x - c(I_{P'})))| dx. \tag{8.2}$$

Since

$$\rho_\alpha(c(I_P) - c(I_{P'})) \leq \rho_\alpha(x - c(I_P)) + \rho_\alpha(x - c(I_{P'})),$$

at least one of $\rho_\alpha(x - c(I_P))$, $\rho_\alpha(x - c(I_{P'}))$ is $\geq \frac{1}{2}\rho_\alpha(c(I_P) - c(I_{P'}))$. Thus, splitting the integral over x accordingly and using rapid decay of ϕ, the right hand side of (8.2) is no greater than a constant times

$$\begin{aligned} &|I_P|^{-\frac{1}{2}} |I_{P'}|^{\frac{1}{2}} (1 + 2^{-k(P)} \rho_\alpha(c(I_P) - c(I_{P'})))^{-\nu} \\ &+ |I_P|^{\frac{1}{2}} |I_{P'}|^{-\frac{1}{2}} (1 + 2^{-k_{P'}} \rho_\alpha(c(I_P) - c(I_{P'})))^{-\nu}. \end{aligned}$$

Recalling that we assumed $|I_P| \geq |I_{P'}|$ we see that the previous display is bounded by a constant times

$$|I_P|^{-\frac{1}{2}} |I_{P'}|^{\frac{1}{2}} (1 + 2^{-k(P)} \rho_\alpha(c(I_P) - c(I_{P'})))^{-\nu}.$$

This proves (4.16). The estimate (4.17) can be proven in the same way, by using the decay estimate (4.15). □

Acknowledgements The author thanks his doctoral advisor Christoph Thiele for encouragement and countless valuable comments and discussions on this project. He is also deeply indebted to Po-Lam Yung for many important and fruitful conversations. This work was carried out while the author was supported by the German Academic Scholarship Foundation.

References

1. Carleson, L.: On convergence and growth of partial sums of Fourier series. Acta Math. **116**, 135–157 (1966)
2. Duoandikoetxea, J.: Fourier Analysis, vol. 29. Graduate Studies in Mathematics. American Mathematical Society, Providence (2001). Translated and revised from the 1995 Spanish original by D. Cruz-Uribe
3. Fabes, E., Rivière, N.: Singular integrals with mixed homogeneity. Stud. Math. **27**(1), 19–38 (1966)
4. Fefferman, C.: Pointwise convergence of Fourier series. Ann. Math. **2**(98), 551–571 (1973)

5. Grafakos, L., Tao, T., Terwilleger, E.: L^p bounds for a maximal dyadic sum operator. Math. Z. **246**(1–2), 321–337 (2004)
6. Guo, S., Pierce, L.B., Roos, J., Yung, P.-L.: Polynomial Carleson operators along monomial curves in the plane. J. Geom. Anal. **27**(4), 2977–3012 (2017)
7. Kenig, C., Tomas, P.A.: Maximal operators defined by Fourier multipliers. Stud. Math. **68**, 79–83 (1980)
8. Lacey, M.T.: Carleson's theorem: proof, complements, variations. Publ. Mat. **48**(2), 251–307 (2004)
9. Lacey, M.T., Thiele, C.: A proof of boundedness of the Carleson operator. Math. Res. Lett. **7**(4), 361–370 (2000)
10. Lie, V.: The (weak-L^2) boundedness of the quadratic Carleson operator. Geom. Funct. Anal. **19**, 457–497 (2009)
11. Pierce, L.B., Yung, P.-L.: A polynomial Carleson operator along the paraboloid (2015). arXiv:1505.03882. Rev. Mat. Iberoam. (To appear)
12. Pramanik, M., Terwilleger, E.: A weak L^2 estimate for a maximal dyadic sum operator on $\mathbb{R}^n$. Ill. J. Math. **47**(3), 775–813 (2003)
13. Sjölin, P.: Convergence almost everywhere of certain singular integrals and multiple Fourier series. Ark. Mat. **9**, 65–90 (1971)
14. Stein, E.M.: Harmonic Analysis: Real-Variable Methods, Orthogonality, and Oscillatory Integrals, vol. 43. Princeton Mathematical Series. Princeton University Press (1993)
15. Thiele, C.: Wave Packet Analysis, vol. 105. CBMS Regional Conference Series in Mathematics. AMS, Providence (2006)

Journal of Fourier Analysis and Applications (2019) 25:2356–2418
https://doi.org/10.1007/s00041-019-09666-0

Harmonic Functions, Conjugate Harmonic Functions and the Hardy Space H^1 in the Rational Dunkl Setting

Jean-Philippe Anker[1] · **Jacek Dziubański**[2] · **Agnieszka Hejna**[2]

Received: 15 February 2018 / Revised: 8 January 2019 / Published online: 14 February 2019
© The Author(s) 2019

Abstract

In this work we extend the theory of the classical Hardy space H^1 to the rational Dunkl setting. Specifically, let Δ be the Dunkl Laplacian on a Euclidean space $\mathbb{R}^N$. On the half-space $\mathbb{R}_+ \times \mathbb{R}^N$, we consider systems of conjugate $(\partial_t^2 + \Delta_{\mathbf{x}})$-harmonic functions satisfying an appropriate uniform L^1 condition. We prove that the boundary values of such harmonic functions, which constitute the real Hardy space H^1_Δ, can be characterized in several different ways, namely by means of atoms, Riesz transforms, maximal functions or Littlewood–Paley square functions.

Keywords Rational Dunkl theory · Hardy spaces · Cauchy–Riemann equations · Riesz transforms · Maximal operators

Mathematics Subject Classification Primary 42B30 · Secondary 33C52, 35J05, 35K08, 42B25, 42B35, 42B37

In memory of Elias M. Stein (1931–2018).

Communicated by Dachun Yang.

✉ Jacek Dziubański
jdziuban@math.uni.wroc.pl

Jean-Philippe Anker
anker@univ-orleans.fr

Agnieszka Hejna
hejna@math.uni.wroc.pl

1 Institut Denis Poisson (UMR 7013), Université d'Orléans, Université de Tours & CNRS, B.P. 6759, 45067 Orléans Cedex 2, France

2 Instytut Matematyczny, Uniwersytet Wrocławski, Pl. Grunwaldzki 2/4, 50-384 Wrocław, Poland

1 Introduction

Real Hardy spaces on $\mathbb{R}^N$ have their origin in the study of holomorphic functions of one variable in the upper half-plane $\mathbb{R}^2_+ = \{z = x + iy \in \mathbb{C} : y > 0\}$. The theorem of Burkholder et al. [5] asserts that a real-valued harmonic function u on $\mathbb{R}^2_+$ is the real part of a holomorphic function $F(z) = u(z) + iv(z)$ satisfying the L^p condition

$$\sup_{y>0} \int_{\mathbb{R}} |F(x+iy)|^p \, dx < \infty, \quad 0 < p < \infty,$$

if and only if the nontangential maximal function $u^*(x) = \sup_{|x-x'|<y} |u(x'+iy)|$ belongs to $L^p(\mathbb{R})$. Here $0 < p < \infty$. The N-dimensional theory was then developed in Stein and Weiss [36] and Fefferman and Stein [19], where the notion of holomorphy was replaced by conjugate harmonic functions. To be more precise, a system of C^2 functions

$$\begin{aligned} &\boldsymbol{u}(x_0, x_1, \dots, x_N) \\ &\quad = (u_0(x_0, x_1, \dots, x_N), u_1(x_0, x_1, \dots, x_N), \dots, u_N(x_0, x_1, \dots, x_N)), \end{aligned}$$

where $x_0 > 0$, satisfies the generalized Cauchy–Riemann equations if

$$\frac{\partial u_j}{\partial x_i} = \frac{\partial u_i}{\partial x_j} \quad \forall\, 0 \le i \ne j \le N \quad \text{and} \quad \sum_{j=0}^{N} \frac{\partial u_j}{\partial x_j} = 0. \tag{1.1}$$

One says that $\boldsymbol{u}$ has the L^p property if

$$\sup_{x_0>0} \int_{\mathbb{R}^N} |\boldsymbol{u}(x_0, x_1, \dots, x_N)|^p \, dx_1 \dots dx_N < \infty. \tag{1.2}$$

As in the case $N = 1$, if $1 - \frac{1}{N} < p < \infty$ and $u_0(x_0, x_1, \dots, x_N)$ is a harmonic function, there is a system $\boldsymbol{u} = (u_0, u_1, \dots, u_N)$ of C^2 functions satisfying (1.1) and (1.2) if and only if

$$u_0^*(\mathbf{x}) = \sup_{\|\mathbf{x}-\mathbf{x}'\|<x_0} |u_0(x_0, \mathbf{x}')|$$

belongs to $L^p(\mathbb{R}^N)$. Here $\mathbf{x} = (x_1, \dots, x_N) \in \mathbb{R}^N$ and similarly $\mathbf{x}' = (x'_1, \dots, x'_N)$. Then u_0 has a limit f_0 in the sense of distributions, as $x_0 \searrow 0$, and u_0 is the Poisson integral of f_0. It turns out that the set of all distributions obtained in this way, which forms the so-called real Hardy space $H^p(\mathbb{R}^N)$, can be equivalently characterized in terms of real analysis (see [19]), namely by means of various maximal functions, square functions or Riesz transforms. Another important contribution to this theory lies in the atomic decomposition introduced by Coifman [7] and extended to spaces of homogeneous type by Coifman and Weiss [8].

The goal of this paper is to study harmonic functions, conjugate harmonic functions, and related Hardy space H^1 for the Dunkl Laplacian Δ (see Sect. 2). We shall prove that these objects have properties analogous to the classical ones. In particular, the related real Hardy space H^1_Δ, which can be defined as the set of boundary values of $(\partial_t^2 + \Delta_{\mathbf{x}})$-harmonic functions satisfying a relevant L^1 property, can be characterized by appropriate maximal functions, square functions, Riesz transforms or atomic decompositions. Apart from the square function characterization, this was achieved previously in [3] and [13] in the one-dimensional case, as well as in the product case.

Hardy spaces associated with semigroups of linear operators have a long history. Let us present a small and selected part of it. Muckenhoupt and Stein [26] introduced a notion of conjugacy for the one–dimensional Bessel operator, which initiated a study of Hardy spaces in the Bessel setting, continued subsequently in [4]. In [20] and [6], the authors developed a theory of real Hardy spaces H^p on homogeneous nilpotent Lie groups, associated either with a sublaplacian (if the group is stratified) or with a Rockland operator (if the group is graded). Another important contribution is the theory of local Hardy spaces in [22], which has several applications, e.g., in the study of Hardy spaces associated with the twisted laplacian [25] or with Schrödinger operators with certain (large) potentials [17]. Hardy spaces associated with semigroups whose kernels satisfy Gaussian bounds were studied in [24]. There, the theory of tent spaces [9,33] was used to produce specific atomic decompositions for Hardy spaces defined by square functions. This theory was further enhanced in [11,37] via characterizations by means of maximal functions.

In the one-dimensional case and in the product case considered in [3,13], the Dunkl kernel can be expressed explicitly in terms of classical special functions (Bessel functions or the confluent hypergeometric function). Thus its behavior is fully understood. In the general case considered in the present paper, no such information is available. Therefore an essential part of our work consists in estimating the Dunkl kernel, the heat kernel, the Poisson kernel, and their derivatives (see Sects. 3–5). As observed in [3], the heat kernel satisfies no Gaussian bound in the Dunkl setting. However, as shown in Sect. 4, some Gaussian-type bounds hold provided the Euclidean distance is replaced by the orbit distance (3.3). Similarly for the Poisson kernel, whose estimates in terms of the orbit distance resemble the analysis on spaces of homogeneous type (see Sect. 5). These crucial observations allow us to adapt the techniques of [11,24,37] in order to obtain atomic, maximal function, and square function characterizations of the Hardy space H^1_Δ. As far as the Riesz transform characterization of H^1_Δ is concerned, we use the maximum principle for Dunkl–Laplace subharmonic functions, together with estimates for the Dunkl and Poisson kernels.

Let us finally mention some further works in the continuation of the present paper. In [14] another atomic decomposition for the Hardy H^1_Δ space is obtained. The article [23] provides characterizations of the Hardy space associated with the Dunkl harmonic oscillator, while [15] is devoted to non-radial multipliers associated with the Dunkl transform.

1.1 Notation

- As usual, $\mathbb{N} = \{0, 1, 2, \dots\}$ denotes the set of nonnegative integers.
- The Euclidean space $\mathbb{R}^N$ is equipped with the standard inner product

$$\langle \mathbf{x}, \mathbf{y} \rangle = \sum\nolimits_{j=1}^{N} x_j \, y_j$$

  and the corresponding norm $\|\mathbf{x}\| = \left(\sum_{j=1}^{N} |x_j|^2\right)^{1/2}$. Throughout the paper,

$$B(\mathbf{x}, r) = \{\mathbf{y} \in \mathbb{R}^N \mid \|\mathbf{x} - \mathbf{y}\| < r\}$$

  stands for the ball with center $\mathbf{x} \in \mathbb{R}^N$ and radius $r > 0$. Finally, $\mathbb{R}_+^{N+1}$ denotes the half-space $(0, \infty) \times \mathbb{R}^N$ in $\mathbb{R}^{N+1}$.
- In $\mathbb{R}^N$, the directional derivative along ξ is denoted by ∂_ξ. As usual, for every multi-index $\alpha = (\alpha_1, \alpha_2, \dots, \alpha_N) \in \mathbb{N}^N$, we set $|\alpha| = \sum_{j=1}^{N} \alpha_j$ and

$$\partial^\alpha = \partial_{e_1}^{\alpha_1} \circ \partial_{e_2}^{\alpha_2} \circ \dots \circ \partial_{e_N}^{\alpha_N},$$

  where $\{e_1, e_2, \dots, e_N\}$ is the canonical basis of $\mathbb{R}^N$. The additional subscript $\mathbf{x}$ in $\partial_{\mathbf{x}}^\alpha$ means that the partial derivative ∂^α is taken with respect to the variable $\mathbf{x} \in \mathbb{R}^N$.
- The symbol $\sim$ between two positive expressions f, g means that their ratio $\frac{f}{g}$ is bounded from above and below by positive constants.
- The symbol $\lesssim$ (respectively $\gtrsim$) between two nonnegative expressions f, g means that there exists a constant $C > 0$ such that $f \le Cg$ (respectively $f \ge Cg$).
- We denote by $C_0(\mathbb{R}^N)$ the space of all continuous functions on $\mathbb{R}^N$ vanishing at infinity, by $C_c^\infty(\mathbb{R}^N)$ the space of all smooth functions on $\mathbb{R}^N$ with compact support, and by $\mathcal{S}(\mathbb{R}^N)$ the Schwartz class on $\mathbb{R}^N$. If $m \in \mathbb{N}$ and Ω is an open subset of $\mathbb{R}^N$, then f is a C^m function on Ω if f and all partial derivatives $\partial^\alpha f$, $|\alpha| \le m$, are continuous functions on Ω.
- If J is a measurable subset of $\mathbb{R}^N$, then χ_J denotes the characteristic function of J, that is, $\chi_J(\mathbf{x}) = 1$ if $\mathbf{x} \in J$ and $\chi_J(\mathbf{x}) = 0$ otherwise.
- Throughout the paper, C, C', c, etc. stand for positive constants, whose values may vary from occurrence to occurrence.

Further notation is defined in the next two sections.

2 Statement of the Results

In this section we first collect basic facts concerning Dunkl operators, the Dunkl Laplacian, and the corresponding heat and Poisson semigroups. For details we refer the reader to [12,30,32]. Next we state our main results.

In the Euclidean space $\mathbb{R}^N$ the reflection σ_α with respect to the hyperplane $\alpha^\perp$ orthogonal to a nonzero vector $\alpha \in \mathbb{R}^N$ is given by

$$\sigma_\alpha(\mathbf{x}) = \mathbf{x} - 2\frac{\langle \mathbf{x}, \alpha\rangle}{\|\alpha\|^2}\alpha.$$

A finite set $R \subset \mathbb{R}^N \setminus \{0\}$ is called a *root system* if $\sigma_\alpha(R) = R$ for every $\alpha \in R$. We shall consider normalized reduced root systems, that is, $\|\alpha\|^2 = 2$ for every $\alpha \in R$. The finite group G generated by the reflections σ_α is called the *Weyl group* (*reflection group*) of the root system. We shall denote by $\mathcal{O}(\mathbf{x})$, resp. $\mathcal{O}(B)$ the G-orbit of a point $\mathbf{x} \in \mathbb{R}^N$, resp. a subset $B \subset \mathbb{R}^N$. A *multiplicity function* is a G-invariant function $k : R \to \mathbb{C}$, which will be fixed and ≥ 0 throughout this paper.

Given a root system R and a multiplicity function k, the *Dunkl operators* T_ξ are the following deformations of directional derivatives ∂_ξ by difference operators:

$$\begin{aligned} T_\xi f(\mathbf{x}) &= \partial_\xi f(\mathbf{x}) + \sum_{\alpha\in R} \frac{k(\alpha)}{2}\langle \alpha, \xi\rangle \frac{f(\mathbf{x}) - f(\sigma_\alpha(\mathbf{x}))}{\langle \alpha, \mathbf{x}\rangle} \\ &= \partial_\xi f(\mathbf{x}) + \sum_{\alpha\in R^+} k(\alpha)\langle \alpha, \xi\rangle \frac{f(\mathbf{x}) - f(\sigma_\alpha(\mathbf{x}))}{\langle \alpha, \mathbf{x}\rangle}. \end{aligned}$$

Here R^+ is any fixed positive subsystem of R. The Dunkl operators T_ξ, which were introduced in [12], commute pairwise and are skew-symmetric with respect to the G-invariant measure $dw(\mathbf{x}) = w(\mathbf{x})\, d\mathbf{x}$, where

$$w(\mathbf{x}) = \prod_{\alpha\in R} |\langle \alpha, \mathbf{x}\rangle|^{k(\alpha)} = \prod_{\alpha\in R^+} |\langle \alpha, \mathbf{x}\rangle|^{2k(\alpha)}.$$

Set $T_j = T_{e_j}$, where $\{e_1, \dots, e_N\}$ is the canonical basis of $\mathbb{R}^N$. The *Dunkl Laplacian* associated with R and k is the differential-difference operator $\Delta = \sum_{j=1}^n T_j^2$, which acts on C^2 functions by

$$\Delta f(\mathbf{x}) = \Delta_{\mathrm{eucl}} f(\mathbf{x}) + \sum_{\alpha\in R} k(\alpha)\delta_\alpha f(\mathbf{x}) = \Delta_{\mathrm{eucl}} f(\mathbf{x}) + 2\sum_{\alpha\in R^+} k(\alpha)\delta_\alpha f(\mathbf{x}),$$

where

$$\delta_\alpha f(\mathbf{x}) = \frac{\partial_\alpha f(\mathbf{x})}{\langle \alpha, \mathbf{x}\rangle} - \frac{f(\mathbf{x}) - f(\sigma_\alpha(\mathbf{x}))}{\langle \alpha, \mathbf{x}\rangle^2}.$$

The operator Δ is essentially self-adjoint on $L^2(dw)$ (see for instance [2, Theorem 3.1]) and generates the heat semigroup

$$H_t f(\mathbf{x}) = e^{t\Delta} f(\mathbf{x}) = \int_{\mathbb{R}^N} h_t(\mathbf{x}, \mathbf{y}) f(\mathbf{y})\, dw(\mathbf{y}). \tag{2.1}$$

Here the heat kernel $h_t(\mathbf{x},\mathbf{y})$ is a C^∞ function in all variables $t>0, \mathbf{x}\in\mathbb{R}^N, \mathbf{y}\in\mathbb{R}^N$, which satisfies

$$h_t(\mathbf{x},\mathbf{y})=h_t(\mathbf{y},\mathbf{x})>0 \quad\text{and}\quad \int_{\mathbb{R}^N} h_t(\mathbf{x},\mathbf{y})\,dw(\mathbf{y})=1.$$

Notice that (2.1) defines a strongly continuous semigroup of linear contractions on $L^p(dw)$, for every $1\le p<\infty$.

The Poisson semigroup $P_t=e^{-t\sqrt{-\Delta}}$ is given by the subordination formula

$$P_t f(\mathbf{x})=\pi^{-1/2}\int_0^\infty e^{-u}\exp\Big(\frac{t^2}{4u}\Delta\Big)f(\mathbf{x})\frac{du}{\sqrt{u}} \tag{2.2}$$

and solves the boundary value problem

$$\begin{cases}(\partial_t^2+\Delta_{\mathbf{x}})\,u(t,\mathbf{x})=0\\ u(0,\mathbf{x})=f(\mathbf{x})\end{cases}$$

in the half-space $\mathbb{R}^{1+N}_+=(0,\infty)\times\mathbb{R}^N\subset\mathbb{R}^{1+N}$ (see [31, Sect. 5]). Let $e_0=(1,0,\ldots,0)$, $e_1=(0,1,\ldots,0)$,..., $e_N=(0,0,\ldots,1)$ be the canonical basis in $\mathbb{R}^{1+N}$. In order to unify our notation we shall denote the variable t by x_0 and set $T_0=\partial_{e_0}$.

Our goal is to study real harmonic functions of the operator

$$\mathcal{L}=T_0^2+\Delta=\sum_{j=0}^N T_j^2. \tag{2.3}$$

The operator $\mathcal{L}$ is the Dunkl Laplacian associated with the root system R, considered as a subset of $\mathbb{R}^{1+N}$ under the embedding $R\subset\mathbb{R}^N\hookrightarrow\mathbb{R}\times\mathbb{R}^N$.

We say that a system

$$\mathbf{u}=(u_0,u_1,\ldots,u_N),\ \text{ where }\ u_j=u_j(x_0,\underbrace{x_1,\ldots,x_N}_{\mathbf{x}})\ \ \forall\, 0\le j\le N,$$

of C^1 real functions on $\mathbb{R}^{1+N}_+$ satisfies the generalized Cauchy–Riemann equations if

$$\begin{cases}T_iu_j=T_ju_i \quad \forall\, 0\le i\ne j\le N,\\ \sum_{j=0}^N T_ju_j=0.\end{cases} \tag{2.4}$$

In this case each component u_j is $\mathcal{L}$-harmonic, i.e., $\mathcal{L}u_j=0$.

We say that a system $\mathbf{u}$ of C^2 real $\mathcal{L}$-harmonic functions on $\mathbb{R}^{1+N}_+$ belongs to the Hardy space $\mathcal{H}^1$ if it satisfies both (2.4) and the L^1 condition

$$\|\mathbf{u}\|_{\mathcal{H}^1}=\sup_{x_0>0}\big\||\mathbf{u}(x_0,\cdot)|\big\|_{L^1(dw)}=\sup_{x_0>0}\int_{\mathbb{R}^N}|\mathbf{u}(x_0,\mathbf{x})|\,dw(\mathbf{x})<\infty,$$

where $|\mathbf{u}(x_0,\mathbf{x})|=\Big(\sum_{j=0}^N|u_j(x_0,\mathbf{x})|^2\Big)^{1/2}$.

We are now ready to state our first main result.

Theorem 2.1 *Let u_0 be a $\mathcal{L}$-harmonic function in the upper half-space $\mathbb{R}^{1+N}_+$. Then there are $\mathcal{L}$-harmonic functions u_j $(j = 1, \ldots, N)$ such that $\mathbf{u} = (u_0, u_1, \ldots, u_N)$ belongs to $\mathcal{H}^1$ if and only if the nontangential maximal function*

$$u_0^*(\mathbf{x}) = \sup_{\|\mathbf{x}'-\mathbf{x}\|<x_0} |u_0(x_0, \mathbf{x}')| \tag{2.5}$$

belongs to $L^1(dw)$. In this case, the norms $\|u_0^*\|_{L^1(dw)}$ and $\|\mathbf{u}\|_{\mathcal{H}^1}$ are moreover equivalent.

If $\mathbf{u} \in \mathcal{H}^1$, we shall prove that the limit $f(\mathbf{x}) = \lim_{x_0 \to 0} u_0(x_0, \mathbf{x})$ exists in $L^1(dw)$ and $u_0(x_0, \mathbf{x}) = P_{x_0} f(\mathbf{x})$. This leads to consider the so-called real Hardy space

$$H^1_\Delta = \{f(\mathbf{x}) = \lim_{x_0 \to 0} u_0(x_0, \mathbf{x}) \mid (u_0, u_1, \ldots, u_N) \in \mathcal{H}^1\},$$

equipped with the norm

$$\|f\|_{H^1_\Delta} = \|(u_0, u_1, \ldots, u_N)\|_{\mathcal{H}^1}.$$

Let us denote by

$$\mathcal{M}_P f(\mathbf{x}) = \sup_{\|\mathbf{x}-\mathbf{x}'\|<t} \left|P_t f(\mathbf{x}')\right| \tag{2.6}$$

the nontangential maximal function associated with the Poisson semigroup $P_t = e^{-t\sqrt{-\Delta}}$. According to Theorem 2.1, H^1_Δ coincides with the following subspace of $L^1(dw)$:

$$H^1_{\max, P} = \{f \in L^1(dw) \mid \|f\|_{H^1_{\max, P}} := \|\mathcal{M}_P f\|_{L^1(dw)} < \infty\}. \tag{2.7}$$

Moreover, the norms $\|f\|_{H^1_\Delta}$ and $\|f\|_{H^1_{\max, P}}$ are equivalent.

Our task is to prove other characterizations of H^1_Δ by means of real analysis.

A Characterization by the Heat Maximal Function

Let

$$\mathcal{M}_H f(\mathbf{x}) = \sup_{\|\mathbf{x}-\mathbf{x}'\|^2<t} |H_t f(\mathbf{x}')|$$

be the nontangential maximal function associated with the heat semigroup $H_t = e^{t\Delta}$ and set

$$H^1_{\max, H} = \{f \in L^1(dw) \mid \|f\|_{H^1_{\max, H}} := \|\mathcal{M}_H f\|_{L^1(dw)} < \infty\}. \tag{2.8}$$

Theorem 2.2 *The spaces H^1_Δ and $H^1_{\max, H}$ coincide and the corresponding norms $\|f\|_{H^1_\Delta}$ and $\|f\|_{H^1_{\max, H}}$ are equivalent.*

B Characterization by Square Functions

For every $1 \leq p \leq \infty$, the operators $Q_t = t\sqrt{-\Delta}e^{-t\sqrt{-\Delta}}$ are uniformly bounded on $L^p(dw)$ (this is a consequence of the estimates (4.3), (5.7) and (5.4)). Consider the square function

$$Sf(\mathbf{x}) = \left(\iint_{\|\mathbf{x}-\mathbf{y}\|<t} |Q_t f(\mathbf{y})|^2 \frac{dt\, dw(\mathbf{y})}{t\, w(B(\mathbf{x},t))} \right)^{1/2} \tag{2.9}$$

and the space

$$H^1_{\text{square}} = \{ f \in L^1(dw) \mid \|Sf\|_{L^1(dw)} < \infty \}.$$

Theorem 2.3 *The spaces H^1_Δ and H^1_{square} coincide and the corresponding norms $\|f\|_{H^1_\Delta}$ and $\|Sf\|_{L^1(dw)}$ are equivalent.*

Remark 2.4 The square function characterization of H^1_Δ is also valid for $Q_t = t^2\Delta\, e^{t^2\Delta}$.

C Characterization by Riesz Transforms

The Riesz transforms, which are defined in the Dunkl setting by

$$R_j f = T_j(-\Delta)^{-1/2} f$$

(see Sect. 8), are bounded operators on $L^p(dw)$, for every $1 < p < \infty$ (cf. [1]). In the limit case $p = 1$, they turn out to be bounded operators from H^1_Δ into $H^1_\Delta \subset L^1(dw)$. This leads to consider the space

$$H^1_{\text{Riesz}} = \{ f \in L^1(dw) \mid \|R_j f\|_{L^1(w)} < \infty,\ \forall\, 1 \leq j \leq N \}.$$

Theorem 2.5 *The spaces H^1_Δ and H^1_{Riesz} coincide and the corresponding norms $\|f\|_{H^1_\Delta}$ and*

$$\|f\|_{H^1_{\text{Riesz}}} := \|f\|_{L^1(dw)} + \sum\nolimits_{j=1}^{N} \|R_j f\|_{L^1(dw)}.$$

are equivalent.

D Characterization by Atomic Decompositions

Let us define atoms in the spirit of [24]. Given a Euclidean ball B in $\mathbb{R}^N$, we shall denote its radius by r_B and its G-orbit by $\mathcal{O}(B)$. For any positive integer M, let $\mathcal{D}(\Delta^M)$ be the domain of Δ^M as an (unbounded) operator on $L^2(dw)$.

Definition 2.6 Let $1 < q \leq \infty$ and let M be a positive integer. A function $a \in L^2(dw)$ is said to be a $(1, q, M)$*–atom* if there exist $b \in \mathcal{D}(\Delta^M)$ and a ball B such that

- $a = \Delta^M b$,
- $\operatorname{supp}(\Delta^\ell b) \subset \mathcal{O}(B)\ \forall\, 0 \leq \ell \leq M$,
- $\|(r_B^2\Delta)^\ell b\|_{L^q(dw)} \leq r_B^{2M} w(B)^{\frac{1}{q}-1}\ \forall\, 0 \leq \ell \leq M$.

Let us remark that $\|a\|_{L^1(dw)} \leq |G|^{1-\frac{1}{q}}$, where $|G|$ denotes the number of elements of G. This follows easily from the above conditions (with $\ell = M$) by using Hölder's inequality.

Definition 2.7 An $L^1(dw)$–function f belongs to $H^1_{(1,q,M)}$ if there exist $(1, q, M)$–atoms a_j and $\lambda_j \in \mathbb{C}$ such that $\sum_j |\lambda_j| < \infty$ and

$$f = \sum\nolimits_j \lambda_j a_j . \tag{2.10}$$

Moreover,

$$\|f\|_{H^1_{(1,q,M)}} = \inf\left\{\sum\nolimits_j |\lambda_j|\right\},$$

where the infimum is taken over all representations (2.10).

Notice that the series (2.10) converges (absolutely) in $L^1(dw)$ and almost everywhere. Moreover, the results in our paper ensure that the convergence of the series (2.10) holds in the Hardy space H^1 as well.

Theorem 2.8 *The spaces* H^1_Δ *and* $H^1_{(1,q,M)}$ *coincide and the corresponding norms are equivalent.*

Let us briefly describe the organization of the proofs of the results. Clearly, $H^1_{(1,q_1,M)} \subset H^1_{(1,q_2,M)}$ for $1 < q_2 \leq q_1 \leq \infty$. The proof $(u_0, u_1, \ldots, u_N) \in \mathcal{H}^1$ implies $u_0^* \in L^1(dw)$, which is actually the inclusion $H^1_\Delta \subset H^1_{\max,P}$, is presented in Sect. 7, see Proposition 7.6. The proof is based on $\mathcal{L}$-subharmonicity of certain function constructed from $\mathbf{u}$ (see Sect. 6). The converse to Proposition 7.6 is proved at the very end of Sect. 11. Inclusions: $H^1_\Delta \subset H^1_{\text{Riesz}} \subset H^1_\Delta$ are shown in Sect. 8. Further, $H^1_{(1,q,M)} \subset H^1_{\text{Riesz}}$ for M large is proved in Sect. 9. Section 10 is devoted to proving $H^1_{\max,H} = H^1_{\max,\text{P}}$. The proofs of $H^1_{\max,H} \subset H^1_{(1,\infty,M)}$ for every $M \geq 1$ are presented in Sect. 11. Inclusion: $H^1_{(1,q,M)} \subset H^1_{\max,H}$ for every $M \geq 1$ is proved in Sect. 12. Finally, $H^1_{(1,2,M)} \subset H^1_{\text{square}} \subset H^1_{(1,2,M)}$ are established in Sect. 13.

3 Dunkl Kernel, Dunkl Transform and Dunkl Translations

The purpose of this section is to collect some facts about the Dunkl kernel, the Dunkl transform and Dunkl translations. General references are [10,12,30,32]. At the end of this section we shall derive estimates for the Dunkl translations of radial functions. These estimates will be used later to obtain bounds for the heat kernel and for the Poisson kernel, as well as for their derivatives, and furthermore upper and lower bounds for the Dunkl kernel.

We begin with some notation. Given a root system R in $\mathbb{R}^N$ and a multiplicity function $k \geq 0$, let

$$\gamma = \sum\nolimits_{\alpha \in R^+} k(\alpha) \quad \text{and} \quad \mathbf{N} = N + 2\gamma. \tag{3.1}$$

The number $\mathbf{N}$ is called the homogeneous dimension, because of the scaling property

$$w(B(t\mathbf{x}, tr)) = t^{\mathbf{N}} w(B(\mathbf{x}, r)).$$

Observe that

$$w(B(\mathbf{x}, r)) \sim r^N \prod_{\alpha \in R} (\,|\langle \alpha, \mathbf{x} \rangle| + r\,)^{k(\alpha)}.$$

Thus the measure w is doubling, that is, there is a constant $C > 0$ such that

$$w(B(\mathbf{x}, 2r)) \leq C\, w(B(\mathbf{x}, r)).$$

Moreover, there exists a constant $C \geq 1$ such that, for every $\mathbf{x} \in \mathbb{R}^N$ and for every $r_2 \geq r_1 > 0$,

$$C^{-1}\Big(\frac{r_2}{r_1}\Big)^N \leq \frac{w(B(\mathbf{x}, r_2))}{w(B(\mathbf{x}, r_1))} \leq C\Big(\frac{r_2}{r_1}\Big)^{\mathbf{N}}. \tag{3.2}$$

Set

$$V(\mathbf{x}, \mathbf{y}, t) = \max\big\{w(B(\mathbf{x}, t)), w(B(\mathbf{y}, t))\big\}.$$

Finally, let

$$d(\mathbf{x}, \mathbf{y}) = \min_{\sigma \in G} \|\mathbf{x} - \sigma(\mathbf{y})\| \tag{3.3}$$

denote the distance between two G-orbits $\mathcal{O}(\mathbf{x})$ and $\mathcal{O}(\mathbf{y})$. Obviously, $\mathcal{O}(B(\mathbf{x}, r)) = \{\mathbf{y} \in \mathbb{R}^N \mid d(\mathbf{y}, \mathbf{x}) < r\}$ and

$$w(B(\mathbf{x}, r)) \leq w(\mathcal{O}(B(\mathbf{x}, r))) \leq |G|\, w(B(\mathbf{x}, r)).$$

3.1 Dunkl Kernel

For fixed $\mathbf{x} \in \mathbb{R}^N$, the *Dunkl kernel* $\mathbf{y} \longmapsto E(\mathbf{x}, \mathbf{y})$ is the unique solution to the system

$$\begin{cases} T_\xi f = \langle \xi, \mathbf{x} \rangle f \quad \forall\, \xi \in \mathbb{R}^N, \\ f(0) = 1. \end{cases}$$

The following integral formula was obtained by Rösler [28]:

$$E(\mathbf{x},\mathbf{y})=\int_{\mathbb{R}^N} e^{\langle\eta,\mathbf{y}\rangle}\,d\mu_{\mathbf{x}}(\eta), \tag{3.4}$$

where $\mu_{\mathbf{x}}$ is a probability measure supported in the convex hull $\operatorname{conv}\mathcal{O}(\mathbf{x})$ of the G-orbit of $\mathbf{x}$. The function $E(\mathbf{x},\mathbf{y})$, which generalizes the exponential function $e^{\langle\mathbf{x},\mathbf{y}\rangle}$, extends holomorphically to $\mathbb{C}^N\times\mathbb{C}^N$ and satisfies the following properties:

- $E(0,\mathbf{y})=1 \quad \forall\,\mathbf{y}\in\mathbb{C}^N$,
- $E(\mathbf{x},\mathbf{y})=E(\mathbf{y},\mathbf{x}) \quad \forall\,\mathbf{x},\mathbf{y}\in\mathbb{C}^N$,
- $E(\lambda\mathbf{x},\mathbf{y})=E(\mathbf{x},\lambda\mathbf{y}) \quad \forall\,\lambda\in\mathbb{C},\ \forall\,\mathbf{x},\mathbf{y}\in\mathbb{C}^N$,
- $E(\sigma(\mathbf{x}),\sigma(\mathbf{y}))=E(\mathbf{x},\mathbf{y}) \quad \forall\,\sigma\in G,\ \forall\,\mathbf{x},\mathbf{y}\in\mathbb{C}^N$,
- $\overline{E(\mathbf{x},\mathbf{y})}=E(\bar{\mathbf{x}},\bar{\mathbf{y}}) \quad \forall\,\mathbf{x},\mathbf{y}\in\mathbb{C}^N$,
- $E(\mathbf{x},\mathbf{y})>0 \quad \forall\,\mathbf{x},\mathbf{y}\in\mathbb{R}^N$,
- $|E(i\mathbf{x},\mathbf{y})|\le 1 \quad \forall\,\mathbf{x},\mathbf{y}\in\mathbb{R}^N$,
- $|\partial_{\mathbf{y}}^{\alpha}E(\mathbf{x},\mathbf{y})|\le\|\mathbf{x}\|^{|\alpha|}\max_{\sigma\in G}e^{\operatorname{Re}\langle\sigma(\mathbf{x}),\mathbf{y}\rangle} \quad \forall\,\alpha\in\mathbb{N}^N,\ \forall\,\mathbf{x}\in\mathbb{R}^N,\ \forall\,\mathbf{y}\in\mathbb{C}^N.$

3.2 Dunkl Transform

The *Dunkl transform* is defined on $L^1(dw)$ by

$$\mathcal{F}f(\xi)=c_k^{-1}\int_{\mathbb{R}^N} f(\mathbf{x})E(\mathbf{x},-i\xi)\,dw(\mathbf{x}),$$

where

$$c_k=\int_{\mathbb{R}^N} e^{-\frac{\|\mathbf{x}\|^2}{2}}\,dw(\mathbf{x})>0\,.$$

In the limit case $k\equiv 0$, the Dunkl transform boils down to the classical Fourier transform

$$\hat{f}(\xi)=(2\pi)^{-N/2}\int_{\mathbb{R}^N} f(\mathbf{x})\,e^{-i\langle\xi,\mathbf{x}\rangle}\,d\mathbf{x}\,.$$

The following properties hold for the Dunkl transform (see [10,32]):

- The Dunkl transform is a topological automorphisms of the Schwartz space $\mathcal{S}(\mathbb{R}^N)$.
- (*Inversion formula*) For every $f\in\mathcal{S}(\mathbb{R}^N)$ and actually for every $f\in L^1(dw)$ such that $\mathcal{F}f\in L^1(dw)$, we have

$$f(\mathbf{x})=(\mathcal{F})^2 f(-\mathbf{x}) \qquad \forall\,\mathbf{x}\in\mathbb{R}^N.$$

- (*Plancherel Theorem*) The Dunkl transform extends to an isometric automorphism of $L^2(dw)$.
- The Dunkl transform of a radial function is again a radial function.

- (*Scaling*) For $\lambda \in \mathbb{R}^*$, we have

$$\mathcal{F}(f_\lambda)(\xi) = \mathcal{F}f(\lambda\xi), \quad \text{where } f_\lambda(\mathbf{x}) = |\lambda|^{-\mathbf{N}} f(\lambda^{-1}\mathbf{x}).$$

- Via the Dunkl transform, the Dunkl operator T_η corresponds to the multiplication by $\pm i\,\langle \eta, \cdot \rangle$. Specifically,

$$\begin{cases} \mathcal{F}(T_\eta f) = i\,\langle \eta, \cdot \rangle\,\mathcal{F}f, \\ T_\eta(\mathcal{F}f) = -i\,\mathcal{F}(\langle \eta, \cdot \rangle f). \end{cases}$$

  In particular, $\mathcal{F}(\Delta f)(\xi) = -\|\xi\|^2 \mathcal{F}f(\xi)$.

3.3 Dunkl Translations and Dunkl Convolution

The *Dunkl translation* $\tau_{\mathbf{x}} f$ of a function $f \in \mathcal{S}(\mathbb{R}^N)$ by $\mathbf{x} \in \mathbb{R}^N$ is defined by

$$\tau_{\mathbf{x}} f(\mathbf{y}) = c_k^{-1} \int_{\mathbb{R}^N} E(i\xi, \mathbf{x})\, E(i\xi, \mathbf{y})\, \mathcal{F}f(\xi)\, dw(\xi). \tag{3.5}$$

Notice the following properties of Dunkl translations:

- each translation $\tau_{\mathbf{x}}$ is a continuous linear map of $\mathcal{S}(\mathbb{R}^N)$ into itself, which extends to a contraction on $L^2(dw)$,
- (*Identity*) $\tau_0 = I$,
- (*Symmetry*) $\tau_{\mathbf{x}} f(\mathbf{y}) = \tau_{\mathbf{y}} f(\mathbf{x}) \quad \forall\, \mathbf{x}, \mathbf{y} \in \mathbb{R}^N, \ \forall\, f \in \mathcal{S}(\mathbb{R}^N)$,
- (*Scaling*) $\tau_{\mathbf{x}}(f_\lambda) = (\tau_{\lambda^{-1}\mathbf{x}} f)_\lambda \quad \forall\, \lambda > 0, \ \forall\, \mathbf{x} \in \mathbb{R}^N, \ \forall\, f \in \mathcal{S}(\mathbb{R}^N)$,
- (*Commutativity*) the Dunkl translations $\tau_{\mathbf{x}}$ and the Dunkl operators T_ξ all commute,
- (*Skew–symmetry*)

$$\int_{\mathbb{R}^N} \tau_{\mathbf{x}} f(\mathbf{y})\, g(\mathbf{y})\, dw(\mathbf{y}) = \int_{\mathbb{R}^N} f(\mathbf{y})\, \tau_{-\mathbf{x}} g(\mathbf{y})\, dw(\mathbf{y}) \forall \mathbf{x} \in \mathbb{R}^N, \forall f, g \in \mathcal{S}(\mathbb{R}^N).$$

The latter formula allows us to define the Dunkl translations $\tau_{\mathbf{x}} f$ in the distributional sense for $f \in L^p(dw)$ with $1 \le p \le \infty$. In particular,

$$\int_{\mathbb{R}^N} \tau_{\mathbf{x}} f(\mathbf{y})\, dw(\mathbf{y}) = \int_{\mathbb{R}^N} f(\mathbf{y})\, dw(\mathbf{y}) \quad \forall\, \mathbf{x} \in \mathbb{R}^N, \ \forall\, f \in \mathcal{S}(\mathbb{R}^N).$$

Finally, notice that $\tau_{\mathbf{x}} f$ is given by (3.5), if $f \in L^1(dw)$ and $\mathcal{F}f \in L^1(dw)$.

The *Dunkl convolution* of two reasonable functions (for instance Schwartz functions) is defined by

$$\begin{aligned}(f * g)(\mathbf{x}) &= c_k\, \mathcal{F}^{-1}[(\mathcal{F}f)(\mathcal{F}g)](\mathbf{x}) \\ &= \int_{\mathbb{R}^N} (\mathcal{F}f)(\xi)\, (\mathcal{F}g)(\xi)\, E(\mathbf{x}, i\xi)\, dw(\xi) \quad \forall\, \mathbf{x} \in \mathbb{R}^N\end{aligned}$$

or, equivalently, by

$$(f*g)(\mathbf{x})=\int_{\mathbb{R}^N} f(\mathbf{y})\,\tau_{\mathbf{x}}g(-\mathbf{y})\,dw(\mathbf{y}) \qquad \forall\,\mathbf{x}\in\mathbb{R}^N.$$

3.4 Dunkl Translations of Radial Functions

The following specific formula was obtained by Rösler [29] for the Dunkl translations of (reasonable) radial functions $f(\mathbf{x})=\tilde{f}(\|\mathbf{x}\|)$:

$$\tau_{\mathbf{x}}f(-\mathbf{y})=\int_{\mathbb{R}^N}(\tilde{f}\circ A)(\mathbf{x},\mathbf{y},\eta)\,d\mu_{\mathbf{x}}(\eta) \qquad \forall\,\mathbf{x},\mathbf{y}\in\mathbb{R}^N. \tag{3.6}$$

Here

$$A(\mathbf{x},\mathbf{y},\eta)=\sqrt{\|\mathbf{x}\|^2+\|\mathbf{y}\|^2-2\langle\mathbf{y},\eta\rangle}=\sqrt{\|\mathbf{x}\|^2-\|\eta\|^2+\|\mathbf{y}-\eta\|^2}$$

and $\mu_{\mathbf{x}}$ is the probability measure occurring in (3.4), which is supported in $\operatorname{conv}\mathcal{O}(\mathbf{x})$.

In the remaining part of this section, we shall derive estimates for the Dunkl translations of certain radial functions. Recall that $d(\mathbf{x},\mathbf{y})$ denotes the distance of the orbits $\mathcal{O}(\mathbf{x})$ and $\mathcal{O}(\mathbf{y})$ (see (3.3)). Let us begin with the following elementary estimates (see, e.g., [1]), which hold for $\mathbf{x},\mathbf{y}\in\mathbb{R}^N$ and $\eta\in\operatorname{conv}\mathcal{O}(\mathbf{x})$:

$$A(\mathbf{x},\mathbf{y},\eta)\ge d(\mathbf{x},\mathbf{y}) \tag{3.7}$$

and

$$\begin{cases}\|\nabla_{\mathbf{y}}\{A(\mathbf{x},\mathbf{y},\eta)^2\}\|\le 2\,A(\mathbf{x},\mathbf{y},\eta),\\ |\,\partial_{\mathbf{y}}^{\beta}\{A(\mathbf{x},\mathbf{y},\eta)^2\}|\le 2 & \text{if } |\beta|=2,\\ \partial_{\mathbf{y}}^{\beta}\{A(\mathbf{x},\mathbf{y},\eta)^2\}=0 & \text{if } |\beta|>2.\end{cases} \tag{3.8}$$

Hence

$$\|\nabla_{\mathbf{y}}A(\mathbf{x},\mathbf{y},\eta)\|\le 1 \tag{3.9}$$

and, more generally,

$$|\partial_{\mathbf{y}}^{\beta}(\theta\circ A)(\mathbf{x},\mathbf{y},\eta)|\le C_\beta\,A(\mathbf{x},\mathbf{y},\eta)^{m-|\beta|} \qquad \forall\,\beta\in\mathbb{N}^N,$$

if $\theta\in C^\infty(\mathbb{R}\smallsetminus\{0\})$ is a homogeneous symbol of order $m\in\mathbb{R}$, i.e.,

$$\left|\left(\tfrac{d}{dx}\right)^{\beta}\theta(x)\right|\le C_\beta\,|x|^{m-\beta} \qquad \forall\,x\in\mathbb{R}\smallsetminus\{0\},\ \forall\,\beta\in\mathbb{N}.$$

Similarly,

$$|\partial_{\mathbf{y}}^{\beta}(\tilde{\theta}\circ A)(\mathbf{x},\mathbf{y},\eta)|\le C_\beta\,\{1+A(\mathbf{x},\mathbf{y},\eta)\}^{m-|\beta|} \qquad \forall\,\beta\in\mathbb{N}^N,$$

if $\tilde{\theta} \in C^\infty(\mathbb{R})$ is an even inhomogeneous symbol of order $m \in \mathbb{R}$, i.e.,

$$\left|\left(\tfrac{d}{dx}\right)^\beta \tilde{\theta}(x)\right| \leq C_\beta\,(1+|x|)^{m-\beta} \qquad \forall\, x \in \mathbb{R}\,,\ \forall\, \beta \in \mathbb{N}\,.$$

Consider the radial function

$$q(\mathbf{x}) = c_M\,(1+\|\mathbf{x}\|^2)^{-M/2}$$

on $\mathbb{R}^N$, where $M > \mathbf{N}$ and $c_M > 0$ is a normalizing constant such that $\int_{\mathbb{R}^N} q(\mathbf{x})\,dw(\mathbf{x}) = 1$. Notice that $\tilde{q}(x) = c_M\,(1+x^2)^{-M/2}$ is an even inhomogeneous symbol of order $-M$. The following estimate holds for the translates $q_t(\mathbf{x},\mathbf{y}) = \tau_{\mathbf{x}} q_t(-\mathbf{y})$ of $q_t(\mathbf{x}) = t^{-\mathbf{N}} q(t^{-1}\mathbf{x})$.

Proposition 3.1 *There exists a constant $C > 0$ (depending on M) such that*

$$0 \leq q_t(\mathbf{x},\mathbf{y}) \leq C\,V(\mathbf{x},\mathbf{y},t)^{-1} \qquad \forall\, t > 0,\ \forall\, \mathbf{x},\mathbf{y} \in \mathbb{R}^N.$$

Proof By scaling we can reduce to $t = 1$. Fix $\mathbf{x},\mathbf{y} \in \mathbb{R}^N$. We shall prove that

$$\begin{aligned}\int_{\mathbb{R}^N} (1 + A(\mathbf{x},\mathbf{y},\eta))^{-M} d\mu_{\mathbf{x}}(\eta) &\sim \int_{\mathbb{R}^N} (1 + A(\mathbf{x},\mathbf{y},\eta)^2)^{-M/2} d\mu_{\mathbf{x}}(\eta) \\ &= q_1(\mathbf{x},\mathbf{y}) \leq C V(\mathbf{x},\mathbf{y},1)^{-1}. \end{aligned} \tag{3.10}$$

Set $\bar{B} = \{\mathbf{y}' \in \mathbb{R}^N \mid \|\mathbf{y}'-\mathbf{y}\| \leq 1\}$. By continuity, the function $\bar{B} \ni \mathbf{y}' \longmapsto q_1(\mathbf{x},\mathbf{y}')$ reaches a maximum $K = q_1(\mathbf{x},\mathbf{y}_0) \geq 0$ on the ball $\bar{B}$ at some point $\mathbf{y}_0 \in \bar{B}$. For every $\mathbf{y}' \in \bar{B}$, we have

$$\begin{aligned} 0 \leq q_1(\mathbf{x},\mathbf{y}_0) - q_1(\mathbf{x},\mathbf{y}') &= \int_{\mathbb{R}^N} \big\{(\tilde{q}\circ A)(\mathbf{x},\mathbf{y}_0,\eta) - (\tilde{q}\circ A)(\mathbf{x},\mathbf{y}',\eta)\big\}\,d\mu_{\mathbf{x}}(\eta) \\ &= \int_{\mathbb{R}^N}\int_0^1 \frac{\partial}{\partial s}\,(\tilde{q}\circ A)(\mathbf{x},\underbrace{\mathbf{y}'+s(\mathbf{y}_0-\mathbf{y}')}_{\mathbf{y}_s},\eta)\,ds\,d\mu_{\mathbf{x}}(\eta) \\ &\leq \|\mathbf{y}_0-\mathbf{y}'\| \int_{\mathbb{R}^N}\int_0^1 |(\tilde{q}'\circ A)(\mathbf{x},\mathbf{y}_s,\eta)|\,ds\,d\mu_{\mathbf{x}}(\eta) \\ &\leq M\,\|\mathbf{y}_0-\mathbf{y}'\| \int_{\mathbb{R}^N}\int_0^1 (\tilde{q}\circ A)(\mathbf{x},\mathbf{y}_s,\eta)\,ds\,d\mu_{\mathbf{x}}(\eta) \\ &= M\,\|\mathbf{y}_0-\mathbf{y}'\| \int_0^1 q_1(\mathbf{x},\mathbf{y}_s)\,ds \\ &\leq M\,\|\mathbf{y}_0-\mathbf{y}'\|\,K\,. \end{aligned}$$

Here we have used (3.9) and the elementary estimate

$$|\tilde{q}'(x)| \leq M\,\tilde{q}(x) \qquad \forall\, x \in \mathbb{R}\,.$$

Hence

$$q_1(\mathbf{x},\mathbf{y}') \geq q_1(\mathbf{x},\mathbf{y}_0) - |q_1(\mathbf{x},\mathbf{y}_0) - q_1(\mathbf{x},\mathbf{y}')| \geq K - \frac{K}{2} = \frac{K}{2},$$

if $\mathbf{y}' \in \bar{B} \cap B(\mathbf{y}_0, r)$ with $r = \frac{1}{2M}$. Moreover, as $w(\bar{B} \cap B(\mathbf{y}_0, r)) \sim w(\bar{B})$, we have

$$\begin{aligned} 1 = \int_{\mathbb{R}^N} q_1(\mathbf{x},\mathbf{y}')dw(\mathbf{y}') &\geq \int_{\bar{B}\cap B(\mathbf{y}_0,r)} q_1(\mathbf{x},\mathbf{y}')\,dw(\mathbf{y}') \\ &\geq \frac{K}{2} w(\bar{B} \cap B(\mathbf{y}_0, r)) \geq \frac{K}{C}\, w(\bar{B})\,. \end{aligned}$$

Therefore

$$0 \leq q_1(\mathbf{x},\mathbf{y}) \leq K \leq C\, w(B(\mathbf{y},1))^{-1}.$$

We deduce (3.10) by using the symmetry $q_1(\mathbf{x},\mathbf{y}) = q_1(\mathbf{y},\mathbf{x})$. □

Consider next a radial function f satisfying

$$|f(\mathbf{x})| \lesssim (1+\|\mathbf{x}\|)^{-M-\kappa} \qquad \forall\, \mathbf{x} \in \mathbb{R}^N$$

with $M > \mathbf{N}$ and $\kappa \geq 0$. Then the following estimate holds for the translates $f_t(\mathbf{x},\mathbf{y}) = \tau_{\mathbf{x}} f_t(-\mathbf{y})$ of $f_t(\mathbf{x}) = t^{-\mathbf{N}} f(t^{-1}\mathbf{x})$.

Corollary 3.2 *There exists a constant $C > 0$ such that*

$$|f_t(\mathbf{x},\mathbf{y})| \leq C\, V(\mathbf{x},\mathbf{y},t)^{-1}\Big(1+\frac{d(\mathbf{x},\mathbf{y})}{t}\Big)^{-\kappa} \qquad \forall\, t>0,\ \forall\, \mathbf{x},\mathbf{y} \in \mathbb{R}^N.$$

Proof By scaling we can reduce to $t = 1$. By using (3.6), (3.7), and (3.10) we get

$$\begin{aligned} |f_1(\mathbf{x},\mathbf{y})| &\lesssim \int_{\mathbb{R}^N} \big(1+A(\mathbf{x},\mathbf{y},\eta)\big)^{-M}\big(1+A(\mathbf{x},\mathbf{y},\eta)\big)^{-\kappa} d\mu_{\mathbf{x}}(\eta) \\ &\leq C\, V(\mathbf{x},\mathbf{y},1)^{-1}\big(1+d(\mathbf{x},\mathbf{y})\big)^{-\kappa}. \end{aligned}$$

□

Notice that the space of radial Schwartz functions f on $\mathbb{R}^N$ identifies with the space of even Schwartz functions $\tilde{f}$ on $\mathbb{R}$, which is equipped with the norms

$$\|\tilde{f}\|_{\mathcal{S}_m} = \max_{0\leq j\leq m} \sup_{x\in\mathbb{R}} (1+|x|)^m \left|\Big(\frac{d}{dx}\Big)^j \tilde{f}(x)\right| \qquad \forall\, m \in \mathbb{N}\,. \tag{3.11}$$

Proposition 3.3 *For every $\kappa \geq 0$, there exist $C \geq 0$ and $m \in \mathbb{N}$ such that, for all even Schwartz functions $\tilde{\psi}^{\{1\}}$, $\tilde{\psi}^{\{2\}}$ and for all even nonnegative integers ℓ_1, ℓ_2, the convolution kernel*

$$\Psi_{s,t}(\mathbf{x},\mathbf{y}) \\ = c_k^{-1} \int_{\mathbb{R}^N} (s\,\|\xi\|)^{\ell_1}\, \psi^{\{1\}}(s\,\|\xi\|)\, (t\,\|\xi\|)^{\ell_2}\, \psi^{\{2\}}(t\,\|\xi\|)\, E(\mathbf{x}, i\xi)\, E(-\mathbf{y}, i\xi)\, dw(\xi)$$

satisfies

$$|\Psi_{s,t}(\mathbf{x},\mathbf{y})| \leq C\,\|\psi^{\{1\}}\|_{\mathcal{S}_{m+\ell_1+\ell_2}}\,\|\psi^{\{2\}}\|_{\mathcal{S}_{m+\ell_1+\ell_2}} \\ \times \min\left\{\left(\tfrac{s}{t}\right)^{\ell_1}, \left(\tfrac{t}{s}\right)^{\ell_2}\right\} V(\mathbf{x},\mathbf{y},s+t)^{-1}\Big(1+\frac{d(\mathbf{x},\mathbf{y})}{s+t}\Big)^{-\kappa},$$

for every $s, t > 0$ and for every $\mathbf{x}, \mathbf{y} \in \mathbb{R}^N$.

Proof By continuity of the inverse Dunkl transform in the Schwartz setting, there exists a positive even integer m and a constant $C > 0$ such that

$$\sup_{\mathbf{z}\in\mathbb{R}^N} (1+\|\mathbf{z}\|)^{M+\kappa}\, |\mathcal{F}^{-1} f(\mathbf{z})| \leq C\,\|\tilde{f}\|_{\mathcal{S}_m},$$

for every even function $\tilde{f} \in C^m(\mathbb{R})$ with $\|\tilde{f}\|_{\mathcal{S}_m} < \infty$. Consider first the case $0 < s \leq t = 1$. Then

$$\|\,(s\xi)^{\ell_1}\,\tilde{\psi}^{\{1\}}(s\xi)\,\xi^{\ell_2}\,\tilde{\psi}^{\{2\}}(\xi)\,\|_{\mathcal{S}_m} \leq C\,\|\psi^{\{1\}}\|_{\mathcal{S}_m}\,\|\psi^{\{2\}}\|_{\mathcal{S}_{m+\ell_1+\ell_2}}\, s^{\ell_1}.$$

According to Corollary 3.2, we deduce that

$$|\Psi_{s,1}(\mathbf{x},\mathbf{y})| \leq C\,\|\psi^{\{1\}}\|_{\mathcal{S}_{m+\ell_1+\ell_2}}\,\|\psi^{\{2\}}\|_{\mathcal{S}_{m+\ell_1+\ell_2}}\, s^{\ell_1}\, V(\mathbf{x},\mathbf{y},1)^{-1}\,\big(1+d(\mathbf{x},\mathbf{y})\big)^{-\kappa} \\ \leq C\,\|\psi^{\{1\}}\|_{\mathcal{S}_{m+\ell_1+\ell_2}}\,\|\psi^{\{2\}}\|_{\mathcal{S}_{m+\ell_1+\ell_2}}\, s^{\ell_1}\, V(\mathbf{x},\mathbf{y},s+1)^{-1}\Big(1+\frac{d(\mathbf{x},\mathbf{y})}{s+1}\Big)^{-\kappa}.$$

In the case $s = 1 \geq t > 0$, we have similarly

$$|\Psi_{1,t}(\mathbf{x},\mathbf{y})| \leq C\,\|\psi^{\{1\}}\|_{\mathcal{S}_{m+\ell_1+\ell_2}}\,\|\psi^{\{2\}}\|_{\mathcal{S}_{m+\ell_1+\ell_2}}\, t^{\ell_2}\, V(\mathbf{x},\mathbf{y},1+t)^{-1}\Big(1+\frac{d(\mathbf{x},\mathbf{y})}{1+t}\Big)^{-\kappa}.$$

The general case is obtained by scaling. □

4 Heat Kernel and Dunkl Kernel

Via the Dunkl transform, the heat semigroup $H_t = e^{t\Delta}$ is given by

$$H_t f(\mathbf{x}) = \mathcal{F}^{-1}\big(e^{-t\|\xi\|^2}\mathcal{F}f(\xi)\big)(\mathbf{x}).$$

Alternately (see, e.g., [32])

$$H_t f(\mathbf{x}) = f * h_t(\mathbf{x}) = \int_{\mathbb{R}^N} h_t(\mathbf{x},\mathbf{y})\, f(\mathbf{y})\, dw(\mathbf{y}),$$

where the heat kernel $h_t(\mathbf{x},\mathbf{y})$ is a smooth positive radial convolution kernel. Specifically, for every $t>0$ and for every $\mathbf{x},\mathbf{y}\in\mathbb{R}^N$,

$$h_t(\mathbf{x},\mathbf{y}) = c_k^{-1}\,(2t)^{-\mathbf{N}/2}\,e^{-\frac{\|\mathbf{x}\|^2+\|\mathbf{y}\|^2}{4t}}\,E\Big(\frac{\mathbf{x}}{\sqrt{2t}},\frac{\mathbf{y}}{\sqrt{2t}}\Big) = \tau_{\mathbf{x}} h_t(-\mathbf{y}), \tag{4.1}$$

where

$$h_t(\mathbf{x}) = \tilde{h}_t(\|\mathbf{x}\|) = c_k^{-1}\,(2t)^{-\mathbf{N}/2}\,e^{-\frac{\|\mathbf{x}\|^2}{4t}}.$$

In particular,

$$\begin{aligned} &h_t(\mathbf{x},\mathbf{y}) = h_t(\mathbf{y},\mathbf{x}) > 0,\\ &\int_{\mathbb{R}^N} h_t(\mathbf{x},\mathbf{y})\, dw(\mathbf{y}) = 1,\\ &h_t(\mathbf{x},\mathbf{y}) \le c_k^{-1}\,(2t)^{-\mathbf{N}/2}\,e^{-\frac{d(\mathbf{x},\mathbf{y})^2}{4t}}. \end{aligned} \tag{4.2}$$

4.1 Upper Heat Kernel Estimates

We prove now Gaussian bounds for the heat kernel and its derivatives, in the spirit of spaces of homogeneous type, except that the metric $\|\mathbf{x}-\mathbf{y}\|$ is replaced by the orbit distance $d(\mathbf{x},\mathbf{y})$ (see (3.3)). In comparison with (4.2), the main difference lies in the factor $t^{\mathbf{N}/2}$, which is replaced by the volume of appropriate balls.

Theorem 4.1 (a) Time derivatives*: for any nonnegative integer m, there are constants $C,c>0$ such that*

$$\big|\partial_t^m h_t(\mathbf{x},\mathbf{y})\big| \le C\,t^{-m}\,V(\mathbf{x},\mathbf{y},\sqrt{t})^{-1}\,e^{-c\,d(\mathbf{x},\mathbf{y})^2/t}, \tag{4.3}$$

for every $t>0$ and for every $\mathbf{x},\mathbf{y}\in\mathbb{R}^N$.

(b) Hölder bounds*: for any nonnegative integer m, there are constants $C,c>0$ such that*

$$\big|\partial_t^m h_t(\mathbf{x},\mathbf{y}) - \partial_t^m h_t(\mathbf{x},\mathbf{y}')\big| \le C\,t^{-m}\Big(\frac{\|\mathbf{y}-\mathbf{y}'\|}{\sqrt{t}}\Big)\,V(\mathbf{x},\mathbf{y},\sqrt{t})^{-1}\,e^{-c\,d(\mathbf{x},\mathbf{y})^2/t}, \tag{4.4}$$

for every $t>0$ and for every $\mathbf{x},\mathbf{y},\mathbf{y}'\in\mathbb{R}^N$ such that $\|\mathbf{y}-\mathbf{y}'\|<\sqrt{t}$.

(c) Dunkl derivative*: for any $\xi\in\mathbb{R}^N$ and for any nonnegative integer m, there are constants $C,c>0$ such that*

$$\Big|T_{\xi,\mathbf{x}}\,\partial_t^m h_t(\mathbf{x},\mathbf{y})\Big| \le C\,t^{-m-1/2}\,V(\mathbf{x},\mathbf{y},\sqrt{t})^{-1}\,e^{-c\,d(\mathbf{x},\mathbf{y})^2/t}, \tag{4.5}$$

for all $t>0$ and $\mathbf{x},\mathbf{y}\in\mathbb{R}^N$.

(d) Mixed derivatives*: for any nonnegative integer m and for any multi-indices α, β, there are constants $C, c>0$ such that, for every $t>0$ and for every $\mathbf{x},\mathbf{y}\in\mathbb{R}^N$,*

$$\left|\partial_t^m \partial_{\mathbf{x}}^{\alpha} \partial_{\mathbf{y}}^{\beta} h_t(\mathbf{x},\mathbf{y})\right| \le C\, t^{-m-\frac{|\alpha|}{2}-\frac{|\beta|}{2}}\, V(\mathbf{x},\mathbf{y},\sqrt{t})^{-1}\, e^{-c\,d(\mathbf{x},\mathbf{y})^2/t}, \tag{4.6}$$

for every $t>0$ and for every $\mathbf{x},\mathbf{y}\in\mathbb{R}^N$.

Proof The proof relies on the expression

$$h_t(\mathbf{x},\mathbf{y}) = \int_{\mathbb{R}^N} \tilde{h}_t\big(A(\mathbf{x},\mathbf{y},\eta)\big)\, d\mu_{\mathbf{x}}(\eta) \tag{4.7}$$

and on the properties of $A(\mathbf{x},\mathbf{y},\eta)$.

(a) Consider first the case $m=0$. By scaling we can reduce to $t=1$. On the one hand, we use (3.7) to estimate

$$\begin{aligned} c_k\, 2^{\mathbf{N}/2}\, h_1(\mathbf{x},\mathbf{y}) &= \int_{\mathbb{R}^N} e^{-A(\mathbf{x},\mathbf{y},\eta)^2/8}\, e^{-A(\mathbf{x},\mathbf{y},\eta)^2/8}\, d\mu_{\mathbf{x}}(\eta) \\ &\le e^{-d(\mathbf{x},\mathbf{y})^2/8} \int_{\mathbb{R}^N} e^{-A(\mathbf{x},\mathbf{y},\eta)^2/8}\, d\mu_{\mathbf{x}}(\eta)\,. \end{aligned}$$

On the other hand, it follows from Proposition 3.1 and Corollary 3.2 that

$$\int_{\mathbb{R}^N} e^{-c\,A(\mathbf{x},\mathbf{y},\eta)^2} d\mu_{\mathbf{x}}(\eta) \lesssim V(\mathbf{x},\mathbf{y},1)^{-1},$$

for any fixed $c>0$. Hence

$$h_1(\mathbf{x},\mathbf{y}) \lesssim V(\mathbf{x},\mathbf{y},1)^{-1}\, e^{-d(\mathbf{x},\mathbf{y})^2/8}.$$

Consider next the case $m>0$. Observe that $\partial_t^m \tilde{h}_t(x)$ is equal to $t^{-m}\tilde{h}_t(x)$ times a polynomial in $\frac{x^2}{t}$. Therefore

$$\left|\partial_t^m \tilde{h}_t(x)\right| \le C_m\, t^{-m}\, \tilde{h}_{2t}(x)\,. \tag{4.8}$$

By differentiating (4.7) and by using (4.8), we deduce that

$$\left|\partial_t^m h_t(\mathbf{x},\mathbf{y})\right| \le C_m\, t^{-m}\, h_{2t}(\mathbf{x},\mathbf{y})\,.$$

We conclude by using the case $m=0$.

(b) Observe now that $\tilde{\mathfrak{h}}_t(x) = \partial_x \partial_t^m \tilde{h}_t(x)$ is equal to $\frac{x}{t^{m+1}}\tilde{h}_t(x)$ times a polynomial in $\frac{x^2}{t}$, hence

$$\left|\tilde{\mathfrak{h}}_t(x)\right| \le C_m\, t^{-m-1/2}\, \tilde{h}_{2t}(x)\,. \tag{4.9}$$

By differentiating (4.7) and by using (3.9) and (4.3), we estimate

$$\begin{aligned}
|\partial_t^m h_t(\mathbf{x},\mathbf{y}) - \partial_t^m h_t(\mathbf{x},\mathbf{y}')| &= \Big|\int_{\mathbb{R}^N} \{\partial_t^m \tilde{h}_t(A(\mathbf{x},\mathbf{y},\eta)) - \partial_t^m \tilde{h}_t(A(\mathbf{x},\mathbf{y}',\eta))\} d\mu_{\mathbf{x}}(\eta)\Big| \\
&= \Big|\int_{\mathbb{R}^N}\int_0^1 \frac{\partial}{\partial s}\partial_t^m \tilde{h}_t(A(\mathbf{x},\underbrace{\mathbf{y}'+s(\mathbf{y}-\mathbf{y}')}_{\mathbf{y}_s},\eta))\, ds\, d\mu_{\mathbf{x}}(\eta)\Big| \\
&\le \|\mathbf{y}-\mathbf{y}'\| \int_0^1 \int_{\mathbb{R}^N} \big|\tilde{\mathfrak{h}}_t(A(\mathbf{x},\mathbf{y}_s,\eta))\big|\, d\mu_{\mathbf{x}}(\eta)\, ds \\
&\le C_m\, t^{-m}\, \frac{\|\mathbf{y}-\mathbf{y}'\|}{\sqrt{t}} \int_0^1 h_{2t}(\mathbf{x},\mathbf{y}_s)\, ds \\
&\le C_m'\, t^{-m}\, \frac{\|\mathbf{y}-\mathbf{y}'\|}{\sqrt{t}} \int_0^1 V(\mathbf{x},\mathbf{y}_s,\sqrt{2t}\,)\, e^{-c\frac{d(\mathbf{x},\mathbf{y}_s)^2}{2t}}\, ds\,.
\end{aligned}$$

In order to conclude, notice that

$$V(\mathbf{x},\mathbf{y}_s,\sqrt{2t}\,) \sim V(\mathbf{x},\mathbf{y},\sqrt{t}\,) \tag{4.10}$$

under the assumption $\|\mathbf{y}-\mathbf{y}'\| < \sqrt{t}$ and let us show that, for every $c>0$, there exists $C \ge 1$ such that

$$C^{-1} e^{-\frac{3}{2}c\,\frac{d(\mathbf{x},\mathbf{y})^2}{t}} \le e^{-c\,\frac{d(\mathbf{x},\mathbf{y}_s)^2}{t}} \le C\, e^{-\frac{1}{2}c\,\frac{d(\mathbf{x},\mathbf{y})^2}{t}}. \tag{4.11}$$

As long as $d(\mathbf{x},\mathbf{y}) \le C\sqrt{t}$, all expressions in (4.11) are indeed comparable to 1. On the other hand, if $d(\mathbf{x},\mathbf{y}) \ge \sqrt{32\,t}$, then

$$\begin{aligned}
|d(\mathbf{x},\mathbf{y})^2 - d(\mathbf{x},\mathbf{y}_s)^2| &= |d(\mathbf{x},\mathbf{y}) - d(\mathbf{x},\mathbf{y}_s)|\, \{d(\mathbf{x},\mathbf{y}) + d(\mathbf{x},\mathbf{y}_s)\} \\
&\le \|\mathbf{y}-\mathbf{y}_s\|\, \{2\,d(\mathbf{x},\mathbf{y}) + \|\mathbf{y}-\mathbf{y}_s\|\} \le \sqrt{2t}\,\{2\,d(\mathbf{x},\mathbf{y}) + \sqrt{2t}\,\} \\
&\le \sqrt{8t}\,d(\mathbf{x},\mathbf{y}) + 2t \le \frac{1}{2}\,d(\mathbf{x},\mathbf{y})^2 + 2t\,.
\end{aligned}$$

Hence

$$\frac{1}{2}\,d(\mathbf{x},\mathbf{y})^2/t - 2 \le d(\mathbf{x},\mathbf{y}_s)^2/t \le \frac{3}{2}\,d(\mathbf{x},\mathbf{y})^2/t + 2\,.$$

(c) By symmetry, we can replace $T_{\xi,\mathbf{x}}$ by $T_{\xi,\mathbf{y}}$. Consider first the contribution of the directional derivative in $T_{\xi,\mathbf{y}}$. By differentiating (4.7) and by using (4.9) and (4.3), we estimate as above

$$\begin{aligned}
|\partial_{\xi,\mathbf{y}}\partial_t^m h_t(\mathbf{x},\mathbf{y})| &\le \|\xi\| \int_{\mathbb{R}^N} |\tilde{\mathfrak{h}}_t(A(\mathbf{x},\mathbf{y},\eta))|\, d\mu_{\mathbf{x}}(\eta) \\
&\le C\, t^{-m-1/2}\, h_{2t}(\mathbf{x},\mathbf{y}) \\
&\le C\, t^{-m-1/2}\, V(\mathbf{x},\mathbf{y},\sqrt{t}\,)^{-1}\, e^{-c\,d(\mathbf{x},\mathbf{y})^2/t}.
\end{aligned}$$

Consider next the contributions

$$\frac{\partial_t^m h_t(\mathbf{x},\mathbf{y}) - \partial_t^m h_t(\mathbf{x},\sigma_\alpha(\mathbf{y}))}{\langle\alpha,\mathbf{y}\rangle} \tag{4.12}$$

of the difference operators in $T_{\xi,\mathbf{y}}$. If $|\langle\alpha,\mathbf{y}\rangle| > \sqrt{t/2}$, we use (4.3) and estimate separately each term in (4.12). If $|\langle\alpha,\mathbf{y}\rangle| \le \sqrt{t/2}$, we estimate again

$$\begin{aligned}\left|\frac{\partial_t^m h_t(\mathbf{x},\mathbf{y}) - \partial_t^m h_t(\mathbf{x},\sigma_\alpha(\mathbf{y}))}{\langle\alpha,\mathbf{y}\rangle}\right| &\le \sqrt{2}\int_{\mathbb{R}^N}\int_0^1 |\tilde{\mathfrak{h}}_t(A(\mathbf{x},\mathbf{y}_s,\eta))|\,ds\,d\mu_{\mathbf{x}}(\eta)\\ &\le C\,t^{-m-1/2}\int_0^1 h_{2t}(\mathbf{x},\mathbf{y}_s)\,ds\\ &\le C\,t^{-m-1/2}\int_0^1 V(\mathbf{x},\mathbf{y}_s,\sqrt{2t})^{-1}\,e^{-c\,\frac{d(\mathbf{x},\mathbf{y}_s)^2}{2t}}\,ds\\ &\le C\,t^{-m-1/2}\,V(\mathbf{x},\mathbf{y},\sqrt{t})^{-1}\,e^{-c\,\frac{d(\mathbf{x},\mathbf{y})^2}{t}}.\end{aligned}$$

In the last step we have used (4.10) and (4.11), which hold as $\|\mathbf{y}_s-\mathbf{y}\|\le\sqrt{t}$.

(d) This time, we use (3.8) to estimate

$$\left|\partial_{\mathbf{y}}^{\beta}\partial_t^m \tilde{h}_t\big(A(\mathbf{x},\mathbf{y},\eta)\big)\right| \le C_{m,\beta}\,t^{-m-\frac{|\beta|}{2}}\,\tilde{h}_{2t}\big(A(\mathbf{x},\mathbf{y},\eta)\big). \tag{4.13}$$

Firstly, by differentiating (4.7) and by using (4.13), we obtain

$$\left|\partial_t^m \partial_{\mathbf{y}}^{\beta} h_t(\mathbf{x},\mathbf{y})\right| \le C_{m,\beta}\,t^{-m-\frac{|\beta|}{2}}\,h_{2t}(\mathbf{x},\mathbf{y}). \tag{4.14}$$

Secondly, by differentiating

$$h_t(\mathbf{x},\mathbf{y}) = \int_{\mathbb{R}^N} h_{t/2}(\mathbf{x},\mathbf{z})\,h_{t/2}(\mathbf{z},\mathbf{y})\,dw(\mathbf{z}),$$

by using (4.14) and by symmetry, we get

$$\left|\partial_t^m \partial_{\mathbf{x}}^{\alpha}\partial_{\mathbf{y}}^{\beta} h_t(\mathbf{x},\mathbf{y})\right| \le C_{m,\alpha,\beta}\,t^{-m-\frac{|\alpha|}{2}-\frac{|\beta|}{2}}\,h_{2t}(\mathbf{x},\mathbf{y}).$$

We conclude by using (4.3). □

4.2 Lower Heat Kernel Estimates

We begin with an auxiliary result.

Lemma 4.2 *Let $\tilde{f}$ be a smooth bump function on $\mathbb{R}$ such that $0\le\tilde{f}\le 1$, $\tilde{f}(x)=1$ if $|x|\le\frac{1}{2}$ and $\tilde{f}(x)=0$ if $|x|\ge 1$. Set as usual*

$$f(\mathbf{x}) = \tilde{f}(\|\mathbf{x}\|) \quad\text{and}\quad f(\mathbf{x},\mathbf{y}) = \tau_{\mathbf{x}} f(-\mathbf{y}).$$

Then $0 \le f(\mathbf{x},\mathbf{y}) \le 1$ *and* $f(\mathbf{x},\mathbf{y}) = 0$ *if* $d(\mathbf{x},\mathbf{y}) \ge 1$. *Moreover, there exists a positive constant* c_1 *such that*

$$\sup_{\mathbf{y}\in\mathcal{O}(B(\mathbf{x},1))} f(\mathbf{x},\mathbf{y}) \ge \frac{c_1}{w(B(\mathbf{x},1))}, \tag{4.15}$$

for every $\mathbf{x} \in \mathbb{R}^N$.

Proof All claims follow from (3.6) and (3.7). Let us prove the last one. On the one hand, by translation invariance,

$$\int_{\mathbb{R}^N} f(\mathbf{x},\mathbf{y})\, dw(\mathbf{y}) = \int_{\mathbb{R}^N} f(\mathbf{y})\, dw(\mathbf{y}) \ge w(B(0,1/2)).$$

On the other hand,

$$\int_{\mathbb{R}^N} f(\mathbf{x},\mathbf{y})\, dw(\mathbf{y}) = \int_{\mathcal{O}(B(\mathbf{x},1))} f(\mathbf{x},\mathbf{y})\, dw(\mathbf{y}) \le |G|\, w(B(\mathbf{x},1)) \sup_{\mathbf{y}\in\mathcal{O}(B(\mathbf{x},1))} f(\mathbf{x},\mathbf{y}).$$

This proves (4.15) with $c_1 = \frac{w(B(0,1/2))}{|G|}$. □

Proposition 4.3 *There exist positive constants* c_2 *and* ε *such that*

$$h_t(\mathbf{x},\mathbf{y}) \ge \frac{c_2}{w(B(\mathbf{x},\sqrt{t}))},$$

for every $t > 0$ *and* $\mathbf{x},\mathbf{y} \in \mathbb{R}^N$ *satisfying* $\|\mathbf{x}-\mathbf{y}\| \le \varepsilon\sqrt{t}$.

Proof By scaling it suffices to prove the proposition for $t = 2$. According to Lemma 4.2, applied to $\tilde{h}_1 \gtrsim \tilde{f}$, there exists $c_3 > 0$ and, for every $\mathbf{x} \in \mathbb{R}^N$, there exists $\mathbf{y}(\mathbf{x}) \in \mathcal{O}(B(\mathbf{x},1))$ such that

$$h_1(\mathbf{x},\mathbf{y}(\mathbf{x})) \ge c_3\, w(B(\mathbf{x},1))^{-1}.$$

This estimate holds true around $\mathbf{y}(\mathbf{x})$, according to (4.4), Specifically, there exists $0 < \varepsilon < 1$ (independent of $\mathbf{x}$) such that

$$h_1(\mathbf{x},\mathbf{y}) \ge \tfrac{c_3}{2}\, w(B(\mathbf{x},1))^{-1} \qquad \forall\, \mathbf{y} \in B(\mathbf{y}(\mathbf{x}),\varepsilon).$$

By using the semigroup property and the symmetry of the heat kernel, we deduce that

$$\begin{aligned} h_2(\mathbf{x},\mathbf{x}) &= \int h_1(\mathbf{x},\mathbf{y})\, h_1(\mathbf{y},\mathbf{x})\, dw(\mathbf{y}) \\ &\ge \int_{B(\mathbf{y}(\mathbf{x}),\varepsilon)} h_1(\mathbf{x},\mathbf{y})^2\, dw(\mathbf{y}) \\ &\ge w(B(\mathbf{y}(\mathbf{x}),\varepsilon))\, (\tfrac{c_3}{2})^2\, w(B(\mathbf{x},1))^{-2}. \end{aligned}$$

By using the fact that the balls $B(\mathbf{y}(\mathbf{x}), \varepsilon)$, $B(\mathbf{x}, 1)$, $B(\mathbf{x}, \sqrt{2})$ have comparable volumes and by using again (4.4), we conclude that

$$h_2(\mathbf{x}, \mathbf{y}) \geq c_4\, w(B(\mathbf{x}, \sqrt{2}))^{-1},$$

for all $\mathbf{x}, \mathbf{y} \in \mathbb{R}^N$ sufficiently close. □

A standard argument, which we include for the reader's convenience, allows us to deduce from such a near on diagonal estimate the following global lower Gaussian bound.

Theorem 4.4 *There exist positive constants C and c such that*

$$h_t(\mathbf{x}, \mathbf{y}) \geq \frac{C}{\min\{w(B(\mathbf{x}, \sqrt{t})), w(B(\mathbf{y}, \sqrt{t}))\}}\, e^{-c\,\|\mathbf{x}-\mathbf{y}\|^2/t}, \tag{4.16}$$

for every $t > 0$ and for every $\mathbf{x}, \mathbf{y} \in \mathbb{R}^N$.

Proof We resume the notation of Proposition 4.3. For $s \in \mathbb{R}$, we define $\lceil s \rceil$ to be the smallest integer larger than or equal to s. Assume that $\|\mathbf{x} - \mathbf{y}\|^2/t \geq 1$ and set $n = \lceil 4\|\mathbf{x} - \mathbf{y}\|^2/(\varepsilon^2 t) \rceil \geq 4$. Let $\mathbf{x}_i = \mathbf{x} + i(\mathbf{y} - \mathbf{x})/n$ $(i = 0, \ldots, n)$, so that $\mathbf{x}_0 = \mathbf{x}$, $\mathbf{x}_n = \mathbf{y}$, and $\|\mathbf{x}_{i+1} - \mathbf{x}_i\| = \|\mathbf{x} - \mathbf{y}\|/n$. Consider the balls $B_i = B(\mathbf{x}_i, \frac{\varepsilon}{4}\sqrt{t/n})$ and observe that

$$\begin{aligned}\|\mathbf{y}_{i+1} - \mathbf{y}_i\| &\leq \|\mathbf{y}_i - \mathbf{x}_i\| + \|\mathbf{x}_i - \mathbf{x}_{i+1}\| + \|\mathbf{x}_{i+1} - \mathbf{y}_{i+1}\| < \frac{\varepsilon}{4}\sqrt{\frac{t}{n}} + \frac{\varepsilon}{2}\sqrt{\frac{t}{n}} + \frac{\varepsilon}{4}\sqrt{\frac{t}{n}} \\ &= \varepsilon\sqrt{\frac{t}{n}}\end{aligned}$$

if $\mathbf{y}_i \in B_i$ and $\mathbf{y}_{i+1} \in B_{i+1}$. By using the semigroup property, Proposition 4.3 and the behavior of the ball volume, we estimate

$$\begin{aligned} h_t(\mathbf{x}, \mathbf{y}) &= \int_{\mathbb{R}^N} \cdots \int_{\mathbb{R}^N} h_{t/n}(\mathbf{x}, \mathbf{y}_1) h_{t/n}(\mathbf{y}_1, \mathbf{y}_2) \ldots h_{t/n}(\mathbf{y}_{n-1}, \mathbf{y})\, dw(\mathbf{y}_1) \ldots dw(\mathbf{y}_{n-1}) \\ &\geq c_2^{n-1} \int_{B_1} \cdots \int_{B_{n-1}} w(B(\mathbf{x}, \sqrt{t/n}))^{-1} \ldots w(B(\mathbf{y}_{n-1}, \sqrt{t/n}))^{-1} \\ &\quad \times dw(\mathbf{y}_1) \ldots dw(\mathbf{y}_{n-1}) \\ &\geq c_3^{n-1} w(B(\mathbf{x}, \sqrt{t/n}))^{-1} \frac{w(B_1) \ldots w(B_{n-1})}{w(B(\mathbf{x}_1, \sqrt{t/n})) \ldots w(B(\mathbf{x}_{n-1}, \sqrt{t/n}))} \\ &\geq c_5^{n-1} w(B(\mathbf{x}, \sqrt{t}))^{-1} = c_5^{-1} w(B(\mathbf{x}, \sqrt{t}))^{-1} e^{-n \ln c_5^{-1}} \\ &\geq C\, w(B(\mathbf{x}, \sqrt{t}))^{-1} e^{-c\frac{\|\mathbf{x}-\mathbf{y}\|^2}{t}}. \end{aligned}$$

We conclude by symmetry. □

By combining (4.3) and (4.16), we obtain in particular the following near on diagonal estimates. Notice that the ball volumes $w(B(\mathbf{x},\sqrt{t}))$ and $w(B(\mathbf{y},\sqrt{t}))$ are comparable under the assumptions below.

Corollary 4.5 *For every $c>0$, there exists $C>0$ such that*

$$\frac{C^{-1}}{w(B(\mathbf{x},\sqrt{t}))}\le h_t(\mathbf{x},\mathbf{y})\le\frac{C}{w(B(\mathbf{x},\sqrt{t}))},$$

for every $t>0$ and $\mathbf{x},\mathbf{y}\in\mathbb{R}^N$ such that $\|\mathbf{x}-\mathbf{y}\|\le c\sqrt{t}$.

4.3 Estimates of the Dunkl Kernel

According to (4.1), the heat kernel estimates (4.3) and (4.16) imply the following results, which partially improve upon known estimates for the Dunkl kernel. Notice that $\mathbf{x}$ can be replaced by $\mathbf{y}$ in the ball volumes below.

Corollary 4.6 *There are constants $c\ge 1$ and $C\ge 1$ such that*

$$\frac{C^{-1}}{w(B(\mathbf{x},1))}\,e^{\frac{\|\mathbf{x}\|^2+\|\mathbf{y}\|^2}{2}}\,e^{-c\,\|\mathbf{x}-\mathbf{y}\|^2}\le E(\mathbf{x},\mathbf{y})\le\frac{C}{w(B(\mathbf{x},1))}\,e^{\frac{\|\mathbf{x}\|^2+\|\mathbf{y}\|^2}{2}}\,e^{-c^{-1}d(\mathbf{x},\mathbf{y})^2},$$

for all $\mathbf{x},\mathbf{y}\in\mathbb{R}^N$. In particular,

- *for every $\varepsilon>0$, there exists $C\ge 1$ such that*

$$\frac{C^{-1}}{w(B(\mathbf{x},1))}\,e^{\frac{\|\mathbf{x}\|^2+\|\mathbf{y}\|^2}{2}}\le E(\mathbf{x},\mathbf{y})\le\frac{C}{w(B(\mathbf{x},1))}\,e^{\frac{\|\mathbf{x}\|^2+\|\mathbf{y}\|^2}{2}},$$

  for all $\mathbf{x},\mathbf{y}\in\mathbb{R}^N$ satisfying $\|\mathbf{x}-\mathbf{y}\|<\varepsilon$;
- *there exist $c>0$ and $C>0$ such that*

$$E(\lambda\mathbf{x},\mathbf{y})\ge\frac{C}{w(B(\sqrt{\lambda}\,\mathbf{x},1))}\,e^{\lambda(1-c\,\|\mathbf{x}-\mathbf{y}\|^2)},$$

  for all $\lambda\ge 1$ and for all $\mathbf{x},\mathbf{y}\in\mathbb{R}^N$ with $\|\mathbf{x}\|=\|\mathbf{y}\|=1$.

5 Poisson Kernel in the Dunkl Setting

The Poisson semigroup $P_t=e^{-t\sqrt{-\Delta}}$ is subordinated to the heat semigroup $H_t=e^{t\Delta}$ by (2.2) and correspondingly for their integral kernels

$$p_t(\mathbf{x},\mathbf{y})=\pi^{-1/2}\int_0^\infty e^{-u}\,h_{\frac{t^2}{4u}}(\mathbf{x},\mathbf{y})\,\frac{du}{\sqrt{u}}\,. \tag{5.1}$$

This subordination formula enables us to transfer properties of the heat kernel $h_t(\mathbf{x},\mathbf{y})$ to the Poisson kernel $p_t(\mathbf{x},\mathbf{y})$. For instance,

$$\begin{aligned} &p_t(\mathbf{x},\mathbf{y}) = p_t(\mathbf{y},\mathbf{x}) > 0, \\ &\int_{\mathbb{R}^N} p_t(\mathbf{x},\mathbf{y})\, dw(\mathbf{y}) = 1, \\ &p_t(\mathbf{x},\mathbf{y}) = \tau_{\mathbf{x}} p_t(-\mathbf{y}), \end{aligned} \tag{5.2}$$

where

$$p_t(\mathbf{x}) = \tilde{p}_t(\|\mathbf{x}\|) = c_k'\, t\, \big(t^2 + \|\mathbf{x}\|^2\big)^{-\frac{\mathbf{N}+1}{2}} \tag{5.3}$$

and

$$c_k' = \frac{2^{\mathbf{N}/2}\,\Gamma(\frac{\mathbf{N}+1}{2})}{\sqrt{\pi}\, c_k} > 0\,.$$

The following global bounds hold for the Poisson kernel and its derivatives.

Proposition 5.1 (a) Upper and lower bounds*: there is a constant* $C \geq 1$ *such that*

$$\frac{C^{-1}}{V(\mathbf{x},\mathbf{y},t+\|\mathbf{x}-\mathbf{y}\|)}\,\frac{t}{t+\|\mathbf{x}-\mathbf{y}\|} \leq p_t(\mathbf{x},\mathbf{y}) \leq \frac{C}{V(\mathbf{x},\mathbf{y},t+d(\mathbf{x},\mathbf{y}))}\,\frac{t}{t+d(\mathbf{x},\mathbf{y})} \tag{5.4}$$

for every $t > 0$ *and for every* $\mathbf{x},\mathbf{y} \in \mathbb{R}^N$.

(b) Dunkl gradient*: for every* $\xi \in \mathbb{R}^N$*, there is a constant* $C > 0$ *such that*

$$\big|T_{\xi,\mathbf{y}}\, p_t(\mathbf{x},\mathbf{y})\big| \leq \frac{C}{V(\mathbf{x},\mathbf{y},t+d(\mathbf{x},\mathbf{y}))}\,\frac{1}{t+d(\mathbf{x},\mathbf{y})} \tag{5.5}$$

for all $t > 0$ *and* $\mathbf{x},\mathbf{y} \in \mathbb{R}^N$.

(c) Mixed derivatives: for any nonnegative integer m *and for any multi-index* β*, there is a constant* $C \geq 0$ *such that, for every* $t > 0$ *and for every* $\mathbf{x},\mathbf{y} \in \mathbb{R}^N$*,*

$$\big|\partial_t^m \partial_{\mathbf{y}}^\beta\, p_t(\mathbf{x},\mathbf{y})\big| \leq C\, p_t(\mathbf{x},\mathbf{y})\big(t+d(\mathbf{x},\mathbf{y})\big)^{-m-|\beta|} \times \begin{cases} 1 & \text{if } m=0, \\ 1+\frac{d(\mathbf{x},\mathbf{y})}{t} & \text{if } m>0. \end{cases} \tag{5.6}$$

Moreover, for any nonnegative integer m *and for any multi-indices* β, β'*, there is a constant* $C \geq 0$ *such that, for every* $t > 0$ *and for every* $\mathbf{x},\mathbf{y} \in \mathbb{R}^N$*,*

$$\big|\partial_t^m \partial_{\mathbf{x}}^\beta \partial_{\mathbf{y}}^{\beta'}\, p_t(\mathbf{x},\mathbf{y})\big| \leq C\, t^{-m-|\beta|-|\beta'|}\, p_t(\mathbf{x},\mathbf{y})\,. \tag{5.7}$$

Notice that, by symmetry, (5.5) holds also with $T_{\xi,\mathbf{x}}$ instead of $T_{\xi,\mathbf{y}}$.

Proof (a) The Poisson kernel bounds (5.4) are obtained by inserting the heat kernel bounds (4.3) and (4.16) in the subordination formula (5.1). For a detailed proof we refer the reader to [16, Proposition 6].

(b) The Dunkl gradient estimate (5.5) is deduced similarly from (4.5).

(c) The estimate (5.6) is proved directly. As $(t,x)\longmapsto(t^2+x^2)^{-(\mathbf{N}+1)/2}$ is a homogeneous symbol of order $-\mathbf{N}-1$ on $\mathbb{R}^2$, we have

$$\begin{cases} |\,\partial_x^{\beta}\,\tilde{p}_t(x)|\leq C_\beta\,(t+|x|)^{-\beta}\,\tilde{p}_t(x)\\ |\,\partial_t^{m}\,\partial_x^{\beta}\,\tilde{p}_t(x)|\leq C_{m,\beta}\,t^{-1}(t+|x|)^{1-m-\beta}\,\tilde{p}_t(x)\end{cases}\qquad\forall\,t>0,\ \forall\,x\in\mathbb{R},\tag{5.8}$$

for every positive integer m and for every nonnegative integer β. By using (3.6), (3.7), (5.2), (5.3) and (5.8), we estimate

$$\begin{aligned}\big|\partial_{\mathbf{y}}^{\beta}\,p_t(\mathbf{x},\mathbf{y})\big|&\leq\int_{\mathbb{R}^N}\big|\,\partial_{\mathbf{y}}^{\beta}\,\tilde{p}_t(A(\mathbf{x},\mathbf{y},\eta))\big|\,d\mu_{\mathbf{x}}(\eta)\\ &\leq C_\beta\int_{\mathbb{R}^N}\big(t+A(\mathbf{x},\mathbf{y},\eta)\big)^{-|\beta|}\,\tilde{p}_t(A(\mathbf{x},\mathbf{y},\eta))\,d\mu_{\mathbf{x}}(\eta)\\ &\leq C_\beta\,\big(t+d(\mathbf{x},\mathbf{y})\big)^{-|\beta|}\,p_t(\mathbf{x},\mathbf{y})\end{aligned}$$

and, similarly,

$$\big|\partial_t^{m}\,\partial_{\mathbf{y}}^{\beta}\,p_t(\mathbf{x},\mathbf{y})\big|\leq C_{m,\beta}\,t^{-1}\big(t+d(\mathbf{x},\mathbf{y})\big)^{1-m-|\beta|}\,p_t(\mathbf{x},\mathbf{y}),$$

for every positive integer m. Finally, (5.7) is deduced from (5.6) by using the semigroup property. More precisely, by differentiating

$$p_t(\mathbf{x},\mathbf{y})=\int_{\mathbb{R}^N}p_{t/2}(\mathbf{x},\mathbf{z})\,p_{t/2}(\mathbf{z},\mathbf{y})\,dw(\mathbf{z})\,,$$

by using (5.6) and by symmetry, we obtain

$$\begin{aligned}\big|\partial_t^{m}\,\partial_{\mathbf{x}}^{\beta}\,\partial_{\mathbf{y}}^{\beta'}\,p_t(\mathbf{x},\mathbf{y})\big|&\lesssim t^{-m-|\beta|-|\beta'|}\int_{\mathbb{R}^N}p_{t/2}(\mathbf{x},\mathbf{z})\,p_{t/2}(\mathbf{z},\mathbf{y})\,dw(\mathbf{z})\\ &=t^{-m-|\beta|-|\beta'|}\,p_t(\mathbf{x},\mathbf{y})\,.\end{aligned}$$

□

Notice the following straightforward consequence of the upper bound in (5.4):

$$\mathcal{M}_P f(\mathbf{x})\lesssim\sum_{\sigma\in G}\mathcal{M}_{HL}f(\sigma(\mathbf{x}))\,,\tag{5.9}$$

where $\mathcal{M}_{HL}$ denotes the Hardy–Littlewood maximal function on the space of homogeneous type $(\mathbb{R}^N,\,\|\mathbf{x}-\mathbf{y}\|,\,dw)$. Likewise, (4.3) yields

$$\mathcal{M}_H f(\mathbf{x})\lesssim\sum_{\sigma\in G}\mathcal{M}_{HL}f(\sigma(\mathbf{x}))\,.$$

Observe that the Poisson kernel is an approximation of the identity in the following sense.

Proposition 5.2 *Given any compact subset $K \subset \mathbb{R}^N$, any $r > 0$ and any $\varepsilon > 0$, there exists $t_0 = t_0(K, r, \varepsilon) > 0$ such that, for every $0 < t < t_0$ and for every $\mathbf{x} \in K$,*

$$\int_{\|\mathbf{x}-\mathbf{y}\|>r} p_t(\mathbf{x}, \mathbf{y})\, dw(\mathbf{y}) < \varepsilon\,.$$

Proof Let K be a compact subset of $\mathbb{R}^N$ and let $r, \varepsilon > 0$. Fix $\mathbf{x}_0 \in K$ and consider $f \in C_c^\infty(\mathbb{R}^N)$ such that $0 \le f \le 1$, $f = 1$ on $B(\mathbf{x}_0, r/4)$ and $\operatorname{supp} f \subset B(\mathbf{x}_0, r/2)$. By the inversion formula,

$$f(\mathbf{x}) - P_t f(\mathbf{x}) = c_k^{-1} \int_{\mathbb{R}^N} (1 - e^{-t\|\xi\|})\, E(i\xi, \mathbf{x})\, \mathcal{F}f(\xi)\, dw(\xi)\,,$$

hence

$$|f(\mathbf{x}) - P_t f(\mathbf{x})| \le c_k^{-1} \int_{\mathbb{R}^N} \left(1 - e^{-t\|\xi\|}\right) |\mathcal{F}f(\xi)|\, dw(\xi)\,. \tag{5.10}$$

As $\mathcal{F}f \in \mathcal{S}(\mathbb{R}^N)$, (5.10) implies that there is $t_0 = t_0(\mathbf{x}_0, r, \varepsilon) > 0$ such that

$$\sup_{\mathbf{x}\in\mathbb{R}^N} |f(\mathbf{x}) - P_t f(\mathbf{x})| < \varepsilon \quad \forall\, 0 < t < t_0\,.$$

In particular, for every $0 < t < t_0$ and for every $\mathbf{x} \in B(\mathbf{x}_0, r/4)$, we have

$$\begin{aligned} 0 \le \int_{\|\mathbf{x}-\mathbf{y}\|>r} p_t(\mathbf{x}, \mathbf{y})\, dw(\mathbf{y}) &= 1 - \int_{\|\mathbf{x}-\mathbf{y}\|\le r} p_t(\mathbf{x}, \mathbf{y})\, dw(\mathbf{y}) \\ &\le f(\mathbf{x}) - \int_{\|\mathbf{x}-\mathbf{y}\|\le r} p_t(\mathbf{x}, \mathbf{y}) f(\mathbf{y})\, dw(\mathbf{y}) \le |f(\mathbf{x}) - P_t f(\mathbf{x})| < \varepsilon\,. \end{aligned}$$

We easily conclude the proof by compactness. □

The following results follow from (5.4), (5.9), and Proposition 5.2.

Corollary 5.3 *Let f be a bounded continuous function on $\mathbb{R}^N$. Then its Poisson integral $u(t, \mathbf{x}) = P_t f(\mathbf{x})$ is also bounded and continuous on $[0, \infty) \times \mathbb{R}^N$.*

Corollary 5.4 *Let $f \in L^p(dw)$ with $1 \le p \le \infty$. Then for almost every $\mathbf{x} \in \mathbb{R}^N$,*

$$\lim_{t\to 0} \sup_{\|\mathbf{y}-\mathbf{x}\|<t} \left| P_t f(\mathbf{y}) - f(\mathbf{x}) \right| = 0.$$

Moreover, for $f \in L^p(dw)$, $1 \le p < \infty$, we have $\lim_{t\to 0} \|P_t f - f\|_{L^p(dw)} = 0$.

Remark 5.5 The assertion of Proposition 5.2 remains valid with the same proof if $p_t(\mathbf{x}, \mathbf{y})$ is replaced by $\Phi_t(\mathbf{x}, \mathbf{y}) = \tau_{\mathbf{x}} \Phi_t(-\mathbf{y})$, where $\Phi \in \mathcal{S}(\mathbb{R}^N)$ is radial, nonnegative, and $\int \Phi(\mathbf{x})\, dw(\mathbf{x}) = 1$.

6 Conjugate Harmonic Functions—Subharmonicity

For $\sigma \in G$, let $f^\sigma(\mathbf{x}) = f(\sigma(\mathbf{x}))$. It is easy to check that

$$T_\xi f^\sigma(\mathbf{x}) = (T_{\sigma\xi} f)^\sigma(\mathbf{x}), \quad \sigma \in G, \ \mathbf{x}, \xi \in \mathbb{R}^N,$$
$$(\Delta f^\sigma)(\mathbf{x}) = (\Delta f)^\sigma(\mathbf{x}). \tag{6.1}$$

Let $\{\sigma_{ij}\}_{i,j=1}^N$ denote the matrix of $\sigma \in G$ written in the canonical basis $e_1, \ldots, e_N$ of $\mathbb{R}^N$. Clearly, $\{\sigma_{ij}\}_{i,j=1}^N$ belongs to the group $\mathbf{O}(N, \mathbb{R})$ of the orthogonal $N \times N$ matrices.

Lemma 6.1 *Assume that* $\mathbf{u}(x_0, \mathbf{x}) = (u_0(x_0, \mathbf{x}), u_1(x_0, \mathbf{x}), \ldots, u_N(x_0, \mathbf{x}))$ *satisfies the Cauchy–Riemann equations* (2.4). *For* $\sigma \in G$, *set*

$$u_{\sigma,0}(x_0, \mathbf{x}) = u_0(x_0, \sigma(\mathbf{x})), \quad u_{\sigma,j}(x_0, \mathbf{x}) = \sum_{i=1}^N \sigma_{ij} u_i(x_0, \sigma(\mathbf{x})), \ j = 1, 2, \ldots, N. \tag{6.2}$$

Then $\mathbf{u}_\sigma(x_0, \mathbf{x}) = (u_{\sigma,0}(x_0, \mathbf{x}), u_{\sigma,1}(x_0, \mathbf{x}), \ldots, u_{\sigma,N}(x_0, \mathbf{x}))$ *satisfies the Cauchy–Riemann equations. Moreover,*

$$|\mathbf{u}_\sigma(x_0, \mathbf{x})| = |\mathbf{u}(x_0, \sigma(\mathbf{x}))|. \tag{6.3}$$

Proof Let $1 \le k, j \le N$. Then

$$T_k u_{\sigma,j}(x_0, \mathbf{x}) = \sum_{i=1}^N \sigma_{ij} T_k(u_i(x_0, \sigma\cdot))(\mathbf{x}) = \sum_{i=1}^N \sigma_{ij} \sum_{\ell=1}^N \sigma_{\ell k}(T_\ell u_i)(x_0, \sigma(\mathbf{x})), \tag{6.4}$$

and, similarly,

$$T_j u_{\sigma,k}(x_0, \mathbf{x}) = \sum_{i=1}^N \sigma_{ik} \sum_{\ell=1}^N \sigma_{\ell j}(T_\ell u_i)(x_0, \sigma(\mathbf{x})). \tag{6.5}$$

Recall that $T_\ell u_i = T_i u_\ell$. Hence, (6.5) becomes

$$T_j u_{\sigma,k}(x_0, \mathbf{x}) = \sum_{i=1}^N \sigma_{ik} \sum_{\ell=1}^N \sigma_{\ell j}(T_i u_\ell)(x_0, \sigma(\mathbf{x})). \tag{6.6}$$

Now we see that (6.4) and (6.6) are equal. The proof that $T_k u_{\sigma,0} = T_0 u_{\sigma,k}$ is straightforward. The second equality of (2.4) follows directly from (6.6) and the fact that $\sigma^{-1} = \sigma^*$.

Since $\{\sigma_{ij}\} \in \mathbf{O}(N, \mathbb{R})$,

$$|u_{\sigma,0}(x_0, \mathbf{x})|^2 + \sum_{j=1}^{N} |u_{\sigma,j}(x_0, \mathbf{x})|^2 = |u_0(x_0, \sigma(\mathbf{x}))|^2 + \sum_{j=1}^{N} \left| \sum_{i=1}^{N} \sigma_{ij} u_i(x_0, \sigma(\mathbf{x})) \right|^2$$

$$= |u_0(x_0, \sigma(\mathbf{x}))|^2 + \sum_{i=1}^{N} |u_i(x_0, \sigma(\mathbf{x}))|^2, \quad (6.7)$$

which proves (6.3). □

Let

$$F(t, \mathbf{x}) = \{\mathbf{u}_\sigma(t, \mathbf{x})\}_{\sigma \in G}. \quad (6.8)$$

We shall always assume that $\mathbf{u}$ and $\mathbf{u}_\sigma$ are related by (6.2). Then, by (6.3),

$$|F(x_0, \mathbf{x})|^2 = \sum_{\sigma \in G} \sum_{\ell=0}^{N} |u_{\sigma,\ell}(x_0, \mathbf{x})|^2 = \sum_{\sigma \in G} |\mathbf{u}_\sigma(x_0, \mathbf{x})|^2 = \sum_{\sigma \in G} |\mathbf{u}(x_0, \sigma(\mathbf{x}))|^2.$$

Observe that $|F(x_0, \mathbf{x})| = |F(x_0, \sigma(\mathbf{x}))|$ for every $\sigma \in G$.

Consequently, for every $\alpha \in R$,

$$\sum_{\sigma \in G} \sum_{\ell=0}^{N} \Big(u_{\sigma,\ell}(x_0, \mathbf{x}) - u_{\sigma,\ell}(x_0, \sigma_\alpha(\mathbf{x})) \Big) \cdot u_{\sigma,\ell}(x_0, \mathbf{x})$$

$$= \frac{1}{2} \sum_{\sigma \in G} \sum_{\ell=0}^{N} \Big| u_{\sigma,\ell}(x_0, \mathbf{x}) - u_{\sigma,\ell}(x_0, \sigma_\alpha(\mathbf{x})) \Big|^2. \quad (6.9)$$

We shall need the following auxiliary lemma.

Lemma 6.2 *For every $\varepsilon > 0$ there is $\delta > 0$ such that for every matrix $A = \{a_{ij}\}_{i,j=0}^{N}$ with real entries a_{ij} one has*

$$\|A\|^2 \le \varepsilon \Big((\,trA)^2 + \sum_{i<j} (a_{ij} - a_{ji})^2 \Big) + (1 - \delta) \|A\|_{\mathrm{HS}}^2,$$

where $\|A\|_{\mathrm{HS}}$ denotes the Hilbert–Schmidt norm of A.

Proof The lemma was proved in [13]. For the convenience of the reader we present a short proof. The inequality is known for trace zero symmetric A (see Stein and Weiss [36, Lemma 2.2]). By homogeneity we may assume that $\|A\|_{\mathrm{HS}} = 1$. Assume that the inequality does not hold. Then there is $\varepsilon > 0$ such that for every $n > 0$ there is $A_n = \{a_{ij}^{\{n\}}\}_{i,j=0}^{N}$, $\|A_n\|_{\mathrm{HS}} = 1$ such that

$$\|A_n\|^2 > \varepsilon \Big((\mathrm{tr} A_n)^2 + \sum_{i<j} (a_{ij}^{\{n\}} - a_{ji}^{\{n\}})^2 \Big) + \left(1 - \frac{1}{n}\right) \|A_n\|_{\mathrm{HS}}^2.$$

Thus there is a subsequence n_s such that $A_{n_s} \to A$, $\|A\|_{\mathrm{HS}} = 1$ and

$$\|A\|^2 \geq \varepsilon \left((\mathrm{tr} A)^2 + \sum_{i<j} (a_{ij} - a_{ji})^2 \right) + \|A\|^2_{\mathrm{HS}}.$$

But then $A = A^*$ and $\mathrm{tr} A = 0$, and so, $\|A\|^2 \geq \|A\|^2_{\mathrm{HS}}$. This contradicts the already known inequality. □

We now state and prove the main theorem of Sect. 6, which is the analog in the Dunkl setting of a Euclidean subharmonicity property (see [34, Chapter VII, Sect. 3.1]) and which was proved in the product case in [13, Proposition 4.1]. Recall (2.3) that $\mathcal{L} = T_0^2 + \Delta$.

Theorem 6.3 *There is an exponent $0 < q < 1$ which depends on k such that if $\mathbf{u} = (u_0, u_1, \ldots, u_N) \in C^2$ satisfies the Cauchy–Riemann equations (2.4), then the function $|F|^q$ is $\mathcal{L}$-subharmonic, that is, $\mathcal{L}(|F|^q)(t, \mathbf{x}) \geq 0$ on the set where $|F| > 0$.*

Proof Observe that $|F|^q$ is C^2 on the set where $|F| > 0$. Let $\cdot$ denote the inner product in $\mathbb{R}^{(N+1)\cdot|G|}$. For $j = 0, 1, \ldots, N$, we have

$$\partial_{e_j} |F|^q = q|F|^{q-2}\Big((\partial_{e_j} F) \cdot F\Big)$$
$$\partial^2_{e_j} |F|^q = q(q-2)|F|^{q-4}\Big((\partial_{e_j} F) \cdot F\Big)^2 + q|F|^{q-2}\Big((\partial^2_{e_j} F) \cdot F + |\partial_{e_j} F|^2\Big).$$

Recall that $|F(x_0, \mathbf{x})| = |F(x_0, \sigma(\mathbf{x}))|$. Hence,

$$\begin{aligned}\mathcal{L}|F|^q &= q(q-2)|F|^{q-4} \left\{ \left(\sum_{j=0}^{N} \Big((\partial_{e_j} F) \cdot F\Big)^2 \right\} \right. \\ &\quad + q|F|^{q-2} \left\{ \left(\sum_{j=0}^{N} \partial^2_{e_j} F + 2 \sum_{\alpha \in R^+} \frac{k(\alpha)}{\langle \alpha, \mathbf{x} \rangle} \partial_\alpha F \right) \cdot F + \sum_{j=0}^{N} |\partial_{e_j} F|^2 \right\}. \end{aligned} \tag{6.10}$$

Since $T_j T_\ell = T_\ell T_j$, we conclude from (2.4) applied to $\mathbf{u}_\sigma$ that for $\ell = 0, 1, \ldots, N$, we have

$$\begin{aligned}&\sum_{j=0}^{N} \partial^2_{e_j} u_{\sigma,\ell}(x_0, \mathbf{x}) + 2 \sum_{\alpha \in R^+} \frac{k(\alpha)}{\langle \alpha, \mathbf{x} \rangle} \partial_\alpha u_{\sigma,\ell}(x_0, \mathbf{x}) \\ &\quad = \sum_{\alpha \in R^+} k(\alpha) \|\alpha\|^2 \frac{u_{\sigma,\ell}(x_0, \mathbf{x}) - u_{\sigma,\ell}(x_0, \sigma_\alpha(\mathbf{x}))}{\langle \alpha, \mathbf{x} \rangle^2}.\end{aligned}$$

Thus,

$$
\begin{aligned}
&\left(\sum_{j=0}^{N} \partial_{e_j}^2 F + 2\sum_{\alpha\in R^+} \frac{k(\alpha)}{\langle\alpha, \mathbf{x}\rangle}\partial_\alpha F\right)\cdot F \\
&= \sum_{\sigma\in G}\sum_{\ell=0}^{N}\left(\sum_{j=0}^{N}\partial_{e_j}^2 u_{\sigma,\ell}(x_0,\mathbf{x}) + 2\sum_{\alpha\in R^+}\frac{k(\alpha)}{\langle\alpha,\mathbf{x}\rangle}\partial_\alpha u_{\sigma,\ell}(x_0,\mathbf{x})\right) u_{\sigma,\ell}(x_0,\mathbf{x}) \\
&= \sum_{\sigma\in G}\sum_{\ell=0}^{N}\sum_{\alpha\in R^+} k(\alpha)\|\alpha\|^2\frac{u_{\sigma,\ell}(x_0,\mathbf{x}) - u_{\sigma,\ell}(x_0,\sigma_\alpha(\mathbf{x}))}{\langle\alpha,\mathbf{x}\rangle^2} u_{\sigma,\ell}(x_0,\mathbf{x}) \\
&= \sum_{\alpha\in R^+}\frac{k(\alpha)\|\alpha\|^2}{\langle\alpha,\mathbf{x}\rangle^2}\sum_{\sigma\in G}\sum_{\ell=0}^{N}\Big(u_{\sigma,\ell}(x_0,\mathbf{x}) - u_{\sigma,\ell}(x_0,\sigma_\alpha(\mathbf{x}))\Big)u_{\sigma,\ell}(x_0,\mathbf{x}) \\
&= \frac{1}{2}\sum_{\alpha\in R^+}\frac{k(\alpha)\|\alpha\|^2}{\langle\alpha,\mathbf{x}\rangle^2}\sum_{\sigma\in G}\sum_{\ell=0}^{N}\Big(u_{\sigma,\ell}(x_0,\mathbf{x}) - u_{\sigma,\ell}(x_0,\sigma_\alpha(\mathbf{x}))\Big)^2 \qquad (6.11)
\end{aligned}
$$

Thanks to (6.10) and (6.11), it suffices to prove that there is $0 < q < 1$ such that

$$
\begin{aligned}
&(2-q)\sum_{j=0}^{N}\Big((\partial_{e_j} F(x_0,\mathbf{x}))\cdot F(x_0,\mathbf{x})\Big)^2 \\
&\le \frac{1}{2}|F(x_0,\mathbf{x})|^2\sum_{\sigma\in G}\sum_{\ell=0}^{N}\sum_{\alpha\in R^+}\frac{k(\alpha)\|\alpha\|^2}{\langle\alpha,\mathbf{x}\rangle^2}\Big(u_{\sigma,\ell}(x_0,\mathbf{x}) - u_{\sigma,\ell}(x_0,\sigma_\alpha(\mathbf{x}))\Big)^2 \\
&\quad + |F(x_0,\mathbf{x})|^2\left(\sum_{j=0}^{N}|\partial_{e_j} F(x_0,\mathbf{x})|^2\right).
\end{aligned} \qquad (6.12)
$$

Set

$$
B_\sigma = \begin{bmatrix} \partial_{e_0} u_{\sigma,0} & \partial_{e_0} u_{\sigma,1} & \dots & \partial_{e_0} u_{\sigma,N} \\ \partial_{e_1} u_{\sigma,0} & \partial_{e_1} u_{\sigma,1} & \dots & \partial_{e_1} u_{\sigma,N} \\ & & \dots & \\ \partial_{e_N} u_{\sigma,0} & \partial_{e_N} u_{\sigma,1} & \dots & \partial_{e_N} u_{\sigma,N} \end{bmatrix}.
$$

Let $\mathbf{B} = \{B_\sigma\}_{\sigma\in G}$ be matrix with $N+1$ rows and $(N+1)\cdot|G|$ columns. It represents a linear operator (denoted by $\mathbf{B}$) from $\mathbb{R}^{(N+1)\cdot|G|}$ into $\mathbb{R}^{1+N}$. Let $\|\mathbf{B}\|$ be its norm.

Observe that for $0 < q < 1$, we have

$$(2-q)\sum_{j=0}^{N}\Big((\partial_{e_j}F)\cdot F\Big)^2 \le (2-q)|F|^2\|\mathbf{B}\|^2,$$
$$|F|^2\sum_{j=0}^{N}|\partial_{e_j}F|^2 = |F|^2\|\mathbf{B}\|_{\mathrm{HS}}^2.$$

Clearly,

$$\|\mathbf{B}\|^2 \le \sum_{\sigma\in G}\|B_\sigma\|^2, \quad \|\mathbf{B}\|_{\mathrm{HS}}^2 = \sum_{\sigma\in G}\|B_\sigma\|_{\mathrm{HS}}^2.$$

Therefore the inequality (6.12) will be proven if we show that

$$(2-q)\sum_{\sigma\in G}\|B_\sigma\|^2 \le \sum_{\sigma\in G}\|B_\sigma\|_{\mathrm{HS}}^2 + \frac{1}{2}\sum_{\sigma\in G}\sum_{\ell=0}^{N}\sum_{\alpha\in R^+}\frac{k(\alpha)\|\alpha\|^2}{\langle\alpha,\mathbf{x}\rangle^2}\Big(u_{\sigma,\ell}(x_0,\mathbf{x}) - u_{\sigma,\ell}(x_0,\sigma_\alpha(\mathbf{x}))\Big)^2. \tag{6.13}$$

Recall that

$$\gamma = \sum_{j=1}^{N}\sum_{\alpha\in R^+}\frac{k(\alpha)\langle\alpha,e_j\rangle^2}{\|\alpha\|^2} = \sum_{j=0}^{N}\sum_{\alpha\in R^+}\frac{k(\alpha)\langle\alpha,e_j\rangle^2}{\|\alpha\|^2}$$

(see (3.1)). By applying first the Cauchy–Riemann equations (2.4) and next the Cauchy–Schwarz inequality, we obtain

$$\begin{aligned}(\operatorname{tr}B_\sigma)^2 &= \left(-\sum_{j=1}^{N}\sum_{\alpha\in R^+}k(\alpha)\langle\alpha,e_j\rangle\frac{u_{\sigma,j}(x_0,\mathbf{x}) - u_{\sigma,j}(x_0,\sigma_\alpha(\mathbf{x}))}{\langle\alpha,\mathbf{x}\rangle}\right)^2\\
&\le \left(\sum_{j=1}^{N}\sum_{\alpha\in R^+}\frac{k(\alpha)\langle\alpha,e_j\rangle^2}{\|\alpha\|^2}\right)\\
&\quad\times\left(\sum_{j=1}^{N}\sum_{\alpha\in R^+}\|\alpha\|^2k(\alpha)\frac{\big(u_{\sigma,j}(x_0,\mathbf{x}) - u_{\sigma,j}(x_0,\sigma_\alpha(\mathbf{x}))\big)^2}{\langle\alpha,\mathbf{x}\rangle^2}\right)\\
&\le \gamma\sum_{j=0}^{N}\sum_{\alpha\in R^+}\|\alpha\|^2k(\alpha)\frac{\big(u_{\sigma,j}(x_0,\mathbf{x}) - u_{\sigma,j}(x_0,\sigma_\alpha(\mathbf{x}))\big)^2}{\langle\alpha,\mathbf{x}\rangle^2}.\end{aligned} \tag{6.14}$$

Utilizing again the Cauchy–Riemann equations (2.4), we get

$$\begin{aligned}
&\sum_{0\le i<j\le N} \left(\partial_{e_i} u_{\sigma,j}(x_0,\mathbf{x}) - \partial_{e_j} u_{\sigma,i}(x_0,\mathbf{x})\right)^2 \\
&= \sum_{j=1}^{N} \left(\sum_{\alpha\in R^+} k(\alpha)\langle \alpha, e_j\rangle \frac{u_{\sigma,0}(x_0,\mathbf{x}) - u_{\sigma,0}(x_0,\sigma_\alpha(\mathbf{x}))}{\langle \alpha,\mathbf{x}\rangle} \right)^2 \\
&\quad + \sum_{1\le i<j\le N} \left(\sum_{\alpha\in R^+} -k(\alpha)\langle \alpha, e_i\rangle \frac{u_{\sigma,j}(x_0,\mathbf{x}) - u_{\sigma,j}(x_0,\sigma_\alpha(\mathbf{x}))}{\langle \alpha,\mathbf{x}\rangle} \right. \\
&\quad \left. + k(\alpha)\langle \alpha, e_j\rangle \frac{u_{\sigma,i}(x_0,\mathbf{x}) - u_{\sigma,i}(x_0,\sigma_\alpha(\mathbf{x}))}{\langle \alpha,\mathbf{x}\rangle} \right)^2 \\
&\le 2 \left(\sum_{j=0}^{N} \sum_{\alpha\in R^+} \frac{k(\alpha)\langle \alpha, e_j\rangle^2}{\|\alpha\|^2} \right) \\
&\quad \times \left(\sum_{j=0}^{N} \sum_{\alpha\in R^+} \|\alpha\|^2 k(\alpha) \frac{\left(u_{\sigma,j}(x_0,\mathbf{x}) - u_{\sigma,j}(x_0,\sigma_\alpha(\mathbf{x}))\right)^2}{\langle \alpha,\mathbf{x}\rangle^2} \right). \qquad (6.15)
\end{aligned}$$

Using the auxiliary Lemma 6.2 together with (6.14) and (6.15) we have that for every $\varepsilon > 0$ there is $0 < \delta < 1$ such that

$$\begin{aligned}
\sum_{\sigma\in G} \|B_\sigma\|^2 &\le (1-\delta) \sum_{\sigma\in G} \|B_\sigma\|_{\mathrm{HS}}^2 \\
&\quad + 3\varepsilon\gamma \sum_{\sigma\in G} \sum_{j=0}^{N} \sum_{\alpha\in R^+} \|\alpha\|^2 k(\alpha) \frac{\left(u_{\sigma,j}(x_0,\mathbf{x}) - u_{\sigma,j}(x_0,\sigma_\alpha(\mathbf{x}))\right)^2}{\langle \alpha,\mathbf{x}\rangle^2}. \qquad (6.16)
\end{aligned}$$

Taking $\varepsilon > 0$ such that $3\varepsilon\gamma \le \frac{1}{4}$ and utilizing (6.16) we deduce that (6.13) holds for q such that $(1-\delta) \le (2-q)^{-1}$. □

7 Harmonic Functions in the Dunkl Setting

In this section we characterize certain $\mathcal{L}$-harmonic functions in the half-space $\mathbb{R}_+^{1+N}$ by adapting the classical proofs (see, e.g., [19,34,36]). Let us first construct an auxiliary barrier function.

7.1 Barrier Function

For fixed $\delta > 0$, let $v_1, \dots, v_s \in \mathbb{R}^N$ be a set of vectors of the unit sphere in $S^{N-1} = \{\mathbf{x} \in \mathbb{R}^N : \|\mathbf{x}\| = 1\}$ which forms a δ–net on S^{N-1}. Let $M, \varepsilon > 0$. Define

$$\mathcal{V}_m(x_0, \mathbf{x}) = 2M\varepsilon x_0 + \varepsilon E\Big(\frac{\varepsilon\pi}{4}\mathbf{x}, v_m\Big) \cos\Big(\frac{\varepsilon\pi}{4}x_0\Big), \quad m = 1, \dots, s, \tag{7.1}$$

(cf. [34, Chapter VII, Sect. 1.2] in the classical setting). The function $\mathcal{V}_m$ is $\mathcal{L}$-harmonic and strictly positive on $[0, \varepsilon^{-1}] \times \mathbb{R}^N$. Set

$$\mathcal{V}(x_0, \mathbf{x}) = \sum_{m=1}^{s} \mathcal{V}_m(x_0, \mathbf{x}).$$

By Corollary 4.6,

$$\lim_{\|\mathbf{x}\| \to \infty} \mathcal{V}(x_0, \mathbf{x}) = \infty \quad \text{uniformly in } x_0 \in [0, \varepsilon^{-1}]. \tag{7.2}$$

7.2 Maximum Principle and the Mean Value Property

As we have already remarked in Sect. 2, the operator $\mathcal{L}$ is the Dunkl–Laplace operator associated with the root system R as a subset of $\mathbb{R}^{1+N} = \mathbb{R} \times \mathbb{R}^N$. We shall denote the element of $\mathbb{R}^{1+N}$ by $\boldsymbol{x} = (x_0, \mathbf{x})$. The associated measure will be denoted by $\boldsymbol{w}$. Clearly, $d\boldsymbol{w}(\boldsymbol{x}) = w(\mathbf{x})\, d\mathbf{x}\, dx_0$. Moreover, $E(\boldsymbol{x}, \boldsymbol{y}) = e^{x_0 y_0} E(\mathbf{x}, \mathbf{y})$. We shall slightly abuse notation and use the same letter σ for the action of the group G in $\mathbb{R}^{1+N}$, so $\sigma(\boldsymbol{x}) = \sigma(x_0, \mathbf{x}) = (x_0, \sigma(\mathbf{x}))$.

The following weak maximum principle for $\mathcal{L}$-subharmonic functions was actually proved in Theorem 4.2 of Rösler [27].

Theorem 7.1 *Let $\Omega \subset \mathbb{R}^{1+N}$ be open, bounded, and $\overline{\Omega} \subset (0, \infty) \times \mathbb{R}^N$. Assume that Ω is G-invariant, that is, $(x_0, \sigma(\mathbf{x})) \in \Omega$ for $(x_0, \mathbf{x}) \in \Omega$ and all $\sigma \in G$. Let $f \in C^2(\Omega) \cap C(\overline{\Omega})$ be real-valued and $\mathcal{L}$-subharmonic. Then*

$$\max_{\overline{\Omega}} f = \max_{\partial\Omega} f.$$

Let $f^{\{r\}}(\boldsymbol{x}) = \chi_{B(0,r)}(\boldsymbol{x})$ be the characteristic function of the ball in $\mathbb{R}^{1+N}$. Set

$$f(r, \boldsymbol{x}, \boldsymbol{y}) = \tau_{\boldsymbol{x}} f^{\{r\}}(-\boldsymbol{y}).$$

Clearly, $0 \le f(r, \boldsymbol{x}, \boldsymbol{y}) \le 1$. The following mean value theorem was proved in [21, Theorem 3.2].

Theorem 7.2 *Let $\Omega \subset \mathbb{R}^{1+N}$ be an open and G-invariant set and let u be a C^2 function in Ω. Then u is $\mathcal{L}$-harmonic if and only if u has the following mean value property:*

for all $\boldsymbol{x} \in \Omega$ *and* $\rho > 0$ *such that* $B(\boldsymbol{x}, \rho) \subset \Omega$, *we have*

$$u(\boldsymbol{x}) = \frac{1}{\boldsymbol{w}(B(0,r))} \int_\Omega f(r, \boldsymbol{x}, \boldsymbol{y}) u(\boldsymbol{y}) d\boldsymbol{w}(\boldsymbol{y}) \quad \textit{for}\ 0 < r < \rho/3.$$

7.3 Characterizations of $\mathcal{L}$-Harmonic Functions in the Upper Half-Space

Theorem 7.3 *Suppose that u is a* C^2 *function on* $\mathbb{R}_+^{1+N}$. *Then u is a Poisson integral of a bounded function on* $\mathbb{R}^N$ *if and only if u is* $\mathcal{L}$*-harmonic and bounded.*

Proof The proof is identical to that of Stein [34]. Clearly, the Poisson integral of a bounded function is bounded and $\mathcal{L}$-harmonic. To prove the converse assume that u is $\mathcal{L}$-harmonic and bounded, so $|u| \le M$. Set $f_n(\mathbf{x}) = u(\frac{1}{n}, \mathbf{x})$ and $u_n(x_0, \mathbf{x}) = P_{x_0} f_n(\mathbf{x})$. Then $U_n(x_0, \mathbf{x}) = u(x_0 + \frac{1}{n}, \mathbf{x}) - u_n(x_0, \mathbf{x})$ is $\mathcal{L}$-harmonic, $|U_n| \le 2M$, continuous on $[0, \infty) \times \mathbb{R}^N$, and $U_n(0, \mathbf{x}) = 0$. We shall prove that $U_n \equiv 0$. Fix $(y_0, \mathbf{y}) \in \mathbb{R}_+^{1+N}$. Set

$$U(x_0, \mathbf{x}) = U_n(x_0, \mathbf{x}) + \mathcal{V}(x_0, \mathbf{x})$$

and consider the function U on the closure of the set $\Omega = (0, \varepsilon^{-1}) \times B(0, R)$, with $\varepsilon > 0$ small and R large enough. Then U is $\mathcal{L}$-harmonic in Ω, continuous on $\bar{\Omega}$, and positive on the boundary of the $\partial\Omega$. Thus, by the maximum principle, U is positive in $\bar{\Omega}$, so

$$U_n(y_0, \mathbf{y}) > -2M\varepsilon y_0 - \sum_{m=1}^{s} \varepsilon E\Big(\frac{\varepsilon\pi}{4}\mathbf{y}, v_m\Big) \cos\Big(\frac{\varepsilon\pi}{4} y_0\Big).$$

Letting $\varepsilon \to 0$ we obtain $U_n(y_0, \mathbf{y}) \ge 0$. The same argument applied to $-u$ gives $-U_n(y_0, \mathbf{y}) \ge 0$, so $U_n \equiv 0$, which can be written as

$$u\Big(x_0 + \frac{1}{n}, \mathbf{x}\Big) = P_{x_0} f_n(\mathbf{x}) = \int p_{x_0}(\mathbf{x}, \mathbf{y}) f_n(\mathbf{y})\, dw(\mathbf{y}). \tag{7.3}$$

Clearly $|f_n| \le M$, so by the *-weak compactness, there is a subsequence n_j and $f \in L^\infty(\mathbb{R}^N)$ such that for $\varphi \in L^1(dw)$, we have

$$\lim_{j\to\infty} \int \varphi(\mathbf{y}) f_{n_j}(\mathbf{y})\, dw(\mathbf{y}) = \int \varphi(\mathbf{y}) f(\mathbf{y})\, dw(\mathbf{y}).$$

So,

$$\begin{aligned} u(x_0, \mathbf{x}) &= \lim_{j\to\infty} u\Big(x_0 + \frac{1}{n_j}, \mathbf{x}\Big) = \lim_{j\to\infty} \int p_{x_0}(\mathbf{x}, \mathbf{y}) f_{n_j}(\mathbf{y})\, dw(\mathbf{y}) \\ &= \int p_{x_0}(\mathbf{x}, \mathbf{y}) f(\mathbf{y})\, dw(\mathbf{y}). \end{aligned}$$

□

Corollary 7.4 *If u is $\mathcal{L}$-harmonic and bounded in $\mathbb{R}_+^{1+N}$ then u has a nontangential limit at almost every point of the boundary.*

Theorem 7.5 *Suppose that u is a C^2-function on $\mathbb{R}_+^{1+N}$. If $1 < p < \infty$ then u is a Poisson integral of an $L^p(dw)$ function if and only if u is $\mathcal{L}$-harmonic and*

$$\sup_{x_0>0} \|u(x_0, \cdot)\|_{L^p(dw)} < \infty. \tag{7.4}$$

If $p = 1$ then u is a Poisson integral of a bounded measure ω if and only if u is $\mathcal{L}$-harmonic and

$$\sup_{x_0>0} \|u(x_0, \cdot)\|_{L^1(dw)} < \infty. \tag{7.5}$$

Moreover, if $u^* \in L^1(dw)$ (see (2.5)), then $d\omega(\mathbf{x}) = f(\mathbf{x})dw(\mathbf{x})$, where $f \in L^1(dw)$.

Proof Assume that either (7.4) or (7.5) holds. Then, by Theorem 7.2, for every $\varepsilon > 0$,

$$\sup_{x_0>0} \sup_{\mathbf{x}\in\mathbb{R}^N} |u(x_0 + \varepsilon, \mathbf{x})| \leq C_\varepsilon < \infty. \tag{7.6}$$

Set $f_n(\mathbf{x}) = u(\frac{1}{n}, \mathbf{x})$. From Theorem 7.3 we conclude that $u(\frac{1}{n} + x_0, \mathbf{x}) = P_{x_0} f_n(\mathbf{x})$. Moreover, there is a subsequence n_j such that f_{n_j} converges weakly-* to $f \in L^p(dw)$ (if $1 < p < \infty$) or to a measure ω (if $p = 1$). In both cases u is the Poisson integral either of f or ω. If additionally $u^* \in L^1(dw)$, then the measure ω is absolutely continuous with respect to dw. □

7.4 Proof of a Part of Theorem 2.1

We are now in a position to prove a part of Theorem 2.1, which is stated in the following proposition. The converse is proven at the very end of Sect. 11 (see Proposition 11.5).

Proposition 7.6 *Assume that $\mathbf{u} \in \mathcal{H}^1$. Then*

$$\|\mathbf{u}^*\|_{L^1(dw)} \leq C\|\mathbf{u}\|_{\mathcal{H}^1}. \tag{7.7}$$

Proof Fix $\varepsilon > 0$. Set $u_{j,\varepsilon}(x_0, \mathbf{x}) = u_j(\varepsilon + x_0, \mathbf{x})$, $f_{j,\varepsilon}(\mathbf{x}) = u_j(\varepsilon, \mathbf{x})$. Then, by Theorem 7.2, the $\mathcal{L}$-harmonic function $u_{j,\varepsilon}(x_0, \mathbf{x})$ is bounded and continuous on the closed set $[0, \infty) \times \mathbb{R}^N$. In particular, $f_{j,\varepsilon} \in L^\infty \cap L^1(dw) \cap C^2$. By Theorem 7.3,

$$u_{j,\varepsilon}(x_0, \mathbf{x}) = P_{x_0} f_{j,\varepsilon}(\mathbf{x}).$$

It is not difficult to conclude using (5.7) (with $m = 0$) that $\lim_{\|(x_0,\mathbf{x})\|\to\infty} |u_{j,\varepsilon}(x_0, \mathbf{x})| = 0$. Thus also $\lim_{\|\mathbf{x}\|\to\infty} f_{j,\varepsilon}(\mathbf{x}) = 0$. Set $\mathbf{u}_\varepsilon = (u_{0,\varepsilon}, u_{1,\varepsilon}, \ldots, u_{N,\varepsilon})$. Clearly, $\mathbf{u}_\varepsilon \in \mathcal{H}^1$. Let $F_\varepsilon(x_0, \mathbf{x}) = F(\varepsilon + x_0, \mathbf{x})$, where $F(x_0, \mathbf{x})$ is defined by (6.8). Set $f_\varepsilon(\mathbf{x}) = |F(\varepsilon, \mathbf{x})|$. Let $0 < q < 1$ be as in Theorem 6.3 and $p = q^{-1} > 1$. Observe that the function $|F_\varepsilon(x_0, \mathbf{x})|^q - P_{x_0}(f_\varepsilon^q)(\mathbf{x})$ vanishes for $x_0 = 0$ and

$$\lim_{\|(x_0,\mathbf{x})\|\to\infty} \Big(|F_\varepsilon(x_0, \mathbf{x})|^q - P_{x_0}(f_\varepsilon^q)(\mathbf{x})\Big) = 0.$$

So, by Theorem 6.3 and the maximum principle (see Theorem 7.1),

$$|\mathbf{u}(\varepsilon + x_0, \mathbf{x})|^q \leq |F_\varepsilon(x_0, \mathbf{x})|^q \leq P_{x_0}(\boldsymbol{f}_\varepsilon^q)(\mathbf{x}). \tag{7.8}$$

Set $\mathbf{u}_\varepsilon^*(\mathbf{x}) = \sup_{\|\mathbf{x}-\mathbf{y}\|<x_0} |\mathbf{u}(\varepsilon + x_0, \mathbf{y})|$. Then, by (7.8) and (5.9),

$$\|\mathbf{u}_\varepsilon^*\|_{L^1(dw)} \leq C_p \|\boldsymbol{f}_\varepsilon^q\|_{L^p(dw)}^p = C_p \|\boldsymbol{f}_\varepsilon\|_{L^1(dw)} \leq C_p \|\mathbf{u}\|_{\mathcal{H}^1}.$$

Since $\mathbf{u}_\varepsilon^*(\mathbf{x}) \to \mathbf{u}^*(\mathbf{x})$ as $\varepsilon \to 0$ and the convergence is monotone, we use the Lebesgue monotone convergence theorem and get (7.7). □

From Theorem 7.5 and Proposition 7.6 we obtain the following corollary.

Corollary 7.7 *If $\mathbf{u} \in \mathcal{H}^1$, then there are $f_j \in L^1(dw)$, $j = 0, 1, \ldots, N$, such that $|f_j(\mathbf{x})| \leq \mathbf{u}^*(\mathbf{x})$ and $u_j(x_0, \mathbf{x}) = P_{x_0} f_j(\mathbf{x})$. Moreover, the limit $\lim_{x_0 \to 0} u_j(x_0, \mathbf{x}) = f_j(\mathbf{x})$ exists in $L^1(dw)$.*

8 Riesz Transform Characterization of H^1_Δ

8.1 Riesz Transforms

The Riesz transforms in the Dunkl setting are defined by

$$\mathcal{F}(R_j f)(\xi) = -i \frac{\xi_j}{\|\xi\|} (\mathcal{F} f)(\xi), \quad j = 1, 2, \ldots, N.$$

They are bounded operators on $L^2(dw)$. Clearly,

$$R_j f = -T_{e_j}(-\Delta)^{-1/2} f = - \lim_{\varepsilon \to 0^+,\, M \to \infty} c \int_\varepsilon^M T_{e_j} e^{t\Delta} f \frac{dt}{\sqrt{t}},$$

and the convergence is in $L^2(dw)$ for $f \in L^2(dw)$. It follows from [1] that R_j are bounded operators on $L^p(dw)$ for $1 < p < \infty$.

Our task is to define $R_j f$ for $f \in L^1(dw)$. To this end we set

$$\mathcal{T} = \{\varphi \in L^2(dw) : (\mathcal{F}\varphi)(\xi)(1 + \|\xi\|)^n \in L^2(dw),\ n = 0, 1, 2, \ldots\}.$$

It is not difficult to check that if $\varphi \in \mathcal{T}$, then $\varphi \in C_0(\mathbb{R}^N)$ and $R_j\varphi \in C_0(\mathbb{R}^N) \cap L^2(dw)$. Moreover, for fixed $\mathbf{y} \in \mathbb{R}^N$ the function $p_t(\mathbf{x}, \mathbf{y})$ belongs to $\mathcal{T}$. Now $R_j f$, for $f \in L^1(dw)$, is defined in a weak sense as a functional on $\mathcal{T}$, by

$$\langle R_j f, \varphi \rangle = - \int_{\mathbb{R}^N} f(\mathbf{x}) R_j \varphi(\mathbf{x})\, dw(\mathbf{x}).$$

8.2 Proof of Theorem 2.5

Assume that $f \in L^1(dw)$ is such that $R_j f$ belong to $L^1(dw)$ for $j = 1, 2, \dots, N$. Set $f_0(\mathbf{x}) = f(\mathbf{x})$, $f_j(\mathbf{x}) = R_j f(\mathbf{x})$, $u_0(x_0, \mathbf{x}) = P_{x_0} f(\mathbf{x})$, $u_j(x_0, \mathbf{x}) = P_{x_0} f_j(\mathbf{x})$. Then $\mathbf{u} = (u_0, u_1, \dots, u_n)$ satisfies (2.4). Moreover,

$$\sup_{x_0>0} \int_{\mathbb{R}^N} |u_j(x_0, \mathbf{x})|\, dw(\mathbf{x}) \leq \|f_j\|_{L^1(dw)} \quad \text{for } j = 0, 1, \dots, N.$$

Thus $\mathbf{u} \in \mathcal{H}^1$ and

$$\|f\|_{H^1_\Delta} = \|\mathbf{u}\|_{\mathcal{H}^1} \leq \|f\|_{L^1(dw)} + \sum_{j=1}^{N} \|R_j f\|_{L^1(dw)}.$$

We turn to prove the converse. Assume that $f_0 \in H^1_\Delta$. By the definition of H^1_Δ there is a system $\mathbf{u} = (u_0, u_1, \dots, u_N) \in \mathcal{H}^1$ such that $f_0(\mathbf{x}) = \lim_{x_0 \to 0} u_0(x_0, \mathbf{x})$ (convergence in $L^1(dw)$). Set $f_j(\mathbf{x}) = \lim_{x_0 \to 0} u_j(x_0, \mathbf{x})$, where limits exist in $L^1(dw)$ (see Corollary 7.7). We have $u_j(x_0, \mathbf{x}) = P_{x_0} f_j(\mathbf{x})$. It suffices to prove that $R_j f_0 = f_j$. To this end, for $\varepsilon > 0$, let $f_{j,\varepsilon}(\mathbf{x}) = u_j(\varepsilon, \mathbf{x})$, $u_{j,\varepsilon}(x_0, \mathbf{x}) = u_j(x_0 + \varepsilon, \mathbf{x})$. Then $f_{j,\varepsilon} \in L^1(dw) \cap C_0(\mathbb{R}^N)$. In particular, $f_{j,\varepsilon} \in L^2(dw)$. Set $g_j = R_j f_{0,\varepsilon}$, $v_j(x_0, \mathbf{x}) = P_{x_0} g_j(\mathbf{x})$. Then $\mathbf{v} = (u_{0,\varepsilon}, v_1, \dots, v_N)$ satisfies the Cauchy–Riemann equations (2.4). Therefore, $T_j u_{0,\varepsilon}(x_0, \mathbf{x}) = T_0 u_{j,\varepsilon}(x_0, \mathbf{x}) = T_0 v_j(x_0, \mathbf{x})$. Hence, $u_{j,\varepsilon}(x_0, \mathbf{x}) - v_j(x_0, \mathbf{x}) = c_j(\mathbf{x})$. But $\lim_{x_0 \to \infty} u_{j,\varepsilon}(x_0, \mathbf{x}) = 0 = \lim_{x_0 \to \infty} v_j(x_0, \mathbf{x})$ for every $\mathbf{x} \in \mathbb{R}^N$. Consequently, $u_{j,\varepsilon}(x_0, \mathbf{x}) = v_j(x_0, \mathbf{x})$. Thus, $f_{j,\varepsilon} = R_j f_{0,\varepsilon}$. Since $\lim_{\varepsilon \to 0} f_{j,\varepsilon} = f_j$ in $L^1(dw)$ and $R_j f_{0,\varepsilon} \to R f_0$ in the sense of distributions, we have $f_j = R_j f_0$.

9 Inclusion $H^1_{(1,q,M)} \subset H^1_\Delta$

In this section we show that the atomic space $H^1_{(1,q,M)}$ with $M > \mathbf{N}$ is contained in the Hardy space H^1_Δ and there exists $C = C_{k,q,M}$ such that

$$\|f\|_{H^1_\Delta} \leq C \|f\|_{H^1_{(1,q,M)}}. \tag{9.1}$$

Let $f \in H^1_{(1,q,M)}$. According to Theorem 2.5, it is enough to show that $R_j f \in L^1(dw)$ and $\|R_j f\|_{L^1(dw)} \leq C \|f\|_{H^1_{(1,q,M)}}$. By the definition of the atomic space, there is a sequence a_j of $(1, q, M)$ atoms and $\lambda_i \in \mathbb{C}$, $(\lambda_i) \in \ell^1$, such that $f = \sum_i \lambda_i a_i$ and $\sum_i |\lambda_i| \leq 2\|f\|_{H^1_{(1,q,M)}}$. Observe that the series converges in $L^1(dw)$, hence $R_j f = \sum_i \lambda_j R_j a_j$ in the sense of distributions. Therefore it suffices to prove that there is a constant $C > 0$ such $\|R_j a\|_{L^1(dw)} \leq C$ for every a being a $(1, q, M)$-atom. Our proof follows ideas of [24]. Let $b \in \mathcal{D}(\Delta^M)$ and $B(\mathbf{y_0}, r)$ be as in the definition of

$(1, q, M)$ atom. Since R_j is bounded on $L^q(dw)$, by the Hölder inequality, we have

$$\|R_j a\|_{L^1(\mathcal{O}(B(\mathbf{y_0},4r)))} \leq C.$$

In order to estimate $R_j a$ on the set $\mathcal{O}(B(\mathbf{y_0}, 4r))^c$ we write

$$\begin{aligned} R_j a &= c_k'' \int_0^\infty T_{j,\mathbf{x}} e^{t\Delta} a \frac{dt}{\sqrt{t}} \\ &= c_k'' \int_0^{r^2} T_{j,\mathbf{x}} e^{t\Delta} a \frac{dt}{\sqrt{t}} + c_k'' \int_{r^2}^\infty T_{j,\mathbf{x}} e^{t\Delta} (\Delta)^M b \frac{dt}{\sqrt{t}} \\ &= c_k'' \int_0^{r^2} T_{j,\mathbf{x}} e^{t\Delta} a \frac{dt}{\sqrt{t}} + c_k'' \int_{r^2}^\infty T_{j,\mathbf{x}} \partial_t^M e^{t\Delta} b \frac{dt}{\sqrt{t}} \\ &= R_{j,0} a + R_{j,\infty} a. \end{aligned}$$

Further, using (4.5) with $m = 0$ together with (3.2), we get

$$\begin{aligned} |R_{j,0} a(\mathbf{x})| &\leq C \int_0^{r^2} \int_{\mathbb{R}^N} t^{-1} w(B(\mathbf{y}, \sqrt{t}))^{-1} e^{-cd(\mathbf{x},\mathbf{y})^2/t} |a(\mathbf{y})|\, dw(\mathbf{y}) dt \\ &\leq C \frac{r^{\mathbf{N}+1}}{d(\mathbf{x}, \mathbf{y})^{\mathbf{N}+1} w(B(\mathbf{y}_0, r))}. \end{aligned} \tag{9.2}$$

To estimate $R_{j,\infty} a$ we recall that $\|b\|_{L^1(dw)} \leq r^{2M}$. Using (4.5) with $m = M$, we obtain

$$\begin{aligned} |R_{j,\infty} a(\mathbf{x})| &\leq C \int_{r^2}^\infty \int_{\mathbb{R}^N} t^{-M-1} w(B(\mathbf{y}, \sqrt{t}))^{-1} e^{-cd(\mathbf{x},\mathbf{y})^2/t} |b(\mathbf{y})|\, dw(\mathbf{y}) dt \\ &\leq C \frac{r^{2M}}{d(\mathbf{x}, \mathbf{y})^{2M} w(B(\mathbf{y}_0, r))}. \end{aligned} \tag{9.3}$$

Obviously, (9.2) and (9.3) combined with (3.2) imply $\|R_j a\|_{L^1(\mathcal{O}(B(\mathbf{y_0},4r))^c)} \leq C$.

10 Maximal Functions

Let $\Phi(\mathbf{x})$ be a radial continuous function such that $|\Phi(\mathbf{x})| \leq C(1 + \|\mathbf{x}\|)^{-\kappa-\beta}$ with $\kappa > \mathbf{N}$. Set $\Phi_t(\mathbf{x}) = t^{-N}\Phi(t^{-1}\mathbf{x})$ and $\Phi_t(\mathbf{x}, \mathbf{y}) = \tau_{\mathbf{x}}\Phi_t(-\mathbf{y})$. Then, by Corollary 3.2,

$$|\Phi_t(\mathbf{x}, \mathbf{y})| \leq C\, V(\mathbf{x}, \mathbf{y}, t)^{-1} \left(1 + \frac{d(\mathbf{x}, \mathbf{y})}{t}\right)^{-\beta}.$$

Set $\mathcal{M}_{\Phi,a} f(\mathbf{x}) = \sup_{\|\mathbf{x}-\mathbf{y}\|<at} |\Phi_t f(\mathbf{y})|$, where

$$\Phi_t f(\mathbf{x}) = \Phi_t * f(\mathbf{x}) = \int_{\mathbb{R}^N} \Phi_t(\mathbf{x}, \mathbf{y})\, f(\mathbf{y})\, dw(\mathbf{y}).$$

If $a = 1$, we write $\mathcal{M}_\Phi$ instead of $\mathcal{M}_{\Phi,1}$. By definition a function $f \in L^1(dw)$ belongs to $H^1_{\max,\Phi}$ if $\mathcal{M}_\Phi f \in L^1(dw)$. In this case $\|f\|_{H^1_{\max,\Phi}} := \|\mathcal{M}_\Phi f\|_{L^1(dw)}$. Recall that if $\Phi(\mathbf{x}) = p_1(\mathbf{x})$ (see (5.3)), then we write $\mathcal{M}_P$, $H^1_{\max,P}$ and $\|\cdot\|_{H^1_{\max,P}}$ for the corresponding maximal function, space and norm respectively (see (2.6) and (2.7)).

10.1 The Space $\mathcal{N}$

The space $H^1_{\max,\Phi}$ is related to the following tent space $\mathcal{N}$.

Definition 10.1 For $a > 0$, $\lambda > \mathbf{N}$ and suitable functions $u(t, \mathbf{x})$, set

$$u_a^*(\mathbf{x}) = \sup_{\|\mathbf{x}-\mathbf{y}\|<at} |u(t,\mathbf{y})| \quad \text{and} \quad u_\lambda^{**}(\mathbf{x}) = \sup_{\mathbf{y}\in\mathbb{R}^N,\, t>0} |u(t,\mathbf{y})| \left(\frac{t}{\|\mathbf{y}-\mathbf{x}\|+t}\right)^\lambda.$$

The *tent space* $\mathcal{N}_a$ is then defined by

$$\mathcal{N}_a = \{u(t,\mathbf{x}) : \|u\|_{\mathcal{N}_a} = \|u_a^*\|_{L^1(dw)} < \infty\}.$$

If $a = 1$, we use the simplified notation $\mathcal{N}$, $\|u\|_{\mathcal{N}}$ and u^* (cf. (2.5)).

Obviously, if $u(t,\mathbf{x}) = \Phi_t f(\mathbf{x})$, then $\|f\|_{H^1_{\max,\Phi}} = \|u\|_{\mathcal{N}}$.

Lemma 10.2 *There are constants $C, c_\lambda, C_\lambda > 0$ such that*

$$\|u\|_{\mathcal{N}_a} \le C\left(\frac{a+b}{b}\right)^{\mathbf{N}} \|u\|_{\mathcal{N}_b}, \tag{10.1}$$

$$c_\lambda \|u\|_{\mathcal{N}} \le \|u_\lambda^{**}\|_{L^1(dw)} \le C_\lambda \|u\|_{\mathcal{N}}. \tag{10.2}$$

Proof Similar to the proofs in [35, Chapter II] and [20, p. 114]. □

If $\Omega \subset \mathbb{R}^N$ is an open set, then the tent over Ω is given by

$$\widehat{\Omega} = \Big((0,\infty)\times\mathbb{R}^N\Big) \setminus \bigcup_{\mathbf{x}\in\Omega^c} \Gamma(\mathbf{x}), \quad \text{where } \Gamma(\mathbf{x}) = \{(t,\mathbf{y}) : \|\mathbf{x}-\mathbf{y}\| < 4t\}.$$

The space $\mathcal{N}$ admits the following atomic decomposition (see [35]).

Definition 10.3 A function $A(t,\mathbf{x})$ is an atom for $\mathcal{N}$ if there is a ball B such that

- $\operatorname{supp} A \subset \hat{B}$,
- $\|A\|_{L^\infty} \le w(B)^{-1}$.

Clearly, $\|A\|_{\mathcal{N}} \le 1$ for every atom A for $\mathcal{N}$. Moreover, every $u \in \mathcal{N}$ can be written as $u = \sum_j \lambda_j A_j$, where the functions A_j are atoms for $\mathcal{N}$, $\lambda_j \in \mathbb{C}$, and $\sum_j |\lambda_j| \le C\,\|u\|_{\mathcal{N}}$.

Proposition 10.4 *Let* $u(t,\mathbf{x}) = P_t f(\mathbf{x})$ *and* $v(t,\mathbf{x}) = t^n \frac{d^n}{dt^n} P_t f(\mathbf{x})$. *Then for* $f \in L^1(dw)$ *we have*

$$\|v\|_{\mathcal{N}} \leq C_n \|u\|_{\mathcal{N}}.$$

Proof Assume that $\|u\|_{\mathcal{N}} < \infty$. Clearly, $v(t,\mathbf{x}) = 2^n Q_{t/2} P_{t/2} f(\mathbf{x})$, where $Q_t = t^n \frac{d^n}{dt^n} P_t$. Set $u^{\{1\}}(t,\mathbf{x}) = u(\frac{t}{2},\mathbf{x})$. Then

$$\|u^{\{1\}}\|_{\mathcal{N}} \leq C\|u\|_{\mathcal{N}}.$$

According to the atomic decomposition, $u^{\{1\}} = \sum_j c_j A_j$, where the functions A_j are atoms for $\mathcal{N}$, $c_j \in \mathbb{C}$, and $\sum |c_j| \lesssim \|u\|_{\mathcal{N}}$, (see Definition 10.3). Thus, by Lemma 10.2, we have

$$v(t,\mathbf{x}) = 2^n \sum_j c_j Q_{t/2} A_j(t,\mathbf{x}),$$

$$Q_{t/2} A_j(t,\mathbf{x}) = \int_{\mathbb{R}^N} Q_{t/2}(\mathbf{x},\mathbf{y}) A_j(t,\mathbf{y}) dw(\mathbf{y}).$$

From Proposition 5.1 and Definition 10.3 we conclude that $\|Q_{t/2} A_j(t,\mathbf{x})\|_{\mathcal{N}} \leq C$. □

10.2 Calderón Reproducing Formula

Fix $m \in \mathbb{N}$ sufficiently large. Let $\tilde{\Theta} \in C^m(\mathbb{R})$ be an even function such that $\|\tilde{\Theta}\|_{\mathcal{S}^m} < \infty$ (see (3.11)). Set $\Theta(\mathbf{x}) = \tilde{\Theta}(\|\mathbf{x}\|)$ and assume that $\int_{\mathbb{R}^N} \Theta(\mathbf{x})\, dw(\mathbf{x}) = 0$. Write

$$L^2\Big(\mathbb{R}_+^{1+N}, \frac{dt}{t}\, dw(\mathbf{x})\Big) = L^2(dw\, dt/t).$$

The Plancherel theorem for the Dunkl transform implies that

$$\|\Theta_t * f(\mathbf{x})\|_{L^2(dw\, dt/t)} \leq C\, \|f\|_{L^2(dw)}\,. \tag{10.3}$$

Thus, $f \mapsto \Theta_t * f(\mathbf{x})$ is a bounded linear operator from $L^2(dw)$ into $L^2(dw\, dt/t)$. By duality, for $F(t,\mathbf{x}) \in L^2(dw\, dt/t)$, the limit

$$\lim_{\varepsilon \to 0^+} \int_{\varepsilon}^{\varepsilon^{-1}} (\Theta_t * F(t,\cdot))(\mathbf{x}) \frac{dt}{t}$$

exists in $L^2(\mathbb{R}^N, dw)$ and defines a bounded linear operator π_Θ from $L^2(dw\, dt/t)$ into $L^2(\mathbb{R}^N, dw)$, that is,

$$\|\pi_\Theta F\|_{L^2(dw)} \leq C\, \|F\|_{L^2(dw\, dt/t)}. \tag{10.4}$$

With the customary abuse of notation, we write

$$\pi_\Theta F(\mathbf{x}) = \int_0^\infty (\Theta_t * F(t,\cdot))(\mathbf{x})\frac{dt}{t} = \int_0^\infty \int_{\mathbb{R}^N} \Theta_t(\mathbf{x},\mathbf{y}) F(t,\mathbf{y})\, dw(\mathbf{y})\frac{dt}{t}.$$

Let Φ be a radial C^∞ function on $\mathbb{R}^N$ such that

$$0 \le \Phi \le 1\,, \quad \operatorname{supp}\Phi \subset B(0,1/4)\,, \quad \Phi = 1 \text{ on } B(0,1/8)\,,$$

and let κ be an integer $> \mathbf{N}/2$. Then

$$\Psi = \Delta^{2\kappa}(\Phi * \Phi) = (\Delta^\kappa \Phi) * (\Delta^\kappa \Phi).$$

is radial, real-valued and

$$\operatorname{supp}\Psi \subset B(0,1/2)\,, \quad \int_{\mathbb{R}^N} \Psi(\mathbf{x})\, dw(\mathbf{x}) = 0\,,$$
$$\mathcal{F}\Psi(\xi) = c_k \,\|\xi\|^{4\kappa}\, (\mathcal{F}\Phi)^2(\xi) = c_k\, \|\xi\|^{4\kappa}\, |\mathcal{F}\Phi(\xi)|^2.$$

Moreover, it follows from (3.6) and (3.7) that

$$\Phi_t(\mathbf{x},\mathbf{y}) = 0 \quad \text{if} \quad d(\mathbf{x},\mathbf{y}) > t/4 \quad \text{and} \quad \Psi_t(\mathbf{x},\mathbf{y}) = 0 \quad \text{if} \quad d(\mathbf{x},\mathbf{y}) > t/2. \qquad (10.5)$$

Furthermore,

$$\int_{\mathbb{R}^N} \Psi_t(\mathbf{x},\mathbf{y})\, dw(\mathbf{y}) = \int_{\mathbb{R}^N} \Psi_t(\mathbf{x},\mathbf{y})\, dw(\mathbf{x}) = 0 \qquad \forall\, t > 0$$

and the following Calderón reproducing formulae hold in $L^2(dw)$, for $n = 1, 2, \ldots$:

$$f = \lim_{\varepsilon\to 0^+} c'_n \int_\varepsilon^{\varepsilon^{-1}} \Psi_t (t\sqrt{-\Delta})^n\, e^{-t\sqrt{-\Delta}} f\, \frac{dt}{t} = \lim_{\varepsilon\to 0^+} c' \int_\varepsilon^{\varepsilon^{-1}} \Psi_t(-t^2\Delta)\, e^{t^2\Delta} f\, \frac{dt}{t}. \qquad (10.6)$$

We may think about the operator in the second equality in (10.6) as the identity operator obtained by the composition of the bounded linear operator

$$L^2(dw) \ni f \mapsto -t^2\Delta e^{t^2\Delta} f(\mathbf{x}) \in L^2(dw\, dt/t)$$

with the linear operator $c'\pi_\Psi$, which is bounded from $L^2(dw\, dt/t)$ into $L^2(dw)$ (see (10.3) and (10.4)).

Recall that m is a large positive integer. Let $\Phi^{\{j\}}(\mathbf{x}) = \tilde{\Phi}^{\{j\}}(\|\mathbf{x}\|)$, $j = 1, 2$, where $\tilde{\Phi}^{\{j\}}$ are even C^m-functions such that

$$\|\tilde{\Phi}^{\{j\}}\|_{\mathcal{S}^m} < \infty \qquad (10.7)$$

and

$$\int_{\mathbb{R}^N} \Phi^{\{j\}}(\mathbf{x})\, dw(\mathbf{x}) = 1 \qquad (j = 1, 2). \tag{10.8}$$

By rescaling possibly $\Phi^{\{j\}}$, we may assume that

$$f = \lim_{\varepsilon\to 0^+} c_j'' \int_\varepsilon^{\varepsilon^{-1}} \Psi_t \Phi_t^{\{j\}} f \frac{dt}{t} \qquad \forall\, f \in L^2(dw), \tag{10.9}$$

where the limit takes place in $L^2(dw)$. Moreover, by Lemma 10.2, there is a constant $C_s > 0$ such that if $u^{\{j\}}(t,\mathbf{x}) = \Phi_t^{\{j\}} f(\mathbf{x})$ and $v^{\{j\}}(t,\mathbf{x}) = \Phi_{ts}^{\{j\}} f(\mathbf{x}) = u^{\{j\}}(st,\mathbf{x})$, then

$$C_s^{-1} \|v^{\{j\}}\|_{\mathcal{N}} \le \|u^{\{j\}}\|_{\mathcal{N}} \le C_s \|v^{\{j\}}\|_{\mathcal{N}}.$$

We are in a position to state the main results of this section.

Proposition 10.5 *For $\Phi^{\{1\}}$ and $\Phi^{\{2\}}$ as above and every $f \in L^2(dw)$, we have*

$$\begin{aligned}\|\Phi_t^{\{1\}} f\|_{\mathcal{N}_\alpha} = \|\mathcal{M}_{\Phi^{\{1\}},\alpha} f\|_{L^1(dw)} &\le C_{\Phi^{\{1\}},\Phi^{\{2\}},\alpha,\alpha'} \|\mathcal{M}_{\Phi^{\{2\}},\alpha'} f\|_{L^1(dw)}\\ &= C_{\Phi^{\{1\}},\Phi^{\{2\}},\alpha,\alpha'} \|\Phi_t^{\{2\}} f\|_{\mathcal{N}_{\alpha'}}.\end{aligned}$$

Proof Let $\Psi^{\{1\}} = \Phi^{\{1\}} - \Phi^{\{2\}}$. Then $\Psi^{\{1\}}$ is radial and thanks to (10.8), we have $|\mathcal{F}\Psi^{\{1\}}(\xi)| \le C\|\xi\|^2$ for $\|\xi\| < 1$. It suffices to prove that

$$\|\Psi_t^{\{1\}} f\|_{\mathcal{N}} \le C \|\Phi_t^{\{2\}} f\|_{\mathcal{N}}.$$

Using (10.9), we obtain

$$\Psi_t^{\{1\}} f = \lim_{\varepsilon\to 0^+} c_2'' \int_\varepsilon^{\varepsilon^{-1}} \Psi_t^{\{1\}} \Psi_s \Phi_s^{\{2\}} f \frac{ds}{s},$$

where the limit takes place both in $L^2(dw)$ and pointwise, because $\Psi_t^{\{1\}}(\mathbf{y},\mathbf{w}) \in L^2(dw(\mathbf{w}))$. According to Proposition 3.3, for any $\eta, \ell > 0$ such that $\ell \le 4\kappa$, the integral kernel $K_{t,s}(\mathbf{y},\mathbf{z})$ of the operator $\Psi_t^{\{1\}}\Psi_s$ satisfies

$$|K_{t,s}(\mathbf{y},\mathbf{z})| \le C_{\eta,\ell} \min\Big(\Big(\frac{t}{s}\Big)^2, \Big(\frac{s}{t}\Big)^\ell\Big) \frac{1}{V(\mathbf{y},\mathbf{z},s+t)} \Big(1 + \frac{d(\mathbf{y},\mathbf{z})}{s+t}\Big)^{-\mathbf{N}-\eta}.$$

We take $\mathbf{N} < \lambda < \eta < \ell$. Then for $\|\mathbf{x}-\mathbf{y}\| < t$, we have

$$\int_{\mathbb{R}^N} |K_{t,s}(\mathbf{y},\mathbf{z})| \Big(1 + \frac{d(\mathbf{x},\mathbf{z})}{s}\Big)^\lambda dw(\mathbf{z}) \le C' \min\Big(\Big(\frac{s}{t}\Big)^{\ell-\lambda}, \Big(\frac{t}{s}\Big)^2\Big). \tag{10.10}$$

Therefore, using (10.10), we obtain

$$\begin{aligned}\sup_{\|\mathbf{x}-\mathbf{y}\|<t} |\Psi_t^{\{1\}} f(\mathbf{y})| &= |c_2''| \sup_{\|\mathbf{x}-\mathbf{y}\|<t} \left| \lim_{\varepsilon\to 0^+} \int_\varepsilon^{\varepsilon^{-1}} \int_{\mathbb{R}^N} K_{t,s}(\mathbf{y},\mathbf{z}) \Phi_s^{\{2\}} f(\mathbf{z})\, dw(\mathbf{z}) \frac{ds}{s} \right| \\ &\leq |c_2''| \sup_{\mathbf{z},s} |\Phi_s^{\{2\}} f(\mathbf{z})| \Big(1 + \frac{d(\mathbf{x},\mathbf{z})}{s}\Big)^{-\lambda} \\ &\quad \times \sup_{\|\mathbf{x}-\mathbf{y}\|<t} \int_0^\infty \int_{\mathbb{R}^N} |K_{t,s}(\mathbf{y},\mathbf{z})| \Big(1 + \frac{d(\mathbf{x},\mathbf{z})}{s}\Big)^{\lambda} dw(\mathbf{z}) \frac{ds}{s} \\ &\leq C \sup_{\mathbf{z},s} |\Phi_s^{\{2\}} f(\mathbf{z})| \Big(1 + \frac{d(\mathbf{x},\mathbf{z})}{s}\Big)^{-\lambda}. \end{aligned}$$

The proof is complete, by applying (10.2). □

Remark 10.6 It follows from the proof of Proposition 10.5 that if $\Theta \in \mathcal{S}(\mathbb{R}^N)$ is radial and $\int_{\mathbb{R}^N} \Theta(\mathbf{x})\, dw(\mathbf{x}) = 0$, and $\Phi^{\{2\}}$ is as above, then for $f \in L^2(dw)$, we have

$$\|\Theta_t f\|_{\mathcal{N}} \leq C \|\Phi_t^{\{2\}} f\|_{\mathcal{N}}.$$

Proposition 10.7 *For a function $\Phi^{\{1\}}$ as above and $\alpha > 0$ there is a constant $C_{\Phi^{\{1\}},\alpha} > 0$ such that*

$$\|\mathcal{M}_{\Phi^{\{1\}},\alpha} f\|_{L^1(dw)} \leq C_{\Phi^{\{1\}},\alpha} \|\mathcal{M}_P f\|_{L^1(dw)}, \quad \text{for } f \in L^1(dw) \cap L^2(dw).$$

Proof For a positive integer n (large), set $\phi(\xi) = e^{-\|\xi\|}\Big(\sum_{j=0}^{n+1} \frac{\|\xi\|^j}{j!}\Big)$. Then

$$|\phi(\xi) - 1| \leq C\, \|\xi\|^{n+1} \quad \text{for} \quad \|\xi\| < 1.$$

So ϕ is a C^n function such that $|\partial^\beta \phi(\xi)| \leq C_\beta \exp(-\|\xi\|/2)$, $|\beta| \leq n$. Put $\Phi^{\{2\}} = c_k^{-1} \mathcal{F}^{-1}\phi$. Applying Proposition 10.5, we have

$$\|\Phi_t^{\{1\}} f\|_{\mathcal{N}} \lesssim \|\Phi_t^{\{2\}} f\|_{\mathcal{N}}.$$

Notice that $\frac{d^j}{dt^j} P_t f(\mathbf{x}) = \mathcal{F}^{-1}(\|\xi\|^j e^{-t\|\xi\|} \mathcal{F} f(\xi))(\mathbf{x})$. Hence, from Proposition 10.4 we conclude,

$$\|\Phi_t^{\{2\}} f\|_{\mathcal{N}} \leq C \sum_{j=0}^{n+1} \Big\| t^j \frac{d^j}{dt^j} P_t f \Big\|_{\mathcal{N}} \leq C' \|P_t f\|_{\mathcal{N}}.$$

□

Lemma 10.8 *$H^1_{\max,H} \subset H^1_{\max,P}$ and there is a constant $C > 0$ such that*

$$\|\mathcal{M}_P f\|_{L^1(dw)} \leq C \|\mathcal{M}_H f\|_{L^1(dw)} \quad \text{for } f \in L^1(dw). \tag{10.11}$$

Proof The proof is standard. Let $f \in L^1(dw)$. Set $u(t, \mathbf{x}) = e^{t^2\Delta} f(\mathbf{x})$. By the subordination formula (2.2) for fixed $t > 0$, we have

$$\begin{aligned}
\sup_{\|\mathbf{x}'-\mathbf{x}\|<t} |P_t f(\mathbf{x}')| &\leq \frac{1}{2\sqrt{\pi}} \int_0^\infty \sup_{\|\mathbf{x}'-\mathbf{x}\|<t} |u(ts, \mathbf{x}')| e^{-\frac{1}{4s^2}} \frac{ds}{s^2} \\
&= \frac{1}{2\sqrt{\pi}} \int_0^\infty \sup_{\|\mathbf{x}'-\mathbf{x}\|<t} |u(ts, \mathbf{x}')| \left(\frac{ts}{\|\mathbf{x} - \mathbf{x}'\| + ts} \right)^\lambda \\
&\quad \times \left(\frac{\|\mathbf{x} - \mathbf{x}'\| + ts}{ts} \right)^\lambda e^{-\frac{1}{4s^2}} \frac{ds}{s^2} \\
&\leq \frac{1}{2\sqrt{\pi}} \int_0^\infty u_\lambda^{**}(\mathbf{x}) \Big(\frac{1+s}{s} \Big)^\lambda e^{-\frac{1}{4s^2}} \frac{ds}{s^2} \\
&\leq C u_\lambda^{**}(\mathbf{x}).
\end{aligned}$$

Now the lemma follows from (10.2). □

Note that Propositions 10.5 and 10.7 together with Lemma 10.8 imply that

$$H^1_{\max, \Phi^{\{1\}}} \cap L^2(dw) = H^1_{\max, H} \cap L^2(dw) = H^1_{\max, P} \cap L^2(dw)$$

and for $f \in L^2(dw)$, we have

$$\|\mathcal{M}_{\Phi^{\{1\}}} f\|_{L^1(dw)} \sim \|\mathcal{M}_H f\|_{L^1(dw)} \sim \|\mathcal{M}_P f\|_{L^1(dw)}. \tag{10.12}$$

Our task is to remove the assumption $f \in L^2(dw)$ from (10.12).

Lemma 10.9 *Assume that $f \in H^1_{\max, P}$. Then $P_t f \in L^2(dw)$ for every $t > 0$ and*

$$\lim_{t \to 0} \|P_t f - f\|_{H^1_{\max, P}} = 0. \tag{10.13}$$

Proof Proposition 5.1 implies that $P_t f \in L^2(dw)$. To prove (10.13) we follow, e.g., [18, proof of (6.5)].

First observe that there is a constant $C > 0$ such that for every $A > 0$ and $t > 0$, we have

$$\Big\| \sup_{s > At,\, \|\mathbf{x}-\mathbf{y}\|<s} |P_{t+s} f(\mathbf{y}) - P_s f(\mathbf{y})| \Big\|_{L^1(dw(\mathbf{x}))} \leq C A^{-1} \|f\|_{L^1(dw)}. \tag{10.14}$$

To see (10.14) fix $\mathbf{z} \in \mathbb{R}^N$. For $s > At$, thanks to (5.4), we have

$$
\begin{aligned}
|p_{s+t}(\mathbf{y},\mathbf{z}) - p_s(\mathbf{y},\mathbf{z})| &= \left| \int_0^t \partial_u p_{s+u}(\mathbf{y},\mathbf{z})\,du \right| \\
&\leq C \int_0^t \frac{1}{u+s+d(\mathbf{y},\mathbf{z})} w(B(\mathbf{z}, s+u+d(\mathbf{y},\mathbf{z})))^{-1}\,du \\
&\leq C \int_0^t \frac{1}{s+d(\mathbf{y},\mathbf{z})} w(B(\mathbf{z}, s+d(\mathbf{y},\mathbf{z})))^{-1}\,du \\
&\leq \frac{C}{A} \frac{s}{s+d(\mathbf{y},\mathbf{z})} w(B(\mathbf{z}, s+d(\mathbf{y},\mathbf{z})))^{-1}.
\end{aligned}
$$

Since $s + d(\mathbf{x},\mathbf{z}) \leq s + d(\mathbf{x},\mathbf{y}) + d(\mathbf{y},\mathbf{z}) \leq s + \|\mathbf{x}-\mathbf{y}\| + d(\mathbf{y},\mathbf{z}) \leq 2(s + d(\mathbf{y},\mathbf{z}))$, we obtain

$$
\sup_{\|\mathbf{x}-\mathbf{y}\|<s} |p_{s+t}(\mathbf{y},\mathbf{z}) - p_s(\mathbf{y},\mathbf{z})| \leq \frac{C}{A} \frac{s}{s+d(\mathbf{x},\mathbf{z})} w(B(\mathbf{z}, s+d(\mathbf{x},\mathbf{z})))^{-1}, \quad (10.15)
$$

which implies (10.14).

In order to finish the proof of (10.13) assume that $f \in H^1_{\max,P}$. Using (10.14), we get

$$
\begin{aligned}
\|P_t f - f\|_{H^1_{\max,P}} \leq & \left\| \sup_{s>At,\ \|\mathbf{x}-\mathbf{y}\|<s} |P_{t+s} f(\mathbf{y}) - P_s f(\mathbf{y})| \right\|_{L^1(dw(\mathbf{x}))} \\
& + \left\| \sup_{s\leq At,\ \|\mathbf{x}-\mathbf{y}\|<s} |P_{t+s} f(\mathbf{y}) - P_s f(\mathbf{y})| \right\|_{L^1(dw(\mathbf{x}))} \\
& \leq CA^{-1} \|f\|_{L^1(dw)} + \left\| \sup_{s\leq At,\ \|\mathbf{x}-\mathbf{y}\|<s} |P_{s+t} f(\mathbf{y}) - f(\mathbf{x})| \right\|_{L^1(dw(\mathbf{x}))} \\
& + \left\| \sup_{s\leq At,\ \|\mathbf{x}-\mathbf{y}\|<s} |P_s f(\mathbf{y}) - f(\mathbf{x})| \right\|_{L^1(dw(\mathbf{x}))} \\
& \leq CA^{-1} \|f\|_{L^1(dw)} \\
& + 2 \left\| \sup_{s\leq (A+1)t,\ \|\mathbf{x}-\mathbf{y}\|<s} |P_s f(\mathbf{y}) - f(\mathbf{x})| \right\|_{L^1(dw(\mathbf{x}))}.
\end{aligned}
$$

Fix $\varepsilon > 0$ and take $A = C\varepsilon^{-1}$. Corollary 5.4 implies

$$
\lim_{t\to 0} \sup_{s\leq (A+1)t,\ \|\mathbf{x}-\mathbf{y}\|<s} |P_s f(\mathbf{y}) - f(\mathbf{x})| = 0 \quad \text{for almost every } \mathbf{x} \in \mathbb{R}^N.
$$

Since $\sup_{s\leq (A+1)t,\ \|\mathbf{x}-\mathbf{y}\|<s} |P_s f(\mathbf{y}) - f(\mathbf{x})| \leq 2\mathcal{M}_P f(\mathbf{x}) \in L^1(dw(\mathbf{x}))$, the proof is complete by applying the Lebesgue dominated convergence theorem. □

Lemma 10.10 *Let $\varphi \in \mathcal{S}(\mathbb{R}^N)$ be a radial function. There is a constant $C > 0$ such that for all $\varepsilon > 0$ and $u(t,\mathbf{x}) \in \mathcal{N}$ if $u_\varepsilon(t,\mathbf{x}) = u(t,\cdot) * \varphi_\varepsilon(\mathbf{x})$, then*

$$
\|u_\varepsilon\|_{\mathcal{N}} \leq C\|u\|_{\mathcal{N}}.
$$

Proof Let $\lambda > \mathbf{N}$ and $M > 0$ be large enough. For fixed $\mathbf{x} \in \mathbb{R}^N$ we have

$$\begin{aligned} u_\varepsilon^*(\mathbf{x}) &\le C \sup_{t \ge \varepsilon,\, \|\mathbf{x}-\mathbf{y}\|<t} \int_{\mathbb{R}^N} |u(t,\mathbf{z})| \Big(1 + \frac{d(\mathbf{z},\mathbf{x})}{t}\Big)^{-\lambda} \Big(1 + \frac{d(\mathbf{z},\mathbf{y})}{t}\Big)^{\lambda} |\varphi_\varepsilon(\mathbf{y},\mathbf{z})|\, dw(\mathbf{z}) \\ &\quad + C \sup_{0<t<\varepsilon} \sup_{d(\mathbf{x},\mathbf{y})<t} \int_{\mathbb{R}^N} |u(t,\mathbf{z})| \Big(1 + \frac{d(\mathbf{x},\mathbf{y})}{\varepsilon}\Big)^{M} V(\mathbf{y},\mathbf{z},\varepsilon)^{-1} \\ &\quad \times \Big(1 + \frac{d(\mathbf{x},\mathbf{z})}{\varepsilon}\Big)^{-M} dw(\mathbf{z}) \\ &\le C'_\lambda \sum_{\sigma \in G} u_\lambda^{**}(\sigma(\mathbf{x})) + C'_M \int_{\mathbb{R}^N} u^*(\mathbf{z}) w(B(\mathbf{z},\varepsilon))^{-1} \Big(1 + \frac{d(\mathbf{x},\mathbf{z})}{\varepsilon}\Big)^{-M} dw(\mathbf{z}). \end{aligned}$$

Integrating the inequality with respect to $dw(\mathbf{x})$ and applying (10.2) we obtain the lemma. □

Theorem 10.11 *Let $\Phi^{\{1\}}$ satisfies (10.7) and (10.8). Then the spaces $H^1_{\max,\Phi^{\{1\}}}$, $H^1_{\max,H}$, and $H^1_{\max,P}$ coincide and the corresponding norms are equivalent (cf. (10.12)).*

Proof Assume that $f \in H^1_{\max,P}$. Using Lemma 10.9 we take a sequence $t_n \to 0$, $n = 0, 1, \ldots$, such that $\|P_{t_0} f\|_{H^1_{\max,P}} \le 2\|f\|_{H^1_{\max,P}}$, $\|P_{t_{n+1}} f - P_{t_n} f\|_{H^1_{\max,P}} \le 2^{-n}\|f\|_{H^1_{\max,P}}$. Then $f = P_{t_0} f + \sum_{n=1}^\infty (P_{t_n} f - P_{t_{n-1}} f) =: g_0 + \sum_{n=1}^\infty g_n$, with the convergence in $L^1(dw)$. The functions $g_n \in L^2(dw) \cap H^1_{\max,P}$, so, by (10.12),

$$\|\mathcal{M}_{\Phi^{\{1\}}} f\|_{L^1(dw)} \le \sum_{j=0}^\infty \|\mathcal{M}_{\Phi^{\{1\}}} g_j\|_{L^1(dw)} \le C \sum_{j=0}^\infty \|\mathcal{M}_P g_j\|_{L^1(dw)} \le 3C\|f\|_{H^1_{\max,P}}.$$

We now turn to prove the converse. Suppose that $f \in H^1_{\max,\Phi^{\{1\}}}$. Then using Lemma 10.10 and the fact that $\|f * h_\varepsilon\|_{L^2(dw)} \le \|f\|_{L^1(dw)}\|h_\varepsilon\|_{L^2(dw)}$ we conclude that $f_\varepsilon = f * h_\varepsilon \in H^1_{\max,\Phi^{\{1\}}} \cap L^2(dw)$ and $\sup_{\varepsilon>0} \|f_\varepsilon\|_{H^1_{\max,\Phi^{\{1\}}}} \le C\|f\|_{H^1_{\max,\Phi^{\{1\}}}}$. Applying (10.12) we get $\sup_{\varepsilon>0} \|f_\varepsilon\|_{H^1_{\max,H}} \le C'\|f\|_{H^1_{\max,\Phi^{\{1\}}}}$. Observe that $\lim_{\varepsilon\to 0} \mathcal{M}_H f_\varepsilon(\mathbf{x}) = \mathcal{M}_H f(\mathbf{x})$ for almost all $\mathbf{x} \in \mathbb{R}^N$ and the convergence is monotone. Hence, by the Lebesgue monotone convergence theorem, we get $\|f\|_{H^1_{\max,H}} \le C'\|f\|_{H^1_{\max,\Phi^{\{1\}}}}$. Finally, the inequality $\|f\|_{H^1_{\max,P}} \le C\|f\|_{H^1_{\max,\Phi^{\{1\}}}}$ is obtained from Lemma 10.8. □

11 Atomic Decompositions; Inclusion $H^1_{\max,H} \subset H^1_{(1,\infty,M)}$

In next theorem we show that all elements in $H^1_{\max,H} \cap L^2(dw) = H^1_{\max,P} \cap L^2(dw)$ admit atomic decompositions into $(1,\infty,M)$-atoms. The $L^2(dw)$ condition is removed afterwards in Theorem 11.4.

Theorem 11.1 *For every positive integer M, there is a constant $C_M > 0$ such that every element $f \in H^1_{\max,H} \cap L^2(dw) = H^1_{\max,P} \cap L^2(dw)$ can be written as*

$$f = \sum \lambda_j a_j,$$

where a_j are $(1,\infty,M)$-atoms, $\sum |\lambda_j| \le C_M \|\mathcal{M}_P f\|_{L^1(dw)}$. Moreover, the convergence takes place in $L^2(dw)$.

Proof This result is known for Hardy spaces associated with semigroups satisfying Gaussian bounds on spaces of homogeneous type (see [11] and [37]). The proof presented here is a straightforward adaptation of [37] with the difference that tents are now constructed with respect to the orbit distance $d(\mathbf{x},\mathbf{y})$. We include details for the convenience of readers unfamiliar with [11] and [37]. More experienced readers may skip the proof and jump to Theorem 11.4.

Without loss of generality, we may assume that M is an even integer $> 2\mathbf{N}$.

Step 1. Reproducing formulae. Let Φ, Ψ be as in the Calderón reproducing formula with $\kappa = M/2$ (see Sect. 10). Set

$$\varphi(\xi) = \mathcal{F}(\Phi)(\xi) = \tilde{\varphi}(\|\xi\|),$$
$$\psi(\xi) = \mathcal{F}(\Psi)(\xi) = c_k\|\xi\|^{2M}|\varphi(\xi)|^2 = \tilde{\psi}(\|\xi\|) = c_k\|\xi\|^{2M}|\tilde{\varphi}(\|\xi\|)|^2.$$

Then there is a constant c such that

$$f = \lim_{\varepsilon\to 0^+} c\int_\varepsilon^{\varepsilon^{-1}} \Psi_t(-t^2\Delta)e^{t^2\Delta} f\frac{dt}{t} = c\,\pi_\Psi(-t^2\Delta e^{t^2\Delta} f)$$

with convergence in $L^2(dw)$ (see Sect. 10.2). We have

$$\mathcal{F}f(\xi) = \lim_{\varepsilon\to 0^+} c_k c\int_\varepsilon^{\varepsilon^{-1}} t^2\|\xi\|^2\tilde{\psi}(t\|\xi\|)e^{-t^2\|\xi\|^2}\mathcal{F}f(\xi)\frac{dt}{t}.$$

For $\xi \neq 0$, set

$$\eta(\xi) = c_k c\int_1^\infty t^2\|\xi\|^2\tilde{\psi}(t\|\xi\|)e^{-t^2\|\xi\|^2}\frac{dt}{t} = c_k c\int_{\|\xi\|}^\infty t^2\tilde{\psi}(t)e^{-t^2}\frac{dt}{t}.$$

Put $\eta(0) = 1$. Then η is a Schwartz class radial real-valued function. Set $\Xi(\mathbf{x}) = c_k^{-1}\mathcal{F}^{-1}\eta(\mathbf{x})$. Then $\Xi \in \mathcal{S}(\mathbb{R}^N)$, $\int \Xi(\mathbf{x})\,dw(\mathbf{x}) = 1$, and

$$c\int_a^b \Psi_t t^2\Delta e^{t^2\Delta} f\frac{dt}{t} = \Xi_a f - \Xi_b f. \tag{11.1}$$

Step 2. Space of orbits. Let $X = \mathbb{R}^N/G$ be the space of orbits equipped with the metric $d(\mathcal{O}(\mathbf{x}), \mathcal{O}(\mathbf{y})) = d(\mathbf{x}, \mathbf{y})$ and the measure $\boldsymbol{m}(A) = w\Big(\bigcup_{\mathcal{O}(\mathbf{x})\in A} \mathcal{O}(\mathbf{x})\Big)$. So $(X, d, \boldsymbol{m})$ is the space of homogeneous type in the sense of Coifman–Weiss. The space X can be identified with a positive Weyl chamber. Any open set in X of finite measure admits the following easily proved Whitney type covering lemma.

Lemma 11.2 *Suppose that $\boldsymbol{\Omega} \subset X$ is an open set with finite measure. Then there is a sequence of balls $B_X(\mathcal{O}(\mathbf{x}_{\{n\}}), r_{\{n\}})$ such that $r_{\{n\}} = d(\mathcal{O}(\mathbf{x}_{\{n\}}), \boldsymbol{\Omega}^c)$,*

$$\bigcup_{n\in\mathbb{N}} B_X(\mathcal{O}(\mathbf{x}_{\{n\}}), r_{\{n\}}/2) = \boldsymbol{\Omega},$$

the balls $B_X(\mathcal{O}(\mathbf{x}_{\{n\}}), r_{\{n\}}/10)$ are disjoint.

Step 3. Decomposition of $\mathbb{R}^{N+1}_+$. Assume that $f \in H^1_{\max,H} \cap L^2(dw)$. Let

$$F(t, \mathbf{x}) = \Big(|t^2\Delta e^{t^2\Delta} f(\mathbf{x})| + |\Xi_t f(\mathbf{x})|\Big),$$
$$\boldsymbol{F}(t, \mathbf{x}) = \sup_{\sigma\in G} F(t, \sigma(\mathbf{x})),$$

and

$$\mathcal{M} f(\mathbf{x}) = \sup_{d(\mathbf{x},\mathbf{y})<5t} F(t, \mathbf{y}) = \sup_{\|\mathbf{x}-\mathbf{y}\|<5t} \boldsymbol{F}(t, \mathbf{y}).$$

Then, by Proposition 10.5 and Remark 10.6, we have $\|\mathcal{M} f\|_{L^1(dw)} \leq C\|f\|_{H^1_{\max,H}}$. Observe that $\mathcal{M} f(\sigma(\mathbf{x})) = \mathcal{M} f(\mathbf{x})$. Therefore $\mathcal{M} f(\mathbf{x})$ can be identified with the function $\boldsymbol{\mathcal{M}} f(\mathcal{O}(\mathbf{x}))$ on X. Moreover, $\|\mathcal{M} f(\mathbf{x})\|_{L^1(dw)} = \|\boldsymbol{\mathcal{M}} f(\mathcal{O}(\mathbf{x}))\|_{L^1(\boldsymbol{m})}$. For an open set $\boldsymbol{\Omega} \subset X$, let

$$\hat{\boldsymbol{\Omega}} = \{(t, \mathcal{O}(\mathbf{x})) : B_X(\mathcal{O}(\mathbf{x}), 4t) \subset \boldsymbol{\Omega}\}$$

be the tent over $\boldsymbol{\Omega}$. For $j \in \mathbb{Z}$ define

$$\boldsymbol{\Omega}_j = \{\mathcal{O}(\mathbf{x}) \in X : \boldsymbol{\mathcal{M}} f(\mathcal{O}(\mathbf{x})) > 2^j\}, \quad \Omega_j = \{\mathbf{x} \in \mathbb{R}^N : \mathcal{M} f(\mathbf{x}) > 2^j\}.$$

Then $\boldsymbol{\Omega}_j$ is open in X, $\Omega_j = \bigcup_{\mathcal{O}(\mathbf{x})\in\boldsymbol{\Omega}_j} \mathcal{O}(\mathbf{x})$, $\boldsymbol{m}(\boldsymbol{\Omega}_j) = w(\Omega_j)$,

$$\sum_j 2^j w(\Omega_j) \sim \|\mathcal{M} f\|_{L^1(dw)} \sim \|f\|_{H^1_{\max,H}}.$$

Clearly, $\widehat{\Omega}_j = \{(t, \mathbf{x}) \in \mathbb{R}^{N+1}_+ : (t, \mathcal{O}(\mathbf{x})) \in \widehat{\boldsymbol{\Omega}}_j\}$. Set $\mathbf{T}_j = \hat{\Omega}_j \setminus \hat{\Omega}_{j+1}$. Then,

$$\operatorname{supp} F(t, \mathbf{x}) \subset \bigcup_{j\in\mathbb{Z}} \hat{\Omega}_j = \bigcup_{j\in\mathbb{Z}} (\hat{\Omega}_j \setminus \hat{\Omega}_{j+1}) = \bigcup_{j\in\mathbb{Z}} \mathbf{T}_j \tag{11.2}$$

Let $B_X(\mathcal{O}(\mathbf{x}_{\{n,j\}}), r_{\{n,j\}}/2))$, $\mathbf{x}_{\{n,j\}} \in \mathbb{R}^N$, $n = 1, 2, \ldots$, be a Whitney covering of $\mathbf{\Omega}_j$. Set

$$Q_{\{n,j\}} = \{\mathbf{x} \in \mathbb{R}^N : \mathcal{O}(\mathbf{x}) \in B_X(\mathcal{O}(\mathbf{x}_{\{n,j\}}), r_{\{n,j\}}/2))\} = \mathcal{O}(B(\mathbf{x}_{\{n,j\}}, r_{\{n,j\}}/2)).$$

Obviously, $w(B(\mathbf{x}_{\{n,j\}}, r_{\{n,j\}}/2)) \leq w(Q_{\{n,j\}}) \leq |G| w(B(\mathbf{x}_{\{n,j\}}, r_{\{n,j\}}/2))$. We define a cone over a G-invariant set E as

$$\mathcal{R}(E) = \{(t, \mathbf{y}) : d(\mathbf{y}, E) < 2t\}.$$

For $n = 1, 2, \ldots$, let

$$\mathbf{T}_{\{n,j\}} = \mathbf{T}_j \cap \left(\mathcal{R}(Q_{\{n,j\}}) \setminus \bigcup_{i=0}^{n-1} \mathcal{R}(Q_{\{i,j\}}) \right), \quad \mathcal{R}(Q_{\{0,j\}}) = \emptyset.$$

Clearly, $\hat{\Omega}_j \subset \bigcup_{n\in\mathbb{N}} \mathcal{R}(Q_{\{n,j\}})$, $\mathbf{T}_{\{n,j\}} \cap \mathbf{T}_{\{n',j'\}} = \emptyset$ if $(j, n) \neq (j', n')$. Thus we have

$$\operatorname{supp} F(t, \mathbf{x}) \subset \bigcup_{j\in\mathbb{Z}} \bigcup_{n\in\mathbb{N}} \mathbf{T}_{\{n,j\}}. \tag{11.3}$$

Step 4. Decomposition of f and $L^2(dw)$-convergence. Set $G(t, \mathbf{x}) = -t^2 \Delta e^{t^2\Delta} f(\mathbf{x})$, $G_{\{n,j\}}(t, \mathbf{x}) = \chi_{\mathbf{T}_{\{n,j\}}}(t, \mathbf{x}) G(t, \mathbf{x})$. By our assumption $f \in L^2(dw) \cap H^1_{\max,H}$, hence $G \in L^2(dw\, dt/t)$ and $G(t, \mathbf{x}) = \sum_{n\in\mathbb{N},\, j\in\mathbb{Z}} G_{\{n,j\}}(t, \mathbf{x})$, where the series converges unconditionally in $L^2(dw\, dt/t)$, because the sets $\mathbf{T}_{\{n,j\}}$ are pairwise disjoint. Since π_Ψ is a bounded linear operator from $L^2(dw\, dt/t)$ into $L^2(dw)$, we get

$$f = \sum_{j\in\mathbb{Z},\, n\in\in\mathbb{N}} c\,\pi_\Psi\big(G_{\{n,j\}}\big) =: \sum_{j\in\mathbb{Z},\, n\in\mathbb{N}} f_{\{n,j\}}, \tag{11.4}$$

where the series converges unconditionally in $L^2(dw)$.

Step 5. What remains to be done. Let $\lambda_{\{n,j\}} = 2^j w(Q_{\{n,j\}})$. Then

$$\sum_{j\in\mathbb{Z},\, n\in\mathbb{N}} |\lambda_{\{n,j\}}| = \sum_{j\in\mathbb{Z},\, n\in\mathbb{N}} 2^j w(Q_{\{n,j\}}) \lesssim \sum_{j\in\mathbb{Z}} 2^j w(\Omega_j) \sim \|f\|_{H^1_{\max,H}}.$$

Our task is to prove (in Steps 6–9) that, thanks to the choice of the sets $\mathbf{T}_{\{n,j\}}$, the functions $a_{\{n,j\}} = \lambda_{\{n,j\}}^{-1} f_{\{n,j\}}$ are proportional to $(1, \infty, M)$-atoms. Once this is done, the series in (11.4) converges (absolutely) in $L^1(dw)$ as well. Moreover,

$$f = \sum_{j\in\mathbb{Z},\, n\in\mathbb{N}} f_{\{n,j\}} = \sum_{j\in\mathbb{Z},\, n\in\mathbb{N}} \lambda_{\{n,j\}} a_{\{n,j\}}$$

will be the desired atomic decomposition.

Step 6. Functions $b_{\{n,j\}}$. Support of $\Delta^m b_{\{n,j\}}$ for $m = 0, 1, \dots, M$. Observe that

$$\begin{aligned} a_{\{n,\,j\}} &= (\lambda_{\{n,\,j\}})^{-1} c \lim_{\varepsilon\to 0^+} \int_{\varepsilon}^{r_{\{n,\,j\}}} \Psi_t\Big(\chi_{\mathbf{T}_{\{n,\,j\}}} t^2(-\Delta)e^{t^2\Delta} f\Big)\frac{dt}{t} \\ &= (\lambda_{\{n,\,j\}})^{-1} c \lim_{\varepsilon\to 0^+} \int_{\varepsilon}^{r_{\{n,\,j\}}} t^{2M}(-\Delta)^M \Phi_t\Phi_t\Big(\chi_{\mathbf{T}_{\{n,\,j\}}} t^2(-\Delta)e^{t^2\Delta} f\Big)\frac{dt}{t}. \end{aligned} \tag{11.5}$$

Indeed, if $t > r_{\{n,\,j\}}$ and $(t, \mathbf{y}) \in \mathcal{R}(Q_{\{n,\,j\}})$ then

$$d(\mathbf{y}, (\Omega_j)^c) \le d(\mathbf{y}, Q_{\{n,\,j\}}) + \frac{1}{2} r_{\{n,\,j\}} + d(\mathbf{x}_{\{n,\,j\}}, (\Omega_j)^c) \le 2t + \frac{1}{2}t + t = \frac{7}{2}t. \tag{11.6}$$

Hence $(t, \mathbf{y}) \notin \mathbf{T}_{\{n,\,j\}}$, which gives (11.5).

As a consequence of (10.5) and (11.5), we have

$$\operatorname{supp} a_{\{n,\,j\}} \subset \Big\{\mathbf{x} \in \mathbb{R}^N : d(\mathbf{x}, \mathbf{x}_{\{n,\,j\}}) \le \frac{7}{2} r_{\{n,\,j\}}\Big\} = \mathcal{O}\Big(B\Big(\mathbf{x}_{\{n,\,j\}}, \frac{7}{2} r_{\{n,\,j\}}\Big)\Big). \tag{11.7}$$

Let

$$b_{\{n,\,j\}} = (\lambda_{\{n,\,j\}})^{-1} c \lim_{\varepsilon\to 0^+} \int_{\varepsilon}^{r_{\{n,\,j\}}} t^{2M} \Phi_t\Phi_t\Big(\chi_{\mathbf{T}_{\{n,\,j\}}} t^2(-\Delta)e^{t^2\Delta} f\Big)\frac{dt}{t}. \tag{11.8}$$

One can prove using the Dunkl transform that $b_{\{n,\,j\}} \in \mathcal{D}(\Delta^M)$ and

$$(-\Delta)^m b_{\{n,\,j\}} = (\lambda_{\{n,\,j\}})^{-1} c \lim_{\varepsilon\to 0^+} \int_{\varepsilon}^{r_{\{n,\,j\}}} t^{2M}(-\Delta)^m \Phi_t\Phi_t\Big(\chi_{\mathbf{T}_{\{n,\,j\}}} t^2(-\Delta)e^{t^2\Delta} f\Big)\frac{dt}{t} \tag{11.9}$$

for $m = 1, 2, \dots M$, because Δ^m is closed on $L^2(dw)$. Taking a sequence $\varepsilon_\ell \to 0^+$ instead of $\varepsilon \to 0^+$ if necessary, we may assume that the convergence in (11.5), (11.8), and (11.9) holds in $L^2(dw)$ and almost everywhere. By the same arguments,

$$\operatorname{supp} \Delta^m b_{\{n,\,j\}} \subset \mathcal{O}\Big(B\Big(\mathbf{x}_{\{n,\,j\}}, \frac{7}{2} r_{\{n,\,j\}}\Big)\Big). \tag{11.10}$$

Note also that $\Delta^m b_{\{n,\,j\}}(\mathbf{x}) \neq 0$ implies that there is $(t, \mathbf{y}) \in \hat{\Omega}_j$ such that $d(\mathbf{x}, \mathbf{y}) < t$. Then $\mathcal{O}(\mathbf{x}) \in B_X(\mathcal{O}(\mathbf{y}), t) \subset B_X(\mathcal{O}(\mathbf{y}), 4t) \subset \mathbf{\Omega}_j$. Hence,

$$\operatorname{supp} \Delta^m b_{\{n,\,j\}} \subset \Omega_j. \tag{11.11}$$

Clearly, $a_{\{n,j\}} = (-\Delta)^M b_{\{n,j\}} = \Delta^M b_{\{n,j\}}$, because M is an even integer.

Step 7. Size of $\Delta^m b_{\{n,j\}}$ for $m = 0, 1, \ldots, M-1$. Suppose that $(t, \mathbf{y})$ is such that $\chi_{\mathbf{T}_{\{n,j\}}}(t, \mathbf{y}) = 1$. Then $(t, \mathbf{y}) \in (\hat{\Omega}_{j+1})^c$, so $|t^2 \Delta e^{t^2\Delta} f(\mathbf{y})| \leq 2^{j+1}$. Consequently,

$$\begin{aligned}
&|\Delta^m b_{\{n,j\}}(\mathbf{x})| \\
&= \frac{c}{\lambda_{\{n,j\}}} \left| \lim_{\ell\to\infty} \int_{\varepsilon_\ell}^{r_{\{n,j\}}} t^{2M-2m} (t^2(-\Delta))^m \Phi_t \Phi_t (\chi_{\mathbf{T}_{\{n,j\}}} t^2(-\Delta) e^{t^2\Delta} f)(\mathbf{x}) \frac{dt}{t} \right| \\
&= \frac{c}{\lambda_{\{n,j\}}} \left| \lim_{\ell\to\infty} \int_{\varepsilon_\ell}^{r_{\{n,j\}}} \int_{\mathbb{R}^N} t^{2M-2m} K_t^m(\mathbf{x}, \mathbf{y}) (\chi_{\mathbf{T}_{\{n,j\}}}(t, \mathbf{y}) t^2(-\Delta) e^{t^2\Delta} f(\mathbf{y})) dw(\mathbf{y}) \frac{dt}{t} \right|,
\end{aligned}$$

where $K_t^m(\mathbf{x}, \mathbf{y})$ is the integral kernel of the operator $(-t^2\Delta)^m \Phi_t \Phi_t$. Recall that

$$|K_t^m(\mathbf{x}, \mathbf{y})| \leq C w(B(\mathbf{x}, t))^{-1}$$

and

$$K_t^m(\mathbf{x}, \mathbf{y}) = 0 \quad \text{for } d(\mathbf{x}, \mathbf{y}) > t/2$$

(see (10.5) and Corollary 3.2). Thus,

$$\begin{aligned}
|\Delta^m b_{\{n,j\}}(\mathbf{x})| &\leq C(\lambda_{\{n,j\}})^{-1} 2^{j+1} \int_0^{r_{\{n,j\}}} \int_{\mathbb{R}^N} t^{2M-2m} |K_t^m(\mathbf{x}, \mathbf{y})| dw(\mathbf{y}) \frac{dt}{t} \\
&\leq C(\lambda_{\{n,j\}})^{-1} 2^{j+1} \int_0^{r_{\{n,j\}}} t^{2M-2m} \frac{dt}{t} \\
&= C(\lambda_{\{n,j\}})^{-1} 2^j (r_{\{n,j\}})^{2M-2m} \\
&= C w(Q_{\{n,j\}})^{-1} (r_{\{n,j\}})^{2M-2m}.
\end{aligned} \tag{11.12}$$

Step 8. Key lemma. It remains to estimate

$$a_{\{n,j\}}(\mathbf{x}) = \frac{c}{\lambda_{\{n,j\}}} \lim_{\ell\to\infty} \int_{\varepsilon_\ell}^{\varepsilon_\ell^{-1}} \int_{\mathbb{R}^N} \Psi_t(\mathbf{x}, \mathbf{y}) \chi_{\mathbf{T}_{\{n,j\}}}(t, \mathbf{y}) (t^2(-\Delta) e^{t^2\Delta} f)(\mathbf{y})\, dw(\mathbf{y}) \frac{dt}{t}.$$

Let $E_{\{n,j\}} = \bigcup_{i=1}^n Q_{\{i,j\}}$. Then

$$\begin{aligned}
\chi_{\mathbf{T}_{\{n,j\}}}(t, \mathbf{y}) &= \chi_{\hat{\Omega}_j}(t, \mathbf{y}) \chi_{(\hat{\Omega}_{j+1})^c}(t, \mathbf{y}) \chi_{\mathcal{R}(E_{\{n,j\}})}(t, \mathbf{y}) \chi_{(\mathcal{R}(E_{\{n-1,j\}}))^c}(t, \mathbf{y}) \\
&= \chi_1(t, \mathbf{y}) \chi_2(t, \mathbf{y}) \chi_3(t, \mathbf{y}) \chi_4(t, \mathbf{y}).
\end{aligned} \tag{11.13}$$

The following lemma (see [37, Lemma 4.2]) plays a crucial role in the remaining part of the proof of Theorem 11.1.

Lemma 11.3 *For every $\mathbf{x} \in \Omega_j$ and every function χ_s, $s = 1, 2, 3, 4$, there are numbers $0 < \delta_s \leq \omega_s$ such that $\omega_s \leq 3\delta_s$ and either $\Psi_t(\mathbf{x}, \mathbf{y}) \chi_s(t, \mathbf{y}) = 0$ for every $0 < t < \delta_s$ or $\Psi_t(\mathbf{x}, \mathbf{y}) \chi_s(t, \mathbf{y}) = \Psi_t(\mathbf{x}, \mathbf{y})$ for every $0 < t < \delta_s$ and either $\Psi_t(\mathbf{x}, \mathbf{y}) \chi_s(t, \mathbf{y}) = 0$ for every $t > \omega_s$ or $\Psi_t(\mathbf{x}, \mathbf{y}) \chi_s(t, \mathbf{y}) = \Psi_t(\mathbf{x}, \mathbf{y})$ for every $t > \omega_s$.*

Proof For the reader's convenience, we include a short proof along the lines of [37]. Fix $t > 0$ and define $\chi_1'(\mathbf{y}) = \chi_{[4t,\infty)}(d(\mathbf{y}, \Omega_j^c))$, $\chi_2'(\mathbf{y}) = \chi_{(-\infty,4t)}(d(\mathbf{y}, \Omega_{j+1}^c))$, $\chi_3'(\mathbf{y}) = \chi_{(-\infty,2t)}(d(\mathbf{y}, E_{\{n,j\}}))$, $\chi_4'(\mathbf{y}) = \chi_{[2t,\infty)}(d(\mathbf{y}, E_{\{n-1,j\}}))$. Clearly, $\chi_s'(\mathbf{y}) = \chi_s(t, \mathbf{y})$ for $s = 1, 2, 3, 4$. If $d(\mathbf{x}, \mathbf{y}) \geq t$, then $\Psi_t(\mathbf{x}, \mathbf{y}) = \Psi_t(\mathbf{x}, \mathbf{y})\chi_s(t, \mathbf{y}) = 0$. Therefore, to finish the proof, we assume that $d(\mathbf{x}, \mathbf{y}) < t$. Then

$$-t + d(A, \mathbf{x}) < d(A, \mathbf{y}) < t + d(A, \mathbf{x}) \text{ for } A = \Omega_j^c, \Omega_{j+1}^c, E_{\{n,j\}}, E_{\{n-1,j\}}.$$

We are in a position to define consecutively δ_s and ω_s.

(1) If $d(\mathbf{x}, \Omega_j^c) < 3t$ or $d(\mathbf{x}, \Omega_j^c) > 5t$, then $\chi_1'(\mathbf{y}) = 0$ and $\chi_1'(\mathbf{y}) = 1$ respectively, so we put $\delta_1 = \frac{1}{5}d(\mathbf{x}, \Omega_j^c)$ and $\omega_1 = \frac{1}{3}d(\mathbf{x}, \Omega_j^c)$.
(2) If $d(\mathbf{x}, \Omega_{j+1}^c) < 3t$ or $d(\mathbf{x}, \Omega_{j+1}^c) > 5t$, then $\chi_2'(\mathbf{y}) = 1$ and $\chi_2'(\mathbf{y}) = 0$ respectively. Hence we set $\delta_2 = \frac{1}{5}d(\mathbf{x}, \Omega_{j+1}^c)$ and $\omega_2 = \frac{1}{3}d(\mathbf{x}, \Omega_{j+1}^c)$ if $d(\mathbf{x}, \Omega_{j+1}^c) \neq 0$, $\delta_2 = \omega_2 = \delta_1$ otherwise.
(3) If $d(\mathbf{x}, E_{\{n,j\}}) < t$ or $d(\mathbf{x}, E_{\{n,j\}}) > 3t$, then $\chi_3'(\mathbf{y}) = 1$ and $\chi_3'(\mathbf{y}) = 0$ respectively. Thus we put $\delta_3 = \frac{1}{3}d(\mathbf{x}, E_{\{n,j\}})$ and $\omega_3 = d(\mathbf{x}, E_{\{n,j\}})$ if $d(\mathbf{x}, E_{\{n,j\}}) \neq 0$, $\delta_3 = \omega_3 = \delta_1$ otherwise.
(4) If $d(\mathbf{x}, E_{\{n-1,j\}}) < t$ or $d(\mathbf{x}, E_{\{n-1,j\}}) > 3t$, then $\chi_4'(\mathbf{y}) = 0$ and $\chi_4'(\mathbf{y}) = 1$ respectively, so we put $\delta_4 = \frac{1}{3}d(\mathbf{x}, E_{\{n-1,j\}})$ and $\omega_4 = d(\mathbf{x}, E_{\{n-1,j\}})$ if $d(\mathbf{x}, E_{\{n-1,j\}}) \neq 0$, $\delta_4 = \omega_4 = \delta_1$ otherwise. □

We finish Step 8 by the remark (see Case 1 of the proof of the lemma) that if $t > \omega_1 > 0$ then

$$\Psi_t(\mathbf{x}, \mathbf{y})\chi_{\mathbf{T}_{\{n,j\}}}(t, \mathbf{y}) = 0.$$

Step 9. Estimates for $a_{\{n,j\}}$. We shall prove that

$$|a_{\{n,j\}}(\mathbf{x})| \leq Cw(Q_{\{n,j\}})^{-1}. \tag{11.14}$$

Fix $\mathbf{x} \in \Omega_j$. Recall that $\operatorname{supp} a_{\{n,j\}} \subset \Omega_j$. Let $J = \bigcup_{s=1}^4 [\delta_s, \omega_s]$, $I = (0, \infty) \setminus J$, where δ_s, ω_s are from Lemma 11.3. Obviously, $I = (a_1, b_1) \cup \ldots \cup (a_m, b_m)$, where $m \leq 5$, $a_1 = 0$, $b_m = \infty$, and (a_l, b_l) are connected disjoint components of I. Clearly,

$$\begin{aligned}
&\left|a_{\{n,j\}}(\mathbf{x})\right| \\
&\leq \sum_{s=1}^{4} (\lambda_{\{n,j\}})^{-1} c \int_{\delta_s}^{\omega_s} \int_{\mathbb{R}^N} \left|\Psi_t(\mathbf{x}, \mathbf{y})\chi_{\mathbf{T}_{\{n,j\}}}(t, \mathbf{y})(t^2(-\Delta)e^{t^2\Delta} f)(\mathbf{y})\right| dw(\mathbf{y})\frac{dt}{t} \\
&\quad + \sum_{s=1}^{m} (\lambda_{\{n,j\}})^{-1} c \left|\int_{a_s}^{b_s} \int_{\mathbb{R}^N} \Psi_t(\mathbf{x}, \mathbf{y})\chi_{\mathbf{T}_{\{n,j\}}}(t, \mathbf{y})(t^2(-\Delta)e^{t^2\Delta} f)(\mathbf{y})\, dw(\mathbf{y})\frac{dt}{t}\right|,
\end{aligned}$$

where some of the integrals are understood as improper ones (see Steps 6 and 7).

Consider the integral over $[\delta_s, \omega_s]$. Take $t \in [\delta_s, \omega_s]$ and $\mathbf{y}$ such that the integrant $\left|\Psi_t(\mathbf{x}, \mathbf{y})\chi_{\mathbf{T}_{\{n,j\}}}(t, \mathbf{y})(t^2(-\Delta)e^{t^2\Delta} f)(\mathbf{y})\right| \neq 0$. Then $(t, \mathbf{y}) \notin \hat{\Omega}_{j+1}$. Thus, there is $\mathbf{x}'$ such that $d(\mathbf{y}, \mathbf{x}') < 4t$ and $\mathbf{x}' \notin \Omega_{j+1}$, which means that $\mathcal{M} f(\mathbf{x}') \leq 2^{j+1}$. Consequently, $|t^2(-\Delta)e^{t^2\Delta} f(\mathbf{y})| \leq 2^{j+1}$. Hence,

$$
\begin{aligned}
&(\lambda_{\{n,j\}})^{-1} c \int_{\delta_s}^{\omega_s} \int_{\mathbb{R}^N} \left|\Psi_t(\mathbf{x}, \mathbf{y})\chi_{\mathbf{T}_{\{n,j\}}}(t, \mathbf{y})(t^2(-\Delta)e^{t^2\Delta} f)(\mathbf{y})\right| dw(\mathbf{y}) \frac{dt}{t} \\
&\leq (\lambda_{\{n,j\}})^{-1} 2^{j+1} c \int_{\delta_s}^{\omega_s} \int_{\mathbb{R}^N} \left|\Psi_t(\mathbf{x}, \mathbf{y})\right| dw(\mathbf{y}) \frac{dt}{t} \\
&\leq C' (\lambda_{\{n,j\}})^{-1} 2^{j+1} c \int_{\delta_s}^{\omega_s} \frac{dt}{t} \\
&\leq C w(Q_{\{n,j\}})^{-1}, \qquad (11.15)
\end{aligned}
$$

because $0 < \omega_s \leq 3\delta_s$.

We turn to estimate the integrals over $[a_s, b_s]$. Assume that

$$
(\lambda_{\{n,j\}})^{-1} c \left| \int_{a_s}^{b_s} \int_{\mathbb{R}^N} \Psi_t(\mathbf{x}, \mathbf{y})\chi_{\mathbf{T}_{\{n,j\}}}(t, \mathbf{y})(t^2(-\Delta)e^{t^2\Delta} f)(\mathbf{y})\, dw(\mathbf{y}) \frac{dt}{t} \right| > 0.
$$

By Lemma 11.3 for fixed $\mathbf{x} \in \Omega_j$ and $s \in \{1, 2, \dots, m\}$, either $\chi_{\mathbf{T}_{\{n,j\}}}(t, \mathbf{y}) \equiv 0$ for all $t \in [a_s, b_s]$ and $d(\mathbf{x}, \mathbf{y}) < t$ or $\chi_{\mathbf{T}_{\{n,j\}}}(t, \mathbf{y}) \equiv 1$ for all $t \in [a_s, b_s]$ and $d(\mathbf{x}, \mathbf{y}) < t$. So the letter holds. This gives that for every $t \in [a_s, b_s]$ and $\mathbf{y}$ such that $d(\mathbf{x}, \mathbf{y}) < t$, we have $(t, \mathbf{y}) \notin \hat{\Omega}_{j+1}$. So there is $\mathbf{x}'$ (which depends on $(t, \mathbf{y})$) such that $d(\mathbf{y}, \mathbf{x}') < 4t$ and $\mathcal{M} f(\mathbf{x}') < 2^{j+1}$. Note that $d(\mathbf{x}, \mathbf{x}') < d(\mathbf{x}, \mathbf{y}) + d(\mathbf{y}, \mathbf{x}') < 5t$. Consequently, for every $t \in [a_s, b_s]$, we have

$$
2^{j+1} \geq \mathcal{M} f(\mathbf{x}') \geq \sup_{d(\mathbf{x}', \mathbf{z}) < 5t} |\Xi_t f(\mathbf{z})| \geq |\Xi_t f(\mathbf{x})|.
$$

Finally, in our case

$$
\begin{aligned}
&(\lambda_{\{n,j\}})^{-1} c \left| \int_{a_s}^{b_s} \int_{\mathbb{R}^N} \Psi_t(\mathbf{x}, \mathbf{y})\chi_{\mathbf{T}_{\{n,j\}}}(t, \mathbf{y})(t^2(-\Delta)e^{t^2\Delta} f)(\mathbf{y})\, dw(\mathbf{y}) \frac{dt}{t} \right| \\
&= (\lambda_{\{n,j\}})^{-1} c \left| \int_{a_s}^{b_s} \int_{\mathbb{R}^N} \Psi_t(\mathbf{x}, \mathbf{y})(t^2(-\Delta)e^{t^2\Delta} f)(\mathbf{y})\, dw(\mathbf{y}) \frac{dt}{t} \right| \\
&= (\lambda_{\{n,j\}})^{-1} \left| \Xi_{a_s} f(\mathbf{x}) - \Xi_{b_s} f(\mathbf{x}) \right| \\
&\leq C w(Q_{\{n,j\}})^{-1}, \qquad (11.16)
\end{aligned}
$$

where in the last equality we have used (11.1). The estimates (11.15) and (11.16) give (11.14). Recall that $w(Q_{\{n,j\}}) \sim w(B(\mathbf{x}_{\{n,j\}}, 7r_{\{n,j\}}/2))$. Hence, from (11.14), (11.12), (11.7), and (11.10) we deduce Step 5. The proof of Theorem 11.1 is complete. □

Having Lemma 10.9 together with Theorems 11.1 and 10.11 we are in a position to complete the proof of the atomic decomposition of $H^1_{\max,H}$ functions. This is stated in the theorem below.

Theorem 11.4 *There is a constant $C > 0$ such that every function $f \in H^1_{\max,H}$ can be written as*

$$f = \sum \lambda_j a_j,$$

where a_j are $(1, \infty, M)$-atoms, $\sum |\lambda_j| \leq C\|\mathcal{M}_H f\|_{L^1(dw)}$.

Proof Recall that $\|f\|_{H^1_{\max,H}} \sim \|f\|_{H^1_{\max,P}}$ (see Theorem 10.11). Take a sequence g_n $(n = 0, 1, \ldots)$ as in the proof of Theorem 10.11. Then $g_n \in H^1_{\max,P} \cap L^2(dw)$, $f = \sum_{n=0}^{\infty} g_n$, and $\sum_{n=0}^{\infty} \|g_n\|_{H^1_{\max,P}} \leq 3\|f\|_{H^1_{\max,P}}$. By Theorem 11.1 the functions g_n admit atomic decompositions into $(1, \infty, M)$-atoms, that is, $g_n = \sum_m \lambda_{n,m} a_{n,m}$, where the functions $a_{n,m}$ are $(1, \infty, M)$-atoms, $\lambda_{n,m} \in \mathbb{C}$, and $\sum_m |\lambda_{n,m}| \leq C\|g_n\|_{\max,P}$. Finally,

$$f = \sum_{n,m} \lambda_{n,m} a_{n,m}$$

is the desired atomic decomposition. □

We are in a position to complete the proof of Theorem 2.1, by proving the following proposition, which is the converse to Proposition 7.6.

Proposition 11.5 *Assume that u_0 is $\mathcal{L}$-harmonic and satisfies $u_0^* \in L^1(dw)$. Then there is a system $\mathbf{u} = (u_0, u_1, \ldots, u_N) \in \mathcal{H}^1$ such that $\|\mathbf{u}\|_{\mathcal{H}^1} \leq C\|u_0^*\|_{L^1(dw)}$.*

Proof By Theorem 7.5 we have $u_0(t, \mathbf{x}) = P_t f_0(\mathbf{x})$, where $f_0 \in L^1(dw)$. So $f_0 \in H^1_{\max,\,P}$ and $\|f_0\|_{H^1_{\max,\,P}} = \|u_0^*\|_{L^1(dw)}$. Using Theorem 11.4 and then (9.1) we obtain that $f_0 \in H^1_\Delta$ and $\|f_0\|_{H^1_\Delta} \leq C\|u_0^*\|_{L^1(dw)}$. □

12 Inclusion $H^1_{(1,q,M)} \subset H^1_{\max,H}$

In this section we shall prove that, for every integer $M \geq 1$ and for every $1 < q \leq \infty$, we have $H^1_{(1,q,M)} \subset H^1_{\max,H}$ and

$$\|f\|_{H^1_{\max,H}} \leq C_{M,q}\|f\|_{H^1_{(1,q,M)}}.$$

It suffices to establish that there is a constant $C_{M,q} > 0$ such that

$$\|a\|_{H^1_{\max,H}} \leq C_{M,q},$$

for every $(1, q, M)$-atom a. Since every $(1, q, M)$-atom is a $(1, q, 1)$-atom, we may reduce to $M = 1$.

Assume that a is a $(1, q, 1)$-atom associated with a set $\mathcal{B} = \bigcup_{\sigma\in G} B(\sigma(\mathbf{y}_0), r)$. Then there is a function $b \in \mathcal{D}(\Delta)$ such that $a = \Delta b$, $\operatorname{supp} \Delta^j b \subset \mathcal{B}$, $\|\Delta^j b\|_{L^q(dw)} \le r^{2-2j} w(\mathcal{B})^{\frac{1}{q}-1}$, $j = 0, 1$. Set $u(t, \mathbf{x}) = e^{t^2\Delta} a(\mathbf{x})$. Observe that

$$\|u^*\|_{L^q(dw)} \le C_q \|a\|_{L^q(dw)} \le w(\mathcal{B})^{\frac{1}{q}-1}$$

(see (2.5) for the definition of u^*). Thus, by the doubling property of the measure $dw(\mathbf{x})\, d\mathbf{x}$ and the Hölder inequality,

$$\int_{d(\mathbf{x},\mathbf{y}_0)\le 8r} u^*(\mathbf{x})\, dw(\mathbf{x}) \le C'_q.$$

We turn to estimate $u^*(\mathbf{x})$ on $d(\mathbf{x}, \mathbf{y}_0) > 8r$. Clearly,

$$\begin{aligned} u^*(\mathbf{x}) &\le \sup_{0<t<d(\mathbf{x},\mathbf{y}_0)/4,\, d(\mathbf{x}',\mathbf{x})<t} |e^{t^2\Delta}\Delta b(\mathbf{x}')| + \sup_{t>d(\mathbf{x},\mathbf{y}_0)/4,\, d(\mathbf{x}',\mathbf{x})<t} |e^{t^2\Delta}\Delta b(\mathbf{x}')| \\ &= J_1(\mathbf{x}) + J_2(\mathbf{x}). \end{aligned} \tag{12.1}$$

Recall that $\|b\|_{L^1(dw)} \le r^2$ and note that

$$e^{t^2\Delta}\Delta = \Delta e^{t^2\Delta} = \frac{d}{ds} e^{s\Delta}\Big|_{s=t^2}.$$

To deal with J_1 we note that if $d(\mathbf{x}', \mathbf{x}) < t \le d(\mathbf{x}, \mathbf{x}_0)/4$, $d(\mathbf{x}, \mathbf{y}_0) > 4r$, and $d(\mathbf{y}, \mathbf{y}_0) < r$, then $d(\mathbf{x}', \mathbf{y}) \sim d(\mathbf{x}, \mathbf{y}_0)$. So, using (4.3), we have

$$\left|\frac{d}{ds} h_s(\mathbf{x}', \mathbf{y})\right|_{s=t^2} \le \frac{C}{t^2 w(B(\mathbf{y}_0, d(\mathbf{y}_0, \mathbf{x})))} e^{-c' d(\mathbf{y}_0,\mathbf{x})^2/t^2}.$$

Hence,

$$J_1(\mathbf{x}) \lesssim w(B(\mathbf{y}_0, d(\mathbf{x}, \mathbf{y}_0)))^{-1} \frac{r^2}{d(\mathbf{x}, \mathbf{y}_0)^2}.$$

In order to estimate J_2, we observe from (4.3) that for $t > d(\mathbf{x}, \mathbf{y})$ and $d(\mathbf{y}, \mathbf{y}_0) < r < t$, we have

$$\left|\frac{d}{ds} h_s(\mathbf{x}', \mathbf{y})\right|_{s=t^2} \le \frac{C}{t^2 w(B(\mathbf{y}_0, d(\mathbf{y}_0, \mathbf{x})))}.$$

Consequently,

$$J_2(\mathbf{x}) \lesssim w(B(\mathbf{y}_0, d(\mathbf{x}, \mathbf{y}_0)))^{-1} \frac{r^2}{d(\mathbf{x}, \mathbf{y}_0)^2}.$$

Now

$$\int_{d(\mathbf{x},\mathbf{y}_0)>8r} u^*(\mathbf{x})\,dw(\mathbf{x}) \lesssim \sum_{j=3}^{\infty} \int_{2^j r<d(\mathbf{x},\mathbf{y}_0)\leq 2^{j+1}r} \frac{r^2}{w(B(\mathbf{y}_0, d(\mathbf{x},\mathbf{y}_0)))d(\mathbf{x},\mathbf{y}_0)^2}\,dw(\mathbf{x})$$
$$\lesssim \sum_{j=3}^{\infty} 2^{-2j} = C.$$

13 Square Function Characterization

In this section we prove Theorem 2.3. More precisely we show that the atomic Hardy space $H^1_{(1,2,M)}$ coincides with the Hardy space defined by the square function (2.9) with $Q_t = t\sqrt{-\Delta}\,e^{-t\sqrt{-\Delta}}$. This is achieved by mimicking arguments in [24]. The proof for $Q_t = t^2(-\Delta)\,e^{t^2\Delta}$ is similar.

13.1 Tent Spaces T_2^p on Spaces of Homogeneous Type

The square function characterization of the Hardy space $H^1_{(1,2,M)}$ can be related with the so called tent space T_2^1. The tent spaces on Euclidean spaces were introduced in [9] and then extended on spaces of homogeneous type (see, e.g. [33]). For more details we refer the reader to [35].

For a measurable function $F(t,\mathbf{x})$ on $(0,\infty)\times\mathbb{R}^N$, let

$$\mathcal{A}F(\mathbf{x}) := \left(\int_0^{\infty}\int_{\|\mathbf{y}-\mathbf{x}\|<t} |F(t,\mathbf{y})|^2 \frac{dw(\mathbf{y})}{w(B(\mathbf{x},t))}\frac{dt}{t}\right)^{1/2}.$$

Definition 13.1 For $1\leq p<\infty$ the tent space T_2^p is defined to be

$$T_2^p = \{F : \|F\|_{T_2^p} := \|\mathcal{A}F\|_{L^p(dw)} < \infty\}.$$

Clearly, by the doubling property,

$$\|F\|^2_{T_2^2} = \|\mathcal{A}F\|^2_{L^2(dw)} \sim \int_0^{\infty}\int_{\mathbb{R}^N} |F(t,\mathbf{y})|^2 \frac{dw(\mathbf{y})dt}{t}. \tag{13.1}$$

Remark 13.2 Let Ψ be as in (10.6). Recall that, by (10.4) and (13.1), the linear operator π_Ψ is bounded from $L^2(dw\,dt/t)$ into $L^2(dw)$. Furthermore, by using the Dunkl transform, one can easily prove that, if $F(t,\mathbf{x}) = Q_t f(\mathbf{x})$ with $f\in L^2(dw)$, then

$$\|F\|_{T_2^2} = \|Sf\|_{L^2(dw)} \sim \|f\|_{L^2(dw)}$$

and $f = c_1\pi_\Psi(F)$.

The tent space T_2^1 on the space of homogeneous type admits the following atomic decomposition (see, e.g., [33]).

Definition 13.3 A measurable function $A(t, \mathbf{x})$ is a T_2^1-*atom* if there is a ball $B \subset \mathbb{R}^N$ such that

- $\operatorname{supp} A \subset \widehat{B}$
- $\iint_{(0,\infty)\times\mathbb{R}^N} |A(t,\mathbf{x})|^2\, dw(\mathbf{x})\, \frac{dt}{t} \le w(B)^{-1}$.

A function F belongs to T_2^1 if and only if there are sequences A_j of T_2^1-atoms and $\lambda_j \in \mathbb{C}$ such that

$$\sum_j \lambda_j A_j = F, \quad \sum_j |\lambda_j| \sim \|F\|_{T_2^1},$$

where the convergence is in T_2^1 norm and almost everywhere.

The Hölder inequality immediately gives that there is a constant $C > 0$ such that for every function $A(t, \mathbf{x})$ being a T_2^1-atom one has

$$\|A\|_{T_2^1} \le C.$$

Observe that for $f \in L^1(dw)$, the function $F(t, \mathbf{x}) = Q_t f(\mathbf{x})$ is well defined. Moreover, $\mathcal{A}F(\mathbf{x}) = Sf(\mathbf{x})$ and $\|Sf\|_{L^1(dw)} = \|F\|_{T_2^1}$.

Remark 13.4 According to the proof of atomic decomposition of T_2^1 presented in [33], the function $\lambda_j A_j$ can be taken of the form $\lambda_j A_j(t,\mathbf{x}) = \chi_{S_j}(t,\mathbf{x})F(t,\mathbf{x})$, where S_j are disjoint, $\mathbb{R}_+^{N+1} = \bigcup S_j$, and S_j is contained in a tent $\widehat{B}_j$.

So, if $F \in T_2^1 \cap T_2^2$, then F can be decomposed into atoms such that $F(t,\mathbf{x}) = \sum_j \lambda_j A_j(t,\mathbf{x})$ and the convergence is in T_2^1, T_2^2, and pointwise.

Lemma 13.5 *The map* $(P_s F)(t,\mathbf{x}) = \int p_s(\mathbf{x},\mathbf{y})F(t,\mathbf{y})\, dw(\mathbf{y})$ *is bounded on* T_2^1. *Moreover, there is a constant* $C > 0$ *independent of* $s > 0$ *such that* $\|P_s F\|_{T_2^1} \le C\|F\|_{T_2^1}$.

Proof Let $F(t,\mathbf{x}) = \sum_j \lambda_j A_j(t,\mathbf{x})$ be an atomic decomposition of $F \in T_2^1$ as described above. Since $p_s(\mathbf{x},\mathbf{y}) \ge 0$, it suffices to prove that there is a constant $C > 0$ such that

$$\Big\| P_s|A| \Big\|_{T_2^1} \le C$$

for every atom A of T_2^1. To this end let $B = B(\mathbf{x}_0, r)$ be a ball associated with A. Obviously, $P_s|A|(t,\mathbf{x}') = 0$ for $t > r$.

Case 1 $s > r$. Then, by (5.4) and the Hölder inequality,

$$P_s|A|(t,\mathbf{x}') \le \frac{Cs}{s + d(\mathbf{x}_0,\mathbf{x}')} \frac{w(B(\mathbf{x}_0,r))^{1/2}}{w(B(\mathbf{x}_0, s + d(\mathbf{x}_0,\mathbf{x}')))} \left(\int |A(t,\mathbf{y})|^2 dw(\mathbf{y}) \right)^{1/2}.$$

If $\|\mathbf{x} - \mathbf{x}'\| < t \leq r$, then $s + d(\mathbf{x}_0, \mathbf{x}') \sim s + d(\mathbf{x}_0, \mathbf{x})$, because, by our assumption, $s > r$. Hence,

$$\begin{aligned} \Big\| P_s|A| \Big\|_{T_2^1} &\leq C \int \frac{s}{s + d(\mathbf{x}_0, \mathbf{x})} \frac{w(B(\mathbf{x}_0, r))^{1/2}}{w(B(\mathbf{x}_0, s + d(\mathbf{x}_0, \mathbf{x})))} \\ &\quad \times \left(\int_0^r \int_{\|\mathbf{x}-\mathbf{x}'\|<t} \int |A(t, \mathbf{y})|^2 dw(\mathbf{y}) \frac{dw(\mathbf{x}')dt}{w(B(\mathbf{x}, t))t} \right)^{1/2} dw(\mathbf{x}) \\ &\leq C \int \frac{s}{s + d(\mathbf{x}_0, \mathbf{x})} \frac{dw(\mathbf{x})}{w(B(\mathbf{x}_0, s + d(\mathbf{x}_0, \mathbf{x})))} \leq C, \end{aligned}$$

where to get the second to last inequality we first integrated with respect to $dw(\mathbf{x}')$ and then used the definition of T_2^1-atom.

Case 2 $s \leq r$. Recall that P_s is a contraction on $L^2(dw)$. Hence,

$$\begin{aligned} \|\mathcal{A}P_s|A|\|_{L^1(\mathcal{O}(B(\mathbf{x}_0,4r)),\, dw)} &\leq Cw(B(\mathbf{x}_0, r))^{1/2} \|\mathcal{A}P_s|A|\|_{L^2(dw)} \\ &\leq Cw(B(\mathbf{x}_0, r))^{1/2} \|P_s|A|\|_{T_2^2} \\ &\leq Cw(B(\mathbf{x}_0, r))^{1/2} \||A|\|_{T_2^2} \leq C. \end{aligned} \tag{13.2}$$

If $d(\mathbf{x}, \mathbf{x}_0) > 4r$, $\|\mathbf{x}' - \mathbf{x}\| < t < r$, and $\|\mathbf{x}_0 - \mathbf{y}\| < r$, then $s + d(\mathbf{x}', \mathbf{y}) \sim s + d(\mathbf{x}, \mathbf{x}_0)$. Now we proceed as in Case 1 to get the required bound on $\mathcal{O}(B(\mathbf{x}_0, 4r))^c$. □

Lemma 13.6 *The family P_s forms an approximation of the identity in T_2^1, that is,*

$$\lim_{s \to 0} \|P_s F - F\|_{T_2^1} = 0.$$

Proof According to Lemma 13.5, it suffices to establish that for every A being a T_2^1-atom, we have

$$\lim_{s \to 0} \|P_s A - A\|_{T_2^1} = \lim_{s \to 0} \|\mathcal{A}(P_s A - A)\|_{L^1(dw)} = 0. \tag{13.3}$$

Let A be such an atom and let $B = B(\mathbf{x}_0, r)$ be its associated ball. To prove (13.3) it suffices to consider $0 < s < r$.

If $d(\mathbf{x}, \mathbf{x}_0) > 4r$, $\|\mathbf{y} - \mathbf{x}_0\| < r$, and $\|\mathbf{x} - \mathbf{x}'\| < t < r$, then $s + d(\mathbf{x}', \mathbf{y}) \sim d(\mathbf{x}, \mathbf{x}_0)$, so

$$|P_s A(t, \mathbf{x}')| \leq \frac{Cs}{s + d(\mathbf{x}_0, \mathbf{x})} \frac{w(B(\mathbf{x}_0, r))^{1/2}}{w(B(\mathbf{x}_0, s + d(\mathbf{x}_0, \mathbf{x})))} \left(\int |A(t, \mathbf{y})|^2 dw(\mathbf{y}) \right)^{1/2}.$$

Since $\operatorname{supp} A \cap \{(t, \mathbf{x}') : \|\mathbf{x}' - \mathbf{x}\| < t < r\} = \emptyset$, we have

$$|\mathcal{A}(P_s A - A)(\mathbf{x})| = |\mathcal{A}(P_s A)(\mathbf{x})| \leq \frac{Cs}{s + d(\mathbf{x}_0, \mathbf{x})} \frac{1}{w(B(\mathbf{x}_0, s + d(\mathbf{x}_0, \mathbf{x})))}.$$

Hence,

$$\lim_{s\to 0}\int_{d(\mathbf{x},\mathbf{x}_0)>4r} |\mathcal{A}(P_sA-A)(\mathbf{x})|\,dw(\mathbf{x})=0.$$

We now turn to estimate $\|\mathcal{A}(P_sA-A)\|_{L^1(\mathcal{O}(B(\mathbf{x}_0,4r)),dw)}$. Observe that

$$\begin{aligned}&|(P_sA-A)(t,\mathbf{x}')|\le 2\mathcal{M}_PA(t,\mathbf{x}') \quad \text{and } \|\mathcal{M}_PA(t,\mathbf{x}')\|_{L^2(dw(\mathbf{x}'))}\\&\le C\|A(t,\mathbf{x}')\|_{L^2(dw(\mathbf{x}'))}.\end{aligned}$$

Moreover, $\lim_{s\to 0}\|P_sA(t,\mathbf{x}')-A(t,\mathbf{x}')\|_{L^2(dw(\mathbf{x}'))}=0$ for almost every $t>0$. Therefore, applying the Hölder inequality and (13.1), we have

$$\begin{aligned}&\limsup_{s\to 0}\|\mathcal{A}(P_sA-A)\|_{L^1(\mathcal{O}(B(\mathbf{x}_0,4r)))}\\&\quad\le \limsup_{s\to 0} Cw(B)^{1/2}\|\mathcal{A}(P_sA-A)\|_{L^2(\mathcal{O}(B(\mathbf{x}_0,4r)))}\\&\quad\le \limsup_{s\to 0} Cw(B)^{1/2}\left(\int_0^r\int |P_sA(t,\mathbf{x})-A(t,\mathbf{x})|^2\frac{dw(\mathbf{x})\,dt}{t}\right)^{1/2}=0,\end{aligned}$$

where in the last equality we have used the Lebesgue dominated convergence theorem. □

13.2 Proof of Theorem 2.3

The inclusion $H^1_{(1,2,M)}\subset H^1_{\text{square}}$ will be established once we prove the following lemma.

Lemma 13.7 *For every positive integer M, there exists a constant $C_M>0$ such that, for every $(1,2,M)$-atom a, we have*

$$\|F(t,\mathbf{x})\|_{T_2^1}\le C_M, \quad \textit{where } F(t,\mathbf{x})=Q_ta(\mathbf{x}).$$

Proof Let a be a $(1,2,M)$–atom, $M\ge 1$, associated with a ball $B=B(\mathbf{x}_0,r)$. By definition $a=\Delta^Mb$ with $\Delta^\ell b$ (for $\ell=0,1,\ldots,M$) satisfying relevant support and size conditions (see Definition 2.6). By the Hölder inequality,

$$\|Sa\|_{L^1(\mathcal{O}(8B))}\lesssim \|Sa\|_{L^2(\mathcal{O}(8B))}w(\mathcal{O}(8B))^{1/2}\lesssim 1.$$

If $d(\mathbf{x},\mathbf{x}_0)>8r$ then choose $n\ge 3$ such that $2^nr\le d(\mathbf{x},\mathbf{x}_0)<2^{n+1}r$ and split the integral as below

$$\begin{aligned}Sa(\mathbf{x})^2&=\int\int_{t>\|\mathbf{x}-\mathbf{y}\|}|Q_ta(\mathbf{y})|^2w(B(\mathbf{y},t))^{-1}\,dw(\mathbf{y})\,\frac{dt}{t}\\&=\int_0^{2^nr/4}\int_{t>\|\mathbf{x}-\mathbf{y}\|}+\int_{2^nr/4}^{\infty}\int_{t>\|\mathbf{x}-\mathbf{y}\|}=I_1+I_2.\end{aligned}$$

Define $a_1 = \Delta^{M-1}b$. Then by the definition of the atom $\|a_1\|_{L^1(w)} \leq r^2$. Note that

$$Q_t(a) = Q_t(\Delta a_1) = (\Delta Q_t)(a_1) = t(\partial_t Q_t)^3(a_1).$$

Estimation for I_1. If $z \in \mathcal{O}(B)$ and $\|\mathbf{x} - \mathbf{y}\| < t \leq 2^n r/4$, then $2^n r \lesssim d(\mathbf{z}, \mathbf{y})$. Therefore, thanks to (5.4) and (5.7) with $m = 3$, we have

$$\begin{aligned}
|Q_t a(\mathbf{y})|^2 &= \left| \int t(\partial_t^3)(p_t(\mathbf{y}, \mathbf{z}))a_1(\mathbf{z})\, dw(\mathbf{z}) \right|^2 \\
&\lesssim \left(\int d(\mathbf{z}, \mathbf{y})^{-2} \frac{t}{t + d(\mathbf{z}, \mathbf{y})} V(\mathbf{z}, \mathbf{y}, t + d(\mathbf{z}, \mathbf{y}))^{-1} |a_1(\mathbf{z})|\, dw(\mathbf{z}) \right)^2 \\
&\lesssim (2^n r)^{-4} \frac{t^2}{(2^n r)^2} w(B(\mathbf{x}_0, 2^n r))^{-2} \|a_1\|^2_{L^1(dw)}.
\end{aligned}$$

Consequently,

$$\begin{aligned}
I_1 &\lesssim \left(\int_0^{2^n r} t\, dt \right) w(B(\mathbf{x}_0, 2^n r))^{-2} \|a_1\|^2_{L^1(dw)} (2^n r)^{-4} (2^n r)^{-2} \\
&\lesssim 2^{-4n} w(B(\mathbf{x}_0, 2^n r))^{-2}.
\end{aligned}$$

Estimation for I_2. In this case $t \geq 2^n r/4$, so thanks to (5.7) with $m = 3$ we have

$$\begin{aligned}
|Q_t a(\mathbf{y})|^2 &= \left(\int t(\partial_t^3)(p_t(\mathbf{y}, \mathbf{z}))a_1(\mathbf{z})\, dw(\mathbf{z}) \right)^2 \\
&\lesssim \left(\int t^{-2} \frac{t}{t + d(\mathbf{z}, \mathbf{y})} V(\mathbf{z}, \mathbf{y}, t + d(\mathbf{z}, \mathbf{y}))^{-1} |a_1(\mathbf{z})|\, dw(\mathbf{z}) \right)^2 \\
&\lesssim t^{-4} w(B(\mathbf{x}_0, 2^n r))^{-2} \|a_1\|^2_{L^1(dw)}.
\end{aligned}$$

Consequently,

$$I_2 \lesssim \left(\int_{2^n r/4}^{\infty} t^{-5}\, dt \right) w(B(\mathbf{x}_0, 2^n r))^{-2} \|a_1\|^2_{L^1(dw)} \lesssim 2^{-4n} w(B(\mathbf{x}_0, 2^n r))^{-2}.$$

Finally,

$$\|Sa\|_{L^1(\mathcal{O}(8B)^c)} \lesssim \sum_{n \geq 3} \int_{2^n r < d(\mathbf{x}, \mathbf{x}_0) \leq 2^{n+1} r} 2^{-2n} w(B(\mathbf{x}_0, 2^n r))^{-1} dw(\mathbf{x}) \lesssim 1.$$

□

The opposite inclusion $H^1_{\text{square}} \subset H^1_{(1,2,M)}$ is contained in the following proposition.

Proposition 13.8 *Let M be a positive integer. Assume that for $f \in L^1(dw)$ the function $F(t,\mathbf{x}) = Q_t f(\mathbf{x})$ belongs to T_2^1. Then there are $\lambda_j \in \mathbb{C}$ and a_j being $(1,2,M)$-atoms such that*

$$f = \sum_j \lambda_j a_j \quad \textit{and} \quad \sum_j |\lambda_j| \leq C\|F\|_{T_2^1}.$$

The constant C depends on M but it is independent of f.

Proof We start our proof under the additional assumption $f \in L^2(dw)$. Then $F(t,\mathbf{x}) = Q_t f(\mathbf{x}) \in T_2^1 \cap T_2^2$. The proof in this case is the same as that of [24, Theorem 4.1]. The only difference is to control support of functions $\Delta^s b_j$. For the convenience of the reader we provide its sketch.

Let $F = \sum_j \lambda_j A_j$ be a T_2^1 atomic decomposition of the function $Q_t f(\mathbf{x})$ as it is described in Remark 13.4. In particular, $\sum_j |\lambda_j| \leq C\|Sf\|_{L^1(dw)}$. Let $\Psi^{\{1\}}$ be a radial C^∞ function supported by $B(0,1/4)$ such that $\int_0^\infty \Psi_t Q_t \frac{dt}{t}$ forms a Calderón reproducing formula, where $\Psi = \Delta^{M+1}\Psi^{\{1\}}$. By (10.4), Remarks 13.2 and 13.4, we have

$$f = \pi_\Psi F = \sum_j \lambda_j \pi_\Psi A_j \tag{13.4}$$

and the series converges unconditionally in $L^2(dw)$. Then $\operatorname{supp} A_j \subset \widehat{B}_j$. Let $B_j = B(\mathbf{y}_j, r_j)$ be a ball associated with A_j. Recall that Δ^m is closed on $L^2(dw)$ for every positive integer m. Set $a_j = \pi_\Psi(A_j)$. We have $a_j = \Delta^M b_j$, where

$$b_j = \lim_{\varepsilon\to 0^+} \int_\varepsilon^{r_j/4} t^{2M}\big(t^2\Delta\Psi_t^{\{1\}} A_j\big)\frac{dt}{t}.$$

Clearly, $\operatorname{supp} b_j \subset \mathcal{O}(B(\mathbf{y}_j, 2r_j))$. The same argument as in the proof of Lemma 4.11. in [24] shows that for every $s = 0, 1, 2, \ldots, M$, the function

$$b_{j,s} = \Delta^s b_j = \lim_{\varepsilon\to 0^+} \int_\varepsilon^{r_j/4} t^{2M}\big(\Delta^s t^2\Delta\Psi_t^{\{1\}} A_j\big)\frac{dt}{t}$$

is supported by $\mathcal{O}(B(\mathbf{y}_j, 2r_j))$ and its $L^2(w)$-norm is bounded by $r^{2M-2s} w(B_j)^{-1/2}$. Thus a_j are proportional to $(1,2,M)$-atoms. In particular, $\|a_j\|_{L^1(dw)} \leq C$ and, consequently, the series (13.4) converges (absolutely) in $L^1(dw)$.

To remove the additional assumption $f \in L^2(dw)$ we recall that the Poisson kernel is an approximation of the identity in $L^1(dw)$ and in H^1_{square} (this is actually Lemma 13.6). Moreover, $P_t f \in L^2(dw)$ for $t > 0$ and $f \in L^1(dw)$. Thus, for $f \in H^1_{\text{square}}$, we can write $f = \sum_n g_n$, where the series converges in $L^1(dw)$, $g_n \in L^2(dw)$, and $\sum_n \|g_n\|_{H^1_{\text{square}}} \leq 3\|f\|_{H^1_{\text{square}}}$. Further we continue as in the proof of Theorem 11.20 to obtain the desired atomic decomposition. □

Acknowledgements The authors want to thank the referees for helpful comments which have improved the presentation of the paper. They are also greatly indebted to Paweł Głowacki for conversations on the subject of the paper. Jean-Philippe Anker and Jacek Dziubański thank their institutions for reciprocal invitations, where part of the present work was carried out. Jacek Dziubański and Agnieszka Hejna supported by the National Science Centre, Poland (Narodowe Centrum Nauki), Grant 2017/25/B/ST1/00599.

Open Access This article is distributed under the terms of the Creative Commons Attribution 4.0 International License (http://creativecommons.org/licenses/by/4.0/), which permits unrestricted use, distribution, and reproduction in any medium, provided you give appropriate credit to the original author(s) and the source, provide a link to the Creative Commons license, and indicate if changes were made.

References

1. Amri, B., Sifi, M.: Riesz transforms for Dunkl transform. Ann. Math. Blaise Pascal **19**(1), 247–262 (2012)
2. Amri, B., Hammi, A.: Dunkl–Schrödinger operators. Complex Anal. Oper. Theory (2018). https://doi.org/10.1007/s11785-018-0834-1
3. Anker, J-Ph, Ben Salem, N., Dziubański, J., Hamda, N.: The Hardy space H^1 in the rational Dunkl setting. Constr. Approx. **42**(1), 93–128 (2015)
4. Betancor, J.J., Dziubański, J., Torrea, J.L.: On Hardy spaces associated with Bessel operators. J. Anal. Math. **107**, 195–219 (2009)
5. Burkholder, D.L., Gundy, R.F., Silverstein, M.L.: A maximal function characterisation of the class H^p. Trans. Am. Math. Soc. **157**, 137–153 (1971)
6. Christ, M., Geller, D.: Singular integral characterizations of Hardy spaces on homogeneous groups. Duke Math. J. **51**, 547–598 (1984)
7. Coifman, R.R.: A real variable characterization of H^p. Studia Math. **51**, 269–274 (1974)
8. Coifman, R.R., Weiss, G.: Extensions of Hardy spaces and their use in analysis. Bull. Am. Math. Soc. **83**(4), 569–615 (1977)
9. Coifman, R.R., Meyer, Y., Stein, E.M.: Some new functions and their applications to harmonic analysis. J. Funct. Anal. **62**(2), 304–315 (1985)
10. de Jeu, M.F.E.: The Dunkl transform. Invent. Math. **113**, 147–162 (1993)
11. Dekel, S., Kerkyacharian, G., Kyriazis, G., Petrushev, P.: Hardy spaces associated with non-negative self-adjoint operators. Studia Math. **239**, 17–54 (2017)
12. Dunkl, C.F.: Differential-difference operators associated to reflection groups. Trans. Am. Math. **311**(1), 167–183 (1989)
13. Dziubański, J.: Riesz transforms characterizations of Hardy spaces H^1 for the rational Dunkl setting and multidimensional Bessel operators. J. Geom. Anal. **26**(4), 2639–2663 (2016)
14. Dziubański, J., Hejna, A.: Remark on atomic decompositions for Hardy space H^1 in the rational Dunkl setting. Studia Math. arXiv:1803.10302
15. Dziubański, J., Hejna, A.: Hörmander's multiplier theorem for the Dunkl transform. arXiv:1807.02640
16. Dziubański, J., Preisner, M.: Hardy spaces for semigroups with Gaussian bounds. Ann. Mat. Pura Appl. **197**, 965–987 (2018)
17. Dziubański, J., Zienkiewicz, J.: Hardy space H^1 associated to Schrödinger operator with potential satisfying reverse Hölder inequality. Rev. Mat. Iberoam. **15**(2), 279–296 (1999)
18. Dziubański, J., Zienkiewicz, J.: A characterization of Hardy spaces associated with certain Schrödinger operators. Potential Anal. **41**(3), 917–930 (2014)
19. Fefferman, C., Stein, E.M.: H^p spaces of several variables. Acta Math. **129**(3–4), 137–195 (1972)
20. Folland, G.B., Stein, E.M.: Hardy Spaces on Homogeneous Groups. Princeton Math. Notes, vol. 28. Princeton Univ. Press, Princeton (1982)
21. Gallardo, L., Rejeb, C.: A new mean value property for harmonic functions relative to the Dunkl–Laplacian operator and applications. Trans. Am. Math. Soc. **368**(5), 3727–3753 (2015)
22. Goldberg, D.: A local version of real Hardy spaces. Duke Math. J. **46**(1), 27–42 (1979)
23. Hejna, A.: Hardy spaces for the Dunkl harmonic oscillator. arXiv:1802.08518
24. Hofmann, S., Lu, G., Mitrea, D., Mitrea, M., Yan, L.: Hardy spaces associated to non-negative self-adjoint operators satisfying Davies–Gaffney estimates. Mem. Am. Math. Soc. **214**(1007), vi+78 (2011)

25. Mauceri, G., Picardello, M., Ricci, F.: A Hardy space associated with twisted convolution. Adv. Math. **39**, 270–288 (1981)
26. Muckenhoupt, B., Stein, E.M.: Classical expansions and their relation to conjugate harmonic functions. Trans. Am. Math. Soc. **118**, 17–92 (1965)
27. Rösler, M.: Generalized Hermite polynomials and the heat equation for Dunkl operators. Commun. Math. Phys. **192**, 519–542 (1998)
28. Rösler, M.: Positivity of Dunkl's intertwining operator. Duke Math. J. **98**(3), 445–463 (1999)
29. Rösler, M.: A positive radial product formula for the Dunkl kernel. Trans. Am. Math. Soc. **355**(6), 2413–2438 (2003)
30. Rösler, M.: Dunkl operators (theory and applications). In: Koelink, E., Van Assche, W. (eds.) Orthogonal polynomials and special functions (Leuven, 2002). Lect. Notes Math., vol. 1817, pp. 93–135. Springer, New York (2003)
31. Rösler, M., Voit, M.: Markov processes related with Dunkl operators. Adv. Appl. Math. **21**, 575–643 (1998)
32. Rösler, M., Voit, M.: Dunkl theory, convolution algebras, and related Markov processes. In: Graczyk, P., Rösler, M., Yor, M. (eds.) Harmonic and Stochastic Analysis of Dunkl Processes, pp. 1–112. Travaux en cours, Hermann (2008)
33. Russ, E.: The atomic decomposition for tent spaces on spaces of homogeneous type. In: McIntosh, A., Portal, P. (eds.) Asymptotic Geometric Analysis, Harmonic Analysis, and Related Topics. Proc. Centre Math. Anal., vol. 42, pp. 125–135. Australian National Univ., Canberra (2007)
34. Stein, E.M.: Singular Integral and Differentiability Properties of Functions. Princeton Math. Series, vol. 30. Princeton Univ. Press, Princeton (1970)
35. Stein, E.M.: Harmonic Analysis (Real Variable Methods, Orthogonality and Oscillatory Integrals). Princeton Math. Series, vol. 43. Princeton Univ. Press, Princeton (1993)
36. Stein, E.M., Weiss, G.: On the theory of harmonic functions of several variables I (the theory of H^p-spaces). Acta Math. **103**, 25–62 (1960)
37. Song, L., Yan, L.: Maximal function characterizations for Hardy spaces associated to nonnegative self-adjoint operators on space of homogeneous type. J. Evol. Equ. **18**, 221–243 (2018)

Publisher's Note Springer Nature remains neutral with regard to jurisdictional claims in published maps and institutional affiliations.

Journal of Fourier Analysis and Applications (2019) 25:2419–2446
https://doi.org/10.1007/s00041-019-09667-z

Sharp Multiplier Theorem for Multidimensional Bessel Operators

Edyta Kania[1] · Marcin Preisner[1]

Received: 5 June 2018 / Revised: 14 January 2019 / Published online: 19 February 2019
© The Author(s) 2019

Abstract

Consider the multidimensional Bessel operator

$$Bf(x) = -\sum_{j=1}^{N} \left(\partial_j^2 f(x) + \frac{\alpha_j}{x_j} \partial_j f(x) \right), \quad x \in (0,\infty)^N.$$

Let $d = \sum_{j=1}^{N} \max(1, \alpha_j + 1)$ be the dimension of the space $(0,\infty)^N$ equipped with the measure $x_1^{\alpha_1} \dots x_N^{\alpha_N} dx_1 \dots dx_N$. In the general case $\alpha_1, \dots, \alpha_N > -1$ we prove multiplier theorems for spectral multipliers $m(B)$ on $L^{1,\infty}$ and the Hardy space H^1. We assume that m satisfies the classical Hörmander condition

$$\sup_{t>0} \|\eta(\cdot) m(t\cdot)\|_{W^{2,\beta}(\mathbb{R})} < \infty$$

with $\beta > d/2$. Furthermore, we investigate imaginary powers B^{ib}, $b \in \mathbb{R}$, and prove some lower estimates on $L^{1,\infty}$ and L^p, $1 < p < 2$. As a consequence, we deduce that our multiplier theorem is sharp.

Keywords Spectral multiplier · Hardy space · Bessel operator · space of homogeneous type

Communicated by Krzysztof Stempak.

The research is supported by the Grant No. 2017/25/B/ST1/00599 from National Science Centre (Narodowe Centrum Nauki), Poland.

✉ Marcin Preisner
marcin.preisner@uwr.edu.pl

Edyta Kania
edyta.kania@uwr.edu.pl

1 Instytut Matematyczny, Uniwersytet Wrocławski, pl. Grunwaldzki 2/4, 50-384 Wroclaw, Poland

Mathematics Subject Classification 42B15 (primary) · 42B30 · 42B20 · 42B25 (secondary)

1 Introduction and Main Results

1.1 The Bessel Operator

Let $N \in \mathbb{N}$ and $\alpha = (\alpha_1, \dots, \alpha_N)$, where $\alpha_j > -1$ for $j = 1, \dots, N$. Consider the space $X = (0, \infty)^N$ equipped with the Euclidean metric and the measure $d\nu(x) = x^\alpha dx = x_1^{\alpha_1} \dots x_N^{\alpha_N} \, dx_1 \dots dx_N$. It is well-known that X satisfies the doubling property, i.e.

$$\nu(B(x, 2r)) \leq C\nu(B(x, r)), \qquad x \in X, \; r > 0,$$

where $B(x, r) = \{y \in X \; : \; |x - y| < r\}$. In other words, there exist $d, C_d > 0$ such that

$$\nu(B(x, \gamma r)) \leq C_d(1 + \gamma)^d \nu(B(x, r)), \qquad x \in X, \; r, \gamma > 0. \tag{D}$$

We choose the constant d (dimension of X) as small as possible. In this case

$$d = \sum_{j=1}^{N} \max(1, \alpha_j + 1). \tag{1.1}$$

The multidimensional Bessel operator is given by $B = B_1 + \dots + B_N$, where

$$B_j f(x) = -\partial_j^2 f(x) - \frac{\alpha_j}{x_j} \partial_j f(x), \qquad x \in X.$$

The operator B, initially defined on, say, $(C_c^2((0, \infty)))^n$, extends to a self-adjoint operator on $L^2(X)$. Slightly abusing notation, we shall denote this extension by the same symbol B. For a precise definition of B we refer the reader to e.g. [6, Sec. 2] (see also [32]). Also, B is the infinitesimal generator of the Bessel semigroup $\mathbf{T}_t f(x) = \int_X T_t(x, y) f(y) \, d\nu(y)$, where $T_t(x, y) = T_t^{[1]}(x_1, y_1) \cdot \ldots \cdot T_t^{[N]}(x_N, y_N)$ and

$$T_t^{[j]}(x_j, y_j) = \frac{1}{2t} (x_j y_j)^{-(\alpha_j - 1)/2} I_{(\alpha_j - 1)/2} \left(\frac{x_j y_j}{2t} \right) \exp\left(-\frac{x_j^2 + y_j^2}{4t} \right), \qquad x_j, y_j, t > 0. \tag{1.2}$$

Here $I_\tau(x) = \sum_{m=0}^{\infty} \frac{1}{m! \Gamma(m + \tau + 1)} \left(\frac{x}{2} \right)^{2m + \tau}$ is the modified Bessel function of the first kind. The kernel $T_t(x, y)$ satisfies the upper and lower gaussian bounds, i.e. there exist constants $c_1, c_2, C_1, C_2 > 0$, such that

$$C_1\nu(B(x,\sqrt{t}))^{-1}\exp\left(-\frac{|x-y|^2}{c_1t}\right)\leq T_t(x,y)$$
$$\leq C_2\nu(B(x,\sqrt{t}))^{-1}\exp\left(-\frac{|x-y|^2}{c_2t}\right). \tag{G}$$

This fact is well known and follows from the asymptotics for $\nu(B(x,\sqrt{t}))$ and I_τ. For details see e.g. [20, Lem. 4.2].

Since B is self-adjoint and nonnegative, for a Borel function $m:(0,\infty)\to\mathbb{C}$ the spectral theorem defines the operator

$$m(B)=\int_0^\infty m(\lambda)\,dE_B(\lambda),$$

where E_B is the spectral resolution of B.

1.2 Multiplier Theorems for *B*

Multiplier theorems for B and other operators are one of the main topics in harmonic analysis. Many authors investigated assumptions on m that guarantee boundedness of $m(B)$ on various function spaces, such as $L^p(X)$, $H^p(X)$, $L^{p,q}(X)$ and others.

For example, in [23] the authors proved weak type (1,1) estimates on $m(B)$ assuming $N=1$, $\alpha>0$ and

$$\left(\int_{R/2}^R |m^{(s)}(\lambda)|^2 d\nu(\lambda)\right)^{1/2}\leq CR^{(\alpha+1)/2-s},\quad R>0,$$

where $s=0,\dots,K$ and K is the least even integer greater than $(\alpha+1)/2=d/2$ (see also [27]). In [18], assuming still $N=1$ and $\alpha>0$, it is proved that if

$$\sup_{t>0}\|\eta(\cdot)m(t\cdot)\|_{W^{2,\beta}(\mathbb{R})}<\infty \tag{S}$$

with some $\beta>d/2$, then $m(B)$ is bounded on the Hardy space $H^1(B)$ related to B. Here and thereafter $W^{2,\beta}(\mathbb{R})$ is the L^2-Sobolev space on $\mathbb{R}$ and η is a fixed nonnegative smooth cut-off function such that $\operatorname{supp}\eta\subseteq(2^{-1},2)$.

In the multidimensional case $N\geq 1$ in [5] the authors prove weak type $(1,1)$ estimates for $m(B)$, where m is of Laplace transform type, i.e. there exists $\phi\in L^\infty(0,\infty)$, such that

$$m(x)=|x|^2\int_0^\infty e^{-t|x|^2}\phi(t)\,dt,\quad x\in(0,\infty)^N.$$

Notice, that if m is of Laplace transform type, then m is radial and (as a function on $(0,\infty)$) satisfies (S) with any $\beta>0$. Another multidimensional result can be found in [20], where it is proved that $m(B)$ is weak type $(1,1)$ and bounded on the Hardy space

$H^1(X)$ provided that $\alpha_j > 1$ for $j = 1, \ldots, N$ and m satisfies (S) with $\beta > d/2$. See also e.g. [21,22,38] for other multiplier results for the Bessel operator.

Our first main goal is to obtain multiplier theorem for B in the most general case $N \geq 1$ and $\alpha_j > -1$, $j = 1, \ldots, N$. Let us notice that many of the results before assumed that $\alpha_j > 0$ and the case $\alpha_j < 0$ is more difficult and less known. One reason for that is the singularity at zero of the measure $x^{\alpha_j} dx_j$ when $\alpha_j < 0$. Also, so-called "generalized translation" operators and convolution structure for B (see, e.g. [7, Sec. 2]), does not help when $\alpha_j < 0$. This is strictly related to the fact, that the generalized eigenfunctions of B are no longer bounded if $\alpha_j < 0$ for some j and, therefore, the generalized translation is not even bounded on L^2. Let us also notice that, we are interested in multiplier results that are sharp in the sense that we assume (S) with β as small as possible. In this case this is expected to be $\beta > d/2$ (we shall discuss this in Sect. 1.3 below).

To state the multiplier result let us recall that the weak L^1 space is given by the semi-norm

$$\|f\|_{L^{1,\infty}(X)} = \sup_{\lambda>0} \lambda \nu \{x \in X \ : \ |f(x)| > \lambda\},$$

and the Hardy space $H^1(B)$ related to B can be defined by the norm

$$\|f\|_{H^1(B)} = \left\| \sup_{t>0} |\mathbf{T}_t f| \right\|_{L^1(X)}.$$

In the case $N = 1$ the space $H^1(B)$ was studied in [7], where $H^1(B)$ was characterized by means of atomic decompositions and the Riesz transforms. In the general case $N \geq 1$ and $\alpha_j > -1$, $j = 1, \ldots, N$ the atomic characterization of $H^1(B)$ can be found in [19] (see also [16,20]). We shall recall this characterization in Sect. 2.3 below.

Theorem A *Let $N \geq 1$ and $\alpha_j > -1$ for $j = 1, \ldots, N$. Assume that $m : (0, \infty) \to \mathbb{C}$ satisfies* (S) *with $\beta > d/2$, see* (1.1). *Then:*

1. *$m(B)$ is bounded from $L^1(X)$ to $L^{1,\infty}(X)$,*
2. *$m(B)$ is bounded from $H^1(B)$ to $H^1(B)$,*
3. *$m(B)$ is bounded from $L^p(X)$ to $L^p(X)$, $1 < p < \infty$.*

Part **1.** of Theorem A will be proved by using results of [36]. More precisely, we shall check the assumptions of [36, Theorem 3.1]. The proof of **2.** will be given in Sect. 2. In fact, in the proof we shall only use general properties of B, such as e.g. (D), (G), and (P_q) below. Thus, the multiplier result in Sect. 2 will be formulated in a more general context. This section can be read independently of the rest of the paper and we shall use different notation. As usual, **3.** is a consequence of either **1.** or **2.** by duality and interpolation, see e.g. [4].

1.3 Imaginary Powers of *B*

Another goal of this paper is to study the imaginary powers B^{ib}, $b \in \mathbb{R}$, of the Bessel operator and establish lower bounds of these operators on some function spaces. We shall concentrate our attention on the dependence of the lower estimates on b for large b. This is related with sharpness of multiplier theorems and may be of independent interest. To state these estimates let us restrict ourselves to the one-dimensional case $N = 1$ ($X = (0, \infty)$, $d\nu(x) = x^\alpha\, dx$, $\alpha > -1$). Motivated by the identity

$$B^{ib} = \Gamma(-ib)^{-1} \int_0^\infty t^{-ib} e^{-tB} \frac{dt}{t} \tag{1.3}$$

let us define for $x \neq y$ the integral kernel

$$K_b(x, y) = \Gamma(-ib)^{-1} \int_0^\infty t^{-ib} T_t(x, y) \frac{dt}{t}. \tag{1.4}$$

Notice, that the integral in (1.3) is not absolutely convergent, thus we have to explain how the kernel $K_b(x, y)$ is related to the operators B^{ib}. Indeed, in Sect. 3.3 we shall prove that for $f \in L^\infty(X)$ with compact support we have

$$B^{ib} f(x) = \int_X K_b(x, y) f(y)\, d\nu(y), \qquad x \notin \operatorname{supp} f \tag{1.5}$$

One of our goals is to provide lower estimates for B^{ib}.

Theorem B *Assume that $\alpha > -1$. Then there exist a constant $C > 0$ and a function f such that $\|f\|_{L^1(X)} = 1$ and for $|b|$ large enough we have*

$$\left\| B^{ib} f \right\|_{L^{1,\infty}(X)} \geq C |b|^{d/2}.$$

Theorem C *Assume that $\alpha > 0$ and $p \in (1, 2)$. Then there exist $C_p > 0$ and f such that $\|f\|_{L^p(X)} = 1$ and for $|b|$ large enough we have*

$$\left\| B^{ib} f \right\|_{L^p(X)} \geq C_p |b|^{\frac{d}{2}\frac{(2-p)}{p}}.$$

The proofs of Theorems B and C are presented in Sect. 3.3. To prove Theorem B we shall carefully analyze the kernels $K_b(x, y)$. More precisely, we prove the following lemma.

Lemma 1.1 *Assume that $\alpha > -1$ and $b \in \mathbb{R}$. Then*

$$\begin{aligned} K_b(x, y) = {} & c_1(b) \left(x^2 + y^2\right)^{-ib-(\alpha+1)/2} \\ & + c_2(b) (xy)^{-\alpha/2} |x - y|^{-2bi-1} \chi_{\{y/2<x<2y\}}(x, y) \\ & + c_3(b) R_b(x, y), \end{aligned} \tag{1.6}$$

where

$$c_1(b) = \frac{2^{2ib+1}}{\Gamma\left((\alpha+1)/4\right)} \frac{\Gamma\left(ib + (\alpha+1)/2\right)}{\Gamma(-ib)},$$

$$c_2(b) = \frac{2^{2ib}}{\sqrt{\pi}} \frac{\Gamma(ib+1/2)}{\Gamma(-ib)}, \quad c_3(b) = \Gamma(-ib)^{-1}.$$

Moreover, there exists $C > 0$ that does not depend on b, such that

$$|R_b(x,y)| \le Cxy(x+y)^{-\alpha-3}.$$

Notice that the kernel $R_b(x,y)$ is related to an operator that is bounded on every $L^p(X)$, $1 \le p \le \infty$, uniformly in $b \in \mathbb{R}$. Thus we may think of $R_b(x,y)$ as of some kind of "error term". However, for $|b| > 1$ the sizes of the constants are the following:

$$|c_1(b)| \simeq |b|^{(\alpha+1)/2}, \quad |c_2(b)| \simeq |b|^{1/2}, \quad |c_3(b)| \simeq |b|^{1/2} \exp\left(\frac{\pi|b|}{2}\right), \tag{1.7}$$

c.f. Lemma 4.1. Thus, $c_3(b)$ grows exponentially when $|b| \to \infty$, while the constants $c_1(b)$ and $c_2(b)$ are much smaller. It appears that the growth of the constant $c_3(b)$ will lead to a problem in deriving lower estimates for B^{ib} (since our goal is to find the exact dependence on b). However, we can overcome this difficulty when analyzing weak $(1,1)$ norm as in Theorem B. The same trick seems not to work in other function spaces (such as $H^1(B)$, $L^p(X)$ and $L^{p,\infty}(X)$ with $p > 1$), thus the proof of Theorem C is different and uses the integral representation of the Bessel function I_τ instead of Lemma 1.1.

As a corollary of Theorems B and C we obtain that Theorem A is sharp (at least for $N = 1$) in the sense that $d/2$ cannot be replaced by a smaller number. The argument is standard, but we shall present it now for the convenience of the reader. One can check that for $m_b(\lambda) = \lambda^{ib}$ we have

$$M_b := \sup_{t>0} \|\eta(\cdot)m_b(t\cdot)\|_{W^{2,\beta}(\mathbb{R})} \le |b|^\beta.$$

Also, Theorem A actually gives that $\|m_b(B)f\|_{L^{1,\infty}(X)} \le CM_b \|f\|_{L^1(X)}$, where C does not depend on b. Combining these estimates with Theorem B for $|b|$ large enough we have

$$|b|^{d/2} \le C\,\|m_b(B)\|_{L^1(X)\to L^{1,\infty}(X)} \le C|b|^\beta.$$

Therefore $\beta \ge d/2$. Actually, one expects that $\beta \ne d/2$, but this question is beyond the scope of this paper.

Similarly, the constant $d/2$ cannot be improved for the Hardy spaces. If $\alpha < 0$ then $d/2 = 1/2$ and (S) with $\beta < 1/2$ would not even guarantee that m is bounded. On the other hand, for $\alpha > 0$ if we could prove multiplier theorem on $H^1(B)$ with a constant lower than $d/2$, then by interpolation we would have better upper bounds for $m_b(B)$ on $L^p(X)$ for $1 < p < 2$, which contradicts Theorem C by an argument similar to the one above.

1.4 Organization of the Paper and Notation

In Sect. 2 we state and prove a "sharp" multiplier theorem on Hardy spaces for self-adjoint operators on spaces of homogeneous type with certain assumptions (Theorem D). This is a slight generalization of Theorem A 2. in the spirit of [36, Theorem 3.1]. In Sect. 2 we shall use different notation, so that it can be read independently of the rest of the paper. In Sect. 3 we prove the results stated above. More precisely, first we check that B satisfies assumption (P_2) (see Sect. 2 below) in the full generality $N \geq 1, \alpha_j > -1$ for $j = 1, \ldots, N$. Thus Theorem D can be applied for B. Then we prove Lemma 1.1 and Theorems B and C. We shall use standard notations, i.e. C and c denote positive constants that may change from line to line.

2 Sharp Multiplier Theorem on Hardy Spaces

2.1 Background and General Assumptions

In this section we consider a space Y with a metric ρ and a nonnegative measure μ. We shall assume that the triple (Y, ρ, μ) is a space of homogeneous type, i.e. there exists $C > 0$ such that $\mu(B(x, 2r)) \leq C\mu(B(x, r))$, for all $x \in Y$ and $r > 0$, where $B(x, r) = \{y \in Y : \rho(x, y) < r\}$, c.f. [14]. It is well-known that this implies the existence of $d, C_d > 0$ such that

$$\mu(B(x, \gamma r)) \leq C_d(1+\gamma)^d \mu(B(x, r)), \qquad x \in Y, \ r, \gamma > 0. \tag{D}$$

As usual, we choose d as small as possible, even at the cost of enlarging C_d.

Let A denote a self-adjoint positive operator and let E_A be its spectral measure, i.e. $A = \int_0^\infty \lambda \, dE_A(\lambda)$. Denote by $\mathbf{P}_t = \exp(-tA)$ the semigroup generated by A. Assume that there exists an integral kernel $P_t(x, y)$ such that $\mathbf{P}_t f(x) = \int_Y P_t(x, y) f(y) \, d\mu(y)$ and that satisfies the upper gaussian bounds, i.e. there exist $c_2, C_2 > 0$ such that

$$P_t(x, y) \leq C_2 \mu(B(x, \sqrt{t}))^{-1} \exp\left(-\frac{\rho(x, y)^2}{c_2 t}\right), \quad t > 0, \ x, y \in Y. \tag{UG}$$

2.2 Multiplier Theorems

By the spectral theorem, for a Borel function m on $(0, \infty)$, we have the operator

$$m(A) = \int_0^\infty m(\lambda) \, dE_A(\lambda).$$

In the classical case $A = -\Delta$, $Y = \mathbb{R}^D$, the Hörmader multiplier theorem states that if m satisfies (S) with $\beta > D/2$, then $m(-\Delta)$ is weak type $(1, 1)$ and bounded on $L^p(\mathbb{R}^D)$ for $1 < p < \infty$. It is well-known that the constant $D/2$ is sharp in the sense that it cannot be replaced by a smaller constant, see e.g. [34].

At this point let us recall one of many multiplier theorems on spaces of homogeneous type. Suppose Y and A are as in Sect. 2.1. Following [36] we introduce additional assumption. Suppose that there exists $C > 0$ and $q \in [2, \infty]$, such that for $R > 0$ and every Borel function m on $\mathbb{R}$ satisfying $\operatorname{supp} m \subseteq [R/2, 2R]$ we have

$$\int_Y \left|K_{m(A)}(x, y)\right|^2 d\mu(x) \leq C\mu\left(B\left(y, R^{-1/2}\right)\right)^{-1} \|m(R\cdot)\|^2_{L^q(\mathbb{R})}. \tag{P_q}$$

Theorem 2.1 [36, Theorem 3.1] *Assume that on a space of homogeneous type (Y, ρ, μ) there is a self-adjoint positive operator A that satisfies* (UG). *Moreover, assume that (P_q) holds with some $q \in [2, \infty]$ and m satisfies*

$$\sup_{t>0} \|\eta(\cdot)m(t\cdot)\|_{W^{q,\beta}(\mathbb{R})} < \infty \tag{S_q}$$

with some $\beta > d/2$. Then $m(A)$ is of weak type $(1, 1)$ *and bounded on $L^p(Y)$ for $p \in (1, \infty)$.*

At this point let us make a few comments.

1. Assuming (UG) the operators $m(A)$ appearing in (P_q) always have integral kernels $K_{m(A)}(x, y)$, c.f. [36, Lemma 2.2].
2. For the Bessel operator we are interested in (S_q) and (P_q) for $q = 2$ only, $(S) = (S_2)$. However, in Sect. 2 the results are stated and proved with an arbitrary $q \in [2, \infty]$.
3. The assumption (P_q) in some sense plays a role of Plancherel theorem in the proof of Theorem 2.1. It is a key to obtain the sharp range $\beta > d/2$. For example, if we would allow m to satisfy (S_q) with $\beta > d/2 + 1/2$, then (P_q) would be superfluous.
4. The assumption (P_q) is written in [36] for m having support in $[0, R]$ not in $[R/2, 2R]$. However, a simple inspection of the proof shows that (P_q) is needed only for m with $\operatorname{supp} m \subseteq [R/2, 2R]$. This makes no difference for many operators. However, it matters e.g. when considering the Bessel operator with negative parameters α_j.
5. Assumption (P_q) in [36] is written for $m(\sqrt{A})$, but we use equivalent version with $m(A)$ (therefore we replace $B(y, R^{-1})$ by $B(y, R^{-1/2})$).

One of the main goals of this paper is to establish a multiplier theorem on Hardy spaces. We shall use the definition of the Hardy space $H^1(A)$ associated with A by means of the maximal operator of the semigroup $\mathbf{P}_t$, namely

$$H^1(A) = \left\{ f \in L^1(Y) \ : \ \|f\|_{H^1(A)} := \left\| \sup_{t>0} |\mathbf{P}_t f| \right\|_{L^1(Y)} < \infty \right\}.$$

To state our result we shall assume additionally that $P_t(x, y)$ satisfies also the the lower Gaussian bounds, namely there exist $c_1, C_1 > 0$, such that

$$P_t(x, y) \geq C_1 \mu(B(x, \sqrt{t}))^{-1} \exp\left(-\frac{\rho(x, y)^2}{c_1 t}\right), \quad t > 0,\ x, y \in Y, \tag{LG}$$

and that the space (Y, ρ, μ) satisfies the following assumption:

$$\text{for all } x \in Y \text{ the function } r \mapsto \mu(B(x, r)) \text{ is a bijection on } (0, \infty). \tag{Y}$$

Notice that (Y) implies that $\mu(Y) = \infty$ and that μ is non-atomic. Now we are ready to state the theorem.

Theorem D *Assume that* (Y, ρ, μ) *is a space of homogeneous type,* d *is as in* (D), *and* (Y) *is satisfied. Suppose that there is a self-adjoint positive operator* A *such that* (UG), (LG), *and* (P_q) *hold with some* $q \in [2, \infty]$. *If* m *satisfies* (S_q) *and* $\beta > d/2$, *then* $m(A)$ *is bounded from* $H^1(A)$ *to* $H^1(A)$, *i.e. there exists* $C > 0$, *such that*

$$\|m(A) f\|_{H^1(A)} \leq C \|f\|_{H^1(A)}.$$

The history of multiplier theorems for spaces of homogeneous type is long and wide. The interested reader is referred to [2,9–11,13,17,20,21,24,26,31,33,35,36] and references therein. Let us concentrate for a moment on the range of parameters β in Theorem D. Obviously, in general, the range $\beta > d/2$ is optimal. However, it may happen that for some particular operators one may obtain multiplier results assuming that $\beta > \widetilde{d}/2$ with $\widetilde{d} < d$, see e.g. [29,30,33]. On the other hand, there are known families of operators for which the constant $d/2$ cannot be lower. One of the methods to prove this is to derive lower estimates for A^{ib} in terms of $b \in \mathbb{R}$, see [12,30,34,36]. Lastly, let us mention that some multiplier results hold also in the non-doubling case, see e.g. [15].

Boundedness of operators on the Hardy space H^1 is a natural counterpart of weak type $(1, 1)$ bound. For example, it is a good end point for the interpolation, see e.g. [4]. However, the Hardy spaces are strictly related to some cancellation conditions and it is usually more involving to study properties of operators on the Hardy space, than on L^p or $L^{p,\infty}$ spaces. Let us also mention that boundedness from H^1 to H^1 obviously implies boundedness from H^1 to L^1, which is usually much easier to prove.

2.3 Hardy Spaces

The Hardy spaces on spaces of homogeneous type are studied extensively from the 60's, see e.g. [14]. In particular, now we have many atomic decompositions for H^p on various spaces and operators acting on this spaces. We refer the reader to e.g. [3,7,19,25] and references therein.

In this subsection we recall some results on Hardy spaces related to A, assuming that (D), (UG), (LG) and (Y) are satisfied. For the proofs and more details we refer the reader to [19]. Firstly, there exists the unique (up to a multiplicative constant) A-harmonic function $\omega : Y \to \mathbb{R}$ such that

$$C^{-1} \leq \omega(x) \leq C, \qquad x \in Y.$$

The function ω plays a special role in the analysis of A and $\mathbf{P}_t$. In particular we have the following Hölder-type estimate.

Theorem 2.2 *Suppose that the semigroup* $\mathbf{P}_t$ *satisfies* (UG), (LG). *Then there exist positive constants* γ, c, C, *such that if* $\rho(y,z) \leq \sqrt{t}$, *then*

$$\left| \frac{P_t(x,y)}{\omega(y)} - \frac{P_t(x,z)}{\omega(z)} \right| \leq C\mu(B(x,\sqrt{t}))^{-1} \left(\frac{\rho(y,z)}{\sqrt{t}} \right)^{\gamma} \exp\left(-\frac{\rho(x,y)^2}{ct} \right).$$

Theorem 2.2 is quite well-known and follows from a general theory. For a short and independent proof see [19, Sec. 4].

Corollary 2.3 *There exist* $\gamma, C > 0$ *such that if* $\rho(y,z) \leq \sqrt{t}$, *then*

$$\int_Y \left| \frac{P_t(x,y)}{\omega(y)} - \frac{P_t(x,z)}{\omega(z)} \right| d\mu(x) \leq C \left(\frac{\rho(y,z)}{\sqrt{t}} \right)^{\gamma}.$$

Using Theorem 2.2 the authors of [19] obtained the following atomic decomposition for the elements of $H^1(A)$. Let us call a function $a : Y \to \mathbb{C}$ an (μ, ω)-atom, if there exists a ball B in Y, such that:

$$\operatorname{supp} a \subseteq B, \qquad \|a\|_\infty \leq \mu(B)^{-1}, \qquad \int_B a(x)\omega(x)\, d\mu(x) = 0.$$

Theorem 2.4 [19, Theorem 1] *There exists a constant* $C > 0$ *such that for each* $f \in H^1(A)$ *there exist* $\lambda_k \in \mathbb{C}$ *and* (μ, ω)*-atoms* a_k $(k \in \mathbb{N})$, *such that*

$$f(x) = \sum_{k \in \mathbb{N}} \lambda_k a_k(x), \quad \text{and} \quad C^{-1} \|f\|_{H^1(A)} \leq \sum_{k \in \mathbb{N}} |\lambda_k| \leq C \|f\|_{H^1(A)}.$$

Let us start by recalling a few consequences of (D) and (UG).

Lemma 2.5 [36, Lemma 2.1] *Suppose that* (D) *and* (UG) *hold. Then*

$$\int_{B(y,r)^c} |P_t(x,y)|^2 \, d\mu(x) \leq C\mu(B(y,\sqrt{t}))^{-1} \exp\left(-\frac{r^2}{c_2 t} \right).$$

In particular

$$\|P_t(x,\cdot)\|_{L^2(Y)}^2 \leq C\mu(B(x,\sqrt{t}))^{-1}.$$

Lemma 2.6 [36, Lemma 4.1] *For $\kappa \geq 0$ there exists a constant $C = C(\kappa) > 0$ such that*

$$\int_Y |P_{(1+i\tau)R^{-1}}(x, y)|^2 (1 + R^{1/2}\rho(x, y))^\kappa \, d\mu(x) \leq C\mu\left(B\left(y, R^{-1/2}\right)\right)^{-1} (1 + |\tau|)^\kappa.$$

Lemma 2.7 [36, Lemma 4.4] *Suppose that* (D) *holds and $\delta > 0$. Then*

$$\int_{B(y,r)^c} (1 + R^{1/2}\rho(x, y))^{-d-2\delta} \, d\mu(x) \leq C\mu\left(B\left(y, R^{-1/2}\right)\right)(1 + rR^{1/2})^{-2\delta}.$$

2.4 Key Kernel Estimates

This subsection is devoted to obtain key estimates needed for the proof of Theorem D. We shall assume (temporarily) that m satisfies $\operatorname{supp} m \subseteq [R/2, 2R]$ with some $R > 0$. Later we shall use a partition of unity for general m. Denote $m_R(\lambda) = m(R\lambda)$, so that $\operatorname{supp} m_R \subseteq [2^{-1}, 2]$. Let us notice that below the letter $q \in [2, \infty]$ is always the exponent related to (P_q) and (S_q). Moreover, all the spectral operators below admit related integral kernels, which can be seen by using an argument identical as in [36, Lem. 2.2]. Let us denote $\widetilde{m}_t(\lambda) = \exp(-t\lambda)m(\lambda)$ and let $M_t(x, y)$ be the kernel associated with $\widetilde{m}_t(A) = \mathbf{P}_t m(A)$.

Proposition 2.8 *Assume that $\operatorname{supp} m \subseteq [R/2, 2R]$ and $m_R \in W^{q,\beta}(\mathbb{R})$ with $\beta > d/2$. Then, there exist $\delta, \gamma, C > 0$ such that for $y, z \in Y$ and $r > 0$ we have*

$$\int_{B(y,r)^c} \sup_{t>0} |M_t(x, y)| \, d\mu(x) \leq C\left(1 + rR^{1/2}\right)^{-\delta} \|m_R\|_{W^{q,\beta}(\mathbb{R})}, \tag{2.1}$$

and, for $\rho(y, z) < R^{-1/2}$,

$$\int_{B(y,r)^c} \sup_{t>0} \left| \frac{M_t(x, y)}{\omega(y)} - \frac{M_t(x, z)}{\omega(z)} \right| d\mu(x) \leq C\left(R^{1/2}\rho(y, z)\right)^\gamma \|m_R\|_{W^{q,\beta}(\mathbb{R})}. \tag{2.2}$$

Let us start by showing the following lemma.

Lemma 2.9 *For $\varepsilon > 0, \kappa \geq 0$ there exists a constant $C = C(\kappa, \varepsilon)$ such that*

$$\int_Y \sup_{t>0} |M_t(x, y)|^2 \left(1 + R^{1/2}\rho(x, y)\right)^\kappa d\mu(x)$$
$$\leq C\mu\left(B\left(y, R^{-1/2}\right)\right)^{-1} \|m_R\|^2_{W^{q,\kappa/2+\varepsilon}(\mathbb{R})}.$$

Proof Fix a cut-off function $\psi \in C_c^\infty(4^{-1}, 4)$, such that $\psi \equiv 1$ on $[2^{-1}, 2]$. Set

$$n_{t,R}(\lambda) = m_R(\lambda) \underbrace{e^{-tR\lambda} e^{\lambda} \psi(\lambda)}_{\lambda_{t,R}(\lambda)}.$$

By the Fourier inversion formula,

$$\widetilde{m}_t(A) = n_{t,R}(AR^{-1})e^{-AR^{-1}} = \frac{1}{2\pi}\int_{\mathbb{R}} \widehat{n}_{t,R}(\tau)\exp\left((i\tau - 1)AR^{-1}\right) d\tau$$

and

$$M_t(x, y) = \frac{1}{2\pi}\int_{\mathbb{R}} \widehat{n}_{t,R}(\tau)P_{(1-i\tau)R^{-1}}(x, y)\, d\tau. \tag{2.3}$$

Notice that $\operatorname{supp}\lambda_{t,R}^{(N)} \subseteq (4^{-1}, 4)$ for arbitrary $N \in \mathbb{N}$. By simple calculus we can find a constant C_N such that

$$\sup_{R>0,t>0} \left|\widehat{\lambda}_{t,R}(\tau)\right| \leq C_N(1 + |\tau|)^{-N}.$$

Since $\widehat{n}_{t,R} = \widehat{m}_R * \widehat{\lambda}_{t,R}$ and $(1 + |\tau|) \leq (1 + |\theta|)(1 + |\tau - \theta|)$, for $\kappa \geq 0$ and $\varepsilon > 0$ we use the Cauchy-Schwarz inequality, getting

$$\begin{aligned}
\int_{\mathbb{R}} \sup_{t>0} |\widehat{n}_{t,R}(\tau)|(1 + |\tau|)^{\kappa/2}\, d\tau &\leq \int_{\mathbb{R}}\int_{\mathbb{R}} \sup_{t>0} |\widehat{m}_R(\theta)||\widehat{\lambda}_{t,R}(\tau - \theta)|(1 + |\tau|)^{\kappa/2}\, d\theta\, d\tau \\
&\leq \int_{\mathbb{R}}\int_{\mathbb{R}} \sup_{t>0} |\widehat{m}_R(\theta)||\widehat{\lambda}_{t,R}(\tau - \theta)| \\
&\quad (1 + |\theta|)^{\kappa/2}\,(1 + |\tau - \theta|)^{\kappa/2}\, d\tau\, d\theta \\
&\leq C\int_{\mathbb{R}} |\widehat{m}_R(\theta)|(1 + |\theta|)^{(\kappa+1)/2+\varepsilon}(1 + |\theta|)^{-1/2-\varepsilon}\, d\theta \\
&\leq C\,\|m_R\|_{W^{2,(1+\kappa)/2+\varepsilon}(\mathbb{R})}\left(\int_{-\infty}^{\infty}(1 + |\theta|)^{-1-\varepsilon}\, d\theta\right)^{1/2} \\
&\leq C\,\|m_R\|_{W^{2,(1+\kappa)/2+\varepsilon}(\mathbb{R})}\,.
\end{aligned}$$

Hence, by (2.3), the Minkowski inequality, and Lemma 2.6 we obtain

$$\begin{aligned}
&\left(\int_Y \sup_{t>0} |M_t(x, y)|^2(1 + R^{1/2}\rho(x, y))^{\kappa}\, d\mu(x)\right)^{1/2} \\
&\leq \int_{\mathbb{R}} \sup_{t>0} |\widehat{n}_{t,R}(\tau)|\left(\int_Y |P_{(1-i\tau)R^{-1}}(x, y)|^2(1 + R^{1/2}\rho(x, y))^{\kappa}\, d\mu(x)\right)^{1/2} d\tau \\
&\leq C\mu\left(B\left(y, R^{-1/2}\right)\right)^{-1/2}\int_{\mathbb{R}} \sup_{t>0} |\widehat{n}_{t,R}(\tau)|(1 + |\tau|)^{\kappa/2}\, d\tau \qquad (2.4) \\
&\leq C\mu\left(B\left(y, R^{-1/2}\right)\right)^{-1/2}\,\|m_R\|_{W^{2,(1+\kappa)/2+\varepsilon}(\mathbb{R})} \\
&\leq C\mu\left(B\left(y, R^{-1/2}\right)\right)^{-1/2}\,\|m_R\|_{W^{q,(1+\kappa)/2+\varepsilon}(\mathbb{R})}\,.
\end{aligned}$$

In the last inequality we have used that $\operatorname{supp} m_R \subseteq [2^{-1}, 2]$ and $q \geq 2$.

Observe that (2.4) is exactly the estimate we look for, but the Sobolev parameter is higher by $1/2$ than we want. To sharpen this estimate, we make use of known interpolation method. Notice, that $M_t(x, y) = \mathbf{P}_t(K_{m(A)}(\cdot, y))(x)$. It is well-known that (UG) implies boundedness on $L^2(Y)$ of the maximal operator $\mathscr{M} f = \sup_{t>0} |\mathbf{P}_t f|$. A second estimate needed for an interpolation is the following

$$\begin{aligned}\left(\int_Y \sup_{t>0} |M_t(x, y)|^2 \, d\mu(x)\right)^{1/2} &= \left\| \mathscr{M} K_{m(A)}(\cdot, y) \right\|_{L^2(Y)} \\ &\leq C \left\| K_{m(A)}(\cdot, y) \right\|_{L^2(Y)} \\ &\leq C\mu\left(B\left(y, R^{-1/2}\right)\right)^{-1/2} \|m_R\|_{L^q(\mathbb{R})} .\end{aligned} \tag{2.5}$$

In the last inequality we have used (P_q) . Now, Lemma 2.9 follows by interpolating (2.4) and (2.5), see e.g. proofs of [36, Lem. 4.3(a)] and [20, Lem. 2.2] for details. □

Proof of (2.1) By the Cauchy-Schwarz inequality and Lemmas 2.7 and 2.9,

$$\begin{aligned}&\int_{B(y,r)^c} \sup_{t>0} |M_t(x, y)| \, d\mu(x) \\ &\leq \left(\int_Y \sup_{t>0} |M_t(x, y)|^2 (1 + R^{1/2}\rho(x, y))^{d+2\delta} \, d\mu(x)\right)^{1/2} \\ &\quad \left(\int_{B(y,r)^c} (1 + R^{1/2}\rho(x, y))^{-d-2\delta} \, d\mu(x)\right)^{1/2} \\ &\leq C\mu\left(B\left(y, R^{-1/2}\right)\right)^{-1/2} \|m_R\|_{W^{q,d/2+\delta+\varepsilon}(\mathbb{R})} \, \mu\left(B\left(y, R^{-1/2}\right)\right)^{1/2} (1 + rR^{1/2})^{-\delta} \\ &\leq C(1 + rR^{1/2})^{-\delta} \|m_R\|_{W^{q,\beta}(\mathbb{R})} ,\end{aligned}$$

where $\delta, \varepsilon > 0$ are such that $d/2 + \delta + \varepsilon \leq \beta$.

□

Consider for a moment the operator $\mathbf{P}_t m(A) \exp(AR^{-1})$ and let $\widetilde{M}_{t,R}(x, y)$ be its kernel. By almost identical arguments as in the proofs of Lemma 2.9 and (2.1), we can show that for $\beta > d/2$ we also have

$$\int_{B(y,r)^c} \sup_{t>0} |\widetilde{M}_{t,R}(x, y)| \, d\mu(x) \leq C \|m_R\|_{W^{q,\beta}(\mathbb{R})} . \tag{2.6}$$

Proof of (2.2) Notice, that $M_t(x,y) = \int_Y \widetilde{M}_{t,R}(x,u) P_{R^{-1}}(u,y)\, d\mu(u)$. For $\rho(y,z) < R^{-1/2}$, by Corollary 2.3 and (2.6),

$$\begin{aligned}
&\int_{B(y,r)^c} \sup_{t>0} \left| \frac{M_t(x,y)}{\omega(y)} - \frac{M_t(x,z)}{\omega(z)} \right| d\mu(x) \\
&= \int_{B(y,r)^c} \sup_{t>0} \left| \int_Y \widetilde{M}_{t,R}(x,u) \left(\frac{P_{R^{-1}}(u,y)}{\omega(y)} - \frac{P_{R^{-1}}(u,z)}{\omega(z)} \right) d\mu(u) \right| d\mu(x) \\
&\leq \int_Y \left| \frac{P_{R^{-1}}(u,y)}{\omega(y)} - \frac{P_{R^{-1}}(u,z)}{\omega(z)} \right| \int_{B(y,r)^c} \sup_{t>0} |\widetilde{M}_{t,R}(x,u)|\, d\mu(x)\, d\mu(u) \\
&\leq C \left(R^{1/2} \rho(y,z) \right)^{\gamma} \|m_R\|_{W^{q,\beta}(\mathbb{R})}.
\end{aligned}$$

□

2.5 Proof of Theorem D

Theorem D follows from Proposition 2.8 by a quite standard argument. We present the details for completeness and convenience of the reader. As usual, by a continuity argument, in order to prove boundedness of the operator $m(A)$ on $H^1(A)$ it is enough to show that there exists $C > 0$ such that

$$\|m(A)a\|_{H^1(A)} = \|\mathcal{M} m(A)a\|_{L^1(Y)} \leq C$$

holds for every (μ,ω)-atom a, see Theorem 2.4. Assume then that: $\operatorname{supp} a \subseteq B(y_0,r) =: B$, $\|a\|_\infty \leq \mu(B)^{-1}$, and $\int a\, \omega d\mu = 0$. As always, the analysis on $2B = B(y_0,2r)$ follows by the Cauchy-Schwarz inequality and boundedness of $\mathcal{M}$ and $m(A)$ on $L^2(Y)$. More precisely,

$$\begin{aligned}
\|\mathcal{M} m(A)a\|_{L^1(2B)} &\leq \mu(2B)^{1/2} \|\mathcal{M} m(A)a\|_{L^2(Y)} \\
&\leq C\mu(B)^{1/2} \|a(x)\|_{L^2(Y)} \leq C.
\end{aligned}$$

Therefore, it is enough to prove that

$$\|\mathcal{M} m(A)a\|_{L^1((2B)^c)} \leq C. \tag{2.7}$$

Let $\eta \in C_c^\infty(2^{-1},2)$ be a fixed function such that $\sum_{j\in\mathbb{Z}} \eta(2^{-j}\lambda) = 1$ for all $\lambda \in (0,\infty)$. By using this partition of unity, we decompose m as

$$m(\lambda) = \sum_{j\in\mathbb{Z}} \eta(2^{-j}\lambda) m(\lambda) = \sum_{j\in\mathbb{Z}} m_j(\lambda).$$

Fix $N \in \mathbb{Z}$ such that $2^{-N} \leq r^2 < 2^{-N+1}$. Then

$$\|\mathcal{M} m(A)a\|_{L^1((2B)^c)} \leq \sum_{j\in\mathbb{Z}} \left\| \mathcal{M} m_j(A)a \right\|_{L^1((2B)^c)} = \sum_{j\geq N} \cdots + \sum_{j<N} \cdots = S_1 + S_2.$$

Denote $m_{j,t}(\lambda) = \exp(-t\lambda)m_j(\lambda)$ and let $M_{j,t}(x, y)$ be the kernel of $m_{j,t}(A) = \mathbf{P}_t m_j(A)$. Obviously, $\operatorname{supp} m_{j,t} \subseteq [2^{j-1}, 2^{j+1}]$ and applying (2.1) we obtain that

$$\begin{aligned} S_1 &\leq \sum_{j\geq N} \int_{(2B)^c} \int_B \sup_{t>0} |M_{j,t}(x, y)||a(y)|d\mu(y)d\mu(x) \\ &\leq \sum_{j\geq N} \int_B |a(y)| \int_{B^c} \sup_{t>0} |M_{j,t}(x, y)|d\mu(x)d\mu(y) \\ &\leq C \|a\|_{L^1(Y)} \sum_{j\geq N} (1 + 2^{j/2}r)^{-\delta} \left\| \eta(\cdot)m(2^j\cdot) \right\|_{W^{q,\beta}(\mathbb{R})} \\ &\leq C \sup_{t>0} \|\eta(\cdot)m(t\cdot)\|_{W^{q,\beta}(\mathbb{R})} \leq C. \end{aligned}$$

If $y \in B$ and $j < N$, then $\rho(y, y_0) < r < 2^{-j/2}$ and we can apply (2.2) for the kernel $M_{j,t}$ with $R = 2^j$. Using the cancellation condition of a,

$$\begin{aligned} S_2 &\leq \sum_{j<N} \int_{(2B)^c} \sup_{t>0} \left| \int_B M_{j,t}(x, y)a(y)d\mu(y) \right| d\mu(x) \\ &= \sum_{j<N} \int_{(2B)^c} \sup_{t>0} \left| \int_B \left(\frac{M_{j,t}(x, y)}{\omega(y)} - \frac{M_{j,t}(x, y_0)}{\omega(y_0)} \right) a(y)\omega(y)d\mu(y) \right| d\mu(x) \\ &\leq \sum_{j<N} \int_B |a(y)| \int_{B(y,r)^c} \sup_{t>0} \left| \frac{M_{j,t}(x, y)}{\omega(y)} - \frac{M_{j,t}(x, y_0)}{\omega(y_0)} \right| d\mu(x)\, \omega(y)d\mu(y) \\ &\leq C \sum_{j<N} 2^{\frac{j\gamma}{2}} \int_B |a(y)|\rho(y, y_0)^\gamma d\mu(y) \left\| \eta(\cdot)m(2^j\cdot) \right\|_{W^{q,\beta}(\mathbb{R})} \\ &\leq C \sup_{t>0} \|\eta(\cdot)m(t\cdot)\|_{W^{q,\beta}(\mathbb{R})} r^\gamma \sum_{j<N} 2^{\frac{j\gamma}{2}} \leq C. \end{aligned}$$

This finishes the proof of (2.7) and Theorem D.

3 The Multidimensional Bessel Operator

In this Section we turn back to the analysis related to B and prove the results stated in Sect. 1.

3.1 The Hankel Transform

Recall that $N \in \mathbb{N}$ and $\alpha_j > -1$ for $j = 1, \ldots, N$. For $x, \xi \in X = (0, \infty)^N$ denote $\varphi_\alpha(x\xi) = \varphi_1(x_1\xi_1) \cdot \ldots \cdot \varphi_N(x_N\xi_N)$, where

$$\varphi_j(z) = 2^{(\alpha_j-1)/2}\Gamma\left((\alpha_j + 1)/2\right) z^{-(\alpha_j-1)/2} J_{(\alpha_j-1)/2}(z), \quad z > 0.$$

Here J_τ denotes the Bessel function of the first kind. By the asymptotics of J_τ one has

$$\left|\varphi_j(z)\right| \leq C(1+z)^{-\alpha_j/2}, \quad z > 0. \tag{3.1}$$

The Hankel transform is defined by

$$H_\alpha f(\xi) = \int_X f(x)\varphi_\alpha(x\xi)d\nu(x), \quad \xi \in X, \tag{3.2}$$

As we have already mentioned, $\varphi_j \in L^\infty$ if and only if $\alpha_j \geq 0$. Nevertheless, it is known that H_α always extends uniquely to an isometric isomorphism on $L^2(X)$, see [5] and [8, Lem. 2.7]. The multipliers $m(B)$ and H_α are related in the same way, as $m(-\Delta)$ and the Fourier transform on $\mathbb{R}^D$. In particular, if

$$n(\lambda) = m(|\lambda|^2), \qquad \lambda \in X,$$

then m is radial and

$$m(B)f = H_\alpha(n \cdot H_\alpha). \tag{3.3}$$

3.2 (P_2) for Multidimensional Bessel Operator

Let us first recall that $\mathbf{T}_t$ satisfies (G) and, obviously, X satisfies (Y). Therefore, Theorem A follows from Theorems 2.1 and D provided that (P_q) holds with $q = 2$, which we now prove. The case $N = 1$ follows by similar and simpler argument, thus we shall concentrate on $N \geq 2$. Let $k \in \{1, \ldots, N-1\}$ and $c_j < 2^{-N}$ for $j = 1, \ldots, k$. Define the sets

$$S_{c_1,\ldots,c_k} = \left\{x \in X \ : \ 1/2 < |x|^2 < 2 \text{ and } x_j < c_j \text{ for } j = 1, \ldots, k\right\}.$$

Lemma 3.1 *Suppose that* $\operatorname{supp} m \subset [R/2, 2R]$, $N \geq 2$, $k \leq N-1$, *and* $c_j < 2^{-N}$ *for* $j = 1, 2, \ldots, k$. *Then there exists* $C > 0$ *such that*

$$\int_{S_{c_1,\ldots,c_k}} \left|m\left(R|x|^2\right)\right|^2 x_1^{\alpha_1} \ldots x_k^{\alpha_k}\, dx_1 \ldots dx_N \leq C c_1^{\alpha_1+1} \ldots c_k^{\alpha_k+1} \, \|m(R\cdot)\|_{L^2(\mathbb{R})}^2 .$$

Proof Introduce the spherical coordinates $(r, \theta_1, \ldots, \theta_{N-1})$ on $\mathbb{R}^N$, namely

$$\begin{cases} x_1 = r\sin(\theta_1), \\ x_i = r\sin(\theta_i)\Pi_{j=1}^{i-1}\cos(\theta_j), \qquad \text{for } i = 2, 3, \ldots, N-1, \\ x_N = r\Pi_{j=1}^{N-1}\cos(\theta_j) \\ dx_1 \ldots dx_N = r^{N-1}\prod_{j=1}^{N-2}\cos^{N-1-j}(\theta_j)dr\, d\theta_1 \ldots d\theta_{N-1}. \end{cases}$$

Since $x \in X$, then $\theta_j \in (0, \pi/2)$ for $j = 1, \dots, N-1$. We claim that if $x \in S_{c_1,\dots,c_k}$, then

$$\sin(\theta_j) < 2^{-N/2} \leq 2^{-1/2} \tag{3.4}$$

for $j = 1, \dots, k$. Obviously, if $\sin(\beta) < 2^{-1/2}$ for $\beta \in (0, \pi/2)$, then $(\cos\beta)^{-1} < 2^{1/2}$. Observe that $2^{-1/2} < r < 2^{1/2}$, since $\operatorname{supp} m \subseteq [R/2, 2R]$. Therefore, (3.4) follows easily by induction, i.e. for $i = 1, \dots, k$,

$$\sin\theta_i = x_i r^{-1}(\cos\theta_1)^{-1} \dots (\cos\theta_{i-1})^{-1} \leq 2^{-N} 2^{i/2} \leq 2^{-N/2}.$$

Denote $S = S_{c_1,\dots,c_k}$. As a consequence of (3.4) we have that $\sin\theta_i \simeq \theta_i$ and $\cos\theta_i \simeq C$ for $i = 1, \dots, k$. Using this, $x_i^{\alpha_i} \simeq r^{\alpha_i}\theta_i^{\alpha_i}$ for $i = 1, \dots, k$ and

$$\begin{aligned}
&\int_S \left| m(R|x|^2) \right|^2 x_1^{\alpha_1} \dots x_k^{\alpha_k} \, dx_1 \dots dx_N \\
&\leq C \int_S \left| m(Rr^2) \right|^2 r^{N-1+\alpha_1+\dots+\alpha_k} \theta_1^{\alpha_1} \dots \theta_k^{\alpha_k} dr d\theta_1 \dots d\theta_{N-1} \\
&\leq C \int_0^{cc_1} \theta_1^{\alpha_1} \, d\theta_1 \cdot \dots \cdot \int_0^{cc_k} \theta_k^{\alpha_k} \, d\theta_k \cdot \int_{1/2<r^2<2} \left| m\left(Rr^2\right) \right|^2 dr \\
&\leq C c_1^{\alpha_1+1} \cdot \dots \cdot c_k^{\alpha_k+1} \cdot \| m(R\cdot) \|_{L^2(\mathbb{R})}^2 .
\end{aligned}$$

□

Proposition 3.2 *Assume that $N \in \mathbb{N}$ and $\alpha_j > -1$ for $j = 1, .., N$. Then (P_q) holds for B with $q = 2$.*

Proof In the proof we consider only the case $N \geq 2$. Let $q = 2$ and suppose that m is supported in $[R/2, 2R]$ for some $R > 0$. Notice that by (3.3) and (3.2) we have

$$\begin{aligned}
m(B) f(x) &= \int_X f(y) \int_X n(\xi) \varphi_\alpha(y\xi) \varphi_\alpha(x\xi) d\nu(\xi) \, d\nu(y) \\
&= \int_X f(y) H_\alpha \left(n(\cdot)\varphi_\alpha(y\cdot) \right)(x) \, d\nu(y)
\end{aligned}$$

and the kernel associated with $m(B)$ has the form

$$K_{m(B)}(x, y) = H_\alpha \left(n(\cdot)\varphi_\alpha(y\cdot) \right)(x).$$

Therefore, by the Plancherel identity for H_α, (P_2) is equivalent to

$$\int_{R/2<|x|^2<2R} \left| m(|x|^2) \right|^2 |\varphi_\alpha(xy)|^2 \, d\nu(x) \leq C \nu \left(B\left(y, R^{-1/2}\right) \right)^{-1} \| m(R\cdot) \|_{L^2(\mathbb{R})}^2 . \tag{3.5}$$

For each $i = 1, \dots, N$ we consider four cases:

C1. $x_i < 2^{-N}, \quad \sqrt{R} y_i > 2^N, \quad \sqrt{R} y_i x_i < 1,$

C2. $x_i < 2^{-N}$, $\sqrt{R}y_i \le 2^N$,
C3. $x_i < 2^{-N}$, $\sqrt{R}y_i > 2^N$, $\sqrt{R}y_i x_i \ge 1$,
C4. $2^{-N} \le x_i \le \sqrt{2}$.

Divide the set $\{x \in X \ : \ 1/2 < |x|^2 < 2\}$ into several disjoint regions using the cases above. Without loss of generality we may consider the set S of points $x \in X$ such that:

- x_i satisfies **C1.** for $i = 1, \ldots, k_1$,
- x_i satisfies **C2.** for $i = k_1 + 1, \ldots, k_2$,
- x_i satisfies **C3.** for $i = k_2 + 1, \ldots, k_3$,
- x_i satisfies **C4.** for $i = k_3 + 1, \ldots, N$,

where $0 \le k_1 \le k_2 \le k_3 < N$. The fact that $k_3 < N$ is implied by $|x|^2 > 1/2$. Notice that it may happen that S is empty. Recall that $\nu(B(y,r)) \simeq \Pi_{j=1}^N \nu_j(B(y_j, r))$, where $d\nu_j(x_j) = x_j^{\alpha_j} dx_j$ is the one-dimensional measure, and

$$\nu_j\left(B\left(y_j, R^{-1/2}\right)\right)^{-1} \simeq R^{(\alpha_j+1)/2}\left(1+\sqrt{R}y_j\right)^{-\alpha_j}.$$

Denote $d_{gl} = N + \alpha_1 + \cdots + \alpha_N$. Using (3.1) and Lemma 3.1 with $k = k_2$, we have

$$\begin{aligned}
&\int_{R/2<|x|^2<2R} \left|m\left(|x|^2\right)\right|^2 |\varphi_\alpha(xy)|^2 \, d\nu(x) \\
&\le C \sum_S R^{\frac{d_{gl}}{2}} \int_S \left|m\left(R|x|^2\right)\right|^2 \prod_{j=1}^N \left(x_j^{-1} + \sqrt{R}y_j\right)^{-\alpha_j} dx \\
&\le C \sum_S R^{\frac{d_{gl}}{2}} \int_S \left|m\left(R|x|^2\right)\right|^2 x_1^{\alpha_1} \ldots x_{k_2}^{\alpha_{k_2}} \prod_{j=k_2+1}^N \left(1+\sqrt{R}y_j\right)^{-\alpha_j} dx_1 \ldots dx_N \\
&\le C \sum_S R^{\frac{d_{gl}}{2}} \prod_{i=1}^{k_1} \left(\sqrt{R}y_i\right)^{-\alpha_i - 1} \prod_{j=k_2+1}^N \left(1+\sqrt{R}y_j\right)^{-\alpha_j} \|m(R\cdot)\|_{L^2(\mathbb{R})}^2 \\
&\le C \prod_{i=1}^{k_1} R^{(\alpha_i+1)/2}\left(\sqrt{R}y_i\right)^{-\alpha_i} \prod_{k=k_1+1}^{k_2} R^{(\alpha_k+1)/2} \prod_{j=k_2+1}^N R^{(\alpha_j+1)/2} \\
&\quad \left(1+\sqrt{R}y_j\right)^{-\alpha_j} \|m(R\cdot)\|_{L^2(\mathbb{R})}^2 \\
&\le C\nu\left(B\left(y, R^{-1/2}\right)\right)^{-1} \|m(R\cdot)\|_{L^2(\mathbb{R})}^2 .
\end{aligned}$$

□

3.3 Imaginary Powers of *B*

In this subsection we prove Lemma 1.1 and Theorems B and C. From now on we consider one-dimensional Bessel operator, i.e. $N = 1$, $X = (0, \infty)$, $\alpha > -1$, and $d\nu(x) = x^\alpha \, dx$.

Let us start this section by recalling well-known asymptotics of the Bessel function I_τ [28,37], i.e.

$$I_\tau(x) = \Gamma\left(\frac{\tau+1}{2}\right)^{-1} \left(\frac{x}{2}\right)^\tau + O\left(x^{\tau+1}\right), x \sim 0, \tag{3.6}$$

$$I_\tau(x) = (2\pi x)^{-1/2} e^x \left(1 + O(x^{-1})\right), x \sim \infty. \tag{3.7}$$

Now we provide a short argument for (1.5). In [6, Sec. 4.3] it is proved that B^{ib} is associated with the kernel

$$-\Gamma(-ib+1)^{-1} \int_0^\infty t^{ib} \partial_t T_t(x, y) \, dt$$

in the sense as in (1.5) (let us notice that in [6] only positive values of α's are considered, but the proof works for $\alpha_j > -1$ as well). By integrating by parts,

$$\begin{aligned} & -\Gamma(-ib+1)^{-1} \int_0^\infty t^{-ib} \partial_t T_t(x, y) \, dt \\ & = -\Gamma(-ib+1)^{-1} \lim_{\varepsilon \to 0} \left(\varepsilon^{ib} T_{\varepsilon^{-1}}(x, y) - \varepsilon^{-ib} T_\varepsilon(x, y)\right) \\ & + \Gamma(-ib)^{-1} \int_0^\infty t^{-ib} T_t(x, y) \frac{dt}{t} = K_b(x, y). \end{aligned}$$

Proof of Lemma 1.1 Let us first notice that for $\kappa \in \mathbb{R}$ and $c, M > 0$, there exists $C = C(\kappa, c, M)$ such that

$$\int_{cz}^\infty t^\kappa \exp\left(-\frac{t}{4}\right) \frac{dt}{t} \le C z^{-M}, \qquad z \ge 1. \tag{3.8}$$

Using (1.4) and (1.2) one obtains

$$\begin{aligned} 2\Gamma(-ib) K_b(x, y) &= \int_0^\infty t^{-ib-1} (xy)^{-(\alpha-1)/2} I_{(\alpha-1)/2}\left(\frac{xy}{2t}\right) \exp\left(-\frac{x^2+y^2}{4t}\right) \frac{dt}{t} \\ &= \int_0^{xy} \cdots + \int_{xy}^\infty \cdots = A_1 + A_2. \end{aligned}$$

Denote $\chi_{loc}(x, y) = \chi_{\{y/2<x<2y\}}(x, y)$ and $\chi_{glob}(x, y) = 1 - \chi_{loc}(x, y), x, y \in X$. In the proof below all expressions denoted by R_k shall be parts of the kernel $R_b(x, y)$.

Using (3.7), we write $A_1 = A_{1,1} + R_1$, where

$$A_{1,1} = \pi^{-1/2} \int_0^{xy} t^{-ib-1/2} (xy)^{-\alpha/2} \exp\left(-\frac{|x-y|^2}{4t}\right) \frac{dt}{t}$$

and

$$\begin{aligned} |R_1| = &\left| \int_0^{xy} t^{-ib-1} (xy)^{-(\alpha-1)/2} \exp\left(-\frac{x^2+y^2}{4t}\right) \right. \\ &\left. \left(I_{(\alpha-1)/2}\left(\frac{xy}{2t}\right) - \left(\frac{\pi xy}{t}\right)^{-1/2} \exp\left(\frac{xy}{2t}\right)\right) \frac{dt}{t} \right| \\ &\le C \int_0^{xy} t^{1/2} (xy)^{-\alpha/2-1} \exp\left(-\frac{|x-y|^2}{4t}\right) \frac{dt}{t} \\ &= C|x-y|(xy)^{-\alpha/2-1} \int_{\frac{|x-y|^2}{xy}}^{\infty} t^{-1/2} e^{-t/4} \frac{dt}{t} \\ &\le Cxy(x+y)^{-\alpha-3}. \end{aligned}$$

In the last inequality we have used (3.8). Denoting

$$\begin{aligned} R_2 &= \chi_{glob}(x,y) A_{1,1}, \\ R_3 &= \pi^{-1/2} \chi_{loc}(x,y) \int_{xy}^{\infty} t^{-ib-1/2} (xy)^{-\alpha/2} \exp\left(-\frac{|x-y|^2}{4t}\right) \frac{dt}{t} \end{aligned}$$

we have

$$\begin{aligned} A_{1,1} - R_2 + R_3 = &\pi^{-1/2} \chi_{loc}(x,y) \left(\int_0^{\infty} t^{-ib-1/2} (xy)^{-\alpha/2} \exp\left(-\frac{|x-y|^2}{4t}\right) \frac{dt}{t} \right) \\ = &\pi^{-1/2} 2^{2bi+1} \Gamma(ib+1/2)\, \chi_{loc}(x,y)(xy)^{-\alpha/2} |x-y|^{-2ib-1}. \end{aligned}$$

Notice that $A_{1,1}$ is one of the terms from (1.6). Next, by (3.8),

$$\begin{aligned} |R_2| &\le C\chi_{glob}(x,y)\, |x-y|^{-1} (xy)^{-\alpha/2} \int_{\frac{|x-y|^2}{xy}}^{\infty} t^{1/2} e^{-t/4} \frac{dt}{t} \\ &\le Cxy(x+y)^{-\alpha-3}, \\ |R_3| &\le C\chi_{loc}(x,y) x^{-\alpha} |x-y|^{-1} \int_0^{\frac{|x-y|^2}{xy}} t^{1/2} \frac{dt}{t} \\ &\simeq C\chi_{loc}(x,y) x^{-\alpha-1}. \end{aligned}$$

Now let us turn to study A_2. Denote $c_\alpha = 4^{-(\alpha-1)/2}\Gamma((\alpha+1)/4)^{-1}$. Then, by using (3.6),

$$A_2 = c_\alpha \int_{xy}^{\infty} t^{-ib-(\alpha+1)/2} \exp\left(-\frac{x^2+y^2}{4t}\right)\frac{dt}{t} + R_4 = A_{2,1} + R_4$$

where, by (3.8),

$$\begin{aligned}|R_4| &= \left| \int_{xy}^{\infty} t^{-ib-1}(xy)^{-(\alpha-1)/2} \exp\left(-\frac{x^2+y^2}{4t}\right) \right.\\ &\quad \left. \left(I_{(\alpha-1)/2}\left(\frac{xy}{2t}\right) - \Gamma\left(\frac{\alpha+1}{4}\right)^{-1}\left(\frac{xy}{4t}\right)^{(\alpha-1)/2}\right)\frac{dt}{t}\right| \\ &\le Cxy \int_{xy}^{\infty} t^{-(\alpha+3)/2} \exp\left(-\frac{x^2+y^2}{4t}\right)\frac{dt}{t} \\ &\le Cxy(x^2+y^2)^{-(\alpha+3)/2} \simeq Cxy(x+y)^{-\alpha-3}.\end{aligned}$$

Moreover,

$$\begin{aligned}A_{2,1} + R_5 &= c_\alpha \int_0^{\infty} t^{-ib-(\alpha+1)/2} \exp\left(-\frac{x^2+y^2}{4t}\right)\frac{dt}{t} \\ &= c_\alpha 4^{ib+(\alpha+1)/2}\Gamma\left(ib+(\alpha+1)/2\right)\left(x^2+y^2\right)^{-ib-(\alpha+1)/2},\end{aligned}$$

where

$$\begin{aligned}|R_5| &= \left| c_\alpha \int_0^{xy} t^{-ib-(\alpha+1)/2} \exp\left(-\frac{x^2+y^2}{4t}\right)\frac{dt}{t}\right| \\ &\le C(x^2+y^2)^{-(\alpha+1)/2} \int_{\frac{x^2+y^2}{xy}}^{\infty} t^{(\alpha+1)/2} e^{-t/4}\frac{dt}{t} \\ &\le Cxy(x+y)^{-\alpha-3}.\end{aligned}$$

□

***Proof of Theorem B* for $\boldsymbol{\alpha < 0}$.** Let $|b| > 1$ and $\varepsilon \in (0, 10^{-1})$ (to be fixed later on). Denote $I = [1, 1+\varepsilon]$ and $S = [1+3\varepsilon, 2]$. Put $f_\varepsilon(x) = \varepsilon^{-1}\chi_I(x)x^{-\alpha}$, so that $\|f_\varepsilon\|_{L^1(X)} = 1$. If $x \in S$, by Lemma 1.1 and the triangle inequality,

$$\begin{aligned}\left|B^{ib} f_\varepsilon(x)\right| &\le |c_2(b)|\, x^{-\alpha/2}|x-1|^{-1}\\ &\quad + |c_2(b)| \left| \int_I \left((xy)^{-\alpha/2}|x-y|^{-2ib-1} - x^{-\alpha/2}|x-1|^{-2ib-1}\right) f_\varepsilon(y)\, d\nu(y)\right|\\ &\quad + |c_1(b)| \left| \int_I \left(x^2+y^2\right)^{-ib-(\alpha+1)/2} f_\varepsilon(y)\, d\nu(y)\right| \\ &\quad + |c_3(b)| \left| \int_I R_b(x,y) f_\varepsilon(y)\, d\nu(y)\right|\\ &= |c_2(b)|\, x^{-\alpha/2}|x-1|^{-1} + \Lambda_1 + \Lambda_2 + \Lambda_3.\end{aligned} \tag{3.9}$$

Observe that for $x \in S$ and $y \in I$ we have $|x-y| \simeq |x-1|$ and $x \simeq y \simeq 1$. By using the Mean Value Theorem for the function $y \mapsto y^{-\alpha/2}|x-y|^{-2ib-1}$,

$$\begin{aligned}\Lambda_1 &\le C\,|c_2(b)|\,\varepsilon^{-1} \int_1^{1+\varepsilon} |b||y-1||x-1|^{-2}\, dy \le C\varepsilon\, |bc_2(b)|\, |x-1|^{-2},\\ \Lambda_2 &\le C\,|c_1(b)|\,\varepsilon^{-1} \int_1^{1+\varepsilon} (x^2+y^2)^{-(1+\alpha)/2}\, dy \simeq C\,|c_1(b)|\,,\\ \Lambda_3 &\le C\,|c_3(b)|\,\varepsilon^{-1} \int_1^{1+\varepsilon} xy(x+y)^{-\alpha-3}\, dy \simeq C\,|c_3(b)|\,.\end{aligned} \tag{3.10}$$

Fix $|b| \ge 1$ and λ such that $\lambda > \max(\Lambda_2, \Lambda_3, |bc_2(b)|)$. Recall that $x^{-\alpha/2} \ge 1$ for $x \in S$, so that for ε small enough

$$\begin{aligned}\nu\left\{x \in S : |c_2(b)|\, x^{-\alpha/2}|x-1|^{-1} > 4\lambda\right\} &\ge \nu\left\{x \in S : |c_2(b)|\, |x-1|^{-1} > 4\lambda\right\}\\ &= \int_{1+3\varepsilon}^{1+|c_2(b)|/(4\lambda)} x^\alpha\, dx\\ &\ge C|c_2(b)|\,/(4\lambda)\end{aligned} \tag{3.11}$$

and

$$\begin{aligned}\nu\left\{x \in S : |\Lambda_1| > \lambda\right\} &\le \nu\left\{x \in S : C\varepsilon\, |bc_2(b)|\, |x-1|^{-2} > \lambda\right\}\\ &\le \int_{1+3\varepsilon}^{1+C(\varepsilon\lambda^{-1}|bc_2(b)|)^{1/2}} x^\alpha\, dx\\ &\le C\left(\varepsilon\lambda^{-1}\,|bc_2(b)|\right)^{1/2}\\ &\le C\varepsilon^{1/2}.\end{aligned} \tag{3.12}$$

Hence, using (3.9)–(3.12) and (1.7) we get

$$\begin{aligned}\left\|B^{ib} f_\varepsilon\right\|_{L^{1,\infty}(X)} &\geq \lambda\nu\left\{x \in S : \left|B^{ib} f_\varepsilon(x)\right| > \lambda\right\}\\ &\geq \lambda\nu\left\{x \in S : |c_2(b)|\, x^{-\alpha/2}|x-1|^{-1} > 4\lambda\right\}\\ &\quad - \lambda\nu\{x \in S : |\Lambda_1| > \lambda\} - \underbrace{\lambda\nu\{x \in S : |\Lambda_2| > \lambda\}}_{=0}\\ &\quad - \underbrace{\lambda\nu\{x \in S : |\Lambda_3| > \lambda\}}_{=0}\\ &\geq C|c_2(b)| - C\lambda\varepsilon^{1/2} \geq C|c_2(b)| \simeq |b|^{1/2} = |b|^{d/2}.\end{aligned}$$

□

Turning to the case $\alpha > 0$ we could also use Lemma 1.1. In this case, the summand with $c_1(b)$ would play the first role. An alternative proof that we shall present here uses integral representation of the modified Bessel function. The same will be used in the proof of Theorem C. It is known that for $\alpha > 0$

$$I_{(\alpha-1)/2}(z) = \left(\Gamma(\alpha/2)^{-1}\sqrt{\pi}\right)^{-1}\left(\frac{z}{2}\right)^{(\alpha-1)/2}\int_{-1}^{1} e^{-zs}(1-s^2)^{\alpha/2-1}ds, \quad z > 0, \tag{3.13}$$

see [37, Ch. 6]. Therefore, for $\alpha > 0$, using (1.4), (1.2), and (3.13) we obtain

$$\begin{aligned}K_b(x,y) &= (2\Gamma(-ib))^{-1}\int_0^\infty t^{-ib-1}(xy)^{-(\alpha-1)/2} I_{(\alpha-1)/2}\left(\frac{xy}{2t}\right)\\ &\quad \exp\left(-\frac{x^2+y^2}{4t}\right)\frac{dt}{t}\\ &= \left(2^\alpha\Gamma(-ib)\Gamma(\alpha/2)\sqrt{\pi}\right)^{-1}\int_{-1}^{1}\int_0^\infty t^{-ib-(\alpha+1)/2}\\ &\quad \exp\left(-\frac{x^2+y^2+2xys}{4t}\right)\frac{dt}{t}(1-s^2)^{\frac{\alpha}{2}-1}ds\\ &= \frac{2^{2ib+1}\Gamma(ib+(\alpha+1)/2)}{\Gamma(-ib)\Gamma(\alpha/2)\sqrt{\pi}}\int_{-1}^{1}\left(x^2+y^2+2xys\right)^{-ib-(\alpha+1)/2}\\ &\quad \left(1-s^2\right)^{\alpha/2-1} ds\\ &= C_\alpha c_1(b)\int_{-1}^{1}\left(x^2+y^2+2xys\right)^{-ib-(\alpha+1)/2}\left(1-s^2\right)^{\alpha/2-1} ds,\end{aligned} \tag{3.14}$$

where $C_\alpha = \pi^{-1/2}\Gamma(\alpha/2)^{-1}\Gamma((\alpha+1)/4)$.

Proof of Theorem B **for $\alpha > 0$.** Let $|b| > 1$, $\varepsilon \in (0, 10^{-1})$, and $f_\varepsilon(x) = x^{-\alpha}\varepsilon^{-1}\chi_{[\varepsilon,2\varepsilon]}(x)$. Similarly as in (3.9), using (3.14) and the Mean Value Theorem, for $x > 3\varepsilon$ we have

$$\begin{aligned}
\left|B^{-ib} f_\varepsilon(x)\right| &\le \left|\int_\varepsilon^{2\varepsilon} K_b(x,0) f_\varepsilon(y)\, d\nu(y)\right| \\
&\quad + \left|\int_\varepsilon^{2\varepsilon} (K_b(x,0) - K_b(x,y))\, f_\varepsilon(y)\, d\nu(y)\right| \\
&\le C|c_1(b)|\varepsilon^{-1} \left\{ \left|\int_\varepsilon^{2\varepsilon}\int_{-1}^{1} \left(1-s^2\right)^{\alpha/2-1} x^{-2ib-(\alpha+1)}\, ds\, dy\right| \right. \\
&\quad + \int_\varepsilon^{2\varepsilon}\int_{-1}^{1} \left(1-s^2\right)^{\alpha/2-1} \\
&\quad \left. \left|\left(x^2+y^2+2sxy\right)^{-ib-\frac{\alpha+1}{2}} - x^{-2ib-(\alpha+1)}\right| ds\, dy \right\} \\
&\le C|c_1(b)| \left(x^{-\alpha-1} + \varepsilon^{-1} \int_{-1}^{1} \left(1-s^2\right)^{\alpha/2-1} \right. \\
&\quad \left. \int_\varepsilon^{2\varepsilon} |b| \left|y^2+2sxy\right| x^{-\alpha-3}\, dy\, ds \right) \\
&\le C|c_1(b)| \left(x^{-\alpha-1} + \varepsilon\, |b|\, x^{-\alpha-2} \right).
\end{aligned} \tag{3.15}$$

Let us fix $|b| > 1$ and $\lambda > |bc_1(b)|$. For all ε small enough, we get

$$\nu\left\{x > 3\varepsilon : C\,|c_1(b)|\, x^{-\alpha-1} > 2\lambda\right\} = \int_{3\varepsilon}^{C(|c_1(b)|/\lambda)^{1/(1+\alpha)}} x^\alpha\, dx \ge C\,|c_1(b)|\,/\lambda,$$

and

$$\begin{aligned}
\nu\left\{x > 3\varepsilon : C\,|bc_1(b)|\,\varepsilon x^{-\alpha-2} > \lambda\right\} &\le \int_0^{C(|bc_1(b)|\varepsilon/\lambda)^{1/(\alpha+2)}} x^\alpha\, dx \\
&= C\left(|bc_1(b)|\,\varepsilon/\lambda\right)^{(\alpha+1)/(\alpha+2)} \\
&\le C\varepsilon^{(1+\alpha)/(\alpha+2)}.
\end{aligned}$$

Therefore, by choosing a proper ε, we obtain

$$\begin{aligned}
\left\|B^{ib} f_\varepsilon\right\|_{L^{1,\infty}(X)} &\ge \lambda\nu\left\{x \in X : \left|B^{ib} f_\varepsilon(x)\right| > \lambda\right\} \\
&\ge \lambda\nu\left\{x > 3\varepsilon : C|c_1(b)|x^{-\alpha-1} > 2\lambda\right\} \\
&\quad - \lambda\nu\left\{x > 3\varepsilon : C|c_1(b)|\varepsilon x^{-\alpha-2} > \lambda\right\} \\
&\ge C|c_1(b)| - C\lambda\varepsilon^{(\alpha+1)/(\alpha+2)} \ge C|c_1(b)| \simeq |b|^{(\alpha+1)/2} = |b|^{d/2}.
\end{aligned}$$

□

Proof of Theorem C Set $\alpha > 0$, $p \in (1,2)$, $|b| > 1$, $\varepsilon \in (0, 10^{-1})$, $\delta > 1$, and a function $f \in L^p(X)$ such that $\operatorname{supp} f \subseteq (0, \varepsilon)$, and $f \geq 0$. Similarly as in (3.15), using (3.14) and Corollary 4.2,

$$\begin{aligned}
\left\|B^{ib} f\right\|_{L^p(X)}^p &\geq \int_\delta^\infty \left|\int_X (K_b(x,0) - (K_b(x,0) - K_b(x,y)) f(y)\, d\nu(y)\right|^p d\nu(x) \\
&\geq C \|f\|_{L^1(X)}^p \int_\delta^\infty |K_b(x,0)|^p \, d\nu(x) \\
&\quad - C \int_\delta^\infty \left|\int_X (K_b(x,y) - K_b(x,0)) f(y)\, d\nu(y)\right|^p d\nu(x) \\
&\geq C \|f\|_{L^1(X)}^p |b|^{p(\alpha+1)/2} \left(\int_\delta^\infty x^{-p(\alpha+1)+\alpha} \, dx - \int_\delta^\infty \varepsilon^p |b|^p x^{-p(\alpha+2)+\alpha} \, dx \right) \\
&\geq C \|f\|_{L^1(X)}^p |b|^{p(\alpha+1)/2} \delta^{(\alpha+1)(1-p)} \left(1 - \varepsilon^p |b|^p \delta^{-p}\right).
\end{aligned}$$

Now we take $\delta = |b|$ and fix ε small enough, independent of b, getting

$$\left\|B^{ib} f\right\|_{L^p(X)} \geq C_p |b|^{\frac{(\alpha+1)(2-p)}{2p}} \|f\|_{L^1(X)} \geq C_{p,\varepsilon} |b|^{\frac{d}{2}\frac{2-p}{p}} \|f\|_{L^p(X)}.$$

□

4 Appendix: Gamma Function Estimate

Lemma 4.1 *Let $a + bi \in \mathbb{C}$. For $a \geq 0$ fixed and all $|b| \geq 1$ we have*

$$|\Gamma(a+bi)| \simeq |b|^{a-1/2} \exp\left(-\frac{\pi b}{2}\right).$$

The result above is known. It is a consequence of the Stirling's Formula, see [1, Ch. 6]. For the convenience of the reader we present a short proof.

Proof Using the reflection formula

$$\Gamma(1-z)\Gamma(z) = \pi / \sin(\pi z), \tag{4.1}$$

and the recursion identity

$$z\Gamma(z) = \Gamma(z+1), \tag{4.2}$$

we have that $|1 - ib|\,|\Gamma(ib)|^2 = |\Gamma(ib)\Gamma(1-ib)| = |\pi/\sin(\pi ib)| \simeq \exp(-\pi|b|)$ for $|b| \geq 1$. Thus,

$$|\Gamma(ib)| \simeq |b|^{-1/2} \exp\left(-\frac{\pi|b|}{2}\right), \qquad |b| \geq 1. \tag{4.3}$$

Denote $S = \{z \in \mathbb{C} : 1 \le \text{Re}(z) \le 2, |\text{Im}(z)| \ge 1\}$ and define a holomorphic function

$$F(z) = \Gamma(z) z^{-z+1/2}, \quad z \in S.$$

Now, we claim that $|F(z)| \le C$ if $z \in \partial S$. This is clear for $z = a \pm i$, $a \in [1, 2]$. For $z = 1 + ib$, $|b| \ge 1$, we use (4.2) and (4.3) getting

$$\begin{aligned} |F(1+ib)| &= |\Gamma(1+ib)| \left|(1+ib)^{-1/2-ib}\right| = |b|\, |\Gamma(ib)|\, (1+b^2)^{-1/4} e^{b \operatorname{arctg} b} \\ &\le C |b|^{1/2} e^{-\pi|b|/2} |b|^{-1/2} e^{b \operatorname{arctg}(b)} \le C. \end{aligned}$$

Similarly we show boundedness of F for $z = 2 + bi$, $|b| \ge 1$.

Observe that $|F(z)| \le |\Gamma(z)||z|^{|-z+1/2|} \le C e^{c|z|^2}$ for $z \in S$. Hence, applying the Phragmén-Lindelöf principle, we obtain that $|F(z)| \le C$ for $z \in S$. Therefore, for a fixed $a \in [1, 2]$ and $|b| \ge 1$ we have

$$\begin{aligned} |\Gamma(a+bi)| &\le C \left|(a+bi)^{a-1/2+bi}\right| = \left(a^2+b^2\right)^{(2a-1)/4} \cdot e^{-b \operatorname{arctg}(b/a)} \\ &\simeq C|b|^{a-1/2} e^{-\pi|b|/2}. \end{aligned} \tag{4.4}$$

This is the desired estimate from above for $a \in [1, 2]$. We extend this for all $a \in [0, \infty)$ by using (4.2). Then, by (4.1), we get estimate from below for $a \in [0, 1]$, and extend this for $a \in [0, \infty)$ using (4.2) once more. □

Corollary 4.2 *For fixed* $a_1, a_2 \ge 0$ *and* $|b| \ge 1$ *we have*

$$\left|\frac{\Gamma(a_1+bi)}{\Gamma(a_2+bi)}\right| \simeq |b|^{a_1-a_2}.$$

Acknowledgements The authors would like to thank Jacek Dziubański, Alessio Martini, Adam Nowak, Błażej Wróbel, and the referees for their helpful comments and suggestions.

Open Access This article is distributed under the terms of the Creative Commons Attribution 4.0 International License (http://creativecommons.org/licenses/by/4.0/), which permits unrestricted use, distribution, and reproduction in any medium, provided you give appropriate credit to the original author(s) and the source, provide a link to the Creative Commons license, and indicate if changes were made.

References

1. Abramowitz, M., Stegun, I.A.: Handbook of mathematical functions with formulas, graphs, and mathematical tables, National Bureau of Standards Applied Mathematics Series, vol. 55. U.S. Government Printing Office, Washington, D.C, For sale by the Superintendent of Documents (1964)
2. Alexopoulos, G.: Spectral multipliers on Lie groups of polynomial growth. Proc. Amer. Math. Soc. **120**(3), 973–979 (1994)
3. Auscher, P., McIntosh, A., Russ, E.: Hardy spaces of differential forms on Riemannian manifolds. J. Geom. Anal. **18**(1), 192–248 (2008)
4. Bernicot, F., Zhao, J.: New abstract Hardy spaces. J. Funct. Anal. **255**(7), 1761–1796 (2008)

5. Betancor, J.J., Castro, A.J., Curbelo, J.: Spectral multipliers for multidimensional Bessel operators. J. Fourier Anal. Appl. **17**(5), 932–975 (2011)
6. Betancor, J.J., Castro, A.J., Nowak, A.: Calderón-Zygmund operators in the Bessel setting. Monatsh. Math. **167**(3–4), 375–403 (2012)
7. Betancor, J.J., Dziubański, J., Torrea, J.L.: On Hardy spaces associated with Bessel operators. J. Anal. Math. **107**, 195–219 (2009)
8. Betancor, J.J., Stempak, K.: Relating multipliers and transplantation for Fourier-Bessel expansions and Hankel transform. Tohoku Math. J. (2) **53**(1), 109–129 (2001)
9. Carbonaro, A., Dragičević, O.: Functional calculus for generators of symmetric contraction semigroups. Duke Math. J. **166**(5), 937–974 (2017)
10. Chen, P.: Spectral multipliers from $H^1_L(X)$ to $L^q(X)$. J. Math. Anal. Appl. **385**(2), 622–633 (2012)
11. Chen, P., Ouhabaz, E.M., Sikora, A., Yan, L.: Restriction estimates, sharp spectral multipliers and endpoint estimates for Bochner-Riesz means. J. Anal. Math. **129**, 219–283 (2016)
12. Chen, P., Sikora, A.: Sharp spectral multipliers for a new class of Grushin type operators. J. Fourier Anal. Appl. **19**(6), 1274–1293 (2013)
13. Christ, M.: L^p bounds for spectral multipliers on nilpotent groups. Trans. Amer. Math. Soc. **328**(1), 73–81 (1991)
14. Coifman, R.R., Weiss, G.: Extensions of Hardy spaces and their use in analysis. Bull. Amer. Math. Soc. **83**(4), 569–645 (1977)
15. Cowling, M.G.: Harmonic analysis on semigroups. Ann. of Math. (2) **117**(2), 267–283 (1983)
16. Dziubański, J.: Riesz transforms characterizations of Hardy spaces H^1 for the rational Dunkl setting and multidimensional Bessel operators. J. Geom. Anal. **26**(4), 2639–2663 (2016)
17. Dziubański, J., Preisner, M.: Remarks on spectral multiplier theorems on Hardy spaces associated with semigroups of operators. Rev. Un. Mat. Argentina **50**(2), 201–215 (2009)
18. Dziubański, J., Preisner, M.: Multiplier theorem for Hankel transform on Hardy spaces. Monatsh. Math. **159**(1–2), 1–12 (2010)
19. Dziubański, J., Preisner, M.: Hardy spaces for semigroups with Gaussian bounds. Ann. Mat. Pur. Appl. (1923-) **197**, 965–987 (2017)
20. Dziubański, J., Preisner, M., Wróbel, B.: Multivariate Hörmander-type multiplier theorem for the Hankel transform. J. Fourier Anal. Appl. **19**(2), 417–437 (2013)
21. Garrigós, G., Seeger, A.: Characterizations of Hankel multipliers. Math. Ann. **342**(1), 31–68 (2008)
22. Gasper, G., Trebels, W.: Multiplier criteria of Hörmander type for Fourier series and applications to Jacobi series and Hankel transforms. Math. Ann. **242**(3), 225–240 (1979)
23. Gosselin, J., Stempak, K.: A weak-type estimate for Fourier–Bessel multipliers. Proc. Amer. Math. Soc. **106**(3), 655–662 (1989)
24. Hebisch, W.: A multiplier theorem for Schrödinger operators. Colloq. Math. **60/61**(2), 659–664 (1990)
25. Hofmann, S., Lu, G., Mitrea, D., Mitrea, M., Yan, L.: Hardy spaces associated to non-negative self-adjoint operators satisfying Davies–Gaffney estimates, Mem. Amer. Math. Soc. **214**, no. 1007, vi+78 (2011)
26. Hörmander, L.: Estimates for translation invariant operators in L^p spaces. Acta Math. **104**, 93–140 (1960)
27. Kapelko, R.: A multiplier theorem for the Hankel transform. Rev. Mat. Complut. **11**(2), 281–288 (1998)
28. Lebedev, N. N.: Special functions and their applications, Dover Publications, Inc., New York, Revised edition, translated from the Russian and edited by Richard A. Silverman, Unabridged and corrected republication. (1972)
29. Martini, A., Müller, D.: A sharp multiplier theorem for Grushin operators in arbitrary dimensions. Rev. Mat. Iberoam. **30**(4), 1265–1280 (2014)
30. Martini, A., Sikora, A.: Weighted Plancherel estimates and sharp spectral multipliers for the Grushin operators. Math. Res. Lett. **19**(5), 1075–1088 (2012)
31. Meda, S.: A general multiplier theorem. Proc. Amer. Math. Soc. **110**(3), 639–647 (1990)
32. Muckenhoupt, B., Stein, E.M.: Classical expansions and their relation to conjugate harmonic functions. Trans. Amer. Math. Soc. **118**, 17–92 (1965)
33. Müller, D., Stein, E.M.: On spectral multipliers for Heisenberg and related groups. J. Math. Pures Appl. **73**(4), 413–440 (1994)
34. Sikora, A., Wright, J.: Imaginary powers of Laplace operators. Proc. Amer. Math. Soc. **129**(6), 1745–1754 (2001)

35. Stein, E.M.: *Harmonic analysis: real-variable methods, orthogonality, and oscillatory integrals*, Princeton Mathematical Series, vol. 43, Princeton University Press, Princeton, NJ, With the assistance of Timothy S. Murphy, Monographs in Harmonic Analysis, III (1993)
36. Thinh Duong, X., Ouhabaz, E.M., Sikora, A.: Plancherel-type estimates and sharp spectral multipliers. J. Funct. Anal. **196**(2), 443–485 (2002)
37. Watson, G.N.: A treatise on the theory of Bessel functions. Cambridge University Press, Cambridge, Cambridge Mathematical Library (1995). Reprint of the second (1944) edition
38. Wróbel, B.: Multivariate spectral multipliers for tensor product orthogonal expansions. Monatsh. Math. **168**(1), 125–149 (2012)

Publisher's Note Springer Nature remains neutral with regard to jurisdictional claims in published maps and institutional affiliations.

Journal of Fourier Analysis and Applications (2019) 25:2447–2462
https://doi.org/10.1007/s00041-019-09668-y

p-Adic Analogue of the Wave Equation

Bo Wu[1] · Andrei Khrennikov[2]

Received: 29 June 2018 / Revised: 16 January 2019 / Published online: 12 February 2019
© Springer Science+Business Media, LLC, part of Springer Nature 2019

Abstract
In this paper, a p-adic analogue of the wave equation with Lipschitz source is considered. Since it is hard to arrive the solution of the problem, we propose a regularized method to solve the problem from a modified p-adic integral equation. Moreover, we give an iterative scheme for numerical computation of the regularlized solution.

Keywords Cauchy problem · p-Adic elliptic equation · Regularized solution

Mathematics Subject Classification 11F85 · 46S10 · 47G30

1 Introduction

p-Adic analysis has attracted many researchers since last century due to its various applications in physics. The definition of derivatives on p-adic field is the most important problem for the study of p-adic equations. In 1960s, Gibbs first gave the definition of logic derivative over dyadic field [8]. Then Butzer, Onneweer, Zheng, Stankovic, Vladimirov, Su, et al. generalized its definition into Vilenkin group and further p-adic field [5,23,26–28,32,38]. Subsequently, Kozyrev, Albeverio, Zuniga-Galindo, Khrennikov et al. have done a lot of work on p-adic pseudo-differential operators [1,13,17,20–22,30,36,37]. Up to now, there are a number of monographs and a huge number of papers that cover various problems involving p-adic pseudo-differential equations since they play important role in p-adic theoretical physics,

Communicated by Hans G. Feichtinger.

✉ Bo Wu
bowu8800@gmail.com

Andrei Khrennikov
andrei.khrennikov@lnu.se

1 Department of Mathematics, Nanjing University of Finance & Economics, Nanjing 210023, China

2 International Center for Mathematical Modelling in Physics and Cognitive Science, Mathematical Institute, Linnnaeus University, Vaxjo, 351 95 Vaxjo, Sweden

see [2–4,6,7,12,14,15,18,19,33–35]. However. there are few results about nonlinear p-adic equations. Andrei Khrennikov et al. [11,16] consider a p-adic analog of the porous medium equation by L^1-theory of the Vladimirov operator D^α. Since nonlinear equations play the important role in mathematical physics, it stimulates us to develop the theory in this field over p-adic field.

In this paper, we consider a p-adic analogue of the wave equation with Lipschitz source

$$\frac{\partial^2 u(t,x)}{\partial t^2} = T^\alpha u(t,x) + f(u), \quad 0 < t < T, \ x \in \mathbb{Q}_p,$$

where f satisfies Lipschitz condition, $\mathbb{Q}_p$ is the field of p-adic numbers (prime number $p > 1$)and T^α, $\alpha > 0$ is pseudo-differential operator defined by Su in [27].

It is worth mentioning that this kind of equation has widely studied in Euclidean space since it linked to elliptic sine-Gordon equations arising in molecular biology [9, 10,24,31]. According to Ostrowski's theorem, there exist two parallel "universe": one is real "universe" based on the field of real numbers $\mathbb{R}$, the other is p-adic "universe" based on the field of p-adic numbers $\mathbb{Q}_p$. So compare to macro study, it is valuable to consider this problem in microscopic view.

The rest of paper is organized as follows. In Sect. 2, we recall some basic notations in p-adic field and introduce the definition of pseudo-differential operator and corresponding orthonormal base. In Sect. 3, we study the mild solution of Cauchy problem for the p-adic analogue of the wave equation. We define a regularized factor to obtain the stable solution. Then, we prove the existence and the stability of the regularized solution. Further, we consider the error estimation.

2 Preliminaries

We will use the notations and results from Taibleson's book [29]. Let $\mathbb{Q}_p$ be the p-adic field, which p is a prime number. It is a nondiscrete, locally compact, totally disconnected and complete topological field. For any $x, y \in \mathbb{Q}_p$, define non-archimedien norm $|\cdot| : \mathbb{Q}_p \to \mathbb{R}^+$ satisfying following properties:

$$\begin{aligned} &\text{(i)} \quad |x| \geq 0; |x| = 0 \Leftrightarrow x = 0; \\ &\text{(ii)} \quad |xy| = |x||y|; \\ &\text{(iii)} \quad |x+y| \leq \max\{|x|, |y|\}. \end{aligned}$$

Denote subset $\mathfrak{D} = \{x \in \mathbb{Q}_p : \ |x| \leq 1\}$ as the ring of integers in $\mathbb{Q}_p$. It is the unique maximal compact subring in $\mathbb{Q}_p$ with the Haar measures $|\mathfrak{D}| = 1$. The prime ideal $B = \{x \in \mathbb{Q}_p : \ |x| \leq p^{-1}\}$ are balls with center 0 in $\mathbb{Q}_p$ satisfying $B = \beta\mathfrak{D}$ where $\beta = p^{-1}$. It is the unique maximal ideal in $\mathfrak{D}$. Then, the fractional ideal can be defined as $B_k = \{x \in \mathbb{Q}_p : \ |x| \leq p^{-k}\}$ with the Haar measures $|B_k| = p^{-k}$, $k \in \mathbb{Z}$.

If $|x| = p^{-t}$, $t \in \mathbb{Z}$, then x admits a (unique) canonical representation

$$x = x_t\beta^t + x_{t+1}\beta^{t+1} + \cdots,$$

where $\beta = p^{-1}$ and $x_t, x_{t+1}, \ldots \in \{0, 1, \ldots, p-1\}$.

For each $l \in \mathbb{Z}$, we choose elements $z_{l,i} \in \mathbb{Q}_p$, $i \in \mathbb{Z}^+$, so that the subsets $B_{l,i} = z_{l,i} + B_l \subset \mathbb{Q}_p$. And if $i \neq j$, then $B_{l,i} \cap B_{l,j} = \emptyset$, $\cup_{i=0}^{\infty} B_{l,i} = \mathbb{Q}_p$.

Define indicative function of Haar measurable subset $E \subset \mathbb{Q}_p$ as

$$\Phi_E(x) = \begin{cases} 1, & x \in E, \\ 0, & x \in E^c, \end{cases}$$

then, the Haar measure of E is $|E| = \int_E dx = \int_{\mathbb{Q}_p} \Phi_E(x)dx$ where dx denote the Haar measure on $\mathbb{Q}_p$ normalized by the condition $\int_{\mathfrak{D}} dx = 1$.

Define translation operator $\tau_h : f \to \tau_h f,\ h \in \mathbb{Q}_p$ as $\tau_h f(x) = f(x-h),\ x \in \mathbb{Q}_p$. Then, the test function space $S = S(\mathbb{Q}_p)$ is defined as

$$\begin{aligned} S(\mathbb{Q}_p) &= \Bigg\{\varphi : \mathbb{Q}_p \to \mathbb{C},\ \varphi(x) \\ &= \sum_{j=1}^{n} c_j \tau_{h_j} \Phi_{B_{k_j}(x)},\ c_j \in \mathbb{C}, h_j \in \mathbb{Q}_{\shortmid},\ k_j \in \mathbb{Z},\ 1 \le j \le n\Bigg\}, \end{aligned}$$

where the element $\varphi(x)$ is called test function.

For the test function space S, we give the following topology: for $\varphi \in S(\mathbb{Q}_{\shortmid})$, there exists unique integers (k, l) such that the function φ is constant on the coset of B_k and with supports in the ball B_l; $\lim_{n\to+\infty} \varphi_n(x) = 0$ converges uniformly for $x \in \mathbb{Q}_{\shortmid}$. Then, S is complete topological linear spaces.

Denote by $S' = S'(\mathbb{Q}_p)$ the distribution space of test function space S. S' is a complete topological linear space under the dual topology.

Let $\chi(x)$ be a fixed non-trivial character of $\mathbb{Q}_p$ which is trivial on $\mathfrak{D}$. For the p-adic field, χ can be constructed by the base value [16] as

$$\chi(\beta^{-j}) = \begin{cases} \exp\left(\frac{2\pi i}{p^j}\right), & \text{for } j \in \mathbb{N}, \\ 1, & \text{otherwise}, \end{cases}$$

Then for $x = x_t\beta_t + x_{t+1}\beta_{t+1} + \cdots$, $\chi(x) = \exp(2\pi i \sum_{j=t}^{-1} x_j p^j)$ and for $\lambda = \lambda_\tau\beta_\tau + \lambda_{\tau+1}\beta_{\tau+1} + \cdots$

$$\chi_\lambda(x) = \chi(\lambda x) = \exp\left(2\pi i \sum_{k=0}^{-(t+\tau+1)} \left(\sum_{j=0}^{k} x_{t+\tau-j}\lambda_{t+\tau}\right) p^{t+\tau+k}\right)$$

For $\varphi \in S(\mathbb{Q}_p)$, we define its Fourier transform $\varphi^\wedge$ by

$$\varphi^{\wedge}(\xi) = \int_{\mathbb{Q}_p} \varphi(x)\overline{\chi_\xi}(x)dx, \quad \xi \in \mathbb{Q}_p,$$

and inverse Fourier transform $\varphi^{\vee}$ by

$$\varphi^{\vee}(x) = \int_{\mathbb{Q}_p} \varphi(\xi)\chi_x(\xi)d\xi, \quad x \in \mathbb{Q}_p,$$

In 1992, Su Weiyi [27] has given definitions of the derivative for the p-adic local fields $\mathbb{Q}_p$, including derivatives of the fractional orders and real orders.

Definition 2.1 Let $< \xi >= \max\{1, |\xi|\}$, $\alpha \geq 0$ if for $\varphi \in S(\mathbb{Q}_p)$, the integral

$$T^{\alpha}\varphi(x) = (< \xi >^{\alpha} \varphi^{\wedge}(\xi))^{\vee}(x)$$

exists at $x \in \mathbb{Q}_p$, where $\chi(x)$ is a fixed non-trivial character of $\mathbb{Q}_p$. Then it is called a pointwise derivative of order α of φ at x.

Note: The defined domain of T^{α} in the definition can be extended to the space $S'(\mathbb{Q}_p)$, where $S'(\mathbb{Q}_p)$ denote the set of all functionals (distributions) on $S(\mathbb{Q}_p)$.

Let $D(T^{\alpha})$ be the domain of T^{α} defined as

$$D(T^{\alpha}) = \{\varphi \in L^2 : < \xi >^{\alpha} \varphi^{\wedge} \in L^2\}.$$

We have

Lemma 2.1 [25] *Let $\psi(x) = \chi(p^{-1}x)\Phi_{\mathfrak{D}}$, then*

$$T^{\alpha}\psi(ax+b) = \begin{cases} p^{\alpha}|a|^{\alpha}\psi(ax+b), & |a| > p^{-1} \\ \psi(ax+b), & |a| \leq p^{-1} \end{cases}$$

where $a, b \in \mathbb{Q}_p$ and $a \neq 0$.

Lemma 2.2 [25] *T^{α} is a positive definite self-adjoint operator on $D(T^{\alpha})$, $\{\psi_{NjI}\}$ is an orthonormal base of L^2 consisting of eigenfunctions of the operator T^{α}, which defined as follows where $\Phi_{\mathfrak{D}}(x)$ is a characteristic function of a unit ball.*

and

$$T^{\alpha}\psi_{NjI}(x) = \begin{cases} p^{(1-N)\alpha}\psi_{NjI}(x), & N < 1 \\ \psi_{NjI}(x), & N \geq 1 \end{cases}$$

Proof We noted that

$$\begin{aligned} \psi_{NjI}(x) &= p^{-\frac{N}{2}}\chi(p^{-1}(p^N jx))\Phi_{\mathfrak{D}}(p^N jx - jz_I) \\ &= p^{-\frac{N}{2}}\chi(p^{-1}jz_I)\chi(p^{-1}(p^N jx - jz_I))\Phi_{\mathfrak{D}}(p^N jx - jz_I) \\ &= p^{-\frac{N}{2}}\chi(p^{-1}jz_I)\psi(p^N jx - jz_I). \end{aligned}$$

For $N < 1$, then $|p^N| > p^{-1}$. So we can obtain

$$\begin{aligned} T^\alpha \psi_{NjI}(x) &= p^{-\frac{N}{2}} \chi(p^{-1} j z_I) T^\alpha \psi(p^N jx - jz_I) \\ &= p^{-\frac{N}{2}} \chi(p^{-1} j z_I) p^\alpha |p^N|^\alpha \psi(p^N jx - jz_I) \\ &= p^{(1-N)\alpha} \psi_{NjI}(x). \end{aligned}$$

For $N \geq 1$, then $|p^N| \leq p^{-1}$. So we can get

$$\begin{aligned} T^\alpha \psi_{NjI}(x) &= p^{-\frac{N}{2}} \chi(p^{-1} j z_I) T^\alpha \psi(p^N jx - jz_I) \\ &= p^{-\frac{N}{2}} \chi(p^{-1} j z_I) \psi(p^N jx - jz_I) \\ &= \psi_{NjI}(x). \end{aligned}$$

To prove the orthogonality of $\{\psi_{NjI}\}$, we consider the inner product $(\psi_{NjI}, \psi_{N'j'I'})$ in L^2.

$$\begin{aligned} (\psi_{NjI}, \psi_{N'j'I'}) &= \int_{p^{-N}I \cap p^{-N'}I'} p^{-\frac{N}{2}} \chi(p^{N-1} jx) p^{-\frac{N'}{2}} \chi(p^{N'-1} j'x) dx \\ &= \delta_{NN'} \int_{p^{-N}(I \cap I')} p^{-N} \chi(p^{N-1} jx) \chi(p^{N-1} j'x) dx \\ &= \delta_{NN'} \delta_{II'} \int_{p^{-N} I} p^{-N} \chi(p^{N-1}(j - j')x) dx \\ &= \delta_{NN'} \delta_{II'} \delta_{jj'}. \end{aligned}$$

To prove the completeness of $\{\psi_{NjI}\}$, we consider the Fourier coefficient of $\Phi_{\mathfrak{D}}$.

$$(\Phi_{\mathfrak{D}}, \psi_{NjI}) = p^{-\frac{N}{2}} \int_{\mathfrak{D} \cap p^{-N} I} \chi(-p^{N-1} jx) dx.$$

For $N \leq 0$, $(\Phi_{\mathfrak{D}}, \psi_{NjI}) = 0$. While for $N > 0$, $(\Phi_{\mathfrak{D}}, \psi_{NjI}) = p^{-\frac{N}{2}} \delta_{I\mathfrak{D}}$. Then

$$\sum_{NjI} |(\Phi_{\mathfrak{D}}, \psi_{NjI})|^2 = (p-1) \sum_{N=1}^{\infty} p^{-N} = 1 = ||\Phi_{\mathfrak{D}}||^2.$$

So by Parserval equality, we proved the completeness of $\{\psi_{NjI}\}$. □

Remark 2.1 There is another mostly used operator over $p-$adic field called Vladimirov operator D^α, $\alpha > 0$. It is defined as a pseudo-differential operator with the symbol $|\xi|^\alpha$:

$$D^\alpha \varphi(x) = (|\xi|^\alpha \varphi^\wedge(\xi))^\vee(x), \quad \varphi \in S(\mathbb{Q}_p).$$

Here, the orthonormal base $\psi_{NjI}(x)$ of D^α is obtained in a different form (see [2]) and $D^\alpha \psi_{NjI}(x) = p^{(1-N)\alpha} \psi_{NjI}(x)$, for any $N \in \mathbb{N}$. In this paper, we use operator

T^α since from the Lemma 2.2, we can find that for any $N \in \mathbb{N}$, the eigenvalues of T^α are no less than 1. This proposition is important for our later estimate.

3 Main Results

3.1 Homogeneous Problem

First, we consider the following homogeneous equation over p-adic field.

$$\begin{cases} \frac{\partial^2 u(t,x)}{\partial t^2} = T_x^\alpha u(t,x), \\ u(0,x) = \phi(x), \\ u_t'(0,x) = g(x). \end{cases} \tag{1}$$

We will use the orthonormal base $\{\psi_{NjI}\}$ constructed in Lemma 2.1 to solve the equation. In the following, we will write $\sum$ instead of $\sum_{N,j,I}$ for simplicity.

Let lacunary series $u(t,x) = \sum u_{NjI}(t)\psi_{NjI}(x)$ be the exact form of problem (1). Then

$$\frac{\partial^2 u(t,x)}{\partial t^2} = \sum \frac{d^2 u_{NjI}(t)}{dt^2}\psi_{NjI}(x),$$
$$T_x^\alpha u(t,x) = \sum_{N<1} p^{\alpha(1-N)} u_{NjI}(t)\psi_{NjI}(x) + \sum_{N\ge 1} u_{NjI}(t)\psi_{NjI}(x).$$

From

$$\frac{\partial^2 u(t,x)}{\partial t^2} = T_x^\alpha u(t,x),$$

we get

$$\sum_{N<1}\{u''_{NjI}(t) - p^{\alpha(1-N)} u_{NjI}(t)\}\psi_{NjI}(x) + \sum_{N\ge 1}\{u''_{NjI}(t) - u_{NjI}(t)\}\psi_{NjI}(x) = 0,$$

Due to the orthogonality of $\{\psi_{NjI}(x)\}$, we have

$$\begin{cases} u''_{NjI}(t) = p^{\alpha(1-N)} u_{NjI}(t), & N < 1, \\ u''_{NjI}(t) = u_{NjI}(t), & N \ge 1. \end{cases}$$

They are the ordinary differential equations of order 2 on $\mathbb{R}$. So the corresponding characteristic equations are given as

$$\begin{cases} \lambda^2 = p^{\alpha(1-N)}, & N < 1, \\ \lambda^2 = 1, & N \ge 1. \end{cases}$$

So, the solution of the equation is obtained as

$$\begin{cases} u_{NjI}(t) = A_{NjI}e^{-tp^{\frac{\alpha}{2}(1-N)}} + B_{NjI}e^{tp^{\frac{\alpha}{2}(1-N)}}, & N < 1, \\ u_{NjI}(t) = C_{NjI}e^{-t} + D_{NjI}e^{t}, & N \geq 1. \end{cases}$$

To determine the coefficients $A_{NjI}, B_{NjI}, C_{NjI}$ and D_{NjI}, we assume that $\phi \in D(T^{\alpha})$ can be expanded as lacunary series $\phi = \sum \phi_{NjI}\psi_{NjI}(x)$, where

$$\phi_{NjI} =< \phi(x), \psi_{NjI}(x) >= \int_{\mathbb{Q}_p} \phi(x)\overline{\psi_{NjI}(x)}dx, \quad \sum |\phi_{NjI}|^2 < +\infty. \tag{2}$$

and

$$\begin{aligned} T^{\alpha}\phi(x) = &\sum_{N<1} p^{\alpha(1-N)} f\phi_{NjI}\psi_{NjI}(x) + \sum_{N\geq 1} \phi_{NjI}\psi_{NjI}(x), \\ &\sum_{N<1} p^{2\alpha(1-N)}|\phi_{NjI}|^2 + \sum_{N\geq 1} |\phi_{NjI}|^2 < +\infty. \end{aligned} \tag{3}$$

With the initial condition $u(0, x) = \phi(x)$, then $u_{NjI}(0) = \phi_{NjI}$, we get

$$\begin{cases} A_{NjI} + B_{NjI} = \phi_{NjI}, & N < 1, \\ C_{NjI} + D_{NjI} = \phi_{NjI}, & N \geq 1. \end{cases}$$

Similarly, we can also expand $g \in D(T^{\alpha})$ as lacunary series $g = \sum g_{NjI}\psi_{NjI}(x)$, where

$$g_{NjI} =< g(x), \psi_{NjI}(x) >= \int_{\mathbb{Q}_p} g(x)\overline{\psi_{NjI}(x)}dx, \quad \sum |g_{NjI}|^2 < +\infty. \tag{4}$$

and

$$\begin{aligned} T^{\alpha}g(x) = &\sum_{N<1} p^{\frac{\alpha(1-N)}{2}} g_{NjI}\psi_{NjI}(x) + \sum_{N\geq 1} g_{NjI}\psi_{NjI}(x), \\ &\sum_{N<1} p^{\alpha(1-N)}|g_{NjI}|^2 + \sum_{N\geq 1} |g_{NjI}|^2 < +\infty. \end{aligned} \tag{5}$$

With the initial condition $u'(0, x) = g(x)$, then $u'_{NjI}(0) = g_{NjI}$, we get

$$\begin{cases} p^{\frac{\alpha(1-N)}{2}}(A_{NjI} - B_{NjI}) = g_{NjI}, & N < 1, \\ C_{NjI} - D_{NjI} = g_{NjI}, & N \geq 1. \end{cases} \tag{6}$$

So, the coefficients are given as

$$\begin{cases} A_{NjI} = \frac{\phi_{NjI}+g_{NjI}p^{-\frac{\alpha(1-N)}{2}}}{2}, \quad B_{NjI}) = \frac{\phi_{NjI}-g_{NjI}p^{-\frac{\alpha(1-N)}{2}}}{2}, & N < 1, \\ C_{NjI} = \frac{\phi_{NjI}+g_{NjI}}{2}, \quad D_{NjI} = \frac{\phi_{NjI}-g_{NjI}}{2}, & N \geq 1. \end{cases} \tag{7}$$

Then the exact solution (or mild solution) of the homogeneous equation is

$$\begin{aligned} u(t,x) = &\sum_{N<1} \left[\phi_{NjI} \cosh\left(tp^{\frac{\alpha(1-N)}{2}}\right) + g_{NjI}p^{-\frac{\alpha(1-N)}{2}} \sinh\left(tp^{\frac{\alpha(1-N)}{2}}\right)\right] \psi_{NjI}(x) \\ &+ \sum_{N\geq 1} [\phi_{NjI} \cosh(t) + g_{NjI} \sinh(t)] \psi_{NjI}(x). \end{aligned}$$

3.2 Inhomogeneous Problem

Next, we consider the case of inhomogeneous equation with nonlinear Lipschitz source.

$$\begin{cases} \frac{\partial^2 u(t,x)}{\partial t^2} = T_x^{\alpha} u(t,x) + f(u), \\ u(0,x) = \phi(x), \\ u_t'(0,x) = g(x), \end{cases} \tag{8}$$

where $f : \mathbb{Q}_p \to \mathbb{Q}_p$ satisfies Lipschitz condition

$$||f(u_1) - f(u_2)||_{L^2(\mathbb{Q}_p)} \leq C||u_1 - u_2||_{L^2(\mathbb{Q}_p)},$$

and C is a positive constant independent of u_1, u_2. By $f(u) = \sum \phi_{NjI}(u)\psi_{NjI}(x)$, the solution u satisfies the following integral equation:

$$\begin{aligned} u(t,x) = &\sum_{N<1} \left[\phi_{NjI} \cosh\left(tp^{\frac{\alpha(1-N)}{2}}\right) + g_{NjI}p^{-\frac{\alpha(1-N)}{2}} \sinh\left(tp^{\frac{\alpha(1-N)}{2}}\right)\right. \\ &\left. + p^{-\frac{\alpha(1-N)}{2}} \int_0^t \sinh\left((t-s)p^{\frac{\alpha(1-N)}{2}}\right) f_{NjI}(u)(s) ds\right] \psi_{NjI}(x) \\ &+ \sum_{N\geq 1} \left[\phi_{NjI} \cosh(t) + g_{NjI} \sinh(t)\right. \\ &\left. + \int_0^t \sinh(t-s) f_{NjI}(u)(s) ds\right] \psi_{NjI}(x). \end{aligned}$$

When $N < 1$, we find that terms $\cosh(tp^{\frac{\alpha(1-N)}{2}})$, $\sinh(tp^{\frac{\alpha(1-N)}{2}})$ and $\sinh((t-s)p^{\frac{\alpha(1-N)}{2}})$ are unstable. So we replace these unstable terms by $\cosh^{\varepsilon}(tp^{\frac{\alpha(1-N)}{2}})$,

$\sinh^{\varepsilon}(tp^{\frac{\alpha(1-N)}{2}})$ and $\sinh^{\varepsilon}((t-s)p^{\frac{\alpha(1-N)}{2}})$, where

$$\cosh^{\varepsilon}\left(tp^{\frac{\alpha(1-N)}{2}}\right) = \frac{E_{\varepsilon}\exp\left(tp^{\frac{\alpha(1-N)}{2}}\right) + \exp\left(-tp^{\frac{\alpha(1-N)}{2}}\right)}{2},$$
$$\sinh^{\varepsilon}\left(tp^{\frac{\alpha(1-N)}{2}}\right) = \frac{E_{\varepsilon}\exp\left(tp^{\frac{\alpha(1-N)}{2}}\right) - \exp\left(-tp^{\frac{\alpha(1-N)}{2}}\right)}{2},$$
$$\sinh^{\varepsilon}\left((t-s)p^{\frac{\alpha(1-N)}{2}}\right) = \frac{E_{\varepsilon}\exp\left((t-s)p^{\frac{\alpha(1-N)}{2}}\right) - \exp\left(-(t-s)p^{\frac{\alpha(1-N)}{2}}\right)}{2},$$
$$E_{\varepsilon} = \frac{\exp\left(-Tp^{\frac{\alpha(1-N)}{2}}\right)}{\varepsilon + \exp\left(-Tp^{\frac{\alpha(1-N)}{2}}\right)}.$$

Then, we could approximate the exact solution u by the regularized solution u_{ε}

$$\begin{aligned} u_{\varepsilon}(t,x) = &\sum_{N<1}\Big[\phi_{NjI}\cosh^{\varepsilon}\left(tp^{\frac{\alpha(1-N)}{2}}\right) + g_{NjI}p^{-\frac{\alpha(1-N)}{2}}\sinh^{\varepsilon}\left(tp^{\frac{\alpha(1-N)}{2}}\right) \\ &+p^{-\frac{\alpha(1-N)}{2}}\int_0^t \sinh^{\varepsilon}\left((t-s)p^{\frac{\alpha(1-N)}{2}}\right) f_{NjI}(u_{\varepsilon})(s)ds\Big]\psi_{NjI}(x) \\ &+\sum_{N\geq 1}\Big[\phi_{NjI}\cosh(t) + g_{NjI}\sinh(t) + \int_0^t \sinh(t-s) f_{NjI}(u_{\varepsilon})(s)ds\Big] \\ &\times \psi_{NjI}(x). \end{aligned} \tag{9}$$

Lemma 3.1

$$\cosh^{\varepsilon}\left(tp^{\frac{\alpha(1-N)}{2}}\right) \leq \varepsilon^{-\frac{t}{T}}, \sinh^{\varepsilon}\left(tp^{\frac{\alpha(1-N)}{2}}\right) \leq \varepsilon^{-\frac{t}{T}}, \sinh^{\varepsilon}\left((t-s)p^{\frac{\alpha(1-N)}{2}}\right) \leq \varepsilon^{\frac{s-t}{T}}.$$

Proof We just prove the first inequality.

$$\begin{aligned} E_{\varepsilon}\exp\left(tp^{\frac{\alpha(1-N)}{2}}\right) &= \frac{\exp\left((t-T)p^{\frac{\alpha(1-N)}{2}}\right)}{\left(\varepsilon + \exp\left(-Tp^{\frac{\alpha(1-N)}{2}}\right)\right)^{\frac{T-t}{T}}\left(\varepsilon + \exp\left(-Tp^{\frac{\alpha(1-N)}{2}}\right)\right)^{\frac{t}{T}}} \\ &\leq \frac{1}{\left(\varepsilon + \exp\left(-Tp^{\frac{\alpha(1-N)}{2}}\right)\right)^{\frac{t}{T}}} \leq \varepsilon^{-\frac{t}{T}}. \end{aligned}$$

From $\varepsilon^{\frac{t}{T}} \leq 1 \leq \exp\left(tp^{\frac{\alpha(1-N)}{2}}\right)$, we obtain

$$\cosh^{\varepsilon}\left(tp^{\frac{\alpha(1-N)}{2}}\right) = \frac{E_{\varepsilon}\exp\left(tp^{\frac{\alpha(1-N)}{2}}\right) + \exp\left(-tp^{\frac{\alpha(1-N)}{2}}\right)}{2} \leq \varepsilon^{-\frac{t}{T}}.$$

□

We shall show the existence and uniqueness of the regularized solution.

Theorem 3.1 *For any given* $\varepsilon > 0$*, the integral Eq. (9) has a unique solution* $u_{\varepsilon} \in C((0, T), \mathbb{Q}_p)$

Proof For $v(t, x) \in C((0, T), \mathbb{Q}_p)$, consider operator F

$$\begin{aligned} Fv = &\sum_{N<1}\left[\phi_{NjI}\cosh^{\varepsilon}\left(tp^{\frac{\alpha(1-N)}{2}}\right) + g_{NjI}p^{-\frac{\alpha(1-N)}{2}}\sinh^{\varepsilon}\left(tp^{\frac{\alpha(1-N)}{2}}\right)\right. \\ &\left. + p^{-\frac{\alpha(1-N)}{2}}\int_0^t \sinh^{\varepsilon}\left((t-s)p^{\frac{\alpha(1-N)}{2}}\right) f_{NjI}(v)(s)ds\right]\psi_{NjI}(x) \\ &+ \sum_{N\geq 1}\left[\phi_{NjI}\cosh(t) + g_{NjI}\sinh(t) + \int_0^t \sinh(t-s)f_{NjI}(v)(s)ds\right]\psi_{NjI}(x). \end{aligned}$$

Then, we have $||F^n v_1 - F^n v_2|| \leq \sqrt{(C^2t^2M_{\varepsilon}^2)^n\frac{1}{n!}}\max_{0\leq t\leq T}||v_1 - v_2||$ for $v_1, v_2 \in C([0, T], \mathbb{Q}_p)$. Indeed,

$$\begin{aligned} ||Fv_1 - Fv_2||^2 = &\sum_{N<1}\left[p^{-\frac{\alpha(1-N)}{2}}\int_0^t \sinh^{\varepsilon}\left((t-s)p^{\frac{\alpha(1-N)}{2}}\right)(f_{NjI}(v_1)(s)\right. \\ &\left. - f_{NjI}(v_2)(s))ds\right]^2 \\ &+ \sum_{N\geq 1}\left[\int_0^t \sinh(t-s)(f_{NjI}(v_1)(s) - f_{NjI}(v_2)(s))ds\right]^2 \\ &\leq M_{\varepsilon}^2 t\sum\int_0^t |f_{NjI}(v_1)(s) - f_{NjI}(v_2)(s)|^2 ds \\ &\leq M_{\varepsilon}^2 t\int_0^t ||f(v_1) - f(v_2)||^2 ds \leq C^2t^2M_{\varepsilon}^2\max_{0\leq t\leq T}||v_1 - v_2||^2 \end{aligned}$$

Here we use the relation $\sinh^{\varepsilon}((t-s)p^{\frac{\alpha(1-N)}{2}}) \leq \varepsilon^{\frac{s-t}{T}} \leq \frac{1}{\varepsilon}$, $\sinh(t-s) \leq \exp(T)$ and for any given ε, let $M_{\varepsilon} = \max\{\frac{1}{\varepsilon}, \exp(T)\}$. Suppose $n = k$ holds, then

$$\begin{aligned} &||F^{k+1}v_1 - F^{k+1}v_2||^2 \\ &= \sum_{N<1}\left[p^{-\frac{\alpha(1-N)}{2}}\int_0^t \sinh^{\varepsilon}\left((t-s)p^{\frac{\alpha(1-N)}{2}}\right)(f_{NjI}(F^k v_1)(s) - f_{NjI}(F^k v_2)(s))ds\right]^2 \\ &+ \sum_{N\geq 1}\left[\int_0^t \sinh(t-s)(f_{NjI}(F^k v_1)(s) - f_{NjI}(F^k v_2)(s))ds\right]^2 \end{aligned}$$

$$
\begin{aligned}
&\leq M_\varepsilon^2 t \sum \int_0^t |f_{NjI}(F^k v_1)(s) - f_{NjI}(F^k v_2)(s)|^2 ds \\
&\leq C^2 t M_\varepsilon^2 \int_0^t ||F^k v_1 - F^k v_2||^2 ds \\
&\leq C^2 t M_\varepsilon^2 \int_0^t (C^2 s^2 M^2)^k \frac{1}{k!} \max_{0\leq t\leq T} ||v_1 - v_2||^2 ds \\
&\leq \frac{(C^2 t M_\varepsilon^2)^{k+1}}{k!} \max_{0\leq t\leq T} ||v_1 - v_2||^2 \int_0^t s^k ds \leq \frac{(C^2 t^2 M_\varepsilon^2)^{k+1}}{(k+1)!} \max_{0\leq t\leq T} ||v_1 - v_2||^2.
\end{aligned}
$$

Since $\lim_{n\to\infty} \sqrt{(C^2 t^2 M_\varepsilon^2)^n \frac{1}{n!}} = 0$ for any given ε, so for some n_0, F^{n_0} is contractive. Then there exists a unique u_ε satisfying $F^{n_0} u_\varepsilon = u_\varepsilon$. From $F u_\varepsilon = F(F^{n_0} u_\varepsilon) = F^{n_0}(F u_\varepsilon)$, so we obtain $F u_\varepsilon = u_\varepsilon$. □

Next, we will consider the stability of regularized solution.

Theorem 3.2 *For any* $0 < \varepsilon < \exp(-T)$, *then*

$$
||u_\varepsilon^1 - u_\varepsilon^2||^2 \leq 3\varepsilon^{-\frac{2t}{T}} \exp(3C^2 T^2)[||\phi^1 - \phi^2||^2 + ||g^1 - g^2||^2],
$$

where the solutions u_ε^1, u_ε^2 *subject to the initial condition* (ϕ^1, g^1), (ϕ^2, g^2) *respectively.*

Proof

$$
\begin{aligned}
&||u_\varepsilon^1 - u_\varepsilon^2||^2 \\
&\quad = \sum_{N<1} \Big[(\phi_{NjI}^1 - \phi_{NjI}^2) \cosh^\varepsilon \left(t p^{\frac{\alpha(1-N)}{2}} \right) + (g_{NjI}^1 - g_{NjI}^2) p^{-\frac{\alpha(1-N)}{2}} \sinh^\varepsilon \left(t p^{\frac{\alpha(1-N)}{2}} \right) \\
&\qquad + p^{-\frac{\alpha(1-N)}{2}} \int_0^t \sinh^\varepsilon \left((t-s) p^{\frac{\alpha(1-N)}{2}} \right) (f_{NjI}(u_\varepsilon^1)(s) - f_{NjI}(u_\varepsilon^2)(s)) ds \Big]^2 \\
&\qquad + \sum_{N\geq 1} [(\phi_{NjI}^1 - \phi_{NjI}^2) \cosh(t) + (g_{NjI}^1 - g_{NjI}^2) \sinh(t) \\
&\qquad + \int_0^t \sinh(t-s)(f_{NjI}(u_\varepsilon^1)(s) - f_{NjI}(u_\varepsilon^2)(s)) ds]^2 \\
&\quad \leq \sum_{N<1} 3 \Big[|\cosh^\varepsilon \left(t p^{\frac{\alpha(1-N)}{2}} \right)|^2 |\phi_{NjI}^1 - \phi_{NjI}^2|^2 + |\sinh^\varepsilon \left(t p^{\frac{\alpha(1-N)}{2}} \right)|^2 |g_{NjI}^1 - g_{NjI}^2|^2 \\
&\qquad + t \int_0^t |\sinh^\varepsilon \left((t-s) p^{\frac{\alpha(1-N)}{2}} \right)|^2 |f_{NjI}(u_\varepsilon^1)(s) - f_{NjI}(u_\varepsilon^2)|^2 ds \Big] \\
&\qquad + \sum_{N\geq 1} 3[\cosh^2(t) |\phi_{NjI}^1 - \phi_{NjI}^2|^2 + \sinh^2(t) |g_{NjI}^1 - g_{NjI}^2|^2 \\
&\qquad + t \int_0^t |\sinh(t-s)|^2 |f_{NjI}(u_\varepsilon^1)(s) - f_{NjI}(u_\varepsilon^2)|^2 ds].
\end{aligned}
$$

For $0 < \varepsilon < \exp(-T)$, we have $\varepsilon^{-\frac{2t}{T}} > \cosh^2(t)$, $\varepsilon^{-\frac{2t}{T}} > \sinh^2(t)$. Further, by using Lemma 3.1,

$$\begin{aligned}||u_\varepsilon^1 - u_\varepsilon^2||^2 &\le 3\varepsilon^{-\frac{2t}{T}}\left[||\phi^1 - \phi^2||^2 + ||g^1 - g^2||^2 + t\int_0^t \varepsilon^{\frac{2s}{T}} ||f(u_\varepsilon^1) - f(u_\varepsilon^2)||^2 ds\right]\\ &\le 3\varepsilon^{-\frac{2t}{T}}\left[||\phi^1 - \phi^2||^2 + ||g^1 - g^2||^2 + C^2 t\int_0^t \varepsilon^{\frac{2s}{T}} ||u_\varepsilon^1 - u_\varepsilon^2||^2 ds\right].\end{aligned}$$

Then, by using Gronwall's inequality, we have

$$||u_\varepsilon^1 - u_\varepsilon^2||^2 \le 3\varepsilon^{-\frac{2t}{T}} \exp(3C^2T^2)\left[||\phi^1 - \phi^2||^2 + ||g^1 - g^2||^2\right].$$

□

The following theorem concerns the error estimate of regularized solution u_ε and the exact solution u.

Theorem 3.3 *For any* $0 < \varepsilon < \exp(-T)$, *denote*

$$\begin{aligned}G_N = &\exp\left(tp^{\frac{\alpha(1-N)}{2}}\right)\phi_{NjI} + g_{NjI}\exp\left(tp^{\frac{\alpha(1-N)}{2}}\right)p^{-\frac{\alpha(1-N)}{2}}\\ &+ \int_0^t \exp\left((t-s)p^{\frac{\alpha(1-N)}{2}}\right) f_{NjI}(u)ds,\end{aligned}$$

and suppose $\sum_{N<1} \exp\left(2(T-t)p^{\frac{\alpha(1-N)}{2}}\right) G_N^2$ *converges and satisfies*

$$\sum_{N<1} \exp\left(2(T-t)p^{\frac{\alpha(1-N)}{2}}\right) G_N^2 \le 2V,$$

where V *is a constant. Then, there exists a regularized solution such that*

$$||u_\varepsilon - u|| \le \sqrt{V\exp(2C^2T^2)\varepsilon^{\frac{T-t}{T}}}.$$

Proof

$$\begin{aligned}||u_\varepsilon - u||^2 = &\sum_{N<1}\Bigg[\frac{E_\varepsilon - 1}{2}\exp\left(tp^{\frac{\alpha(1-N)}{2}}\right)\left(\phi_{NjI} + g_{NjI}p^{-\frac{\alpha(1-N)}{2}}\right)\\ &+ p^{-\frac{\alpha(1-N)}{2}}\left(\int_0^t \sinh^\varepsilon\left((t-s)p^{\frac{\alpha(1-N)}{2}}\right) f_{NjI}(u_\varepsilon)ds\right.\\ &\left.- \int_0^t \sinh\left((t-s)p^{\frac{\alpha(1-N)}{2}}\right) f_{NjI}(u)ds\right)\Bigg]^2\\ &+ \sum_{N\ge 1}\left[\int_0^t \sinh(t-s)(f_{NjI}(u_\varepsilon)(s) - f_{NjI}(u))ds\right]^2\end{aligned}$$

$$= \sum_{N<1} \Bigg[\frac{E_\varepsilon - 1}{2} (\exp\left(tp^{\frac{\alpha(1-N)}{2}}\right) \phi_{NjI} + g_{NjI} \exp\left(tp^{\frac{\alpha(1-N)}{2}}\right) p^{-\frac{\alpha(1-N)}{2}}$$
$$+ \int_0^t \exp\left((t-s)p^{\frac{\alpha(1-N)}{2}}\right) f_{NjI}(u)ds) + p^{-\frac{\alpha(1-N)}{2}}$$
$$\left(\int_0^t \sinh^\varepsilon\left((t-s)p^{\frac{\alpha(1-N)}{2}}\right)(f_{NjI}(u_\varepsilon)ds - f_{NjI}(u)ds) \Bigg]^2\right.$$
$$+ \sum_{N\geq 1} \left[\int_0^t \sinh(t-s)(f_{NjI}(u_\varepsilon)(s) - f_{NjI}(u))ds \right]^2.$$

With inequality $(a+b)^2 \leq 2a^2 + 2b^2$ and

$$(E_\varepsilon - 1)^2 = \Big(\frac{\varepsilon}{\varepsilon + \exp\left(-Tp^{\frac{\alpha(1-N)}{2}}\right)}\Big)^2$$
$$= \varepsilon^2 \Big(\frac{\exp\left((t-T)p^{\frac{\alpha(1-N)}{2}}\right)}{\varepsilon + \exp\left(-Tp^{\frac{\alpha(1-N)}{2}}\right)}\Big)^2 \exp\left(2(T-t)p^{\frac{\alpha(1-N)}{2}}\right)$$
$$\leq \varepsilon^{2-\frac{2t}{T}} \exp\left(2(T-t)p^{\frac{\alpha(1-N)}{2}}\right),$$

we have

$$||u_\varepsilon - u||^2 \leq \sum_{N<1} \Bigg[\frac{(E_\varepsilon - 1)^2}{2} G_N^2 + 2t \int_0^t \left(\sinh^\varepsilon\left((t-s)p^{\frac{\alpha(1-N)}{2}}\right)\right)^2 (f_{NjI}(u_\varepsilon)ds$$
$$-f_{NjI}(u))^2 ds \Bigg]$$
$$+ \sum_{N\geq 1} \left[t \int_0^t (\sinh(t-s))^2 (f_{NjI}(u_\varepsilon)(s) - f_{NjI}(u)))^2 ds \right]$$
$$\leq \frac{1}{2} \varepsilon^{2-\frac{2t}{T}} \sum_{N<1} \exp\left(2(T-t)p^{\frac{\alpha(1-N)}{2}}\right) G_N^2$$
$$+ 2t \int_0^t \varepsilon^{\frac{2s-2t}{T}} \sum (f_{NjI}(u_\varepsilon)ds - f_{NjI}(u))^2 ds$$
$$\leq \varepsilon^{\frac{-2t}{T}} \left\{ \varepsilon^2 V + 2C^2 T \int_0^t \varepsilon^{\frac{2s}{T}} ||u_\varepsilon - u||^2 ds \right\}.$$

Then, by using Gronwall's inequality, we have

$$||u_\varepsilon - u||^2 \leq V \exp(2C^2T^2) \varepsilon^{2-\frac{2t}{T}}.$$

So,

$$||u_\varepsilon - u|| \le \sqrt{V\exp(2C^2T^2)\varepsilon^{\frac{T-t}{T}}}.$$

□

Remark 3.1 To show the numerical computation of the regularlized solution u_ε, we need to divide the time interval into equal subintervals $t_i = \frac{i}{n}t$, $i = 0, 1, \ldots, n$, $n \in \mathbb{N}$. Then, the nonlinear term $\int_0^t \sinh^\varepsilon\left((t-s)p^{\frac{\alpha(1-N)}{2}}\right) f_{NjI}(u_\varepsilon)(s)ds$ and $\int_0^t \sinh(t-s) f_{NjI}(u_\varepsilon)(s)ds$ in (9) can be approximated by following iterative scheme respectively

$$\sum_{j=1}^{i}\int_{t_{j-1}}^{t_j} \sinh^\varepsilon\left((t_i - s)p^{\frac{\alpha(1-N)}{2}}\right) f_{NjI}(u_\varepsilon)(t_{j-1})ds$$

$$\sum_{j=1}^{i}\int_{t_{j-1}}^{t_j} \sinh(t_i - s) f_{NjI}(u_\varepsilon)(t_{j-1})ds.$$

Then, the regularlized solution u_ε can be approximated by

$$\begin{aligned} u_\varepsilon(t_i, x) = &\sum_{N<1}\Big[\phi_{NjI}\cosh^\varepsilon\left(t_i p^{\frac{\alpha(1-N)}{2}}\right) + g_{NjI}p^{-\frac{\alpha(1-N)}{2}}\sinh^\varepsilon\left(t_i p^{\frac{\alpha(1-N)}{2}}\right) \\ &+ p^{-\frac{\alpha(1-N)}{2}}\sum_{j=1}^{i}\int_{t_{j-1}}^{t_j}\sinh^\varepsilon\left((t_i - s)p^{\frac{\alpha(1-N)}{2}}\right) f_{NjI}(u_\varepsilon)(t_{j-1})ds\Big]\psi_{NjI}(x) \\ &+ \sum_{N\ge 1}[\phi_{NjI}\cosh(t_i) + g_{NjI}\sinh(t_i) \\ &+ \sum_{j=1}^{i}\int_{t_{j-1}}^{t_j}\sinh(t_i - s) f_{NjI}(u_\varepsilon)(t_{j-1})ds]\psi_{NjI}(x), \end{aligned}$$

where $u_\varepsilon(t_0, x) = u_\varepsilon(0, x)$ determined by the initial condition.

Acknowledgements The first author was supported by National Natural Science Foundation of China (Grant No.11701270), Natural Science Foundation of the Jiangsu Higher Education Institutions of China (Grant No. 17KJB110003) and the Jiangsu Government Scholarship for Overseas Studies. We are grateful to Prof. Weiyi Su for fruitful discussions and valuable comments.

References

1. Albeverio, S., Khrennikov, A.Yu., Shelkovich, V.M.: Harmonic analysis in the p-adic Lizorkin spaces: fractional operators, pseudo-differential equations, p-adic wavelets. Tauberian theorems. J. Fourier Anal. Appl. **12**(4), 393–425 (2006)

2. Albeverio, S., Khrennikov, A.Yu., Shelkovich, V.M.: Theory of p-adic distributions. In: Linear and Nonlinear Models. Cambridge University Press, Cambridge (2010)
3. Albeverio, S., Khrennikov, A.Yu., Shelkovich, V.M.: The Cauchy problem for evolutionary pseudo-differential equations and the wavelet theory. J. Math. Anal. Appl. **375**, 82–98 (2011)
4. Aref'eva, I.Ya., Dragovich, B.G., Volovich, I.V.: On the p-adic summability of the an harmonic oscillator. Phys. Lett. B **200**, 512–514 (1988)
5. Butzer, P.L., Wagner, H.J.: Walsh-Fourier series and the concept of a derivative. Appl. Anal. **3**, 29–46 (1973)
6. Chuong, N.M., Nguyen, V.C.: The Cauchy problem for a class of pseudo-differential equations over p-adic field. J. Math. Anal. Appl. **340**, 629–645 (2008)
7. Dragovich, B.G.: Adelic harmonic oscillator. Int. J. Modern Phys. A **10**, 2349–2359 (1995)
8. Gibbs, J.E., Millard, M.J.: Walsh functions as a solution of logical differential equations. NPL DES Rept. (1969)
9. Khoa, V.A., Hung, T.T.: Regularity bounds for a Gevrey criterion in a kernel-based regularization of the Cauchy problem of elliptic equations. Appl. Math. Lett. **69**, 75–81 (2017)
10. Khoa, V.A., Truong, M.T.N., Duy, N.H.M., Tuan, N.H.: The Cauchy problem of coupled elliptic sine-Gordon equations with noise: analysis of a general kernel-based regularization and reliable tools of computing. Comput. Math. Appl. **73**, 141–162 (2017)
11. Khrennikov, A.Yu., Kochubei, A.N.: p-Adic analogue of the porous medium equation. arXiv:1611.08863
12. Khrennikov, A.Yu., Shelkovich, V.M.: Non-haar p-adic wavelets and their application to pseudo-differential operators and equations. Appl. Comp. Harmon. Anal. **28**, 1–23 (2010)
13. Khrennikov, A.Yu., Shelkovich, V.M., Van Der Walt, J.H.: Adelic multiresolution analysis, construction of wavelet bases and pseudo-differential operators. J. Fourier Anal. Appl. **19**, 1323–1358 (2013)
14. Khrennikov, A.Yu., Oleschko, K., Correa Lopez, M.J.: Application of p-adic wavelets to model reaction-diffusion dynamics in random porous media. J. Fourier Anal. Appl. **22**, 809–822 (2016)
15. Khrennikov, A.Yu., Oleschko, K., Correa Lopez, M.J.: Modeling fluid's dynamics with master equations in ultrametric spaces representing the treelike structure of capillary networks. Entropy **18**, 249 (2016)
16. Khrennikov, A.Yu., Kozyrev, S.V., Zuniga-Galindo, W.A.: Ultrametric Pseudodifferential Equations and Applications. Cambridge University Press, Cambridge (2018)
17. Kochubei, A.N.: Pseudo-Differential Equations and Stochastics Over Non-archimedean Field. CRC Press, Boca Raton (2001)
18. Kochubei, A.N.: Pseudo-Differential Equations and Stochastics Over Non-archimedean Fields. Marcel Dekker, New York (2001)
19. Kochubei, A.N.: A non-archimedean wave equation. Pac. J. Math. **235**, 245–261 (2008)
20. Kozyrev, S.V.: Wavelet theory as p-adic spectral analysis. Izv. Math. **66**, 367–376 (2002)
21. Kozyrev, S.V.: P-adic pseudodifferential operators and p-adic wavelets. Theor. Math. Phys. **138**, 322–332 (2004)
22. Kozyrev, S.V.: Wavelets and spectral analysis of ultrametric pseudodifferential operators. Sb. Math. **198**, 97–116 (2007)
23. Onneweer, C.W.: Differentiation on a p-adic or p-series field. In: Linear Spaces and Approximation, pp. 187–198. Birkhauser Verlag, Basel (1978)
24. Qian, Z., Fu, C.L., Li, Z.P.: Two regularization methods for a Cauchy problem for the Laplace equation. J. Math. Anal. Appl. **338**, 479–489 (2008)
25. Qiu, H., Su, W.Y.: Pseudo-differential operators over p-adic fields. Sci. China Ser. A **41**(4), 323–336 (2011)
26. Stankovic, R.S.: A note on differential operators on finite non-Abelian groups. Appl. Anal. **21**, 31–41 (1986)
27. Su, W.Y.: Psuedo-differential operators and derivatives on locally compact Vilenkin groups. Sci. China Ser. A **35**(7), 826–836 (1992)
28. Su, W.Y.: Harmonic Analysis and Fractal Analysis Over Local Fields and Applications. World Scientific, Singapore (2017)
29. Taibleson, M.H.: Fourier Analysis on Local Fields. Princeton University Press, Princeton (1975)
30. Torba, S.M., Zuniga-Galindo, W.A.: Parabolic type equations and Markov stochastic processes on Adeles. J. Fourier Anal. Appl. **19**(4), 792–835 (2013)

31. Tuan, N.H., Trong, D.D., Quan, P.H.: A note on a Cauchy problem for the Laplace equation: regularization and error estimates. Appl. Math. Comput. **217**, 2913–2922 (2010)
32. Vladimirov, V.S.: Generalized functions over p-adic number field. Uspekhi Mat. Nauk. **43**, 17–53 (1988)
33. Vladimirov, V.S., Volovich, I.V., Zelenov, E.I.: p-Adic Analysis and Mathematical Physics. WSP, Singapore (1994)
34. Volovich, I.V.: p-Adic string. Class. Quant. Grav. **4**, 83–87 (1987)
35. Volovich, I.V.: p-Adic space-time and string theory. Theor. Math. Phys. **71**, 574–576 (1987)
36. Zuniga-Galindo, W.A.: Fundamental solutions of pseudo-differential operators over p-adic fields. Rend. Sem. Mat. Univ. Padova **109**, 241–245 (2003)
37. Zuniga-Galindo, W.A.: Parabolic equations and Markov processes over p-adic fields. Potential Anal. **28**, 185–200 (2008)
38. Zheng, W.X.: Derivatives and approximation theorems on local fields. Rocky Mt. J. Math. **15**, 803–817 (1985)

Publisher's Note Springer Nature remains neutral with regard to jurisdictional claims in published maps and institutional affiliations.

Journal of Fourier Analysis and Applications (2019) 25:2463–2491
https://doi.org/10.1007/s00041-019-09670-4

Diaconis–Shahshahani Upper Bound Lemma for Finite Quantum Groups

J. P. McCarthy[1]

Received: 19 October 2018 / Published online: 14 February 2019
© Springer Science+Business Media, LLC, part of Springer Nature 2019

Abstract
A central tool in the study of ergodic random walks on finite groups is the Upper Bound Lemma of Diaconis and Shahshahani. The Upper Bound Lemma uses Fourier analysis on the group to generate upper bounds for the distance to random and thus can be used to determine convergence rates for ergodic walks. The Fourier analysis of quantum groups is remarkably similar to that of classical groups. This allows for a generalisation of the Upper Bound Lemma to an Upper Bound Lemma for finite quantum groups. The Upper Bound Lemma is used to study the convergence of ergodic random walks on the dual group $\widehat{S_n}$ as well as on the truly quantum groups of Sekine.

Keywords Random walks · Finite quantum groups · Representation theory

Mathematics Subject Classification 46L53 (60J05, 20G42)

Introduction

Let $\sigma_1, \sigma_2, \ldots, \sigma_k$ be a sequence of shuffles of a deck of cards. If the deck starts in some known order, the order of the deck after these k shuffles is given by

$$\Sigma_k = \sigma_k \cdots \sigma_2 \cdot \sigma_1.$$

Suppose the shuffles are random variables independently and identically distributed as $\sigma_i \sim \nu \in M_p(S_{52})$, the set of probability distributions on S_{52}, then $\Sigma_k \sim \nu^{\star k}$ where $\nu^{\star k}$ is defined inductively as

Communicated by Hans G. Feichtinger.

✉ J. P. McCarthy
jeremiah.mccarthy@cit.ie

1 Department of Mathematics, Cork Institute of Technology, Cork, Ireland

$$\nu^{\star(k+1)}(\{s\}) = \sum_{t\in G} \nu(\{st^{-1}\})\nu^{\star k}(\{t\}).$$

If ν is not concentrated on a subgroup, nor a coset of a normal subgroup of S_{52}, then as $k \to \infty$, $\nu^{\star k} \to \pi_{S_{52}}$, the *uniform* or *random* distribution on S_{52}.

This generalises to arbitrary finite groups. Given independent and identically distributed $s_i \sim \nu \in M_p(G)$, not concentrated on a subgroup, nor a coset of a normal subgroup, as $k \to \infty$:

$$\xi_k = s_k \cdots s_2 \cdot s_1 \sim \nu^{\star k} \to \pi,$$

the random distribution on G. In this case the random walk is said to be *ergodic*.

In the previous statement, convergence is with respect to the *total variation distance*:

$$\|\nu^{\star k} - \pi\| := \sup_{S\subset G} |\nu^{\star k}(S) - \pi(S)|.$$

The inspiration for this work, which is the subject of the author's Ph.D. thesis [10], is the following result of Diaconis and Shahshahani [4]:

Theorem (Upper Bound Lemma) *Let ν be a probability on a finite group G and $k \in \mathbb{N}$. Then*

$$\|\nu^{\star k} - \pi\|^2 \leq \frac{1}{4} \sum_{\alpha\in \text{Irr}(G)\backslash\{\tau\}} d_\alpha \operatorname{Tr}\left(\left(\widehat{\nu}(\alpha)^*\right)^k \widehat{\nu}(\alpha)^k\right) \qquad \bullet$$

The sum is over non-trivial irreducible representations and $\widehat{\nu}(\alpha)$ is the *Fourier transform of ν at the representation ρ^α*

$$\widehat{\nu}(\alpha) = \sum_{t\in G} \nu(\{t\})\rho^\alpha(t).$$

As an application, Diaconis and Shahshahani show that $k_n = \frac{1}{2} n \ln n$ random transpositions are necessary and sufficient to 'mix up' a deck of n cards. This result comprises two elements: an upper bound at $k_n + c\,n$ derived using the Upper Bound Lemma, as well as a lower bound at $k_n - c\,n$ derived using a carefully selected test set. A phenomenon occurs in this example that as $n \to \infty$, the lower bound for $k_n - c\,n$ tends to one for large c, and the upper bound for $k_n + c\,n$ approaches zero for large c. Note that $\frac{2cn}{k_n} \to 0$ so that as $n \to \infty$, the convergence from far from random to close to random is abrupt. This behaviour, called the *cut-off phenomenon*, occurs in a number of examples of random walks on finite groups. Saloff-Coste [15] is an excellent survey that includes more ideas even than the excellent monograph of Diaconis [3]. The author's MSc thesis [11] also comprises a leisurely introduction to the subject.

Finite quantum groups are a noncommutative or 'quantum' generalisation of finite groups, and the central result of this work is that the Diaconis–Shahshahani Upper Bound Lemma also holds for finite quantum groups.

Considered as a research programme, quantum probability is concerned with generalising, where possible, objects in the study of classical probability to quantum objects in the study of quantum probability theory. It is under this programme that this work lies: the study of random walks on finite groups uses classical probability theory—a study of random walks on *quantum* groups *should* be the corresponding area of study in quantum probability.

Therefore, this work is concerned with a generalisation of a generalisation of card shuffling: generalising, where possible, the ideas and results of Diaconis and Shahshahani to the case of finite quantum groups. While the central object of card shuffling—the set of shuffles—is generalised to that of a set of elements of a group, the generalisation to *quantum* groups moves away from a 'set of points' interpretation.

However, in the study of random walks on finite quantum groups at least, the quantum theory generalises so nicely from the classical setup that it can be fruitful to refer to the *virtual* object that is a quantum group as if it really exists. At the very least this approach gives a most pleasing notation for quantum objects. At its very best it can inspire the noncommutative geometer to make good choices in terms of definitions, etc.

One of the most exciting and potentially lucrative aspects of quantum probability, or rather more specifically quantum group theory, is that theorems about finite groups may in fact be true for quantum groups also. For example, the classical finite Peter–Weyl Theorem concerning the matrix elements of representations of classical groups is exactly the same as the finite Peter–Weyl Theorem 3.3 for finite *quantum* groups in the sense that replacing in the classical statement 'finite group' with 'finite quantum group' yields the quantum statement. What this really means is that the classical finite Peter–Weyl Theorem is actually just a special case of the quantum finite Peter–Weyl Theorem.

This is rather comforting on the conceptional level—these quantum objects behave so much like their 'set of points' classical counterparts—but in this study it is on the pragmatic level of proving results about these quantum objects that this principle really comes to the fore. On the one hand, some theorems concerning the theory of finite groups are just corollaries to results about quantum groups. On this other, pragmatic, hand, there is a translation principle: any proof of a classical group theorem, written without regard to any of the points in the 'set of points', may be directly translatable into a proof of the corresponding quantum group theorem.

For example, the Haar state, h, can allow classical 'sum over points' arguments and statements about the algebra of functions on a finite classical group to be translated:

$$\underbrace{\frac{1}{|G|}\sum_{t\in G} f(t)}_{\text{classical: references points } t\in G} \qquad = \qquad \underbrace{h(f)}_{\text{quantum: no reference to points}} \qquad .$$

Representation Theory and 'sum over points' arguments, therefore, are translatable and it is precisely these ideas that play a central role in the representation-theoretic approach of Diaconis and Shahshahani to analysing the rate of convergence of random walks on finite groups [4]. Once the quantum versions of the various objects and maps used by Diaconis and Shahshahani are established, it is largely straightforward to derive and prove the quantum Diaconis–Shahshahani Upper Bound Lemma.

The restriction to *finite* quantum groups is for two reasons. Firstly the classical work that this work is building upon is the Diaconis–Shahshahani Theory approach to random walks on *finite* groups. Secondly, the approach to finding the quantum versions of classical objects in this work—extolled in Sect. 1.1—requires the isomorphism $\mathbb{C}(X \times Y) \cong \mathbb{C}X \otimes \mathbb{C}Y$ for sets X and Y, and this holds only when X and Y are finite.

However, the classical Diaconis–Shahshahani Theory has been extended to possibly-infinite compact groups by Rosenthal [14]. Freslon [7,8] has also extended this work on finite quantum groups, developed during the author's Ph.D. studies, to the far more technically-involved case of compact quantum groups, where the algebras are no longer necessarily finite dimensional. Freslon presents many interesting examples of random walks on compact quantum groups—with the cut-off phenomenon—and also addresses, particularly in [7], a number of concerns not answered or addressed in the author's Ph.D. thesis. As there is a generalisation of the classical finite symmetric group that is infinite dimensional for $n \geq 4$, it would be remiss not to point out that the restriction to *finite* quantum groups is more than a little unnatural, and also brushes many technical difficulties under the carpet.

Therefore this work should be considered a first, pioneering, attempt at a quantum generalisation of the work of Diaconis and Shahshahani, with Freslon's papers a far more accomplished and reaching generalisation.

The paper is organised as follows. In Sect. 1, category theory language is used to motivate the definition of a finite quantum group. Results concerning the dual of a finite quantum group that are used in the sequel are presented at this point. Finally the definition of a random walk on a finite quantum group is given. In Sect. 2, the total variation distance, used for measuring the 'distance to random' of a random walk, is introduced. The central result is contained in Sect. 3; this leans so heavily on the work of Van Daele, that the author calls it by Diaconis–Van Daele Theory. In Sect. 4, two applications of the Upper Bound Lemma are presented: to a random walk on the dual of S_n; and to a random walk on the truly quantum groups of Sekine. In neither example is the cut-off phenomenon exhibited, and in the case of the random walk on the quantum groups of Sekine it is shown the cut-off phenomenon does *not* occur.

1 Preliminaries

1.1 Finite Quantum Groups

The following approach to introducing finite quantum groups is a brief summary of the approach outlined in detail in Sect. 2.2 of [10].

A finite classical group G is an object in the category of finite sets, and the group multiplication $G \times G \to G$, inclusion of the unit, and inverse $G \to G$, are morphisms in this category. The three group axioms are given by three commutative diagrams.

Let $\mathbb{C}G$ be the complex vector space with basis elements $\{\delta^s : s \in G\}$. The map $G \mapsto \mathbb{C}G$ is a covariant functor, the $\mathbb{C}$ functor, from the category of finite sets to the category of finite dimensional complex vector spaces. The image of the group multiplication is a multiplication, $\nabla : \mathbb{C}G \otimes \mathbb{C}G \to \mathbb{C}G$ given by $\nabla(\delta^s \otimes \delta^t) = \delta^{st}$. The presence of the tensor product comes via the isomorphism $\mathbb{C}(G \times G) \cong \mathbb{C}G \otimes \mathbb{C}G$ which does not hold if G is infinite. This gives $\mathbb{C}G$ the structure of a complex associative algebra called the *group algebra of* G. The image of the three commutative diagrams (which comprise the group axioms) under the functor $\mathbb{C}$ give relations that hold in $\mathbb{C}G$, and if $\varphi = \sum_{t \in G} \alpha_t \delta^t$, then the involution

$$\varphi^* = \sum_{t \in G} \overline{\alpha_t} \delta^{t^{-1}}$$

gives the group algebra the structure of a *-algebra.

Now apply the contravariant dual endofunctor to the group algebra. The dual endofunctor maps a finite dimensional complex vector space U to its dual U^*, and a morphism $T : U \to V$ to a morphism $V^* \to U^*$ given by $f \mapsto f \circ T$. The dual of the group algebra is the algebra of functions on G, $F(G)$, with dual basis $\{\delta_s : s \in G\}$ such that $\delta_s(\delta^t) = \delta_{s,t}$, and involution $f^*(\delta^s) = \overline{f(\delta^s)}$. The composition of the $\mathbb{C}$ functor and the dual endofunctor gives a functor $G \mapsto F(G)$ and the images of the group multiplication, inverse, and inclusion of the unit, under this functor composition give structure maps on $F(G)$. These structure maps are called the *comultiplication*, an algebra homomorphism $\Delta : F(G) \to F(G) \otimes F(G)$, $\delta_s \mapsto \mathbb{1}_{m^{-1}(s)} \equiv \sum_{t \in G} \delta_{st^{-1}} \otimes \delta_t$ (for finite G); the counit, $\varepsilon : F(G) \to \mathbb{C}$, $\delta_s \mapsto \delta_{s,e}$; and the antipode, $S : F(G) \to F(G)$, $\delta_s \mapsto \delta_{s^{-1}}$. The associativity, inverse, and identity group axiom commutative diagrams, under this functor, give *coassociativity*, the *counital property*, and the *antipodal property*:

$$\begin{aligned}(\Delta \otimes I_{F(G)}) \circ \Delta &= (I_{F(G)} \otimes \Delta) \circ \Delta \\ (\varepsilon \otimes I_{F(G)}) \circ \Delta &= I_{F(G)} = (I_{F(G)} \otimes \varepsilon) \circ \Delta \\ M \circ (S \otimes I_{F(G)}) \circ \Delta &= \eta_{F(G)} \circ \varepsilon = M \circ (I_{F(G)} \otimes S) \circ \Delta \end{aligned} \tag{1}$$

Here $M : F(G) \otimes F(G) \to F(G)$ is pointwise multiplication and $\eta_{F(G)}$ is the inclusion of the unit, $\mathbb{1}_G$. With the fact that $f^* f = 0$ if and only if $f = 0$, $F(G)$ can be given the structure of a finite dimensional C*-algebra. It is a unital commutative algebra and the group axioms are encoded by these relations. Note furthermore that Δ satisfies $\Delta(f^*) = \Delta(f)^*$, where the involution in $F(G) \otimes F(G)$ is given by $(f \otimes g)^* = f^* \otimes g^*$.

A finite dimensional C*-algebra A with a *-homomorphism $\Delta : A \to A \otimes A$, and maps $\varepsilon : A \to \mathbb{C}$, and $S : A \to A$, that satisfies the above relations, but is not necessarily commutative is thus considered the *algebra of functions on a finite quantum group*. Such an algebra is called a C*-Hopf algebra.

Definition 1.1 An *algebra of functions on a finite quantum group*, is a unital C*-Hopf algebra A; that is a C*-algebra A with a *-homomorphism Δ, a counit ε, and an antipode S, satisfying the relations (1).

Denote an algebra of functions on a finite quantum group by $A =: F(G)$ and refer to G as a finite quantum group, with unit denoted by $\mathbb{1}_G$, and note that S is an involution (not necessarily true for infinite G, see remark to Theorem 2.2, [21]). Timmermann presents in Chapter 1 of his book [18] further properties of Hopf algebras, for example the fact that the counit is a *-homomorphism (used in the sequel). The simplest noncommutative example of an algebra of functions on a finite quantum group is $\mathbb{C}S_3$, where S_3 is the classical symmetric group on three elements, where the comultiplication is given by $\Delta_{\mathbb{C}S_3}(\delta^\sigma) = \delta^\sigma \otimes \delta^\sigma$, and is the dual of the pointwise multiplication in $F(S_3)$. Where τ is the flip map $a \otimes b \mapsto b \otimes a$, this comultiplication has the property that $\tau \circ \Delta_{\mathbb{C}S_3} = \Delta_{\mathbb{C}S_3}$. Algebras of functions on finite quantum groups, $F(C)$, whose comultiplications have this property, $\tau \circ \Delta = \Delta$, are said to be *cocommutative*, and are of the form $F(C) = \mathbb{C}G$ for G a classical finite group. Similarly a commutative algebra of functions on a finite quantum group is the algebra of functions on some classical finite group (can be inferred from Theorem 3.3, [19]).

For the remainder of this work, unless explicitly stated otherwise, G is a quantum group, and assumed finite. There are well established notions of compact and locally compact quantum groups. See, for example, Timmermann [18] for more.

1.2 The Dual of a Quantum Group

Where π is the random/uniform distribution on a classical group G, consider the functional on $F(G)$:

$$f \mapsto \int_G f(t)\, d\pi(t) = \frac{1}{|G|} \sum_{t \in G} f(t).$$

This is a normalised, positive functional, invariant in the sense that for each $s \in G$,

$$\begin{aligned} &\int_G f(t)\, d\pi(t) = \int_G f(st)\, d\pi(t) = \int_G f(ts)\, d\pi(t) \\ &\Rightarrow \mathbb{1}_G \cdot \int_G f = \left(I_{F(G)} \otimes \int_G \right) \Delta(f) = \left(\int_G \otimes I_{F(G)} \right) \Delta(f) . \qquad (f \in F(G)) \end{aligned} \tag{2}$$

This normalised, invariant, positive functional is called a *Haar state*, it is unique, and is denoted by $\int_G$ (earlier h was used).

A finite *quantum* group also has a unique Haar state [21, Theorem 1.3], denoted by $\int_G$. Invariance of the Haar state is given by (2) and furthermore it is tracial: $\int_G ab = \int_G ba$ [21, Theorem 2.2].

Consider the space, $\widehat{A}$, of linear functionals on $A := F(G)$ of the form

$$b \mapsto \int_G ba . \qquad (a, b \in F(G))$$

As $F(G)$ is finite dimensional, the continuous and algebraic duals coincide. Furthermore, the Haar state is faithful and so

$$\langle a, b\rangle := \int_G a^* b$$

defines an inner product making $F(G)$ a Hilbert space. Via the Riesz Representation Theorem for Hilbert spaces, for every element $\varphi \in F(G)'$, there exists a density $a_\varphi^* \in F(G)$ such that:

$$\varphi(b) = \langle a_\varphi^*, b\rangle = \int_G a_\varphi b\,, \qquad (b \in F(G))$$

so that $F(G)' = \widehat{A}$. This space can be given the structure of an algebra of functions on a finite quantum group by employing the contravariant dual functor to $F(G)$ and its structure maps. The dual functor maps the object $F(G)$ to $\widehat{A}$; the comultiplication $\Delta : F(G) \to F(G) \otimes F(G)$ to an associative multiplication $\nabla : \widehat{A} \otimes \widehat{A} \to \widehat{A}$; and the multiplication $M : F(G) \otimes F(G) \to F(G)$ to a comultiplication $\widehat{\Delta} : \widehat{A} \to \widehat{A} \otimes \widehat{A}$ (see [10], Section 2.5 for more). The *-involution on $\widehat{A}$ is given by:

$$\varphi^*(a) = \overline{\varphi(S(a)^*)}\,. \qquad (\varphi \in \widehat{A},\ a \in F(G))$$

The finite quantum group formed in this way is called the *dual* of the quantum group G, and is denoted by $\widehat{G}$. An invariant, positive functional on $F(\widehat{G})$ is given by $\widehat{h}(\varphi) = \varepsilon(a_\varphi)$. It is not normalised and so not the Haar state of $F(\widehat{G})$. In the case of a classical group G, $F(\widehat{G}) = \mathbb{C}G$.

The analogue of a probability distribution on a classical group is a state on the algebra of functions on a quantum group, that is a positive functional of norm one. Denote the states on $F(G)$ by $M_p(G)$ and note that $M_p(G)$ is a subset of $F(\widehat{G})$. A density a_ν defines a state $\nu \in M_p(G)$ if and only if a_ν is positive and $\int_G a_\nu = 1$. There are Convolution and Plancherel Theorems for $F(\widehat{G})$ that will be used to prove the Upper Bound Lemma. The Convolution Theorem is well presented by Bhowmick et al. [2, Sect. 1.1], which heavily cites Van Daele [20–23]; while the Plancherel Theorem is well presented by Timmermann [18, Sect. 2.3], which also cites Van Daele [21].

The *convolution product* on $F(G)$ is given by:

$$a \circledast b := \left(\int_G \otimes I_{F(G)}\right) \left(\left((S \otimes I_{F(G)})\Delta(b)\right)(a \otimes \mathbb{1}_G)\right).$$

There is a Convolution Theorem relating this convolution to a convolution in $F(\widehat{G})$:

$$\varphi_1 \star \varphi_2 := (\varphi_1 \otimes \varphi_2)\Delta\,. \qquad (\varphi_1, \varphi_2 \in F(\widehat{G})).$$

This is the natural multiplication on $F(\widehat{G})$, and therefore briefly write $\varphi_1\varphi_2$ for $\varphi_1 \star \varphi_2$.

Within the proof of Proposition 3.11 of [22], Van Daele proves the following identity:

$$S\left(\left(I_{F(G)}\otimes\int_G\right)(\Delta(a)(\mathbb{1}_G\otimes b))\right)=\left(I_{F(G)}\otimes\int_G\right)((\mathbb{1}_G\otimes a)\Delta(b))$$
$$(a,\ b\in F(G))$$

This identity can be used to prove the following (see Sect. 1.1 of [2]):

Theorem 1.2 (Van Daele's Convolution Theorem) *For* $\varphi_1,\varphi_2\in F(\widehat{G})$

$$a_{\varphi_1\varphi_2}=a_{\varphi_1}\circledast a_{\varphi_2}\qquad\bullet$$

Timmerman [18, Lemma 2.3.8] presents a proof of the following lemma.

Lemma 1.3 *Let* $\varphi_1,\ \varphi_2\in F(\widehat{G})$. *Then*

$$\widehat{h}(\varphi_1\varphi_2)=\varphi_1(S^{-1}(a_{\varphi_2}))\qquad\bullet$$

Theorem 1.4 (Plancherel Theorem) *Let* G *be a quantum group and* $\varphi\in F(\widehat{G})$. *Then*

$$\widehat{h}\left(\varphi^*\varphi\right)=\int_G a_\varphi^*a_\varphi.$$

Proof Applying the previous lemma to the left-hand side:

$$\begin{aligned}\widehat{h}(\varphi^*\varphi)&=\varphi^*\left(S^{-1}(a_\varphi)\right)=\overline{\varphi\left(S(S^{-1}(a_\varphi))^*\right)}\\&=\overline{\varphi(a_\varphi^*)}=\overline{\int_G a_\varphi^*a_\varphi}=\int a_\varphi^*a_\varphi,\end{aligned}$$

where the traciality of $\int_G$ was used • □

1.3 Random Walks on Finite Quantum Groups

Early work on quantum stochastic processes by various authors led to random walks on duals of compact groups (particularly Biane, see [5] for references), and other examples, but Franz and Gohm [5] defined with clarity a *random walk on a finite quantum group*. Given a random walk on a finite classical group G, by applying the functor composition referenced in Sect. 1.1 to the random variables $\xi_k : G^k\to G$, random variables on the level of $F(G)$, $j_k : F(G)\to\bigotimes_{k\text{ copies}}F(G)$ can be defined by:

$$j_k(f)=f\circ\delta^{\xi_k},$$

and the generalisation to random walks on finite quantum groups can be made in this way. It is also possible to generalise from classical to quantum random walks using stochastic operators, and also $F(G)$ analogues of the random variables s_i (denoted by k_i by Franz and Gohm and z_i by the author in [10]).

In Sect. 3.2 of [10], this generalisation of Franz and Gohm from random walks on finite classical groups to random walks on finite quantum groups is explored in detail, and concludes with Franz and Gohm's assertion that a random walk on a finite quantum group G driven by a state $\nu \in M_p(G)$ is essentially the same thing as the semigroup of convolution powers, $\{\nu^{\star k}\}$, defined inductively through

$$\nu^{\star(k+1)} = (\nu \otimes \nu^{\star k}) \circ \Delta.$$

Indeed, Amaury Freslon defines a random walk on a (compact) quantum group simply as a state on $C(G)$, and proceeds to study the semigroup of convolution powers without mentioning the random variables j_k nor z_i, only mentioning the stochastic operator $P_\nu = (\nu \otimes I_{F(G)}) \circ \Delta$ for analytical reasons (see [7, Sect. 4.2]).

2 Total Variation Distance

In the classical case, the distance between ν and $\mu \in M_p(G)$ is given by the *total variation distance*:

$$\|\nu - \mu\| := \sup_{S \subset G} |\nu(\mathbb{1}_S) - \mu(\mathbb{1}_S)|.$$

Given a subset $S \subset G$, the indicator function on S, $\mathbb{1}_S = \sum_{s \in S} \delta_s$, is a projection, and all projections in $F(G)$ are formed in this way, and so there is a one-to-one correspondence between projections in $F(G)$ and subsets of G. Using this correspondence, this definition can be extended to quantum groups:

$$\|\nu - \mu\| := \sup_{p \in F(G) \text{ a projection}} |\nu(p) - \mu(p)|.$$

The Haar state is a normal, faithful trace, therefore non-commutative $\mathcal{L}^p$ machinery [13] can be used to put p-norms on $F(G)$:

$$\|a\|_p := \left(\int_G |a|^p \right)^{1/p}. \qquad (a \in F(G))$$

Set the infinity norm equal to the operator norm.

Proposition 2.1 *Let G be a quantum group and ν, $\mu \in M_p(G)$:*

$$\|\nu - \mu\| = \frac{1}{2} \left\| a_\nu - a_\mu \right\|_1.$$

Proof Standard non-commutative $\mathcal{L}^p$ machinery from [13] shows that:

$$\left\| a_\nu - a_\mu \right\|_1 = \sup_{\phi \in F(G) : \|\phi\|_\infty \leq 1} \left| \int_G ((a_\nu - a_\mu)\phi) \right| = \sup_{\|\phi\|_\infty \leq 1} |\nu(\phi) - \mu(\phi)|.$$

Embed the algebra of functions $F(G)$ into a matrix algebra. As everything is in finite dimensions, the supremum on the right is actually a maximum. Suppose that the two states ν and μ are given by density matrices: $\nu = \operatorname{Tr}(\rho_\nu \cdot)$ and $\mu = \operatorname{Tr}(\rho_\mu \cdot)$, and $\theta := \rho_\nu - \rho_\mu$ so that

$$\|a_\nu - a_\mu\|_1 = \max_{\|\phi\|_\infty \leq 1} |\operatorname{Tr}((\rho_\nu - \rho_\mu)\phi)| = \max_{\|\phi\|_\infty \leq 1} |\operatorname{Tr}(\theta\phi)|.$$

Using the $M_n(\mathbb{C})$ inequality ($\|A\|_1 = \operatorname{Tr}(|A|)$, $|A| := \sqrt{A^*A}$), $|\operatorname{Tr}(AB)| \leq \|A\|_1 \|B\|_\infty$,

$$\begin{aligned}\|a_\nu - a_\mu\|_1 &= \max_{\|\phi\|_\infty \leq 1} |\operatorname{Tr}(\theta\phi)| \\ &\leq \max_{\|\phi\|_\infty \leq 1} \|\theta\|_1 \|\phi\|_\infty = \|\theta\|_1.\end{aligned}$$

Note that θ is self-adjoint so that there exists a polar decomposition $\theta = U|\theta|$ such that U is self-adjoint. Consider $\phi = U$, the phase of θ, so that $U\theta = |\theta|$, and so

$$\operatorname{Tr}(\theta U) = \operatorname{Tr}(|\theta|) = \|\theta\|_1,$$

that is equality is achieved at the self-adjoint unitary U:

$$\|a_\mu - a_\nu\|_1 = |\nu(U) - \mu(U)|.$$

Consider $q = \frac{\mathbb{1}_G + U}{2}$. The self-adjointness of U gives $q = q^*$, and the fact U is also a unitary implies that U is an involution. This implies that $q = q^2$ and so q is a projection. The projection distance satisfies

$$\|\nu - \mu\| \geq |\nu(q) - \mu(q)| = \frac{1}{2}|\nu(U) - \mu(U)| = \frac{1}{2}\|a_\nu - a_\mu\|_1.$$

Now suppose the projection distance is attained at p and consider the infinity-norm-one element $\phi = 2p - \mathbb{1}_G \in F(G)$:

$$\begin{aligned}&\|a_\nu - a_\mu\|_1 \geq |\nu(2p - \mathbb{1}_G) - \mu(2p - \mathbb{1}_G)| = 2|\nu(p) - \mu(p)| = 2\|\mu - \nu\| \\ &\Rightarrow \|\nu - \mu\| \leq \frac{1}{2}\|a_\nu - a_\mu\|_1 \quad \bullet\end{aligned}$$

□

This proof takes advantage of the finite dimensionality of $F(G)$. Amaury Freslon [7, Lemma 2.6] states and proves a similar result for $\left\|\nu - \int_G\right\|$ which holds for not-necessarily finite-dimensional compact quantum groups whenever ν is a state on $\mathcal{L}^\infty(G)$ that has an $\mathcal{L}^1(G)$-density a_ν, which is not always the case in infinite dimensions.

The total variation distance defined using projections has three properties that suggest it is a suitable generalisation of the classical variation distance.

Firstly, in the case of a classical group, G, the one-to-one correspondence between projections and subsets, and between probability measures on G and states on $F(G)$, means that there is agreement in the classical case. Furthermore, there is a one-to-one correspondence between representations of a classical group G and *co*representations of $F(G)$ and this implies that estimates of the distance to random are the same whether the classical or quantum Upper Bound Lemmas are used.

The Upper Bound Lemma is a formula for a two norm, and thus requires a Cauchy–Schwarz inequality to relate back to the total variation distance. Such an inequality for total variation distance is provided for by non-commutative $\mathcal{L}^p$ machinery (see [13]):

$$\|ab\|_1 \leq \|a\|_2\|b\|_2 \,, \qquad (a,\, b \in F(G))$$

applied to

$$\begin{aligned}\|\nu-\mu\|^2 &= \frac{1}{4}\|a_\nu - a_\mu\|_1^2 = \frac{1}{4}\|(a_\nu - a_\mu)\mathbb{1}_G\|_1^2 \\ &\leq \frac{1}{4}\|a_\nu - a_\mu\|_2^2\|\mathbb{1}_G\|_2^2 = \frac{1}{4}\widehat{h}\left((\nu-\mu)^*(\nu-\mu)\right)\end{aligned}$$

The final equality here is the Plancherel Theorem 1.4.

Finally, to establish lower bounds, the total variation distance has a presentation(s) as a supremum:

$$\|\nu-\mu\| = \sup_{p\in F(G) \text{ a projection}} |\nu(p)-\mu(p)| = \frac{1}{2}\sup_{\|\phi\|_\infty\leq 1}|\nu(\phi)-\mu(\phi)| \qquad (3)$$

3 Diaconis–Van Daele Theory

3.1 Corepresentation Theory

Definition 3.1 A *corepresentation* of the algebra of functions on a quantum group G on a complex vector space V is a linear map $\kappa : V \to V \otimes F(G)$ that satisfies:

$$(\kappa \otimes I_{F(G)}) \circ \kappa = (I_V \otimes \Delta) \circ \kappa \qquad \text{and} \qquad (I_V \otimes \varepsilon) \circ \kappa = I_V .$$

The *dimension* of a corepresentation κ is given by the dimension of the vector space V. If V is finite dimensional, with basis $\{e_i\}_{i=1,\ldots,\dim V}$, the *matrix elements of* κ are the $(\dim V)^2$ elements of $F(G)$ given by

$$\kappa(e_j) = \sum_{i=1}^{\dim V} e_i \otimes \rho_{ij} .$$

If V is equipped with a Hermitian inner product, $\langle , \rangle_V$, define for $v, w \in V$, and $a, b \in F(G)$

$$\langle v \otimes a, w \otimes b \rangle_{F(G)} = \langle v, w \rangle_V \cdot a^* b.$$

A corepresentation κ on such a vector space is said to be *unitary* if $\langle \kappa(v), \kappa(w) \rangle_{F(G)} = \langle v, w \rangle_V \cdot \mathbb{1}_G$.

Let κ be a corepresentation on a vector space V. Denote by $\overline{V}$ the conjugate vector space of V and by $v \mapsto \overline{v}$ the canonical conjugate-linear isomorphism. Since Δ and ε are $*$-homomorphisms, the map

$$\overline{\kappa} : \overline{V} \to \overline{V} \otimes F(G), \quad \overline{e_j} \mapsto \sum \overline{e_i} \otimes \rho_{ij}^*,$$

is a corepresentation again, called the *conjugate* of κ.

As there is a one-to-one correspondence between representations of a classical group G and *co*representations of (commutative) $F(G)$, treating a quantum group G as an object, via an abuse of notation, corepresentations κ of a (noncommutative) $F(G)$ on V will be referred to as representations of (quantum) G on V.

Denoting the map $a_\varphi \mapsto \varphi$ by $\mathcal{F}$, so that $\mathcal{F}(a_\varphi) = \varphi$, for each $a_\varphi \in F(G)$ define the map $\widehat{a_\varphi} \in L(\overline{V})$:

$$\widehat{a_\varphi}(\kappa) = (I_{\overline{V}} \otimes \mathcal{F}(a_\varphi)) \circ \overline{\kappa} = (I_{\overline{V}} \otimes \varphi) \circ \overline{\kappa}.$$

Proposition 3.2 *For $a_{\varphi_1}, a_{\varphi_2} \in F(G)$ and κ a representation of G on V*

$$\widehat{a_{\varphi_1 \varphi_2}}(\kappa) = \widehat{a_{\varphi_1}}(\kappa) \circ \widehat{a_{\varphi_2}}(\kappa).$$

Furthermore, if κ is unitary then

$$\widehat{a_{\varphi^*}}(\kappa) = \widehat{a_\varphi}(\kappa)^*. \tag{4}$$

Note the first involution is in $F(\widehat{G})$ and the second is in $L(\overline{V})$.

Proof The proposition is that the map $F(\widehat{G}) \to L(\overline{V})$, $\varphi \mapsto \widehat{a_\varphi}(\kappa)$ is a $*$-homomorphism. See Timmermann for a proof (Proposition 3.1.7 ii. and v., [18]) • □

The following result is presented by Timmermann [18, Proposition 3.29 and Theorem 3.2.12], but is due to Woronowicz [25]

Theorem 3.3 (Finite Quantum Peter–Weyl Theorem) *Let $\mathcal{I} = \mathrm{Irr}(G)$ be an index set for a family of pairwise-inequivalent irreducible unitary representations of G. If d_α is the dimension of the vector space on which ρ^α acts ($\alpha \in \mathcal{I}$),*

$$\left\{ \rho_{ij}^\alpha \,|\, i, j = 1 \ldots d_\alpha, \ \alpha \in \mathcal{I} \right\},$$

the set of matrix elements of a family of pairwise-inequivalent irreducible unitary representations of G, is an orthogonal basis of F(G) with respect to the inner product

$$\langle a, b\rangle := \int_G a^* b\,, \qquad (a,\, b \in F(G))$$

such that $\langle \rho_{ij}^{\alpha}, \rho_{ij}^{\alpha}\rangle = 1/d_\alpha$ •

3.2 Fourier Transform

The following definition is similar to that of Simeng Wang ((2.5), [24]) save for a choice of left-right. As remarked upon by Simeng Wang, his definition is similar to earlier definitions of Kahng and also Caspers save for the presence of the conjugate representation $\overline{\kappa}_\alpha$ rather than κ_α itself. As Wang explains, the conjugate representation is used to be compatible with standard definitions in classical analysis on compact groups and hence most welcome for this work.

Definition 3.4 (The Fourier Transform) Let G be a quantum group with representation notation as before. Then the *Fourier transform* is a map:

$$F(G) \to \bigoplus_{\alpha\in \mathrm{Irr}(G)} L(\overline{V}_\alpha)\,,\ a_\varphi \mapsto \widehat{a_\varphi},$$

defined for each $\alpha \in \mathrm{Irr}(G)$:

$$\widehat{a_\varphi}(\alpha) = (I_{\overline{V}} \otimes \mathcal{F}(a_\varphi)) \circ \overline{\kappa_\alpha} = \left(I_{\overline{V_\alpha}} \otimes \varphi\right) \circ \overline{\kappa_\alpha}.$$

Each $\widehat{a_\varphi}(\alpha)$ is called the *Fourier transform* of a_φ at the representation κ_α. For $\varphi \in F(\widehat{G})$:

$$\widehat{\varphi}(\alpha) := \widehat{a_\varphi}(\alpha) = (I_{\overline{V_\alpha}} \otimes \varphi) \circ \overline{\kappa_\alpha},$$

so $\widehat{\varphi}$ is identified with $\widehat{a_\varphi}$.

The maps $\{\widehat{a_\varphi}(\alpha)\ :\ \alpha \in \mathrm{Irr}(G)\}$ play a key role in the sequel.

Theorem 3.5 (Diaconis–Van Daele Inversion Theorem) *Let ε be the counit of a quantum group G and* $\varphi \in F(\widehat{G})$*. Then*

$$\widehat{h}(\varphi) := \varepsilon\left(a_\varphi\right) = \sum_{\alpha\in Irr(G)} d_\alpha \operatorname{Tr}\left(\widehat{a_\varphi}\,(\alpha)\right).$$

where the sum is over the irreducible unitary representations of G.

Proof Both sides are linear in a_φ so it suffices to check $a_\varphi = \rho^\beta_{kl}$ for $\beta \in \mathrm{Irr}(G)$. The left-hand side reads

$$\varepsilon\left(\rho^\beta_{kl}\right) = \delta_{k,l}.$$

To calculate the right-hand-side, calculate for a given representation the trace of $\widehat{\rho^\beta_{kl}}(\alpha)$. Let $\overline{e_j} \in \overline{V_\alpha}$ and calculate

$$\begin{aligned}\widehat{\rho^\beta_{kl}}(\alpha)\,\overline{e_j} &= \left(I_{\overline{V_\alpha}} \otimes \mathcal{F}\left(\rho^\beta_{kl}\right)\right)\sum_i \overline{e_i} \otimes \left(\rho^\alpha_{ij}\right)^* \\ &= \sum_i \mathcal{F}\left(\rho^\beta_{kl}\right)\left(\rho^\alpha_{ij}\right)^* \overline{e_i} = \sum_i \int_G \left(\left(\rho^\alpha_{ij}\right)^* \rho^\beta_{kl}\right)\overline{e_i}\end{aligned}$$

This is zero unless $\alpha \equiv \beta$. If $\alpha \equiv \beta$ then

$$\begin{aligned}\widehat{\rho^\beta_{kl}}(\alpha)\,\overline{e_j} &= \int_G \left(\left(\rho^\beta_{kj}\right)^* \rho^\beta_{kl}\right)\cdot \overline{e_k} = \frac{1}{d_\beta}\delta_{j,l}\overline{e_k} \\ \Rightarrow \mathrm{Tr}\left(\widehat{\rho^\beta_{kl}}(\beta)\right) &= \sum_j \langle \overline{e_j}, \widehat{\rho^\beta_{kl}}(\beta)\overline{e_j}\rangle_{\overline{V_\beta}} = \sum_j \left\langle \overline{e_j}, \frac{1}{d_\beta}\delta_{j,l}\overline{e_k}\right\rangle = \frac{1}{d_\beta}\delta_{k,l}.\end{aligned}$$

Multiply this by d_β to get $\delta_{k,l}$ • □

Theorem 3.6 (Diaconis–Van Daele Convolution Theorem) *For a representation κ_α of G and $a_{\varphi_1}, a_{\varphi_2} \in F(G)$*

$$\widehat{a_{\varphi_1} \circledast a_{\varphi_2}}(\alpha) = \widehat{a_{\varphi_1}}(\alpha) \circ \widehat{a_{\varphi_2}}(\alpha).$$

Proof Recalling the identifications $\widehat{a_\varphi}(\alpha) =: \widehat{\mathcal{F}(a_\varphi)}(\alpha) = \widehat{\varphi}(\alpha)$, and that

$$\begin{aligned}\left(\widehat{a_{\varphi_1} \circledast a_{\varphi_2}}\right)(\alpha) &= \left(I_{\overline{V_\alpha}} \otimes \mathcal{F}\left(a_{\varphi_1} \circledast a_{\varphi_2}\right)\right)\overline{\kappa_\alpha} = \left(I_{\overline{V_\alpha}} \otimes \mathcal{F}\left(a_{\varphi_1\varphi_2}\right)\right)\overline{\kappa_\alpha} \\ &= \widehat{a_{\varphi_1\varphi_2}}(\alpha) = \widehat{a_{\varphi_1}}(\alpha) \circ \widehat{a_{\varphi_2}}(\alpha).\end{aligned}$$

The second equality uses Van Daele's Convolution Theorem 1.2 and final equality is Proposition 3.2 • □

Lemma 3.7 *Where the sum is over unitary irreducible representations,*

$$\widehat{h}(\varphi_1\varphi_2) = \sum_{\alpha \in Irr(G)} d_\alpha \,\mathrm{Tr}\left(\widehat{a_{\varphi_1}}(\alpha)\,\widehat{a_{\varphi_2}}(\alpha)\right). \qquad (\varphi_1, \varphi_2 \in F(\widehat{G}))$$

Proof The proof uses the convolution theorem of Van Daele and the definition of $\widehat{h}$ to find

$$\widehat{h}\left(\mathcal{F}\left(a_{\varphi_1}\right)\mathcal{F}\left(a_{\varphi_2}\right)\right) = \widehat{h}\left(\mathcal{F}\left(a_{\varphi_1} \circledast a_{\varphi_2}\right)\right) = \varepsilon\left(a_{\varphi_1} \circledast a_{\varphi_2}\right).$$

Now use the Diaconis–Van Daele Inversion Theorem 3.5 and the Diaconis–Van Daele Convolution Theorem 3.2

$$\varepsilon\left(a_{\varphi_1} \circledast a_{\varphi_2}\right) = \sum_{\alpha\in\mathrm{Irr}(G)} d_\alpha \operatorname{Tr}\left(\widehat{a_{\varphi_1} \circledast a_{\varphi_2}}(\alpha)\right) = \sum_{\alpha\in\mathrm{Irr}(G)} d_\alpha \operatorname{Tr}\left(\widehat{a_{\varphi_1}}(\alpha)\,\widehat{a_{\varphi_2}}(\alpha)\right) \quad \bullet$$

□

Proposition 3.8 *Suppose that* $\nu \in M_p(G)$. *If* κ_τ *is the trivial representation,* $\lambda \mapsto \lambda \otimes \mathbb{1}_G$, *then* $\widehat{a_\nu}(\tau) = I_{\overline{\mathbb{C}}}$.

Proof A calculation:

$$\widehat{a_\nu}(\tau)\,\lambda = \left(I_{\overline{\mathbb{C}}} \otimes \nu\right)\overline{\kappa_\tau}(\lambda) = \left(I_{\overline{\mathbb{C}}} \otimes \nu\right)(\lambda \otimes \mathbb{1}_G^*) = \lambda \otimes 1 \equiv \lambda \quad \bullet$$

□

Proposition 3.9 *Suppose that* κ_α *is a non-trivial and irreducible unitary representation, then* $\widehat{\mathbb{1}_G}(\alpha) = 0$.

Proof Another calculation:

$$\begin{aligned}\widehat{\mathbb{1}_G}(\alpha)\,\overline{e_j} &= (I \otimes \mathcal{F}(\mathbb{1}_G))\sum_i \overline{e_i} \otimes \left(\rho_{ij}^\alpha\right)^* = \sum_i \overline{e_i}\,\mathcal{F}(\mathbb{1}_G)\left(\rho_{ij}^\alpha\right)^* \\ &= \sum_i \int_G \left(\left(\rho_{ij}^\alpha\right)^* \mathbb{1}_G\right) \cdot \overline{e_i}.\end{aligned}$$

Note that $\mathbb{1}_G$ is the matrix element of the trivial representation and α is not equivalent to the trivial representation. Therefore, by the orthogonality relation $\widehat{\mathbb{1}_G}(\alpha) = 0$ • □

3.3 The Upper Bound Lemma

At this point, to mirror the classical notation, use π for $\int_G$. Recall, from Sect. 2, that

$$\|\nu - \pi\|^2 \le \frac{1}{4}\widehat{h}\left((\nu-\pi)^*(\nu-\pi)\right). \qquad (\nu,\ \mu \in F(\widehat{G}))$$

Lemma 3.10 (Diaconis–Shahshahani Upper Bound Lemma) *Let* $\nu \in M_p(G)$. *Then*

$$\|\nu^{\star k} - \pi\|^2 \le \frac{1}{4}\sum_{\alpha\in Irr(G)\setminus\{\tau\}} d_\alpha \operatorname{Tr}\left[\left(\widehat{\nu}(\alpha)^*\right)^k \left(\widehat{\nu}(\alpha)\right)^k\right],$$

where the sum is over all non-trivial irreducible unitary representations.

Proof Noting that $a_\pi = \mathbb{1}_G$ write

$$\widehat{h}\left((\nu-\pi)^*(\nu-\pi)\right) = \widehat{h}\left(\mathcal{F}(a_\nu - \mathbb{1}_G)^*\,\mathcal{F}(a_\nu - \mathbb{1}_G)\right).$$

Now using Lemma 3.7 and (4), this is equal to

$$\widehat{h}\left((\nu-\pi)^*(\nu-\pi)\right)=\sum_{\alpha\in\mathrm{Irr(G)}} d_\alpha \operatorname{Tr}\left[\widehat{(a_\nu-\mathbb{1}_G)}(\alpha)^*\,\widehat{(a_\nu-\mathbb{1}_G)}(\alpha)\right].$$

Now note that

$$\widehat{(a_\nu-\mathbb{1}_G)}(\alpha)=\widehat{a_\nu}(\alpha)-\widehat{\mathbb{1}_G}(\alpha).$$

If $\alpha=\tau$, the trivial representation, then this yields zero as both terms are the identity on $\mathbb{C}$. If α is non-trivial, then $\widehat{\mathbb{1}_G}(\alpha)=0$ and thus (using the notation $\widehat{a_\nu}(\alpha)=\widehat{\nu}(\alpha)$):

$$\|\nu-\pi\|^2\leq\frac{1}{4}\widehat{h}\left((\nu-\pi)^*(\nu-\pi)\right)=\frac{1}{4}\sum_{\alpha\in\mathrm{Irr}(G)\setminus\{\tau\}} d_\alpha \operatorname{Tr}\left[\widehat{\nu}(\alpha)^*\widehat{\nu}(\alpha)\right].$$

Apply Proposition 3.2 k times • □

Note this is *exactly* the same as the classical Diaconis–Shahshahani Upper Bound Lemma.

3.4 Lower Bounds

Lemma 3.11 (Lower Bound Lemma) *Suppose that $\nu\in M_p(G)$ and ρ the matrix element of a non-trivial one dimensional representation. Then*

$$\|\nu^{\star k}-\pi\|\geq\frac{1}{2}|\nu(\rho)|^k.$$

Proof Starting with (3), note that the argument from Proposition 3.9 shows that ρ has zero expectation under the random distribution. Note also that ρ is unitary (Proposition 3.1.7 v., [18]), thus norm one and thus a suitable test function.

Note that for a one dimensional representation, by Proposition 3.2:

$$|\nu^{\star k}(\rho)|=\left|\overline{\nu^{\star k}(\rho)}\right|=\left|\widehat{\nu^{\star k}}(\rho)\right|=|\widehat{\nu}(\rho)^k|=\left|\overline{\nu(\rho)}\right|^k=|\nu(\rho)|^k \quad \bullet$$

□

4 Examples

4.1 Random Walks on Dual Groups

Consider the dual quantum group, $\widehat{G}$, of a not-necessarily abelian group, G, given by $F(\widehat{G}):=\mathbb{C}G$. States on $F(\widehat{G})$ are given by positive definite functions $u\in F(G)=\mathbb{C}\widehat{G}$ (see Bekka, de la Harpe and Valette [1, Proposition C.4.2]). Furthermore, there is a correspondence between positive definite functions and unitary representations on G together with a vector. In particular, for each positive definite function u there exists a unitary representation $\rho : G \to \mathrm{GL}(V)$ and a vector $\xi \in V$ such that

$$u(s) = \langle \rho(s)\xi, \xi \rangle, \tag{5}$$

and for each unitary representation and vector (5) defines a positive definite function on G. For u to be a state it is required that $u(e) = 1$ and so $\langle \xi, \xi \rangle = 1$; i.e. ξ is a unit vector. Therefore probabilities on $\widehat{G}$ can be chosen by selecting a given representation and unit vector.

Since $\Delta(\delta^s) = \delta^s \otimes \delta^s$, it follows that $\kappa_s(\lambda) = \lambda \otimes \delta^s$ defines a (co)representation (with $\tau = \kappa_e$), and thus all irreducible representations of $\widehat{S_n}$ are of this form by counting. This makes the application of the upper bound lemma straightforward. Let $u \in M_p(\widehat{G})$ so that $\widehat{u}(\kappa_s) = \overline{u(s)}$ and so

$$\widehat{u}(\kappa_s)^* \widehat{u}(\kappa_s) = |u(s)|^2.$$

Therefore the upper bound lemma yields:

$$\|u^{\star k} - \pi\|^2 \leq \frac{1}{4} \sum_{t \in G \setminus \{e\}} |u(t)|^{2k}.$$

4.1.1 A Walk on $\widehat{S_n}$

Consider the quantum group $\widehat{S_n}$ (given by $F(\widehat{S_n}) := \mathbb{C}S_n$) with a state $u \in M_p(\widehat{S_n})$ given by the permutation representation on $\mathbb{C}^n$ given by $\pi(\sigma)(e_i) = e_{\sigma(i)}$ together with the unit vector ξ with components

$$\alpha_i = \sqrt{n^{n-i} \frac{n-1}{n^n - 1}}.$$

This vector is such that the α_i are in geometric progression with common ratio $\dfrac{1}{\sqrt{n}}$. For large n, this vector is approximately given by:

$$\xi \approx \left(1, \frac{1}{\sqrt{n}}, \frac{1}{\sqrt{n^2}}, \dots\right) \approx (1, 0, 0, \dots).$$

Following this through

$$u(\sigma) \approx \begin{cases} 1 & \text{if } \sigma(1) = 1, \\ 0 & \text{otherwise.} \end{cases}$$

To establish the upper bound some elementary inequalities will be used.

Lemma 4.1 *The following hold for $n \geq 5$:*

$$\left(\frac{4}{n}\right)^k (n-1) \leq 1, \ \textit{for } k > n^{n-1}\ln(n)/2, \tag{6}$$

$$\sqrt{\ln(n-1)} - n + 2 \leq 0, \tag{7}$$

$$A_n = \frac{(n-1)(\sqrt{n}-1)^2}{n^n - 1} n^{n-2} \to 1^- \ \textit{monotonically}, \tag{8}$$

$$g(n) = \frac{(\sqrt{n}-1)^2}{n^n - 1} n^{n-1} \to 1^- \ \textit{monotonically}. \tag{9}$$

Proof For (6), note that $k > n^2 \ln(n)/2$ and thus $\left(\frac{4}{n}\right)^k (n-1) \leq n \left(\frac{4}{n}\right)^{n^2\ln(n)/2} =:$ $h(n)$. Note that $h(5) < 1$ and

$$\frac{d}{dn}(\ln(h(n))) = n^2 \ln(n)\left[\ln\left(\frac{4}{n}\right)\right] + n^2\left[\ln\left(\frac{2e^{1/n^2}}{n}\right)\right] < 0,$$

while (7) is trivial. Note that $A_n \to 1$ and

$$\begin{aligned}\frac{(n^n-1)^2}{n^{n-3}(\sqrt{n}-1)}\frac{dA_n}{dn} &= n^n(n-2) + n^2\sqrt{n}(n^{n-2} - (\ln n + 1)) + n\ln n(n-1)\\ &\quad + n(n-2) + \sqrt{n}(n\ln n - 1) + n\sqrt{n} + 2 > 0\end{aligned}$$

for $n \geq 3$, and so the convergence is monotonic. Also $g(n) \to 1$ and

$$\frac{(n^n-1)^2}{n^{n-5/2}(\sqrt{n}-1)} g'(n) = n^2(n^{n-2}\sqrt{n} - (\ln n + 1)) + \sqrt{n}(n\ \ln n + n - 1) > 0$$

for $n \geq 3$ • □

Upper Bounds

For $k = \alpha\, n^{n-1}\ln(n)/2 + c\, n^{n-2}$, $\alpha > 1$, $c > 0$, and n large

$$\|u^{\star k} - \pi\|^2 \leq \frac{1}{2} e^{-2Ac}$$

where A can be chosen to be any $A < 1$ for n sufficiently large.

Proof Note that

$$u(\sigma) = \langle \xi, \pi(\sigma)\xi \rangle = \frac{n^{n+1} - n^n}{n^n - 1} \sum_{i=1}^{n} \frac{1}{\sqrt{n^{i+\sigma(i)}}}.$$

Therefore, using the Upper Bound Lemma,

$$\|u^{\star k} - \pi\|^2 \leq \frac{1}{4}\left(\frac{n^{n+1} - n^n}{n^n - 1}\right)^{2k} \sum_{\sigma \in S_n \backslash \{e\}} \left[\sum_{i=1}^{n} \frac{1}{\sqrt{n^{i+\sigma(i)}}}\right]^{2k}.$$

Define for $a_i = 1/\sqrt{n^i}$

$$S(\sigma) = \sum_{i=1}^{n} a_i a_{\sigma(i)}.$$

An *inversion for* σ is an ordered pair (j, k) with $j, k \in \{1, \ldots, n\}$ with $j < k$ and $\sigma(j) > \sigma(k)$. All non-identity permutations have at least one inversion.

Taking the approach of Steele [17, P. 79], for any inversion for σ, (j, k), define a new permutation by

$$\tau_1(i) := (j \quad k)\sigma(i) = \begin{cases} \sigma(i) & \text{if} \quad i \neq j, k, \\ \sigma(j) & \text{if} \quad i = k, \\ \sigma(k) & \text{if} \quad i = j. \end{cases}$$

Steele shows that

$$S(\sigma) \leq S(\tau_1).$$

That is, multiplying by $(j \quad k)$, whenever (j, k) is an inversion for σ, in this fashion, increases the number of fixed points, reduces the number of inversions (for τ_1), and increases S. This can be done iteratively until $\tau_r = e$. If the maximising $\sigma \in S_n/\{e\}$ were not a transposition, then it would be the product of at least two transpositions. Take one of the transpositions $(j \quad k)$: note that (j, k) is certainly an inversion for $(j \quad k)$. By the referenced calculation, $\tau_i := (j \quad k)\sigma$ has $S(\tau_i) \geq S(\sigma)$. Therefore, no matter what the starting permutation σ, τ_{r-1} is a transposition and therefore to maximise S on $S_n \backslash \{e\}$ one just maximises over transpositions.

The one-step differences between the a_i are decreasing in i and so the smallest one-step difference is between a_{n-1} and a_n. Let $(j \quad k)$ be a transposition with $j < k$. Note that

$$\begin{aligned} S((n-1 \quad n)) - S((j \quad k)) &= a_j^2 + a_k^2 + 2a_{n-1}a_n \\ &\quad - a_{n-1}^2 - a_n^2 - 2a_j a_k \\ &= (a_j - a_k)^2 - (a_n - a_{n-1})^2 \geq 0, \end{aligned}$$

so that S is maximised at $(n-1 \quad n)$.

Partition $S_n \backslash \{e\}$ into F_1 and F_1^C where F_1 is the set of permutations with $\sigma(1) = 1$. On F_1,

$$S(\sigma) \leq S((n-1 \quad n)) = \frac{1}{n^n}\frac{n^n - n^2}{n-1} + \frac{2\sqrt{n}}{n^n} =: f_0.$$

Now consider the maximum of S on F_1^C. From Steele's argument [17], it is known that strictly increasing the number of fixed points (by multiplying by suitably chosen transpositions), increases S. Also, if written in the disjoint cycle notation, elements of F_1^C must contain a cycle of the form $(1 \quad i_2 \quad \dots \quad i_N)$. By multiplying by suitably chosen transpositions, any disjoint cycle not containing 1 may be factored out whilst increasing S. Then write

$$(1 \quad i_2 \quad \dots \quad i_N) = (1 \quad i_N)(1 \quad i_{N-1}) \cdots (1 \quad i_2),$$

so that the maximum of S on F_1^C occurs at an element of the form

$$\sigma = \prod_{k=N}^{2} (1 \quad i_k).$$

All transpositions $(1 \quad i_k)$ give $(1, i_k)$ an inversions for $(1 \quad i_k)$, therefore can be removed—all the time increasing S—until one gets a transposition of the form $(1 \quad i)$. The maximum must occur at such a transposition. Note that

$$\begin{aligned} S((1 \quad 2)) - S((1 \quad i)) &= 2a_1a_2 + a_i^2 - 2a_1a_i - a_2^2 \\ &= (a_1 - a_i)^2 - (a_1 - a_2)^2 \geq 0. \end{aligned}$$

Therefore $S(1 \quad i) \leq S(1 \quad 2)$ and for any $\sigma \in F_1^C$:

$$S(\sigma) \leq S(1 \quad 2) = \frac{2\sqrt{n}}{n^2} + \frac{1}{n^2}\frac{1}{n^n}\frac{n^n - n^2}{n-1} =: f_1.$$

For $n \geq 4$, $f_1 \leq 2f_0/\sqrt{n}$ as

$$\begin{aligned} \frac{2}{\sqrt{n}}f_0 - f_1 = \frac{1}{n^2n^n(n-1)}&[n^3\sqrt{n}((n^{n-3} - 2) + n^{n-4}(n - \sqrt{n})) \\ &+ n^2(4n-3)] \geq 0. \end{aligned}$$

Therefore the Upper Bound Lemma yields:

$$\|u^{\star k} - \pi\|^2 \leq \frac{1}{4}\left(\frac{n^{n+1} - n^n}{n^n - 1}\right)^{2k}\left(\sum_{\sigma \in F_1} S(\sigma)^{2k} + \sum_{\sigma \in F_1^C} S(\sigma)^{2k}\right).$$

It is shown in the author's Ph.D. thesis [10, p. 93] that this yields:

$$\|u^{\star k} - \pi\|^2 \leq \frac{1}{4}\left(1 - \frac{(n-1)(\sqrt{n}-1)^2}{n^n - 1}\right)^{2k} \sqrt{n-1}(n-1)^{n-1}e^{2-n} \times \left[1 + \left(\frac{4}{n}\right)^k (n-1)\right]. \tag{10}$$

Using (6), and $(1-x) \leq \exp(-x)$, a rewriting yields:

$$\|u^{\star k} - \pi\|^2 \leq \frac{1}{2}\exp\left(-2k\,\frac{(n-1)(\sqrt{n}-1)^2}{n^n-1} + (n-1)\ln(n-1)\right) \times \exp\left(\ln(\sqrt{n-1}) - n + 2\right)$$

By (7) the second exponential is less than one. Writing $k = \alpha\, n^{n-1}\ln(n)/2 + c\, n^{n-2}$ and rewriting again

$$\|u^{\star k} - \pi\| \leq \frac{1}{2}\exp\left(-2c\,\overbrace{\frac{(n-1)(\sqrt{n}-1)^2}{n^n-1}n^{n-2}}^{=A_n}\right) \times \exp\left((n-1)\ln(n-1) - \alpha(n-1)\ln(n)\underbrace{\frac{(\sqrt{n}-1)^2}{n^n-1}n^{n-1}}_{=g(n)}\right)$$

Note that for $\alpha > 1$ and n sufficiently large by (9)

$$\begin{aligned}(n-1)\ln(n-1)\,(1-\alpha\, g(n)) &\leq 0\\ \Rightarrow (n-1)\ln(n-1) &\leq \alpha\,(n-1)\ln(n-1)\,g(n)\\ &\leq \alpha\,(n-1)\ln(n)\,g(n)\\ \Rightarrow (n-1)\ln(n-1) - \alpha\,(n-1)\ln(n)\,g(n) &\leq 0,\end{aligned}$$

and the result follows • □

If there were an effective lower bound for the total variation distance for $k = n^{n-1}\ln(n)/2 - c\, n^{n-2}$, and if such a bound approached one for large c, then the cut-off phenomenon would be exhibited for this random walk. Unfortunately no such result is at hand, but the following gives a partial result, showing that the random walk is still far from random for small multiples of n^{n-2}. The following will be used.

Lemma 4.2 *The following hold for, respectively,* $0 < x < 1/2$ *and* $n \geq 2$*:*

$$1 - x \geq e^{-x^2-x}, \tag{11}$$

$$B_n = \frac{n^n}{n^n - 1}\left(\frac{n^2}{n^n - 1} + 1\right) \to 1^+ \textit{ monotonically}. \tag{12}$$

Proof The bound (11) is [10, Lemma 5.5.1] with $n = 1$. Note that $B_n \to 1$, $B_2 > 1$, and note

$$\frac{(n^n - 1)^3}{n^n}\frac{dB_n}{dn} = n^n[2n - (1 + \ln n)(n^2 + 1)] - (1 + \ln n)n^2 \\ -(2n - (1 + \ln n)) < 0 \quad \bullet$$

□

Lower Bounds

For $k = \alpha\, n^{n-2}$, $\alpha > 0$, and n large

$$\|u^{\star k} - \pi\| \geq \frac{1}{2}e^{-B\alpha}$$

where B can be chosen to be any $B > 1$ for sufficiently large n.

Proof Using the Lower Bound Lemma 3.11 with the matrix element $\delta^{(n-1 \quad n)}$, using a calculation from the Ph.D. thesis [10, p. 95], together with (11),

$$\begin{aligned}\|u^{\star k} - \pi\| &\geq \frac{1}{2}\left|1 - \frac{(n-1)(\sqrt{n}-1)^2}{n^n - 1}\right|^{\alpha n^{n-2}} \\ &\geq \frac{1}{2}\exp\left[\left(-\frac{(n-1)^2(\sqrt{n}-1)^4}{(n^n-1)^2} - \frac{(n-1)(\sqrt{n}-1)^2}{n^n-1}\right)\alpha n^{n-2}\right] \\ &\geq \frac{1}{2}\exp\left[\left(-\frac{n^2 n^n}{(n^n-1)^2} - \frac{n^n}{n^n-1}\right)\alpha\right] \\ &= \frac{1}{2}\exp\left[\underbrace{\left(\frac{n^n}{n^n-1}\left(\frac{n^2}{n^n-1}+1\right)\right)}_{=B_n}\alpha\right] \quad \bullet\end{aligned}$$

□

4.2 A Family of Walks on the Sekine Quantum Groups

4.2.1 Sekine Quantum Groups

Sekine [16] introduced a family a finite quantum groups of order $2n^2$ that are neither commutative nor cocommutative.

The following follows the presentation of Franz and Skalski [6] rather than of Sekine. Let $n \geq 3$ be fixed and $\zeta_n = e^{2\pi i/n}$ and

$$\mathbb{Z}_n = \{0, 1, \ldots, n-1\},$$

with addition modulo n.

Consider n^2 one-dimensional spaces $\mathbb{C}e_{(i,j)}$ spanned by elements indexed by $\mathbb{Z}_n \times \mathbb{Z}_n$, $\{e_{(i,j)} : i, j \in \mathbb{Z}_n\}$. Together with a copy of $M_n(\mathbb{C})$, spanned by elements E_{ij} indexed by $\{(i, j) : i, j = 1, \ldots, n, 0 \equiv n\}$, a direct sum of these $n^2 + 1$ spaces, the $2n^2$ dimensional space

$$A_n = \left(\bigoplus_{i,j \in \mathbb{Z}_n} \mathbb{C}e_{(i,j)} \right) \oplus M_n(\mathbb{C}),$$

can be given the structure of the algebra of functions on a finite quantum group denoted by Y_n (so that $A_n = F(Y_n)$). On the one dimensional elements the comultiplication is given by, for $i, j \in \mathbb{Z}_n$:

$$\Delta(e_{(i,j)}) = \sum_{\ell,m \in \mathbb{Z}_n} (e_{(\ell,m)} \otimes e_{(i-\ell,j-m)}) + \frac{1}{n} \sum_{\ell,m=1}^{n} \left(\zeta_n^{i(\ell-m)} E_{\ell,m} \otimes E_{\ell+j,m+j} \right). \quad (13)$$

On the matrix elements in the $M_n(\mathbb{C})$ factor:

$$\Delta(E_{i,j}) = \sum_{\ell,m \in \mathbb{Z}_n} (e_{(-\ell,-m)} \otimes \zeta_n^{\ell(i-j)} E_{i-m,j-m}) + \sum_{\ell,m \in \mathbb{Z}_n} \left(\zeta_n^{\ell(j-i)} E_{i-m,j-m} \otimes e_{(\ell,m)} \right) \quad (14)$$

The antipode is given by $S(e_{(i,j)}) = e_{(-i,-j)}$ on the one dimensional factors and the transpose for the $M_n(\mathbb{C})$ factor. Sekine does not give the counit but by noting that $u_{(0,0)} = I_n$ (where $U \in M_n(M_n(\mathbb{C}))$ is defined in Sekine's original paper), it can be seen that the coefficient of the $e_{(0,0)}$ one-dimensional factor satisfies the counital property. The Haar state $\int_{Y_n} \in M_p(Y_n)$ is given by:

$$\int_{Y_n} \left(\sum_{i,j \in \mathbb{Z}_n} x_{(i,j)} e_{(i,j)} + a \right) = \frac{1}{2n^2} \left(\sum_{i,j \in \mathbb{Z}_n} x_{(i,j)} + n \cdot \mathrm{Tr}(a) \right).$$

Although Sekine restricts the construction to $n \geq 3$, for $n = 1$ and $n = 2$ the construction still satisfies the conditions of Kac and Paljutkin [9] and so are algebras of functions of quantum groups. Sekine does not clarify but the construction for $n = 2$ does not give the celebrated Kac–Paljutkin quantum group of order eight and indeed Y_2 is commonly mistaken for the Kac–Paljutkin quantum group in the literature. In fact, Y_2 is the dual group $\widehat{D_4}$ while Y_1 is the classical group $\mathbb{Z}_2$ [10].

To use the quantum Diaconis–Shahshahani Upper Bound Lemma, the representation theory of the quantum group must be well understood. The representation theory of the Sekine quantum groups changes according to the parity of the parameter n and the below restricts to n odd.

4.2.2 Representation Theory for n Odd

For n odd there are $2n$ one dimensional representations and $\binom{n}{2}$ two dimensional representations. Consider the convolution algebra $(F(Y_n), \circledast_{F(Y_n)})$. Sekine gives $2n$ minimal one-dimensional central projections, $\binom{n}{2}$ minimal two-dimensional central projections, and matrix units in the two-dimensional subspaces. Sébastian Palcoux (private communication, March 2016) suggests a connection between projections and matrix units in the convolution algebra and the comultiplication in the algebra of functions. Palcoux's approach uses slightly different Fourier transforms and convolutions—and the language of planar algebras (see [12])—therefore his results could not be used directly. However it was possible to show that the one-dimensional central projections in $(F(Y_n), \circledast_{F(Y_n)})$ were the matrix elements of the one dimensional representations, while the two-dimensional central projections were, up to a factor of two, the matrix elements of the irreducible two-dimensional representations (see the appendix to [10] for the brute-force verification). As far as the author knows this was not explicitly stated in the existing literature.

Let $\ell \in \{0, 1, \dots, n-1\}$. Then

$$\rho_\ell^{\pm} = \sum_{i,j\in\mathbb{Z}_n} \zeta_n^{i\ell} e_{(i,j)} \pm \sum_{m=1}^{n} E_{m,m+\ell},$$

are the $2n$ matrix elements of the one dimensional representations so that

$$\kappa_\ell^{\pm}(\lambda) = \lambda \otimes \rho_\ell^{\pm} \text{ and } \Delta(\rho_\ell^{\pm}) = \rho_\ell^{\pm} \otimes \rho_\ell^{\pm}.$$

Note that $\rho_0^{+} = \mathbb{1}_{Y_n}$ is the matrix element of the trivial representation.

Now let $u \in \{0, 1, \dots, n-1\}$ and $v \in \{1, 2, \dots, (n-1)/2\}$. Each pair gives a two dimensional representation $\kappa^{u,v} : \mathbb{C}^2 \to \mathbb{C}^2 \otimes F(Y_n)$ with matrix elements:

$$\begin{pmatrix} \rho_{11}^{u,v} & \rho_{12}^{u,v} \\ \rho_{21}^{u,v} & \rho_{22}^{u,v} \end{pmatrix} = \begin{pmatrix} \sum_{i,j\in\mathbb{Z}_n} \zeta_n^{iu+jv} e_{(i,j)} & \sum_{m=1}^{n} \zeta_n^{-mv} E_{m,m+u} \\ \sum_{m=1}^{n} \zeta_n^{mv} E_{m,m+u} & \sum_{i,j\in\mathbb{Z}_n} \zeta_n^{iu-jv} e_{(i,j)} \end{pmatrix}.$$

4.2.3 States on the Sekine Quantum Groups

Consider the basis of $F(\widehat{Y_n})$ dual to $\{e_{(i,j)} : i, j \in \mathbb{Z}_n\} \cup \{E_{i,j} : i, j = 1, 2, \dots, n\}$ given by

$$\begin{array}{ccc} e^{(i,j)} e_{(r,s)} = \delta_{i,r}\delta_{j,s} & \text{and} & e^{(i,j)} E_{r,s} = 0, \\ E^{i,j} e_{(r,s)} = 0 & \text{and} & E^{i,j} E_{r,s} = \delta_{i,r}\delta_{j,s} \end{array}.$$

Let $\mu \in F(\widehat{Y_n})$:

$$\mu = \sum_{i,j \in \mathbb{Z}_n} x_{(i,j)} e^{(i,j)} + \sum_{p,q=1}^{n} a_{pq} E^{p,q}.$$

Franz and Skalski [6] show that $\mu \in M_p(Y_n)$ if and only if

- $x_{(i,j)} \geq 0$ for all $i, j \in \mathbb{Z}_n$,
- the matrix $A = (a_{pq})$ is positive,
- $\mathrm{Tr}(\mu) := \sum_{i,j \in \mathbb{Z}_n} x_{(i,j)} + \sum_{p=1}^{n} a_{pp} = 1$.

4.2.4 A Random Walk

Where n is odd, and $J_n = \sum_{p,q=1}^{n} E^{p,q} \in M_n(\mathbb{C})$ the matrix of all ones, consider the state

$$\nu = \frac{1}{8}(e^{(0,1)} + e^{(1,0)} + e^{(-1,0)} + e^{(0,-1)}) + \frac{1}{2n} J_n \in M_p(Y_n).$$

Note that ν is not supported on a commutative subalgebra and so the associated random walk is not a classical Markov chain. The Diaconis–Shahshahani Upper Bound Lemma gives:

$$\|\nu^{\star k} - \pi\|^2 \leq \frac{1}{4} \sum_{\alpha \in \mathrm{Irr}(Y_n) \setminus \{\tau\}} d_\alpha \, \mathrm{Tr}\left[\left(\widehat{\nu}(\alpha)^*\right)^k \widehat{\nu}(\alpha)^k \right] \tag{15}$$

Proposition 4.3 *For $k = \alpha n^2$, with $\alpha > 1/20$ and $n \geq 3$*

$$\frac{1}{2} e^{-\alpha \pi^2/2} \leq \|\nu^{\star k} - \pi\| \leq c_n e^{-\alpha \pi^2/4},$$

with $c_n \to 1$ as $n \to \infty$.

Therefore, as Fig. 1 shows, the cut-off phenomenon does *not* occur.

Proof For the Upper Bound, consider first the one-dimensional representations $\kappa_\ell^{\pm}$ (except the trivial representation κ_0^{+}).

$$\begin{aligned} \nu(\rho_\ell^{\pm}) &= \frac{1}{8}\left(\zeta_n^\ell + \zeta_n^{-\ell} + 2\right) \pm 2 \\ &= \begin{cases} \frac{1}{2}\left(\cos^2\left(\frac{\pi \ell}{n}\right) + 1\right) & \text{if } \pm = + \\ -\frac{1}{2}\sin^2\left(\frac{\pi \ell}{n}\right) & \text{if } \pm = - \\ 0 & \text{for } \kappa_\ell^{\pm} = \kappa_0^{-} \end{cases}. \end{aligned}$$

Note that the κ_ℓ^{+} will dominate to such an extent that very trivial bounds may be used on the other terms.

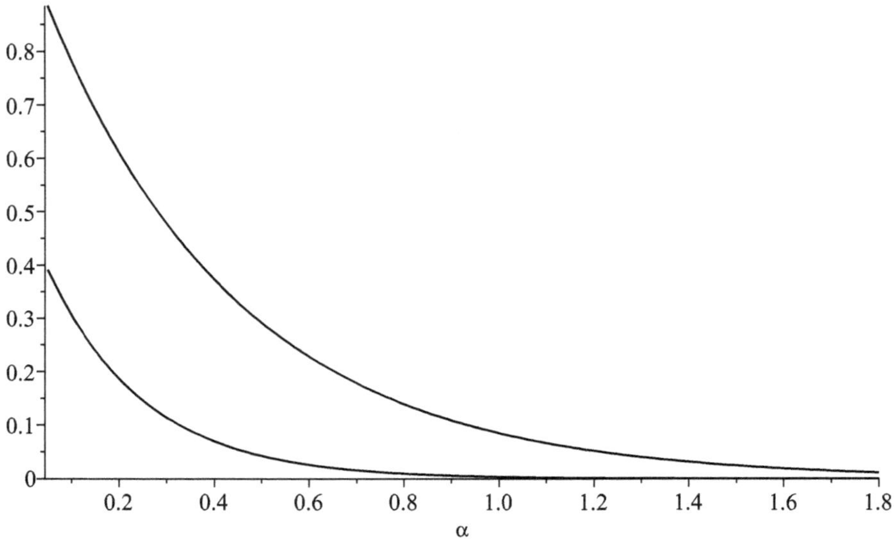

Fig. 1 Note $c_n \to 1$ very rapidly. Plotted here are the bounds for $c_n = 1$. Note that for $k = n^2/5$ the distance to random is bounded above by 3/5 and bounded below away from zero. However at $k = 3n^2/5$ the distance to random is still bounded away from zero

The contribution to (15) from the κ_ℓ^+ is given by

$$\frac{1}{4}\sum_{\alpha=\kappa_\ell^+} d_\alpha \operatorname{Tr}\left[\left(\widehat{\nu}(\alpha)^*\right)^k \widehat{\nu}(\alpha)^k\right] = \frac{1}{4}\sum_{\ell=1}^{n-1} |\nu(\rho_\ell^+)|^{2k} = \frac{1}{4}\sum_{\ell=1}^{n-1}\left(\frac{1}{2}\left(\cos^2\left(\frac{\pi\ell}{n}\right)+1\right)\right)^{2k}.$$

Using the symmetry of $\cos^2(x)$ about $x = \pi/2$, $\cos x \leq e^{-x^2/2}$ on $[0, \pi/2]$ [3, Theorem 2, p. 26], and employing the following inequality:

$$\frac{1}{2}(\cos^2 x + 1) \leq \frac{1}{2}(\cos x + 1) = \left(\cos\left(\frac{x}{2}\right)\right)^2 \leq \left(e^{-(x/2)^2/2}\right)^2 = e^{-x^2/4},$$

$$\Rightarrow \frac{1}{4}\sum_{\alpha=\kappa_\ell^+} d_\alpha \operatorname{Tr}\left[\left(\widehat{\nu}(\alpha)^*\right)^k \widehat{\nu}(\alpha)^k\right] \leq \frac{1}{2}\sum_{\ell=1}^{\frac{n-1}{2}} e^{-\pi^2\ell^2\alpha/2}$$

A similar sum occurs in the analysis of the classical walk on $\mathbb{Z}_n$. See p.26 of Diaconis [3] to see how such sums are handled (details teased out in Sect. 3.4 [11]):

$$\frac{1}{4}\sum_{\alpha=\kappa_\ell^+} d_\alpha \operatorname{Tr}\left[\left(\widehat{\nu}(\alpha)^*\right)^k \widehat{\nu}(\alpha)^k\right] \leq e^{-\alpha\pi^2/2}$$

for $\alpha \geq 1/20$ □

With the dominant term identified, the other terms can be bound crudely. In particular,

$$
\begin{aligned}
\frac{1}{4}\sum_{\alpha=\kappa_\ell^-} d_\alpha \operatorname{Tr}\left[\left(\widehat{\nu}(\alpha)^*\right)^k \widehat{\nu}(\alpha)^k\right] &= \frac{1}{4}\sum_{\ell=1}^{n-1}\frac{1}{4^k}\sin^{4k}\left(\frac{\pi\ell}{n}\right) \le \frac{1}{4}\sum_{\ell=1}^{n-1}\frac{1}{4^k} = \frac{1}{4}\frac{n-1}{4^{\alpha n^2}}\\
&\le \left(\frac{n}{4}\cdot\frac{e^{\alpha\pi^2/2}}{4^{\alpha n^2}}\right)e^{-\alpha\pi^2/2} \le \left(\frac{n}{4}\cdot\frac{e^{\alpha\pi^2/2}}{e^{\alpha n^2}}\right)e^{-\alpha\pi^2/2}\\
&\le \left(\frac{n}{4}\cdot e^{-\alpha(n^2-\pi^2/2)}\right)e^{-\alpha\pi^2/2}\\
&\underset{\alpha\ge 1/20}{\le} \left(\frac{n}{4}\cdot e^{-(n^2-\pi^2/2)/20}\right)e^{-\alpha\pi^2/2} =: d(n)\cdot e^{-\alpha\pi^2/2}.
\end{aligned}
$$

Note that $\widehat{\nu}(\kappa^{u,v})$ is real and diagonal:

$$
\begin{aligned}
\widehat{\nu}(\kappa^{u,v}) &= \frac{1}{2}\left(\cos\left(\frac{2\pi u}{n}\right)+\cos\left(\frac{2\pi v}{n}\right)\right)I_2\\
\Rightarrow \operatorname{Tr}[(\widehat{\nu}(\kappa^{u,v})^*)^k\widehat{\nu}(\kappa^{u,v})^k] &= \frac{2}{4^k}\left(\cos\left(\frac{2\pi u}{n}\right)+\cos\left(\frac{2\pi v}{n}\right)\right)^{2k}
\end{aligned}
$$

Using the trivial $|\cos(x)+\cos(y)|\le 2$ bound,

$$
\begin{aligned}
\frac{1}{4}\sum_{\alpha=\kappa^{u,v}} d_\alpha \operatorname{Tr}\left[\left(\widehat{\nu}(\alpha)^*\right)^k \widehat{\nu}(\alpha)^k\right] &\le \frac{1}{4^k}\sum_{\substack{u=0,\dots,n-1\\ v=1,\dots,\frac{n-1}{2}}} 2^{2k} = \frac{2^{2k}}{4^{2k}}\frac{n(n-1)}{2}\\
&\le \frac{1}{2}\frac{n^2}{4^{\alpha n^2}} \le \left(\frac{1}{2}n^2 e^{-\alpha(n^2-\pi^2/2)}\right)e^{-\alpha\pi^2/2}\\
&\le 2n\cdot d(n)\cdot e^{-\alpha\pi^2/2}.
\end{aligned}
$$

Putting these bounds together gives the result with

$$
c_n = \sqrt{1+d(n)+2n\cdot d(n)}
$$

Consider the Lower Bound Lemma 3.11 with ρ_1^+:

$$
\|\nu^{\star k}-\pi\| \ge \frac{1}{2}\left|\frac{\cos^2\left(\frac{\pi}{n}\right)+1}{2}\right|^k.
$$

Consider $h(x) = \ln\left(\frac{1}{2}(\cos^2 x + 1)\, e^{x^2/2}\right)$. Note $h(0) = h'(0) = h''(0) = 0$ but

$$h''(x) = \frac{(1 - \cos^2 x)(3 - \cos^2 x)}{(\cos^2 x + 1)^2} > 0,$$

for $x \in (0, \pi)$. Therefore $\frac{1}{2}(\cos^2 x + 1) \geq e^{-x^2/2}$ and

$$\|\nu^{\star k} - \pi\| \geq \frac{1}{2}e^{-\pi^2 k/2n^2} = \frac{1}{2}e^{-\alpha\pi^2/2} \quad \bullet$$

Acknowledgements I would like to thank Adam Skalski; Example 4.2 was developed during a visit to Adam at the Institute of Mathematics of the Polish Academy of Sciences (IMPAN), Warsaw, Poland. This trip was financially supported by IMPAN and also Cork Institute of Technology. I would like to thank Uwe Franz for assisting with Proposition 2.1. I would like to thank Amaury Freslon for encouragement and helpful comments; in particular for help in greatly improving the presentation of the bounds for the random walk on $\widehat{S_n}$. The rest of the paper was developed during the author's Ph.D. study at University College Cork, under the supervision of Stephen Wills.

References

1. Bekka, B., de la Harpe, P., Valette, A.: Kazhdans Property (T), New Mathematical Monographs, vol. 11. Cambridge University Press, Cambridge (2008)
2. Bhowmick, J., Skalski, A., Sołtan, P.M.: Quantum group of automorphisms of a finite quantum group. J. Algebra **423**, 514–537 (2015)
3. Diaconis, P.: Group Representations in Probability and Statistics. IMS, Hayward (1988)
4. Diaconis, P., Shahshahani, M.: Generating a random permutation with random transpositions. Z. Wahrscheinlichkeitstheorie Verw. Gebiete **57**, 159–179 (1981)
5. Franz, U., Gohm, R.: Random Walks on Finite Quantum Groups, Quantum Independent Increment Processes II. Lecture Notes in Mathematics, pp. 1–32. Springer, Berlin (2006)
6. Franz, U., Skalski, A.: On idempotent states on quantum groups. J. Algebra **322**(5), 1774–1802 (2009)
7. Freslon, A.: Cut-off phenomenon for random walks on free orthogonal free groups, to appear in Probab. Theory Relat. Fields (2018)
8. Freslon, A.: Quantum reflections, random walks and cut-off (2018). arXiv:1802.09402
9. Kac, G.I., Paljutkin, V.G.: Finite group rings. Trudy Moskov. Mat. Obšč. **15**, 224–261 (1966). Translated in Trans. Moscow Math. Soc. (1967), 251–284, 1996
10. McCarthy, J.P.: Random walks on finite quantum groups: Diaconis–Shahshahani theory for quantum groups. PhD thesis (2017). arXiv:1709.09357
11. McCarthy, J.P.: The cut-off phenomenon in random walks on finite groups. MSc thesis (2010). arXiv:1504.05387
12. Palcoux, S.: Ore's theorem for cyclic subfactor planar algebras and beyond. Pac. J. Math. **292–1**, 203–221 (2018)
13. Pisier, G.: Non-commutative $\mathcal{L}^p$-spaces. In: Johnson, W.B., Lindenstrauss, J. (eds.) Handbook of the Geometry of Banach Spaces, vol. 2, pp. 1459–1517. North Holland, Amsterdam (2003)
14. Rosenthal, J.S.: Random rotation: characters and random walks on $SO(N)$. Ann. Probab. **22**(1), 398–423 (1997)
15. Saloff-Coste, L.: Random Walks on Finite Groups (Probability on Discrete Structures), pp. 263–346. Springer, Berlin (2004)
16. Sekine, Y.: An example of finite-dimensional Kac algebras of Kac–Paljutkin type. Proc. Am. Math. Soc. **124**(4), 1139–1147 (1996)
17. Steele, J.M.: The Cauchy–Schwarz Master Class: An Introduction to the Art of Mathematical Inequalities. Cambridge University Press, New York (2004)

18. Timmermann, T.: An invitation to quantum groups and duality. Eur. Math. Soc. (2008)
19. Vainermann, L.I., Kac, G.I.: Nonunimodular ring groups and Hopf–von Neumann algebras. Dokl. Akad. Nauk SSSR **211**, 194–225; English transl. Math. USSR Sb. **23** (1974), 170–181 (1973)
20. Van Daele, A.: Multiplier Hopf algebras. Trans. Am. Math. Soc. **342**(2), 917–932 (1994)
21. Van Daele, A.: The Haar measure on finite quantum groups. Proc. Am. Math. Soc. **125**(12), 3489–3500 (1997)
22. Van Daele, A.: An algebraic framework for group duality. Adv. Math. **140**(2), 323–366 (1998)
23. Van Daele, A.: The Fourier transform for group duality (2007). Arxiv preprint, arXiv:math/0609502
24. Wang, S.: L_p-improving convolution operators on finite quantum groups. Indiana Univ. Math. J. **65**(5), 1609–1637 (2016)
25. Woronowicz, S.L.: Compact matrix pseudogroups. Commun. Math. Phys. **111**, 613–665 (1987)

Publisher's Note Springer Nature remains neutral with regard to jurisdictional claims in published maps and institutional affiliations.

Journal of Fourier Analysis and Applications (2019) 25:2492–2502
https://doi.org/10.1007/s00041-019-09671-3

A Paley–Wiener Type Theorem for Singular Measures on $(-1/2, 1/2)$

Eric S. Weber[1]

Received: 29 March 2018 / Revised: 27 January 2019 / Published online: 22 February 2019
© Springer Science+Business Media, LLC, part of Springer Nature 2019

Abstract
For a fixed singular Borel probability measure μ on $(-1/2, 1/2)$, we give several characterizations of when an entire function is the Fourier transform of some $f \in L^2(\mu)$. The first characterization is given in terms of criteria for sampling functions of the form $\hat{f}$ when $f \in L^2(\mu)$. The second characterization is given in terms of criteria for interpolation of bounded sequences on $\mathbb{N}_0$ by $\hat{f}$. Both characterizations use the construction of Fourier series for $f \in L^2(\mu)$ demonstrated by Herr and Weber via the Kaczmarz algorithm and classical results concerning the Cauchy transform of μ.

Keywords Entire function · Fourier transform · Paley–Wiener space · Singular measure

Mathematics Subject Classification Primary 30D10 · 42A38; Secondary 30E05 · 42A70 · 94A20

1 Introduction

The classical Paley–Wiener theorem states that an entire function F is the Fourier Transform for some $f \in L^2(-1/2, 1/2)$ if and only if F is of exponential type at most π and the restriction of F to $\mathbb{R}$ is square-integrable. An equivalent description of such entire functions is that F satisfy the following two conditions: (i) the values of F on the integer lattice are square-summable and (ii) for every $z \in \mathbb{C}$, $F(z)$ can be recovered from the values of F on the integer lattice via cardinal interpolation, i.e. the Shannon–Whittaker–Kotelnikov Sampling Theorem [3]. While this latter description of the inhabitants of the Paley–Wiener space is less elegant than the former, we will

---

Communicated by Yura Lyubarskii.

✉ Eric S. Weber
esweber@iastate.edu

1 Department of Mathematics, Iowa State University, 396 Carver Hall, Ames, IA 50011, USA

demonstrate that it is amenable to generalization to singular measures on $(-1/2, 1/2)$. Indeed, here is the question we shall answer: given a fixed singular Borel probability measure μ on $(-1/2, 1/2)$, when can an entire function F be written as

$$F(z) = \int_{-1/2}^{1/2} f(x) e^{-2\pi i x z}\, d\mu(x) \tag{1}$$

for some $f \in L^2(\mu)$? Note that we are *a priori* fixing the measure μ.

There are numerous results—in addition to the Paley–Wiener theorem—on when an entire function is the Fourier transform of a function, measure or distribution. The classical results in this regard include the Plancheral–Pólya theory [21], the "Bochner–Schoenberg–Eberlein conditions" contained collectively in [6,12,24], and the Beurling–Malliavin theory [4].

We consider the question from the viewpoint of the sampling and interpolating problems for bandlimited functions as presented in [14,29]. In particular, in [29] Strichartz poses a more difficult question than the one we address in the present paper: for a compact set K, when is an entire function F the Fourier transform of a (complex) measure supported on K?

2 Main Results

For our characterization of those functions which admit the representation in Eq. (1), we require both the Fourier-Stieltjes transform $\widehat{\mu}$ and the Cauchy transform μ_+ of μ:

$$\widehat{\mu}(z) := \int_{-1/2}^{1/2} e^{-2\pi i z x}\, d\mu(x); \qquad \mu_+(z) := \int_{-1/2}^{1/2} \frac{1}{1 - z e^{2\pi i x}}\, d\mu(x).$$

Note that μ_+ is nonvanishing on $\mathbb{D}$ since $Re(1/1 - z) > 1/2$ for $|z| < 1$; it is in fact the reciprocal of μ_+ that we require. We also need a description of the Fourier series expansions for $f \in L^2(\mu)$; this is given in the following theorem, which uses the Kaczmarz algorithm [17], as well as the main result of [18]. See [15] for details; see also [22] for the original proof of the existence of Fourier series for $f \in L^2(\mu)$.

The Kaczmarz algorithm is an iterative method for reconstructing a vector x in a Hilbert space H given the data $\{\langle x, \varphi_n \rangle\}_{n=0}^{\infty}$. It is defined as follows (assume $\|\varphi_n\| = 1$):

$$x_0 = \langle x, \varphi_0 \rangle \varphi_0; \qquad x_k = x_{k-1} + \langle x - x_{k-1}, \varphi_k \rangle \varphi_k.$$

Only certain sequences $\{\varphi_n\}_n$ have the property that $\|x_k - x\| \to 0$, those that do are called *effective*. Kaczmarz proves [18] that if $\{\varphi_n\}_n$ is a periodic sequence that spans H, then $\|x_k - x\| \to 0$. A complete characterization of which sequences are effective is given in [13]. The main result of [18] says that if $\{\varphi_n\}_n$ is a spanning, stationary sequence with singular spectral measure, then it is effective. Also in [18], if $\{\varphi_n\}_n$ is an effective sequence, then there is a second sequence, called the *auxiliary sequence* and denoted by $\{g_n\}_n$, such that for every $x \in H$,

$$x = \sum_{n=0}^{\infty} \langle x, g_n \rangle \varphi_n,$$

with convergence in the norm. The auxiliary sequence is obtained by the following recursion:

$$g_0 = \varphi_0; \qquad g_k = \varphi_k - \sum_{j=0}^{k-1} \langle \varphi_k, \varphi_j \rangle g_j.$$

Here we denote the exponential functions by $e_n(x) := e^{2\pi i n x}$; for a singular measure μ, $\{e_n\}_n$ is a stationary sequence which is effective.

Theorem A *Suppose μ is a singular Borel probability measure on $(-1/2, 1/2)$, and let $\{\alpha_n\}$ be the sequence of Taylor coefficients of $\dfrac{1}{\mu_+(z)}$. Define the sequence of functions $g_n(x) = \sum_{j=0}^{n} \overline{\alpha_{n-j}} e^{2\pi i j x}$. Then the sequence $\{g_n\}_{n=0}^{\infty} \subset L^2(\mu)$ has the property that for all $f \in L^2(\mu)$,*

$$f = \sum_{n=0}^{\infty} \langle f, g_n \rangle e_n = \sum_{n=0}^{\infty} \langle f, g_n \rangle g_n \tag{2}$$

with the convergence of both series occuring in the norm. Moreover, Parseval's identity holds: $\|f\|^2 = \sum_{n=0}^{\infty} |\langle f, g_n \rangle|^2$.

The $\{g_n\}_n$ as defined in Theorem A is the auxiliary sequence of the exponential functions in $L^2(\mu)$ [15]. As a consequence of Parseval's identity, the sequence $\{g_n\}_{n=0}^{\infty}$ is a Bessel sequence and hence for any square-summable sequence $\{c_n\} \in \ell^2(\mathbb{N}_0)$, the series $\sum_{n=0}^{\infty} c_n g_n$ also converges in norm.

2.1 Characterization Using Sampling Criteria

As noted previously, the Paley–Wiener theorem can be reformulated in terms of the Sampling Theorem. Our first characterization of which entire functions $F = \hat{f}$ for some $f \in L^2(\mu)$ is analogous.

Theorem 1 *Suppose μ is a singular Borel probability measure on $(-1/2, 1/2)$, and let $\{\alpha_n\}_{n=0}^{\infty}$ be the Taylor coefficients for $\dfrac{1}{\mu_+(z)}$. An entire function F admits the representation in Eq. (1) for some $f \in L^2(\mu)$ if and only if the following conditions hold:*

(i)

$$\sum_{n=0}^{\infty} \left| \sum_{j=0}^{n} \alpha_{n-j} F(j) \right|^2 < \infty;$$

(ii) for all $z \in \mathbb{C}$,

$$F(z) = \sum_{n=0}^{\infty} \left(\sum_{j=0}^{n} \alpha_{n-j} F(j) \right) \left(\sum_{k=0}^{n} \overline{\alpha_{n-k}} \widehat{\mu}(z-k) \right). \tag{3}$$

Proof For $F = \hat{f}$, note that $\sum_{j=0}^{n} \alpha_{n-j} F(j) = \langle f, g_n \rangle$, so the necessity of (i) follows by the Parseval identity. The necessity of (ii) follows by the previous observation and applying the Fourier transform to the second series expansion of f in Eq. (2).

We turn now to the sufficiency. Combining (ii) with the fact that the sequence $\{g_n\}_{n=0}^{\infty} \subset L^2(\mu)$ is a Bessel sequence, we define the function

$$f = \sum_{n=0}^{\infty} \left(\sum_{j=0}^{n} \alpha_{n-j} F(j) \right) g_n.$$

As this series converges in $L^2(\mu)$, we obtain

$$\begin{aligned} \hat{f}(z) &= \sum_{n=0}^{\infty} \left(\sum_{j=0}^{n} \alpha_{n-j} F(j) \right) \widehat{g_n}(z) \\ &= \sum_{n=0}^{\infty} \left(\sum_{j=0}^{n} \alpha_{n-j} F(j) \right) \left(\sum_{k=0}^{n} \overline{\alpha_{n-k}} \widehat{\mu}(z-k) \right) \\ &= F(z) \end{aligned}$$

by Item (iii). □

2.2 Characterization Using Interpolation Criteria

We consider now whether the sampling condition in Eq. (3) of Theorem 1 can be replaced by a different criteria. Our approach here is to view the characterization from an interpolation viewpoint rather than a sampling viewpoint. The question then becomes the following: when is $\{F(n)\}_{n=0}^{\infty}$ the sequence of Fourier moments of some $f \in L^2(\mu)$? In other words: given an entire function F does there exist some $f \in L^2(\mu)$ such that for all $n \in \mathbb{N}_0$,

$$F(n) = \int_{-1/2}^{1/2} f(x) e^{-2\pi i n x} \, d\mu(x)? \tag{4}$$

Certainly this is a necessary condition for $F = \hat{f}$, and in all (Theorem 2) but the extremal case (Theorem 3) concerning the support of μ this is sufficient.

The interpolation problem can be decided using the model subspaces of $H^2(\mathbb{D})$, where $\mathbb{D}$ denotes the open unit disc in $\mathbb{C}$. For a singular measure μ on $(-1/2, 1/2)$, there exists a unique inner function b on $\mathbb{D}$ given by the Herglotz Representation [20], i.e. there exists a unique inner function b such that:

$$Re\left(\frac{1+b(z)}{1-b(z)}\right) = \int_{-1/2}^{1/2} \frac{1-|z|^2}{|z-e^{2\pi i x}|^2} d\mu(x).$$

This inner function defines a backwards shift invariant space $\mathcal{H}(b) = H^2 \ominus bH^2$ as a consequence of Beurling's theorem [2]. Observe that $G \in \mathcal{H}(b)$ if and only if $T_{\overline{b}}G = 0$, where T_φ is the Toeplitz operator on $H^2(\mathbb{D})$ with symbol φ. Clark proves [7] that the Normalized Cauchy transform V_μ defined as

$$V_\mu : L^2(\mu) \to \mathcal{H}(b) : f \mapsto \frac{\displaystyle\int_{-1/2}^{1/2} \frac{f(x)}{1 - ze^{-2\pi i x}}\, d\mu(x)}{\mu_+(z)}. \tag{5}$$

is a unitary operator. If $G \in \mathcal{H}(b)$, then there exists a unique $g \in L^2(\mu)$ such that $G = V_\mu g$. We denote it by $g := G^\star$, and call $G^\star$ the $L^2(\mu)$-boundary of G because

$$\lim_{r \to 1^-} \|g(x) - G(re^{2\pi i x})\|_\mu = 0. \tag{6}$$

This limit was demonstrated by Poltoratskiĭ [22] (see also [1]).

The Normalized Cauchy Transform can be expressed in terms of the sequence $\{g_n\}$ appearing in Theorem A [15]:

$$V_\mu f(z) = \sum_{n=0}^{\infty} \langle f, g_n \rangle z^n, \qquad f \in L^2(\mu). \tag{7}$$

Therefore, we have the following characterization of the interpolation problem posed in Eq. (4).

Lemma 1 *Suppose μ is a singular Borel probability measure on $(-1/2, 1/2)$, b is the inner function on $\mathbb{D}$ associated to μ via the Herglotz representation, and suppose $\vec{a} := \{a_n\}_{n=0}^{\infty} \subset \mathbb{C}$. The following conditions are equivalent:*

(i) there exists a function $f \in L^2(\mu)$ with the property that

$$a_n = \int_{-1/2}^{1/2} f(x) e^{-2\pi i n x}\, d\mu(x); \tag{8}$$

(ii) the series $\sum_{n=0}^{\infty} a_n z^n$ has a radius of convergence of at least 1, and the following inclusion holds:

$$G_{\vec{a}}(z) := \frac{\sum_{n=0}^{\infty} a_n z^n}{\mu_+(z)} \in \mathcal{H}(b).$$

Proof Observe that

$$G_{\vec{a}}(z) = \left(\sum_{m=0}^{\infty} \alpha_m z^m \right) \left(\sum_{n=0}^{\infty} a_n z^n \right) = \sum_{n=0}^{\infty} \left(\sum_{j=0}^{n} \alpha_{n-j} a_j \right) z^n. \tag{9}$$

Suppose that the moment problem in Eq. (8) has a solution for some $f \in L^2(\mu)$. Combining Eqs. (9) and (7) demonstrates that $G_{\vec{a}} = V_\mu f$. Therefore, by (5) we obtain that $G_{\vec{a}} \in \mathcal{H}(b)$.

Conversely, if $G_{\vec{a}} \in \mathcal{H}(b)$, then reversing the previous argument yields the existence of a function $f \in L^2(\mu)$ such that $G_{\vec{a}}(z) = \sum_{n=0}^{\infty} \langle f, g_n \rangle z^n$. Since we have for every n

$$\sum_{j=0}^{n} \alpha_{n-j} a_j = \langle f, g_n \rangle = \sum_{j=0}^{n} \alpha_{n-j} \int_{-1/2}^{1/2} f(x) e^{-2\pi i j x} \, d\mu(x),$$

it now follows that Eq. (8) holds. □

For an entire function F of exponential type, we use h_F to denote the Phragmén-Lindelöf indicator function.

Theorem 2 *Suppose μ is a singular Borel probability measure with support in $[\alpha, \beta] \subset (-1/2, 1/2)$. Let b be the inner function associated to μ via the Herglotz Representation. The entire function F admits the representation in Eq. (1) if and only if*

(i) F is of exponential type;
(ii) the indicator function of F satisfies $h_F(\frac{\pi}{2}) \le 2\pi\beta$ and $h_F(-\frac{\pi}{2}) \le -2\pi\alpha$;
(iii) the following inclusion holds:

$$G_F(z) := \frac{\sum_{n=0}^{\infty} F(n) z^n}{\mu_+(z)} \in \mathcal{H}(b)$$

i.e. the function G_F is in the kernel of the Toeplitz operator $T_{\overline{b}}$.

Proof If F admits the representation in Eq. (1), then F satisfies (i) and (ii) using standard estimates (see, e.g. [19]; see also Lemma 2 below). Additionally, (iii) follows from Lemma 1.

Conversely, if F satisfies (i), (ii), and (iii), then by Lemma 1, there exists a $f \in L^2(\mu)$ such that $\hat{f}(n) = F(n)$ for $n \in \mathbb{N}_0$. Moreover, $\hat{f}$ also satisfies conditions (i) and (ii), i.e. $\hat{f}$ is of exponential type and the indicator function of $\hat{f}$ satisfies $h_{\hat{f}}(\frac{\pi}{2}) \le 2\pi\beta$ and $h_{\hat{f}}(-\frac{\pi}{2}) \le -2\pi\alpha$. Therefore, we must have $F = \hat{f}$ by Carlson's Theorem ([5, Theorem 9.2.1]). □

Corollary 1 *Suppose μ is a singular Borel probability measure with support in $[-1/2+\epsilon, 1/2-\epsilon]$, where $0 < \epsilon < 1/2$. Let b be the inner function associated to μ via the Herglotz Representation. The entire function F admits the representation in Eq. (1) if and only if*

(i) F is of exponential type at most $2\pi(1/2 - \epsilon)$;
(ii) the following inclusion holds:

$$G_F(z) := \frac{\sum_{n=0}^{\infty} F(n) z^n}{\mu_+(z)} \in \mathcal{H}(b)$$

i.e. the function G_F is in the kernel of the Toeplitz operator $T_{\overline{b}}$.

As noted previously, for $f \in L^2(\mu)$, standard arguments can be used to show that $|\hat{f}(z)| \le Ae^{\pi|z|}$ (see also Theorem 7.23 in [23]). We require a stronger growth estimate on $\hat{f}$.

Lemma 2 *If μ is a Borel measure on $(-1/2, 1/2)$, then for $f \in L^2(\mu)$, $\hat{f}$ satisfies the estimate*

$$|\hat{f}(z)| \le \varepsilon(|z|)e^{\pi|z|}, \qquad \varepsilon(r) = o(1). \tag{10}$$

Proof For $z = x + iy$ with $y > 0$, we estimate $\|e^{2\pi itz}\|_\mu^2$ as follows: let $\{x_n\}$ be an increasing sequence with $x_0 = -1/2$, $x_n < 1/2$, $x_n \to 1/2$ and $\gamma_n = \mu((x_{n-1}, x_n])$. We have

$$\int_{-1/2}^{1/2} |e^{2\pi itz}|^2 d\mu(t) = \sum_n \int_{(x_{n-1},x_n]} e^{4\pi ty} d\mu(t) \le O(1) + \sum_n \gamma_n e^{4\pi x_n y}.$$

Since $\sum \gamma_n < \infty$, we obtain

$$\frac{\int_{-1/2}^{1/2} |e^{2\pi itz}|^2 d\mu(t)}{e^{2\pi y}} \lesssim \sum_n \gamma_n e^{2\pi(2x_n-1)y} = o(1)$$

by Lebesgue's Convergence Theorem. The same estimate holds for $y < 0$. Equation (10) now follows from the Cauchy-Schwarz inequality. □

Theorem 3 *Suppose μ is a singular Borel probability measure on $(-1/2, 1/2)$, and let b be the inner function associated to μ by the Herglotz Representation. The entire function F admits the representation in Eq. (1) if and only if*

(i) $|F(z)| \le \varepsilon(|z|)e^{\pi|z|}$ with $\varepsilon(r) = o(1)$;
(ii) the following inclusions hold:

$$G_+(z) := \frac{\sum_{n=0}^\infty F(n)z^n}{\mu_+(z)} \in \mathcal{H}(b), \qquad G_-(z) := \frac{\sum_{n=0}^\infty \overline{F(-n)}z^n}{\mu_+(z)} \in \mathcal{H}(b);$$

(iii) the $L^2(\mu)$-boundaries of G_+ and G_- satisfy the relationship

$$\overline{G_+^\star} = G_-^\star.$$

Proof The necessity of Item (i) follows from Lemma 2; the necessity of Item (ii) follows from Lemma 1; and Item (iii) follows from a routine calculation.

For the converse, the issue again is when can the sequence $\{F(n)\}_{n=-\infty}^\infty$ be interpolated by a function of the form $\hat{f}$ for some $f \in L^2(\mu)$. If Item (ii) holds, then by Lemma 1, there exist $f_+, f_- \in L^2(\mu)$ such that for all $n \in \mathbb{N}_0$,

$$F(n) = \int_{-1/2}^{1/2} f_+(x)e^{-2\pi inx}\, d\mu(x); \qquad \overline{F(-n)} = \int_{-1/2}^{1/2} f_-(x)e^{-2\pi inx}\, d\mu(x). \tag{11}$$

Since $V_\mu f_+ = G_+$ and $V_\mu f_- = G_-$, we have $f_+ = G_+^\star$ and $f_- = G_-^\star$. Therefore, if in addition Item (iii) holds, then we have for all $n \in \mathbb{Z}$

$$F(n) = \int_{-1/2}^{1/2} f_+(x) e^{-2\pi i n x} \, d\mu(x). \tag{12}$$

Consequently, if Item (i) holds, F and $\hat{f}_+$ are both entire functions satisfying the same estimate by Lemma 2 and agree on $\mathbb{Z}$. Therefore, Carlson's Theorem in the form given in [5, Corollary 9.4.4] guarantees that $F = \hat{f}$. □

3 The Paley–Wiener Theorem and Exponential Frames

The power and beauty of the Paley–Wiener theorem is that the Paley–Wiener space is the intersection of two simple collections: entire functions of exponential type at most π and $L^2(\mathbb{R})$. In general, this simplicity cannot be replicated. There can be no integrability characterization for entire functions F that admit a representation as in Eq. (1), at least in the sense we make precise presently. We note that in [26] (see also [25,27]), Strichartz demonstrated that for measures μ which are "uniformly β-dimensional", the function $\hat{f}$ satisfies the following integrability/Plancherel identity condition:

$$\limsup_{R\to\infty} \frac{1}{R^{1-\beta}} \int_{-R}^{R} |\hat{f}(t)|^2 dt \simeq \|f\|_\mu^2. \tag{13}$$

However, this condition does not characterize the functions $\hat{f}$.

For our purposes here, let us denote $PW(\mu) = \{\hat{f} \mid f \in L^2(\mu)\}$, and let us denote by $\mathcal{C}_\tau$ the collection of all entire functions of exponential type at most τ and bounded on $\mathbb{R}$ $(0 < \tau \leq \pi)$. By a weight w on $\mathbb{R}$ we mean a nonnegative measurable function; we denote $L^2(w) := \{f \mid \int_{\mathbb{R}} |f(x)|^2 w(x) dx < \infty\}$. The following is a folklore result, which we include here for completeness.

Theorem B *For a singular probability measure μ on $(-1/2, 1/2)$, there is no weight or measure w such that $PW(\mu) = \mathcal{C}_\tau \cap L^2(w)$.*

Proof For convenience, assume that $L^2(\mu)$ is at least 3 dimensional. If so, there exists a function $F \in PW(\mu)$ that has at least two zeros, say z_1 and z_2. If $F \notin \mathcal{C}_\tau \cap L^2(w)$ we are done; so suppose $F \in \mathcal{C}_\tau \cap L^2(w)$. The function

$$G(z) := \frac{F(z)}{(z - z_1)(z - z_2)} \in \mathcal{C}_\tau \cap L^2(w).$$

We now have $G(x) \in L^2(\mathbb{R})$, so by the classical Paley–Wiener theorem, there exists $g \in L^2(-\tau, \tau)$ such that $G = \hat{g}$. Thus, $G \notin PW(\mu)$. □

We can quantify this discrepancy further by considering the problem of the existence of exponential bases and frames in $L^2(\mu)$.

Definition 1 An exponential frame for $L^2(\mu)$ is a sequence $\{\omega_n e^{2\pi i\lambda_n x}\}_{n\in\mathbb{Z}} \subset L^2(\mu)$ such that there exist constants $0 < A \leq B < \infty$ such that for all $f \in L^2(\mu)$,

$$A\|f\|_\mu^2 \leq \sum_n |\langle f, \omega_n e^{2\pi i\lambda_n \cdot}\rangle|^2 \leq B\|f\|_\mu^2.$$

An exponential Riesz basis is an exponential frame that ceases to be a frame if any of its elements are removed.

Jorgensen and Pedersen [16] prove that the uniform measure μ_3 on the middle-third Cantor set is not spectral, meaning it does not have an orthonormal basis of exponentials. Strichartz posed the problem of whether μ_3 has an exponential frame in [28]; this problem is still unresolved. We illustrate that a description of those functions F that admit the representation in Eq. (1) in terms similar to the classical Paley–Wiener theorem is related to the question of when the measure μ possesses an exponential frame or Riesz basis.

Theorem 4 *Suppose $PW(\mu) = \mathcal{C}_\tau \cap L^2(w)$ for some $\tau \in (0, \pi]$ and some weight or measure w on $\mathbb{R}$ with*

$$\|f\|_\mu \simeq \|\hat{f}\|_w. \tag{14}$$

Then there exists a Riesz basis of the form

$$\{\omega_n e^{2\pi i\lambda_n x}\}_{n\in\mathbb{Z}} \subset L^2(\mu) \tag{15}$$

for some sequence $\{\lambda_n\} \subset \mathbb{R}$ and $\omega_n > 0$. More generally, if $PW(\mu) \subset \mathcal{C}_\tau \cap L^2(w)$ is a closed subspace satisfying Eq. (14), then there exists a frame of the form (15).

Proof If $PW(\mu) = \mathcal{C}_\tau \cap L^2(w)$, then $PW(\mu)$ endowed with the $\|\cdot\|_w$ norm is a de Branges space. Indeed, by [8, Theorem 23], we need to verify the conditions identified as (H1), (H2), and (H3). Clearly $\mathcal{C}_\tau \cap L^2(w)$ satisfies (H1) and (H3). For (H2), note that if $w \in \mathbb{C}$ and $F \in PW(\mu)$, then by our assumption that the norms are equivalent, we obtain

$$|F(w)| \leq \|f\|_\mu \|e^{2\pi i w\cdot}\|_\mu \lesssim \|F\|_w. \tag{16}$$

It follows by Eq. (16) that $PW(\mu)$ is complete.

Consequently, by [8, Theorem 22] there exists a sequence $\{\lambda_n\} \subset \mathbb{R}$ such that for all $f \in L^2(\mu)$,

$$\|\hat{f}\|_w^2 = \sum_{n\in\mathbb{Z}} \frac{|\hat{f}(\lambda_n)|^2}{K(\lambda_n, \lambda_n)} \simeq \|f\|_\mu^2. \tag{17}$$

Here K is the reproducing kernel for the space; the sequence of kernels $\{K(\lambda_n, \cdot)\}$ form a complete orthogonal set in the space. This combined with Eq. (17) proves the claim, where $\omega_n = K(\lambda_n, \lambda_n)^{-1/2}$. □

Measures which possess a Riesz basis or frame of exponentials are rare (see e.g. [9,11]). An alternate proof of the second statement in Theorem 4 is found in [10].

Remark 1 We point out that our techniques in the present paper are quite similar to the alternative proof of the Paley–Wiener theorem given in [5, p. 106].

Acknowledgements We thank the anonymous referees for numerous suggestions that improved the presentation of this paper. We also thank the referees for bringing to our attention the folklore result Theorem B and the proof presented there. The results of this paper were inspired while the author attended the workshop "Hilbert Spaces of Entire Functions and their Applications" at the Institute of Mathematics, Polish Academy of Sciences (IM PAN). The author thanks the organizers of the workshop for the invitation to participate and IM PAN for their hospitality.

References

1. Aleksandrov, A.B.: Inner functions and related spaces of pseudocontinuable functions, Zap. Nauchn. Sem. Leningrad. Otdel. Mat. Inst. Steklov. (LOMI) **170** (1989), no. Issled. Lineĭn. Oper. Teorii Funktsiĭ. 17, 7–33, 321
2. Beurling, A.: On two problems concerning linear transformations in Hilbert space. Acta Math. **81**, 17 (1948)
3. Benedetto, J., Ferriera, P.J.S.G. (eds.): Modern Sampling Theory. Birkhauser, Basel (2001)
4. Beurling, A., Malliavin, P.: On Fourier transforms of measures with compact support. Acta Math. **107**, 291–309 (1962)
5. Boas Jr., R.P.: Entire Functions. Academic Press Inc., New York (1954)
6. Bochner, S.: A theorem on Fourier-Stieltjes integrals. Bull. Am. Math. Soc. **40**(4), 271–276 (1934)
7. Clark, D.N.: One dimensional perturbations of restricted shifts. J. Anal. Math. **25**, 169–191 (1972)
8. de Branges, L.: Hilbert Spaces of Entire Functions. Prentice-Hall Inc., Englewood Cliffs, NJ (1968)
9. Dutkay, D.E., Han, D., Sun, Q., Weber, E.: On the Beurling dimension of exponential frames. Adv. Math. **226**(1), 285–297 (2011)
10. Dutkay, D.E., Han, D., Weber, E.: Continuous and discrete Fourier frames for fractal measures. Trans. Am. Math. Soc. **366**(3), 1213–1235 (2014)
11. Dutkay, D.E., Lai, C.-K.: Uniformity of measures with Fourier frames. Adv. Math. **252**, 684–707 (2014)
12. Eberlein, W.F.: Characterizations of Fourier-Stieltjes transforms. Duke Math. J. **22**, 465–468 (1955)
13. Haller, R., Szwarc, R.: Kaczmarz algorithm in Hilbert space. Stud. Math. **169**, 123–132 (2005)
14. Huang, N.N., Strichartz, R.S.: Sampling theory for functions with fractal spectrum. Exp. Math. **10**(4), 619–638 (2001)
15. Herr, J.E., Weber, E.S.: Fourier series for singular measures. Axioms **6**(2:7), 13 (2017). https://doi.org/10.3390/axioms6020007
16. Jorgensen, P., Pedersen, S.: Dense analytic subspaces in fractal L^2-spaces. J. Anal. Math. **75**, 185–228 (1998)
17. Kaczmarz, Stefan: Angenäherte auflösung von systemen linearer gleichungen. Bull. Int. Acad. Pol. Sic. Let., Cl. Sci. Math. Nat. **35**, 355–357 (1937)
18. Kwapień, S., Mycielski, J.: On the Kaczmarz algorithm of approximation in infinite-dimensional spaces. Stud. Math. **148**(1), 75–86 (2001)
19. Paley, R.E., Wiener, N.: Fourier Transforms in the Complex Domain, vol. 19. American Mathematical Society Colloquium Publications, Providence (1987). Reprint of the 1934 original
20. Peter, L.: Duren, Theory of H^p Spaces, Pure and Applied Mathematics, vol. 38. Academic Press, New York (1970)
21. Plancherel, M., Pólya, G.: Fonctions entières et intégrales de fourier multiples. Comment. Math. Helv. **10**(1), 110–163 (1937)
22. Poltoratskiĭ, A.G.: Boundary behavior of pseudocontinuable functions. Algebra i Analiz **5**(2), 189–210 (1993). English translation in St. Petersburg Math. 5:2 (1994): 389–406
23. Rudin, W.: Functional Analysis. International Series in Pure and Applied Mathematics, 2nd edn. McGraw-Hill Inc., New York (1991)
24. Schoenberg, I.J.: A remark on the preceding note by Bochner. Bull. Am. Math. Soc. **40**(4), 277–278 (1934)
25. Strichartz, R.S.: Self-similar measures and their Fourier transforms. I. Indiana Univ. Math. J. **39**(3), 797–817 (1990)
26. Strichartz, R.S.: Self-similar measures and their Fourier transforms. II. Trans. Am. Math. Soc. **336**(1), 335–361 (1993)

27. Strichartz, R.S.: Self-similar measures and their Fourier transforms. III. Indiana Univ. Math. J. **42**(2), 367–411 (1993)
28. Strichartz, R.S.: Remarks on: "Dense analytic subspaces in fractal L^2-spaces" [J. Anal. Math. **75** (1998), 185–228; MR1655831 (2000a:46045)] by P. E. T. Jorgensen and S. Pedersen. J. Anal. Math. **75**, 229–231 (1998). MR 1655832
29. Strichartz, R.S.: Mock Fourier series and transforms associated with certain Cantor measures. J. Anal. Math. **81**, 209–238 (2000)

Publisher's Note Springer Nature remains neutral with regard to jurisdictional claims in published maps and institutional affiliations.

Journal of Fourier Analysis and Applications (2019) 25:2503–2520
https://doi.org/10.1007/s00041-019-09672-2

Markov Semi-groups Associated with the Complex Unimodular Group $Sl(2, \mathbb{C})$

Nizar Demni[1]

Received: 24 September 2018 / Published online: 21 February 2019
© Springer Science+Business Media, LLC, part of Springer Nature 2019

Abstract
In this paper, we derive the explicit expressions of the Markov semi-groups constructed by Biane (ESAIM Probab Stat 15:S2–S10, 2011) from the restriction of a particular positive definite function on the complex unimodular group $SL(2, \mathbb{C})$ to two commutative subalgebras of its universal $C^{\star}$-algebra. Our computations use Euclidean Fourier analysis together with the generating function of Laguerre polynomials with index -1, and yield absolutely-convergent double series representations of the semi-group densities. We also supply some arguments supporting the coincidence, noticed by Biane as well, occurring between the heat kernel on the Heisenberg group and the semi-group corresponding to the intersection of the principal and the complementary series. To this end, we appeal to the metaplectic representation $Mp(4, \mathbb{R})$ and to the Landau operator in the complex plane.

Keywords Positive definite functions · Intertwining operators · Gelfand pairs · Metaplectic representation · Landau Laplacian

Mathematics Subject Classification 60B15 · 60G51 · 42A82 · 22E46

1 Reminder: Intertwining Operators Arising from Gelfand Pairs

Two Markov semigroups $(P_t)_{t\geq 0}$ and $(Q_t)_{t\geq 0}$ defined on measurables spaces $(E, \mathscr{E})$ and $(F, \mathscr{F})$ respectively are said to be intertwined by a Markov kernel

$$\Lambda : (E, \mathscr{E}) \to (F, \mathscr{F})$$

Communicated by Hans G. Feichtinger.

✉ Nizar Demni
nizar.demni@univ-rennes1.fr

[1] IRMAR, Université de Rennes 1, Campus de Beaulieu, 35042 Rennes Cedex, France

if

$$Q_t \Lambda = \Lambda P_t. \tag{1}$$

When (1) holds, it allows for instance to transfer (under regularity assumptions on Λ) analytical and probabilistic properties of the semigroup $(P_t)_{t\geq 0}$ to $(Q_t)_{t\geq 0}$ as it is the case of the Brownian and the Dunkl semi-groups (see e.g. [11], CH. II). However, proving the existence of an intertwining relation and/or constructing the Markov kernel Λ is in general not obvious. For instance, numerous examples of intertwining operators were explicitly computed in [10] using the filtering procedure (see also [18] for further developments). In [7], it was shown that the semigroups of the Brownian motion in a finite-dimensional Euclidean space and of the Brownian motion conditioned to stay in the interior of the Weyl chamber of a finite Coxeter group are intertwined by means of the so-called Duistermaat–Heckman measure. Though this intertwining uses the filtering procedure as well, it relies heavily on the action of the Coxeter group on the underlying Euclidean space which allows in this setting for the construction of the Brownian motion from its conditioned process through the so-called generalized Pitman transforms. In the same vein and as explained in [6], the existence of a coupling between two Markov processes together with a suitable group action provides an intertwining relation between their semigroups. A noncommutative version of this construction was used in [5] to produce a Markov kernel intertwining the so-called noncommutative Bessel semigroup and the heat semi-group on $\mathbb{R}^2$. In that version, the noncommutativity is only concerned with the coupling process which corresponds in this new picture to a completely-positive contraction semigroup $(T_t)_{t\geq 0}$ on a noncommutative $C^\star$-algebra A. As to the intertwining relation (1), it holds between the restrictions of $(T_t)_{t\geq 0}$ to two commutative subalgebras B and C of A. Moreover, the substitute of the aforementioned group action is now a completely positive projection $\pi : A \to C$ which intertwines T_t and its restriction to C at any time $t > 0$. Further examples of intertwinings arising from these considerations were further provided and analyzed in [6]. One of them arises from the diagonal subgroup of the complex unimodular group $SL(2, \mathbb{C})$ and from the Gelfand pair $(SL(2, \mathbb{C}), SU(2))$, yet lacked the explicit expressions of the corresponding Markov semigroups and in turn of the intertwining kernel. More precisely, let

$$T := \left\{ \begin{pmatrix} e^x & 0 \\ 0 & e^{-x} \end{pmatrix}, \quad x \in \mathbb{R} \right\},$$

be the diagonal subgroup of $SL(2, \mathbb{C})$. Then, by the virtue of the Cartan decomposition, a $SU(2)$-bi-invariant function on $SL(2, \mathbb{C})$ depends only on the real variable x. Besides, the map

$$\varphi : x \mapsto x\coth(x) - 1 = x\frac{1 + e^{-2x}}{1 - e^{-2x}} - 1$$

defines a $SU(2)$-biinvariant conditionally positive definite function [8] or equivalently (by Schoenberg Theorem),

$$\psi_t : x \mapsto e^{t(1 - x\coth(x))}$$

is a continuous $SU(2)$-biinvariant positive definite function. The function φ is obtained from a suitable limit of positive definite spherical functions along a path in the complementary series tending to the trivial representation. In the representation-theoretical realm, such a function belongs to the so-called Lie cone and corresponds by the GNS construction to an infinitesimal small representation in a neighborhood of the trivial one together with an unbounded cocycle [25].

On the other hand, ϕ is the Lévy exponent of a background driving Lévy process generating the self-decomposable random variable whose characteristic function is $x/\sinh(x)$ [19]. In particular, the following Lévy–Kintchine formula holds:

$$1 - x\coth(x) = \int_{\mathbb{R}\setminus\{0\}} (e^{iux} - 1)\frac{\pi}{4}\frac{du}{\sinh^2(\pi u/2)}, \tag{2}$$

whence one deduces the action of the corresponding infinitesimal generator on sufficiently regular functions (see e.g. [2], Theorem 3.3.3., [8], Corollary 2.9):

$$\begin{aligned}\mathscr{L}(f)(u) &= \frac{1}{2\pi}\int_{\mathbb{R}} e^{iux}[1 - x\coth(x)]\mathcal{F}(f)(x)dx \\ &= \frac{\pi}{4}\int_{\mathbb{R}\setminus\{0\}} (f(u+v) - f(u))\frac{dv}{\sinh^2(\pi v/2)},\end{aligned}$$

where $\mathcal{F}$ stands for the Euclidean Fourier transform. A non commutative approach to this Lévy process stems from the previous considerations. Indeed, let ψ_t act by multiplication on the convolution algebra $L^1(Sl(2,\mathbb{C}))$, then this action defines a completely positive contraction semigroup on the universal $C^\star$-algebra and leaves invariant the commutative subalgebras $L^1(T)$. By completing $L^1(T)$ to the universal commutative $C^\star$-algebra $C^\star(T)$, the Lévy semi-group $(Q_t)_{t\geq 0}$ associated with $(\psi_t)_{t\geq 0}$ arises from the restriction of the aforementioned action to $C^\star(T)$. More concretely, the Gelfand spectrum of $C^\star(T)$ is isomorphic to $\mathbb{R}$ through the Euclidean Fourier transform and one has:

$$\mathcal{F}(f\psi_t)(-x) = \int_{\mathbb{R}} \mathscr{F}(f)(u)q_t(x-u)du,$$

for suitable $f \in L^1(T)$.

In [6], another new Markov semi-group was introduced by Biane using the fact that $(SL(2,\mathbb{C}), SU(2))$ is a Gelfand pair. In this case, the Gelfand spectrum of the convolution algebra of $SU(2)$-biinvariant functions is given by the set of bounded spherical functions:

$$x \mapsto \phi_\omega(x) = \frac{\sinh(\omega x)}{\omega\sinh(x)}, \quad \omega \in \Omega,$$

where $\Omega = i\mathbb{R} \cup [-1, 1]$. In particular, for any $\omega \in \Omega$, one obtains through the Gelfand–Fourier transform a Markov semi-group $(P_t)_{t\geq 0}$ such that:

$$\phi_\omega(x)\psi_t(x) = \int_\Omega \phi_\xi(x)P_t(\omega, d\xi). \tag{3}$$

Since ϕ_ω is also a positive definite function, then the RHS of (3) is nothing else but the decomposition of the positive definite function $\phi_\omega \psi_t$ into extreme ones (see e.g. [15]).

In this paper, we shall give explicit expressions of semi-groups of $(Q_t)_{t\geq 0}$ and $(P_t)_{t\geq 0}$. Our computations use the euclidean Fourier transform and appeal to the generating function of Laguerre polynomials with index -1 which are related to the so-called Lah numbers [9]. By standard arguments from Fourier analysis, $Q_t, t > 0$, is absolutely continuous with respect to Lebesgue measure on $\mathbb{R}$. As noticed in [6], the same holds true for $P_t, t > 0$, when either $\omega \in i\mathbb{R}$ (the principal series) or $\omega \in [-1, 1], t \leq |w|$ (the complementary series). Otherwise, an extra atom shows up in the Lebesgue decomposition of $P_t(w, d\xi), |w| > 0, 0 < t < |w|$. In the last part of the paper, we supply some arguments supporting the occurrence (noticed also by Biane in the same paper) of $\phi_0 \psi_t$ in the subelliptic heat kernel of the Heisenberg group or equivalently in the Lévy stochastic area formula [12]. To this end, we appeal to the Schrödinger operator with a constant magnetic field in the plane (sometimes referred to as the Euclidean Landau laplacian [3]) and its realization by means of the metaplectic representation of $Mp(4, \mathbb{R})$ [23]. In particular, the heat semi-group of the Landau Laplacian may be interpreted via the GNS construction as an average of a continuous family of unitary representations with respect to the two-dimensional Gaussian distribution. However, we still do not know whether or not these representations occur in the metaplectic representation though the answer seems to be positive. Indeed, it turns out that the radial part of the symplectic matrix representing this Schrödinger operator is the tensor product of a 2×2 matrix with the identity matrix. Since the maximal compact subgroup $O(4) \cap Sp(4, \mathbb{R})$ of the symplectic group may be identified with the unitary group $U(2)$, then we can turn the Cartan decomposition of this matrix in $Sp(4, \mathbb{R})$ into a Cartan decomposition of another one in $Sl(2, \mathbb{C})$.

The paper is organized as follows. In Sect. 2, we write down the semi-group density of $(Q_t)_{t\geq 0}$ while Sect. 3 contains the (three) different expressions of the kernel $P_t(\omega, d\xi)$ according to the values of the spectral parameter ω. The last section is devoted to the special value $\omega = 0$ in relation with the Heisenberg group and the the metaplectic representation.

2 Explicit Expression of the Lévy Semigroup $(Q_t)_{t\geq 0}$

It is clear from (2) that ψ_1 is the characteristic function of an infinitely divisible distribution on the real line generating the Lévy semi-group $(Q_t)_{t\geq 0}$ (see Corollary 1 in [19]). Moreover, since ψ_t is integrable for any $t > 0$, then the corresponding semi-group density q_t is smooth [20, Theorem 2.1] and may be computed using the Fourier inversion formula:

$$q_t(\xi) = \frac{1}{2\pi} \int_{\mathbb{R}} e^{-i\xi x} \psi_t(x) dx = \frac{e^t}{2\pi} \int_{\mathbb{R}} e^{-i\xi x} e^{-tx \coth(x)} dx, \quad \xi \in \mathbb{R}.$$

The following theorem provides a double-series representation for $q_t(\xi)$:

Theorem 1 *For any $t > 0$ and any $u \in \mathbb{R}$,*

$$q_t(\xi) = \frac{e^t}{\pi} \sum_{m \geq 0} \sum_{j \geq 0} \frac{(-2t)^m (m)_j}{j![(2j+2m+t)^2 + \xi^2]^{(m+1)/2}} T_{m+1} \times \left(\frac{2j+2m+t}{\sqrt{(2j+2m+t)^2+\xi^2}} \right),$$

where T_{m+1} is the $(m+1)$-th Tchebycheff polynomial of the first kind [1]*:*

$$T_{m+1}(\cos a) = \cos((m+1)a), \quad a \in \mathbb{R}.$$

Moreover, the double series is absolutely convergent.

Proof Expand

$$e^{-tx\coth(x)} = e^{-tx} \sum_{j \geq 0} L_j^{(-1)}(2tx) e^{-2jx}, x \geq 0, \tag{4}$$

where

$$L_j^{(-1)}(x) = \frac{1}{j!} \sum_{m=0}^{j} (-1)^m \binom{j}{m} (m)_{j-m} x^m, \quad j \geq 0, \tag{5}$$

are the Laguerre polynomials of index -1 [1] and for a real number a,

$$(a)_j = a(a+1)\cdots(a+j-1),$$

is the Pochhammer symbol with the convention $(0)_0 = 1$, $(0)_j = 0$, $j \geq 1$. Note that there is no constant term in these polynomials provided that $j \geq 1$ and that they are connected to Lah numbers [9]. Now, the following bound (see e.g. [22, p. 530]):

$$|L_j^{(-1)}(2tx)| \leq \frac{C}{j^{1/4}} e^{tx}, \tag{6}$$

where C is an absolute constant, allows to apply Fubini Theorem to compute

$$\begin{aligned}
\int_{\mathbb{R}} e^{-i\xi x} e^{-tx\coth(x)} dx &= 2 \int_0^\infty \cos(\xi x) \sum_{j \geq 0} L_j^{(-1)}(2x) e^{-(2j+t)x} dx \\
&= 2 \sum_{j \geq 0} \int_0^\infty \cos(\xi x) e^{-(2j+t)x} L_j^{(-1)}(2tx) dx \\
&= 2 \sum_{j \geq 0} \sum_{m=0}^{j} \frac{(-2t)^m}{(j-m)!m!} (m)_{j-m} \\
&\times \int_0^\infty \cos(\xi x) e^{-(2j+t)x} x^m dx.
\end{aligned}$$

Using formula 3.944. 6., in [16], we further get:

$$\int_{\mathbb{R}} e^{-i\xi x} e^{-tx\coth(x)} dx = 2\sum_{j\geq 0}\sum_{m=0}^{j} \frac{(-2t)^m}{(j-m)!} \frac{(m)_{j-m}}{[(2j+t)^2+\xi^2]^{(m+1)/2}} \times \cos\left\{(m+1)\arctan\left(\frac{\xi}{2j+t}\right)\right\}$$
$$= 2\sum_{j\geq 0}\sum_{m=0}^{j} \frac{(-2t)^m}{(j-m)!} \frac{(m)_{j-m}}{[(2j+t)^2+\xi^2]^{(m+1)/2}} T_{m+1} \times \left\{\cos\arctan\left(\frac{\xi}{2j+t}\right)\right\},$$

which together with the identity

$$\cos(y) = \frac{1}{\sqrt{1+\tan^2(y)}}, \quad y \in \left[-\frac{\pi}{2}, \frac{\pi}{2}\right],$$

yield

$$q_t(\xi) = \frac{e^t}{\pi}\sum_{j\geq 0}\sum_{m=0}^{j} \frac{(-2t)^m}{(j-m)!} \frac{(m)_{j-m}}{[(2j+t)^2+\xi^2]^{(m+1)/2}} T_{m+1}\left(\frac{2j+t}{\sqrt{(2j+t)^2+\xi^2}}\right). \tag{7}$$

Finally, use the bound

$$(2j+t+2m)^2 \geq 4(j+m)^2,$$

together with the Gamma integral:

$$\frac{1}{(j+m)^{m+1}} = \frac{1}{m!}\int_0^{\infty} v^m e^{-(j+m)v} dv$$

to see that

$$\sum_{j\geq 1}\sum_{m=1}^{j} \frac{(2t)^m}{(j-m)!} \frac{(m)_{j-m}}{[(2j+t)^2+\xi^2]^{(m+1)/2}}$$
$$= \sum_{m\geq 1}\sum_{j\geq m} \frac{(2t)^m}{(j-m)!} \frac{(m)_{j-m}}{[(2j+t)^2+\xi^2]^{(m+1)/2}}$$
$$= \sum_{m\geq 1}\sum_{j\geq 0} \frac{(2t)^m}{j!} \frac{(m)_j}{[(2j+2m+t)^2+\xi^2]^{(m+1)/2}}$$
$$\leq \sum_{m\geq 1} t^m \sum_{j\geq 0} \frac{(m)_j}{j!} \frac{1}{(j+m)^{m+1}}$$

$$= \int_0^\infty \sum_{m\geq 1} \frac{(tv)^m}{m!(e^v-1)^m} dv$$
$$= \int_0^\infty \left(e^{tv/(e^v-1)} - 1\right) dv \quad < \infty$$

uniformly in ξ. As a matter of fact, the double series in the RHS of (7) converges absolutely and the sought expression for q_t follows after inverting the summation order in (7). □

Remark 1 Recall from (2) that the Lévy measure of the semi-group $(q_t)_{t\geq 0}$ is given by:

$$\frac{\pi}{4} \frac{du}{\sinh^2(\pi u/2)} du, \quad u \neq 0.$$

Recall also from Lemma 2.16 in [8] that this measure is uniquely determined by the vague limit:

$$\lim_{t\to 0^+} \frac{1}{t} \int_{\mathbb{R}\setminus\{0\}} f(\xi) q_t(\xi) d\xi = \frac{\pi}{4} \int_{\mathbb{R}\setminus\{0\}} f(\xi) \frac{d\xi}{\sinh^2(\pi\xi/2)},$$

for compactly-supported functions f in $\mathbb{R} \setminus \{0\}$. Since $(0)_j = \delta_{j0}$ and since

$$T_1(u) = u, \quad T_2(u) = 2u^2 - 1,$$

then quick computations show:

$$\lim_{t\to 0^+} \frac{1}{t} q_t(\xi) = \frac{1}{\pi\xi^2} - \frac{1}{2\pi} \sum_{j\geq 1} \frac{j^2 - (\xi/2)^2}{[(j^2 + (\xi/2)^2]^2}.$$

Consequently, the Lebesgue convergence Theorem leads to the following identity:

$$\frac{1}{\sinh^2(\pi\xi/2)} = \frac{4}{\pi^2\xi^2} - \frac{2}{\pi^2} \sum_{j\geq 1} \frac{j^2 - (\xi/2)^2}{[(j^2 + (\xi/2)^2]^2},$$

which may be derived from

$$\pi \cot(\pi x) = \frac{1}{x} + \sum_{j\geq 1} \frac{2x}{x^2 - j^2}$$

after complexification and differentiation.

Remark 2 The Tchebycheff polynomial may be expanded as [1]:

$$T_{m+1}(y) = \sum_{k=0}^{[(m+1)/2]} a_{k,m} y^{m+1-2k},$$

for some real coefficients $a_{k,m}$. Plugging this expansion in (7), we get

$$q_t(\xi) = \frac{e^t}{\pi}\sum_{j\geq 0}\sum_{m=0}^{j}\frac{(-2t)^m (m)_{j-m}}{(j-m)!}\sum_{k=0}^{[(m+1)/2]} a_{k,m}\frac{(2j+t)^{m+1-2k}}{[(2j+t)^2+\xi^2]^{m+1-k}}.$$

On the other hand, the semi-group density of the Cauchy process in $\mathbb{R}^d$, $d \geq 1$, starting at the origin reads [2]:

$$\mathscr{C}_{s,d}(y) = \frac{1}{\pi^{(d+1)/2}}\Gamma\left(\frac{d+1}{2}\right)\frac{s}{(s^2+|y|^2)^{(d+1)/2}}, \quad y \in \mathbb{R}^d, s > 0.$$

Hence, q_t may be seen as a superposition of radial parts of Cauchy semi-group densities with varying even dimensions.

Remark 3 Let $(R_t)_{t\geq 0}$ be a Bessel process of dimension $\delta > 0$ and starting at $R_0 = 0$. In [14], Theorem 2.1, the author derives an explicit expression for the density f_a^δ of the conditional distribution of (see also [24]):

$$\int_0^1 R_s^2 ds$$

given $R_1 = a$. In particular, the Laplace transform of this random integral at $x^2/2$ is given by

$$\left(\frac{x}{\sinh(x)}\right)^{\delta/2} e^{(a^2/2)(1-x\coth(x))},$$

which tends to $\psi_{a^2/2}(x)$ as $\delta \to 0^+$. However,

$$\lim_{\delta\to 0^+} f_a^\delta \neq q_{a^2/2},$$

which may be checked directly from the double series representation of f_a^δ.

3 Explicit Expression of the Kernel $P_t(w, d\xi)$, $w \in \Omega$

The Lebesgue decomposition of the kernel $P_t(w, d\xi)$ depends on the decay of $\phi_\omega\psi_t$ at infinity and more precisely on the square-integrability of this function with respect to the radial part of Haar measure. Since the latter is given by $\sinh^2(x)dx$, we distinguish the three cases according to whether ω belongs to the principal series or to the complementary one with either $t \geq |\omega|$ or $t < |\omega|$, $\omega \neq 0$.

3.1 The Principal Series

Let $\omega \in i\mathbb{R}$, then $\phi_\omega\psi_t$ is square integrable with respect to the Haar measure. As a matter of fact, the decomposition (3) is in this case only over the principal series since the complementary one does not appear in the decomposition of the left regular representation into irreducible ones (see e.g. [4,13]). In this respect, we shall prove the following:

Theorem 2 *Let $\omega \in \mathbb{R}$ and $x > 0$, then*

$$\phi_{i\omega}(x)e^{t(1-x\coth(x))} = \int_{\mathbb{R}} \phi_{i\xi}(x)\frac{q_t(\xi-\omega)-q_t(\xi+\omega)}{2\omega}\xi d\xi.$$

In particular, $P_t(\omega, d\xi)$ is absolutely continuous with respect to Lebesgue measure $d\xi$.

Proof Since $\phi_{i\omega} = \phi_{-i\omega}$, then we shall assume $w \in \mathbb{R}_+$. Then

$$\frac{\sin(\omega x)}{\omega} = \frac{x}{2\omega}\int e^{ix\xi}\mathbf{1}_{[-\omega,\omega]}(\xi)d\xi,$$

where for $\omega = 0$, one uses the weak limit

$$\lim_{\omega\to 0^+}\frac{1}{2\omega}\mathbf{1}_{[-\omega,\omega]} = \delta_0.$$

It follows that

$$\begin{aligned}\frac{\sin(\omega x)}{\omega}e^{t(1-x\coth(x))} &= \frac{x}{2\omega}\int e^{ix\xi}\left[q_t\star\mathbf{1}_{[-\omega,\omega]}\right](\xi)d\xi,\\ &= \frac{x}{2\omega}\int_{\mathbb{R}}\cos(x\xi)\left[q_t\star\mathbf{1}_{[-\omega,\omega]}\right](\xi)d\xi,\end{aligned}$$

where

$$\left[q_t\star\mathbf{1}_{[-\omega,\omega]}\right](\xi) = \int_{\mathbb{R}} q_t(\xi-u)\mathbf{1}_{[-\omega,\omega]}(u)du = \int_{-\omega}^{\omega} q_t(\xi-u)du.$$

Now, it is straightforward from the double series representation of q_t (or from the Riemann-Lebesgue Lemma) that

$$\lim_{u\to\pm\infty} q_t(u) = 0,$$

so that an integration by parts yields:

$$\frac{\sin(\omega x)}{\omega}e^{t(1-x\coth(x))} = -\frac{1}{2\omega}\int_{\mathbb{R}}\sin(x\xi)\partial_\xi\left[q_t\star\mathbf{1}_{[-\omega,\omega]}\right](\xi)d\xi.$$

But

$$\partial_\xi\left[q_t\star\mathbf{1}_{[-\omega,\omega]}\right](\xi) = \partial_\xi\int_{-\omega}^{\omega} q_t(\xi-y)dy$$

$$= \int_{-\omega}^{\omega} -\partial_y (q_t(\xi - \cdot))(y) dy$$
$$= q_t(\xi + \omega) - q_t(\xi - \omega).$$

As a result,

$$\frac{\sin(\omega x)}{\omega} e^{t(1-x\coth(x))} = \int_{\mathbb{R}} \frac{\sin(x\xi)}{\xi} \frac{q_t(\xi - \omega) - q_t(\xi + \omega)}{2\omega} \xi d\xi,$$

as desired. □

Remark 4 A slightly different form of the previous result is:

$$\phi_{i\omega}(x) e^{t(1-x\coth(x))} = \int_{\mathbb{R}} \phi_{i\xi}(x) \frac{q_t(\xi - \omega) - q_t(\xi + \omega)}{2\omega\xi} \xi^2 d\xi.$$

Since $\xi^2 d\xi$ is the Plancherel measure of the group $Sl(2, C)$ (see for instance [21], p. 8 and use the fact that $Sl(2, C)$ is isomorphic to $SO(3, 1)$), Theorem 2 gives the inverse spherical Fourier transform of $\phi_\omega \psi_t$ when ω belongs to the principal series [21, p. 9]. Moreover, since q_t is even then the kernel

$$\frac{q_t(\xi - \omega) - q_t(\xi + \omega)}{2\omega\xi}$$

is so in both variables (ω, ξ).

Remark 5 The Markov property of the classical Markov process associated with the semi-group density $P_t(\omega, d\xi)$ may be deduced in this case directly from that of the Lévy process. Indeed, it is equivalent to the Chapman–Kolmogorov equation: for any $\xi, \omega, \gamma, \in \mathbb{R}$,

$$\gamma \int_{\mathbb{R}} \frac{q_t(\xi - \omega) - q_t(\xi + \omega)}{2\omega} \frac{q_s(\gamma - \xi) - q_s(\gamma + \xi)}{2\xi} \xi d\xi$$
$$= \gamma \frac{q_{t+s}(\gamma - \omega) - q_{t+s}(\gamma + \omega)}{2\omega}.$$

Remark 6 The fact that $P_t(\omega, d\xi)$ is a probability measure is equivalent to the harmonicity of the identity function with respect to Q_t:

$$\omega = \int_{\mathbb{R}} \frac{q_t(\xi - \omega) - q_t(\xi + \omega)}{2} \xi d\xi = \int_{\mathbb{R}} q_t(\omega - \xi) \xi d\xi.$$

In this respect, the action of the semi-group $(P_t)_{t \geq 0}$ on even functions in $C_0(\Omega)$ may be written as:

$$P_t(f)(\omega) = \int_{\mathbb{R}} f(\xi) \frac{\xi}{\omega} q_t(\xi - \omega) d\xi,$$

which allows to think of the corresponding Markov process as a Doob-transform of the Lévy process. Furthermore, we can see by direct computations that the intertwining operator is given by the kernel:

$$(\omega, x) \mapsto \int_{\mathbb{R}} \phi_{i\omega}(u) e^{iux} du = \frac{\pi}{\omega} \frac{\sinh(\pi\omega)}{\cosh(\pi\omega) + \cosh(\pi x)}$$

where the equality follows from formula 3.986. 2, in [16].

3.2 The Complementary Series

Now, let $\omega \in (-1, 1)$ be in the complementary series and assume $t \geq |\omega|$. Then, $\phi_\omega \psi_t$ is still square integrable therefore we have a similar decomposition as in the previous theorem:

$$\phi_\omega(x)\psi_t(x) = \int_{\mathbb{R}} \frac{\sin(\xi x)}{\xi \sinh(x)} P_t(\omega, d\xi),$$

or equivalently

$$\frac{\sinh(\omega x)}{\omega} \psi_t(x) = \int_{\mathbb{R}} \frac{\sin(\xi x)}{\xi} P_t(\omega, d\xi).$$

More precisely,

Theorem 3 *Let $\omega \in (-1, 1)$ and $t \geq |w|$. Then*

$$\phi_\omega(x)\psi_t(x) = \frac{e^t}{2\pi} \int_{\mathbb{R}} \phi_{i\xi}(x) \frac{I_t^-(\omega, \xi) - I_t^+(\omega, \xi)}{\omega\xi} \xi^2 d\xi,$$

where $I_t^{\mp}(\omega, \xi)$ are absolutely convergent double series defined below.

Proof Since $\phi_\omega(x) = \phi_\omega(-x)$ and from the uniqueness of the integral representation (3), we deduce that the probability measure $P(\cdot, d\xi)$ is symmetric. It is also absolutely continuous with respect to Lebesgue measure on $\mathbb{R}$: indeed,

$$\partial_x \left(\frac{\sinh(\omega \cdot)}{\omega} \psi_t \right)(x) = \int_{\mathbb{R}} e^{i\xi x} P_t(\omega, d\xi)$$

and the LHS of this equality is integrable as a function of the variable x. Hence, for any $\omega \in [0, 1]$ and any $t \geq \omega$,

$$\begin{aligned} P_t(\omega, \xi) &= \frac{1}{2\pi} \int_{\mathbb{R}} e^{-ix\xi} \partial_x \left(\frac{\sinh(\omega \cdot)}{\omega} \psi_t \right)(x) dx \\ &= i \frac{\xi}{2\pi\omega} \int_{\mathbb{R}} e^{-ix\xi} \sinh(\omega x) \psi_t(x) dx \\ &= \xi \frac{e^t}{2\pi\omega} \int_0^\infty \sin(x\xi)(e^{-(t-\omega)x} - e^{-(t+\omega)x}) \sum_{j \geq 0} L_j^{(-1)}(2tx) e^{-2jx} dx. \end{aligned}$$

Using the bound (6), we can split the above integral into the difference of the following series:

$$I_t^{\mp}(\omega,\xi) := \sum_{j\geq 0}\int_0^{\infty} \sin(x\xi)L_j^{(-1)}(2tx)e^{-(2j+t\mp\omega)x}dx,$$

which, by the virtue of (5) together with formula 3.944. 5., in [16], may be expanded as

$$\begin{aligned} I_t^{\mp}(\omega,\xi) &= \sum_{j\geq 0}\sum_{m=0}^{j}(m)_{j-m}\frac{(-2t)^m}{(j-m)!m!}\int_{\mathbb{R}} x^m \sin(x\xi)e^{-(2j+t\mp\omega)x}dx \\ &= \sum_{j\geq 0}\sum_{m=0}^{j}(m)_{j-m}\frac{(-2t)^m}{(j-m)![(2j+t\mp\omega)^2+\xi^2]^{(m+1)/2}} \\ &\quad \times \sin\left\{(m+1)\arctan\left(\frac{\xi}{2j+t\mp\omega}\right)\right\}. \end{aligned}$$

Moreover, the same arguments used in the proof of Theorem 1 show that we can intertwine the summation order in such a way that

$$\begin{aligned} I_t^{\mp}(\omega,\xi) = \sum_{m\geq 0}\sum_{j\geq 0}(m)_j \frac{(-2t)^m}{j![(2j+2m+t\mp\omega)^2+\xi^2]^{(m+1)/2}} \\ \times \sin\left\{(m+1)\arctan\left(\frac{\xi}{2j+2m+t\mp\omega}\right)\right\}. \end{aligned}$$

Since $\phi_\omega = \phi_{-\omega}$ and $I_t^-(\omega,\xi) = I_t^+(-\omega,\xi)$, then the proof is complete. □

Finally, we deal with the case $t < |\omega|, \omega \in (-1,1)\setminus\{0\}$. Then, the spherical function ϕ_ω decays as $e^{(|w|-t-1)x}$, $x \to +\infty$ so that a residual part shows up in the decomposition (3). Actually, straightforward computations show that the map

$$x \mapsto \phi_\omega(x)\psi_t(x) - e^t\frac{|\omega|-t}{|\omega|}\phi_{|\omega|-t}(x) = \frac{e^t}{|\omega|\sinh(x)}[\sinh(|\omega|x)e^{-tx\coth(x)} - \sinh((|\omega|-t)x)]$$

is square integrable with respect to the radial part of the Haar measure.[1] As a matter of fact, it may be decomposed as:

$$\phi_\omega(x)\psi_t(x) - e^t\frac{|\omega|-t}{|\omega|}\phi_{|\omega|-t}(x) = \int_{\mathbb{R}} \phi_{i\xi}(x)G_t(\omega,d\xi), \tag{8}$$

for some (submarmovian) kernel $G_t(\omega,d\xi)$ and the latter may be computed along similar, yet more complicated, lines as those written in the proof of Theorem 3. For sake of simplicity, we shall assume (without loss of generality) that $\omega \in (0,1)$ and prove that:

[1] There is missed factor e^t in [6].

Theorem 4 *The kernel of $G_t(\omega, d\xi)$ is absolutely continuous with respect to Lebsegue measure:*

$$G_t(\omega, d\xi) = \xi \frac{e^t}{2\pi\omega}\left[J_t(\omega,\xi) - I_t^+(\omega,\xi) + \frac{1}{\sqrt{\xi^2+(\omega-t)^2)}} \times \sin\left(\arctan\left(\frac{\xi}{\omega-t}\right)\right)\right] d\xi,$$

where $J_t(\omega, \xi)$ is is displayed in (9).

Proof We start by writing (8) as

$$\frac{e^t}{\omega}\partial_x\left[\sinh(\omega x)e^{-tx\coth(x)} - \sinh((\omega-t)x)\right] = \int_{\mathbb{R}} e^{ix\xi}G_t(\omega, d\xi),$$

and noting that the derivative of the function

$$x \mapsto \sinh(\omega x)e^{-tx\coth(x)} - \sinh((\omega-t)x)$$

is integrable. Consequently,

$$\begin{aligned} G_t(\omega, d\xi) &= \frac{e^t}{2\pi\omega}\int_{\mathbb{R}} e^{-ix\xi}\partial_x\left[\sinh(\omega x)e^{-tx\coth(x)} - \sinh((\omega-t)x)\right]dx \\ &= i\xi\frac{e^t}{2\pi\omega}\int_{\mathbb{R}} e^{-ix\xi}\left[\sinh(\omega x)e^{-tx\coth(x)} - \sinh((\omega-t)x)\right]dx \\ &= \xi\frac{e^t}{\pi\omega}\int_{\mathbb{R}} \sin(x\xi)\left[\sinh(\omega x)e^{-tx\coth(x)} - \sinh((\omega-t)x)\right]dx. \end{aligned}$$

Now, split

$$\begin{aligned} \sinh(\omega x)e^{-tx\coth(x))} - \sinh((\omega-t)x) =& \frac{e^{\omega x}}{2}\left[e^{-tx\coth(x)} - e^{-tx}\right] \\ &- \frac{1}{2}\left[e^{-(\omega+t\coth(x))x} - e^{-(\omega-t)x}\right], \end{aligned}$$

and use the generating function (4) to write:

$$e^{-tx\coth(x)} - e^{-tx} = \sum_{j\geq 1} L_j^{(-1)}(2tx)e^{-(2j+t)x}.$$

Together with the estimate (6) lead to the following absolutely-convergent integrals:

$$J_t(\omega,\xi) := \int_0^\infty \sin(x\xi)\left\{\sum_{j\geq 1} L_j^{(-1)}(2tx)e^{-(2j+t-\omega)x}\right\}dx$$

$$= \sum_{j\geq 1} \int_0^\infty \sin(x\xi) L_j^{(-1)}(2tx) e^{-(2j+t-\omega)x} dx$$
$$= \sum_{j\geq 1} \sum_{m=0}^{j} (m)_{j-m} \frac{(-2t)^m}{(j-m)!m!} \int_0^\infty x^m \sin(x\xi) e^{-(2j+t-\omega)x} dx,$$

and

$$K_t(\omega, \xi) := \int_0^\infty \sin(x\xi) \left\{ e^{-(\omega+t)x} \sum_{j\geq 0} L_j^{(-1)}(2tx) e^{-2jx} - e^{-(\omega-t)x} \right\} dx$$
$$= \sum_{j\geq 0} \int_0^\infty \sin(x\xi) e^{-(\omega+t+2j)x} L_j^{(-1)}(2tx) dx - \int_0^\infty \sin(x\xi) e^{-(\omega-t)x} dx$$
$$= I_t^+(\omega, \xi) - \int_0^\infty \sin(x\xi) e^{-(\omega-t)x} dx.$$

Finally, formula 3.944. 5., in [16] provides the expressions:

$$J_t(\omega, \xi) = \sum_{j\geq 1} \sum_{m=0}^{j} (m)_{j-m} \frac{(-2t)^m}{(j-m)![(2j+t-\omega)^2 + \xi^2]^{(m+1)/2}}$$
$$\times \sin\left\{ (m+1) \arctan\left(\frac{\xi}{2j+t-\omega} \right) \right\}, \tag{9}$$

and

$$\int_0^\infty \sin(x\xi) e^{-(\omega-t)x} dx = \frac{1}{\sqrt{\xi^2 + (\omega-t)^2)}} \sin\left(\arctan\left(\frac{\xi}{\omega - t} \right) \right),$$

which finish the proof. □

4 $\omega = 0$: The Lévy Stochastic Area

For $w = 0$, Theorem 2 specializes to

$$\phi_0(x)\psi_t(x) = \frac{x}{\sinh(x)} \psi_t(x) = \lim_{\omega\to 0} \int_{\mathbb{R}} \phi_{i\xi}(x) \frac{q_t(\xi-\omega) - q_t(\xi+\omega)}{2\omega\xi} \xi^2 d\xi$$
$$= \frac{1}{\sinh(x)} \int_{\mathbb{R}} \sin(\xi x) \lim_{\omega\to 0} \frac{q_t(\xi-\omega) - q_t(\xi+\omega)}{2\omega} d\xi$$
$$= \frac{1}{\sinh(x)} \int_{\mathbb{R}} \sin(\xi x) \partial_\xi q_t(\xi) d\xi.$$

This result may be directly derived from an integration by parts, namely:

$$x\psi_t(x) = \int_{\mathbb{R}} x\cos(\xi x)q_t(\xi)d\xi = -\int_{\mathbb{R}} \sin(\xi x)\partial_\xi q_t(\xi)d\xi.$$

On the other hand, $\phi_0\psi_t$ is also the Euclidean Fourier transform of the Lévy stochastic area at unit time conditional on the planar Brownian motion defining it being on the circle of radius $\sqrt{2t}$ [12, Lemma 2]. In [6], the author asks for an explanation of this coincidence and we supply in the sequel some arguments supporting it. Of course, this is far from being a definitive answer.

Our key observation is that the Lévy stochastic area rather encodes the heat kernel of the so-called Landau Laplacian in the plane (the Schrödinger operator in the plane with a perpendicular constant magnetic field [17]):

$$H_b := -\frac{1}{2}\left[(\partial_x + iby)^2 + (\partial_y - ibx)^2\right]$$

where $b > 0$ is the strength of the magnetic field. This fact is not surprising and stems from the fact that this Hamiltonian may be obtained from the sublaplacian of the Heisenberg group after performing a partial Fourier transform with respect to the vertical variable. Now, H_b may be realized as the image of a matrix $\mathcal{A}$ in the Lie algebra $sp(4, \mathbb{R})$ by the derived metaplectic representation $Mp(4, \mathbb{R})$ (see [23] for further details). Indeed, let $\alpha := b/(2\pi)$ and consider the matrix $\mathcal{A} \in sp(4, \mathbb{R})$:

$$\mathcal{A}_\alpha = \begin{pmatrix} \alpha A & I_2 \\ -\alpha^2 I_2 & \alpha A \end{pmatrix}, \qquad A := \begin{pmatrix} 0 & -1 \\ 1 & 0 \end{pmatrix},$$

where I_2 is the 2×2 identity matrix. Then, the spectrum of H_b is the countable set $\{b(2m+1), m \geq 0\}$ and with π denoting the metaplectic representation, we have

$$H_b = \frac{d}{dt}(\pi[e^{-t\mathcal{A}_\alpha}])_{t=0} \equiv d\pi(\mathcal{A}_\alpha)$$

where $d\pi$ is the derived representation. Besides, its heat semi-group at unit time is given by (see e.g. [17,23], Eq. (6.1)):

$$\pi[e^{-\mathcal{A}_\alpha}](f)(0) = e^{-d\pi(\mathcal{A}_\alpha)}(f)(0) = \int_{\mathbb{R}^2} \phi_0(b)\psi_{|y|^2/2}(b)f(y)e^{-|y|^2/2}\frac{dy}{2\pi}. \qquad (10)$$

for any square integrable function f. Thus, if $(\rho_y, v_y, \mathcal{H}_y)$, $y \in \mathbb{R}^2$, denotes the unitary representation of $Sl(2, \mathbb{C})$ associated with the positive definite function $\phi_0\psi_{|y|^2/2}$ via the GNS construction,[2] then (10) may be written as an average of the continuous family of representations $(\rho_y, v_y, \mathcal{H}_y)_{y\in\mathbb{R}^2}$ with respect to the Gaussian distribution $e^{-|y|^2/2}/(2\pi)$:

[2] $\mathcal{H}_y$ is the representation Hilbert space endowed with an inner product $\langle\cdot,\cdot\rangle_{\mathcal{H}_y}$ and $v_y \in \mathcal{H}_y$ is a cyclic $SU(2)$-invariant unit vector.

$$\pi[e^{-\mathcal{A}_\alpha}](f)(0) = \int_{\mathbb{R}^2} \langle \rho_y(\mathrm{diag}(e^b, e^{-b}))v_y, v_y\rangle_{\mathcal{H}_y} e^{-|y|^2/2} \frac{dy}{2\pi}.$$

In this respect, it seems like the representations $(\rho_y)_{y\in\mathbb{R}^2}$ occur in the decomposition of π into irreducible ones.

This possible guess is indeed strengthened by the fact that the Cartan decomposition of $e^{-\mathcal{A}_\alpha}$ is somehow special in the sense that it may be turned into a Cartan decomposition in $Sl(2,\mathbb{C})$. More precisely, straightforward computations show that $\mathcal{A}_\alpha^2 = 2\alpha\mathcal{A}\mathcal{J}_\alpha$ where

$$\mathcal{J}_\alpha := \begin{pmatrix} 0_2 & (1/\alpha)I_2 \\ -\alpha I_2 & 0_2 \end{pmatrix}$$

satisfies $\mathcal{J}_\alpha^2 = -I_4$. As a result, one gets for any $j \geq 1$,

$$\mathcal{A}_\alpha^{2j} = (-1)^{j-1}(2\alpha)^{2j-1}\mathcal{A}_\alpha\mathcal{J}_\alpha, \quad \mathcal{A}_\alpha^{2j-1} = (-1)^{j-1}(2\alpha)^{2j-2}\mathcal{A}_\alpha,$$

whence it follows that for any $t \geq 0$,

$$\begin{aligned} e^{-t\mathcal{A}_\alpha} &= I_4 + \frac{1-\cos(2\alpha t)}{2\alpha}\mathcal{A}_\alpha\mathcal{J}_\alpha + \frac{\sin(2\alpha t)}{2\alpha}\mathcal{A}_\alpha \\ &= I_4 + \frac{\sin^2(\alpha t)}{\alpha}\mathcal{A}_\alpha\mathcal{J}_\alpha + \frac{\sin(\alpha t)\cos(\alpha t)}{\alpha}\mathcal{A}_\alpha. \end{aligned}$$

or in block form:

$$e^{-t\mathcal{A}_\alpha} = \begin{pmatrix} \cos^2(\alpha t)I_2 + \sin(\alpha t)\cos(\alpha t)A & [\sin^2(\alpha t)A + \sin(\alpha t)\cos(\alpha t)I_2]/\alpha \\ -\alpha[\sin(\alpha t)\cos(\alpha t)I_2 + \sin^2(\alpha t)A] & \cos^2(\alpha t)I_2 + \sin(\alpha t)\cos(\alpha t)A \end{pmatrix}.$$

Besides, setting

$$\begin{aligned} U_\alpha(t) &:= \cos^2(\alpha t)I_2 + \sin(\alpha t)\cos(\alpha t)A \\ V_\alpha(t) &:= \sin^2(\alpha t)A + \sin(\alpha t)\cos(\alpha t)I_2, \end{aligned}$$

and noting that $A^2 = -I_2$, we readily get the relations

$$\begin{aligned} U_\alpha(t)U_\alpha^T(t) &= \cos^2(\alpha t)I_2 \\ V_\alpha(t)V_\alpha^T(t) &= \sin^2(\alpha t)I_2, \\ U_\alpha(t)V_\alpha(t)^T &= \sin(\alpha t)\cos(\alpha t)I_2 = U_\alpha(t)^T V_\alpha(t), \end{aligned}$$

which in turn show that:

$$e^{-t\mathcal{A}_\alpha}e^{-t\mathcal{A}_\alpha^T} = \begin{pmatrix} \cos^2(\alpha t) + \dfrac{\sin^2(\alpha t)}{\alpha^2} & \sin(\alpha t)\cos(\alpha t)\left(\dfrac{1}{\alpha} - \alpha\right) \\ \sin(\alpha t)\cos(\alpha t)\left(\dfrac{1}{\alpha} - \alpha\right) & \cos^2(\alpha t) + \alpha^2\sin^2(\alpha t) \end{pmatrix} \otimes I_2,$$

and similarly

$$e^{-t\mathcal{A}_\alpha^T} e^{-t\mathcal{A}_\alpha} = \begin{pmatrix} \cos^2(\alpha t) + \alpha^2 \sin^2(\alpha t) & \sin(\alpha t)\cos(\alpha t)\left(\dfrac{1}{\alpha} - \alpha\right) \\ \sin(\alpha t)\cos(\alpha t)\left(\dfrac{1}{\alpha} - \alpha\right) & \cos^2(\alpha t) + \dfrac{\sin^2(\alpha t)}{\alpha^2} \end{pmatrix} \otimes \quad I_2.$$

Consequently, the Cartan decomposition of $e^{-\mathcal{A}_\alpha}$ admits the block form:

$$e^{-\mathcal{A}_\alpha} = M_\alpha \begin{pmatrix} \lambda_1 I_2 & \\ & \lambda_2 I_2 \end{pmatrix} R_\alpha,$$

where $M_\alpha, R_\alpha \in O(4) \cap Sp(4, \mathbb{R})$ are orthogonal symplectic matrices and $\lambda_1, \lambda_2 > 0, \lambda_1\lambda_2 = 1$. Since the subgroup $O(4) \cap Sp(4, \mathbb{R}) \subset SO(4)$ consists of 2×2 block matrices

$$\begin{pmatrix} G & F \\ -F & G \end{pmatrix},$$

then it may be identified with $U(2)$ through the map $(G, F) \mapsto G + iF$. As a matter of fact, we may assign to $e^{-\mathcal{A}_\alpha}$ the following matrix in $SL(2, \mathbb{C})$:

$$M_\alpha \begin{pmatrix} \lambda_1 & \\ & 1/\lambda_1 \end{pmatrix} R_\alpha.$$

Acknowledgements We would like to thank Jacques Faraut, Philippe Biane, Bachir Bekka and René Schilling for stimulating discussions and remarks.

References

1. Andrews, G.E., Askey, R., Roy, R.: Special Functions. Cambridge University Press, Cambridge (1999)
2. Applebaum, D.: Lévy Processes and Stochastic Calculus. Cambridge Studies in Advanced Mathematics, p. xxx+460. Cambridge University Press, Cambridge (2009)
3. Askour, N., Intissar, A., Mouayn, Z.: Espaces de Bargmann généralisés et formules explicites pour leurs noyaux reproduisants, C. R. A. S. Paris, t.**325**, Série I, pp. 707–712 (1997)
4. Bernstein, J.N.: On the support of Plancherel measure. J. Geom. Phys. **5**(4), 663–710 (1988)
5. Biane, P.: Intertwining of Markov semi-groups, some examples. Séminaire de probabilités de Strasbourg **29**, 30–36 (1995)
6. Biane, P.: Entrelacements de semi-groupes provenant de paires de Gelfand. ESAIM Probab. Stat. **15**, S2–S10 (2011)
7. Biane, P., Bougerol, P., O'Connell, N.: Continuous crystal and Duistermaat–Heckman measure for Coxeter groups. Adv. Math. **221**, 1522–1583 (2009)
8. Böttcher, B., Schilling, R., Wang, J.: Lévy Matters III. Lévy-type processes: construction, approximation and simple path properties. With a short biography of Paul Lévy by Jean Jacod. Lecture Notes in Mathematics, 2099. Lévy Matters. Springer, Cham (2013)
9. Boyadzhiev, K.N.: Lah numbers, Laguerre polynomials of order negative one, and the nth derivative of $\exp(1/x)$. Acta Univ. Sapientiae Math. **8**(1), 22–31 (2016)
10. Carmona, P., Petit, F., Yor, M.: Beta-gamma random variables and intertwining relations between certain Markov processes. Rev. Mat. Iberoam. **14**(2), 311–367 (1998)
11. Chybiryakov, O., Demni, N., Gallardo, L., Rösler, M., Voit, M., Yor, M.: Harmonic and Stochastic Analysis of Dunkl Processes. Travaux en Cours, Hermann (2008)

12. Gaveau, B.: Principe de moindre action, propagation de la chaleur et estimées sous-elliptiques sur certains groupes nilpotents. Acta Math. **139**(1–2), 95–153 (1977)
13. Gelfand, I.M., Graev, M.I., Pyatetskii-Shapiro, I.I.: Representation Theory and Automorphic Functions. Translated from the Russian by K. A. Hirsch. Reprint of the 1969 edition. Generalized Functions, 6. Academic Press, Inc., Boston (1990)
14. Ghomrasni, R.: On distributions associated with the generalized Lévy's stochastic area formula. Studia Sci. Math. Hung. **41**(1), 93–100 (2004)
15. Godement, R.: A theory of spherical functions I. Trans. Am. Math. Soc. **73**(3), 496–556 (1952)
16. Gradshteyn, I.S., Ryzhik, I.M.: Table of Integrals, Series and Products, 7th edn. Academic Press, Boston (1994)
17. Helmes, K.: On Lévy's Area Process. Stochastic Differential Systems (Bad Honnef, 1985), 187–194, Lecture Notes in Control and Information Science, 78, Springer, Berlin (1986)
18. Hirsch, F., Yor, M.: Fractional intertwinings between two Markov semigroups. Potential Anal. **31**, 133–146 (2009)
19. Jurek, Z.J., Yor, M., Selfdecomposable laws associated with hyperbolic functions. Probab. Math. Stat. 24 (2004), no. 1. Acta Univ. Wratislav. No. 2646, 181–190 (2004)
20. Knopova, V., Schilling, R.: A note on the existence of transition probability densities. Forum Math. **25**(1), 125–149 (2013)
21. Koornwinder, T.: Jacobi functions and analysis on non compact semi simple Lie groups. Special Functions: Group Theoretical Aspects and Applications, pp. 1–85, Math. Appl., Reidel, Dordrecht (1984)
22. Lewandowski, Z., Szynal, J.: An upper bound for the Laguerre polynomials. Proceedings of the VIIIth Symposium on Orthogonal Polynomials and Their Applications (Seville, 1997). J. Comput. Appl. Math. **99**(1–2), 529–533 (1998)
23. Matsumoto, H., Ueki, N.: Applications of the theory of the metaplectic representation to quadratic Hamiltonians on the two dimensional Euclidean space. J. Math. Soc. Jpn. **52**(2), 269–292 (2000)
24. Pitman, J., Yor, M.: A decomposition of Bessel bridges. Z. Wahrsch. Verw. Gebiete. **59**(4), 425–457 (1982)
25. Vershik, A.M., Karpushev, S.I.: Cohomology of groups in unitary representations, the neighborhood of the identity, and conditionally positive definite functions. Math. USSR-Sbornik. **47**(2), 513–526 (1984)

Publisher's Note Springer Nature remains neutral with regard to jurisdictional claims in published maps and institutional affiliations.

Journal of Fourier Analysis and Applications (2019) 25:2521–2569
https://doi.org/10.1007/s00041-019-09673-1

Dual-Root Lattice Discretization of Weyl Orbit Functions

Jiří Hrivnák[1] · Lenka Motlochová[1]

Received: 13 September 2017 / Revised: 12 February 2019 / Published online: 4 March 2019
© Springer Science+Business Media, LLC, part of Springer Nature 2019

Abstract
Four types of discrete transforms of Weyl orbit functions on finite point sets are developed. The point sets are formed by intersections of dual-root lattices with fundamental domains of affine Weyl groups. The finite sets of weights, labelling the orbit functions, obey symmetries of the dual extended affine Weyl groups. Fundamental domains of the dual extended affine Weyl groups are detailed in full generality. Identical cardinality of the point and weight sets is proved and explicit counting formulas for these cardinalities are derived. Discrete orthogonality of complex-valued Weyl and real-valued Hartley orbit functions over the point sets is established and the corresponding discrete Fourier–Weyl and Hartley–Weyl transforms are formulated. The unitary transform matrices of the discrete transforms are exemplified.

Keywords Weyl orbit function · Root lattice · Discrete Fourier transform · Hartley transform

Mathematics Subject Classification 33C52 · 65T50 · 20F55

1 Introduction

The purpose of this article is to extend the collection of discrete Fourier transforms of Weyl orbit functions on Weyl group invariant lattices [15,17,18]. A finite fragment of a refinement of the classical dual-root lattice [2] serves as the starting set of points over which the discrete orthogonality of four types of complex Weyl orbit functions [21,22,28] is developed. The entire resulting transform formalism produces a real-

Communicated by Hartmut Führ.

✉ Lenka Motlochová
lenka.motlochova@fjfi.cvut.cz

Jiří Hrivnák
jiri.hrivnak@fjfi.cvut.cz

1 Department of Physics, Faculty of Nuclear Sciences and Physical Engineering, Czech Technical University in Prague, Břehová 7, 115 19 Prague, Czech Republic

valued multidimensional Weyl group invariant generalizations of the one-dimensional discrete Hartley transform [3].

The antisymmetric and symmetric exponential orbit sums over Weyl groups form a standard part of the theory of Lie algebras and their representations [2]. From the viewpoint of the Coxeter groups theory, Weyl groups cover all finite crystallographic reflection groups [19]. Depending on the type of the underlying crystallographic root system, two or four sign homomorphisms exist [28]. Each sign homomorphism determines signs in the exponential sums and thus generates for each Weyl group two or four types of complex special functions. Lattice shifts and Weyl group invariance properties of the resulting Weyl orbit functions generalize periodicity and boundary behaviour of the standard cosine and sine functions of one variable. Investigating Weyl orbit functions as special functions, the results range from generalizations of continuous multivariate Fourier transforms in [21,22] to generalized Chebyshev polynomial methods [25,30]. Discrete Fourier methods are comprehensively studied for Weyl orbit functions [15,17,18] as well as for their multivariate Chebyshev polynomial generalizations [16,28,30,36]. The refinement of the dual weight lattice intersected with the fundamental domain of the affine Weyl group forms a finite point set on which the majority of the discrete Fourier and Chebyshev methods is developed [15,17,25,28]. This choice of the point set generates symmetries of labels of orbit functions governed by the dual affine Weyl group [17]. Choosing as the starting point set the refinement of the weight lattice produced different argument and label symmetries, both controlled by the same affine Weyl group [18].

The dual root and root lattices constitute the last classical Weyl group invariant lattices for which the inherent Fourier methods have not yet been studied. Several apparent relative difficulties, surmounted in the present paper for the dual-root lattice, stem from the fact that the label symmetries are in this case determined by the dual extended affine Weyl group. Firstly, even though the structure of the extended affine Weyl group and its dual version is detailed already in [2], relevant rigorous results about their fundamental domains appeared much later in [23]. Moreover, these fundamental domains, essential for selecting the labels of orbit functions in discrete transforms, are determined in [23] only up to their boundaries. A uniform description of the fundamental domains from [23], including a unique layout of the boundary points, is achieved in the present paper by introducing lexicographical ordering on the Kac coordinates [20]. Secondly, the main challenge poses linking the number of weights, found in the fundamental domain of the dual extended affine Weyl group, with the number of points from the refined dual-root lattice, lying in the fundamental domain of the affine Weyl group. Both sets form topologically distant finite subsets of the underlying Euclidean space and their common cardinality constitutes a novel invariant characteristic of the crystallographic root systems and corresponding simple Lie algebras. Determining the cardinality of these sets in full generality requires invoking and extending concepts from the theory of invariant polynomials [19,34]. The common cardinality of the point and weight sets guarantees in turn, the existence of novel complex-valued Fourier–Weyl and real-valued Hartley–Weyl dual-root lattice discrete transforms.

Both complex and real types of developed Fourier-like transforms significantly enhance the collection of available Weyl group invariant discrete transforms [15,17,

25,28]. The topology of the point sets of the dual-root lattice discretization and their relative position with respect to the fundamental domain of the affine Weyl group substantially diverge from the locations of the original dual weight lattice points. Moreover, fundamentally novel options are generated by combining both dual root and dual weight discretizations to produce subtractive discrete transforms on generalized and composed grids. These options encompass transforms on the refined dual weight lattice with points from the dual-root lattice omitted, such as two-variable discrete transforms on the honeycomb lattice [14]. Assuming Weyl symmetric or antisymmetric armchair boundary conditions in mechanical graphene eigenvibrations model [5], triangular quantum dot model [9,33] or in quantum field lattice models [7] yields novel Weyl invariant solutions and dispersion relations in terms of four types of extended Weyl orbit functions [14]. Moreover, the subtractive discrete transforms on the zig-zag boundary triangular graphene dot potentially yield exact solutions and analytical dispersion relations, which are not attainable by standard methods [9]. Real-valued multivariate Hartley–Weyl versions of the transforms augment the application potential of discrete Hartley transforms in pattern recognition [4], geophysics [24], signal processing [31,32], optics [26] and measurement [35]. The unitary matrices of the derived transforms also permit the construction of novel analogues of the Kac-Peterson matrices from the conformal field theory [18].

This paper is organized as follows. In Sect. 2, the necessary facts concerning root systems and invariant lattices of Weyl groups are recalled. Section 3 contains a description of infinite extensions of Weyl groups. In Sect. 4, fundamental domains of infinite extensions of Weyl groups and the corresponding invariant polynomials are detailed. Section 5 is devoted to the study of finite sets of points and weights. The identical cardinality of these sets is proven and explicit counting formulas for the cardinalities are listed in full generality. Section 6 describes the Weyl and Hartley orbit functions together with their discrete orthogonality and discrete Fourier transforms. Comments and follow-up questions are contained in the last section.

2 Invariant Lattices of Weyl Groups

The goal of Sect. 2 is to recall and establish the notation related to the four classical Weyl group invariant lattices Q, $Q^\vee$ and P, $P^\vee$, introduce the dual marks $m_i^\vee$ together with the crucial index set J and summarize definitions and facts concerning sign homomorphisms σ.

2.1 Root Systems and Weyl Groups

The notation used in this article is based on papers [15,17]. The purpose of this subsection is to extend this notation and recall other pertinent details [2]. Each simple Lie algebra from the classical four series $A_n\,(n \geq 1)$, $B_n\,(n \geq 3)$, $C_n\,(n \geq 2)$, $D_n\,(n \geq 4)$ and from the five exceptional cases E_6, E_7, E_8, F_4, G_2 determines its set $\Delta = \{\alpha_1, \ldots, \alpha_n\}$ of the simple roots [2,37]. For the cases of simple Lie algebras with two different root-lengths, the set Δ is disjointly decomposed into Δ_s of the short simple roots and Δ_l of the long simple roots,

$$\Delta = \Delta_s \cup \Delta_l .$$

The set Δ forms a non-orthogonal basis of the Euclidean space $\mathbb{R}^n$ with the standard scalar product $\langle \cdot, \cdot \rangle$. To every simple root $\alpha_i \in \Delta$ corresponds a reflection r_i given by the formula

$$r_i a \equiv a - \frac{2\,\langle a, \alpha_i \rangle}{\langle \alpha_i, \alpha_i \rangle}\alpha_i, \quad a \in \mathbb{R}^n .$$

Reflections r_i, $i \in \{1, \dots, n\}$ generate an irreducible Weyl group

$$W \equiv \langle r_1, \dots, r_n \rangle$$

and W in turn, generates the entire root system $\Pi \equiv W\Delta$. The set of the simple roots Δ induces a partial ordering $\leq$ on $\mathbb{R}^n$ such that for $\lambda, \nu \in \mathbb{R}^n$ it holds $\nu \leq \lambda$ if and only if $\lambda - \nu = k_1\alpha_1 + \dots + k_n\alpha_n$ with $k_i \in \mathbb{Z}^{\geq 0}$ for all $i \in \{1, \dots, n\}$. There exists a unique root $\xi \in \Pi$, highest with respect to this ordering, of the form

$$\xi = m_1\alpha_1 + \dots + m_n\alpha_n .$$

To each simple root $\alpha_i \in \Delta$ relates the dual simple root given by

$$\alpha_i^\vee = \frac{2\alpha_i}{\langle \alpha_i, \alpha_i \rangle}.$$

The set of dual simple roots $\Delta^\vee = \{\alpha_1^\vee, \dots, \alpha_n^\vee\}$ generates the entire dual root system $\Pi^\vee \equiv W\Delta^\vee$. The dual root system $\Pi^\vee$ contains the highest dual root with respect to the ordering induced by $\Delta^\vee$ of the form

$$\eta = m_1^\vee\alpha_1^\vee + \dots + m_n^\vee\alpha_n^\vee .$$

The expansion coefficients of the highest root $m_1, \dots, m_n$ and of the highest dual root $m_1^\vee, \dots, m_n^\vee$, named the marks and the dual marks, respectively, are listed in Table 1 in [17]. Setting additionally $m_0 = m_0^\vee = 1$, the Coxeter number m is given by

$$m = \sum_{i=0}^{n} m_i = \sum_{i=0}^{n} m_i^\vee . \tag{1}$$

The dual marks with unitary values determine an important index set $J \subset \{1, \dots, n\}$,

$$J \equiv \left\{1 \leq i \leq n \mid m_i^\vee = 1\right\}.$$

Note that J is empty for the simple Lie algebras E_8, F_4 and G_2.

The minimal number of generators r_i, necessary to generate an element $w \in W$, is called the length $l(w)$ of w. There is a unique longest element $w_0 \in W$, also called the opposite involution, and for its length it holds that

$$l(w_0) = \frac{|\Pi|}{2}.$$

For root systems with even Coxeter numbers, the opposite involutions are of the form

$$w_0 = (r_1 \dots r_n)^{\frac{m}{2}}.$$

An important parabolic subgroup W_i of the Weyl group W is obtained by omitting r_i from the set of generators of W,

$$W_i \equiv \langle r_1, \dots, r_{i-1}, r_{i+1}, \dots, r_n \rangle .$$

The subgroup W_i forms the Weyl group corresponding to the root system $\Pi_i \equiv W_i(\Delta \backslash \alpha_i)$. The opposite involution in W_i with respect to the root system Π_i is denoted by w_i.

2.2 Invariant Lattices

Four classical Weyl group invariant lattices [2] comprise the root lattice, the dual weight lattice, the dual-root lattice and the weight lattice. The root lattice Q is the $\mathbb{Z}$-span of the set of simple roots Δ,

$$Q = \mathbb{Z}\alpha_1 + \dots + \mathbb{Z}\alpha_n.$$

The dual weight lattice is $\mathbb{Z}$-dual to the root lattice Q,

$$P^\vee = \left\{\omega^\vee \in \mathbb{R}^n \mid \left\langle \omega^\vee, \alpha_i \right\rangle \in \mathbb{Z},\ \forall \alpha_i \in \Delta\right\} = \mathbb{Z}\omega_1^\vee + \dots + \mathbb{Z}\omega_n^\vee,$$

with the dual fundamental weights $\omega_i^\vee$ given by

$$\left\langle \omega_i^\vee, \alpha_j \right\rangle = \delta_{ij}.$$

The dual-root lattice is the $\mathbb{Z}$-span of the set of dual simple roots $\Delta^\vee$,

$$Q^\vee = \mathbb{Z}\alpha_1^\vee + \dots + \mathbb{Z}\alpha_n^\vee.$$

The weight lattice is $\mathbb{Z}$-dual to the dual-root lattice $Q^\vee$,

$$P = \left\{\omega \in \mathbb{R}^n \mid \left\langle \omega, \alpha_i^\vee \right\rangle \in \mathbb{Z},\ \forall \alpha_i^\vee \in \Delta^\vee\right\} = \mathbb{Z}\omega_1 + \dots + \mathbb{Z}\omega_n, \tag{2}$$

with the fundamental weights ω_i given by

$$\langle \omega_i, \alpha_j^\vee \rangle = \delta_{ij}.$$

The weight lattice P is partitioned into $|J| + 1$ components and this decomposition consists of the root lattice Q and its $|J|$ shifted copies,

$$P = Q \cup \bigcup_{i \in J} (\omega_i + Q).$$

The Cartan matrix C with entries given by

$$C_{ij} = \left\langle \alpha_i, \alpha_j^\vee \right\rangle \tag{3}$$

relates the simple roots and fundamental weights as well as their dual versions via formulas

$$\omega_i = \sum_{j=1}^{n} C_{ij}^{-1} \alpha_j, \quad \omega_i^\vee = \sum_{j=1}^{n} \left(C_{ij}^T \right)^{-1} \alpha_j^\vee. \tag{4}$$

The determinant c of the Cartan matrix C determines the index of connection of the root system Π and the order of the quotient group P/Q,

$$c = \det C = |P/Q| = |J| + 1.$$

2.3 Sign Homomorphisms

Recall from [15] that a homomorphism $\sigma : W \mapsto \{\pm 1\}$ is called a sign homomorphism. The identity $\mathbb{1}$ and the determinant σ^e sign homomorphisms, which exist for any Weyl group W, are given on the generating reflections r_i as

$$\begin{aligned} \mathbb{1}(r_i) &= 1, \\ \sigma^e(r_i) &= -1. \end{aligned}$$

For the root systems with two lengths of roots, the short and long sign homomorphisms σ^s and σ^l are defined as

$$\sigma^s(r_i) = \begin{cases} -1 & \text{if } \alpha_i \in \Delta_s, \\ 1 & \text{otherwise,} \end{cases}$$

$$\sigma^l(r_i) = \begin{cases} -1 & \text{if } \alpha_i \in \Delta_l, \\ 1 & \text{otherwise.} \end{cases}$$

To each sign homomorphism σ is attached a vector $\varrho^\sigma = \varrho_1^\sigma \omega_1 + \cdots + \varrho_n^\sigma \omega_n \in P$ defined by

$$\begin{aligned} &\varrho_i^{\mathbb{1}} \equiv 0 \quad i = 1, \ldots, n, \\ &\varrho_i^{\sigma^e} \equiv 1 \quad i = 1, \ldots, n, \\ &\varrho_i^{\sigma^s} \equiv 1 \quad \alpha_i \in \Delta_s, \qquad \varrho_i^{\sigma^s} \equiv 0 \quad \alpha_i \in \Delta_l, \\ &\varrho_i^{\sigma^l} \equiv 0 \quad \alpha_i \in \Delta_s, \qquad \varrho_i^{\sigma^l} \equiv 1 \quad \alpha_i \in \Delta_l. \end{aligned} \tag{5}$$

The vector ϱ^{σ^e} becomes the standard ϱ vector defined as the half-sum of the positive roots. Zero coordinates ϱ_0^σ of the vectors ϱ^σ are for convenience defined by

$$\varrho_0^{\mathbb{1}} \equiv 0, \quad \varrho_0^{\sigma^e} \equiv 1, \quad \varrho_0^{\sigma^s} \equiv 1, \quad \varrho_0^{\sigma^l} \equiv 0.$$

Furthermore, to each sign homomorphism σ is associated a generalized Coxeter number m^σ by defining relation

$$m^\sigma = \sum_{i=0}^{n} m_i^\vee \varrho_i^\sigma . \tag{6}$$

Note that $m^{\mathbb{1}} = 0$ and the number $m^{\sigma^e} = m$ coincides with the standard Coxeter number. The short m^{σ^s} and long m^{σ^l} Coxeter numbers are tabulated in [15].

3 Extensions of Weyl Groups

The goal of Sect. 3 is to unify the notation for affine and dual affine Weyl groups W^{aff}, W_Q^{aff} together with their fundamental domains F and F_Q. The introduced ε-function enters as a weight coefficient the discrete Fourier–Weyl transforms. Explicit lists of actions of the subgroups Γ_M of the extended dual affine Weyl groups W_P^{aff}, tabulated in Table 1, serve for calculation of the normalization coefficients in the discrete transforms.

3.1 Affine Weyl Groups

The affine Weyl group is a semidirect product of the group of translations $Q^\vee$ and W,

$$W^{\mathrm{aff}} = Q^\vee \rtimes W. \tag{7}$$

For $q^\vee \in Q^\vee$ and $w \in W$, any element $T(q^\vee)w \in W^{\mathrm{aff}}$ acts on $\mathbb{R}^n$ as

$$T(q^\vee)w \cdot a \equiv wa + q^\vee, \quad a \in \mathbb{R}^n.$$

The standard retraction homomorphism $\psi : W^{\mathrm{aff}} \to W$ of the semidirect product (7) is given by

$$\psi(T(q^\vee)w) = w.$$

The fundamental domain $F \subset \mathbb{R}^n$ of W^{aff} is a simplex explicitly given by

$$F = \left\{a_1\omega_1^\vee + \cdots + a_n\omega_n^\vee \mid a_0 + m_1a_1 + \cdots + m_na_n = 1, a_i \geq 0, i = 0, \ldots, n\right\}.$$

The stabilizer $\mathrm{Stab}_{W^{\mathrm{aff}}}(a)$ is the subgroup of W^{aff} of elements stabilizing $a \in \mathbb{R}^n$,

$$\mathrm{Stab}_{W^{\mathrm{aff}}}(a) = \left\{w^{\mathrm{aff}} \in W^{\mathrm{aff}} \mid w^{\mathrm{aff}}a = a\right\},$$

and the function $\varepsilon : \mathbb{R}^n \to \mathbb{N}$ is defined by

$$\varepsilon(a) = \frac{|W|}{\left|\mathrm{Stab}_{W^{\mathrm{aff}}}(a)\right|}. \tag{8}$$

The stabilizers $\mathrm{Stab}_{W^{\mathrm{aff}}}(a)$ and $\mathrm{Stab}_{W^{\mathrm{aff}}}(w^{\mathrm{aff}}a)$ are conjugated and therefore, the function ε is W^{aff}-invariant

$$\varepsilon(a) = \varepsilon(w^{\mathrm{aff}}a), \quad w^{\mathrm{aff}} \in W^{\mathrm{aff}}. \tag{9}$$

The standard action of W on the torus $\mathbb{R}^n/Q^\vee$ generates for $x \in \mathbb{R}^n/Q^\vee$ its isotropy groups $\mathrm{Stab}(x)$ and orbits Wx of orders

$$\widetilde{\varepsilon}(x) \equiv |Wx|, \quad x \in \mathbb{R}^n/Q^\vee.$$

The following three properties from Proposition 2.2 in [17] of the action of W on the torus $\mathbb{R}^n/Q^\vee$ are essential,

(1) For any $x \in \mathbb{R}^n/Q^\vee$, there exists $x' \in F \cap \mathbb{R}^n/Q^\vee$ and $w \in W$ such that

$$x = wx'. \tag{10}$$

(2) If $x, x' \in F \cap \mathbb{R}^n/Q^\vee$ and $x' = wx$, $w \in W$, then

$$x' = x = wx. \tag{11}$$

(3) If $x \in F \cap \mathbb{R}^n/Q^\vee$, i.e. $x = a + Q^\vee$, $a \in F$, then $\psi(\mathrm{Stab}_{W^{\mathrm{aff}}}(a)) = \mathrm{Stab}(x)$ and

$$\mathrm{Stab}(x) \cong \mathrm{Stab}_{W^{\mathrm{aff}}}(a). \tag{12}$$

Relation (12) grants that for $x = a + Q^\vee$, $a \in F$ it holds that

$$\varepsilon(a) = \widetilde{\varepsilon}(x). \tag{13}$$

Note that instead of $\widetilde{\varepsilon}(x)$, the symbol $\varepsilon(x)$ is used for $|Wx|$, $x \in F \cap \mathbb{R}^n/Q^\vee$ in [15,17]. The algorithm for the calculation of the coefficients $\varepsilon(x)$ is described in [17, §3.7].

To each sign homomorphism σ is associated a subset $F^\sigma \subset F$ of the form

$$F^\sigma \equiv \left\{a \in F \mid \sigma \circ \psi\left(\mathrm{Stab}_{W^{\mathrm{aff}}}(a)\right) = \{1\}\right\}. \tag{14}$$

The sets F^σ are detailed in [15,17] and described via non-negative symbols a_i^σ defined by

$$\begin{aligned}
&a_0^{\mathfrak{1}}, a_i^{\mathfrak{1}} \geq 0 && \text{if } \alpha_i \in \Delta, \\
&a_0^{\sigma^e}, a_i^{\sigma^e} > 0 && \text{if } \alpha_i \in \Delta, \\
&a_0^{\sigma^s}, a_i^{\sigma^s} \geq 0 && \text{if } \alpha_i \in \Delta_l, \qquad a_i^{\sigma^s} > 0 \quad \text{if } \alpha_i \in \Delta_s, \\
&a_0^{\sigma^l}, a_i^{\sigma^l} > 0 && \text{if } \alpha_i \in \Delta_l, \qquad a_i^{\sigma^l} \geq 0 \quad \text{if } \alpha_i \in \Delta_s.
\end{aligned}$$

The explicit formulas for F^σ are thus of the form

$$F^\sigma = \left\{a_1^\sigma \omega_1^\vee + \cdots + a_n^\sigma \omega_n^\vee \mid a_0^\sigma + m_1 a_1^\sigma + \cdots + m_n a_n^\sigma = 1\right\}. \tag{15}$$

Note that $F^1 = F$ and $F^{\sigma^e} = \mathrm{int}(F)$.

3.2 Dual Affine Weyl Groups

The dual affine Weyl group W_Q^{aff} is a semidirect product of the group of translations in the root lattice Q and W,

$$W_Q^{\mathrm{aff}} \equiv Q \rtimes W. \tag{16}$$

For $q \in Q$ and $w \in W$, any element $T(q)w \in W_Q^{\mathrm{aff}}$ acts on $\mathbb{R}^n$ as

$$T(q)w \cdot b \equiv wb + q, \quad b \in \mathbb{R}^n. \tag{17}$$

The fundamental domain $F_Q \subset \mathbb{R}^n$ of the dual affine Weyl group W_Q^{aff}, denoted by $F^\vee$ in [15,17], is explicitly given by

$$F_Q \equiv \left\{b_1\omega_1 + \cdots + b_n\omega_n \mid b_0 + m_1^\vee b_1 + \cdots + m_n^\vee b_n = 1, b_i \geq 0, i = 0, \ldots, n\right\}. \tag{18}$$

By identifying each $b = b_1\omega_1 + \cdots + b_n\omega_n \in F_Q$ with its Kac coordinates $[b_0, b_1, \ldots, b_n]$ from (18),

$$b \equiv [b_0, b_1, \ldots, b_n],$$

the lexicographic ordering on F_Q is introduced in the following way. An element $b = [b_0, b_1, \ldots, b_n]$ is lexicographically higher than $b' = [b'_0, b'_1, \ldots, b'_n]$,

$$b >_{\mathrm{lex}} b', \tag{19}$$

if and only if $b_i > b'_i$ for the first $i = 0, 1, \ldots, n$ where b_i differs from b'_i.

The standard dual retraction homomorphism $\widehat{\psi} : W_Q^{\mathrm{aff}} \to W$ of the semidirect product (16) is given by

$$\widehat{\psi}(T(q)w) = w. \tag{20}$$

Similar to (14), four subsets of $F_Q^\sigma \subset F_Q$ are introduced by

$$F_Q^\sigma \equiv \left\{b \in F_Q \mid \sigma \circ \widehat{\psi}\left(\mathrm{Stab}_{W_Q^{\mathrm{aff}}}(b)\right) = \{1\}\right\}.$$

The domains F_Q^σ are explicitly described by

$$F_Q^\sigma = \left\{b_1^\sigma \omega_1 + \ldots + b_n^\sigma \omega_n \mid b_0^\sigma + m_1^\vee b_1^\sigma + \cdots + m_n^\vee b_n^\sigma = 1\right\} \tag{21}$$

with the symbols $b_0^\sigma, \dots, b_n^\sigma$ satisfying

$$\begin{aligned} & b_0^1, b_i^1 \geq 0 && \text{if } \alpha_i \in \Delta, \\ & b_0^{\sigma^e}, b_i^{\sigma^e} > 0 && \text{if } \alpha_i \in \Delta, \\ & b_i^{\sigma^s} \geq 0 && \text{if } \alpha_i \in \Delta_l, && b_0^{\sigma^s}, b_i^{\sigma^s} > 0 && \text{if } \alpha_i \in \Delta_s, \\ & b_i^{\sigma^l} > 0 && \text{if } \alpha_i \in \Delta_l, && b_0^{\sigma^l}, b_i^{\sigma^l} \geq 0 && \text{if } \alpha_i \in \Delta_s. \end{aligned} \tag{22}$$

3.3 Extended Dual Affine Weyl Groups

The extended dual affine Weyl group W_P^{aff} is a semidirect product of the group of translations P and the Weyl group W,

$$W_P^{\text{aff}} \equiv P \rtimes W. \tag{23}$$

For $p \in P$ and $w \in W$, extending the action (17) on $\mathbb{R}^n$ to elements $T(p)w \in W_P^{\text{aff}}$ yields

$$T(p)w \cdot b \equiv wb + p, \quad b \in \mathbb{R}^n.$$

The extended dual retraction homomorphism $\widehat{\psi} : W_P^{\text{aff}} \to W$ of the semidirect product (23) is a natural extension of (20) given by

$$\widehat{\psi}(T(p)w) = w.$$

Introducing the subgroup $\Gamma \subset W_P^{\text{aff}}$ of all elements of W_P^{aff} leaving the fundamental domain F_Q of W_Q^{aff} invariant

$$\Gamma = \left\{ \gamma \in W_P^{\text{aff}} \mid \gamma \cdot F_Q = F_Q \right\} \tag{24}$$

allows to represent W_P^{aff} as a semidirect product

$$W_P^{\text{aff}} = W_Q^{\text{aff}} \rtimes \Gamma. \tag{25}$$

Explicit structure of the abelian group Γ is detailed in [2, Ch. VI,§2] as

$$\Gamma = \left\{ 1, \gamma_i \in W_P^{\text{aff}} \mid i \in J \right\}, \quad \gamma_i \equiv T(\omega_i) w_i w_0, \quad |\Gamma| = c. \tag{26}$$

The action of Γ on F_Q assigns to each fixed $\gamma \in \Gamma$ a bijection on F_Q given on the Kac coordinates $[b_0, \dots, b_n]$ by

$$\gamma \cdot b = \gamma \cdot [b_0, \dots, b_n] = [b_{\pi_\gamma(0)}, \dots, b_{\pi_\gamma(n)}] \tag{27}$$

where π_γ denotes a permutation of the index set $\{0, 1, \dots, n\}$. These permutations of the Kac coordinates $[b_0, \dots, b_n]$, which determine the group Γ, are specified for every simple Lie algebra in Table 1. Direct analysis of the permutations $\gamma \in \Gamma$ in Table 1 on the ϱ^σ vector (5) yields for its coordinates ϱ^σ_i that

$$\varrho^\sigma_{\pi_\gamma(i)} = \varrho^\sigma_i, \quad i \in \{0, 1, \dots, n\}. \tag{28}$$

Considering a resolution factor $M \in \mathbb{N}$, an important class of subgroups Γ_M of W^{aff}_P is given by

$$\Gamma_M = \left\{1, \gamma_{M,i} \in W^{\mathrm{aff}}_P \mid i \in J\right\}, \quad \gamma_{M,i} \equiv T(M\omega_i)w_i w_0, \quad |\Gamma_M| = c. \tag{29}$$

For each M, the subgroup $\Gamma_M \subset W^{\mathrm{aff}}_P$ is isomorphic to Γ by an isomorphism

$$\Gamma_M \ni \gamma_M \mapsto \gamma \in \Gamma \tag{30}$$

defined by assigning $\gamma_i \in \Gamma$ to $\gamma_{M,i} \in \Gamma_M$. The action of $\gamma_M \in \Gamma_M$ on $b \in \mathbb{R}^n$ is directly related to action of $\gamma \in \Gamma$ by

$$\gamma_M \cdot b = M\left(\gamma \cdot \frac{b}{M}\right). \tag{31}$$

Consequently, Γ_M acts naturally on the magnified fundamental domain MF_Q. Setting for $b = b_1\omega_1 + \dots + b_n\omega_n \in MF_Q$ the magnified Kac coordinates $b_0 \equiv M - \left(m^\vee_1 b_1 + \dots + m^\vee_n b_n\right)$, this action is described by

$$\gamma_M \cdot b = \gamma_M \cdot [b_0, \dots, b_n] = [b_{\pi_\gamma(0)}, \dots, b_{\pi_\gamma(n)}]. \tag{32}$$

To each element $\gamma_M \in \Gamma_M$ is thus assigned the same permutation π_γ of $\{0, 1, \dots, n\}$ from Table 1 as to the corresponding $\gamma \in \Gamma$ in (27).

4 Fundamental Domains and Invariant Polynomials

The purpose of Sect. 4 is to describe the orders of the stabilizers $h_{P,M}$ and σ-dependent fundamental domains F^σ_P of extended dual affine Weyl groups. The calculation procedure for the stabilizer function $h_{P,M}$ allows explicit evaluation of the normalization coefficients of the discrete transforms and the magnified domains F^σ_P subsequently contain the labels of the discretely orthogonal orbit functions and supersede the role of the Brillouin zone intersections. Subsection 4.3, committed to the exposition of the spaces of the multivariate invariant polynomials, culminates in isomorphism Theorem 4.6 that afterwards guarantees existence of the discrete transforms.

Table 1 Non-trivial abelian groups Γ, isomorphic to the groups in the second column, are listed

	Γ	γ	$\gamma \cdot [b_0, \ldots, b_n]$	$\sigma \circ \widehat{\psi}(\gamma)$		
				σ^e	σ^s	σ^l
A_n	$\mathbb{Z}_{n+1}$	γ_i	$[b_{n-i+1}, b_{n-i+2}, \ldots, b_n, b_0, b_1, \ldots, b_{n-i}]$	$(-1)^{ni}$	–	–
B_n	$\mathbb{Z}_2$	γ_n	$[b_n, b_{n-1}, \ldots, b_1, b_0]$	$(-1)^{\frac{n(n+1)}{2}}$	$(-1)^n$	$(-1)^{\frac{(n-1)n}{2}}$
C_n	$\mathbb{Z}_2$	γ_1	$[b_1, b_0, b_2, \ldots, b_n]$	-1	1	-1
D_{2k}	$\mathbb{Z}_2 \times \mathbb{Z}_2$	γ_1	$[b_1, b_0, b_2, \ldots, b_{2k-2}, b_{2k}, b_{2k-1}]$	1	–	–
		γ_{2k-1}	$[b_{2k-1}, b_{2k}, b_{2k-2}, \ldots, b_2, b_0, b_1]$	$(-1)^k$	–	–
		γ_{2k}	$[b_{2k}, b_{2k-1}, \ldots, b_1, b_0]$	$(-1)^k$	–	–
D_{2k+1}	$\mathbb{Z}_4$	γ_1	$[b_1, b_0, b_2, \ldots, b_{2k-1}, b_{2k+1}, b_{2k}]$	1	–	–
		γ_{2k}	$[b_{2k+1}, b_{2k}, \ldots, b_2, b_0, b_1]$	$(-1)^k$	–	–
		γ_{2k+1}	$[b_{2k}, b_{2k+1}, b_{2k-1}, \ldots, b_1, b_0]$	$(-1)^k$	–	–
E_6	$\mathbb{Z}_3$	γ_1	$[b_1, b_5, b_4, b_3, b_6, b_0, b_2]$	1	–	–
		γ_5	$[b_5, b_0, b_6, b_3, b_2, b_1, b_4]$	1	–	–
E_7	$\mathbb{Z}_2$	γ_6	$[b_6, b_5, b_4, b_3, b_2, b_1, b_0, b_7]$	-1	–	–

The action of each $\gamma \in \Gamma$, different from the identity, on the Kac coordinates is specified and the corresponding values of sign homomorphisms on $\widehat{\psi}(\gamma)$ are given

4.1 Stabilizers

The stabilizer $\mathrm{Stab}_{W_P^{\mathrm{aff}}}(b)$ is a subgroup of W_P^{aff} stabilizing $b \in \mathbb{R}^n$ and the related counting discrete function $h_{P,M} : \mathbb{R}^n \to \mathbb{N}$ is for any $M \in \mathbb{N}$ defined by

$$h_{P,M}(b) = \left|\mathrm{Stab}_{W_P^{\mathrm{aff}}}\left(\frac{b}{M}\right)\right|. \tag{33}$$

Characterizing the structure of $\mathrm{Stab}_{W_P^{\mathrm{aff}}}(b)$ in the following proposition subsequently allows the calculation of the function $h_{P,M}$.

Proposition 4.1 *For any $b \in F_Q$ it holds that*

$$\mathrm{Stab}_{W_P^{\mathrm{aff}}}(b) = \mathrm{Stab}_{W_Q^{\mathrm{aff}}}(b) \rtimes \mathrm{Stab}_{\Gamma}(b). \tag{34}$$

Proof The semidirect decomposition (25), where W_Q^{aff} is normal in W_P^{aff}, directly guarantees that $\mathrm{Stab}_{W_Q^{\mathrm{aff}}}(b)$ is a normal subgroup of $\mathrm{Stab}_{W_P^{\mathrm{aff}}}(b)$. Moreover, for any $b \in F_Q$ stabilized by both $w^{\mathrm{aff}} \in W_Q^{\mathrm{aff}}$ and $\gamma \in \Gamma$, $w^{\mathrm{aff}}\gamma$ is in $\mathrm{Stab}_{W_P^{\mathrm{aff}}}(b)$. Conversely, for any $w_P \in \mathrm{Stab}_{W_P^{\mathrm{aff}}}(b)$ exists from (25) a unique decomposition $w_P = w_Q\gamma$ with $w_Q \in W_Q^{\mathrm{aff}}$ and $\gamma \in \Gamma$. Invariance (24) of F_Q under the action of Γ implies that $\gamma \cdot b \in F_Q$. Since the fundamental domain F_Q contains only one point from each W_Q^{aff}-orbit, the relation

$$w_P \cdot b = w_Q \cdot (\gamma \cdot b) = b$$

forces $\gamma \cdot b = b$. □

The isomorphism (30) and relation (31) imply for the stabilizers that

$$\mathrm{Stab}_{\Gamma_M}(b) \simeq \mathrm{Stab}_{\Gamma}\left(\frac{b}{M}\right). \tag{35}$$

The resulting counting formula for $h_{P,M}$ is deduced for $b \in MF_Q$ from relations (34) and (35) as

$$h_{P,M}(b) = \left|\mathrm{Stab}_{W_Q^{\mathrm{aff}}}\left(\frac{b}{M}\right)\right| \cdot \left|\mathrm{Stab}_{\Gamma_M}(b)\right|. \tag{36}$$

The calculation procedure for $|\mathrm{Stab}_{W_Q^{\mathrm{aff}}}(b/M)|$ is described in [17, §3.7]. The orders of stabilizers $\mathrm{Stab}_{\Gamma_M}(b)$ are directly derived from the explicit form of permutations given by (27) in Table 1.

Note that since $\gamma_M \in \Gamma_M$ and $\gamma \in \Gamma$ from (26) and (29) differ only in the translation part, their retractions coincide

$$\widehat{\psi}(\gamma_M) = \widehat{\psi}(\gamma). \tag{37}$$

Consequently, the retraction of the stabilizers are also identical

$$\widehat{\psi}\left(\mathrm{Stab}_{\Gamma_M}(b)\right) = \widehat{\psi}\left(\mathrm{Stab}_{\Gamma}\left(\frac{b}{M}\right)\right). \tag{38}$$

4.2 Fundamental Domains

Significant results concerning the structure of the fundamental domain F_P of the group W_P^{aff} are achieved in [23]. Firstly, from the semidirect decomposition (25) follows that the fundamental domain F_P coincides with a fundamental domain of the group Γ acting on F_Q. Secondly, the interior of F_P is determined as [23]

$$\mathrm{int}(F_P) = \left\{b \in \mathrm{int}(F_Q) \mid \langle \eta + \alpha_i^\vee, b\rangle < 1,\ i \in J\right\}.$$

It is also asserted that the extended interior $\mathrm{ent}(F_P)$ of F_P defined by

$$\mathrm{ent}(F_P) \equiv \left\{b \in F_Q \mid \langle \eta + \alpha_i^\vee, b\rangle < 1,\ i \in J\right\} \tag{39}$$

forms a subset of F_P, $\mathrm{ent}(F_P) \subset F_P$.

In order to uniquely determine the remaining boundary points from $F_P \setminus \mathrm{ent}(F_P)$, note that the defining relation in (39)

$$b_i + b_1 m_1^\vee + \cdots + m_n^\vee b_n = \langle \eta + \alpha_i^\vee, b\rangle < 1 = b_0 + m_1^\vee b_1 + \cdots + m_n^\vee b_n$$

implies that $b_0 > b_i$, $i \in J$. Taking into account the explicit forms of Γ-permutations of $[b_0, \ldots, b_n]$ from Table 1 and the lexicographic ordering (19), the inequality $b_0 > b_i$, $i \in J$ grants that the point $[b_0, \ldots, b_n]$ is the lexicographically highest among the points lying in its Γ-orbit. Consequently, the fundamental domain F_P is taken as such a subset of F_Q which contains the lexicographically highest point from each Γ-orbit of $b \in F_Q$. This resulting form of the fundamental domain F_P, which contains exactly one point from each W_P^{aff}-orbit of $\mathbb{R}^n$, is summarized in the following theorem.

Theorem 4.2 *The set $F_P \subset \mathbb{R}^n$ defined by*

$$F_P = \left\{b \in F_Q \mid b = \max_{>_{\mathrm{lex}}} \Gamma b\right\} \tag{40}$$

forms a fundamental domain of the extended dual affine Weyl group W_P^{aff}.

Four crucial subsets F_P^σ of F_P are for each sign homomorphism σ defined by

$$F_P^\sigma \equiv \left\{b \in F_P \mid \sigma \circ \widehat{\psi}(\mathrm{Stab}_{W_P^{\mathrm{aff}}}(b)) = \{1\}\right\}. \tag{41}$$

Recall also from [23] that the stabilizer $\mathrm{Stab}_{W_P^{\mathrm{aff}}}(b)$ of any interior point $b \in \mathrm{int}(F_P)$ is trivial, therefore, it holds that $\mathrm{int}(F_P) \subset F_P^\sigma$. The form of the sets F_P^σ is simplified in the following proposition.

Proposition 4.3 *The sets F_P^σ, defined by* (41)*, are of the form*

$$F_P^\sigma = \left\{ b \in F_P \cap F_Q^\sigma \,|\, \sigma \circ \widehat{\psi}\,(\mathrm{Stab}_\Gamma(b)) = \{1\} \right\}. \tag{42}$$

Proof Since both stabilizers $\mathrm{Stab}_{W_Q^{\mathrm{aff}}}(b)$ and $\mathrm{Stab}_\Gamma(b)$ are subgroups of $\mathrm{Stab}_{W_P^{\mathrm{aff}}}(b)$, it holds for any point $b \in F_P^\sigma$ that

$$\sigma \circ \widehat{\psi}(\mathrm{Stab}_{W_Q^{\mathrm{aff}}}(b)) = \sigma \circ \widehat{\psi}\,(\mathrm{Stab}_\Gamma(b)) = \{1\}, \tag{43}$$

and thus $b \in F_Q^\sigma$. Conversely, for any $b \in F_P$ follows from (43) and the semidirect decomposition (34) that $\sigma \circ \widehat{\psi}(\mathrm{Stab}_{W_P^{\mathrm{aff}}}(b)) = \{1\}$ and thus $b \in F_P^\sigma$. □

Theorem 4.2 and relation (31) guarantee that the magnified domain MF_P is a fundamental domain of the action of Γ_M on MF_Q. Relations (38) and (42) allow to express the magnified domains MF_P^σ,

$$MF_P^\sigma = \left\{ b \in M(F_P \cap F_Q^\sigma) \;|\; \sigma \circ \widehat{\psi}\left(\mathrm{Stab}_{\Gamma_M}(b)\right) = \{1\} \right\}. \tag{44}$$

Example 4.1 For the simple Lie algebra A_3, the fundamental domain of the affine Weyl group F_Q is according to (18) of the form

$$F_Q = \{b_1\omega_1 + \cdots + b_3\omega_3 \,|\, b_0 + b_1 + b_2 + b_3 = 1, b_i \geq 0, i = 0, 1, 2, 3\},$$

and the fundamental domain of the extended dual affine Weyl group $F_P \subset F_Q$ is by (40) of the explicit form

$$\begin{aligned} F_P = &\left\{ b \in F_Q \,|\, b_0 > b_1, b_0 > b_2, b_0 > b_3 \right\} \cup \left\{ b \in F_Q \,|\, b_0 = b_1 > b_2, b_0 > b_3 \right\} \\ &\cup \left\{ b \in F_Q \,|\, b_0 = b_2 > b_1 \geq b_3 \right\} \cup \left\{ b \in F_Q \,|\, b_0 = b_1 = b_2 \geq b_3 \right\}. \end{aligned}$$

The sets (21) coincide for the identity sign homomorphism with the original sets F_Q and F_P,

$$F_Q^{\mathfrak{1}} = F_Q, \quad F_P^{\mathfrak{1}} = F_P.$$

For the sign homomorphism σ^e, the set $F_Q^{\sigma^e}$ is by (21) and (22) of the form

$$\begin{aligned} F_Q^{\sigma^e} = &\left\{ b_1^{\sigma^e}\omega_1 + b_2^{\sigma^e}\omega_2 + b_3^{\sigma^e}\omega_3 | b_0^{\sigma^e} + b_1^{\sigma^e} \right. \\ &\left. + b_2^{\sigma^e} + b_3^{\sigma^e} = 1, b_i^{\sigma^e} > 0, i = 0, 1, 2, 3 \right\}, \end{aligned}$$

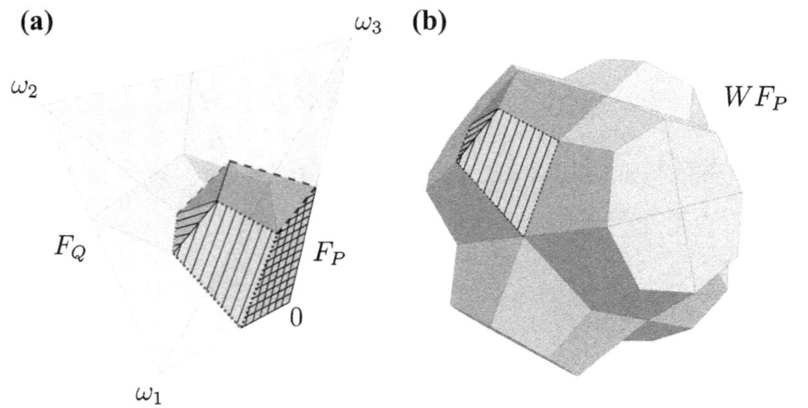

Fig. 1 **a** The domain F_P of A_3, bounded by six kite-shaped surfaces, contains its entire boundary except for the points lying on the hatched surfaces and the dotted lines. The domain $F_P^{\sigma^e}$ is obtained from F_P by omitting the crosshatched areas together with the dashed lines and the point in the intersection of the top three kite-shaped surfaces. The enveloping tetrahedron represents the domain F_Q, tiled by Γ-images of F_P. **b** The polyhedron is formed by images of the Weyl group W acting on F_P

and relation (42) together with Table 1 implies the following explicit description of $F_P^{\sigma^e}$,

$$\begin{aligned} F_P^{\sigma^e} =& \left\{ b^{\sigma^e} \in F_Q^{\sigma^e} \mid b_0^{\sigma^e} > b_1^{\sigma^e}, b_0^{\sigma^e} > b_2^{\sigma^e}, b_0^{\sigma^e} > b_3^{\sigma^e} \right\} \\ &\cup \left\{ b^{\sigma^e} \in F_Q^{\sigma^e} \mid b_0^{\sigma^e} = b_1^{\sigma^e} > b_2^{\sigma^e}, b_0^{\sigma^e} > b_3^{\sigma^e} \right\} \\ &\cup \left\{ b^{\sigma^e} \in F_Q^{\sigma^e} \mid b_0^{\sigma^e} = b_2^{\sigma^e} > b_1^{\sigma^e} \geq b_3^{\sigma^e} \right\} \\ &\cup \left\{ b^{\sigma^e} \in F_Q^{\sigma^e} \mid b_0^{\sigma^e} = b_1^{\sigma^e} = b_2^{\sigma^e} > b_3^{\sigma^e} \right\}. \end{aligned}$$

The domains F_Q, F_P, $F_Q^{\sigma^e}$, $F_P^{\sigma^e}$ and WF_P of A_3 are depicted in Fig. 1.

Setting $x = x_1\omega_1 + x_2\omega_2 + x_3\omega_3$, $x_1, x_2, x_3 \in \mathbb{R}^{\geq 0}$, another geometric shape for the choice of the fundamental domain of the extended dual affine Weyl group exists. The domain $\mathcal{B}$, bounded by the planes orthogonal to ω_i and passing through $\omega_i/2$,

$$\begin{aligned} \mathcal{B} =& \left\{ x \in \mathbb{R}^3 \mid 3x_1 + 2x_2 + x_3 < \tfrac{3}{2}, x_1 + 2x_2 + x_3 < 1, x_1 + 2x_2 + 3x_3 < \tfrac{3}{2} \right\} \\ &\cup \left\{ x \in \mathbb{R}^3 \mid 3x_1 + 2x_2 + x_3 = \tfrac{3}{2}, x_1 + 2x_2 + x_3 \leq 1, x_1 + 2x_2 + 3x_3 < \tfrac{3}{2} \right\} \\ &\cup \left\{ x \in \mathbb{R}^3 | 3x_1 + 2x_2 + x_3 < \tfrac{3}{2}, x_1 + 2x_2 + x_3 = 1, \right. \\ &\quad \left. x_1 + 2x_2 + 3x_3 < \tfrac{3}{2}, x_1 \geq x_3 \right\} \\ &\cup \left\{ x \in \mathbb{R}^3 \mid x_1 = x_2 = x_3 = 1/4 \right\}, \end{aligned}$$

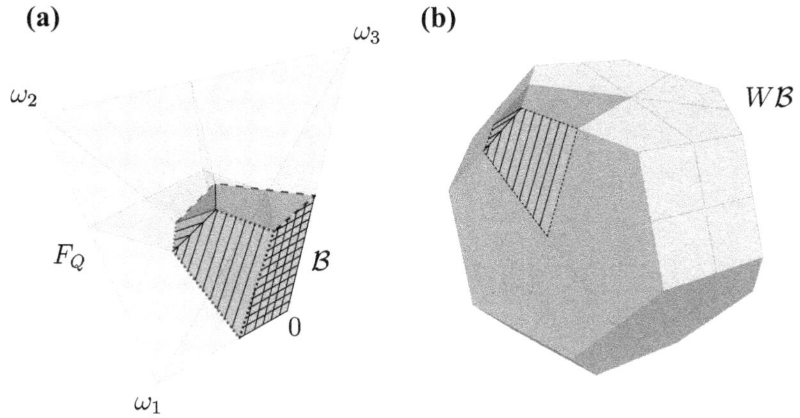

Fig. 2 **a** The domain $\mathcal{B}$ of A_3, bounded by one square-shaped and five kite-shaped surfaces, contains its entire boundary except for the points on the hatched areas and the dotted lines. The enveloping tetrahedron represents the domain F_Q, tiled by Γ-images of $\mathcal{B}$. **b** The polyhedron is formed by the images of the Weyl group W acting on $\mathcal{B}$

is also a fundamental domain of W_P^{aff}. The closure of the domain $W\mathcal{B}$ coincides, up to a factor 2π, with the Brillouin zone [27] of the dual-root lattice of A_3. The domains $\mathcal{B}$ and $W\mathcal{B}$ are depicted in Fig. 2.

4.3 $(\mathcal{R}, \sigma)$-Invariant Polynomials

The vector space $\mathbb{C}[x]$ comprises polynomials of $n+1$ variables $x \equiv [x_0, \dots, x_n]^T$ over $\mathbb{C}$. The extended m-degree $\text{edg}_m\, x^\lambda$ of a monomial $x^\lambda \equiv x_0^{\lambda_0} x_1^{\lambda_1} \dots x_n^{\lambda_n}$, $\lambda \equiv [\lambda_0, \dots, \lambda_n]$ is defined as

$$\text{edg}_m\, x^\lambda \equiv \lambda_0 + m_1^\vee \lambda_1 + \dots + m_n^\vee \lambda_n.$$

The extended m-degree of any polynomial $f \in \mathbb{C}[x]$ is then the maximum extended m-degree of homogeneous parts of f. Recall from [17] that the finite sets $\Lambda_M \subset P$ consist for any $M \in \mathbb{N}$ of the weights contained in the set MF_Q,

$$\Lambda_M = P \cap MF_Q, \tag{45}$$

and is of the explicit form

$$\Lambda_M = \Big\{ \lambda_1 \omega_1 + \dots + \lambda_n \omega_n \,\Big|\, \lambda_0 + m_1^\vee \lambda_1 + \dots + m_n^\vee \lambda_n = M,\ \lambda_i \in \mathbb{Z}^{\geq 0}, i = 0, \dots, n \Big\}. \tag{46}$$

Identifying each element of Λ_M with its magnified Kac coordinates,

$$\lambda \equiv [\lambda_0, \dots, \lambda_n], \tag{47}$$

results in the conclusion that $\mathrm{edg}_m\, x^\lambda = M$ if and only if $\lambda \in \Lambda_M$. All linear combinations of monomials of extended m-degree equal to M form a vector subspace of $\mathbb{C}[x]$ denoted by Π_M,

$$\Pi_M \equiv \left\{ \sum_{\lambda \in \Lambda_M} c_\lambda x^\lambda \mid c_\lambda \in \mathbb{C},\, \lambda \in \Lambda_M \right\}.$$

The standard action [19,34] of an operator $\mathbb{G} \in \mathrm{GL}_{n+1}(\mathbb{C})$ on $\mathbb{C}[x]$ is given by

$$\mathbb{G} \cdot f(x) \equiv f(\mathbb{G}^{-1}x), \quad f \in \mathbb{C}[x]. \tag{48}$$

For any representation $\mathcal{R} : \Gamma \to \mathrm{GL}_{n+1}(\mathbb{C})$ of the abelian group (26) and a sign homomorphism σ, a polynomial $f \in \mathbb{C}[x]$ is called $(\mathcal{R}, \sigma)$-invariant if it for all $\gamma \in \Gamma$ satisfies

$$\mathcal{R}(\gamma) \cdot f = \sigma \circ \widehat{\psi}(\gamma)\, f. \tag{49}$$

A vector subspace $\Pi_M^{\mathcal{R},\sigma} \subset \Pi_M$ contains all $(\mathcal{R}, \sigma)$-invariant polynomials from Π_M,

$$\Pi_M^{\mathcal{R},\sigma} \equiv \left\{ f \in \Pi_M \mid \mathcal{R}(\gamma) \cdot f = \sigma \circ \widehat{\psi}(\gamma)\, f,\ \gamma \in \Gamma \right\}.$$

Proposition 4.4 *Let $\mathcal{R}_1, \mathcal{R}_2$ be representations of Γ to $\mathrm{GL}_{n+1}(\mathbb{C})$ for which there exists $\mathbb{P} \in \mathrm{GL}_{n+1}(\mathbb{C})$ such that*

(i) *$\mathcal{R}_2(\gamma) = \mathbb{P}^{-1}\mathcal{R}_1(\gamma)\mathbb{P}$ for all $\gamma \in \Gamma$, i.e. $\mathcal{R}_1$ and $\mathcal{R}_2$ are equivalent,*
(ii) *$\mathbb{P} \cdot f \in \Pi_M \Leftrightarrow f \in \Pi_M$.*

Then the spaces $\Pi_M^{\mathcal{R}_1,\sigma}$ and $\Pi_M^{\mathcal{R}_2,\sigma}$ are isomorphic, for any sign homomorphism σ,

$$\Pi_M^{\mathcal{R}_1,\sigma} \simeq \Pi_M^{\mathcal{R}_2,\sigma}.$$

Proof Assumption (ii) implies for any $f \in \Pi_M^{\mathcal{R}_2,\sigma}$ that $\mathbb{P} \cdot f \in \Pi_M$. Definitions (48) and (49) and assumption (i) yield $(\mathcal{R}_1, \sigma)$-invariance of $\mathbb{P} \cdot f$,

$$\begin{aligned}\mathcal{R}_1(\gamma) \cdot \mathbb{P} \cdot f(x) &= f(\mathbb{P}^{-1}\mathcal{R}_1(\gamma)^{-1}x) = f(\mathcal{R}_2(\gamma)^{-1}\mathbb{P}^{-1}x)\\ &= \mathbb{P} \cdot \mathcal{R}_2(\gamma) \cdot f(x) = \sigma \circ \widehat{\psi}(\gamma)\, \mathbb{P} \cdot f(x)\end{aligned}$$

and therefore $\mathbb{P} \cdot f \in \Pi_M^{\mathcal{R}_1,\sigma}$. Conversely, it holds that if $f \in \Pi_M^{\mathcal{R}_1,\sigma}$ then $\mathbb{P}^{-1} \cdot f \in \Pi_M^{\mathcal{R}_2,\sigma}$. The map $\Pi_M^{\mathcal{R}_2,\sigma} \ni f \mapsto \mathbb{P} \cdot f \in \Pi_M^{\mathcal{R}_1,\sigma}$ is linear and its inverse is the map $\Pi_M^{\mathcal{R}_1,\sigma} \ni f \mapsto \mathbb{P}^{-1} \cdot f \in \Pi_M^{\mathcal{R}_2,\sigma}$.

Proposition 4.5 *Let $\mathbb{P} \in \mathrm{GL}_{n+1}(\mathbb{C})$ be such that $(\mathbb{P}^{-1}x)_i$ is a linear combination of monomials of the extended m-degree $m_i^\vee$ for every $i \in \{0, 1, \dots, n\}$. Then $f \in \Pi_M$ if and only if $\mathbb{P} \cdot f \in \Pi_M$.*

Proof For any monomial $x^\lambda \in \Pi_M$, the assumption implies that the factors of $\mathbb{P} \cdot x^\lambda$,

$$\mathbb{P} \cdot x^\lambda = (\mathbb{P}^{-1}x)_0^{\lambda_0} (\mathbb{P}^{-1}x)_1^{\lambda_1} \dots (\mathbb{P}^{-1}x)_n^{\lambda_n}, \tag{50}$$

satisfy for every $i \in \{0, 1, \dots, n\}$ that

$$(\mathbb{P}^{-1}x)_i^{\lambda_i} = \left(\sum_{k=1}^{l} p_{j_k} x_{j_k} \right)^{\lambda_i}, \quad p_{j_k} \in \mathbb{C} \backslash \{0\}, \quad \text{edg}_m\, x_{j_k} = m_i^\vee,\ k = 1, \dots, l. \tag{51}$$

Multinomial expansion of (51) provides the relation

$$(\mathbb{P}^{-1}x)_i^{\lambda_i} = \sum_{\substack{r_1, r_2, \dots, r_l \geq 0 \\ r_1 + r_2 + \dots + r_l = \lambda_i}} \left(\frac{\lambda_i!}{r_1! r_2! \dots r_l!} p_{j_1}^{r_1} p_{j_2}^{r_2} \dots p_{j_l}^{r_l} \right) x_{j_1}^{r_1} x_{j_2}^{r_2} \dots x_{j_l}^{r_l},$$

which guarantees that the polynomial $(\mathbb{P}^{-1}x)_i^{\lambda_i}$ is a linear combination of monomials of extended m-degree $\lambda_i m_i^\vee$ and thus

$$\text{edg}_m\, (\mathbb{P}^{-1}x)_i^{\lambda_i} = \lambda_i m_i^\vee.$$

Since $\mathbb{P} \cdot x^\lambda$ is a product of non-zero polynomials $(\mathbb{P}^{-1}x)_i^{\lambda_i}$ in (50), it is also non-zero and a linear combination of monomials of extended m-degree M,

$$\text{edg}_m\, \mathbb{P} \cdot x^\lambda = \text{edg}_m\, x^\lambda.$$

The resulting polynomial $\mathbb{P} \cdot f$ is thus a linear combination of monomials of extended m-degree M and therefore $\mathbb{P} \cdot f \in \Pi_M$. Conversely, the assumption on $\mathbb{P}$ is equivalent to $\mathbb{P}^{-1}$ being a block diagonal matrix in a suitable permutation of the standard basis with its blocks determined by the same values of $m_i^\vee$. Taking into account that the matrix $\mathbb{P}$ retains the same form, $f \in \Pi_M$ forces also $\mathbb{P}^{-1} \cdot f \in \Pi_M$. □

The action (27) of Γ on F_Q via permutations of the Kac coordinates $[b_0, \dots, b_n]$ induces a faithful representation $\mathcal{A} : \Gamma \to \text{GL}_{n+1}(\mathbb{C})$ by assigning each element $\gamma \in \Gamma$ its permutation matrix $\mathcal{A}(\gamma) \in \text{GL}_{n+1}(\mathbb{C})$,

$$\mathcal{A}(\gamma)[b_0, \dots, b_n]^T = [b_{\pi_\gamma(0)}, \dots, b_{\pi_\gamma(n)}]^T. \tag{52}$$

Similarly, the action of Γ_M on MF_Q, described by (32), assigns to each element $\gamma_M \in \Gamma_M$ its permutation matrix corresponding to $\gamma \in \Gamma$, provided by isomorphism (30). Consequently, the action of Γ_M also induces, independently on M, the identical representation matrices $\mathcal{A}(\gamma)$,

$$\mathcal{A}(\gamma) = \mathcal{A}(\gamma_M). \tag{53}$$

Since the finite group Γ is abelian, the commuting diagonalizable matrices $\mathcal{A}(\gamma)$ can be simultaneously diagonalized, i.e. there exists a unitary matrix $\mathbb{P} \in \text{GL}_{n+1}(\mathbb{C})$

such that $\mathbb{P}^{\dagger}\mathcal{A}(\gamma)\mathbb{P}$ is a diagonal matrix for each $\gamma \in \Gamma$. Therefore, the diagonal representation $\mathcal{D} : \Gamma \to \mathrm{GL}_{n+1}(\mathbb{C})$, equivalent to $\mathcal{A}$, is given by

$$\mathcal{D}(\gamma) = \mathbb{P}^{\dagger}\mathcal{A}(\gamma)\mathbb{P}. \tag{54}$$

The diagonal matrices $\mathcal{D}(\gamma)$ are for the generators of the non-trivial groups Γ of the following form

$$\begin{aligned}
A_n : \quad & \mathcal{D}(\gamma_1) = \mathrm{diag}\left(1, e^{\frac{2\pi \mathrm{i}}{n+1}}, e^{\frac{4\pi \mathrm{i}}{n+1}}, \dots, e^{\frac{2n\pi \mathrm{i}}{n+1}}\right), \\
B_{2k+1} : \quad & \mathcal{D}(\gamma_{2k+1}) = \mathrm{diag}\,(1, -1, 1, -1, \dots, 1, -1), \\
B_{2k} : \quad & \mathcal{D}(\gamma_{2k}) = \mathrm{diag}\,(1, 1, 1, -1, 1, -1 \dots, 1, -1, -1), \\
C_n : \quad & \mathcal{D}(\gamma_1) = \mathrm{diag}\,(1, -1, 1, 1, \dots, 1, 1), \\
D_{2k+1} : \quad & \mathcal{D}(\gamma_{2k+1}) = \mathrm{diag}\,(1, -1, 1, -1, \dots, 1, -1, \mathrm{i}, -\mathrm{i}), \\
D_{2k} : \quad & \mathcal{D}(\gamma_1) = \mathrm{diag}\,(1, 1, \dots, 1, -1, -1), \\
& \mathcal{D}(\gamma_{2k}) = \mathrm{diag}\,(1, -1, 1, -1, \dots, 1, -1, 1), \\
E_6 : \quad & \mathcal{D}(\gamma_5) = \mathrm{diag}\left(1, e^{\frac{2\pi \mathrm{i}}{3}}, e^{\frac{4\pi \mathrm{i}}{3}}, 1, e^{\frac{2\pi \mathrm{i}}{3}}, e^{\frac{4\pi \mathrm{i}}{3}}, 1\right), \\
E_7 : \quad & \mathcal{D}(\gamma_6) = \mathrm{diag}\,(1, 1, 1, 1, -1, 1, -1, -1)
\end{aligned} \tag{55}$$

and the unitary conjugation matrices $\mathbb{P}$ are listed in Table 2.

Theorem 4.6 *Let $\mathcal{A}$ be the permutation representation* (52) *of Γ and $\mathcal{D}$ the corresponding diagonal representation* (54). *Then the spaces $\Pi_M^{\mathcal{A},\sigma}$ and $\Pi_M^{\mathcal{D},\sigma}$ are for any sign homomorphism σ isomorphic,*

$$\Pi_M^{\mathcal{A},\sigma} \simeq \Pi_M^{\mathcal{D},\sigma}.$$

Proof Explicit forms of the unitary conjugation matrices $\mathbb{P}$ in Table 2 guarantee that each $\mathbb{P}$ satisfies the assumption of Proposition 4.5. This in turn, results in the validity of assumption (ii) in Proposition 4.4. □

5 Sets of Points and Weights

Section 5 defines the σ-dependent point and label sets $F^{\sigma}_{Q^{\vee},M}$ and $\Lambda^{\sigma}_{P,M}$ of the discrete transforms. Deduced from the theory of invariant polynomial spaces, crucial Theorem 5.4 guarantees the common cardinality of the sets $F^{\sigma}_{Q^{\vee},M}$ and $\Lambda^{\sigma}_{P,M}$. The counting formulas from Theorem 5.5 provide the explicit numbers of elements of the point and label sets.

Table 2 The non-zero entries of the unitary conjugation matrices $\mathbb{P}$, which for non-trivial groups Γ diagonalize their permutation representations $\mathcal{A}$

	$\mathbb{P} = (p_{ij}) \in \mathbb{C}^{n+1,n+1}, \quad i, j = 0, \ldots, n$		$\mathbb{P} = (p_{ij}) \in \mathbb{C}^{n+1,n+1}, \quad i, j = 0, \ldots, n$
A_n	$p_{ij} = \frac{1}{\sqrt{n+1}} \alpha^{-ij}, \quad i, j = 0, \ldots, n, \quad \alpha = e^{\frac{2\pi \mathrm{i}}{n+1}}$	C_n	$p_{00} = p_{10} = p_{01} = \frac{1}{\sqrt{2}}, \quad p_{11} = -\frac{1}{\sqrt{2}},$ $p_{ii} = 1, \quad i = 2, \ldots, n$
B_{2k+1}	$p_{ii} = p_{i,2k+1-i} = \frac{1}{\sqrt{2}},$ $p_{2k+1-i,i} = \frac{(-1)^i}{\sqrt{2}},$ $p_{2k+1-i,2k+1-i} = \frac{(-1)^{i+1}}{\sqrt{2}}, \quad i = 0, \ldots, k$	B_{2k}	$p_{00} = p_{2k,0} = p_{0,2k} = \frac{1}{\sqrt{2}},$ $p_{k,1} = 1, \quad p_{2k,2k} = -\frac{1}{\sqrt{2}},$ $p_{i-1,i} = p_{i-1,2k+1-i} = \frac{1}{\sqrt{2}},$ $p_{2k+1-i,i} = \frac{(-1)^i}{\sqrt{2}},$ $p_{2k+1-i,2k+1-i} = \frac{(-1)^{i+1}}{\sqrt{2}}, \quad i = 2, \ldots, k$
D_{2k+1}	$p_{00} = p_{01} = p_{10} = p_{11} = \frac{1}{2},$ $p_{2k,0} = p_{2k+1,0} = p_{2k+1,2k} = p_{2k+1,2k+1} = \frac{1}{2},$ $p_{2k,1} = p_{2k+1,1} = p_{2k,2k} = p_{2k,2k+1} = -\frac{1}{2},$ $p_{1,2k} = p_{0,2k+1} = -\frac{\mathrm{i}}{2}, \quad p_{0,2k} = p_{1,2k+1} = \frac{\mathrm{i}}{2},$ $p_{i,2i-2} = p_{i,2i-1} = p_{2k+1-i,2i-2} = \frac{1}{\sqrt{2}},$ $p_{2k+1-i,2i-1} = -\frac{1}{\sqrt{2}}, \quad i = 2, \ldots, k$	D_{2k}	$p_{00} = p_{01} = p_{10} = p_{11} = p_{2k-1,0} = p_{2k,0} = \frac{1}{2},$ $p_{0,2k-1} = p_{0,2k} = p_{2k-1,2k-1} = p_{2k,2k} = \frac{1}{2}$ $p_{k,2} = 1, \quad p_{2k-1,1} = p_{2k,1} = p_{1,2k-1} = -\frac{1}{2},$ $p_{1,2k} = p_{2k-1,2k} = p_{2k,2k-1} = -\frac{1}{2},$ $p_{i-1,i} = p_{i-1,2k+1-i} = \frac{1}{\sqrt{2}}, \quad p_{2k+1-i,i} = \frac{(-1)^i}{\sqrt{2}},$ $p_{2k+1-i,2k+1-i} = \frac{(-1)^{i+1}}{\sqrt{2}}, \quad i = 3, \ldots, k$
E_6	$\frac{1}{\sqrt{3}} \begin{pmatrix} 1 & e^{\frac{4\pi \mathrm{i}}{3}} & 0 & 0 & 0 & e^{\frac{2\pi \mathrm{i}}{3}} & 0 \\ 1 & e^{\frac{2\pi \mathrm{i}}{3}} & 0 & 0 & 0 & e^{\frac{4\pi \mathrm{i}}{3}} & 0 \\ 0 & 0 & e^{\frac{2\pi \mathrm{i}}{3}} & 0 & e^{\frac{4\pi \mathrm{i}}{3}} & 0 & 1 \\ 0 & 0 & 0 & \sqrt{3} & 0 & 0 & 0 \\ 0 & 0 & e^{\frac{4\pi \mathrm{i}}{3}} & 0 & e^{\frac{2\pi \mathrm{i}}{3}} & 0 & 1 \\ 1 & 1 & 0 & 0 & 0 & 1 & 0 \\ 0 & 0 & 1 & 0 & 1 & 0 & 1 \end{pmatrix}$	E_7	$\frac{1}{\sqrt{2}} \begin{pmatrix} 1 & 0 & 0 & 0 & 0 & 0 & 1 & 0 \\ 0 & 1 & 0 & 0 & 0 & 0 & 0 & 1 \\ 0 & 0 & 1 & 0 & 1 & 0 & 0 & 0 \\ 0 & 0 & 0 & \sqrt{2} & 0 & 0 & 0 & 0 \\ 0 & 0 & 1 & 0 & -1 & 0 & 0 & 0 \\ 0 & 1 & 0 & 0 & 0 & 0 & 0 & -1 \\ 1 & 0 & 0 & 0 & 0 & 0 & -1 & 0 \\ 0 & 0 & 0 & 0 & 0 & \sqrt{2} & 0 & 0 \end{pmatrix}$

5.1 Sets $\Lambda^\sigma_{P,M}$ and $F^\sigma_{Q^\vee,M}$

The Fourier calculus of Weyl orbit functions is commonly formulated [15,17,36] on points $F^\sigma_{P^\vee,M}$, $M\in\mathbb{N}$ from the refined dual weight lattice $\frac{1}{M}P^\vee$ contained in F^σ,

$$F^\sigma_{P^\vee,M}=\tfrac{1}{M}P^\vee\cap F^\sigma, \tag{56}$$

while weights $\Lambda^\sigma_{Q,M}$ labelling the functions are taken from the magnified dual fundamental domain MF^σ_Q,

$$\Lambda^\sigma_{Q,M}=P\cap MF^\sigma_Q. \tag{57}$$

This work develops the Fourier calculus on points $F^\sigma_{Q^\vee,M}$, $M>m^\sigma$ from the refined dual-root lattice $\frac{1}{M}Q^\vee$ contained in F^σ,

$$F^\sigma_{Q^\vee,M}\equiv\frac{1}{M}Q^\vee\cap F^\sigma, \tag{58}$$

with weights $\Lambda^\sigma_{P,M}$ labelling the functions taken from the magnified fundamental domain MF^σ_P,

$$\Lambda^\sigma_{P,M}\equiv P\cap MF^\sigma_P. \tag{59}$$

The set of points $F^\sigma_{Q^\vee,M}\subset F^\sigma_{P^\vee,M}$ contains all points of $F^\sigma_{P^\vee,M}$ belonging to $\frac{1}{M}Q^\vee$ and the set of weights $\Lambda^\sigma_{P,M}\subset\Lambda^\sigma_{Q,M}$ is a suitable fragment of $\Lambda^\sigma_{Q,M}$.

In order to describe explicitly the point sets $F^\sigma_{Q^\vee,M}$ for any sign homomorphism σ, the symbols s^σ_i, $i\in\{0,1,\dots,n\}$ are introduced via the relations

$$\begin{array}{llll}
s^1_0,\, s^1_i\in\mathbb{Z}^{\geq 0} & \text{if } \alpha_i\in\Delta, & & \\
s^{\sigma^e}_0,\, s^{\sigma^e}_i\in\mathbb{N} & \text{if } \alpha_i\in\Delta, & & \\
s^{\sigma^s}_0,\, s^{\sigma^s}_i\in\mathbb{Z}^{\geq 0} & \text{if } \alpha_i\in\Delta_l, & s^{\sigma^s}_i\in\mathbb{N} & \text{if } \alpha_i\in\Delta_s, \\
s^{\sigma^l}_0,\, s^{\sigma^l}_i\in\mathbb{N} & \text{if } \alpha_i\in\Delta_l, & s^{\sigma^l}_i\in\mathbb{Z}^{\geq 0} & \text{if } \alpha_i\in\Delta_s.
\end{array}$$

The explicit description of the point set $F^\sigma_{Q^\vee,M}$ follows from relations (4) and (15),

$$\begin{aligned}F^\sigma_{Q^\vee,M}=\Bigg\{&\frac{s^\sigma_1}{M}\omega^\vee_1+\cdots+\frac{s^\sigma_n}{M}\omega^\vee_n \,\Big|\, \sum_{i=0}^n m_i s^\sigma_i\\ &=M,\ \sum_{i=1}^n (C^T)^{-1}_{ij}s^\sigma_i\in\mathbb{Z},\ j=1,\dots,n\Bigg\}\end{aligned} \tag{60}$$

where C^T denotes the transposed Cartan matrix (3). The set of n equations

$$\sum_{i=1}^n (C^T)^{-1}_{ij}s^\sigma_i\in\mathbb{Z},\quad j=1,\dots,n \tag{61}$$

selects such points from $F^{\sigma}_{P^\vee,M}$ which belong to $\frac{1}{M}Q^\vee$. Standard reduction of the sets of equations (61) yields their simplified form for each non-trivial case as

$$
\begin{aligned}
&A_n: && s_1^\sigma + 2s_2^\sigma + 3s_3^\sigma + \dots + ns_n^\sigma \equiv 0 \mod n+1,\\
&B_{2k+1}: && s_1^\sigma + s_3^\sigma + s_5^\sigma + \dots + s_{2k+1}^\sigma \equiv 0 \mod 2,\\
&B_{2k}: && s_1^\sigma + s_3^\sigma + s_5^\sigma + \dots + s_{2k-1}^\sigma \equiv 0 \mod 2,\\
&C_n: && s_n^\sigma \equiv 0 \mod 2,\\
&D_{4k}: && s_1^\sigma + s_3^\sigma + \dots + s_{4k-3}^\sigma + s_{4k-1}^\sigma \equiv 0 \mod 2,\\
& && s_{4k-1}^\sigma + s_{4k}^\sigma \equiv 0 \mod 2,\\
&D_{4k+1}: && 2s_1^\sigma + 2s_3^\sigma + \dots + 2s_{4k-1}^\sigma + 3s_{4k}^\sigma + s_{4k+1}^\sigma \equiv 0 \mod 4,\\
&D_{4k+2}: && s_1^\sigma + s_3^\sigma + \dots + s_{4k-1}^\sigma + s_{4k+2}^\sigma \equiv 0 \mod 2,\\
& && s_{4k+1}^\sigma + s_{4k+2}^\sigma \equiv 0 \mod 2,\\
&D_{4k+3}: && 2s_1^\sigma + 2s_3^\sigma + \dots + 2s_{4k+1}^\sigma + s_{4k+2}^\sigma + 3s_{4k+3}^\sigma \equiv 0 \mod 4,\\
&E_6: && s_1^\sigma + 2s_2^\sigma + s_4^\sigma + 2s_5^\sigma \equiv 0 \mod 3,\\
&E_7: && s_4^\sigma + s_6^\sigma + s_7^\sigma \equiv 0 \mod 2.
\end{aligned}
\tag{62}
$$

Example 5.1 The point and label sets $F^{\sigma}_{Q^\vee,M}$ and $\Lambda^{\sigma}_{P,M}$ are explicitly constructed for the Lie algebra A_2. The marks and dual marks of A_2 are listed in [17] as

$$m_1 = m_2 = m_1^\vee = m_2^\vee = 1.$$

Firstly, the explicit formulas for the dual weight point sets $F^{\sigma}_{P^\vee,M}$, that depend on the identity and determinant sign homomorphisms $\mathbb{1}$ and σ^e, are given from defining relation (56) and explicit formula (15) as

$$
\begin{aligned}
F^{\mathbb{1}}_{P^\vee,M} &= \left\{ \frac{s_1^{\mathbb{1}}}{M}\omega_1^\vee + \frac{s_2^{\mathbb{1}}}{M}\omega_2^\vee \mid s_0^{\mathbb{1}} + s_1^{\mathbb{1}} + s_2^{\mathbb{1}} = M,\ s_0^{\mathbb{1}}, s_1^{\mathbb{1}}, s_2^{\mathbb{1}} \in \mathbb{Z}^{\geq 0} \right\},\\
F^{\sigma^e}_{P^\vee,M} &= \left\{ \frac{s_1^{\sigma^e}}{M}\omega_1^\vee + \frac{s_2^{\sigma^e}}{M}\omega_2^\vee \mid s_0^{\sigma^e} + s_1^{\sigma^e} + s_2^{\sigma^e} = M,\ s_0^{\sigma^e}, s_1^{\sigma^e}, s_2^{\sigma^e} \in \mathbb{N} \right\}.
\end{aligned}
$$

The dual-root lattice point set $F^{\sigma}_{Q^\vee,M}$ contains only such points from the dual weight point set $F^{\sigma}_{P^\vee,M}$ which belong to the refined dual-root lattice $\frac{1}{M}Q^\vee$. These points are determined from relation (62) by the condition $s_1^\sigma + 2s_2^\sigma = 0 \bmod 3$,

$$F^{\sigma}_{Q^\vee,M} = \left\{ \frac{s_1^\sigma}{M}\omega_1^\vee + \frac{s_2^\sigma}{M}\omega_2^\vee \in F^{\sigma}_{P^\vee,M} \mid s_1^\sigma + 2s_2^\sigma = 0 \bmod 3 \right\}.$$

Fixing the scaling factor $M = 7$, the points of the set $F^{\mathbb{1}}_{P^\vee,7}$ are determined by their coordinates $\left(s_1^{\mathbb{1}}/M, s_2^{\mathbb{1}}/M\right)$ in the $\omega^\vee$-basis,

$$
\begin{aligned}
F^{1}_{P^\vee,7} = &\Big\{(0,0), \left(0,\frac{3}{7}\right), \left(\frac{3}{7},0\right), \left(0,\frac{6}{7}\right), \left(\frac{6}{7},0\right), \left(\frac{1}{7},\frac{1}{7}\right), \left(\frac{1}{7},\frac{4}{7}\right),\\
&\left(\frac{4}{7},\frac{1}{7}\right), \left(\frac{2}{7},\frac{2}{7}\right), \left(\frac{2}{7},\frac{5}{7}\right), \left(\frac{5}{7},\frac{2}{7}\right), \left(\frac{3}{7},\frac{3}{7}\right)\Big\}\\
&\cup\Big\{\left(\frac{1}{7},0\right), \left(\frac{1}{7},\frac{3}{7}\right), \left(\frac{4}{7},0\right), \left(\frac{1}{7},\frac{6}{7}\right), (1,0), \left(\frac{2}{7},\frac{1}{7}\right), \left(\frac{2}{7},\frac{4}{7}\right),\\
&\left(\frac{5}{7},\frac{1}{7}\right), \left(\frac{3}{7},\frac{2}{7}\right), \left(0,\frac{5}{7}\right), \left(0,\frac{2}{7}\right), \left(\frac{4}{7},\frac{3}{7}\right)\Big\}\\
&\cup\Big\{\left(0,\frac{1}{7}\right), \left(0,\frac{4}{7}\right), \left(\frac{3}{7},\frac{1}{7}\right), (0,1), \left(\frac{6}{7},\frac{1}{7}\right), \left(\frac{1}{7},\frac{2}{7}\right), \left(\frac{1}{7},\frac{5}{7}\right),\\
&\left(\frac{4}{7},\frac{2}{7}\right), \left(\frac{2}{7},\frac{3}{7}\right), \left(\frac{2}{7},0\right), \left(\frac{5}{7},0\right), \left(\frac{3}{7},\frac{4}{7}\right)\Big\}.
\end{aligned}
$$

The first 12 points of $F^{1}_{P^\vee,7}$ satisfy the condition $s^{1}_1 + 2s^{1}_2 = 0 \bmod 3$ and therefore, they form the dual-root lattice point set $F^{1}_{Q^\vee,7}$.

The point set $F^{\sigma^e}_{P^\vee,7}$ comprises the points of the set $F^{1}_{P^\vee,7}$ that are situated in the interior of the fundamental domain F. Thus, the points of the point set $F^{\sigma^e}_{P^\vee,7}$ are determined by their coordinates $\left(s^{\sigma^e}_1/M, s^{\sigma^e}_2/M\right)$ in $\omega^\vee$-basis,

$$
\begin{aligned}
F^{\sigma^e}_{P^\vee,7} = &\Big\{\left(\frac{1}{7},\frac{1}{7}\right), \left(\frac{1}{7},\frac{4}{7}\right), \left(\frac{4}{7},\frac{1}{7}\right), \left(\frac{2}{7},\frac{2}{7}\right), \left(\frac{3}{7},\frac{3}{7}\right)\Big\}\\
&\cup\Big\{\left(\frac{1}{7},\frac{3}{7}\right), \left(\frac{2}{7},\frac{1}{7}\right), \left(\frac{2}{7},\frac{4}{7}\right), \left(\frac{5}{7},\frac{1}{7}\right), \left(\frac{3}{7},\frac{2}{7}\right)\Big\}\\
&\cup\Big\{\left(\frac{3}{7},\frac{1}{7}\right), \left(\frac{1}{7},\frac{2}{7}\right), \left(\frac{1}{7},\frac{5}{7}\right), \left(\frac{4}{7},\frac{2}{7}\right), \left(\frac{2}{7},\frac{3}{7}\right)\Big\}.
\end{aligned}
$$

The first 5 points of $F^{\sigma^e}_{P^\vee,7}$ satisfy the condition $s^{\sigma^e}_1 + 2s^{\sigma^e}_2 = 0 \bmod 3$ and therefore, they form the dual-root lattice point set $F^{\sigma^e}_{Q^\vee,7}$.

Secondly, the explicit formulas for the label sets $\Lambda^{\sigma}_{Q,M}$ are given from defining relation (57) and explicit formula for the fundamental domain (18) by

$$
\begin{aligned}
\Lambda^{1}_{Q,M} &= \left\{\lambda^{1}_1\omega_1 + \lambda^{1}_2\omega_2 \mid \lambda^{1}_0 + \lambda^{1}_1 + \lambda^{1}_2 = M,\ \lambda^{1}_0, \lambda^{1}_1, \lambda^{1}_2 \in \mathbb{Z}^{\geq 0}\right\},\\
\Lambda^{\sigma^e}_{Q,M} &= \left\{\lambda^{\sigma^e}_1\omega_1 + \lambda^{\sigma^e}_2\omega_2 \mid \lambda^{\sigma^e}_0 + \lambda^{\sigma^e}_1 + \lambda^{\sigma^e}_2 = M,\ \lambda^{\sigma^e}_0, \lambda^{\sigma^e}_1, \lambda^{\sigma^e}_2 \in \mathbb{N}\right\}.
\end{aligned}
$$

Fixing again the scaling factor $M = 7$, the magnified Kac coordinates (47) of the labels from $\Lambda^{1}_{Q,7}$ are of the form $\left[\lambda^{1}_0, \lambda^{1}_1, \lambda^{1}_2\right]$ and the set $\Lambda^{1}_{Q,7}$ is expressed as

$$
\begin{aligned}
\Lambda^{1}_{Q,7} = &\{[7,0,0],[0,7,0],[0,0,7]\} \cup \{[6,1,0],[0,6,1],[1,0,6]\}\\
&\cup \{[5,2,0],[0,5,2],[2,0,5]\} \cup \{[4,3,0],[0,4,3],[3,0,4]\}\\
&\cup \{[4,0,3],[3,4,0],[0,3,4]\} \cup \{[5,0,2],[2,5,0],[0,2,5]\}
\end{aligned}
$$

$$\cup \{[6,0,1],[1,6,0],[0,1,6]\} \cup \{[5,1,1],[1,5,1],[1,1,5]\}$$
$$\cup \{[4,1,2],[2,4,1],[1,2,4]\} \cup \{[3,3,1],[3,1,3],[1,3,3]\}$$
$$\cup \{[4,2,1],[1,4,2],[2,1,4]\} \cup \{[3,2,2],[2,3,2],[2,2,3]\}.$$

The points of the label set $\Lambda^{1}_{Q,7}$ are divided into the orbits of the group Γ_7 which is for any scaling factor M formed by the elements $\gamma_{M,i}$, $i = 0, 1, 2$ and acts according to Table 1 on the magnified Kac coordinates $[\lambda_0, \lambda_1, \lambda_2]$ as

$$\begin{aligned} \gamma_{M,0}[\lambda_0,\lambda_1,\lambda_2] &= [\lambda_0,\lambda_1,\lambda_2], \\ \gamma_{M,1}[\lambda_0,\lambda_1,\lambda_2] &= [\lambda_2,\lambda_0,\lambda_1], \\ \gamma_{M,2}[\lambda_0,\lambda_1,\lambda_2] &= [\lambda_1,\lambda_2,\lambda_0]. \end{aligned} \tag{63}$$

The label set of weights $\Lambda^{1}_{P,M}$ contains the lexicographically highest label from each Γ_M-orbit of $\Lambda^{1}_{Q,M}$ and therefore, the label set $\Lambda^{1}_{P,7}$ is given by

$$\Lambda^{1}_{P,7} = \{[7,0,0],[6,1,0],[5,2,0],[4,3,0],[4,0,3],[5,0,2],[6,0,1],$$
$$[5,1,1],[4,1,2],[3,3,1],[4,2,1],[3,2,2]\}.$$

The label set $\Lambda^{\sigma^e}_{Q,7}$ contains the points of $\Lambda^{1}_{Q,7}$ which are from the interior of $7F_Q$, i.e.

$$\Lambda^{\sigma^e}_{Q,7} = \{[5,1,1],[1,5,1],[1,1,5]\} \cup \{[4,1,2],[2,4,1],[1,2,4]\}$$
$$\cup \{[3,3,1],[3,1,3],[1,3,3]\} \cup \{[4,2,1],[1,4,2],[2,1,4]\}$$
$$\cup \{[3,2,2],[2,3,2],[2,2,3]\}.$$

Since for A_2 it holds for any $b \in M(F_P \cap F^{\sigma^e}_Q)$ that $\sigma^e \circ \widehat{\psi}(\mathrm{Stab}_{\Gamma_M}(b)) = \{1\}$, the label set $\Lambda^{\sigma^e}_{P,M}$ consists of the points from $\Lambda^{1}_{P,M}$ that are situated in the interior of MF_Q. Equivalently, $\Lambda^{\sigma^e}_{P,M}$ contains the lexicographically highest point from each Γ_M-orbit of $\Lambda^{\sigma^e}_{Q,M}$. In particular, the label set $\Lambda^{\sigma^e}_{P,7}$ is of the following form,

$$\Lambda^{\sigma^e}_{P,7} = \{[5,1,1],[4,1,2],[3,3,1],[4,2,1],[3,2,2]\}.$$

The point sets $F^{\sigma}_{Q^\vee,7}$ and $F^{\sigma}_{P^\vee,7}$ together with the label sets $\Lambda^{\sigma}_{P,7}$ and $\Lambda^{\sigma}_{Q,7}$ are depicted in Fig. 3.

5.2 Cardinality of $\Lambda^{\sigma}_{P,M}$ and $F^{\sigma}_{Q^\vee,M}$

In order to prove that the point sets $F^{\sigma}_{Q^\vee,M}$ and the sets of weights $\Lambda^{\sigma}_{P,M}$ have for $M > m^{\sigma}$ the same cardinality, the isomorphism of polynomial spaces in Theorem 4.6 is employed. The first step is to introduce an auxiliary finite set of weights $\widetilde{\Lambda}^{\sigma}_{Q,M}$ by the relation

$$\widetilde{\Lambda}^{\sigma}_{Q,M} \equiv \left\{\lambda \in \Lambda_M \mid \sigma \circ \widehat{\psi}(\mathrm{Stab}_{\Gamma_M}(\lambda)) = \{1\}\right\} \tag{64}$$

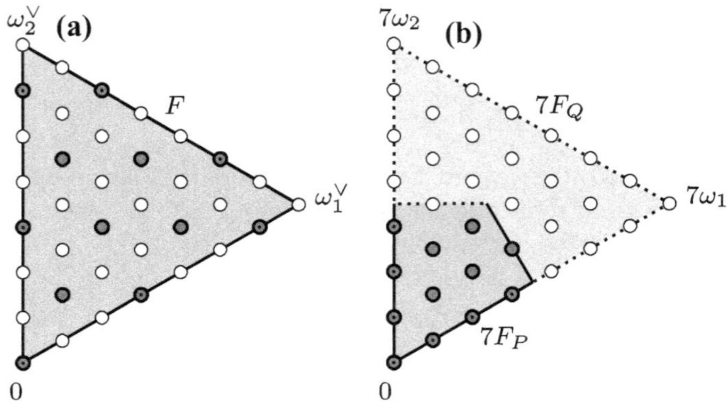

Fig. 3 **a** The fundamental domain F of A_2, depicted as the triangle, contains 24 white nodes and 12 dark nodes forming together the set of points $F^1_{P^\vee,7}$. The 12 dark nodes represent the points of $F^1_{Q^\vee,7}$. The dark nodes without the dotted ones on the boundary of F depict the 5 points of $F^{\sigma^e}_{Q^\vee,7}$. **b** The domain $7F_Q$ of A_2, depicted as the triangle, contains 24 white nodes and 12 dark nodes that form the set of weights $\Lambda^1_{Q,7}$. The 12 dark nodes in the kite-shaped domain $7F_P$ belong to the weight set $\Lambda^1_{P,7}$. The dark nodes without the dotted ones depict the 5 points of the set $\Lambda^{\sigma^e}_{P,7}$

together with its complementary set $\widetilde{H}^\sigma_{Q,M}$,

$$\widetilde{H}^\sigma_{Q,M} \equiv \left\{\lambda \in \Lambda_M \mid \sigma \circ \widehat{\psi}(\mathrm{Stab}_{\Gamma_M}(\lambda)) = \{\pm 1\}\right\} \tag{65}$$

and their corresponding sets $\widetilde{\Lambda}^\sigma_{P,M}$ and $\widetilde{H}^\sigma_{P,M}$ of representative weights in Γ_M-orbits,

$$\widetilde{\Lambda}^\sigma_{P,M} \equiv MF_P \cap \widetilde{\Lambda}^\sigma_{Q,M}, \tag{66}$$
$$\widetilde{H}^\sigma_{P,M} \equiv MF_P \cap \widetilde{H}^\sigma_{Q,M}. \tag{67}$$

Indeed, since MF_P is a fundamental domain of the action of Γ_M on MF_Q, it holds that

$$\widetilde{\Lambda}^\sigma_{Q,M} = \Gamma_M \widetilde{\Lambda}^\sigma_{P,M}, \tag{68}$$
$$\widetilde{H}^\sigma_{Q,M} = \Gamma_M \widetilde{H}^\sigma_{P,M}. \tag{69}$$

The following disjoint decomposition of the weight set (45) and the set $\Lambda^1_{P,M}$ are thus obtained,

$$\Lambda_M = \widetilde{\Lambda}^\sigma_{Q,M} \cup \widetilde{H}^\sigma_{Q,M}, \tag{70}$$
$$\Lambda^1_{P,M} = \widetilde{\Lambda}^\sigma_{P,M} \cup \widetilde{H}^\sigma_{P,M}. \tag{71}$$

The set $\widetilde{\Lambda}^\sigma_{P,M}$ is related by ϱ^σ shifts (5) to the set $\Lambda^\sigma_{P,M+m^\sigma}$ in the following proposition.

Proposition 5.1 *The set of weights* $\Lambda^{\sigma}_{P,M+m^{\sigma}}$ *coincides for any* $M \in \mathbb{N}$ *with the* ϱ^{σ}*-shifted set* $\widetilde{\Lambda}^{\sigma}_{P,M}$,

$$\Lambda^{\sigma}_{P,M+m^{\sigma}} = \varrho^{\sigma} + \widetilde{\Lambda}^{\sigma}_{P,M}.$$

Proof Firstly, for any coordinates $[\lambda_0, \dots, \lambda_n]$ of $\lambda = \lambda_1\omega_1 + \ldots + \lambda_n\omega_n \in P$, the corresponding coordinates $[\lambda^{\sigma}_0, \dots, \lambda^{\sigma}_n]$ of $\lambda^{\sigma} = \lambda^{\sigma}_1\omega_1 + \cdots + \lambda^{\sigma}_n\omega_n \in P$ are defined by

$$\lambda^{\sigma}_i = \lambda_i + \varrho^{\sigma}_i, \quad i \in \{0, 1, \dots, n\}.$$

According to (59) and (44), any weight $\lambda^{\sigma} \in \Lambda^{\sigma}_{P,M+m^{\sigma}}$ satisfies the conditions

$$\lambda^{\sigma} \in P \cap (M + m^{\sigma})F_P, \tag{72}$$

$$\lambda^{\sigma} \in P \cap (M + m^{\sigma})F^{\sigma}_Q, \tag{73}$$

together with

$$\sigma \circ \widehat{\psi}\left(\mathrm{Stab}_{\Gamma_{M+m^{\sigma}}}(\lambda^{\sigma})\right) = \{1\}. \tag{74}$$

Defining relations (5), (6) and (21), (22) imply that $\varrho^{\sigma} \in P \cap m^{\sigma} F^{\sigma}_Q$ and consequently (73) grants that

$$\lambda \in P \cap M F_Q. \tag{75}$$

Taking into account (28) and comparing the action of $\gamma_{M+m^{\sigma}} \in \Gamma_{M+m^{\sigma}}$ on $\lambda^{\sigma} \in P \cap (M + m^{\sigma})F^{\sigma}_Q$,

$$\gamma_{M+m^{\sigma}} \cdot \lambda^{\sigma} = [\lambda^{\sigma}_{\pi_{\gamma}(0)}, \dots, \lambda^{\sigma}_{\pi_{\gamma}(n)}] = [\lambda_{\pi_{\gamma}(0)} + \varrho_0, \dots, \lambda_{\pi_{\gamma}(n)} + \varrho_n], \tag{76}$$

to the action of $\gamma_M \in \Gamma_M$ on $\lambda \in P \cap M F_Q$,

$$\gamma_M \cdot \lambda = [\lambda_{\pi_{\gamma}(0)}, \dots, \lambda_{\pi_{\gamma}(n)}], \tag{77}$$

yields that the point λ^{σ} is the lexicographically highest in its $\Gamma_{M+m^{\sigma}}$-orbit whenever λ is the lexicographically highest in its Γ_M-orbit. Thus, properties (72), (75) and (40) ensure that

$$\lambda \in P \cap M F_P. \tag{78}$$

Connecting action (76) for the stabilizing $\gamma_{M+m^{\sigma}} \in \mathrm{Stab}_{\Gamma_{M+m^{\sigma}}}(\lambda^{\sigma})$ and action (77) of $\gamma_M \in \Gamma_M$ on $\lambda \in P \cap M F_P$ forces γ_M to stabilize λ, i.e. $\gamma_M \in \mathrm{Stab}_{\Gamma_M}(\lambda)$. Relation (37) thus gives

$$\sigma \circ \widehat{\psi}\left(\mathrm{Stab}_{\Gamma_M}(\lambda)\right) = \sigma \circ \widehat{\psi}\left(\mathrm{Stab}_{\Gamma_{M+m^{\sigma}}}(\lambda^{\sigma})\right), \tag{79}$$

moreover (74) guarantees in turn that

$$\sigma \circ \widehat{\psi}\left(\mathrm{Stab}_{\Gamma_M}(\lambda)\right) = \{1\}, \tag{80}$$

and therefore $\lambda \in \widetilde{\Lambda}^{\sigma}_{P,M}$.

Conversely, $\lambda \in \widetilde{\Lambda}^{\sigma}_{P,M}$ implies validity of (80) and (78). Defining relations (5), (6) and (21), (22) then imply that $\lambda \in P \cap MF_Q$ grants (73). By comparing actions (76) and (77), property (78) forces relation (72), properties (80) and (79) guarantee property(74) and therefore $\lambda^{\sigma} \in \Lambda^{\sigma}_{P,M+m^{\sigma}}$. □

Proposition 5.2 *The dimension of $\Pi^{\mathcal{A},\sigma}_M$ is for any $M \in \mathbb{N}$ equal to the number of points in $\Lambda^{\sigma}_{P,M+m^{\sigma}}$,*

$$\dim \Pi^{\mathcal{A},\sigma}_M = \left|\Lambda^{\sigma}_{P,M+m^{\sigma}}\right|.$$

Proof Utilizing the group action (32), the polynomial $f^{\sigma}_{\lambda} \in \Pi_M$ is for any sign homomorphism σ and any $\lambda \in \widetilde{\Lambda}^{\sigma}_{P,M}$ introduced by

$$f^{\sigma}_{\lambda}(x) \equiv \sum_{\gamma\in\Gamma_M} \sigma \circ \widehat{\psi}(\gamma)\, x^{\gamma\cdot\lambda}. \tag{81}$$

Any $\gamma \in \Gamma_M$ is represented by (53) as a permutation matrix $\mathcal{A}(\gamma)$ and thus the polynomial action (48) on the monomials takes the form

$$\mathcal{A}(\gamma) \cdot x^{\lambda} = x^{\gamma\cdot\lambda}. \tag{82}$$

Substituting (82) into (81) and taking into account relations (37) and (53), the polynomials f^{σ}_{λ}, $\lambda \in \widetilde{\Lambda}^{\sigma}_{P,M}$ are deduced to be $(\mathcal{A}, \sigma)$-invariant,

$$f^{\sigma}_{\lambda} \in \Pi^{\mathcal{A},\sigma}_M.$$

Since any $\gamma \in \mathrm{Stab}_{\Gamma_M}(\lambda)$ satisfies from (66) that $\sigma \circ \widehat{\psi}(\gamma) = 1$, a general linear combination of polynomials (81) is of the form

$$\begin{aligned}
\sum_{\lambda\in\widetilde{\Lambda}^{\sigma}_{P,M}} c_{\lambda} f^{\sigma}_{\lambda}(x) &= \sum_{\lambda\in\widetilde{\Lambda}^{\sigma}_{P,M}} \sum_{\gamma\in\Gamma_M} c_{\lambda}\, \sigma \circ \widehat{\psi}(\gamma)\, x^{\gamma\cdot\lambda} \\
&= \sum_{\lambda\in\widetilde{\Lambda}^{\sigma}_{P,M}} \left(\sum_{\gamma\in\mathrm{Stab}_{\Gamma_M}(\lambda)} c_{\lambda}\, x^{\lambda} + \sum_{\gamma\in\Gamma_M\setminus\mathrm{Stab}_{\Gamma_M}(\lambda)} c_{\lambda}\, \sigma \circ \widehat{\psi}(\gamma)\, x^{\gamma\cdot\lambda} \right) \\
&= \sum_{\lambda\in\widetilde{\Lambda}^{\sigma}_{P,M}} |\mathrm{Stab}_{\Gamma_M}(\lambda)|\, c_{\lambda}\, x^{\lambda} \\
&\quad + \sum_{\lambda\in\widetilde{\Lambda}^{\sigma}_{P,M}} \sum_{\gamma\in\Gamma_M\setminus\mathrm{Stab}_{\Gamma_M}(\lambda)} c_{\lambda}\, \sigma \circ \widehat{\psi}(\gamma)\, x^{\gamma\cdot\lambda}.
\end{aligned} \tag{83}$$

Since MF_P is a fundamental domain of Γ_M acting on MF_Q, the second term of (83) is a linear combination of monomials x^{λ} with $\lambda \notin \widetilde{\Lambda}^{\sigma}_{P,M}$. Since the monomials x^{λ}, $\lambda \in \widetilde{\Lambda}^{\sigma}_{P,M}$ in the first term of (83) are linearly independent, setting (83) equal to zero grants that all $c_\lambda \in \mathbb{C}$ with $\lambda \in \widetilde{\Lambda}^\sigma_{P,M}$ are also zero. The polynomials (81) are thus linearly independent.

For any $(\mathcal{A}, \sigma)$-invariant polynomial $f \in \Pi^{\mathcal{A},\sigma}_M$ of the form

$$f(x) = \sum_{\lambda \in \Lambda_M} c_\lambda \, x^\lambda$$

and for any $\gamma \in \Gamma_M$, properties (49), (37), (53) and (82) imply that

$$\sigma \circ \widehat{\psi}(\gamma) \sum_{\lambda \in \Lambda_M} c_\lambda \, x^\lambda = \sum_{\lambda \in \Lambda_M} c_\lambda \, x^{\gamma \cdot \lambda}. \tag{84}$$

The set of weights Λ_M is invariant under Γ_M and therefore, comparing the coefficients in (84) yields that $\sigma \circ \widehat{\psi}(\gamma) c_\lambda = c_{\gamma^{-1} \cdot \lambda}$. Thus, relation $\sigma \circ \widehat{\psi}(\gamma^{-1}) = \sigma \circ \widehat{\psi}(\gamma)$ guarantees for all $\gamma \in \Gamma_M$ that

$$\sigma \circ \widehat{\psi}(\gamma) c_\lambda = c_{\gamma \cdot \lambda}. \tag{85}$$

The disjoint decomposition (70) of the weight set Λ_M together with relations (68), (69) and (85) ensures that

$$\begin{aligned} f(x) &= \sum_{\gamma \in \Gamma_M} \sum_{\lambda \in \widetilde{\Lambda}^\sigma_{P,M}} c_{\gamma \cdot \lambda} \, x^{\gamma \cdot \lambda} + \sum_{\gamma \in \Gamma_M} \sum_{\lambda \in \widetilde{H}^\sigma_{P,M}} c_{\gamma \cdot \lambda} \, x^{\gamma \cdot \lambda} \\ &= \sum_{\lambda \in \widetilde{\Lambda}^\sigma_{P,M}} c_\lambda f^\sigma_\lambda(x) + \sum_{\lambda \in \widetilde{H}^\sigma_{P,M}} c_\lambda \sum_{\gamma \in \Gamma_M} \sigma \circ \widehat{\psi}(\gamma) \, x^{\gamma \cdot \lambda}. \end{aligned} \tag{86}$$

Defining relation (67) guarantees that for any $\lambda \in \widetilde{H}^\sigma_{P,M}$ there exists $\widetilde{\gamma} \in \mathrm{Stab}_{\Gamma_M}(\lambda)$ such that $\sigma \circ \widehat{\psi}(\widetilde{\gamma}) = -1$ and thus the second term in (86) vanishes,

$$\sum_{\gamma \in \Gamma_M} \sigma \circ \widehat{\psi}(\gamma) \, x^{\gamma \cdot \lambda} = \sum_{\gamma \in \Gamma_M} \sigma \circ \widehat{\psi}(\gamma) \, x^{\gamma \cdot (\widetilde{\gamma} \cdot \lambda)} = - \sum_{\gamma \in \Gamma_M} \sigma \circ \widehat{\psi}(\gamma) \, x^{\gamma \cdot \lambda} = 0,$$

and the polynomials (81) generate the space $\Pi^{\mathcal{A},\sigma}_M$. The constructed basis f^σ_λ, $\lambda \in \widetilde{\Lambda}^\sigma_{P,M}$ provides together with Proposition 5.1 the resulting relation for dimension of $\Pi^{\mathcal{A},\sigma}_M$,

$$\dim \Pi^{\mathcal{A},\sigma}_M = \left| \widetilde{\Lambda}^\sigma_{P,M} \right| = \left| \Lambda^\sigma_{P,M+m^\sigma} \right|.$$

□

Proposition 5.3 *The dimension of $\Pi^{\mathcal{D},\sigma}_M$ is for any $M \in \mathbb{N}$ equal to the number of points in $F^\sigma_{Q^\vee,M+m^\sigma}$,*

$$\dim \Pi^{\mathcal{D},\sigma}_M = \left| F^\sigma_{Q^\vee,M+m^\sigma} \right|.$$

Proof Invariance property (49) of the diagonal representation (54) grants that any polynomial from $\Pi_M^{\mathcal{D},\sigma}$ is a linear combination of $(\mathcal{D},\sigma)$-invariant monomials. Therefore, the set of $(\mathcal{D},\sigma)$-invariant monomials x^λ, $\lambda \in \Lambda_M$ satisfying for all $\gamma \in \Gamma$

$$\mathcal{D}(\gamma) \cdot x^\lambda = \sigma \circ \widehat{\psi}(\gamma)\, x^\lambda, \tag{87}$$

forms a basis of $\Pi_M^{\mathcal{D},\sigma}$. Since for any $\gamma_1, \gamma_2 \in \Gamma$ it holds that

$$\mathcal{D}(\gamma_1\gamma_2) \cdot x^\lambda = \mathcal{D}(\gamma_1) \cdot (\mathcal{D}(\gamma_2) \cdot x^\lambda) = \sigma \circ \widehat{\psi}(\gamma_2)\mathcal{D}(\gamma_1) \cdot x^\lambda = \sigma \circ \widehat{\psi}(\gamma_1\gamma_2)\, x^\lambda,$$

verifying property (87) only for the generators of Γ yields its validity for all $\gamma \in \Gamma$. For any $\lambda = [\lambda_0, \ldots, \lambda_n] \in \Lambda_M$, with its Kac coordinates $\lambda_i \in \mathbb{Z}^{\geq 0}$, $i \in \{0, 1, \ldots, n\}$ satisfying the defining relation in (46)

$$\lambda_0 + m_1^\vee \lambda_1 + \cdots + m_n^\vee \lambda_n = M, \tag{88}$$

and for any diagonal generator $\mathcal{D}(\gamma) \equiv \operatorname{diag}(d_0, \ldots, d_n)$ of Γ is condition (87) equivalent to the relation

$$d_0^{\lambda_0} \ldots d_n^{\lambda_n} = \sigma \circ \widehat{\psi}(\gamma). \tag{89}$$

Explicit forms (55) of the diagonal generators yield the following explicit reformulations of (89) for the non-trivial cases,

$$\begin{aligned}
&A_n: && \lambda_1 + 2\lambda_2 + 3\lambda_3 + \cdots + n\lambda_n \equiv h^\sigma \mod n+1,\\
&B_{2k+1}: && \lambda_1 + \lambda_3 + \cdots + \lambda_{2k+1} \equiv h^\sigma \mod 2,\\
&B_{2k}: && \lambda_3 + \lambda_5 + \cdots + \lambda_{2k-1} + \lambda_{2k} \equiv h^\sigma \mod 2,\\
&C_n: && \lambda_1 \equiv h^\sigma \mod 2,\\
&D_{2k+1}: && 2\lambda_1 + 2\lambda_3 + \cdots + 2\lambda_{2k-1} + \lambda_{2k} + 3\lambda_{2k+1} \equiv h^\sigma \mod 4,\\
&D_{2k}: && \lambda_{2k-1} + \lambda_{2k} \equiv 0 \mod 2,\\
& && \lambda_1 + \lambda_3 + \cdots + \lambda_{2k-1} \equiv h^\sigma \mod 2,\\
&E_6: && \lambda_1 + 2\lambda_2 + \lambda_4 + 2\lambda_5 \equiv 0 \mod 3,\\
&E_7: && \lambda_4 + \lambda_6 + \lambda_7 \equiv h^\sigma \mod 2,
\end{aligned} \tag{90}$$

where non-zero values of h^σ are listed in Table 3.

Recall from Proposition 2.1 in [15] that the short and long Coxeter numbers (6) are of the form

$$m^{\sigma^s} = \sum_{\alpha_i \in \Delta_s} m_i, \quad m^{\sigma^l} = \sum_{\alpha_i \in \Delta_l} m_i + 1. \tag{91}$$

Introducing the symbols $s_i \in \mathbb{Z}^{\geq 0}$ by relations

$$\begin{aligned}
&s_i^{\mathbb{1}} = s_i && i = 0, \ldots, n,\\
&s_i^{\sigma^e} = 1 + s_i && i = 0, \ldots, n,\\
&s_i^{\sigma^s} = 1 + s_i, && \alpha_i \in \Delta_s, \qquad s_0^{\sigma^s} = s_0,\ s_i^{\sigma^s} = s_i \qquad \alpha_i \in \Delta_l,
\end{aligned}$$

Table 3 The non-zero values of symbols h^{σ}

	A_{2k+1}	B_{4k+1}	B_{4k+2}	B_{4k+3}	C_n	D_{4k+2}	D_{4k+3}	E_7
h^{σ^e}	$k+1$	1	1	0	1	1	2	1
h^{σ^s}	–	1	0	1	0	–	–	–
h^{σ^l}	–	0	1	1	1	–	–	–

$$s_i^{\sigma^l} = s_i, \qquad \alpha_i \in \Delta_s, \qquad s_0^{\sigma^l} = 1 + s_0,\ s_i^{\sigma^l} = 1 + s_i \qquad \alpha_i \in \Delta_l,$$

and substituting them into the expressions (60) and (62) together with formulas (1) and (91) implies, that the number of points in $F^{\sigma}_{Q^{\vee}, M+m^{\sigma}}$ is equal to the number of solutions of the equations

$$s_0 + m_1 s_1 + \cdots + m_n s_n = M \tag{92}$$

and

$$\begin{aligned}
A_n :&\quad s_1 + 2s_2 + 3s_3 + \cdots + ns_n \equiv h^{\sigma} \mod n+1,\\
B_{2k+1} :&\quad s_1 + s_3 + s_5 + \cdots + s_{2k+1} \equiv h^{\sigma} \mod 2,\\
B_{2k} :&\quad s_1 + s_3 + s_5 + \cdots + s_{2k-1} \equiv h^{\sigma} \mod 2,\\
C_n :&\quad s_n \equiv h^{\sigma} \mod 2,\\
D_{4k}:&\quad s_1 + s_3 + \cdots + s_{4k-3} + s_{4k-1} \equiv 0 \mod 2,\\
&\quad s_{4k-1} + s_{4k} \equiv 0 \mod 2,\\
D_{4k+1} :&\quad 2s_1 + 2s_3 + \cdots + 2s_{4k-1} + 3s_{4k} + s_{4k+1} \equiv 0 \mod 4,\\
D_{4k+2} :&\quad s_1 + s_3 + \cdots + s_{4k-1} + s_{4k+2} \equiv h^{\sigma} \mod 2,\\
&\quad s_{4k+1} + s_{4k+2} \equiv 0 \mod 2,\\
D_{4k+3} :&\quad 2s_1 + 2s_3 + \cdots + 2s_{4k+1} + s_{4k+2} + 3s_{4k+3} \equiv h^{\sigma} \mod 4,\\
E_6 :&\quad s_1 + 2s_2 + s_4 + 2s_5 \equiv 0 \mod 3,\\
E_7 :&\quad s_4 + s_6 + s_7 \equiv h^{\sigma} \mod 2.
\end{aligned} \tag{93}$$

The solutions $s_i \in \mathbb{Z}^{\geq 0}$ of the equations (92) and (93) and solutions $\lambda_i \in \mathbb{Z}^{\geq 0}$ of the equations (88) and (90) coincide for each case up to a permutation and therefore, their numbers are identical. □

Theorem 5.4 *For any $M \in \mathbb{N}$, $M > m^{\sigma}$ it holds that*

$$\left| F^{\sigma}_{Q^{\vee}, M} \right| = \left| \Lambda^{\sigma}_{P,M} \right|.$$

Proof Combining Propositions 5.2, Proposition 5.3 and Theorem 4.6 grants for any $M \in \mathbb{N}$ that

$$\left| F^{\sigma}_{Q^{\vee}, M+m^{\sigma}} \right| = \dim \Pi_M^{\mathcal{D},\sigma} = \dim \Pi_M^{\mathcal{A},\sigma} = \left| \Lambda^{\sigma}_{P, M+m^{\sigma}} \right|.$$

5.3 Counting Formulas

In order to obtain explicit counting formulas for cardinalities of all $\Lambda^{\sigma}_{P,M}$ and $M > m^{\sigma}$, the following crucial identity stemming from Proposition 5.1 is used,

$$\left|\Lambda^{\sigma}_{P,M}\right| = \left|\widetilde{\Lambda}^{\sigma}_{P,M-m^{\sigma}}\right|. \tag{94}$$

The calculation of the cardinality of $\Lambda^{\sigma}_{P,M}$ is thus reverted to counting the weights in $\widetilde{\Lambda}^{\sigma}_{P,M}$ for all $M \in \mathbb{N}$. The group Γ_M partitions the sets of weights (64) and (65) into Γ_M-orbits and the sets $\widetilde{\Lambda}^{\sigma}_{P,M}$ and $\widetilde{H}^{\sigma}_{P,M}$ consist of exactly one point from each Γ_M-orbit. Therefore, the number of points in $\widetilde{\Lambda}^{\sigma}_{P,M}$ and $\widetilde{H}^{\sigma}_{P,M}$ is equal to the number of Γ_M-orbits in $\widetilde{\Lambda}^{\sigma}_{Q,M}$ and $\widetilde{H}^{\sigma}_{Q,M}$, respectively.

Starting with the identity sign homomorphism set $\widetilde{\Lambda}^{\mathfrak{1}}_{P,M}$ and introducing a set of points in Λ_M fixed by a given $\gamma \in \Gamma_M$,

$$\mathrm{Fix}_M(\gamma) \equiv \{\lambda \in \Lambda_M \mid \gamma \cdot \lambda = \lambda\},$$

the Burnside's lemma applied to the set $\widetilde{\Lambda}^{\mathfrak{1}}_{Q,M} = \Lambda_M$ provides the relation

$$\left|\Lambda^{\mathfrak{1}}_{P,M}\right| = \left|\widetilde{\Lambda}^{\mathfrak{1}}_{P,M}\right| = \frac{1}{c} \sum_{\gamma \in \Gamma_M} |\mathrm{Fix}_M(\gamma)|. \tag{95}$$

Moreover, the group Γ_M is partitioned into disjoint sets containing the elements of the same order $|\gamma| = d$ which implies that (95) is reformulated as

$$\left|\Lambda^{\mathfrak{1}}_{P,M}\right| = \frac{1}{c} \sum_{d|c} \sum_{\substack{|\gamma|=d \\ \gamma \in \Gamma_M}} |\mathrm{Fix}_M(\gamma)|. \tag{96}$$

For a general sign homomorphism, disjoint decomposition (71) yields the relation

$$\left|\widetilde{\Lambda}^{\sigma}_{P,M}\right| = \left|\Lambda^{\mathfrak{1}}_{P,M}\right| - \left|\widetilde{H}^{\sigma}_{P,M}\right|, \tag{97}$$

and employing the Burnside's lemma to the set $\widetilde{H}^{\sigma}_{Q,M}$ produces the identity

$$\left|\widetilde{H}^{\sigma}_{P,M}\right| = \frac{1}{c} \sum_{d|c} \sum_{\substack{|\gamma|=d \\ \gamma \in \Gamma_M}} \left|\mathrm{Fix}_M(\gamma) \cap \widetilde{H}^{\sigma}_{Q,M}\right|. \tag{98}$$

Using the Euler's totient function φ together with Eqs. (94), (96), (97) and (98), the explicit counting formulas for the numbers of points in the weight sets $\Lambda^{\sigma}_{P,M}$ are for all cases listed in the following theorem.

Theorem 5.5 *The numbers of elements in $\Lambda^{\sigma}_{P,M}$ are for any $M \in \mathbb{N}$, $M > m^{\sigma}$ determined by the following formulas.*

(1) $A_n (n \geq 1)$:

$$\left|\Lambda^{\mathbb{1}}_{P,M}(A_n)\right| = \frac{1}{n+M+1} \sum_{d \mid \gcd(n+1,M)} \varphi(d) \binom{\frac{n+M+1}{d}}{\frac{n+1}{d}}, \tag{99}$$

$$\left|\Lambda^{\sigma^e}_{P,M}(A_n)\right| = \frac{(-1)^{n+1}}{M} \sum_{d \mid \gcd(n+1,M)} (-1)^{\frac{n+1}{d}} \varphi(d) \binom{\frac{M}{d}}{\frac{n+1}{d}}, \tag{100}$$

(2) $B_n (n \geq 3)$:

$$\left|\Lambda^{\mathbb{1}}_{P,M}(B_n)\right| = \begin{cases} \binom{n+k}{n} & M = 2k+1, \\ \frac{1}{2}\left[\binom{2m+2k+l+1}{2m+1} + \binom{2m+2k+l}{2m+1} + \binom{m+k}{m}\right] & M = 4k+2l, n = 2m+1, \\ \frac{1}{2}\left[\binom{2m+2k+l}{2m} + \binom{2m+2k+l-1}{2m} + \binom{m+k}{m} + \binom{m+k+l-1}{m}\right] & M = 4k+2l, n = 2m, \end{cases}$$

$$\left|\Lambda^{\sigma^e}_{P,M}(B_n)\right| = \begin{cases} \frac{1}{2}\left[\binom{2k+l}{4m+1} + \binom{2k+l-1}{4m+1} - \binom{k+l-1}{2m}\right] & M = 4k+2l, n = 4m+1, \\ \frac{1}{2}\left[\binom{2k+l}{4m+2} + \binom{2k+l-1}{4m+2} - \binom{k}{2m+1} - \binom{k+l-1}{2m+1}\right] & M = 4k+2l, n = 4m+2, \\ \left|\Lambda^{\mathbb{1}}_{P,M-2n}(B_n)\right| & \text{otherwise}, \end{cases}$$

$$\left|\Lambda^{\sigma^s}_{P,M}(B_n)\right| = \begin{cases} \frac{1}{2}\left[\binom{2m+2k+l}{2m+1} + \binom{2m+2k+l-1}{2m+1} - \binom{m+k+l-1}{m}\right] & M = 4k+2l, n = 2m+1, \\ \left|\Lambda^{\mathbb{1}}_{P,M-2}(B_n)\right| & \text{otherwise}, \end{cases}$$

$$\left|\Lambda^{\sigma^l}_{P,M}(B_n)\right| = \begin{cases} \frac{1}{2}\left[\binom{2k+l+1}{4m+2} + \binom{2k+l}{4m+2} - \binom{k}{2m+1} - \binom{k+l}{2m+1}\right] & M = 4k+2l, n = 4m+2, \\ \frac{1}{2}\left[\binom{2k+l+1}{4m+3} + \binom{2k+l}{4m+3} - \binom{k}{2m+1}\right] & M = 4k+2l, n = 4m+3, \\ \left|\Lambda^{\mathbb{1}}_{P,M-2n+2}(B_n)\right| & \text{otherwise}, \end{cases}$$

where $l \in \{0, 1\}$.

(3) $C_n (n \geq 2)$:

$$\left|\Lambda^{\mathbb{1}}_{P,M}(C_n)\right| = \binom{n+k}{n}, \quad \text{where } k = \left\lfloor \frac{M}{2} \right\rfloor,$$

$$\left|\Lambda^{\sigma^e}_{P,M}(C_n)\right| = \left|\Lambda^{\mathbb{1}}_{P,M-2n-1}(C_n)\right|,$$

$$\left|\Lambda^{\sigma^s}_{P,M}(C_n)\right| = \left|\Lambda^{\mathbb{1}}_{P,M-2n+2}(C_n)\right|,$$

$$\left|\Lambda^{\sigma^l}_{P,M}(C_n)\right| = \left|\Lambda^{\mathbb{1}}_{P,M-3}(C_n)\right|.$$

(4) $D_n (n \geq 4)$:

$$\left|\Lambda^{\mathbb{1}}_{P,M}(D_n)\right| = \begin{cases} \binom{n+k}{n} + \binom{n+k-1}{n} & M = 2k+1, \\ \frac{1}{4}\left[\binom{2m+2k+1}{2m+1} + 6\binom{2m+2k}{2m+1} + \binom{2m-1+2k}{2m+1} + \binom{2m-1+2k}{2m-1} + 2\binom{m+k-1}{m-1}\right] & M = 4k, n = 2m+1, \\ \frac{1}{4}\left[\binom{2m+2k+2}{2m+1} + 6\binom{2m+2k+1}{2m+1} + \binom{2m+2k}{2m+1} + \binom{2m+2k}{2m-1}\right] & M = 4k+2, n = 2m+1, \\ \frac{1}{4}\left[\binom{2m+2k}{2m} + 6\binom{2m+2k-1}{2m} + \binom{2m+2k-2}{2m} + \binom{2m+2k-2}{2m-2}\right. & \\ \quad \left. +2\binom{m+k}{m} + 6\binom{m+k-1}{m}\right] & M = 4k, n = 2m, \\ \frac{1}{4}\left[\binom{2m+2k+1}{2m} + 6\binom{2m+2k}{2m} + \binom{2m+2k-1}{2m} + \binom{2m+2k-1}{2m-2}\right. & \\ \quad \left. +6\binom{m+k}{m} + 2\binom{m+k-1}{m}\right] & M = 4k+2, n = 2m, \end{cases}$$

$$\left|\Lambda^{\sigma^e}_{P,M}(D_n)\right| = \begin{cases} \frac{1}{4}\left[\binom{2k+1}{4m+3} + 6\binom{2k}{4m+3} + \binom{2k-1}{4m+3} + \binom{2k-1}{4m+1} - 2\binom{k-1}{2m}\right] & M = 4k, n = 4m+3, \\ \frac{1}{4}\left[\binom{2k+2}{4m+2} + 6\binom{2k+1}{4m+2} + \binom{2k}{4m+2} + \binom{2k}{4m} - 2\binom{k+1}{2m+1} - 6\binom{k}{2m+1}\right] & M = 4k+2, n = 4m+2, \\ \frac{1}{4}\left[\binom{2k+1}{4m+2} + 6\binom{2k}{4m+2} + \binom{2k-1}{4m+2} + \binom{2k-1}{4m} - 6\binom{k}{2m+1} - 2\binom{k-1}{2m+1}\right] & M = 4k, n = 4m+2, \\ \left|\Lambda^{\mathbb{1}}_{P,M-2n+2}(D_n)\right| & \textit{otherwise.} \end{cases}$$

(5) E_6:

$$\left|\Lambda^{\mathbb{1}}_{P,M}(E_6)\right| = \begin{cases} \frac{1}{3}\left[|\Lambda_{6k}(E_6)| + 2\binom{k+2}{2} + 2\binom{k+1}{2}\right] & M = 6k, \\ \frac{1}{3}\left[|\Lambda_{6k+3}(E_6)| + 4\binom{k+2}{2}\right] & M = 6k+3, \\ \frac{1}{3}\left|\Lambda_M(E_6)\right| & \textit{otherwise,} \end{cases}$$

$$\left|\Lambda^{\sigma^e}_{P,M}(E_6)\right| = \left|\Lambda^{\mathbb{1}}_{P,M-12}(E_6)\right|.$$

(6) E_7:

$$\left|\Lambda^{\mathbb{1}}_{P,M}(E_7)\right| = \begin{cases} \frac{1}{2}\left|\Lambda_{2k+1}(E_7)\right| & M = 2k+1, \\ \frac{1}{2}\left[|\Lambda_{12k+2l}(E_7)| + \sum_{i=0}^{3} d_{li}\binom{4-i+k}{4}\right] & M = 12k+2l, l \in \{0, \ldots, 5\}, \end{cases}$$

$$\left|\Lambda^{\sigma^e}_{P,M}(E_7)\right| = \begin{cases} \frac{1}{2}\left[|\Lambda_{12k+2l-18}(E_7)| - \sum_{i=0}^{3} d_{l+3,i}\binom{2-i+k}{4}\right] & M = 12k+2l, l \in \{0, 1, 2\}, \\ \frac{1}{2}\left[|\Lambda_{12k+2l-18}(E_7)| - \sum_{i=0}^{3} d_{l-3,i}\binom{3-i+k}{4}\right] & M = 12k+2l, l \in \{3, 4, 5\}, \\ \left|\Lambda^{\mathbb{1}}_{P,M-18}(E_7)\right| & \textit{otherwise,} \end{cases}$$

where

$$(d_{li}) = \begin{pmatrix} 1 & 34 & 64 & 9 \\ 2 & 46 & 55 & 5 \\ 5 & 55 & 46 & 2 \\ 9 & 64 & 34 & 1 \\ 16 & 67 & 25 & 0 \\ 25 & 67 & 16 & 0 \end{pmatrix}.$$

(7) E_8, F_4, G_2:

$$\left|\Lambda^{\sigma}_{P,M}\right| = \left|\Lambda^{\sigma}_{Q,M}\right|.$$

Proof A detailed calculation is presented for the infinite series of groups Γ_M of A_n. Since the remaining infinite series B_n, C_n and D_n share common groups Γ_M listed in Table 1, the proof of the corresponding counting formulas is less complex.

The group Γ_M of A_n is a cyclic group of order $n+1$ generated by the permutation γ_1 satisfying $\gamma_1^{n+1} = 1$,

$$\Gamma_M = \{\gamma_1, \gamma_1^2, \dots, \gamma_1^{n+1}\}.$$

For each $k \in \{1, \dots, n+1\}$ such that order $\left|\gamma_1^k\right| = d$ it holds that

$$d = \frac{n+1}{l}, \quad l = \gcd(n+1, k).$$

Since for any weight $\lambda \in \mathrm{Fix}_M\left(\gamma_1^k\right)$, its cyclic isotropy subgroup $\mathrm{Stab}_{\Gamma_M}(\lambda)$ contains both γ_1^k and γ_1^l, it holds that $\lambda \in \mathrm{Fix}_M\left(\gamma_1^l\right)$. Conversely, as l divides k, the set of fixed points $\mathrm{Fix}_M\left(\gamma_1^l\right)$ is contained in the set $\mathrm{Fix}_M\left(\gamma_1^k\right)$ and thus

$$\mathrm{Fix}_M\left(\gamma_1^k\right) = \mathrm{Fix}_M\left(\gamma_1^l\right). \tag{101}$$

Since the number of elements of order d in a cyclic group Γ_M is the value of the Euler's totient function $\varphi(d)$ and all elements of order d satisfy (101), counting relation (96) specializes to

$$\left|\Lambda^{1}_{P,M}\right| = \frac{1}{n+1} \sum_{d|n+1} \varphi(d) \left|\mathrm{Fix}_M(\gamma_1^l)\right|. \tag{102}$$

The Kac coordinates $\lambda_0, \dots, \lambda_n \in \mathbb{Z}^{\geq 0}$ of a point $\lambda \in \Lambda_M$ satisfy equation (88) specialized to A_n,

$$\lambda_0 + \lambda_1 + \cdots + \lambda_n = M. \tag{103}$$

Employing explicit expression for the cyclic permutation listed in Table 1, a weight $\lambda \in \Lambda_M$ is stabilized by $\gamma_1^l \in \Gamma_M$, i.e. $\lambda \in \mathrm{Fix}_M\left(\gamma_1^l\right)$, if and only if

$$\lambda_i = \lambda_{i+l} = \cdots = \lambda_{i+(d-1)l}, \quad i = 0, \dots, l-1. \tag{104}$$

Substituting (104) into relation (103) yields

$$d(\lambda_0 + \cdots + \lambda_{l-1}) = M, \tag{105}$$

and thus, if M is not divisible by d, then

$$\left|\mathrm{Fix}_M\left(\gamma_1^l\right)\right| = 0, \quad d \nmid M. \tag{106}$$

If d divides M, then the number $\left|\mathrm{Fix}_M(\gamma_1^l)\right|$ coincides with the number of non-negative integer solutions of equation (105). This number is determined by Proposition 3.1 in [17] as

$$\left|\mathrm{Fix}_M\left(\gamma_1^l\right)\right| = \binom{\frac{n+1+M}{d} - 1}{\frac{n+1}{d} - 1} = \frac{n+1}{n+M+1}\binom{\frac{n+1+M}{d}}{\frac{n+1}{d}}, \quad d \mid M. \tag{107}$$

Substituting relations (107) and (106) into (102) results in counting formula (99).

Taking into account (37), the values of the composition of homomorphisms $\sigma^e \circ \widehat{\psi}$ on the elements of the group Γ_M follow for the case A_n from Table 1 as

$$\sigma^e \circ \widehat{\psi}(\gamma_1^k) = (-1)^{nk}. \tag{108}$$

For n even, definition (65) and relation (108) immediately guarantee that $\widetilde{H}_{Q,M}^{\sigma^e} = \emptyset$ and thus

$$\left|\widetilde{H}_{P,M}^{\sigma^e}\right| = 0, \quad n \text{ even}. \tag{109}$$

Continuing with n odd, equality (101) again grants that counting formula (98) simplifies as

$$\begin{aligned}\left|\widetilde{H}_{P,M}^{\sigma^e}\right| &= \frac{1}{n+1}\sum_{d|n+1} \varphi(d)\left|\mathrm{Fix}_M(\gamma_1^l) \cap \widetilde{H}_{Q,M}^{\sigma^e}\right| \\ &= \frac{1}{n+1}\sum_{l|n+1} \varphi\left(\frac{n+1}{l}\right)\left|\mathrm{Fix}_M(\gamma_1^l) \cap \widetilde{H}_{Q,M}^{\sigma^e}\right|.\end{aligned} \tag{110}$$

In order to determine the structure of the intersection sets in (110), note that for an odd n there exist an odd number m and $j \in \mathbb{N}$ such that

$$n+1 = 2^j m,$$

and any divisor l of $n+1$ is of the form

$$l = 2^i p, \quad p \mid m,\ i \in \{0, 1, \dots, j\}. \tag{111}$$

Taking any $\lambda \in \mathrm{Fix}_M(\gamma_1^l) \cap \widetilde{H}_{Q,M}^{\sigma^e}$, relations (108) and (65) yield that there exists an odd number $u \in \{1, \dots, n+1\}$ such that $\gamma_1^u \in \mathrm{Stab}_{\Gamma_M}(\lambda)$. The stabilizer subgroup $\mathrm{Stab}_{\Gamma_M}(\lambda)$ therefore, contains a subgroup generated by $\gamma_1^u, \gamma_1^l \in \mathrm{Stab}_{\Gamma_M}(\lambda)$,

$$\left\{(\gamma_1^u)^r(\gamma_1^l)^t \mid r,t\in\mathbb{Z}\right\} = \left\{\gamma_1^{ur+lt} \mid r,t\in\mathbb{Z}\right\} \subset \mathrm{Stab}_{\Gamma_M}(\lambda),$$

and thus $\gamma_1^{\gcd(u,l)} \in \mathrm{Stab}_{\Gamma_M}(\lambda)$. Since u is odd, the greatest common divisor $\gcd(u,l)$ is odd and divides p and therefore $\lambda \in \mathrm{Fix}_M(\gamma_1^{\gcd(u,l)}) \subset \mathrm{Fix}_M(\gamma_1^p)$. Conversely, taking any $\lambda \in \mathrm{Fix}_M(\gamma_1^p) \subset \mathrm{Fix}_M(\gamma_1^l)$, the divisor p, being odd, grants that $\lambda \in \widetilde{H}^{\sigma^e}_{Q,M}$ and hence

$$\mathrm{Fix}_M(\gamma_1^l) \cap \widetilde{H}^{\sigma^e}_{Q,M} = \mathrm{Fix}_M(\gamma_1^p). \tag{112}$$

Relations (112) and (111) allow to further evaluate the counting formula (110) as

$$\begin{aligned}\left|\widetilde{H}^{\sigma^e}_{P,M}\right| &= \frac{1}{n+1}\sum_{l|n+1}\varphi\left(\frac{n+1}{l}\right)\left|\mathrm{Fix}_M(\gamma_1^p)\right| \\ &= \frac{1}{n+1}\sum_{p|m}\sum_{i=0}^{j}\varphi\left(\frac{n+1}{2^i p}\right)\left|\mathrm{Fix}_M(\gamma_1^p)\right|.\end{aligned} \tag{113}$$

Since the Euler's totient function φ is multiplicative and $\varphi(2^i) = 2^{i-1}$ holds for any $i \in \mathbb{N}$, the following identities are obtained,

$$\begin{aligned}\sum_{i=0}^{j}\varphi\left(\frac{n+1}{2^i p}\right) &= \sum_{i=0}^{j}\varphi\left(2^i\frac{m}{p}\right) = \varphi\left(\frac{m}{p}\right)\sum_{i=0}^{j}\varphi\left(2^i\right) \\ &= 2^j\varphi\left(\frac{m}{p}\right) = 2\varphi\left(\frac{n+1}{p}\right),\end{aligned}$$

and thus (113) is simplified as

$$\begin{aligned}\left|\widetilde{H}^{\sigma^e}_{P,M}\right| &= \frac{2}{n+1}\sum_{p|m}\varphi\left(\frac{n+1}{p}\right)\left|\mathrm{Fix}_M(\gamma_1^p)\right| \\ &= \frac{1}{n+1}\sum_{l|n+1}\left(1-(-1)^l\right)\varphi\left(\frac{n+1}{l}\right)\left|\mathrm{Fix}_M(\gamma_1^l)\right| \\ &= \frac{1}{n+1}\sum_{d|n+1}\left(1-(-1)^l\right)\varphi(d)\left|\mathrm{Fix}_M(\gamma_1^l)\right|.\end{aligned} \tag{114}$$

The resulting counting formula

$$\left|\widetilde{H}^{\sigma^e}_{P,M}\right| = \frac{1}{n+1}\sum_{d|n+1}\left(1-(-1)^{l+n+1}\right)\varphi(d)\left|\mathrm{Fix}_M(\gamma_1^l)\right| \tag{115}$$

specializes for n odd to (114) and for n even to (109). Substituting relations (106) and (107) into the counting formula (115) and the result into relations (97) and (94), while taking into account that the Coxeter number of A_n is $m^{\sigma^e} = n+1$, yields the final counting relation (100). □

6 Discrete Transforms of Weyl Orbit Functions

Section 6 firstly describes the argument and label symmetries which are induced by the dual-root lattice discretization of Weyl orbit functions. Fundamental Theorems 6.4 and 6.6 explicitly state the discrete orthogonality relations of Weyl and Hartley orbit functions sampled on the point sets $F^{\sigma}_{Q^{\vee},M}$ and labelled by the weights from $\Lambda^{\sigma}_{P,M}$. The discrete forward and backward Fourier–Weyl and Hartley–Weyl transforms together with the interpolation and Plancherel formulas are deduced.

6.1 Weyl and Hartley Orbit Functions

The sign homomorphisms σ of the Weyl group W induce up to four families of Weyl orbit functions. The Weyl orbit functions $\varphi^{\sigma}_{b} : \mathbb{R}^n \to \mathbb{C}$, labelled by a parameter $b \in \mathbb{R}^n$, are defined for any $a \in \mathbb{R}^n$ by

$$\varphi^{\sigma}_{b}(a) \equiv \sum_{w \in W} \sigma(w)\, e^{2\pi \mathrm{i} \langle wb, a\rangle}. \tag{116}$$

The functions φ^{1}_{b} and $\varphi^{\sigma^e}_{b}$ are called C-functions and S-functions, respectively. In the cases of simple Lie algebras with two different root lengths, the functions $\varphi^{\sigma^s}_{b}$ and $\varphi^{\sigma^l}_{b}$ are termed the S^s- and S^l-functions in [28], respectively.

Recall from Proposition 3.1 in [6] that while restricting the label to the the weight lattice $\lambda \in P$, the argument invariance of Weyl orbit functions with respect to the action of the element of the affine Weyl group $w^{\mathrm{aff}} \in W^{\mathrm{aff}}$ is for any $a \in \mathbb{R}^n$ of the form

$$\varphi^{\sigma}_{\lambda}(w^{\mathrm{aff}} a) = \sigma \circ \psi(w^{\mathrm{aff}}) \cdot \varphi^{\sigma}_{\lambda}(a). \tag{117}$$

Moreover, the functions $\varphi^{\sigma}_{\lambda}$ are zero for the boundary points $a' \in F \backslash F^{\sigma}$ of the sets (14),

$$\varphi^{\sigma}_{\lambda}(a') = 0, \quad a' \in F \backslash F^{\sigma}. \tag{118}$$

Restricting the arguments of Weyl orbit functions to the refined dual-root lattice $\frac{1}{M} Q^{\vee}$, additional label symmetries with respect to the extended dual affine Weyl group (23) are generated.

Proposition 6.1 *Let $s \in \frac{1}{M} Q^{\vee}$, then for any $w_P \in W^{\mathrm{aff}}_{P}$ and any $b \in \mathbb{R}^n$ it holds that*

$$\varphi^{\sigma}_{M w_P\left(\frac{b}{M}\right)}(s) = \sigma \circ \widehat{\psi}(w_P) \cdot \varphi^{\sigma}_{b}(s). \tag{119}$$

Additionally, the Weyl orbit function $\varphi^{\sigma}_{b'}$ vanishes for any $b' \in M(F_P \backslash F^{\sigma}_{P})$,

$$\varphi^{\sigma}_{b'}(s) = 0, \quad b' \in M(F_P \backslash F^{\sigma}_{P}). \tag{120}$$

Proof The $\mathbb{Z}$-duality (2) of the Weyl group invariant lattices P and $Q^\vee$ ensures that the relation

$$\langle Mwp, s\rangle = \langle wp, Ms\rangle \in \mathbb{Z}$$

is valid for any $p \in P$ and $w \in W$ and thus, for any element of the extended dual affine Weyl group $w_P = T(p)w \in W_P^{\mathrm{aff}}$ it holds that

$$\begin{aligned}\varphi^\sigma_{Mw_P\left(\frac{b}{M}\right)}(s) &= \sum_{w'\in W} \sigma(w')\, e^{2\pi \mathrm{i}\langle w'wb+Mw'p,s\rangle}\\ &= \sum_{w'\in W} \sigma(w')\, e^{2\pi \mathrm{i}\langle w'wb,s\rangle} = \sigma(w)\varphi^\sigma_b(s).\end{aligned}$$

For points from the boundary set $b' \in M(F_P \setminus F_P^\sigma)$ there exists by (41) an element $w_P = \mathrm{Stab}_{W_P^{\mathrm{aff}}}(b'/M)$ such that $\sigma \circ \widehat{\psi}(w_P) = -1$ and hence

$$\varphi^\sigma_{b'}(s) = \varphi^\sigma_{Mw_P\left(\frac{b'}{M}\right)}(s) = \sigma \circ \widehat{\psi}(w_P) \cdot \varphi^\sigma_{b'}(s) = -\varphi^\sigma_{b'}(s).$$

□

A real-valued version of the Weyl orbit functions employs the Hartley kernel functions

$$\mathrm{cas}\,\alpha = \cos\alpha + \sin\alpha, \quad \alpha \in \mathbb{R}. \tag{121}$$

The sign homomorphisms σ of the Weyl group W induce up to four families of the Hartley orbit functions $\zeta^\sigma_b : \mathbb{R}^n \to \mathbb{R}$, labelled by parameter $b \in \mathbb{R}^n$,

$$\zeta^\sigma_b(a) = \sum_{w\in W} \sigma(w)\mathrm{cas}\,(2\pi\langle wb, a\rangle).$$

The relation between the complex exponential function in (116) and the Hartley kernel function (121) provides the following expression,

$$\zeta^\sigma_b = \mathrm{Re}\,\varphi^\sigma_b + \mathrm{Im}\,\varphi^\sigma_b. \tag{122}$$

Equation (122) straightforwardly extracts from (117) that restricting the label to the weight lattice $\lambda \in P$, the argument invariance of Hartley orbit functions with respect to the action of the element of the affine Weyl group $w^{\mathrm{aff}} \in W^{\mathrm{aff}}$ is for any $a \in \mathbb{R}^n$ of the form

$$\zeta^\sigma_\lambda(w^{\mathrm{aff}}a) = \sigma \circ \psi(w^{\mathrm{aff}}) \cdot \zeta^\sigma_\lambda(a). \tag{123}$$

Moreover, the functions ζ^σ_λ are zero for the boundary points $a' \in F\setminus F^\sigma$ of the sets (14),

$$\zeta^\sigma_\lambda(a') = 0, \quad a' \in F\setminus F^\sigma. \tag{124}$$

Using (122) again, a real-valued analogue of Proposition 6.1 is derived.

Proposition 6.2 *Let $s \in \frac{1}{M}Q^\vee$, then for any $w_P \in W_P^{\mathrm{aff}}$ and any $b \in \mathbb{R}^n$ it holds that*

$$\zeta^\sigma_{Mw_P\left(\frac{b}{M}\right)}(s) = \sigma \circ \widehat{\psi}(w_P) \cdot \zeta^\sigma_b(s).$$

Additionally, the Hartley orbit function $\zeta^\sigma_{b'}$ vanishes for any $b' \in M(F_P \setminus F_P^\sigma)$,

$$\zeta^\sigma_{b'}(s) = 0, \quad b' \in M(F_P \setminus F_P^\sigma).$$

6.2 Discrete Orthogonality

The argument symmetries of Weyl orbit functions (117), (118) grant that the functions φ^σ_b, $\lambda \in P$ discretized to the grid $\frac{1}{M}Q^\vee$ are fully determined by their non-zero values in the sets (58). Label symmetries, induced by the selection of the discrete grid $\frac{1}{M}Q^\vee$ and determined by Proposition 6.1, restrict the labels to the weight sets (59). The same argument and label symmetries are guaranteed for the Hartley orbit functions by relations (123), (124) and Proposition 6.2. A scalar product of two discrete complex valued functions $f, g : F^\sigma_{Q^\vee,M} \to \mathbb{C}$ is introduced via relation

$$\langle f, g \rangle^\sigma_M \equiv \sum_{s \in F^\sigma_{Q^\vee,M}} \varepsilon(s) f(s) \overline{g(s)}, \tag{125}$$

with the function ε defined by (8), and the resulting finite-dimensional Hilbert space is denoted by $\mathcal{H}^\sigma_{Q^\vee,M}$. Note that for any $s \in \frac{1}{M}Q^\vee$ and $\lambda \in P$ it holds

$$e^{2\pi \mathrm{i}\langle \lambda, s+Q^\vee \rangle} = e^{2\pi \mathrm{i}\langle \lambda, s \rangle},$$

and the mapping $e^{2\pi \mathrm{i}\langle \lambda, y \rangle}$ with $y \in \frac{1}{M}Q^\vee/Q^\vee$ is well defined. In order to derive the discrete orthogonality of Weyl orbit functions (116) with respect to the scalar product (125), the following basic orthogonality relations of exponential functions are essential.

Proposition 6.3 *Let $M \in \mathbb{N}$ and $\lambda, \lambda' \in P$, then*

$$\sum_{y \in \frac{1}{M}Q^\vee/Q^\vee} e^{2\pi \mathrm{i}\langle \lambda - \lambda', y \rangle} = \begin{cases} M^n & \lambda - \lambda' \in MP, \\ 0 & \textit{otherwise}. \end{cases} \tag{126}$$

Proof For any $\lambda \in P$, such that for all $y \in \frac{1}{M}Q^\vee/Q^\vee$ is satisfied $\langle \lambda, y \rangle \in \mathbb{Z}$, the $\mathbb{Z}$-duality of P and $Q^\vee$ yields that $\lambda \in MP$. Thus for all $\lambda \in P \setminus MP$, there exists $y' \in \frac{1}{M}Q^\vee/Q^\vee$ such that $\langle \lambda, y' \rangle \notin \mathbb{Z}$ and $e^{2\pi \mathrm{i}\langle \lambda, y' \rangle} \neq 1$. Therefore, the following calculation from [29], specialized for the finite quotient group $\frac{1}{M}Q^\vee/Q^\vee$ of order M^n,

$$e^{2\pi \mathrm{i}\langle \lambda, y' \rangle} \sum_{y \in \frac{1}{M}Q^\vee/Q^\vee} e^{2\pi \mathrm{i}\langle \lambda, y \rangle} = \sum_{y \in \frac{1}{M}Q^\vee/Q^\vee} e^{2\pi \mathrm{i}\langle \lambda, y+y' \rangle} = \sum_{y \in \frac{1}{M}Q^\vee/Q^\vee} e^{2\pi \mathrm{i}\langle \lambda, y \rangle},$$

forces basic orthogonality relations (126). □

Theorem 6.4 *For any* $\lambda, \lambda' \in \Lambda^{\sigma}_{P,M}$*, it holds that*

$$\langle \varphi^{\sigma}_{\lambda}, \varphi^{\sigma}_{\lambda'} \rangle^{\sigma}_{M} = |W|\, M^n h_{P,M}(\lambda)\, \delta_{\lambda\lambda'}, \tag{127}$$

where the coefficients $h_{P,M}$ *are defined by* (33).

Proof Since the functions $\varphi^{\sigma}_{\lambda}$ vanish by (118) on the sets $F \setminus F^{\sigma}$, the scalar product is evaluated as

$$\langle \varphi^{\sigma}_{\lambda}, \varphi^{\sigma}_{\lambda'} \rangle^{\sigma}_{M} = \sum_{s \in F^{\sigma}_{Q^\vee, M}} \varepsilon(s) \varphi^{\sigma}_{\lambda}(s) \overline{\varphi^{\sigma}_{\lambda'}(s)} = \sum_{s \in \frac{1}{M} Q^\vee \cap F} \varepsilon(s) \varphi^{\sigma}_{\lambda}(s) \overline{\varphi^{\sigma}_{\lambda'}(s)}. \tag{128}$$

The W^{aff}-invariance properties (9) and (117) grant that the summands in (128) are W^{aff}-invariant, i.e. for all $w^{\mathrm{aff}} \in W^{\mathrm{aff}}$ it holds that

$$\varepsilon(s) \varphi^{\sigma}_{\lambda}(s) \overline{\varphi^{\sigma}_{\lambda'}(s)} = \varepsilon(w^{\mathrm{aff}} s) \varphi^{\sigma}_{\lambda}(w^{\mathrm{aff}} s) \overline{\varphi^{\sigma}_{\lambda'}(w^{\mathrm{aff}} s)}. \tag{129}$$

The $Q^\vee$-shift invariance in (129) and relation (13) imply

$$\sum_{s \in \frac{1}{M} Q^\vee \cap F} \varepsilon(s) \varphi^{\sigma}_{\lambda}(s) \overline{\varphi^{\sigma}_{\lambda'}(s)} = \sum_{x \in \left[\frac{1}{M} Q^\vee / Q^\vee\right] \cap F} \widetilde{\varepsilon}(x) \varphi^{\sigma}_{\lambda}(x) \overline{\varphi^{\sigma}_{\lambda'}(x)}$$

and the W-invariance in (129) and relations (10), (11) produce the identity

$$\sum_{x \in \left[\frac{1}{M} Q^\vee / Q^\vee\right] \cap F} \widetilde{\varepsilon}(x) \varphi^{\sigma}_{\lambda}(x) \overline{\varphi^{\sigma}_{\lambda'}(x)} = \sum_{y \in \frac{1}{M} Q^\vee / Q^\vee} \varphi^{\sigma}_{\lambda}(y) \overline{\varphi^{\sigma}_{\lambda'}(y)}.$$

The W-invariance of the quotient group $\frac{1}{M} Q^\vee / Q^\vee$ is used to further simplify the result,

$$\begin{aligned}\sum_{y \in \frac{1}{M} Q^\vee / Q^\vee} \varphi^{\sigma}_{\lambda}(y) \overline{\varphi^{\sigma}_{\lambda'}(y)} &= \sum_{w' \in W} \sum_{w \in W} \sum_{y \in \frac{1}{M} Q^\vee / Q^\vee} \sigma(ww') e^{2\pi \mathrm{i} \langle w\lambda - w'\lambda', y \rangle} \\ &= |W| \sum_{w' \in W} \sigma(w') \sum_{y \in \frac{1}{M} Q^\vee / Q^\vee} e^{2\pi \mathrm{i} \langle \lambda - w'\lambda', y \rangle}. \end{aligned} \tag{130}$$

If $\lambda - w'\lambda' \in MP$, then $\lambda/M = w'\lambda'/M + p$ holds for some $p \in P$ and $w' \in W$ and thus λ/M and λ'/M are in the same W^{aff}_{P}-orbit. Definition (59) of the weight set $\Lambda^{\sigma}_{P,M}$ guarantees that both λ/M and λ'/M are in the fundamental domain F_P and therefore $\lambda = \lambda'$. Contrapositive implication yields that if $\lambda \neq \lambda'$, then for all $w' \in W$ it holds that $\lambda - w'\lambda' \notin MP$ and hence basic orthogonality relations (126) grant zero value $\langle \varphi^{\sigma}_{\lambda}, \varphi^{\sigma}_{\lambda'} \rangle^{\sigma}_{M} = 0$.

If $\lambda = \lambda'$, then basic orthogonality relations (126) guarantee that summands in (130) do not vanish only if $\lambda - w'\lambda \in MP$, or equivalently $w' \in \widehat{\psi}\left(\mathrm{Stab}_{W_P^{\mathrm{aff}}}(\lambda/M)\right)$ and thus

$$|W| \sum_{w' \in W} \sigma(w') \sum_{y \in \frac{1}{M} Q^\vee / Q^\vee} e^{2\pi \mathrm{i} \langle \lambda - w'\lambda, y \rangle} = |W| \, M^n \sum_{w' \in \widehat{\psi}\left(\mathrm{Stab}_{W_P^{\mathrm{aff}}}\left(\frac{\lambda}{M}\right)\right)} \sigma(w').$$

Since for $w_P \in \mathrm{Stab}_{W_P^{\mathrm{aff}}}(\lambda/M)$ the property $\widehat{\psi}(w_P) = 1$ forces $w_P = 1$, the subgroups $\mathrm{Stab}_{W_P^{\mathrm{aff}}}(\lambda/M)$ and $\widehat{\psi}\left(\mathrm{Stab}_{W_P^{\mathrm{aff}}}(\lambda/M)\right)$ are isomorphic and hence

$$\sum_{w' \in \widehat{\psi}\left(\mathrm{Stab}_{W_P^{\mathrm{aff}}}\left(\frac{\lambda}{M}\right)\right)} \sigma(w') = \sum_{w_P \in \mathrm{Stab}_{W_P^{\mathrm{aff}}}\left(\frac{\lambda}{M}\right)} \sigma \circ \widehat{\psi}(w_P).$$

Definition (59) of the weight set $\Lambda^\sigma_{P,M}$ also ensures that for any $\lambda \in \Lambda^\sigma_{P,M}$ it holds that $\lambda/M \in F^\sigma_P$ and, taking into account defining relation (41) and notation (33), the final identity yielding relations (127) is derived,

$$\sum_{w_P \in \mathrm{Stab}_{W_P^{\mathrm{aff}}}\left(\frac{\lambda}{M}\right)} \sigma \circ \widehat{\psi}(w_P) = \sum_{w_P \in \mathrm{Stab}_{W_P^{\mathrm{aff}}}\left(\frac{\lambda}{M}\right)} 1 = h_{P,M}(\lambda).$$

□

Theorem 6.5 *The functions φ^σ_λ, $\lambda \in \Lambda^\sigma_{P,M}$ form for any $M \in \mathbb{N}$, $M > m^\sigma$ an orthogonal basis of the Hilbert space $\mathcal{H}^\sigma_{Q^\vee,M}$.*

Proof Theorem 6.4 grants linear independence of the set of discretized functions $\varphi^\sigma_\lambda : F^\sigma_{Q^\vee,M} \to \mathbb{C}$, $\lambda \in \Lambda^\sigma_{P,M}$, and Theorem 5.4 guarantees that this set of orthogonal functions has the cardinality

$$\left|\Lambda^\sigma_{P,M}\right| = \left|F^\sigma_{Q^\vee,M}\right| = \dim \mathcal{H}^\sigma_{Q^\vee,M}.$$

□

As a consequence of the discrete orthogonality of Weyl orbit functions, the discrete orthogonality of Hartley orbit functions is derived in the following theorem.

Theorem 6.6 *For any $\lambda, \lambda' \in \Lambda^\sigma_{P,M}$, it holds that*

$$\langle \zeta^\sigma_\lambda, \zeta^\sigma_{\lambda'} \rangle^\sigma_M = |W| \, M^n h_{P,M}(\lambda) \, \delta_{\lambda\lambda'},$$

where the coefficients $h_{P,M}$ are defined by (33).

Proof The following trigonometric identity,

$$(\cos\alpha+\sin\alpha)(\cos\alpha'+\sin\alpha')=\operatorname{Re}\left(e^{\mathrm{i}\alpha}e^{-\mathrm{i}\alpha'}\right)+\operatorname{Im}\left(e^{\mathrm{i}\alpha}e^{\mathrm{i}\alpha'}\right),$$

valid for any $\alpha,\alpha'\in\mathbb{R}$, implies together with discrete orthogonality relations (127) that

$$\begin{aligned}\langle\zeta^\sigma_\lambda,\zeta^\sigma_{\lambda'}\rangle^\sigma_M&=\operatorname{Re}\langle\varphi^\sigma_\lambda,\varphi^\sigma_{\lambda'}\rangle^\sigma_M+\operatorname{Im}\langle\varphi^\sigma_\lambda,\overline{\varphi^\sigma_{\lambda'}}\rangle^\sigma_M\\&=|W|\,M^n h_{P,M}(\lambda)\,\delta_{\lambda\lambda'}+\operatorname{Im}\langle\varphi^\sigma_\lambda,\overline{\varphi^\sigma_{\lambda'}}\rangle^\sigma_M.\end{aligned}$$

Definition (116) immediately provides the following relation for complex conjugated function $\varphi^\sigma_{\lambda'}$,

$$\overline{\varphi^\sigma_{\lambda'}}=\varphi^\sigma_{-\lambda'},\tag{131}$$

and lattice property (2) of the weight lattice P ensures that $-\lambda'\in P$. Since the lattice $\frac{1}{M}P$ is W^{aff}_P-invariant, there exist $\mu\in P\cap MF_P$ and $w_P\in W^{\mathrm{aff}}_P$ such that

$$-\lambda'=Mw_P\left(\frac{\mu}{M}\right).\tag{132}$$

Relations (131), (132) and label symmetry (119) allow to calculate

$$\operatorname{Im}\langle\varphi^\sigma_\lambda,\overline{\varphi^\sigma_{\lambda'}}\rangle^\sigma_M=\operatorname{Im}\langle\varphi^\sigma_\lambda,\varphi^\sigma_{-\lambda'}\rangle^\sigma_M=\sigma\circ\widehat{\psi}(w_P)\cdot\operatorname{Im}\langle\varphi^\sigma_\lambda,\varphi^\sigma_\mu\rangle^\sigma_M.$$

If $\mu\in\Lambda^\sigma_{P,M}$, then discrete orthogonality relations (127) ensure that $\operatorname{Im}\langle\varphi^\sigma_\lambda,\varphi^\sigma_\mu\rangle^\sigma_M=0$. If, on the other hand, $\mu\in M(F_P\setminus F^\sigma_P)$, then vanishing property (120) grants directly that $\langle\varphi^\sigma_\lambda,\varphi^\sigma_\mu\rangle^\sigma_M=0$. □

The discrete orthogonality of Hartley orbit functions also generates Hartley version of Theorem 6.5.

Theorem 6.7 *The functions ζ^σ_λ, $\lambda\in\Lambda^\sigma_{P,M}$ form for any $M\in\mathbb{N}$, $M>m^\sigma$ an orthogonal basis of the Hilbert space $\mathcal{H}^\sigma_{Q^\vee,M}$.*

6.3 Discrete Transforms

The interpolating function $I[f]^\sigma_M:\mathbb{R}^n\to\mathbb{C}$ of any function $f\in\mathcal{H}^\sigma_{Q^\vee,M}$ is defined as a linear combination of the basis functions φ^σ_λ,

$$I[f]^\sigma_M(a)\equiv\sum_{\lambda\in\Lambda^\sigma_{P,M}}c^\sigma_\lambda\varphi^\sigma_\lambda(a),\tag{133}$$

satisfying the condition

$$I[f]^\sigma_M(s)=f(s),\quad s\in F^\sigma_{Q^\vee,M}.$$

The frequency spectrum coefficients c^σ_λ in (133) are uniquely determined by Theorem 6.5 and calculated as standard Fourier coefficients,

$$c^\sigma_\lambda = \frac{\langle f, \varphi^\sigma_\lambda \rangle^\sigma_M}{\langle \varphi^\sigma_\lambda, \varphi^\sigma_\lambda \rangle^\sigma_M} = (|W|\, M^n h_{P,M}(\lambda))^{-1} \sum_{s \in F^\sigma_{Q^\vee,M}} \varepsilon(s) f(s) \overline{\varphi^\sigma_\lambda(s)}, \tag{134}$$

and the corresponding Plancherel formulas also hold

$$\sum_{s \in F^\sigma_{Q^\vee,M}} \varepsilon(s)\, |f(s)|^2 = |W|\, M^n \sum_{\lambda \in \Lambda^\sigma_{P,M}} h_{P,M}(\lambda)\, |c_\lambda|^2 .$$

Equations (134) and (133) establish the forward and backward discrete Fourier–Weyl transforms, respectively, of the function $f \in \mathcal{H}^\sigma_{Q^\vee,M}$.

Similarly, the Hartley interpolating function $H[f]^\sigma_M : \mathbb{R}^n \to \mathbb{C}$ of any function $f \in \mathcal{H}^\sigma_{Q^\vee,M}$ is defined as a linear combination of the Hartley basis functions ζ^σ_λ,

$$H[f]^\sigma_M(a) \equiv \sum_{\lambda \in \Lambda^\sigma_{P,M}} d^\sigma_\lambda\, \zeta^\sigma_\lambda(a), \tag{135}$$

satisfying the condition

$$H[f]^\sigma_M(s) = f(s), \quad s \in F^\sigma_{Q^\vee,M}.$$

The frequency spectrum coefficients d^σ_λ in (135) are uniquely determined by Theorem 6.7 and calculated as standard Fourier coefficients,

$$d^\sigma_\lambda = \frac{\langle f, \zeta^\sigma_\lambda \rangle^\sigma_M}{\langle \zeta^\sigma_\lambda, \zeta^\sigma_\lambda \rangle^\sigma_M} = (|W|\, M^n h_{P,M}(\lambda))^{-1} \sum_{s \in F^\sigma_{Q^\vee,M}} \varepsilon(s) f(s) \zeta^\sigma_\lambda(s), \tag{136}$$

and the corresponding Plancherel formulas also hold

$$\sum_{s \in F^\sigma_{Q^\vee,M}} \varepsilon(s)\, |f(s)|^2 = |W|\, M^n \sum_{\lambda \in \Lambda^\sigma_{P,M}} h_{P,M}(\lambda)\, \left|d^\sigma_\lambda\right|^2 .$$

Equations (136) and (135) establish the forward and backward discrete Hartley–Weyl transforms, respectively, of the function $f \in \mathcal{H}^\sigma_{Q^\vee,M}$.

For any fixed ordering of the points and labels in the sets $\Lambda^\sigma_{P,M}$ and $F^\sigma_{Q^\vee,M}$, the unitary transform matrices $\mathbb{I}^\sigma_M$ and $\mathbb{H}^\sigma_M$, that correspond to the discrete Fourier–Weyl and Hartley–Weyl transforms, are given by

$$\begin{aligned} \left(\mathbb{I}^\sigma_M\right)_{\lambda s} &= \sqrt{\varepsilon(s)(|W|\, M^n h_{P,M}(\lambda))^{-1}}\, \overline{\varphi^\sigma_\lambda(s)}, \quad \lambda \in \Lambda^\sigma_{P,M},\ s \in F^\sigma_{Q^\vee,M}, \\ \left(\mathbb{H}^\sigma_M\right)_{\lambda s} &= \sqrt{\varepsilon(s)(|W|\, M^n h_{P,M}(\lambda))^{-1}}\, \zeta^\sigma_\lambda(s), \quad \lambda \in \Lambda^\sigma_{P,M},\ s \in F^\sigma_{Q^\vee,M}. \end{aligned} \tag{137}$$

Table 4 The values of the function ε of $s \in F^{1}_{Q^\vee,M}$ with the Kac coordinates $[s^1_0, s^1_1, s^1_2]$ and $s^1_0, s^1_1, s^1_2 \neq 0$

	$[s^1_0, s^1_1, s^1_2]$	$[0, s^1_1, s^1_2]$	$[s^1_0, 0, s^1_2]$	$[s^1_0, s^1_1, 0]$	$[0, 0, s^1_2]$	$[0, s^1_1, 0]$	$[s^1_0, 0, 0]$
$\varepsilon(s)$	6	3	3	3	1	1	1

As demonstrated in the following example, once the discrete functions ε and $h_{P,M}$ are tabulated and the orbit functions explicitly evaluated, computation of the unitary transforms matrices $\mathbb{I}^\sigma_M$ and $\mathbb{H}^\sigma_M$ is straightforward.

Example 6.1 The explicit formulas for the Weyl orbit functions of A_2, with the label $b = b_1\omega_1 + b_2\omega_2$ and the point $a = a_1\omega^\vee_1 + a_2\omega^\vee_2$, are calculated as

$$\begin{aligned}
\varphi^1_b(a) &= e^{\frac{2}{3}\pi \mathrm{i}((2b_1+b_2)a_1+(b_1+2b_2)a_2)} + e^{\frac{2}{3}\pi \mathrm{i}(-b_1+b_2)a_1+(b_1+2b_2)a_2)} \\
&\quad + e^{\frac{2}{3}\pi \mathrm{i}((-b_1-2b_2)a_1+(b_1-b_2)a_2)} \\
&\quad + e^{\frac{2}{3}\pi \mathrm{i}((-b_1-2b_2)a_1+(-2b_1-b_2)a_2)} + e^{\frac{2}{3}\pi \mathrm{i}((-b_1+b_2)a_1+(-2b_1-b_2)a_2)} \\
&\quad + e^{\frac{2}{3}\pi \mathrm{i}((2b_1+b_2)a_1+(b_1-b_2)a_2)}, \\
\varphi^{\sigma^e}_b(a) &= e^{\frac{2}{3}\pi \mathrm{i}((2b_1+b_2)a_1+(b_1+2b_2)a_2)} - e^{\frac{2}{3}\pi \mathrm{i}(-b_1+b_2)a_1+(b_1+2b_2)a_2)} \\
&\quad + e^{\frac{2}{3}\pi \mathrm{i}((-b_1-2b_2)a_1+(b_1-b_2)a_2)} \\
&\quad - e^{\frac{2}{3}\pi \mathrm{i}((-b_1-2b_2)a_1+(-2b_1-b_2)a_2)} + e^{\frac{2}{3}\pi \mathrm{i}((-b_1+b_2)a_1+(-2b_1-b_2)a_2)} \\
&\quad - e^{\frac{2}{3}\pi \mathrm{i}((2b_1+b_2)a_1+(b_1-b_2)a_2)}.
\end{aligned}$$

The real-valued Hartley orbit functions are obtained merely by replacing the exponential functions with the Hartley kernel functions (121).

The Weyl group of A_2 contains six elements, $|W| = 6$, and the coefficients $\varepsilon(s)$ and the orders of the stabilizers $\left|\mathrm{Stab}_{W^{\mathrm{aff}}_Q}(\lambda/M)\right|$ are tabulated in Tables 4 and 5, respectively.

Analysis of the action (63) yields that the only element stabilized by $\gamma_{M,1}$ or $\gamma_{M,2}$ has the magnified Kac coordinates of the form $\left[\frac{M}{3}, \frac{M}{3}, \frac{M}{3}\right]$. Thus, the order of the stabilizer $\mathrm{Stab}_{\Gamma_M}(\lambda)$ is given by

$$\left|\mathrm{Stab}_{\Gamma_M}(\lambda)\right| = \begin{cases} 3 & \lambda = \left[\frac{M}{3}, \frac{M}{3}, \frac{M}{3}\right], \\ 1 & \text{otherwise}, \end{cases}$$

and the function $h_{P,M}$ is directly calculated from Table 5 and relation (36).

Fixing the scaling factor $M = 7$, the unitary matrices $\mathbb{I}^\sigma_7$ and $\mathbb{H}^\sigma_7$, that correspond to the ordering of point and label sets $F^\sigma_{Q^\vee,7}$ and $\Lambda^\sigma_{P,7}$ in Example 5.1, are explicitly computed from (137). The Fourier–Weyl identity homomorphism unitary matrix $\mathbb{I}^1_7 = \mathrm{Re}\,\mathbb{I}^1_7 + \mathrm{i}\,\mathrm{Im}\,\mathbb{I}^1_7$, with its entries rounded, is determined by

Table 5 The orders of the stabilizer $\mathrm{Stab}_{W_Q^{\mathrm{aff}}}(\lambda/M)$ of $\lambda \in \Lambda^{\mathbb{1}}_{P,M}$ with the magnified Kac coordinates $[\lambda_0^{\mathbb{1}}, \lambda_1^{\mathbb{1}}, \lambda_2^{\mathbb{1}}]$ and $\lambda_0^{\mathbb{1}}, \lambda_1^{\mathbb{1}}, \lambda_2^{\mathbb{1}} \neq 0$

	$[\lambda_0^{\mathbb{1}}, \lambda_1^{\mathbb{1}}, \lambda_2^{\mathbb{1}}]$	$[0, \lambda_1^{\mathbb{1}}, \lambda_2^{\mathbb{1}}]$	$[\lambda_0^{\mathbb{1}}, 0, \lambda_2^{\mathbb{1}}]$	$[\lambda_0^{\mathbb{1}}, \lambda_1^{\mathbb{1}}, 0]$	$[0, 0, \lambda_2^{\mathbb{1}}]$	$[0, \lambda_1^{\mathbb{1}}, 0]$	$[\lambda_0^{\mathbb{1}}, 0, 0]$
$\lvert \mathrm{Stab}_{W_Q^{\mathrm{aff}}}(\lambda/M) \rvert$	1	2	2	2	6	6	6

$$
\mathrm{Re}\,\mathbb{I}_7^{\mathbb{1}} = \begin{pmatrix}
0.143 & 0.247 & 0.247 & 0.247 & 0.247 & 0.35 & 0.35 & 0.35 & 0.35 & 0.247 & 0.247 & 0.35 \\
0.247 & 0.146 & 0.146 & -0.192 & -0.192 & 0.454 & -0.101 & -0.101 & 0.112 & -0.168 & -0.168 & -0.162 \\
0.247 & -0.192 & -0.192 & -0.168 & -0.168 & 0.112 & -0.101 & -0.101 & -0.162 & 0.146 & 0.146 & 0.454 \\
0.247 & -0.168 & -0.168 & 0.146 & 0.146 & -0.162 & -0.101 & -0.101 & 0.454 & -0.192 & -0.192 & 0.112 \\
0.247 & -0.168 & -0.168 & 0.146 & 0.146 & -0.162 & -0.101 & -0.101 & 0.454 & -0.192 & -0.192 & 0.112 \\
0.247 & -0.192 & -0.192 & -0.168 & -0.168 & 0.112 & -0.101 & -0.101 & -0.162 & 0.146 & 0.146 & 0.454 \\
0.247 & 0.146 & 0.146 & -0.192 & -0.192 & 0.454 & -0.101 & -0.101 & 0.112 & -0.168 & -0.168 & -0.162 \\
0.35 & -0.162 & -0.162 & 0.454 & 0.454 & 0.293 & -0.143 & -0.143 & -0.385 & 0.112 & 0.112 & -0.337 \\
0.35 & -0.101 & -0.101 & -0.101 & -0.101 & -0.143 & 0.357 & 0.357 & -0.143 & -0.101 & -0.101 & -0.143 \\
0.35 & 0.112 & 0.112 & -0.162 & -0.162 & -0.337 & -0.143 & -0.143 & 0.293 & 0.454 & 0.454 & -0.385 \\
0.35 & -0.101 & -0.101 & -0.101 & -0.101 & -0.143 & 0.357 & 0.357 & -0.143 & -0.101 & -0.101 & -0.143 \\
0.35 & 0.454 & 0.454 & 0.112 & 0.112 & -0.385 & -0.143 & -0.143 & -0.337 & -0.162 & -0.162 & 0.293
\end{pmatrix},
$$

and

$$
\mathrm{Im}\,\mathbb{I}_7^{\mathbb{1}} = \begin{pmatrix}
0 & 0 & 0 & 0 & 0 & 0 & 0 & 0 & 0 & 0 & 0 & 0 \\
0 & -0.084 & 0.084 & -0.341 & 0.341 & 0 & -0.267 & 0.267 & 0 & -0.236 & 0.236 & 0 \\
0 & -0.341 & 0.341 & 0.236 & -0.236 & 0 & -0.267 & 0.267 & 0 & 0.084 & -0.084 & 0 \\
0 & -0.236 & 0.236 & 0.084 & -0.084 & 0 & 0.267 & -0.267 & 0 & -0.341 & 0.341 & 0 \\
0 & 0.236 & -0.236 & -0.084 & 0.084 & 0 & -0.267 & 0.267 & 0 & 0.341 & -0.341 & 0 \\
0 & 0.341 & -0.341 & -0.236 & 0.236 & 0 & 0.267 & -0.267 & 0 & -0.084 & 0.084 & 0 \\
0 & 0.084 & -0.084 & 0.341 & -0.341 & 0 & 0.267 & -0.267 & 0 & 0.236 & -0.236 & 0 \\
0 & 0 & 0 & 0 & 0 & 0 & 0 & 0 & 0 & 0 & 0 & 0 \\
0 & 0.267 & -0.267 & 0.267 & -0.267 & 0 & -0.189 & 0.189 & 0 & -0.267 & 0.267 & 0 \\
0 & 0 & 0 & 0 & 0 & 0 & 0 & 0 & 0 & 0 & 0 & 0 \\
0 & -0.267 & 0.267 & -0.267 & 0.267 & 0 & 0.189 & -0.189 & 0 & 0.267 & -0.267 & 0 \\
0 & 0 & 0 & 0 & 0 & 0 & 0 & 0 & 0 & 0 & 0 & 0
\end{pmatrix}.
$$

The Hartley–Weyl identity homomorphism orthogonal matrix $\mathbb{H}_7^{\mathbb{1}}$ is given by

$$
\mathbb{H}_7^{\mathbb{1}} = \begin{pmatrix}
0.143 & 0.248 & 0.248 & 0.248 & 0.248 & 0.35 & 0.35 & 0.35 & 0.35 & 0.248 & 0.248 & 0.35 \\
0.248 & 0.231 & 0.062 & 0.148 & -0.533 & 0.454 & 0.166 & -0.368 & 0.112 & 0.067 & -0.404 & -0.162 \\
0.248 & 0.148 & -0.533 & -0.404 & 0.067 & 0.112 & 0.166 & -0.368 & -0.162 & 0.062 & 0.231 & 0.454 \\
0.248 & 0.067 & -0.404 & 0.062 & 0.231 & -0.162 & -0.368 & 0.166 & 0.454 & 0.148 & -0.533 & 0.112 \\
0.248 & -0.404 & 0.067 & 0.231 & 0.062 & -0.162 & 0.166 & -0.368 & 0.454 & -0.533 & 0.148 & 0.112 \\
0.248 & -0.533 & 0.148 & 0.067 & -0.404 & 0.112 & -0.368 & 0.166 & -0.162 & 0.231 & 0.062 & 0.454 \\
0.248 & 0.062 & 0.231 & -0.533 & 0.148 & 0.454 & -0.368 & 0.166 & 0.112 & -0.404 & 0.067 & -0.162 \\
0.35 & -0.162 & -0.162 & 0.454 & 0.454 & 0.293 & -0.143 & -0.143 & -0.384 & 0.112 & 0.112 & -0.337 \\
0.35 & -0.368 & 0.166 & -0.368 & 0.166 & -0.143 & 0.546 & 0.168 & -0.143 & 0.166 & -0.368 & -0.143 \\
0.35 & 0.112 & 0.112 & -0.162 & -0.162 & -0.337 & -0.143 & -0.143 & 0.293 & 0.454 & 0.454 & -0.384 \\
0.35 & 0.166 & -0.368 & 0.166 & -0.368 & -0.143 & 0.168 & 0.546 & -0.143 & -0.368 & 0.166 & -0.143 \\
0.35 & 0.454 & 0.454 & 0.112 & 0.112 & -0.384 & -0.143 & -0.143 & -0.337 & -0.162 & -0.162 & 0.293
\end{pmatrix}.
$$

The Fourier–Weyl and Hartley–Weyl determinant homomorphism unitary matrices $\mathbb{I}_7^{\sigma^e}$ and $\mathbb{H}_7^{\sigma^e}$ are computed as

$$
\mathbb{I}_7^{\sigma^e} = \begin{pmatrix}
0.168\mathrm{i} & 0.378\mathrm{i} & 0.378\mathrm{i} & 0.681\mathrm{i} & 0.471\mathrm{i} \\
0.378\mathrm{i} & -0.5-0.189\mathrm{i} & 0.5-0.189\mathrm{i} & 0.378\mathrm{i} & -0.378\mathrm{i} \\
0.471\mathrm{i} & -0.378\mathrm{i} & -0.378\mathrm{i} & -0.168\mathrm{i} & 0.681\mathrm{i} \\
0.378\mathrm{i} & 0.5-0.189\mathrm{i} & -0.5-0.189\mathrm{i} & 0.378\mathrm{i} & -0.378\mathrm{i} \\
0.681\mathrm{i} & 0.378\mathrm{i} & 0.378\mathrm{i} & -0.471\mathrm{i} & -0.168\mathrm{i}
\end{pmatrix},
$$

and

$$\mathbb{H}_7^{\sigma^e} = \begin{pmatrix} -0.168 & -0.378 & -0.378 & -0.681 & -0.471 \\ -0.378 & -0.311 & 0.689 & -0.378 & 0.378 \\ -0.471 & 0.378 & 0.378 & 0.168 & -0.681 \\ -0.378 & 0.689 & -0.311 & -0.378 & 0.378 \\ -0.681 & -0.378 & -0.378 & 0.471 & 0.168 \end{pmatrix}.$$

Concluding Remarks

- The choice of the fundamental domain F_P of the dual extended affine Weyl group is not unique. As demonstrated in Example 4.1, the Brillouin zone [27] of A_3, intersected with the dominant Weyl chamber and with certain boundaries omitted, produces another viable fundamental domain for the A_3 case. The significant advantage of the presented Kac coordinates approach stems from existence of an effective algorithm for constructing the weight sets $\Lambda^{\sigma}_{P,M}$ from formula (44) for any case. Indeed, explicit relations (21) and (22) immediately produce the weight sets $P \cap MF^{\sigma}_{Q}$. Sorting the weights from $P \cap MF^{\sigma}_{Q}$ into Γ_M-orbits, selecting the lexicographically highest in each orbit, while excluding those with negative sign homomorphism values of stabilizing $\gamma \in \Gamma_M$, yields directly the set $\Lambda^{\sigma}_{P,M}$ for any fixed $M \in \mathbb{N}$.
- The group theoretical and combinatorial methods, used to develop the presented dual-root lattice discrete Fourier–Weyl and Hartley–Weyl transforms, demonstrate interconnectedness of the generalized Fourier analysis with the theory of invariant polynomials, combinatorics and the group theory. Indeed, the counting formulas for the case A_n are due to their link to combinatorics of necklaces already present in various contexts in the mathematical literature. The action of the cyclic group Γ_M on the set Λ_M implies that the number $|\Lambda^{\mathbb{1}}_{P,M}(A_n)|$ coincides with the number of necklaces with $n+1$ white and M black beads. Moreover, conditions (90) for the point sets $F^{\sigma}_{Q^\vee,M}(A_n)$ are in fact special cases of equation (2) in [8] and thus, preserving the notation from [8], it holds that

$$\begin{aligned} \left|F^{\mathbb{1}}_{Q^\vee,M}(A_n)\right| &= a_0(n+1, M), \\ \left|F^{\sigma^e}_{Q^\vee,M}(A_{2k})\right| &= a_0(2k+1, M-2k-1), \\ \left|F^{\sigma^e}_{Q^\vee,M}(A_{2k+1})\right| &= a_{k+1}(2k+2, M-2k-2). \end{aligned} \tag{138}$$

  Using the Hölder's identity [11] to evaluate the Ramanujan sums in the explicit expressions for $a_k(m,n)$ in [8, Thm. 1] allows to bring equations (138) to the form of $|\Lambda^{\sigma}_{P,M}(A_n)|$ counting formulas (99) and (100). Such direct comparison provides an alternative proof of the A_n case of Theorem 5.4. Note also that the counting formulas studied in conjunction with the perfect forms in Lemma 5.1 in [1] coincide with the counting formula for $|\Lambda^{\sigma^e}_{P,M}(A_3)|$.
- The good behaviour of the dual-weight discretization of Weyl orbit functions in interpolation estimates [10] indicates similar viability of the discrete transforms (134) and (136) in various applications related to digital data processing. The

interpolation performance of the dual-root lattice discretization and existence of interpolation convergence criteria [25] pose open problems. As a direct application of the presented dual-root lattice transforms of A_2, the subtractive armchair honeycomb transforms and related interpolation tests are successfully achieved in [14]. The four families of Weyl orbit functions induce four families of orthogonal polynomials [28,30] which are special cases of multivariate Jacobi and Macdonald polynomials [13]. Existence of the generalization of the dual-weight lattice orthogonality of the selected subset of Macdonald polynomials [36] to the dual-root lattice poses an open problem. The related polynomial interpolation and approximation methods, cubature formulas [16] and their comparison to the weight and dual weight versions deserve further study.

- Besides developing novel discrete transforms on generalized and composed grids, other fundamentally different options are provided by existence of functions invariant with respect to some subgroups of the given Weyl group. These even normal subgroups of index 2 of the Weyl groups are generated as kernels of each of the sign homomorphisms. There exist six more types of special functions induced by the even subgroups, called E-functions, for root systems with two lengths of the roots [12] and one type for root systems with one length of the roots [29]. The dual-weight lattice discretization of all ten types of Weyl and Hartley orbit functions is derived in a unified manner and full generality in [12]. Extending the present dual-root lattice Fourier calculus to all ten types of Weyl and Hartley orbit functions poses a deep unsolved problem.

Acknowledgements This work was supported by the Grant Agency of the Czech Technical University in Prague, grant number SGS16/239/OHK4/3T/14. LM and JH gratefully acknowledge the support of this work by RVO14000.

References

1. Barnes, E.S.: The construction of perfect and extreme forms. The non-zero entries of the unitary. I, Acta Arith. **5**, 57–79 (1958)
2. Bourbaki, N.: Groupes et algèbres de Lie, Chapiters IV, V, VI. Hermann, Paris (1968)
3. Bracewell, R.N.: Discrete Hartley transform. J. Opt. Soc. Am. **73**, 1832–1835 (1983). https://doi.org/10.1364/JOSA.73.001832
4. Bharkad, S.D., Kokare, M.: Hartley transform based fingerprint matching. J. Inf. Process. Syst. **8**, 85–100 (2012). https://doi.org/10.3745/JIPS.2012.8.1.085
5. Cserti, J., Tichy, G.: A simple model for the vibrational modes in honeycomb lattices. Eur. J. Phys. **25**, 723–736 (2004). https://doi.org/10.1088/0143-0807/25/6/004
6. Czyżycki, T., Hrivnák, J.: Generalized discrete orbit function transforms of affine Weyl groups. J. Math. Phys. **55**, 113508 (2014). https://doi.org/10.1063/1.4901230
7. Drissi, L.B., Saidi, E.H., Bousmina, M.: Graphene, lattice field theory and symmetries. J. Math. Phys. **52**, 022306 (2011). https://doi.org/10.1063/1.3546030
8. Elashvili, A., Jibladze, M., Pataraia, D.: Combinatorics of necklaces and hermite reciprocity. J. Alg. Comb. **10**, 173–188 (1999). https://doi.org/10.1023/A:1018727630642
9. Güçlü, A.D., Potasz, P., Korkusinski, M., Hawrylak, P.: Graphene Quantum Dots. Springer, Berlin (2014)
10. Háková, L., Hrivnák, J., Patera, J.: Four families of Weyl group orbit functions of B_3 and C_3. J. Math. Phys. **54**, 083501 (2013). https://doi.org/10.1063/1.4817340
11. Hölder, O.: Zur Theorie der Kreisteilungsgleichung $K_m(x) = 0$. Prace Matematyczno Fizyczne **43**, 13–23 (1936)

12. Hrivnák, J., Juránek, M.: On E-discretization of tori of compact simple Lie groups. II. J. Math. Phys. **58**, 103504 (2017). https://doi.org/10.1063/1.4997520
13. Hrivnák, J., Motlochová, L.: On connecting Weyl-orbit functions to Jacobi polynomials and multivariate (anti)symmetric trigonometric functions. Acta Polytech. **56**, 283–290 (2016). https://doi.org/10.14311//AP.2016.56.0282
14. Hrivnák, J., Motlochová, L.: Discrete cosine and sine transforms generalized to honeycomb lattice. J. Math. Phys. **59**, 063503 (2018). https://doi.org/10.1063/1.5027101
15. Hrivnák, J., Motlochová, L., Patera, J.: On discretization of tori of compact simple Lie groups II. J. Phys. A **45**, 255201 (2012). https://doi.org/10.1088/1751-8113/45/25/255201
16. Hrivnák, J., Motlochová, L., Patera, J.: Cubature formulas of multivariate polynomials arising from symmetric orbit functions. Symmetry **8**, 63 (2016). https://doi.org/10.3390/sym8070063
17. Hrivnák, J., Patera, J.: On discretization of tori of compact simple Lie groups. J. Phys. A: Math. Theor. **42**, 385208 (2009). https://doi.org/10.1088/1751-8113/42/38/385208
18. Hrivnák, J., Walton, M.A.: Weight-lattice discretization of Weyl-orbit functions. J. Math. Phys. **57**, 083512 (2016). https://doi.org/10.1063/1.4961154
19. Humphreys, J.E.: Reflection Groups and Coxeter Groups. Cambridge Studies in Advanced Mathematics, vol. 29. Cambridge University Press, Cambridge (1990). https://doi.org/10.1017/CBO9780511623646
20. Kac, V.: Automorphisms of finite order of semi-simple Lie algebras. Funct. Anal. Appl. **3**, 252–254 (1969). https://doi.org/10.1007/BF01676631
21. Klimyk, A.U., Patera, J.: Orbit functions. SIGMA **2**, 006 (2006). https://doi.org/10.3842/SIGMA.2006.006
22. Klimyk, A.U., Patera, J.: Antisymmetric orbit functions. SIGMA **3**, 023 (2007). https://doi.org/10.3842/SIGMA.2007.023
23. Komrakov, B.P., Premet, A.A.: The fundamental domain of an extended affine Weyl group (in Russian). Vestsi Akad. Navuk BSSR Ser. Fiz.-Mat. Navuk **3**, 18–22 (1984)
24. Kühl, H., Sacchi, M.D., Fertig, J.: The Hartley transform in seismic imaging. Geophysics **66**, 1251–1257 (2001). https://doi.org/10.1190/1.1487072
25. Li, H., Xu, Y.: Discrete Fourier analysis on fundamental domain and simplex of A_d lattice in d-variables. J. Fourier Anal. Appl. **16**, 383–433 (2010). https://doi.org/10.1007/s00041-009-9106-9
26. Liu, S., Guo, C., Sheridan, J.T.: A review of optical image encryption techniques. Opt. Laser Technol. **57**, 327–342 (2014). https://doi.org/10.1016/j.optlastec.2013.05.023
27. Michel, L.: Fundamental concepts for the study of crystal symmetry. Phys. Rep. **341**, 265–336 (2001). https://doi.org/10.1016/S0370-1573(00)00091-0
28. Moody, R.V., Motlochová, L., Patera, J.: Gaussian cubature arising from hybrid characters of simple Lie groups. J. Fourier Anal. Appl. **20**, 1257–1290 (2014). https://doi.org/10.1007/s00041-014-9355-0
29. Moody, R.V., Patera, J.: Orthogonality within the families of C-, S-, and E-functions of any compact semisimple Lie group. SIGMA **2**, 076 (2006). https://doi.org/10.3842/SIGMA.2006.076
30. Moody, R.V., Patera, J.: Cubature formulae for orthogonal polynomials in terms of elements of finite order of compact simple Lie groups. Adv. Appl. Math. **47**, 509–535 (2011). https://doi.org/10.1016/j.aam.2010.11.005
31. Paraskevas, I., Barbarosou, M., Chilton, E.: Hartley transform and the use of the Whitened Hartley spectrum as a tool for phase spectral processing. J. Eng. (2015). https://doi.org/10.1049/joe.2014.0350
32. Püschel, M., Moura, J.M.F.: Algebraic signal processing theory: foundation and 1-D time. IEEE Trans. Signal Process. **56**, 3572–3585 (2008). https://doi.org/10.1109/TSP.2008.925261
33. Rozhkov, A.V., Nori, F.: Exact wave functions for an electron on a graphene triangular quantum dot. Phys. Rev. B **81**, 155401 (2010). https://doi.org/10.1103/PhysRevB.81.155401
34. Springer, T.A.: Invariant Theory. Lecture Notes in Mathematics, vol. 585. Springer, Berlin (1977)
35. Sun, H.: The Hartley transform applied to particle image velocimetry. Meas. Sci. Technol. **13**, 1996–2000 (2002). https://doi.org/10.1088/0957-0233/13/12/326
36. van Diejen, J.F., Emsiz, E.: Orthogonality of Macdonald polynomials with unitary parameters. Math. Z. **276**, 517–542 (2014). https://doi.org/10.1007/s00209-013-1211-4
37. Vinberg, E.B., Onishchik, A.L.: Lie Groups and Lie Algebras. Springer, New York (1994)

Publisher's Note Springer Nature remains neutral with regard to jurisdictional claims in published maps and institutional affiliations.

Journal of Fourier Analysis and Applications (2019) 25:2570–2607
https://doi.org/10.1007/s00041-019-09674-0

RESEARCH ARTICLE

Gabor Frames for Model Sets

Ewa Matusiak[1]

Received: 8 February 2018 / Revised: 19 February 2019 / Published online: 4 March 2019
© The Author(s) 2019

Abstract

We generalize three main concepts of Gabor analysis for lattices to the setting of model sets: fundamental identity of Gabor analysis, Janssen's representation of the frame operator and Wexler–Raz biorthogonality relations. Utilizing the connection between model sets and almost periodic functions, as well as Poisson's summations formula for model sets we develop a form of a bracket product that plays a central role in our approach. Furthermore, we show that, if a Gabor system for a model set admits a dual which is of Gabor type, then the density of the model set has to be greater than one.

Keywords Gabor frames · Model sets · Almost periodic functions · Poisson's summation formula

Mathematics Subject Classification 42B05 · 42B35 · 42C15 · 43A60

1 Introduction

One of the central themes within Gabor analysis for lattices is a duality theory for Gabor frames, including Wexler–Raz biorthogonality relations [28] and Janssen's representation of a Gabor frame operator [17]. These results are closely connected with the so-called fundamental identity of Gabor analysis, that can be derived from an application of Poisson's summation formula for the symplectic Fourier transform [8]. These duality conditions allow us for example to specify whether a given Gabor system is a tight frame or whether two Gabor systems are dual to each other. An immediate consequence of the duality theory one can obtain necessary density conditions on a lattice so that a given Gabor system forms a frame. In this exposition we leave the

Communicated by Massimo Fornasier.

✉ Ewa Matusiak
ewa.matusiak@univie.ac.at

1 Department of Mathematics, University of Vienna, Vienna, Austria

setting of a lattice and consider a certain type of irregular sets of time-frequency shifts, namely model sets.

The first examples of model sets were studied by Meyer in [22]. Meyer thought of model sets as generalizations of lattices which retain enough lattice-like structure to be useful for studying sampling problems in harmonic analysis [20,23]. The crucial property of model sets is that there exists a form of Poisson's summation formula, which in turn will allow us to derive analogous duality theory as in the case of Gabor analysis for lattices.

First work on Gabor frames for model sets was done by Kreisel [18], where he showed how Gabor frames for a simple model set can be made compatible with its topological dynamics and derived existence conditions for multiwindow Gabor frames for model sets. We will use some of his results here. A constructive approach, that is a characterization of tight and dual frames of semi-regular Gabor systems (where time shifts come from a lattice, and frequency shifts come from a model set, or vice versa), was recently obtained in [21].

For general irregular sets of time-frequency shifts, that are however discrete and relatively separated, it is difficult to provide any constructive results as the tools to deal with such sets are missing. However, certain extensions into irregular Gabor frames were undertaken, for example in [9] or [12]. In [9], the author gave a characterization of the weighted irregular Gabor tight frame and dual systems in $L^2(\mathbb{R}^d)$ in terms of the distributional symplectic Fourier transform of a positive Borel measure on $\mathbb{R}^{2d}$ in the case where the window belongs to the Schwartz class. More recently in [13] the authors study nonuniform sampling in shift invariant spaces and construct semi-regular Gabor frames with respect to the class of totally positive functions. Their results are Beurling type results, expressed by means of density of the sampling sets.

We utilize the connection between model sets and almost periodic functions and use harmonic analysis of the latter to develop a certain form of duality theory for Gabor frames for model sets. We rely strongly on Poisson's summation formula for model sets to introduce the so-called bracket product, in analogy to the bracket product for lattices introduced in [6] to study shift-invariant spaces, or later in multi-dimensional setting to study frames of translates [14,19].

Almost periodic functions were recently investigated in the connection with Gabor frames in [4,10,26]. As the space of almost periodic functions is non-separable, it can not admit countable frames, and the problem arises in which sense frame-type inequalities are still possible for norm estimation in this space [4,10,26]. In [4] the authors also provide Gabor frames for a suitable separable subspaces of the space of almost periodic functions. We, on the other hand, use almost periodic functions as a tool to develop existence results for irregular Gabor frames for the space of square integrable functions.

The article is organized as follows. In Sect. 2 we establish some notations and definitions that we will use throughout the article. In Sect. 3 we derive main identities of Gabor analysis for lattices using a different approach than the one presented in the literature, namely by constructing a certain bracket product. We introduce model sets in Sect. 4, and we also shortly present main facts from the theory of almost periodic functions and point out some connections between the two. Section 5 is devoted to developing a technical tool, that is a bracket product for model sets, that we later use in Sect. 6 to obtain fundamental identity of Gabor analysis, Janssen representation and Wexler–Raz biorthogonality relations for Gabor systems for model sets.

2 Notation and Preliminaries

We will work with the Hilbert space of square integrable functions on $L^2(\mathbb{R}^d)$. The key element in time-frequency analysis is the *time-frequency shift operator* $\pi(z)$, $z = (x, \omega) \in \mathbb{R}^{2d}$, which acts on $L^2(\mathbb{R}^d)$ by

$$\pi(z) f(t) = M_\omega T_x f(t) = e^{2\pi i \omega \cdot t} f(t - x) .$$

Here M_ω denotes the modulation operator and T_x the translation operator which are defined as

$$T_x f(t) = f(t - x) \quad \text{and } M_\omega f(f) = e^{2\pi i \omega \cdot t} f(t) .$$

We define a Fourier transform on $L^2(\mathbb{R}^d)$ as

$$\widehat{f}(\zeta) = \int_{\mathbb{R}^d} f(t) e^{-2\pi i t \cdot \zeta} \, dt ,$$

and denote the inverse Fourier transform of f by $\check{f}$. In the sequel we will distinguish between Fourier transform on $L^2(\mathbb{R}^d)$ and Fourier transform on $L^2(\mathbb{R}^{2d})$ by writing $\mathcal{F} f$ for the latter, $f \in L^2(\mathbb{R}^{2d})$. The Fourier transform has a property of interchanging translation and modulation, that is

$$\widehat{T_x f} = M_{-x} \widehat{f} \quad \text{and} \quad \widehat{M_\omega f} = T_\omega \widehat{f} .$$

The translation and modulation operators obey the following commutation relation

$$M_\omega T_x = e^{2\pi i x \cdot \omega} T_x M_\omega .$$

Combining the last two properties, we have that

$$\widehat{\pi(x, \omega) f} = e^{2\pi i x \cdot \omega} \pi(\omega, -x) \widehat{f} \quad \text{and} \quad \pi(x, \check{\omega}) f = e^{2\pi i x \cdot \omega} \pi(-\omega, x) \check{f} .$$

Given a non-zero function $g \in L^2(\mathbb{R}^d)$, the *short-time Fourier transform* of $f \in L^2(\mathbb{R}^d)$ with respect to the window g, is defined as

$$\mathcal{V}_g f(x, \omega) := \int_{\mathbb{R}^d} f(t) \overline{g(t - x)} e^{-2\pi i \omega \cdot t} \, dt = \langle f, \pi(z) g \rangle , \quad z = (x, \omega) \in \mathbb{R}^{2d} .$$

We define the *modulation spaces* as follows: fix a non-zero Schwartz function $g \in \mathcal{S}(\mathbb{R}^d)$, and let

$$M^p(\mathbb{R}^d) := \left\{ f \in \mathcal{S}'(\mathbb{R}^d) : \mathcal{V}_g f \in L^p(\mathbb{R}^d) \right\} , \quad 1 \le p \le \infty$$

with the norm $\|f\|_{M^p} = \|\mathcal{V}_g f\|_p$. Different choices for g give rise to equivalent norms on $M^p(\mathbb{R}^d)$. For $p = 2$, we have $M^2(\mathbb{R}^d) = L^2(\mathbb{R}^d)$. The space $M^1(\mathbb{R}^d)$, known as Feichtinger's algebra, can also be characterized as

$$M^1(\mathbb{R}^d) := \{f \in L^2(\mathbb{R}^d) : \|\mathcal{V}_f f\|_1 < \infty\}.$$

Proposition 2.1 [11] *The space $M^1(\mathbb{R}^d)$ has the following properties:*

(i) *$M^1(\mathbb{R}^d)$ is a Banach algebra under pointwise multiplication.*
(ii) *$M^1(\mathbb{R}^d)$ is a Banach algebra under convolution.*
(iii) *$M^1(\mathbb{R}^d)$ is invariant under time-frequency shifts.*
(vi) *$M^1(\mathbb{R}^d)$ is invariant under Fourier transform.*

$M^1(\mathbb{R}^d)$ contains the Schwartz space $\mathcal{S}(\mathbb{R}^d)$ and it is dense in $L^p(\mathbb{R}^d)$, $1 \le p < \infty$, therefore it is a very useful space in the time-frequency analysis. This will be the space of windows g.

Another collection of function spaces that will be useful in our calculations are the amalgam spaces $\mathbf{W}(L^p, \ell^q)(\mathbb{R}^d)$, $1 \le p, q \le \infty$. The amalgam space $\mathbf{W}(L^p, \ell^q)(\mathbb{R}^d)$ is the space of functions f such that

$$\|f\|_{\mathbf{W}(L^p,\ell^q)(\mathbb{R}^d)} := \left(\sum_{k\in\mathbb{Z}^d} \|f\mathbb{1}_k\|_p^q\right)^{1/q},$$

where $\mathbb{1}_k$ is the characteristic function of the cube $[0, 1]^d + k$, $k \in \mathbb{Z}^d$. Different partitions of $\mathbb{R}^d$ give equivalent norms. The space $\mathbf{W}(L^\infty, \ell^1)(\mathbb{R}^d)$, which is a subspace of $L^1(\mathbb{R}^d)$, is referred to as Wiener's algebra, and will be denoted by $\mathbf{W}(\mathbb{R}^d)$. It consists of all $f \in L^\infty(\mathbb{R}^d)$ such that

$$\|f\|_{\mathbf{W}(\mathbb{R}^d)} = \|f\|_{\mathbf{W}(L^\infty,\ell^1)(\mathbb{R}^d)} := \sum_{k\in\mathbb{Z}^d} \|f\mathbb{1}_k\|_\infty .$$

Moreover, we have $\mathbf{W}(\mathbb{R}^d) \subseteq \mathbf{W}(L^\infty, \ell^2)(\mathbb{R}^d)$. The following convolution relation will be often used:

$$L^1 * \mathbf{W} \subseteq \mathbf{W}.$$

For a function $f \in M^1(\mathbb{R}^d)$ it follows that $f \in \mathbf{W}(L^\infty, \ell^1)(\mathbb{R}^d)$ and $\widehat{f} \in \mathbf{W}(L^\infty, \ell^1)(\mathbb{R}^d)$, [11].

Let $C_0(\mathbb{R}^d)$ be the space of all continuous functions that vanish at infinity. Then the closed subspace of $\mathbf{W}(\mathbb{R}^d)$ consisting of continuous functions is $\mathbf{W}(C_0, \ell^1)(\mathbb{R}^d)$. Continuity of elements in $\mathbf{W}(L^\infty, \ell^1)(\mathbb{R}^d)$ allows for pointwise evaluations, and we have (see Proposition 11.1.4 in [11]): if $F \in \mathbf{W}(C_0, \ell^1)(\mathbb{R}^d)$, then $F|\Lambda \in \ell^1(\Lambda)$, for Λ any discrete relatively separated set in $\mathbb{R}^d$ with the norm estimate

$$\sum_{\lambda\in\Lambda} |F(\lambda)| \le \mathrm{rel}(\Lambda)\|F\|_{\mathbf{W}(\mathbb{R}^d)} .$$

A discrete subset Λ of $\mathbb{R}^d$ is relatively separated if

$$\mathrm{rel}(\Lambda) := \sup\{\#\{\Lambda \cap B(x,1)\} : x \in \mathbb{R}^d\} < \infty\,,$$

where $B(x,1)$ is a ball of radius 1 in $\mathbb{R}^d$ centered at x.

There are two more time frequency representations of f that we will be using. For $f, g \in L^2(\mathbb{R}^d)$, the *cross-ambiguity function* of f and g

$$\mathcal{A}(f,g)(x,\omega) = \int_{\mathbb{R}^d} f\Big(t+\frac{x}{2}\Big)\overline{g\Big(t-\frac{x}{2}\Big)}e^{-2\pi i t\cdot\omega}\,dt\,,$$

and the *cross-Wigner distribution* of f and g

$$\mathcal{W}(f,g)(x,\omega) = \int_{\mathbb{R}^d} f\Big(x+\frac{t}{2}\Big)\overline{g\Big(x-\frac{t}{2}\Big)}e^{-2\pi i t\cdot\omega}\,dt\,.$$

The three time-frequency representations, $\mathcal{V}_g f$, $\mathcal{A}(f,g)$ and $\mathcal{W}(f,g)$ are related to each other. Before we state the relationships, we define the rotation $\mathcal{U} = \mathcal{U}_J$ of a function F on $\mathbb{R}^{2d}$ as

$$\mathcal{U}F(x,\omega) = \mathcal{U}_J F(x,\omega) = F(J(x,\omega)^T) = F(\omega,-x) \quad \text{with} \quad J = \begin{pmatrix} 0 & I \\ -I & 0 \end{pmatrix},$$

where I is the $d \times d$ identity matrix. Then $\mathcal{U}^{-1} = \mathcal{U}_J^{-1} = \mathcal{U}_{-J}$. In the following two propositions we list the properties of the cross-ambiguity function and the cross-Wigner distribution that we will be using throughout the exposition. For the proofs we refer the reader to [11].

Proposition 2.2 *For $f, g \in L^2(\mathbb{R}^d)$ the cross-ambiguity function has the following properties.*

(a) *$\mathcal{A}(f,g)$ is uniformly continuous on $\mathbb{R}^{2d}$.*
(b) *$\mathcal{A}(f,g)(x,\omega) = e^{\pi i x\cdot\omega}\mathcal{V}_g f(x,\omega)$.*
(c) *$\mathcal{A}(f,g)^* = \mathcal{A}(g,f)$, where $\mathcal{A}(f,g)^*(x,\omega) = \overline{\mathcal{A}(f,g)(-x,-\omega)}$ is the involution.*
(d) *$\mathcal{A}(\pi(x,\omega)f,g)(t,\zeta) = e^{\pi i t\cdot\omega}e^{-\pi i x\cdot(\zeta-\omega)}\mathcal{A}(f,g)(t-x,\zeta-\omega)$ for $(x,\omega) \in \mathbb{R}^{2d}$.*
(e) *$\mathcal{A}(f,g) = \mathcal{F}^{-1}\mathcal{U}^{-1}\mathcal{W}(f,g)$.*

Moreover, if $f, g \in M^1(\mathbb{R}^d)$, then $\mathcal{A}(f,g) \in \mathbf{W}(C_0,\ell^1)(\mathbb{R}^{2d})$, and by b), also $\mathcal{V}_g f \in \mathbf{W}(C_0,\ell^1)(\mathbb{R}^{2d})$.

Proposition 2.3 *For $f, g \in L^2(\mathbb{R}^d)$ the cross-Wigner distribution has the following properties.*

(a) *$\mathcal{W}(f,g)$ is uniformly continuous on $\mathbb{R}^{2d}$.*
(b) *$\mathcal{W}(f,g) = \overline{\mathcal{W}(g,f)}$.*
(c) *$\mathcal{W}(\widehat{f},\widehat{g}) = \mathcal{U}^{-1}W(f,g)$.*

(d) $\mathcal{W}(f,g) = \mathcal{F}\mathcal{U}\mathcal{A}(f,g)$.
(e) *For* $(x,\omega) \in \mathbb{R}^{2d}$ *and* $(x',\omega') \in \mathbb{R}^{2d}$, *we have*

$$\begin{aligned}\mathcal{W}(T_x M_\omega f, T_{x'} M_{\omega'} g)(t,\zeta) \\ &= e^{-\pi i(x+x')\cdot(\omega-\omega')} e^{2\pi i t\cdot(\omega-\omega')} e^{-2\pi i \zeta\cdot(x-x')} \\ &\quad \times \mathcal{W}(f,g)\Big(t - \frac{x+x'}{2}, \zeta - \frac{\omega+\omega'}{2}\Big)\end{aligned}$$

f) Moyal's formula: for $f_1, g_1 \in L^2(\mathbb{R}^d)$,

$$\langle W(f,g), W(f_1,g_1)\rangle_{L^2(\mathbb{R}^{2d})} = \langle f, f_1\rangle \overline{\langle g, g_1\rangle}\,.$$

Moreover, if $f, g \in M^1(\mathbb{R}^d)$, *then* $\mathcal{W}(f,g) \in \mathbf{W}(C_0, \ell^1)(\mathbb{R}^{2d})$.

We note here that Moyal's formula also holds for the cross-ambiguity function and the short-time Fourier transform. We also mention the following tensor-product properties: if $Z = (z, \tilde{z})$ with $z \in \mathbb{R}^{2n}$, $\tilde{z} \in \mathbb{R}^{2d-2n}$ and $\psi_1, \phi_1 \in M^1(\mathbb{R}^n)$, $\psi_2, \phi_2 \in M^1(\mathbb{R}^{d-n})$, then

$$\begin{aligned}\mathcal{A}(\psi_1 \otimes \psi_2, \phi_1 \otimes \phi_2)(Z) &= \big(\mathcal{A}(\psi_1,\phi_1) \otimes \mathcal{A}(\psi_2,\phi_2)\big)(z,\tilde{z}) \\ &= \mathcal{A}(\psi_1,\phi_1)(z)\mathcal{A}(\psi_2,\phi_2)(\tilde{z})\,.\end{aligned}$$

Similarly for the cross-Wigner distribution and short-time Fourier transform.

Before we turn to Gabor systems, we need one more result. It is originally stated for the short-time Fourier transform, but we state it here for the cross-ambiguity function. The proof is the same.

Lemma 2.4 [7] *Let* $f, g \in M^1(\mathbb{R}^d)$. *Then* $\mathcal{A}(f,g) \in M^1(\mathbb{R}^{2d})$.

The main object of our study here are Gabor systems. Let Λ be a relatively separated subset of $\mathbb{R}^{2d}$. A *Gabor system* is a collection of time-frequency shifts of one or more window functions $g_i \in M^1(\mathbb{R}^d)$, $i = 1, \ldots, M$, with respect to $\Lambda \subset \mathbb{R}^{2d}$, and it is denoted by

$$\mathcal{G}(g_1, \ldots, g_M; \Lambda) = \{\pi(\lambda) g_i : i = 1, \ldots, M,\ \lambda \in \Lambda\}\,.$$

The analysis operator and the synthesis operator for a Gabor system $\mathcal{G}(g_1, \ldots, g_M; \Lambda)$ are defined as

$$C_{g,\Lambda} f := (\langle f, \pi(\lambda) g_i\rangle)_{\lambda\in\Lambda; i=1,\ldots,M}$$
$$D_{g,\Lambda} c := \sum_{i=1}^{M} \sum_{\lambda\in\Lambda} c_\lambda \pi(\lambda) g_i$$

are bounded between $M^p(\mathbb{R}^d)$ and $\ell^p(\Lambda)$ spaces, with estimates

$$\|C_{g,\Lambda} f\|_{\ell^p} \le \mathrm{rel}(\Lambda)\, \|f\|_{M^p} \max_{i=1,\ldots,M} \|g_i\|_{M^1}$$

$$\|D_{g,\Lambda}c\|_{M^p} \le \mathrm{rel}(\Lambda)\,\|c\|_{\ell^p} \max_{i=1,\dots,M} \|g_i\|_{M^1}$$

A Gabor system $\mathcal{G}(g_1, \dots, g_M; \Lambda)$ with $g_i \in M^1(\mathbb{R}^d)$, $i = 1, \dots, M$, will be called an *M^p-frame* if $C_{g,\Lambda}$ is bounded below on $M^p(\mathbb{R}^d)$. This is equivalent to having constants $A_g, B_g > 0$ so that for all $f \in M^p(\mathbb{R}^d)$

$$\sqrt{A_g}\|f\|_{M^p} \le \|S_g^\Lambda f\|_{M^p} \le \sqrt{B_g}\|f\|_{M^p},$$

where S_g^Λ is a *frame operator* given by

$$S_g^\Lambda f = \sum_{i=1}^{M}\sum_{\lambda\in\Lambda}\langle f, \pi(\lambda)g_i\rangle \pi(\lambda)g_i\,. \tag{1}$$

In this case the frame operator is invertible on $M^p(\mathbb{R}^d)$. Theorem 3.2 in [12] states that when each $g_i \in M^1(\mathbb{R}^d)$, $i = 1, \dots, M$, then $\mathcal{G}(g_1, \dots, g_M; \Lambda)$ is an M^p-frame for some $p \in [1, \infty]$, if and only if it is an M^p-frame for all p. The constants A_g and B_g are called lower and upper frame bounds, respectively. If $A_g = B_g$ then the frame is called a *tight Gabor frame*, and if $A_g = B_g = 1$, a *normalized tight Gabor frame*.

When $\mathcal{G}(g_1, \dots, g_M; \Lambda)$ is an M^p-frame for some $p \in [1, \infty]$, then we have a frame decomposition

$$f = \sum_{i=1}^{M}\sum_{\lambda\in\Lambda}\langle f, \left(S_g^\Lambda\right)^{-1}\pi(\lambda)g_i\rangle \pi(\lambda)g_i\,, \quad \text{for all } f \in M^p(\mathbb{R}^d). \tag{2}$$

The sequence $\{\left(S_g^\Lambda\right)^{-1}\pi(\lambda)g_i : i = 1, \dots, M,\ \lambda \in \Lambda\}$ is also a frame for $M^p(\mathbb{R}^d)$, called the *canonical dual frame* of $\mathcal{G}(g_1, \dots, g_M; \Lambda)$, and has upper and lower frame bounds B_g^{-1} and A_g^{-1}, respectively. If the frame is tight, then $\left(S_g^\Lambda\right)^{-1} = A_g^{-1}I$, where I is the identity operator, and the frame decomposition becomes

$$f = A_g^{-1}\sum_{i=1}^{M}\sum_{\lambda\in\Lambda}\langle f, \pi(\lambda)g_i\rangle \pi(\lambda)g_i\,, \quad \text{for all } f \in M^p(\mathbb{R}^d)\,.$$

In order to use the representation (2) in practice, we need to be able to calculate $\left(S_g^\Lambda\right)^{-1}$. While the existence of $\left(S_g^\Lambda\right)^{-1}$ is guaranteed by the frame condition, it is usually tedious to find this operator explicitly. Moreover, if Λ is not a lattice in $\mathbb{R}^d$, then the frame operator of S_g^Λ does not commute with time-frequency shifts, that is $\pi(\beta)S_g^\Lambda \ne S_g^\Lambda\pi(\beta)$ for $\beta \in \Lambda$. Indeed, let $\beta \in \Lambda$ and $f \in L^2(\mathbb{R}^d)$, then

$$S_g^\Lambda \pi(\beta) f = \sum_{i=1}^{M} \sum_{\lambda\in\Lambda} \langle \pi(\beta) f, \pi(\lambda) g_i\rangle \pi(\lambda) g_i$$
$$= \sum_{i=1}^{M} \sum_{\lambda\in\Lambda} e^{-2\pi i \beta_1\cdot(\lambda_2-\beta_2)} \langle f, \pi(\lambda-\beta) g_i\rangle \pi(\lambda) g_i \, ,$$

where $\beta = (\beta_1, \beta_2)$ and $\lambda = (\lambda_1, \lambda_2)$. On the other hand,

$$\pi(\beta) S_g^\Lambda f = \sum_{i=1}^{M} \sum_{\lambda\in\Lambda} \langle f, \pi(\lambda) g_i\rangle \pi(\beta)\pi(\lambda) g_i = \sum_{i=1}^{M} \sum_{\lambda\in\Lambda} e^{-2\pi i \beta_1\cdot\lambda_2} \langle f, \pi(\lambda) g_i\rangle \pi(\lambda+\beta) g_i$$
$$= \sum_{i=1}^{M} \sum_{\lambda\in\Lambda+\beta} e^{-2\pi i \beta_1\cdot(\lambda_2-\beta_2)} \langle f, \pi(\lambda-\beta) g_i\rangle \pi(\lambda) g_i \, ,$$

and the two expressions are not equal since $\Lambda + \beta \neq \Lambda$. Therefore, the canonical dual frame $\{(S_g^\Lambda)^{-1}\pi(\lambda) g_i : i = 1, \ldots, M, \lambda \in \Lambda\}$ does not have the same structure as $\mathcal{G}(g_1, \ldots, g_M; \Lambda)$, that is it is not a Gabor frame, and, in order to compute the canonical dual frame we would have to apply $(S_g^\Lambda)^{-1}$ to $\pi(\lambda) g_i$, for all $i = 1, \ldots, M$ and all $\lambda \in \Lambda$. Hence, we search for a pair of dual frames, rather than just one frame. Let $\mathcal{G}(g_1, \ldots, g_M; \Lambda)$ and $\mathcal{G}(h_1, \ldots, h_M; \Lambda)$ be Gabor systems, then we can define a mixed frame operator

$$S_{g,h}^\Lambda f = \sum_{i=1}^{M} \sum_{\lambda\in\Lambda} \langle f, \pi(\lambda) g_i\rangle \pi(\lambda) h_i$$

which is a bounded linear operator on $M^p(\mathbb{R}^d)$. If $S_{g,h}^\Lambda f = f$ for every $f \in M^p(\mathbb{R}^d)$, then we call $\mathcal{G}(h_1, \ldots, h_M; \Lambda)$ a *generalized dual Gabor frame* of $\mathcal{G}(g_1, \ldots, g_M; \Lambda)$.

3 Gabor Frames for Lattices: Revised

Before we turn our attention to Gabor frames for model sets, we revisit here known results for regular Gabor frames, that is where time-frequency shifts come from a lattice. We present a different approach then the one presented in the literature, by constructing a bracket product for the time-frequency plane. Alternative, but in some sense similar approaches, were also developed in [8] and more recently in [16].

Throughout this section Λ will be a lattice, that is a discrete subgroup of $\mathbb{R}^{2d}$. A lattice can be represented by an invertible matrix $A \in GL(2d, \mathbb{R})$ and is then given by $\Lambda = A\mathbb{Z}^{2d}$. We define the volume of a lattice $\Lambda = A\mathbb{Z}^{2d}$ by $\mathrm{vol}(\Lambda) = |\det(A)|$. The density of a lattice is given by the reciprocal of the volume, that is $D(\Lambda) = \mathrm{vol}(\Lambda)^{-1}$. A dual lattice is defined as $\Lambda^* = A^{-T}\mathbb{Z}^{2d}$.

A tool that is heavily utilized in time-frequency analysis is the Poisson summation formula for functions on $\mathbb{R}^d$. However, we will use here Poisson summation formula for functions on $\mathbb{R}^{2d}$.

Theorem 3.1 (Poisson Summation Formula for Lattices) *Let Λ be a lattice in $\mathbb{R}^{2d}$. Then, for every $F \in M^1(\mathbb{R}^{2d})$ we have*

$$\sum_{\lambda\in\Lambda} F(\lambda)e^{-2\pi i\lambda\cdot z} = vol(\Lambda)^{-1} \sum_{\lambda^*\in\Lambda^*} \mathcal{F}F(z-\lambda^*),$$

where Λ^* is a dual lattice. The identity holds pointwise for all $z \in \mathbb{R}^{2d}$, and both sums converge uniformly and absolutely $z \in \mathbb{R}^{2d}$.

Let $f, g, h \in M^1(\mathbb{R}^d)$. Then, by Lemma 2.4 $\mathcal{A}(f,g)$ and $\mathcal{W}(f,g)$ belong to $M^1(\mathbb{R}^{2d})$, and by Poisson summation formula,

$$\sum_{\lambda\in\Lambda} \mathcal{A}(f,g)(\lambda)e^{-2\pi i\lambda\cdot z} = \mathrm{vol}(\Lambda)^{-1} \sum_{\lambda^*\in\Lambda^*} \mathcal{W}(\widehat{f},\widehat{g})(z-\lambda^*), \tag{3}$$

where we used the relation $\mathcal{A}(f,g) = \mathcal{F}^{-1}\mathcal{W}(\widehat{f},\widehat{g})$ derived from Proposition 2.2 and Proposition 2.3. Assume that $\Lambda = A\mathbb{Z}^{2d}$, then we can write (3) more explicitly as

$$\sum_{n\in\mathbb{Z}^{2d}} \mathcal{A}(f,g)(An)e^{-2\pi i An\cdot z} = (\det A)^{-1} \sum_{n\in\mathbb{Z}^{2d}} \mathcal{W}(\widehat{f},\widehat{g})(z-A^{-T}n). \tag{4}$$

We are now in the position to define a main ingredient in our approach, the bracket product. Let $f \in L^2(\mathbb{R}^d)$ and $g \in M^1(\mathbb{R}^d)$. For a fixed $z = (x,\omega) \in \mathbb{R}^{2d}$, the *generalized $\Lambda-$bracket product* of $\pi(z)f$ and g is defined as

$$\left[\widehat{\pi(z)f},\widehat{g}\right]_\Lambda(\tilde{z}) = \mathrm{vol}(\Lambda)^{-1} \sum_{\lambda^*\in\Lambda^*} M_{-z}\mathcal{W}(\widehat{f},\widehat{g})(\tilde{z}-\lambda^*), \tag{5}$$

where M_z denotes the $\mathbb{R}^{2d}$ modulation by z. It follows from Monotone Convergence Theorem and the fact that $\mathcal{W}(\widehat{f},\widehat{g}) \in L^1(\mathbb{R}^{2d})$, that the series (5) converges absolutely to a function $L^1(\mathbb{T}_{\Lambda^*})$, $\mathbb{T}_{\Lambda^*} = \mathbb{R}^{2d}/\Lambda^*$. When Λ is represented by a matrix A, we have $\mathbb{T}_{\Lambda^*} = A^{-1}\mathbb{T}$ where $\mathbb{T}$ is the torus in $\mathbb{R}^{2d}$. Since $\left[\widehat{\pi(z)f},\widehat{g}\right]_\Lambda \in L^1(\mathbb{T}_{\Lambda^*})$, we can compute the Fourier coefficients

$$\begin{aligned}
&\int_{\mathbb{T}_{\Lambda^*}} \left[\widehat{\pi(z)f},\widehat{g}\right]_\Lambda(\tilde{z})e^{2\pi i\tilde{z}\cdot\lambda}\,d\tilde{z}\\
&= \mathrm{vol}(\Lambda)^{-1}\int_{\mathbb{T}_{\Lambda^*}} \sum_{\lambda^*\in\Lambda^*} M_{-z}\mathcal{W}(\widehat{f},\widehat{g})(\tilde{z}-\lambda^*)e^{2\pi i\tilde{z}\cdot\lambda}\,d\tilde{z}\\
&= \int_{\mathbb{R}^{2d}} M_{-z}\mathcal{W}(\widehat{f},\widehat{g})(\tilde{z})e^{2\pi i\tilde{z}\cdot\lambda}\,d\tilde{z} = \mathcal{F}^{-1}M_{-z}\mathcal{W}(\widehat{f},\widehat{g})(\lambda)\\
&= T_z\mathcal{F}^{-1}\mathcal{W}(\widehat{f},\widehat{g})(\lambda) = T_z\mathcal{A}(f,g)(\lambda)\\
&= \mathcal{A}(f,g)(\lambda-z).
\end{aligned}$$

Then the application of the Plancherel theorem for Fourier series, gives us the following proposition.

Proposition 3.2 *Let Λ be a lattice in $\mathbb{R}^{2d}$. Fix $z \in \mathbb{R}^{2d}$. Then for all $g, h \in M^1(\mathbb{R}^{2d})$ and $f_1, f_2 \in L^2(\mathbb{R}^{2d})$, we have*

$$\sum_{\lambda\in\Lambda} \mathcal{A}(f_1, g)(\lambda - z)\overline{\mathcal{A}(f_2, h)(\lambda - z)} = \int_{\mathbb{T}_{\Lambda^*}} \left[\widehat{\pi(z)f_1}, \widehat{g}\right]_\Lambda(\tilde{z}) \overline{\left[\widehat{\pi(z)f_2}, \widehat{h}\right]_\Lambda(\tilde{z})}\, d\tilde{z},$$

with the absolute convergence of the integral.

The following two results are the main ingredients in deriving Janssen representation of the Gabor frame operator $S^\Lambda_{g,h}$ and successive characterization of tight and dual Gabor frames.

Proposition 3.3 *Let Λ be a lattice in $\mathbb{R}^{2d}$. Assume that $f_1, f_2, g, h \in M^1(\mathbb{R}^d)$. Then the function*

$$F(z, \tilde{z}) = \left[\widehat{\pi(z)f_1}, \widehat{g}\right]_\Lambda(\tilde{z}) \overline{\left[\widehat{\pi(z)f_2}, \widehat{h}\right]_\Lambda(\tilde{z})} \tag{6}$$

is continuous and periodic, and coincides pointwise with its Fourier series

$$vol(\Lambda)^{-1} \sum_{\beta^*\in\Lambda^*}\sum_{\lambda\in\Lambda} \left\langle \mathcal{A}(f_1, g), T_\lambda M_{\beta^*}\mathcal{A}(f_2, h)\right\rangle_{L^2(\mathbb{R}^{2d})} e^{-2\pi i\lambda\cdot\tilde{z}} e^{-2\pi i z\cdot\beta^*}. \tag{7}$$

Proof Let Λ be a lattice and $f_1, f_2, g, h \in M^1(\mathbb{R}^d)$. Using Proposition 3.2 and a change of index, we write explicitely

$$\begin{aligned}
F(z, \tilde{z}) &= \left[\widehat{\pi(z)f_1}, \widehat{g}\right]_\Lambda(\tilde{z}) \overline{\left[\widehat{\pi(z)f_2}, \widehat{h}\right]_\Lambda(\tilde{z})}\, d\tilde{z}\\
&= \mathrm{vol}(\Lambda)^{-2} \sum_{\lambda^*,\beta^*\in\Lambda^*} M_{-z}\mathcal{W}(\widehat{f_1}, \widehat{g})(\tilde{z} - \lambda^*)\overline{M_{-z}\mathcal{W}(\widehat{f_2}, \widehat{h})(\tilde{z} - \beta^*)}\\
&= \mathrm{vol}(\Lambda)^{-2} \sum_{\lambda^*,\beta^*\in\Lambda^*} \mathcal{W}(\widehat{f_1}, \widehat{g})(\tilde{z} - \lambda^*)\overline{\mathcal{W}(\widehat{f_2}, \widehat{h})(\tilde{z} - \beta^*)}\, e^{-2\pi i z\cdot(\beta^*-\lambda^*)}\\
&= \mathrm{vol}(\Lambda)^{-2} \sum_{\lambda^*,\beta^*\in\Lambda^*} \mathcal{W}(\widehat{f_1}, \widehat{g})(\tilde{z} - \lambda^*)\overline{\mathcal{W}(\widehat{f_2}, \widehat{h})(\tilde{z} - \lambda^* - \beta^*)}\, e^{-2\pi i z\cdot\beta^*}\\
&= \mathrm{vol}(\Lambda)^{-1} \sum_{\beta^*\in\Lambda^*} \left[\mathrm{vol}(\Lambda)^{-1} \sum_{\lambda^*\in\Lambda^*} \mathcal{W}(\widehat{f_1}, \widehat{g})(\tilde{z} - \lambda^*)\overline{T_{\beta^*}\mathcal{W}(\widehat{f_2}, \widehat{h})(\tilde{z} - \lambda^*)}\right] e^{-2\pi i z\cdot\beta^*}.
\end{aligned}$$

For a fixed $\beta^* \in \Lambda^*$, consider the series

$$F_{\beta^*}(\tilde{z}) = \mathrm{vol}(\Lambda)^{-1} \sum_{\lambda^*\in\Lambda^*} \mathcal{W}(\widehat{f_1}, \widehat{g})(\tilde{z} - \lambda^*)\overline{T_{\beta^*}\mathcal{W}(\widehat{f_2}, \widehat{h})(\tilde{z} - \lambda^*)}.$$

Since $\mathcal{W}(\widehat{f_1}, \widehat{g})$ and $\mathcal{W}(\widehat{f_2}, \widehat{h})$ lie in $M^1(\mathbb{R}^{2d})$, their product as well. Moreover,

$$\mathcal{F}^{-1}\left(\mathcal{W}(\widehat{f_1}, \widehat{g})\overline{T_{\beta^*}\mathcal{W}(\widehat{f_2}, \widehat{h})}\right) = \mathcal{A}(f_1, g) * (M_{\beta^*}\mathcal{A}(f_2, h))^* \in M^1(\mathbb{R}^{2d}),$$

since $\mathcal{A}(f_1, g), \mathcal{A}(f_2, h) \in M^1(\mathbb{R}^{2d})$. Therefore, by Poisson summation formula we can write $F_{\beta^*}(\tilde{z})$ as

$$F_{\beta^*}(\tilde{z}) = \sum_{\lambda\in\Lambda} \big(\mathcal{A}(f_1, g) * (M_{\beta^*}\mathcal{A}(f_2, h))^*\big)(\lambda)\, e^{-2\pi i\lambda\cdot\tilde{z}}$$
$$= \sum_{\lambda\in\Lambda} \Big\langle \mathcal{A}(f_1, g), T_\lambda M_{\beta^*}\mathcal{A}(f_2, h)\Big\rangle_{L^2(\mathbb{R}^{2d})} e^{-2\pi i\lambda\cdot\tilde{z}},$$

and $F(z, \tilde{z})$ becomes

$$F(z, \tilde{z}) = \mathrm{vol}(\Lambda)^{-1} \sum_{\beta^*\in\Lambda^*}\sum_{\lambda\in\Lambda} \Big\langle \mathcal{A}(f_1, g), T_\lambda M_{\beta^*}\mathcal{A}(f_2, h)\Big\rangle_{L^2(\mathbb{R}^{2d})} e^{-2\pi i\lambda\cdot\tilde{z}}\, e^{-2\pi i z\cdot\beta^*}. \tag{8}$$

By Lemma 2.4, $\mathcal{A}(f_1, g), \mathcal{A}(f_2, h) \in M^1(\mathbb{R}^{2d})$, and therefore,

$$\mathcal{V}_{\mathcal{A}(f_2,h)}(\mathcal{A}(f_1, g))(w, \tilde{w}) = e^{2\pi i w\cdot\tilde{w}}\Big\langle \mathcal{A}(f_1, g), \pi(w, \tilde{w})\mathcal{A}(f_2, h)\Big\rangle$$
$$\in \mathbf{W}(C_0, \ell^1)(\mathbb{R}^{2d} \times \mathbb{R}^{2d}).$$

Hence, $\mathcal{V}_{\mathcal{A}(f_2,h)}(\mathcal{A}(f_1, g))$ restricted to $\Lambda \times \Lambda^*$ belongs to $\ell^1(\Lambda \times \Lambda^*)$, and as a consequence, the series (8) defining $F(z, \tilde{z})$ is absolutely convergent. By the uniqueness of the Fourier series, (8) is the Fourier series of F. □

Proposition 3.4 *Let Λ be a lattice in $\mathbb{R}^{2d}$. Assume that $g_i, h_i \in M^1(\mathbb{R}^{2d})$, for every $i = 1, \ldots, M$. Then for every $f_1, f_2 \in M^1(\mathbb{R}^{2d})$, the function*

$$\mathcal{N}(z) = \sum_{i=1}^{M}\sum_{\lambda\in\Lambda} \langle\pi(z)f_1, \pi(\lambda)g_i\rangle\overline{\langle\pi(z)f_2, \pi(\lambda)h_i\rangle} \tag{9}$$

is continuous and periodic, and coincides pointwise with its Fourier series $\sum_{\lambda^\circ\in\Lambda^\circ} \widehat{\mathcal{N}}(\lambda^\circ)e^{-2\pi i\sigma(\lambda^\circ,z)}$, with

$$\widehat{\mathcal{N}}(\lambda^\circ) = vol(\Lambda)^{-1}\sum_{i=1}^{M}\langle h_i, \pi(\lambda^\circ)g_i\rangle\langle\pi(\lambda^\circ)f_1, f_2\rangle, \tag{10}$$

where $\Lambda^\circ = J\Lambda^*$ is the adjoint lattice and σ a symplectic form defined as $\sigma(\theta, z) = \theta \cdot Jz$, with θ and z seen as column vectors in $\mathbb{R}^{2d}$.

Proof Let Λ be a lattice in $\mathbb{R}^{2d}$ and choose $f_1, f_2 \in M^1(\mathbb{R}^{2d})$. Then the function $\mathcal{N}(z)$ is well defined and can be expressed using the mixed frame operator $S^\Lambda_{g,h}$ as $\mathcal{N}(z) = \langle S^\Lambda_{g,h}\pi(z)f_1, \pi(z)f_2\rangle$. Moreover, using the relations between short time Fourier transform and cross-ambiguity function, we can write $\mathcal{N}$ as

$$
\begin{aligned}
\mathcal{N}(z) &= \sum_{i=1}^{M} \sum_{\lambda\in\Lambda} \mathcal{V}_{g_i}(\pi(z)f_1)(\lambda)\overline{\mathcal{V}_{h_i}(\pi(z)f_2)(\lambda)} \\
&= \sum_{i=1}^{M} \sum_{\lambda\in\Lambda} \mathcal{A}(\pi(z)f_1, g_i)(\lambda)\overline{\mathcal{A}(\pi(z)f_2, h_i)(\lambda)} \\
&= \sum_{i=1}^{M} \sum_{\lambda\in\Lambda} \mathcal{A}(f_1, g_i)(\lambda - z)\overline{\mathcal{A}(f_2, h_i)(\lambda - z)}\,.
\end{aligned}
$$

For $i = 1, \dots, M$ fixed, let

$$
\mathcal{N}_i(z) = \sum_{\lambda\in\Lambda} \mathcal{A}(f_1, g_i)(\lambda - z)\overline{\mathcal{A}(f_2, h_i)(\lambda - z)}\,,
$$

which is a periodic function. Using Proposition 3.2 and Proposition 3.3, we have

$$
\begin{aligned}
\mathcal{N}_i(z) &= \int_{\mathbb{T}_{\Lambda^*}} \left[\widehat{\pi(z)f_1}, \widehat{g_i}\right]_\Lambda(\tilde{z})\,\overline{\left[\widehat{\pi(z)f_2}, \widehat{h_i}\right]_\Lambda(\tilde{z})}\,d\tilde{z} \\
&= \mathrm{vol}(\Lambda)^{-1} \int_{\mathbb{T}_{\Lambda^*}} \sum_{\beta^*\in\Lambda^*} \sum_{\lambda\in\Lambda} \left\langle \mathcal{A}(f_1, g), T_\lambda M_{\beta^*}\mathcal{A}(f_2, h)\right\rangle_{L^2(\mathbb{R}^{2d})} e^{-2\pi i\lambda\cdot\tilde{z}} e^{-2\pi i z\cdot\beta^*}\,d\tilde{z} \\
&= \mathrm{vol}(\Lambda)^{-1} \sum_{\beta^*\in\Lambda^*} \left[\int_{\mathbb{T}_{\Lambda^*}} \sum_{\lambda\in\Lambda} \left\langle \mathcal{A}(f_1, g), T_\lambda M_{\beta^*}\mathcal{A}(f_2, h)\right\rangle_{L^2(\mathbb{R}^{2d})} e^{-2\pi i\lambda\cdot\tilde{z}}\,d\tilde{z}\right] e^{-2\pi i z\cdot\beta^*} \\
&= \mathrm{vol}(\Lambda)^{-1} \sum_{\beta^*\in\Lambda^*} \left\langle \mathcal{A}(f_1, g), M_{\beta^*}\mathcal{A}(f_2, h)\right\rangle_{L^2(\mathbb{R}^{2d})} e^{-2\pi i z\cdot\beta^*} \\
&= \mathrm{vol}(\Lambda)^{-1} \sum_{\beta^*\in\Lambda^*} \left\langle \mathcal{W}(\widehat{f_1}, \widehat{g_i}), T_{\beta^*}\mathcal{W}(\widehat{f_2}, \widehat{h_i})\right\rangle_{L^2(\mathbb{R}^{2d})} e^{-2\pi i z\cdot\beta^*}\,.
\end{aligned}
$$

The interchange of the integral and the sum is possible due to the Fubini's Theorem. Now, since $\left\langle \mathcal{A}(f_1, g), T_\lambda M_{\beta^*}\mathcal{A}(f_2, h)\right\rangle_{L^2(\mathbb{R}^{2d})}$ are in $\ell^1(\Lambda \times \Lambda^*)$ by the proof of Proposition 3.3, the coefficients $\left\langle \mathcal{W}(\widehat{f_1}, \widehat{g_i}), T_{\beta^*}\mathcal{W}(\widehat{f_2}, \widehat{h_i})\right\rangle_{L^2(\mathbb{R}^{2d})}$ are in $\ell^1(\Lambda^*)$. Hence, for each $i = 1, \dots, M$, the function $\mathcal{N}_i$ is continuous, as it equals the absolutely convergent trigonometric series

$$
\mathcal{N}_i(z) = \mathrm{vol}(\Lambda)^{-1} \sum_{\beta^*\in\Lambda^*} \left\langle \mathcal{W}(\widehat{f_1}, \widehat{g_i}), T_{\beta^*}\mathcal{W}(\widehat{f_2}, \widehat{h_i})\right\rangle_{L^2(\mathbb{R}^{2d})} e^{-2\pi i z\cdot\beta^*}\,.
$$

The coefficients $\left\langle \mathcal{W}(\widehat{f_1}, \widehat{g_i}), T_{\beta^*}\mathcal{W}(\widehat{f_2}, \widehat{h_i})\right\rangle_{L^2(\mathbb{R}^{2d})}$ can be simplified. Let $\frac{\beta^*}{2} = (\frac{\beta_1^*}{2}, \frac{\beta_2^*}{2})$, then using Proposition 2.3 *e*), we have

$$
T_{\frac{-\beta^*}{2}}\mathcal{W}(\widehat{f_1}, \widehat{g_i})(t, \zeta) = e^{-\pi i\beta_1^*\cdot\beta_2^*} e^{-2\pi i t\cdot\beta_2^*} e^{2\pi i\zeta\cdot\beta_1^*}\mathcal{W}(\widehat{f_1}, T_{-\beta_1^*}M_{-\beta_2^*}\widehat{g_i})(t, \zeta)\,,
$$

$$T_{\frac{\beta^*}{2}}\mathcal{W}(\widehat{f_2},\widehat{h_i})(t,\zeta)=e^{\pi i\beta_1^*\cdot\beta_2^*}e^{-2\pi it\cdot\beta_2^*}e^{2\pi i\zeta\cdot\beta_1^*}\mathcal{W}(T_{\beta_1^*}M_{\beta_2^*}\widehat{f_2},\widehat{h_i})(t,\zeta),$$

and applying Moyal's formula we obtain

$$\begin{aligned}&\Big\langle\mathcal{W}(\widehat{f_1},\widehat{g_i}),T_{\beta^*}\mathcal{W}(\widehat{f_2},\widehat{h_i})\Big\rangle_{L^2(\mathbb{R}^{2d})}\\&=\Big\langle T_{\frac{-\beta^*}{2}}\mathcal{W}(\widehat{f_1},\widehat{g_i}),T_{\frac{\beta^*}{2}}\mathcal{W}(\widehat{f_2},\widehat{h_i})\Big\rangle_{L^2(\mathbb{R}^{2d})}\\&=e^{-2\pi i\beta_1^*\cdot\beta_2^*}\Big\langle\mathcal{W}(\widehat{f_1},T_{-\beta_1^*}M_{-\beta_2^*}\widehat{g_i}),\mathcal{W}(T_{\beta_1^*}M_{\beta_2^*}\widehat{f_2},\widehat{h_i})\Big\rangle_{L^2(\mathbb{R}^{2d})}\\&=e^{-2\pi i\beta_1^*\cdot\beta_2^*}\langle\widehat{f_1},T_{\beta_1^*}M_{\beta_2^*}\widehat{f_2}\rangle\overline{\langle T_{-\beta_1^*}M_{-\beta_2^*}\widehat{g_i},\widehat{h_i}\rangle}\\&=\langle M_{-\beta_1^*}T_{\beta_2^*}f_1,f_2\rangle\langle h_i,M_{-\beta_1^*}T_{\beta_2^*}g_i\rangle\\&=\langle\pi(\lambda^\circ)f_1,f_2\rangle\langle h_i,\pi(\lambda^\circ)g_i\rangle,\end{aligned}$$

where $\lambda^\circ=J\beta^*\in J\Lambda^*=\Lambda^\circ$. We can then express $\mathcal{N}_i(z)$ as

$$\mathcal{N}_i(z)=\mathrm{vol}(\Lambda)^{-1}\sum_{\lambda^\circ\in\Lambda^\circ}\langle\pi(\lambda^\circ)f_1,f_2\rangle\langle h_i,\pi(\lambda^\circ)g_i\rangle\,e^{-2\pi i\sigma(\lambda^\circ,z)},$$

and we have

$$\mathcal{N}(z)=\sum_{i=1}^M\mathcal{N}_i(z)=\mathrm{vol}(\Lambda)^{-1}\sum_{\lambda^\circ\in\Lambda^\circ}\left(\sum_{i=1}^M\langle h_i,\pi(\lambda^\circ)g_i\rangle\langle\pi(\lambda^\circ)f_1,f_2\rangle\right)e^{-2\pi i\sigma(\lambda^\circ,z)}. \tag{11}$$

The function $\mathcal{N}$ is continuous since it is a finite sum of continuous functions. By the uniqueness of the Fourier series, (26) is the Fourier series of $\mathcal{N}$. □

We are now in the position to state the three main identities in Gabor analysis.

Theorem 3.5 *Let Λ be a lattice in $\mathbb{R}^{2d}$ with adjoint lattice Λ°. Then for $g_i,h_i\in M^1(\mathbb{R}^d)$, $i=1,\ldots,M$, the following hold.*

(i) *Fundamental identity of Gabor analysis:*

$$\sum_{i=1}^M\sum_{\lambda\in\Lambda}\langle f_1,\pi(\lambda)g_i\rangle\langle\pi(\lambda)h_i,f_2\rangle=vol(\Lambda)^{-1}\sum_{i=1}^M\sum_{\lambda^\circ\in\Lambda^\circ}\langle h_i,\pi(\lambda^\circ)g_i\rangle\langle\pi(\lambda^\circ)f_1,f_2\rangle \tag{12}$$

for all $f_1,f_2\in L^2(\mathbb{R}^d)$.

(ii) *Janssen Representation:*

$$S^\Lambda_{g,h}=vol(\Lambda)^{-1}\sum_{i=1}^M\sum_{\lambda^\circ\in\Lambda^\circ}\langle h_i,\pi(\lambda^\circ)g_i\rangle\pi(\lambda^\circ), \tag{13}$$

where the series converges unconditionally in the strong operator sense.

(iii) *Wexler–Raz Biorthogonality Relations:*

$$S^{\Lambda}_{g,h} = I \text{ on } L^2(\mathbb{R}^d) \iff vol(\Lambda)^{-1} \sum_{i=1}^{M} \langle h_i, \pi(\lambda^\circ) g_i \rangle = \delta_{\lambda^\circ,0} \text{ for } \lambda^\circ \in \Lambda^\circ . \tag{14}$$

Proof The Fundamental Identity of Gabor Analysis and Janssen representation follow directly from Proposition 3.4. It suffices to prove the statements on a dense subspace of $L^2(\mathbb{R}^d)$. Let $f_1, f_2 \in M^1(\mathbb{R}^d)$, then the left hand side of (12) equals the function $\mathcal{N}$ of Proposition 3.4 evaluated at $z = 0$. Since $\mathcal{N}$ equals its Fourier series expansion, we have

$$\mathcal{N}(0) = \text{vol}(\Lambda)^{-1} \sum_{\lambda^\circ \in \Lambda^\circ} \sum_{i=1}^{M} \langle h_i, \pi(\lambda^\circ) g_i \rangle \langle \pi(\lambda^\circ) f_1, f_2 \rangle ,$$

which gives (12).

As for Janssen representation, we observed in the proof of Proposition 3.4 that $\mathcal{N}(z) = \langle S^{\Lambda}_{g,h} \pi(z) f_1, \pi(z) f_2 \rangle$ for fixed $f_1, f_2 \in M^1(\mathbb{R}^d)$. Evaluating $\mathcal{N}$ at $z = 0$ and using the Fourier series representation of $\mathcal{N}$, we obtain

$$\begin{aligned} \langle S^{\Lambda}_{g,h} f_1, f_2 \rangle &= \text{vol}(\Lambda)^{-1} \sum_{i=1}^{M} \sum_{\lambda^\circ \in \Lambda^\circ} \langle h_i, \pi(\lambda^\circ) g_i \rangle \langle \pi(\lambda^\circ) f_1, f_2 \rangle \\ &= \Big\langle \text{vol}(\Lambda)^{-1} \sum_{i=1}^{M} \sum_{\lambda^\circ \in \Lambda^\circ} \langle h_i, \pi(\lambda^\circ) g_i \rangle \pi(\lambda^\circ) f_1, f_2 \Big\rangle , \end{aligned}$$

which is the Janssen representation of the frame operator $S^{\Lambda}_{g,h}$.

The implication $\Longleftarrow$ of $iii)$ follows trivially from the Janssen representation of $S^{\Lambda}_{g,h}$. For the converse, assume that $S^{\Lambda}_{g,h} = I$. Let $f_1, f_2 \in M^1(\mathbb{R}^d)$, then $\mathcal{N}$ of Proposition 3.4 is a constant function. Indeed,

$$\langle f_1, f_2 \rangle = \langle \pi(z) f_1, \pi(z) f_2 \rangle = \langle S^{\Lambda}_{g,h} \pi(z) f_1, \pi(z) f_2 \rangle = \mathcal{N}(z)$$

for every $z \in \mathbb{R}^{2d}$. Since $\mathcal{N}$ is a constant function, it equals its 0th Fourier coefficient $\widehat{\mathcal{N}}(0) = \langle f_1, f_2 \rangle$. By Proposition 3.4, we have

$$\text{vol}(\Lambda)^{-1} \sum_{i=1}^{M} \langle h_i, \pi(\lambda^\circ) g_i \rangle \langle \pi(\lambda^\circ) f_1, f_2 \rangle = \delta_{\lambda^\circ,0} \langle f_1, f_2 \rangle . \tag{15}$$

Fix $\lambda^\circ = (\lambda^\circ_1, \lambda^\circ_2) \in \Lambda^\circ$ and let $f \in M^1(\mathbb{R}^d)$ be a nonzero function. By letting $f_1 = T_{-\lambda^\circ_1} M_{-\lambda^\circ_2} f$ and $f_2 = f$, (15) becomes

$$\text{vol}(\Lambda)^{-1} \sum_{i=1}^{M} \langle h_i, \pi(\lambda^\circ) g_i \rangle \langle f, f \rangle = \delta_{\lambda^\circ,0} \langle T_{-\lambda^\circ_1} M_{-\lambda^\circ_2} f, f \rangle ,$$

and the right hand side of (14) holds. □

In the subsequent sections we will turn our attention to Gabor frames for model sets by generalizing the construction we have just presented.

4 Almost Periodic Functions, Model Sets and Local Functions

The main object of our investigation are Gabor frames for model sets, and in the treatment of such frames, we naturally come across almost periodic functions. Therefore, we begin with the review of some basic facts about almost periodic functions and finish with a definition and some properties of model sets. For a detailed exposition on almost periodic functions we refer to [2,3,5].

We say that a bounded and continuous function $f : \mathbb{R}^{2d} \to \mathbb{C}$ is *almost periodic*, if to every $\epsilon > 0$ there corresponds a relatively dense set $E(f,\epsilon) \subseteq \mathbb{R}^{2d}$, such that for every $\tau \in E(f,\epsilon)$,

$$\sup_{z\in\mathbb{R}^{2d}} |f(z+\tau) - f(z)| \leq \epsilon\,.$$

A subset D is called relatively dense in $\mathbb{R}^{2d}$ when there exists $r > 0$, such that for all $z \in \mathbb{R}^{2d}$, $D \cap B(z,r) \neq \emptyset$, where $B(z,r)$ is a ball of radius r in $\mathbb{R}^{2d}$ centered at z. Each $\tau \in E(f,\epsilon)$ is called an ϵ-period of f. Let $AP(\mathbb{R}^{2d})$ denote the space of almost periodic functions. Each almost periodic function is uniformly continuous and admits a formal Fourier series

$$f(z) \sim \sum_{\lambda\in\sigma(f)} a(\lambda, f)e^{-2\pi i\lambda\cdot z}\,, \tag{16}$$

where

$$a(\lambda, f) := \mathcal{M}_z\{f(z)e^{2\pi i\lambda\cdot z}\} = \lim_{R\to\infty} \frac{1}{R^{2d}} \int_{B(0,R)} f(z)e^{2\pi i\lambda\cdot z}\,dz\,,$$

are the Fourier coefficients of f, and $\sigma(f)$ is the so-called Bohr spectrum of f,

$$\sigma(f) := \left\{\lambda \in \mathbb{R}^m \,:\, a(\lambda, f) \neq 0\right\},$$

and it forms a countable set, that is only for a countable number of $\lambda \in \mathbb{R}^m$, $a(\lambda, f)$ is nonzero ([5]). The 0-th Fourier coefficient of f,

$$a(0, f) := \mathcal{M}_z\{f(z)\}\,,$$

is called *the mean value* of f. If $f(z,\tilde{z})$, with $(z,\tilde{z}) \in \mathbb{R}^{2d} \times \mathbb{R}^{2d}$, is almost periodic, then it is almost periodic with respect to each of the variables z and $\tilde{z}$. Moreover, $\mathcal{M}_{\tilde{z}}\{f(z,\tilde{z})\}$ is an almost periodic function of z.

We gather the important identities of almost periodic functions, that we will use throughout the article, in the following theorem.

Theorem 4.1 ([5]) *Let $f \in AP(\mathbb{R}^{2d})$ with the Fourier series given by* (16). *Then the following hold.*

(i) *Bohr's Fundamental Theorem:* $\mathcal{M}_z\{|f(z)|^2\} = \sum_{\lambda\in\sigma(f)}|a(\lambda, f)|^2$.
(ii) *Plancherel's Theorem:* $\mathcal{M}_z\{f(z)\overline{g(z)}\} = \sum_{\lambda\in\sigma(f)} a(\lambda, f)\overline{a(\lambda, g)}$ *for all* $g \in AP(\mathbb{R}^{2d})$ *with* $\sigma(f) = \sigma(g)$.
(iii) *If all the coefficients* $a(\lambda, f)$ *of* $f \in AP(\mathbb{R}^m)$ *are zero, then the function* $f \equiv 0$.
(iv) *When* f *is non-negative, then* $\mathcal{M}\{f\} = 0$ *if and only if* $f \equiv 0$.

Throughout the exposition we will be encountering almost periodic functions whose spectrums lie in model set. We state the basic definitions and theorems for even dimensional model sets since only those we will use, however the same definitions and properties apply in any dimension.

We begin with a lattice in $\Gamma \subset \mathbb{R}^{2d} \times \mathbb{R}^n$, where $\mathbb{R}^{2d}$ and $\mathbb{R}^n$ are equipped with Euclidean metrics and $\mathbb{R}^{2d} \times \mathbb{R}^n$ is the orthogonal sum of the two spaces. Let $p_1 : \mathbb{R}^{2d} \times \mathbb{R}^n \to \mathbb{R}^{2d}$ and $p_2 : \mathbb{R}^{2d} \times \mathbb{R}^n \to \mathbb{R}^n$ be projection maps such that $p_1|\Gamma$ is injective and $L = p_1(\Gamma)$ is a dense subgroup of $\mathbb{R}^{2d}$. We impose the same properties on p_2. For the dual lattice of Γ, denoted by Γ^*, let p_1^*, p_2^* be defined as p_1, p_2. It holds then, that $p_1^*|\Gamma^*$ is injective and $p_1^*(\Gamma^*)$ is a dense subgroup of $\mathbb{R}^{2d}$, and the same holds for p_2^*. Moreover, for $\gamma \in \Gamma$ and $\gamma^* \in \Gamma^*$,

$$\mathbb{Z} \ni \gamma \cdot \gamma^* = (p_1(\gamma), p_2(\gamma)) \cdot (p_1^*(\gamma^*), p_2^*(\gamma^*)) = p_1(\gamma) \cdot p_1^*(\gamma^*) + p_2(\gamma) \cdot p_2^*(\gamma^*)$$

Let $\Omega \subset \mathbb{R}^n$ be compact, equal to the closure of its interior and have boundary of measure 0. We call Ω a window. Then a model set $\Lambda(\Omega)$ is defined as

$$\Lambda(\Omega) := \{p_1(\gamma) : \gamma \in \Gamma,\ p_2(\gamma) \in \Omega\} \subset L \subset \mathbb{R}^{2d}.$$

If Ω is symmetric around the origin then $0 \in \Lambda(\Omega)$. Model set is *generic* if the boundary of Ω has no common points with $p_2(\Gamma)$. A model set is *simple* if $n = 1$. We will be working only with simple model sets. We assume, without loss of generality, that from now on Ω is symmetric around the origin.

Model sets are a very natural generalizations of lattices, and for $n = 0$ they reduce to a lattice and, thus, the results that we develop later on in the article reduce to the known ones for lattices. If $\Lambda(\Omega)$ is a model set, then it is uniformly discrete, relatively dense, and has a well defined density

$$D(\Lambda(\Omega)) = \lim_{R\to\infty} \frac{\#\{\Lambda(\Omega) \cap B(x, R)\}}{R^{2d}},$$

where $\#S$ denotes the cardinality of the set S and $B(x, R)$ is a ball of radius R in $\mathbb{R}^{2d}$ centered at x . The limit is independent of $x \in \mathbb{R}^{2d}$. For a simple model set $\Lambda(\Omega)$, we have $D(\Lambda(\Omega)) = \text{vol}(\Gamma)^{-1}|\Omega|$, see [1].

Due to the underlying lattice structure of a model set, there exists a Poisson summation formula for $\Lambda(\Omega)$. Let $C_0^\infty(\Omega)$ be the space of all smooth, real valued functions on $\mathbb{R}$ with support in Ω. Via the mapping $p_2 \circ (p_1|_\Gamma)^{-1} : L \to \mathbb{R}$ we obtain a space $\mathcal{C}(\Lambda(\Omega))$ of functions on L, vanishing off $\Lambda(\Omega)$: for $\psi \in C_0^\infty(\Omega)$, we define $w_\psi \in \mathcal{C}(\Lambda(\Omega))$ by

$$w_\psi : L \to \mathbb{R}, \qquad w_\psi(\lambda) := \psi(p_2(\gamma)) \quad \text{for} \quad \lambda = p_1(\gamma) \in \Lambda(\Omega), \tag{17}$$

and $w_\psi(\lambda) = 0$ for $\lambda \notin \Lambda(\Omega)$. If ψ were the indicator function of Ω, we would have $w_\psi(\lambda) = 1$ on $\Lambda(\Omega)$ and $w_\psi(\lambda) = 0$ if $\lambda \notin \Lambda(\Omega)$. However, the indicator function is not smooth. The Poisson summation formula for model sets was originally stated for the class of Schwartz functions in [23]. However, since it relies on the original Poisson summation formula, we can state it for a bigger space.

Theorem 4.2 (Poisson Summation Formula for Model Sets) *Let $\Lambda(\Omega)$ be a simple model set defined by a relatively compact set $\Omega \subseteq \mathbb{R}$ of non-empty interior and a lattice $\Gamma \subseteq \mathbb{R}^{2d} \times \mathbb{R}$. Let $\psi \in C_0^\infty(\Omega)$, and the weight factors $w_\psi(\lambda)$ on $\Lambda(\Omega)$ be defined as in* (17). *Then, for every $F \in M^1(\mathbb{R}^{2d})$, the following holds*

$$\sum_{\lambda \in \Lambda(\Omega)} w_\psi(\lambda) F(\lambda) e^{-2\pi i \lambda \cdot z} = \sum_{\gamma^* \in \Gamma^*} \widetilde{w}_\psi\big(-p_2^*(\gamma^*)\big) \widehat{F}\big(z - p_1^*(\gamma^*)\big), \tag{18}$$

where

$$\widetilde{w}_\psi\big(p_2^*(\gamma^*)\big) := vol(\Gamma)^{-1} \widehat{\psi}\big(p_2^*(\gamma^*)\big) \quad for\ \gamma^* \in \Gamma^*. \tag{19}$$

The identity holds pointwise for all $z \in \mathbb{R}^{2d}$, and both sums converge uniformly and absolutely for all $t \in \mathbb{R}^{2d}$.

Meyer, in [23], originally stated the Poisson summation formula for model sets for functions in the Schwartz class $\mathcal{S}(\mathbb{R}^{2d})$. As the Poisson summation formula for model sets follows from the ordinary Poisson summation formula, which holds not only for the elements from $\mathcal{S}(\mathbb{R}^{2d})$ but also for functions in $M^1(\mathbb{R}^{2d})$, we were able to extend the former one to a bigger class of functions.

Proof Since $w_\psi(\lambda) = \psi(p_2(\gamma))$ for $\lambda = p_1(\gamma)$, one can forget about the restriction $\lambda \in \Lambda$ which is given for free by the support of ψ, and consider a function $f = F \otimes \psi$. Since $\psi \in C_0^\infty(\Omega) \subset M^1(\mathbb{R})$, $f \in M^1(\mathbb{R}^{2d} \times \mathbb{R})$ by the tensor product property of Feichtinger's algebra M^1. We can now apply the ordinary Poisson summation formula, Theorem 3.1, to the lattice Γ, its dual lattice Γ^* and the function f and obtain

$$\sum_{\gamma \in \Gamma} f(\gamma) e^{-2\pi i \gamma \cdot z} = \mathrm{vol}(\Gamma)^{-1} \sum_{\gamma^* \in \Gamma^*} \mathcal{F} f(z - \gamma^*),$$

Then (18) follows by taking $z = (x, 0) \in \mathbb{R}^{2d} \times \mathbb{R}$. □

Remark 1 Poisson summation formula for model sets gives a method for constructing almost periodic functions with desired spectrum. Indeed, the function on the right hand side of (18) is almost periodic since it equals an absolutely convergent trigonometric series. By the property of almost periodic functions, the Fourier series of this function coincides with this trigonometric series. That means that the Fourier coefficients of the right hand side of (18) equal $w_\psi(\lambda) F(\lambda)$.

On the collection of point sets in $\mathbb{R}^{2d}$ that are relatively dense and uniformly separated, with minimal separation greater than r, denoted by $D_r(\mathbb{R}^{2d})$, we can put a topology, called local topology: two sets Λ and Λ' of $D_r(\mathbb{R}^{2d})$ are close if, for some large R and some small ϵ, one has

$$\Lambda' \cap B(0, R) = (\Lambda + v) \cap B(0, R) \quad \text{for some } v \in B(0, \epsilon). \tag{20}$$

Thus for each point of Λ within the ball $B(0, R)$, there is a point of Λ' within the distance ϵ of that point, and vice versa. The pairs (Λ, Λ') satisfying (20) are called (R, ϵ)-close. More formally, for $\epsilon > 0$ and a ball $B(x, R)$, define

$$U(\epsilon, B(x, R)) := \Big\{(\Lambda, \Lambda') \in D_r(\mathbb{R}^{2d}) \times D_r(\mathbb{R}^{2d}) : \\ (\Lambda + v) \cap B(x, R) = \Lambda' \cap B(x, R), \text{ for some } v \in B(0, \epsilon)\Big\}.$$

These sets form a fundamental system for a uniform structure on $D_r(\mathbb{R}^{2d})$ whose topology has the sets

$$U(\epsilon, B(x, R))[\Lambda] := \Big\{\Lambda' \in D_r(\mathbb{R}^{2d}) : (\Lambda, \Lambda') \in U(\epsilon, B(x, R))\Big\}$$

as a neighbourhood basis of Λ. Note, all the point sets Λ from $D_r(\mathbb{R}^{2d})$ have the same relative separation $\mathrm{rel}(\Lambda)$.

On the set $D_r(\mathbb{R}^{2d})$ we can put a metric. Let $\Lambda, \Lambda' \in D_r(\mathbb{R}^{2d})$, then

$$d(\Lambda, \Lambda') := \limsup_{R\to\infty} \frac{\#\{((\Lambda \cup \Lambda') \setminus (\Lambda \cap \Lambda')) \cap B(0, R)\}}{R^{2d}}$$

is a pseudometric on $D_r(\mathbb{R}^{2d})$. We obtain a metric by defining the equivalence relation

$$\Lambda \equiv \Lambda' \iff d(\Lambda, \Lambda') = 0\,.$$

Later in the article, we will work with a collection of model sets. Let Ω be a window, then for each $(s, t) \in \mathbb{R}^{2d} \times \mathbb{R}$ we may define

$$\Lambda^{\Omega}_{(s,t)} = s + \Lambda(\Omega - t)$$

Note that $\Lambda(\Omega)$ and all its shifts have the same relative separation $\mathrm{rel}(\Lambda(\Omega))$.

If $(s, t) \equiv (s', t') \mod \Gamma$, then $\Lambda^{\Omega}_{(s,t)} = \Lambda^{\Omega}_{(s',t')}$, however the inverse is not necessarily true. In the sequel we will write $(s, t)_L$ for the congruence class $(s, t) \mod \Gamma$. These model sets are parametrized by the torus $\mathbb{T} := (\mathbb{R}^{2d} \times \mathbb{R})/\Gamma = (\mathbb{R}/\mathbb{Z})^{2d+1}$. There is a natural measure, Haar measure, θ on $\mathbb{T}$. It is invariant under the action of $\mathbb{R}^{2d}$ on $(\mathbb{R}^{2d} \times \mathbb{R})/\Gamma$ and it acts by

$$z + (s, t)_L = (z + s, t)_L\,.$$

We can define an embedding $\mathbb{R}^{2d} \to \mathbb{T}$, $z \mapsto (z, 0)_L$. The image of this embedding is dense in $\mathbb{T}$.

Now, let $\Lambda(\Omega)$ be a model set, and we translate it by elements $z \in \mathbb{R}^{2d}$

$$z + \Lambda(\Omega) = z + \Lambda(\Omega + 0) = \Lambda^{\Omega}_{(z,0)} \,.$$

The closure of the set of all translates $\Lambda^{\Omega}_{(z,0)}$ of $\Lambda(\Omega)$ under the local topology (20) forms the so-called local hull $X(\Lambda(\Omega))$ of $\Lambda(\Omega)$, $X(\Lambda(\Omega)) = \overline{\{z + \Lambda(\Omega) : z \in \mathbb{R}^{2d}\}}$, ([25,27]).

Proposition 4.3 [27] *Let $\Lambda(\Omega)$ be a model set. There is a continuous mapping*

$$\beta : X(\Lambda(\Omega)) \to \mathbb{T} \,,$$

called the torus parametrization, such that (i) β is onto; (ii) β is injective almost everywhere with respect to the Haar measure θ; (iii) $\beta(z + \Lambda') = z + \beta(\Lambda')$ for all $z \in \mathbb{R}^{2d}$ and all $\Lambda' \in X(\Lambda)$; and (iv) $\beta(z + \Lambda(\Omega)) = (z, 0)_L$ for all $z \in \mathbb{R}^{2d}$.

By injective almost everywhere, we mean that the set P of points $z \in \mathbb{T}$, for which there is more than one point set of $X(\lambda(\Omega))$ over z, satisfies $\theta(P) = 0$.

There is a unique $\mathbb{R}^{2d}$-invariant measure μ on $X(\Lambda(\Omega))$, with $\mu(X(\Lambda(\Omega))) = 1$, and β relates the Haar measure θ and μ through: $\theta(P) = \mu(\beta^{-1}P)$ for all measurable subsets P of $\mathbb{T}$. Having μ we can introduce the space $L^2(X(\Lambda(\Omega)), \mu)$ of square integrable functions on $X(\Lambda(\Omega))$. Square integrable functions on $X(\Lambda(\Omega))$ and square integrable functions on $\mathbb{T}$ can be identified,

$$L^2(X(\Lambda(\Omega)), \mu) \simeq L^2(\mathbb{T}, \theta) \,. \tag{21}$$

The mapping takes a function $\mathfrak{N} \in L^2(\mathbb{T}, \theta)$ and creates $\widetilde{\mathcal{N}} = \mathfrak{N} \circ \beta \in L^2(X(\Lambda(\Omega)), \mu)$, and since β is almost everywhere injective, the map is a bijection. This allows us to analyze functions on $X(\Lambda(\Omega))$ by treating them as functions on $\mathbb{T}$.

Consider a function $\widetilde{\mathcal{N}} : X(\Lambda(\Omega)) \to \mathbb{C}$. We can define from it a function $\mathcal{N} : \mathbb{R}^{2d} \to \mathbb{C}$ by

$$\mathcal{N}(z) = \widetilde{\mathcal{N}}(z + \Lambda(\Omega)) \,.$$

If $\widetilde{\mathcal{N}}$ is continuous, then for all $z_1, z_2 \in \mathbb{R}^{2d}$, if $z_1 + \Lambda(\Omega)$ and $z_2 + \Lambda(\Omega)$ are close, then $\widetilde{\mathcal{N}}(z_1 + \Lambda(\Omega))$ and $\widetilde{\mathcal{N}}(z_2 + \Lambda(\Omega))$ are close, and as a consequence, $\mathcal{N}(z_1)$ and $\mathcal{N}(z_2)$ are close. Thus continuity of $\widetilde{\mathcal{N}}$ implies continuity of $\mathcal{N}$, or a certain locality. More formally, a function $\mathcal{N} : \mathbb{R}^{2d} \to \mathbb{C}$ is called *local with respect to* $\Lambda(\Omega)$, if for all $\delta > 0$ there exists $R > 0$ and $\epsilon > 0$ so that whenever $z_1 + \Lambda(\Omega)$ and $z_2 + \Lambda(\Omega)$, for $z_1, z_2 \in \mathbb{R}^{2d}$, are (R, ϵ)-close, then

$$|\mathcal{N}(z_1) - \mathcal{N}(z_2)| < \delta \,.$$

Intuitively, $\mathcal{N}$ looks very much the same at places where the local environment looks the same. It can be easily verified that local functions are continuous on $\mathbb{R}^{2d}$ and almost periodic.

Using locality, we can go in the opposite direction. Let $\mathcal{N}$ be a local function with respect to $\Lambda(\Omega)$. Define a function $\widetilde{\mathcal{N}}$ on the orbit of $\Lambda(\Omega)$:

$$\widetilde{\mathcal{N}} : \{z + \Lambda(\Omega) : z \in \mathbb{R}^{2d}\} \to \mathbb{C} \quad \text{by} \quad \widetilde{\mathcal{N}}(z + \Lambda(\Omega)) = \mathcal{N}(z)\,.$$

Then $\widetilde{\mathcal{N}}$ is uniformly continuous on $\{z + \Lambda(\Omega) : z \in \mathbb{R}^{2d}\}$ with respect to the local topology. The reason for this is that the continuity condition which defines the localness of $\mathcal{N}$ is based on the uniformity defining the local topology on $\{z+\Lambda(\Omega) : z \in \mathbb{R}^{2d}\}$. It follows that $\widetilde{\mathcal{N}}$ lifts uniquely to a continuous function on a local hull $X(\Lambda(\Omega))$.

Proposition 4.4 [24] *For each local function $\mathcal{N}$ with respect to $\Lambda(\Omega)$ there is a unique continuous function $\widetilde{\mathcal{N}}$ on a local hull $X(\Lambda(\Omega))$, whose restriction to the orbit of $\Lambda(\Omega)$ is $\mathcal{N}$. Every continuous function on the local hull of $\Lambda(\Omega)$ arises in this way.*

The spectral theory of $L^2(X(\Lambda(\Omega)), \mu)$ allows us to analyze $\mathcal{N}$ by analyzing its corresponding function $\widetilde{\mathcal{N}}$ on $L^2(X(\Lambda(\Omega)), \mu)$. Suppose $\mathcal{N}$ is a local function with respect to the model set $\Lambda(\Omega)$. From the locality of $\mathcal{N}$ we have its extension $\widetilde{\mathcal{N}} \in L^2(X(\Lambda(\Omega)), \mu)$ which is continuous. Then we obtain $\mathfrak{N} \in L^2(\mathbb{T}, \theta)$, where

$$\mathfrak{N}((z, 0)_L) = \mathfrak{N}(\beta(z + \Lambda(\Omega))) = \widetilde{\mathcal{N}}(z + \Lambda(\Omega)) = \mathcal{N}(z)\,,$$

and since functions in $L^2(\mathbb{T}, \theta)$ have Fourier expansions, we can write

$$\begin{aligned}\mathcal{N}(z) = \widetilde{\mathcal{N}}(z + \Lambda(\Omega)) = \mathfrak{N}((z, 0)_L) &= \sum_{\eta\in\Gamma^*} \widehat{\mathfrak{N}}(\eta)e^{-2\pi i(z,0)\cdot\eta} \\ &= \sum_{\eta\in\Gamma^*} \widehat{\mathfrak{N}}(\eta)e^{-2\pi iz\cdot p_1^*(\eta)}\,, \end{aligned} \tag{22}$$

almost everywhere, with

$$\widehat{\mathfrak{N}}(\eta) = \int_{\mathbb{T}} \mathfrak{N}((s, t)_L)e^{2\pi i(s,t)\cdot\eta}\, d\theta(s, t)\,.$$

However, we know $\mathfrak{N}$ only on $(\mathbb{R}^{2d}, 0)_L$. To compute the coefficients $\widehat{\mathfrak{N}}(\eta)$ out of $\mathcal{N}$ alone, we can use the Birkhoff ergodic theorem

$$\begin{aligned}\widehat{\mathfrak{N}}(\eta) &= \int_{\mathbb{T}} \mathfrak{N}((s, t)_L)e^{2\pi i(s,t)\cdot\eta}\, d\theta(s, t) = \lim_{R\to\infty} \frac{1}{R^{2d}} \int_{B(0,R)} \mathfrak{N}((z, 0)_L)e^{2\pi i(z,0)\cdot\eta}\, dz \\ &= \lim_{R\to\infty} \frac{1}{R^{2d}} \int_{B(0,R)} \mathcal{N}(z)e^{2\pi iz\cdot p_1^*(\eta)}\, dz\,,\end{aligned}$$

where we used $\mathfrak{N}((z, 0)_L) = \mathcal{N}(z)$ and $\eta = (p_1^*(\eta), p_2^*(\eta))$, so

$$(z, 0)\cdot \eta = z \cdot p_1^*(\eta) + 0 \cdot p_2^*(\eta)\,.$$

If $\sum_{\eta\in\Gamma^*} |\widehat{\mathfrak{N}}(\eta)| < \infty$, then the Fourier series (22) converges absolutely to $\mathcal{N}(z)$ for all $z \in \mathbb{R}^{2d}$.

5 Bracket Product on Model Sets

As described in the introduction, we are interested in the charaterization of tight and dual Gabor frames for simple model sets. We are going to imitate the approach presented in Sect. 3 for model sets, and like in the previous Section, the Poisson summation formula will play a crucial role.

We assume from now on that Ω is symmetric around the origin and that $p_2(\Gamma)$ and $p_2^*(\Gamma^*)$ have no common points with the boundary of Ω. Let $\Lambda(\Omega)$ be a simple model set and $\psi \in C_0^\infty(\Omega)$. Let $\widetilde{w}_\psi$ be a function defined as in Theorem 4.2. Then for a fixed $z = (x, \omega) \in \mathbb{R}^{2d}$ the generalized ψ*-bracket product* of f and g is defined as

$$\big[\widehat{\pi(z)f}, \widehat{g}\big]^{\psi}_{\Lambda(\Omega)}(\tilde{z}) := \sum_{\gamma^* \in \Gamma^*} \widetilde{w}_\psi\big(-p_2^*(\gamma^*)\big) M_{-z}\mathcal{W}(\widehat{f}, \widehat{g})\big(\tilde{z} - p_1^*(\gamma^*)\big). \tag{23}$$

For $f, g \in M^1(\mathbb{R}^d)$, we have $\mathcal{W}(\widehat{f}, \widehat{g}) \in M^1(\mathbb{R}^{2d})$ and the bracket product is well defined. Moreover, $\mathcal{F}^{-1}M_{-z}\mathcal{W}(\widehat{f}, \widehat{g}) = T_z\mathcal{A}(f, g)$ and is also an element of $M^1(\mathbb{R}^{2d})$, and by Remark 1, $\big[\widehat{\pi(z)f}, \widehat{g}\big]^{\psi}_{\Lambda(\Omega)}$ is an almost periodic function represented by the trigonometric series

$$\big[\widehat{\pi(z)f}, \widehat{g}\big]^{\psi}_{\Lambda(\Omega)}(\tilde{z}) = \sum_{\lambda \in \Lambda} w_\psi(\lambda)\mathcal{A}(f, g)(\lambda - z)e^{-2\pi i \lambda \cdot \tilde{z}}.$$

The Fourier coefficients are given by

$$\mathcal{M}_{\tilde{z}}\Big\{\big[\widehat{\pi(z)f}, \widehat{g}\big]^{\psi}_{\Lambda(\Omega)}(\tilde{z})e^{2\pi i \lambda \cdot \tilde{z}}\Big\} = w_\psi(\lambda)\mathcal{A}(f, g)(\lambda - z). \tag{24}$$

We make the following useful observation that is in analogy with regular shifts.

Lemma 5.1 *Let $\Lambda(\Omega)$ be a simple model set and $\psi \in C_0^\infty(\Omega)$. For all functions $f_1, f_2, g, h \in M^1(\mathbb{R}^d)$, we have*

$$\sum_{\lambda \in \Lambda(\Omega)} w_\psi(\lambda)^2\, \mathcal{A}(f_1, g)(\lambda - z)\overline{\mathcal{A}(f_2, h)(\lambda - z)}$$
$$= \mathcal{M}\Big\{\big[\widehat{\pi(z)f_1}, \widehat{g}\big]^{\psi}_{\Lambda(\Omega)} \cdot \overline{\big[\widehat{\pi(z)f_2}, \widehat{h}\big]^{\psi}_{\Lambda(\Omega)}}\Big\}.$$

Proof The bracket products $\big[\widehat{\pi(z)f_1}, \widehat{g}\big]^{\psi}_{\Lambda(\Omega)}$ and $\big[\widehat{\pi(z)f_2}, \widehat{h}\big]^{\psi}_{\Lambda(\Omega)}$ are almost periodic function with Fourier coefficients given by $w_\psi(\lambda)\mathcal{A}(f_1, g)(\lambda - z)$ and $w_\psi(\lambda)\mathcal{A}(f_2, h)(\lambda - z)$, respectively. Using Plancherel Theorem for Fourier series of almost periodic functions and (24), we obtain

$$\sum_{\lambda\in\Lambda(\Omega)} w_\psi(\lambda)^2\,\mathcal{A}(f_1,g)(\lambda-z)\overline{\mathcal{A}(f_2,h)(\lambda-z)}$$
$$=\sum_{\lambda\in\Lambda(\Omega)}\mathcal{M}_{\tilde z}\Big\{\big[\widehat{\pi(z)f_1},\widehat{g}\big]^\psi_{\Lambda(\Omega)}(\tilde z)e^{2\pi i\lambda\cdot\tilde z}\Big\}\overline{\mathcal{M}_{\tilde z}\Big\{\big[\widehat{\pi(z)f_2},\widehat{h}\big]^\psi_{\Lambda(\Omega)}(\tilde z)e^{2\pi i\lambda\cdot\tilde z}\Big\}}$$
$$=\mathcal{M}\Big\{\big[\widehat{\pi(z)f_1},\widehat{g}\big]^\psi_{\Lambda(\Omega)}\cdot\overline{\big[\widehat{\pi(z)f_2},\widehat{h}\big]^\psi_{\Lambda(\Omega)}}\Big\}.$$

□

The following result concerning the bracket product will be important in many calculations to follow.

Proposition 5.2 *Let $\Lambda(\Omega)$ be a simple model set and $\psi\in C_0^\infty(\mathbb{R})$ be non-negative. Assume that $g,h\in M^1(\mathbb{R}^d)$. Then, for $f_1,f_2\in M^1(\mathbb{R}^d)$,*

$$F(z,\tilde z)=\big[\widehat{\pi(z)f_1},\widehat{g}\big]^\psi_{\Lambda(\Omega)}(\tilde z)\cdot\overline{\big[\widehat{\pi(z)f_2},\widehat{h}\big]^\psi_{\Lambda(\Omega)}(\tilde z)},\quad (z,\tilde z)\in\mathbb{R}^{2d}\times\mathbb{R}^{2d}$$

is an almost periodic function.

Proof Let $f_1,f_2\in M^1(\mathbb{R}^d)$. Moreover, for $\eta\in\Gamma^*$, we define Ψ_η such that $\widehat{\Psi_\eta}=\widehat{\psi}\cdot T_{p_2^*(\eta)}\widehat{\psi}$. Then each Ψ_η belongs to $C_0^\infty(\mathbb{R})$ and is compactly supported on $\Omega+\Omega$, and we can define $\widetilde{w}_{\Psi_\eta}$ as $\widetilde{w}_{\Psi_\eta}(p_2^*(\gamma^*))=\mathrm{vol}(\Gamma)^{-1}\widehat{\Psi_\eta}(p_2^*(\gamma^*))$, for all $\gamma^*\in\Gamma^*$, as in (19). Then, by the change of index, we have

$$\begin{aligned}F(z,\tilde z)&=\sum_{\mu,\theta\in\Gamma^*}\widetilde{w}_\psi\Big(-p_2^*(\mu)\widetilde{w}_\psi\big(-p_2^*(\theta)\big)M_{-z}\mathcal{W}(\widehat{f_1},\widehat{g})(\tilde z-p_1^*(\mu)\Big)\\&\quad\times\overline{M_{-z}\mathcal{W}(\widehat{f_2},\widehat{h})(\tilde z-p_1^*(\theta))}\\&=\sum_{\mu,\theta\in\Gamma^*}\widetilde{w}_\psi\big(-p_2^*(\mu)\big)\widetilde{w}_\psi\big(-p_2^*(\theta)\big)\mathcal{W}(\widehat{f_1},\widehat{g})(\tilde z-p_1^*(\mu))\\&\quad\times\overline{\mathcal{W}(\widehat{f_2},\widehat{h})(\tilde z-p_1^*(\theta))}e^{-2\pi i\left(p_1^*(\theta)-p_1^*(\mu)\right)\cdot z}\\&=\sum_{\eta,\mu\in\Gamma^*}\widetilde{w}_\psi\big(-p_2^*(\mu)\big)\widetilde{w}_\psi\big(-p_2^*(\mu)-p_2^*(\eta)\big)\mathcal{W}(\widehat{f_1},\widehat{g})(\tilde z-p_1^*(\mu))\\&\quad\times\overline{T_{p_1^*(\eta)}\mathcal{W}(\widehat{f_2},\widehat{h})(\tilde z-p_1^*(\mu))}e^{-2\pi i p_1^*(\eta)\cdot z}\\&=\sum_{\eta,\mu\in\Gamma^*}\mathrm{vol}^{-2}(\Gamma)\widehat{\Psi_\eta}\big(-p_2^*(\mu)\big)\mathcal{W}(\widehat{f_1},\widehat{g})(\tilde z-p_1^*(\mu))\\&\quad\times\overline{T_{p_1^*(\eta)}\mathcal{W}(\widehat{f_2},\widehat{h})(\tilde z-p_1^*(\mu))}e^{-2\pi i p_1^*(\eta)\cdot z}\\&=\mathrm{vol}^{-1}(\Gamma)\sum_{\eta\in\Gamma^*}\\&\quad\times\left[\sum_{\mu\in\Gamma^*}\widetilde{w}_{\Psi_\eta}\big(-p_2^*(\mu)\big)\mathcal{W}(\widehat{f_1},\widehat{g})(\tilde z-p_1^*(\mu))\overline{T_{p_1^*(\eta)}\mathcal{W}(\widehat{f_2},\widehat{h})(\tilde z-p_1^*(\mu))}\right]e^{-2\pi i p_1^*(\eta)\cdot z}.\end{aligned}$$

For a fixed $\eta\in\Gamma^*$, consider the series

$$F_\eta(\tilde z)=\sum_{\mu\in\Gamma^*}\widetilde{w}_{\Psi_\eta}\big(-p_2^*(\mu)\big)\Big(\mathcal{W}(\widehat{f_1},\widehat{g})\cdot\overline{T_{p_1^*(\eta)}\mathcal{W}(\widehat{f_2},\widehat{h})}\Big)(\tilde z-p_1^*(\mu)).$$

Since $\mathcal{W}(\widehat{f_1}, \widehat{g})$ and $\mathcal{W}(\widehat{f_2}, \widehat{h})$ lie in $M^1(\mathbb{R}^{2d})$, their product as well. Moreover,

$$\mathcal{F}^{-1}\big(\mathcal{W}(\widehat{f_1}, \widehat{g})\overline{T_{p_1^*(\eta)}\mathcal{W}(\widehat{f_2}, \widehat{h})}\big) = \mathcal{A}(f_1, g) * (M_{p_1^*(\eta)}\mathcal{A}(f_2, h))^*$$

which is in $M^1(\mathbb{R}^{2d})$ since $\mathcal{A}(f_1, g), \mathcal{A}(f_2, h) \in M^1(\mathbb{R}^{2d})$. Therefore, by Poisson summation formula for model sets we can write $F_\eta(\tilde{z})$ as

$$\begin{aligned} F_\eta(\tilde{z}) &= \sum_{\gamma\in\Gamma} \Psi_\eta(p_2(\gamma))\big(\mathcal{A}(f_1, g) * (M_{p_1^*(\eta)}\mathcal{A}(f_2, h))^*\big)\big(p_1(\gamma)\big)\, e^{-2\pi i p_1(\gamma)\cdot\tilde{z}} \\ &= \sum_{\gamma\in\Gamma} \Psi_\eta(p_2(\gamma))\Big\langle \mathcal{A}(f_1, g), T_{p_1(\gamma)}M_{p_1^*(\eta)}\mathcal{A}(f_2, h)\Big\rangle_{L^2(\mathbb{R}^{2d})} e^{-2\pi i p_1(\gamma)\cdot\tilde{z}}, \end{aligned}$$

and $F(z, \tilde{z})$ becomes

$$\begin{aligned} &F(z, \tilde{z}) \\ &= \mathrm{vol}^{-1}(\Gamma) \sum_{\eta\in\Gamma^*}\sum_{\gamma\in\Gamma} \Psi_\eta(p_2(\gamma)) \\ &\quad\times \Big\langle \mathcal{A}(f_1, g), T_{p_1(\gamma)}M_{p_1^*(\eta)}\mathcal{A}(f_2, h)\Big\rangle_{L^2(\mathbb{R}^{2d})} e^{-2\pi i p_1(\gamma)\cdot\tilde{z}}\, e^{-2\pi i p_1^*(\eta)\cdot z} \\ &= \mathrm{vol}^{-1}(\Gamma) \sum_{\eta\in\Gamma^*}\sum_{\gamma\in\Gamma} \langle \psi, T_{p_2(\gamma)}M_{p_2^*(\eta)}\psi\rangle \\ &\quad\times \Big\langle \mathcal{A}(f_1, g), T_{p_1(\gamma)}M_{p_1^*(\eta)}\mathcal{A}(f_2, h)\Big\rangle_{L^2(\mathbb{R}^{2d})} e^{-2\pi i p_1(\gamma)\cdot\tilde{z}}\, e^{-2\pi i p_1^*(\eta)\cdot z} \\ &= \mathrm{vol}^{-1}(\Gamma) \sum_{\eta\in\Gamma^*}\sum_{\gamma\in\Gamma} \\ &\quad\times \Big\langle \psi\otimes\mathcal{A}(f_1, g), T_\gamma M_\eta\big(\psi\otimes\mathcal{A}(f_2, h)\big)\Big\rangle_{L^2(\mathbb{R}\times\mathbb{R}^{2d})} e^{-2\pi i p_1(\gamma)\cdot\tilde{z}}\, e^{-2\pi i p_1^*(\eta)\cdot z}. \end{aligned}$$

The coefficients in the series defining $F(z, \tilde{z})$ are in $\ell^1(\Gamma\times\Gamma^*)$, because $\psi\otimes\mathcal{A}(f_1, g)$ and $\psi\otimes\mathcal{A}(f_2, h)$ lie in $M^1(\mathbb{R}\times\mathbb{R}^{2d})$, and hence

$$\mathcal{V}_{(\psi\otimes\mathcal{A}(f_2,h))}\big(\psi\otimes\mathcal{A}(f_1, g)\big) \in \mathbf{W}(C_0, \ell^1)\big(\big(\mathbb{R}\times\mathbb{R}^{2d}\big)\times\big(\mathbb{R}\times\mathbb{R}^{2d}\big)\big).$$

That means that F equals a generalized trigonometric polynomial, and therefore is almost periodic. □

We need one more result that will be an important tool in the characterization of tight frames and dual frames.

Proposition 5.3 *Let $\Lambda(\Omega)$ be a simple model set and $\psi \in C_0^\infty(\Omega)$ non-negative function. Assume that $g_i, h_i \in M^1(\mathbb{R}^d)$ for every $i = 1, \ldots, M$. Then, for every $f_1, f_2 \in M^1(\mathbb{R}^d)$, the function*

$$\mathcal{N}^{\psi}(z) = \sum_{i=1}^{M} \sum_{\lambda \in \Lambda(\Omega)} w_{\psi}(\lambda)^2 \langle \pi(z) f_1, \pi(\lambda) g_i \rangle \langle \pi(\lambda) h_i, \pi(z) f_2 \rangle \tag{25}$$

is almost periodic and coincides pointwise with its Fourier series $\sum_{\eta \in \Gamma^*} \widehat{\mathcal{N}}^{\psi}(\eta) e^{-2\pi i p_1^*(\eta)\cdot z}$, *with*

$$\widehat{\mathcal{N}}^{\psi}(\eta) = vol(\Gamma)^{-1}\, \widehat{\psi^2}(-p_2^*(\eta)) \left\langle \pi\left(J p_1^*(\eta)\right) f_1, f_2 \right\rangle \sum_{i=1}^{M} \left\langle h_i, \pi\left(J p_1^*(\eta)\right) g_i \right\rangle,$$

where $\eta \in \Gamma^*$ *and* J *is the symplectic matrix.*

Proof Let $f_1, f_2 \in M^1(\mathbb{R}^d)$. Using the relationship between short time Fourier transform and cross-ambiguity function, we can express $\mathcal{N}^{\psi}$ as

$$\begin{aligned}
\mathcal{N}^{\psi}(z) &= \sum_{i=1}^{M} \sum_{\lambda \in \Lambda(\Omega)} w_{\psi}(\lambda)^2 \langle \pi(z) f_1, \pi(\lambda) g_i \rangle \langle \pi(\lambda) h_i, \pi(z) f_2 \rangle \\
&= \sum_{i=1}^{M} \sum_{\lambda \in \Lambda(\Omega)} w_{\psi}(\lambda)^2\, \mathcal{A}(f_1, g_i)(\lambda - z) \overline{\mathcal{A}(f_2, h_i)(\lambda - z)}\,.
\end{aligned}$$

For $i = 1, \ldots, M$ fixed, let

$$\mathcal{N}_i^{\psi}(z) = \sum_{\lambda \in \Lambda(\Omega)} w_{\psi}(\lambda)^2\, \mathcal{A}(f_1, g_i)(\lambda - z) \overline{\mathcal{A}(f_2, h_i)(\lambda - z)}\,.$$

By Lemma 5.1, we have

$$\mathcal{N}_i^{\psi}(z) = \mathcal{M}_{\tilde{z}} \left\{ \left[\widehat{\pi(z) f_1}, \widehat{g_i}\right]^{\psi}_{\Lambda(\Omega)}(\tilde{z}) \cdot \overline{\left[\widehat{\pi(z) f_2}, \widehat{h_i}\right]^{\psi}_{\Lambda(\Omega)}(\tilde{z})} \right\}.,$$

and by Proposition 5.2 and properties of almost periodic functions of two variables, $\mathcal{N}_i^{\psi}(z)$ is almost periodic. Moreover, using functions F_{η} and Ψ_{η} defined in the proof of Proposition 5.2, we have

$$\begin{aligned}
\mathcal{N}_i^{\psi}(z) &= \mathrm{vol}(\Gamma)^{-1} \sum_{\eta \in \Gamma^*} \mathcal{M}_{\tilde{z}} \left\{ F_{\eta}(\tilde{z}) \right\} e^{-2\pi i z \cdot p_1^*(\eta)} \\
&= \mathrm{vol}(\Gamma)^{-1} \sum_{\eta \in \Gamma^*} \Psi_{\eta}(0) \left\langle \mathcal{A}(f_1, g_i), M_{p_1^*(\eta)} \mathcal{A}(f_2, h_i) \right\rangle_{L^2(\mathbb{R}^{2d})} e^{-2\pi i z \cdot p_1^*(\eta)} \\
&= \mathrm{vol}(\Gamma)^{-1} \sum_{\eta \in \Gamma^*} \widehat{\psi^2}(-p_2^*(\eta)) \left\langle \pi\left(J p_1^*(\eta)\right) f_1, f_2 \right\rangle \left\langle h_i, \pi\left(J p_1^*(\eta)\right) g_i \right\rangle e^{-2\pi i z \cdot p_1^*(\eta)},
\end{aligned}$$

where the last equality follows from relations between cross-ambiguity function, cross-Wigner distribution and Moyal's formula, as in the proof of Proposition 3.4.

Now, since $\mathcal{N}$ is a finite sum of almost periodic functions, it is almost periodic and it equals a generalized trigonometric series

$$\mathcal{N}^{\psi}(z) = \mathrm{vol}(\Gamma)^{-1} \times \sum_{\eta \in \Gamma^*} \left(\widehat{\psi^2}(-p_2^*(\eta)) \left\langle \pi\left(Jp_1^*(\eta)\right) f_1, f_2 \right\rangle \sum_{i=1}^{M} \left\langle h_i, \pi\left(Jp_1^*(\eta)\right) g_i \right\rangle \right) e^{-2\pi i z \cdot p_1^*(\eta)}. \tag{26}$$

By the uniqueness of the Fourier series, (26) is the Fourier series of $\mathcal{N}^{\psi}(z)$. □

6 Gabor Analysis for Model Sets

We first begin with weighted Gabor frames and characterize normalized tight and dual weighted Gabor frames. The characterization follows directly from the bracket product defined in the previous section. Next, we move to the non-weighted scenario, where we develop Fundamental Identity of Gabor Analysis for model sets, Janssen representation and Wexler–Raz orthogonality relations.

6.1 Weighted Gabor Systems

Let $\Lambda(\Omega)$ be a simple model set, $\psi \in C_0^{\infty}(\Omega)$ non-negative function and the windows $g_1, \ldots, g_M \in M^1(\mathbb{R}^d)$. Then, a *weighted Gabor system*, denoted by $\mathcal{G}_{\psi}(g_1, \ldots, g_M; \Lambda(\Omega))$, is a collection of elements $w_{\psi}(\lambda)\pi(\lambda)g_i$, where $\lambda \in \Lambda(\Omega)$, $i = 1, \ldots, M$ and w_{ψ} as defined in (17). The system $\mathcal{G}_{\psi}(g_1, \ldots, g_M; \Lambda(\Omega))$ is a frame for $L^2(\mathbb{R}^d)$ if there exist constants $A_g, B_g > 0$ such that

$$A_g \|f\|_2^2 \leq \sum_{i=1}^{M} \sum_{\lambda \in \Lambda(\Omega)} w_{\psi}(\lambda)^2 \, |\langle f, \pi(\lambda) g_i \rangle|^2 \leq B_g \|f\|_2^2 \tag{27}$$

holds for all $f \in L^2(\mathbb{R}^d)$. $\mathcal{G}_{\psi}(g_1, \ldots, g_M; \Lambda(\Omega))$ is a *normalized weighted tight Gabor frame* if $A_g = B_g = 1$. As in the case of non-weighted Gabor systems, $\mathcal{G}_{\psi}(g_1, \ldots, g_M; \Lambda(\Omega))$ is a Bessel sequence (only the right hand side of (27) holds) if all $g_i \in M^1(\mathbb{R}^d)$, $i = 1, \ldots, M$. Equipped with this notions, we can now characterize weighted tight Gabor frames.

Proposition 6.1 *Let $\Lambda(\Omega)$ be a simple model set and $\psi \in C_0^{\infty}(\Omega)$ non-negative function. Then the family $\mathcal{G}_{\psi}(g_1, \ldots, g_M; \Lambda(\Omega))$, with $g_i \in M^1(\mathbb{R}^d)$ for every $i = 1, \ldots, M$, is a normalized weighted tight Gabor frame for $L^2(\mathbb{R}^m)$, that is*

$$\sum_{i=1}^{M} \sum_{\lambda \in \Lambda(\Omega)} w_{\psi}(\lambda)^2 \, |\langle f, \pi(\lambda) g_i \rangle|^2 = \|f\|_2^2 \quad \textit{for all } f \in L^2(\mathbb{R}^m) \tag{28}$$

if and only if

$$vol(\Gamma)^{-1}\,\widehat{\psi^2}\big(p_2^*(\eta)\big)\sum_{i=1}^{M}\Big\langle g_i,\pi\big(Jp_1^*(\eta)\big)g_i\Big\rangle=\delta_{\eta,0}\,, \tag{29}$$

for each $\eta\in\Gamma^*$, where δ is the Kronecker delta.

Proof By [15], it is sufficient to prove the theorem when $f\in M^1(\mathbb{R}^d)$. Assume that $\mathcal{G}_\psi(g_1,\ldots,g_M;\Lambda(\Omega))$ is a normalized weighted tight Gabor frame. Since $g_i\in M^1(\mathbb{R}^d)$ for every $i=1,\ldots,M$, by Proposition 5.3 with $g_i=h_i$ for all $i=1,\ldots,M$ and $f_1=f_2=f$, we can define a function $\mathcal{O}^\psi(z)$ as

$$\mathcal{O}^\psi(z)=\sum_{i=1}^{M}\sum_{\lambda\in\Lambda(\Omega)}w_\psi(\lambda)^2\,|\langle\pi(z)f,\pi(\lambda)g_i\rangle|^2\,. \tag{30}$$

By (28), this function is constant and equals $\|f\|_2^2$. Let $\mathcal{E}^\psi(z)=\mathcal{O}^\psi(z)-\|f\|_2^2$. Then, $\mathcal{E}^\psi(z)$ is almost periodic and $\mathcal{E}^\psi=0$. By the property of almost periodic function, it implies that the Fourier coefficients of $\mathcal{E}^\psi(z)$,

$$\widehat{\mathcal{E}^\psi}(\eta)=\begin{cases}\widehat{\mathcal{O}}^\psi(\eta)-\|f\|_2^2, & \eta=0\\ \widehat{\mathcal{O}}^\psi(\eta), & \eta\neq 0\end{cases}$$

are zero. By Proposition 5.3, we have then

$$\mathrm{vol}(\Gamma)^{-1}\,\widehat{\psi^2}\big(p_2^*(\eta)\big)\Big\langle\pi\big(Jp_1^*(\eta)\big)f,f\Big\rangle\sum_{i=1}^{M}\Big\langle g_i,\pi\big(Jp_1^*(\eta)\big)g_i\Big\rangle=\delta_{\eta,0}\|f\|_2^2\,, \tag{31}$$

with $\eta\in\Gamma^*$, for every $f\in M^1(\mathbb{R}^d)$.

Let $f\in M^1(\mathbb{R}^d)\setminus\{0\}$. Consider first $\eta=0$. Then (31) becomes

$$\mathrm{vol}(\Gamma)^{-1}\,\widehat{\psi^2}(0)\,\langle f,f\rangle\sum_{i=1}^{M}\langle g_i,g_i\rangle=\delta_{\eta,0}\|f\|_2^2\,,$$

and (29) follows. Now, let $\eta\neq 0$ be fixed and take $f(x)=e^{-\pi x^2}$, a Gausian. Then (31) implies

$$\mathrm{vol}(\Gamma)^{-1}\,\widehat{\psi^2}\big(p_2^*(\eta)\big)\Big\langle\pi\big(Jp_1^*(\eta)\big)f,f\Big\rangle\sum_{i=1}^{M}\Big\langle g_i,\pi\big(Jp_1^*(\eta)\big)g_i\Big\rangle=0,$$

and since $\Big\langle\pi\big(Jp_1^*(\eta)\big)f,f\Big\rangle\neq 0$, (29) is satisfied.

Conversely, assume that (29) holds. Since $g_i\in M^1(\mathbb{R}^d)$ for every $i=1,\ldots,M$, by Proposition 5.3 with $f_1=f_2=f$ and $g_i=h_i$ for every $i=1,\ldots,M$, we can define function $\mathcal{O}^{\psi}(z)$ as in (30). Then, by Proposition 5.3 and relation (29), the function $\mathcal{O}^{\psi}(z)$ is given by the trigonometric series

$$\sum_{\eta\in\Gamma^*} \widehat{\mathcal{O}}^{\psi}(\eta)e^{-2\pi i z\cdot p_1^*(\eta)} = \mathcal{O}^{\psi}(z),$$

for every $z \in \mathbb{R}^d$, with

$$\widehat{\mathcal{O}}^{\psi}(\eta) = \delta_{\eta,0}\Big\langle \pi\left(Jp_1^*(\eta)\right)f, f\Big\rangle .$$

Hence, $\mathcal{O}^{\psi}(z)$ is constant and $\mathcal{O}^{\psi}(z) = \|f\|_2^2$. Evaluating $\mathcal{O}^{\psi}(z)$ at $z = 0$, gives the claim. □

We know state conditions for a weighted Gabor system $\mathcal{G}_{\psi}(h_1, \ldots, h_M; \Lambda(\Omega))$ to be a dual system of $\mathcal{G}_{\psi}(g_1, \ldots, g_M; \Lambda(\Omega))$.

Proposition 6.2 *Let $\Lambda(\Omega)$ be a simple model set and $\psi \in C_0^{\infty}(\Omega)$ non-negative function. Let $g_i \in M^1(\mathbb{R}^d)$ and $h_i \in M^1(\mathbb{R}^d)$, for every $i = 1, \ldots, M$. Then $\mathcal{G}_{\psi}(g_1, \ldots, g_M; \Lambda(\Omega))$ and $\mathcal{G}_{\psi}(h_1, \ldots, h_M; \Lambda(\Omega))$ are weighted dual Gabor frames, that is*

$$\sum_{i=1}^{M} \sum_{\lambda\in\Lambda(\Omega)} w_{\psi}(\lambda)^2 \langle f, \pi(\lambda)g_i\rangle\langle \pi(\lambda)h_i, f\rangle = \|f\|_2^2 \quad \textit{for all } f \in L^2(\mathbb{R}^m), \tag{32}$$

if and only if

$$vol(\Gamma)^{-1}\, \widehat{\psi^2}(p_2^*(\eta)) \sum_{i=1}^{M} \Big\langle h_i, \pi\left(Jp_1^*(\eta)\right)g_i\Big\rangle = \delta_{\eta,0}, \tag{33}$$

for each $\eta \in \Gamma^*$, where δ is the Kronecker delta.

Proof The proof is analogous to the proof of Proposition 6.1 with obvious adjustments. □

Based on the last proposition we can also derive density condition for weighted Gabor frames.

Proposition 6.3 *Let $\Lambda(\Omega)$ be a simple model set and $\psi \in C_0^{\infty}(\Omega)$ a non-negative function such that $\|\psi\|_2 = 1$. If the Gabor frame $\mathcal{G}_{\psi}(g; \Lambda(\Omega))$, with $g \in M^1(\Lambda(\Omega))$ admits a weighted dual that is also a Gabor system, then $D(\Lambda(\Omega)) \geq 1$.*

Proof Let $g \in M^1(\mathbb{R}^d)$ and assume that the Gabor frame $\mathcal{G}_{\psi}(g; \Lambda(\Omega))$ admits a dual $\mathcal{G}_{\psi}(h; \Lambda(\Omega))$ with $h \in M^1(\mathbb{R}^d)$. Let B_g be the upper frame bound of $\mathcal{G}_{\psi}(g; \Lambda(\Omega))$, and we can assume without loss of generality that $\|h\|_2^2 = |\Omega| B_g^{-1}$. Then, by polarization, we have the frame decomposition

$$\langle f_1, f_2 \rangle = \sum_{\lambda \in \Lambda(\Omega)} \langle f_1, \pi(\lambda) g \rangle w_\psi(\lambda)^2 \langle \pi(\lambda) h, f_2 \rangle \quad \text{for all } f_1, f_2 \in L^2(\mathbb{R}^m),$$

If we set $f_1 = h$ and $f_2 = g$, by the Bessel property of $\mathcal{G}_\psi(g; \Lambda(\Omega))$, we obtain

$$\langle h, g \rangle = \sum_{\lambda \in \Lambda(\Omega)} w_\psi(\lambda)^2 |\langle h, \pi(\lambda) g \rangle|^2 \le B_g \|h\|_2^2 = |\Omega| .$$

On the other hand, by Proposition 6.2 with $M = 1$, $\mathrm{vol}(\Gamma)^{-1} \widehat{\psi^2}(0) \langle h, g \rangle = 1$, and therefore $D(\Lambda(\Omega)) = \mathrm{vol}(\Gamma)^{-1} |\Omega| \ge 1$ because $\widehat{\psi^2}(0) = \|\psi\|_2^2 = 1$. □

Gabardo, in [9], gave a characterization of the weighted irregular Gabor tight frames and dual systems in terms of the distributional symplectic Fourier transform of a positive Borel measure where the windows belong to the Schwartz class. It is possible to derive his results in the setting of model sets, using the characterization just presented.

6.2 (Nonweighted) Gabor Systems

Let $g_i, h_i \in M^1(\mathbb{R}^d)$, $i = 1, \ldots, M$, and $\Lambda(\Omega)$ a simple model set. At the beginning of Sect. 2, we showed that the frame operator $S_g^{\Lambda(\Omega)}$ of $\mathcal{G}(g_1, \ldots, g_M; \Lambda)$ does not commute with the time-frequency shifts taken from $\Lambda(\Omega)$. The same holds in particular for any $\Lambda \in X(\Lambda(\Omega))$ and a time-frequency shift by $z \in \mathbb{R}^{2d}$. Let $S_{g,h}^{\Lambda - z}$ denote the mixed frame operator associated to $\mathcal{G}(g_1, \ldots, g_M; \Lambda - z)$ and $\mathcal{G}(h_1, \ldots, h_M; \Lambda - z)$. Then there is a covariance relation relating $S_{g,h}^{\Lambda}$ and $S_{g,h}^{\Lambda - z}$. The following result was obtained by Kreisel in [18]. We state it here for the mixed frame operators.

Proposition 6.4 [18] *If $\mathcal{G}(g_1, \ldots, g_M; \Lambda)$ and $\mathcal{G}(h_1, \ldots, h_M; \Lambda)$,are Gabor systems for Λ, and, $\mathcal{G}(g_1, \ldots, g_M; \Lambda - z)$ and $\mathcal{G}(h_1, \ldots, h_M; \Lambda - z)$ are Gabor systems for $\Lambda - z$, then the mixed frame operators $S_{g,h}^{\Lambda}$ and $S_{g,h}^{\Lambda - z}$ satisfy*

$$S_{g,h}^{\Lambda} \pi(z) = \pi(z) S_{g,h}^{\Lambda - z} .$$

Moreover, the following continuity property holds.

Proposition 6.5 [18] *Suppose $\Lambda_n \to \Lambda$ in $X(\Lambda(\Omega))$ and the window functions g_i, h_i lie in $M^1(\mathbb{R}^d)$, for each $i = 1, \ldots, M$. Then $S_{g,h}^{\Lambda_n} \to S_{g,h}^{\Lambda}$ in the strong operator topology on $B(M^1(\mathbb{R}^d))$.*

Even though the mapping $\Lambda \to S_g^{\Lambda}$, $\Lambda \in X(\Lambda(\Omega))$ is not continuous when $B(M^1(\mathbb{R}^d))$ is given the norm topology, all the frames $\mathcal{G}(g_1, \ldots, g_M; \Lambda)$ have the same optimal frame bounds.

Proposition 6.6 [18] *Suppose $\mathcal{G}(g_1, \ldots, g_M; \Lambda)$ is a frame for each $\Lambda \in X(\Lambda(\Omega))$ and each $g_i \in M^1(\mathbb{R}^d)$. For any $\Lambda \in X(\Lambda(\Omega))$ the optimal upper and lower frame bounds for $\mathcal{G}(g_1, \ldots, g_M; \Lambda)$ are the same as those for $\mathcal{G}(g_1, \ldots, g_M; \Lambda(\Omega))$. As a result, $\|S_g^{\Lambda}\|_{M^1} = \|S_g^{\Lambda(\Omega)}\|_{M^1}$ and $\|(S_g^{\Lambda})^{-1}\|_{M^1} = \|(S_g^{\Lambda(\Omega)})^{-1}\|_{M^1}$, where $\|\cdot\|_{M^1}$ denotes the operator norm on $B(M^1(\mathbb{R}^d))$.*

As a result of the continuity property, we have the following Corollary.

Corollary 6.7 [18] *Suppose $g_1, \ldots, g_M \in M^1(\mathbb{R}^d)$ and $\mathcal{G}(g_1, \ldots, g_M; \Lambda(\Omega))$ is an M^1-frame. Then for any $\Lambda \in X(\Lambda(\Omega))$, $\mathcal{G}(g_1, \ldots, g_M; \Lambda)$ is also an M^1-frame.*

Now, let $f_1, f_2 \in M^1(\mathbb{R}^d)$ be fixed and let $g_i, h_i \in M^1(\mathbb{R}^d)$ for $i = 1, \ldots, M$. We define a function $\widetilde{\mathcal{N}} : X(\Lambda(\Omega)) \to \mathbb{C}$ through the mixed frame operator, as

$$\widetilde{\mathcal{N}}(\Lambda) = \langle S_{g,h}^{\Lambda} f_1, f_2 \rangle.$$

Since, by Proposition 6.5, $S_{g,h}^{\Lambda}$ is continuous, in the strong operator topology, over $X(\Lambda(\Omega))$, the function $\widetilde{\mathcal{N}}$ is continuous. As was presented in Sect. 4, we can define from $\widetilde{\mathcal{N}}$ a function $\mathcal{N} : \mathbb{R}^m \to \mathbb{C}$, by

$$\mathcal{N}(z) = \widetilde{\mathcal{N}}(\Lambda(\Omega) - z)\,, \tag{34}$$

and since $\widetilde{\mathcal{N}}$ is continuous, $\mathcal{N}$ is local with respect to $\Lambda(\Omega)$. As was shown in Sect. 4, it has a Fourier expansion

$$\mathcal{N}(z) = \sum_{\eta \in \Gamma^*} \widehat{\mathfrak{N}}(\eta) e^{-2\pi i p_1^*(\eta)\cdot z}$$

where

$$\widehat{\mathfrak{N}}(\eta) = \lim_{R\to\infty} \frac{1}{R^{2d}} \int_{B(0,R)} \mathcal{N}(z) e^{2\pi i z \cdot p_1^*(\eta)}\, dz\,. \tag{35}$$

Applying the tools developed in Sect. 5, we will be able to compute the Fourier coefficients $\widehat{\mathfrak{N}}(\eta)$ of $\mathcal{N}$.

Before we proceed further we introduce a sequence of auxiliary functions that will be crucial in proving our results. The following Lemma is a particular case of Prop 3.6 in [1].

Lemma 6.8 *Let $0 < \epsilon < 1$, Ω be a compact subset of $\mathbb{R}$, $\widetilde{\Omega} = (1-\epsilon)\Omega$ and $f_s = \frac{\mathbb{1}_{\epsilon^s \widetilde{\Omega}}}{|\epsilon^s \widetilde{\Omega}|}$ for $s \in \mathbb{N}$. Then,*

(i) *the infinite convolution product $\mathop{\ast}_{s=0}^{\infty} f_s$ converges in $L^1(\mathbb{R})$ and defines a non-negative smooth function $\psi = \mathop{\ast}_{s=0}^{\infty} f_s$ compactly supported on Ω, with $\|\psi\|_1 = 1$;*
(ii) *the Fourier transform of ψ is the smooth function $\widehat{\psi} = \prod_{s=0}^{\infty} \widehat{f_s}$, with uniform convergence of the product, where $\widehat{f_s}(t) = sinc(t\,|\epsilon^s \widetilde{\Omega}|)$ and $sinc(t) = \frac{\sin(\pi t)}{\pi t}$.*

Now, for $0 < \epsilon < 1$ and Ω a compact subset of $\mathbb{R}$, we define a sequence of compact sets $\Omega_n = (1-\epsilon^n)\Omega$. The sets Ω_n are increasing and $\overline{\bigcup_n \Omega_n} = \Omega$. Let ψ_n be an infinite convolution product

$$\psi_n = \mathop{\huge *}_{s=0}^{\infty} \frac{\mathbb{1}_{\epsilon^{ns}\Omega_n}}{|\epsilon^{ns}\Omega_n|} = \frac{\mathbb{1}_{\Omega_n}}{|\Omega_n|} * \left(\mathop{\huge *}_{s=1}^{\infty} \frac{\mathbb{1}_{\epsilon^{ns}\Omega_n}}{|\epsilon^{ns}\Omega_n|} \right). \tag{36}$$

Then, by Lemma 6.8, each ψ_n is well defined and forms a sequence of $C_0^\infty(\Omega)$ non-negative functions, with Fourier transform of ψ_n being

$$\widehat{\psi_n}(t) = \prod_{s=0}^{\infty} \operatorname{sinc}\big(t\,|\epsilon^{ns}\Omega_n|\big) = \operatorname{sinc}(t\,|\Omega_n|) \cdot \prod_{s=1}^{\infty} \operatorname{sinc}\big(t\,|\epsilon^{ns}\Omega_n|\big). \tag{37}$$

Lemma 6.9 *With the above notation,*

(i) *the sequence $\{\psi_n\}_{n=1}^\infty$ converges pointwise to $\frac{\mathbb{1}_\Omega}{|\Omega|}$ on $\Omega \setminus \partial\Omega$, where $\partial\Omega$ is the boundary of Ω;*

(ii) *the sequence of Fourier transforms, $\{\widehat{\psi_n}\}_{n=1}^\infty$, converges uniformly.*

Proof Let $t_0 \in \Omega \setminus \partial\Omega$ and $\delta > 0$. By the properties of the sets Ω_n we have: $\Omega_m \subset \Omega_n \subset \Omega$ and $\epsilon^{ns}\Omega_n \subset \epsilon^{ms}\Omega_m$ for $n \geq m$ and $\epsilon^{ns}\Omega_n \subset \epsilon^{n(s+1)}\Omega_n$ for all n and s. Then, there exists $N \geq 0$, such that for all $n \geq N$, $|\Omega \setminus \Omega_n| < \delta|\Omega|^2$ and $t_0 \in \Omega_n$. That means, for $n \geq N$

$$\left[\left(\sum_{s=1}^{\infty} \operatorname{supp} \frac{\mathbb{1}_{\epsilon^{ns}\Omega_n}}{|\epsilon^{ns}\Omega_n|}\right) - t_0\right] \cap \Omega_n = \left[\left(\sum_{s=1}^{\infty} \epsilon^{ns}\Omega_n - t_0\right)\right] \cap \Omega_n \neq \emptyset\,.$$

Then, for all $n \geq N$,

$$\begin{aligned}
\left|\psi_n(t_0) - \frac{\mathbb{1}_\Omega(t_0)}{|\Omega|}\right| &= \left|\left[\frac{\mathbb{1}_{\Omega_n}}{|\Omega_n|} * \left(\mathop{\huge *}_{s=1}^{\infty} \frac{\mathbb{1}_{\epsilon^{ns}\Omega_n}}{|\epsilon^{ns}\Omega_n|}\right)\right](t_0) - \frac{1}{|\Omega|}\right| \\
&= \left|\frac{1}{|\Omega_n|}\int_{\Omega_n} \left(\mathop{\huge *}_{s=1}^{\infty} \frac{\mathbb{1}_{\epsilon^{ns}\Omega_n}}{|\epsilon^{ns}\Omega_n|}\right)(t_0 - x)\,dx - \frac{1}{|\Omega|}\right| \\
&\leq \left|\frac{1}{|\Omega_n|}\left\|\mathop{\huge *}_{s=1}^{\infty} \frac{\mathbb{1}_{\epsilon^{ns}\Omega_n}}{|\epsilon^{ns}\Omega_n|}\right\|_1 - \frac{1}{|\Omega|}\right| \\
&= \left|\frac{1}{|\Omega_n|} - \frac{1}{|\Omega|}\right| = \frac{|\Omega \setminus \Omega_n|}{|\Omega_n||\Omega|} \\
&< \frac{\delta}{(1-\epsilon^N)}\,,
\end{aligned}$$

and claim (i) follows.

For (ii) it suffices to show that the sequence of Fourier transforms $\widehat{\psi_n}$ is uniformly Cauchy. By (i) and Lemma 6.8, $\{\psi_n\}_{n=1}^\infty$ is a sequence of L^1 functions converging pointwise almost everywhere to $\frac{\mathbb{1}_\Omega}{|\Omega|}$. Then by the L^1 Dominated Convergence Theorem, $\{\psi_n\}_{n=1}^{\infty}$ converges to $\frac{\mathbb{1}_\Omega}{|\Omega|}$ in L^1, and it follows that $\{\psi_n\}_{n=1}^{\infty}$ is an L^1 Cauchy sequence. Meaning, for $\delta > 0$ there exists $N > 0$ such that $\|\psi_n - \psi_m\|_1 < \delta$ for all $n, m \geq N$. Let $n, m \geq N$, then

$$\|\widehat{\psi_m} - \widehat{\psi_n}\|_\infty \leq \|\psi_m - \psi_n\|_1 < \delta\,,$$

and $\{\widehat{\psi_n}\}_{n=1}^{\infty}$ is uniformly Cauchy. By the completeness of $L^\infty(\mathbb{R})$, it converges uniformly. □

Let, Ψ denote the uniform limit of the sequence defined in (37). To make things more convenient later, we normalize Ψ, and define a new function

$$\Phi = |\Omega| \cdot (\Psi * \Psi)\,. \tag{38}$$

Note that $\Phi(0) = 1$.

The following observation will be the main ingredient in our approach. It is analogous to the results for lattices developed in Sect. 3. With the above notation we have

Proposition 6.10 *Let $\Lambda(\Omega)$ be a simple generic model set. Assume that $g_i, h_i \in M^1(\mathbb{R}^d)$ for every $i = 1, \ldots, M$. Then, for every $f_1, f_2 \in M^1(\mathbb{R}^d)$, the function*

$$\mathcal{N}(z) = \sum_{i=1}^{M} \sum_{\lambda \in \Lambda(\Omega)} \Big\langle \pi(z) f_1, \pi(\lambda) g_i \Big\rangle \Big\langle \pi(\lambda) h_i, \pi(z) f_2 \Big\rangle$$

is continuous and coincides pointwise with its Fourier series $\sum_{\eta\in\Gamma^*} \widehat{\mathcal{N}}(\eta) e^{-2\pi i p_1^*(\eta)\cdot z}$, with

$$\widehat{\mathcal{N}}(\eta) = D(\Lambda(\Omega))\, \Phi\big(p_2^*(\eta)\big) \Big\langle \pi\big(J p_1^*(\eta)\big) f_1, f_2 \Big\rangle \sum_{i=1}^{M} \Big\langle h_i, \pi\big(J p_1^*(\eta)\big) g_i \Big\rangle.$$

where $\eta \in \Gamma^*$, Φ defined in (38).

We can approximate function $\mathcal{N}$ with the desired accuracy by an almost periodic function whose spectrum lies in a 'dual' model set. Let $\epsilon > 0$ and $C = \max \|\mathcal{V}_{g_i} h_i\|_{\mathbf{W}(L^\infty, \ell^2)}$. The function Φ decays rapidly outside its essential support, hence, for the essential support, we can choose a compact interval $\widetilde{\Omega}_\epsilon$, depending on the windows g_i, h_i, such that

$$\Big\| \Phi \mathbb{1}_{\widetilde{\Omega}_\epsilon{}^c} \Big\|_{\mathbf{W}(\mathbb{R})} < \frac{\epsilon}{D(\Lambda(\Omega))\, \mathrm{rel}(\Gamma^*)\, C\, M} \tag{39}$$

where $\big|\Phi\big|^*(x) = \big|\Phi\big|(-x)$ is the involution and $\widetilde{\Omega}_\epsilon^c$ a complement of $\widetilde{\Omega}_\epsilon$. Let us define an $\epsilon-$dual model set $\Lambda^*(\widetilde{\Omega}_\epsilon)$ originating from Γ^* and $\widetilde{\Omega}_\epsilon$ by

$$\Lambda^*(\widetilde{\Omega}_\epsilon) = \big\{\beta = p_1^*(\eta) \,:\, \eta \in \Gamma^*,\ p_2^*(\eta) \in \widetilde{\Omega}_\epsilon\big\} \tag{40}$$

We note here, that the concept of an ϵ-dual model set defined here differs from the original ϵ-dual model set definition by Meyer. Then, the function $\mathcal{N}_\epsilon$, given by the series

$$\mathcal{N}_\epsilon(z) := \sum_{\beta\in\Lambda^*(\widetilde{\Omega}_\epsilon)} \widehat{\mathcal{N}}(\beta)e^{-2\pi i\beta\cdot z},$$

defines an almost periodic function with spectrum in $\Lambda^*(\widetilde{\Omega}_\epsilon)$. Moreover, by Cauchy-Schwarz inequality, we have

$$\begin{aligned}
\|\mathcal{N}-\mathcal{N}_\epsilon\|_\infty &\le D(\Lambda(\Omega))\sum_{i=1}^{M}\sum_{\eta\in\Gamma^*;\ p_2^*(\eta)\notin\widetilde{\Omega}_\epsilon}\left|\Phi(-p_2^*(\eta))\left\langle\pi\left(Jp_1^*(\eta)\right)f_1, f_2\right\rangle\left\langle h_i,\pi\left(Jp_1^*(\eta)\right)g_i\right\rangle\right| \\
&\le D(\Lambda(\Omega))\sum_{i=1}^{M}\left[\sum_{\eta\in\Gamma^*}\left(\left(|\Phi|^*\mathbb{1}_{\widetilde{\Omega}_\epsilon^c}\right)\otimes\mathcal{U}|\mathcal{V}_{g_i}h_i|^2\right)(\eta)\right]^{1/2} \\
&\quad\times\left[\sum_{\eta\in\Gamma^*}\left(\left(|\Phi|^*\mathbb{1}_{\widetilde{\Omega}_\epsilon^c}\right)\otimes\mathcal{U}|\mathcal{V}_{f_1}f_2|^2\right)(\eta)\right]^{1/2} \\
&\le D(\Lambda(\Omega))\,\mathrm{rel}(\Gamma^*)\sum_{i=1}^{M}\left\|\left(|\Phi|^*\mathbb{1}_{\widetilde{\Omega}_\epsilon^c}\right)\otimes\mathcal{U}|\mathcal{V}_{g_i}h_i|^2\right\|_{\mathbf{W}(\mathbb{R}\times\mathbb{R}^{2d})}^{1/2} \\
&\quad\left\|\left(|\Phi|^*\mathbb{1}_{\widetilde{\Omega}_\epsilon^c}\right)\otimes\mathcal{U}|\mathcal{V}_{f_1}f_2|^2\right\|_{\mathbf{W}(\mathbb{R}\times\mathbb{R}^{2d})}^{1/2}.
\end{aligned}$$

By the property of tensor product and (39), we obtain

$$\begin{aligned}
\|\mathcal{N}-\mathcal{N}_\epsilon\|_\infty &\le D(\Lambda(\Omega))\,\mathrm{rel}(\Gamma^*)\left\|\Phi\mathbb{1}_{\widetilde{\Omega}_\epsilon^c}\right\|_{\mathbf{W}(\mathbb{R})}\sum_{i=1}^{M}\left\||V_{g_i}h_i|^2\right\|_{\mathbf{W}(\mathbb{R}^{2d})}\left\||V_{f_1}f_2|^2\right\|_{\mathbf{W}(\mathbb{R}^{2d})} \\
&\le D(\Lambda(\Omega))\,\mathrm{rel}(\Gamma^*)\left\|\Phi\mathbb{1}_{\widetilde{\Omega}_\epsilon^c}\right\|_{\mathbf{W}(\mathbb{R})}\sum_{i=1}^{M}\left\|V_{g_i}h_i\right\|_{\mathbf{W}(L^\infty,\ell^2)(\mathbb{R}^{2d})}\left\|V_{f_1}f_2\right\|_{\mathbf{W}(L^\infty,\ell^2)(\mathbb{R}^{2d})} \\
&< \epsilon\left\|V_{f_1}f_2\right\|_{\mathbf{W}(L^\infty,\ell^2)(\mathbb{R}^{2d})}.
\end{aligned}$$

Proof of Proposition 6.10 Let $f_1, f_2 \in M^1(\mathbb{R}^d)$. Let ψ_n be a sequence of $C_0^\infty(\Omega)$ non-negative functions defined in (36). By Proposition 5.3, for each $n\in\mathbb{N}$, the functions

$$\mathcal{N}^{\psi_n}(z) = \sum_{i=1}^{M}\sum_{\lambda\in\Lambda(\Omega)} w_{\psi_n}(\lambda)^2\left\langle\pi(z)f_1,\pi(\lambda)g_i\right\rangle\left\langle\pi(\lambda)h_i,\pi(z)f_2\right\rangle$$

are well defined almost periodic functions that are pointwise equal to their Fourier series $\sum_{\eta\in\Gamma^*}\widehat{\mathcal{N}}^{\psi_n}(\eta)e^{-2\pi i p_1^*(\eta)\cdot z}$, where $\widehat{\mathcal{N}}^{\psi_n}(\eta)$ are given by

$$\widehat{\mathcal{N}}^{\psi_n}(\eta) = \mathrm{vol}(\Gamma)^{-1}\,\widehat{\psi_n^2}\left(p_2^*(\eta)\right)\left\langle\pi\left(Jp_1^*(\eta)\right)f_1, f_2\right\rangle\sum_{i=1}^{M}\left\langle h_i,\pi\left(Jp_1^*(\eta)\right)g_i\right\rangle.$$

On the other hand, the series $\sum_{\eta\in\Gamma^*}\widehat{\mathcal{N}}(\eta)e^{-2\pi i p_1^*(\eta)\cdot z}$ converges absolutely (by a similar argument as in the proof of Proposition 5.3) and gives rise to a uniformly continuous function. By the uniform convergence of $\widehat{\psi_n^2}$ to $\Psi * \Psi$, it can be easily verified that $\mathcal{N}^{\psi_n}$ converges uniformly to $|\Omega|^{-2}\sum_{\eta\in\Gamma^*}\widehat{\mathcal{N}}(\eta)e^{-2\pi i p_1^*(\eta)\cdot z}$, since $D(\Lambda(\Omega)) = \mathrm{vol}(\Gamma)^{-1}|\Omega|$ and $\Phi = |\Omega|\cdot(\Psi * \Psi)$.

Now, since $\Lambda(\Omega)$ is genereic, that is the boundary $\partial\Omega$ of Ω has no common points with $p_2(\Gamma)$, and ψ_n converges pointwise to $\frac{\mathbb{1}_\Omega}{|\Omega|}$ on $\Omega\setminus\partial\Omega$, by Lemma 6.8, we show that $\mathcal{N}(z)$ is a pointwise limit of $|\Omega|^2\mathcal{N}^{\psi_n}(z)$. Indeed, by the Lebesgue Dominated Convergence Theorem, we can move the limit inside the sum, and for every $z\in\mathbb{R}^{2d}$, we have

$$\begin{aligned}
\lim_{n\to\infty}\mathcal{N}^{\psi_n}(z) &= \sum_{i=1}^{M}\sum_{\lambda\in\Lambda(\Omega)}\lim_{n\to\infty} w_{\psi_n}^2(\lambda)\left\langle\pi(z)f_1,\pi(\lambda)g_i\right\rangle\left\langle\pi(\lambda)h_i,\pi(z)f_2\right\rangle\\
&= \sum_{i=1}^{M}\sum_{\lambda\in\Lambda(\Omega)}\lim_{n\to\infty}\psi_n^2(p_2(\gamma))\left\langle\pi(z)f_1,\pi(\lambda)g_i\right\rangle\left\langle\pi(\lambda)h_i,\pi(z)f_2\right\rangle\\
&\quad(\lambda = p_1(\gamma),\ \gamma\in\Gamma)\\
&= \sum_{i=1}^{M}\sum_{\lambda\in\Lambda(\Omega)}|\Omega|^{-2}\left[\mathbb{1}_\Omega(p_2(\gamma))\right]^2\left\langle\pi(z)f_1,\pi(\lambda)g_i\right\rangle\left\langle\pi(\lambda)h_i,\pi(z)f_2\right\rangle\\
&= |\Omega|^{-2}\mathcal{N}(z)\,.
\end{aligned}$$

By the uniqueness of the the limits, we must have $\mathcal{N}(z) = \sum_{\eta\in\Gamma^*}\widehat{\mathcal{N}}(\eta)e^{-2\pi i p_1^*(\eta)\cdot z}$, and by the uniqueness of the Fourier series, (26) is the Fourier series of $\mathcal{N}(z)$. □

For $g_i, h_i\in M^1(\mathbb{R}^d)$ with $i = 1,\ldots,M$, the function $\mathcal{N}(z)$ of Proposition 6.10 coincides with the function $\mathcal{N}(z)$ defined in (34). Indeed, using Proposition 6.4, we can write $\mathcal{N}(z)$ from (34) explicitly as

$$\begin{aligned}
\mathcal{N}(z) = \widetilde{\mathcal{N}}(\Lambda(\Omega)-z) &= \left\langle S_{g,h}^{\Lambda(\Omega)-z}f_1, f_2\right\rangle = \left\langle S_{g,h}^{\Lambda(\Omega)}\pi(z)f_1,\pi(z)f_2\right\rangle\\
&= \sum_{i=1}^{M}\sum_{\lambda\in\Lambda(\Omega)}\left\langle\pi(z)f_1,\pi(\lambda)g_i\right\rangle\left\langle\pi(\lambda)h_i,\pi(z)f_2\right\rangle.
\end{aligned}$$

By the uniqueness of the Fourier coefficients, $\widehat{\mathfrak{N}}(\eta)$ in (35) equal $\widehat{\mathcal{N}}(\eta)$ from Proposition 6.10, for all $\eta\in\Gamma^*$.

We are now in the position to state the main results.

Theorem 6.11 *Let $\Lambda(\Omega)$ be a simple generic model set. Then for $g_i, h_i\in M^1(\mathbb{R}^d)$, $i = 1,\ldots,M$ the following hold.*

(i) *Fundamental Identity of Gabor Analysis for Model Sets:*

$$\sum_{i=1}^{M} \sum_{\lambda\in\Lambda(\Omega)} \langle f_1, \pi(\lambda)g_i\rangle\langle\pi(\lambda)h_i, f_2\rangle$$
$$= D(\Lambda(\Omega)) \sum_{\eta\in\Gamma^*} \Phi\big(p_2^*(\eta)\big)\Big\langle\pi\big(Jp_1^*(\eta)\big)f_1, f_2\Big\rangle \sum_{i=1}^{M}\Big\langle h_i, \pi\big(Jp_1^*(\eta)\big)g_i\Big\rangle \tag{41}$$

for all $f_1, f_2 \in M^1(\mathbb{R}^d)$.

(ii) *Janssen representation:*

$$S_{g,h}^{\Lambda(\Omega)} = D(\Lambda(\Omega))\sum_{i=1}^{M}\sum_{\eta\in\Gamma^*} \Phi\big(p_2^*(\eta)\big)\Big\langle h_i, \pi\big(Jp_1^*(\eta)\big)g_i\Big\rangle\pi\big(Jp_1^*(\eta)\big), \tag{42}$$

where the series converges unconditionally in the strong operator topology.

(iii) *Wexler–Raz Biorthogonality Relations:*

$$S_{g,h}^{\Lambda(\Omega)} = I \text{ on } M^1(\mathbb{R}^d) \iff D(\Lambda(\Omega))\sum_{i=1}^{M}\Phi\big(p_2^*(\eta)\big)\Big\langle h_i, \pi\big(Jp_1^*(\eta)\big)g_i\Big\rangle = \delta_{\eta,0} \tag{43}$$

for $\eta \in \Gamma^*$.

The relation (41), as well as (42), can be written using dual model sets, and giving a better understanding of the above relations to the ones for regular lattices. Let $\widetilde{\Omega}$ be a compact subset of $\mathbb{R}$, equal closure of its interior and with measure of the boundary equal to zero. We define a sequence of dual model sets as

$$\Lambda_m^*(\widetilde{\Omega}) = \left\{\beta = p_1^*(\eta)\ :\ \eta\in\Gamma^*,\ p_2^*(\eta)\in m\widetilde{\Omega}\setminus(m-1)\widetilde{\Omega}\right\}.$$

Then, the right hand side of (41) equals

$$D(\Lambda(\Omega)) \sum_{\eta\in\Gamma^*} \Phi\big(p_2^*(\eta)\big)\Big\langle\pi\big(Jp_1^*(\eta)\big)f_1, f_2\Big\rangle \sum_{i=1}^{M}\Big\langle h_i, \pi\big(Jp_1^*(\eta)\big)g_i\Big\rangle$$
$$= D(\Lambda(\Omega))\sum_{m=1}^{\infty}\sum_{\beta\in\Lambda_m^*(\widetilde{\Omega})} w_\Phi(\beta)\Big\langle\pi\big(J\beta\big)f_1, f_2\Big\rangle\sum_{i=1}^{M}\Big\langle h_i, \pi\big(J\beta\big)g_i\Big\rangle,$$

where w_Φ is defined as in (17) for the function Φ. Since Φ is well concentrated, with rapid decay outside its essential support, the sum over m has only a finite number of relevant terms, and we can approximate (41) as

$$\sum_{i=1}^{M} \sum_{\lambda \in \Lambda(\Omega)} \langle f_1, \pi(\lambda) g_i \rangle \langle \pi(\lambda) h_i, f_2 \rangle$$
$$\approx D(\Lambda(\Omega)) \sum_{m=1}^{M} \sum_{\beta \in \Lambda_m^*(\widetilde{\Omega})} w_\Phi(\beta) \Big\langle \pi\big(J\beta\big) f_1, f_2 \Big\rangle \sum_{i=1}^{M} \Big\langle h_i, \pi\big(J\beta\big) g_i \Big\rangle .$$

Approximation depends on M and $\widetilde{\Omega}$. We can choose $\widetilde{\Omega}$ to be an $\epsilon-$essential support of Φ in the sense of (39), that is $\widetilde{\Omega} = \widetilde{\Omega}_\epsilon$ and we have $\Lambda_1^*(\widetilde{\Omega}) = \Lambda^*(\widetilde{\Omega}_\epsilon)$ as in (40). Then we obtain a good approximation already for $M = 1$

(i) Fundamental Identity of Gabor Analysis for Model Sets becomes

$$\sum_{i=1}^{M} \sum_{\lambda \in \Lambda(\Omega)} \langle f_1, \pi(\lambda) g_i \rangle \langle \pi(\lambda) h_i, f_2 \rangle$$
$$\approx D(\Lambda(\Omega)) \sum_{\beta \in \Lambda^*(\widetilde{\Omega}_\epsilon)} w_\Phi(\beta) \Big\langle \pi(J\beta) f_1, f_2 \Big\rangle \sum_{i=1}^{M} \Big\langle h_i, \pi(J\beta) g_i \Big\rangle .$$

(ii) Janssen Representation gives us an approximation of the frame operator in the form of

$$S_{g,h}^{\Lambda(\Omega)} \approx D(\Lambda(\Omega)) \sum_{i=1}^{M} \sum_{\beta \in \Lambda^*(\widetilde{\Omega}_\epsilon)} w_\Phi(\beta) \Big\langle h_i, \pi(J\beta) g_i \Big\rangle \pi(J\beta) .$$

(iii) Wexler–Raz Biorthogonality Relations provide an approximation to the identity operator:

$$S_{g,h}^{\Lambda(\Omega)} \approx I \text{ on } M^1(\mathbb{R}^d) \iff D(\Lambda(\Omega)) \sum_{i=1}^{M} w_\Phi(\beta) \Big\langle h_i, \pi(J\beta) g_i \Big\rangle = \delta_{\beta,0}$$

for $\beta \in \Lambda^*(\widetilde{\Omega}_\epsilon)$.

These relations resemble the relations of Gabor systems for lattices, where there is a connection between a lattice and its dual (or symplectic dual). Here $\Lambda(\Omega)$ takes the place of a lattice, and an $\epsilon-$dual model set $\Lambda^*(\widetilde{\Omega}_\epsilon)$ takes the place of a dual lattice, and depends on the desired accuracy of the approximation and window functions g_i, h_i.

Proof The Fundamental Identity of Gabor Analysis and Janssen representation follow directly from Proposition 6.10. Let $f_1, f_2 \in M^1(\mathbb{R}^d)$, then the left hand side of (41) equals the function $\mathcal{N}$ of Proposition 6.10 evaluated at $z = 0$. Since $\mathcal{N}$ equals its Fourier series expansion, we have

$$\mathcal{N}(0) = D(\Lambda(\Omega)) \sum_{\eta \in \Gamma^*} \Phi\big(p_2^*(\eta)\big) \Big\langle \pi\big(J p_1^*(\eta)\big) f_1, f_2 \Big\rangle \sum_{i=1}^{M} \Big\langle h_i, \pi\big(J p_1^*(\eta)\big) g_i \Big\rangle .$$

which gives (41).

As for Janssen representation, we observe that $\mathcal{N}(z) = \langle S_{g,h}^{\Lambda(\Omega)}\pi(z)f_1, \pi(z)f_2\rangle$ for fixed $f_1, f_2 \in M^1(\mathbb{R}^d)$. Evaluating $\mathcal{N}$ at $z = 0$ and using the Fourier series representation of $\mathcal{N}$, we obtain

$$\langle S_{g,h}^{\Lambda(\Omega)} f_1, f_2\rangle = D(\Lambda(\Omega)) \sum_{\eta\in\Gamma^*} \Phi(p_2^*(\eta))\Big\langle\pi\big(Jp_1^*(\eta)\big)f_1, f_2\Big\rangle \sum_{i=1}^{M}\Big\langle h_i, \pi\big(Jp_1^*(\eta)\big)g_i\Big\rangle$$

$$= \Big\langle D(\Lambda(\Omega)) \sum_{\eta\in\Gamma^*} \Phi(p_2^*(\eta)) \sum_{i=1}^{M}\Big\langle h_i, \pi\big(Jp_1^*(\eta)\big)g_i\Big\rangle\pi\big(Jp_1^*(\eta)\big)f_1, f_2\Big\rangle,$$

which is the Janssen representation of the frame operator $S_{g,h}^{\Lambda(\Omega)}$.

The implication $\Longleftarrow$ of $iii)$ follows trivially from the Janssen representation of $S_{g,h}^{\Lambda(\Omega)}$. For the converse, assume that $S_{g,h}^{\Lambda(\Omega)} = I$. Let $f_1, f_2 \in M^1(\mathbb{R}^d)$, then $\mathcal{N}$ of Proposition 3.4 is a constant function. Indeed,

$$\langle f_1, f_2\rangle = \langle\pi(z)f_1, \pi(z)f_2\rangle = \langle S_{g,h}^{\Lambda(\Omega)}\pi(z)f_1, \pi(z)f_2\rangle = \mathcal{N}(z)$$

for every $z \in \mathbb{R}^{2d}$. Since $\mathcal{O} = \mathcal{N} - \langle f_1, f_2\rangle$ is a zero function, all its Fourier coefficients

$$\widehat{\mathcal{O}}(\eta) = \begin{cases} \widehat{\mathcal{N}}(\eta) - \langle f_1, f_2\rangle, & \eta = 0 \\ \widehat{\mathcal{N}}(\eta), & \eta \neq 0 \end{cases}$$

are zero. By Proposition 6.10, we have then

$$D(\Lambda(\Omega)) \sum_{i=1}^{M} \Phi(p_2^*(\eta))\Big\langle h_i, \pi\big(Jp_1^*(\eta)\big)g_i\Big\rangle\Big\langle\pi\big(Jp_1^*(\eta)\big)f_1, f_2\Big\rangle = \delta_{\eta,0}\langle f_1, f_2\rangle\,. \tag{44}$$

Fix $Jp_1^*(\eta) \in \mathbb{R}^{2d}$ and let f be a Gausian, that is $f(x) = e^{-\pi x^2}$. Then $\Big\langle\pi\big(Jp_1^*(\eta)\big)f_1, f_2\Big\rangle = \Big\langle\pi\big(Jp_1^*(\eta)\big)f, f\Big\rangle \neq 0$ and the right hand side of (43) holds. □

As a consequence of the Wexler–Raz biorthogonality relations we obtain a density result for Gabor systems for model sets.

Proposition 6.12 *Let $\Lambda(\Omega)$ be a simple generic model set. If the Gabor frame $\mathcal{G}(g; \Lambda(\Omega))$, with $g \in M^1(\Lambda(\Omega))$ admits a dual that is also a Gabor system, then* $D(\Lambda(\Omega)) \geq 1$.

Proof Let $g \in M^1(\mathbb{R}^d)$ and assume that the Gabor frame $\mathcal{G}(g; \Lambda(\Omega))$ admits a dual $\mathcal{G}(h; \Lambda(\Omega))$ with $h \in M^1(\mathbb{R}^d)$. Let B_g be the upper frame bound of $\mathcal{G}(g; \Lambda(\Omega))$, and we can assume without loss of generality that $\|h\|_2^2 = B_g^{-1}$. Then we have the frame decomposition

$$\langle f_1, f_2\rangle = \sum_{\lambda\in\Lambda(\Omega)} \langle f_1, \pi(\lambda)g\rangle\langle\pi(\lambda)h, f_2\rangle \quad \text{for all } f_1, f_2 \in L^2(\mathbb{R}^m),$$

If we set $f_1 = h$ and $f_2 = g$, by the Bessel property of $\mathcal{G}(g; \Lambda(\Omega))$, we obtain

$$\langle h, g \rangle = \sum_{\lambda \in \Lambda(\Omega)} |\langle h, \pi(\lambda) g \rangle|^2 \leq B_g \|h\|_2^2 = 1 \,.$$

On the other hand, by Theorem 6.11 $iii)$ with $L = 1$, $D(\Lambda(\Omega))\langle h, g \rangle = 1$, and therefore $D(\Lambda(\Omega)) \geq 1$. □

Acknowledgements Open access funding provided by University of Vienna. This research was supported by the Austrian Science Fund, FWF Project (V312-N25).

Open Access This article is distributed under the terms of the Creative Commons Attribution 4.0 International License (http://creativecommons.org/licenses/by/4.0/), which permits unrestricted use, distribution, and reproduction in any medium, provided you give appropriate credit to the original author(s) and the source, provide a link to the Creative Commons license, and indicate if changes were made.

References

1. Baake, M., Moody, R.V.: Self-similar measures for quasicrystals. In: Baake, M. (ed.) Directions in Mathematical Quasicrystals. AMS, American Mathematical Society, Providence, RI (2000)
2. Besicovitch, A.: Almost Periodic Functions. Neudruck, vol. XIII. Dover Publications, Inc., New York (1955)
3. Besicovitch, A., Bohr, H.: Almost periodicity and general trigonometric series. Acta Math. **57**, 203–292 (1931)
4. Boggiatto, P., Fernández, C., Galbis, A.: Gabor systems and almost periodic functions. Appl. Comput. Harmon. Anal. **42**(1), 65–87 (2017)
5. Bohr, H.: Almost Periodic Functions. Chelsea Publishing Company, New York (1947)
6. de Boor, C., DeVore, R.A., Ron, A.: The structure of finitely generated shift-invariant spaces in $L_2(\mathbb{R}^d)$. J. Funct. Anal. **119**(1), 37–78 (1994)
7. Feichtinger, H.G.: On a new Segal algebra. Monatsh. Math. **92**, 269–289 (1981)
8. Feichtinger, H.G., Luef, F.: Wiener amalgam spaces for the fundamental identity of Gabor analysis. Collect. Math. **57**(Extra Volume (2006)), 233–253 (2006)
9. Gabardo, J.-P.: Weighted irregular Gabor tight frames and dual systems using windows in the Schwartz class. J. Funct. Anal. **256**(3), 635–672 (2009)
10. Galindo, F.: Some remarks on "On the windowed Fourier transform and wavelet transform of almost periodic functions," by J. R. Partington and B. Ünalmış. Appl. Comput. Harmon. Anal., **16**(3), 174–181 (2004)
11. Gröchenig, K.: Foundations of Time-Frequency Analysis. Birkhäuser, Boston (2001)
12. Gröchenig, K., Ortega-Cerdà, J., Romero, J.L.: Deformation of Gabor systems. Adv. Math. **277**, 388–425 (2015)
13. Gröchenig, K., Romero, J.L., Stöckler, J.: Sampling theorems for shift-invariant spaces, Gabor frames, and totally positive functions. Invent. Math. **211**(3), 1119–11148 (2017)
14. Hernández, E., Labate, D., Weiss, G.: A unified characterization of reproducing systems generated by a finite family. II. J. Geom. Anal. **12**(4), 615–662 (2002)
15. Hernández, E., Weiss, G.: A First Course on Wavelets. CRC Press, Boca Raton, FL (1996)
16. Jakobsen, M., Lemvig, J.: Co-compact Gabor systems on locally compact Abelian groups. J. Fourier Anal. Appl. **22**(1), 36–70 (2016)
17. Janssen, A.: Duality and biorthogonality for Weyl-Heisenberg frames. J. Fourier Anal. Appl. **1**(4), 403–436 (1995)
18. Kreisel, M.: Gabor frames for quasicrystals, K-theory, and twisted gap labeling. J. Funct. Anal. **270**(3), 1001–1030 (2016)
19. Labate, D.: A unified characterization of reproducing systems generated by a finite family. J. Geom. Anal. **12**(3), 469–491 (2002)

20. Matei, B., Meyer, Y.: Simple quasicrystals are sets of stable sampling. Complex Var. Elliptic Equ. **55**(8–10), 947–964 (2010)
21. Matusiak, E.: Frames of translates for model sets. preprint (2018)
22. Meyer, Y.: Algebraic Numbers and Harmonic Analysis. North-Holland Mathematical Library, vol. 2. North-Holland Publishing Company, Amsterdam (1972)
23. Meyer, Y.: Quasicrystals, almost periodic patterns, mean-periodic functions and irregular sampling. Afr. Diaspora J. Math. **13**(1), 1–45 (2012)
24. Moody, R., Nesterenko, M., Patera, J.: Computing with almost periodic functions. Acta Crystallogr. Sect. A **64**(6), 654–669 (2008)
25. Moody, R.V.: Mathematical quasicrystals: a tale of two topologies. In: XIVth international congress on mathematical physics (ICMP 2003), Lisbon, Portugal, 28 July–2 August 2003. Selected papers based on the presentation at the conference., pp. 68–77. Hackensack, NJ: World Scientific (2005)
26. Partington, J., Ünalmış, B.: On the windowed Fourier transform and wavelet transform of almost periodic functions. Appl. Comput. Harmon. Anal. **10**(1), 45–60 (2001)
27. Schlottmann, M.: Generalized model sets and dynamical systems. In: Baake, M. (ed.) Directions in Mathematical Quasicrystals, pp. 143–159. AMS, American Mathematical Society, Providence, RI (2000)
28. Wexler, J., Raz, S.: Discrete Gabor expansions. Signal Process. **21**, 207–220 (1990)

Publisher's Note Springer Nature remains neutral with regard to jurisdictional claims in published maps and institutional affiliations.

Journal of Fourier Analysis and Applications (2019) 25:2608–2634
https://doi.org/10.1007/s00041-019-09675-z

RESEARCH ARTICLE

Vector-Valued Extensions of Operators Through Multilinear Limited Range Extrapolation

Emiel Lorist[1] · **Bas Nieraeth**[1]

Received: 15 January 2018 / Revised: 11 January 2019 / Published online: 1 March 2019
© The Author(s) 2019

Abstract

We give an extension of Rubio de Francia's extrapolation theorem for functions taking values in UMD Banach function spaces to the multilinear limited range setting. In particular we show how boundedness of an m-(sub)linear operator

$$T : L^{p_1}(w_1^{p_1}) \times \cdots \times L^{p_m}(w_m^{p_m}) \to L^p(w^p)$$

for a certain class of Muckenhoupt weights yields an extension of the operator to Bochner spaces $L^p(w^p; X)$ for a wide class of Banach function spaces X, which includes certain Lebesgue, Lorentz and Orlicz spaces. We apply the extrapolation result to various operators, which yields new vector-valued bounds. Our examples include the bilinear Hilbert transform, certain Fourier multipliers and various operators satisfying sparse domination results.

Keywords Extrapolation · Limited range · Multilinear · UMD · Muckenhoupt weights · Banach function space · p-convexity · Bilinear Hilbert transform · Fourier multipliers · Sparse domination

Mathematics Subject Classification Primary: 42B25; Secondary: 42B15 · 46E30

Communicated by Rodolfo H. Torres.

The first author is supported by the VIDI subsidy 639.032.427 of the Netherlands Organisation for Scientific Research (NWO).

✉ Emiel Lorist
e.lorist@tudelft.nl

Bas Nieraeth
b.nieraeth@tudelft.nl

[1] Delft Institute of Applied Mathematics, Delft University of Technology, P.O. Box 5031, 2600 GA Delft, The Netherlands

1 Introduction

Scalar-valued extrapolation, using the theory of Muckenhoupt weights, has proven to be an essential tool in harmonic analysis. The classical extrapolation result (see [59] and [29, Chap. IV]) says that if a (sub)linear operator T satisfies for a fixed $p_0 \in (1,\infty)$ and all weights w in the *Muckenhoupt* class A_{p_0} the norm inequality

$$\|Tf\|_{L^{p_0}(w)} \leq C\,\|f\|_{L^{p_0}(w)} \tag{1.1}$$

for all $f \in L^{p_0}(w)$, then we have for all $p \in (1,\infty)$ and all weights $w \in A_p$

$$\|Tf\|_{L^{p}(w)} \leq C\,\|f\|_{L^{p}(w)} \tag{1.2}$$

for all $f \in L^p(w)$. Numerous generalizations of this result have appeared, see for example [4,21,22,32,35]. We mention several of them.

It was shown by Grafakos and Martell [32] that extrapolation extends to the *multilinear* setting. Indeed, they showed that given fixed exponents $p_1,\ldots,p_m \in (1,\infty)$, if for an m-(sub)linear operator T and all weights $w_j^{p_j} \in A_{p_j}$ we have

$$\|T(f_1,\ldots,f_m)\|_{L^p(w^p)} \leq C \prod_{j=1}^m \|f_j\|_{L^{p_j}(w_j^{p_j})},$$

where $w = \prod_{j=1}^m w_j$ and $\frac{1}{p} = \sum_{j=1}^m \frac{1}{p_j}$, then the same estimate holds for all $p_j \in (1,\infty)$, weights $w_j^{p_j} \in A_{p_j}$ and w and p as before.

Considering a different kind of generalization, it was shown by Auscher and Martell [4] that a *limited range* version of the extrapolation result holds: if there are exponents $0 < p_- < p_+ \leq \infty$ such that the estimate (1.1) is valid for a fixed $p_0 \in (p_-,p_+)$ and all weights w in the Muckenhoupt and *Reverse Hölder* class $A_{p_0/p_-} \cap RH_{(p_+/p_0)'}$, then (1.2) is valid for all $p \in (p_-,p_+)$ and all weights $w \in A_{p/p_-} \cap RH_{(p_+/p)'}$.

Vector-valued extensions of the extrapolation theory have also been considered. Through an argument using Fubini's Theorem, the initial estimate (1.1) immediately implies not only the estimate (1.2) for all $p \in (1,\infty)$, but also for extensions of the operator T to functions taking values in the sequence spaces ℓ^s or more generally Lebesgue spaces L^s for $s \in (1,\infty)$. Moreover, Rubio de Francia showed in [60, Theorem 5] that one can take this even further. Indeed, this result states that assuming (1.1) holds for some $p_0 \in (1,\infty)$ and for all weights $w \in A_{p_0}$, then for each Banach function space X with the UMD property, T extends to an operator $\widetilde{T}$ on the Bochner space $L^p(X)$ which satisfies

$$\left\|\widetilde{T}f\right\|_{L^p(X)} \leq C\,\|f\|_{L^p(X)}$$

for all $p \in (1,\infty)$ and all $f \in L^p(X)$. Recently, it was shown by Amenta, Veraar, and the first author in [2] that given $p_- \in (0,\infty)$, if (1.1) holds for $p_0 \in (p_-,\infty)$ and all weights $w \in A_{p_0/p_-}$, then for each Banach function space X such that X^{p_-}

has the UMD property, T extends to an operator $\widetilde{T}$ on the Bochner space $L^p(w; X)$ and satisfies

$$\|\widetilde{T} f\|_{L^p(w;X)} \leq C \|f\|_{L^p(w;X)}$$

for all $p \in (p_-, \infty)$, all weights $w \in A_{p/p_-}$ and all $f \in L^p(w; X)$. Here X^{p_-} is the p_--concavification of X, see Sect. 2 for the definition.

Vector-valued estimates in harmonic analysis have been actively developed in the past decades. Important for the mentioned vector-valued extrapolation are the equivalence of the boundedness of the vector-valued Hilbert transform on $L^p(X)$ and the UMD property of X for a Banach space X (see [9,13]) and the fact that for a Banach function space X the UMD property implies the boundedness of the lattice Hardy–Littlewood maximal operator on $L^p(X)$ (see [10,61]). For recent results in vector-valued harmonic analysis in UMD Banach function spaces, see for example [8,25,36,38,66].

In the recent work [21] by Cruz-Uribe and Martell both the limited range and the multilinear extrapolation result were combined, yielding a unified multilinear limited range version of the extrapolation result in the scalar-valued case. This result also covers vector-valued extensions to ℓ^s for certain $s \in (0, \infty)$. This opened the question whether a unified multilinear limited range extrapolation theorem also holds for more general Banach function spaces. In this work, we give a positive answer to this question.

We now state our main result, in which we denote $X \in \mathrm{UMD}_{p_-,p_+}$ for the technical assumption that $\big((X^{p_-})^*\big)^{(p_+/p_-)'}$ has the UMD property, see Sect. 3 for a thorough discussion of this assumption. A more general version of this theorem can be found in Theorem 4.3 below.

Theorem 1.1 *Let $m \in \mathbb{N}$ and fix $0 < p_j^- < p_j^+ \leq \infty$ for $j \in \{1, \dots, m\}$. Let T be an operator defined on m-tuples of functions and suppose there exist $p_j \in (p_j^-, p_j^+)$ such that for all weights $w_j^{p_j} \in A_{p_j/p_j^-} \cap RH_{(p_j^+/p_j)'}$ and $f_j \in L^{p_j}(w_j^{p_j})$ we have*

$$\big\|T(f_1, \dots, f_m)\big\|_{L^p(w^p)} \leq C \prod_{j=1}^{m} \|f_j\|_{L^{p_j}(w_j^{p_j})},$$

with $w = \prod_{j=1}^m w_j$, $\frac{1}{p} = \sum_{j=1}^m \frac{1}{p_j}$, and where $C > 0$ depends only on the characteristic constants of the weights. Moreover, assume that T satisfies one of the following conditions:

(i) *T is m-linear.*
(ii) *T is m-sublinear and positive valued.*

Let $X_1, \dots, X_m$ be quasi-Banach function spaces over a σ-finite measure space (S, μ) and define $X = X_1 \dots X_m$. Assume that for all simple functions $f_j : \mathbb{R}^d \to X_j$ the function $\widetilde{T} f : \mathbb{R}^d \to X$ given by

$$\widetilde{T}(f_1, \dots, f_m)(x, s) := T(f_1(\cdot, s), \dots, f_m(\cdot, s))(x), \quad x \in \mathbb{R}^d, \quad s \in S$$

is well-defined and strongly measurable. If $X_j \in \mathrm{UMD}_{p_j^-,p_j^+}$*, then for all* $p_j \in (p_j^-, p_j^+)$ *and weights* $w_j^{p_j} \in A_{p_j/p_j^-} \cap RH_{(p_j^+/p_j)'}$*,* $\widetilde{T}$ *extends to a bounded operator on* $L^{p_1}(w_1^{p_1}; X_1) \times \cdots \times L^{p_m}(w_m^{p_m}; X_m)$ *with*

$$\left\|\widetilde{T}(f_1, \ldots, f_m)\right\|_{L^p(w^p;X)} \leq C' \prod_{j=1}^{m} \|f_j\|_{L^{p_j}(w_j^{p_j};X_j)},$$

for all $f_j \in L^{p_j}(w_j^{p_j}; X_j)$*, with* w *and* p *are as before, and where* $C' > 0$ *depends on the* p_j*,* p_j^-*,* p_j^+*, the characteristic constants of the weights, and the spaces* X_j*.*

Remark 1.2

- If T is a linear operator as in Theorem 1.1, we have for $f_j \otimes \xi_j \in L^{p_j}(w_j^{p_j}) \otimes X_j$ that

$$\widetilde{T}(f_1 \otimes \xi_1, \ldots, f_m \otimes \xi_m) = T(f_1, \ldots, f_m) \otimes \xi_1 \ldots \xi_m \in L^p(w^p) \otimes X.$$

  So in this case $\widetilde{T}$ is automatically well-defined and strongly measurable for all simple functions $f_j : \mathbb{R}^d \to X$.
- Although we state Theorem 1.1 for Banach function spaces, it extends to spaces isomorphic to a closed subspace of a Banach function space and by standard representation techniques also to certain Banach lattices, see [54,57] for the details.
- In [21] scalar-valued multilinear limited range extrapolation is proven through off-diagonal extrapolation. Relying on this result, in this paper we prove the vector-valued multilinear limited range result. Our method does not directly generalize to the off-diagonal setting, which leaves vector-valued off-diagonal extrapolation as an open problem.
- In Theorem 1.1 one could allow for $p_j^- = 0$. In this case one would have to interpret $X_j \in \mathrm{UMD}_{0,p_j^+}$ as $X_j \in \mathrm{UMD}_{p,p_j^+}$ for some $p \in (0, p_j^+)$.

Even in the linear case $m = 1$ our result is new in the sense that it extends the main result of [2] to allow for finite p_j^+, which yields many new applications. We are now able to consider, for example, Riesz transforms associated to elliptic operators through the weighted estimates obtained in [4]. Many more examples of such operators can also be considered through recent advances in the theory of sparse dominations. Indeed, for example for certain Fourier multipliers such as Bochner-Riesz multipliers as well as for spherical maximal operators, sparse bounds have been found. Sparse bounds naturally imply weighted norm estimates which, through our result, yield bounded vector-valued extensions for such operators. For a more elaborate discussion as well as for references we refer the reader to Sect. 5.

Our result is also new in the full range multilinear case, i.e., if $p_j^- = 1$, $p_j^+ = \infty$ for all $j \in \{1, \ldots, m\}$. This can, for example, be applied to multilinear Calderón–Zygmund operators, as these satisfy the appropriate weighted bounds to apply our result. We elaborate on this in Sect. 5.

Finally, for the case $m = 2$ our result yields new results for boundedness of the vector-valued bilinear Hilbert transform $\widetilde{\mathrm{BHT}}$, due to known scalar-valued weighted bounds as were first established by Culiuc et al. [24]. Bounds for the vector-valued bilinear Hilbert transform $\widetilde{\mathrm{BHT}}$ have useful applications in PDEs, see [5] and references therein. The precise result we obtain can be found in Theorem 5.1.

Remark 1.3 In the recent work [53] of Li et al., and the recent work [58] of the second author, scalar-valued extrapolation results were obtained using the multilinear weight classes from [52], which were made public after this paper first appeared. Rather than considering a condition for each weight individually, these weight classes allow for an interaction between the various weights, making them more appropriately adapted to the multilinear setting. This gives rise to the problem of extending these results to the vector-valued case. To facilitate this, it seems that an appropriate multilinear UMD condition on tuples of Banach function spaces is required. We leave this as a basis for future research.

This article is organized as follows:

- In Sect. 2 we summarize the preliminaries on Muckenhoupt weights, product quasi-Banach function spaces and the UMD property.
- In Sect. 3 we discus the UMD_{p_-,p_+} property and give examples of quasi-Banach function spaces satisfying the UMD_{p_-,p_+} property.
- In Sect. 4 we prove our main result in terms of $(m+1)$-tuples of functions, proving Theorem 1.1 as a corollary.
- In Sect. 5 we prove new vector-valued bounds for various operators.

2 Preliminaries

2.1 Muckenhoupt Weights

A locally integrable function $w : \mathbb{R}^d \to (0, \infty)$ is called a *weight*. For $p \in [1, \infty)$ and a weight w the space $L^p(w)$ is the subspace of all measurable functions $f : \mathbb{R}^d \to \mathbb{C}$, which we denote by $f \in L^0(\mathbb{R}^d)$, such that

$$\|f\|_{L^p(w)} := \left(\int_{\mathbb{R}^d} |f(x)|^p w(x)\,\mathrm{d}x\right)^{1/p} < \infty.$$

By a cube $Q \subseteq \mathbb{R}^d$ we will mean a half-open cube whose sides are parallel to the coordinate axes and for a locally integrable function $f \in L^0(\mathbb{R}^d)$ we will write $\langle f\rangle_Q := \frac{1}{|Q|}\int_Q f\,\mathrm{d}x$.

For $p \in [1, \infty)$ we will say that a weight w lies in the *Muckenhoupt class* A_p and write $w \in A_p$ if it satisfies

$$[w]_{A_p} := \sup_Q \langle w\rangle_Q \langle w^{1-p'}\rangle_Q^{p-1} < \infty,$$

where the supremum is taken over all cubes $Q \subseteq \mathbb{R}^d$ and the second factor is replaced by $(\operatorname{ess\,inf}_Q w)^{-1}$ if $p = 1$. We define $A_\infty := \bigcup_{p\in[1,\infty)} A_p$.

Lemma 2.1 *Let $p \in [1, \infty)$ and $w \in A_p$.*

(i) *$w \in A_q$ for all $q \in [p, \infty)$ with $[w]_{A_q} \le [w]_{A_p}$.*
(ii) *If $p > 1$, $w^{1-p'} \in A_{p'}$ with $[w]_{A_p}^{\frac{1}{p}} = [w^{1-p'}]_{A_{p'}}^{\frac{1}{p'}}$.*
(iii) *If $p > 1$, there exists an $\varepsilon > 0$ such that $w \in A_{p-\varepsilon}$ and $[w]_{A_{p-\varepsilon}} \le C_p [w]_{A_p}$.*

The first two properties of Lemma 2.1 are immediate from the definition. For the third see [31, Corollary 7.2.6]. The linear estimate of $[w]_{A_{p-\varepsilon}}$ in terms of $[w]_{A_p}$ can be found in [39, Theorem 1.2]. Note that self-improvement properties for A_p weights are classical. We opt to use this quantitative version of the result for clarity in the proof of our main theorem.

For $s \in [1, \infty)$ we say that $w \in A_\infty$ satisfies a *reverse Hölder* property and write $w \in RH_s$ if

$$[w]_{RH_s} := \sup_Q \langle w^s \rangle_Q^{\frac{1}{s}} \langle w \rangle_Q^{-1} < \infty.$$

We will require the following properties of the reverse Hölder classes, see [42].

Lemma 2.2 *Let $r \in (1, \infty)$, $s \in [1, \infty)$ and define $p = s(r-1)+1$. For $w \in A_\infty$ the following are equivalent*

(i) $w \in A_r \cap RH_s$.
(ii) $w^s \in A_p$.
(iii) $w^{1-r'} \in A_{p'}$.

Moreover we have

$$\max\{[w]_{RH_s}^s, [w]_{A_r}^s\} \le [w^s]_{A_p} \le \big([w]_{A_r}[w]_{RH_s}\big)^s.$$

For $n \in \mathbb{N}$ we will write $\phi_{a,b,\ldots}$ for a non-decreasing function $[1, \infty)^n \to [1, \infty)$, depending on the parameters $a, b, \ldots$ and the dimension d. This function may change from line to line. We need non-decreasing dependence on the Muckenhoupt characteristics in our proofs. In [2, Appendix A] it is shown how to deduce non-decreasing dependence from a more general estimate in terms of the Muckenhoupt characteristics.

2.2 Banach Function Spaces

Let (S, μ) be a σ-finite measure space. A subspace X of $L^0(S)$ equipped with a quasi-norm $\|\cdot\|_X$ is called a *quasi-Banach function space* if it satisfies the following properties:

(i) If $\xi \in L^0(S)$ and $\eta \in X$ with $|\xi| \le |\eta|$, then $\xi \in X$ and $\|\xi\|_X \le \|\eta\|_X$.
(ii) There is an $\xi \in X$ with $\xi > 0$.
(iii) If $0 \le \xi_n \uparrow \xi$ with $(\xi_n)_{n=1}^\infty$ a sequence in X, $\xi \in L^0(S)$ and $\sup_{n\in\mathbb{N}} \|\xi_n\|_X < \infty$, then $\xi \in X$ and $\|\xi\|_X = \sup_{n\in\mathbb{N}} \|\xi_n\|_X$.

It is called a *Banach function space* if $\|\cdot\|_X$ is a norm. A Banach function space X is called *order continuous* if for any sequence $0 \le \xi_n \uparrow \xi \in X$ it holds that

$\|\xi_n - \xi\|_X \to 0$. Order continuity of a Banach function space X ensures that its dual X^* is again a Banach function space (see [54, Sect. 1.b]), and that the Bochner space $L^p(S'; X)$ is a Banach function space over $(S \times S', \mu \times \mu')$ for any σ-finite measure space (S', μ'). As an example we note that any reflexive Banach function space is order-continuous.

A quasi-Banach function space X is said to be *p-convex* for $p \in (0, \infty]$ if

$$\Big\| \Big(\sum_{k=1}^{n} |\xi_k|^p \Big)^{\frac{1}{p}} \Big\|_X \leq \Big(\sum_{k=1}^{n} \|\xi_k\|_X^p \Big)^{\frac{1}{p}}$$

for all $\xi_1, \ldots, \xi_n \in X$ with the usual modification when $p = \infty$. It is said to be *p-concave* when the reverse inequality holds. Usually the defining inequalities for p-convexity and p-concavity include a constant depending on p and X, but as shown in [54, Theorem 1.d.8], X can be renormed equivalently such that these constants equal 1. See [54, Sect. 1.d] for a thorough introduction of p-convexity and concavity in Banach function spaces and see [43] for the quasi-Banach function space case.

We define the *p-concavification* of a quasi-Banach function space X for $p \in (0, \infty)$ by

$$X^p := \{\xi \in L^0(S) : |\xi|^{\frac{1}{p}} \in X\} = \{|\xi|^p \operatorname{sgn} \xi : \xi \in X\},$$

equipped with the quasi-norm $\|\xi\|_{X^p} := \big\| |\xi|^{\frac{1}{p}} \big\|_X^p$. Note that X^p is a Banach function space if and only if X is p-convex. In particular, X is a Banach function space if and only if it is 1-convex.

For two quasi-Banach function spaces X_0, X_1 over the same measure space (S, μ) we define the vector space $X_0 \cdot X_1$ as

$$X_0 \cdot X_1 := \{\xi_0 \cdot \xi_1 : \xi_0 \in X_0, \xi_1 \in X_1\}$$

and for $\xi \in X_0 \cdot X_1$ we define

$$\|\xi\|_{X_0 \cdot X_1} := \inf\big\{ \|\xi_0\|_{X_0} \|\xi_1\|_{X_1} : |\xi| = \xi_0 \cdot \xi_1, 0 \leq \xi_0 \in X_0, 0 \leq \xi_1 \in X_1 \big\}$$

We call $X_0 \cdot X_1$ a *product quasi-Banach function space* if $\| \cdot \|_{X_0 \cdot X_1}$ defines a complete quasi-norm on $X_0 \cdot X_1$. We will mostly be working with so called *Calderón–Lozanovskii* products. These are product quasi-Banach function spaces of the form $X_0^{1-\theta} \cdot X_1^{\theta}$ for some $\theta \in (0, 1)$, see [15,55]. Of course the definition of product quasi-Banach function spaces and Calderón–Lozanovskii products can be canonically extended to m quasi-Banach function spaces over the same measure space for any $m \in \mathbb{N}$. We give a few examples of product Banach function spaces, see also [12].

Example 2.3 Fix $m \in \mathbb{N}$ and let (S, μ) be an atomless or atomic σ-finite measure space.

(i) Lebesgue spaces: $L^p(S) = L^{p_1}(S) \ldots L^{p_m}(S)$ for $p_j \in (0, \infty)$ and $\frac{1}{p} = \sum_{j=1}^{m} \frac{1}{p_j}$.

(ii) Lorentz spaces: $L^{p,q}(S) = L^{p_1,q_1}(S)\dots L^{p_m,q_m}(S)$ for $p_j, q_j \in (0,\infty)$, $\frac{1}{p} = \sum_{j=1}^m \frac{1}{p_j}$ and $\frac{1}{q} = \sum_{j=1}^m \frac{1}{q_j}$.
(iii) Orlicz spaces: $L^{\Phi}(S) = L^{\Phi_1}(S)\dots L^{\Phi_m}(S)$ for Young functions Φ_j and $\Phi^{-1} = \Phi_1^{-1}\dots\Phi_m^{-1}$.

We will use the following properties of product Banach function spaces:

Proposition 2.4 *Let X, X_0, X_1 be Banach function spaces over a σ-finite measure space (S,μ) and let $\theta \in (0,1)$.*

(i) *If X_0 or X_1 is reflexive, then $X_0^{1-\theta}\cdot X_1^{\theta} = [X_0, X_1]_\theta$.*
(ii) *If X_0 or X_1 is reflexive, then $X_0^{1-\theta}\cdot X_1^{\theta}$ is reflexive.*
(iii) $\left(X_0^{1-\theta}\cdot X_1^{\theta}\right)^* = \left(X_0^*\right)^{1-\theta}\cdot\left(X_1^*\right)^{\theta}$.
(iv) $(X^\theta)^* = (X^*)^\theta \cdot L^{1/(1-\theta)}(S)$.
(v) *If X_0 and X_1 have the* UMD *property, then $X_0^{1-\theta}\cdot X_1^{\theta}$ has the* UMD *property.*

Part (i) follows from [15], it has been extended to the product quasi-Banach function space setting in [44,45]. Part (ii) is proven in [55, Theorem 3]. It also follows from [15] through complex interpolation. Part (iii) is proven in [55, Theorem 2] and for (iv) see [62, Theorem 2.9]. Finally part (v) follows from part (i) and [41, Proposition 4.2.17], see also the next section on the UMD property.

2.3 The UMD Property

We say that a Banach space X has the UMD property if the martingale difference sequence of any finite martingale in $L^p(\Omega; X)$ is unconditional for some (equivalently all) $p \in (1,\infty)$. The UMD property implies reflexivity and if X has the UMD property, then X^* has the UMD property as well. Standard examples of Banach spaces with the UMD property include reflexive L^p-spaces, Lorentz spaces, Orlicz spaces, Sobolev spaces, Besov spaces and Schatten classes. For a thorough introduction to the theory of UMD spaces we refer the reader to [14,41].

Throughout this paper we will consider Banach function spaces with the UMD property. In this case we have a characterisation of the UMD property in terms of the lattice Hardy-Littlewood maximal operator, which for simple $f : \mathbb{R}^d \to X$ is defined by

$$\widetilde{M} f(x) := \sup_Q \langle |f| \rangle_Q \mathbf{1}_Q(x),$$

where the supremum is taken over all cubes $Q \subseteq \mathbb{R}^d$ (see [30] for the details). The boundedness of $\widetilde{M}$ on both $L^p(\mathbb{R}^d; X)$ and $L^p(\mathbb{R}^d; X^*)$ for some (equivalently all) $p \in (1,\infty)$ is equivalent to X having the UMD property by a result of Bourgain [10] and Rubio de Francia [61]. Moreover, if X has the UMD property we have the following weighted bound for all $p \in (1,\infty)$, $w \in A_p$ and $f \in L^p(w; X)$

$$\|\widetilde{M} f\|_{L^p(w;X)} \le \phi_{X,p}([w]_{A_p}) \|f\|_{L^p(w;X)}, \tag{2.1}$$

see [30]. A more precise dependence on the weight characteristic can be found in [34].

The UMD property of a Banach function space X implies that certain q-concavifications of X also have the UMD property, see [61, Theorem 4].

Proposition 2.5 (Rubio de Francia) *Let X be a Banach function space over a σ-finite measure space (S, μ) such that X has the* UMD *property. Then there exists an $\varepsilon > 0$ such that such that X^q has the* UMD *property for all $0 < q < 1 + \varepsilon$.*

Note that the difficult part of Proposition 2.5 is the claim that X^q has the UMD property for $1 < q < 1 + \varepsilon$.

3 The UMD_{p_-,p_+} Property of Quasi-Banach Function Spaces

For our main result we need an extension of the UMD property, as we will often consider quasi-Banach function spaces of which a concavification has the UMD property. In particular, we will use the following notion:

Definition 3.1 Let X be a quasi-Banach function space and let $0 < p_- < p_+ \leq \infty$. Then we say X has the UMD_{p_-,p_+} property if and only if X is p_--convex, p_+-concave and $\big((X^{p_-})^*\big)^{(p_+/p_-)'}$ has the UMD property. We denote this by $X \in \mathrm{UMD}_{p_-,p_+}$.

Note that X is a Banach function space with the UMD property if and only if $X \in \mathrm{UMD}_{1,\infty}$ and we denote this by $X \in \mathrm{UMD}$.

Remark 3.2

- The p_--convexity in Definition 3.1 implies that X^{p_-} is a Banach function space, so its dual $(X^{p_-})^*$ is non-trivial. Moreover $(X^{p_-})^*$ is a Banach function space, since it has the UMD property by Proposition 2.5 and is therefore reflexive, which implies that X^{p_-} is order-continuous.
- The p_+-concavity assumption in Definition 3.1 is not restrictive, as any quasi-Banach function space with the UMD property is actually isomorphic to a Banach function space (see [19]), which implies that $(X^{p_-})^*$ is $(p_+/p_-)'$-convex and thus that X is p_+-concave by [54, Sect. 1.d]

We first show some basic results for the UMD_{p_-,p_+} property.

Proposition 3.3 *Fix $0 < p_- < p_+ \leq \infty$ and let X be a quasi-Banach function space over a σ-finite measure space (S, μ) such that $X \in \mathrm{UMD}_{p_-,p_+}$.*

(i) $X^{p_-} \in \mathrm{UMD}_{1,p_+/p_-}$.
(ii) $X^* \in \mathrm{UMD}_{p_+',p_-'}$ *if* $p_- \geq 1$.
(iii) $X \in \mathrm{UMD}_{\tilde{p}_-,\tilde{p}_+}$ *for all* $\tilde{p}_- \in (0, p_-]$ *and* $\tilde{p}_+ \in [p_+, \infty]$.
(iv) *If* $1 < p_- < p_+ < \infty$, *then* $X = [Y, L^2(S)]_\theta$ *for a Banach function space* $Y \in \mathrm{UMD}$ *and* $\theta = 2/\max\{p_-', p_+\}$.
(v) $L^p(S'; X) \in \mathrm{UMD}_{p_-,p_+}$ *for all* $p \in (p_-, p_+)$ *and any σ-finite measure space* (S', μ').

Proof Part (i) follows directly from the definition. For part (ii) the p'_+-convexity and p'_--concavity follow from [54, Sect. 1.d]. If $p_- = 1$, the claim is that $\big((X^*)^{p'_+}\big)^* \in$ UMD, which is clear. Assuming $p_- > 1$, we have by Proposition 2.4

$$\begin{aligned}
(((X^*)^{p'_+})^*)^{(p'_-/p'_+)'} &= \Big(\Big(((X^{p_-})^*)^{p'_+/p_-} \cdot L^{p'_-/p'_+}(S)\Big)^*\Big)^{(p'_-/p'_+)'} \\
&= \Big(\Big(((X^{p_-})^*)^{(p_+/p_-)'\theta} \cdot L^1(S)^{1-\theta}\Big)^*\Big)^{1/\theta} \\
&= \Big(((X^{p_-})^*)^{(p_+/p_-)'}\Big)^* \cdot (L^\infty(S))^{(1-\theta)/\theta} \\
&= \Big(((X^{p_-})^*)^{(p_+/p_-)'}\Big)^*
\end{aligned}$$

with $\theta := \frac{1}{(p'_-/p'_+)'} < 1$, since taking a product with $L^\infty(S)^{(1-\theta)/\theta} = L^\infty(S)$ has no effect on the space. Thus we conclude that $X^* \in \mathrm{UMD}_{p'_+,p'_-}$.

For part (iii) the p'_+-convexity and p'_--concavity follow from [56, Theorem 4.2]. First assume that $p_- = 1$ and let $\tilde{p}_- \in (0, 1)$. By Proposition 2.4(iv) we have

$$\big((X^{\tilde{p}_-})^*\big)^{(p_+/\tilde{p}_-)'} = (X^*)^{\tilde{p}_-(p_+/\tilde{p}_-)'} \cdot L^{\frac{p_+-\tilde{p}_-}{(1-\tilde{p}_-)p_+}}(S) = (X^*)^{p'_+\theta} \cdot L^1(S)^{1-\theta} \tag{3.1}$$

with

$$\theta := \frac{\tilde{p}_-(p_+ - 1)}{p_+ - \tilde{p}_-} < 1.$$

By assumption $(X^*)^{p'_+} \in$ UMD, so

$$\big[(X^*)^{p_+}, L^1(S)\big]^*_\theta = \big[((X^*)^{p'_+})^*, L^\infty(S)\big]_\theta = \Big(((X^*)^{p'_+})^*\Big)^\theta \in \mathrm{UMD}$$

by Propositions 2.4 and 2.5. Using Proposition 2.4(i), we obtain from (3.1) that $X \in \mathrm{UMD}_{\tilde{p}_-,p_+}$. For arbitrary $0 < p_- < p_+ \le \infty$ we know that $X \in \mathrm{UMD}_{\tilde{p}_-,p_+}$ for all $\tilde{p}_- \in (0, p_-]$ by (i) and Proposition 2.5 yields that $X \in \mathrm{UMD}_{\tilde{p}_-,\tilde{p}_+}$ for all $\tilde{p}_+ \in [p_+, \infty]$.

For part (iv) note that $X \in \mathrm{UMD}_{p',p}$ with $p = \max\{p'_-, p_+\}$ by part (iii). Therefore

$$Y := \Big(((X^{p'})^*)^{(p/p')'}\Big)^* \in \mathrm{UMD}.$$

Then using Proposition 2.4 we have

$$X = \Big(((X^{p'})^*)^{1/p'} \cdot L^p(S)\Big)^* = \big[((X^{p'})^*)^{(p/p')'}, L^2(S)\big]^*_{2/p} = [Y, H]_\theta.$$

Finally part (v) follows from [41, Proposition 4.2.15] as

$$\big((L^p(S';X)^{p_-})^*\big)^{(p_+/p_-)'} = L^{\frac{(p_+-p_-)p}{p_+(p-p_-)}}\big(S';((X^{p_-})^*)^{(p_+/p_-)'}\big).$$

□

Next we note how product quasi-Banach function spaces work under the UMD_{p_-,p_+} property. In particular the following result describes some properties of the space X in our main theorem, Theorem 1.1.

Proposition 3.4 *Let $X_1,\dots,X_m$ be quasi-Banach function spaces. For $j=1,\dots,m$ let $0<p_j^-<p_j^+\le\infty$ and assume that $X_j\in\mathrm{UMD}_{p_j^-,p_j^+}$. Let $X=X_1\dots X_m$, then $X\in\mathrm{UMD}_{p_-,p_+}$, where $\frac{1}{p_-}:=\sum_{j=1}^m\frac{1}{p_j^-}$ and $\frac{1}{p_+}:=\sum_{j=1}^m\frac{1}{p_j^+}$.*

Proof We will prove the proposition for $m=2$. The general case can be proven by induction, cf. the proof of Lemma 4.2. First note that $X^{p_-}=X_1^{p_1^-(p_-/p_1^-)}\cdot X_2^{p_2^-(p_-/p_2^-)}$ is a Banach function space by assumption, so X is p_--convex. By Proposition 2.4 we have

$$\begin{aligned}\big((X^{p_-})^*\big)^{(p_+/p_-)'} &= \big((X_1^{p_-}\cdot X_2^{p_-})^*\big)^{(p_+/p_-)'}\\ &= \big((X_1^{p_1^-})^*\big)^{(p_1^+/p_1^-)'(1-\theta)}\cdot\big((X_2^{p_2^-})^*\big)^{(p_2^+/p_2^-)'\theta}\end{aligned}$$

with

$$\theta=\frac{\frac{1}{p_2^-}-\frac{1}{p_2^+}}{\frac{1}{p_-}-\frac{1}{p_+}}.$$

Thus by Proposition 2.4(v) and Remark 3.2 we know that $X\in\mathrm{UMD}_{p_-,p_+}$. □

The UMD_{p_-,p_+} property of a quasi-Banach function space X looks quite technical. However, as we will see in the next example, this abstract assumption is quite natural for concrete examples of Banach function spaces.

Example 3.5 Let $0<p_-<p_+\le\infty$ and let X be a quasi-Banach function space over an atomless or atomic σ-finite measure space (S,μ). Then $X\in\mathrm{UMD}_{p_-,p_+}$ in each of the following cases:

(i) The Lebesgue spaces $X=L^p(S)$ for $p\in(p_-,p_+)$.
(ii) The Lorentz spaces $X=L^{p,q}(S)$ with $p,q\in(p_-,p_+)$.
(iii) The Orlicz spaces $X=L^\Phi(S)$ for which $t\mapsto\Phi(t^{1/p})$ is a convex function and $t\mapsto\Phi(t^{1/q})$ is a concave function with $p,q\in(p_-,p_+)$.

Note that Theorem 1.1 for the Lebesgue spaces described in Example 3.5(i) follows directly from scalar-valued limited range extrapolation using Fubini's theorem, see also [23].

Proof Note that (i) is a special case of (ii). For (ii) the p_--convexity and p_+-concavity follow from [56, Theorems 4.4 and 5.1]. Furthermore by the definition of $L^{p,q}(S)$ and the duality of Lorentz spaces (see [37]) we have that

$$\big((X^{p_-})^*\big)^{(p_+/p_-)'} = \big(\big(L^{p/p_-,q/p_-}(S)\big)^*\big)^{(p_+/p_-)'} = L^{\frac{(p_+-p_-)p}{p_+(p-p_-)},\frac{(p_+-p_-)q}{p_+(q-p_-)}}(S).$$

Since $L^{r,s}(S) \in \mathrm{UMD}$ for $r, s \in (1,\infty)$ (see [41]), this proves (ii).

For (iii) note that $L^\Phi(S)$ is p-convex and q-concave by [46]. So $Y := \big((X^{p_-})^*\big)^{(p_+/p_-)'}$ is $\frac{(p_+-p_-)p}{p_+(p-p_-)}$-convex and $\frac{(p_+-p_-)q}{p_+(q-p_-)}$-concave. By [54, Theorem 1.f.1] this implies that both Y and Y^* are uniformly convex. Note that Y is an Orlicz space with Young function $\Psi(t) = \varphi^*(t^{(p_+-p_-)/p_-})$, where $\varphi(t) = \Phi(t^{1/p_-})$, and Y^* is an Orlicz space with Young function Ψ^*. Therefore we know by [46, Proposition 1] that both Ψ and its conjugate function Ψ^* satisfy the Δ_2-condition. Thus it follows from [28, Theorem 6.2] that $Y \in \mathrm{UMD}$. □

We end our discussion of the UMD_{p_-,p_+} property by extending the result of Rubio de Francia for the UMD property of Banach function spaces in Proposition 2.5 to the UMD_{p_-,p_+} property of quasi-Banach function spaces.

Theorem 3.6 *Let $0 < p_- < p_+ \le \infty$ and let X be a quasi-Banach function space over a σ-finite measure space (S,μ) such that $X \in \mathrm{UMD}_{p_-,p_+}$. Then there exists an $\varepsilon > 0$ such that such that $X \in \mathrm{UMD}_{p_-q_-,p_+/q_+}$ for all $0 < q_-, q_+ < 1+\varepsilon$.*

Proof By Proposition 3.3(i) we may assume $p_- = 1$ without loss of generality. Note that the case $p_+ = \infty$ was already included in Proposition 2.5, so we restrict our attention to $p_+ < \infty$.

Applying Proposition 2.5 to $(X^*)^{p'}$ yields an $r_1 > 1$ such that $(X^*)^{p'_+r_1} \in \mathrm{UMD}$. Furthermore since $p'_+ > 1$ we know that $X^* \in \mathrm{UMD}$ and thus also $X \in \mathrm{UMD}$. So by Proposition 2.5 applied to X there exists an $r_2 > 1$ such that $X^{r_2} \in \mathrm{UMD}$. Define $r = \min\{r_1, r_2, 1+\frac{1}{p'_+}\}$.

Let $\theta = \frac{r'}{p_++r'} \in (0,1)$ and define the complex interpolation space

$$Y := [(X^r)^*, (X^*)^{p'_+r}]_\theta.$$

Note that since $(X^r)^*, (X^*)^{p'_+r} \in \mathrm{UMD}$, we know by Proposition 2.4(v) that $Y \in \mathrm{UMD}$ as well. Moreover using Proposition 2.4 we have

$$\begin{aligned}
Y &= \big((X^r)^*\big)^{1-\theta} \cdot \big((X^*)^{p'_+r}\big)^\theta \\
&= \big((X^r)^*\big)^{1-\theta} \cdot \Big(\big((X^r)^{1/r}\big)^*\Big)^{p'_+r\theta} \\
&= \big((X^r)^*\big)^{\frac{p_+}{p_++r'}} \cdot \Big(\big((X^r)^*\big)^{1/r} \cdot L^{r'}(S)\Big)^{\frac{p'_+rr'}{p_++r'}} \\
&= \big((X^r)^*\big)^{\frac{p_++p'_+r'}{p_++r'}} \cdot L^{\frac{p_++r'}{p'_+r}}(S)
\end{aligned}$$

Define

$$\alpha := \frac{p_+ + p_+' r'}{p_+ + r'}, \quad \beta := \frac{p_+' r}{p_+ + r'} < p_+'(r-1) < 1.$$

Again by Proposition 2.4 we have

$$Y = \big((X^r)^*\big)^\alpha \cdot L^{\frac{1}{\beta}}(S) = \Big(\big((X^r)^*\big)^{\frac{\alpha}{\alpha+\beta}} \cdot L^{\frac{\alpha+\beta}{\beta}}(S)\Big)^{\alpha+\beta} = \big((X^{\frac{r\alpha}{\alpha+\beta}})^*\big)^{\alpha+\beta}.$$

Take $q_- = \frac{r\alpha}{\alpha+\beta}$. Then we have

$$q_- = \frac{r\alpha}{\alpha+\beta} = \frac{r(p_+ + p_+' r')}{p_+ + p_+' r' + p_+' r} = \frac{p_+ r + p_+' r' + p_+' r}{p_+ + p_+' r' + p_+' r} > 1. \tag{3.2}$$

Moreover

$$\alpha + \beta - \frac{r\alpha}{p_+} = \frac{p_+ + p_+' r' + p_+' r - r - (p_+' - 1)(r + r')}{p_+ + r'} = 1$$

and therefore

$$(p_+/q_-)' = \frac{\alpha+\beta}{\alpha+\beta-\frac{r\alpha}{p_+}} = \alpha+\beta.$$

So $Y = \big((X^{q_-})^*\big)^{(p_+/q_-)'}$ and since $Y \in \mathrm{UMD}$, this implies that $X \in \mathrm{UMD}_{q_-,p_+}$. By applying Proposition 2.5 once more, we can find a $q_+ > 1$ such that $X \in \mathrm{UMD}_{q_-,p_+/q_+}$. By Proposition 3.3(iii), this completes the proof with $\varepsilon = \min\{q_- - 1, q_+ - 1\} > 0$. □

4 Proof of the Main Result

In this section we will prove our main result, Theorem 1.1. The proof of Theorem 1.1 consists of following ingredients:

- The extension of Rubio de Francia's result for the UMD property to the setting of the UMD_{p_-,p_+} property, proven in Theorem 3.6.
- A vector-valued Rubio de Francia iteration algorithm, see Lemma 4.1.
- A result for the product of weighted Bochner spaces, proven below in Lemma 4.2.

We start with the Rubio de Francia iteration algorithm lemma. We remark that Rubio de Francia iteration algorithms also play a key role in scalar-valued extrapolation, see for example [23]. Recall that we write $\phi_{a,b,\ldots}$ for a non-decreasing function $[1,\infty)^2 \to [1,\infty)$, depending on the parameters $a, b, \ldots$ and the dimension d.

Lemma 4.1 *Fix $1 < r < r_+ \leq \infty$ and let Y be a Banach function space over a σ-finite measure space (S,μ) with $Y \in \mathrm{UMD}_{r_+',\infty}$. For all $w \in A_r \cap RH_{(r_+/r)'}$ and* *nonnegative* $u \in L^{r'}(w;Y)$ *there is a nonnegative* $v \in L^{r'}(w;Y)$ *such that* $u \leq v$, $\|v\|_{L^{r'}(w;Y)} \leq 2\|u\|_{L^{r'}(w;Y)}$, *and* $v(\cdot,s)w \in A_1 \cap RH_{r'_+}$ *with*

$$\max\left\{[v(\cdot,s)w]_{A_1}, [v(\cdot,s)w]_{RH_{r'_+}}\right\} \leq \phi_{Y,r,r_+}\left([w]_{A_r}, [w]_{RH_{(r_+/r)'}}\right)$$

for μ*-a.e* $s \in S$.

Proof Fix $w \in A_r \cap RH_{(r_+/r)'}$ and $u \in L^{r'}(w;Y)$. Define

$$u_w := (uw)^{r'_+} \quad \text{and} \quad X := L^{r'/r'_+}\left(w^{1-r'}; Y^{r'_+}\right).$$

Then $u_w \in X$. By Lemma 2.2(iii) we know that for $p := (*)r_+/r'(r-1)+1$ we have $w^{1-r'} \in A_{p'}$ with

$$[w^{1-r'}]_{A_{p'}}^{\frac{1}{p'}} = \left[w^{(r_+/r)'}\right]_{A_p}^{\frac{1}{p}} \leq \left([w]_{A_r}[w]_{RH_{(r_+/r)'}}\right)^{(r_+/r)'}$$

So since

$$p' = 1 + \frac{1}{(*)r_+/r'(r-1)} = \frac{r'}{r'_+}$$

we know that $\widetilde{M}$ is bounded on X by (2.1) with

$$\|\widetilde{M}\|_{X\to X} \leq \phi_{Y,r,r_+}\left([w]_{A_r}, [w]_{RH_{(r_+/r)'}}\right). \tag{4.1}$$

Define

$$v := w^{-1} \cdot \left(\sum_{n=0}^{\infty} \frac{\widetilde{M}^n u_w}{\left(2\|\widetilde{M}\|_{X\to X}\right)^n}\right)^{1/r'_+}$$

where $\widetilde{M}^n$ is given by n iterations of $\widetilde{M}$. As $\widetilde{M}^n u_w$ is nonnegative we know that $u \leq v$. Furthermore $v \in L^{r'}(w;Y)$ with

$$\|v\|_{L^{r'}(w;Y)} = \left\|\sum_{n=0}^{\infty} \frac{\widetilde{M}^n u_w}{\left(2\|\widetilde{M}\|_{X\to X}\right)^n}\right\|_X^{1/r'_+} \leq 2\|u_w\|_X^{1/r'_+} = 2\|u\|_{L^{r'}(w;Y)}.$$

Moreover, since

$$\widetilde{M}\left((vw)^{r'_+}\right)(\cdot,s) \leq 2\|\widetilde{M}\|_{X\to X}(v(\cdot,s)w)^{r'_+},$$

we know that $(v(\cdot,s)w)^{r'_+} \in A_1$ for μ-a.e $s \in S$. Thus it follows from (4.1) and Lemma 2.2 that $v(\cdot,s)w \in A_1 \cap RH_{r'_+}$ with

$$\max\left\{[v(\cdot,s)w]_{A_1}, [v(\cdot,s)w]_{RH_{r'_+}}\right\} \leq \phi_{Y,r,r_+}\left([w]_{A_r}, [w]_{RH_{(r_+/r)'}}\right)$$

for μ-a.e $s \in S$ □

Next we prove the result for the product of weighted Bochner spaces, which follows from the properties of product Banach function spaces in Proposition 2.4 and complex interpolation of weighted Bochner spaces.

Lemma 4.2 *Fix $m \in \mathbb{N}$ and $r \in (1, \infty)$. Let $Y_1, \dots, Y_m$ be reflexive Banach function spaces over a σ-finite measure space (S, μ), let $w_1, \dots, w_m$ be weights and take $\theta_1, \dots, \theta_m \in (0, 1)$ such that $\sum_{j=1}^m \theta_j = 1$. Define $Y = Y_1^{\theta_1} \dots Y_m^{\theta_m}$ and $w = \prod_{j=1}^m w_j^{\theta_j}$. Then we have*

$$L^r(w; Y) = L^r\left(w_1; Y_1\right)^{\theta_1} \dots L^r\left(w_m; Y_m\right)^{\theta_m}$$

Proof We will prove the lemma by induction. For $m = 1$ the result is trivial. Now assume that the statement holds for $m = k - 1$ for some $k \in \mathbb{N}$. We will show the statement for $m = k$.

Let $\tilde{\theta}_j = \frac{\theta_j}{1-\theta_k}$ for $j = 1, \dots, k-1$ and define

$$X = Y_1^{\tilde{\theta}_1} \dots Y_{k-1}^{\tilde{\theta}_{k-1}}, \quad v = \prod_{j=1}^{m} w_j^{\tilde{\theta}_j}.$$

Using Proposition 2.4(i) twice and complex interpolation of weighted Bochner spaces (see [65, Theorem 1.18.5] and [12]) we get

$$\begin{aligned} L^r(w; Y) &= L^r\left(w; [X, Y_k]_{\theta_k}\right) = \left[L^r(v; X), L^r(w_k; Y_k)\right]_{\theta_k} \\ &= L^r\left(v; X\right)^{1-\theta_k} \cdot L^r\left(w_k; Y_k\right)^{\theta_k} \\ &= \left(L^r\left(w_1; Y_1\right)^{\tilde{\theta}_1} \dots L^r\left(w_{k-1}; Y_{k-1}\right)^{\tilde{\theta}_{k-1}}\right)^{1-\theta_k} \cdot \left(L^r\left(w_k; Y_k\right)\right)^{\theta_k} \\ &= L^r\left(w_1; Y_1\right)^{\theta_1} \dots L^r\left(w_k; Y_k\right)^{\theta_k}, \end{aligned}$$

which proves the lemma. □

With these preparatory lemmata we are now ready to prove our main theorem. We first state and prove the result in terms of $(m+1)$-tuples of functions. Afterwards, we present the main result, Theorem 1.1, as a corollary. We write $\phi_{a_j, b_j, \dots}^{j=1,\dots,m}$ for a non-decreasing function $[1, \infty)^{2m} \to [1, \infty)$ depending on the parameters $a_j, b_j, \dots$ for $j = 1, \dots, m$ and the dimension d.

Theorem 4.3 *(Multilinear limited range extrapolation for vector-valued functions) Fix $m \in \mathbb{N}$, let $X_1, \dots, X_m$ be quasi-Banach function spaces over a σ-finite measure space (S, μ) and define $X = X_1 \dots X_m$. Let*

$$\mathcal{F} \subseteq L_+^0(\mathbb{R}^d; X) \times L_+^0(\mathbb{R}^d; X_1) \times \dots \times L_+^0(\mathbb{R}^d; X_m).$$

For $j = 1, \dots, m$ fix $0 < p_j^- < p_j^+ \le \infty$ and assume that $X_j \in \mathrm{UMD}_{p_j^-, p_j^+}$. Moreover assume that for all $p_j \in (p_j^-, p_j^+)$, weights $w_j^{p_j} \in A_{p_j/p_j^-} \cap RH_{(p_j^+/p_j)'}$ and $(f, f_1, \dots, f_m) \in \mathcal{F}$, we have

$$\|f(\cdot, s)\|_{L^p(w^p)} \le \phi^{j=1,\dots,m}_{p_j, p_j^-, p_j^+}\left([w_j^{p_j}]_{A_{p_j/p_j^-}}, [w_j^{p_j}]_{RH_{(p_j^+/p_j)'}}\right) \prod_{j=1}^m \|f_j(\cdot, s)\|_{L^{p_j}(w_j^{p_j})} \tag{4.2}$$

for μ-a.e. $s \in S$, where $\frac{1}{p} = \sum_{j=1}^m \frac{1}{p_j}$ and $w = \prod_{j=1}^m w_j$.

Then for all $p_j \in (p_j^-, p_j^+)$, weights $w_j^{p_j} \in A_{p_j/p_j^-} \cap RH_{(p_j^+/p_j)'}$, and $(f, f_1, \dots, f_m) \in \mathcal{F}$, we have

$$\|f\|_{L^p(w^p;X)} \le \phi^{j=1,\dots,m}_{X_j, p_j, p_j^-, p_j^+}\left([w_j^{p_j}]_{A_{p_j/p_j^-}}, [w_j^{p_j}]_{RH_{(p_j^+/p_j)'}}\right) \prod_{j=1}^m \|f_j\|_{L^{p_j}(w_j^{p_j};X_j)}, \tag{4.3}$$

with w and p as before.

Proof We split the proof in two steps. In the first step we show that the conclusion of the theorem holds for specific choices of $p_j \in (p_j^-, p_j^+)$. In the second step we conclude that the result holds for all $p_j \in (p_j^-, p_j^+)$ through scalar-valued extrapolation.

Step 1 Let $1 < \beta < \min_j \frac{p_j^+}{p_j^-}$. We will first prove the theorem for $p_j := \beta \cdot p_j^-$. Let $(f, f_1, \dots, f_m) \in \mathcal{F}$ and take weights $w_j^{p_j} \in A_{p_j/p_j^-} \cap RH_{(p_j^+/p_j)'}$ for $j = 1, \dots, m$. From Theorem 3.6 and Lemma 2.1(iii) it follows that there exists an $1 < \alpha < \beta$ such that

$$X_j \in \mathrm{UMD}_{\alpha p_j^-, p_j^+} \quad \text{and} \quad w_j^{p_j} \in A_{p_j/(\alpha p_j^-)} \cap RH_{(*) p_j^+/p_j{}'} \tag{4.4}$$

with $[w_j^{p_j}]_{A_{p_j/(\alpha p_j^-)}} \le C_{p_j, p_j^-}\, [w_j^{p_j}]_{A_{p_j/p_j^-}}$ for all $j \in \{1, \dots, m\}$. We define

$$q_j := \alpha p_j^- \quad \text{and} \quad q := \frac{\alpha}{\beta} p.$$

Note that

$$\frac{p}{q} = \frac{\beta}{\alpha} = \frac{p_j}{q_j} \quad \text{and} \quad \frac{1}{q} = \frac{\beta}{\alpha} \sum_{j=1}^m \frac{1}{p_j} = \sum_{j=1}^m \frac{1}{q_j}.$$

Let $u_j \in L^{(p_j/q_j)'}(w_j^{p_j}; (X_j^{q_j})^*)$. By Proposition 3.3 and (4.4) we may apply Lemma 4.1 for $j = 1, \dots, m$ with

$$r = p_j/q_j, \qquad r_+ = p_j^+/q_j, \qquad Y = (X_j^{q_j})^*,$$

and weight $w_j^{p_j}$ to find nonnegative $v_j \in L^{(p_j/q_j)'}(w_j^{p_j}; (X^{q_j})^*)$ such that

- $u_j \leq v_j$.
- $\|v_j\|_{L^{(p_j/q_j)'}(w_j^{p_j};(X_j^{q_j})^*)} \leq 2$.
- $v_j(\cdot, s)w_j^{p_j} \in A_{q_j/p_j^-} \cap RH_{(p_j^+/q_j)'}$ with for μ-a.e. $s \in S$

$$\max\{[v_j(\cdot, s)w_j^{p_j}]_{A_{q_j/p_j^-}}, [v_j(\cdot, s)w_j^{p_j}]_{RH_{(p_j^+/q_j)'}}\}$$
$$\leq \phi_{X_j,p_j,p_j^-,p_j^+}([w_j^{p_j}]_{A_{p_j/p_j^-}}, [w_j^{p_j}]_{RH_{(p_j^+/p_j)'}}).$$

We set $v = \prod_{j=1}^m v_j^{1/q_j} w_j^{p_j/q_j}$ so that

$$\left(\prod_{j=1}^m u_j^{p/p_j}\right) w^p \leq \prod_{j=1}^m v_j^{q/q_j} w_j^{qp_j/q_j} = v^q.$$

Let $(f, f_1, \ldots, f_m) \in \mathcal{F}$. By Fubini's theorem, Hölder's inequality, the assumption (4.2), and the properties of the v_j we have

$$\begin{aligned}
&\int_{\mathbb{R}^d}\int_S f^q \prod_{j=1}^m u_j^{p/p_j} \,\mathrm{d}\mu\, w^p \,\mathrm{d}x \leq \int_S\int_{\mathbb{R}^d} f^q v^q \,\mathrm{d}x\,\mathrm{d}\mu \\
&\leq \int_S \phi_{q_j,p_j^-,p_j^+}^{j=1,\ldots,m}([v_j(\cdot, s)w_j^{p_j}]_{A_{q_j/p_j^-}}, [v_j(\cdot, s)w_j^{p_j}]_{RH_{(p_j^+/q_j)'}}) \\
&\quad\times \prod_{j=1}^m \|f_j(\cdot, s)\|^q_{L^{q_j}(v_j w_j^{p_j})} \,\mathrm{d}\mu(s) \\
&\leq \phi_{X_j,p_j,p_j^-,p_j^+}^{j=1,\ldots,m}([w_j^{p_j}]_{A_{p_j/p_j^-}}, [w_j^{p_j}]_{RH_{(p_j^+/p_j)'}}) \\
&\quad\times \prod_{j=1}^m \left(\int_S\int_{\mathbb{R}^d} f_j^{q_j} v_j w_j^{p_j} \,\mathrm{d}x\,\mathrm{d}\mu\right)^{q/q_j} \qquad (4.5)\\
&\leq \phi_{X_j,p_j,p_j^-,p_j^+}^{j=1,\ldots,m}([w_j^{p_j}]_{A_{p_j/p_j^-}}, [w_j^{p_j}]_{RH_{(p_j^+/p_j)'}}) \\
&\quad\times \prod_{j=1}^m \|f_j^{q_j}\|^{q/q_j}_{L^{p_j/q_j}(w_j^{p_j};X_j^{q_j})} \|v_j\|^{q/q_j}_{L^{(p_j/q_j)'}(w_j^{p_j};(X_j^{q_j})^*)} \\
&\leq \phi_{X_j,p_j,p_j^-,p_j^+}^{j=1,\ldots,m}([w_j^{p_j}]_{A_{p_j/p_j^-}}, [w_j^{p_j}]_{RH_{(p_j^+/p_j)'}}) \\
&\quad\times \left(\prod_{j=1}^m \|f_j\|_{L^{p_j}(w_j^{p_j};X_j)} \|u_j\|^{1/q_j}_{L^{(p_j/q_j)'}(w_j^{p_j};(X_j^{q_j})^*)}\right)^q.
\end{aligned}$$

Now by Lemma 4.2 with

$$r = p/q, \qquad Y_j = X_j^{q_j}, \qquad \theta_j = q/q_j$$

and weights $w_j^{p_j}$, Proposition 2.4(iii) and the duality of Bochner spaces (see [41, Corollary 1.3.22]), we have

$$L^{p/q}(w^p; X^q)^* = L^{(p_1/q_1)'}\big(w_1^{p_1}; (X_1^{q_1})^*\big)^{q/q_1} \dots L^{(p_m/q_m)'}\big(w_m^{p_m}; (X_m^{q_m})^*\big)^{q/q_m}.$$

Thus, picking $u \in L^{p/q}(w^p; X^q)^*$ of norm 1, by taking an infimum over all decompositions $u = \prod_{j=1}^m u_j^{p/p_j}$ with $u_j \in L^{(p_j/q_j)'}\big(w_j^{p_j}; (X_j^{q_j})^*\big)$, we may conclude from (4.5) that

$$\int_{\mathbb{R}^d}\int_S f^q u \,\mathrm{d}\mu\, w^p \,\mathrm{d}x \le \phi^{j=1,\dots,m}_{X_j,p_j,p_j^-,p_j^+}\big([w_j^{p_j}]_{A_{p_j/p_j^-}}, [w_j^{p_j}]_{RH_{(p_j^+/p_j)'}}\big) \times \Big(\prod_{j=1}^m \|f_j\|_{L^{p_j}(w_j^{p_j};X_j)}\Big)^q.$$

Thus, the result for these specific p_j's follows from

$$\|f\|^q_{L^p(w^p;X)} = \|f^q\|_{L^{p/q}(w^p;X^q)} = \sup_{\|u\|_{L^{p/q}(w^p;X^q)^*}=1} \int_{\mathbb{R}^d}\int_S f^q u \,\mathrm{d}\mu\, w^p\,\mathrm{d}x.$$

Step 2: We may finish the proof for general p_j's by appealing to the scalar-valued limited range multilinear extrapolation result by Cruz-Uribe and Martell [21]. Indeed, we define a new family

$$\widetilde{\mathcal{F}} := \big\{\big(\|f\|_X, \|f_1\|_{X_1}, \dots, \|f_m\|_{X_m}\big) : (f, f_1, \dots, f_m) \in \mathcal{F}\big\}.$$

Then $\widetilde{\mathcal{F}} \subset L^0_+(\mathbb{R}^d)^{m+1}$ and by Step 1 we have

$$\|\tilde f\|_{L^p(w^p)} \le \phi^{j=1,\dots,m}_{X,p_j^-,p_j^+}\big([w_j^{p_j}]_{A_{p_j/p_j^-}}, [w_j^{p_j}]_{RH_{(p_j^+/p_j)'}}\big) \prod_{j=1}^m \|\tilde f_j\|_{L^{p_j}(w_j^{p_j})}$$

for certain $p_j \in (p_j^-, p_j^+)$, all $(\tilde f, \tilde f_1, \dots, \tilde f_m) \in \widetilde{\mathcal{F}}$, and all weights $w_j^{p_j} \in A_{p_j/p_j^-} \cap RH_{(p_j^+/p_j)'}$. The result for general $p_j \in (p_j^-, p_j^+)$ then follows directly from [21, Theorem 1.3 and Corollary 1.11], proving the assertion. □

Finally, we will prove the main result from the introduction, which is a direct corollary of Theorem 4.3.

Proof of Theorem 1.1 We wish to apply Theorem 4.3 to the collection

$$\mathcal{F} = \big\{(|\widetilde{T}(f_1, \dots, f_m)|, |f_1|, \dots, |f_m|) : \ f_j : \mathbb{R}^d \to X_j \text{ simple}\big\}.$$

Our assumption implies that there are $p_j \in (p_j^-, p_j^+)$ so that for all weights $w_j^{p_j} \in A_{p_j/p_j^-} \cap RH_{(p_j^+/p_j)'}$ the a priori estimate (4.2) in Theorem 4.3 holds. By appealing to the scalar-valued limited range multilinear extrapolation result [21] we may conclude that (4.2) in fact holds for all $p_j \in (p_j^-, p_j^+)$ and weights $w_j^{p_j} \in A_{p_j/p_j^-} \cap RH_{(p_j^+/p_j)'}$. Thus, Theorem 4.3 implies that

$$\|\widetilde{T}(f_1, \ldots, f_m)\|_{L^p(w^p;X)} \leq C \prod_{j=1}^m \|f_j\|_{L^{p_j}(w_j^{p_j};X_j)} \tag{4.6}$$

for all simple functions $f_j \colon \mathbb{R}^d \to X_j$, where C depends only on the X_j, p_j, and the characteristic constants of the weights. If T is m-linear, then (4.6) extends directly to all $f_j \in L^{p_j}(w_j^{p_j}; X_j)$ by density. If T is m-sublinear and positive valued, then we fix simple functions $f_j \colon \ \colon \mathbb{R}^d \to X_j$ for $j \in \{2, \ldots, m\}$. For any pair of simple functions $f_1, g_1 : \mathbb{R}^d \to X_1$ we have

$$\begin{aligned}\widetilde{T}(f_1, \ldots, f_m) &= \widetilde{T}(f_1 - g_1 + g_1, f_2 \ldots, f_m)\\ &\leq \widetilde{T}(f_1 - g_1, f_2 \ldots, f_m) + \widetilde{T}(g_1, f_2 \ldots, f_m)\end{aligned}$$

so that

$$\begin{aligned}&\|\widetilde{T}(f_1, \ldots, f_m) - \widetilde{T}(g_1, f_2 \ldots, f_m)\|_{L^p(w^p;X)}\\ &\quad\leq \|\widetilde{T}(f_1 - g_1, f_2 \ldots, f_m)\|_{L^p(w^p;X)}\\ &\quad\leq C\,\|f_1 - g_1\|_{L^{p_1}(w_1^{p_1};X_1)} \prod_{j=2}^m \|f_j\|_{L^{p_j}(w_j^{p_j};X_j)}.\end{aligned}$$

Thus, (4.6) extends to arbitrary $f_1 \in L^{p_1}(w_1^{p_1}; X_1)$ by density. Iterating this argument for $j = 2, \ldots m$ proves the result. □

5 Applications

In this section we apply our main result to various operators, for which we obtain new vector-valued bounds.

5.1 The Bilinear Hilbert Transform

For $d = 1$, The bilinear Hilbert transform BHT is defined by

$$\mathrm{BHT}(f, g)(x) = \mathrm{p.\,v.} \int_{\mathbb{R}} f(x-t)g(x+t)\frac{\mathrm{d}t}{t}.$$

After its initial introduction by Calderón, it took thirty years until L^p estimates were established by Lacey and Thiele [50]. They showed that for $p_1, p_2 \in (1, \infty]$ with $\frac{1}{p} = \frac{1}{p_1} + \frac{1}{p_2} < \frac{3}{2}$ one has

$$\|\mathrm{BHT}(f, g)\|_{L^p} \leq C\|f\|_{L^{p_1}}\|g\|_{L^{p_2}}. \tag{5.1}$$

As for weighted bounds, the first results were obtained by Culiuc et al. [24], and through the extrapolation result of Cruz-Uribe and Martell the range of exponents was increased [21], in particular recovering the full range of exponents for the unweighted result (5.1). It was already shown in [21] that this result implies corresponding vector-valued bounds for BHT for certain ℓ^s-spaces. Moreover, vector-valued bounds for BHT have also been considered by Benea and Muscalu [5]. In particular, they consider functions taking values in iterated L^s-spaces, see [5, Theorem 8] including the case $s = \infty$.

Through our main result we are able to obtain a new bounded vector-valued extension of the bilinear Hilbert transform. By combining the weighted estimates in [21, Theorem 1.18] with Theorem 1.1, we get:

Theorem 5.1 *Let $q_1, q_2 \in (1, \infty)$ so that $\frac{1}{q_1} + \frac{1}{q_2} < 1$. For $j \in \{1, 2\}$, define*

$$p_j^- := \frac{2q_j}{1 + q_j}, \quad p_j^+ := 2q_j.$$

Let $X = X_1 \cdot X_2$, where X_1, X_2 are quasi-Banach function spaces over a σ-finite measure space (S, μ) satisfying $X_j \in \mathrm{UMD}_{p_j^-, p_j^+}$. Then for all p_1, p_2 with $p_j \in (p_j^-, p_j^+)$ and all weights w_1, w_2 satisfying $w_j^{p_j} \in A_{p_j/p_j^-} \cap RH_{(p_j^+/p_j)'}$ we have

$$\|\widetilde{\mathrm{BHT}}(f, g)\|_{L^p(w^p; X)} \leq C' \|f\|_{L^{p_1}(w_1^{p_1}; X_1)} \|g\|_{L^{p_2}(w_2^{p_2}; X_2)}$$

for all $f \in L^{p_1}(w_1^{p_1}; X_1)$, $g \in L^{p_2}(w_2^{p_2}; X_2)$, where $\frac{1}{p} = \frac{1}{p_1} + \frac{1}{p_2}$, $w = w_1 w_2$, and where $C' > 0$ depends only on the X_j, p_j, q_j, and the characteristic constants of the weights.

By Example 2.3 we have $\ell^s = \ell^{s_1} \cdot \ell^{s_2}$ for $s_1, s_2 \in (0, \infty)$ and $\frac{1}{s} = \frac{1}{s_1} + \frac{1}{s_2}$. Thus, we recover [21, Theorem 1.29] by Example 3.5. It is implicit from the arguments in [24] that there are more general weighted estimates for BHT leading to a wider range of vector-valued extensions. For a technical discussion on this, we refer the reader to [21, Sect. 5].

Furthermore by Proposition 3.3(v) we can also handle iterated L^s-spaces as considered by Benea and Muscalu [5], but our results do not overlap as we do not obtain bounds involving L^∞-spaces. Such spaces might be in the scope of a generalized version of our main theorem using multilinear weight classes combined with a multilinear UMD condition, see also Remark 1.3 and [58].

Finally, we mention the vector-valued bounds obtained by Hytönen, Lacey, and Parissis [40] for the related bilinear quartile operator (the Fourier-Walsh model of BHT). They consider estimates involving triples of more general UMD Banach spaces with so called *quartile type q*. It is unknown whether these estimates hold for BHT itself. Note that a Banach function space $X \in \mathrm{UMD}_{p_-, p_+}$ has quartile type $\max\{p_-', p_+\}$ by Proposition 3.3(iv) and [40, Proposition 4.1].

5.2 Multilinear Calderón–Zygmund Operators

Let T be an m-linear operator, initially defined for m-tuples $f_1, \ldots, f_m \in C_c^\infty(\mathbb{R}^d)$, that satisfies

$$T(f_1, \ldots, f_m)(x) = \int_{(\mathbb{R}^d)^m} K(x, y_1, \ldots, y_m) \prod_{j=1}^{m} f_j(y_j)\,\mathrm{d}y,$$

whenever $x \notin \cap_{j=1}^m \operatorname{supp} f_j$, where K is a kernel defined in $(\mathbb{R}^d)^{m+1}$ outside of the diagonal $y_0 = y_1 = \cdots = y_m$. If K satisfies the estimate

$$|\partial_{y_0}^{\alpha_0} \ldots \partial_{y_m}^{\alpha_m} K(y_0, \ldots, y_m)| \le \frac{C}{\left(\sum_{j,k=0}^m |y_j - y_k|\right)^{md+|\alpha_1|+\cdots+|\alpha_m|}}$$

for all multi-indices α_j so that $\sum_{j=1}^m |\alpha_j| \le 1$ and if there exist $p_1, \ldots, p_m$ so that T extends to a bounded operator $L^{p_1} \times \cdots L^{p_m} \to L^p$ with $\frac{1}{p} = \sum_{j=1}^m \frac{1}{p_j}$, then T is called an m-linear Calderón-Zygmund operator.

Multilinear Calderón-Zygmund operators first appeared in the work [20] by Coifman and Meyer. Weighted estimates for these operators have been considered for example by Grafakos and Torres in [33] and subsequently by Grafakos and Martell in [32], where it was shown that for all $p_j \in (1, \infty)$, all weights $w_j^{p_j} \in A_{p_j}$, and all $f_j \in L^{p_j}(w_j^{p_j})$ we have

$$\left\| T(f_1, \ldots, f_m) \right\|_{L^p(w^p)} \le C \prod_{j=1}^{m} \|f_j\|_{L^{p_j}(w_j^{p_j})},$$

where $w = \prod_{j=1}^m w_j$ and $\frac{1}{p} = \sum_{j=1}^m \frac{1}{p_j}$, and where C depends only on the characteristic constants of the weights. Thus, by Theorem 1.1 we obtain the following result:

Theorem 5.2 *Let T be an m-linear Calderón-Zygmund operator and suppose $X_1, \ldots, X_m \in$ UMD. Then for all $p_j \in (1, \infty)$, all weights $w_j^{p_j} \in A_{p_j}$, and all $f_j \in L^{p_j}(w_j^{p_j}; X_j)$ we have*

$$\left\| \widetilde{T}(f_1, \ldots, f_m) \right\|_{L^p(w^p; X)} \le C' \prod_{j=1}^{m} \|f_j\|_{L^{p_j}(w_j^{p_j}; X_j)},$$

where $X = X_1 \ldots X_m$, $w = \prod_{j=1}^m w_j$, $\frac{1}{p} = \sum_{j=1}^m \frac{1}{p_j}$, and where C' depends only on the X_j, p_j, and the characteristic constants of the weights.

This result is new, as previously only ℓ^s-valued extensions had been considered in [32].

We wish to point out that, using a more appropriate multilinear weight condition, more general weighted bounds for multilinear Calderón-Zygmund operators have been found in [52].

5.3 Limited Range Extrapolation

The remaining examples are for the linear case $m = 1$.

Example 5.3 (*Fourier multipliers I*) For $a < b$, $q \in [1, \infty)$, and a function $m : [a, b] \to \mathbb{C}$ we define the q-variation norm

$$\|m\|_{V^q([a,b])} := \|m\|_{L^\infty([a,b])} + \sup \left(\sum_{j=0}^{n-1} |m(t_{j+1}) - m(t_j)|^q \right)^{\frac{1}{q}},$$

where the supremum is taken over all partitions $a = t_0 < \ldots < t_n = b$ of the interval $[a, b]$. Let $\mathscr{D} := \{ \pm (2^k, 2^{k+1}] : k \in \mathbb{Z} \}$ be the dyadic decomposition of $\mathbb{R}$. Then we define a class of multipliers

$$V^q(\mathscr{D}) := \{ m : \mathbb{R} \to \mathbb{C} : \sup_{I \in \mathscr{D}} \|m|_I\|_{V^q(I)} < \infty \}.$$

For $q > 2$ and $p_+ := 2\left(\frac{q}{2}\right)'$ it was shown by Król [48, Theorem A(ii)] that for all $p \in [2, p_+)$, $w \in A_{p/2} \cap RH_{(p_+/p)'}$ and $m \in V^q(\mathscr{D})$ the Fourier multiplier T_m defined by $\mathscr{F}(T_m f) = m\mathscr{F} f$ satisfies

$$\|T_m\|_{L^p(w) \to L^p(w)} < \infty.$$

Therefore one may readily apply Theorem 1.1 with $p_- = 2$ to the linear operator T_m. So for any Banach function space X such that $X \in \mathrm{UMD}_{2,p_+}$ we find for all $p \in (2, p_+)$, all $w \in A_{p/2} \cap RH_{(p_+/p)'}$ and $m \in V^q(\mathscr{D})$ that

$$\|\widetilde{T}_m\|_{L^p(w;X) \to L^p(w;X)} < \infty.$$

Note that [48, Theorem A(i)] was already extrapolated to the vector-valued setting by Amenta, Veraar, and the first author in [2], proving that for $m \in V^q(\mathscr{D})$ with $q \in [1, 2]$ the Fourier multiplier T_m has a bounded vector-valued extension for Banach function spaces $X \in \mathrm{UMD}_{q,\infty}$. Furthermore extensions of [48, Theorem A] for operator-valued Fourier multipliers have been obtained in [1].

Example 5.4 (*Riesz transforms associated with elliptic operators*) Let $A \in L^\infty(\mathbb{R}^d; \mathbb{C}^{d\times d})$ satisfy an ellipticity condition $\mathrm{Re}(A(x)\xi \cdot \overline{\xi}) \geq \lambda|\xi|^2$ for a.e. $x \in \mathbb{R}^d$, and all $\xi \in \mathbb{C}^d$. Then we may consider a second order divergence form operator

$$L := -\operatorname{div}(A\nabla f),$$

defined on L^2, which due to the ellipticity condition on A generates an analytic semigroup $(e^{-tL})_{t>0}$ in L^2. Let $1 \leq p_- < p_+ \leq \infty$. If both the semigroup and the gradient family $(\sqrt{t}\nabla e^{-tL})_{t>0}$ satisfy L^{p_-}–L^{p_+} off-diagonal estimates, then the Riesz transform $R := \nabla L^{-1/2}$ is a bounded operator in $L^p(w)$ for all $p \in (p_-, p_+)$ and all weights $w \in A_{p/p_-} \cap RH_{(p_+/p)'}$, see [4,7]. The values of p_- and p_+ for which such off-diagonal estimates hold depend on the dimension d and on the matrix-valued function A and are studied in detail in [3]. The result we obtain is that if a Banach function space X satisfies $X \in \mathrm{UMD}_{p_-,p_+}$, then for all $p \in (p_-, p_+)$ and all weights $w \in A_{p/p_-} \cap RH_{(p_+/p)'}$ we have

$$\|\widetilde{R}\|_{L^p(w;X)\to L^p(w;X)} < \infty.$$

This result is new in the sense that previously such bounds were previously only known for $X = \ell^s$ through the limited range extrapolation result in [4].

Next, we consider a class of operators satisfying a certain sparse domination property. A collection $\mathcal{S}$ of cubes in $\mathbb{R}^d$ is called *sparse* if there is a pairwise disjoint collection of sets $(E_Q)_{Q\in\mathcal{S}}$ so that for each $Q \in \mathcal{S}$ we have $E_Q \subseteq Q$ and $|Q| \le 2|E_Q|$. We say that a (sub)linear operator T satisfies the sparse domination property with parameters $1 \le p_- < p_+ \le \infty$ if there is a $C > 0$ so that for all compactly supported smooth functions $f, g : \mathbb{R}^d \to \mathbb{C}$ we have

$$|\langle Tf, g\rangle| \le C \sup_{\mathcal{S}\text{ sparse}} \sum_{Q\in\mathcal{S}} \langle |f|^{p_-}\rangle_Q^{\frac{1}{p_-}} \langle |g|^{p_+}\rangle_Q^{\frac{1}{p_+}} |Q|, \tag{5.2}$$

where the supremum runs over all sparse collections of cubes $\mathcal{S}$. For an operator T we denote the optimal constant C appearing in (5.2) by $\|T\|_{S(p_-,p_+)}$. Estimates in the form (5.2) were first considered in [7] where it was shown that

$$\|T\|_{S(p_-,p_+)} < \infty \Rightarrow \|T\|_{L^p(w)\to L^p(w)} < \infty$$

for $p \in (p_-, p_+)$ and $w \in A_{p/p_-} \cap RH_{(p_+/p)'}$ by giving a quantitative estimate in terms of the characteristic constants of the weight. Thus, we may readily apply Theorem 1.1 with $m = 1$ to any linear or positive-valued sublinear T such that $\|T\|_{S(p_-,p_+)} < \infty$. This yields the following result:

Theorem 5.5 *Let T be a linear or a positive-valued sublinear operator and let X be a Banach function space over a σ-finite measure space (S, μ). Assume that for all simple functions $f : \mathbb{R}^d \to X$ the function $\widetilde{T} f(x, s) := T(f(\cdot, s))(x)$ is well-defined and strongly measurable.*

If there are $1 \le p_- < p_+ \le \infty$ such that

$$X \in \mathrm{UMD}_{p_-,p_+}, \qquad \|T\|_{S(p_-,p_+)} < \infty,$$

then for all $p \in (p_-, p_+)$, all weights $w \in A_{p/p_-} \cap RH_{(p_+/p)'}$, and all $f \in L^p(w; X)$, we have

$$\|\widetilde{T} f\|_{L^p(w;X)} \le C\|f\|_{L^p(w;X)}, \tag{5.3}$$

where C depends only on X, p, p_-, p_+, and the characteristic constants of w.

We emphasize again that if T is linear, then $\widetilde{T} f$ is automatically well-defined and strongly measurable for any simple function $f : \mathbb{R}^d \to X$, see also Remark 1.2.

We conclude this section by giving several examples of operators satisfying sparse bounds.

Example 5.6 (*Fourier multipliers II*) For each $\delta \geq 0$, the Bochner-Riesz multiplier B_δ is defined as the Fourier multiplier $\mathscr{F}(B_\delta f) = (1-|\xi|^2)_+^\delta \mathscr{F} f$, where $t_+ = \max(t, 0)$. For $\delta \geq (d-1)/2$, B_δ satisfies weighted bounds $\|B_\delta\|_{L^p(w)\to L^p(w)} < \infty$ for any $p \in (1, \infty)$ and any $w \in A_p$, see [11,26,63].

The situation is more complicated when $0 < \delta < (d-1)/2$ and weighted bounds for such δ have, for example, been considered in [17,18,27]. The idea to quantify weighted bounds for B_δ for $0 < \delta < (d-1)/2$ through sparse domination was initiated by Benea, Bernicot, and Luque [6]. It was shown by Lacey, Mena, and Reguera that for this range of δ there are explicit subsets $R_{\delta,d}$ of the plane so that

$$\|B_\delta\|_{S(p_-,p_+)} < \infty$$

for $(p_-, p_+) \in R_{\delta,d}$, see [51]. We also refer the reader to the recent work by Kesler and Lacey [47] containing certain sparse endpoint bounds in dimension $d = 2$.

As far as we know, the only vector-valued estimates that have been shown for B_δ have been for $X = \ell^s$, see [6]. For any p_-, p_+ and δ for which $\|B_\delta\|_{S(p_-,p_+)} < \infty$, we obtain by Theorem 5.5 that inequality (5.3) with $\widetilde{T} = \widetilde{B}_\delta$ holds for any Banach function space X satisfying $X \in \mathrm{UMD}_{p_-,p_+}$, yielding new vector-valued estimates.

Example 5.7 (*Spherical maximal operators*) Let (S^{d-1}, σ) denote the unit sphere in $\mathbb{R}^d$ equipped with its normalized Euclidean surface measure σ. For a smooth function f on $\mathbb{R}^d$ we denote by $A_r f(x)$ the average of f over the sphere centered at x of radius $r > 0$, i.e.,

$$A_r f(x) := \int_{S^{d-1}} f(x - r\omega)\, \mathrm{d}\sigma(\omega).$$

We respectively define the lacunary spherical maximal operator and the full spherical maximal operator by

$$M_{\mathrm{lac}} f := \sup_{k\in\mathbb{Z}} |A_{2^k} f|, \quad M_{\mathrm{full}} f := \sup_{r>0} |A_r f|,$$

the latter having been introduced by Stein [64] and the former having been studied by Calderón [16]. It was shown by Lacey [49] that for explicit subsets L_d, F_d of the plane we have

$$\begin{aligned} \|M_{\mathrm{lac}}\|_{S(p_-,p_+)} &< \infty, \quad \text{for } (p_-, p_+) \in L_d, \\ \|M_{\mathrm{full}}\|_{S(p_-,p_+)} &< \infty, \quad \text{for } (p_-, p_+) \in F_d. \end{aligned}$$

These results recover the previous known L^p-bounds for these operators and yield weighted bounds.

To apply Theorem 5.5 to M_{lac} and M_{full}, one needs to check that these operators have well-defined and strongly measurable extensions to X-valued simple functions with X a Banach function space over (S, μ). This can be checked as in [34, Lemma 3.1]. Therefore it follows from Theorem 5.5 that if $(p_-, p_+) \in L_d$ or $(p_-, p_+) \in F_d$, then for any Banach function space $X \in \mathrm{UMD}_{p_-,p_+}$ we obtain the bound (5.3) for $\widetilde{T} = \widetilde{M}_{\mathrm{lac}}$ or $\widetilde{T} = \widetilde{M}_{\mathrm{full}}$ respectively. As far as we know, this is the first instance that vector-valued extensions have been considered for these operators.

Acknowledgements The authors thank Mark Veraar and Dorothee Frey for their helpful comments on the draft. Moreover the authors thank Alex Amenta for his suggestion to consider the multilinear setting. Finally the authors would like to thank the anonymous referees for their suggestions.

Open Access This article is distributed under the terms of the Creative Commons Attribution 4.0 International License (http://creativecommons.org/licenses/by/4.0/), which permits unrestricted use, distribution, and reproduction in any medium, provided you give appropriate credit to the original author(s) and the source, provide a link to the Creative Commons license, and indicate if changes were made.

References

1. Amenta, A., Lorist, E., Veraar, M.C.: Fourier multipliers in Banach function spaces with UMD concavifications. Published electronically in Trans. Am. Soc. (2018)
2. Amenta, A., Lorist, E., Veraar, M.C.: Rescaled extrapolation for vector-valued functions. Publ. Mat. **63**(1), 155–182 (2019)
3. Auscher, P.: On necessary and sufficient conditions for L^p-estimates of Riesz transforms associated to elliptic operators on $\mathbb{R}^n$ and related estimates. Mem. Am. Math. Soc. **186**(871), xviii+75 (2007)
4. Auscher, P., Martell, J.M.: Weighted norm inequalities, off-diagonal estimates and elliptic operators. I. General operator theory and weights. Adv. Math. **212**(1), 225–276 (2007)
5. Benea, C., Muscalu, C.: Multiple vector-valued inequalities via the helicoidal method. Anal. PDE **9**(8), 1931–1988 (2016)
6. Benea, C., Bernicot, F., Luque, T.: Sparse bilinear forms for Bochner Riesz multipliers and applications. Trans. Lond. Math. Soc. **4**(1), 110–128 (2017)
7. Bernicot, F., Frey, D., Petermichl, S.: Sharp weighted norm estimates beyond Calderón–Zygmund theory. Anal. PDE **9**(5), 1079–1113 (2016)
8. Betancor, J.J., Fariña, J.C., Rodríguez-Mesa, L.: Hardy–Littlewood and UMD Banach lattices via Bessel convolution operators. J. Oper. Theory **67**(2), 349–368 (2012)
9. Bourgain, J.: Some remarks on Banach spaces in which martingale difference sequences are unconditional. Ark. Mat. **21**(2), 163–168 (1983)
10. Bourgain, J.: Extension of a result of Benedek, Calderón and Panzone. Ark. Mat. **22**(1), 91–95 (1984)
11. Buckley, S.M.: Estimates for operator norms on weighted spaces and reverse Jensen inequalities. Trans. Am. Math. Soc. **340**(1), 253–272 (1993)
12. Bukhvalov, A.V.: Interpolation of linear operators in spaces of vector functions and with a mixed norm. Sibirsk. Mat. Zh. **28**(1), i, 37–51 (1987)
13. Burkholder, D.L.: A geometric condition that implies the existence of certain singular integrals of Banach-space-valued functions. In: Conference on harmonic analysis in honor of Antoni Zygmund, vols. I, II (Chicago, Ill., 1981). Wadsworth Mathematics Series, pp. 270–286. Wadsworth, Belmont, CA (1983)
14. Burkholder, D.L.: Martingales and singular integrals in Banach spaces. In: Handbook of the Geometry of Banach Spaces, vol. I, pp. 233–269. North-Holland, Amsterdam (2001)
15. Calderón, A.P.: Intermediate spaces and interpolation, the complex method. Stud. Math. **24**, 113–190 (1964)
16. Calderón, C.P.: Lacunary spherical means. Ill. J. Math. **23**(3), 476–484 (1979)
17. Carro, M.J., Duoandikoetxea, J., Lorente, M.: Weighted estimates in a limited range with applications to the Bochner–Riesz operators. Indiana Univ. Math. J. **61**(4), 1485–1511 (2012)
18. Christ, M.: On almost everywhere convergence of Bochner–Riesz means in higher dimensions. Proc. Am. Math. Soc. **95**(1), 16–20 (1985)

19. Cioica, P., Cox, S.G., Veraar, M.C.: Stochastic integration in quasi-Banach spaces (2018). arXiv:1804.08947
20. Coifman, R.R., Meyer, Y.: On commutators of singular integrals and bilinear singular integrals. Trans. Am. Math. Soc. **212**, 315–331 (1975)
21. Cruz-Uribe, D., Martell, J.M.: Limited range multilinear extrapolation with applications to the bilinear Hilbert transform. Math. Ann. **371**(1–2), 615–653 (2018)
22. Cruz-Uribe, D., Martell, J.M., Pérez, C.: Extrapolation from A_∞ weights and applications. J. Funct. Anal. **213**(2), 412–439 (2004)
23. Cruz-Uribe, D.V., Martell, J.M., Pérez, C.: Weights, extrapolation and the theory of Rubio de Francia. In: Operator Theory: Advances and Applications, vol. 215. Birkhäuser/Springer Basel AG, Basel (2011)
24. Culiuc, A., Di Plinio, F., Ou, Y.: Domination of multilinear singular integrals by positive sparse forms. J. Lond. Math. Soc. (2) **98**(2), 369–392 (2018)
25. Deleaval, L., Kriegler, C.: Dunkl spectral multipliers with values in UMD lattices. J. Funct. Anal. **272**(5), 2132–2175 (2017)
26. Duoandikoetxea, J., Rubio de Francia, J.L.: Maximal and singular integral operators via Fourier transform estimates. Invent. Math. **84**(3), 541–561 (1986)
27. Duoandikoetxea, J., Moyua, A., Oruetxebarria, O., Seijo, E.: Radial A_p weights with applications to the disc multiplier and the Bochner–Riesz operators. Indiana Univ. Math. J. **57**(3), 1261–1281 (2008)
28. Fernandez, D .L., Garcia, J .B.: Interpolation of Orlicz-valued function spaces and U.M.D. property. Studia Math. **99**(1), 23–40 (1991)
29. García-Cuerva, J., Rubio de Francia, J.L.: Weighted Norm Inequalities and Related Topics. North-Holland Mathematics Studies. vol. 116. Notas de Matemática, 104. North-Holland, Amsterdam (1985)
30. García-Cuerva, J., Macías, R., Torrea, J.L.: The Hardy–Littlewood property of Banach lattices. Isr. J. Math. **83**(1–2), 177–201 (1993)
31. Grafakos, L.: Classical Fourier Analysis. Graduate Texts in Mathematics, 3rd edn, vol. 249. Springer, New York (2014)
32. Grafakos, L., Martell, J.M.: Extrapolation of weighted norm inequalities for multivariable operators and applications. J. Geom. Anal. **14**(1), 19–46 (2004)
33. Grafakos, L., Torres, R.H.: Maximal operator and weighted norm inequalities for multilinear singular integrals. Indiana Univ. Math. J. **51**(5), 1261–1276 (2002)
34. Hänninen, T.S., Lorist, E.: Sparse domination for the lattice Hardy–Littlewood maximal operator. Proc. Am. Math. Soc. **147**(1), 271–284 (2019)
35. Harboure, E., Macías, R., Segovia, C.: Extrapolation results for classes of weights. Am. J. Math. **110**(3), 383–397 (1988)
36. Hong, G.: Banach lattice-valued q-variation and convexity (2014). arXiv:1410.1575
37. Hunt, R.A.: On $L(p, q)$ spaces. Enseignement Math. (2) **12**, 249–276 (1966)
38. Hytönen, T.P., Vähäkangas, A.V.: The local non-homogeneous Tb theorem for vector-valued functions. Glasg. Math. J. **57**(1), 17–82 (2015)
39. Hytönen, T.P., Pérez, C., Rela, E.: Sharp reverse Hölder property for A_∞ weights on spaces of homogeneous type. J. Funct. Anal. **263**(12), 3883–3899 (2012)
40. Hytönen, T.P., Lacey, M.T., Parissis, I.: The vector valued quartile operator. Collect. Math. **64**(3), 427–454 (2013)
41. Hytönen, T.P., van Neerven, J.M.A.M., Veraar, M.C., Weis, L.: Analysis in Banach Spaces. Volume I: Martingales and Littlewood–Paley Theory. Ergebnisse der Mathematik und ihrer Grenzgebiete, vol. 63. Springer, Cham (2016)
42. Johnson, R., Neugebauer, C.J.: Change of variable results for A_p- and reverse Hölder RH_r-classes. Trans. Am. Math. Soc. **328**(2), 639–666 (1991)
43. Kalton, N.J.: Convexity conditions for non-locally convex lattices. Glasg. Math. J. **25**(2), 141–152 (1984)
44. Kalton, N.J., Mitrea, M.: Stability results on interpolation scales of quasi-Banach spaces and applications. Trans. Am. Math. Soc. **350**(10), 3903–3922 (1998)
45. Kalton, N.J., Mayboroda, S., Mitrea, M.: Interpolation of Hardy-Sobolev–Besov–Triebel–Lizorkin spaces and applications to problems in partial differential equations. In: Interpolation Theory and Applications, vol. 445. Contemporary Mathematics, pp. 121–177. American Mathematical Society, Providence (2007)

46. Kamińska, A.: Indices, convexity and concavity in Musielak–Orlicz spaces. Funct. Approx. Comment. Math. **26**, 67–84 (1998) (Dedicated to Julian Musielak)
47. Kesler, R., Lacey, M.T.: Sparse endpoint estimates for Bochner–Riesz multipliers on the plane. Collect. Math. **69**(3), 427–435 (2018)
48. Król, S.: Fourier multipliers on weighted L^p spaces. Math. Res. Lett. **21**(4), 807–830 (2014)
49. Lacey, M.T.: Sparse bounds for spherical maximal functions (2017). arXiv:1702.08594
50. Lacey, M.T., Thiele, C.: On Calderón's conjecture. Ann. Math. (2) **149**(2), 475–496 (1999)
51. Lacey, M.T., Mena, D., Reguera, M.C.: Sparse bounds for Bochner–Riesz multipliers. J. Fourier Anal. Appl. (2017). https://doi.org/10.1007/s13348-018-0214-1
52. Lerner, A.K., Ombrosi, S., Pérez, C., Torres, R.H., Trujillo-González, R.: New maximal functions and multiple weights for the multilinear Calderón-Zygmund theory. Adv. Math. **220**(4), 1222–1264 (2009)
53. Li, K., Martell, J.M., Ombrosi, S.: Extrapolation for multilinear Muckenhoupt classes and applications to the bilinear Hilbert transform (2018). arXiv:1802.03338
54. Lindenstrauss, J., Tzafriri, L.: Classical Banach spaces. II. Ergebnisse der Mathematik und ihrer Grenzgebiete, vol. 97. Springer, Berlin (1979)
55. Lozanovskii, G.Ya.: On some Banach lattices. Sib. Math. J. **10**(3), 419–431 (1969)
56. Maligranda, L.: Type, cotype and convexity properties of quasi-Banach spaces. In: Banach and Function Spaces, pp. 83–120. Yokohama Publishers, Yokohama (2004)
57. Meyer-Nieberg, P.: Banach Lattices. Universitext. Springer, Berlin (1991)
58. Nieraeth, B.: Quantitative estimates and extrapolation for multilinear weight classes (2018). arXiv:1808.01985
59. Rubio de Francia, J.L.: Factorization and extrapolation of weights. Bull. Am. Math. Soc. (N.S.) **7**(2), 393–395 (1982)
60. Rubio de Francia, J.L.: A Littlewood–Paley inequality for arbitrary intervals. Rev. Mat. Iberoam. **1**(2), 1–14 (1985)
61. Rubio de Francia, J.L.: Martingale and integral transforms of Banach space valued functions. In: Probability and Banach Spaces (Zaragoza, 1985). Lecture Notes in Mathematics, vol. 1221, pp. 195–222. Springer, Berlin (1986)
62. Schep, A.: Products and factors of Banach function spaces. Positivity **14**(2), 301–319 (2010)
63. Shi, X.L., Sun, Q.Y.: Weighted norm inequalities for Bochner–Riesz operators and singular integral operators. Proc. Am. Math. Soc. **116**(3), 665–673 (1992)
64. Stein, E .M.: Maximal functions. I. Spherical means. Proc. Natl Acad. Sci. U.S.A. **73**(7), 2174–2175 (1976)
65. Triebel, H.: Interpolation Theory, Function Spaces, Differential Operators, vol. 18. North-Holland Mathematical Library, North-Holland, Amsterdam (1978)
66. Xu, Q.: H^∞ functional calculus and maximal inequalities for semigroups of contractions on vector-valued L_p-spaces. Int. Math. Res. Not. IMRN **14**, 5715–5732 (2015)

Publisher's Note Springer Nature remains neutral with regard to jurisdictional claims in published maps and institutional affiliations.

Journal of Fourier Analysis and Applications (2019) 25:2635–2652
https://doi.org/10.1007/s00041-019-09676-y

RESEARCH ARTICLE

An Endpoint Weak-Type Estimate for Multilinear Calderón–Zygmund Operators

Cody B. Stockdale[1] · Brett D. Wick[1]

Received: 26 September 2018 / Revised: 8 February 2019 / Published online: 4 March 2019
© Springer Science+Business Media, LLC, part of Springer Nature 2019

Abstract
The purpose of this article is to provide an alternative proof of the weak-type $\left(1, \ldots, 1; \frac{1}{m}\right)$ estimate for m-multilinear Calderón–Zygmund operators on $\mathbb{R}^n$ first proved by Grafakos and Torres. Subsequent proofs in the bilinear setting have been given by Maldonado and Naibo and also by Pérez and Torres. The proof given here is motivated by the proof of the weak-type $(1, 1)$ estimate for Calderón–Zygmund operators in the nonhomogeneous setting by Nazarov, Treil, and Volberg.

Keywords Singular integrals · Multilinear operators · Weak-type estimates

Mathematics Subject Classification 42B20

1 Introduction

The Calderón–Zygmund theory of singular integral operators is central in the study of harmonic analysis. A key property of Calderón–Zygmund operators in the linear setting is their boundedness from $L^p(\mathbb{R}^n)$ to $L^p(\mathbb{R}^n)$ for all $p \in (1, \infty)$, assuming a priori that the operators are bounded from $L^2(\mathbb{R}^n)$ to $L^2(\mathbb{R}^n)$. The well-known method of proof is as follows:

(1) establish a weak-type $(1, 1)$ estimate for the operator,

---

Communicated by Loukas Grafakos.

B. D. Wick's research supported in part by National Science Foundation Grant DMS #1560955 and #1800057.

✉ Cody B. Stockdale
codystockdale@wustl.edu

Brett D. Wick
wick@math.wustl.edu

1 Department of Mathematics and Statistics, Washington University in St. Louis, One Brookings Drive, St. Louis, MO 63130, USA

(2) use the Marcinkiewicz interpolation theorem to obtain a strong type (p, p) bound for all $p \in (1, 2)$, and
(3) use duality to deduce the strong type (p, p) estimate for all $p \in (1, \infty)$.

For a detailed treatment of the classical Calderón–Zygmund theory, see [2,10].

The classical proof of the weak-type $(1, 1)$ bound utilizes the Calderón–Zygmund decomposition for $L^1(\mathbb{R}^n)$ functions. This technique readily extends to handle Calderón–Zygmund operators on spaces that have an underlying measure possessing the doubling property. Such spaces are called spaces of homogeneous type. Recall that a measure μ possesses the doubling property if there exists a constant $C > 0$ such that

$$\mu(B(x, 2r)) \leq C\mu(B(x, r))$$

for all $r > 0$ and all x in the space. The classical technique, however, does not generalize as easily to spaces of nonhomogeneous type, which are spaces whose underlying measures instead have a polynomial growth condition. To address this setting, Tolsa developed a version of the Calderón–Zygmund decomposition adapted to nonhomogeneous measures to prove the weak-type estimate in a similar manner to the classical case in [11]. In [8], Nazarov, Treil, and Volberg provided a proof of the weak-type $(1, 1)$ bound of Calderón–Zygmund operators in the nonhomogeneous setting without using the Calderón–Zygmund decomposition. The proof in [8] also works in the classical setting on $\mathbb{R}^n$.

More recently, attention has been given to the study of multilinear Calderón–Zygmund operators (see [1,3–7,9]). To describe the setting, let m be a positive integer. We say $K : \mathbb{R}^{n(m+1)} \to \mathbb{C}$ is an *m-multilinear Calderón–Zygmund kernel* if there exist $C_K, \delta > 0$ such that the following conditions hold:

(1) (*size*)

$$|K(x, y_1, \ldots, y_m)| \leq \frac{C_K}{\left(\sum_{i=1}^m |x - y_i|\right)^{nm}}$$

for all $x, y_1, \ldots, y_m \in \mathbb{R}^n$ with $x \neq y_j$ for some j,

(2) (*smoothness*)

$$\left|K(x, y_1, \ldots, y_m) - K\left(x', y_1, \ldots, y_m\right)\right| \leq \frac{C_K |x - x'|^\delta}{(\sum_{i=1}^m |x - y_i|)^{nm+\delta}}$$

whenever $|x - x'| \leq \dfrac{1}{2} \max\limits_{1 \leq i \leq m} |x - y_i|$, and

$$\left|K(x, y_1, \ldots, y_j, \ldots, y_m) - K\left(x, y_1, \ldots, y_j', \ldots, y_m\right)\right| \leq \frac{C_K |y_j - y_j'|^\delta}{(\sum_{i=1}^m |x - y_i|)^{nm+\delta}}$$

for each $j \in \{1, \ldots, m\}$ whenever $|y_j - y_j'| \leq \dfrac{1}{2} \max\limits_{1 \leq i \leq m} |x - y_i|$.

We say a bounded multilinear operator $T : (L^2(\mathbb{R}^n))^m \to L^{\frac{2}{m}}(\mathbb{R}^n)$ is an *m-multilinear Calderón–Zygmund operator associated to a kernel K* if K is an m-multilinear Calderón–Zygmund kernel and if

$$T(f_1, \ldots, f_m)(x) = \int_{(\mathbb{R}^n)^m} K(x, y_1, \ldots, y_m) f_1(y_1) \cdots f_m(y_m) dy_1 \cdots dy_m$$

for almost every $x \in \mathbb{R}^n \setminus \bigcap_{i=1}^m \operatorname{supp} f_i$. Throughout the remainder of this paper, T will denote a multilinear Calderón–Zygmund operator.

Let $\mathcal{M}(\mathbb{R}^n)$ denote the space of $\mathbb{R}$-valued Borel measures on $\mathbb{R}^n$. For $\nu_1, \ldots, \nu_m \in \mathcal{M}(\mathbb{R}^n)$ and $x \in \mathbb{R}^n \setminus \bigcap_{i=1}^m \operatorname{supp} \nu_i$, define

$$T(\nu_1, \ldots, \nu_m)(x) := \int_{(\mathbb{R}^n)^m} K(x, y_1, \ldots, y_m) d\nu(y_1) \cdots d\nu(y_m).$$

We will denote the total variation of $\nu \in \mathcal{M}(\mathbb{R}^n)$ by $\|\nu\|$. Notice that if f is a Borel measurable function, then $\||f|dm\| = \|f\|_{L^1(\mathbb{R}^n)}$ and

$$T(f_1 dm, \ldots, f_m dm) = T(f_1, \ldots, f_m).$$

Here $f dm \in \mathcal{M}(\mathbb{R}^n)$ is defined for Borel subsets of $\mathbb{R}^n$ by $f dm(A) = \int_A f(x)dx$. Also, if $\nu_i = \sum_{j=1}^N a_{i,j}\delta_{x_{i,j}}$ for $i \in \{1, \ldots, m\}$, then

$$T(\nu_1, \ldots, \nu_m)(x) = \sum_{j_1, \ldots, j_m=1}^{N} \left(\prod_{i=1}^m a_{i,j_i} \right) K(x, x_{1,j_1}, \ldots, x_{m,j_m}).$$

Analogous properties to the classical case were established for multilinear operators in [4] (see also [7,9]). In particular, a weak-type estimate is proved and used to establish strong type estimates via interpolation. The appropriate weak-type estimate for multilinear Calderón–Zygmund operators is of type $\left(1, \ldots, 1; \frac{1}{m}\right)$. It is stated as the following:

Theorem 1 *Let T be a multilinear Calderón–Zygmund operator. If $f_1, \ldots, f_m \in L^1(\mathbb{R}^n)$, then*

$$\|T(f_1, \ldots, f_m)\|_{L^{\frac{1}{m},\infty}(\mathbb{R}^n)} \le A_2 \prod_{i=1}^m \|f_i\|_{L^1(\mathbb{R}^n)}$$

where A_2 depends on C_K, n, and m.

As in the classical situation, Grafakos and Torres [4] prove Theorem 1 using the Calderón–Zygmund decomposition. An alternative proof is presented in Sect. 3.

The new proof is modeled after the argument for the weak-type estimate in [8]. Instead of obtaining cancellation by means of the Calderón–Zygmund decomposition, we do so by subtracting terms involving certain point mass measures. The argument is then completed by establishing a weak-type estimate on a mixture of linear combinations of point mass measures and of $L^1(\mathbb{R}^n) \cap L^\infty(\mathbb{R}^n)$ functions with appropriate $L^\infty(\mathbb{R}^n)$ norm. This is stated precisely as the following:

Theorem 2 *Let T be a multilinear Calderón–Zygmund operator, $t > 0$, and $l \in \{1, \ldots, m\}$ be given. If $\nu_1, \ldots, \nu_l \in \mathcal{M}(\mathbb{R}^n)$ are of the form $\nu_i = \sum_{j=1}^N a_{i,j}\delta_{x_{i,j}}$ where $x_{i,j} \in \mathbb{R}^n$ and $a_{i,j} \in \mathbb{R}$ and if $f_{l+1}, \ldots, f_m \in L^1(\mathbb{R}^n) \cap L^\infty(\mathbb{R}^n)$ satisfy $\|f_i\|_{L^\infty(\mathbb{R}^n)} \le t^{\frac{1}{m}}$ for all i, then*

$$|\{|T(\nu_1, \ldots, \nu_l, f_{l+1}, \ldots, f_m)| > t\}| \le A_3 t^{-\frac{1}{m}} \left(\prod_{i=1}^{l} \|\nu_i\|^{\frac{1}{m}}\right) \left(\prod_{i=l+1}^{m} \|f_i\|^{\frac{1}{m}}_{L^1(\mathbb{R}^n)}\right)$$

where A_3 depends on C_K, n, and m.

It is not important that the ν_i are applied in the first l slots of T – an identical proof yields the theorem whenever the set of indices of the ν_i is a nonempty subset of $\{1, \ldots, m\}$.

Since the proof of Theorem 1 still requires a decomposition of the arbitrary $L^1(\mathbb{R}^n)$ functions into bounded and unbounded pieces ("good" and "bad" pieces), there is an analogy to be made between our proof and the proof in [4]. First, the term where the operator is only being applied to the "good" functions is handled identically – both proofs use Chebyshev's inequality, the a priori boundedness of T, and the $L^\infty(\mathbb{R}^n)$ norms of the good functions to obtain the appropriate estimate. However, the terms where the operator has at least one "bad" function as an input are treated differently.

First we describe the Calderón–Zygmund decomposition approach to handling these terms. Because of the nature of this decomposition, each "bad" function, b_i, can be written as the sum $b_i = \sum_{j=1}^\infty b_{i,j}$ where each $b_{i,j}$ has mean value zero, is supported on a cube of appropriate measure, and has useful $L^1(\mathbb{R}^n)$ control. The cancellation involved in the $b_{i,j}$ allows one to introduce a term with the kernel evaluated at the center of the cube on which $b_{i,j}$ is supported, then one can use the smoothness assumption of the kernel to obtain the desired estimate. The disjointness of supp$(b_{i,j})$ over the j allows one to recover the estimate for the original term.

Without the Calderón–Zygmund decomposition, there is no immediate cancellation that may be exploited in the "bad" functions. Instead, we apply a Whitney decomposition to write the support of each "bad" function, b_i, as a union of dyadic cubes with disjoint interiors and restrict b_i to each cube given by the Whitney decomposition. Call these restrictions $b_{i,j}$. It suffices to approximate $b_i \approx \sum_{j=1}^N b_{i,j}$ and get an appropriate estimate with these sums, uniform in N. With this goal in mind, denote the center of supp$(b_{i,j})$ by $c_{i,j}$ and define measures ν_i^N by $\nu_i^N := \sum_{j=1}^N a_{i,j}\delta_{c_{i,j}}$ where $a_{i,j} := \int_{\mathbb{R}^n} b_{i,j}(x)dx$. Adding and subtracting terms involving these ν_i^N introduces cancellation. We then subtract a term involving the kernel evaluated at the $c_{i,j}$ and use the regularity discussed in Sect. 2 to get the desired control. It is then left to control a term involving a mixture of linear combinations of point mass measures and "good" functions. This can be done with Theorem 2.

Section 2 includes Lemma 1, a regularity condition first described in [9] for bilinear kernels, which is the key use of cancellation. Section 3 contains the main results. The proof of Theorem 1 assuming Theorem 2 is given first. The proof of Theorem 2 is given at the end.

We would like to acknowledge Rodolfo Torres for kindly providing comments and references.

2 Preliminaries

The proofs of Theorems 1 and 2 use the multilinear geometric Hörmander condition given in Lemma 1 below. This type of regularity was first introduced in the bilinear setting by Pérez and Torres in [9]. Throughout the rest of this paper, we use the notation $\vec{y}_{i,k} = (y_i, y_{i+1}, \ldots, y_k)$, $\vec{f}_{i,k} = (f_i, f_{i+1}, \ldots, f_k)$, $\vec{\nu}_{i,k} = (\nu_i, \nu_{i+1}, \ldots, \nu_k)$, $\vec{c}_{(i,j_i),(k,j_k)} = (c_{i,j_i}, c_{i+1,j_{i+1}}, \ldots, c_{k,j_k})$, and $\vec{\nu}_{(i,j_i),(k,j_k)} = (\nu_{i,j_i}, \nu_{i+1,j_{i+1}}, \ldots, \nu_{k,j_k})$. We apologize for further complicating the notation; however, this is necessary to compactly describe many expressions that follow.

Lemma 1 *There exists $A_1 > 0$ such that if $l \in \{1, \ldots, m\}$ and $\mathcal{S}_1, \ldots, \mathcal{S}_l$ are countable collections of sets satisfying either*

(1) *each $\mathcal{S}_i = \{S_{i,1}, S_{i,2}, \ldots\}$ consists of dyadic cubes with disjoint interiors or*
(2) *each $\mathcal{S}_i = \{S_{i,1}, S_{i,2}, \ldots\}$ consists of sets satisfying:*

- *$S_{i,j}$ have disjoint interiors,*
- *$S_{i,j} \subseteq B(c_{i,j}, r_{i,j})$, and*
- $\Omega_i := \bigcup_{j=1}^{\infty} S_{i,j} = \bigcup_{j=1}^{\infty} B(c_{i,j}, r_{i,j})$,

then

$$\sum_{j_1,\ldots,j_l=1}^{\infty} \prod_{i=1}^{l} |S_{i,j_i}| \int_{\mathbb{R}^{n(m-l)}} \sup_{\substack{(y_1,\ldots,y_l) \\ \in \prod_{i=1}^{l} S_{i,j_i}}} \times \int_{\mathbb{R}^n \setminus \left(\bigcup_{i=1}^{l} \Omega_i^*\right)} \left| K\left(x, \vec{y}_{1,m}\right) - K\left(x, \vec{c}_{(1,j_1),(l,j_l)}, \vec{y}_{l+1,m}\right) \right| dx d\vec{y}_{l+1,m}$$
$$\leq A_1 \sum_{i=1}^{l} |\Omega_i|$$

where $\Omega_i^* := \bigcup_{j=1}^{\infty} 2S_{i,j} = \bigcup_{j=1}^{\infty} B(c_{i,j}, 2r_{i,j})$.

It is not important that the indices of the $\mathcal{S}_i$ range from 1 to l – an identical proof yields the lemma whenever the set of indices is a nonempty subset of $\{1, \ldots, m\}$.

This regularity was considered in [9] when the $\mathcal{S}_i$ are collections of dyadic cubes with disjoint interiors. We will use the lemma when the collections $\mathcal{S}_i$ consist of dyadic cubes in the proof of Theorem 1 and when they are of the second type in the proof of Theorem 2.

Proof We only prove the statement when the collections $\mathcal{S}_i$ are of the second type. The proof for collections of dyadic cubes is similar and is addressed in the bilinear setting in [9]. For $i = 1, \dots, l$, fix $S_{i,j_i} \in \mathcal{S}_i$. Use the smoothness condition of K and the fact that $S_{i,j_i} \subseteq B(c_{i,j_i}, r_{i,j_i}) \subseteq \overline{B(c_{i,j_i}, r_{i,j_i})}$ to see

$$\begin{aligned}
&\sup_{\substack{(y_1,\dots,y_l)\\ \in\prod_{i=1}^l S_{i,j_i}}} \int_{\mathbb{R}^n\setminus\left(\bigcup_{i=1}^l \Omega_i^*\right)} \left|K\left(x, \vec{y}_{1,m}\right) - K\left(x, \vec{c}_{(1,j_1),(l,j_l)}, \vec{y}_{l+1,m}\right)\right| dx \\
&\quad\lesssim \sup_{\substack{(y_1,\dots,y_l)\\ \in\prod_{i=1}^l S_{i,j_i}}} \int_{\mathbb{R}^n\setminus\left(\bigcup_{i=1}^l \Omega_i^*\right)} \frac{\sum_{i=1}^l |y_i - c_{i,j_i}|^\delta}{(\sum_{i=1}^m |x-y_i|)^{nm+\delta}} dx \\
&\quad\leq \sup_{\substack{(y_1,\dots,y_l)\\ \in\prod_{i=1}^l \overline{B(c_{i,j_i}, r_{i,j_i})}}} \int_{\mathbb{R}^n\setminus\left(\bigcup_{i=1}^l \Omega_i^*\right)} \frac{\sum_{i=1}^l r_{i,j_i}^\delta}{(\sum_{i=1}^m |x-y_i|)^{nm+\delta}} dx.
\end{aligned}$$

Since for fixed $y_i \in \overline{B(c_{i,j_i}, r_{i,j_i})}$, $i = l+1, \dots, m$, the function $\int_{\mathbb{R}^n\setminus\left(\bigcup_{i=1}^l \Omega_i^*\right)} \frac{\sum_{i=1}^l r_{i,j_i}^\delta}{(\sum_{i=1}^m |x-y_i|)^{nm+\delta}} dx$ is continuous in the variables $y_i \in \overline{B(c_{i,j_i}, r_{i,j_i})}$, $i = 1, \dots, l$, and since $\overline{B(c_{i,j_i}, r_{i,j_i})}$ is a compact set, we may write

$$\begin{aligned}
&\sup_{\substack{(y_1,\dots,y_l)\\ \in\prod_{i=1}^l \overline{B(c_{i,j_i}, r_{i,j_i})}}} \int_{\mathbb{R}^n\setminus\left(\bigcup_{i=1}^l \Omega_i^*\right)} \frac{\sum_{i=1}^l r_{i,j_i}^\delta}{(\sum_{i=1}^m |x-y_i|)^{nm+\delta}} dx \\
&\quad= \int_{\mathbb{R}^n\setminus\left(\bigcup_{i=1}^l \Omega_i^*\right)} \frac{\sum_{i=1}^l r_{i,j_i}^\delta}{(\sum_{i=1}^l |x-y_i^*| + \sum_{i=l+1}^m |x-y_i|)^{nm+\delta}} dx \quad \text{and}
\end{aligned}$$

$$\begin{aligned}
&\inf_{\substack{(y_1,\dots,y_l)\\ \in\prod_{i=1}^l \overline{B(c_{i,j_i}, r_{i,j_i})}}} \int_{\mathbb{R}^n\setminus\left(\bigcup_{i=1}^l \Omega_i^*\right)} \frac{\sum_{i=1}^l r_{i,j_i}^\delta}{(\sum_{i=1}^m |x-y_i|)^{nm+\delta}} dx \\
&\quad= \int_{\mathbb{R}^n\setminus\left(\bigcup_{i=1}^l \Omega_i^*\right)} \frac{\sum_{i=1}^l r_{i,j_i}^\delta}{(\sum_{i=1}^l |x-y_{i_*}| + \sum_{i=l+1}^m |x-y_i|)^{nm+\delta}} dx.
\end{aligned}$$

Note that for $x \in \mathbb{R}^n \setminus \left(\bigcup_{i=1}^l \Omega_i^*\right)$, $|x - y_{i_*}| \leq 2r_i + |x - y_i^*|$ and $|x - y_i^*| \geq r_i$, so

$$\frac{\sum_{i=1}^l |x-y_{i_*}| + \sum_{i=l+1}^m |x-y_i|}{\sum_{i=1}^l |x-y_i^*| + \sum_{i=l+1}^m |x-y_i|} \leq \frac{\sum_{i=1}^l 2r_i}{\sum_{i=1}^l |x-y_i^*| + \sum_{i=l+1}^m |x-y_i|} + 1 \leq 3.$$

Then

$$\sup_{\substack{(y_1,\ldots,y_l)\\ \in\prod_{i=1}^l S_{i,j_i}}} \int_{\mathbb{R}^n\setminus\left(\bigcup_{i=1}^l \Omega_i^*\right)} \left|K\left(x,\vec{y}_{1,m}\right)-K\left(x,\vec{c}_{(1,j_1),(l,j_l)},\vec{y}_{l+1,m}\right)\right|dx$$

$$\lesssim \int_{\mathbb{R}^n\setminus\left(\bigcup_{i=1}^l \Omega_i^*\right)} \frac{\sum_{i=1}^l r_{i,j_i}^\delta}{\left(\sum_{i=1}^l |x-y_{i_*}|+\sum_{i=l+1}^m |x-y_i|\right)^{nm+\delta}}$$

$$\times\left(\frac{\sum_{i=1}^l |x-y_{i_*}|+\sum_{i=l+1}^m |x-y_i|}{\sum_{i=1}^l |x-y_i^*|+\sum_{i=l+1}^m |x-y_i|}\right)^{nm+\delta} dx$$

$$\lesssim \int_{\mathbb{R}^n\setminus\left(\bigcup_{i=1}^l \Omega_i^*\right)} \frac{\sum_{i=1}^l r_{i,j_i}^\delta}{\left(\sum_{i=1}^l |x-y_{i_*}|+\sum_{i=l+1}^m |x-y_i|\right)^{nm+\delta}}dx$$

$$=\inf_{\substack{(y_1,\ldots,y_l)\\ \in\prod_{i=1}^l \overline{B(c_{i,j_i},r_{i,j_i})}}} \int_{\mathbb{R}^n\setminus\left(\bigcup_{i=1}^l \Omega_i^*\right)} \frac{\sum_{i=1}^l r_{i,j_i}^\delta}{\left(\sum_{i=1}^m |x-y_i|\right)^{nm+\delta}}dx$$

$$\le \inf_{\substack{(y_1,\ldots,y_l)\\ \in\prod_{i=1}^l S_{i,j_i}}} \int_{\mathbb{R}^n\setminus\left(\bigcup_{i=1}^l \Omega_i^*\right)} \frac{\sum_{i=1}^l r_{i,j_i}^\delta}{\left(\sum_{i=1}^m |x-y_i|\right)^{nm+\delta}}dx.$$

Using the previous estimate, trivial estimates to pull the infimum outside of the integral, Fubini, and integral estimates, we get the bound

$$\int_{\mathbb{R}^{n(m-l)}} \sup_{\substack{(y_1,\ldots,y_l)\\ \in\prod_{i=1}^l S_{i,j_i}}} \int_{\mathbb{R}^n\setminus\left(\bigcup_{i=1}^l \Omega_i^*\right)} \left|K\left(x,\vec{y}_{1,m}\right)-K\left(x,\vec{c}_{(1,j_1),(l,j_l)},\vec{y}_{l+1,m}\right)\right|dx\,d\vec{y}_{l+1,m}$$

$$\lesssim \int_{\mathbb{R}^{n(m-l)}} \inf_{\substack{(y_1,\ldots,y_l)\\ \in\prod_{i=1}^l S_{i,j_i}}} \int_{\mathbb{R}^n\setminus\left(\bigcup_{i=1}^l \Omega_i^*\right)} \frac{\sum_{i=1}^l r_{i,j_i}^\delta}{\left(\sum_{i=1}^m |x-y_i|\right)^{nm+\delta}}dx\,d\vec{y}_{l+1,m}$$

$$\le \inf_{\substack{(y_1,\ldots,y_l)\\ \in\prod_{i=1}^l S_{i,j_i}}} \int_{\mathbb{R}^n\setminus\left(\bigcup_{i=1}^l \Omega_i^*\right)} \int_{\mathbb{R}^{n(m-l)}} \frac{\sum_{i=1}^l r_{i,j_i}^\delta}{\left(\sum_{i=1}^m |x-y_i|\right)^{nm+\delta}}d\vec{y}_{l+1,m}\,dx$$

$$\lesssim \inf_{\substack{(y_1,\ldots,y_l)\\ \in\prod_{i=1}^l S_{i,j_i}}} \int_{\mathbb{R}^n\setminus\left(\bigcup_{i=1}^l \Omega_i^*\right)} \frac{\sum_{i=1}^l r_{i,j_i}^\delta}{\left(\sum_{i=1}^l |x-y_i|\right)^{nl+\delta}}dx.$$

Therefore

$$\sum_{j_1,\ldots,j_l=1}^\infty \prod_{i=1}^l |S_{i,j_i}|$$

$$\times\int_{\mathbb{R}^{n(m-l)}} \sup_{\substack{(y_1,\ldots,y_l)\\ \in\prod_{i=1}^l S_{i,j_i}}} \int_{\mathbb{R}^n\setminus\left(\bigcup_{i=1}^l \Omega_i^*\right)} \left|K\left(x,\vec{y}_{1,m}\right)-K\left(x,\vec{c}_{(1,j_1),(l,j_l)},\vec{y}_{l+1,m}\right)\right|dx\,d\vec{y}_{l+1,m}$$

$$
\begin{aligned}
&\lesssim C_K \sum_{j_1,\ldots,j_l=1}^{\infty} \prod_{i=1}^{l} |S_{i,j_i}| \inf_{\substack{(y_1,\ldots,y_l)\\ \in \prod_{i=1}^{l} S_{i,j_i}}} \int_{\mathbb{R}^n \setminus \left(\bigcup_{i=1}^{l} \Omega_i^*\right)} \frac{\sum_{i=1}^{l} r_{i,j_i}^{\delta}}{(\sum_{i=1}^{l} |x-y_i|)^{nl+\delta}} dx \\
&\le C_K \sum_{j_1,\ldots,j_l=1}^{\infty} \int_{S_{l,j_l}} \cdots \int_{S_{1,j_1}} \int_{\mathbb{R}^n \setminus \left(\bigcup_{i=1}^{l} \Omega_i^*\right)} \frac{\sum_{i=1}^{l} r_{i,j_i}^{\delta}}{(\sum_{i=1}^{l} |x-y_i|)^{nl+\delta}} dx d\vec{y}_{1,l} \\
&= C_K \sum_{k=1}^{l} \left(\sum_{\substack{j_1,\ldots,j_l=1\\ r_{k,j_k} \ge r_{i,j_i} \text{ all } i}}^{\infty} \int_{S_{l,j_l}} \cdots \int_{S_{1,j_1}} \int_{\mathbb{R}^n \setminus \left(\bigcup_{i=1}^{l} \Omega_i^*\right)} \frac{\sum_{i=1}^{l} r_{i,j_i}^{\delta}}{(\sum_{i=1}^{l} |x-y_i|)^{nl+\delta}} dx d\vec{y}_{1,l} \right).
\end{aligned}
$$

We will control the term of the summation above with $k = 1$; the other terms are handled identically. Using trivial estimates, Fubini's theorem, the fact that the S_{i,j_i} have disjoint interiors, and integral estimates, we obtain

$$
\begin{aligned}
&\sum_{\substack{j_1,\ldots,j_l=1\\ r_{1,j_1} \ge r_{i,j_i} \text{ all } i}}^{\infty} \int_{S_{l,j_l}} \cdots \int_{S_{1,j_1}} \int_{\mathbb{R}^n \setminus \left(\bigcup_{i=1}^{l} \Omega_i^*\right)} \frac{\sum_{i=1}^{l} r_{i,j_i}^{\delta}}{(\sum_{i=1}^{l} |x-y_i|)^{nl+\delta}} dx d\vec{y}_{1,l} \\
&\lesssim \sum_{j_1,\ldots,j_l=1}^{\infty} \int_{S_{l,j_l}} \cdots \int_{S_{1,j_1}} \int_{\mathbb{R}^n \setminus \left(\bigcup_{i=1}^{l} \Omega_i^*\right)} \frac{r_{1,j_1}^{\delta}}{(\sum_{i=1}^{l} |x-y_i|)^{nl+\delta}} dx d\vec{y}_{1,l} \\
&\lesssim \sum_{j_1=1}^{\infty} \int_{S_{1,j_1}} \int_{\mathbb{R}^n \setminus \left(\bigcup_{i=1}^{l} \Omega_i^*\right)} \sum_{j_2,\ldots,j_l=1}^{\infty} \int_{S_{l,j_l}} \cdots \int_{S_{2,j_2}} \frac{1}{(\sum_{i=1}^{l} |x-y_i|)^{nl}} d\vec{y}_{2,l} \frac{r_{1,j_1}^{\delta}}{|x-c_{1,j_1}|^{\delta}} dx dy_1 \\
&\le \int_{\Omega_1} \int_{\mathbb{R}^n \setminus \left(\bigcup_{i=1}^{l} \Omega_i^*\right)} \int_{\mathbb{R}^{n(l-1)}} \frac{1}{(\sum_{i=1}^{l} |x-y_i|)^{nl}} d\vec{y}_{2,l} \frac{r_{1,j_1}^{\delta}}{|x-c_{1,j_1}|^{\delta}} dx dy_1 \\
&\lesssim \int_{\Omega_1} \int_{\mathbb{R}^n \setminus \left(\bigcup_{i=1}^{l} \Omega_i^*\right)} \frac{r_{1,j_1}^{\delta}}{|x-y_1|^n |x-c_{1,j_1}|^{\delta}} dx dy_1 \\
&\lesssim |\Omega_1| \int_{\mathbb{R}^n \setminus \Omega_1^*} \frac{r_{1,j_1}^{\delta}}{|x-c_{1,j_1}|^{n+\delta}} dx \\
&\le |\Omega_1| \int_{|x|>2r_{1,j_1}} \frac{r_{1,j_1}^{\delta}}{|x|^{n+\delta}} dx \\
&\le |\Omega_1|.
\end{aligned}
$$

Similarly, for $k = 2, \ldots, l$,

$$
\sum_{\substack{j_1,\ldots,j_l=1\\ r_{k,j_k} \ge r_{i,j_i} \text{ all } i}}^{\infty} \int_{S_{l,j_l}} \cdots \int_{S_{1,j_1}} \int_{\mathbb{R}^n \setminus \left(\bigcup_{i=1}^{l} \Omega_i^*\right)} \frac{\sum_{i=1}^{l} r_{i,j_i}^{\delta}}{(\sum_{i=1}^{l} |x-y_i|)^{nl+\delta}} dx d\vec{y}_{1,l} \lesssim |\Omega_k|.
$$

This completes the proof. □

3 Main Results

We now turn to proving the main result of Grafakos and Torres [4].

Theorem 1 *If $f_1, \ldots, f_m \in L^1(\mathbb{R}^n)$, then*

$$\|T(f_1, \ldots, f_m)\|_{L^{\frac{1}{m},\infty}(\mathbb{R}^n)} \leq A_2 \prod_{i=1}^{m} \|f_i\|_{L^1(\mathbb{R}^n)}$$

where A_2 depends on K, n, m, and $\|T\|_{(L^2(\mathbb{R}^n))^m \to L^{\frac{2}{m}}(\mathbb{R}^n)}$. That is, for every $t > 0$, it holds that

$$|\{|T(f_1, \ldots, f_m)| > t\}| \leq A_2^{\frac{1}{m}} t^{-\frac{1}{m}} \prod_{i=1}^{m} \|f_i\|_{L^1(\mathbb{R}^n)}^{\frac{1}{m}}.$$

Our contribution is the following.

Theorem 2 *Let $t > 0$ and $l \in \{1, \ldots, m\}$ be given. If $\nu_1, \ldots, \nu_l \in \mathcal{M}(\mathbb{R}^n)$ are of the form $\nu_i = \sum_{j=1}^{N} a_{i,j}\delta_{x_{i,j}}$ where $x_{i,j} \in \mathbb{R}^n$ and $a_{i,j} \in \mathbb{R}$ and if $f_{l+1}, \ldots, f_m \in L^1(\mathbb{R}^n) \cap L^\infty(\mathbb{R}^n)$ satisfy $\|f_i\|_{L^\infty(\mathbb{R}^n)} \leq t^{\frac{1}{m}}$ for all i, then*

$$|\{|T(\nu_1, \ldots, \nu_l, f_{l+1}, \ldots, f_m)| > t\}| \leq A_3 t^{-\frac{1}{m}} \left(\prod_{i=1}^{l} \|\nu_i\|^{\frac{1}{m}}\right)\left(\prod_{i=l+1}^{m} \|f_i\|_{L^1(\mathbb{R}^n)}^{\frac{1}{m}}\right)$$

where A_3 depends on K, n, m, and $\|T\|_{(L^2(\mathbb{R}^n))^m \to L^{\frac{2}{m}}(\mathbb{R}^n)}$.

Note that Theorem 2 holds whenever the set of indices of the ν_i is a nonempty subset of $\{1, \ldots, m\}$. We will first prove Theorem 1 assuming Theorem 2. We will then prove Theorem 2. Let M denote the uncentered Hardy-Littlewood maximal function and recall its formula

$$Mf(x) := \sup_{Q \ni x} \frac{1}{|Q|} \int_Q |f(y)|dy.$$

Write $\|M\|$ for $\|M\|_{L^1(\mathbb{R}^n) \to L^{1,\infty}(\mathbb{R}^n)}$.

Proof of Theorem 1 Let $t > 0$ be given. By density, we may assume $f_1, \ldots, f_m$ are continuous functions with compact support. Normalize to assume $\|f_1\|_{L^1(\mathbb{R}^n)} = \cdots = \|f_m\|_{L^1(\mathbb{R}^n)} = 1$. Set

$$G_i := \left\{Mf_i > t^{\frac{1}{m}}\right\} \quad \text{and } G := \bigcup_{i=1}^{m} G_i.$$

Notice that

$$|G| \le \sum_{i=1}^{m} |G_i| \le m\|M\| t^{-\frac{1}{m}}.$$

Put

$$b_i := f_i \mathbb{1}_{G_i} \quad \text{and } g_i := f_i \mathbb{1}_{\mathbb{R}^n \setminus G_i}.$$

Set

$$\begin{aligned} E_1 &:= \left\{ |T(g_1, g_2, \ldots, g_m)| > \frac{t}{2^m} \right\}, \\ E_2 &:= \left\{ |T(b_1, g_2, \ldots, g_m)| > \frac{t}{2^m} \right\}, \\ E_3 &:= \left\{ |T(g_1, b_2, \ldots, g_m)| > \frac{t}{2^m} \right\}, \\ &\cdots \\ E_{2^m} &:= \left\{ |T(b_1, b_2, \ldots, b_m)| > \frac{t}{2^m} \right\}; \end{aligned}$$

where each $E_s = \left\{ |T(h_1, \ldots, h_m)| > \frac{t}{2^m} \right\}$ with $h_i \in \{b_i, g_i\}$ and all the sets E_s are distinct. Since

$$|\{|T(f_1, \ldots, f_m)| > t\}| \le \sum_{s=1}^{2^m} |E_s|,$$

it suffices to control each $|E_s|$ by a constant multiplied by $t^{-\frac{1}{m}}$.

We will first estimate $|E_1|$. Note that since $|f_i(x)| \le Mf_i(x) \le t^{\frac{1}{m}}$ for almost every $x \in \mathbb{R}^n \setminus G_i$, it is true that $\|g_i\|_{L^\infty(\mathbb{R}^n)} \le t^{\frac{1}{m}}$. Use Chebyshev's inequality, the boundedness of T from $(L^2(\mathbb{R}^n))^m$ to $L^{\frac{2}{m}}(\mathbb{R}^n)$, and the fact that $\|g_i\|_{L^\infty(\mathbb{R}^n)} \le t^{\frac{1}{m}}$ to see

$$\begin{aligned} |E_1| &\le 4t^{-\frac{2}{m}} \int_{\mathbb{R}^n} |T(g_1, \ldots, g_m)(x)|^{\frac{2}{m}} dx \\ &\le 4\|T\|^{\frac{2}{m}}_{(L^2(\mathbb{R}^n))^m \to L^{\frac{2}{m}}(\mathbb{R}^n)} t^{-\frac{2}{m}} \prod_{i=1}^{m} \left(\int_{\mathbb{R}^n} |g_i(x)|^2 dx \right)^{\frac{1}{m}} \\ &\le 4\|T\|^{\frac{2}{m}}_{(L^2(\mathbb{R}^n))^m \to L^{\frac{2}{m}}(\mathbb{R}^n)} t^{-\frac{1}{m}} \prod_{i=1}^{m} \|g_i\|^{\frac{1}{m}}_{L^1(\mathbb{R}^n)} \\ &\le 4\|T\|^{\frac{2}{m}}_{(L^2(\mathbb{R}^n))^m \to L^{\frac{2}{m}}(\mathbb{R}^n)} t^{-\frac{1}{m}} \prod_{i=1}^{m} \|f_i\|^{\frac{1}{m}}_{L^1(\mathbb{R}^n)} \end{aligned}$$

$$
\begin{aligned}
&= 4\|T\|^{\frac{2}{m}}_{(L^2(\mathbb{R}^n))^m \to L^{\frac{2}{m}}(\mathbb{R}^n)} t^{-\frac{1}{m}} \\
&= B_1 t^{-\frac{1}{m}}
\end{aligned}
$$

where $B_1 := 4\|T\|^{\frac{2}{m}}_{(L^2(\mathbb{R}^n))^m \to L^{\frac{2}{m}}(\mathbb{R}^n)}$.

Consider the set E_s for a fixed $2 \le s \le 2^m$. Suppose that there are l functions of the form b_i and $m-l$ functions of the form g_i appearing as entries in the $T(h_1, \ldots, h_m)$ involved in the definition of E_s. For notational simplicity, assume that the b_i are in the first l entries and the g_i are in the remaining $m-l$ entries (analogous arguments hold in the other cases). Apply a Whitney decomposition to write each G_i as a union of dyadic cubes with disjoint interiors,

$$
G_i = \bigcup_{j=1}^{\infty} Q_{i,j},
$$

where

$$
2\mathrm{diam}(Q_{i,j}) \le d\big(Q_{i,j}, \mathbb{R}^n \setminus G_i\big) \le 8\mathrm{diam}(Q_{i,j}).
$$

Put

$$
b_{i,j} = b_i \mathbb{1}_{Q_{i,j}} \quad \text{and } b_i^N = \sum_{j=1}^{N} b_{i,j}.
$$

It suffices to control (uniformly in N) the measure of E_s with b_i replaced by b_i^N. Denote this set by $\widetilde{E}_s$.

Let $c_{i,j}$ denote the center of $Q_{i,j}$ and set

$$
a_{i,j} = \int_{Q_{i,j}} b_{i,j}(x)dx, \quad \nu_{i,j} = a_{i,j}\delta_{c_{i,j}}, \text{ and } \nu_i^N = \sum_{j=1}^{N} \nu_{i,j}.
$$

Notice, by adding and subtracting $T\left(\vec{\nu^N}_{1,k}, \vec{b^N}_{k+1,l}, \vec{g}_{l+1,m}\right)$ for $1 \le k \le l$, we have

$$
\begin{aligned}
|\widetilde{E}_s| \le{} & \sum_{k=1}^{l} \left| \left\{ \left| T\left(\vec{\nu^N}_{1,k-1}, b_k^N - \nu_k^N, \vec{b^N}_{k+1,l}, \vec{g}_{l+1,m}\right) \right| > \frac{t}{(l+1)2^m} \right\} \right| \\
& + \left| \left\{ \left| T\left(\vec{\nu^N}_{1,l}, \vec{g}_{l+1,m}\right) \right| > \frac{t}{(l+1)2^m} \right\} \right| \\
\le{} & \sum_{k=1}^{l} \left(|G| + \left| \left\{ \mathbb{R}^n \setminus G : \left| T\left(\vec{\nu^N}_{1,k-1}, b_k^N - \nu_k^N, \vec{b^N}_{k+1,l}, \vec{g}_{l+1,m}\right) \right| > \frac{t}{(l+1)2^m} \right\} \right| \right) \\
& + \left| \left\{ \left| T\left(\vec{\nu^N}_{1,l}, \vec{g}_{l+1,m}\right) \right| > \frac{t}{(l+1)2^m} \right\} \right| \\
\le{} & m^2 \|M\| t^{-\frac{1}{m}} + \sum_{k=1}^{l} |S_k| + |S|,
\end{aligned}
$$

where

$$S_k := \left\{ \mathbb{R}^n \setminus G : \left| T\left(\vec{\nu^N}_{1,k-1}, b_k^N - \nu_k^N, \vec{b^N}_{k+1,l}, \vec{g}_{l+1,m} \right) \right| > \frac{t}{(l+1)2^m} \right\}, \quad \text{and}$$

$$S := \left\{ \left| T\left(\vec{\nu^N}_{1,l}, \vec{g}_{l+1,m} \right) \right| > \frac{t}{(l+1)2^m} \right\}.$$

We will control each $|S_k|$ and $|S|$ individually.

We will first control $|S_k|$. Begin by using Chebyshev's inequality, the fact that $(b_{k,j_k}^N dm - \nu_{k,j_k}^N)(Q_{k,j_k}) = 0$, Fubini, and trivial estimates to see

$$\begin{aligned}
|S_k| &\le (l+1)2^m t^{-1} \int_{\mathbb{R}^n \setminus G} \left| T\left(\vec{\nu^N}_{1,k-1}, b_k^N - \nu_k^N, \vec{b^N}_{k+1,l}, \vec{g}_{l+1,m} \right)(x) \right| dx \\
&\le (m+1)2^m t^{-1} \sum_{j_1,\ldots,j_l=1}^{N} \int_{\mathbb{R}^n \setminus G} \left| \int_{\mathbb{R}^{n(m-l)}} \int_{Q_{l,j_l}} \cdots \int_{Q_{1,j_1}} K(x, \vec{y}_{1,m}) \right. \\
&\quad \times \left(\prod_{i=k+1}^{l} b_{i,j_i}(y_i) \right) \left(\prod_{i=l+1}^{m} g_i(y_i) \right) \\
&\quad \times d\vec{\nu}_{(1,j_1),(k-1,j_{k-1})}(\vec{y}_{1,k-1}) d(b_{k,j_k} dm - \nu_{k,j_k})(y_k) d\vec{y}_{k+1,m} \Big| dx \\
&\le (m+1)2^m t^{-1} \sum_{j_1,\ldots,j_l=1}^{N} \int_{\mathbb{R}^{n(m-l)}} \int_{Q_{l,j_l}} \cdots \int_{Q_{1,j_1}} \int_{\mathbb{R}^n \setminus G} \\
&\quad \times |K(x, \vec{y}_{1,m}) - K(x, \vec{c}_{(1,j_1),(l,j_l)}, \vec{y}_{l+1,m})| \left(\prod_{i=k+1}^{l} |b_{i,j_i}(y_i)| \right) \left(\prod_{i=l+1}^{m} |g_i(y_i)| \right) \\
&\quad \times dx d|\vec{\nu}|_{(1,j_1),(k-1,j_{k-1})}(\vec{y}_{1,k-1}) d|b_{k,j_k} dm - \nu_{k,j_k}|(y_k) d\vec{y}_{k+1,m} \\
&\le (m+1)2^m t^{-1} \sum_{j_1,\ldots,j_l=1}^{N} \left(\prod_{i=1}^{k-1} |a_{i,j_i}| \right) |b_{k,j_k} dm - \nu_{k,j_k}|(Q_{k,j_k}) \\
&\quad \times \left(\prod_{i=k+1}^{l} \|b_{i,j_i}\|_{L^1(\mathbb{R}^n)} \right) \left(\prod_{i=l+1}^{m} \|g_i\|_{L^\infty(\mathbb{R}^n)} \right) \\
&\quad \times \int_{\mathbb{R}^{n(m-l)}} \sup_{\substack{(y_1,\ldots,y_l) \\ \in \prod_{i=1}^{l} Q_{i,j_i}}} \int_{\mathbb{R}^n \setminus G} |K(x, \vec{y}_{1,m}) - K(x, \vec{c}_{(1,j_1),(l,j_l)}, \vec{y}_{l+1,m})| dx d\vec{y}_{l+1,m}.
\end{aligned}$$

Now, note that $\|b_{i,j}\|_{L^1(\mathbb{R}^n)} \le (17\sqrt{n})^n t^{\frac{1}{m}} |Q_{i,j}|$. Indeed, for a fixed $Q_{i,j}$, let $Q_{i,j}^*$ be the cube with the same center, but diameter $17\sqrt{n}$ times as large. Then $Q_{i,j}^* \cap (\mathbb{R}^n \setminus G_i) \neq \emptyset$. So there is a point $x \in Q_{i,j}^*$ such that $Mf_i(x) \le t^{\frac{1}{m}}$. In particular, $\fint_{Q_{i,j}^*} |f_i(y)| dy \le t^{\frac{1}{m}} |Q_{i,j}^*|$. Since $|Q_{i,j}^*| = (17\sqrt{n})^n |Q_{i,j}|$, we have

$$\|b_{i,j}\|_{L^1(\mathbb{R}^n)} = \int_{Q_{i,j}} |f_i(y)| dy \le \int_{Q_{i,j}^*} |f_i(y)| dy \le t^{\frac{1}{m}} |Q_{i,j}^*| = (17\sqrt{n})^n t^{\frac{1}{m}} |Q_{i,j}|.$$

Use the fact that $|b_{k,j_k}^N dm - v_{k,j_k}^N|(Q_{k,j_k}) \leq 2\|b_{k,j_k}\|_{L^1(\mathbb{R}^n)}$, the L^∞ control of the g_i, the L^1 control of the $b_{i,j}$, and Lemma 1 (which applies since $2\mathrm{diam}(Q_{i,j}) \leq d(Q_{i,j}, \mathbb{R}^n \setminus G_i)$) to continue the estimate

$$
\begin{aligned}
|S_k| &\leq (m+1)2^{m+1}t^{-1} \sum_{j_1,\ldots,j_l=1}^{N} \left(\prod_{i=1}^{l} \|b_{i,j_i}\|_{L^1(\mathbb{R}^n)}\right)\left(\prod_{i=l+1}^{m} \|g_i\|_{L^\infty(\mathbb{R}^n)}\right) \\
&\quad \times \int_{\mathbb{R}^{n(m-l)}} \sup_{\substack{(y_1,\ldots,y_l) \\ \in \prod_{i=1}^{l} Q_{i,j_i}}} \int_{\mathbb{R}^n \setminus G} |K(x, \vec{y}_{1,m}) - K(x, \vec{c}_{(1,j_1),(l,j_l)}, \vec{y}_{l+1,m})| dx d\vec{y}_{l+1,m} \\
&\leq (m+1)2^{m+1}(17\sqrt{n})^{nl} \sum_{j_1,\ldots,j_l=1}^{N} \left(\prod_{i=1}^{l} |Q_{i,j_i}|\right) \\
&\quad \times \int_{\mathbb{R}^{n(m-l)}} \sup_{\substack{(y_1,\ldots,y_l) \\ \in \prod_{i=1}^{l} Q_{i,j_i}}} \int_{\mathbb{R}^n \setminus G} |K(x, \vec{y}_{1,m}) - K(x, \vec{c}_{(1,j_1),(l,j_l)}, \vec{y}_{l+1,m})| dx d\vec{y}_{l+1,m} \\
&\leq A_1(m+1)2^{m+1}(17\sqrt{n})^{nm} \sum_{i=1}^{l} |G_i| \\
&\leq A_1 m(m+1)2^{m+1}(17\sqrt{n})^{nm} \|M\| t^{-\frac{1}{m}}.
\end{aligned}
$$

The control of $|S|$ follows from applying Theorem 2 below

$$
\begin{aligned}
|S| &\leq 2(l+1)^{\frac{1}{m}} A_3 t^{-\frac{1}{m}} \left(\prod_{i=1}^{l} \left\| v_i^N \right\|^{\frac{1}{m}}\right)\left(\prod_{i=l+1}^{m} \|g_i\|_{L^1(\mathbb{R}^n)}^{\frac{1}{m}}\right) \\
&\leq 2(m+1)^{\frac{1}{m}} A_3 t^{-\frac{1}{m}} \left(\prod_{i=1}^{l} \left\| b_i^N \right\|_{L^1(\mathbb{R}^n)}^{\frac{1}{m}}\right)\left(\prod_{i=l+1}^{m} \|g_i\|_{L^1(\mathbb{R}^n)}^{\frac{1}{m}}\right) \\
&\leq 2(m+1)^{\frac{1}{m}} A_3 t^{-\frac{1}{m}} \left(\prod_{i=1}^{m} \|f_i\|_{L^1(\mathbb{R}^n)}^{\frac{1}{m}}\right) \\
&= 2(m+1)^{\frac{1}{m}} A_3 t^{-\frac{1}{m}}.
\end{aligned}
$$

Put the estimates of $|S_k|$ and $|S|$ together to get

$$
\begin{aligned}
|\widetilde{E}_s| &\leq \left(m^2\|M\| + \sum_{k=1}^{l} A_1 m(m+1)2^{m+1}(17\sqrt{n})^{nm}\|M\| + 2(m+1)^{\frac{1}{m}} A_3\right) t^{-\frac{1}{m}} \\
&\leq \left(m^2\|M\| + A_1 m^2(m+1)2^{m+1}(17\sqrt{n})^{nm}\|M\| + 2(m+1)^{\frac{1}{m}} A_3\right) t^{-\frac{1}{m}} \\
&= B_2 t^{-\frac{1}{m}}
\end{aligned}
$$

where $B_2 := m^2\|M\| + A_1 m^2(m+1)2^{m+1}(17\sqrt{n})^{nm}\|M\| + 2(m+1)^{\frac{1}{m}} A_3$. Since the above estimate is independent of N, letting $N \to \infty$ yields

$$|E_s| \le B_2 t^{-\frac{1}{m}}$$

Finally, use the estimates of $|E_s|$, $1 \le s \le 2^m$ to observe

$$|\{|T(f_1, \ldots, f_m)| > t\}| \le |E_1| + \sum_{s=2}^{2^m} |E_s| \le \left(B_1 + \left(2^m - 1\right) B_2\right) t^{-\frac{1}{m}}.$$

Take $A_2 = (B_1 + (2^m - 1) B_2)^m$ to complete the proof. □

We now prove Theorem 2.

Proof of Theorem 2 Assume without loss of generality that each $a_{i,j} > 0$ and that $\|\nu_1\| = \cdots = \|\nu_l\| = \|f_{l+1}\|_{L^1(\mathbb{R}^n)} = \cdots = \|f_m\|_{L^1(\mathbb{R}^n)} = 1$. For $i = 1, \ldots, l$, set

$$E_{i,1} := B(x_{i,1}, r_{i,1})$$

where $r_{i,1} > 0$ is chosen so that $|E_{i,1}| = a_{i,1} t^{-\frac{1}{m}}$. Subsequently, set

$$E_{i,2} := B(x_{i,2}, r_{i,2}) \setminus E_{i,1}$$

where $r_{i,2} > 0$ is chosen so that $|E_{i,2}| = a_{i,2} t^{-\frac{1}{m}}$. In general, for $j = 3, \ldots, N$, set

$$E_{i,j} := B(x_{i,j}, r_{i,j}) \setminus \bigcup_{k=1}^{j-1} E_{i,k}$$

where $r_{i,j} > 0$ is chosen so that $|E_{i,j}| = a_{i,j} t^{-\frac{1}{m}}$. Set

$$E_i := \bigcup_{j=1}^{N} E_{i,j}$$

and notice that by construction

$$|E_i| = \sum_{j=1}^{N} |E_{i,j}| = \sum_{j=1}^{N} a_{i,j} t^{-\frac{1}{m}} = \|\nu_i\| t^{-\frac{1}{m}} = t^{-\frac{1}{m}}.$$

Similarly, set

$$E^*_{i,1} := B(x_{i,1}, 2r_{i,1}),$$

and subsequently for $j = 2, \ldots, N$,

$$E^*_{i,j} := B(x_{i,j}, 2r_{i,j}) \setminus \bigcup_{k=1}^{j-1} E^*_{i,k}.$$

Set

$$E^*_i := \bigcup_{j=1}^{N} E^*_{i,j}, \quad \text{and } E^* := \bigcup_{i=1}^{l} E^*_i.$$

By the doubling property of Lebesgue measure,

$$|E^*| \le \sum_{i=1}^{l} |E^*_i| \le 2^n \sum_{i=1}^{l} |E_i| \le m 2^n t^{-\frac{1}{m}}.$$

For $k \in \{0, \ldots, l\}$, set

$$\sigma_k := T\left(t^{\frac{1}{m}} \mathbb{1}_{E_1}, \ldots, t^{\frac{1}{m}} \mathbb{1}_{E_k}, \nu_{k+1} \ldots, \nu_l, f_{l+1}, \ldots, f_m\right),$$

noticing that $\sigma_0 = T(\nu_1, \ldots, \nu_l, f_{l+1}, \ldots, f_m)$. Then, by adding and subtracting σ_k for $1 \le k \le l$, we have

$$\begin{aligned}
|\{|T(\nu_1, \ldots, \nu_l, f_{l+1}, \ldots, f_m)| > t\}| &\le \sum_{k=1}^{l} \left|\left\{|\sigma_{k-1} - \sigma_k| > \frac{t}{l+1}\right\}\right| + \left|\left\{|\sigma_l| > \frac{t}{l+1}\right\}\right| \\
&= \sum_{k=1}^{l} |E^*| + \sum_{k=1}^{l} \left|\left\{\mathbb{R}^n \setminus E^* : |\sigma_{k-1} - \sigma_k| > \frac{t}{l+1}\right\}\right| + \left|\left\{|\sigma_l| > \frac{t}{l+1}\right\}\right| \\
&\le m^2 2^n t^{-\frac{1}{m}} + \sum_{k=1}^{l} |P_k| + |P|,
\end{aligned}$$

where

$$\begin{aligned}
P_k &:= \left\{\mathbb{R}^n \setminus E^* : |\sigma_{k-1} - \sigma_k| > \frac{t}{l+1}\right\}, \text{ and} \\
P &:= \left\{|\sigma_l| > \frac{t}{l+1}\right\}.
\end{aligned}$$

The remainder of the proof will focus on bounding $|P|$ and each $|P_k|$ by a constant multiplied by $t^{-\frac{1}{m}}$.

To control $|P|$, use Chebyshev's inequality, the boundedness of T from $(L^2(\mathbb{R}^n))^m$ to $L^{\frac{2}{m}}(\mathbb{R}^n)$, and the L^∞ control of the f_i to observe

$$
\begin{aligned}
|P| &\leq \frac{(l+1)^{\frac{2}{m}}}{t^{\frac{2}{m}-\frac{2l}{m^2}}} \int_{\mathbb{R}^n} \left|T\left(\mathbb{1}_{E_1}, \ldots, \mathbb{1}_{E_l}, f_{l+1}, \ldots, f_m\right)(x)\right|^{\frac{2}{m}} dx \\
&\leq \frac{(m+1)^{\frac{2}{m}}}{t^{\frac{2}{m}-\frac{2l}{m^2}}} \|T\|^{\frac{2}{m}}_{(L^2(\mathbb{R}^n))^m \to L^{\frac{2}{m}}(\mathbb{R}^n)} \left(\prod_{i=1}^{l} \|\mathbb{1}_{E_i}\|^{\frac{2}{m}}_{L^2(\mathbb{R}^n)}\right)\left(\prod_{i=l+1}^{m} \|f_i\|^{\frac{2}{m}}_{L^2(\mathbb{R}^n)}\right) \\
&\leq \frac{(m+1)^{\frac{2}{m}}}{t^{\frac{2}{m}-\frac{2l}{m^2}}} \|T\|^{\frac{2}{m}}_{(L^2(\mathbb{R}^n))^m \to L^{\frac{2}{m}}(\mathbb{R}^n)} \left(t^{-\frac{l}{m^2}}\right)\left(t^{\frac{m-l}{m^2}}\right) \prod_{i=l+1}^{m} \|f_i\|^{\frac{1}{m}}_{L^1(\mathbb{R}^n)} \\
&\leq (m+1)^{\frac{2}{m}} \|T\|^{\frac{2}{m}}_{(L^2(\mathbb{R}^n))^m \to L^{\frac{2}{m}}(\mathbb{R}^n)} t^{-\frac{1}{m}}.
\end{aligned}
$$

We will now control $|P_k|$. Notice

$$
\sigma_{k-1} - \sigma_k = T\left(t^{\frac{1}{m}}\mathbb{1}_{E_1}, \ldots, t^{\frac{1}{m}}\mathbb{1}_{E_{k-1}}, \nu_k - t^{\frac{1}{m}}\mathbb{1}_{E_k}, \nu_{k+1}, \ldots, \nu_l, f_{l+1}, \ldots, f_m\right).
$$

Use Chebyshev's inequality, the fact that $(\nu_{k,j_k} - t^{\frac{1}{m}}\mathbb{1}_{E_{k,j_k}}dm)(E_k) = 0$, Fubini, and trivial estimates to see

$$
\begin{aligned}
|P_k| &\leq \frac{l+1}{t^{\frac{m-k+1}{m}}} \int_{\mathbb{R}^n \setminus E^*} \left|T\left(\mathbb{1}_{E_1}, \ldots, \mathbb{1}_{E_{k-1}}, \nu_k - t^{\frac{1}{m}}\mathbb{1}_{E_k}, \nu_{k+1}, \ldots, \nu_l, f_{l+1}, \ldots, f_m\right)(x)\right| dx \\
&\leq \frac{m+1}{t^{\frac{m-k+1}{m}}} \sum_{j_1,\ldots,j_l=1}^{N} \int_{\mathbb{R}^n \setminus E^*} \left| \int_{\mathbb{R}^{n(m-l)}} \int_{E_{l,j_l}} \cdots \int_{E_{1,j_1}} K(x, \vec{y}_{1,m}) \right. \\
&\quad \times \left. \left(\prod_{i=l+1}^{m} f_i(y_i)\right) d\vec{y}_{1,k-1} d(\nu_{k,j_k} - t^{\frac{1}{m}}\mathbb{1}_{E_{k,j_k}}dm)(y_k) d\vec{\nu}_{(k+1,j_{k+1}),(l,j_l)}(\vec{y}_{k+1,l}) d\vec{y}_{l+1,m} \right| dx \\
&\leq \frac{m+1}{t^{\frac{m-k+1}{m}}} \sum_{j_1,\ldots,j_l=1}^{N} \int_{\mathbb{R}^{n(m-l)}} \int_{E_{l,j_l}} \cdots \int_{E_{1,j_1}} \int_{\mathbb{R}^n \setminus E^*} |K(x, \vec{y}_{1,m}) - K(x, \vec{x}_{(1,j_1),(l,j_l)}, \vec{y}_{l+1,m})| \\
&\quad \times \left(\prod_{i=l+1}^{m} |f_i(y_i)|\right) dx d\vec{y}_{1,k-1} d\left|\nu_{k,j_k} - t^{\frac{1}{m}}\mathbb{1}_{E_{k,j_k}}dm\right|(y_k) d|\vec{\nu}|_{(k+1,j_{k+1}),(l,j_l)}(\vec{y}_{k+1,l}) d\vec{y}_{l+1,m} \\
&\leq \frac{m+1}{t^{\frac{m-k+1}{m}}} \sum_{j_1,\ldots,j_l=1}^{N} \left(\prod_{i=1}^{k-1} |E_{i,j_i}|\right) \left|\nu_{k,j_k} - t^{\frac{1}{m}}\mathbb{1}_{E_{k,j_k}}dm\right|(E_{k,j_k}) \left(\prod_{i=k+1}^{l} a_{i,j_i}\right) \\
&\quad \times \left(\prod_{i=l+1}^{m} \|f_i\|_{L^\infty(\mathbb{R}^n)}\right) \\
&\quad \times \int_{\mathbb{R}^{n(m-l)}} \sup_{\substack{(y_1,\ldots,y_l) \\ \in \prod_{i=1}^{l} E_{i,j_i}}} \int_{\mathbb{R}^n \setminus E^*} |K(x, \vec{y}_{1,m}) - K(x, \vec{x}_{(1,j_1),(l,j_l)}, \vec{y}_{l+1,m})| dx d\vec{y}_{l+1,m}.
\end{aligned}
$$

Use the fact that $|\nu_{k,j_k} - t^{\frac{1}{m}} \mathbb{1}_{E_{k,j_k}} dm|(E_k) \leq 2t^{\frac{1}{m}} |E_{k,j_k}|$, the $L^\infty(\mathbb{R}^n)$ control of the f_i, and Lemma 1 to continue the estimate

$$
\begin{aligned}
|P_k| &\leq 2(m+1) \sum_{j_1,\ldots,j_l=1}^{N} \left(\prod_{i=1}^{l} |E_{i,j_i}| \right) \\
&\quad \times \int_{\mathbb{R}^{n(m-l)}} \sup_{\substack{(y_1,\ldots,y_l) \\ \in \prod_{i=1}^{l} E_{i,j_i}}} \int_{\mathbb{R}^n \setminus E^*} |K(x, \vec{y}_{1,m}) - K(x, \vec{x}_{(1,j_1),(l,j_l)}, \vec{y}_{l+1,m})| dx d\vec{y}_{l+1,m} \\
&\leq 2(m+1) A_1 \sum_{i=1}^{l} |E_i| \\
&\leq 2m(m+1) A_1 t^{-\frac{1}{m}}.
\end{aligned}
$$

Using these estimates of $|P|$ and $|P_k|$, we have

$$
\begin{aligned}
|\{|T(\nu_1, \ldots, \nu_l, f_1, \ldots, f_{m-l})| > t\}| &\leq m^2 2^n t^{-\frac{1}{m}} + \sum_{k=1}^{l} |P_k| + |P| \\
&\leq \left(m^2 2^n + \left(\sum_{k=1}^{l} 2m(m+1) A_1 \right) + (m+1)^{\frac{2}{m}} \|T\|^{\frac{2}{m}}_{(L^2(\mathbb{R}^n))^m \to L^{\frac{2}{m}}(\mathbb{R}^n)} \right) t^{-\frac{1}{m}} \\
&\leq \left(m^2 2^n + 2m^2(m+1) A_1 + (m+1)^{\frac{2}{m}} \|T\|^{\frac{2}{m}}_{(L^2(\mathbb{R}^n))^m \to L^{\frac{2}{m}}(\mathbb{R}^n)} \right) t^{-\frac{1}{m}}.
\end{aligned}
$$

Take $A_3 = \left(m^2 2^n + 2m^2(m+1) A_1 + (m+1)^{\frac{2}{m}} \|T\|^{\frac{2}{m}}_{(L^2(\mathbb{R}^n))^m \to L^{\frac{2}{m}}(\mathbb{R}^n)} \right)$ to complete the proof. □

References

1. Damián, W., Lerner, A.K., Pérez, C.: Sharp weighted bounds for multilinear maximal functions and Calderón–Zygmund operators. J. Fourier Anal. Appl **21**, 161–181 (2015)
2. Grafakos, L.: Classical and Modern Fourier Analysis. Pearson/Prentice Hall, Upper Saddle River, NJ (2004)
3. Grafakos, L., Kalton, N.: Multilinear Calderón–Zygmund operators on Hardy spaces. Collect. Math **52**, 169–179 (2001)
4. Grafakos, L., Torres, R.H.: Multilinear Calderón–Zygmund theory. Adv. Math **165**, 124–164 (2002)
5. Lerner, A.K., Ombrosi, S., Pérez, C., Torres, R.H., Trujillo-González, R.: New maximal functions and multiple weights for the multilinear Calderón–Zygmund theory. Adv. Math. **220**, 1222–1264 (2009)
6. Lin, Y.: Endpoint estimates for multilinear singular integral operators. Georgian Math. J. **23**, 559–570 (2016)
7. Maldonado, D., Naibo, V.: Weighted norm inequalities for paraproducts and bilinear pseudodifferential operators with mild regularity. J. Fourier Anal. Appl. **15**, 218–261 (2009)
8. Nazarov, F., Treil, S., Volberg, A.: Weak-type estimates and Cotlar inequalities for Calderón–Zygmund operators on nonhomogeneous spaces. Int. Math. Res. Notices. **9**, 463–487 (1998)
9. Pérez, C., Torres, R.H.: Minimal regularity conditions for the end-point estimate of bilinear Calderón–Zygmund operators. Proc. Am. Math. Soc. **1**, 1–13 (2014)

10. Stein, E.M.: Harmonic Analysis: Real-Variable Methods, Orthogonality, and Oscillatory Integrals. Princeton University Press, Princeton, NJ (1993)
11. Tolsa, X.: A proof of the weak $(1, 1)$ inequality for singular integrals with non doubling measures based on a Calderón–Zygmund decomposition. Publ. Mat. **45**, 163–174 (2001)

Publisher's Note Springer Nature remains neutral with regard to jurisdictional claims in published maps and institutional affiliations.

Journal of Fourier Analysis and Applications (2019) 25:2653–2689
https://doi.org/10.1007/s00041-019-09677-x

RESEARCH ARTICLE

Existence of Extremizers for a Model Convolution Operator

Chandan Biswas[1]

Received: 2 May 2018 / Revised: 4 January 2019 / Published online: 14 March 2019
© Springer Science+Business Media, LLC, part of Springer Nature 2019

Abstract
The operator T, defined by convolution with the affine arc length measure on the moment curve parametrized by $h(t) = (t, t^2, \dots, t^d)$ is a bounded operator from L^p to L^q if $(\frac{1}{p}, \frac{1}{q})$ lies on a line segment. In this article we prove that at non-end points there exist functions which extremize the associated inequality and any extremizing sequence is pre compact modulo the action of the symmetry of T. We also establish a relation between extremizers for T at the end points and the extremizers of an X-ray transform restricted to directions along the moment curve. Our proof is based on the ideas of Michael Christ on convolution with the surface measure on the paraboloid.

Keywords Extremizers · Precompactness · Radon transform

Mathematics Subject Classification Primary 42B10 · 44A35; Secondary 44A12

1 Introduction and Statement of Results

Let X and Y be two Banach spaces and T be a linear operator from X into Y. Beyond the immediate question of boundedness of T or the value of the operator norm of T, it is natural to investigate the operator in more details, such as the various properties of the operator—both qualitative and quantitative—of the existence of extremizers, the properties of extremizers, the extremizing sequences, the near extremizers of the operator.

This kind of detailed study of a bounded operator has a long and rich history. One of the most celebrated examples is the work of Beckner [1], where he studied the existence of extremizers for the Hausdorff-Young inequality on $\mathbb{R}^d$. Recently there has been a series of works on the above questions for different operators, such

Communicated by Hans G. Feichtinger.

This work is supported in part by NSF Grants DMS-1266336 and DMS-1600458.

✉ Chandan Biswas
cbiswas@wisc.edu

1 Department of Mathematics, University of Wisconsin, Madison, USA

as the Stein-Tomas Inequality [8,9], an improved Hausdorff-Young inequality [5], convolution with the surface measure on the paraboloid [3,10], for k-plane transforms [13,15], convolution with the surface measure on the sphere [24] to name a few. One motivation to investigate such questions is to produce improved inequalities and thus an inverse result concerning the stability of the inequality near an extremizer (see [5], for example) and to make qualitative studies of PDEs [25].

In this paper we investigate the above questions for a generalized Radon transform along the moment curve. Let T be the linear operator which acts on the continuous functions on $\mathbb{R}^d$ by convolution with affine arc length measure on the moment curve $(t, t^2, \ldots, t^d)$. That is, for a continuous function f on $\mathbb{R}^d$, Tf is defined by

$$Tf(x) = \int_{\mathbb{R}} f(x + (t, t^2, \ldots, t^d))dt. \tag{1.1}$$

Theorem 1.1 (Christ, Littman, Oberlin, Stovall) *T maps $L^p(\mathbb{R}^d)$ into $L^q(\mathbb{R}^d)$ as a bounded operator with $1 \le p, q \le \infty$ if and only if*

$$\frac{1}{p} = \frac{1}{p_\theta} = \frac{1-\theta}{p_0} + \frac{\theta}{p_1} \quad \text{and} \quad \frac{1}{q} = \frac{1}{q_\theta} = \frac{1-\theta}{q_0} + \frac{\theta}{q_1} \tag{1.2}$$

for some $\theta \in [0, 1]$ where $p_0 = \frac{d+1}{2}$, $q_0 = \frac{d(d+1)}{2(d-1)}$ and $p_1 = q_0'$, $q_1 = p_0'$.

The above theorem was proved for $d = 2$ by Littman in [21]. Oberlin [22] showed that it is sufficient to satisfy the above condition when $d = 3$. The theorem was proved up to the end point for any dimension by Christ in [2]. Extending ideas from [2], Stovall proved the strong type end point bound [23].

By using methods based on several observations and results of Christ (see [2–4,23]), such as the method of refinement to get quantitative restricted weak type inequality and an almost orthogonality argument using a distance-like function between two near extremizers, together with some refinements by Dendrinos and Stovall (using the increasing structure in the method of refinement), [12], we have been able to improve the associated inequality for this operator for $\theta \in (0, 1)$, such as establishing existence of functions that optimize the inequality and a qualitative information about functions that nearly optimize the corresponding inequality. At the endpoint i.e. for $\theta = 0, 1$ we apply a scaling, not adapted to the moment curve, to establish existence of functions that optimize the inequality corresponding to the bounds of an Xray transform. To state our results more precisely, first we need to introduce some definitions.

Let (p, q) be as above. Let A_p be the operator norm of T. That is

$$A_p = \sup_{\|f\|_{L^p}=1} \|Tf\|_{L^q}.$$

Note that although A_p depends on p, we shall write it simply as A when it is clear what value of p is under consideration.

Definition 1.2 **Extremizer.** Let $f \in L^p$. We say that f is an extremizer if

$$\|Tf\|_{L^q} = A\|f\|_{L^p} \neq 0. \tag{1.3}$$

Definition 1.3 **δ-Quasiextremizer.** For any $\delta > 0$, $f \in L^p$ is a δ-quasiextremizer if

$$\|Tf\|_{L^q} \geq \delta \|f\|_{L^p} \neq 0. \tag{1.4}$$

Definition 1.4 **δ-Quasiextremizer pair.** Let $\delta > 0$. We say that an ordered pair (f, g) of measurable functions on $\mathbb{R}^d$ is an δ-quasiextremizer pair if

$$\langle T(f), g\rangle \geq \delta \|g\|_{L^{q'}} \|f\|_{L^p} \neq 0.$$

Definition 1.5 **Extremizing sequence.** An extremizing sequence is any sequence $f_n \in L^p$ such that

$$\|f_n\|_{L^p} = 1,$$

$$\|Tf_n\|_{L^q} \to A.$$

Let us denote the moment curve by $h(t) = (t, t^2, \ldots, t^d)$. Our main theorems are as follows.

Theorem 1.6 *For every $\theta \in (0, 1)$ there exists an extremizer for $T : L^{p_\theta} \to L^{q_\theta}$ for the inequality* (1.3) *when $d \geq 2$. Furthermore for any nonnegative extremizing sequence $\{f_n\}$ there exists a sequence of symmetries (diffeomorphisms of $\mathbb{R}^d$, preserving L^p norm of f), $\{\phi_n^*\}$, of $\mathbb{R}^d$, such that there is a subsequence of $\{\phi_n^*(f_n)\}$ which converges in L^{p_θ} to some extremizer for $T : L^{p_\theta} \to L^{q_\theta}$.*

To state the second theorem we need a few more definitions.

Let X be the X-ray transform restricted to directions along the moment curve defined on continuous functions on $\mathbb{R}^d = \mathbb{R} \times \mathbb{R}^{d-1}$ by

$$Xf(t, y) = \int_{\mathbb{R}} f(s, y + s(2t, 3t^2, \ldots, dt^{d-1}))\, ds. \tag{1.5}$$

Theorem 1.7 (Christ and Erdogan [6], Dendrinos and Stovall [11], Erdogan [14], Laghi [19]) *X maps L^p into $L^q(L^r, dt)$ if for some $\theta \in [0, 1)$*

$$\left(\frac{1}{p}, \frac{1}{q}, \frac{1}{r}\right) = \left(\frac{1}{p_\theta}, \frac{1}{q_\theta}, \frac{1}{r_\theta}\right) = \left(1 - \theta + \frac{\theta d}{d+2}, \frac{\theta d}{d+2}, 1 - \theta + \frac{\theta(d^2 - d - 2)}{d^2 + d - 2}\right)$$

and the restricted weak type bound holds for X at the end point i.e., for $\theta = 1$.

The X-ray transform has been studied by many authors for its connection to many other parts of mathematics. It was first studied by Gelfand in [16]. There has been a lot of work done to investigate the boundedness properties of the X-ray like transforms, such as [17] and [18], to name a few.

It has been proved by Michael Christ, that extremizers of T exist in the case $d = 2$, and they have been identified and shown to be unique up to symmetries. Although it is still not known whether for $d > 2$, there exists an extremizer for $T : L^{\frac{d+1}{2}} \to L^{\frac{d(d+1)}{2(d-1)}}$ (corresponding to $\theta = 0$ in Theorem 1.1), we have been able to prove the following.

Theorem 1.8 *Let T be defined as in 1.1 and $d > 2$.*

- *Every extremizing sequence for $T : L^{p_0} \to L^{q_0}$ has a subsequence that either converges modulo symmetries of T to an extremizer for $T : L^{p_0} \to L^{q_0}$, or that converges modulo the nonsymmetry, $f_n \to r_n^{\frac{2(d-1)}{d+1}} f_n((0, r_n x') + h(x_1))$, to an extremizer for $X^* : L^{p_0} \to L^{q_0}$ corresponding to $\theta = \frac{(d+2)(d-1)}{d^2+d}$ in Theorem 1.7.*
- *Likewise, every extremizing sequence for $T : L^{p_1} \to L^{q_1}$ has a subsequence that either converges modulo symmetries of T to an extremizer for $T : L^{p_1} \to L^{q_1}$, or that converges modulo nonsymmetry, $f_n \to r_n^{\frac{d^2-d+2}{d+1}} f_n(r_n x)$, to an extremizer for $X : L^{p_1} \to L^{q_1}$.*
- *$\|T\|_{L^{p_0}\to L^{q_0}} \geq \|X^*\|_{L^{p_0}\to L^{q_0}}$ and if there exists an extremizing sequence for $T : L^{p_0} \to L^{q_0}$ that does not have a subsequence converging to an extremizer modulo symmetries of T, then $\|T\|_{L^{p_0}\to L^{q_0}} = \|X^*\|_{L^{p_0}\to L^{q_0}}$.*

Corollary 1.9 *At least one of the following must hold:*

- *(A) There exists an extremizer for $T : L^{p_0} \to L^{q_0}$; or*
- *(B) There exists an extremizer for $X : L^{p_1} \to L^{q_1}$.*

2 Outline of the Proof

A simplified outline of the argument is as follows. Given a function f which is very close to being an extremizer, we consider a dyadic decomposition of the range: $f = \sum_j 2^j f_j$ where $\frac{1}{2}\chi_{E_j} \leq f_j < \chi_{E_j}$ for pairwise disjoint measurable sets E_j in $\mathbb{R}^d$. Following [3] we prove that for f to be a near extremizer (say $\|Tf\|_q \geq A(1-\delta)\|f\|_p$ for small δ) the set of indices $\{j\}$ essentially lies in an interval around $J \in \mathbb{Z}$ in $\mathbb{Z}$ of length depending only on δ but independent of f. This is done by applying a certain"trilinear" bound for T (see Lemma 9.2 and Theorem 10.2) using Christ's method of refinement [2] and the increasing structure of the refinement due to Dendrinos and Stovall [12]. The trilinear bound has already been established in Lemma 5.2 in [23]. Our proof is much simpler in comparison due to the increasing structure in the method of refinement. We show that E_j are very close to being curved "parallelepipeds", and these are "almost" pairwise disjoint. More precisely these parallelepipeds are projections of balls in the incidence manifold. This is accomplished by introducing a mock distance on the set of all natural extremizers and proving that any two distant natural extremizers are almost orthogonal in the L^q space, see Lemma 9.1 (this is similar to the argument due to Christ in [3]). One significant difference in the structure of these parallelepipeds corresponding to when $d = 2$ (in this case the paraboloid in [3] is the same as the moment curve) from $d > 2$ is when $d = 2$ the symmetries of the operator act transitively on this set of parallelepipeds. For $d > 2$ the situation is quite different. For $\theta \in (0, 1)$ the symmetries of the operator act transitively on this set of parallelepipeds (near extremizers for $L^{p_\theta} \to L^{q_\theta}$ bound). So after applying symmetries we can assume that all these near extremizers are well adapted to the unit ball in $\mathbb{R}^d$. In other words $E_J \sim B(0,1)$, the unit ball in $\mathbb{R}^d$. So we can proceed as in the case of Paraboloid [3] using Christ's argument to prove the existence of extremizers. On the other hand when $d > 2$, the symmetry group does not act transitively on the set of near extremizers for $L^{p_0} \to L^{q_0}$ bound. As a consequence one has to allow the thickness of these parallelepipeds to become arbitrarily small as f becomes closer to being an extremizer. We overcome this obstruction by applying a non-symmetric "scaling" to the function f, so that we can avoid an extremizing sequence converging to 0 pointwise while simultaneously preserving the L^p norm of f. This is essentially the only new part of our analysis. This enables us to make the now rescaled near extremizers for $T : L^{p_0} \to L^{q_0}$ well adapted to the unit ball in $\mathbb{R}^d$ and thus proving existence of extremizers for the Xray tranform, $X : L^{p_0} \to L^{q_0}$.

3 Notation

Most of the notation we will use is fairly standard. In this note c, C denote implicit small and large positive constants respectively, which are allowed to change from one line to another. If $1 \le p \le \infty$, we denote by p' the exponent dual to p. We use $|E|$ to indicate Lebesgue measure. When A and B are non-negative real numbers, we write $A \lesssim B$ to mean $A \le CB$ for an implicit constant C, and $A \sim B$ when $A \lesssim B$ and $B \lesssim A$. We will also employ the somewhat less standard notation $\mathcal{T}(E, F) := \langle T(\chi_E), \chi_F \rangle$ when E and F are Borel sets and T is a linear operator. We will also use (E, F) to denote the pair of functions (χ_E, χ_F). We say that the sequence $\{f_n\} \subset L^p$ converges weakly to f in L^p if for any function $\psi \in L^{p'}$, $\int f_n\psi$ converges to $\int f\psi$ and $\{f_n\}$ converges strongly to f in L^p if $\int |f_n - f|^p$ converges to 0. Since $T(f) \le T(|f|)$ for all $f \in L^p$ and we are interested in only all those f for which $|T(f)|$ is large, in this paper all the functions f will be assumed to be nonnegative.

4 Symmetries

In this section we study the symmetries of the operator T.

Definition 4.1 A symmetry of $T : L^p \to L^q$ is an L^p isometry ϕ^* for which $T \circ \phi^* = \psi^* \circ T$ for some L^q isometry ψ^*.

The operator T has many symmetries. Let $\Theta : \mathbb{R}^{d+d} \to \mathbb{R}^{d-1}$ be the function defined by

$$\Theta(x, y) = (y_2 - x_2 - (y_1 - x_1)^2, y_3 - x_3 - (y_1 - x_1)^3, \ldots, y_d - x_d - (y_1 - x_1)^d)$$

and Let Σ be the incidence manifold $\Sigma = \{(x, y) : \Theta(x, y) = 0\}$. Let us denote the set of all diffeomorphisms of $\mathbb{R}^d$ by $\mathrm{Diff}(\mathbb{R}^d)$.

Definition 4.2 Let $G_{d,d}$ denote the set of all $(\phi, \psi) \in \mathrm{Diff}(\mathbb{R}^d) \times \mathrm{Diff}(\mathbb{R}^d)$ such that

$$\Theta(\phi(x), \psi(y)) = 0 \quad \text{if and only if} \quad \Theta(x, y) = 0 \quad \text{for all} \quad (x, y) \in \mathbb{R}^{d+d}.$$

In other words, $G_{d,d}$ denotes the set of all ordered pairs of diffeomorphisms of $\mathbb{R}^d$ which preserve the incidence manifold Σ. We also let G_d denote the set of all $\phi \in \mathrm{Diff}(\mathbb{R}^d)$ such that there exists $\psi \in \mathrm{Diff}(\mathbb{R}^d)$ such that $(\phi, \psi) \in G_{d,d}$.

The followings are examples of elements of $G_{d,d}$.

- Translation: $(\phi(x), \psi(y)) = (x + v, y + v)$ for some $v \in \mathbb{R}^d$.
- Scaling: $(\phi(x), \psi(y)) = \big(S_r(x), S_r(y)\big) = \big((rx_1, r^2x_2, \ldots, r^d x_d), (ry_1, r^2y_2, \ldots, r^d y_d)\big)$ for some $r \in \mathbb{R} - \{0\}$.
- *Gliding along h*: $(\phi(x), \psi(y)) = (G_{t_0}(x), G_{t_0}(y) + h(t_0))$ for some $t_0 \in \mathbb{R}$, where G_{t_0} is the linear operator defined on $\mathbb{R}^d$ associated to the $(d \times d)$ matrix

$$G_{t_0} = \begin{bmatrix} 1 & 0 & 0 \ldots 0 \\ 2t_0 & 1 & 0 \ldots 0 \\ 3t_0^2 & 3t_0 & 1 \ldots 0 \\ \vdots & \vdots & \ldots \; 0 \\ \binom{m}{1} t_0^{m-1} & \binom{m}{2} t_0^{m-2} & \ldots \; 0 \\ \vdots & \vdots & \ldots \; 0 \\ \binom{d}{1} t_0^{d-1} & \binom{d}{2} t_0^{d-2} & \ldots \; 1 \end{bmatrix}.$$

Note that $h(t + t_0) = G_{t_0}(h(t)) + h(t_0)$ for all $t, t_0 \in \mathbb{R}$.

The elements of G_d play a central role in our analysis. There might be more elements in G_d than the ones in the above examples but as we shall see these are enough for our analysis. For each of the three types of symmetries described above the associated diffeomorphism has constant Jacobian. For each ϕ we define the associated operators $\phi^* : L^p \to L^p$ by $\phi^* f(x) = J_\phi^{\frac{1}{p}} f(\phi(x))$. Then

$$\|\phi^*(f)\|_{L^p} = \|f\|_{L^p}, \quad \langle T(\phi^* f), \psi^* g\rangle = \langle T(f), g\rangle.$$

5 Paraballs

In this section we shall study an essentially exhaustive list of quasiextremal pairs. They are natural in the sense that every quasi-extremal pair is close, in a sense that degrades as the constant of quasiextremality decreases, to one of these pairs, see Theorem 6.1. It is elementary to show that the characteristic function of the set $\{x \in \mathbb{R}^d : \|x\| < \delta\}$ is an δ^C-quasiextremal for $T : L^{p_\theta} \to L^{q_\theta}$ for each $\theta \in [0, 1]$. For $\theta = 0$, we have, in addition, that the characteristic function of the set $\{x \in \mathbb{R}^d : \|x\| < \delta\}$, and for $\theta = 1$, the δ-tubular neighborhood of the set $\{(t, t^2, \ldots, t^d) : t \in [-1, 1]\}$ i.e. $\{x \in \mathbb{R}^d : \|x - (t, t^2, \ldots, t^d)\| < \delta$ for some $t \in [-1, 1]\}$ are c-quasiextremal where c is a small positive number that depends only on d and independent of δ, see

Proposition 5.7. The set of all paraballs is the collection of sets that we produce by applying the elements of G_d to these sets. Below is a more detailed description of the "paraballs".

Definition 5.1 For $0 < \alpha, \beta \leq 1$, we define

- $B(0, 0, \alpha, 1) = \{y \in \mathbb{R}^d : |y_1| < 1 \text{ and } \|y - h(y_1)\| < \alpha\}$ and $B^*(0, 0, \alpha, 1) = \{x \in \mathbb{R}^d : |x_i| < \alpha \text{ for all } 1 \leq i \leq d\}$.
- $B(0, 0, 1, \beta) = \{y \in \mathbb{R}^d : |y_i| < \beta \text{ for all } 1 \leq i \leq d\}$ and $B^*(0, 0, 1, \beta) = \{x \in \mathbb{R}^d : |x_1| < 1 \text{ and } \|x + h(-x_1)\| < \beta\}$.
- $$\begin{aligned} B(\bar{x}, t_0, \lambda\alpha, \lambda) &= G_{t_0} S_\lambda B(0, 0, \alpha, 1) + \bar{x} + h(t_0) \\ B^*(\bar{x}, t_0, \lambda\alpha, \lambda) &= G_{t_0} S_\lambda B^*(0, 0, \alpha, 1) + \bar{x}. \end{aligned} \tag{5.1}$$

We also define a scaling of a paraball by

$$\lambda B(\bar{x}, t_0, \alpha, \beta) = B(\bar{x}, t_0, \lambda\alpha, \lambda\beta). \tag{5.2}$$

Note that this does not correspond to a symmetry of the operator.

As an example in the special case when $0 < \alpha \leq \beta$ the paraball $B = B(\bar{x}, t_0, \alpha, \beta)$ is the set of all $y \in \mathbb{R}^d$ satisfying all of

- $|y_1 - \bar{y}_1| \leq \beta$;
- $|\sum_{i=1}^m \binom{m}{i}(-t_0)^{m-i}(y_i - \bar{y}_i) - (y_1 - \bar{y}_1)^m| \leq \beta^{m-1}\alpha$ for all $1 < m \leq d$

where $\bar{y} = \bar{x} + h(t_0)$.

For $0 < \alpha \leq \beta$, the dual paraball, denoted by $B^* = B^*(\bar{x}, t_0, \alpha, \beta)$, is the set of all $x \in \mathbb{R}^d$ such that

- $|x_1 - \bar{x}_1| \leq \alpha$;
- $|\sum_{i=1}^m \binom{m}{i}(-t_0)^{m-i}(x_i - \bar{x}_i)| \leq \beta^{m-1}\alpha$ for all $1 < m \leq d$.

Similarly when $0 < \beta < \alpha$ the paraball $B = B(\bar{x}, t_0, \alpha, \beta)$ is the set of all $y \in \mathbb{R}^d$ satisfying all of

- $|y_1 - \bar{y}_1| \leq \beta$;
- $|\sum_{i=1}^m \binom{m}{i}(-t_0)^{m-i}(y_i - \bar{y}_i)| \leq \alpha^{m-1}\beta$ for all $1 < m \leq d$

where $\bar{y} = \bar{x} + h(t_0)$.

The dual paraball, denoted by $B^* = B^*(\bar{x}, t_0, \alpha, \beta)$ is the set of all $x \in \mathbb{R}^d$ such that

- $|x_1 - \bar{x}_1| \leq \alpha$;
- $|\sum_{i=1}^m \binom{m}{i}(-t_0)^{m-i}(x_i - \bar{x}_i) + (\bar{x}_1 - x_1)^m| \leq \alpha^{m-1}\beta$ for all $1 < m \leq d$.

For our analysis of an extremizing sequence it is important to measure how two paraballs interact with each other. We define a mock-distance on the set of all paraballs to measure the interaction between any two distant paraballs similar to the distance defined in [3].

Definition 5.2 Let $B^a = B(\bar{x}^a, t_a, \alpha_a, \beta_a)$ and $B^b = B(\bar{x}^b, t_b, \alpha_b, \beta_b)$ be two paraballs. Let $\bar{y}^a = \bar{x}^a + h(t_a)$ and $\bar{y}^b = \bar{x}^b + h(t_b)$ be the centers of the dual paraballs of B^a and B^b respectively. We define:

- If $\alpha_a \le \beta_a$ and $\alpha_b \le \beta_b$,

$$\begin{aligned} d(B^a, B^b) =& \frac{\max\left(\alpha_a\beta_a\alpha_a{\beta_a}^2 \ldots \alpha_a\beta_a^{d-1}, \;\; \alpha_b\beta_b\alpha_b{\beta_b}^2 \ldots \alpha_b\beta_b^{d-1}\right)}{\min\left(\alpha_a\beta_a\alpha_a{\beta_a}^2 \ldots \alpha_a\beta_a^{d-1}, \;\; \alpha_b\beta_b\alpha_b{\beta_b}^2 \ldots \alpha_b\beta_b^{d-1}\right)} + \left(\frac{\beta_a}{\beta_b} + \frac{\beta_b}{\beta_a}\right) \\ &+ \left(\frac{\alpha_a}{\alpha_b} + \frac{\alpha_b}{\alpha_a}\right) + |\bar{y}_1^a - \bar{y}_1^b|\left(\frac{1}{\beta_a} + \frac{1}{\beta_b}\right) + |\bar{x}_1^a - \bar{x}_1^b|\left(\frac{1}{\alpha_a} + \frac{1}{\alpha_b}\right) \\ &+ \sum_{m=2}^{d} \frac{|\sum_{i=1}^m \binom{m}{i}(-t_a)^{m-i}(\bar{y}_i^b - \bar{y}_i^a) - (\bar{y}_1^b - \bar{y}_1^a)^m|}{\alpha_a\beta_a^{m-1}} \\ &+ \sum_{m=2}^{d} \frac{|\sum_{i=1}^m \binom{m}{i}(-t_b)^{m-i}(\bar{y}_i^a - \bar{y}_i^b) - (\bar{y}_1^a - \bar{y}_1^b)^m|}{\alpha_b\beta_b^{m-1}} \\ &+ \sum_{m=2}^{d} \frac{|\sum_{i=1}^m \binom{m}{i}(-t_a)^{m-i}(\bar{x}_i^b - \bar{x}_i^a)|}{\alpha_a\beta_a^{m-1}} \\ &+ \sum_{m=2}^{d} \frac{|\sum_{i=1}^m \binom{m}{i}(-t_b)^{m-i}(\bar{x}_i^a - \bar{x}_i^b)|}{\alpha_b\beta_b^{m-1}}; \end{aligned} \tag{5.3}$$

- If $\beta_a < \alpha_a$ and $\beta_b < \alpha_b$,

$$\begin{aligned} d(B^a, B^b) =& \frac{\max\left(\alpha_a\beta_a\alpha_a^2\beta_a \ldots {\alpha_a}^{d-1}\beta_a, \;\; \alpha_b\beta_b\beta_b{\alpha_b}^2 \ldots \beta_b\alpha_b^{d-1}\right)}{\min\left(\alpha_a\beta_a\beta_a{\alpha_a}^2 \ldots \beta_a\alpha_a^{d-1}, \;\; \alpha_b\beta_b\beta_b{\alpha_b}^2 \ldots \beta_b\alpha_b^{d-1}\right)} + \left(\frac{\beta_a}{\beta_b} + \frac{\beta_b}{\beta_a}\right) \\ &+ \left(\frac{\alpha_a}{\alpha_b} + \frac{\alpha_b}{\alpha_a}\right) + |\bar{y}_1^a - \bar{y}_1^b|\left(\frac{1}{\beta_a} + \frac{1}{\beta_b}\right) + |\bar{x}_1^a - \bar{x}_1^b|\left(\frac{1}{\alpha_a} + \frac{1}{\alpha_b}\right) \\ &+ \sum_{m=2}^{d} \frac{|\sum_{i=1}^m \binom{m}{i}(-t_a)^{m-i}(\bar{y}_i^b - \bar{y}_i^a)|}{\beta_a\alpha_a^{m-1}} \\ &+ \sum_{m=2}^{d} \frac{|\sum_{i=1}^m \binom{m}{i}(-t_b)^{m-i}(\bar{y}_i^a - \bar{y}_i^b)|}{\beta_b\alpha_b^{m-1}} \\ &+ \sum_{m=2}^{d} \frac{|\sum_{i=1}^m \binom{m}{i}(-t_a)^{m-i}(\bar{x}_i^b - \bar{x}_i^a) + (\bar{x}_1^a - \bar{x}_1^b)^m|}{\beta_a\alpha_a^{m-1}} \\ &+ \sum_{m=2}^{d} \frac{|\sum_{i=1}^m \binom{m}{i}(-t_b)^{m-i}(\bar{x}_i^a - \bar{x}_i^b) + (\bar{x}_1^b - \bar{x}_1^a)^m|}{\beta_b\alpha_b^{m-1}}. \end{aligned} \tag{5.4}$$

A few comments are in order. Note that we have defined this distance d for only two of the four possible cases, only for $(\alpha_a \le \beta_a, \alpha_b \le \beta_b)$ and $(\beta_a < \alpha_a, \beta_b < \alpha_b)$. This is because in our analysis we need to use the distance between only these two types of paraballs. So from now on when when we talk about distance between paraballs it would be one of these two cases. Note that d is not a distance on the set of all paraballs, simply because for any paraball B, $d(B, B) = 5$. But as we shall see that this is not of any significance to our analysis, for we shall use the properties of d only when the distance between two paraballs is large. Note that our "distance" function, d is not a pseudo-distance either, as it does not satisfy the properties of a pseudo-distance, So for the lack of a better term we shall call it a mock-distance.

In the first term in the expression we compare the $(d-1)$-dimensional volume of the cross sections of the paraballs. The second and the third terms measure the ratio between the lengths of the bases of the paraballs and the dual paraballs respectively. The fourth term measures the distance between the first coordinates of the centers of the paraballs and the fifth term for the centers of the dual paraballs. The sixth and the seventh term measure how far are the centers of each paraball from the other paraball. Likewise the eighth and ninth terms measure how far are the centers of the dual paraball from the other dual paraball.

We shall see in the proof of Proposition 5.4 that the third, eighth and the ninth terms are redundant, in the sense that these are essentially dominated by first and second, sixth and the seventh term respectively. But we include these terms to make the mock-distance symmetric i.e. $d(B^a, B^b) = d(B^{a*}, B^{b*})$. In addition we have the following property of this mock-distance.

Lemma 5.3 *For every pair of paraballs B^a, B^b and $\phi \in G_d$ we have*

$$d(B^a, B^b) = d(\phi^*(B^a), \phi^*(B^b)).$$

Proof It is enough to prove the stated equality when ϕ is either a translation or scaling or gliding along h. Let $B^a = B(\bar{x}^a, t_a, \alpha_a, \beta_a)$ and $B^b = B(\bar{x}^b, t_b, \alpha_b, \beta_b)$ be two paraballs. When ϕ is a translation i.e. $\phi(x) = x + v$ for some $v \in \mathbb{R}^d$, then $\phi^*(B^a) = B(\bar{x}^a + v, t_a, \alpha_a, \beta_a)$ and $\phi^*(B^b) = B(\bar{x}^b + v, t_b, \alpha_b, \beta_b)$. When ϕ is a scaling i.e. $\phi = S_r$ for some $r > 0$, then $\phi^*(B^a) = B(S_r(\bar{x}^a), rt_a, r\alpha_a, r\beta_a)$ and $\phi^*(B^b) = B(S_r(\bar{x}^b), rt_b, r\alpha_b, r\beta_b)$. In both these cases each term in the Definition 5.2 remains unchanged. Therefore $d(B^a, B^b) = d(\phi^*(B^a), \phi^*(B^b))$.

Let us now assume that ϕ is a gliding along h i.e. $\phi(x) = G_{t_0}(x)$ for some $t_0 \in \mathbb{R}$. Then $\phi^*(B^a) = B(G_{t_0}(\bar{x}^a), t_a + t_0, \alpha_a, \beta_a)$ and $\phi^*(B^b) = B(G_{t_0}(\bar{x}^b), t_b + t_0, \alpha_b, \beta_b)$. So the first five terms in $d(\phi^*(B^a), \phi^*(B^b))$ remain same as the first five terms in $d(B^a), (B^b))$. If $\alpha_a \le \beta_a$ then the sixth term in $d(\phi^*(B^a), \phi^*(B^b))$ is

$$\begin{aligned}
&\sum_{m=2}^{d} \frac{|\sum_{i=1}^{m} \binom{m}{i} (-t_a - t_0)^{m-i} (G_{t_0}(\bar{y}^b - \bar{y}^a))_i - (\bar{y}_1^b - \bar{y}_1^a)^m|}{\alpha_a \beta_a^{m-1}} \\
&\quad = \sum_{m=2}^{d} \frac{|(G_{-t_0-t_a} G_{t_0}(\bar{y}^b - \bar{y}^a))_m - (\bar{y}_1^b - \bar{y}_1^a)^m|}{\alpha_a \beta_a^{m-1}} \\
&\quad = \sum_{m=2}^{d} \frac{|(G_{-t_a}(\bar{y}^b - \bar{y}^a))_m - (\bar{y}_1^b - \bar{y}_1^a)^m|}{\alpha_a \beta_a^{m-1}} \\
&\quad = \sum_{m=2}^{d} \frac{|\sum_{i=1}^{m} \binom{m}{i} (-t_a)^{m-i} (\bar{y}_i^b - \bar{y}_i^a) - (\bar{y}_1^b - \bar{y}_1^a)^m|}{\alpha_a \beta_a^{m-1}},
\end{aligned} \tag{5.5}$$

which is the also sixth term in $d(B^a, B^b)$. Similarly when $\beta_a < \alpha_a$ the sixth term remains unchanged. By similar arguments one can prove that all other terms remain unchanged in $d(\phi^*(B^a), \phi^*(B^b))$. This finishes the proof. □

Proposition 5.4 *There exists a constant $C < \infty$ which depends only on the dimension d, such that for any two paraballs B^a, B^b*

$$d(B^a, B^b) \le C\left(\frac{\max\left(|B^a|, |B^b|\right)}{|B^a \cap B^b|}\right)^C.$$

Proof The proof of this lemma will be an adaptation of the proof of Lemma 3.7 in [3]. We shall give the proof for the case $\alpha_a \le \beta_a$ and $\alpha_b \le \beta_b$ for the paraballs $B^a = B(\bar{x}^a, t_a, \alpha_a, \beta_a)$ and $B^b = B(\bar{x}^b, t_b, \alpha_b, \beta_b)$ respectively, the other case being identical. Without loss of generality we may assume that $d(B^a, B^b)$ is large. Otherwise $d(B^a, B^b)$ would be bounded by a large constant C. For any paraball $B(\bar{x}, t_0, \alpha, \beta)$, $y \in B$ implies $|y_1 - \bar{y}_1| \le \beta$ where $\bar{y} = \bar{x} + h(t_0)$. This implies that

$$\begin{aligned} |B^a \cap B^b| &\le \min(\alpha_a\beta_a\alpha_a{\beta_a}^2 \ldots \alpha_a\beta_a^{d-1}, \alpha_b\beta_b\alpha_b{\beta_b}^2 \ldots \alpha_b\beta_b^{d-1})\min(\beta_a, \beta_b) \\ &\le \frac{\min(\alpha_a\beta_a\alpha_a{\beta_a}^2 \ldots \alpha_a\beta_a^{d-1}, \alpha_b\beta_b\alpha_b{\beta_b}^2 \ldots \alpha_b\beta_b^{d-1})}{\max(\alpha_a\beta_a\alpha_a{\beta_a}^2 \ldots \alpha_a\beta_a^{d-1}, \alpha_b\beta_b\alpha_b{\beta_b}^2 \ldots \alpha_b\beta_b^{d-1})}\max(|B^a|, |B^b|). \end{aligned}$$

If $\frac{\max\left(\alpha_a\beta_a\alpha_a{\beta_a}^2 \ldots \alpha_a\beta_a^{d-1}, \alpha_b\beta_b\alpha_b{\beta_b}^2 \ldots \alpha_b\beta_b^{d-1}\right)}{\min\left(\alpha_a\beta_a\alpha_a{\beta_a}^2 \ldots \alpha_a\beta_a^{d-1}, \alpha_b\beta_b\alpha_b{\beta_b}^2 \ldots \alpha_b\beta_b^{d-1}\right)} \ge c\, d(B^a, B^b)$, this concludes the proof.

For the paraball $B^a = B(\bar{x}^a, t_a, \alpha_a, \beta_a)$, we define

$$S^a = (-\beta_a, \beta_a).$$

Similarly, we define S^b for B^b. Now

$$\begin{aligned} |B^a \cap B^b| &\le \frac{|(\bar{y}_1^a + S^a) \cap (\bar{y}_1^b + S^b)|}{\max(|S^a|, |S^b|)}\max(|B^a|, |B^b|) \\ &\le \frac{\min(|S^a|, |S^b|)}{\max(|S^a|, |S^b|)}\max(|B^a|, |B^b|) \\ &= \frac{\min(\beta_a, \beta_b)}{\max(\beta_a, \beta_b)}\max(|B^a|, |B^b|) \\ &\sim \left(\frac{\beta_a}{\beta_b} + \frac{\beta_b}{\beta_a}\right)^{-1}\max(|B^a|, |B^b|). \end{aligned}$$

Therefore the desired inequality follows if $\frac{\beta_a}{\beta_b} + \frac{\beta_b}{\beta_a} \ge c\, d(B^a, B^b)$. In addition, for some absolute constant C,

$$|(\bar{y}_1^a + S^a) \cap (\bar{y}_1^b + S^b)| \le C\left[|\bar{y}_1^a - \bar{y}_1^b|\left(\frac{1}{\beta_a} + \frac{1}{\beta_b}\right)\right]^{-1}\max(|S^a|, |S^b|).$$

Hence the desired inequality follows if the fourth term is $\geq c\, d(B^a, B^b)$.

Let us now consider the third term. We can assume that the first two terms are small i.e.

- $\frac{\beta_a}{\beta_b} + \frac{\beta_b}{\beta_a} \leq c' d(B^a, B^b)^{c'}$;
- $\frac{\max\left(\alpha_a\beta_a\alpha_a\beta_a{}^2...\alpha_a\beta_a^{d-1},\quad \alpha_b\beta_b\alpha_b\beta_b{}^2...\alpha_b\beta_b^{d-1}\right)}{\min\left(\alpha_a\beta_a\alpha_a\beta_a{}^2...\alpha_a\beta_a^{d-1},\quad \alpha_b\beta_b\alpha_b\beta_b{}^2...\alpha_b\beta_b^{d-1}\right)} \leq c' d(B^a, B^b)^{c'}$.

We shall prove that this implies the third term is also small i.e.

$$\frac{\alpha_a}{\alpha_b} + \frac{\alpha_b}{\alpha_a} \leq c d(B^a, B^b)^c.$$

WLOG let us assume that $\beta_a \leq \beta_b$. Since $\frac{\beta_a}{\beta_b} + \frac{\beta_b}{\beta_a} \leq c' d(B^a, B^b)^{c'}$, this implies

$$\beta_a \leq \beta_b \leq c' d(B^a, B^b)^{c'} \beta_a.$$

Therefore

$$\begin{aligned}
\left(\frac{\alpha_a}{\alpha_b} + \frac{\alpha_b}{\alpha_a}\right)^{d-1} &\sim \frac{\max(\alpha_a{}^{d-1}, \alpha_b{}^{d-1})}{\min(\alpha_a{}^{d-1}, \alpha_b{}^{d-1})} \\
&\leq \frac{\max\left(\alpha_a\beta_b\alpha_a\beta_b{}^2\ldots\alpha_a\beta_b^{d-1},\quad \alpha_b\beta_b\alpha_b\beta_b{}^2\ldots\alpha_b\beta_b^{d-1}\right)}{\min\left(\alpha_a\beta_a\alpha_a\beta_a{}^2\ldots\alpha_a\beta_a^{d-1},\quad \alpha_b\beta_a\alpha_a\beta_a{}^2\ldots\alpha_b\beta_a{}^{d-1}\right)} \\
&\sim \left(c' d(B^a, B^b)^{c'}\right)^C \frac{\max\left(\alpha_a\beta_a\alpha_a\beta_a{}^2\ldots\alpha_a\beta_a^{d-1},\quad \alpha_b\beta_b\alpha_b\beta_b{}^2\ldots\alpha_b\beta_b^{d-1}\right)}{\min\left(\alpha_a\beta_a\alpha_a\beta_a{}^2\ldots\alpha_a\beta_a^{d-1}\quad \alpha_b\beta_b\alpha_b\beta_b{}^2\ldots\alpha_b\beta_b^{d-1}\right)}.
\end{aligned} \tag{5.6}$$

We choose c' such that $\left(c' d(B^a, B^b)^{c'}\right)^C < c d(B^a, B^b)^c$.

Now we consider the fifth term. Suppose that $|\bar{x}_1^a - \bar{x}_1^b| \geq c d(B^a, B^b)\alpha_a$. If $(w, s_1, \ldots, s_{d-1}) \in \mathbb{R}\times\mathbb{R}^{d-1}$ belongs to $B^a \cap B^b$, then one has $|s_1 - \bar{x}_2^a - (w - \bar{x}_1^a)^2| \leq \alpha_a\beta_a$ and $|s_1 - \bar{x}_2^b - (w - \bar{x}_1^b)^2| \leq \alpha_b\beta_b$. Subtracting gives us

$$|2w\,(\bar{x}_1^a - \bar{x}_1^b) - d| \leq 2\max(\alpha_a\beta_a, \alpha_b\beta_b),$$

where $d = 2(\bar{x}_1^a)^2 - 2\bar{x}_1^a\bar{y}_1^b$. Since $|\bar{x}_1^a - \bar{x}_1^b| \geq c d(B^a, B^b)\alpha_a$, this implies

$$\left|\{w \in S^a : |2w(\bar{x}_1^a - \bar{x}_1^b) - d| \leq 2\max(\alpha_a\beta_a, \alpha_b\beta_b)\}\right| \leq C d(B^a, B^b)^{-1}|S^a|$$

uniformly for all $d \in \mathbb{R}$. This implies the required upper bound on $|B^a \cap B^b|$.

Next let us assume that for some m with $2 \leq m \leq d$, we have

$$\frac{|\sum_{i=1}^m \binom{m}{i}(-t_a)^{m-i}(\bar{y}_i^b - \bar{y}_i^a) - (\bar{y}_1^b - \bar{y}_1^a)|}{\alpha_a\beta_a^{m-1}} \geq c d(B^a, B^b).$$

We define polynomials Q_j^a and Q_j^b on $\mathbb{R}^d$ by

$$Q_j^a(y) = \sum_{i=1}^{j} \binom{j}{i} (-t_a)^{j-i}(y_i - \bar{y}_i^a) - (y_1 - \bar{y}_1^a)^j$$

$$Q_j^b(y) = \sum_{i=1}^{j} \binom{j}{i} (-t_b)^{j-i}(y_i - \bar{y}_i^b) - (y_1 - \bar{y}_1^b)^j$$

for each $2 \le j \le d$. For every $z \in \mathbb{R}$ we define $t(z) \in \mathbb{R}^{d-1}$ so that $Q_j^b(z, t(z)) = 0$ for each j. Now we define a one variable polynomial $P(z) = Q_m^a(z, t(z))$. Observe that

$$|P(z)| > \alpha_a \beta_a^{m-1} + \alpha_b \beta_b^{m-1} \quad \text{implies} \quad B^a \cap B^b \cap (\{z\} \times \mathbb{R}^{d-1}) = \emptyset.$$

Note that $P(\bar{x}_1^b) \ge cd(B^a, B^b)\alpha_a \beta_a^{m-1}$. Let

$$\epsilon = \frac{3\max(\alpha_a \beta_a^{m-1}, \alpha_b \beta_b^{m-1})}{d(B^a, B^b)\, \alpha_a \beta_a^{m-1}} \le d(B^a, B^b)^{-\frac{1}{2}}.$$

Then for all $z \in \bar{x}_1^b + S^b$ we have

$$P(z) \ge \epsilon d(B^a, B^b)\alpha_a \beta_a^{m-1} = 3\max(\alpha_a \beta_a^{m-1}, \alpha_b \beta_b^{m-1}) \ge \alpha_a \beta_a^{m-1} + \alpha_b \beta_b^{m-1}$$

except on a set of measure smaller than $C\epsilon^c|S^b|$. Therefore one has

$$|B^a \cap B^b| \le C\epsilon^c |B^b| \le Cd(B^a, B^b)^{-c}|B^b|.$$

Similar arguments give the required inequality if $\frac{\left|\sum_{i=1}^m \binom{m}{i}(-t_b)^{m-i}(\bar{y}_i^a - \bar{y}_i^b) - (\bar{y}_1^a - \bar{y}_1^b)^m\right|}{\alpha_b \beta_b^{m-1}} \ge cd(B^a, B^b)$.

Now let us assume that for some $2 \le m \le d$ we have

$$\frac{|\sum_{i=1}^m \binom{m}{i}(-t_a)^{m-i}(\bar{x}_i^b - \bar{x}_i^a)|}{\alpha_a \beta_a^{m-1}} \ge cd(B^a, B^b)$$

and that all the previous terms are less than $c'd(B^a, B^b)$ where c' is a small positive number to be chosen precisely in a moment. Since both sides of the equation are invariant if we replace (B^a, B^b) by $(\phi^*(B^a), \phi^*(B^b))$, we can assume that $B^a = B(0, 0, \alpha_a, 1)$ and $B^b = B(\bar{x}^b, t, \alpha_b, \beta_b)$. This implies

$$|\bar{x}_m^b| = |\bar{y}_m^b - t^m| > cd(B^a, B^b)\alpha_a \tag{5.7}$$

and

- $\frac{\alpha_a}{\alpha_b} + \frac{\alpha_b}{\alpha_a} < c'd(B^a, B^b)$;
- $|\bar{y}_1^b - t| < c'd(B^a, B^b)\max(\alpha_a, \alpha_b)$;
- $|\bar{y}_m^b - (\bar{y}_1{}^b)^m| < c'd(B^a, B^b)\alpha_a$.

This implies $t = \bar{y}_1^b + \mathcal{O}(c'd(B^a, B^b))\alpha_a$ which in turn implies

$$|\bar{x}_m^b| = |\bar{y}_m^b - t^m| = |\bar{y}_m^b - (\bar{y}_1^b)^m + \mathcal{O}(c'd(B^a, B^b))\alpha_a| < \mathcal{O}(c'd(B^a, B^b))\alpha_a.$$

We choose c' small enough such that this contradicts 5.7. □

We also have an almost triangle inequality.

Lemma 5.5 *There exists a constant $C < \infty$ depending only on the dimension d such that for any three paraballs B^a, B^b, B^d we have*

$$d(B^a, B^b) \le C\left(d(B^a, B^d)^C + d(B^d, B^b)^C\right).$$

Proof Without loss of generality we may assume that $B^d = B(0, 0, \alpha, 1)$ or $B(0, 0, 1, \beta)$. The parameters specifying B^a and B^b are controlled by $\eta_1 = d(B^a, B^d)$ and $\eta_2 = d(B^b, B^d)$ respectively. Also for any three positive numbers β_1, β_2, β, one has $\frac{\beta_1}{\beta_2} + \frac{\beta_2}{\beta_1} \le C\left(\left(\frac{\beta_1}{\beta} + \frac{\beta}{\beta_1}\right)^C + \left(\frac{\beta_2}{\beta} + \frac{\beta}{\beta_2}\right)^C\right)$. Therefore $d(B^a, B^b)$ is bounded from above by $C\eta^C$ where $\eta = \max\left(d(B^a, B^d), d(B^d, B^b)\right)$. □

We also have the following covering property.

Lemma 5.6 *There exists a constant $C < \infty$ depending only on the dimension d such that for any two paraballs B^a, B^b we have*

$$B^a \subset C(d(B^a, B^b))^C B^b$$

where $C(d(B^a, B^b))^C B^b$ is defined as in (5.2).

Proof Without loss of generality we may assume that $B^b = B(0, 0, \alpha, 1)$. Let $B^a = B(\bar{x}, t_0, \alpha_1, \beta_1)$ and let $\eta = d(B^a, B^b)$. Then there is a constant C such that the parameters corresponding to B^a are controlled by $C\eta^C$. After some elementary algebra it follows that $B^a \subset C\eta^C B^b$. □

In the next Proposition we shall prove that $(B(0, 0, 1, 1), B^*(0, 0, 1, 1))$ [and hence $(B(0, 0, \alpha, \alpha), B^*(0, 0, \alpha, \alpha))$] are quasiextremal pairs for $T : L^{p_\theta} \to L^{q_\theta}$ for every $\theta \in [0, 1]$. In Lemma 7.3 we shall prove that for $0 < \theta < 1$ these are essentially the only quasi-extremal pairs. For $\theta = 0$, in addition to the above we also have $\big(B(0, 0, \alpha, 1), B^*(0, 0, \alpha, 1)\big)$ for every $0 < \alpha < 1$ and for $\theta = 1$, we have $\big(B(0, 0, 1, \beta), B^*(0, 0, 1, \beta)\big)$ for every $0 < \beta < 1$ which are quasiextremal pairs for T.

Proposition 5.7 *There exists $c > 0$ which depends only on the dimension d with the following property.*

- $\big(B(0,0,\alpha,1), B^*(0,0,\alpha,1)\big)$ *is a c-quasi-extremal pair for* $T : L^{p_0} \to L^{q_0}$ *for all* $0 < \alpha < 1$;
- $\big(B(0,0,1,\beta), B^*(0,0,1,\beta)\big)$ *is a c-quasi-extremal pair for* $T : L^{p_1} \to L^{q_1}$ *for all* $0 < \beta < 1$;
- $\big(B(0,0,\alpha,\alpha), B^*(0,0,\alpha,\alpha)\big)$ *is a c-quasi-extremal pair for* $T : L^{p_\theta} \to L^{q_\theta}$ *for all* $0 \le \theta \le 1$ *and for all* $\alpha > 0$.

Proof We shall write the proof of the first claim, the others being identical. Let $B = B(0,0,\alpha,1)$ with $0 < \alpha < 1$. We claim that

$$T(B, B^*) \ge c\alpha^d, \qquad |B| \le \alpha^{d-1}, \qquad |B^*| \le \alpha^d \tag{5.8}$$

which after some elementary calculations implies that

$$\frac{T(B, B^*)}{|B|^{\frac{2}{d+1}} |B^*|^{1-\frac{2(d-1)}{d(d+1)}}} \ge c.$$

The upper bounds on the sizes of B and B^* follow directly from the definition. Let us fix a small number $r > 0$ (to be chosen precisely later) which depends only on d. Define B_r to be the set of all $y \in \mathbb{R}^d$ such that

- $|y_1| \le r$;
- $|y_m - y_1^m| \le r\alpha$ for all $1 < m \le d$.

Then $|B_r| \ge r^d |B|$.

We want to show that if r is sufficiently small then for all $y \in B_r$, the set of all $x \in \mathbb{R}^d$ such that $x \in B^*$ has measure at least $r\alpha$. Therefore $T(B, B^*) \ge r^{d+1}\alpha|B| \ge r^{d+1}|B|^{\frac{2}{d+1}}|B^*|^{1-\frac{2(d-1)}{d(d+1)}}$.

Let us fix $y \in B_r$. For each $x_1 \in \mathbb{R}$ with $|x_1| < r\alpha$, we define $x' = (x_2, x_3, \ldots, x_d) \in \mathbb{R}^{d-1}$ by $x_m = y_m - (y_1 - x_1)^m$ for $2 \le m \le d$, so that $(x, y) \in \Sigma$. Now $x \in B^*$ if and only if

$$|x_m| = |y_m - (y_1 - x_1)^m| < \alpha. \tag{5.9}$$

Now,

$$\begin{aligned} x_m &= y_m - (y_1 - x_1)^m \\ &= y_1^m - (y_1 - x_1)^m + \mathcal{O}(r)\alpha \\ &\le \mathcal{O}(r)\alpha < \alpha, \end{aligned} \tag{5.10}$$

if we choose r to be sufficiently small. □

6 Quasiextremal Pairs and Paraballs

Let E and F be subsets of $\mathbb{R}^d$ with finite positive Lebesgue measure. Write $\mathcal{T}(E, F) = \langle T(\chi_E), \chi_F\rangle$ and $\mathcal{T}(f, g) = \langle T(f), g\rangle$. Define α and β by

$$\alpha|E| = \beta|F| = \langle T(\chi_E), \chi_F\rangle.$$

Then T being restricted weak type $(p_0, q_0) = \left(\frac{d+1}{2}, \frac{d(d+1)}{2(d-1)}\right)$ is equivalent to

$$|E| \geq c\beta^{\frac{d(d+1)}{2}} \left(\frac{\alpha}{\beta}\right)^{d-1}.$$

In addition, if (E, F) is an ϵ-quasiextremal pair then by Definition 1.4 we also have

$$|E| \leq c\epsilon^{-\mathbf{C}}\beta^{\frac{d(d+1)}{2}} \left(\frac{\alpha}{\beta}\right)^{d-1}.$$

for some $C > 0$. We aim to exploit these two inequalities simultaneously to obtain information about ϵ-quasiextremal pairs and prove the following theorem.

Theorem 6.1 *Let $d > 2$. There exists an absolute constant C, depending only on d such that for any ϵ-quasiextremal pair (E, F), there exists a paraball B such that*

$$T(E \cap B, F \cap B^*) \geq C^{-1}\epsilon^{C}\mathcal{T}(E, F)$$

and

$$|B| \leq |E| \quad \text{and} \quad |B^*| \leq |F|.$$

7 Parametrization of Subsets of E and F

The following Lemma is proved in Lemma 3.7 in [12].

Lemma 7.1 *If d is even there exists a point $\bar{y}$ in E, a measurable subset $\Omega \subset \mathbb{R}^{d+1}$ such that*

- $|\Omega| = c\alpha^{\frac{d+2}{2}}\beta^{\frac{d}{2}}$;
- $\bar{y} - h(t_1) + h(t_2) - h(t_3) - \cdots + h(t_j) \in E$ *for every* $t = (t_1, \ldots, t_{d+1}) \in \Omega$ *and for every even* j;
- $\bar{x} - h(t_1) + h(t_2) - h(t_3) - \cdots - h(t_j) \in F$ *for every* $t = (t_1, \ldots, t_{d+1}) \in \Omega$ *and for every odd* j;
- $t_1 < t_2 < \cdots < t_d$ *for every* $t = (t_1, t_2, \ldots, t_{d+1}) \in \Omega$;
- $t_i - t_{i-1} \geq c\beta$ *for every even* i;
- $t_i - t_{i-1} \geq c\alpha$ *for every odd* i;

and if d is odd, there exists a point $\bar{x}$ in F, a measurable subset $\Omega \subset \mathbb{R}^{d+1}$ such that

- $|\Omega| = c\alpha^{\frac{d+1}{2}}\beta^{\frac{d+1}{2}}$;
- $\bar{x} + h(t_1) - h(t_2) + h(t_3) - \cdots + h(t_j) \in E$ *for every* $t = (t_1, .., t_{d+1}) \in \Omega$ *and for every odd* j;
- $\bar{x} + h(t_1) - h(t_2) + h(t_3) - \cdots - h(t_j) \in F$ *for every* $t = (t_1, \ldots, t_{d+1}) \in \Omega$ *and for every even* j;
- $t_1 < t_2 < \cdots < t_d$ *for every* $t = (t_1, t_2, \ldots, t_{d+1}) \in \Omega$.

- $t_i - t_{i-1} \geq c\beta$ *for every odd* i;
- $t_i - t_{i-1} \geq c\alpha$ *for every even* i.

Here c *is a small positive constant independent of* E, F, α, β *and depends only on* d.

Now as in [2] if we consider the map $(t_1, \ldots, t_d)$ goes to $\bar{y} - h(t_1) + h(t_2) - h(t_3) - \cdots + (-1)^d h(t_d) \in E$ we have

$$|E| \geq c \int_{\Omega} \prod_{1 \leq i < j \leq d} (t_j - t_i).$$

Since (E, F) is a ϵ-quasiextremal in addition to Lemma 7.1 we also have for each $t \in \Omega$,

- $t_i \leq t_{i-1} + \epsilon^{-C}\alpha$ for every odd $1 < i \leq d+1$;
- $t_i \leq t_{i-1} + \epsilon^{-C}\beta$ for every even $1 < i \leq d+1$;

where C is an absolute constant depending only on d.

Lemma 7.2 *There exists* $C < \infty$, *depending only on* d, *with the following properties. If* (E, F) *is an* ϵ*-quasiextremal with* $\alpha|E| = \beta|F| = \mathcal{T}(E, F)$ *then there exists* $t_0 \in \mathbb{R}$ *and a point* $\bar{y}$ *in* E, *a measurable subset* $\Omega \subset \mathbb{R}^{d+1}$ *such that if* d *is even*

- $|\Omega| = c\alpha^{\frac{d+2}{2}}\beta^{\frac{d}{2}}$;
- $\bar{y} - h(t_1) + h(t_2) - h(t_3) + \cdots + h(t_j) \in E$ *for every* $t = (t_1, \ldots, t_{d+1}) \in \Omega$ *and for every even* j;
- $\bar{y} - h(t_1) + h(t_2) - h(t_3) - \cdots - h(t_j) \in F$ *for every* $t = (t_1, \ldots, t_{d+1}) \in \Omega$ *and for every odd* j;
- $t_i \leq t_{i-1} + \epsilon^{-C}\alpha$ *for every odd* $1 < i < d$;
- $t_i \leq t_{i-1} + \epsilon^{-C}\beta$ *for every even* $1 < i \leq d$;
- $t_1 < t_2 < \cdots < t_d$ *for every* $t \in \Omega$;
- $|t_1 - t_0| \leq \epsilon^{-C}\alpha$ *for all* $t \in \Omega$:
- $t_i - t_{i-1} \geq c\beta$ *for every even* i;
- $t_i - t_{i-1} \geq c\alpha$ *for every odd* i.

and if d *is odd, there exists a point* $\bar{x}$ *in* F, *a measurable subset* $\Omega \subset \mathbb{R}^{d+1}$ *such that*

- $|\Omega| = c\alpha^{\frac{d+1}{2}}\beta^{\frac{d+1}{2}}$;
- $\bar{x} + h(t_1) - h(t_2) - h(t_3) - \cdots + h(t_j) \in E$ *for every* $t = (t_1, .., t_{d+1}) \in \Omega$ *and for every odd* j;
- $\bar{x} + h(t_1) - h(t_2) + h(t_3) - \cdots - h(t_j) \in F$ *for every* $t = (t_1, \ldots, t_{d+1}) \in \Omega$ *and for every even* j;
- $t_1 < t_2 < \cdots < t_d$ *for every* $t = (t_1, t_2, \ldots, t_{d+1}) \in \Omega$.
- $|t_1 - t_0| \leq \epsilon^{-C}\beta$ *for all* $t \in \Omega$:
- $t_i - t_{i-1} \geq c\beta$ *for every odd* i;
- $t_i - t_{i-1} \geq c\alpha$ *for every even* i.

Proof The proof is quite straightforward. If the Ω from Lemma 7.1 does not satisfy the property that for some t_0, $|t_1 - t_0| \leq \epsilon^{-C}\alpha$ for all $t \in \Omega$, then in the proof of Lemma 7.1 we iterate the construction of the sets Ω_k upto (d+3) times. Now we fix a point $s \in \Omega_{d+3}$ and apply Lemma 7.1 with $\bar{y}$ replaced by $\bar{y} - h(s_1) + h(s_2)$. □

Lemma 7.3 *There exist $c, C < \infty$ with the following properties. Let (E, F) be an ϵ-quasiextremal pair for $T : L^{p_\theta} \to L^{q_\theta}$ and $\alpha|E| = \beta|F| = \mathcal{T}(E, F)$. Then*

- *If $\theta = 0$ then $\alpha \le C\epsilon^{-C}\beta$;*
- *If $\theta = 1$ then $\beta \le C\epsilon^{-C}\alpha$;*
- *If $0 < \theta < 1$ then $c\epsilon^{\frac{C}{1-\theta}}\alpha \le \beta \le C\epsilon^{\frac{-C}{\theta}}\alpha$.*

Proof We shall first consider the case when $\theta = 0$ and d is even. The proof when d is odd is identical. If (E, F) is an ϵ-quasiextremal for $T : L^{p_0} \to L^{q_0}$, by Lemma 7.2 one has for all $t = (t_1, t_2, \ldots, t_d) \in \Omega$,

$$t_2 + c\alpha < t_3 < t_1 + C\epsilon^{-C}\beta < t_2 + C\epsilon^{-C}\beta.$$

This implies $\alpha < C\epsilon^{-C}\beta$.

Next Let us consider the case when $\theta = 1$. If (E, F) is an ϵ-quasiextremal for $T : L^{p_1} \to L^{q_1}$. Then one has

$$\langle T^*(\chi_F), \chi_E\rangle = \langle T(\chi_E), \chi_F\rangle \ge \epsilon|E|^{\frac{1}{p_1}}|F|^{\frac{1}{q_1'}}$$

and

$$\beta|F| = \alpha|E| = \langle T^*(\chi_F), \chi_E)\rangle.$$

This implies (F, E) is an ϵ-quasiextremal pair of $T^* : L^{q_1'} = L^{p_0} \to L^{p_1'} = L^{q_0}$. Since T^* is the convolution with the affine arclength measure of $-h(t)$, one has

$$\beta \le C\epsilon^{-C}\alpha.$$

Let us now fix a $\theta \in (0, 1)$. Let (E, F) be such that

$$\langle T(\chi_E), \chi_F\rangle < \epsilon^{\frac{C}{1-\theta}}|E|^{\frac{1}{p_0}}|F|^{\frac{1}{q_0'}}.$$

This implies

$$\begin{aligned}\langle T(\chi_E), \chi_F\rangle &= (\langle T(\chi_E), \chi_F\rangle)^{1-\theta}(\langle T(\chi_E), \chi_F\rangle)^{\theta}\\ &< \epsilon^C\big(|E|^{\frac{1}{p_0}}|F|^{\frac{1}{q_0'}}\big)^{1-\theta}\big(A_{p_1}|E|^{\frac{1}{p_1}}|F|^{\frac{1}{q_1'}}\big)^{\theta} \\ &< \epsilon^C A_{p_1}^{\theta}|E|^{\frac{1}{p_\theta}}|F|^{\frac{1}{q_\theta'}}.\end{aligned} \tag{7.1}$$

Similarly $\langle T(\chi_E), \chi_F\rangle < \epsilon^C A_{p_0}^{1-\theta}|E|^{\frac{1}{p_\theta}}|F|^{\frac{1}{q_\theta'}}$ if $\langle T(\chi_E), \chi_F\rangle < \epsilon^{\frac{C}{\theta}}|E|^{\frac{1}{p_1}}|F|^{\frac{1}{q_1'}}$. This implies if (E, F) is an ϵ-quasiextremal pair of $T : L^{p_\theta} \to L^{q_\theta}$ then (E, F) is an $\epsilon^{\frac{C}{1-\theta}}$-quasiextremal pair of $T : L^{p_0} \to L^{q_0}$ and $\epsilon^{\frac{C}{\theta}}$-quasiextremal pair of $T : L^{p_1} \to L^{q_1}$.

Thus by the results for $\theta = 0$ and $\theta = 1$, we have

$$c\epsilon^{\frac{C}{1-\theta}}\alpha \le \beta \le C\epsilon^{\frac{-C}{\theta}}\alpha.$$

□

Let us consider the paraball $B = B(\bar{x} = \bar{y} - h(t_0), t_0, C\epsilon^{-C}\alpha, C\epsilon^{-C}\beta)$.

Lemma 7.4 *If B is as above then if d is even,*

$$E \cap B \supset \bar{y} - h(t_1) + h(t_2) - h(t_3) - \cdots + h(t_d),$$
$$F \cap B^* \supset \bar{y} - h(t_1) + h(t_2) + h(t_3) - \cdots - h(t_{d+1})$$

for every $t \in \Omega$, and when d is odd

$$E \cap B \supset \bar{y} - h(t_1) + h(t_2) - h(t_3) - \cdots + h(t_{d+1}),$$
$$F \cap B^* \supset \bar{y} - h(t_1) + h(t_2) - h(t_3) - \cdots - h(t_d)$$

for every $t \in \Omega$.

Proof We will give the details when d is even, $\epsilon = 1$ and $\theta = 0$; the proof for other cases are essentially the same. By Lemma 7.2 it is enough to prove that $\bar{y} - h(t_1) + h(t_2) + h(t_3) - \cdots + h(t_d) \in B$ and $\bar{y} - h(t_1) + h(t_2) - h(t_3) - \cdots + h(t_d) - h(t_{d+1}) \in B^*$ for every $t \in \Omega$.

Let us first prove that $\bar{y} - h(t_1) + h(t_2) + h(t_3) - \cdots + h(t_d) \in B$. We can assume, by applying suitable symmetry, if necessary, that $\bar{x} = \bar{y} = 0$ with $t_0 = 0$. If $y = (y_1, y_2, \ldots, y_d) = -h(t_1) + h(t_2) - h(t_3) - \cdots + h(t_d)$ then Lemma 7.2 implies

$$|y_1| = |t_1 - t_2 + \cdots - t_d| \le C\beta,$$

and

$$\begin{aligned} y_m - y_1^m =& \big(-t_1^m + t_2^m - \cdots + t_d^m\big) - (-t_1 + t_2 - \cdots + t_d)^m \\ =& -t_1^m + (t_2^m - t_3^m) + \cdots + (t_{d-2}^m - t_{d-1}^m) + t_d^m \\ & - \big[-t_1 + (t_2 - t_3) + \cdots + (t_{d-2} - t_{d-1}) + t_d\big]^m \\ =& -t_1^m - (-t_1)^m + [t_2^m - t_3^m - (t_2 - t_3)^m] \\ & + \cdots + [t_{d-2}^m - t_{d-1}^m - (t_{d-2} - t_{d-1})^m] \\ & - \sum_{r_1 + \cdots + r_{\frac{d}{2}} = m, r_i \ne m} (-t_1)^{r_1}(t_2 - t_3)^{r_2} \ldots (t_{d-2} - t_{d-1})^{r_{\frac{d}{2}-1}} t_d^{\frac{d}{2}} \\ \le& C\alpha^m + C\beta^{m-1}|t_2 - t_3| + C\beta^{m-1}|t_4 - t_5| + \cdots + C\beta^{m-1}|t_{d-2} - t_{d-1}|. \end{aligned} \tag{7.2}$$

By Lemma 7.2 we have $|t_i - t_{i+1}| < C\alpha$ for all even $2 \le i \le d$. By Lemma 7.3 we have $\alpha \le C\beta$. Therefore we have $|y_m - y_1^m| < C\alpha\beta^{m-1}$. This proves our claim.

Now we shall prove that $\bar{y} - h(t_1) + h(t_2) - h(t_3) - \cdots + h(t_d) - h(t_{d+1}) \in B^*$ for every $t \in \Omega$. WLOG $\bar{x} = \bar{y} = 0$ with $t_0 = 0$. Let $x = (x_1, x_2, \ldots, x_d) = -h(t_1) + h(t_2) - h(t_3) - \cdots + h(t_d) - h(t_{d+1})$. Then $x \in B^*$ if and only if $|x_1| < C\alpha$ and $|x_m| < C\alpha\beta^{m-1}$ for all $1 < m \leq d$. By Lemma 7.2

$$|x_1| = |-t_1 + (t_2 - t_3) + \cdots + (t_d - t_{d+1})| \leq C\alpha$$

and

$$\begin{aligned}
|x_m| &= |-t_1^m + t_2^m - \cdots + t_d^m - t_{d+1}^m| \\
&= |-t_1^m + (t_2^m - t_3^m) + \cdots + (t_d^m - t_{d+1}^m)| \\
&\leq |-t_1^m| + C\beta^{m-1}|t_2 - t_3| + \cdots + C\beta^{m-1}|t_d - t_{d+1}| \\
&\leq C\beta^{m-1}\alpha.
\end{aligned} \tag{7.3}$$

□

Therefore we have $|E \cap B(\bar{x}, t_0, C\epsilon^{-C}\alpha, C\epsilon^{-C}\beta)| \geq |\{\bar{y} - h(t_1) + h(t_2) + h(t_3) - \cdots + h(t_d) : t \in \Omega\}| \geq c\epsilon^C|E|$. Similarly $|F \cap B^*(\bar{x}, t_0, C\epsilon^{-C}\alpha, C\epsilon^{-C}\beta)| \geq |\{\bar{y} - h(t_1) + h(t_2) - h(t_3) - \cdots - h(t_{d+1}) : t \in \Omega\}| \geq c\epsilon^C|F|$. This implies

- $\mathcal{T}(E \cap B, F \cap B^*) \geq c\epsilon^C \mathcal{T}(E, F)$
- $|B| \leq C\epsilon^{-C}|E|$
- $|B^*| \leq C\epsilon^{-C}|F|$

This is stronger than the conclusion of Theorem 6.1 which will be proved in the following lemma.

Lemma 7.5 *There exist absolute constants $N, C < \infty$ with the following property. For each paraball B and given any $0 < \delta \leq 1$, there exists a family of paraballs $\{B_l : l \in L\}$ with the following properties,*

- $B \subset \cup_{l \in L} B_l$;
- $B^* \subset \cup_{l \in L} B_l^*$;
- $|L| \leq N\delta^{-C}$;
- $|B_l| \sim \delta|B|$ *for all l;*
- $|B_l^*| \sim \delta|B^*|$ *for all l.*

Proof The proof of this lemma will be similar to the proof of Lemma 7.2 in [4]. WLOG we can assume that $B = B(0, 0, \alpha, 1)$ or $B(0, 0, 1, \beta)$. Let $B = B(0, 0, \alpha, 1)$, the proof for $B(0, 0, 1, \beta)$ follows from a similar argument. Then $|B| \sim \alpha^{d-1}$ and $|B^*| \sim \alpha^d$. Let $\eta = \delta^{\frac{2}{d(d+1)}}$. Let us select a maximal $\eta^d\alpha$-separated subset of $B^*(0, 0, \alpha, 1)$ with respect to the regular Euclidean distance. Let us denote this set by $\{z^l : l \in L\}$. Then $|L| \leq C\eta^{-C}$. Now we choose a maximal η-seperated of $[-1, 1]$. Let us denote this set by $\{t_k : k \in K\}$. Then $|K| \leq C\eta^{-C}$.

Now we define $B_{l,k} = B(z^l, t_k, C\eta\alpha, C\eta)$. Then $|B_{l,k}| \sim \eta^{\frac{d(d+1)}{2}}\alpha^{d-1} \sim \delta|B|$ and $|B_{l,k}^*| \sim \eta^{\frac{d(d+1)}{2}}\alpha^d \sim \delta|B^*|$. By the definition of $B_{l,k}$ it directly follows that

$$B^* \subset \cup_{l,k} B_{l,k}^*.$$

We shall now prove that $B \subset \cup_{l,k} B_{l,k}$. It is enough to prove that

$$B(0, 0, \eta^d \alpha, 1) \subset \cup_k B(0, t_k, C\eta\alpha, C\alpha) \tag{7.4}$$

since this implies

$$B(0, 0, 1, 1) = \cup_l B(z^l, 0, \eta^d \alpha, 1) \subset \cup_l \cup_k B(z^l, t_k, C\eta\alpha, C\alpha).$$

To prove 7.4, let $y \in B(0, 0, \eta^d \alpha, 1)$. Now choose t_k such that $t_k \leq y_1 < t_{k+1}$. We claim that $y \in B(0, t_k, C\eta\alpha, C\eta)$. Our claim is true if and only if for each $2 \leq m \leq d$

$$\left| \sum_{i=1}^{m} \binom{m}{i} (-t_k)^{m-i} (y_i - t_k^i) - (y_1 - t_k)^m \right| \leq C^m \eta^m \alpha.$$

Now

$$\begin{aligned} &\left| \sum_{i=1}^{m} \binom{m}{i} (-t_k)^{m-i} (y_i - t_k^i) - (y_1 - t_k)^m \right| \\ &= \left| \sum_{i=1}^{m} \binom{m}{i} (-t_k)^{m-i} (y_i - y_1^i) \right| \leq C\eta^d \alpha \leq C\eta^m \alpha \end{aligned}$$

as $|t_k| \leq 1$ and $y \in B(0, 0, \eta^d \alpha, 1)$.

□

To complete the proof of the Theorem 6.1 we choose C sufficiently large such that with $\delta = \epsilon^C$ we apply Lemma 7.5 to obtain paraballs $\{B_l\}_{l \in L}$ such that for each $l \in L$, $|B_l| \leq |E|$ and $|B_l^*| \leq |F|$. We now have

$$T(E \cap B, F \cap B^*) \leq \sum_{l \in L} T(E \cap B_l, F \cap B_l^*). \tag{7.5}$$

In addition we have $|L| \leq C\epsilon^{-C}$. Therefore there exists a paraball B_l such that

$$T(E \cap B_l, F \cap B_l^*) \geq C^{-1} \epsilon^C T(E \cap B, F \cap B^*). \geq c\epsilon^C \mathbb{T}(E, F)$$

and

$$|B_l| \leq |E| \qquad |B_l^*| \leq |F|.$$

8 Lorentz Spaces and ϵ-Quasiextremal Function

Definition 8.1 Let f be a nonnegative function which is finite almost everywhere. By a rough level set decomposition of f we mean a representation of f as $f =$

$\sum_{j=-\infty}^{\infty} 2^j f_j$ where $\chi_{E_j} \leq f_j \leq 2\chi_{E_j}$ with the sets E_j pairwise disjoint and measurable.

We may approximate the Lorentz norms of f by,

$$\|f\|_{p,r} \sim \begin{cases} \left(\sum_j (2^j |E_j|^{\frac{1}{p}})^r\right)^{\frac{1}{r}}, & \text{if} \quad r < \infty \\ \sup_j 2^j |E_j|^{\frac{1}{p}}, & \text{if} \quad r = \infty \end{cases}$$

where $f = \sum_{j=-\infty}^{\infty} 2^j f_j, f_j \sim \chi_{E_j}$ is a rough level set decomposition of f. In particular, $L^{p,r}(\mathbb{R}^d) = \{f : \|f\|_{p,r} < \infty\}$.

The following lemma is Theorem 4.1 in [23].

Lemma 8.2 *T maps $L^{p,r}$ boundedly to L^q for every $r \in (p, q)$ for every (p, q) as in* 1.2.

The following lemma is also proved in the proof of Theorem 4.1 in [23].

Lemma 8.3 *There exist $C, c > 0$ with the following property. Let $\epsilon > 0$. Let $f = \sum_j 2^j f_j, f_j \sim \chi_{E_j}$ and $g = \sum_k 2^k g_k, g_k \sim \chi_{F_k}$ be such that either $\mathcal{T}(E_j, F_k) \leq \epsilon |E_j|^{\frac{1}{p}} |F_k|^{1-\frac{1}{q}}$ or $2^j |E_j|^{\frac{1}{p}} \leq \epsilon \|f\|_p$ for each j and k. Then $\mathcal{T}(f, g) \leq C\epsilon^c \|f\|_p \|g\|_{q'}$.*

Lemma 8.4 *There exist c, $C < \infty$ with the following property. For each $\epsilon > 0$, if f is a nonnegative function with rough level set decomposition $f = \sum_{j=-\infty}^{\infty} 2^j f_j, f_j \sim \chi_{E_j}$ and if f is a ϵ-quasiextremal then there exists $j \in \mathbb{Z}$ and a paraball B such that*

$$\|2^j \chi_{E_j \cap B}\|_p \geq c\epsilon^C \|f\|_p$$

and

$$|B| \leq |E_j|.$$

Proof The proof of this lemma follows directly from Theorem 6.1 and the previous lemma. If $f = \sum_j 2^j f_j, f_j \sim \chi_{E_j}$ by the previous lemma there exists $j \in \mathbb{Z}$ such that $\|2^j \chi_{E_j \cap B}\|_p \geq c\epsilon^C \|f\|_p$ and E_j is an ϵ^c-quasietremal. Now we apply Theorem (6.1) to E_j to get the desired conclusion. □

9 Two Key Lemmas

In this section we shall prove two lemmas that will be used in the later sections. The first lemma is about how paraballs interact with each other when they are distant from each other. It shows in some sense when we have a collection of paraballs which are at a large distance from each other, then their image under T act on nearly disjoint portions of any given set. The precise statement is given below.

Lemma 9.1 *Let $d > 2$ and let $(\frac{1}{p}, \frac{1}{q})$ be on the line segment joining the points $(\frac{2}{d+1}, \frac{2(d-1)}{d(d+1)})$ and $(1 - \frac{2(d-1)}{d(d+1)}, 1 - \frac{2}{d+1})$. Then there exists a positive finite constant C depending only on d with the following property. Let $\{B_i\}_{i\in S}$ be a collection of paraballs such that for any $i \neq j$ with $i, j \in S$ we have $d(B_i, B_j) \geq C\eta^{-C}$ for some $\eta > 0$. Then for any F subset of $\mathbb{R}^d$ with positive finite Lebesgue measure, we can write $F = \sqcup F_i$ so that*

$$T(B_i, F_j) \leq \eta |B_i|^{\frac{1}{p}} |F_j|^{\frac{1}{q'}}, \textit{for all} \quad i \neq j.$$

Proof The proof of this lemma will be a straightforward adaptation of the proof of Lemma 4.1 in [3]. For the sake of completeness we give a sketch of the proof here. Define

$$\gamma_i = \frac{1}{3}\eta |F|^{\frac{1}{q'}-1} |B_i|^{\frac{1}{p}}$$

and

$$\tilde{F}_i = \{x \in F : T(B_i) > \gamma_i\}. \tag{9.1}$$

We note that

$$T(B_i, F \setminus \tilde{F}_i) \leq \gamma_i |F| \leq \frac{1}{3}\eta |B_i|^{\frac{1}{p}} |F|^{\frac{1}{q'}}. \tag{9.2}$$

Now choose $F_i \subset \tilde{F}_i$ such that $\cup_i \tilde{F}_i = \sqcup F_i$. Note that the there are many choices of F_i. We just choose one such collection. Also, there might be elements in F which do not belong to $\tilde{F}_i$ for all $i \in S$. We pick one F_i and include these points to this particular set. Since by (9.2) we have $T(B_i, F \setminus \sqcup F_i) \leq \frac{1}{3}\eta |B_i|^{\frac{1}{p}} |F|^{\frac{1}{q'}}$, it is enough to prove that for $i \neq j$

$$T(B_i, F_j) \leq \frac{2}{3}\eta |B_i|^{\frac{1}{p}} |F_j|^{\frac{1}{q'}}. \tag{9.3}$$

Suppose (9.3) does not hold. Then there exists $i \neq j$ such that $T(B_i, F_j) > \frac{2}{3}\eta |B_i|^{\frac{1}{p}} |F_j|^{\frac{1}{q'}}$. For the rest of this proof we fix these two indices i, j. Define $\mathcal{F} = F_j \cap \tilde{F}_i$. By (9.2) $T(B_i, F_j \setminus \tilde{F}_i) \leq \frac{1}{3}\eta |B_i|^{\frac{1}{p}} |F|^{\frac{1}{q'}}$, so we have

$$\frac{1}{3}\eta |B_i|^{\frac{1}{p}} |F|^{\frac{1}{q'}} \leq T(B_i, \mathcal{F}) \leq A |B_i|^{\frac{1}{p}} |\mathcal{F}|^{\frac{1}{q'}}. \tag{9.4}$$

This implies

$$|\mathcal{F}| \geq \left(\frac{1}{3}\right)^{q'} \eta^{q'} A^{-q'} |F|.$$

Now we apply Theorem 6.1 to the pair $(B_j, \mathcal{F})$ to obtain a paraball $\tilde{B}_j$ such that

$$|\tilde{B}_j| \leq |B_j|, \quad |\tilde{B}_j^*| \leq |\mathcal{F}| \leq |F|, \qquad |\tilde{B}_j \cap B_j| \geq c\eta^\gamma |B_j|, \quad |\tilde{B}_j^* \cap \mathcal{F}| \geq c\eta^\gamma |F|. \tag{9.5}$$

Now we replace $\mathcal{F}$ by $\tilde{\mathcal{F}} = \mathcal{F} \cap \tilde{B_j}^*$. Since $|\tilde{\mathcal{F}}| \geq c\eta^\gamma |F|$ and $T(B_i)(x) > \gamma_i$ for all $x \in F_i \supset \mathcal{F} \supset \tilde{\mathcal{F}}$, consequently

$$T(B_i, \tilde{\mathcal{F}}) \geq \gamma_i |\tilde{\mathcal{F}}| \geq c\eta^\gamma |\tilde{\mathcal{F}}|^{\frac{1}{q'}} |B_i|^{\frac{1}{p}}.$$

This means the pair $(B_i, \tilde{\mathcal{F}})$ is a $c\eta^\gamma$-quasietxremal. Therefore by applying Theorem 6.1 once more we get another paraball $\tilde{B}_i$ such that

$$|\tilde{B}_i| \leq |B_i|, \quad |\tilde{B}_i^*| \leq |\tilde{\mathcal{F}}| \leq |F|, \qquad |\tilde{B}_i \cap B_i| \geq c\eta^\gamma |B_i|, \quad |\tilde{B}_i^* \cap \tilde{\mathcal{F}}| \geq c\eta^\gamma |F|. \tag{9.6}$$

Since $\tilde{B_i}^* \cap \tilde{B_j}^* \supset \tilde{B_i}^* \cap \tilde{B_j}^* \cap \mathcal{F} \supset \tilde{B_i}^* \cap \tilde{\mathcal{F}}$, we have

$$|\tilde{B_i}^* \cap \tilde{B_j}^*| \geq |\tilde{B_i}^* \cap \tilde{\mathcal{F}}| \geq c\eta^\gamma |F| \geq c\eta^\gamma \max(|\tilde{B_i}^*|, |\tilde{B_j}^*|).$$

Now by applying Proposition 5.4 to the pair of dual paraballs $(\tilde{B_i}^*, \tilde{B_j}^*)$ we get $d(\tilde{B_i}^*, \tilde{B_j}^*) \leq C\eta^{-C}$. This implies

$$d(\tilde{B}_i, \tilde{B}_j) \leq C\eta^{-C}.$$

Since $|\tilde{B}_i| \leq |B_i|$ and $|\tilde{B}_i \cap B_i| \geq c\eta^\gamma |B_i|$, we have

$$d(\tilde{B}_i, B_i) \leq C\eta^{-C}.$$

Similarly

$$d(\tilde{B}_j, B_j) \leq C\eta^{-C}.$$

By applying Lemma 5.5 we get $d(B_i, B_j) \leq C\eta^{-C}$, which contradicts our hypothesis. □

Lemma 9.2 *Let $d > 2$ and $(\frac{1}{p}, \frac{1}{q})$ be a point on the line segment joining the points $(\frac{2}{d+1}, \frac{2(d-1)}{d(d+1)})$ and $(1 - \frac{2(d-1)}{d(d+1)}, 1 - \frac{2}{d+1})$. There exists C, C' positive finite constants depending only on d with the following property. Let E_1, E_2, F be subsets of $\mathbb{R}^d$ with positive finite Lebesgue measure such that $T(\chi_{E_1}) \geq \eta |E_1|^{\frac{1}{p}} |F|^{\frac{1}{q'}-1}$ and $T(\chi_{E_2}) \geq \eta |E_2|^{\frac{1}{p}} |F|^{\frac{1}{q'}-1}$ on F, then if $|E_2| \geq |E_1|$ we have $|E_2| \leq C' \eta^{-C} |E_1|$.*

Proof This lemma is essentially proved in the proof of Theorem 4.1 in [23] by applying extrapolation method of Christ. Here we give a simplified proof using the increasing structure, $(t_1 < t_2 < \cdots < t_d)$ of Ω in Lemma 7.2. We shall give the proof when d is even for the other case being similar.

Let $p_0 = \frac{d+1}{2}$ and $q_0 = \frac{d(d+1)}{2(d-1)}$. Let us first consider the case when $p = p_0$ and $q = q_0$. Define

$$\alpha = \eta|E_1|^{\frac{1}{p_0}-1}|F|^{\frac{1}{q_0'}}, \quad \beta = \eta|E_1|^{\frac{1}{p_0}}|F|^{\frac{1}{q_0'}-1} \text{ and } \quad \gamma = \eta|E_2|^{\frac{1}{p_0}}|F|^{\frac{1}{q_0'}-1}.$$

Since $T(\chi_{E_1}) \geq \eta|E_1|^{\frac{1}{p_0}}|F|^{\frac{1}{q_0'}-1}$ on F, we have

$$\langle \chi_{E_1}, T^*(\chi_F)\rangle = T(E_1, F) \geq \eta|E_1|^{\frac{1}{p_0}}|F|^{\frac{1}{q_0'}}.$$

Therefore on a large subset of E_1, $T^*(\chi_F) \geq \alpha$. Similarly on a large subset of F, $T(\chi_{E_1}) \geq \beta$ and $T(\chi_{E_2}) \geq \gamma$.

Similar to the proof of Theorem 6.1 there exists a point $\bar{y} \in E_1$ such that we can travel along the curve shifted to $\bar{y}$ inside F for a length of α. Then for each of these points on this travelled path we can travel back inside E_1 for a length of β. We continue this process $d-1$ times. At the dth step we move into E_2 along the curve for a length γ.

As a result we get a $\Omega \subset \mathbb{R}^d$ such that

- $|\Omega| = c\alpha^{\frac{d}{2}}\beta^{\frac{d}{2}-1}\gamma$;
- $\bar{y} - h(t_1) + h(t_2) - h(t_3) - \cdots + h(t_j) \in E_1$ for every $t = (t_1, \ldots, t_d) \in \Omega$ and for every even $j \leq d-2$;
- $\bar{y} - h(t_1) + h(t_2) - h(t_3) - \cdots - h(t_j) \in F$ for every $t = (t_1, \ldots, t_d) \in \Omega$ and for every odd $j \leq d-1$;
- $\bar{y} - h(t_1) + h(t_2) - h(t_3) - \cdots + h(t_d) \in E_2$;
- $t_1 < t_2 < \cdots < t_d$ for every $t = (t_1, t_2, \ldots, t_d) \in \Omega$.

Now we consider the Jacobian, $J(t)$, of the map $(t_1, t_2, \ldots, t_d) \mapsto \bar{x} + h(t_1) - h(t_2) + h(t_3) - \cdots - h(t_d)$. We have

$$J \geq \prod_{1\leq i<j\leq d} |t_i - t_j|.$$

Since $t_1 < t_2 < \cdots < t_d$ for every $t \in \Omega$, we have

- $\prod_{1\leq i<d}(t_i - t_d) \geq \gamma^{d-1}$;
- $\prod_{1\leq i<j}(t_i - t_j) \geq \beta^{j-1}$ for every even $j < d$;
- $\prod_{1\leq i<j}(t_i - t_j) \geq \alpha\beta^{j-2}$ for every odd $j \leq d-1$.

Therefore as in the proof of Theorem 6.1 we get

$$|E_2| \geq |\Omega| \min_t J(t) \geq c\alpha^{\frac{d}{2}}\beta^{\frac{d}{2}-1}\gamma \;\; \alpha^{\frac{d}{2}-1}\beta^{\frac{d-2^2}{2}}\gamma^{d-1}.$$

After substituting the values of α, β and γ in terms of $|E_1|$, $|E_2|$ and $|F|$ we get

$$|E_2| \geq c\eta^{\frac{d(d-1)}{2}} |E_2|^{\frac{d}{p_0}} |E_1|^{(d-1)\left(\frac{1}{p_0}-1\right)+\frac{\left(\frac{d}{2}-1\right)(d-1)}{p_0}} |F|^{d\left(\frac{1}{q_0}-1\right)+\frac{d-1}{q_0}+\left(\frac{d}{2}-1\right)(d-1)\left(\frac{1}{q_0}-1\right)}$$

which implies

$$|E_2|^{-\frac{d-1}{d+1}} \geq c\eta^{\frac{d(d-1)}{2}} |E_1|^{-\frac{d-1}{d+1}}.$$

This is equivalent to

$$|E_2| \leq C\eta^{-\frac{d(d+1)}{2}} |E_1|.$$

Now let us consider the case when $\frac{1}{p} = \frac{1}{p_1} = 1 - \frac{2(d-1)}{d(d+1)}$ and $\frac{1}{q} = \frac{1}{q_1} = 1 - \frac{2}{d+1}$. The argument in this case is similar to the above case. Let α, β, γ, Ω, $\bar{y}$ and J be as before.

Since $t_1 < t_2 < \cdots < t_d$ for every $t \in \Omega$, we have

- $\prod_{1 \leq i < d}(t_i - t_d) \geq \gamma\alpha^{d-2}$;
- $\prod_{1 \leq i < j}(t_i - t_j) \geq \beta\alpha^{j-2}$ for every even $j < d$;
- $\prod_{1 \leq i < j}(t_i - t_j) \geq \alpha^{j-1}$ for every odd $j \leq d - 1$.

Therefore we have

$$\begin{aligned}|E_2| &\geq |\Omega| \min_t J(t) \geq c\alpha^{\frac{d}{2}}\beta^{\frac{d}{2}-1}\gamma \quad \alpha^{d(\frac{d}{2}-1)}\beta^{\frac{d}{2}-1}\gamma \\ &= c\eta^{\frac{d(d-1)}{2}} |E_2|^{\frac{2}{p_2}} \; |E_1|^{\frac{d-2}{p_2}+\left(\frac{1}{p_2}-1\right)\frac{d}{2}(d-1)} \; |F|^{d\left(\frac{1}{q_2}-1\right)+\frac{1}{q_2}\frac{d}{2}(d-1)}.\end{aligned}$$

This implies

$$|E_2|^{-\frac{d^2-3d+4}{d(d+1)}} \geq c\eta^{\frac{d(d-1)}{2}} |E_1|^{-\frac{d^2-3d+4}{d(d+1)}}.$$

This is equivalent to

$$|E_2| \leq C\eta^{-\frac{d^2(d^2-1)}{d^2-3d+4}} |E_1|.$$

We shall now consider the case when

$$\frac{1}{p} = \frac{\theta}{p_1} + \frac{1-\theta}{p_0} \quad \text{and} \quad \frac{1}{q} = \frac{\theta}{q_1} + \frac{1-\theta}{q_0}$$

for some $\theta \in (0, 1)$ and p_0, p_1, q_0, q_1 as mentioned earlier. By the hypothesis of the theorem we have

$$\begin{aligned} T(\chi_{E_1}) &\geq \eta |E_1|^{\frac{\theta}{p_1}+\frac{1-\theta}{p_0}} |F|^{\frac{\theta}{q_1'}+\frac{1-\theta}{q_0'}-1} \\ &= (\eta |E_1|^{\frac{1}{p_0}} |F|^{\frac{1}{q_0'}-1})^{1-\theta} (\eta |E_1|^{\frac{1}{p_1}} |F|^{\frac{1}{q_1'}-1})^{\theta}. \end{aligned}$$

and

$$\begin{aligned} T(\chi_{E_2}) &\geq \eta |E_2|^{\frac{\theta}{p_1}+\frac{1-\theta}{p_0}} |F|^{\frac{\theta}{q_1'}+\frac{1-\theta}{q_0'}-1} \\ &= (\eta |E_2|^{\frac{1}{p_0}} |F|^{\frac{1}{q_0'}-1})^{1-\theta} (\eta |E_2|^{\frac{1}{p_1}} |F|^{\frac{1}{q_1'}-1})^{\theta}. \end{aligned}$$

Now let us consider the case when

$$|E_2|^{\frac{1}{p_0}} |F|^{\frac{1}{q_0'}-1} \geq |E_2|^{\frac{1}{p_1}} |F|^{\frac{1}{q_1'}-1}.$$

This is equivalent to $|E_2|^{\frac{1}{p_0}-\frac{1}{p_1}} \leq |F|^{\frac{1}{q_1'}-\frac{1}{q_1'}}$. Since $|E_2| \geq |E_1|$, we also have $|E_1|^{\frac{1}{p_0}-\frac{1}{p_1}} \leq |E_2|^{\frac{1}{p_0}-\frac{1}{p_1}} \leq |F|^{\frac{1}{q_1'}-\frac{1}{q_0'}}$. This implies

$$|E_1|^{\frac{1}{p_0}} |F|^{\frac{1}{q_0'}-1} \geq |E_1|^{\frac{1}{p_1}} |F|^{\frac{1}{q_1'}-1}.$$

Therefore we have for all $x \in F$

- $T(\chi_{E_1})(x) \geq \eta |E_1|^{\frac{1}{p_0}} |F|^{\frac{1}{q_0'}-1}$;
- $T(\chi_{E_2})(x) \geq \eta |E_2|^{\frac{1}{p_0}} |F|^{\frac{1}{q_0'}-1}$.

Now we apply the proof for the case $(p, q) = (p_0, q_0)$ to get the desired inequality. For the other case we have

$$|E_2|^{\frac{1}{p_0}} |F|^{\frac{1}{q_0'}-1} \leq |E_2|^{\frac{1}{p_1}} |F|^{\frac{1}{q_1'}-1}.$$

In this case we apply the proof for $(p, q) = (p_1, q_1)$ to get the desired inequality. □

10 Entropy Refinement

The following lemma is proved for the paraboloid in Lemma 5.3 in [3]. The proof for the moment curve is almost identical and so we omit the proof.

Lemma 10.1 *Let $d \geq 2$. There exist $c, C < \infty$ with the following property. Let $\delta > 0$. Let f be any function in $L^p(\mathbb{R}^d)$ satisfying $\|Tf\|_q \geq (1-\delta)A\|f\|_p$ that has rough level set decomposition $f = \sum_{j \in \mathbb{Z}} 2^j f_j$, $\chi_{E_j} \leq f_j \leq 2\chi_{E_j}$. Then for any $\eta \in (0, 1]$,*

$$\Big\| \sum_{j:2^j|E_j|^{\frac{1}{p}}<\eta\|f\|_p} 2^j f_j \Big\|_p \le C(\delta^{\frac{1}{p}} + \eta^c)\|f\|_p.$$

Lemma 10.2 *Let $d \ge 2$. There exist $c, C, \tilde{C} < \infty$ with the following property. Let $\rho \in (0,1)$. Let f be a $(1-\delta)$-quasiextremal for T. If $\delta \le C\rho^C$, then there exists a function $\tilde{f}$ satisfying $\|f - \tilde{f}\| \le C\rho^C$ with a rough level set decomposition $\tilde{f} = \sum_{j\in\mathbb{Z}} 2^j f_j$, $f_j \sim \chi_{E_j}$ such that if both $\|2^i \chi_{E_i}\| \ge \rho$ and $\|2^j \chi_{E_j}\| \ge \rho$, then*

$$|i - j| \le \tilde{C}\rho^{-\tilde{C}}.$$

Proof This lemma is an improvement over the previous lemma, in the sense that the indices $\{j\}$ in the sum for which $\|2^j \chi_{E_j}\|_p \ge \eta$, can not be too far from each other. All of them are inside an interval of $\mathbb{Z}$ of length at most $C\eta^{-C}$. This lemma is proved corresponding to the paraboloid in Lemma 6.1 in [3]. The proof of this lemma will be an application of Lemma 9.2 together with the previous lemma.The proof is almost identical in our case. Therefore we omit the proof. □

Following Corollary 6.3 in [3] the above lemma immediately implies the following.

Corollary 10.3 *There exist a finite constant C and a function $\Psi : (0,\infty) \to (0,\infty)$ satisfying $\frac{\Psi(t)}{t^p} \to 0$ as $t \to 0, \infty$ with the following property. For any $\epsilon > 0$ there exists a $\delta > 0$ such that for any nonnegative function f with $\|f\|_p = 1$ and $\|T(f)\|_q \ge (1-\delta)A$, there exists $\phi \in G_d$ and a decomposition $\phi^*(f) = g + h$ with $g, h \ge 0$ satisfying $\|h\|_p < \epsilon$ and*

$$\int \Psi(g) \le C. \tag{10.1}$$

11 Uniform Decay and Extremizers at Non-end Points

In this section we show that any extremizing sequence behaves in a uniform manner. Using this we prove that at non end points any extremizing sequence, after applying symmetries if necessary, converges to a non zero function in L^p. By continuity, the limit must be an extremizer for the corresponding L^p bound.

11.1 Spatial Localization

Lemma 11.1 *There exists $C < \infty$ such that for any $\epsilon > 0$ there exists $\delta > 0$ with the following property. Let f be a nonnegative function with $\|f\|_p = 1$ and $\|T(f)\|_q \ge (1-\delta)A$. Then there exists F with rough level set decomposition $F = \sum_{j\in S} 2^j f_j$, $f_j \sim \chi_{E_j}$ satisfying*

$$0 \le F \le f,$$
$$\|T(F)\|_q \ge (1-\epsilon)A,$$

$$|i-j| \le C\epsilon^{-C} \quad \textit{for all} \quad i, j \in S$$

and for each $j \in S$ *there exist* $N(\sim C\epsilon^{-C})$ *paraballs* $B_{j,i}$ *such that*

$$E_j \subset \bigcup_{i=1}^{N} B_{j,i};$$
$$\sum_{i=1}^{N} |B_{j,i}| \le C\epsilon^{-C}|E_j|.$$

Proof Let $\epsilon > 0$. Fix $\delta > 0$ sufficiently small for later purposes, and suppose $\|f\|_p = 1$, $\|Tf\|_q \ge (1-\delta)A$. By using Lemma 10.2, by losing ϵ amount of L^p norm we can assume that f has a finite level set decomposition. In other words, $f = \sum_S 2^j f_j$ with $S \subset (J - C\epsilon^{-C}, J + C\epsilon^{-C}) \cap \mathbb{Z}$ for some $J \in \mathbb{Z}$. Let $0 < \eta (\le c\epsilon^C)$ be a small quantity to be chosen later. Then $\|T(f)\|_q \ge (1-2\delta)\, A \ge \eta$. Now we apply Lemma 8.4 to f to get a paraball B_1 and $i_1 \in S$ such that $\|2^{i_1}\chi_{E_{i_1}\cap B_1}\|_p \ge c\eta^C$ and $|B_1| \le |E_{i_1}|$.

At the next step we set $g_1 = f\chi_{E_{i_1}\cap B_1}$ and write $f = g_1 + h_1$. So $h_1 = \sum_{j\ne i_1} f(\chi_{E_j} + \chi_{E_{i_1}\setminus B_1})$. Now we look at $\|T(h_1)\|_q$. If $\|T(h_1)\|_q \ge \eta$ then by applying Lemma 8.4 to h_1 we get another paraball B_2 and $i_2 \in \mathbb{Z}$ such that

$$\|2^{i_2}\chi_{E_{i_2}\cap B_2\setminus\left(E_{i_1}\cap B_1\right)}\| \ge c\eta^C \quad \text{and} \quad |B_2| \le |E_{i_2}|.$$

Now define $g_2 = f\chi_{E_{i_2}\cap B_2\setminus\left(E_{i_1}\cap B_1\right)}$ and $f = g_1 + g_2 + h_2$.

We continue this process. Now suppose we are at the $(n-1)$-th step. So we have a collection of paraballs $\{B_j\}_{1\le j\le n-1}$ and indices $\{i_j\}_{1\le j\le n-1}$ such that

- $|B_j| \le |E_{i_j}|$;
- $g_m = f\chi_{E_{i_m}\cap B_m\setminus\cup_{1\le j\le m-1}(E_{i_j}\cap B_j)}$;
- $\|g_m\|_p \ge c\,\eta^C$;
- $f = \sum_{j=1}^{n-1} g_j + h_{n-1}$.

Now if $\|T(h_{n-1})\|_q < \eta$ we stop. Otherwise after applying Lemma 8.4 one more time we get another paraball B_n and $i_n \in \mathbb{Z}$ such that $\|2^{i_n}\chi_{E_{i_n}\cap B_n}\|_p \ge c\eta^C$ and $|B_n| \le |E_{i_n}|$.

Since the g_j have disjoint support, $\sum_{1\le j\le n}\|g_j\|_p^p \le \|f\|_p^p \le 1$. So this process must stop after at most $C\eta^{-C}$ steps. Let the process stops at the n-th step. Then we define $F = \sum_{1\le j\le n} g_j$, so that $\|T(f-F)\|_q < \eta$. This means $\|T(F)\|_q \ge (1-2\delta-\eta)A \ge (1-\epsilon)A$ provided η is sufficiently small compared to ϵ. At the same time since $\|T(F)\|_q \le A\|F\|_p$, we have $\|F\|_p \ge 1-\epsilon$. This implies $\|f-F\|_p \le \epsilon$. Now we set $\{B_{j,i}\} = \{B_i : i_j = j\}$. □

Lemma 11.2 *There exists* $C < \infty$ *such that for any* $\epsilon > 0$ *there exists* $\delta > 0$ *with the following property. Let* f *be a nonnegative function with* $\|f\|_p = 1$ *and* $\|T(f)\|_q \ge$

$(1-\delta)\,A$. Then there exists $\tilde{f}$ with rough level set decomposition $\tilde{f}=\sum_{j\in S}2^j f_j$, $f_j\sim\chi_{E_j}$ satisfying

$$
\begin{aligned}
&0\le\tilde{f}\le f,\\
&\|\tilde{f}\|_p\ge(1-\epsilon),\\
&\|T(\tilde{f})\|_q\ge(1-\epsilon)\,A
\end{aligned}
$$

and there exists a distinguished $J\in S$ and a paraball B_J such that

$$
\begin{aligned}
&|J-j|\le C\epsilon^{-C}\quad\text{for all}\quad j\in S,\\
&E_j\subset C\epsilon^{-C^{C\epsilon^{-C}}}B_J\quad\text{for all}\quad j\in S,\\
&\|2^J\chi_{B_J}\|_p\le C.
\end{aligned}
$$

Proof This lemma is an improvement over the previous lemma, in the sense that the paraballs $\{B_{j,i}\}$ have been replaced by a single paraball B, after scaling it with a factor of $(C\epsilon^{-C})^{C\epsilon^{-C}}$. The proof of this lemma will be an application of Lemma 9.1 together with the previous lemma.

By the previous lemma we can assume that $f=\sum_{j\in S}2^j f_j$, $f_j\sim\chi_{E_j}$ where $E_j\subset\cup_{i=1}^{C\epsilon^{-C}}B_{j,i}$ and $|S|\le C\epsilon^{-C}$. Let $0<\eta<\epsilon^C$ be a small quantity to be chosen later. Let us write the collection of paraballs $B_{j,i}$ as $\{B_l:1\le l\le N\}$. Then $N\le C\epsilon^{-C}$. Let if possible $\{1,2,\ldots,N\}=S^a\cup S^b$ be a partition of $\{1,2,\ldots,N\}$ such that for each $(i,j)\in S^a\times S^b$, $d(B_i,B_j)>C\eta^{-C}$. We continue as in the proof of Lemma 10.2. Let $g=\sum_{k\in\tilde{S}}2^k g_k$, $g_k\sim\chi_{F_k}$, be an arbitrary $L^{q'}$ function with $\|g\|_{q'}=1$. Now partition each of the sets F_k measurably as $F_k=F_k^a\cup F_k^b\cup F_k^c$ with the following property:

- For each $x\in F_k^a$ there exists $j\in S^a$ so that $T\chi_{B_j}(x)\ge\eta|B_j|^{\frac{1}{p}}\,|F_k|^{-\frac{1}{q}}$;
- For each $x\in F_k^b$ there exists $j\in S^b$ so that $T\chi_{B_j}(x)\ge\eta|B_j|^{\frac{1}{p}}\,|F_k|^{-\frac{1}{q}}$.

Write

$$
\begin{aligned}
h^a&=g\sum_{k\in\tilde{S}}2^k\chi_{F_k^a},\\
h^b&=g\sum_{k\in\tilde{S}}2^k\chi_{F_k^b},\\
f^a&=f\sum_{j\in S^a}2^j\chi_{E_j},\\
f^b&=f\sum_{j\in S^b}2^j\chi_{E_j},
\end{aligned}
$$

As in the proof of Lemma 10.2, if η is sufficiently small, there is $i\in S^a$ and $k\in\tilde{S}$ with

$$T(\chi_{B_i}, \chi_{F_k^b}) \geq \eta |B_i|^{\frac{1}{p}} |F_k|^{1-\frac{1}{q}}.$$

But by the proof of Lemma 9.1 this implies there is $j \in S^b$, such that $d(B_i, B_j) \leq C\epsilon^{-C}$ which contradicts our hypothesis. Therefore it is not possible to decompose the collection of paraballs corresponding to f into a disjoint union of two sets such that any two elements belonging to different sets are at least $C\epsilon^{-C}$ far with respect to the mock-distance d. Now let us fix a paraball B_{j_0} corresponding to f_0. Now we construct inductively a sequence of collection of paraballs by

- $\{B_1, \ldots, B_N\} = \cup_{j=0}^{N} \mathbb{B}_j$;
- $\mathbb{B}_0 = \{B_{j_0}\}$;
- $B \in \mathbb{B}_j$ and $B' \in \mathbb{B}_{j+1}$ implies $d(B, B') \leq C\epsilon^{-C}$.

By quasi-tirangle inequality this implies $d(B_{j_0}, B_j) \leq C\epsilon^{-C^{C\epsilon^{-C}}}$ for all $j = 1, 2, \ldots, N$. □

12 Weak Convergence and Extremizers for $\theta \in (0, 1)$

Lemma 12.1 *There exists a constant C (depending only on d) and positive functions $\Psi_1, \Psi_2 : (0, \infty) \to (0, \infty)$ and $\rho_1, \rho_2 : (1, \infty) \to (0, \infty)$ satisfying $\frac{\Psi_1(t)}{t^p} \to 0$ and $\frac{\Psi_2(t)}{t^{q'}} \to 0$ as $t \to 0, \infty$ and $\rho_i(R) \to \infty$ as $R \to \infty$ with the following property. For any $\epsilon > 0$ there exists a $\delta > 0$ such that for any nonnegative function f with $\|f\|_p = 1$ and $\|T(f)\|_q \geq (1-\delta)\, A$, there exists $\phi \in G_d$ and a decomposition*

$$\phi^*(f) = g + h$$

with $g, h \geq 0$ satisfying for large R,

$$\|h\|_p < \epsilon, \quad \int_{\mathbb{R}^d} \Psi_1(g) \leq C, \quad support(g) \subset B(0, \rho_1(R)).$$

In addition, there exists $F \geq 0$ satisfying $\|F\|_{q'} = 1$ and

$$\langle T(g), F\rangle \geq (1-\epsilon)\, A, \quad \int_{\mathbb{R}^d} \Psi_2(F) \leq C, \quad support(F) \subset B(0, \rho_2(R)).$$

Proof Let $\epsilon > 0$. Fix $\delta > 0$ sufficiently small. Let $\|f\|_p = 1$ with $\|Tf\|_q \geq A(1-\delta)$. By applying Lemma 11.2 to f we can assume that there is a $\phi \in G_d$ such that

$$\phi^*(f) = g + h$$

with $\|h\|_p < \epsilon$ and support$(g) \subset B_J = B(\bar{x}_J, t_J, \alpha_J, \beta_J)$. In addition, by Lemma 7.3, we can assume that in Lemma 11.2 the distinguished index $J = 0$ and the paraball $B_J = \{y \in \mathbb{R}^d;\ \|y\| < 1\}$. Therefore $\|h\|_p < \epsilon$ and support$(g) \subset B(0, C\epsilon^{-C^{C\epsilon^{-C}}})$.

We set $\rho(R) = CR^{C^{CR^C}}$. The second part of the conclusion follows similarly by applying the same proof to the operator T^* since T^* is the convolution with the affine arc length measure on the curve $-h(t)$. □

12.1 Proof of Existence of Extremizers for $\theta \in (0, 1)$

The proof of the existence of extremizers is along the lines of the proof for the corresponding result in [3]. For the sake of completeness we give the details of the proof. Let $\{f_n\}$ be any extremizing sequence. By the previous Lemma there exist $\{\phi_n \in G_d\}$, such that $\phi_n^*(f_n) = g_n + h_n$, while the functions g_n and h_n satisfy all the conclusions of the previous Lemma corresponding to $\epsilon_n = \frac{1}{n}$. Also there exists a sequence $\{F_n\}$ with $\|F_n\|_{q'} = 1$ and $T(g_n, F_n) \to A$.

By applying the Banach-Alaoglu Theorem, after passing through a subsequence we can assume that $g_n \rightharpoonup g$ and $F_n \rightharpoonup F$ weakly. Therefore $\|F\|_{q'} \leq 1$ and $\|g\|_p \leq 1$. Let us fix a large R. Now we set

$$g_{n,R}(x) = g_n(x)\,\chi_{\|x\|\leq R}(x)\,\chi_{g_n \leq R}(x), \qquad g_R(x) = g(x)\,\chi_{\|x\|\leq R}(x)\,\chi_{g\leq R}(x); \tag{12.1}$$

$$F_{n,R}(x) = F_n(x)\,\chi_{\|x\|\leq R}(x)\,\chi_{F_n\leq R}(x),. \qquad F_R(x) = F(x)\,\chi_{\|x\|\leq R}(x)\,\chi_{F\leq R}(x). \tag{12.2}$$

By stationary phase argument for any fixed $\psi \in C_0^1(\mathbb{R}^d)$, the operator $f \mapsto \psi T(\psi f)$ maps $L^2(\mathbb{R}^d)$ boundedly to the Sobolev space $H^{\frac{1}{d}}$, which in turn embeds into L^s where $\frac{1}{s} = \frac{1}{2} - \frac{1}{d^2}$. Thus the weak convergence of $g_{n,R}$ to g_R as $n \to \infty$ implies the L^s norm convergence of $T(g_{n,R})$ to $T(g_R)$ as $n \to \infty$, for every fixed R. Therefore $T(G_{n,R}, F_{n,R}) \to T(g_R, F_R)$ as $n \to \infty$ for a fixed R.

By Lemma 12.1 the integral of $g_n^{p_\theta}$ outside the ball of radius R goes to zero as R goes to infinity uniformly in n. This implies $g_{n,R}$ converges to g_n in L^{p_θ} as R goes to infinity uniformly in n. Similarly $F_{n,R}$ converges to F_n in $L^{q'_\theta}$ uniformly in n. This together with the conclusion from previous paragraph implies

$$A = \lim_{n\to\infty} T(g_n, F_n) = T(g, F). \tag{12.3}$$

So g is an extremizer.

13 L^p Convergence of Extremizing Subsequence

The main result of this section is the L^p convergence of a subsequence of any extremizing sequence after applying suitable symmetries. The proof is similar to the corresponding result in [3]. For the convenience of the reader we give the details of the proof.

Euler-Lagrange Identity Let f be a nonnegative extremizer with $\|f\|_p = 1$. Then by Holder's inequality

$$\begin{aligned} A^q &= \|T(f)\|_q^q = \langle T(f), T(f)^{q-1}\rangle = \langle f, T^*(T(f)^{q-1})\rangle \\ &\leq \|f\|_p \|T^*(T(f)^{q-1})\|_{p'} = A\|(Tf)^{q-1}\|_{q'} = A\|T(f)\|_q^{\frac{q}{q'}} = AA^{\frac{q}{q'}} = A^q. \end{aligned}$$

Since the equality holds for the above chain of inequalities, we have $T^*(T(f)^{q-1})$ agrees with a constant multiple of $f^{\frac{p}{p'}}$ almost everywhere on $\mathbb{R}^d$. The above equality implies this constant is A^q. So finally, we have for any nonnegative extremizer f with $\|f\|_p = 1$,

$$T^*\big((T(f))^{q-1}\big) = A^q f^{p-1} \quad \text{almost everywere on} \quad \mathbb{R}^d. \tag{13.1}$$

Lemma 13.1 *Let f_n be an extremizing sequence. Then there exist a sequence of symmetries $\{\phi_n\} \subset G_d$ and an extremal F such that $\{\phi_{n_k}^*(f_{n_k})\}$ converges to F in L^p for some subsequence $\{f_{n_k}\}$.*

Proof After passing through a subsequence, if necessary and applying suitable symmetry ϕ_n^* we have that $\phi_n^*(f_n) = g_n + h_n$ with $\|h_n\|_p \to 0$ and $\|T(g_n)\|_q \to A$. Therefore it is enough to prove that $\{\phi_n^*(f_n)\}$ converges to g in L^p where g is as in (12.3).

By (13.1) there exists $H \geq 0$ such that $T^*(H) = A^q g^{p-1}$ almost everywhere on $\mathbb{R}^d$. By (12.3) we have

$$\begin{aligned} A^q &= A^q\langle g, g^{p-1}\rangle = \langle g, T^*(H)\rangle = \lim_{n\to\infty}\langle g_n, T^*(H)\rangle \\ &= A^q \lim_{n\to\infty}\langle g_n, g^{p-1}\rangle = A^q \lim_{n\to\infty}\langle \phi_n^*(f_n), g^{p-1}\rangle. \end{aligned}$$

At this point we apply Theorem 2.11 in [20] to the pair $(\{\phi_n^*(f_n)\}, g)$ to get the desired conclusion. □

14 Extremizers at End Points

In this section we shall prove Theorem 1.8, which describes a relation between extremizers for $T : L^{p_0} \to L^{q_0}$ and extremizers for $X^* : L^{p_0} \to L^{q_0}$, adjoint of X as defined in 1.5. Simultaneously we prove a relation between extremizers for $T : L^{p_1} \to L^{q_1}$ and extremizers for $X : L^{p_1} \to L^{q_1}$. We would like to thank Michael Christ for his suggestion to look at the restricted X-ray transform, X, for the endpoint cases.

Lemma 14.1 *Let T and X be defined as in* 1.1 *and* 1.5 *respectively and $d > 2$. Then $\|T\|_{L^{p_0}\to L^{q_0}} \geq \|X^*\|_{L^{p_0}\to L^{q_0}}$ and $\|T\|_{L^{p_1}\to L^{q_1}} \geq \|X\|_{L^{p_1}\to L^{q_1}}$.*

Proof It suffices to show that $\|T\|_{L^{p_0}\to L^{q_0}} \geq \|X^*\|_{L^{p_0}\to L^{q_0}}$ since $(p_1, q_1) = (q_0', p_0')$ and T^* is the same operator as T with the curve $h(t)$ replaced by $-h(t)$. Let $\epsilon > 0$.

Let $f \in L^{p_0}$ and $g \in L^{q_0'}$. Let $\gamma(t) = (t^2, t^3, \dots, t^d) \in \mathbb{R}^{d-1}$. Define

$$\tilde{f}_\epsilon(x) := \epsilon^{\frac{d-1}{p_0}} f((0, \epsilon(x_2, \dots, x_d) + h(x_1))$$

and

$$\tilde{g}_\epsilon(x) = \epsilon^{\frac{d}{q_0'}} g(\epsilon x).$$

Let f and g be compactly supported smooth functions. Then

$$\begin{aligned} \langle X^* f, g\rangle &= \lim_{\epsilon\to 0} \int_{(t,y)\in\mathbb{R}\times\mathbb{R}^d} f\big(\epsilon y_1 + t, y - \frac{\gamma(\epsilon y_1 + t) - \gamma(t)}{\epsilon}\big) g(y)\, dy\, dt \\ &= \lim_{\epsilon\to 0} \langle T f_\epsilon, g_\epsilon\rangle \le \|T\| \end{aligned} \tag{14.1}$$

where f_ϵ, g_ϵ are the functions such that $\tilde{f}_\epsilon = f$ and $\tilde{g}_\epsilon = g$. □

Lemma 14.2 *Let T and X be as above and $d > 2$.*

- *If there exists an extremizing sequence for $T : L^{p_0} \to L^{q_0}$ that does not have a subsequence converging to an extremizer modulo symmetries of T, then $\|T\|_{L^{p_0}\to L^{q_0}} = \|X^*\|_{L^{p_0}\to L^{q_0}}$.*
- *If there exists an extremizing sequence for $T : L^{p_1} \to L^{q_1}$ that does not have a subsequence converging to an extremizer modulo symmetries of T, then $\|T\|_{L^{p_1}\to L^{q_1}} = \|X\|_{L^{p_1}\to L^{q_1}}$.*

Proof We shall prove the lemma only for $T : L^{p_0} \to L^{q_0}$, the other case being identical. By the previous Lemma it suffices to show that $\|T\|_{L^{p_0}\to L^{q_0}} \le \|X^*\|_{L^{p_0}\to L^{q_0}}$. By hypothesis, there exists an extremizing sequence $\{f_n\}$ for $T : L^{p_0} \to L^{q_0}$ such that for any sequence of symmetries $\{\phi_n^*\}$, the sequence $\{\phi_n^*(f_n)\}$ has no subsequence which converges to a non-zero limit in L^{p_0}. Let us start with such an extremizing sequence $\{f_n\}$ such that $\|Tf_n\|_{L^{q_0}} \ge (1 - \frac{1}{n})A$ for each n. Then there exists another sequence $\{g_n\}$ such that $\langle Tf_n, g_n\rangle$ converges to $A = \|T\|_{L^{p_0}\to L^{q_0}}$ and $\|g_n\|_{L^{q_0'}} = 1$. By Lemma 11.2 there is a sequence of symmetries $\{\phi_n^*\}$ such that after changing f_n to $\phi_n^*(f_n)$ and g_n to $\psi_n^*(g_n)$, if necessary, we have

- $f_n = \sum_{j\in S_n} 2^j f_{n,j}$, $f_{n,j} \sim \chi_{E_{n,j}}$;
- $|j - J_n| \le Cn^{-C}$ for all $j \in S_n$;
- $E_{n,j} \subset Cn^{-C^{Cn^{-C}}} B(0, 0, \alpha_n, 1)$ for all $j \in S_n$;
- $\|2^{J_n} \chi_{B(0,0,\alpha_n,1)}\|_{p_0} \le C$.

and

- $g_n = \sum_{k\in\tilde{S}_n} 2^k g_{n,k}$, $g_{n,k} \sim \chi_{F_{n,k}}$;
- $|k - K_n| \le Cn^{-C}$ for all $k \in \tilde{S}_n$;
- $F_{n,k} \subset Cn^{-C^{Cn^{-C}}} B^*(0, 0, \alpha_n, 1)$ for all $k \in \tilde{S}_n$;
- $\|2^{K_n} \chi_{B^*(0,0,\alpha_n,1)}\|_{q_0'} \le C$.

Since f_n has no subsequence converging to a nonzero limit in L^{p_0}, by the proof of Lemma 12.1, $\alpha_n \to 0$.

We define

$$\tilde{f}_n(x) := \alpha_n^{\frac{d-1}{p_0}} f_n((0, \alpha_n(x_2, \dots, x_d) + h(x_1))$$

and

$$\tilde{g}_n(x) = \alpha_n^{\frac{d}{q_0'}} g_n(\alpha_n x).$$

Let $c_n = Cn^{-C^{Cn^{-C}}}$. This implies for each $\tilde{f}_n$, we now have the following,

$$\tilde{f}_n = \sum_{|j|<c_n} 2^j f_{n,j}((0, 2^{\frac{-p_0 J_n}{d-1}}(x_2, \dots, x_d) + h(x_1))).$$

Let $f^*_{n,k}(x) = f_{n,k}((0, 2^{\frac{-p_0 J_n}{d-1}}(x_2, \dots, x_d) + h(x_1)))$. Then $f^*_{n,j} \sim \chi_{E^*_{n,j}}$ with the following properties:

- $\chi_{E^*_{n,0}} \sim \chi_{B(0,1)}$;
- $E^*_{n,j} \subset B(0, c_n)$ for all $|j| < c_n$.

Similarly we have

$$\tilde{g}_n(x) = \sum_{|k|<c_n} 2^k g_{n,k}(2^{\frac{-q_0' K_n}{d}} x).$$

If $g^*_{n,k}(x) = g_{n,k}(2^{\frac{-q_0' K_n}{d}} x)$ then $g^*_{n,k} \sim \chi_{F^*_{n,k}}$ with

- $\chi_{F^*_{n,0}} \sim \chi_{B(0,1)}$;
- $F^*_{n,k} \subset B(0, c_n)$ for all $|j| < c_n$.

By a simple change of variable we see that for each n, we have $\|\tilde{g}_n\|_{L^{q_0'}} = 1$ and $\|\tilde{f}_n\|_{L^{p_0}} = 1$. Furthermore, we now apply the proof of Lemma 12.1 to show that there is an $f \in L^{p_0}$ such that $\{\tilde{f}_n\}$ has a subsequence that converges weakly to f as n goes to infinity and there is an $g \in L^{q_0'}$ such that $\{\tilde{g}_n\}$ has a subsequence weakly converges to g as n goes to infinity. WLOG we assume that the sequence $\{\tilde{f}_n\}$ it self converges weakly to f and likewise for $\{\tilde{g}_n\}$. We shall now prove that

$$\begin{aligned}
\|T\|_{L^{p_0}\to L^{q_0}} &= \lim_{n\to\infty} \langle Tf_n, g_n\rangle \\
&= \lim_{n\to\infty} \int_{(t,y)\in\mathbb{R}\times\mathbb{R}^d} \tilde{f}_n(\epsilon y_1 + t, y - \frac{\gamma(\epsilon y_1 + t) - \gamma(t)}{\epsilon})\tilde{g}_n(y)\, dy\, dt \\
&= \lim_{n\to\infty} \langle X^*(\tilde{f}_n), \tilde{g}_n\rangle \\
&= \langle X^* f, g\rangle \\
&\le \|X^*\|_{L^{p_0}\to L^{q_0}}.
\end{aligned} \tag{14.2}$$

where $\epsilon = \alpha_n$.

The proof of this claim is similar to the proof of 12.1. We define for a large R,

$$\tilde{f}_{n,R}(x) = \tilde{f}_n(x)\,\chi_{\|x\|\le R}(x)\,\chi_{\tilde{f}_n \le R}(x), \qquad f_R(x) = f(x)\,\chi_{\|x\|\le R}(x)\,\chi_{f\le R}(x); \tag{14.3}$$

$$\tilde{g}_{n,R}(x) = \tilde{g}_n(x)\,\chi_{\|x\|\le R}(x)\,\chi_{\tilde{g}_n \le R}(x),. \qquad g_R(x) = g(x)\,\chi_{\|x\|\le R}(x)\,\chi_{g\le R}(x). \tag{14.4}$$

For any compactly supported smooth function ψ, the operator $f \mapsto \psi X^*(\psi f)$ maps $L^2(\mathbb{R}^d)$ boundedly to the Sobolev space H^s, which in turn embeds into L^q where $q > 2$, see [7]. Thus the weak convergence of $\tilde{f}_{n,R}$ to f_R as $n \to \infty$ implies the L^q norm convergence of $X^*(\tilde{f}_{n,R})$ to $X^*(f_R)$ as $n \to \infty$, for every fixed R. Therefore $\langle X^* \tilde{f}_{n,R}, \tilde{g}_{n,R}\rangle \to \langle X^* f_R, g_R\rangle$ as $n \to \infty$ for a fixed R.

By Lemma 12.1 we know that the $L^{q_0'}$ norms of $\tilde{g}_n$ and the L^{p_0} norms of $\tilde{f}_n$ decreases uniformly in n outside the ball of radius R centered at 0 as R goes to infinity. This together with the previous paragraph imply that $\|T\| \le \|X^*\|$. □

We now have the following immediate corollary.

Corollary 14.3 *Let T and X be as above and $d > 2$.*

- *If there exists an extremizing sequence, $\{f_n\}$, for $T : L^{p_0} \to L^{q_0}$ that does not have a subsequence converging to an extremizer modulo symmetries of T, then after applying the nonsymmetry, $f_n \to r_n^{\frac{d-1}{p_0}} f_n((0, r_n x') + h(x_1))$, it has a subsequence that converges weakly to an extremizer for $X^* : L^{p_0} \to L^{q_0}$.*
- *If there exists an extremizing sequence, $\{f_n\}$, for $T : L^{p_1} \to L^{q_1}$ that does not have a subsequence converging to an extremizer modulo symmetries of T, then after applying the nonsymmetry, $f_n \to r_n^{\frac{d}{q_0'}} f_n(r_n x)$, it has a subsequence that converges weakly to an extremizer for $X : L^{p_1} \to L^{q_1}$.*

14.1 Proof of Theorem 1.8

To complete the proof of Theorem 1.8, it suffices to prove that the weak convergence in Corollary 14.3 is in fact L^p convergence. The proof is along the same line of proof of Lemma 13.1. Let $\{\tilde{f}_n\}$ converges weakly to f in L^{p_0} and $\{\tilde{g}_n\}$ converges weakly to g in L^{q_0}. Since g is an extremizer for $X : L^{q_0'} \to L^{p_0'}$, By the Euler Lagrange equation for X we have

$$X^*\left((X(g))^{p_0'-1}\right) = A^{p_0'} g^{q_0'-1}.$$

Now we apply Theorem 2.11 in [20] to the tuple $(p, g_n, g) = (q_0', \tilde{g}_n, g)$ to prove that $\tilde{g}_n$ converges to g in $L^{q_0'}$. Similarly $\tilde{f}_n$ converges to f in L^{p_0}.

Acknowledgements I would like to thank my Ph.D. adviser, Betsy Stovall, for suggesting this problem and for many insightful comments and numerous discussions during the course of this work. This work would have been impossible without her guidance. I also like to thank Almut Burchard and Michael Christ for their many valuable remarks, in particular Christ's suggestion to look at the X-ray transform for the end point case. I would like to thank Andreas Seeger for pointing out few references and many suggestions that improved the exposition of this article. This work was supported in part by NSF grants DMS-1266336 and DMS-1600458. Part of this work was carried out when the author visited the Mathematical Sciences Research Institute as a program associate in the Harmonic Analysis program in Spring 2017, supported by DMS-1440140. A version of this article formed part of the author's Ph.D. thesis. I would like to thank the referees for their comments and suggestions which significantly improved the exposition of this article.

References

1. Beckner, W.: Inequalities in Fourier analysis. Ann. Math. **102**(1), 159–182 (1975)
2. Christ, M.: Convolution, curvature, and combinatorics: a case study. Int. Math. Res. Not. **1998**(19), 1033–1048 (1998)
3. Christ, M.: On extremals for a Radon-like transform. Preprint. arXiv:1106.0728 (2011)
4. Christ, M.: Quasiextremals for a Radon-like transform. Preprint. arXiv:1106.0722 (2011)
5. Christ, M.: A sharpened Hausdorff Young inequality. Preprint arXiv:1406.1210 (2014)
6. Christ, M., Erdoğan, M.B.: Mixed norm estimates for a restricted x-ray transform. J. Anal. Math. **87**, 187–198 (2002)
7. Christ, M., Nagel, A., Stein, E.M., Wainger, S.: Singular and maximal radon transforms: analysis and geometry. Ann. Math. **150**(2), 489–577 (1999)
8. Christ, M., Shao, S.: Existence of extremals for a Fourier restriction inequality. Anal. PDE **5**, 261–312 (2012)
9. Christ, M., Shao, S.: On the extremizers of an adjoint Fourier restriction inequality. Adv. Math. **230**, 957–977 (2012)
10. Christ, M., Xue, Q.: Smoothness of extremizers of a convolution inequality. J. Math. Pures Appl. **97**, 120–141 (2012)
11. Dendrinos, S., Stovall, B.: Uniform estimates for the x-ray transform restricted to polynomial curves. J. Funct. Anal. **262**, 4986–5020 (2012)
12. Dendrinos, S., Stovall, B.: Uniform bounds for convolution and restricted x-ray transforms along degenerate curves. J. Funct. Anal. **268**, 585–633 (2015)
13. Drouot, A.: Sharp constant for a k-plane transform inequality. Anal. PDE **7**, 1237–1252 (2014)
14. Erdoğan, M.B.: Mixed-norm estimates for a restricted x-ray transform in $\mathbb{R}^4$ and $\mathbb{R}^5$. Int. Math. Res. Not. **200**(11), 575–600 (2001)
15. Flock, T.C.: Uniqueness of extremizers for an endpoint inequality of the k-plane transform. J. Geom. Anal. **26**(1), 570–602 (2016)
16. Gelfand, I.M., Graev, M.I.: Integral transformations connected with straight line complexes in complex affine space. Dokl. Acad. Nauk SSSR **2**, 809–812 (1961)
17. Gelfand, I.M., Graev, M.I., Vilenkin, N.: Generalized Functions, vol. 5. Academic Press, New York (1966)
18. Greenleaf, A., Seeger, A.: Fourier integral operators with fold singularities. J. Reine Angew. Math. **455**, 35–56 (1994)
19. Laghi, N.: A note on restricted x-ray transforms. Math. Proc. Camb. Philos. Soc. **146**(3), 719–729 (2009)
20. Lieb, E.H., Loss, M., Loss, M.A.: American Mathematical Society. Analysis. Crm Proceedings & Lecture Notes. American Mathematical Society (2001)
21. Littman, W.: L^p -L^q estimates for singular integral operators arising from hyperbolic equations in Partial differential equations (Berkeley, Calif., 1971). Proc. Symp. Pure Math. **23**, 479–481 (1973)
22. Oberlin, D.M.: Convolution estimates for some measures on curves. Proc. Am. Math. Soc. **99**(1), 56–60 (1987)
23. Stovall, B.: Endpoint bounds for a generalized Radon transform. J. Lond. Math. Soc. **80**(2), 357–374 (2009)

24. Stovall, B.: Quasi-extremals for convolution with surface measure on the sphere. Ill. J. Math. **53**(2), 391–412 (2009)
25. Weinstein, M.I.: Nonlinear Schrödinger equations and sharp interpolation estimates. Commun. Math. Phys. **87**(4), 567–576 (1982)

Publisher's Note Springer Nature remains neutral with regard to jurisdictional claims in published maps and institutional affiliations.

Journal of Fourier Analysis and Applications (2019) 25:2690–2696
https://doi.org/10.1007/s00041-019-09678-w

A Remark on the Arcsine Distribution and the Hilbert transform

Ronald R. Coifman[1] · **Stefan Steinerberger**[1]

Received: 29 October 2018 / Revised: 26 February 2019 / Published online: 21 March 2019
© Springer Science+Business Media, LLC, part of Springer Nature 2019

Abstract

It is known that if $(p_n)_{n\in\mathbb{N}}$ is a sequence of orthogonal polynomials in $L^2([-1,1], w(x)dx)$, then the roots are distributed according to an arcsine distribution $\pi^{-1}(1-x^2)^{-1}dx$ for a wide variety of weights $w(x)$. We connect this to a result of the Hilbert transform due to Tricomi: if $f(x)(1-x^2)^{1/4} \in L^2(-1,1)$ and its Hilbert transform Hf vanishes on $(-1,1)$, then the function f is a multiple of the arcsine distribution

$$f(x) = \frac{c}{\sqrt{1-x^2}}\chi_{(-1,1)} \qquad \text{where } c \in \mathbb{R}.$$

We also prove a localized Parseval-type identity that seems to be new: if $f(x)(1-x^2)^{1/4} \in L^2(-1,1)$ and $f(x)\sqrt{1-x^2}$ has mean value 0 on $(-1,1)$, then

$$\int_{-1}^{1} (Hf)(x)^2\sqrt{1-x^2}dx = \int_{-1}^{1} f(x)^2\sqrt{1-x^2}dx.$$

Keywords Hilbert transform · Arcsine distribution · Integral operators

Mathematics Subject Classification Primary 44A15 · 60E99; Secondary 33C45

Communicated by Thomas Strohmer.

C. R. is partially supported by the NIH. S.S. is partially supported by the NSF (DMS-1763179) and the Alfred P. Sloan Foundation.

✉ Stefan Steinerberger
stefan.steinerberger@yale.edu

Ronald R. Coifman
ronald.coifman@yale.edu

1 Department of Mathematics, Yale University, New Haven, CT 06511, USA

1 Introduction

This short paper is concerned with properties of the Hilbert transform

$$(Hf)(x) = \frac{1}{\pi} \int_{\mathbb{R}} \frac{f(y)}{x-y} dy$$

when interpreted as an operator acting on $L^2([-1,1])$ (this is sometimes called the 'finite' Hilbert transform H_T). One of the first people who seems to have studied this problem is Tricomi [18,19] who was interested in solving the so-called airfoil equation $H_t f = g$. He observed the following fundamental result.

Theorem 1 (Tricomi [18]) *Let $f(x)(1-x^2)^{1/4} \in L^2(-1,1)$. If the Hilbert transform Hf vanishes identically on $(-1,1)$, then*

$$f(x) = \frac{c}{\sqrt{1-x^2}} \chi_{(-1,1)} \qquad \text{where } c \in \mathbb{R}.$$

We will give a particularly simple proof of Theorem 1 which has the advantage of also establishing a Plancherel-type theorem for H_T that seems to be new (see Theorem 2 in §1.3.). We also discuss a connection to orthogonal polynomials where Theorem 1 has a particularly nice interpretation and inverse problems for integral operators.

1.1 Orthogonal Polynomials

It is well understood that, given a nonnegative weight on $(-1,1)$, the associated family of orthogonal polynomials has roots whose distribution tend to the arcsine distribution (this dates back to Erdős and Turan [5] in 1940, see also Erdős and Freud [6], Ullman [20] and Van Assche [21]). Our main result provides a fairly natural way to explain why that this is indeed the only natural smooth distribution that could have that property. If p_n is a polynomial having roots $\{x_1, \dots, x_n\} \subset (-1,1)$, then

$$\frac{p_n'(x)}{p_n(x)} = \sum_{k=1}^{n} \frac{1}{x - x_k}$$

can be used to derive information about the distribution of the roots of p_n'. If the quantity was very big (in absolute values) in a region, then it would repel any roots from locating there. This suggests that the only way to have the roots of p_n' follow the same distribution as the roots of p_n is for the sum to somewhat cancel out; conversely, the sum is approximated by the Hilbert transform of the probability measure of the distribution of roots which we would thus like to see vanish everywhere. It was this consideration, motivated by a recent dynamical interpretation [16] of roots, that originally led us to Theorem 1.

A specific recent application of Theorem 1 (partially inspired by this interpretation) in this context is as follows: suppose $p_n(x)$ is a polynomial of degree n on the real line having only real roots. Suppose furthermore that the roots are approximately distributed according to a smooth (possibly compactly supported) probability distribution $u_0(x)$. What can one say about the distribution of roots of p_n'? Rolle's theorem implies that p_n' has $n-1$ roots on the real line and that these roots interlace with the roots of p_n. In particular, we can expect, as $n \to \infty$, that the roots of p_n' are also distributed according to $u_0(x)$ and, more generally, the same is true for the $k-$th derivative $p_n^{(k)}$. However, this is no longer true when k is allowed to move with n: what can be said about the distribution of roots of $p_n^{(0.1n)}$ or, more generally the distribution $u(t,x)$ of $p_n^{(tn)}$ for $0 < t < 1$ depending on $u_0(x)$? The second author [17] recently proposed a nonlinear, nonlocal transport equation

$$\frac{\partial u}{\partial t} + \frac{1}{\pi}\frac{\partial}{\partial x}\left(\arctan\left(\frac{Hu}{u}\right)\right) = 0,$$

where H is the Hilbert transform and the equation acts on $\mathrm{supp}(u(t,x))$. The derivation is not rigorous but recovers the correct results for orthogonal polynomials on $(-1,1)$, the family of Hermite polynomials (where the equation turns into a one-parameter family of shrinking semicircle distributions) and the family of Laguerre polynomials (where the equation turns into a one-parameter flow within the family of Marchenko-Pastur distribution). Theorem 1 becomes significant in the first case since $u_t = 0$ locally requires $Hu = 0$ leading naturally to the arcsine distribution. A naturally variant of the equation on the torus has since been studied by Granero-Belinchon [7] who established well-posed in certain spaces.

1.2 Integral Operators

There is a secondary motivation: while upper bounds on integral operators are well understood, there is no such corresponding theory for lower bounds. One simple question one could ask is the following: let $f \in C_c^\infty(-1,1)$, how big does the Hilbert transform have to be on, say, the interval $(2,3)$? A sharp result was given by Alaifari, Pierce and the second author in [1] (see also Rüland [15]) and reads

$$\|Hf\|_{L^2(2,3)} \geq c_1 \exp\left(-c_2\frac{\|f'\|_{L^2(-1,1)}}{\|f\|_{L^2(-1,1)}}\right)\|f\|_{L^2(-1,1)},$$

The proof is far from stable (in the sense that it is not clear how to establish arguments of this type for more general integral operators). The problem seems to be completely open for general integral operators (we refer to [11] for sharp results for the Laplace transform and the Fourier transform and to [8] for a dyadic model). When dealing with the Hilbert transform, the identity $\|Hf\|_{L^2(\mathbb{R})} = \|f\|_{L^2(\mathbb{R})}$ suggests a rephrasing of the question: how does the Hilbert transform move the L^2-mass of a function around? Our main result in that direction reads as follows.

Theorem 2 *Let* $f(x)(1-x^2)^{1/4} \in L^2(-1,1)$. *If* $f(x)\sqrt{1-x^2}$ *has mean value 0, then*

$$\int_{-1}^{1} (Hf)(x)^2 \sqrt{1-x^2}\,dx = \int_{-1}^{1} f(x)^2 \sqrt{1-x^2}\,dx.$$

This is a nice addition to the classical global L^2–isometry $\|Hf\|_{L^2(\mathbb{R})} = \|f\|_{L^2(\mathbb{R})}$ for compactly supported functions. We were surprised to not find this result in the literature, it seems so nice that one would expect that it should be known and not difficult to find. Are there similar statements (perhaps not identities but maybe inequalities) for more general singular integral operators of convolution type? We refer to papers of Astalaet al. [2], Bertola et al. [3], Katsevich [9,10] and Okada and Elliott [12–14] and references therein for an overview of existing results.

2 Proofs

2.1 A Real-Variable Approach to Theorems 1 and 2

Before giving a rigorous proof in § 2.2, we sketch an argument that is not entirely rigorous (we do not specify the regularity of the function and freely interchange summation and integration). However, the argument does motivate the rigorous proof in the sense that it can be interpreted as the 'correct' argument (happening on the upper half disk) projected down onto the interval (this also explains the origin of several of the underlying identities).

Sketch of an argument We introduce the function $f(x)\sqrt{1-x^2}$ which reduces the problem to showing that if the Hilbert transform of that function vanishes on $(-1,1)$, then that function is constant. We expand it into Chebychev polynomials

$$f(x)\sqrt{1-x^2} = \sum_{k=0}^{\infty} a_k T_k(x)$$

and write

$$\begin{aligned} \frac{1}{\pi}\int_{-1}^{1} \frac{f(y)}{x-y}dy &= \frac{1}{\pi}\int_{-1}^{1} \frac{f(y)\sqrt{1-y^2}}{(x-y)\sqrt{1-y^2}}dy \\ &= \frac{1}{\pi}\int_{-1}^{1} \sum_{k=0}^{\infty} \frac{a_k T_k(y)}{(x-y)\sqrt{1-y^2}}dy. \end{aligned}$$

We want to show that this quantity vanishing implies that $f(x)\sqrt{1-x^2} = a_0$. The key ingredient is the identity

$$\frac{1}{\pi}\int_{-1}^{1} \frac{a_k T_k(y)}{(x-y)\sqrt{1-y^2}}dy = a_k U_{k-1}(x)$$

where U_ℓ denotes the Chebychev polynomials of the second kind given by

$$U_0(x) = 1, U_1(x) = 2x \quad \text{and} \quad U_{\ell+1}(x) = 2xU_\ell(x) - U_{\ell-1}(x).$$

Writing

$$f(x)\sqrt{1-x^2} = \sum_{k=0}^{\infty} a_k T_k(x),$$

we have that

$$\begin{aligned}
\int_{-1}^{1} f(x)^2\sqrt{1-x^2}dx &= \int_{-1}^{1} \left(f(x)\sqrt{1-x^2}\right)^2 \frac{dx}{\sqrt{1-x^2}} \\
&= \int_{-1}^{1} \sum_{k,\ell=0}^{\infty} a_k T_k(x) a_\ell T_\ell(x) \frac{dx}{\sqrt{1-x^2}} \\
&= \sum_{k,\ell=0}^{\infty} \int_{-1}^{1} a_k T_k(x) a_\ell T_\ell(x) \frac{dx}{\sqrt{1-x^2}} \\
&= \sum_{k=0}^{\infty} \int_{-1}^{1} a_k^2 T_k(x)^2 \frac{dx}{\sqrt{1-x^2}} \\
&= \frac{\pi}{2} \sum_{k=0}^{\infty} a_k^2
\end{aligned}$$

and the identity above shows that

$$\begin{aligned}
\int_{-1}^{1} (Hf)(x)^2\sqrt{1-x^2}dx &= \int_{-1}^{1} \left(\sum_{k=1}^{\infty} a_k U_{k-1}(x)\right)^2 \sqrt{1-x^2}dx \\
&= \int_{-1}^{1} \sum_{k,\ell=1}^{\infty} a_k U_{k-1}(x) a_\ell U_{\ell-1}(x) \sqrt{1-x^2}dx \\
&= \sum_{k=1}^{\infty} \int_{-1}^{1} a_k^2 U_{k-1}(x)^2 \sqrt{1-x^2}dx \\
&= \frac{\pi}{2} \sum_{k=1}^{\infty} a_k^2
\end{aligned}$$

resulting in the desired identity

$$\int_{-1}^{1} f(x)^2\sqrt{1-x^2}dx = \int_{-1}^{1} (Hf)(x)^2\sqrt{1-x^2}dx$$

whenever $f(x)\sqrt{1-x^2}$ has mean value 0. □

The sketch of the argument has perhaps a somewhat uneasy feel to it: it is not a priori clear that integration and summation can be exchanged like that and the algebraic identities seem to be coming out of nowhere: in truth, we are observing a fairly natural proof in a different space projected down on the unit interval. The next section contains a more illuminating proof (which, simultaneously, explains the origin of the identities used above).

2.2 Proof of Theorems 1 and 2

The main idea is a substitution that can be found in a paper of Coifman and Weiss [4] relating the Hilbert transform on $(-1, 1)$ to the conjugate function on the boundary of the disk (the trick is surely older than that); as such, the proof is algebraic in nature and unlikely to generalize to other integral operators.

Proof Let $f(x)(1-x^2)^{1/4} \in L^2(-1, 1)$ be given. We define a function $g : [-\pi, \pi] \to \mathbb{R}$ via

$$g(\theta) = f(\cos\theta)\sqrt{1-(\cos\theta)^2}.$$

We want to make sure that $g \in L^2(-\pi, \pi)$, a simple substitution $x = \cos\psi$ shows that

$$\int_{-\pi}^{\pi} g(\theta)^2 d\theta = \int_{-\pi}^{\pi} f(\cos\theta)^2(1-(\cos\theta)^2)d\theta = \int_{-1}^{1} f(x)^2\sqrt{1-x^2}dx$$

which is finite by assumption. The, however, a simple substitution (carried out in [4]) shows that with $x = \cos\psi$

$$\text{p.v.}\frac{1}{\pi}\int_{-1}^{1}\frac{f(y)}{x-y}dy = -\frac{1}{2\sin\psi}\text{p.v.}\frac{1}{\pi}\int_{-\pi}^{\pi} g(\theta)\cot\left(\frac{\psi-\theta}{2}\right)d\theta.$$

However, that last principial value is merely the formula for the conjugate function of g. If g has mean value 0 (this being the orthogonality to the arcsine distribution), then the conjugate function has the same L^2-norm as the function g. This implies the result. □

Acknowledgements We are grateful to the referee for refering us to the work of Tricomi and several helpful suggestions.

References

1. Alaifari, R., Pierce, L.B., Steinerberger, S.: Lower bounds on the truncated Hilbert transform. Rev. Mat. Iberoam. **32**(1), 23–56 (2016)
2. Astala, K., Päivärinta, L., Saksman, E.: The finite Hilbert transform in weighted spaces. Proc. R. Soc. Edinb. Sect. A **126**(6), 1157–1167 (1996)
3. Bertola, M., Katsevich, A., Tovbis, A.: Inversion formulae for the cosh-weighted Hilbert transform. Proc. Am. Math. Soc. **141**(8), 2703–2718 (2013)

4. Coifman, R., Weiss, G.: Extensions of Hardy spaces and their use in analysis. Bull. Am. Math. Soc. **83**(4), 569–645 (1977)
5. Erdős, P., Turán, P.: On interpolation. III. Interpolatory theory of polynomials. Ann. Math. **41**, 510–553 (1940)
6. Erdős, P., Freud, G.: On orthogonal polynomials with regularly distributed zeros. Proc. Lond. Math. Soc. **29**, 521–537 (1974)
7. Granero-Belinchon, R.: On a nonlocal differential equation describing roots of polynomials under differentiation, arXiv:1812.00082
8. Jaming, P., Pozzi, E., Wick, B.: Lower bounds for the dyadic Hilbert transform. Ann. Fac. Sci. Toulouse Math. **27**(1), 265–284 (2018)
9. Katsevich, A.: Singular value decomposition for the truncated Hilbert transform. Inverse Probl. **26**, 115011 (2010)
10. Katsevich, A.: Singular value decomposition for the truncated Hilbert transform: part II. Inverse Probl. **27**(7), 075006 (2011)
11. Lederman, R., Steinerberger, S.: Stability estimates for truncated fourier and laplace transforms. Integral Equ. Oper. Theory **87**, 529–543 (2017)
12. Okada, S., Elliott, D.: The finite Hilbert transform in L^2. Math. Nachr. **153**, 43–56 (1991)
13. Okada, S., Elliott, D.: Hölder continuous functions and the finite Hilbert transform. Math. Nachr. **169**, 219–233 (1994)
14. Okada, S., Elliot, D.: The finite Hilbert transform and weighted Sobolev spaces. Math. Nachr. **266**, 34–47 (2004)
15. Ruland, A.: Quantitative Invertibility and Approximation for the Truncated Hilbert and Riesz Transforms, arXiv:1708.04285
16. Steinerberger, S.: Electrostatic interpretation of zeros of orthogonal polynomials. Proc. Am. Math. Soc. **146**, 5323–5331 (2018)
17. Steinerberger, S.: A nonlocal transport equation describing roots of polynomials under differentiation, arXiv:1811.04844
18. Tricomi, F.: On the finite Hilbert transformation. Quart. J. Math. Oxf Ser. **2**, 199–211 (1951)
19. Tricomi, F.: Integral Equations. Pure and Applied Mathematics, vol. 5. Interscience Publishers, Inc., New York (1957)
20. Ullman, J.L.: On the regular behaviour of orthogonal polynomials. Proc. Lond. Math. Soc. **24**, 119–148 (1972)
21. Van Assche, W.: Asymptotics for orthogonal polynomials. Lecture Notes in Mathematics, vol. 1265. Springer, Berlin (1987)

Publisher's Note Springer Nature remains neutral with regard to jurisdictional claims in published maps and institutional affiliations.

Journal of Fourier Analysis and Applications (2019) 25:2697–2702
https://doi.org/10.1007/s00041-019-09679-9

Letter to the Editor: A Short Complex-Variable Proof of the Titchmarsh Convolution Theorem

Thomas Ransford[1]

Received: 18 January 2019 / Published online: 27 March 2019
© Springer Science+Business Media, LLC, part of Springer Nature 2019

Abstract
The Titchmarsh convolution theorem is a celebrated result about the support of the convolution of two functions. We present a simple proof based on the canonical factorization theorem for bounded holomorphic functions on the unit disk.

Keywords Convolution · Laplace transform · Singular inner function

Mathematics Subject Classification Primary 42A85 · Secondary 30JXX

1 Introduction

Let $f, g : \mathbb{R} \to \mathbb{C}$ be integrable functions, and let $f * g$ be their convolution product, namely

$$(f * g)(t) := \int_{-\infty}^{\infty} f(t-s)g(s)\,ds \quad (t \in \mathbb{R}).$$

It is clear that if $f = 0$ a.e. on $(-\infty, a)$ and if $g = 0$ a.e. on $(-\infty, b)$, then $f * g = 0$ a.e. on $(-\infty, a+b)$. Hence, defining

$$\alpha(f) := \sup\{a \in \mathbb{R} : f = 0 \text{ a.e. on } (-\infty, a)\},$$

Communicated by Yura Lyubarskii.

Research supported by Grants from NSERC and the Canada Research Chairs program.

✉ Thomas Ransford
thomas.ransford@mat.ulaval.ca

1 Département de mathématiques et de statistique, Université Laval, Québec City, QC G1V 0A6, Canada

and likewise for g and $f * g$, we have $\alpha(f * g) \geq \alpha(f) + \alpha(g)$. Much less obvious is the following theorem.

Theorem 1.1 *If $\alpha(f) > -\infty$ and $\alpha(g) > -\infty$, then*

$$\alpha(f * g) = \alpha(f) + \alpha(g).$$

Theorem 1.1 is a version of the Titchmarsh convolution theorem [13]. This celebrated result has applications in a number of fields, including harmonic analysis, Banach algebras, operator theory and partial differential equations.

There are several proofs of the Titchmarsh convolution theorem, none of them easy. Broadly speaking, they divide into two groups: those using real-variable methods (see e.g. [4,7,11,12]), and those based on complex-variable techniques (Titchmarsh's original proof was of this kind, see also [1,2,5,8,9]). The complex-variable proofs are maybe more natural, but tend to require more advanced results from function theory. In this paper we present a simple proof that uses nothing beyond the well-known canonical factorization theorem for bounded holomorphic functions on the unit disk.

2 Laplace Transforms

Complex-variable proofs of Titchmarsh's theorem usually proceed via the Laplace transform, so we begin by briefly reviewing this notion. In what follows, we write $\mathbb{H}$ for the right half-plane, namely $\mathbb{H} := \{z \in \mathbb{C} : \operatorname{Re} z > 0\}$.

Let $f : \mathbb{R} \to \mathbb{C}$ be an integrable function with $\alpha(f) > -\infty$. Its Laplace transform $\mathcal{L}f : \mathbb{H} \to \mathbb{C}$ is defined by

$$\mathcal{L}f(z) := \int_{-\infty}^{\infty} f(t)e^{-zt}\,dt \qquad (z \in \mathbb{H}).$$

It is easy to see that $\mathcal{L}f$ is holomorphic on $\mathbb{H}$. Furthermore, a straightforward estimate shows that

$$|\mathcal{L}f(z)| \leq Ae^{-\alpha(f)\operatorname{Re} z} \qquad (z \in \mathbb{H}),$$

where $A := \int_{-\infty}^{\infty} |f(t)|\,dt$. In particular, if $\alpha(f) \geq 0$, then $\mathcal{L}f$ is bounded on $\mathbb{H}$. The converse is also true:

Lemma 2.1 *If $\mathcal{L}f$ is bounded on $\mathbb{H}$, then $\alpha(f) \geq 0$.*

Proof Set $f_1 := f1_{(-\infty,0]}$ and $f_2 := f1_{(0,\infty)}$. As f_1 is supported on a compact subset of $(-\infty, 0]$, its Laplace transform $\mathcal{L}f_1$ is an entire function which is bounded on the left half-plane $\operatorname{Re} z \leq 0$. On the right half-plane $\operatorname{Re} z > 0$, we have $\mathcal{L}f_1 = \mathcal{L}f - \mathcal{L}f_2$, where $\mathcal{L}f_2$ is clearly bounded, and $\mathcal{L}f$ is bounded by assumption. So in fact $\mathcal{L}f_1$ is bounded on the whole complex plane. By Liouville's theorem $\mathcal{L}f_1$ is constant. It is easily seen that $\lim_{x\to-\infty} \mathcal{L}f_1(x) = 0$, so the constant must be zero, i.e. $\mathcal{L}f_1 \equiv 0$. By the uniqueness theorem for Laplace transforms, $f_1 = 0$ a.e. Hence $f = f_2$ a.e., and so, finally, $\alpha(f) = \alpha(f_2) \geq 0$. □

Lastly in this section, we remark that, if f, g are integrable functions such that $\alpha(f), \alpha(g) > -\infty$, then $\mathcal{L}(f * g) = \mathcal{L}f.\mathcal{L}g$, the pointwise product on $\mathbb{H}$. The proof is a routine argument using Fubini's theorem.

3 Beginning of the Proof of Theorem 1.1

We argue by contradiction. Suppose, if possible, that there are integrable functions f, g with $\alpha(f) > -\infty$ and $\alpha(g) > -\infty$ such that $\alpha(f*g) > \alpha(f)+\alpha(g)$. Replacing f and g by appropriate translates, we may further assume that $\alpha(f) < 0$ and $\alpha(g) < 0$ and $\alpha(f * g) > 0$. By Lemma 2.1, the Laplace transforms $\mathcal{L}f, \mathcal{L}g$ are unbounded on $\mathbb{H}$, whereas their product $\mathcal{L}f.\mathcal{L}g$ is bounded on $\mathbb{H}$. So, to obtain a contradiction, and thereby complete the proof of Theorem 1.1, it suffices to establish the following general result about holomorphic functions.

Proposition 3.1 *Let F and G be holomorphic functions on $\mathbb{H}$ satisfying $|F(z)| \le Ae^{a\mathrm{Re}\, z}$ and $|G(z)| \le Be^{b\mathrm{Re}\, z}$ for some positive constants A, a, B, b. If both F and G are unbounded on $\mathbb{H}$, then so is their product FG.*

Up till now, we have followed a fairly standard route. The novelty in our approach lies in our proof Proposition 3.1, for which we shall use the canonical factorization theorem for bounded holomorphic functions on the unit disk. We pause briefly to review this theorem.

4 Canonical Factorization Theorem

Let $\mathbb{D}$ denote the open unit disk and $\mathbb{T}$ denote the unit circle.

A *Blaschke product* on $\mathbb{D}$ is a function of the form

$$B(z) := cz^m \prod \frac{|a_n|}{a_n} \frac{a_n - z}{1 - \overline{a}_n z} \quad (z \in \mathbb{D}),$$

where c is a unimodular constant, m is a non-negative integer, and (a_n) is a (finite or infinite) sequence of points in $\mathbb{D}$ such that $\sum_n (1 - |a_n|) < \infty$.

A *bounded outer function* on $\mathbb{D}$ is a function of the form

$$O(z) := \exp\Bigl(\int_{\mathbb{T}} \frac{e^{i\theta} + z}{e^{i\theta} - z} \log \omega(\theta) \frac{d\theta}{2\pi}\Bigr) \quad (z \in \mathbb{D}),$$

where $\omega : \mathbb{T} \to (0, \infty)$ is a bounded positive function such that $\log \omega \in L^1(\mathbb{T})$.

A *singular inner function* on $\mathbb{D}$ is a function of the form

$$S(z) := \exp\Bigl(-\int_{\mathbb{T}} \frac{e^{i\theta} + z}{e^{i\theta} - z} d\sigma(e^{i\theta})\Bigr) \quad (z \in \mathbb{D}),$$

where σ is finite positive Borel measure on $\mathbb{T}$ that is singular with respect to Lebesgue measure. The measure σ is uniquely determined by S. We shall often write $S = S_\sigma$ to indicate the dependence of S upon σ.

The following result is known as the canonical factorization theorem. A proof can be found for example in [6, Theorem 2.8].

Theorem 4.1 *Every bounded holomorphic function H on $\mathbb{D}$ with $H \not\equiv 0$ has a unique factorization of the form $H = BOS$, where B is a Blaschke product, O is a bounded outer function, and S is a singular inner function.*

The following corollary will prove crucial in what follows.

Corollary 4.2 *Let B be a Blaschke product, let O be a bounded outer function, and let S_{σ_1} and S_{σ_2} be singular inner functions. Then $BOS_{\sigma_1}/S_{\sigma_2}$ is bounded on $\mathbb{D}$ if and only if $\sigma_1 - \sigma_2$ is a positive measure.*

Proof The 'if' is obvious, since $BOS_{\sigma_1}/S_{\sigma_2} = BOS_{\sigma_1-\sigma_2}$. For the 'only if', suppose that $BOS_{\sigma_1}/S_{\sigma_2}$ is bounded. Then, by Theorem 4.1, we can write it as $\widetilde{B}\widetilde{O}\widetilde{S}$, where $\widetilde{B}$ is a Blaschke product, $\widetilde{O}$ is a bounded outer function, and $\tilde{S}$ is a singular inner function, say $\widetilde{S} = S_\sigma$. Multiplying up by S_{σ_2}, we then have $BOS_{\sigma_1} = \widetilde{B}\widetilde{O}S_{\sigma_2+\sigma}$. By the uniqueness part of Theorem 4.1, it follows that $\sigma_1 = \sigma_2 + \sigma$. Thus $\sigma_1 - \sigma_2$ is a positive measure. □

5 Completion of the Proof of Theorem 1.1

Proof of Proposition 3.1 By assumption, the function $F(z)e^{-az}$ is bounded on $\mathbb{H}$. Composing with $\phi(z) := (1+z)/(1-z)$, which maps $\mathbb{D}$ conformally onto $\mathbb{H}$, we obtain a bounded holomorphic function on $\mathbb{D}$. Hence, by Theorem 4.1,

$$F(\phi(z))e^{-a\phi(z)} = B(z)O(z)S_\sigma(z) \quad (z \in \mathbb{D}),$$

where B is a Blaschke product, O is a bounded outer function, and S_σ is a singular inner function. Notice also that $e^{-a\phi(z)} = S_{a\delta_1}(z)$, where δ_1 denotes the Dirac mass at the point 1. Thus we have

$$F \circ \phi = BOS_\sigma / S_{a\delta_1}.$$

If F is unbounded on $\mathbb{H}$, then $F \circ \phi$ is unbounded on $\mathbb{D}$, so by Corollary 4.2 $\sigma - a\delta_1$ is not a positive measure, in other words $\sigma(\{1\}) < a$.

Likewise, we can write

$$G \circ \phi = \widetilde{B}\widetilde{O}S_\tau / S_{b\delta_1},$$

where $\widetilde{B}$ is a Blaschke product, $\widetilde{O}$ is a bounded outer function, S_τ is a singular inner function, and, if G is unbounded, then $\tau(\{1\}) < b$.

Multiplying these equations together, we obtain

$$(FG) \circ \phi = (B\widetilde{B})(O\tilde{O})S_{\sigma+\tau}/S_{(a+b)\delta_1}.$$

If F and G are both unbounded, then $(\sigma+\tau)(\{1\}) < a+b$, so the measure $(\sigma+\tau) - (a+b)\delta_1$ is not positive. Corollary 4.2 then shows that $(FG) \circ \phi$ is unbounded on $\mathbb{D}$, whence FG is unbounded on $\mathbb{H}$. This completes the proof of Proposition 3.1, and therefore also that of Theorem 1.1. □

6 Concluding Remarks

The Titchmarsh convolution theorem can be expressed in terms of supports. The *support* an integrable function $f : \mathbb{R} \to \mathbb{C}$ is the smallest closed subset of $\mathbb{R}$ such that $f = 0$ a.e. on the complement. Writing $\operatorname{supp}(f)$ for this support, it is easy to see that, for all integrable functions f, g, we have

$$\operatorname{supp}(f * g) \subset \operatorname{supp}(f) + \operatorname{supp}(g). \tag{6.1}$$

If further $\operatorname{supp}(f)$ and $\operatorname{supp}(g)$ are both compact, then Theorem 1.1 tells us that the two sides of (6.1) have the same minimum and (after reflection) the same maximum. Thus, writing $\operatorname{conv}(\cdot)$ for the convex hull, we have

$$\operatorname{conv}(\operatorname{supp}(f * g)) = \operatorname{conv}(\operatorname{supp}(f)) + \operatorname{conv}(\operatorname{supp}(g)). \tag{6.2}$$

In this form, the result generalizes to higher dimensions. If $f, g : \mathbb{R}^n \to \mathbb{C}$ are integrable functions with compact supports, then (6.2) holds. More generally still, (6.2) holds whenever f and g are distributions on $\mathbb{R}^n$ with compact supports. This generalization is due to Lions [10]. It can be deduced quite easily from the basic version, Theorem 1.1 (see e.g. [3, Sect. 45]).

References

1. Boas Jr., R.P.: Entire Functions. Academic Press Inc., New York (1954)
2. Crum, M.M.: On the resultant of two functions. Q. J. Math. Oxford Ser. **12**, 108–111 (1941)
3. Donoghue Jr., W.F.: Distributions and Fourier Transforms. Pure and Applied Mathematics, vol. 32. Academic Press, New York (1969)
4. Doss, R.: An elementary proof of Titchmarsh's convolution theorem. Proc. Am. Math. Soc. **104**(1), 181–184 (1988)
5. Dufresnoy, J.: Sur le produit de composition de deux fonctions. C. R. Acad. Sci. Paris **225**, 857–859 (1947)
6. Duren, P.L.: Theory of H^p Spaces. Pure and Applied Mathematics, vol. 38. Academic Press, New York (1970)
7. Kalisch, G.K.: A functional analysis proof of Titchmarsh's theorem on convolution. J. Math. Anal. Appl. **5**, 176–183 (1962)
8. Koosis, P.: On functions which are mean periodic on a half-line. Commun. Pure Appl. Math. **10**, 133–149 (1957)
9. Lax, P.D.: Translation invariant spaces. Acta Math. **101**, 163–178 (1959)

10. Lions, J.L.: Supports dans la transformation de Laplace. J. Anal. Math. **2**, 369–380 (1953)
11. Mikusiński, J.G.: A new proof of Titchmarsh's theorem on convolution. Stud. Math. **13**, 56–58 (1953)
12. Mikusiński, J.G.: Operational Calculus. International Series of Monographs on Pure and Applied Mathematics, vol. 8. Pergamon Press, Państwowe Wydawnictwo Naukowe, New York, Warsaw (1959)
13. Titchmarsh, E.C.: The zeros of certain integral functions. Proc. Lond. Math. Soc. (2) **25**, 283–302 (1926)

Publisher's Note Springer Nature remains neutral with regard to jurisdictional claims in published maps and institutional affiliations.

Journal of Fourier Analysis and Applications (2019) 25:2703–2732
https://doi.org/10.1007/s00041-019-09680-2

RESEARCH ARTICLE

On Some Spectral Properties of Pseudo-differential Operators on $\mathbb{T}$

Juan Pablo Velasquez-Rodriguez[1]

Received: 17 August 2018 / Revised: 13 March 2019 / Published online: 27 March 2019
© Springer Science+Business Media, LLC, part of Springer Nature 2019

Abstract

In this paper we use Riesz spectral Theory and Gershgorin Theory to obtain explicit information concerning the spectrum of pseudo-differential operators defined on the unit circle $\mathbb{T} := \mathbb{R}/2\pi\mathbb{Z}$. For symbols in the Hörmander class $S^m_{1,0}(\mathbb{T}\times\mathbb{Z})$, we provide a sufficient and necessary condition to ensure that the corresponding pseudo-differential operator is a Riesz operator in $L^p(\mathbb{T})$, $1 < p < \infty$, extending in this way compact operators characterisation in Molahajloo (Pseudo-Differ Oper Anal Appl Comput 213:25–29, 2011) and Ghoberg's lemma in Molahajloo and Wong (J Pseudo-Differ Oper Appl 1(2):183–205, 2011) to $L^p(\mathbb{T})$. We provide an example of a non-compact Riesz pseudo-differential operator in $L^p(\mathbb{T})$, $1 < p < 2$. Also, for pseudo-differential operators with symbol satisfying some integrability condition, it is defined its associated matrix in terms of the Fourier coefficients of the symbol, and this matrix is used to give necessary and sufficient conditions for L^2-boundedness without assuming any regularity on the symbol, and to locate the spectrum of some operators.

Keywords Spectral theory · Pseudo-differential operators · Riesz operators · Operator ideals · Gershgorin theory · Fourier analysis

Mathematics Subject Classification Primary 58J40 · Secondary 47A10

1 Introduction

Pseudo-differential operators acting on functions defined on smooth manifolds are an important generalisation of differential operators, and their study is a very active branch of contemporary mathematics, among other reasons, because of its applications

---

Communicated by Michael Ruzhansky.

✉ Juan Pablo Velasquez-Rodriguez
velasquez.juan@correounivalle.edu.co

1 Department of Mathematics, Universidad del Valle, Cali, Colombia

in various areas of physics, such as black-hole physics, quantum electrodynamics and quantum field theory [6,14]. Pseudo-differential operators considered when the manifold is a compact Lie group have arisen as in interesting arena from the point of view of Ruzhansky–Turunen Theory, in which the representations of the group allow to define a global symbol instead of symbols depending on charts.

The simplest example in this direction is the case of the one dimensional torus, on which we have the concept of periodic pseudo-differential operators acting on periodic functions. These operators have been widely studied [4,5,9,12,16–18,23,27] and remarkable results have been obtained, being one of the most important, perhaps the most, the global definition of pseudo-differential operators on the unit circle in [1], proposed by Agranovich in 1979 crediting L.R. Volevich. The definition is illuminating and readily generalisable for any torus $\mathbb{T}^d$, but it is a non-trivial fact that it is equivalent to the original one given by Hörmander. However, time after Agranovich published a definition, McLean in [21] proved the equivalence for operators with symbols in the Hörmander (ρ, δ)- classes, and since then other proofs have been published.

In this paper we will be working within the framework of the global symbols introduced by Ruzhansky and Turunen with extension and periodisation techniques [29] since their approach is useful and elegant and that is why their book [30] will be one of our main references. This work is organised as follows:

- In Sect. 2 we recall some of the basics on periodic pseudo-differential operators from the point of view of Ruzhansky and Turunen Theory.
- In Sect. 3, we treat Riez operators in the sense of Pietsch, which are those operators acting on a Banach space with spectrum behaviour identical to that of compact operators. Inspired by Molahajloo in [22], we give the main theorem of the section, a necessary and sufficient condition for a pseudo-differential operator with symbol in the Hörmander class $S^0_{1,0}(\mathbb{T} \times \mathbb{Z})$ to be a Riesz operator in $L^p(\mathbb{T})$, $1 < p < \infty$ (Theorem 3.4), together with an example of a non-compact Riesz pseudo-differential operator.
- In Sect. 4 the interest is focused in the study of global pseudo-differential operators acting on $L^2(\mathbb{T})$. It is defined the infinite matrix associated to the operator, and it is used to study operators in $L^2(\mathbb{T})$ factorising in terms of operators in $L^2(\mathbb{Z})$, which are generally easier to deal with. As a result we obtain necessary and sufficient conditions for boundedness (Theorems 4.1 and 4.2) without assuming any regularity on the symbol, and sufficient conditions for invertibility, and spectrum localisation (Theorems 4.4 and 4.5).
- In Sect. 5 we discuss a proof of Ghoberg's lemma in $L^p(\mathbb{T})$ as an extension of [24, Theorem 3.3].

2 Periodic Pseudo-differential Operators

In this section we recall the definition of periodic global pseudo-differential operator, the periodic version of Hörmander classes and the basics for the corresponding pseudo-differential calculus in the sense of Ruzhansky-Turunen theory [30]. We remark that it is possible to consider global pseudo-differential operators associated to full symbols which are p-integrable in the toroidal variable.

As usual for an integrable function $f : \mathbb{T} \to \mathbb{C}$ we denote its k-th Fourier coefficient by $\widehat{f}(k)$, which is defined by

$$\widehat{f}(k) := \int_{\mathbb{T}} f(x)e^{-ix\cdot k}dx,$$

where $k \in \mathbb{Z}$, and dx denotes the normalized Haar measure on $\mathbb{T}$. Let $\mathcal{P}(\mathbb{T}) := Span\{e^{ix\cdot k} : k \in \mathbb{Z}\}$ denote the collection of trigonometric polynomials on the unit circle. Consider a given linear operator $A : \mathcal{P}(\mathbb{T}) \to L^p(\mathbb{T})$ for $1 < p < \infty$. Following [30, Theorem 10.4.6] , we associate to the linear operator A the *full symbol* $\sigma_A : \mathbb{T} \times \mathbb{Z} \to \mathbb{C}$, defined by

$$\sigma_A(x, k) := e^{-ix\cdot k}(Ae_k)(x),$$

where $e_k(x) := e^{ix\cdot k}$. This full symbol satisfies that

$$(A_\sigma f)(x) = \sum_{k\in\mathbb{Z}} \sigma_A(x, k)\widehat{f}(k)e^{ix\cdot k},$$

for all $f \in \mathcal{P}(\mathbb{T})$. On the other hand, given a measurable function $\sigma : \mathbb{T} \times \mathbb{Z} \to \mathbb{C}$ we can define using the equality above a linear operator acting, initially, on some subspace of $L^1(\mathbb{T})$. Our objective is to find suitable conditions on the function σ in order to ensure that the domain of the corresponding pseudo-differential operator A_σ can be extended to $L^p(\mathbb{T})$.

Let $m \in \mathbb{R}$, and $0 \le \rho, \delta \le 1$ given. The periodic Hörmander class $S^m_{\rho,\delta}(\mathbb{T} \times \mathbb{Z})$ of symbols $\sigma \in C^\infty(\mathbb{T} \times \mathbb{Z})$ is the set of functions satisfying the estimates

$$|\Delta_k^t D_x^r \sigma(x, k)| \le C_{t,r,\rho,\delta,m}\langle k\rangle^{m-\rho t+\delta r},$$

for each $t, r \in \mathbb{N}$, $k \in \mathbb{Z}$, $x \in \mathbb{T}$, $C_{t,r,\rho,\delta,m} \in \mathbb{R}^+$, and $\langle k\rangle := (1 + k^2)^{1/2}$. Here the difference operator $\Delta_k^t \varphi$ acts on functions $\varphi : \mathbb{Z} \to \mathbb{C}$ as

$$\Delta_k^t \varphi(k) := \sum_{h=0}^{t} (-1)^{t-h}\binom{t}{h}\varphi(k + h),$$

and for functions $f \in C^r(\mathbb{T})$ the operator $D_x^r f$ is defined by

$$D_x^r f(x) := (-i)^r \partial_x^r f(x).$$

Considering symbols in Hörmander classes allow us to treat with a graded C^*-algebra of pseudo-differential operators as a consequence of the pseudo-differential calculus summarised in [30].

We introduce the object of study in this paper, the periodic pseudo-differential operators with p-integrable symbols for $1 < p < \infty$ fixed. This definition comes from the fact that, in order to obtain a well defined operator from $\mathcal{P}(\mathbb{T})$ with the L^p-norm to $L^p(\mathbb{T})$, the linear operator defined by formula

$$(A_\sigma f)(x) = \sum_{k\in\mathbb{Z}} \sigma_A(x,k)\widehat{f}(k)e^{ix\cdot k},$$

should have symbol $\sigma_A(x,k)$ in $L^p(\mathbb{T})$ for every fixed k.

Definition 2.1 Let $1 < p < \infty$ be given. Consider $\sigma : \mathbb{T}\times\mathbb{Z} \to \mathbb{C}$ be a measurable function such that $\sigma(\cdot,k) \in L^p(\mathbb{T})$ for each $k \in \mathbb{Z}$. The *associated periodic pseudo-differential operator* $T_\sigma : \mathcal{P}(\mathbb{T}) \to L^p(\mathbb{T})$ is defined by

$$(T_\sigma f)(x) = \sum_{k\in\mathbb{Z}} \sigma(x,k)\widehat{f}(k)e^{ix\cdot k}.$$

In particular, if $\sigma(x,k) := \sigma(k)$ is a function depending only on k, the pseudo-differential operator T_σ is called a Fourier multiplier.

Recall that for symbols in Hörmander classes, it is possible to provide a symbolic calculus to periodic pseudo-differential operators. This is stated formally in the following propositions [30].

Proposition 2.1 (Toroidal Composition Formula) *Let T_α, T_β be periodic pseudo-differential operators with symbols $\alpha \in S^m_{\rho,\delta}(\mathbb{T}\times\mathbb{Z})$, $\beta \in S^l_{\rho,\delta}(\mathbb{T}\times\mathbb{Z})$. Then $T_\alpha T_\beta$ is a pseudo-differential operator with symbol σ in the Hörmander class $S^{m+l}_{\rho,\delta}(\mathbb{T}\times\mathbb{Z})$ and the toroidal symbol σ has the following asymptotic expansion:*

$$\sigma(x,k) \sim \sum_h \frac{1}{h!}\Delta_k^h\alpha(x,k)D_x^h\beta(x,k),$$

where the above means that for each $N \in \mathbb{N}$ we have

$$\sigma(x,k) - \sum_{h<N} \frac{1}{h!}\Delta_k^h\alpha(x,k)D_x^h\beta(x,k) \in S^{m-N}_{\rho,p\delta}(\mathbb{T}\times\mathbb{Z}).$$

Proposition 2.2 (Adjoint Operator Formula) *Let T_σ be a pseudo-differential operator with symbol $\sigma \in S^m_{\rho,\delta}(\mathbb{T}\times\mathbb{Z})$. Then the adjoint operator $(T_\sigma)^*$ is a pseudo-differential operator with symbol $\sigma^* \in S^m_{\rho,\delta}(\mathbb{T}\times\mathbb{Z})$ and it has the following asymptotic expansion*

$$\sigma^* \sim \sum_h \frac{1}{h!}\Delta_k^h\partial_x^h\overline{\sigma(x,k)}.$$

3 Riesz Pseudo-differential Operators

In this section we extend the main result of [22] on the compactness of periodic pseudo-differential operators. We will provide an alternative proof of main Theorem in that paper, which characterises the compact pseudo-differential operators on $L^2(\mathbb{T})$, using different arguments not requiring the existence of an inner product, implying a result, the main result of this section, for the $L^p(\mathbb{T})$ case, where $1 < p < \infty$. For this purpose we review the already known Gohberg's Lemma, wich we extend to $L^p(\mathbb{T})$ in Sect. 5. In what follows, $\mathcal{L}(E, F)$ will denote the collection of all bounded linear operators from E to F normed spaces, and $\mathfrak{K}(E, F)$ denotes the collection of compact operators in $\mathcal{L}(E, F)$. For $E = F$ we write $\mathcal{L}(E)$ instead of $\mathcal{L}(E, E)$. For $T \in \mathcal{L}(E)$ the resolvent set of T will be denoted by

$$Res(T) : \{\lambda \in \mathbb{C} := (T - \lambda I)^{-1} \in \mathcal{L}(E)\},$$

and the spectrum by $Spec(T) := \mathbb{C} \setminus Res(T)$.

Theorem 3.1 (Gohberg's Lemma) *Let T_σ be a pseudo-differential operator with symbol $\sigma \in S^0_{1,0}(\mathbb{T} \times \mathbb{Z})$, and take $1 < p < \infty$ fixed. Then $\|T_\sigma - K\|_{\mathcal{L}(L^p(\mathbb{T}))} \geq d_\sigma$ for any compact operator $K \in \mathfrak{K}(L^p(\mathbb{T}))$, where*

$$d_\sigma := \limsup_{|k| \to \infty} \{\sup_{x \in \mathbb{T}} |\sigma(x, k)|\}.$$

A proof of the theorem above can be found in [24], where the authors study the case $p = 2$. We assert that the proof also works if the L^2-norm is replaced by L^p-norm in each step. This statement will be proved in Sect. 5.

Theorem 3.2 (Molahajloo) *Let T_σ be a pseudo-differential operator with symbol $\sigma \in S^0_{1,0}(\mathbb{T} \times \mathbb{Z})$. Then T_σ is a compact operator on $L^2(\mathbb{T})$ if and only if*

$$d_\sigma := \limsup_{|k| \to \infty} \{\sup_{x \in \mathbb{T}} |\sigma(x, k)|\} = \lim_{|k| \to \infty} \{\sup_{x \in \mathbb{T}} |\sigma(x, k)|\} = 0.$$

Proof Assume that $d_\sigma = 0$ and let $f \in C^2(\mathbb{T})$. For all $x \in \mathbb{T}$

$$\begin{aligned}
(T_\sigma f)(x) &= \sum_{k \in \mathbb{Z}} \sigma(x, k) \widehat{f}(k) e^{ix \cdot k} \\
&= \sum_{k \in \mathbb{Z}} \left(\sum_{m \in \mathbb{Z}} \widehat{\sigma}(m, k) e^{ix \cdot m} \right) \widehat{f}(k) e^{ix \cdot k} \\
&= \sum_{m \in \mathbb{Z}} e^{ix \cdot m} \left(\sum_{k \in \mathbb{Z}} \widehat{\sigma}(m, k) \widehat{f}(k) e^{ix \cdot k} \right) \\
&= \sum_{m \in \mathbb{Z}} e^{ix \cdot m} (T_{\widehat{\sigma}_m} f)(x),
\end{aligned}$$

where $\widehat{\sigma}_m(k) := \widehat{\sigma}(m, k)$ and the change in the order of summation is justified by Fubini–Tonelli's theorem since

$$\sum_{k \in \mathbb{Z}} \sum_{m \in \mathbb{Z}} |\widehat{\sigma}(m, k)| |\widehat{f}(k)| = \sum_{k \in \mathbb{Z}} \|\widehat{\sigma}(\cdot, k)\|_{L^1(\mathbb{Z})} \cdot |\widehat{f}(k)|$$

$$\leq C_{0,2,0,0,0} \sum_{m \in \mathbb{Z}} \langle m \rangle^{-2} \cdot \sum_{k \in \mathbb{Z}} |\widehat{f}(k)| < \infty.$$

By defining the operator $(A_m f)(x) := e^{ix \cdot m} f(x)$, a multiplication operator, we have

$$(T_\sigma f)(x) = \sum_{m \in \mathbb{Z}} (A_m T_{\widehat{\sigma}_m} f)(x),$$

and clearly $A_m \in \mathcal{L}(L^2(\mathbb{T}))$. Now, for each $m \in \mathbb{Z}$, the operator $T_{\widehat{\sigma}_m}$ is a Fourier multiplier. As it is well known, a pseudo-differential operator with symbol depending only on the Fourier variable is a Fourier multiplier which is a compact operator in $L^2(\mathbb{T})$ if and only if its symbol η satisfy

$$\lim_{|k| \to \infty} |\eta(k)| = 0.$$

Now, for each $m \in \mathbb{Z}$ we have that

$$\lim_{|k| \to \infty} |\widehat{\sigma}(m,k)| \leq \lim_{|k| \to \infty} \sup_{x \in \mathbb{T}} |\sigma(x,k)| \leq \limsup_{|k| \to \infty} \sup_{x \in \mathbb{T}} |\sigma(x,k)| = 0.$$

This implies that each operator $T_{\widehat{\sigma}_m}$ is a compact operator. As a consequence each $A_m T_{\widehat{\sigma}_m}$ is compact and for all $N \in \mathbb{N}$, the operator

$$\sum_{|m| \leq N} A_m T_{\widehat{\sigma}_m},$$

is also compact since the set of compact operators $\mathfrak{K}(L^2(\mathbb{T}))$ form a two sided ideal in $\mathcal{L}(L^2(\mathbb{T}))$ (see [13, Proposition 4.3.4]) and this ideal of compact operators is a closed subset of $\mathcal{L}(L^2(\mathbb{T}))$ in the operator norm topology. For this reason, if the series

$$\sum_{m \in \mathbb{Z}} A_m T_{\widehat{\sigma}_m},$$

converge in the operator norm topology then

$$T_\sigma = \lim_{N \to \infty} \sum_{|k| \leq N} A_m T_{\widehat{\sigma}_m},$$

is compact as it is the limit of a sequence of compact operators. □

Remark 3.1 This argument is the most important part of the proof because it shows that it is not necessary to appeal to inner product or Calkin's algebra to characterise compact operators. Moreover, it exposes, similar to Schatten classes [10,11], a relationship between operator ideals, approximation properties and pseudo-differential operators. In fact Theorem 3.1, seen from the perspective of operator ideals theory, is a characterisation of belonging to a certain closed ideal in terms of the symbol. This idea will be exploited later.

To conclude our proof, it is well known (See [24, Proposition 2.4]) that if $\sigma \in S^0_{1,0}(\mathbb{T} \times \mathbb{Z})$ then

$$\sum_{m \in \mathbb{Z}} \| A_m T_{\widehat{\sigma}_m} \|_{\mathcal{L}(L^2(\mathbb{T}))} = \sum_{m \in \mathbb{Z}} \| T_{\widehat{\sigma}_m} \|_{\mathcal{L}(L^2(\mathbb{T}))} < \infty.$$

In summary, T_σ is a compact operator. Now, assume that $d_\sigma \neq 0$. We need only to show that T_σ is not compact on $L^2(\mathbb{T})$. Suppose that T_σ is compact. If we set $T_\sigma = K$ in Theorem 3.1 then it contradicts our assumption that $d_\sigma \neq 0$. □

3.1 Riesz Periodic Fourier Multipliers

The previous proof of Theorem 3.2 essentially uses two facts: First, Gohber's lemma which allow us to proof one implication. Second, the set of compact operators in $\mathcal{L}(L^2(\mathbb{T}))$ is a closed operator ideal. In the case $L^p(\mathbb{T})$ $1 < p < \infty$ we will use the same arguments as in Theorem 3.2 but, instead of the ideal of compact operators, we will use its radical.

We want to recall some basic definitions for the reader.

Definition 3.1 Let $\mathcal{L}$ be the collection of all bounded operators defined on Banach spaces. That is

$$\mathcal{L} := \bigcup_{E, F \text{ Banach Spaces}} \mathcal{L}(E, F).$$

An operator ideal is a sub-collection $\mathfrak{I}$ of $\mathcal{L}$ such that each one of its components

$$\mathfrak{I}(E, F) := \mathfrak{I} \cap \mathcal{L},$$

satisfy the following conditions:

(i) For each one-dimensional Banach space E the identity map I_E belongs to $\mathfrak{I}(E)$.
(ii) If $T_1, T_2 \in \mathfrak{I}(E, F)$ then $T_1 + T_2 \in \mathfrak{I}(E, F)$.
(iii) If $T \in \mathfrak{I}(E, F)$ and $X \in \mathcal{L}(F, F_0), Y \in \mathcal{L}(E_0, E)$ then their composition $XTY \in \mathfrak{I}(E_0, F_0)$, where E_0 and F_0 are Banach spaces.

If additionally each component $\mathfrak{I}(E, F)$ is closed in the operator norm topology of $\mathcal{L}(E, F)$ then it is said that $\mathfrak{I}$ is a closed operator ideal.

Given a operator ideal $\mathfrak{I}$ it is said that $S \in \mathcal{L}(E, F)$ belongs to the radical of $\mathfrak{I}$ (see [25, 4.3.1]) if for all $L \in \mathcal{L}(F, E)$ there exist $U \in \mathcal{L}(E)$ and $X \in \mathfrak{I}(E)$ such that

$$U(I_E - LS) = I_E - X.$$

The collection of operators that satisfy this condition is denoted by $\mathfrak{I}^{rad}$. It can be proved that $\mathfrak{I}^{rad}$ is a closed operator ideal so, to take advantage of this fact, we will work from now on with bounded operators in the radical of an special ideal. The radical of compact operators ideal is called the ideal of Gohberg or inessential operators, and is completely characterised by [25, Theorem 26.7.2].

Theorem 3.3 (Pietsch) $\mathfrak{R} := \mathfrak{K}^{rad}$ *is the largest operator ideal such that all components* $\mathfrak{R}(E)$ *consists of Riesz operators only.*

Riesz operators are a generalisation of compact operators that has been widely studied [3,26,28,32] and for that reason there are many equivalent definitions. One can use, for example, the definition given in [26].

Definition 3.2 Let E be a Banach space. It is said that $T \in \mathcal{L}(E)$ is a *Riesz operator* if for all $\varepsilon > 0$ exist an exponent s and points $v_1, \ldots, v_h \in E$ which depend on ε such that

$$T^s(B_E) \subseteq \bigcup_{j=1}^{h} v_j + \varepsilon^s B_E,$$

where B_E is the unit ball of E.

This definition does not provides too much information at first sight, but fortunately one can think of Riesz operators in a more intuitive way. In simple terms a bounded linear operator is a Riesz operator if its spectrum is countable, it contains only eigenvalues, each one of these has finite-dimensional associated eigen-space, and the number zero is the only possible accumulation point. In other words, Riesz operators are those bounded linear operators whose spectrum behaves like the spectrum of compact operators. They are generally easier to study than the compact operators because of their different characterisations. Several necessary and sufficient conditions to be a Riesz operator are listed in the following lemma, which is the collection of Propositions 3.2.13 and 3.2.24 in [26], 26.5.1 in [25], and the corollary of Theorem 3.4.3 in [3].

Lemma 3.1 *Let E be a Banach space and suppose* $T \in \mathcal{L}(E)$. *The following statements are equivalent*

- (i) T *is a Riesz operator.*
- (ii) T^m *is Riesz for some (for all) exponent* m.
- (iii) λT *is iterative compact for all* $\lambda \in \mathbb{C}$.
- (iv) *Non-zero points in* $Spec(T)$ *are isolated eigenvalues with finite-dimensional associated eigenspace and zero as the only cluster point. If E is infinite dimensional then* $0 \in Spec(T)$.
- (v) $T - \lambda I$ *is a Fredholm operator for all* $\lambda \in \mathbb{C}$.

In general, Riesz operators do not form an operator ideal. However, what we want to show now is that in certain subalgebras of $\mathcal{L}(L^p(\mathbb{T}))$, $1 < p < \infty$, there are ideals that only contain Riesz operators, and this will be a crucial fact in this work. The following two propositions are the first step in that direction.

Proposition 3.1 *Let $\sigma : \mathbb{Z} \to \mathbb{C}$ be a function such that*

$$|\Delta_k^t \sigma(k)| \leq C \langle k \rangle^{-t} \ , \ k \in \mathbb{Z} \tag{1}$$

for $0 \leq t \leq [1/2] + 1$. Then for $1 < p < \infty$ the spectrum of its associated Fourier multiplier T_σ as an operator in $\mathcal{L}(L^p(\mathbb{T}))$ is:

$$Spec(T_\sigma) = \overline{\{\sigma(k) \ : \ k \in \mathbb{Z}\}}$$

Proof First, the condition (1) guarantees that T_σ satisfy the conditions of the toroidal version of Mikhlin multiplier Theorem (see [33], Lemma 22.12) then $T_\sigma \in \mathcal{L}(L^p(\mathbb{T}))$. Define

$$\sigma_\lambda(k) := \sigma(k) - \lambda \ \text{ and } \ \sigma_\lambda^{-1}(k) := \frac{1}{\sigma_\lambda(k)}$$

for $\lambda \in \mathbb{C} \setminus \overline{\{\sigma(k) \ : \ k \in \mathbb{Z}\}}$. Since $\mathbb{C}$ is a regular topological space, there exist a $d > 0$ such that

$$\inf_{k \in \mathbb{Z}} |\sigma(k) - \lambda| = d.$$

Then $T_{\sigma_\lambda^{-1}}$ also satisfy the conditions of Mikhlin multiplier theorem because

$$\begin{aligned} |\Delta_k \sigma_\lambda^{-1}(k)| &= \left| \frac{1}{\sigma_\lambda(k+1)} - \frac{1}{\sigma_\lambda(k)} \right| \\ &= \left| \frac{\sigma(k+1) - \sigma(k)}{\sigma_\lambda(k+1)\sigma_\lambda(k)} \right| \leq \frac{C}{d^2} \langle k \rangle^{-1}. \end{aligned}$$

Finally, it is clear that

$$(T_\sigma - \lambda I)(T_{\sigma_\lambda^{-1}}) = (T_{\sigma_\lambda})(T_{\sigma_\lambda^{-1}}) = I,$$

which proves that $Spec(T) \subseteq \overline{\{\sigma(k) \ : \ k \in \mathbb{Z}\}}$. Now its easy to see that $\overline{\{\sigma(k) \ : \ k \in \mathbb{Z}\}} \subseteq Spec(T)$ because the spectrum is closed and each number $\sigma(k)$ is an eigenvalue of T_σ with associated eigenvector $e^{ix \cdot k}$, thus

$$\{\sigma(k) \ : \ k \in \mathbb{Z}\} \subseteq Spec(T).$$

□

As an straightforward consequence we obtain the following.

Proposition 3.2 *Let $T_\sigma : L^p(\mathbb{T}) \to L^p(\mathbb{T})$ $1 < p < \infty$ be a Fourier multiplier whose symbol satisfy the condition (1). Then T_σ is a Riesz operator if and only if*

$$\lim_{|k| \to \infty} |\sigma(k)| = 0.$$

Proof By the previous proposition and (iv) in Lemma 3.1. □

Remark 3.2 (Algebraic considerations, part 1)

(i) The collection of Fourier multipliers

$$M_p := \{T_\sigma : L^p(\mathbb{T}) \to L^p(\mathbb{T}) \ : \ \sigma \text{ satisfy (1) } \},$$

is a sub-algebra of $\mathcal{L}(L^p(\mathbb{T}))$, $1 < p < \infty$. This is a consequence of discrete Leibniz rule.

(ii) If $\mathcal{A}$ is an algebra and $\mathcal{B}$ is a sub-algebra of $\mathcal{A}$ then each ideal $\mathcal{I}$ of $\mathcal{A}$ defines an ideal of $\mathcal{B}$, this is $\mathcal{I} \cap \mathcal{B}$.

(iii) The component $\mathfrak{R}(L^p(\mathbb{T}))$ is the largest ideal in $\mathcal{L}(L^p(\mathbb{T}))$ which consists only of Riesz operators. As an immediate consequence $\mathfrak{R}(L^p(\mathbb{T})) \cap M_p$ is the largest ideal on M_p that consists only of Riesz operators.

(iv) The set

$$\mathfrak{R}(M_p) := \{T_\sigma \in M_p \ : \ \lim_{|k|\to\infty} |\sigma(k)| = 0\},$$

is an ideal of M_p, which, in virtue of Proposition 3.2, contains only Riesz operators, so $\mathfrak{R}(M_p) \subseteq \mathfrak{R} \cap M_p$.

(v) Again, by Proposition 3.2, the symbol of every Fourier multiplier $T_\sigma \in \mathfrak{R} \cap M_p$ satisfy $\lim_{|k|\to\infty} |\sigma(k)| = 0$, so $\mathfrak{R} \cap M_p \subseteq \mathfrak{R}(M_p)$ and in conclusion $\mathfrak{R} \cap M_p = \mathfrak{R}(M_p)$, that is, all Riesz Fourier multipliers are inessential operators.

3.2 Riesz Periodic Pseudo-differential Operators

We have constructed enough tools to prove the main theorem of this section. We will use the same techniques as in Theorem 3.2 but instead of compact operators we will use inessential operators.

Theorem 3.4 *Let T_σ be a pseudo-differential operator with symbol $\sigma \in S^0_{1,0}(\mathbb{T} \times \mathbb{Z})$ and $1 < p < \infty$. Then T_σ is a Riesz operator on $L^p(\mathbb{T})$ if and only if*

$$d'_\sigma := \lim_{|k|\to\infty} \{\sup_{x\in\mathbb{T}} |\sigma(x, k)|\} = 0.$$

Proof We assert that a pseudo-differential operator with symbol in $S^0_{1,0}(\mathbb{T} \times \mathbb{Z})$ is a Riesz operator if and only if $d'_{\sigma^n} = 0$ for some $n \in \mathbb{N}$, where σ^n denotes the symbol of $(T_\sigma)^n$. Then to finish we will show that $d'_{\sigma^n} = 0$ if and only if $d'_\sigma = 0$.

First, suppose that T_σ is a Riesz operator and $d'_{\sigma^n} > 0$ for all n. Using (iii) in Lemma 3.1 we know that, for some $h \in \mathbb{N}$, $(T_\sigma)^h$ is a compact operator. Then, by Gohberg's lemma, we have

$$0 < d'_{\sigma^h} \le d_{\sigma^h} \le \left\| (T_\sigma)^h - K \right\|_{\mathcal{L}(L^p(\mathbb{T}))},$$

for all compact operators K. If we set $K = (T_\sigma)^h$, the above chain of inequalities contradicts our assumption that $d'_{\sigma^h} \neq 0$.

Now suppose $d'_{\sigma^n} = 0$ for some $n \in \mathbb{N}$. Then for $f \in C^2(\mathbb{T})$

$$\begin{aligned}(T_{\sigma^n} f)(x) &= \sum_{k\in\mathbb{Z}} \sigma^n(x,k)\widehat{f}(k)e^{ix\cdot k} \\ &= \sum_{k\in\mathbb{Z}}\left(\sum_{m\in\mathbb{Z}} \widehat{\sigma}^n(m,k)e^{ix\cdot m}\right)\widehat{f}(k)e^{ix\cdot k} \\ &= \sum_{m\in\mathbb{Z}}(A_m T_{\widehat{\sigma}^n_m} f)(x),\end{aligned}$$

and

$$\lim_{|k|\to\infty} |\widehat{\sigma}^n(m,k)| \leq \lim_{|k|\to\infty} \sup_{x\in\mathbb{T}} |\sigma^n(x,k)| = 0,$$

so, for each $m \in \mathbb{Z}$, $T_{\sigma^n_m}$ is an operator in the ideal $\mathfrak{R}(M_p)$ which is a subset of the closed ideal $\mathfrak{R}(L^p(\mathbb{T}))$ of inessential operators. As a consequence each operator $A_m T_{\widehat{\sigma}^n_m}$ belongs to $\mathfrak{R}(L^p(\mathbb{T}))$ and for all $N \in \mathbb{N}$ the operator

$$\sum_{|m|\leq N} A_m T_{\widehat{\sigma}^n_m},$$

belongs to $\mathfrak{R}(L^p(\mathbb{T}))$ as well. As already said, the series

$$\sum_{m\in\mathbb{Z}} A_m T_{\widehat{\sigma}^n_m},$$

converge in the operator norm topology. In summary T_{σ^n} belongs to $\mathfrak{R}(L^p(\mathbb{T}))$ as the limit of a sequence of inessential operators

$$T_{\sigma^n} = \lim_{N\to\infty} \sum_{|m|\leq N} A_m T_{\widehat{\sigma}^n_m}.$$

Finally, let us see that $d'_{\sigma^n} = 0$ if and only if $d'_\sigma = 0$. This is a simple consequence of toroidal composition formula. As we know, by toroidal composition formula, taking the first two terms in the asymptotic expansion of $T_{\sigma^n} = T_\sigma T_{\sigma^{n-1}}$ we have

$$\sigma^n(x,k) = (\sigma(x,k))(\sigma^{n-1}(x,k)) + r_n(x,k),$$

with $r_n(x,k) \in S^{-1}_{1,0}(\mathbb{T}\times\mathbb{Z})$. Repeating this process we obtain

$$\begin{aligned}\sigma^n(x,k) &= (\sigma(x,k))^2(\sigma^{n-2}(x,k)) + (\sigma(x,k))(r_{n-1}(x,k)) + r_n(x,k) \\ &= \ldots\end{aligned}$$

$$= (\sigma(x,k))^n + \sum_{h=0}^{n-1} (\sigma(x,k))^h (r_{n-h}(x,k)),$$

with each $r_j(x,k) \in S^{-1}_{1,0}(\mathbb{T} \times \mathbb{Z})$. From this we have

$$(\sigma(x,k))^n = \sigma^n(x,k) - \sum_{h=0}^{n-1} (\sigma(x,k))^h (r_{n-h}(x,k)),$$

which implies

$$|\sigma(x,k)|^n \leq |\sigma^n(x,k)| + \sum_{h=0}^{n-1} |\sigma(x,k)|^h |r_{n-h}(x,k)|,$$

and

$$\sup_{x\in\mathbb{T}} |\sigma(x,k)|^n \leq \sup_{x\in\mathbb{T}} |\sigma^n(x,k)| + \sum_{k=0}^{n-1} C^h_\sigma \sup_{x\in\mathbb{T}} |r_{n-k}(x,k)|,$$

where C_σ is a constant that bounds uniformly $|\sigma(x,k)|$ and whose existence is clear because $\sigma \in S^0_{1,0}(\mathbb{T} \times \mathbb{Z})$. In conclusion

$$\lim_{|k|\to\infty} \sup_{x\in\mathbb{T}} |\sigma(x,k)|^n \leq \lim_{|k|\to\infty} \sup_{x\in\mathbb{T}} |\sigma^n(x,k)| + \sum_{h=0}^{n-1} \lim_{|k|\to\infty} C^h_\sigma \sup_{x\in\mathbb{T}} |r_{n-h}(x,k)|.$$

This finish the proof. □

Remark 3.3 (Algebraic considerations, part 2.)

(i) The collection of pseudo-differential operators

$$\Psi^m_p := \{T_\sigma : L^p(\mathbb{T}) \to L^p(\mathbb{T}) \; : \; \sigma \in S^m_{1,0}(\mathbb{T} \times \mathbb{Z})\},$$

is a sub-algebra of $\mathcal{L}(L^p(\mathbb{T}))$, this as a consequence of the toroidal composition formula.

(ii) The component $\mathfrak{R}(L^p(\mathbb{T}))$ is the largest ideal in the algebra $\mathcal{L}(L^p(\mathbb{T}))$ consisting only of Riesz operators. As an immediate consequence $\mathfrak{R} \cap \Psi^m_p$ is the largest ideal of Ψ^m_p consisting only of Riesz operators

(iii) The set

$$\mathfrak{R}(\Psi^m_p) := \{T_\sigma : L^p(\mathbb{T}) \to L^p(\mathbb{T}) \; : \; \lim_{|k|\to\infty} \{\sup_{x\in\mathbb{T}} |\sigma(x,k)|\} = 0\},$$

is an ideal of Ψ^m_p that, by the previous theorem , consist only of Riesz operators. Then $\mathfrak{R}(\Psi^m_p) \subseteq \mathfrak{R} \cap \Psi^m_p$.

(iv). Again, by the previous theorem, the symbol of any linear operator $T_\sigma \in \mathfrak{R} \cap \Psi_p^m$ satisfy

$$\lim_{|k|\to\infty} \{\sup_{x\in\mathbb{T}} |\sigma(x,k)|\} = 0.$$

In conclusion $\mathfrak{R}(\Psi_p^m) = \mathfrak{R} \cap \Psi_p^m$.

We finish this section with a remark about compact operators on $L^p(\mathbb{T})$. The statements of Proposition 3.2 and Theorem 3.4 could be more general. With sufficient conditions on the symbol for the boundedness of Fourier multipliers or pseudo-differential operators, let us call this conditions (BF) and (BP) respectively, and similar proofs as in Proposition 3.2 and Theorem 3.4, one can state:

Let $T_\sigma : L^p(\mathbb{T}) \to L^p(\mathbb{T})$ $1 < p < \infty$ be a Fourier multiplier whose symbol σ satisfy the condition (BF). Then T_σ is a Riesz operator if and only if

$$\lim_{|k|\to\infty} |\sigma(k)| = 0.$$

Let $T_\sigma : L^p(\mathbb{T}) \to L^p(\mathbb{T})$ be a pseudo-differential operator whose symbol σ satisfy the condition (BP). Then T_σ is a Riesz operator in $L^p(\mathbb{T})$ if and only if

$$d'_\sigma := \lim_{|k|\to\infty} \{\sup_{x\in\mathbb{T}} |\sigma(x,k)|\} = 0.$$

Unfortunately conditions (BF) and (BP) are not easy to find, but we emphasise this more general form of Proposition 3.2 and Theorem 3.4 to give an example of a Riesz non-compact pseudo-differential operators, showing then that the problem of find necessary an sufficient conditions on the symbol for the compactness of the operator in $L^p(\mathbb{T})$ is still open. For this, we will exhibit a bounded pseudo-differential operator whose symbol tends to zero uniformly, and thus is a Riesz operator, but not a compact operator on $L^p(\mathbb{T})$.

As it is shown in [19], there exist strictly singular operators on $L^p[0, 1]$, and then on $L^p(\mathbb{T})$, which are non-compact operators. Strictly singular operators are bounded linear operators between normed spaces which are not bounded below on any infinite-dimensional subspace, and the class of strictly singular operators is contained in the class of Riesz operators on any Banach space. For $L^p(\mathbb{T})$, $1 < p < 2$, one can construct an example of a strictly singular non-compact operator through the following process: consider a complemented subspace F_p of $L^p(\mathbb{T})$ isomorphic to $\ell^p(\mathbb{N})$, generated by disjointly supported functions, and denote by P_p a projection from $L^p(\mathbb{T})$ onto F_p. Take the inclusion $i_{p,2} : \ell^p(\mathbb{N}) \to \ell^2(\mathbb{N})$, and define the operator $Q : \ell^2(\mathbb{N}) \to L^p(\mathbb{T})$ by

$$Q[(t_n)](x) = \sum_{n\in\mathbb{N}} t_n r_n(x),$$

where r_n are the Rademacher functions $r_n(x) = sign(\sin(2^{k-1}x))$. By Khintchine's inequality, the operator Q is an isomorphic embedding of $\ell^2(\mathbb{N})$ into $L^p(\mathbb{T})$ for every $1 < p < \infty$, and the operator $A_p : L^p(\mathbb{T}) \to L^p(\mathbb{T})$ given by

$$A = Qi_{p,2}P_p,$$

is a non-compact strictly singular operator [19]. To construct our example let us define

$$E_n := \left(\frac{2\pi}{2^{n+2}}, \frac{2\pi}{2^{n+1}}\right), \quad n \in \mathbb{N},$$

and let $\chi_n(x)$ be the characteristic function on E_n. Take the closed subspace F_p as

$$F_p := Span\{2^{(n+2)/2}\chi_n\}_{n\in\mathbb{N}},$$

which, as it is discussed in [20, Chapter 3, p.p. 126], is isomorphic to $\ell^p(\mathbb{N})$ and 1-complemented with the projection P_p given by

$$P_p[f](x) = \sum_{n\in\mathbb{N}} \left(\int_{\mathbb{T}} 2^{(n+2)(p-1)/2}\chi_n(x)\overline{f(x)}dx\right)2^{(n+2)/2}\chi_n(x).$$

Thus the linear operator $A = Qi_{p,2}P_p$ is a pseudo-differential operator with symbol

$$\begin{aligned}
\sigma_A(x,k) = e^{-ix\cdot k}A(e^{ix\cdot k}) &= e^{-ix\cdot k}\sum_{n\in\mathbb{N}}\left(\int_{\mathbb{T}} 2^{(n+2)(p-1)/2}\chi_n(x)e^{-ix\cdot k}dx\right)r_n(x) \\
&= e^{-ix\cdot k}\sum_{n\in\mathbb{N}}\left(\int_{2\pi/2^{n+2}}^{2\pi/2^{n+1}} 2^{(n+2)(p-1)/2}e^{-ix\cdot k}dx\right)r_n(x) \\
&= \frac{e^{-ix\cdot k}}{ik}\sum_{n\in\mathbb{N}} 2^{(n+2)(p-1)/2}e^{-ik\frac{2\pi}{2^{n+2}}}(e^{-ik\frac{2\pi}{2^{n+2}}} - 1)r_n(x).
\end{aligned}$$

We note that each $k \in \mathbb{Z}$ fixed, and $n \in \mathbb{N}$ large enough, we get

$$\begin{aligned}
\left|e^{-ik\frac{2\pi}{2^{n+2}}} - 1\right|^2 = 2\left|\cos\frac{2\pi k}{2^{n+2}} - 1\right| &= 2\left|\cos\frac{2\pi k + 0}{2^{n+2}} - \cos 0\right| \\
&= 4\left|\sin\frac{2\pi k - 0}{2^{n+3}}\sin\frac{2\pi k + 0}{2^{n+3}}\right| \\
&\leq \frac{\pi|k|}{2^{n+2}},
\end{aligned}$$

which proves that there exist a constant $C > 0$ such that

$$S_k := \sum_{n\in\mathbb{N}} 2^{(n+2)(p-1)/2}|e^{ik\frac{2\pi}{2^{n+2}}} - 1| \leq C|k|^{1/2},$$

and then

$$\lim_{|k|\to\infty} \sup_{x\in\mathbb{T}} |\sigma_A(x,k)| \leq \lim_{|k|\to\infty} \frac{S_k}{|k|} \leq \lim_{|k|\to\infty} \frac{C}{|k|^{1/2}} = 0,$$

so A is a non-compact Riesz operator.

4 Gershgorin Theory for Periodic Pseudo-differential Operators

In this section we will leave aside the spaces $L^p(\mathbb{T})$, $p \neq 2$, and focus in the Hilbert space $L^2(\mathbb{T})$. The main reason is that the Fourier transform is an isomorphism between $L^2(\mathbb{T})$ and $L^2(\mathbb{Z})$ which allows to factorise operators in $L^2(\mathbb{T})$ through $L^2(\mathbb{Z})$ a sequence space that, in some cases, it is easier to study. This factorisation allows one to obtain necessary and sufficient conditions for boundedness (Theorems 4.1 and 4.2) without assuming any regularity on the symbol, and sufficient conditions for invertibility, and spectrum localisation (Theorems 4.4 and 4.5). To start with , it is necessary to mention some details about the infinite matrices theory.

4.1 Infinite Matrices

Definition 4.1 An infinite matrix is a function $M : \mathbb{Z} \times \mathbb{Z} \to \mathbb{C}$ with matrix entries defined by $M_{jk} := M(j,k)$. If M is an infinite matrix and φ an infinite vector (or a function from $\mathbb{Z}$ to $\mathbb{C}$) then the product of the vector φ an the matrix M is defined as

$$M\varphi(j) := \sum_{k\in\mathbb{Z}} M_{jk}\varphi(k).$$

For infinite matrices P and Q their product is defined as the infinite matrix with entries

$$PQ_{jk} = \sum_{h\in\mathbb{Z}} P_{jh}Q_{hk},$$

and as usual, the adjoint of the infinite matrix M is the infinite matrix M^* with entries

$$M^*_{jk} := \overline{(M_{kj})}.$$

It is easy to see that with the above definition for any pair of infinite vectors (functions $\varphi_1, \varphi_2 : \mathbb{Z} \to \mathbb{C}$) and complex numbers λ_1, λ_2 one has

$$M(\lambda_1\varphi_1 + \lambda_2\varphi_2) = \lambda_1 M\varphi_1 + \lambda_2 M\varphi_2,$$

So it is reasonable to think that an infinite matrix M can define a linear operator on some sequence space. However, not all infinite matrices define linear operators, and some conditions should be imposed on the matrix to be sufficiently well behaved to define a linear operator. In this case we are interested in linear operators on $L^2(\mathbb{Z})$.

Fortunately infinite matrices that define operators in $L^2(\mathbb{Z})$ have already been studied and many results have been obtained. We state the most relevant for our work below. They can be found in [8].

Lemma 4.1 (Crone) *Let M be an infinite matrix with rows and columns in $L^2(\mathbb{Z})$ and define $P_n(x) := \sum_{|k|\leq n}\langle x, e_k\rangle e_k$, where e_k is the k-th coordinate vector of $L^2(\mathbb{Z})$. Then M define a bounded operator in $L^2(\mathbb{Z})$ if and only if*

$$\sup_{n\in\mathbb{N}} \left\| P_n M^* M P_n \right\|_{\mathcal{L}(L^2(\mathbb{Z}))} < \infty,$$

when this happens

$$\sup_{n\in\mathbb{N}} \left\| P_n M^* M P_n \right\|_{\mathcal{L}(L^2(\mathbb{Z}))} = \|M\|^2_{\mathcal{L}(L^2(\mathbb{Z}))}.$$

Lemma 4.2 (Crone) *Let M be a infinite matrix. Then M defines a bounded linear operator on $L^2(\mathbb{Z})$ if and only if satisfy the following conditions*

(i) *The rows of M are functions in $L^2(\mathbb{Z})$.*
(ii) *$(M^*M)^n$ is defined for all $n \in \mathbb{N}$.*
(iii) $\sup_{n\in\mathbb{N}}\sup_{k\in\mathbb{Z}} |(M^*M)^n_{kk}|^{1/n} < \infty$

when this happens we get

$$\sup_{n\in\mathbb{N}}\sup_{k\in\mathbb{Z}} |(M^*M)^n_{kk}|^{1/n} = \|M\|^2_{\mathcal{L}(L^2(\mathbb{Z}))}.$$

Next we will adapt these theorems to periodic pseudo-differential operators.

4.2 L^2-Boundedness

Let $\sigma : \mathbb{T}\times\mathbb{Z} \to \mathbb{C}$ be a measurable function such that $\sigma(\cdot, k) \in L^2(\mathbb{T})$ for each $k \in \mathbb{Z}$ and T_σ be its associated pseudo-differential operator. Then, for $f \in L^2(\mathbb{T})$, we can write

$$\begin{aligned}(T_\sigma f)(x) &= \sum_{k\in\mathbb{Z}}\left(\sum_{m\in\mathbb{Z}}\widehat{\sigma}(m,k)e^{ix\cdot m}\right)\widehat{f}(k)e^{ix\cdot k}\\ &= \sum_{j\in\mathbb{Z}}\left(\sum_{k\in\mathbb{Z}}\widehat{\sigma}(j-k,k)\widehat{f}(k)\right)e^{ix\cdot j}.\end{aligned}$$

From this, it is straightforward that the j-th Fourier coefficient of $T_\sigma f$ is

$$\sum_{k\in\mathbb{Z}}\widehat{\sigma}(j-k,k)\widehat{f}(k),$$

which is similar to the j-th entry of the matrix-vector product

$$\sum_{k\in\mathbb{Z}} M_{jk}\varphi(k).$$

This observation is the key fact of this section and is the motivation for the following definition.

Definition 4.2 (Associated matrix.) Let $\sigma : \mathbb{T}\times\mathbb{Z}\to\mathbb{C}$ be a measurable function such that $\sigma(\cdot,k)\in L^2(\mathbb{T})$ for each $k\in\mathbb{Z}$ and T_σ be its associated pseudo-differential operator. Then its associated matrix M_σ is defined as the infinite matrix with entries

$$(M_\sigma)_{jk} := \widehat{\sigma}(j-k,k) = \int_{\mathbb{T}} \sigma(x,k)e^{-ix\cdot(j-k)}dx.$$

With this definition in mind, the operator T_σ can be factored through $L^2(\mathbb{Z})$ as the following diagram shows

$$\begin{array}{ccc} L^2(\mathbb{T}) & \xrightarrow{T_\sigma} & L^2(\mathbb{T}) \\ \mathcal{F}_{\mathbb{T}}\downarrow & & \mathcal{F}_{\mathbb{T}}^{-1}\uparrow \\ L^2(\mathbb{Z}) & \xrightarrow{M_\sigma} & L^2(\mathbb{Z}) \end{array}$$

where $\mathcal{F}_{\mathbb{T}}$ and $\mathcal{F}_{\mathbb{T}}^{-1}$ are the toroidal Fourier transform and toroidal inverse Fourier transform, defined by

$$\mathcal{F}_{\mathbb{T}}(f)(k) := \int_{\mathbb{T}} f(x)e^{-ix\cdot k}dx, \quad \text{and} \quad \mathcal{F}_{\mathbb{T}}^{-1}(\widehat{f})(x) := \sum_{k\in\mathbb{Z}} \widehat{f}(k)e^{ix\cdot k}.$$

These linear operators extend to a unitary operator. For this reason the operator T_σ is bounded in $L^2(\mathbb{T})$ if and only if the infinite matrix M_σ defines a bounded operator in $L^2(\mathbb{Z})$ and $\|T_\sigma\|_{\mathcal{L}(L^2(\mathbb{T}))} = \|M_\sigma\|_{\mathcal{L}(L^2(\mathbb{Z}))}$. This allow us to apply Lemmas 4.1 and 4.2 to give necessary and sufficient conditions for L^2-Boundedness of pseudo-differential operators.

Theorem 4.1 *Let $\sigma : \mathbb{T}\times\mathbb{Z}\to\mathbb{C}$ be a measurable function such that $\sigma(\cdot,k)\in L^2(\mathbb{T})$ for each $k\in\mathbb{Z}$ and T_σ be its associated pseudo-differential operator. Then T_σ defines a bounded linear operator on $L^2(\mathbb{T})$ if and only if rows of the associated matrix M_σ are in $L^2(\mathbb{Z})$, $(M_\sigma^* M_\sigma)^n$ is defined for all $n\in\mathbb{N}$, and*

$$\sup_{n\in\mathbb{N}}\sup_{k\in\mathbb{Z}}\left|\int_{\mathbb{T}} \tau^n(x,k)dx\right|^{1/n} < \infty,$$

where τ is the symbol of $(T_\sigma)^*T_\sigma = T_{\sigma^*}T_\sigma$ and τ^n is the symbol of $(T_\tau)^n$. When this happens then

$$\sup_{n\in\mathbb{N}}\sup_{k\in\mathbb{Z}}\left|\int_{\mathbb{T}} \tau^n(x,k)dx\right|^{1/n} = \|T_\sigma\|^2_{\mathcal{L}(L^2(\mathbb{T}))}.$$

Proof Let M_σ be the associated matrix of T_σ. Then, since rows of M_σ are $L^2(\mathbb{Z})$, the adjoint matrix is well defined as a densely defined operator in $L^2(\mathbb{Z})$, and is a pseudo-differential operator with symbol

$$\sigma^*(x, j) = \sum_{k\in\mathbb{Z}} \widehat{\sigma}(j-k, k)e^{ix\cdot k}.$$

Now consider the operator $T_{\sigma^*}T_\sigma$. Since $(M_\sigma^* M_\sigma)^n$ is defined for all n, it is a pseudo-differential operator with symbol $\tau(\cdot, k) \in L^2(\mathbb{T})$ for each k and, again by the existence of $(M_\sigma^* M_\sigma)^n$ for all n, T_τ^n is a pseudo-differential operator with symbol $\tau^n(\cdot, k) \in L^2(\mathbb{T})$ for each k and for all $n \in \mathbb{N}$. We have already cover satisfy hypothesis (i) and (ii) of Lemma 4.2 and consequently M_σ defines a bounded linear operator in $L^2(\mathbb{Z})$ if and only if aditionally we have

$$\sup_{n\in\mathbb{N}}\sup_{k\in\mathbb{Z}} |(M_{\sigma^*}M_\sigma)^n_{kk}|^{1/n} = \sup_{n\in\mathbb{N}}\sup_{k\in\mathbb{Z}} \left|\int_{\mathbb{T}} \tau^n(x,k)dx\right|^{1/n} < \infty.$$

□

Corollary 4.1 *Let T_σ be a pseudo-differential operator with symbol $\sigma \in S^m_{\rho,\delta}(\mathbb{T}\times\mathbb{Z})$. Then T_σ defines a bounded operator on $L^2(\mathbb{T})$ if and only if*

$$\sup_{n\in\mathbb{N}}\sup_{k\in\mathbb{Z}} \left|\int_{\mathbb{T}} \tau^n(x,k)dx\right|^{1/n} < \infty,$$

where τ is the symbol of $(T_\sigma)^*T_\sigma = T_{\sigma^*}T_\sigma$ and τ^n is the symbol of $(T_\tau)^n$. When this happens then

$$\sup_{n\in\mathbb{N}}\sup_{k\in\mathbb{Z}} \left|\int_{\mathbb{T}} \tau^n(x,k)dx\right|^{1/n} = \|T_\sigma\|^2_{\mathcal{L}(L^2(\mathbb{T}))}.$$

Proof The existence of $(M_\sigma^* M_\sigma)^n$ for all $n \in \mathbb{N}$ follows from toroidal composition formula, an the fact that $(M_\sigma^* M_\sigma)^n$ is the matrix associated to the symbol τ^n of $((T_\sigma)^*T_\sigma)^n = (T_{\sigma^*}T_\sigma)^n$. Also, the $L^2(\mathbb{Z})$-norm of the j-th row of M_σ is

$$\begin{aligned}\left(\sum_{k\in\mathbb{Z}} |\widehat{\sigma}(j-k,k)|^2\right)^{1/2} &= \left(\sum_{k\in\mathbb{Z}} |\overline{\widehat{\sigma}^*(k-j,j)}|^2\right)^{1/2}\\ &= \left(\sum_{k\in\mathbb{Z}} |\widehat{\sigma}^*(k-j,j)|^2\right)^{1/2}\\ &= \left\|\sigma^*(\cdot,j)\right\|_{L^2(\mathbb{T})} < \infty.\end{aligned}$$

It follows from Theorem 4.1 that T_σ is bounded if and only if

$$\sup_{n\in\mathbb{N}}\sup_{k\in\mathbb{Z}} \left|\int_{\mathbb{T}} \tau^n(x,k)dx\right|^{1/n} < \infty.$$

□

Theorem 4.2 *Let $\sigma : \mathbb{T} \times \mathbb{Z} \to \mathbb{C}$ be a measurable function such that $\sigma(\cdot, k) \in L^2(\mathbb{T})$ for each $k \in \mathbb{Z}$ and T_σ be its associated pseudo-differential operator. Then T_σ defines a bounded operator on $L^2(\mathbb{T})$ if and only if rows of associated matrix M_σ are in $L^2(\mathbb{Z})$ and*

$$\sup_{n\in\mathbb{N}} \left\| M_{\sigma,n} \right\|_{\mathcal{L}(\ell^2(\mathbb{C}^{2n+1}))} < \infty,$$

where $M_{\sigma,n}$ is the $(2n+1) \times (2n+1)$ matrix with entries

$$(M_{\sigma,n})_{jk} := \overline{\langle \sigma(x, j)e^{ix\cdot j} | \sigma(x, k)e^{ix\cdot k} \rangle}_{L^2(\mathbb{T})}\ , \ |j|, |k| \le 2n+1,$$

when this happens

$$\sup_{n\in\mathbb{N}} \left\| M_{\sigma,n} \right\|_{\mathcal{L}(\ell^2(\mathbb{C}^{2n+1}))} = \| T_\sigma \|^2_{\mathcal{L}(L^2(\mathbb{T}))} \, .$$

Proof We just have to see that

$$\begin{aligned}(M_{\sigma^*} M_\sigma)_{jk} : &= \sum_{h\in\mathbb{Z}} \widehat{\sigma^*}(j-h, h) \cdot \widehat{\sigma}(h-k, k) \\ &= \sum_{h\in\mathbb{Z}} \overline{\widehat{\sigma}(h-j, j)} \cdot \widehat{\sigma}(h-k, k) \\ &= \overline{\langle \sigma(x, j)e^{ix\cdot j} | \sigma(x, k)e^{ix\cdot k} \rangle}_{L^2(\mathbb{T})},\end{aligned}$$

and

$$\| P_n M_{\sigma^*} M_\sigma P_n \|_{\mathcal{L}(L^2(\mathbb{Z}))} = \left\| M_{\sigma,n} \right\|_{\mathcal{L}(\ell^2(\mathbb{C}^{2n+1}))} \, .$$

with this the result follows from Lemma 4.1. □

Remark 4.1 Usually, sufficient conditions for boundedness of pseudo-differential operators are given appealing to the existence of a certain number of derivatives in the toroidal variable of the symbol [5,9,30]. However, Theorems 4.1 and 4.2 provide necessary and sufficient conditions for boundedness without any derivative. This allows one, for example, to recover the classical result about boundedness of a multiplication operators by a fixed function, which is the same as an operator with associated Toeplitz matrix, using Theorem 4.1.

Let us define $T_\phi f := \phi f$. Then T_ϕ is a pseudo-differential operator with symbol $\sigma(x, k) = \phi(x)$, and consequently T_ϕ^*, $T_\phi^* T_\phi$ and $(T_\phi^* T_\phi)^n$ are pseudo-differential operators with symbols $\overline{\phi}$, $|\phi|^2$ and $|\phi|^{2n}$, respectively. This proves that rows of the associated matrix M_σ are in $L^2(\mathbb{Z})$, and that $(M_\sigma^* M_\sigma)^n$ is defined for all $n \in \mathbb{N}$ so, T_ϕ is bounded if and only if

$$\sup_{n\in\mathbb{N}} \sup_{k\in\mathbb{Z}} \left| \int_{\mathbb{T}} |\phi|^{2n}(x)dx \right|^{1/n} < \infty,$$

but this equivalent to

$$|\phi|^2 \in L^p(\mathbb{T}) \text{ for all } 1 \le p < \infty, \text{ and } \sup_{p\in[1,\infty)} ||(|\phi|^2)||_{L^p(\mathbb{T})} < \infty,$$

which is equivalent to $\phi \in L^\infty(\mathbb{T})$. Another interesting example is given by the pseudo-differential operator associated to the symbol

$$\sigma(x,k) = \begin{cases} 0 & \text{if } \frac{1}{2^{|k|}} \le x < 2\pi, \\ 1 & \text{if } 0 \le x < \frac{1}{2^{|k|}}. \end{cases}$$

For this operator one has

$$\begin{aligned} \overline{\langle \sigma(x,j)e^{ix\cdot j} | \sigma(x,k)e^{ix\cdot k}\rangle}_{L^2(\mathbb{T})} &= \int_0^{1/2^{\max(|j|,|k|)}} e^{-ix\cdot(j-k)}dx \\ &= \frac{-1}{i(j-k)}(e^{-i(j-k)/2^{\max(|j|,|k|)}} - 1), \end{aligned}$$

and for $|j|, |k|$ large enough we have the inequality

$$\begin{aligned} |\overline{\langle \sigma(x,j)e^{ix\cdot j} | \sigma(x,k)e^{ix\cdot k}\rangle}_{L^2(\mathbb{T})}| &= \frac{1}{|j-k|}|e^{-i(j-k)/2^{\max(|j|,|k|)}} - 1| \\ &\le \frac{1}{2^{(\max(|j|,|k|)-1)/2}}, \end{aligned}$$

proving that the entries of the matrix $M_\sigma^* M_\sigma$ decay rapidly, and thus Theorem 4.2 ensures that T_σ defines a bounded operator, even when $\sigma(x,k)$ has no derivatives in the toroidal variable.

4.3 Gershgorin Theory

A beautiful and useful result about the spectrum of a matrix is the Gershgorin's circle theorem which we enunciate below.

Theorem 4.3 (Gershgorin) *Let M be a $n \times n$ matrix with entries a_{jk} and define $r_j := \sum_{k\neq j} |a_{jk}|$. Then each eigenvalue λ of M lies in one of the disks $\overline{B_{\mathbb{C}}(a_{jj}, r_j)}$.*

This theorem can be extended to operators that act on a infinite dimensional space, particularly to infinite matrices. There are many papers on the subject and indeed the Gershgorin theorem gives rise to an entire theory, called the Gershgorin theory. We enunciate below a theorem pertaining to this theory which is a subtle variation Theorem 2 in [31]. See also [2].

Lemma 4.3 *Let M be a infinite matrix. Define two new matrices D and F by*

$$D_{jk} := \delta_{jk} M_{jk}, \quad \textit{and} \quad F_{jk} := (1-\delta_{jk})M_{jk},$$

where δ_{jk} is the Kronecker delta. If the following conditions are met

(i). $M_{kk} \neq 0$ *for all* $k \in \mathbb{Z}$ *and* $\inf_{k\in\mathbb{Z}} |M_{kk}| > 0$.
(ii). $I + FD^{-1}$ *defines a bounded operator in* $L^2(\mathbb{Z})$ *with bounded inverse.*

then M is a invertible densely defined linear operator in $L^2(\mathbb{Z})$ with bounded inverse. If in addition

$$\lim_{|k|\to\infty} |M_{kk}| = \infty,$$

then the inverse of M is a compact operator.

Proof We just have to write $M = D + F$. If M satisfy condition (i) then D^{-1} exist and defines a bounded operator in $L^2(\mathbb{Z})$. Clearly

$$M = D + F = (I + FD^{-1})D,$$

so the inverse of M, if it exists, is

$$M^{-1} = D^{-1}(I + FD^{-1})^{-1}.$$

Applying the hypothesis (ii) we conclude that M is invertible with continuous inverse. In addition if

$$\lim_{|k|\to\infty} |M_{kk}| = \infty,$$

then

$$\lim_{|k|\to\infty} |M_{kk}|^{-1} = 0,$$

so D^{-1} is a compact operator, and in consequence M^{-1} is compact as well. □

We will now apply this theorem to pseudo-differential operators to provide a way to locate the espectrum. For this the following proposition is needed.

Proposition 4.1 *Let M be a infinite matrix such that M and M^* both define a bounded operator on $L^\infty(\mathbb{Z})$. Then M defines a bounded operator in $L^2(\mathbb{Z})$ and*

$$\|M\|_{\mathcal{L}(L^2(\mathbb{Z}))} \leq \sqrt{\|M\|_{\mathcal{L}(L^\infty(\mathbb{Z}))}\, \|M^*\|_{\mathcal{L}(L^\infty(\mathbb{Z}))}}.$$

Theorem 4.4 *Let $\sigma : \mathbb{T}\times\mathbb{Z} \to \mathbb{C}$ be a measurable function such that $\sigma(\cdot, k) \in L^2(\mathbb{T})$ for each $k \in \mathbb{Z}$ and T_σ be its associated pseudo-differential operator. If σ satisfy the following three properties*

(i)

$$\int_{\mathbb{T}} \sigma(x,k)dx \neq 0 \;\; \textit{for all} \;\; k \in \mathbb{Z}\,, \;\; \inf_{k\in\mathbb{Z}} \left| \int_{\mathbb{T}} \sigma(x,j)dx \right| > 0,$$

(ii)

$$\sup_{k\in\mathbb{Z}}\left|\int_{\mathbb{T}}\sigma(x,k)dx\right|^{-1}\cdot\sum_{j\neq k}|\widehat{\sigma}(j-k,k)|<1,$$

(iii)

$$\sup_{j\in\mathbb{Z}}\left|\int_{\mathbb{T}}\sigma(x,j)dx\right|^{-1}\cdot\sum_{k\neq j}|\widehat{\sigma}(j-k,k)|<1,$$

or equivalently,

(i')

$$\int_{\mathbb{T}}\sigma(x,k)dx\neq 0 \ \textit{for all}\ k\in\mathbb{Z}\ ,\ \inf_{k\in\mathbb{Z}}\left|\int_{\mathbb{T}}\sigma(x,j)dx\right|>0,$$

(ii')

$$\sup_{k\in\mathbb{Z}}\left|\int_{\mathbb{T}}\sigma(x,k)dx\right|^{-1}\cdot\|\widehat{\sigma}(\cdot,k)\|_{L^1(\mathbb{Z})}<2,$$

(iii')

$$\sup_{j\in\mathbb{Z}}\left|\int_{\mathbb{T}}\sigma(x,j)dx\right|^{-1}\cdot\left\|\widehat{\sigma^*}(\cdot,j)\right\|_{L^1(\mathbb{Z})}<2,$$

where $\sigma^*(x,k)$ is the symbol of adjoint operator, then T_σ is an invertible linear operator with bounded inverse. In particular if

$$\lim_{|k|\to\infty}\left|\int_{\mathbb{T}}\sigma(x,k)dx\right|=\infty$$

the inverse is a compact operator.

Proof Let M_σ be the associated matrix of T_σ. We will show that this infinite matrix satisfy the hypothesis of Lemma 4.3. First its easy to see that (i) in previous statement is equivalent to i in Lemma 4.3. For the remaining hypothesis define D_σ and F_σ as

$$(D_\sigma)_{jk}:=\delta_{jk}(M_\sigma)_{jk},\ \text{ and }\ (F_\sigma)_{jk}:=(1-\delta_{jk})(M_\sigma)_{jk},$$

then by Proposition 4.1

$$\begin{aligned}\left\|F_\sigma D_\sigma^{-1}\right\|^2_{\mathcal{L}(L^2(\mathbb{Z}))}&=\left\|I-(I+F_\sigma D_\sigma^{-1})\right\|^2_{\mathcal{L}(L^2(\mathbb{Z}))}\\&\leq\left\|F_\sigma D_\sigma^{-1}\right\|_{\mathcal{L}(L^\infty(\mathbb{T}))}\cdot\left\|(F_\sigma D_\sigma^{-1})^*\right\|_{\mathcal{L}(L^\infty(\mathbb{T}))},\end{aligned}$$

but clearly

$$\left\| F_\sigma D_\sigma^{-1} \right\|_{\mathcal{L}(L^\infty(\mathbb{T}))} = \sup_{j\in\mathbb{Z}} \left| \int_{\mathbb{T}} \sigma(x,j)dx \right|^{-1} \cdot \sum_{k\neq j} |\widehat{\sigma}(j-k,k)| < 1,$$

and

$$\left\| (F_\sigma D_\sigma^{-1})^* \right\|_{\mathcal{L}(L^\infty(\mathbb{T}))} = \sup_{k\in\mathbb{Z}} \left| \int_{\mathbb{T}} \sigma(x,k)dx \right|^{-1} \cdot \sum_{j\neq k} |\widehat{\sigma}(j-k,k)| < 1,$$

so $\| F_\sigma D_\sigma^{-1} \|_{\mathcal{L}(L^2(\mathbb{Z}))} < 1$ and by Lemma 2.1 in [7] the operator defined by $I + F_\sigma D_\sigma{}^{-1}$ is invertible with bounded inverse in $L^2(\mathbb{Z})$. □

Example 4.1 Let $\alpha : \mathbb{Z} \to \mathbb{C}$ a measurable function that satisfy $\alpha(k) \neq 0$ for all k and $V \in \mathcal{F}_{\mathbb{T}}^{-1}(L^1(\mathbb{Z}))$ such that $\int_{\mathbb{T}} V(x)dx \neq 0$. Then for the symbol $\sigma(x,k) := \alpha(k)V(x)$ one has

$$\widehat{\sigma}(j-k,k) = \alpha(k)\widehat{V}(j-k)$$

and from this

$$\sup_{k\in\mathbb{Z}} \left| \int_{\mathbb{T}} \sigma(x,k)dx \right|^{-1} \cdot \|\widehat{\sigma}(\cdot,k)\|_{L^1(\mathbb{Z})} = \left| \int_{\mathbb{T}} V(x)dx \right|^{-1} \cdot \|\widehat{V}\|_{L^1(\mathbb{Z})},$$

$$\sup_{j\in\mathbb{Z}} \left| \int_{\mathbb{T}} \sigma(x,j)dx \right|^{-1} \cdot \left\|\widehat{\sigma^*}(\cdot,j)\right\|_{L^1(\mathbb{Z})}$$

$$= \sup_{j\in\mathbb{Z}} \left| \alpha(j) \int_{\mathbb{T}} V(x)dx \right|^{-1} \cdot \sum_{k\in\mathbb{Z}} \alpha(k)\widehat{V}(j-k).$$

so if V satisfy the conditions

$$\|\widehat{V}\|_{L^1(\mathbb{Z})} < 2\left| \int_{\mathbb{T}} V(x)dx \right|,$$

and

$$\left| \alpha(j) \int_{\mathbb{T}} V(x)dx \right|^{-1} \cdot \sum_{k\in\mathbb{Z}} \alpha(k)\widehat{V}(j-k) < 1 \quad \text{for all} \quad j \in \mathbb{Z},$$

then T_σ is a invertible linear operator with bounded inverse. If a additionally

$$\lim_{|k|\to\infty} |\alpha(k)| = \infty,$$

then the inverse is a compact operator.

Corollary 4.2 *Let $\sigma : \mathbb{T}\times\mathbb{Z} \to \mathbb{C}$ be a measurable function such that $\sigma(\cdot, k) \in L^2(\mathbb{T})$ for each $k \in \mathbb{Z}$ and T_σ be its associated pseudo-differential operator. If the following conditions hold*

(i)

$$\int_{\mathbb{T}} \sigma(x,k)dx - \lambda \neq 0, \quad \forall k \in \mathbb{Z} \ , \inf_{k\in\mathbb{Z}} \left| \int_{\mathbb{T}} \sigma(x,j)dx - \lambda \right| > 0,$$

(ii)

$$\sup_{k\in\mathbb{Z}} \left| \int_{\mathbb{T}} \sigma(x,k)dx - \lambda \right|^{-1} \cdot \sum_{j\neq k} |\widehat{\sigma}(j-k,k)| < 1,$$

(iii)

$$\sup_{j\in\mathbb{Z}} \left| \int_{\mathbb{T}} \sigma(x,j)dx - \lambda \right|^{-1} \cdot \sum_{k\neq j} |\widehat{\sigma}(j-k,k)| < 1,$$

then $\lambda \in Res(T_\sigma)$.

Proof We just have to see that for all $\lambda \in \mathbb{C}$ $T_\sigma - \lambda I$ is a pseudo-differential operator with symbol $\sigma(x,k) - \lambda$. □

Example 4.2 Let $\alpha : \mathbb{Z} \to \mathbb{C}$ a measurable function and $V \in \mathcal{F}_{\mathbb{T}}^{-1}(L^1(\mathbb{Z}))$. Then for the symbol $\sigma(x,k) := \alpha(k) + V(x)$ one has

$$\widehat{\sigma}(j-k,k) = \begin{cases} \widehat{V}(j-k) & j \neq k \\ \alpha(k) + \int_{\mathbb{T}} V(x)dx & j = k \end{cases}$$

and then if

$$\sum_{k\neq j} |\widehat{V}(j-k)| < \inf_{k\in\mathbb{Z}} \left| \alpha(k) + \int_{\mathbb{T}} V(x)dx - \lambda \right|$$

$\lambda \in Res(T_\sigma)$ which proves that

$$Spec(T_\sigma) \subseteq \bigcup_{k\in\mathbb{Z}} \overline{B_{\mathbb{C}}(c_k, r)}$$

where

$$c_k = \alpha(k) + \int_{\mathbb{T}} V(x)dx \quad \text{and} \quad r = \sum_{j\neq k} |\widehat{V}(j-k)|$$

To finish this section we enunciate an improved version of Lemma 4.3, which can be found as the Theorem 3.1 in [15]. We use it to prove an improved version of Theorem 4.4.

Lemma 4.4 (Farid and Lancaster) *Let M be an infinite matrix, considered as a linear operator on $L^p(\mathbb{Z})$ for $1 \leq p < \infty$, with columns in $L^1(\mathbb{Z})$. Define $r_k := \sum_{j\in\mathbb{Z}, j\neq k} |M_{jk}|$ and assume that*

(i) *$M_{kk} \neq 0$, for all $k \in \mathbb{Z}$ and $|M_{kk}| \to \infty$ as $|k| \to \infty$.*
(ii) *There exits $s \in [0, 1)$ such that for all $k \in \mathbb{Z}$*

$$r_k = s_k|M_{kk}|, \quad s_k \in [0, s].$$

(iii) *Either FD^{-1} and $(I + \mu FD^{-1})^{-1}$ exists and are in $\mathcal{L}(L^p(\mathbb{Z}))$ for every $\mu \in (0, 1]$, or $D^{-1}F$ and $(I + \mu D^{-1}F)^{-1}$ exists and are in $\mathcal{L}(L^p(\mathbb{T}))$.*

Then M is a closed operator and the spectrum $Spec(M)$ is nonempty and consists of discrete nonzero eigenvalues, lying in the set

$$\bigcup_{k\in\mathbb{Z}} \overline{B_{\mathbb{C}}(M_{kk}, r_k)},$$

where the closed balls $\overline{B_{\mathbb{C}}(M_{kk}, r_k)}$ are called the Gershgorin discs. Furthermore, any set consisting of n Gershgorin discs whose union is disjoint from all other Gersgorin discs intersects $Spec(M)$ in a finite set of eigenvalues of M with total algebraic multiplicity n.

As an immediate consequence we have:

Theorem 4.5 *Let $\sigma : \mathbb{T}\times\mathbb{Z} \to \mathbb{C}$ be a measurable function such that $\sigma(\cdot, k) \in L^2(\mathbb{T})$ for each $k \in \mathbb{Z}$ and T_σ be its associated pseudo-differential operator. Let M_σ be the associated matrix. Assume that*

(i) *$\int_{\mathbb{T}} \sigma(x, k)dx \neq 0$, for all $k \in \mathbb{Z}$ and $\left|\int_{\mathbb{T}} \sigma(x, k)dx\right| \to \infty$ as $|k| \to \infty$.*
(ii) *Rows and columns of M_σ are in $L^1(\mathbb{Z})$.*
(iii)

$$\sup_{k\in\mathbb{Z}} \left|\int_{\mathbb{T}} \sigma(x, k)dx\right|^{-1} \cdot \sum_{j\neq k} |\widehat{\sigma}(j-k, k)| < 1,$$

and

$$\sup_{j\in\mathbb{Z}} \left|\int_{\mathbb{T}} \sigma(x, j)dx\right|^{-1} \cdot \sum_{k\neq j} |\widehat{\sigma}(j-k, k)| < 1.$$

Then T_σ is a closed operator and the spectrum $Spec(T_\sigma)$ is nonempty and consists of discrete nonzero eigenvalues, lying in the set

$$\bigcup_{k\in\mathbb{Z}} \overline{B_{\mathbb{C}}(a_{kk}, r_k)} \text{ where } a_{kk} = \int_{\mathbb{T}} \sigma(x, k)dx, \; r_k = \sum_{j\neq k} |\widehat{\sigma}(j-k, k)|.$$

Furthermore, any set of n Gershgorin discs whose union is disjoint from all other Gersgorin discs intersects $Spec(T_\sigma)$ in a finite set of eigenvalues of T_σ with total algebraic multiplicity n.

Proof Condition (i) in Theorem 4.5. is the condition (i) of Lemma 4.3. Proposition 4.1. and conditions (ii),(iii) of Theorem 4.5 implies conditions (ii) and (iii) of Lemma 4.3. □

5 Gohberg's Lemma in $L^p(\mathbb{T})$

In this section we discuss a proof of Gohberg's Lemma in $L^p(\mathbb{T})$.

Proof of Gohberg's lemma: If u is a nonzero function in $C^\infty(\mathbb{T})$ then

$$\begin{aligned} T_\sigma u(x) &= \sum_{k\in\mathbb{Z}} \sigma(x,k)\widehat{u}(k)e^{ix\cdot k} \\ &= \sum_{k\in\mathbb{Z}} \int_{\mathbb{T}} \sigma(x,k)u(y)e^{i(x-y)k}dy \\ &= \int_{\mathbb{T}} (\mathcal{F}_{\mathbb{T}}^{-1}\sigma)(x,x-y)u(y)dy \\ &= \int_{\mathbb{T}} (\mathcal{F}_{\mathbb{T}}^{-1}\sigma)(x,y)u(x-y)dy, \end{aligned}$$

where $\mathcal{F}_{\mathbb{T}}^{-1}\sigma$ is the inverse Fourier transform of σ with respect to the second variable in a distributional sense. Hence in a distributional sense

$$T_\sigma u(x) = \int_{\mathbb{T}} (\mathcal{F}_{\mathbb{T}}^{-1}\sigma)(x,y)u(x-y)dy.$$

Since σ is continuous it follows that for all $k\in\mathbb{Z}$ exists a $x_k\in[-\pi,\pi]$ such that

$$|\sigma(x_k,k)| = \sup_{x\in\mathbb{T}} |\sigma(x,k)|.$$

By definition of d_σ, there exists a subsequence $\{(x_{k_l},k_l)\}_{l\in\mathbb{N}}$ such that

$$|k_l|\to\infty \text{ and } |\sigma(x_{k_l},k_l)|\to d_\sigma,$$

as $l\to\infty$. Then, since $\mathbb{T}$ is compact, the sequence $\{x_{k_l}\}_{l\in\mathbb{N}}$ must have an cluster point x_0 in $\mathbb{T}$ and a subsequence $\{x_{k_{l_m}}\}_{m\in\mathbb{N}}$ such that $x_{k_{l_m}}\to x_0$ as $m\to\infty$. For simplicity we will rename the sequence $\{(x_{k_{l_m}},k_{l_m})\}_{m\in\mathbb{N}}$ as the original one $\{(x_{k_l},k_l)\}_{l\in\mathbb{N}}$. Define the functions u_{k_l} on $\mathbb{T}$ by

$$u_{k_l}(x) = u(x-x_{k_l})e^{ix\cdot k_l},$$

then wee see that $\|u_{k_l}\|_{L^p(\mathbb{T})} = \|u\|_{L^p(\mathbb{T})}$ for $l \in \mathbb{N}$. Let's see that the sequence $\{u_{k_l}\}_{l\in\mathbb{N}}$ converges weakly to zero. First, for any $v \in L^q(\mathbb{T})$

$$\int_{\mathbb{T}} u_{k_l}(x)v(x)dx = \int_{\mathbb{T}} (u(x-x_{k_l}) - u(x-x_0))v(x)e^{ix\cdot k_l}dx + \int_{\mathbb{T}} u(x-x_0)v(x)e^{ix\cdot k_l}dx.$$

Since by Hölder inequality $u(x-x_0)v(x) \in L^1(\mathbb{T})$ then we can apply the Riemann-Lebesgue lemma and consequently, for any $\varepsilon > 0$, there is a natural number N_1 such that

$$l \geq N_1 \implies \left|\int_{\mathbb{T}} u(x-x_0)v(x)e^{ix\cdot k_l}dx\right| < \frac{\varepsilon}{2},$$

and second, since u is continuous, for any $\varepsilon > 0$ there exist $N(\varepsilon) \in \mathbb{N}$, $\delta(\varepsilon) > 0$ such that

$$\begin{aligned} l \geq N(\varepsilon) &\implies |x_{k_l} - x_0| < \delta(\varepsilon) \\ &\implies |u(x-x_{k_l}) - u(x-x_0)| < \frac{\varepsilon}{2}\cdot\|v\|^{-1}_{L^q(\mathbb{T})}, \end{aligned}$$

thus for any $\varepsilon > 0$

$$l \geq \max(N_1, N(\varepsilon)) \implies \left|\int_{\mathbb{T}} u_{k_l}(x)v(x)dx\right| < \varepsilon,$$

as an immediate consequence

$$\|Ku_{k_l}\|_{L^p(\mathbb{T})} \to 0 \text{ as } l \to \infty,$$

for every $K \in \mathfrak{K}(L^p(\mathbb{T}))$. Then for an arbitrary positive number ε and sufficiently large l

$$\|Ku_{k_l}\|_{L^p(\mathbb{T})} \leq \varepsilon \|u_{k_l}\|_{L^p(\mathbb{T})}. \tag{G1}$$

On the other hand, we have the following lemma, whose proof can be found in [24].

Lemma 5.1 $\|\sigma(\cdot,k_l) - T_\sigma u_{k_l}\|_{L^p(\mathbb{T})} \to 0$ *as* $l \to \infty$.

Then for sufficiently large l,

$$\|\sigma(\cdot,k_l)\|_{L^p(\mathbb{T})} - \|T_\sigma u_{k_l}\|_{L^p(\mathbb{T})} \leq \varepsilon \|u\|_{L^p(\mathbb{T})}. \tag{G2}$$

By using again the continuity of $\sigma(\cdot, k_l)$ there exists a positive number δ such that for all x in $[-\pi + x_{k_l}, \pi + x_{k_l}]$ with $|x - x_{k_l}| < \delta$, we have

$$|\sigma(x, k_l) - \sigma(x_{k_l}, k_l)| < \varepsilon. \tag{G3}$$

Let $u \in C^\infty(\mathbb{T})$ such that $u(x) = 0$ for $|x| \geq \delta$. Then $u_{k_l}(x) = 0$ for all x in $[-\pi + x_{k_l}, \pi + x_{k_l}]$ with $|x - x_{k_l}| \geq \delta$. So

$$\begin{aligned}
\left\|\sigma(x_{k_l}, k_l)u_{k_l}\right\|_{L^p(\mathbb{T})} - \left\|\sigma(\cdot, k_l)u_{k_l}\right\|_{L^p(\mathbb{T})} &\leq \left\|\sigma(x_{k_l})u_{k_l} - \sigma(\cdot, k_l)u_{k_l}\right\|_{L^p(\mathbb{T})} \\
&= \left(\int_{-\pi+x_{k_l}}^{\pi+x_{k_l}} |\sigma(x_{k_l}, k_l) - \sigma(x, k_l)|^p |u_{k_l}(x)|^p dx\right)^{1/p} \\
&= \left(\int_{|x-x_{k_l}|<\delta} |\sigma(x_{k_l}, k_l) - \sigma(x, k_l)|^p |u_{k_l}(x)|^p dx\right)^{1/p},
\end{aligned}$$

then by (G3)

$$|\sigma(x_{k_l}, k_l)| \,\|u\|_{L^p(\mathbb{T})} - \left\|\sigma(\cdot, k_l)u_{k_l}\right\|_{L^p(\mathbb{T})} < \varepsilon \,\|u\|_{L^p(\mathbb{T})}. \tag{G4}$$

Combining (G1), (G2) and (G4), we get for sufficiently large k,

$$\begin{aligned}
\|u\|_{L^p(\mathbb{T})} \|T_\sigma - K\|_{\mathcal{L}(L^p(\mathbb{T}))} &\geq \|(T_\sigma - K)u\|_{L^p(\mathbb{T})} \\
&\geq \|T_\sigma u\|_{L^p(\mathbb{T})} - \|Ku\|_{L^p(\mathbb{T})} \\
&\geq \|T_\sigma u\|_{L^p(\mathbb{T})} - \varepsilon \,\|u\|_{L^p(\mathbb{T})} \\
&\geq \left\|\sigma(\cdot, k_l)u_{k_l}\right\|_{L^p(\mathbb{T})} - 2\varepsilon \,\|u\|_{L^p(\mathbb{T})} \\
&\geq |\sigma(x_{k_l}, k_l)| \,\|u\|_{L^p(\mathbb{T})} - 3\varepsilon \,\|u\|_{L^p(\mathbb{T})} \\
&= (|\sigma(x_{k_l}, k_l)| - 3\varepsilon) \,\|u\|_{L^p(\mathbb{T})}.
\end{aligned}$$

Letting $l \to \infty$, we get $\|T_\sigma - K\|_{\mathcal{L}(L^p(\mathbb{T}))} \geq d_\sigma - 3\varepsilon$, and finally, using the fact that ε is an arbitrary positive number, we have

$$\|T_\sigma - K\|_{\mathcal{L}(L^p(\mathbb{T}))} \geq d_\sigma.$$

Acknowledgements I sincerely thank the guidance of Carlos Andres Rodriguez Torijano who proposed me this research as the first step in my career as a mathematical researcher. I also want to thank professor Michael Ruzhansky and the anonymous referees for their comments.

References

1. Agranovich, M.S.: Spectral properties of elliptic pseudodifferential operators on a closed curve. Funct. Anal. Appl. **13**(4), 279–281 (1979)
2. Aleksić, J., Kostić, V., Žigić, M.: Spectrum localizations for matrix operators on lp spaces. Appl. Math. Comput. **249**, 541–553 (2014)

3. Caradus, S.R., Pfaffenberger, W.E.: Calkin Algebras and Algebras of Operators on Banach Spaces, 1st edn. Lecture Notes in Pure and Applied Mathematics. Marcel Dekker, New York (1974)
4. Cardona, D.: Hölder-Besov boundedness for periodic pseudo-differential operators. J. Pseudo-Differ. Oper. Appl. **8**(1), 13–34 (2017)
5. Cardona, D.: On the boundedness of periodic pseudo-differential operators. arXiv:1701.08184 [math.AP] (2017)
6. Carey, A., Ellwood, D., Paycha, S., Rosenberg S.: Quantum field theory, and pseudodifferential operators. In: Clay Mathematics Proceedings, vol. 12, pp. 37 – 72. AMS, New York (2010)
7. Conway, J.B.: A Course in Functional Analysis, 2 edn. (Corrected Fourth Printing) ed. Graduate Texts in Mathematics 096. Springer, New York (1997)
8. Crone, L.: A characterization of matrix operators on l2. Math. Z. **123**, 315–317 (1971)
9. Delgado, J.: Lp-bounds for pseudo-differential operators on the torus. In: Molahajloo, S., et al. (eds.) Pseudo-Differential Operators, Generalized Functions and Asymptotics, pp. 103–116. Birkhauser, Basel (2013)
10. Delgado, J., Ruzhansky, M.: Kernel and symbol criteria for schatten classes and r-nuclearity on compact manifolds. Comptes Rendus Math. **352**(10), 779–784 (2014)
11. Delgado, J., Ruzhansky, M.: Schatten classes and traces on compact Lie groups. Math. Res. Lett. arXiv:1303.3914 [math.FA] (2016)
12. Delgado, J., Ruzhansky, M.: L^p-bounds for pseudo-differential operators on compact lie groups. J. Inst. Math. Jussieu (2017). https://doi.org/10.1017/S1474748017000123
13. Eidelman, Y., Milman, V., Tsolomitis, A.: Functional Analysis: An introduction. Graduate Studies in Mathematics. American Mathematical Society, New York (2004)
14. Esposito, G., Napolitano, G.: Geometry and physics of pseudodifferential operators on manifolds. Il Nuovo Cimento C **38**, 158–172 (2015)
15. Farid, F., Lancaster, P.: Spectral properties of diagonally dominant infinite matrices. II. Linear Algebr. Appl. **143**, 7–17 (1991)
16. Ghaemi, M.B., Birgani, M., Morsalfard, E.N.: A study on pseudo-differential operators on $\mathbb{S}^1$ and $\mathbb{Z}$. J. Pseudo-Differ. Oper. Appl. **7**, 237–247 (2016)
17. Ghaemi, M.B., Birgani, M.J., Wong, M.W.: Characterizations of nuclear pseudo-differential operators on $\mathbb{S}^1$ with applications to adjoints and products. J. Pseudo-Differ. Oper. Appl. **8**(2), 191–201 (2017)
18. Ghaemi, M. B., Nabizadeh Morsalfard, E., Birgani, M.: Some criteria for boundedness and compactness of pseudo-differential operators on $l^p(\mathbb{S}^1)$, $1<p<\infty$
19. Hernández, F.L., Semenov, E.M., Tradacete, P.: Strictly singular operators on L^p spaces and interpolation. Proc. Am. Math. Soc. **138**(2), 675–686 (2010)
20. Lindenstrauss, J., Johnson, W.: Handbook of the geometry of Banach spaces, 1st, vol. 1, ed edn. Elsevier, Amsterdam (2001)
21. McLean, W.: Local and global descriptions of periodic pseudodifferential operators. Math. Nachr. **150**(1), 151–161 (1991)
22. Molahajloo, S.: A characterization of compact pseudo-differential operators on $\mathbb{S}^1$. Pseudo-Differ. Oper. Anal. Appl. Comput. **213**, 25–29 (2011)
23. Molahajloo, S., Wong, M.W.: Pseudo-differential operators on $\mathbb{S}^1$. New Developments in Pseudo-Differential Operators: ISAAC Group in Pseudo-Differential Operators (IGPDO), pp. 297–306
24. Molahajloo, S., Wong, M.W.: Ellipticity, fredholmness and spectral invariance of pseudo-differential operators on $\mathbb{S}^1$. J. Pseudo-Differ. Oper. Appl. **1**(2), 183–205 (2010)
25. Pietsch, A.: Operator Ideals. North-Holland Mathematical Library, vol. 20. Elsevier, Amterdam (1980)
26. Pietsch, A.: Eigenvalues and S-Numbers. Cambridge Studies in Advanced Mathematics, vol. 13. Cambridge University Press, Cambridge (1987)
27. Pirhayati, M.: Spectral Theory of Pseudo-Differential Operators on $\mathbb{S}^1$, pp. 15–23. Springer, Basel (2011)
28. Ruston, A.F.: Operators with a fredholm theory. J. Lond. Math. Soc. **3**, 318–326 (1964)
29. Ruzhansky, M., Turunen, V.: Quantization of pseudo-differential operators on the torus. arXiv:0805.2892 [math.FA] (2008)
30. Ruzhansky, M., Turunen, V.: Pseudo-Differential Operators and Symmetries. Background Analysis and Advanced Topics, 1st edn. Birkhäuser, Basel (2009)
31. Shivakumar, P., Williams, J., Rudraiah, N.: Eigenvalues for infinite matrices. Linear Algebr. Appl. **96**, 35–63 (1987)

32. West, T.T.: Riesz operators in banach spaces. Proc. Lond. Math. Soc. **16**(1), 131–140 (1996)
33. Wong, M.W.: Discrete Fourier Analysis, 1 edn. Pseudo-Differential Operators 5. Birkhäuser Basel (2011)

Publisher's Note Springer Nature remains neutral with regard to jurisdictional claims in published maps and institutional affiliations.

Journal of Fourier Analysis and Applications (2019) 25:2733–2781
https://doi.org/10.1007/s00041-019-09681-1

Interpolation Theory for Sobolev Functions with Partially Vanishing Trace on Irregular Open Sets

Sebastian Bechtel[1] · Moritz Egert[2]

Received: 3 September 2018 / Revised: 26 March 2019 / Published online: 12 April 2019
© Springer Science+Business Media, LLC, part of Springer Nature 2019

Abstract
A full interpolation theory for Sobolev functions with smoothness between 0 and 1 and vanishing trace on a part of the boundary of an open set is established. Geometric assumptions are mostly of measure theoretic nature and reach beyond Lipschitz regular domains. Previous results were limited to regular geometric configurations or Hilbertian Sobolev spaces. Sets with porous boundary and their characteristic multipliers on smoothness spaces play a major role in the arguments.

Keywords Interpolation of Banach spaces · (fractional) Sobolev spaces · Traces and extensions of Sobolev functions · Porous sets · Measure density conditions · Hardy's inequality

Mathematics Subject Classification Primary: 46B70 · Secondary: 46E35

1 Introduction and Main Results

Recent years have witnessed an ever-growing interest in the treatment of quasilinear equations of parabolic type through maximal regularity techniques [10]. To a large extend this stems from the flexibility of the approach and its applicability to rough geometric configurations that arise in applications, for example reaction-diffusion models

Communicated by Mieczysław Mastyło.

✉ Sebastian Bechtel
bechtel@mathematik.tu-darmstadt.de

Moritz Egert
moritz.egert@math.u-psud.fr

1 Fachbereich Mathematik, Technische Universität Darmstadt, Schlossgartenstr. 7, 64289 Darmstadt, Germany

2 Laboratoire de Mathématiques d'Orsay, Univ. Paris-Sud, CNRS, Université Paris-Saclay, 91405 Orsay, France

related to differential operators or systems in divergence form with mixed boundary conditions [7,11–13,22,23,31,32,39]. There, the linear second order operator usually admits a proper theory of weak solutions on Sobolev spaces of type $\mathrm{W}^{1,p}$ carrying the boundary conditions, but in controlling non-linear terms of high order and reaction processes on lower-dimensional substructures of the boundary simultaneously, interpolation spaces of order $s \in (0,1)$ are most appropriate. We refer to [12, Sect. 4.1] for a detailed account on this paradigm in the context of dynamics in a semiconductor device, see also [22, Sect. 6]. This leads to the problem of identifying such interpolation spaces in the presence of rough boundaries to the fractional Sobolev- and Bessel potential spaces, in analogy with what is long known for function spaces on $\mathbb{R}^d$ or on smooth domains with pure Dirichet boundary condition [6,35,40].

At the heart of the matter lies the following question. Given an open set $O \subseteq \mathbb{R}^d$ and a piece $D \subseteq \partial O$ of its boundary, define the Sobolev space $\mathrm{W}^{1,p}_D(O)$ as the $\mathrm{W}^{1,p}(O)$-closure of smooth functions whose support stays away from D. Under which geometric assumptions can one determine explicitly the interpolation spaces

$$[\mathrm{L}^p(O), \mathrm{W}^{1,p}_D(O)]_s \quad \text{and} \quad (\mathrm{L}^p(O), \mathrm{W}^{1,p}_D(O))_{s,p}$$

defined through either Calderón–Lions' complex method or Peetre's real method? The space $\mathrm{W}^{1,p}_D(O)$ should be thought of the collection of $\mathrm{W}^{1,p}(O)$-functions with homogeneous Dirichlet boundary condition on D.

Interpolation theory related to the spaces $\mathrm{W}^{1,p}_D(O)$ has recently been studied in [3,4,8,16,18,23], but mostly with a focus on interpolating with respect to integrability. Interpolation in differentiability appears in [4,16] for $p = 2$ and in [18] for general p on certain model sets. The main difficulty lies in that taking the boundary trace on D can, if at all, be defined in a meaningful way only on the Sobolev space [6, Sect. 6.6]. This forbids to treat the question via purely functorial techniques.

We close this gap by establishing a full interpolation theory under geometric assumptions in the spirit of what has become standard for treating mixed boundary value problems [3,8,23,39]. In particular, we confirm the formula for the complex interpolation spaces that was conjectured in connection with fractional powers of divergence form operators in [3, Rem. 10.5] and listed as an open problem in [12, Sect. 5.3]. We also treat interpolation simultaneously in differentiability and integrability. Some of our results appear to be new even on much more regular domains since we do not require that the interface of D with the complementary boundary part $\partial O \setminus D$ can be parametrized by coordinate charts in any sense.

1.1 Geometric Setting

We shall work on open sets $O \subseteq \mathbb{R}^d$, not necessarily connected or bounded, satisfying the thickness condition

$$c \leq \frac{|B \cap O|}{|B|} \leq C \tag{1}$$

Fig. 1 The domain $O \subseteq \mathbb{R}^3$ is obtained by transforming a cylinder such that one lateral boundary part degenerates to a line segment touching the opposed side from outside. The dark-shaded boundary parts carry the Dirichlet condition

for some constants $0 < c \leq C < 1$ and all balls B of radius $\mathrm{r}(B) \leq 1$ centered at the boundary ∂O. This excludes that O have interior or exterior cusps. We assume that the Dirichlet part $D \subseteq \partial O$ is a $(d-1)$-regular, not necessarily closed set in the sense of Jonsson–Wallin [27]. Only around the complementary boundary part $\overline{\partial O \setminus D}$ we demand Lipschitz coordinate charts with uniformly controlled bi-Lipschitz constants, which on domains with compact boundary reduces to the usual weak Lipschitz condition. Finally, the interface ∂D between the two boundary parts should be a *porous* subset of the full boundary. This means that there should exist some $\kappa \in (0, 1)$ with the property that every ball B of radius $\mathrm{r}(B) \leq 1$ centered in ∂D contains a ball of radius κr centered in ∂O that avoids ∂D. A related condition appeared in [39].

Porosity plays a fundamental role in our considerations and for the reader's convenience we have collect some folklore on this concept in Appendix A. We often take advantage of it in form of equivalent but less transparent conditions related to Aikawa- and Assouad dimension. In particular, all our results hold if ∂D is $(d-2)$-regular as in Fig. 1. We believe that this setting is rather common in applications. It includes the Gröger regular sets [20] used for instance in [7,22,31,32]. Our interpolation results are new even in this context since compared to earlier work [18] we remove the requirement that the Lipschitz coordinate charts should be measure preserving.

1.2 Main Results

Fractional Sobolev spaces $\mathrm{W}^{s,p}$ and Bessel potential spaces $\mathrm{H}^{s,p}$ on O, with and without Dirichlet boundary conditions on D indicated by a subscript, are properly defined in Sects. 2.4–2.5 through restriction from $\mathbb{R}^d$. The Dirichlet condition is understood for $s > 1/p$ in virtue of Jonsson–Wallin's trace theory [27]. For integer s we have $\mathrm{W}^{s,p} = \mathrm{H}^{s,p}$ up to equivalent norms. In particular, we retrieve from $\mathrm{L}^p = \mathrm{H}^{0,p}$ and $\mathrm{W}^{1,p} = \mathrm{H}^{1,p}$ answers to the interpolation questions raised above.

In most interpolation results we shall have possibly different Lebesgue exponents $p_0, p_1 \in (1, \infty)$ and smoothness parameters $s_0, s_1 \in \mathbb{R}$ and we interpolate in both scales simultaneously. In order to straighten the presentation, we introduce here, given $\theta \in (0, 1)$, the interpolating parameters $p \in (1, \infty)$ and $s \in \mathbb{R}$ through

$$\frac{1}{p}:=\frac{1-\theta}{p_0}+\frac{\theta}{p_1}, \qquad s:=(1-\theta)s_0+\theta s_1. \tag{2}$$

In the presence of p_i, s_i and θ as above we shall *exclusively* use the symbols p and s in that very sense, sometimes without further mentioning.

Let us state the central result of our paper. The proof is given in Sect. 4, including an informal outline in Sect. 4.1.

Theorem 1.1 *Assume the geometric setting of Sect. 1.1. Let $p_0, p_1 \in (1,\infty)$, $s_0 \in [0, 1/p_0)$, $s_1 \in (1/p_1, 1]$, and for $\theta \in (0,1)$ define p and s as in* (2). *If* X *denotes either* H *or* W, *then the complex interpolation identity*

(a)

$$[\mathrm{X}^{s_0,p_0}(O), \mathrm{X}_D^{s_1,p_1}(O)]_\theta = \begin{cases} \mathrm{X}_D^{s,p}(O) & (\textit{if } s > 1/p) \\ \mathrm{X}^{s,p}(O) & (\textit{if } s < 1/p) \end{cases}$$

holds up to equivalent norms as well as the real interpolation identity
(b)

$$(\mathrm{X}^{s_0,p_0}(O), \mathrm{X}_D^{s_1,p_1}(O))_{\theta,p} = \begin{cases} \mathrm{W}_D^{s,p}(O) & (\textit{if } s > 1/p) \\ \mathrm{W}^{s,p}(O) & (\textit{if } s < 1/p) \end{cases}$$

with the exception that $s_0 \neq 0$ and $s_1 \neq 1$ are required in (a) for $\mathrm{X} = \mathrm{W}$.

Interpolation theory for the spaces $\mathrm{X}^{s,p}(O)$ without boundary conditions becomes apparent from an extension result of Rychkov [34] that we shall review in Sect. 2.5 and make extensive use of. We establish further structural properties of our function spaces in Sect. 3. Abstract techniques then lead us in Sect. 3.2 to the following interpolation results for two function spaces with Dirichlet condition. This only requires O and D to be regular in the sense of Jonsson–Wallin [27].

Theorem 1.2 *Let $O \subseteq \mathbb{R}^d$ be an open, d-regular set, and let $D \subseteq \overline{O}$ be $(d-1)$-regular. Let $p_0, p_1 \in (1,\infty)$, $s_0 \in (1/p_0, 1+1/p_0)$, $s_1 \in (1/p_1, 1+1/p_1)$, and for $\theta \in (0,1)$ define p and s as in* (2). *Let* X *denote either* H *or* W. *Up to equivalent norms it follows that*

(c)

$$[\mathrm{X}_D^{s_0,p_0}(O), \mathrm{X}_D^{s_1,p_1}(O)]_\theta = \mathrm{X}_D^{s,p}(O),$$

(d)

$$(\mathrm{X}_D^{s_0,p_0}(O), \mathrm{X}_D^{s_1,p_1}(O))_{\theta,p} = \mathrm{W}_D^{s,p}(O),$$

with the two exceptions that in (c) for $\mathrm{X} = \mathrm{W}$ *either all or none of s_0, s_1, s have to be 1 and that in (d) the value $s = 1$ is only permitted when $s_0 = s_1 = 1$.*

As a cautionary tale, let us remark that *a priori* all function spaces are defined by restrictions. In particular, $\mathrm{W}^{1,p}(O) = \mathrm{H}^{1,p}(O)$ might be smaller than the collection of $\mathrm{L}^p(O)$-functions whose first-order distributional derivatives are in $\mathrm{L}^p(O)$ under the assumptions of Theorem 1.2. Under the full set of geometric assumptions in Theorem 1.1, however, there is no such ambiguity. We have collected further background on this issue in Appendix B.

1.3 Extensions and Generalizations

Abstract reiteration and duality theorems [6,25,40] imply numerous further interpolation results that invoke our Theorems 1.1 and 1.2 "off-the-shelf". We leave the care of writing them down to the interested readers. Here, we only present one such result that turned out useful in the $\mathrm{W}^{-1,p}$-theory of divergence form operators [11,31,32] and previously was available only in the restrictive setup of [18, Lemma 3.4]. The proof of this result will be given in Sect. 5. We write $\mathrm{W}_D^{-1,p}(O)$ for the space of conjugate linear functionals on $\mathrm{W}_D^{1,p'}(O)$, where $1/p + 1/p' = 1$.

Theorem 1.3 *Assume the geometric setting of Sect. 1.1 and let $p \in (1, \infty)$. Up to equivalent norms it follows that*

(e)

$$[\mathrm{W}_D^{-1,p}(O), \mathrm{W}_D^{1,p}(O)]_{1/2} = \mathrm{L}^p(O).$$

In Sect. 6 we present a method tailored for real interpolation of fractional Sobolev spaces with the same integrability that is in some sense dual to the proof of Theorem 1.1. It bears the advantage of a more general geometric setting.

Theorem 1.4 *Let $O \subseteq \mathbb{R}^d$ be an open, d-regular set with $(d-1)$-regular boundary, and let $D \subseteq \overline{O}$ be uniformly $(d-1)$-regular. Let $p \in (1, \infty)$, $s_0 \in [0, 1/p)$, $s_1 \in (1/p, 1]$, and $\theta \in (0, 1)$. Up to equivalent norms it follows that*

(f)

$$(\mathrm{W}^{s_0,p}(O), \mathrm{W}_D^{s_1,p}(O))_{\theta,p} = \begin{cases} \mathrm{W}_D^{s,p}(O) & (\text{if } s > 1/p) \\ \mathrm{W}^{s,p}(O) & (\text{if } s < 1/p) \end{cases},$$

where $s := (1-\theta)s_0 + \theta s_1$.

Our proof simplifies [16, Sect. 7], where the case $p = 2$ was treated on bounded domains with a Lipschitz assumption around $\overline{\partial O \setminus D}$. Uniform $(d-1)$-regularity is defined in the next section. For bounded sets there is no difference with $(d-1)$-regularity.

2 Notation and Background

2.1 Geometry

We work in Euclidean space $\mathbb{R}^d$ of dimension $d \geq 2$. We write $B = \mathrm{B}(x, r)$ for the open ball of radius $\mathrm{r}(B) = r$ centered at $x \in \mathbb{R}^d$ and cB for the concentric ball of radius $c\mathrm{r}(B)$. By $\mathcal{H}^\ell$ we denote the ℓ-dimensional *Hausdorff measure* in $\mathbb{R}^d$ [44, Ch. 7] and by $|\cdot|$ the Lebesgue measure in $\mathbb{R}^d$. We write $\mathrm{diam}(\cdot)$ and $\mathrm{d}(\cdot, \cdot)$ for the *diameter* and the *(semi) distance* of sets induced by the Euclidean distance on $\mathbb{R}^d$.

Definition 2.1 An open set $O \subseteq \mathbb{R}^d$ satisfies a *uniform Lipschitz condition* around a closed boundary part $F \subseteq \partial O$ if the following holds. For every $x \in F$ there is an open neighborhood $U_x \ni x$ and a bi-Lipschitz transformation $\Phi_x : U_x \to (-1, 1)^d$ such that $\Phi_x(x) = 0$ and

$$\Phi_x(U_x \cap O) = (0, 1) \times (-1, 1)^{d-1}, \qquad \Phi_x(U_x \cap \partial O) = \{0\} \times (-1, 1)^{d-1}, \quad (3)$$

and there exists a number L that bounds the bi-Lipschitz constants of all Φ_x. *Bi-Lipschitz constant* refers to the maximum of the Lipschitz constants of Φ_x and Φ_x^{-1}.

Remark 2.2 If F is compact, then the uniform Lipschitz condition is equivalent to requiring existence of bi-Lipschitz maps Φ_x with (3) for every $x \in F$. This so-called *weak Lipschitz condition* is weaker than requiring that O has a Lipschitz boundary near x, see [22, Sect. 7.3] for a relevant example.

To see the equivalence, we cover F by finitely many corresponding neighborhoods. Let L_0 be the maximal bi-Lipschitz constant and fix a Lebesgue number $\delta > 0$, that is, for $x \in F$ we have $B(x, \delta) \subseteq U$ for one pre-selected neighborhood U with bi-Lipschitz transformation Φ. Suitable dilations yield a bi-Lipschitz map T_x of the unit cube onto itself that preserves the upper and lower half and maps $\Phi(x)$ to 0. Due to $B(\Phi(x), \delta/L_0) \subseteq (-1, 1)^d$ its bi-Lipschitz constant is controlled by δ and L. So, we can take $U_x := U$ and $\Phi_x := T_x \circ \Phi$.

Definition 2.3 A set $E \subseteq \mathbb{R}^d$ is called *ℓ-Ahlfors regular* or simply *ℓ-regular*, if there is comparability

$$\mathcal{H}^\ell(B \cap E) \approx \mathrm{r}(B)^\ell$$

uniformly for all open balls B of radius $\mathrm{r}(B) \leq 1$ centered in E. If comparability holds for $\mathrm{r}(B) \leq \mathrm{diam}(E)$, then E is called *uniformly* ℓ-regular.

We remark that ℓ-regular sets are ℓ-sets in the sense of Jonsson–Wallin [27, Theorem II.1.1]. Uniformly ℓ-regular sets are ℓ-regular and the converse holds for bounded sets, see Lemma A.4. Many authors consider only *closed* regular sets, but most considerations adapt *verbatim* since in the situation above the closure $\overline{E}$ is still ℓ-regular and $\overline{E} \setminus E$ is an $\mathcal{H}^\ell$ null set [27, Prop. VIII.1]. We shall frequently use this result without further reference.

Example 2.4 If $O \subseteq \mathbb{R}^d$ satisfies the thickness condition (1), then O and ${}^cO$ are both d-regular. Indeed, by symmetry of (1) it suffices to check d-regularity of O. For B a ball centered in O we distinguish whether or not the concentric ball of half the radius intersects ∂O. If it does not, $\mathcal{H}^d(B \cap O) \approx \mathrm{r}(B)^d$ is obvious and if it does, the same is guaranteed by (1) since B contains a ball of radius $\mathrm{r}(B)/2$ centered in ∂O.

Example 2.5 If, as in Sect. 1.1, D is a $(d-1)$-regular part of the boundary of an open set $O \subseteq \mathbb{R}^d$ that satisfies the uniform Lipschitz condition around $\overline{\partial O \setminus D}$, then the full boundary ∂O is also $(d-1)$-regular. Indeed, we can use that bi-Lipschitz images have comparable $\mathcal{H}^{d-1}$-measure [44, Theorem 28.10 a)] and that the bi-Lipschitz constants are uniformly bounded to show that $\overline{\partial O \setminus D}$ is $(d-1)$-regular. We conclude by the observation that the class of $(d-1)$-regular sets is closed under finite unions.

We recall with slight modification the notion of porous sets introduced by Vaisälä [42].

Definition 2.6 Let $E \subseteq F \subseteq \mathbb{R}^d$. Then E is *porous in* F if there exists a constant $\kappa \in (0, 1]$ with the following property:

$$\forall x \in E, r \leq 1 \quad \exists y \in \mathrm{B}(x, r) \cap F \ : \ \mathrm{B}(y, \kappa r) \cap E = \emptyset. \tag{4}$$

If this holds for all $r \leq \mathrm{diam}(E)$, then E is called *uniformly porous in* F. If $F = \mathbb{R}^d$, then E is simply called *(uniformly) porous*.

Remark 2.7 If E is uniformly porous with constant κ, then it is porous with constant $\min\{\kappa, \kappa \,\mathrm{diam}(E)\}$. Condition (4) implies the seemingly stronger statement

$$\forall x \in F, r \leq 1 \quad \exists y \in \mathrm{B}(x, r) \cap F \ : \ \mathrm{B}(y, \kappa r/4) \subseteq \mathrm{B}(x, r) \setminus E.$$

This is seen by distinguishing whether or not $B(x, r/2)$ intersects E. An analogous remark applies to uniformly porous sets.

Whenever necessary, the reader can refer to Appendix A for further background on porous sets. It is instructive to think of them as lower dimensional than the ambient space. This is made precise in Proposition A.9 via the following notions of Assouad dimension.

Definition 2.8 Let $E \subseteq \mathbb{R}^d$. Let $\overline{\mathcal{AS}}(E)$ denote the set of $\lambda > 0$ for which there exists $C \geq 0$ such that, if $0 < r < R < 2\,\mathrm{diam}(E)$ and $x \in E$, then at most $C(R/r)^\lambda$ balls of radius r centered in E are needed to cover $E \cap \mathrm{B}(x, R)$. The number $\overline{\dim}_{\mathcal{AS}}(E) := \inf \overline{\mathcal{AS}}(E)$ is called *upper Assouad dimension* of E. The corresponding *lower Assouad dimension* is defined as $\underline{\dim}_{\mathcal{AS}}(E) := \sup \underline{\mathcal{AS}}(E)$ with $\underline{\mathcal{AS}}(E)$ the set of $\lambda > 0$ for which there exists $C \geq 0$ such that in the former situation at least $C(R/r)^\lambda$ balls are needed.

There is no ambiguity with uniformly ℓ-regular sets since their dimension is ℓ for any of these concepts, see Proposition A.6. The reader can readily check that Definitions 2.3, 2.6, and 2.8 are purely topological in that they do not change when replacing open by closed balls and/or balls by axis-aligned cubes.

2.2 Banach Spaces and Interpolation

All Banach spaces are over the complex numbers. We assume some familiarity with real and complex interpolation of Banach spaces and refer to the textbooks [6,40] for background. However, understanding this paper does not require the precise construction of interpolation spaces. We shall only need the general methodology, their fundamental properties, and standard results on interpolation of function spaces on $\mathbb{R}^d$ measuring smoothness to be recalled further below.

Let (X_0, X_1) be an *interpolation couple*, that is, a pair of Banach spaces that are included in a common linear Hausdorff space. Then the following Banach spaces can be defined between $X_0 \cap X_1$ and $X_0 + X_1$ with respect to continuous inclusion: For $\theta \in [0, 1]$ the *complex interpolation spaces* $[X_0, X_1]_\theta$ of Calderón–Lions [6, Sect. 4.1] and for $\theta \in (0, 1)$ and $p \in [1, \infty]$ the *real interpolation spaces* $(X_0, X_1)_{\theta,p}$ obtained from Peetre's K-method [6, Sect. 3.1]. In any of these spaces $X_0 \cap X_1$ is dense [6, Theorem 3.4.2 & 4.4.2]. In particular, the endpoints $[X_0, X_1]_j$, $j \in \{0, 1\}$, coincide with X_j only if $X_0 \cap X_1$ is dense in X_j.

2.3 Function Spaces Measuring Smoothness

We define the relevant function spaces of smoothness $s \in \mathbb{R}$ and integrability $p \in (1, \infty)$. The *Bessel potential space* $\mathrm{H}^{s,p}(\mathbb{R}^d)$ consists of those tempered distributions $f \in \mathcal{S}'(\mathbb{R}^d)$ for which

$$\|f\|_{\mathrm{H}^{s,p}} := \|\mathcal{F}^{-1}(1 + |\cdot|^2)^{s/2}\mathcal{F}f\|_{\mathrm{L}^p} < \infty.$$

Here, $\mathcal{F}$ denotes the Fourier transform. With $1/p' := 1 - 1/p$ the spaces $\mathrm{H}^{-s,p}(\mathbb{R}^d)$ and $\mathrm{H}^{s,p'}(\mathbb{R}^d)$ are in a sesquilinear duality extending the L^2 inner product [40, Sect. 2.4.2].

If $k \geq 0$ is an integer, then $\mathrm{H}^{k,p}(\mathbb{R}^d)$ coincides up to equivalent norms with the *Sobolev space* $\mathrm{W}^{k,p}(\mathbb{R}^d)$ of tempered distributions such that

$$\|f\|_{\mathrm{W}^{k,p}} := \left(\|f\|_{\mathrm{L}^p}^p + \sum_{j=1}^d \|\partial_j^k f\|_{\mathrm{L}^p}^p\right)^{1/p} < \infty,$$

where we could have equivalently taken all derivatives up to order k into account, see [40, Sect. 2.3.3]. Note that we have $\mathrm{H}^{0,p}(\mathbb{R}^d) = \mathrm{W}^{0,p}(\mathbb{R}^d) = \mathrm{L}^p(\mathbb{R}^d)$.

For $s = k + \sigma$ with $k \geq 0$ an integer and $\sigma \in (0, 1)$ there is a *fractional Sobolev space* $\mathrm{W}^{s,p}(\mathbb{R}^d)$ defined through requiring

$$\|f\|_{\mathrm{W}^{s,p}(\mathbb{R}^d)} := \|f\|_{\mathrm{W}^{k,p}(\mathbb{R}^d)} + \left(\sum_{j=1}^d \iint_{\mathbb{R}^d \times \mathbb{R}^d} \frac{|\partial_j^k f(x) - \partial_j^k f(y)|^p}{|x - y|^{d+\sigma p}} \,\mathrm{d}x\,\mathrm{d}y\right)^{1/p} < \infty.$$

We could also have restricted integration to $|x - y| < 1$. Further common equivalent norms on these spaces exist [40, Sect. 2.5.1]. We define $\mathrm{W}^{-s,p}(\mathbb{R}^d)$ as the space of conjugate-linear functionals on $\mathrm{W}^{s,p'}(\mathbb{R}^d)$ in accordance with what we have seen for

Bessel potential spaces. We could have also given an equivalent intrinsic definition using the scale of Besov spaces [40, Sect. 2.3.2/2.6.1] but the view point of dual spaces is better suited to our circumstances.

We collect all interpolation properties proved in [40, Sect. 2.4.2] that shall be used "off-the-shelf" in the further course. In [40] the nomenclature is $\mathrm{H}^{s,p} = \mathrm{F}^s_{p,2}$ and $\mathrm{W}^{s,p} = \mathrm{F}^s_{p,p}$ for non-integer s.

Proposition 2.9 *Let $p_0, p_1 \in (1, \infty)$, $s_0, s_1 \in \mathbb{R}$, and $\theta \in (0, 1)$. Let* X *denote either* H *or* W. *Up to equivalent norms it follows that*

(i)

$$[\mathrm{X}^{s_0,p_0}(\mathbb{R}^d), \mathrm{X}^{s_1,p_1}(\mathbb{R}^d)]_\theta = \mathrm{X}^{s,p}(\mathbb{R}^d),$$

(ii)

$$(\mathrm{X}^{s_0,p_0}(\mathbb{R}^d), \mathrm{X}^{s_1,p_1}(\mathbb{R}^d))_{\theta,p} = \mathrm{W}^{s,p}(\mathbb{R}^d),$$

with the two exceptions that in (i) for $\mathrm{X} = \mathrm{W}$ *either all or none of s_0, s_1, s have to be integers and that in (ii) integer s is only permitted when $s_0 = s_1 (= s)$.*

2.4 Function Spaces on $\mathbb{R}^d$ Incorporating a Dirichlet Condition

We define analogous spaces of functions with positive smoothness on $\mathbb{R}^d$ that vanish on some $(d-1)$-regular set $E \subseteq \mathbb{R}^d$. All this is based on celebrated results of Jonsson–Wallin [27].

We need the notion of fractional Sobolev spaces on E. They are denoted $\mathrm{B}^{p,p}_s(E)$ in [27] but to keep the analogy with the previous section we shall write $\mathrm{W}^{s,p}(E)$ instead. Having equipped E with $(d-1)$-dimensional Hausdorff measure, we define for $s \in (0, 1)$ and $p \in (1, \infty)$ this space as the Banach space of those $f \in \mathrm{L}^p(E)$ for which

$$\|f\|_{\mathrm{W}^{s,p}(E)} := \Big(\int_E |f(x)|^p \,\mathcal{H}^{d-1}(\mathrm{d}x) \Big)^{1/p} + \Big(\iint_{\substack{x,y\in E\\ |x-y|<1}} \frac{|f(x) - f(y)|^p}{|x - y|^{d-1+sp}} \,\mathcal{H}^{d-1}(\mathrm{d}x)\,\mathcal{H}^{d-1}(\mathrm{d}y) \Big)^{1/p} < \infty.$$

If E is closed, the following is proved in [27, Theorems VI.1 & VII.1]. The general case follows from the discussion in Sect. 2.1.

Proposition 2.10 *Suppose $E \subseteq \mathbb{R}^d$ is $(d-1)$-regular. Let $p \in (1, \infty)$, $s \in (1/p, 1 + 1/p)$, and let* X *denote either* H *or* W.

(i) If $f \in \mathrm{X}^{s,p}(\mathbb{R}^d)$, then for $\mathcal{H}^{d-1}$-almost every $x \in E$ the limit

$$(\mathcal{R}_E f)(x) := \lim_{r\to 0} \frac{1}{|\mathrm{B}(x, r)|} \int_{\mathrm{B}(x,r)} f(y)\,\mathrm{d}y$$

exists. The restriction operator $\mathcal{R}_E$ *maps* $\mathrm{X}^{s,p}(\mathbb{R}^d)$ *boundedly into* $\mathrm{W}^{s-1/p,p}(E)$.

(ii) *Conversely, there exists a bounded extension operator* $\mathcal{E}_E : \mathrm{W}^{s-1/p,p}(E) \to \mathrm{X}^{s,p}(\mathbb{R}^d)$ *that serves as a right inverse for* $\mathcal{R}_E$. *It does not depend on* p, s.

We often refer to $\mathcal{R}_E$ and $\mathcal{E}_E$ as the *Jonsson–Wallin operators for* E.

Definition 2.11 Let $E \subseteq \mathbb{R}^d$ be $(d-1)$-regular. Given $p \in (1, \infty)$ and $s \in (1/p, 1+1/p)$, define

$$\mathrm{X}^{s,p}_E(\mathbb{R}^d) := \{f \in \mathrm{X}^{s,p}(\mathbb{R}^d) : \mathcal{R}_E f = 0\},$$

where X denotes either H or W, and equip it with the norm inherited from $\mathrm{X}^{s,p}(\mathbb{R}^d)$.

2.5 Function Spaces on Open Sets With and Without Partially Vanishing Trace

As usual, let X denote either H or W. Since for $s \geq 0$ we have $\mathrm{X}^{s,p}(\mathbb{R}^d) \subseteq \mathrm{L}^p(\mathbb{R}^d)$, the *pointwise restriction* $|_O$ of functions to O is defined on $\mathrm{X}^{s,p}(\mathbb{R}^d)$.

Definition 2.12 Let $O \subseteq \mathbb{R}^d$ be an open set and let $s \geq 0$, $p \in (1, \infty)$. Define $\mathrm{X}^{s,p}(O) := \{f|_O : f \in \mathrm{X}^{s,p}(\mathbb{R}^d)\}$ with quotient norm

$$\|f\|_{\mathrm{X}^{s,p}(O)} := \inf\big\{\|F\|_{\mathrm{X}^{s,p}} : F \in \mathrm{X}^{s,p}(\mathbb{R}^d) \text{ and } F|_O = f\big\}.$$

If in addition $E \subseteq \overline{O}$ is $(d-1)$-regular, define $\mathrm{X}^{s,p}_E(O) := \{f|_O : f \in \mathrm{X}^{s,p}_E(\mathbb{R}^d)\}$ for $s \in (1/p, 1+1/p)$ with quotient norm

$$\|f\|_{\mathrm{X}^{s,p}_E(O)} := \inf\big\{\|F\|_{\mathrm{X}^{s,p}} : F \in \mathrm{X}^{s,p}_E(\mathbb{R}^d) \text{ and } F|_O = f\big\}.$$

By construction $|_O : \mathrm{X}^{s,p}(\mathbb{R}^d) \to \mathrm{X}^{s,p}(O)$ is bounded. To let $\mathrm{X}^{s,p}(O)$ inherit non-trivial properties of its whole space analogue, a bounded linear right inverse is needed. If O is d-regular, this has been constructed in a beautiful paper of Rychkov [34, Theorem 5.1].

Proposition 2.13 *(Rychkov) Let* $O \subseteq \mathbb{R}^d$ *be an open,* d*-regular set. Let* X *denote either* H *or* W. *For any* $s > 0$ *and* $p \in (1, \infty)$ *there exists a bounded linear extension operator* $\mathcal{E} : \mathrm{X}^{s,p}(O) \to \mathrm{X}^{s,p}(\mathbb{R}^d)$ *that serves as a right inverse for* $|_O$. *Moreover, if* $m \geq 1$ *is an integer, then* $\mathcal{E}$ *can be taken the same for all* $p \in (1, \infty)$ *and all* $s \in (0, m)$.

Remark 2.14 Indeed, though not stated explicitly in [34], the consistency of the extension operator becomes apparent from the construction of the operator Λ on [34, p. 155].

Let us stress that Rychkov's operator is not defined on $\mathrm{L}^p(O)$. But in the low-regularity regime $s < 1/p$ we can simply extend $\mathrm{X}^{s,p}(O) \to \mathrm{X}^{s,p}(\mathbb{R}^d)$ by zero as we shall see soon. The following definition goes back to Sickel [37] and Jawerth–Frazier [26].

Definition 2.15 Let $t \in (0, 1)$. An open set $O \subseteq \mathbb{R}^d$ belongs to the class $\mathcal{D}^t$ if

$$\sup_{x \in \partial O} \sup_{0<r \leq 1} r^{t-d} \int_{\mathrm{B}(x,r)\setminus \partial O} \mathrm{d}(y, \partial O)^{-t} \, \mathrm{d}y < \infty.$$

The relevant examples for us are as follows. For a proof we refer to Proposition A.10 in the appendix.

Example 2.16 An open set with $(d-1)$-regular boundary is of class $\mathcal{D}^t$ for any $t \in (0, 1)$. An open set with porous boundary is of class $\mathcal{D}^t$ for some $t \in (0, 1)$.

We cite the following multiplier theorem for characteristic functions [37, Theorem 4.4].

Proposition 2.17 *(Sickel) Let $O \subseteq \mathbb{R}^d$ be of class $\mathcal{D}^t$ for some $t \in (0, 1)$. Let $p \in (1, \infty)$ and $0 \leq s < t/p$. If* X *denotes either* H *or* W, *then pointwise multiplication by $\mathbb{1}_O$ is a bounded operator on $\mathrm{X}^{s,p}(\mathbb{R}^d)$. For $t(1/p-1) < s < 0$ the dual operator $\mathbb{1}_O \varphi := \varphi \circ \mathbb{1}_O$ is also bounded on $\mathrm{X}^{s,p}(\mathbb{R}^d)$.*

Corollary 2.18 *Let $O \subseteq \mathbb{R}^d$ be an open set with $(d-1)$-regular boundary. Let* X *denote either* H *or* W. *Then the zero extension operator*

$$\mathcal{E}_0 : \mathrm{X}^{s,p}(O) \to \mathrm{X}^{s,p}(\mathbb{R}^d), \quad \mathcal{E}_0 f(x) := \begin{cases} f(x) & (\textit{if } x \in O) \\ 0 & (\textit{if } x \in {}^c O) \end{cases}$$

is bounded provided $p \in (1, \infty)$ and $s \in [0, 1/p)$.

3 First Properties of Function Spaces with Partially Vanishing Trace

We establish first properties and techniques dealing with the spaces introduced in Sect. 2.5. They will frequently be used in the bulk of the paper.

3.1 Structural Properties

Throughout, X denotes either H or W. We begin by showing that incorporating boundary conditions leads to complemented subspaces.

Lemma 3.1 *Let $E \subseteq \mathbb{R}^d$ be $(d-1)$-regular and let $\mathcal{R}$ and $\mathcal{E}$ be the corresponding Jonsson–Wallin operators. Then $\mathcal{P} := 1 - \mathcal{E}\mathcal{R}$ is a bounded projection from $\mathrm{X}^{s,p}(\mathbb{R}^d)$ onto $\mathrm{X}^{s,p}_E(\mathbb{R}^d)$ for any $p \in (1, \infty)$ and $s \in (1/p, 1 + 1/p)$. In particular, $\mathrm{X}^{s,p}_E(\mathbb{R}^d)$ is a closed subspace of $\mathrm{X}^{s,p}(\mathbb{R}^d)$.*

Proof The operator $\mathcal{E}\mathcal{R}$ is bounded on $\mathrm{X}^{s,p}(\mathbb{R}^d)$ by Proposition 2.10. Since $\mathcal{E}$ is a right inverse for $\mathcal{R}$, we have $(\mathcal{E}\mathcal{R})^2 = \mathcal{E}\mathcal{R}$, that is to say, $\mathcal{E}\mathcal{R}$ is a projection with the same nullspace as $\mathcal{R}$. Now, on the one hand, the nullspace of $\mathcal{R}$ is $\mathrm{X}^{s,p}_E(\mathbb{R}^d)$ and on the other hand, the nullspace of $\mathcal{E}\mathcal{R}$ equals the range of $\mathcal{P}$. The conclusion follows. □

Next, we introduce test functions with *support* $\operatorname{supp}(\cdot)$ away from a given set E.

Definition 3.2 Given $E \subseteq \mathbb{R}^d$, define

$$\mathrm{C}_E^\infty(\mathbb{R}^d) := \left\{ f \in \mathrm{C}_0^\infty(\mathbb{R}^d) : \mathrm{d}(\operatorname{supp}(f), E) > 0 \right\}$$

and if $O \subseteq \mathbb{R}^d$ is any open set, let $\mathrm{C}_E^\infty(O) := \{ f|_O : f \in \mathrm{C}_E^\infty(\mathbb{R}^d)\}$.

Lemma 3.3 *Let $O \subseteq \mathbb{R}^d$ be open and $E \subseteq \overline{O}$ be $(d-1)$-regular. For $p \in (1, \infty)$ and $s \in (1/p, 1]$ the set $\mathrm{C}_E^\infty(O)$ is dense in $\mathrm{X}_E^{s,p}(O)$.*

Proof Since the restriction $|_O : \mathrm{X}_E^{s,p}(\mathbb{R}^d) \to \mathrm{X}_E^{s,p}(O)$ is bounded and onto, it suffices to treat $O = \mathbb{R}^d$. In this case we shall reduce the claim to the fact that any continuous function $f \in \mathrm{W}^{1,p}(\mathbb{R}^d)$ that vanishes everywhere on a closed set $F \subseteq \mathbb{R}^d$ can be approximated by $\mathrm{C}_F^\infty(\mathbb{R}^d)$-functions in $\mathrm{W}^{1,p}(\mathbb{R}^d)$-norm. This is easily proved on using that $\mathrm{W}^{1,p}(\mathbb{R}^d)$ is closed under truncation, see [1, Sect. 9.2].

Let $\mathcal{P} : \mathrm{X}^{s,p}(\mathbb{R}^d) \to \mathrm{X}_E^{s,p}(\mathbb{R}^d)$ be the bounded projection provided by Lemma 3.1. Since $\mathrm{C}_0^\infty(\mathbb{R}^d)$ is dense in $\mathrm{X}^{s,p}(\mathbb{R}^d)$, it suffices to approximate elements in $\mathcal{P}(\mathrm{C}_0^\infty(\mathbb{R}^d))$ by test functions from $\mathrm{C}_E^\infty(\mathbb{R}^d)$. Moreover, it suffices to achieve this for the $\mathrm{W}^{1,p}(\mathbb{R}^d)$-norm, which is stronger than the $\mathrm{X}^{s,p}(\mathbb{R}^d)$-norm since we have $s \leq 1$. Since the projection $\mathcal{P}$ in Lemma 3.1 is the same for all admissible values of s and p, we have in particular

$$\mathcal{P}(\mathrm{C}_0^\infty(\mathbb{R}^d)) \subseteq \mathcal{P}((\mathrm{W}^{1,d+1} \cap \mathrm{W}^{1,p})(\mathbb{R}^d)) = (\mathrm{W}_E^{1,d+1} \cap \mathrm{W}_E^{1,p})(\mathbb{R}^d).$$

Sobolev embeddings yield for every function in the right-hand space a continuous representative f that vanishes $\mathcal{H}^{d-1}$-almost everywhere on E. By Ahlfors-regularity the intersection of E with arbitrarily small balls centered in E still has positive $\mathcal{H}^{d-1}$-measure. Thus every point on $F := \overline{E}$ is an accumulation point of zeros of f. It follows that f vanishes everywhere on F and the above-mentioned approximation result kicks in. □

By a similar argument we prove the surprising feature that Rychkov's extension operator automatically preserves Dirichlet conditions on $(d-1)$-regular sets. Once again, this comes as a byproduct of consistency of the extension operator and Sobolev embeddings and has nothing to do with the particular construction.

Lemma 3.4 *Let $O \subseteq \mathbb{R}^d$ be an open, d-regular set, and let $E \subseteq \overline{O}$ be $(d-1)$-regular. Suppose $p \in (1, \infty)$ and $s \in (1/p, 1 + 1/p)$. If $\mathcal{E} : \mathrm{X}^{s,p}(O) \to \mathrm{X}^{s,p}(\mathbb{R}^d)$ is the extension operator of Proposition 2.13 constructed with $m \geq 2$, then*

$$\mathcal{E} : \mathrm{X}_E^{s,p}(O) \to \mathrm{X}_E^{s,p}(\mathbb{R}^d)$$

is bounded for the $\mathrm{X}^{s,p}(O) \to \mathrm{X}^{s,p}(\mathbb{R}^d)$-norm. In particular, $\mathrm{X}_E^{s,p}(O)$ is a closed subspace of $\mathrm{X}^{s,p}(O)$.

Proof By definition of the quotient norm we obtain $\mathrm{X}^{s,p}_E(O) \subseteq \mathrm{X}^{s,p}(O)$ with continuous inclusion of Banach spaces from the fact that $\mathrm{X}^{s,p}_E(\mathbb{R}^d)$ is a closed subspace of $\mathrm{X}^{s,p}(\mathbb{R}^d)$.

We begin with the case $s \leq 1$. Since $\mathcal{E} : \mathrm{X}^{s,p}(O) \to \mathrm{X}^{s,p}(\mathbb{R}^d)$ is bounded, it suffices to check that $\mathcal{E}$ maps a dense subset of $\mathrm{X}^{s,p}_E(O)$ into $\mathrm{X}^{s,p}_E(\mathbb{R}^d)$. Owing to Lemma 3.3 we can take this subset to be $\mathrm{C}^\infty_E(O) = \mathrm{C}^\infty_E(\mathbb{R}^d)|_O$. So, let $f \in \mathrm{C}^\infty_E(\mathbb{R}^d)$. Since $\mathcal{E}$ acts consistently, we obtain $\mathcal{E}(f|_O) \in (\mathrm{W}^{1,d+1} \cap \mathrm{X}^{s,p})(\mathbb{R}^d)$. Due to Sobolev embeddings $\mathcal{E}(f|_O)$ admits a continuous representative and we need to check that it vanishes everywhere on E. To this end, we let $B \subseteq \mathbb{R}^d$ be an arbitrary open ball centered in $E \subseteq \overline{O}$ with radius $\mathrm{r}(B) < \mathrm{d}(\mathrm{supp}(f), E)$. Since O is d-regular, $B \cap O$ has positive Lebesgue measure but on this set we have $\mathcal{E}(f|_O) = f = 0$ almost everywhere. The conclusion follows.

If $s \in (1, 1+1/p)$ and $f \in \mathrm{X}^{s,p}_E(O)$, then we can use Proposition 2.13 to get $\mathcal{E}f \in \mathrm{X}^{s,p}(\mathbb{R}^d)$ and from the inclusion $\mathrm{X}^{s,p}_E(O) \subseteq \mathrm{X}^{1,p}_E(O)$ and the first part of the proof we get $\mathcal{E}f \in \mathrm{X}^{1,p}_E(\mathbb{R}^d)$. According to Definition 2.11 this implies $\mathcal{E}f \in \mathrm{X}^{s,p}_E(\mathbb{R}^d)$.

As for the final statement, given $f \in \mathrm{X}^{s,p}_E(O)$ we have already seen $\|f\|_{\mathrm{X}^{s,p}_E(O)} \geq \|f\|_{\mathrm{X}^{s,p}(O)}$ and we have just proved $\|f\|_{\mathrm{X}^{s,p}_E(O)} \leq \|\mathcal{E}f\|_{\mathrm{X}^{s,p}_E(\mathbb{R}^d)} \lesssim \|f\|_{\mathrm{X}^{s,p}(O)}$. □

3.2 Symmetric Interpolation Results

We establish symmetric interpolation results for the spaces $\mathrm{X}^{s,p}(O)$ and $\mathrm{X}^{s,p}_E(O)$. *Symmetric* means that either both or none of the spaces are with vanishing trace on E. In particular, we prove Theorem 1.2.

The whole theory relies on the retraction-coretraction principle. Given two Banach spaces X and Y, a bounded linear operator $\mathcal{R} : X \to Y$ is called *retraction* if it has a bounded left-inverse $\mathcal{E} : Y \to X$ such that $\mathcal{RE} = 1$ is the identity on Y. In this case $\mathcal{E}$ is called the associated *coretraction*. It is instructive to think of $\mathcal{R}$ as a restriction and $\mathcal{E}$ a compatible extension operator. The following is proved in [40, Sect. 1.2.4].

Proposition 3.5 *Let (X_0, X_1) and (Y_0, Y_1) be interpolation couples and $\mathcal{R} : X_0 + X_1 \to Y_0 + Y_1$, $\mathcal{E} : Y_0 + Y_1 \to X_0 + X_1$ be linear operators such that $\mathcal{R} : X_j \to Y_j$ is a retraction with associated coretraction $\mathcal{E} : Y_j \to X_j$ for $j = 0, 1$. Let $\langle \cdot, \cdot \rangle$ denote either a complex or a real interpolation bracket and put $X = \langle X_1, X_2 \rangle$ and $Y = \langle Y_1, Y_2 \rangle$. Then $\mathcal{ER}$ restricts to a bounded projection in X and*

$$\mathcal{E} : Y \to \mathcal{ER}(X)$$

is an isomorphism of Banach spaces, where $\mathcal{ER}(X)$ carries the norm of X.

Remark 3.6 Above we may apply $\mathcal{R}$ to the equality of sets $\mathcal{E}(Y) = \mathcal{ER}(X)$ provided by the invertibility of $\mathcal{E} : Y \to \mathcal{ER}(X)$. This yields $Y = \mathcal{R}(X)$ as sets along with comparability $\|y\|_Y \approx \|\mathcal{E}y\|_X$ for $y \in Y$. In particular, if $\mathcal{R}(X)$ carries the quotient norm inherited from $X/\mathsf{R}(\mathcal{R})$, then the inclusion $Y \subseteq \mathcal{R}(X)$ is continuous and the open mapping theorem yields $Y = \mathcal{R}(X)$ as Banach spaces with equivalent norms.

An important special case arises when $\mathcal{R} = \mathcal{P}$ is a projection and $\mathcal{E} = 1$ is the identity, compare with [40, Sect. 1.17.1].

Corollary 3.7 *Let (X_0, X_1) be an interpolation couple and $\mathcal{P}$ a bounded projection in $X_0 + X_1$ with range Z. Then $(Z \cap X_0, Z \cap X_1)$ is an interpolation couple and if $\langle \cdot, \cdot \rangle$ denotes either a complex or a real interpolation bracket, then up to equivalent norms*

$$\langle Z \cap X_0, Z \cap X_1 \rangle = Z \cap \langle X_0, X_1 \rangle.$$

As a first application, we obtain a result similar to Proposition 2.9 for spaces on d-regular sets. We repeat the well-known argument since it will be re-used several times.

Proposition 3.8 *Let $O \subseteq \mathbb{R}^d$ be open and d-regular. Let $p_0, p_1 \in (1, \infty)$, $s_0, s_1 \in (0, \infty)$, and $\theta \in (0, 1)$. Let* X *denote either* H *or* W. *Up to equivalent norms it follows that*

(i)

$$[\mathrm{X}^{s_0,p_0}(O), \mathrm{X}^{s_1,p_1}(O)]_\theta = \mathrm{X}^{s,p}(O),$$

(ii)

$$(\mathrm{X}^{s_0,p_0}(O), \mathrm{X}^{s_1,p_1}(O))_{\theta,p} = \mathrm{W}^{s,p}(O),$$

with the two exceptions that in (i) for $\mathrm{X} = \mathrm{W}$ either all or none of s_0, s_1, s have to be integers and in (ii) integer s is only permitted when $s_0 = s_1 (= s)$.

Proof We apply Proposition 3.5 with $X_j := \mathrm{X}^{s_j,p_j}(\mathbb{R}^d)$, $Y_j := \mathrm{X}^{s_j,p_j}(O)$, $\mathcal{R} := |_O$ the pointwise restriction, and $\mathcal{E}$ Rychkov's extension operator from Proposition 2.13 constructed with an integer $m > \max\{s_0, s_1\}$.

Let us prove (i). According to Proposition 2.9 we have $X := [X_0, X_1]_\theta = \mathrm{X}^{s,p}(\mathbb{R}^d)$. By definition, $\mathcal{R}(X) = \mathrm{X}^{s,p}(O)$ carries the quotient norm inherited from $X / \mathsf{R}(\mathcal{R})$. Hence, Remark 3.6 yields $Y := [Y_0, Y_1]_\theta = \mathrm{X}^{s,p}(O)$ with equivalent norms. The proof of (ii) follows *verbatim* from the identity $(X_0, X_1)_{\theta,p} = \mathrm{W}^{s,p}(\mathbb{R}^d)$ also provided by Proposition 2.9. □

Remark 3.9 Suppose that in addition O has $(d-1)$-regular boundary. In the proof above we could then replace Rychkov's extension operator $\mathcal{E}$ with the zero extension operator $\mathcal{E}_0$ discussed in Corollary 2.18. Consequently, Proposition 3.8 remains valid for parameters $s_j \in [0, 1/p_j)$, which includes the case of Lebesgue spaces.

The same technique yields the

Proof of Theorem 1.2 First, we assume $O = \mathbb{R}^d$. Proposition 2.9 provides the identities analogous to (c) and (d) for the spaces without Dirichlet conditions. Hence, the claim follows from Corollary 3.7 applied to the projection $\mathcal{P}$ provided by Lemma 3.1.

Having established the interpolation identities on $\mathbb{R}^d$, we can now pass to the spaces on O via Proposition 3.5 as in the proof of Proposition 3.8. Indeed, if we take $X_j := \mathrm{X}_E^{s_j,p_j}(\mathbb{R}^d)$, $Y_j := \mathrm{X}_E^{s_j,p_j}(O) = (X_j)|_O$, $\mathcal{R} := |_O$, and $\mathcal{E}$ as Rychkov's extension operator, then the only property that needs to be checked is that $\mathcal{E}$ maps Y_j boundedly into X_j. But the latter is precisely the statement of Lemma 3.4. □

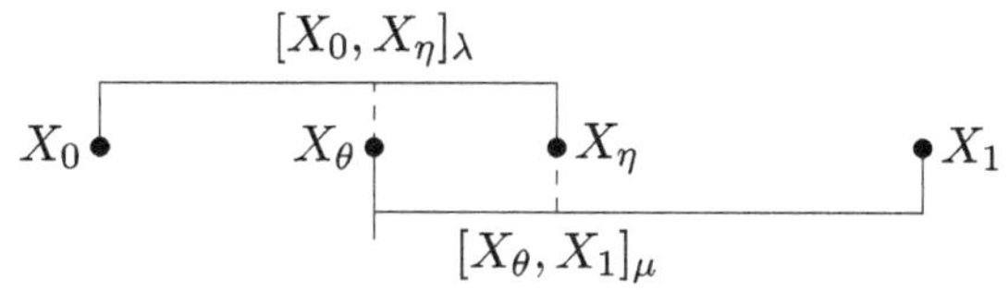

Fig. 2 Assuming the interpolation identities indicated by dashed lines, Wolff's result recovers X_θ and X_η as interpolation spaces associated with the couple (X_0, X_1) for the correct convex combination parameters θ and η, respectively

3.3 Gluing Interpolation Scales

We recall a general interpolation technique due to Wolff [43]. Here, we cite (with adapted notation) the refined version proved in [25, Theorem 1&2]. The statement is visualized in Fig. 2 for complex interpolation.

Proposition 3.10 *(Wolff) Let $X_0, X_\theta, X_\eta, X_1$ be Banach spaces included in a common linear Hausdorff space. Suppose $\theta, \eta, \lambda, \mu \in (0, 1)$ satisfy $\theta = \lambda\eta$ and $\eta = (1-\mu)\theta + \mu$, and let $p_\theta, p_\eta \in [1, \infty]$.*

(i) If $X_\theta = [X_0, X_\eta]_\lambda$ and $X_\eta = [X_\theta, X_1]_\mu$, then also $X_\theta = [X_0, X_1]_\theta$ and $X_\eta = [X_0, X_1]_\eta$.
(ii) If $X_\theta = (X_0, X_\eta)_{\lambda, p_\theta}$ and $X_\eta = (X_\theta, X_1)_{\mu, p_\eta}$, then also $X_\theta = (X_0, X_1)_{\theta, p_\theta}$ and $X_\eta = (X_0, X_1)_{\eta, p_\eta}$.

All equalities above are in the sense of equal sets with equivalent norms.

For further reference we demonstrate once in detail how the results of Proposition 3.8 and Remark 3.9 can be patched together using Wolff's result.

Proposition 3.11 *If in the setting of Proposition 3.8 the boundary ∂O is $(d-1)$-regular, then the conclusion remains valid for $s_0, s_1 \in [0, \infty)$.*

Proof In view of Proposition 3.8, Remark 3.9, and symmetry of the assumption, we only have to treat the case $s_0 = 0$ and $s_1 > 0$. For any $\eta \in (0, 1)$ we abbreviate the relevant convex combinations by $s_\eta := \eta s_1$ and $1/p_\eta := (1 - \eta)/p_0 + \eta/p_1$.

We begin with (i). Since s_0 is an integer, we are only claiming something new in the case $\mathrm{X} = \mathrm{H}$. We have to prove for all $\eta \in (0, 1)$ the equality

$$[\mathrm{H}^{s_0, p_0}(O), \mathrm{H}^{s_1, p_1}(O)]_\eta = \mathrm{H}^{s_\eta, p_\eta}(O). \tag{5}$$

Throughout, the reader should keep in mind Fig. 2. Let us first suppose $s_\eta < 1/p_\eta$ so that $\mathrm{H}^{s_\eta, p_\eta}(\mathbb{R}^d)$ belongs to the regime covered by Remark 3.9. We pick $\theta \in (0, \eta)$ and $\lambda, \mu \in (0, 1)$ such that $\theta = \lambda\eta$ and $\eta = (1 - \mu)\theta + \mu$. The quadruple of spaces $(X_i)_i := (\mathrm{H}^{s_i, p_i}(O))_i$ satisfies the assumption in part (i) of Wolff's result owing to Remark 3.9 and Proposition 3.8. Hence, we obtain (5). Now, suppose $s_\eta \geq 1/p_\eta$. Due to $s_0 = 0$ we can pick $\theta \in (0, \eta)$ to arrange $s_\theta < 1/p_\theta$. The first part of the proof applies to s_η in place of s_1 and yields $[X_0, X_\eta]_\lambda = X_\theta$. Consequently, we can apply Wolff's result with the same numerology as before to obtain (5).

As for (ii), the claim for W-spaces follows *verbatim* on using part (ii) of Wolff's result with p_θ, p_η corresponding to θ, η as above and systematicaly replacing H by W.

Real interpolation of H-spaces requires a different argument since the result will be a W-space. We rely on the one-sided reiteration theorem in Proposition 3.12 below. Indeed, given $\theta \in (0, 1)$ we pick $\eta \in (0, \theta)$ and write $\theta = (1-\lambda)\eta+\lambda$ with $\lambda \in (0, 1)$. Then we use in succession one-sided reiteration, complex interpolation of H-spaces established above, and Proposition 3.8, to give

$$\begin{aligned}\big(\mathrm{H}^{s_0,p_0}(O), \mathrm{H}^{s_1,p_1}(O)\big)_{\theta,p_\theta} &= \big([\mathrm{H}^{s_0,p_0}(O), \mathrm{H}^{s_1,p_1}(O)]_\eta, \mathrm{H}^{s_1,p_1}(O)\big)_{\lambda,p_\theta}\\ &= \big(\mathrm{H}^{s_\eta,p_\eta}(O), \mathrm{H}^{s_1,p_1}(O)\big)_{\lambda,p_\theta}\\ &= \mathrm{W}^{s_\theta,p_\theta}(O).\end{aligned}$$

Concerning the last line we remark that $(1-\lambda)s_\eta+\lambda s_1 = s_\theta$ and $(1-\lambda)/p_\eta+\lambda/p_1 = 1/p_\theta$ hold by construction. □

The reiteration result that we have invoked above is as follows. We refer to [40, Sect. 1.10.3, Theorem 2] for real interpolation and to [9] for complex interpolation, noting that in the latter case the density of $X_0 \cap X_1$ in X_1 guarantees $[X_0, X_1]_1 = X_1$.

Proposition 3.12 *Let* (X_0, X_1) *be an interpolation couple. Let* $\eta, \lambda \in (0, 1)$ *and put* $\theta = (1-\lambda)\eta + \lambda$. *The interpolation identity*

$$\langle [X_0, X_1]_\eta, X_1\rangle_\lambda = \langle X_0, X_1\rangle_\theta$$

holds up to equivalent norms in the following cases. If $\langle\cdot\,,\cdot\rangle$ *is a* $(\cdot\,, p)$*-real interpolation bracket with* $p \in [1,\infty]$ *fixed or if* $\langle\cdot\,,\cdot\rangle$ *is the complex interpolation bracket and* $X_0 \cap X_1$ *is dense in* X_1.

3.4 Non Symmetric Interpolation: The Easy Inclusion

The main difficulty in Theorem 1.1 lies in proving the inclusion "$\supseteq$". Indeed, here we can already prove

Proposition 3.13 *Let* $O \subseteq \mathbb{R}^d$ *be an open, d-regular set with* $(d-1)$*-regular boundary, and let* $D \subseteq \overline{O}$ *be* $(d-1)$*-regular. Let* $p_0, p_1 \in (1,\infty)$, $s_0 \in [0, 1/p_0)$, $s_1 \in (1/p_1, 1]$, *and* $\theta \in (0, 1)$. *Define p and s as in* (2). *Then there are continuous inclusions and*

(i)

$$[\mathrm{X}^{s_0,p_0}(O), \mathrm{X}^{s_1,p_1}_D(O)]_\theta \subseteq \begin{cases}\mathrm{X}^{s,p}_D(O) & (\text{if } s > 1/p)\\ \mathrm{X}^{s,p}(O) & (\text{if } s < 1/p)\end{cases}$$

and

(ii)

$$(\mathrm{X}^{s_0,p_0}(O), \mathrm{X}_D^{s_1,p_1}(O))_{\theta,p} \subseteq \begin{cases} \mathrm{W}_D^{s,p}(O) & (\textit{if } s > 1/p) \\ \mathrm{W}^{s,p}(O) & (\textit{if } s < 1/p) \end{cases}$$

with the exception that $s_0 \neq 0$ and $s_1 \neq 1$ are required in (i) for $\mathrm{X} = \mathrm{W}$. If $p_0 = p_1$, then the result remains true for all $s_1 \in (1/p_1, 1+1/p_1)$ with the additional exception that only in (i) for $\mathrm{X} = \mathrm{H}$ the value $s = 1$ is permitted.

Proof First, we check that Proposition 3.12 applies in its real and its complex version to the couple $(\mathrm{X}^{s_0,p_0}(O), \mathrm{X}^{s_1,p_1}(O))$. If $s_1 \leq 1$ then $\mathrm{X}^{s_0,p_0}(O) \cap \mathrm{X}_D^{s_1,p_1}(O) \supseteq \mathrm{C}_D^\infty(O)$ is dense in $\mathrm{X}_D^{s_1,p_1}(O)$ by Lemma 3.3 and if $p_0 = p_1$ then $\mathrm{X}^{s_0,p_0}(O) \cap \mathrm{X}_D^{s_1,p_1}(O) = \mathrm{X}_D^{s_1,p_1}(O)$ for all $s_1 \in (1/p_1, 1+1/p_1)$. This being said, we denote by $\langle \cdot\,, \cdot \rangle$ either the $(\cdot\,, p)$-real or the complex interpolation bracket and treat all assertions simultaneously.

By definition we have $\mathrm{X}_D^{s_1,p_1}(O) \subseteq \mathrm{X}^{s_1,p_1}(O)$ and hence we get

$$\langle \mathrm{X}^{s_0,p_0}(O), \mathrm{X}_D^{s_1,p_1}(O)\rangle_\theta \subseteq \langle \mathrm{X}^{s_0,p_0}(O), \mathrm{X}^{s_1,p_1}(O)\rangle_\theta$$

with continuous inclusion. The interpolation space on the right has been determined in Proposition 3.11. It coincides (up to equivalent norms) with $\mathrm{W}^{s,p}(O)$ in case of real interpolation and with $\mathrm{X}^{s,p}(O)$ in case of complex interpolation. In the case $s < 1/p$ this already is the desired conclusion.

Let now $s > 1/p$. We fix $\eta \in (0, \theta)$ sufficiently close to θ, so to arrange $1/p_\eta := (1-\eta)/p_0 + \eta/p_1$ and $s_\eta := (1-\eta)s_0 + \eta s_1$ satisfying $s_\eta > 1/p_\eta$. We write $\theta = (1-\lambda)\eta + \lambda$ with $\lambda \in (0, 1)$. From Proposition 3.12 and the reasoning in the first case we obtain

$$\begin{aligned} \langle \mathrm{X}^{s_0,p_0}(O), \mathrm{X}_D^{s_1,p_1}(O)\rangle_\theta &= \langle [\mathrm{X}^{s_0,p_0}(O), \mathrm{X}_D^{s_1,p_1}(O)]_\eta, \mathrm{X}_D^{s_1,p_1}(O)\rangle_\lambda \\ &\subseteq \langle \mathrm{X}^{s_\eta,p_\eta}(O), \mathrm{X}_D^{s_1,p_1}(O)\rangle_\lambda \end{aligned}$$

with continuous inclusion. Let $\mathcal{E}$ be Rychkov's extension operator for O. From Lemma 3.4 and the above we can infer by interpolation that

$$\mathcal{E} : \langle \mathrm{X}^{s_0,p_0}(O), \mathrm{X}_D^{s_1,p_1}(O)\rangle_\theta \to \langle \mathrm{X}^{s_\eta,p_\eta}(\mathbb{R}^d), \mathrm{X}_D^{s_1,p_1}(\mathbb{R}^d)\rangle_\lambda =: Y \tag{6}$$

is bounded. As before, we see that Y is continuously included into $\mathrm{W}^{s,p}(\mathbb{R}^d)$ in case of real interpolation and into $\mathrm{X}^{s,p}(\mathbb{R}^d)$ in case of complex interpolation.

Consider the Jonsson–Wallin restriction operator to D, see Proposition 2.10. It maps $\mathrm{X}^{s_\eta,p_\eta}(\mathbb{R}^d)$ boundedly into $\mathrm{W}^{s_\eta-1/p_\eta,p_\eta}(D)$ since we have $s_\eta > 1/p_\eta$ and it maps $\mathrm{X}_D^{s_1,p_1}(\mathbb{R}^d)$ into $\{0\}$ by definition. By interpolation it maps Y into $\langle \mathrm{W}^{s_\eta-1/p_\eta,p_\eta}(D), \{0\}\rangle_\lambda$. This interpolation space equals $\{0\}$ since it contains $\{0\}$ as a dense subspace. Hence, we have continuous inclusion of Y into $\mathrm{W}_D^{s,p}(\mathbb{R}^d)$ in case of real interpolation and into $\mathrm{X}_D^{s,p}(\mathbb{R}^d)$ in case of complex interpolation. By (6) every function in $\langle \mathrm{X}^{s_0,p_0}(O), \mathrm{X}_D^{s_1,p_1}(O)\rangle_\theta$ has an extension in Y in virtue of a bounded extension operator. The required continuous inclusion follows. □

4 Proof of Theorem 1.1

It will be convenient to reformulate the assumptions of Theorem 1.1 as follows, relying on Examples 2.4 and 2.5.

Assumption 4.1 The set $O \subseteq \mathbb{R}^d$ is open, d-regular, has d-regular complement, and $(d-1)$-regular boundary. The Dirichlet part $D \subseteq \partial O$ is $(d-1)$-regular and ∂D is porous in ∂O. Moreover, O satisfies a uniform Lipschitz condition around $\overline{\partial O \setminus D}$.

Throughout the whole section let X denote either H or W. We are given $p_0, p_1 \in (1, \infty)$, $s_0 \in [0, 1/p_0)$, $s_1 \in (1/p_1, 1]$, and $\theta \in (0, 1)$. When concerned with complex interpolation for X = W, we implictly restrict ourselves to $s_0 \neq 0$ and $s_1 \neq 1$. Our goal is to establish set inclusions

$$[\mathrm{X}^{s_0,p_0}(O), \mathrm{X}_D^{s_1,p_1}(O)]_\theta \supseteq \begin{cases} \mathrm{X}_D^{s,p}(O) & (\textit{if } s > 1/p) \\ \mathrm{X}^{s,p}(O) & (\textit{if } s < 1/p) \end{cases} \tag{7}$$

and

$$(\mathrm{X}^{s_0,p_0}(O), \mathrm{X}_D^{s_1,p_1}(O))_{\theta,p} \supseteq \begin{cases} \mathrm{W}_D^{s,p}(O) & (\textit{if } s > 1/p) \\ \mathrm{W}^{s,p}(O) & (\textit{if } s < 1/p) \end{cases}. \tag{8}$$

This will complete the proof of Theorem 1.1 since under Assumption 4.1 the converse inclusions are continuous due to Proposition 3.13 and hence become equalities with equivalent norms thanks to the bounded inverse theorem.

4.1 Road Map to the Proof

We give the outline for complex interpolation. The real case will be treated in the same way up to replacing the complex interpolation bracket with the $(\cdot\,, p)$-real interpolation bracket and keeping in mind that real interpolation spaces of X-spaces are always W-spaces.

First, we show in Sect. 4.3 that (7) and (8) hold in the case $D = \partial O$ of pure Dirichlet boundary condition. Then the inclusion with general D and $s \in (0, 1/p)$ follows readily:

$$\mathrm{X}^{s,p}(O) \subseteq \big[\mathrm{X}^{s_0,p_0}(O), \mathrm{X}_{\partial O}^{s_1,p_1}(O)\big]_\theta \subseteq \big[\mathrm{X}^{s_0,p_0}(O), \mathrm{X}_D^{s_1,p_1}(O)\big]_\theta.$$

In the case $s \in (1/p, 1)$ we localize in order to reduce the problem to pure Dirichlet interpolation and interpolation with mixed boundary conditions, but for a simpler geometry. Precisely, we will have $O = \mathbb{R}^d_+$ the upper half-space and E_i a transformed version of a portion of D with a security area for good measure that is still $(d-1)$-regular and has porous boundary in $\partial\mathbb{R}^d_+ \cong \mathbb{R}^{d-1}$. Then we have to show that

$$\mathrm{X}_{E_i}^{s,p}(\mathbb{R}^d_+) \subseteq \big[\mathrm{X}^{s_0,p_0}(\mathbb{R}^d_+), \mathrm{X}_{E_i}^{s_1,p_1}(\mathbb{R}^d_+)\big]_\theta. \tag{9}$$

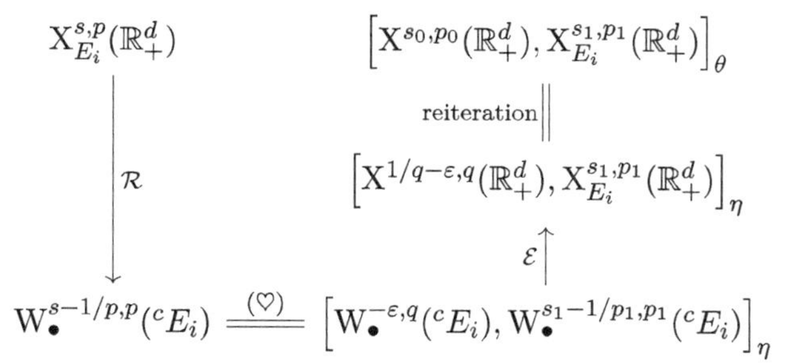

Fig. 3 Schematic presentation of the main argument to prove the inclusion "⊇" in part (a) of Theorem 1.1

This reduction will be done in Sect. 4.4.

The heart of the matter lies in showing (9) in Sect. 4.6. To do so, we decompose $f \in X^{s,p}_{E_i}(\mathbb{R}^d_+)$ as $f = (f - \mathcal{E}\mathcal{R}f) + \mathcal{E}\mathcal{R}f$, where $\mathcal{R}$ is the restriction to $\partial\mathbb{R}^d_+$ and $\mathcal{E}$ is a corresponding extension operator. The term $f - \mathcal{E}\mathcal{R}f$ will be in $[X^{s_0,p_0}(\mathbb{R}^d_+), X^{s_1,p_1}_{E_i}(\mathbb{R}^d_+)]_\theta$ because it satisfies pure Dirichlet boundary conditions on $\partial\mathbb{R}^d_+$. The argument for $\mathcal{E}\mathcal{R}f$ happens completely at the boundary and is displayed in Fig. 3.

Here, $W^{s,p}_\bullet({}^cE_i)$ is a subspace of $W^{s,p}(\mathbb{R}^{d-1})$ with zero condition on the full dimensional set $E_i \subseteq \mathbb{R}^{d-1}$ and q, ε and η are parameters yet to be determined. We need to establish

- the construction of an extension operator $\mathcal{E}$ from $\partial\mathbb{R}^d_+$ to $\mathbb{R}^d_+$ which is consistent in $s \in \mathbb{R} \setminus \mathbb{Z}$ and $p \in (1, \infty)$ and
- the precise definition of the spaces $W^{s,p}_\bullet({}^cE_i)$ for a suitable range of s including verification of the interpolation identity (♡).

The passage through spaces of negative order in (♡) is inevitable and can be implemented in virtue of Proposition 2.17 only because ∂D is porous in ∂O.

4.2 Spaces of Functions Vanishing on a Full-Dimensional Subset

For this part we work with a d-regular set $U \subseteq \mathbb{R}^d$ whose boundary is a Lebesgue null set and whose interior $\mathring{U}$ is of class $\mathcal{D}^t$ for some $t \in (0, 1)$, compare with Definition 2.15.

We remark that most results stated in Sect. 2.5 for open sets still apply in this context. Pointwise multiplication by the characteristic functions of U and $\mathring{U}$ coincide on $L^p(\mathbb{R}^d)$. Moreover, $\mathring{U}$ is d-regular and the corresponding Rychkov's extension operators can also be regarded as extension operators for functions defined on U.

Let $\mathcal{R}$ denote the pointwise restriction operator $|_U$ and let $\mathcal{E}$ denote *some* extension operator $X^{s,p}(U) \to X^{s,p}(\mathbb{R}^d)$. We will specify consistency requirements later on. For $p \in (1, \infty)$ and $s \in (0, \infty)$ we define

$$X^{s,p}_\bullet({}^cU) := \{f \in X^{s,p}(\mathbb{R}^d) : \mathcal{R}f = 0\}$$

with subspace topology. This subspace is complemented in virtue of the projection $1 - \mathcal{E}\mathcal{R}$.

The pointwise multiplier $\mathbb{1}_U$ is bounded on $\mathrm{X}^{s,p}(\mathbb{R}^d)$ for $t(1/p-1) < s < t/p$ due to Proposition 2.17. This allows us to extend the definition of $\mathrm{X}^{s,p}_{\bullet}({}^cU)$ to such s by

$$\mathrm{X}^{s,p}_{\bullet}({}^cU):=\{f - \mathbb{1}_U f : f \in \mathrm{X}^{s,p}(\mathbb{R}^d)\},$$

where the topology is again the subspace topology. Note that for $s \in (0, t/p)$ this gives the same space as before and that now $1 - \mathbb{1}_U$ becomes the complementing projection.

The following lemma captures the interpolation behavior of these spaces.

Lemma 4.2 *Let $p_0, p_1 \in (1,\infty)$, $s_0 \in (t(1/p_0 - 1), \infty)$, $s_1 \in (t(1/p_1 - 1), \infty)$, and $\theta \in (0,1)$. Up to equivalent norms it follows that*

(i)

$$[\mathrm{X}^{s_0,p_0}_{\bullet}({}^cU), \mathrm{X}^{s_1,p_1}_{\bullet}({}^cU)]_\theta = \mathrm{X}^{s,p}_{\bullet}({}^cU),$$

(ii)

$$(\mathrm{X}^{s_0,p_0}_{\bullet}({}^cU), \mathrm{X}^{s_1,p_1}_{\bullet}({}^cU))_{\theta,p} = \mathrm{W}^{s,p}_{\bullet}({}^cU),$$

with the two exceptions that in (i) for $\mathrm{X} = \mathrm{W}$ either all or none of s_0, s_1, s have to be integers and that in (ii) integer s is only permitted when $s_0 = s_1 (= s)$.

Proof By symmetry we may assume $s_0 \leq s_1$. In virtue of Corollary 3.7 we shall transfer the interpolation identities of Proposition 2.9 for the $\mathrm{X}^{s,p}(\mathbb{R}^d)$-spaces to the $\mathrm{X}^{s,p}_{\bullet}({}^cU)$-spaces. We only have to identify suitable projections $\mathcal{P}$.

If $s_0 > 0$, then we pick a Rychkov's extension operator $\mathcal{E}$ that is consistent up to a positive integer greater s_1 and use $\mathcal{P}:=1 - \mathcal{E}\mathcal{R}$.

Now assume $s_0 \leq 0$. If $s_1 < t/p_1$, then we can directly use $\mathcal{P}:=1-\mathbb{1}_U$. Otherwise, there are $p \in (1,\infty)$ and $s \in (0, t/p)$ such that $(s, 1/p)^\top$ lies on the segment connecting $(s_0, 1/p_0)^\top$ and $(s_1, 1/p_1)^\top$ in the $(s, 1/p)$-plane. If necessary, we can arrange that s is not an integer. We have just obtained interpolation for the spaces on the segment connecting $(s_0, 1/p_0)^\top$ and $(s, 1/p)^\top$ and in order to conclude, we patch this interpolation scale together with the one for positive differentiability by the technique illustrated in the proof of Proposition 3.11. □

4.3 The Case of Pure Dirichlet Conditions

For this part we strengthen our previous requirements on $U \subseteq \mathbb{R}^d$ to the effect that it should be a closed d-regular set with $(d-1)$-regular boundary.

It follows that ∂U is a Lebesgue null set and we claim that $\mathring{U}$ is of class $\mathcal{D}^t$ for all $t \in (0,1)$. Indeed, by Example 2.16 the open set ${}^cU$ has this property and since we have $\partial\mathring{U} \subseteq \partial U = \partial({}^cU)$ with set difference of zero Lebesgue measure, we see by the very definition that if ${}^cU$ is of class $\mathcal{D}^t$, then so is $\mathring{U}$.

We start out with a reformulation of Corollary 2.18.

Lemma 4.3 *If* $p \in (1, \infty)$ *and* $s \in [0, 1/p)$, *then* $\mathrm{X}^{s,p}({}^cU) = \mathrm{X}^{s,p}_{\bullet}({}^cU)|_{{}^cU}$ *with equivalent norms.*

Proof The inclusion $\mathrm{X}^{s,p}_{\bullet}({}^cU)|_{{}^cU} \subseteq \mathrm{X}^{s,p}({}^cU)$ is clear. For the converse let $f \in \mathrm{X}^{s,p}({}^cU)$ and F an extension of f in $\mathrm{X}^{s,p}(\mathbb{R}^d)$. We get $\mathbb{1}_{{}^cU} F \in \mathrm{X}^{s,p}(\mathbb{R}^d)$ owing to Corollary 2.18. Hence, we have $\mathbb{1}_{{}^cU} F \in \mathrm{X}^{s,p}_{\bullet}({}^cU)$ and $f = (\mathbb{1}_{{}^cU} F)|_{{}^cU} \in \mathrm{X}^{s,p}_{\bullet}({}^cU)|_{{}^cU}$ follows. For the boundedness, we calculate

$$\|f\|_{\mathrm{X}^{s,p}_{\bullet}({}^cU)|_{{}^cU}} \leq \|\mathbb{1}_{{}^cU} F\|_{\mathrm{X}^{s,p}(\mathbb{R}^d)} \lesssim \|F\|_{\mathrm{X}^{s,p}(\mathbb{R}^d)}$$

and take the infimum over all such extensions F. □

In order to proceed, we need a generic re-norming lemma and its consequence for the pointwise multiplication by $\mathbb{1}_{{}^cU}$. To fix ideas for the following, we decided to include a proof even though the result is known in the literature.

Lemma 4.4 *If* $p \in (1, \infty)$ *and* $s \in \mathbb{R}$, *then*

$$\|f\|_{\mathrm{X}^{s,p}} \approx \|f\|_{\mathrm{X}^{s-1,p}} + \|\nabla f\|_{\mathrm{X}^{s-1,p}} \qquad (f \in \mathcal{S}'(\mathbb{R}^n)). \tag{10}$$

Proof In the following all function spaces are on $\mathbb{R}^d$ and we omit the dependence. The operator $\mathcal{I}_{-1} f := \mathcal{F}^{-1}(1 + |\xi|^2)^{1/2} \mathcal{F} f$ is invertible from $\mathcal{S}'$ into itself. By definition it restricts to an isomorphism $\mathcal{I}_{-1} : \mathrm{H}^{s,p} \to \mathrm{H}^{s-1,p}$. By interpolation the same holds for $\mathcal{I}_{-1} : \mathrm{W}^{s,p} \to \mathrm{W}^{s-1,p}$, see Proposition 2.9. Hence, we find for all $f \in \mathcal{S}'$,

$$\|f\|_{\mathrm{X}^{s,p}} \approx \|\mathcal{F}^{-1}(1 + |\xi|^2)^{1/2} \mathcal{F} f\|_{\mathrm{X}^{s-1,p}}.$$

Comparing with (10), we see that it remains to prove

$$\|f\|_{\mathrm{X}^{s-1,p}} + \sum_{j=1}^{d} \|\mathcal{F}^{-1} \xi_j \mathcal{F} f\|_{\mathrm{X}^{s-1,p}} \approx \|\mathcal{F}^{-1}(1 + |\xi|^2)^{1/2} \mathcal{F} f\|_{\mathrm{X}^{s-1,p}}. \tag{11}$$

To this end we consider Fourier multipliers $f \mapsto \mathcal{F}^{-1} m \mathcal{F} f$, defined on $\mathcal{S}'$ via a smooth and bounded function $m : \mathbb{R}^d \to \mathbb{C}$, to pass from one side to the other. If such multiplier is bounded on L^p, then it is bounded on $\mathrm{H}^{k,p}$ for all integers k since it commutes with $\mathcal{I}_{-1}$ and its inverse. Hence, it is bounded on $\mathrm{X}^{s,p}$ for all $s \in \mathbb{R}$ by interpolation. This being said, we obtain "$\lesssim$" in (11) by considering the Fourier multipliers associated with $(1+|\xi|^2)^{-1/2}$ and $\xi_j(1+|\xi|^2)^{-1/2}$. Their L^p boundedness follows easily from the Mihlin multiplier theorem [6, Theorem 6.1.6]. Next, we pick a smooth function $\chi : \mathbb{R} \to [0, 1]$ that vanishes on $(-1, 1)$ and is identically 1 outside of $[-2, 2]$ in order to write

$$(1 + |\xi|^2)^{1/2} = \left(\frac{(1 + |\xi|^2)^{1/2}}{1 + \sum_{j=1}^{d} \chi(\xi_j)|\xi_j|} \right) + \sum_{j=1}^{d} \left(\frac{(1 + |\xi|^2)^{1/2}}{1 + \sum_{j=1}^{d} \chi(\xi_j)|\xi_j|} \right) \left(\frac{\chi(\xi_j)|\xi_j|}{\xi_j} \right) \xi_j.$$

Again by Mihlin's theorem each bracket corresponds to an L^p bounded Fourier multiplier. This yields the converse estimate "$\gtrsim$". □

Lemma 4.5 *For $p \in (1, \infty)$ and $s \in (1/p, 1 + 1/p)$ pointwise multiplication by $\mathbb{1}_{^cU}$ is $\mathrm{X}^{s,p}_{\partial U}(\mathbb{R}^d) \to \mathrm{X}^{s,p}_{\bullet}(^cU)$ bounded.*

Proof For $f \in \mathrm{C}^{\infty}_{\partial U}(\mathbb{R}^d)$ we have that $\nabla(\mathbb{1}_{^cU} f) = \mathbb{1}_{^cU} \nabla f$. Hence, we can combine Lemma 4.4 and Proposition 2.17 to the effect that

$$\begin{aligned} \|\mathbb{1}_{^cU} f\|_{\mathrm{X}^{s,p}} &\approx \|\mathbb{1}_{^cU} f\|_{\mathrm{X}^{s-1,p}} + \|\mathbb{1}_{^cU} \nabla f\|_{\mathrm{X}^{s-1,p}} \\ &\lesssim \|f\|_{\mathrm{X}^{s-1,p}} + \|\nabla f\|_{\mathrm{X}^{s-1,p}} \\ &\approx \|f\|_{\mathrm{X}^{s,p}}. \end{aligned} \tag{12}$$

For $s \in (1/p, 1]$ we can use that $\mathrm{C}^{\infty}_{\partial U}(\mathbb{R}^d)$ is dense in $\mathrm{X}^{s,p}_{\partial U}(\mathbb{R}^d)$ by Lemma 3.3 to conclude that $\mathbb{1}_{^cU} : \mathrm{X}^{s,p}_{\partial U}(\mathbb{R}^d) \to \mathrm{X}^{s,p}(\mathbb{R}^d)$ is bounded. That it actually maps into the closed subspace $\mathrm{X}^{s,p}_{\bullet}(^cU)$ follows by construction. Suppose now $s \in (1, 1+1/p)$. The commutation $\nabla(\mathbb{1}_{^cU} \cdot) = \mathbb{1}_{^cU}\nabla(\cdot)$ extends by density to all $f \in \mathrm{X}^{1,p}_{\partial U}(\mathbb{R}^d)$. Hence, it holds in particular on $\mathrm{X}^{s,p}_{\partial U}(\mathbb{R}^d)$ and the calculation (12) re-applies. □

We get the analogue of Lemma 4.3 in the case $s \in (1/p, 1 + 1/p)$.

Lemma 4.6 *If $p \in (1, \infty)$ and $s \in (1/p, 1 + 1/p)$, then $\mathrm{X}^{s,p}_{\partial U}(^cU) = \mathrm{X}^{s,p}_{\bullet}(^cU)|_{^cU}$ with equivalent norms.*

Proof The inclusion $\mathrm{X}^{s,p}_{\partial U}(^cU) \subseteq \mathrm{X}^{s,p}_{\bullet}(^cU)|_{^cU}$ works as in the proof of Lemma 4.3, on using Lemma 4.5 instead of Corollary 2.18.

Conversely, let $f \in \mathrm{X}^{s,p}_{\bullet}(^cU)$. Since f is in particular a member of $\mathrm{X}^{s,p}(\mathbb{R}^d)$, we find a sequence $(f_n)_n \subseteq \mathrm{C}^{\infty}_0(\mathbb{R}^d)$ that approximates f in the topology of $\mathrm{X}^{s,p}(\mathbb{R}^d)$. Let $\mathcal{E}$ be Rychkov's extension operator for U, which, as we recall, acts consistently on $\mathrm{W}^{1,d+1}(\mathbb{R}^d)$. We apply the projection $\mathcal{P} = 1 - \mathcal{E}\mathcal{R}$ to that sequence. Since $\mathcal{P}$ projects onto $\mathrm{X}^{s,p}_{\bullet}(^cU)$, we get $\mathcal{P} f_n = 0$ almost everywhere on U on the one hand and $\mathcal{P} f_n \in \mathrm{C}(\mathbb{R}^d)$ by Sobolev embeddings on the other hand. By d-regularity, the intersection of U with balls centered in U has positive Lebesgue measure. Hence, the $\mathcal{P} f_n$ vanish everywhere on U. In particular they vanish on ∂U, which means $\mathcal{P} f_n \in \mathrm{X}^{s,p}_{\partial U}(\mathbb{R}^d)$. Now, since $f_n \to f$ in $\mathrm{X}^{s,p}(\mathbb{R}^d)$, also $\mathcal{P} f_n \to \mathcal{P} f = f$ in $\mathrm{X}^{s,p}(\mathbb{R}^d)$, which gives $f \in \mathrm{X}^{s,p}_{\partial U}(\mathbb{R}^d)$. □

Eventually, we can transfer the interpolation settled in Lemma 4.2 to the spaces incorporating pure Dirichlet boundary conditions. Since we can take $U = {}^cO$, this gives the full claim of Theorem 1.1 for pure Dirichlet conditions.

Proposition 4.7 *Let $p_0, p_1 \in (1, \infty)$, $s_0 \in [0, 1/p_0)$, $s_1 \in (1/p_1, 1]$, and $\theta \in (0, 1)$. There are continuous inclusions*

(i)

$$[\mathrm{X}^{s_0,p_0}(^cU), \mathrm{X}^{s_1,p_1}_{\partial U}(^cU)]_\theta \supseteq \begin{cases} \mathrm{X}^{s,p}_{\partial U}(^cU) & (\textit{if } s > 1/p) \\ \mathrm{X}^{s,p}(^cU) & (\textit{if } s < 1/p) \end{cases},$$

(ii)

$$(\mathrm{X}^{s_0,p_0}({}^cU), \mathrm{X}^{s_1,p_1}_{\partial U}({}^cU))_{\theta,p} \supseteq \begin{cases} \mathrm{W}^{s,p}_{\partial U}({}^cU) & (\textit{if } s > 1/p) \\ \mathrm{W}^{s,p}({}^cU) & (\textit{if } s < 1/p) \end{cases},$$

with the exception that $s_0 \neq 0$ *and* $s_1 \neq 1$ *are required in (i) for* $\mathrm{X} = \mathrm{W}$*. If in addition* ${}^cU$ *is d-regular, then both inclusions become equalities with equivalent norms.*

Proof Let $\langle \cdot, \cdot \rangle$ denote either the θ-complex or (θ, p)-real interpolation bracket. Using Lemmas 4.3 and 4.6, we get

$$\begin{aligned} \langle \mathrm{X}^{s_0,p_0}_{\bullet}({}^cU), \mathrm{X}^{s_1,p_1}_{\bullet}({}^cU)\rangle|_{{}^cU} &\subseteq \langle \mathrm{X}^{s_0,p_0}_{\bullet}({}^cU)|_{{}^cU}, \mathrm{X}^{s_1,p_1}_{\bullet}({}^cU)|_{{}^cU}\rangle \\ &= \langle \mathrm{X}^{s_0,p_0}({}^cU), \mathrm{X}^{s_1,p_1}_{\partial U}({}^cU)\rangle. \end{aligned}$$

Lemma 4.2 identifies the space on the left-hand side with either $\mathrm{X}^{s,p}_{\bullet}({}^cU)|_{{}^cU}$ or $\mathrm{W}^{s,p}_{\bullet}({}^cU)|_{{}^cU}$. The claim follows from Lemma 4.3 in the case $s < 1/p$ and from Lemma 4.6 in the case $s > 1/p$.

The final statement on equalities in these inclusions follows from Proposition 3.13. □

4.4 Localization

We recall that O satisfies a uniform Lipschitz condition around $N := \overline{\partial O \setminus D}$ with bi-Lipschitz constant L as in Definition 2.1. We claim that we can select countably many points $x_i \in \overline{\partial O \setminus D}$, $i \in I \subseteq \mathbb{N} \setminus \{0\}$, with corresponding coordinate charts $(U_{x_i}, \Phi_{x_i}) =: (U_i, \Phi_i)$, and an open set U_0 that does not intersect N, with the following properties. With $J := \{0\} \cup I$, the covering

$$\overline{O} \subseteq \dot{\bigcup_{j\in J}} U_j \tag{13}$$

admits a smooth partition of unity by functions $\eta_j \in \mathrm{C}^\infty(\mathbb{R}^d)$ satisfying

(i) $\operatorname{supp}(\eta_j) \subseteq U_j$, (ii) $\sum_{j\in J} \eta_j = 1$ on $\mathbb{R}^d$,

(iii) $\sum_{j\in J} \mathbb{1}_{U_j} \leq C$ on $\mathbb{R}^d$, (iv) $\|\eta_j\|_{\mathrm{L}^\infty} + \|\nabla \eta_j\|_{\mathrm{L}^\infty} \leq C'$,

and there are auxiliary functions $\chi_i \in \mathrm{C}^\infty(\mathbb{R}^d)$ with $\|\chi_i\|_{\mathrm{L}^\infty} + \|\nabla \chi_i\|_{\mathrm{L}^\infty} \leq C'$ such that χ_i is 1 on $\Phi_i(\operatorname{supp} \eta_i)$ and supported in $(-1, 1)^d$, whereas χ_0 is 1 on $\operatorname{supp} \eta_0$ and supported in U_0. Here, C and C' are constants that depend only on L and d.

The construction is as follows. For any $x \in N$ we extend Φ_x to a bi-Lipschitz map $\overline{U_x} \to [-1, 1]^d$ with the same Lipschitz constant not larger than L. From $\Phi_x(x) = 0$

we conclude that $\Phi_x(\overline{U_x} \cap \mathrm{B}(x, \frac{1}{2L}))$ is contained in $B(0, \frac{1}{2})$ and hence does not intersect the boundary of the unit cube. The inclusion $B_x := B(x, \frac{1}{2L}) \subseteq U_x$ then follows from the fact that bi-Lipschitz mappings between closed sets preserve the boundaries. Starting from $\bigcup_{x \in N} \frac{1}{8} B_x \supseteq N$, we use the Vitali covering lemma (Lemma A.2) to extract a countable collection $(x_i)_{i \in I} \subseteq N$ such that $\bigcup_{i \in I} \frac{5}{8} B_i \supseteq N$ with the $\frac{1}{8} B_i$ mutually disjoint. We have abbreviated as usual $B_i := B_{x_i}$.

If $x \in \mathbb{R}^d$ is contained in U_i, then $U_i \subseteq \mathrm{B}(x_i, L\sqrt{d}) \subseteq \mathrm{B}(x, 2L\sqrt{d})$ by the Lipschitz property. Due to $B(x_i, \frac{1}{16L}) \subseteq U_i$ and mutual disjointness there are at most $(32L^2\sqrt{d})^d$ such i. Finite overlap guarantees that $U_0 := \mathbb{R}^d \setminus \bigcup_{i \in I} \frac{5}{8} \overline{B}_i$ is an open set that pays for (13) and we can take $C := 1 + (32L^2\sqrt{d})^d$ in (iii).

For $i \in I$ we pick $\varphi_i \in \mathrm{C}_0^\infty(B_i)$ with range in $[0, 1]$, equal to 1 on $\frac{7}{8} B_i$, and $\|\nabla \varphi_i\|_\infty \leq cL$ for a dimensional constant c. We also pick a smooth φ_0 with range in $[0, 1]$, support in $\mathbb{R}^d \setminus \bigcup_{i \in I} \frac{6}{8} B_i$, and equal to 1 on $\mathbb{R}^d \setminus \bigcup_{i \in I} \frac{7}{8} B_i$. For any $x \in \mathbb{R}^d$ the sum $\sum_{j \in J} \varphi_j(x)$ contains at most C non-zero terms, one of which is equal to 1. Hence, functions η_j with the properties specified in (i), (ii), (iv) are given by $\eta_j := \varphi_j / \sum_{j \in J} \varphi_j$. For $i \in I$ we can take the χ_i all the same since $\Phi_i(\mathrm{supp}(\eta_i))$ is contained in $B(0, \frac{1}{2})$. We pick $\chi_0 \in \mathrm{C}^\infty(U_0)$ equal to 1 on $\mathbb{R}^d \setminus \bigcup_{i \in I} \frac{6}{8} B_i$ to complete the construction.

With this formalism at hand, we define the retraction-coretraction pair

$$\mathcal{E} : f \longmapsto \big(\chi_0 f, (\chi_i (f \circ \Phi_i^{-1}))_{i \in I}\big), \tag{14}$$

$$\mathcal{R} : (g_j)_{j \in J} \longmapsto \eta_0 g_0 + \sum_{i \in I} \eta_i (g_i \circ \Phi_i). \tag{15}$$

Indeed, we find $\mathcal{R}\mathcal{E} f = f$ for $f \in \mathrm{L}^p(O)$. It is implicitly understood that functions with compact support are extended by zero and domains of definitions are appropriately restricted to make these definitions meaningful. We introduce natural function spaces for these mappings.

Definition 4.8 For $p \in (1, \infty)$ and $s \in [0, 1]$ define the Banach space

$$\mathbb{X}^{s,p}(O) := \mathrm{X}^{s,p}(O) \times \ell^p(I; \mathrm{X}^{s,p}(\mathbb{R}^d_+)), \quad \|g\|_{\mathbb{X}^{s,p}(O)} := \Big(\sum_{j \in J} \|g_j\|_{\mathrm{X}^{s,p}}^p\Big)^{1/p}.$$

Remark 4.9 The space $\mathbb{X}^{s,p}(O)$ is constructed by ℓ^p-superposition from $\mathrm{X}^{s,p}(O)$ and $\mathrm{X}^{s,p}(\mathbb{R}^d_+)$. Real and complex interpolation behaves in the best possible (componentwise) way under this operation [40, Sect. 1.18.1]. Precisely, the spaces $\mathbb{X}^{s,p}(O)$ interpolate according to the same rules as do $\mathrm{X}^{s,p}(O)$ and $\mathrm{X}^{s,p}(\mathbb{R}^d_+)$ according to Proposition 3.11.

Lemma 4.10 *For $p \in (1, \infty)$ and $s \in [0, 1]$ the maps $\mathcal{E} : \mathrm{X}^{s,p}(O) \to \mathbb{X}^{s,p}(O)$ and $\mathcal{R} : \mathbb{X}^{s,p}(O) \to \mathrm{X}^{s,p}(O)$ are bounded.*

Proof In view of Remark 4.9 we only have to treat the extremal cases $s = 0$ and $s = 1$. For convenience we write $\mathbb{L}^p(O)$ and $\mathbb{W}^{1,p}(O)$ instead of $\mathbb{X}^{s,p}(O)$, respectively.

Given $f \in \mathrm{L}^p(O)$, we use the uniformity and support properties of the partition of unity along with the uniform bi-Lipschitz property of the Φ_i when applying the transformation formula [33, Sect. 2.3.1], to give

$$\|\mathcal{E} f\|_{\mathbb{L}^p(O)}^p = \int_O |\chi_0 f|^p \,\mathrm{d}x + \sum_{i\in I} \int_{\mathbb{R}^d_+} |\chi_i(f \circ \Phi_i^{-1})| \,\mathrm{d}x \lesssim \int_O \sum_{j\in J} \mathbb{1}_{U_j} |f|^p \,\mathrm{d}x. \tag{16}$$

The right-hand side is bounded by $C\|f\|_{\mathrm{L}^p}^p$ due to the finite overlap property (iii). Similarly, given $g \in \mathbb{L}^p(O)$, we can estimate

$$\begin{aligned} \|\mathcal{R} g\|_{\mathrm{L}^p(O)}^p &\leq \int_O \Big(|\eta_0 g_0| + \sum_{i\in I} |\eta_i(g_i \circ \Phi_i)|\Big)^p \mathrm{d}x \\ &\approx \int_O |\eta_0 g_0|^p + \sum_{i\in I} |\eta_i(g_i \circ \Phi_i)|^p \,\mathrm{d}x \\ &\lesssim \int_O |g_0|^p \,\mathrm{d}x + \sum_{i\in I} \int_{\mathbb{R}^d_+} |g_i|^p \,\mathrm{d}x = \|g\|_{\mathbb{L}^p(O)}, \end{aligned} \tag{17}$$

where in the second step we have used again that for fixed x the sum contains at most C non-zero terms and hence the ℓ^1-norm can be replaced by an ℓ^p-norm at the expense of a constant depending on C. The previous two estimates yield the claim in case $s = 0$.

We turn to the case $s = 1$ and recall that $\mathrm{W}^{1,p}$-spaces are defined by restriction. Let $f \in \mathrm{W}^{1,p}(O)$ and let $F \in \mathrm{W}^{1,p}(\mathbb{R}^d)$ be any extension. Calculating $\nabla(\mathcal{E} F)$ by the product rule and chain rules [33, Sect. 2.3.1], we can use the same argument as in (16) to get

$$\sum_{j\in J} \|(\mathcal{E} F)_j\|_{\mathrm{L}^p(\mathbb{R}^d)}^p + \|\nabla(\mathcal{E} F)_j\|_{\mathrm{L}^p(\mathbb{R}^d)}^p \lesssim \|F\|_{\mathrm{L}^p(\mathbb{R}^d)}^p + \|\nabla F\|_{\mathrm{L}^p(\mathbb{R}^d)}^p.$$

Since each $(\mathcal{E} F)_j$ extends $(\mathcal{E} f)_j$, the left-hand side controls $\|\mathcal{E} f\|_{\mathbb{W}^{1,p}(O)}$ from above and we can pass to the infimum over F to obtain the required boundedness of $\mathcal{E}$. Likewise, given $G \in \mathrm{X}^{s,p}(\mathbb{R}^d) \times \ell^p(I; \mathrm{X}^{s,p}(\mathbb{R}^d))$ we can recycle (17) to the effect that

$$\|\mathcal{R} G\|_{\mathrm{L}^p(\mathbb{R}^d)}^p + \|\nabla \mathcal{R} G\|_{\mathrm{L}^p(\mathbb{R}^d)}^p \lesssim \sum_{j\in J} \|G_j\|_{\mathrm{L}^p(\mathbb{R}^d)}^p + \|\nabla G_j\|_{\mathrm{L}^p(\mathbb{R}^d)}^p$$

and we conclude as before. □

To bring the boundary conditions into play, we introduce a modified version of $\mathbb{X}^{s,p}(O)$. We set

$$E_i := \Phi_i(D) \cup \big(\mathbb{R}^{d-1} \setminus (-1,1)^{d-1}\big) \qquad (i \in I) \tag{18}$$

and define

$$\mathbb{X}^{s,p}_E(O) := \mathrm{X}^{s,p}_{\partial O}(O) \times \bigtimes_{i \in I} \mathrm{X}^{s,p}_{E_i}(\mathbb{R}^d_+) \qquad (s > 1/p), \tag{19}$$

which we consider as a closed subspace of $\mathbb{X}^{s,p}(O)$ in virtue of Lemma 3.4. Let us make sure that these transformed Dirichlet parts are of the same geometric quality as D.

Lemma 4.11 *The set E_i defined in* (18) *is $(d-1)$-regular in $\mathbb{R}^{d-1}$ and has porous boundary.*

Proof Let $B \subseteq \mathbb{R}^{d-1}$ be a ball of radius $\mathrm{r}(B) \leq 1$ centered in $\overline{E_i}$. There are two cases. The first one is that $\frac{1}{2}B$ intersects the complement of $(-1,1)^{d-1}$. Then there is a ball B' of radius $\mathrm{r}(B)/4$ contained in $B \setminus [-1,1]^{d-1}$. The second one is that $\frac{1}{2}B$ is properly contained in the domain of Φ_i^{-1} and thus there is a ball $B'' \subseteq \mathbb{R}^d$ centered in ∂O such that $\mathrm{r}(B) \approx \mathrm{r}(B'')$ and $\Phi_i^{-1}(B) \supseteq B'' \cap \partial O$.

(i) We pick the center of B in E_i and show that E_i is $(d-1)$-regular. In the first case we have $|B \cap E_i| \gtrsim (\mathrm{r}(B)/4)^{d-1}$. In the second case we use that bi-Lipschitz images have comparable Hausdorff measure [44, Theorem 28.10 a)] and that D is $(d-1)$-regular to conclude

$$|B \cap E_i| \approx \mathcal{H}^{d-1}(\Phi_i^{-1}(B \cap E_i)) \geq \mathcal{H}^{d-1}(B'' \cap D) \gtrsim \mathrm{r}(B)^{d-1}.$$

(ii) We pick the center of B in ∂E_i and show that ∂E_i is porous. Again, in the first case, already B' does not intersect ∂E_i. Otherwise, we use porosity of ∂D in ∂O, taking Remark 2.7 into account, to find a ball centered in ∂O and contained in B'' which avoids ∂D. Transforming this ball back using Φ_i, we find a ball centered in B with comparably smaller radius that does not intersect ∂E_i. □

The next lemma shows that $\mathcal{E}$ and $\mathcal{R}$ defined in (14) and (15) are well-behaved with respect to the Dirichlet conditions defined in (18) and (19).

Lemma 4.12 *For $p \in (1,\infty)$ and $s \in (1/p, 1]$, the operators $\mathcal{E} : \mathrm{X}^{s,p}_D(O) \to \mathbb{X}^{s,p}_E(O)$ and $\mathcal{R} : \mathbb{X}^{s,p}_E(O) \to \mathrm{X}^{s,p}_D(O)$ are bounded.*

Proof As for $\mathcal{E}$, it suffices to consider $f \in \mathrm{C}^\infty_D(O)$ since the general case follows by density, see Lemma 3.3. Since χ_0 is smooth with support away from $\overline{\partial O \setminus D}$, we get that $\chi_0 f$ is smooth with compact support away from ∂O. In particular, we have $\chi_0 f \in \mathrm{X}^{s,p}_{\partial O}(O)$. We conclude from the bi-Lipschitz property of Φ_i that

$$\mathrm{d}(E_i, \mathrm{supp}(f \circ \Phi_i^{-1})) = \mathrm{d}(\Phi_i(D), \Phi_i(\mathrm{supp}\, f)) \approx \mathrm{d}(D, \mathrm{supp}\, f) > 0.$$

Hence, $\chi_i(f \circ \Phi_i^{-1})$ is a Lipschitz continuous function on $\mathbb{R}^d_+$ whose compact support has positive distance to E_i. Thus, it is contained in $\mathrm{W}^{1,p}_{E_i}(\mathbb{R}^d_+) \subseteq \mathrm{X}^{s,p}_{E_i}(\mathbb{R}^d_+)$.

As for $\mathcal{R}$, we take $g = (g_j)_{j \in J}$ from $\mathrm{C}^\infty_{\partial O}(O) \times \bigtimes_{i \in I} \mathrm{C}^\infty_{E_i}(\mathbb{R}^d_+)$, which is dense in $\mathbb{X}^{s,p}_E(O)$ due to Lemmas 3.3 and 4.11. As before, we only have to show that the support of $\mathcal{R}g$ has positive distance to D. But $\mathrm{supp}(\eta_0\, g_0) \subseteq \mathrm{supp}(g_0)$ has positive distance to D by construction and for $\mathrm{supp}(\eta_i(g_i \circ \Phi_i))$ we can argue as above. □

We formulate a reduction result based on this localization.

Proposition 4.13 *The set inclusions* (7) *and* (8) *follow from the set inclusions*

$$[\mathrm{X}^{s_0,p_0}(\mathbb{R}^d_+), \mathrm{X}^{s_1,p_1}_{E_i}(\mathbb{R}^d_+)]_\theta \supseteq \begin{cases} \mathrm{X}^{s,p}_{E_i}(\mathbb{R}^d_+) & (\textit{if } s > 1/p) \\ \mathrm{X}^{s,p}(\mathbb{R}^d_+) & (\textit{if } s < 1/p) \end{cases} \tag{20}$$

and

$$(\mathrm{X}^{s_0,p_0}(\mathbb{R}^d_+), \mathrm{X}^{s_1,p_1}_{E_i}(\mathbb{R}^d_+))_{\theta,p} \supseteq \begin{cases} \mathrm{W}^{s,p}_{E_i}(\mathbb{R}^d_+) & (\textit{if } s > 1/p) \\ \mathrm{W}^{s,p}(\mathbb{R}^d_+) & (\textit{if } s < 1/p) \end{cases}. \tag{21}$$

Proof We apply Proposition 3.5 with the pair $(\mathcal{E}, \mathcal{R})$ defined in (14) and (15). Owing to the mapping properties derived in Lemmas 4.10 and 4.12, we get equal sets

$$[\mathrm{X}^{s_0,p_0}(O), \mathrm{X}^{s_1,p_1}_{D}(O)]_\theta = \mathcal{R}[\mathbb{X}^{s_0,p_0}(O), \mathbb{X}^{s_1,p_1}_{E}(O)]_\theta.$$

Lemma 4.12 asserts that the inclusion (7) holds provided that we can prove

$$[\mathbb{X}^{s_0,p_0}(O), \mathbb{X}^{s_1,p_1}_{E}(O)]_\theta \supseteq \begin{cases} \mathbb{X}^{s,p}_{E}(O) & (\text{if } s > 1/p) \\ \mathbb{X}^{s,p}(O) & (\text{if } s < 1/p) \end{cases}. \tag{22}$$

The ℓ^p-superpositions of spaces $\mathrm{X}^{s,p}(O)$ and $\mathrm{X}^{s,p}_{E_i}(\mathbb{R}^d_+)$ on the left interpolate componentwise, see Remark 4.9. This being said, the above follows from the assumption (20) for the components on $\mathbb{R}^d_+$ and Proposition 4.7 for the component on O.

The real case is the same upon using $\mathbb{W}$-spaces on the right of (22) and appealing to assumption (21) instead. □

Remark 4.14 It stems from the interpolation on the left-hand side of (22) that at least at this stage of the proof we prefer talking about set inclusions only. Continuity of (22) would require continuity of (20) (which we could obtain) – but with uniform bounds in I (which we believe to be rather painful).

4.5 Extension and Restriction Operators for the Half-Space

We introduce the extension and restriction operators appearing in Fig. 3. As usual, X denotes either H or W.

The Restriction Operator $\mathcal{R}$

Let $F \subseteq \partial\mathbb{R}^d_+$ be $(d-1)$-regular. We identify $\partial\mathbb{R}^d_+$ with $\mathbb{R}^{d-1}$ whenever convenient. Proposition 2.10 yields a restriction operator $\mathcal{R}_F : \mathrm{X}^{s,p}(\mathbb{R}^d) \to \mathrm{W}^{s-1/p,p}(F)$ for $p \in (1,\infty)$ and $s \in (1/p, 1+1/p)$. By construction, we have for $u \in \mathrm{X}^{s,p}(\mathbb{R}^d) \cap \mathrm{C}(\mathbb{R}^d)$,

$$\mathcal{R}_F u(x') = u(x', 0) \qquad (\text{a.e. } x' \in F). \tag{23}$$

In virtue of this formula $\mathcal{R}_F$ is well-defined on the quotient space $\mathrm{X}^{s,p}(\mathbb{R}^d_+) \cap \mathrm{C}(\overline{\mathbb{R}^d_+})$. The inclusion chain

$$\mathrm{C}_0^\infty(\mathbb{R}^d)|_{\mathbb{R}^d_+} \subseteq \mathrm{X}^{s,p}(\mathbb{R}^d_+) \cap \mathrm{C}(\overline{\mathbb{R}^d_+}) \subseteq \mathrm{X}^{s,p}(\mathbb{R}^d_+) = \mathrm{X}^{s,p}(\mathbb{R}^d)|_{\mathbb{R}^d_+}$$

and the density of the first space in the last space shows that we can extend $\mathcal{R}_F$ to $\mathrm{X}^{s,p}(\mathbb{R}^d_+)$ by continuity. We abbreviate $\mathcal{R}:=\mathcal{R}_{\partial\mathbb{R}^d_+}$.

The Extension Operator $\mathcal{E}$

For the extension operator we also need to consider spaces of negative smoothness. They have been defined on the whole space in Sect. 2.3. We set $\mathrm{X}^{s,p}(\mathbb{R}^d_+):= \mathrm{X}^{s,p}(\mathbb{R}^d)|_{\mathbb{R}^d_+}$, where the restriction of distributions $|_{\mathbb{R}^d_+} : \mathcal{S}'(\mathbb{R}^d) \to \mathcal{D}'(\mathbb{R}^d_+)$ coincides with the pointwise restriction when s is non-negative.

We construct $\mathcal{E}$ via the *bounded analytic* C_0*-semigroup* $(\mathrm{e}^{-\Lambda t})_{t\geq 0}$ generated by $\Lambda:=(1-\Delta_{x'})^{1/2}$ in $\mathrm{L}^p(\mathbb{R}^{d-1})$. Here, $\Delta_{x'}$ denotes the *Laplacian* in $\mathbb{R}^{d-1}$. A reader who is not familiar with these notions may consult the textbook [2], in particular Example 3.7.6 and Theorem 3.8.3. By means of the Fourier transform $\mathcal{F}$ in $\mathbb{R}^{d-1}$ the operators $\mathrm{e}^{-\Lambda t}$ are unambiguously defined on all of $\mathcal{S}'(\mathbb{R}^{d-1})$ through

$$\mathrm{e}^{-\Lambda t} : \mathcal{S}'(\mathbb{R}^{d-1}) \to \mathcal{D}'(\mathbb{R}^d_+), \quad u \mapsto \mathcal{F}^{-1}\big(\mathrm{e}^{-t\sqrt{1+|\xi'|^2}}\mathcal{F}u(\xi')\big).$$

We write $\mathsf{D}_p(\Lambda^k)$ for the maximal domain of Λ^k in $\mathrm{L}^p(\mathbb{R}^d)$ and equip it with the graph norm $\|\cdot\|_{\mathrm{L}^p} + \|\Lambda^k \cdot\|_{\mathrm{L}^p}$. By definition of Bessel potential spaces, we have for $k \in \mathbb{N}$ up to equivalent norms,

$$\mathsf{D}_p(\Lambda^k) = \mathrm{H}^{k,p}(\mathbb{R}^{d-1}) = \mathrm{W}^{k,p}(\mathbb{R}^{d-1}). \tag{24}$$

Abstract semigroup theory [40, Theorem 1.14.5] provides an equivalent norm on the real interpolation space $\mathrm{W}^{k-1/p,p} = (\mathrm{L}^p, \mathrm{H}^{k,p})_{1-\frac{1}{kp},p}$:

$$\|\partial_t^k \mathrm{e}^{-\Lambda t} u\|_{\mathrm{L}^p(\mathbb{R}_+;\mathrm{L}^p(\mathbb{R}^{d-1}))} \approx \|u\|_{\mathrm{W}^{k-1/p,p}(\mathbb{R}^{d-1})} \qquad (u \in \mathrm{W}^{k-1/p,p}(\mathbb{R}^{d-1})). \tag{25}$$

With this at hand, we fix $\chi \in \mathrm{C}_0^\infty(\overline{\mathbb{R}_+})$ with $\chi(0) = 1$ and define the operator

$$\mathcal{E}u(x', x_d):=\chi(x_d)\mathrm{e}^{-\Lambda x_d}u(x') \qquad (x' \in \mathbb{R}^{d-1}, x_d \geq 0). \tag{26}$$

Proposition 4.15 *The operator* $\mathcal{E}$ *defined in* (26) *is* $\mathrm{W}^{s,p}(\mathbb{R}^{d-1}) \to \mathrm{X}^{s+1/p,p}(\mathbb{R}^d_+)$ *bounded for all* $p \in (1,\infty)$ *and* $s \in \mathbb{R} \setminus \mathbb{Z}$.

Proof Our argument is an adaption of [40, Sect. 2.9.3] and divides into six steps.

Step 1: *$\mathcal{E} : \mathrm{W}^{1-1/p,p}(\mathbb{R}^{d-1}) \to \mathrm{W}^{1,p}(\mathbb{R}^d_+)$ is bounded.* First, we note that for $\varphi \in \mathrm{C}_0^\infty(\overline{\mathbb{R}_+})$ the multiplication operator

$$\varphi(x_d) : \mathrm{L}^p(\mathbb{R}_+; \mathrm{L}^p(\mathbb{R}^{d-1})) \to \mathrm{L}^p(\mathbb{R}_+; \mathrm{L}^p(\mathbb{R}^{d-1})) \tag{27}$$

is bounded. By boundedness of the semigroup the same is true for

$$\varphi(x_d)\mathrm{e}^{-\Lambda x_d} : \mathrm{L}^p(\mathbb{R}^{d-1}) \to \mathrm{L}^p(\mathbb{R}_+; \mathrm{L}^p(\mathbb{R}^{d-1})). \tag{28}$$

In particular, we get $\mathcal{E} : \mathrm{L}^p(\mathbb{R}^{d-1}) \to \mathrm{L}^p(\mathbb{R}^d_+)$ if we choose $\varphi = \chi$ in (28). Using the product rule and (27), we deduce from (25) that for $k \in \mathbb{N}$ we have

$$\|\partial_d^k \mathcal{E} u\|_{\mathrm{L}^p(\mathbb{R}^d_+)} \lesssim \|u\|_{\mathrm{W}^{k-1/p}(\mathbb{R}^{d-1})}. \tag{29}$$

Using (24), (27), the identity $\Lambda \mathrm{e}^{-\Lambda x_d} = -\partial_d \mathrm{e}^{-\Lambda x_d}$ and (29), we obtain

$$\begin{aligned}
\|\mathcal{E}u\|_{\mathrm{W}^{1,p}(\mathbb{R}^d_+)} &\approx \|\mathcal{E}u\|_{\mathrm{L}^p(\mathbb{R}_+;\mathrm{W}^{1,p}(\mathbb{R}^{d-1}))} + \|\mathcal{E}u\|_{\mathrm{W}^{1,p}(\mathbb{R}_+;\mathrm{L}^p(\mathbb{R}^{d-1}))} \\
&\approx \|\mathcal{E}u\|_{\mathrm{L}^p(\mathbb{R}^d_+)} + \|\chi(x_d)\Lambda \mathrm{e}^{-\Lambda x_d} u(x')\|_{\mathrm{L}^p(\mathbb{R}^d_+)} \\
&\quad + \|\partial_d \mathcal{E} u\|_{\mathrm{L}^p(\mathbb{R}^d_+)} \\
&\lesssim \|u\|_{\mathrm{W}^{1-1/p,p}(\mathbb{R}^{d-1})}.
\end{aligned}$$

Step 2: *$\mathcal{E} : \mathrm{W}^{k-1/p,p}(\mathbb{R}^{d-1}) \to \mathrm{W}^{k,p}(\mathbb{R}^d_+)$ is bounded for $k \in \mathbb{N}$.* We argue by induction. The case $k = 1$ was treated in Step 1. Moreover, the derivatives in x_d-direction are under control owing to (29). We fix $1 \leq j \leq d-1$. As $\mathrm{e}^{-\Lambda x_d}$ and ∂_j both are Fourier multipliers on $\mathcal{S}'(\mathbb{R}^{d-1})$, they commute. Assume the claimed boundedness holds for $k \in \mathbb{N}$. Then $\partial_j^k \mathcal{E} : \mathrm{W}^{k-1/p,p}(\mathbb{R}^{d-1}) \to \mathrm{L}^p(\mathbb{R}^d_+)$ is bounded and we conclude from Lemma 4.4 that

$$\begin{aligned}
\|\partial_j^{k+1} \mathcal{E} u\|_{\mathrm{L}^p(\mathbb{R}^d_+)} = \|\partial_j^k \mathcal{E} \partial_j u\|_{\mathrm{L}^p(\mathbb{R}^d_+)} &\lesssim \|\partial_j u\|_{\mathrm{W}^{k-1/p,p}(\mathbb{R}^{d-1})} \\
&\lesssim \|u\|_{\mathrm{W}^{k+1-1/p,p}(\mathbb{R}^{d-1})}.
\end{aligned}$$

Step 3: *Lifting property.* To bring negative orders of differentiability into play, we introduce for $m \in \mathbb{N}$ the lift operator $\mathcal{I}_{2m} := \mathcal{F}^{-1}(1 + |\xi'|^2)^{-m}\mathcal{F}$ defined on $\mathcal{S}'(\mathbb{R}^{d-1})$. It is invertible and we write $\mathcal{I}_{-2m} := \mathcal{I}_{2m}^{-1}$. For $s \in \mathbb{R}$ the operator $\mathcal{I}_{2m}$ is an isomorphism $\mathrm{H}^{s,p}(\mathbb{R}^{d-1}) \to \mathrm{H}^{s+2m,p}(\mathbb{R}^{d-1})$ by definition of the norms on Bessel potential spaces. Since the Fourier multipliers $\mathcal{E}$ and $\mathcal{I}_{-2m}$ commute, we can decompose

$$\mathcal{E} = \mathcal{I}_{-2m} \circ \mathcal{E} \circ \mathcal{I}_{2m}, \tag{30}$$

in order to lift the argument of $\mathcal{E}$ into a space with positive order of differentiability.

Step 4: *$\mathcal{I}_{-2m}$ in d-dimensional space.* Since $\mathcal{I}_{-2m} = (1 - \Delta_{x'})^m$ is a differential operator of order $2m$ acting only in $d-1$ coordinates, we have $\mathcal{I}_{-2m} : \mathrm{H}^{s+2m,p}(\mathbb{R}^d) \to \mathrm{H}^{s,p}(\mathbb{R}^d)$ for integer s. Interpolation by means of Proposition 2.9 yields $\mathcal{I}_{-2m} : \mathrm{X}^{s+2m,p}(\mathbb{R}^d) \to \mathrm{X}^{s,p}(\mathbb{R}^d)$. The differential operator $\mathcal{I}_{-2m}$ is local in the sense that it commutes with the distributional restriction. Hence, its restriction to the upper half-space is well-defined and we get

$$\mathcal{I}_{-2m} : \mathrm{X}^{s+2m,p}(\mathbb{R}^d_+) \to \mathrm{X}^{s,p}(\mathbb{R}^d_+). \tag{31}$$

Step 5: *Interpolation of $\mathcal{I}_{2m}$ and $\mathcal{E}$.* As before, we interpolate $\mathcal{I}_{2m} : \mathrm{H}^{s,p}(\mathbb{R}^{d-1}) \to \mathrm{H}^{s+2m,p}(\mathbb{R}^{d-1})$ from Step 3 to obtain for all $s \in \mathbb{R}$ boundedness of

$$\mathcal{I}_{2m} : \mathrm{W}^{s,p}(\mathbb{R}^{d-1}) \to \mathrm{W}^{s+2m,p}(\mathbb{R}^{d-1}). \tag{32}$$

Similarly, real and complex interpolation of the outcome of Step 2 with the aid of Proposition 3.8 yields

$$\mathcal{E} : \mathrm{W}^{s,p}(\mathbb{R}^{d-1}) \to \mathrm{X}^{s+1/p,p}(\mathbb{R}^d_+) \tag{33}$$

if $s \geq 1 - 1/p$ is not an integer.

Step 6: *Patching everything together.* Let $s \in \mathbb{R} \setminus \mathbb{Z}$. If $s \geq 1 - 1/p$, then $\mathcal{E} : \mathrm{W}^{s,p}(\mathbb{R}^{d-1}) \to \mathrm{X}^{s+1/p,p}(\mathbb{R}^d_+)$ follows by (33). Otherwise, we choose $m \in \mathbb{N}$ such that $2m + s \geq 1 - 1/p$. We use the decomposition (30) to conclude $\mathcal{E} : \mathrm{W}^{s,p}(\mathbb{R}^{d-1}) \to \mathrm{X}^{s+1/p,p}(\mathbb{R}^d_+)$ from (32), (33) and (31). □

The next lemma justifies calling $\mathcal{E}$ an extension operator.

Lemma 4.16 *Let $F \subseteq \partial\mathbb{R}^d_+$ be $(d-1)$-regular and $\mathcal{R}_F$ the corresponding restriction operator. Let $p \in (1, \infty)$ and suppose that $s > 0$ is not an integer. If $u \in \mathrm{W}^{s,p}(\mathbb{R}^{d-1})$, then $\mathcal{R}_F \mathcal{E} u = u$ holds almost everywhere on F*

Proof By density it suffices to prove the claim for $u \in \mathrm{C}_0^\infty(\mathbb{R}^{d-1})$. Due to (24) we have $u \in \mathsf{D}_p(\Lambda^k)$ for all $k \in \mathbb{N}$ and $p \in (1, \infty)$. We pick k and p such that $\mathsf{D}_p(\Lambda^k)$ is continuously included into $\mathrm{C}(\mathbb{R}^{d-1})$ in virtue of Sobolev embeddings. Since we have $\Lambda^k \mathrm{e}^{-t\Lambda} u = \mathrm{e}^{-t\Lambda} \Lambda^k u$ for $t \geq 0$, the strong continuity of the semigroup on $\mathrm{L}^p(\mathbb{R}^{d-1})$ implies $\mathcal{E} u \in \mathrm{C}(\overline{\mathbb{R}^+}; \mathrm{C}(\mathbb{R}^{d-1})) = \mathrm{C}(\overline{\mathbb{R}^d_+})$ and $\mathcal{E} u(x', 0) = u(x', 0)$ for almost every $x' \in \mathbb{R}^{d-1}$. Proposition 4.15 guarantees $\mathcal{E} u \in \mathrm{X}^{s+1/p,p}(\mathbb{R}^d_+)$ and we conclude from (23) that $\mathcal{R}_F \mathcal{E} u = u$ holds almost everywhere on F. □

4.6 Conclusion of the Proof

Here, we will verify the set inclusions (20) and (21). Thereby we complete the proof of Theorem 1.1.

We start out with the interpolation in the case $s \in (0, 1/p)$, which we treat slightly more generally for a later use.

Proposition 4.17 *Let $p_0, p_1 \in (1,\infty)$, $s_0 \in [0, 1/p_0)$, $s_1 \in (1/p_1, 1]$, and for $\theta \in (0,1)$ define p and s as in (2). Suppose $s < 1/p$. Assume that $U \subseteq \mathbb{R}^d$ is a closed d-regular set with $(d-1)$-regular boundary. Moreover, assume that ${}^cU$ is also d-regular and that $D \subseteq \partial U$ is $(d-1)$-regular. Then it follows up to equivalent norms that*

(i)

$$[\mathrm{X}^{s_0,p_0}({}^cU), \mathrm{X}_D^{s_1,p_1}({}^cU)]_\theta = \mathrm{X}^{s,p}({}^cU),$$

(ii)

$$(\mathrm{X}^{s_0,p_0}({}^cU), \mathrm{X}_D^{s_1,p_1}({}^cU))_{\theta,p} = \mathrm{W}^{s,p}({}^cU),$$

with the exception that $s_0 \neq 0$ and $s_1 \neq 1$ are required in (i) for $\mathrm{X} = \mathrm{W}$.

Proof The "$\subseteq$"-inclusions follow from Proposition 3.13. For the converse let $\langle \cdot\,, \cdot \rangle$ denote either the θ-complex or (θ, p)-real interpolation bracket. Using the inclusion $\mathrm{X}_{\partial U}^{s_1,p_1}({}^cU) \subseteq \mathrm{X}_D^{s_1,p_1}({}^cU)$, we get

$$\langle \mathrm{X}^{s_0,p_0}({}^cU), \mathrm{X}_{\partial U}^{s_1,p_1}({}^cU)\rangle \subseteq \langle \mathrm{X}^{s_0,p_0}({}^cU), \mathrm{X}_D^{s_1,p_1}({}^cU)\rangle.$$

We identify the space on the left-hand side according to Proposition 4.7 to conclude. □

Since this proposition applies to $U = {}^cO$, we get (20) and (21) in case $s < 1/p$.

In a next step we establish the rest of Fig. 3. To this end, we shall appeal to the theory of Sect. 4.2 with $U = E_i$ in $\mathbb{R}^{d-1}$. This requires E_i to be $(d-1)$-regular in $\mathbb{R}^{d-1}$, its boundary to be a Lebesgue null set, and its interior to be of some class $\mathcal{D}^t$. The first requirement is met by Lemma 4.11, which also guarantees that ∂E_i is porous. Hence, so is its subset ∂E_i°. In view of Example 2.16 the interior of E_i is of class $\mathcal{D}^t$ for some $t \in (0,1)$. Finally, the boundary of a porous set is a null set by Lemma A.1.

Due to Lemma 4.2 the spaces $\mathrm{W}_\bullet^{s,p}({}^cE_i)$ interpolate as expected. Next, we check that the extension operator constructed in the previous section preserves the zero condition when restricted to $\mathrm{W}_\bullet^{s,p}({}^cE_i)$.

Lemma 4.18 *Let $p \in (1,\infty)$ and let $s > t(1/p - 1)$ not be an integer. If X denotes either H or W, then $\mathcal{E} : \mathrm{W}_\bullet^{s,p}({}^cE_i) \to \mathrm{X}_{E_i}^{s+1/p,p}(\mathbb{R}^d_+)$.*

Proof Let $u \in \mathrm{W}_\bullet^{s,p}({}^cE_i)$. Due to Proposition 4.15 we have $\mathcal{E}u \in \mathrm{X}^{s+1/p,p}(\mathbb{R}^d_+)$. By Lemma 4.16 and the definition of $\mathrm{W}_\bullet^{s,p}({}^cE_i)$ we know that $\mathcal{R}_{E_i}\mathcal{E}u = u = 0$ holds. This means $\mathcal{E}u \in \mathrm{X}_{E_i}^{s+1/p,p}(\mathbb{R}^d_+)$. □

Let now $p_0, p_1 \in (1,\infty)$, $s_0 \in [0, 1/p_0)$,$s_1 \in (1/p_1, 1]$, and $\theta \in (0,1)$. Let us recall

$$\frac{1}{p} = \frac{1-\theta}{p_0} + \frac{\theta}{p_1}, \quad s = (1-\theta)s_0 + \theta s_1$$

and that we assume $s > 1/p$. By these restrictions on the parameters there are $q \in (1, \infty)$ and $\varepsilon \in (0, \min\{1/q, t(1-1/q)\})$ such that the point $(1/q, -\varepsilon)^\top$ lies on the segment connecting $(1/p_1, s_1 - 1/p_1)^\top$ and $(1/p_0, s_0 - 1/p_0)^\top$ in the $(1/p, s)$-plane. Since we have by construction

$$\begin{pmatrix} 1/p \\ s - 1/p \end{pmatrix} = (1-\theta) \begin{pmatrix} 1/p_0 \\ s_0 - 1/p_0 \end{pmatrix} + \theta \begin{pmatrix} 1/p_1 \\ s_1 - 1/p_1 \end{pmatrix},$$

we can fix $\eta \in (0, \theta)$ such that

$$\begin{pmatrix} 1/p \\ s - 1/p \end{pmatrix} = (1-\eta) \begin{pmatrix} 1/q \\ -\varepsilon \end{pmatrix} + \eta \begin{pmatrix} 1/p_1 \\ s_1 - 1/p_1 \end{pmatrix}.$$

This yields identity (♡) in Fig. 3. Adding both lines of the previous equation gives

$$s = (1-\eta)(1/q - \varepsilon) + \eta s_1.$$

We deduce

$$\Big(1 - \frac{\theta - \eta}{1-\eta}\Big) s_0 + \frac{\theta - \eta}{1 - \eta} s_1 = 1/q - \varepsilon.$$

In the following all function spaces are on $\mathbb{R}^d_+$ and we omit the dependence. Let $\langle \cdot\,, \cdot \rangle$ denote either the complex or the $(\cdot\,, p)$-real interpolation bracket. From Proposition 4.17 and Proposition 3.12 we deduce

$$\Big\langle \mathrm{X}^{1/q-\varepsilon,q}, \mathrm{X}^{s_1,p_1}_{E_i} \Big\rangle_\eta = \Big\langle \Big[\mathrm{X}^{s_0,p_0}, \mathrm{X}^{s_1,p_1}_{E_i} \Big]_{\frac{\theta-\eta}{1-\eta}}, \mathrm{X}^{s_1,p_1}_{E_i} \Big\rangle_\eta = \Big\langle \mathrm{X}^{s_0,p_0}, \mathrm{X}^{s_1,p_1}_{E_i} \Big\rangle_\theta,$$

where $s_0 \neq 0$ and $s_1 \neq 1$ are required in case $\mathrm{X} = \mathrm{W}$. This establishes Fig. 3 in case of complex interpolation. It also establishes the analogue that corresponds to real interpolation of H-spaces. As for real interpolation of W-spaces, we invoke the following reiteration theorem [6, Theorem 3.5.3].

Proposition 4.19 *Let (X_0, X_1) be an interpolation couple. Let $p \in [1, \infty]$, $\theta_0, \theta_1 \in [0, 1]$ with $\theta_0 \neq \theta_1$, and $\lambda \in (0, 1)$. With $\theta := (1-\lambda)\theta_0 + \lambda\theta_1$ it follows that up to equivalent norms*

$$((X_0, X_1)_{\theta_0,p}, (X_0, X_1)_{\theta_1,p})_{\lambda,p} = (X_0, X_1)_{\theta,p},$$

subject to the interpretation $(X_0, X_1)_{j,p} := X_j$ in the endpoint cases $j \in \{0, 1\}$.

Indeed, in combination with Proposition 4.17 we can give

$$\Big(\mathrm{W}^{1/q-\varepsilon,q}, \mathrm{W}^{s_1,p_1}_{E_i}\Big)_{\eta,p} = \Big(\Big(\mathrm{W}^{s_0,p_0}, \mathrm{W}^{s_1,p_1}_{E_i}\Big)_{\frac{\theta-\eta}{1-\eta},p}, \mathrm{W}^{s_1,p_1}_{E_i}\Big)_{\eta,p} = \Big(\mathrm{W}^{s_0,p_0}, \mathrm{W}^{s_1,p_1}_{E_i}\Big)_{\theta,p}$$

without requiring $s_0 \neq 0$ or $s_1 \neq 1$. This completes Fig. 3 in the remaining case.

With this at hand, we complete the proof. Let $\langle \cdot\,, \cdot \rangle$ denote either the complex or the $(\cdot\,, p)$-real interpolation bracket. With Lemma 4.16 we derive $\mathcal{R}(f - \mathcal{E}\mathcal{R}f) = 0$ for $f \in \mathrm{X}^{s,p}$, which means $f - \mathcal{E}\mathcal{R}f \in \mathrm{X}^{s,p}_{\partial\mathbb{R}^d_+}$. We have

$$\langle \mathrm{X}^{s_0,p_0}, \mathrm{X}^{s_1,p_1}_{\partial\mathbb{R}^d_+} \rangle_\theta \subseteq \langle \mathrm{X}^{s_0,p_0}, \mathrm{X}^{s_1,p_1}_{E_i} \rangle_\theta,$$

where Proposition 4.7 identifies the left-hand space as $\mathrm{X}^{s,p}_{\partial\mathbb{R}^d_+}$ for complex interpolation and as $\mathrm{W}^{s,p}_{\partial\mathbb{R}^d_+}$ for real interpolation. From the decomposition

$$f = (f - \mathcal{E}\mathcal{R}f) + \mathcal{E}\mathcal{R}f$$

we conclude $f \in [\mathrm{X}^{s_0,p_0}, \mathrm{X}^{s_1,p_1}_{E_i}]_\theta$ for $f \in \mathrm{X}^{s,p}_{E_i}$ in case of complex interpolation, which completes the proof of (7), and $f \in (\mathrm{X}^{s_0,p_0}, \mathrm{X}^{s_1,p_1}_{E_i})_{\theta,p}$ for $f \in \mathrm{W}^{s,p}_{E_i}$, which shows (8).

5 A Complex $(\mathrm{W}^{-1,p}_D, \mathrm{W}^{1,p}_D)$ Interpolation Formula

In this section we prove Theorem 1.3. We begin by defining spaces of negative smoothness with boundary conditions on an open set.

Definition 5.1 Let $O \subseteq \mathbb{R}^d$ be open and $D \subseteq \overline{O}$ be $(d-1)$-regular. Let $p \in (1, \infty)$ and $s \in [0, 1]$. For X either H or W define

$$\mathrm{X}^{-s,p}(O) := (\mathrm{X}^{s,p'}(O))^*$$

and if $s > 1 - 1/p$ define

$$\mathrm{X}^{-s,p}_D(O) := (\mathrm{X}^{s,p'}_D(O))^*.$$

The second part of the definition is consistent with the case $O = \mathbb{R}^d$, see Sect. 2.3.

We are concerned with interpolation spaces between $\mathrm{W}^{-1,p}_D(O)$ and $\mathrm{W}^{1,p}_D(O)$. These two spaces form an interpolation couple since we can naturally view $\mathrm{W}^{1,p}_D(O)$ as a subspace of $\mathrm{W}^{-1,p}_D(O)$ by extending the $\mathrm{L}^p(O) - \mathrm{L}^{p'}(O)$ duality. We also recall that as a consequence of Lemma 3.3 the inclusion $\mathrm{W}^{1,p}_D(O) \subseteq \mathrm{L}^p(O)$ is dense.

As for interpolation of dual spaces, we have the following principle [6, Cor. 4.5.2], see also [5, Cor. 2.15] for a proper treatment of the spaces of conjugate-linear functionals indicated by a superscript asterisk.

Proposition 5.2 *Let (X_0, X_1) be an interpolation couple such that $X_0 \cap X_1$ is dense in both X_0 and X_1 and assume that X_0 is reflexive. For $\theta \in (0, 1)$ it follows that with equal norms*

$$[X_1^*, X_0^*]_{1-\theta} = ([X_0, X_1]_\theta)^*.$$

The idea of proof is to patch together the interpolation scale provided by Theorem 1.1 with its dual scale. This requires some overlap of interpolation scales. The following lemmas use some notions introduced in Sect. 3.2.

Lemma 5.3 *Let $O \subseteq \mathbb{R}^d$ be an open set with $(d-1)$-regular boundary. Let $p \in (1, \infty)$, $s \in (1/p - 1, 1/p)$, and let* X *denote either* H *or* W. *There is a retraction $\mathcal{R} : \mathrm{X}^{s,p}(\mathbb{R}^d) \to \mathrm{X}^{s,p}(O)$ with corresponding coretraction $\mathcal{E} : \mathrm{X}^{s,p}(O) \to \mathrm{X}^{s,p}(\mathbb{R}^d)$. These operators are the same for all p and s.*

Proof If $s \in [0, 1/p)$, then due to Corollary 2.18 we can take $\mathcal{R} := |_O$ and $\mathcal{E} := \mathcal{E}_0$ the extension by 0. By the usual identification of functions with distributions, these operators consistently act on $\mathrm{X}^{s,p}$ also when $s \in (1/p - 1, 0]$. Indeed, if $f \in \mathrm{X}^{s,p}(\mathbb{R}^d)$ then for all $\varphi \in \mathrm{X}^{-s,p'}(O)$ we set

$$\langle f|_O, \varphi\rangle_{\mathrm{X}^{s,p}(O),\mathrm{X}^{-s,p'}(O)} := \langle f, \mathcal{E}_0\varphi\rangle_{\mathrm{X}^{s,p}(\mathbb{R}^d),\mathrm{X}^{-s,p'}(\mathbb{R}^d)},$$

where $\langle\cdot\,,\cdot\rangle$ denotes the respective duality pairing. Well-definedness and boundedness of $|_O : \mathrm{X}^{s,p}(\mathbb{R}^d) \to \mathrm{X}^{s,p}(O)$ follows again from Corollary 2.18. Conversely, given $g \in \mathrm{X}^{s,p}(O)$, we let the zero extension $\mathcal{E}_0 g$ act on $\psi \in \mathrm{X}^{-s,p'}(\mathbb{R}^d)$ via

$$\langle \mathcal{E}_0 g, \psi\rangle_{\mathrm{X}^{s,p}(\mathbb{R}^d),\mathrm{X}^{-s,p'}(\mathbb{R}^d)} := \langle g, \psi|_O\rangle_{\mathrm{X}^{s,p}(O),\mathrm{X}^{-s,p'}(O)}.$$

It is bounded since $|_O : \mathrm{X}^{-s,p'}(\mathbb{R}^d) \to \mathrm{X}^{-s,p'}(O)$ is bounded by definition of the quotient norm. Finally, $(\mathcal{E}_0 g)|_O = g$ follows by concatenating the two identities above. □

Lemma 5.4 *Let $O \subseteq \mathbb{R}^d$ be an open set with $(d-1)$-regular boundary. Let $p \in (1, \infty)$, $s_0, s_1 \in (1/p - 1, 1/p)$, $\theta \in (0, 1)$, and set $s := (1-\theta)s_0 + \theta s_1$. If* X *denotes either* H *or* W, *then up to equivalent norms*

$$[\mathrm{X}^{s_0,p}(O), \mathrm{X}^{s_1,p}(O)]_\theta = \mathrm{X}^{s,p}(O)$$

with the exception that $s = 0$ is only allowed if X = H.

Proof The corresponding identities on $O = \mathbb{R}^d$ are due to Proposition 2.9. The conclusion follows from Proposition 3.5 applied with the retraction-coretraction pair from Lemma 5.3. □

With these tools at hand, we can give the

Proof of Theorem 1.3 We appeal to Wolff's result, Proposition 3.10. All function spaces will be on O and we omit the dependence. We fix some $s \in (0, \min\{1/p, 1 - 1/p\})$ and consider the following diagram.

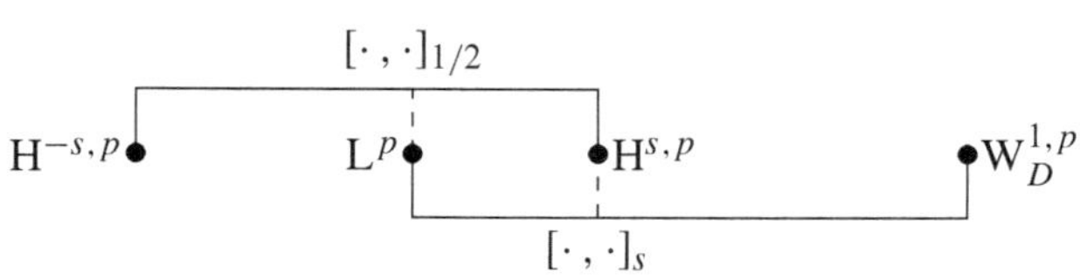

The $1/2$-interpolation is due to Lemma 5.4 and s-interpolation is due to Theorem 1.1. Proposition 3.10 yields $\mathrm{L}^p = [\mathrm{H}^{-s,p}, \mathrm{W}_D^{1,p}]_{s/(1+s)}$. Therefore we can consider the diagram

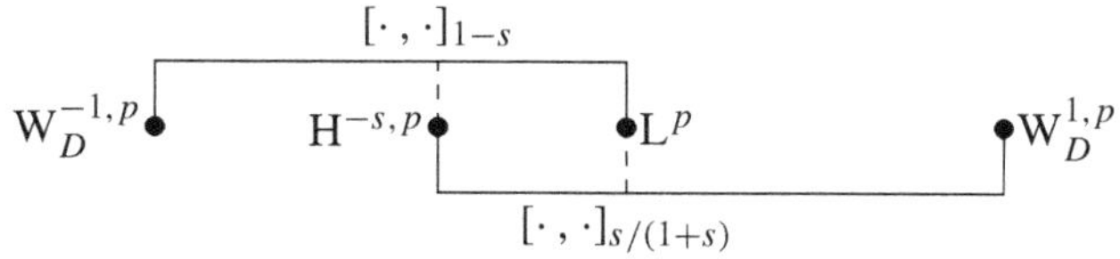

where the $(1-s)$-interpolation follows from Theorem 1.1 by means of the duality principle of Proposition 5.2. Another application of Proposition 3.10 completes the proof. □

6 Real Interpolation Via the Trace Method

Here, we present the proof of Theorem 1.4.

6.1 Road Map

The main new ingredient is Grisvard's trace characterization of real interpolation spaces [19, Theorem 5.12] stated in Proposition 6.1 below.

For X a Banach space we need the usual Bochner–Lebesgue space $\mathrm{L}^p(\mathbb{R}; X)$ of X-valued p-integrable functions on the real line and for $s > 0$ the respective (fractional) Sobolev spaces $\mathrm{W}^{s,p}(\mathbb{R}; X)$ that are defined as in the scalar case upon replacing absolute values by norms on X. For $s > 1/p$ such functions have a continuous representative and in that sense $\mathrm{W}^{s,p}(\mathbb{R}; X) \subseteq \mathrm{C}(\mathbb{R}; X)$ holds with continuous inclusion [38, Cor. 26]. In particular, the pointwise evaluation $|_{t=0} : \mathrm{W}^{s,p}(\mathbb{R}; X) \to X$ is well-defined and bounded. All this was already used in [19] and known at the time by different proofs.

Proposition 6.1 *(Grisvard) Let X_0, X_1 be Banach spaces such that $X_1 \subseteq X_0$ with dense and continuous inclusion. Let $p \in (1, \infty)$ and $s > 1/p$. Then*

$$\big(\mathrm{L}^p(\mathbb{R}; X_1) \cap \mathrm{W}^{s,p}(\mathbb{R}; X_0)\big)|_{t=0} \subseteq \big(X_0, X_1\big)_{1-\frac{1}{sp},p}.$$

The strategy to obtain Theorem 1.4 is schematically displayed in Fig. 4. Owing to Proposition 3.13 and the bounded inverse theorem, we only need to prove the set inclusion "$\supseteq$" in (f).

The key observation is that functions in the starting space of Fig. 4 can be extended by zero to the set

$$O \perp D := \big(O \times \{0\}\big) \cup \big(D \times \mathbb{R}\big),$$

without losing Sobolev regularity. We shall see that $O \perp D$ is, as expected, a d-regular subset of $\mathbb{R}^{d+1}$. By means of the Jonsson–Wallin operator $\mathcal{E}_{O \perp D}$ we can then extend to

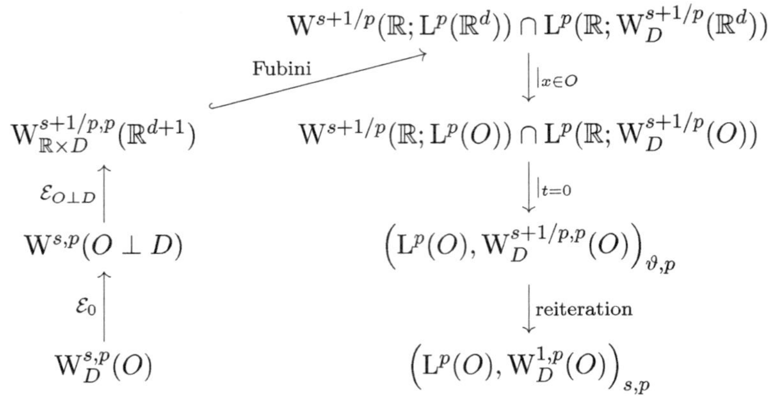

Fig. 4 Schematic presentation of the argument for obtaining the inclusion "$\supseteq$" in Theorem 1.4 for $s > 1/p$. For $s < 1/p$ the diagram would start with $W^{s,p}(O)$ instead

all of $\mathbb{R}^{d+1}$ and via a Fubini property we end up in a space suitable for Grisvard's result. Taking the trace yields the desired inclusion, up to applying reiteration techniques from Proposition 4.19 in the final step.

Unless otherwise stated, we make the following

Assumption 6.2 The set $O \subseteq \mathbb{R}^d$ is open and d-regular. The Dirichlet part $D \subseteq \overline{O}$ is uniformly $(d-1)$-regular.

Only the final step will use the $(d-1)$-regularity of the full boundary ∂O additionally assumed in Theorem 1.4.

6.2 Hardy's Inequality

In order to obtain the mapping property of the zero extension $\mathcal{E}_0$ in Fig. 4, we establish a fractional Hardy inequality adapted to mixed boundary conditions that might be of independent interest. In contrast to related inequalities in [16] we completely avoid the use of capacities.

Definition 6.3 A set $U \subseteq \mathbb{R}^d$ is *plump* if there exists $\kappa \in (0, 1)$ with the property:

$$\forall x \in \overline{U}, r \leq \operatorname{diam}(U) \quad \exists y \in B(x, r) \; : \; B(y, \kappa r) \subseteq U.$$

Remark 6.4 A comparison with Definition 2.6 yields first examples of plump sets. Namely, if $E \subseteq \mathbb{R}^d$ is uniformly porous, then ${}^cE$ is plump. This example can be modified to the effect that E is bounded and (uniformly) porous and $Q \subseteq \mathbb{R}^d$ is an open cube containing $\overline{E}$: Still we have that $Q \setminus E$ is plump.

We cite a result of Dyda–Vähäkangas [14, Theorem 1].

Proposition 6.5 *[Dyda–Vähäkangas] Let* $p \in (1, \infty)$ *and* $s \in (0, 1)$. *Suppose that* $U \subseteq \mathbb{R}^d$ *is a proper, plump, open set in* $\mathbb{R}^d$. *Assume one of the following conditions:*

(i) $\overline{\dim}_{\mathcal{AS}}(\partial U) < d - sp$ *and* U *is unbounded.*
(ii) $\underline{\dim}_{\mathcal{AS}}(\partial U) > d - sp$ *and either* U *is bounded or* ∂U *is unbounded.*

Then there exists a constant c *such that the inequality*

$$\int_U \frac{|f(x)|^p}{\mathrm{d}(x, \partial U)^{sp}} \,\mathrm{d}x \leq c \int_U \int_U \frac{|f(x) - f(y)|^p}{|x - y|^{sp+d}} \,\mathrm{d}x \,\mathrm{d}y$$

holds for all measurable functions f *for which the left-hand side is finite.*

Upper and lower Assouad dimension have been introduced in Definition 2.8. For uniformly ℓ-regular sets they both equal ℓ, see Proposition A.6. With this at hand, we can state and prove the Hardy inequality alluded to above.

Proposition 6.6 *Let* $p \in (1, \infty)$ *and* $s \in (0, 1)$, $s \neq 1/p$. *Under Assumption 6.2 there is a constant* $C > 0$ *such that the fractional Hardy inequality*

$$\int_O \frac{|f(x)|^p}{\mathrm{d}(x, D)^{sp}} \,\mathrm{d}x \leq C \|f\|^p_{\mathrm{W}^{s,p}(O)} \tag{34}$$

holds for all $f \in \mathrm{W}^{s,p}(O)$ *if* $s < 1/p$ *and for all* $f \in \mathrm{W}^{s,p}_D(O)$ *if* $s > 1/p$.

Proof In both cases we shall reduce the claim to Proposition 6.5 on some auxiliary set.

Case 1: $s < 1/p$. Since D is uniformly $(d-1)$-regular, so is $\overline{D}$. Due to Propositions A.6 and A.9 we have that $\overline{D}$ is uniformly porous and hence $U := \mathbb{R}^d \setminus \overline{D}$ is plump. Since $(d-1)$-regular sets have empty interior, we conclude $\partial U = \overline{D}$, which has upper and lower Assouad dimension $d-1$. Part (i) of Proposition 6.5 yields for all measurable f for which the left-hand side is finite

$$\int_{\mathbb{R}^d} \frac{|f(x)|^p}{\mathrm{d}(x, D)^{sp}} \,\mathrm{d}x \leq c \int_{\mathbb{R}^d} \int_{\mathbb{R}^d} \frac{|f(x) - f(y)|^p}{|x - y|^{sp+d}} \,\mathrm{d}x \,\mathrm{d}y \leq c \|f\|^p_{\mathrm{W}^{s,p}(\mathbb{R}^d)}. \tag{35}$$

Example 2.16 guarantees that $x \mapsto \mathrm{d}(x, D)^{-sp}$ is locally integrable. Thus, (35) applies to every $f \in \mathrm{C}_0^\infty(\mathbb{R}^d)$, a dense subspace of $\mathrm{W}^{s,p}(\mathbb{R}^d)$, and we can use Fatou's lemma to extend (35) to all $f \in \mathrm{W}^{s,p}(\mathbb{R}^d)$. Restriction to O yields (34) for $f \in \mathrm{W}^{s,p}(O)$.

Case 2: $s > 1/p$ *and* D *is unbounded.* Part (ii) of Proposition 6.5 applies to $U := \mathbb{R}^d \setminus \overline{D}$ and we can argue as before, except that now we have (35) *a priori* for $f \in \mathrm{C}_D^\infty(\mathbb{R}^d)$, a dense class of $f \in \mathrm{W}^{s,p}_D(\mathbb{R}^d)$ in view of Lemma 3.3. Hence we get (34) for $f \in \mathrm{W}^{s,p}_D(O)$.

Case 3: $s > 1/p$ *and* D *is bounded.* Let Q be an open cube that contains $\overline{D}$. As before we obtain that $U := 2Q \setminus \overline{D}$ is plump, where $2Q$ denotes the concentric cube with twice the sidelength. Moreover, $\partial U = \partial(2Q) \cup \overline{D}$ is uniformly $(d-1)$-regular as a finite union of sets with that property. Hence, it has lower

Assouad dimension $d-1$. Part (ii) of Proposition 6.5 yields for all measurable g for which the left-hand side is finite

$$\int_U \frac{|g(x)|^p}{\mathrm{d}(x, \partial U)^{sp}} \, \mathrm{d}x \leq c \int_U \int_U \frac{|g(x) - g(y)|^p}{|x-y|^{sp+d}} \, \mathrm{d}x \, \mathrm{d}y \leq c \|g\|^p_{\mathrm{W}^{s,p}(U)}. \tag{36}$$

This applies to $g \in \mathrm{C}^\infty_{\partial U}(U)$ and extends to $g \in \mathrm{W}^{s,p}_{\partial U}(U)$ as in Case 2.

Let now $f \in \mathrm{W}^{s,p}_D(O)$. Let us fix $\eta \in \mathrm{C}^\infty_0(2Q)$ equal to 1 on Q and let $\mathcal{E}$ be an extension operator for O as in Proposition 2.13. We can bound

$$\begin{aligned}\int_O \frac{|f(x)|^p}{\mathrm{d}(x, D)^{sp}} \, \mathrm{d}x &\leq \int_{O \cap Q} \frac{|\eta(x)\mathcal{E} f(x)|^p}{\mathrm{d}(x, D)^{sp}} \, \mathrm{d}x + \int_{O \setminus Q} \frac{|f(x)|^p}{\mathrm{d}(x, D)^{sp}} \, \mathrm{d}x \\ &\leq \int_U \frac{|\eta(x)\mathcal{E} f(x)|^p}{\mathrm{d}(x, \partial U)^{sp}} \, \mathrm{d}x + \int_{O \setminus Q} \frac{|f(x)|^p}{\mathrm{d}(x, D)^{sp}} \, \mathrm{d}x =: I_1 + I_2,\end{aligned}$$

where we have used $O \cap Q \subseteq U$ and $D \subseteq \partial U$ to obtain I_1. Since on ${}^cQ$ we have $\mathrm{d}(\cdot\,, D) \geq \mathrm{d}(D, \partial Q) > 0$, we control I_2 by $\|f\|^p_{\mathrm{L}^p(O)}$. Next, $\mathcal{E} f \in \mathrm{W}^{s,p}_D(\mathbb{R}^d)$ follows from Lemma 3.4. Pointwise multiplication by η is bounded on $\mathrm{W}^{s,p}(\mathbb{R}^d)$ and maps $\mathrm{W}^{s,p}_D(\mathbb{R}^d) \to \mathrm{W}^{s,p}_{\partial U}(\mathbb{R}^d)$ since this is true for the respective dense subsets provided by Lemma 3.3. In conclusion, we have $\eta \mathcal{E} f \in \mathrm{W}^{s,p}_{\partial U}(\mathbb{R}^d)$. Hence, (36) gives control on I_1 by $\|\eta \mathcal{E} f\|^p_{\mathrm{W}^{s,p}(\mathbb{R}^d)}$. The boundedness of $\mathcal{E}$ leads us to a desirable bound for I_1. □

6.3 Details of the Proof

We are in a position to give a precise meaning to Fig. 4. We begin with the zero extension part on the left.

The following lemma is a straightforward consequence of a product formula for Hausdorff measure [17, Theorem 2.10.45]. Full details are written out in [16, Cor. 7.6].

Lemma 6.7 *If $O \subseteq \mathbb{R}^d$ is d-regular and $D \subseteq \mathbb{R}^d$ is $(d-1)$-regular, then $D \times \mathbb{R}$, $O \times \{0\}$, and $O \perp D \subseteq \mathbb{R}^d$ are d-regular.*

Since $O \perp D$ is a d-regular subset of $\mathbb{R}^{d+1}$, the fractional Sobolev spaces $\mathrm{W}^{s,p}(O \perp D)$ can be defined as in Sect. 2.4 and there is a corresponding Jonsson–Wallin theory in $\mathbb{R}^{d+1}$. In the following, we systematically use bold face to distinguish geometric objects such as points, balls, and Hausdorff measures in $\mathbb{R}^{d+1}$ from their counterparts in $\mathbb{R}^d$.

Proposition 6.8 *Let $p \in (1, \infty)$ and $s \in (0, 1)$, $s \neq 1/p$. Under Assumption 6.2 the zero extension operator*

$$(\mathcal{E}_0 f)(x, t) := \begin{cases} f(x) & (\text{if } x \in O,\, t = 0) \\ 0 & (\text{if } x \in D,\, t \in \mathbb{R}) \end{cases}$$

is $\mathrm{W}^{s,p}(O) \to \mathrm{W}^{s,p}(O \perp D)$ *bounded if* $s < 1/p$ *and* $\mathrm{W}^{s,p}_D(O) \to \mathrm{W}^{s,p}(O \perp D)$ *bounded if* $s > 1/p$.

Proof Since the outer measure $E \mapsto \mathcal{H}^d(E \times \{0\})$ on $\mathbb{R}^d$ is a translation invariant Borel measure that assigns finite measure to the unit cube [44, §27], the induced measure coincides up to a norming constant $c_d > 0$ with the d-dimensional Lebesgue measure. Thus, $\mathcal{E}_0 f \in \mathrm{L}^p(O \perp D)$ is a consequence of $f \in \mathrm{L}^p(O)$.

We use Tonelli's theorem to bound the remaining part of the $\mathrm{W}^{s,p}(O \perp D)$-norm by

$$\begin{aligned}
&\iint_{\substack{\boldsymbol{x},\boldsymbol{y} \in O \perp D \\ |\boldsymbol{x}-\boldsymbol{y}|<1}} \frac{|\mathcal{E}_0 f(\boldsymbol{x}) - \mathcal{E}_0 f(\boldsymbol{y})|^p}{|\boldsymbol{x}-\boldsymbol{y}|^{d+sp}} \, \mathcal{H}^d(\mathrm{d}\boldsymbol{x}) \, \mathcal{H}^d(\mathrm{d}\boldsymbol{y}) \\
&\leq c_d \iint_{\substack{x,y \in O \\ |x-y|<1}} \frac{|f(x) - f(y)|^p}{|x-y|^{d+sp}} \, \mathrm{d}x \, \mathrm{d}y \\
&\quad + 2 \int_O \int_{\substack{\boldsymbol{y} \in D \times \mathbb{R} \\ |\boldsymbol{y}-(x,0)|<1}} \frac{|f(x)|^p}{|\boldsymbol{y}-(x,0)|^{d+sp}} \, \mathcal{H}^d(\mathrm{d}\boldsymbol{y}) \, \mathrm{d}x.
\end{aligned} \tag{37}$$

The first integral on the right is bounded by $\|f\|^p_{\mathrm{W}^{s,p}(O)}$. If the inner domain of integration in the second integral is non-empty, then there exists an integer $n_0 \geq 0$ such that $2^{-n_0-1} < \mathrm{d}(x, D) \leq 2^{-n_0}$. We then split the integral into dyadic annuli

$$\boldsymbol{C}_n := \big(D \times \mathbb{R}\big) \cap \big((\mathbf{B}((x,0), 2^{-n}) \setminus \mathbf{B}((x,0), 2^{-n-1})\big),$$

each of which satisfies $\mathcal{H}^d(\boldsymbol{C}_n) \lesssim 2^{-dn}$ since $D \times \mathbb{R}$ is d-regular, to give

$$\begin{aligned}
\int_{\substack{\boldsymbol{y} \in D \times \mathbb{R} \\ |\boldsymbol{y}-(x,0)|<1}} \frac{1}{|\boldsymbol{y}-(x,0)|^{d+sp}} \, \mathcal{H}^d(\mathrm{d}\boldsymbol{y}) &\lesssim \sum_{n=0}^{n_0} 2^{(n+1)(d+sp)} 2^{-dn} \\
&= \frac{2^{d+sp}}{2^{sp}-1}(2^{sp(n_0+1)} - 1).
\end{aligned}$$

By choice of n_0, the right-hand side is controlled by $\mathrm{d}(x, D)^{-sp}$. In conclusion, the second integral on the right of (37) is bounded by

$$\int_O \int_{\substack{\boldsymbol{y} \in D \times \mathbb{R} \\ |\boldsymbol{y}-(x,0)|<1}} \frac{|f(x)|^p}{|\boldsymbol{y}-(x,0)|^{d+sp}} \, \mathcal{H}^d(\mathrm{d}\boldsymbol{y}) \, \mathrm{d}x \lesssim \int_O \frac{|f(x)|^p}{\mathrm{d}(x, D)^{sp}} \, \mathrm{d}x.$$

The claim follows by Proposition 6.6. □

The Fubini property appearing in Fig. 4 is as follows. Throughout, we canonically identify $\mathrm{L}^p(\mathbb{R}^{d+1})$ with $\mathrm{L}^p(\mathbb{R}; \mathrm{L}^p(\mathbb{R}^d))$ by means of Fubini's theorem.

Lemma 6.9 *If* $p \in (1, \infty)$ *and* $s \geq 0$, *then up to equivalent norms*

$$\mathrm{W}^{s,p}(\mathbb{R}^{d+1}) = \mathrm{L}^p(\mathbb{R}; \mathrm{W}^{s,p}(\mathbb{R}^d)) \cap \mathrm{W}^{s,p}(\mathbb{R}; \mathrm{L}^p(\mathbb{R}^d)). \tag{38}$$

Proof For $s \geq 0$ an integer, the claim follows directly from Fubini's theorem. Let now $s = k + \sigma$, where $k \geq 0$ is an integer and $\sigma \in (0, 1)$. According to [40, Sect. 2.5.1] we can equivalently norm $\mathrm{W}^{s,p}(\mathbb{R}^d)$ by

$$\|f\| := \|f\|_{\mathrm{L}^p} + \sum_{j=1}^{d} \left(\int_{\mathbb{R}} \int_{\mathbb{R}^d} \left| \frac{\partial^k f}{\partial x_j^k}(x + he_j) - \frac{\partial^k f}{\partial x_j^k}(x) \right|^p \mathrm{d}x \, \frac{\mathrm{d}h}{|h|^{1+sp}} \right)^{1/p},$$

where $(e_j)_j$ denote the standard unit vectors in $\mathbb{R}^d$. This equivalent norm only takes into account differences of f along the coordinate axes. Therefore we obtain (38) from Fubini's theorem if we equivalently norm all appearing spaces as described before. □

The next lemma makes Fig. 4 precise, except for the final step.

Lemma 6.10 *Let $p \in (1, \infty)$ and $s \in (0, 1)$, $s \neq 1/p$. Under Assumption 6.2 the set inclusion*

$$(\mathrm{L}^p(O), \mathrm{W}_D^{s+1/p,p}(O))_{\vartheta,p} \supseteq \begin{cases} \mathrm{W}_D^{s,p}(O) & (\text{if } s > 1/p) \\ \mathrm{W}^{s,p}(O) & (\text{if } s < 1/p) \end{cases},$$

holds for $\vartheta \in (0, 1)$ satisfying $\vartheta(s + 1/p) = s$.

Proof We fix f, which is a function in $\mathrm{W}^{s,p}(O)$ if $s < 1/p$ and in $\mathrm{W}_D^{s,p}(O)$ if $s > 1/p$, respectively.

The inclusion $\mathrm{W}_D^{s+1/p,p}(O) \subseteq \mathrm{L}^p(O)$ is continuous and it is dense since already $\mathrm{C}_{\partial O}^\infty(O)$ is dense in $\mathrm{L}^p(O)$. In view of Proposition 6.1 it suffices to construct a function

$$F \in \mathrm{L}^p(\mathbb{R}; \mathrm{W}_D^{s+1/p,p}(O)) \cap \mathrm{W}^{s+1/p,p}(\mathbb{R}; \mathrm{L}^p(O)) \quad \text{such that} \quad F|_{t=0} = f.$$

For the construction we start by extending f to $O \perp D$ by zero. This extension $\mathcal{E}_0 f$ is in $\mathrm{W}^{s,p}(O \perp D)$ due to Proposition 6.8. Since $O \perp D$ is a d-regular subset of $\mathbb{R}^{d+1}$ according to Lemma 6.7, we can use Proposition 2.10 to extend $\mathcal{E}_0 f$ to a function $G \in \mathrm{W}^{s+1/p,p}(\mathbb{R}^{d+1})$ in virtue of the corresponding Jonsson–Wallin operator. In view of Lemma 6.9 we have by canonical identification

$$G \in \mathrm{L}^p(\mathbb{R}; \mathrm{W}^{s+1/p,p}(\mathbb{R}^d)) \cap \mathrm{W}^{s+1/p,p}(\mathbb{R}; \mathrm{L}^p(\mathbb{R}^d)). \tag{39}$$

A closer inspection reveals the following.

(i) Let $\mathcal{R}$ be the Jonsson–Wallin restriction to the d-set $D \times \mathbb{R}$ in $\mathbb{R}^{d+1}$. We have $\mathcal{R}G = 0$ by construction and therefore $G \in \mathrm{W}_{D\times\mathbb{R}}^{s+1/p,p}(\mathbb{R}^{d+1})$. We introduce $t := \min\{s + 1/p, 1\}$ to have Lemma 3.3 at our disposal and approximate $G \in \mathrm{W}_{D\times\mathbb{R}}^{t,p}(\mathbb{R}^{d+1})$ in that space by test functions $G_n \in \mathrm{C}_{D\times\mathbb{R}}^\infty(\mathbb{R}^{d+1})$. By slicing

$$G_n \in \mathrm{L}^p(\mathbb{R}; \mathrm{W}_D^{t,p}(\mathbb{R}^d)) \cap \mathrm{W}^{t,p}(\mathbb{R}; \mathrm{L}^p(\mathbb{R}^d))$$

and due to the Fubini property of Lemma 6.9 the limit G is contained in the same space. From consistency of the restriction operator $\mathcal{R}$ on fractional Sobolev spaces we can infer $\mathrm{W}_D^{t,p}(\mathbb{R}^d) \cap \mathrm{W}^{s+1/p,p}(\mathbb{R}^d) = \mathrm{W}_D^{s+1/p,p}(\mathbb{R}^d)$. This being said, it follows from (39) that we have

$$G \in \mathrm{L}^p(\mathbb{R}; \mathrm{W}_D^{s+1/p,p}(\mathbb{R}^d)) \cap \mathrm{W}^{s+1/p,p}(\mathbb{R}; \mathrm{L}^p(\mathbb{R}^d)).$$

(ii) Let $\mathcal{R}$ be the Jonsson–Wallin restriction to the d-set $\mathbb{R}^d \times \{0\}$ in $\mathbb{R}^{d+1}$. This operator is bounded from $\mathrm{W}^{s+1/p}(\mathbb{R}^{d+1})$ into $\mathrm{L}^p(\mathbb{R}^d \times \{0\})$ by Proposition 2.10. On the other hand, we can look at the restriction $|_{t=0}$ defined on $\mathrm{L}^p(\mathbb{R}; \mathrm{W}^{s+1/p,p}(\mathbb{R}^d)) \cap \mathrm{W}^{s+1/p,p}(\mathbb{R}; \mathrm{L}^p(\mathbb{R}^d))$ and bounded into $\mathrm{L}^p(\mathbb{R}^d)$. Identifying corresponding objects via Fubini's theorem as before, it turns out that these two restrictions are the same since they obviously agree on a dense class of continuous functions. Since $\mathcal{R}G$ and f coincide $\mathcal{H}^d$-almost everywhere on $O \times \{0\}$ by construction, we can record

$$G|_{t=0} = f \quad \text{almost everywhere on } O.$$

The outcome of observations (i) and (ii) shows that $F := G|_{O \times \mathbb{R}}$ verifies (39). □

Together with Proposition 3.13 we obtain

Corollary 6.11 *If in addition to Assumption 6.2 the set O has $(d-1)$-regular boundary, then the set inclusion in Lemma 6.10 is an equality with equivalent norms.*

Eventually, we can complete the

Proof of Theorem 1.4 In the following all function spaces will be on O and we omit the dependence on O for clarity. In view of the reiteration theorem above it suffices to treat the case $s_0 = 0$ and $s_1 = 1$ and prove for $s \in (0, 1)$ that up to equivalent norms it follows that

$$(\mathrm{L}^p(O), \mathrm{W}_D^{1,p})_{s,p} = \begin{cases} \mathrm{W}_D^{s,p} & (\text{if } s > 1/p) \\ \mathrm{W}^{s,p} & (\text{if } s < 1/p) \end{cases}. \tag{40}$$

If $s + 1/p = 1$, then the claim follows from Corollary 6.11. The proof for $s + 1/p \neq 1$ divides into four cases.

Case 1: $s > 1/p$ and $s + 1/p < 1$. We have a diagram suitable for Wolff interpolation:

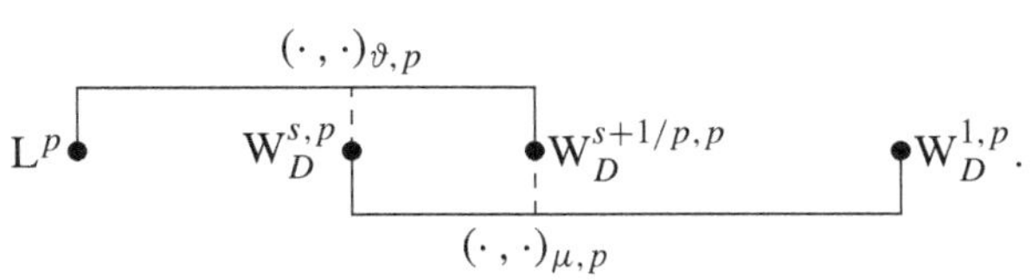

Indeed, the (ϑ, p)-interpolation is due to Corollary 6.11 and the (μ, p)-interpolation with suitable $\mu \in (0, 1)$ is due to Theorem 1.2. The claim follows by Proposition 3.10.

Case 2: $s > 1/p$ and $s + 1/p > 1$. We fix any $t \in (s, 1)$ and let $\lambda \in (0, 1)$ satisfy $(1-\lambda)s + \lambda(s + 1/p) = t$. Applying one after the other Theorem 1.2, Corollary 6.11, and Proposition 4.19, we obtain

$$\begin{aligned}\mathrm{W}_D^{t,p} &= (\mathrm{W}_D^{s,p}, \mathrm{W}_D^{s+1/p,p})_{\lambda,p} = ((\mathrm{L}^p, \mathrm{W}_D^{s+1/p,p})_{\vartheta,p}, \mathrm{W}_D^{s+1/p,p})_{\lambda,p} \\ &= (\mathrm{L}^p, \mathrm{W}^{s+1/p,p})_{\theta,p},\end{aligned}$$

with $\theta = t/(s + 1/p)$. Once again by Proposition 4.19 and Corollary 6.11 we find

$$(\mathrm{L}^p, \mathrm{W}_D^{t,p})_{s/t,p} = (\mathrm{L}^p, (\mathrm{L}^p, \mathrm{W}_D^{s+1/p,p})_{\theta,p})_{s/t,p} = (\mathrm{L}^p, \mathrm{W}_D^{s+1/p,p})_{\vartheta,p} = \mathrm{W}_D^{s,p}.$$

Thus we obtain the desired result (40) from Proposition 3.10 applied as follows:

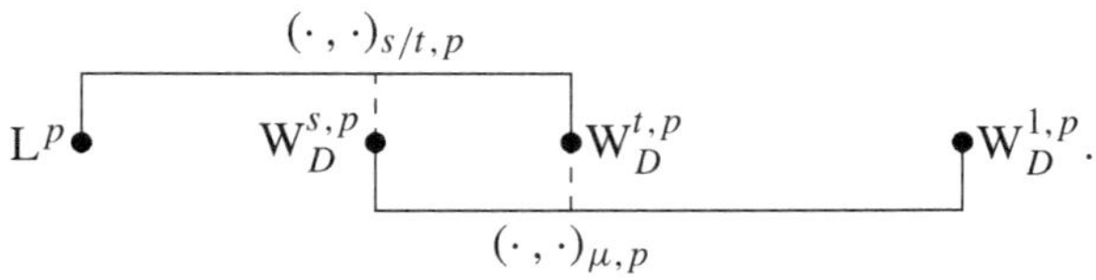

Indeed, we have obtained the $(s/t, p)$-interpolation above and the (μ, p)-interpolation for appropriately chosen μ is due to Theorem 1.2. Note that because of the exceptional case for real interpolation of Sobolev spaces we cannot pick $t = 1$ right away.

Case 3: $s < 1/p$ and $s + 1/p < 1$. We can apply one of the previous two cases with $s + 1/p$ in place of s to obtain $(\mathrm{L}^p, \mathrm{W}_D^{1,p})_{s+1/p,p} = \mathrm{W}_D^{s+1/p,p}$. Together with Corollary 6.11 in the first and reiteration in the third step, we are led to the desired result

$$\mathrm{W}^{s,p} = (\mathrm{L}^p, \mathrm{W}_D^{s+1/p,p})_{\vartheta,p} = (\mathrm{L}^p, (\mathrm{L}^p, \mathrm{W}_D^{1,p})_{s+1/p,p})_{\vartheta,p} = (\mathrm{L}^p, \mathrm{W}_D^{1,p})_{s,p}.$$

Case 4: $s < 1/p$ and $s + 1/p > 1$. We pick $1/p < \lambda < \kappa < 1$. By one of the first two cases along with Proposition 4.19, we find

$$(\mathrm{L}^p, \mathrm{W}_D^{\kappa,p})_{\lambda/\kappa,p} = (\mathrm{L}^p, (\mathrm{L}^p, \mathrm{W}_D^{1,p})_{\kappa,p})_{\lambda/\kappa,p} = (\mathrm{L}^p, \mathrm{W}_D^{1,p})_{\lambda,p} = \mathrm{W}_D^{\lambda,p}.$$

Together with Theorem 1.2 this establishes for suitable μ the diagram

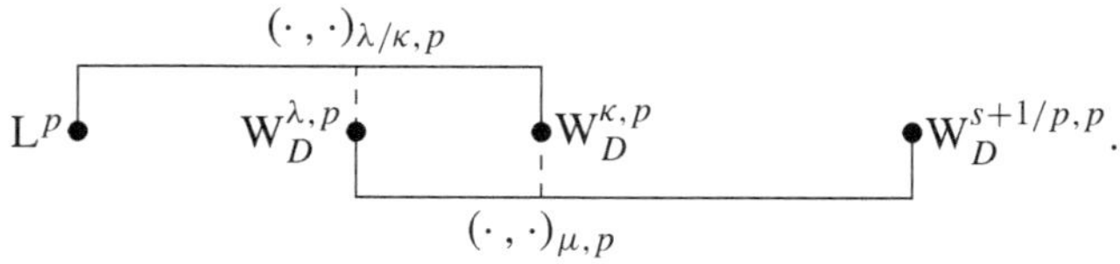

Proposition 3.10 yields $\mathrm{W}_D^{\kappa,p} = (\mathrm{L}^p, \mathrm{W}_D^{s+1/p,p})_{\theta,p}$ with $\theta = \kappa/(s + 1/p)$. We conclude by using one after the other Corollary 6.11, reiteration, one of the first two cases, and again reiteration:

$$\begin{aligned}\mathrm{W}^{s,p} &= (\mathrm{L}^p, \mathrm{W}^{s+1/p,p}_D)_{\vartheta,p} = (\mathrm{L}^p, (\mathrm{L}^p, \mathrm{W}^{s+1/p,p}_D)_{\theta,p})_{s/\kappa,p} = (\mathrm{L}^p, \mathrm{W}^{\kappa,p}_D)_{s/\kappa,p} \\ &= (\mathrm{L}^p, (\mathrm{L}^p, \mathrm{W}^{1,p}_D)_{\kappa,p})_{s/\kappa,p} = (\mathrm{L}^p, \mathrm{W}^{1,p}_D)_{s,p}.\end{aligned}$$

□

Acknowledgements Both authors are grateful to Joachim Rehberg for many fruitful discussions on and around the topic. The first named author thanks his Ph.D. advisor Robert Haller-Dintelmann for his support and the Laboratoire de Mathématiques d'Orsay for hospitality during a stay in March 2018 where this project got started.

Appendix A: Porous Sets

We provide a streamlined approach to the geometry of porous sets. All this is known to the experts but some results require going through existing literature in a rather opaque way. The reader may look up relevant definitions in Sect. 2.1.

Lemma A.1 *Every porous set $E \subseteq \mathbb{R}^d$ is a Lebesgue null set.*

Proof By Remark 2.7, each ball B centered in E contains a ball of comparable radius that does not intersect E. Hence, there is $\delta \in (0, 1)$ depending only on E such that

$$\frac{|B \cap E|}{|B|} \leq 1 - \delta.$$

By Lebesgue's differentiation theorem this implies $\mathbb{1}_E = 0$ almost everywhere. □

We recall the Vitali covering lemma that will be used frequently in the following, see [24, Theorem 1.2].

Lemma A.2 *Let $\{B_i\}_{i \in I}$ be a family of open balls with uniformly bounded radii. Then there exists a subfamily $\{B_j\}_{j \in J}$ of disjoint balls such that*

$$\bigcup_i B_i \subseteq \bigcup_j 5B_j.$$

Corollary A.3 *Let $E \subseteq \mathbb{R}^d$ and $0 < r \leq R < \infty$. For any ball B of radius R the set $E \cap B$ can be covered by $10^d(R/r)^d$ ball of radius r centered in $E \cap B$.*

Proof Consider the covering $\{\mathrm{B}(x, r/5)\}_{x \in B \cap E}$ of $B \cap E$. We find a disjoint subfamily $\{B_i\}_{i \in I}$ such that $B \cap E \subseteq \cup_{i \in I} 5B_i$. We denote by $\#_i$ the cardinality of I and calculate

$$\#_i c_d (r/5)^d = | \cup_{i \in I} B_i| \leq |2B| = c_d 2^d R^d,$$

where c_d is the measure of the unit ball. This shows $\#_i \leq 10^d(R/r)^d$. □

We continue with the simple observation that the radius bound by 1 in the definition of ℓ-regularity is arbitrary.

Lemma A.4 *Let $E \subseteq \mathbb{R}^d$ and $0 < \ell \le d$. If for some $M \in (0, \infty)$ there is comparability $\mathcal{H}^\ell(B \cap E) \approx \mathrm{r}(B)^\ell$ uniformly for all open balls B of radius $\mathrm{r}(B) \le M$ centered in E, then the same is true for any $M \in (0, \infty)$.*

Proof Suppose we have uniform comparability for balls up to radius $\mathrm{r}(B) \le m$. Given $M > m$, we need to extend it to balls B centered in E of radius $\mathrm{r}(B) \le M$. Let $c{:=}m/M$. The calculation

$$\frac{m^\ell \mathrm{r}(B)^\ell}{M^\ell} \lesssim \mathcal{H}^l(cB \cap E) \le \mathcal{H}^\ell(B \cap E)$$

gives the lower estimate. For the upper one, we cover $B \cap E$ by $10^d/c^d$ balls of radius $c\mathrm{r}(B)$ centered in $B \cap E$ according to Corollary A.3 and conclude $\mathcal{H}^\ell(B \cap E) \lesssim \mathrm{r}(B)^\ell$. □

We come to computing the Assouad dimensions of Ahlfors-regular sets.

Lemma A.5 *Let $E \subseteq \mathbb{R}^d$ be ℓ-regular for some $0 < \ell \le d$ and let $M < \infty$. There exist constants $c, C > 0$ such that, if $x \in E$ and $0 < r \le R < M$, then in order to cover $E \cap \mathrm{B}(x, R)$ by balls of radius r centered in E, at least $c(R/r)^\ell$ and at most $C(R/r)^\ell$ balls are needed. If E is unbounded and uniformly ℓ-regular, then this also holds for $M = \infty$.*

Proof Let $\{B_i\}_{i \in I}$ be *some* cover of $E \cap \mathrm{B}(x, R)$ by balls of radius r. We use Lemma A.4 to calculate

$$R^\ell \lesssim \mathcal{H}^\ell(B(x, R) \cap E) \le \mathcal{H}^\ell(\cup_{i \in I} B_i \cap E) \le \sum_{i \in I} \mathcal{H}^\ell(B_i \cap E) \lesssim \#_i r^\ell,$$

which shows $\#_i \gtrsim (R/r)^\ell$ and gives the constant c. As for C, we select a subfamily of disjoint balls B_j from the covering $\{\mathrm{B}(x, r/5)\}_{x \in B \cap E}$ of $B \cap E$. Then we estimate, using Lemma A.4,

$$\#_j (r/5)^\ell \lesssim \sum_{j \in J} \mathcal{H}^\ell(B_j \cap E) \le \mathcal{H}^\ell(2B \cap E) \lesssim (2R)^\ell$$

and conclude $\#_j \lesssim (R/r)^\ell$. □

Proposition A.6 *Let $E \subseteq \mathbb{R}^d$ be uniformly ℓ-regular. It follows that $\underline{\dim}_{\mathcal{A}}\mathcal{S}(E) = \overline{\dim}_{\mathcal{A}}\mathcal{S}(E) = \ell$.*

Proof We can rephrase Lemma A.5 in the language of Definition 2.8. It precisely asserts that $\ell \in \underline{\mathcal{AS}}(E) \cap \overline{\mathcal{AS}}(E)$. Hence, we get $\underline{\dim}_{\mathcal{AS}}(E) \ge \ell$ and $\overline{\dim}_{\mathcal{AS}}(E) \le \ell$. The claim follows since $\underline{\dim}_{\mathcal{AS}}(E) \le \overline{\dim}_{\mathcal{AS}}(E)$ holds for any set E. Indeed, given $\lambda \in \underline{\mathcal{AS}}(E)$ and $\mu \in \overline{\mathcal{AS}}(E)$ we have $(R/r)^\lambda \lesssim (R/r)^\mu$ for all $0 < r < R < \mathrm{diam}(E)$ and hence $\lambda \le \mu$. □

We turn to porosity. The following result was already mentioned in Sect. 1.1.

Lemma A.7 *Let $E \subseteq F \subseteq \mathbb{R}^d$. If F is ℓ-regular and E is m-regular with $0 < m < \ell \leq d$, then E is porous in F. Likewise, if $\overline{\dim}_{\mathcal{AS}}(E) < \underline{\dim}_{\mathcal{AS}}(F)$, then E is uniformly porous in F.*

Proof We begin with the first claim. Lemma A.5 yields some $C \geq 1$ such that, if $x \in E$ and $0 < r \leq R \leq 1$, then at most $C(2R/r)^m$ balls of radius r centered in E are needed to cover $E \cap \mathrm{B}(x, 2R)$. It also yields some $c > 0$ such that at least $c(R/(2r))^\ell$ balls of radius $2r$ centered in F are needed to cover $F \cap \mathrm{B}(x, R)$. We use this observation with $r = \kappa R$, where $\kappa \in (0, 1)$ satisfies $c/(2\kappa)^\ell > C(2/\kappa)^m$. This is possible due to $m < \ell$.

Let $\{B_i\}_{i \in I}$ be a family of $\#_i \leq C(2/\kappa)^m$ balls of radius r centered in E that cover $E \cap \mathrm{B}(x, 2R)$. By choice of κ the balls $\{2B_i\}_{i \in I}$ cannot cover $F \cap \mathrm{B}(x, R)$. Pick $y \in F \cap \mathrm{B}(x, R)$ that is not contained in any of the $2B_i$. By construction we have $\mathrm{B}(y, r) \subseteq \mathbb{R}^d \setminus \cup_{i \in I} B_i$ but due to $r < R$ we also have $\mathrm{B}(y, r) \subseteq \mathrm{B}(x, 2R)$ and hence $E \cap \mathrm{B}(y, r) \subseteq \cup_i B_i$. Thus, we must have $E \cap \mathrm{B}(y, r) = \emptyset$ and conclude that E is porous in F.

The proof of the second claim is identical, but we do not assume $R \leq 1$ and have the covering properties for some $m \in \overline{\mathcal{AS}}(E)$ and $\ell \in \underline{\mathcal{AS}}(F)$ with $m < \ell$ by assumption. □

Lemma A.8 *If $E \subseteq \mathbb{R}^d$ is porous, then there exist $C \geq 1$ and $0 < s < d$ such that, given $x \in E$ and $0 < r < R \leq 1$, there is a covering of $E \cap B(x, R)$ by $C(R/r)^s$ balls of radius r centered in E. Moreover, if E is uniformly porous, then $\overline{\dim}_{\mathcal{A}}\mathcal{S}(E) < d$.*

Proof We only show the porous case since the uniform case again just follows by dropping all restrictions on the radii. In the following all cubes are closed and axis-aligned. As indicated in Sect. 2.1, we can equivalently replace balls by cubes and radii by side lengths in the definition of porosity and Assouad dimension. Likewise, it suffices to establish the claim of the lemma with cubes.

In view of Remark 2.7 we can fix $n \in \mathbb{N}$ such that for every cube $Q \subseteq \mathbb{R}^d$ there is a cube $Q' \subseteq Q \setminus E$ of sidelength $\ell(Q') = \ell(Q)/n$. We fix a cube Q centered in E of side length $R \leq 1$. Let $0 < r \leq R$ and fix $k \in \mathbb{N}$ such that $R/(2n)^{k+1} \leq r < R/(2n)^k$. We claim that we can cover Q by $((2n)^d - 1)^{k+1}$ closed cubes of side length $R/(2n)^{k+1}$. Put $s := \log((2n)^d - 1)/\log(2n) < d$. Then

$$((2n)^d - 1)^{k+1} = (2n)^s (2n)^{ks} < (2n)^s (R/r)^s$$

shows the assertion.

For the claim we start with $k = 1$. There is a cube $Q' \subseteq Q \setminus E$ of side length R/n. Then there is a cube Q'' in the grid of $(2n)^d$ cubes with sidelength $R/(2n)$ covering Q that is contained in Q'. This means that we only need $(2n)^d - 1$ cubes of side length $R/(2n)$ to cover E. We conclude by applying this argument inductively on each cube of the previous covering. □

Combining the uniform cases of the two preceding lemmas lets us re-obtain a result of Luukkainen [30, Thm 5.2]. Note that $\underline{\dim}_{\mathcal{AS}}(\mathbb{R}^d) = d$ due to Proposition A.6.

Proposition A.9 *A set $E \subseteq \mathbb{R}^d$ is uniformly porous if and only if its upper Assouad dimension is strictly less than d.*

We can use the non-uniform cases to show that some open sets are of class $\mathcal{D}^t$. The argument is a slight adaption of [29, Theorem 4.2].

Proposition A.10 *Let $O \subseteq \mathbb{R}^d$ be open. If ∂O is porous, then $O \in \mathcal{D}^t$ for some $t \in (0, 1)$. If ∂O is ℓ-regular for some $0 < \ell < d$, then $O \in \mathcal{D}^t$ for all $t \in (0, \max\{1, d - \ell\})$.*

Proof If ∂O is porous, then we pick $C \geq 1$ and $0 < s < d$ according to Lemma A.8 such that for each $j \geq 0$ and for any ball B with radius $r \leq 1$ centered in ∂O we can cover $B \cap \partial O$ by at most $C2^{js}$ balls of radius $r2^{-j}$. If ∂O is ℓ-regular, then Lemma A.5 guarantees that we can take $s = \ell$. In any case, fix $\max(s, d-1) < u < d$. Put $E_j := \{x \in B : \mathrm{d}(x, \partial O) \leq r2^{-j}\}$ and $A_j := E_j \setminus E_{j+1}$. By construction, the covering property for $B \cap \partial O$ implies that we can cover E_j by at most $C2^{js}$ balls of radius $r2^{-(j-1)}$. The d-regularity of Lebesgue measure then implies

$$|A_j| \leq |E_j| \lesssim 2^{js} r^d 2^{d-jd}. \tag{41}$$

We use that $\{A_j\}_{j\geq 0}$ is a disjoint cover of $B \setminus \partial O$, comparability $\mathrm{d}(x, \partial O) \approx r2^{-j}$ on A_j, estimate (41), and $s < u$ to calculate

$$\begin{aligned}\int_{B\setminus\partial O} \mathrm{d}(y, \partial E)^{u-d} \mathrm{d}y &\leq \sum_j \int_{A_j} \mathrm{d}(y, \partial O)^{u-d} \mathrm{d}y \lesssim \sum_j |A_j| 2^{dj-uj} r^{u-d} \\ &\lesssim \sum_j r^u 2^{j(s-u)} \lesssim r^u.\end{aligned}$$

Setting $t := d - u \in (0, 1)$, we write this in the form

$$\sup_{x\in\partial O} \sup_{0<r\leq 1} r^{t-d} \int_{\mathrm{B}(x,r)\setminus\partial O} \mathrm{d}(y, \partial O)^{-t} < \infty,$$

which just means that $O \in \mathcal{D}^t$. In the case of ℓ-regular boundary, every $u \in (\max\{\ell, d-1\}, 1)$ and thus every $t \in (0, \max\{1, d-\ell\})$ was admissible in the proof. □

Appendix B: Intrinsic Characterizations

Although perfectly suited for interpolation questions, the function spaces $\mathrm{X}^{s,p}(O) = \mathrm{X}^{s,p}(\mathbb{R}^d)|_O$ lack an *intrinsic characterization* through a norm that only uses information on O. The problem of finding such characterizations has a long history and we refer for instance to [15,21,27,34,36,40,41,45] and references therein. Here, we only mention two results that are of particular importance for putting our paper into context of work on mixed boundary value problems [3,8,11,13,22,23,39].

The following is the full-dimensional case in [27, Theorem VI.1].

Proposition B.1 *Let $O \subseteq \mathbb{R}^d$ be an open d-regular set and let $p \in (1, \infty)$ and $s \in (0, 1)$. Then $\mathrm{W}^{s,p}(O)$ is up to equivalent norms the space of those $f \in \mathrm{L}^p(O)$ for which*

$$\|f\| := \|f\|_{\mathrm{L}^p(O)} + \left(\iint_{O \times O} \frac{|f(x) - f(y)|^p}{|x-y|^{d+sp}} \, \mathrm{d}x \, \mathrm{d}y \right)^{1/p} < \infty.$$

Remark B.2 If in the setting above $D \subseteq \overline{O}$ is $(d-1)$-regular, then $\mathrm{W}^{s,p}_D(O)$ is the closure of $\mathrm{C}^\infty_D(O)$ for the intrinsic norm $\|\cdot\|$. This follows from Lemma 3.3.

Probably most important result of the above type concerns $\mathrm{W}^{1,p}_D(O)$.

Proposition B.3 *Suppose $O \subseteq \mathbb{R}^d$ is an open set, $D \subseteq \overline{O}$ is $(d-1)$-regular, and O satisfies a uniform Lipschitz condition around $\overline{\partial O \setminus D}$. Then $\mathrm{W}^{1,p}_D(O)$ can equivalently be normed by*

$$\|f\| := \left(\|f\|^p_{\mathrm{L}^p(O)} + \|\nabla f\|^p_{\mathrm{L}^p(O)} \right)^{1/p}$$

and in fact it is the closure of $\mathrm{C}^\infty_D(O)$ for this norm.

Proof Let X be the closure of $\mathrm{C}^\infty_D(O)$ for the norm $\|\cdot\|$. Clearly we have $\mathrm{W}^{1,p}_D(O) \subseteq X$ with continuous inclusion. For the converse inclusion we argue as in [3, Lemma 3.2], using the localization formalism of Sect. 4.4 and in particular the maps $\mathcal{E}$ and $\mathcal{R}$ defined in (14) and (15). We let σ be the extension of functions from $(0, 1) \times (-1, 1)^{d-1}$ to $(-1, 1)^d$ by even reflection and consider the operator

$$\mathcal{T} : f \mapsto \mathcal{R}((\mathcal{E}f)_0, (\sigma(\mathcal{E}f)_i)_{i \in I}).$$

By construction $\mathcal{T}f|_O = f$ holds for $f \in X$. It follows from the proof of Lemma 4.10 that $\mathcal{T} : X \to \mathrm{W}^{1,p}(\mathbb{R}^d)$ is bounded. The proof of Lemma 4.12 reveals that $\mathcal{T}$ in fact maps X into $\mathrm{W}^{1,p}_D(\mathbb{R}^d)$, which implies $X \subseteq \mathrm{W}^{1,p}_D(O)$. □

References

1. Adams, D.R., Hedberg, L.I.: Function Spaces and Potential Theory. Grundlehren der Mathematischen Wissenschaften, vol. 314. Springer, Berlin (1996)
2. Arendt, W., Batty, C.J.K., Hieber, M., Neubrander, F.: Vector-valued Laplace Transforms and Cauchy Problems. Monographs in Mathematics, vol. 96. Birkhäuser, Basel (2001)
3. Auscher, P., Badr, N., Haller-Dintelmann, R., Rehberg, J.: The square root problem for second order divergence form operators with mixed boundary conditions on L^p. J. Evol. Equ. **15**(1), 165–208 (2015)
4. Axelsson, A., Keith, S., McIntosh, A.: The Kato square root problem for mixed boundary value problems. J. Lond. Math. Soc. **74**, 113–130 (2006)
5. Bechtel, S.: The Kato Square Root Property for Mixed Boundary Conditions. Master's thesis, TU Darmstadt, (2017). http://www3.mathematik.tu-darmstadt.de/index.php?id=3305
6. Bergh, J., Löfström, J.: Interpolation Spaces. An Introduction. Grundlehren der Mathematischen Wissenschaften, vol. 223. Springer, Berlin (1976)

7. Bonifacius, L., Neitzel, I.: Second order optimality conditions for optimal control of quasilinear parabolic equations. Math. Control Relat. Fields **8**(1), 1–34 (2018)
8. Brewster, K., Mitrea, D., Mitrea, I., Mitrea, M.: Extending Sobolev functions with partially vanishing traces from locally (ε, δ)-domains and applications to mixed boundary problems. J. Funct. Anal. **266**(7), 4314–4421 (2014)
9. Cwikel, M.: Complex interpolation spaces, a discrete definition and reiteration. Indiana Univ. Math. J. **27**(6), 1005–1009 (1978)
10. Denk, R., Hieber, M., Prüss, J.: $\mathcal{R}$-boundedness, Fourier multipliers and problems of elliptic and parabolic type. Mem. Am. Math. Soc. **166**, 788 (2003)
11. Disser, K.: Well-posedness for coupled bulk-interface diffusion with mixed boundary conditions. Analysis **35**(4), 309–317 (2015)
12. Disser, K., Rehberg, J.: The 3D transient semiconductor equations with gradient-dependent and interfacial recombination. arXiv preprint, available at https://arxiv.org/abs/1805.01348
13. Disser, K., Meyries, M., Rehberg, J.: A unified framework for parabolic equations with mixed boundary conditions and diffusion on interfaces. J. Math. Anal. Appl. **430**(2), 1102–1123 (2015)
14. Dyda, B., Vähäkangas, A.: A framework for fractional Hardy inequalities. Ann. Acad. Sci. Fenn. Math. **39**(2), 675–689 (2014)
15. Egert, M., Tolksdorf, P.: Characterizations of Sobolev functions that vanish on a part of the boundary. Discret. Contin. Dyn. Syst. Ser. S **10**(4), 729–743 (2017)
16. Egert, M., Haller-Dintelmann, R., Tolksdorf, P.: The Kato Square Root Problem for mixed boundary conditions. J. Funct. Anal. **267**(5), 1419–1461 (2014)
17. Federer, H.: Geometric Measure Theory. Die Grundlehren der mathematischen Wissenschaften, vol. 153. Springer, New York (1969)
18. Griepentrog, J.A., Gröger, K., Kaiser, H.-C., Rehberg, J.: Interpolation for function spaces related to mixed boundary value problems. Math. Nachr. **241**, 110–120 (2002)
19. Grisvard, P.: Équations différentielles abstraites. Ann. Sci. École. Norm. Supér. **2**, 311–395 (1969)
20. Gröger, K.: A $W^{1,p}$-estimate for solutions to mixed boundary value problems for second order elliptic differential equations. Math. Ann. **283**(4), 679–687 (1989)
21. Hajłasz, P., Koskela, P., Tuominen, H.: Sobolev embeddings, extensions and measure density condition. J. Funct. Anal. **254**(5), 1217–1234 (2008)
22. Haller-Dintelmann, R., Rehberg, J.: Maximal parabolic regularity for divergence operators including mixed boundary conditions. J. Differ. Equ. **247**(5), 1354–1396 (2009)
23. Haller-Dintelmann, R., Jonsson, A., Knees, D., Rehberg, J.: Elliptic and parabolic regularity for second-order divergence operators with mixed boundary conditions. Math. Methods Appl. Sci. **39**(17), 5007–5026 (2016)
24. Heinonen, J.: Lectures on Analysis on Metric Spaces Universitext. Springer, New York (2001)
25. Janson, S., Nilsson, P., Peetre, J.: Notes on Wolff's note on interpolation spaces. Proc. Lond. Math. Soc. **48**(2), 283–299 (1984)
26. Jawerth, B., Frazier, M.: A discrete transform and decompositions of distribution spaces. J. Funct. Anal. **93**(1), 34–170 (1990)
27. Jonsson, A., Wallin, H.: Function spaces on subsets of $\mathbb{R}^n$. Math. Rep. **2**(1), (1984)
28. Lehrbäck, J.: Weighted Hardy inequalities and the size of the boundary. Manuscr. Math. **127**(2), 249–273 (2008)
29. Lehrbäck, J., Tuominen, H.: A note on the dimensions of Assouad and Aikawa. J. Math. Soc. Jpn. **65**(2), 343–356 (2013)
30. Luukkainen, J.: Assouad dimension: antifractal metrization, porous sets, and homogeneous measures. J. Korean Math. Soc. **35**(1), 23–76 (1998)
31. Meinlschmidt, H., Meyer, C., Rehberg, J.: Optimal control of the thermistor problem in three spatial dimensions, Part 1: Existence of optimal solutions. SIAM J. Control Optim. **55**(5), 2876–2904 (2017)
32. Meinlschmidt, H., Meyer, C., Rehberg, J.: Optimal control of the thermistor problem in three spatial dimensions, Part 2: Optimality conditions. SIAM J. Control Optim. **55**(4), 2368–2392 (2017)
33. Nečas, J.: Direct Methods in the Theory of Elliptic Equations. Springer Monographs in Mathematics. Springer, Berlin (2012)
34. Rychkov, V.S.: Linear extension operators for restrictions of function spaces to irregular open sets. Stud. Math. **140**(2), 141–162 (2000)
35. Seeley, R.: Interpolation in L^p with boundary conditions. Stud. Math. **44**, 44–60 (1972)

36. Shvartsman, P.: Local approximations and intrinsic characterization of spaces of smooth functions on regular subsets of $\mathbb{R}^n$. Math. Nachr. **279**(11), 1212–1241 (2006)
37. Sickel, W.: Pointwise multipliers of Lizorkin–Triebel spaces. The Maz'ya Anniversary Collection. Operator Theory: Advances and Applications, vol. 110. Birkhäuser, Basel (1999)
38. Simon, J.: Sobolev, Besov and Nikol'skii fractional spaces: imbeddings and comparisons for vector valued spaces on an interval. Ann. Mater. Pura Appl. **157**, 117–148 (1990)
39. ter Elst, A.F.M., Rehberg, J.: Hölder estimates for second-order operators on domains with rough boundary. Adv. Differ. Equ. **20**(3–4), 299–360 (2015)
40. Triebel, H.: Interpolation Theory, Function Spaces, Differential Operators. North-Holland Mathematical Library, vol. 18. North-Holland, Amsterdam (1978)
41. Triebel, H.: A note on function spaces in rough domains. Tr. Mat. Inst. Steklova **293**, 346–351 (2016)
42. Väisälä, J.: Porous sets and quasisymmetric maps. Trans. Am. Math. Soc. **299**(2), 525–533 (1987)
43. Wolff, T.H.: A note on interpolation spaces. Lecture Notes in Mathematics. Harmonic analysis (Minneapolis, Minn., 1981), vol. 908. Springer, Berlin (1982)
44. Yeh, J.: Real Analysis. World Scientific Publishing, Hackensack (2006)
45. Zhou, Y.: Fractional Sobolev extension and imbedding. Trans. Am. Math. Soc. **367**(2), 959–979 (2015)

Publisher's Note Springer Nature remains neutral with regard to jurisdictional claims in published maps and institutional affiliations.

Journal of Fourier Analysis and Applications (2019) 25:2782–2800
https://doi.org/10.1007/s00041-019-09682-0

RESEARCH ARTICLE

Generalized Fourier Transforms Associated with the Oscillator Representation

Jing-Song Huang[1] · **Ling Zhou**[2]

Received: 28 September 2018 / Revised: 20 March 2019 / Published online: 11 April 2019
© Springer Science+Business Media, LLC, part of Springer Nature 2019

Abstract

By identifying the Fourier transform $\mathcal{F}$ with an operator in the oscillator representation of the metaplectic group $\widetilde{Sp}(2n, \mathbb{R})$, the twofold cover of the symplectic group $Sp(2n, \mathbb{R})$, we study generalized Fourier transforms inspired by the work of De Bie, Oste and Van der Jeugt. We obtain several families of operators $\mathcal{T}$'s that have the important properties similar to $\mathcal{F}$ using various dual pair correspondences.

Keywords Fourier transform · Oscillator representation · Dual pair correspondence

1 Introduction

The Fourier transform is a powerful tool in harmonic analysis with wide applications. For instance, it plays a pivotal role in the recent progress on the spherical packing problem as it can simultaneously diagonalize translations and make periodic structures easier to understand [1]. There are several conventions to define the Fourier transform over $\mathbb{R}^n$. Let $L^p(\mathbb{R}^n)$ denotes the space of all p-th power Lebesgue integrable functions on $\mathbb{R}^n$ for $1 \leq p < \infty$. Throughout this paper, the Fourier transform of $f \in L^1(\mathbb{R}^n) \cap L^2(\mathbb{R}^n)$ is given by

---

Communicated by Hans G. Feichtinger.

Jing-Song Huang: research is supported by research Grants from the Research Grant Council of Hong Kong SAR in China.

✉ Jing-Song Huang
mahuang@ust.hk

Ling Zhou
zhou.2568@osu.edu

1 Department of Mathematics, Hong Kong University of Science and Technology, Clear Water Bay, Kowloon, Hong Kong SAR, China

2 Department of Mathematics, Ohio State University, Columbus, OH 43210, USA

$$(\mathcal{F}f)(y) = \frac{1}{(2\pi)^{n/2}} \int_{\mathbb{R}^n} e^{-i\langle x,y\rangle} f(x)dx,$$

where $\langle x, y\rangle$ is the standard inner product on $\mathbb{R}^n$. The inverse Fourier transform $\mathcal{F}^{-1}$ of $f \in L^1(\mathbb{R}^n) \cap L^2(\mathbb{R}^n)$ is given by

$$\left(\mathcal{F}^{-1} f\right)(y) = \frac{1}{(2\pi)^{n/2}} \int_{\mathbb{R}^n} e^{i\langle x,y\rangle} f(x)dx,$$

which is indeed the inverse of $\mathcal{F}$ as a consequence of the Fourier inverse theorem. It follows that $\mathcal{F}$ is a unitary operator on $L^2(\mathbb{R}^n)$, with respect to the inner product $\langle f, g\rangle = \int_{\mathbb{R}^n} f(x)\overline{g(x)}dx$. The Schwartz space $\mathcal{S}(\mathbb{R}^n)$ is a dense subspace of $L^2(\mathbb{R}^n)$ and the action of $\mathcal{F}$ on $L^2(\mathbb{R}^n)$ is extended from its action on $\mathcal{S}(\mathbb{R}^n)$. In addition, $\mathcal{F}$ is a homeomorphism of $\mathcal{S}(\mathbb{R}^n)$ onto itself (cf. [11]). For convenience, we specify the underlying space of $\mathcal{F}$, as well as the generalized operators we will construct in this paper, to be $\mathcal{S}(\mathbb{R}^n)$.

In the framework of the oscillator representation, the Fourier transform $\mathcal{F}$ is expressed by

$$\mathcal{F} = i^{\frac{n}{2}} e^{-\frac{\pi i}{2}\mathfrak{k}} \tag{1.1}$$

with $\mathfrak{k}$ an element in $\mathfrak{sl}(2, \mathbb{R})$ that is embedded into $\mathfrak{sp}(2n, \mathbb{R})$ in a dual pair $(O(n), \mathfrak{sl}(2, \mathbb{R}))$. De Bie, Oste and Van der Jeugt [2] have obtained a class of generalized Fourier transforms as exponential of elements in the universal enveloping algebra $\mathcal{U}(\mathfrak{sl}(2, \mathbb{R}))$ by determining their series expansions. In this paper, we construct generalized Fourier transforms by a different approach.

To illuminate our idea, we recall the basic concepts of the dual pair correspondence. Following Howe, we define a *reductive dual pair* in $Sp(2n, \mathbb{R})$ to be a pair of closed reductive subgroups G, G' of $Sp(2n, \mathbb{R})$ such that G and G' are centralizers of each other. The dual pair correspondence can be obtained by the following consideration. Denote by $\mathbb{D}(\mathbb{R}^n)$ the *Weyl algebra* of polynomial coefficient differential operators on $\mathbb{R}^n$ (cf. [3, p. 278]). Let G be a reductive algebraic subgroup of $GL(n, \mathbb{R})$ (cf. [7, p. 446]). Then G acts on $\mathcal{S}(\mathbb{R}^n)$ by

$$\rho(g)f(x) = f(g^{-1}x).$$

Let $\mathbb{D}(\mathbb{R}^n)^G$ be the commutant of G in $\mathbb{D}(\mathbb{R}^n)$. If the generators of $\mathbb{D}(\mathbb{R}^n)^G$ span a Lie algebra $\mathfrak{g}'$, then it is the Lie algebra of G' in a dual pair (G, G'). In this case there is a correspondence of representations between G and $\mathfrak{g}'$ (cf. [3], Sect. 9.2.1).

We can easily construct the following four families of dual pairs with m, n, p, q positive integers:

(i) $(O(n), Sp(2m, \mathbb{R})) \subset Sp(2nm, \mathbb{R})$,
(ii) $(U(n), U(m)) \subset Sp(2nm, \mathbb{R})$,
(iii) $(O(p, q), Sp(2m, \mathbb{R})) \subset Sp(2(p + q)m, \mathbb{R})$,
(iv) $(U(p, q), U(m)) \subset Sp(2(p + q)m, \mathbb{R})$.

In this paper we are only concerned with the case $m = 1$ for the construction of generalized Fourier transforms.

A very important property of Fourier transform $\mathcal{F}$ is that it commutes with the standard action of the orthogonal group $O(n)$ on $\mathcal{S}(\mathbb{R}^n)$. In other words, $\mathcal{F}$ is $O(n)$-equivariant. In case $G = O(n)$ the other party in the dual pair is $\mathfrak{g}' = \mathfrak{sl}(2,\mathbb{R})$. The Fourier transform $\mathcal{F}$ in the oscillator representation has the following $O(n)\times\mathfrak{sl}(2,\mathbb{R})$-module decomposition [9]

$$\mathcal{F} \simeq \sum_{k=0}^{\infty} Id|_{\mathcal{H}_k} \otimes T_k, \tag{1.2}$$

where $\mathcal{H}_k$ is the space of harmonic polynomials homogeneous of degree k and T_k is the Hankel transform. This leads us to consider the generalized Fourier transforms in the dual pair $(O(n), \mathfrak{sl}(2,\mathbb{R}))$.

We note that a key technique in our approach is Schur's lemma. The discrete decomposition as above (1.2) is crucial. It is natural to consider the dual pair $(O(p,q), \mathfrak{sl}(2,\mathbb{R}))$ $(p+q=n)$. However, our approach fails in this case, since the decomposition of the oscillator representation is not discrete [4,10]. Nevertheless, we are able to deal with another two pairs $(U(n), U(1))$ and $(U(p,q), U(1))$ due to the discrete decompositions of the oscillator representation.

We are very grateful to Hendrik De Bie for his insightful comments. We also thank the referee for very helpful remarks.

2 Fourier Transform and the Oscillator Representation

In this section, we recall the basic properties of the Fourier transform $\mathcal{F}$ in the framework of the oscillator representation.

Proposition 2.1 *$\mathcal{F}$ has the following properties:*

(i) $\mathcal{F}$ is $O(n)-equivariant$, i.e.

$$\mathcal{F}(gf) = g(\mathcal{F}f), \forall f \in \mathcal{S}(\mathbb{R}^n), g \in O(n).$$

(ii) Write ∂_j for $\frac{\partial}{\partial x_j}$. Then

$$\mathcal{F}\circ x_j = i\partial_j \circ \mathcal{F},\ \ \mathcal{F}\circ \partial_j = ix_j \circ \mathcal{F}$$
$$\mathcal{F}^{-1}\circ x_j = -i\partial_j \circ \mathcal{F}^{-1},\ \ \mathcal{F}^{-1}\circ \partial_j = -ix_j \circ \mathcal{F}^{-1}.$$

(iii) Let $\tilde{f}$ be the reflection of f, i.e. $\tilde{f}(x) = f(-x), \forall x \in \mathbb{R}^n$. Then

$$\mathcal{F}^2 f = \mathcal{F}^{-2} f = \tilde{f}, \forall f \in \mathcal{S}(\mathbb{R}^n).$$

These properties are obtained by straightforward calculation [11, Chapter I §2]. Recall that the Schwartz space on $\mathbb{R}^n$ is defined by is

$$\mathcal{S}(\mathbb{R}^n) = \left\{ f \in C^\infty(\mathbb{R}^n) : \sup_{x\in\mathbb{R}^n} |x^\alpha D^\beta f(x)| < \infty, \forall \alpha, \beta \in \mathbb{N}^n \right\}$$

where $x^\alpha D^\beta = x_1^{\alpha_1} \ldots x_n^{\alpha_n} \frac{\partial^{|\beta|}}{\partial x_1^{\alpha_1} \ldots \partial x_n^{\alpha_n}}$. When $n = 1$, $\mathcal{S}(\mathbb{R})$ is space of the oscillator representation of $\mathfrak{sl}(2, \mathbb{R})$ and is denoted by ω. Set

$$SL(2, \mathbb{R}) = \left\{ \begin{bmatrix} a & b \\ c & d \end{bmatrix} \middle| a, b, c, d \in \mathbb{R} \text{ satisfying } ad - bc = 1 \right\}.$$

The group $SL(2, \mathbb{R})$ has fundamental group $\mathbb{Z}$ and thus has a unique (up to isomorphism) connected twofold cover, denoted by $\widetilde{SL}(2, \mathbb{R})$. The Lie algebra $\mathfrak{sl}(2, \mathbb{R})$ has a natural basis:

$$e^+ = \begin{bmatrix} 0 & 1 \\ 0 & 0 \end{bmatrix}, \; e^- = \begin{bmatrix} 0 & 0 \\ -1 & 0 \end{bmatrix}, \; h = \begin{bmatrix} 1 & 0 \\ 0 & -1 \end{bmatrix}$$

with commutator relations

$$[h, e^+] = 2e^+, [h, e^-] = -2e^-, [e^+, e^-] = h.$$

The group $SO(2)$ is a maximal compact subgroup of $SL(2, \mathbb{R})$, and its Lie algebra is $\mathfrak{k}_0 = \mathfrak{so}(2, \mathbb{R})$. The complexification $\mathfrak{k} = (\mathfrak{k}_0)_{\mathbb{C}}$ is spanned by

$$k = i(e^- - e^+) = \begin{bmatrix} 0 & -i \\ i & 0 \end{bmatrix}.$$

Then k together with

$$n^+ = \frac{1}{2}\left(h + i\left(e^- + e^+\right)\right), n^- = \frac{1}{2}\left(h - i\left(e^- + e^+\right)\right),$$

form another basis for $\mathfrak{sl}(2, \mathbb{R})$, with the similar commutation relations as $\{h, e^+, e^-\}$.

The oscillator representation $(\omega, \mathcal{S}(\mathbb{R}))$ of $\mathfrak{sl}(2, \mathbb{R})$ is given by

$$\omega(e^+) = \frac{i}{2}x^2, \; \omega(e^-) = \frac{i}{2}\frac{d^2}{dx^2}, \; \omega(h) = x\frac{d}{dx} + \frac{1}{2}.$$

The Lie algebra action exponentiates to a unitary representation of $\widetilde{SL}(2, \mathbb{R})$ on $L^2(\mathbb{R})$ [5, Theorem 2.1.2].

Set $a = x + \frac{\partial}{\partial x}$, $a^+ = x - \frac{\partial}{\partial x}$. Then k acts on $\mathcal{S}(\mathbb{R})$ diagonally by

$$\omega(k) = i\left(\omega\left(e^- - e^+\right)\right) = \frac{1}{2}\left(x^2 - \frac{d^2}{dx^2}\right) = \frac{1}{4}(aa^+ + a^+a),$$

which is known as the *Hermite operator* (see [5, p. 103]). Set

$$v_j := (a^+)^j e^{-\frac{r^2}{2}}, \text{ for } j \in \mathbb{Z}_{\geq 0}.$$

Then $\{v_j\}$ forms an orthogonal k-eigenbasis.

Lemma 2.2 [5, Chapter III §2.1] *Set*

$$V_{\frac{1}{2}} = Span\{v_{2j} | j \in \mathbb{N}\} \text{ and } V_{\frac{3}{2}} = Span\{v_{2j+1} | j \in \mathbb{N}\}.$$

Then $V_{\frac{1}{2}}$ and $V_{\frac{3}{2}}$ are irreducible lowest weight modules of $\mathfrak{sl}(2, \mathbb{R})$ with lowest weights $\frac{1}{2}$ and $\frac{3}{2}$ respectively, and the oscillator representation is decomposed as

$$\mathcal{S}(\mathbb{R}) \simeq V_{\frac{1}{2}} \oplus V_{\frac{3}{2}}.$$

Let $v_j^* \in \mathcal{S}(\mathbb{R})^*$ be the dual basis of v_j. Then $\mathcal{S}(\mathbb{R})^*_{adm} = \text{Span}\left\{v_j^* | j \in \mathbb{Z}\right\}$, the admissible dual of $(\omega, \mathcal{S}(\mathbb{R}))$, can be identified with the $\mathfrak{k}$-finite vectors in $\mathcal{S}(\mathbb{R})$ with $\mathfrak{sl}(2, \mathbb{R})$-action given by

$$\omega^*(e^+) = -\frac{i}{2}x^2, \ \omega^*(e^-) = -\frac{i}{2}\frac{d^2}{dx^2}, \ \omega^*(h) = x\frac{d}{dx} + \frac{1}{2}.$$

The representation $(\omega^n, \mathcal{S}(\mathbb{R}^n))$ of $\mathfrak{sl}(2, \mathbb{R})$ is the $n-$fold tensor power of $(\omega, \mathcal{S}(\mathbb{R}))$, where $\mathcal{S}(\mathbb{R})^{\otimes n}$ is identified with $\mathcal{S}(\mathbb{R}^n)$ and the action is given by

$$\omega^n(e^+) = \frac{i}{2}r^2, \ \omega^n(e^-) = \frac{i}{2}\Delta, \ \omega^n(h) = E + \frac{n}{2}.$$

Here $r^2 = \sum_{j=1}^n x_j^2$ is the multiplication by r^2, $\Delta = \sum_{j=1}^n \left(\frac{\partial}{\partial x_j}\right)^2$ is the Laplace operator and $E = \sum_{j=1}^n x_j \frac{\partial}{\partial x_j}$. When there is no confusion on the action, we omit ω^n.

Set $a_j = x_j + \frac{\partial}{\partial x_j}$, $a_j^+ = x_j - \frac{\partial}{\partial x_j}$. Then

$$\omega^n(\mathfrak{k}) = \frac{1}{2}(r^2 - \Delta) = \frac{1}{4}\sum_{j=1}^n \left(a_j a_j^+ + a_j^+ a_j\right).$$

For $\beta = (\beta_1, \ldots, \beta_n) \in \mathbb{N}_+^n$, set

$$v_\beta = \left(a_1^+\right)^{\beta_1} \ldots \left(a_n^+\right)^{\beta_n} v_0 \text{ with } v_0 = e^{-\frac{1}{2}r^2}.$$

By computation, $v_\beta = P_\beta(x_1, \ldots, x_n)v_0$, where P_β are the well-known Hermite polynomials. And v_β are known as *Hermite functions*. Moreover, $\{v_\beta | \beta \in \mathbb{N}^n\}$ forms an orthogonal $\mathfrak{k}$ eigenbasis for $\mathcal{S}(\mathbb{R}^n)$, because

$$\omega^n(\mathfrak{k})v_\beta = \sum_{j=1}^n \left(\beta_j + \frac{1}{2}\right) v_\beta = \left(|\beta| + \frac{n}{2}\right) v_\beta$$

where $|\beta| = \sum_{j=1}^{n} \beta_j$, for all $\beta \in \mathbb{N}^n$. It follows that

$$\omega^n\left(e^{-\frac{\pi i}{2}\mathfrak{k}}\right)v_\beta = (-i)^{\left(|\beta|+\frac{n}{2}\right)}v_\beta.$$

In addition, by comparing their actions on the basis (cf. [5], p. 121), we conclude that

$$\mathcal{F} = i^{\frac{n}{2}}e^{-\frac{\pi i}{2}\mathfrak{k}}. \tag{2.1}$$

The action of $\mathcal{F}$ on the basis of $\mathcal{S}(\mathbb{R}^n)$ spanned by v_β's indicates that the intertwining relation in (ii) of Proposition 2.1 characterize $\mathcal{F}$. We make the precise statement in the following proposition [9, Propositions 1.2.1 and 1.2.2].

Proposition 2.3 *Suppose $\mathcal{T}$ is a continuous operator on $\mathcal{S}(\mathbb{R}^n)$.*

(i) *The condition*

$$\mathcal{T} \circ x_j = i\partial_j \circ \mathcal{T} \text{ and } \mathcal{T} \circ \partial_j = ix_j \circ \mathcal{T}\ (1 \le i \le n) \tag{2.2}$$

implies that $\mathcal{T}$ is a multiple of $\mathcal{F}$.

(ii) *The condition*

$$\mathcal{T} \circ x_i x_j = -\partial_i\partial_j \circ \mathcal{T} \text{ and } \mathcal{T} \circ \partial_i\partial_j = -x_i x_j \circ \mathcal{T}(1 \le i, j \le n) \tag{2.3}$$

implies that $\mathcal{T} = a\mathcal{F} + b\mathcal{F}^{-1}$ for some $a, b \in \mathbb{C}$.

Proof These two facts are consequences of straightforward calculations. We include these calculations, since this is the simplest case for our approach of using the dual pair correspondence and Schur's lemma.

(i) The condition $\mathcal{T} \circ x_j = i\partial_j \circ \mathcal{T}$ and $\mathcal{T} \circ \partial_j = ix_j \circ \mathcal{T}$ $(1 \le j \le n)$ implies that

$$\mathcal{T} \circ a_j^+ = \mathcal{T} \circ (x_j - \partial_j) = (i\partial_j - ix_j) \circ \mathcal{T} = -ia_j^+ \circ \mathcal{T}.$$

We have $\mathcal{T}v_0 = cv_0$ for some $c \in \mathbb{C}$, because for $1 \le j \le n$

$$\begin{aligned}\partial_j \frac{\mathcal{T}v_0}{v_0} &= \partial_j\left(\mathcal{T}\left(e^{-\frac{r^2}{2}}\right)e^{\frac{r^2}{2}}\right)\\ &= (\partial_j\mathcal{T})\left(e^{-\frac{r^2}{2}}\right)e^{\frac{r^2}{2}} + x_j\mathcal{T}\left(e^{-\frac{r^2}{2}}\right)e^{\frac{r^2}{2}}\\ &= 0,\end{aligned}$$

since $\partial_j\mathcal{T}\big(e^{-\frac{r^2}{2}}\big) = -i\mathcal{T}\big(x_je^{-\frac{r^2}{2}}\big) = i\mathcal{T}\big(\partial_j\big(e^{-\frac{r^2}{2}}\big)\big) = -x_j\mathcal{T}(e^{-\frac{r^2}{2}})$. It follows that

$$\begin{aligned}\mathcal{T}v_\beta &= \mathcal{T}\left(\left(a_1^+\right)^{\beta_1}\dots\left(a_n^+\right)^{\beta_n}v_0\right)\\ &= (-i)^{|\beta|}\left(a_1^+\right)^{\beta_1}\dots\left(a_n^+\right)^{\beta_n}\mathcal{T}v_0\\ &= (-i)^{|\beta|}cv_\beta\end{aligned}$$

for all $\beta \in \mathbb{N}^n$. Therefore, $\mathcal{T} = ci^{\frac{n}{2}}\omega^n(e^{-\frac{\pi i}{2}k}) = c\mathcal{F}$.

(ii) Note that $\{x_i\partial_j + \frac{1}{2}\delta_i^j, x_ix_j, \partial_i\partial_j | 1 \le i, j \le n\}$ forms a basis of the Lie algebra isomorphic to $\mathfrak{sp}(2n, \mathbb{R})$ via the map ψ

$$\begin{aligned} x_i\partial_j + \frac{1}{2}\delta_i^j &\mapsto e_{i,j} - e_{n+j,n+i}, \\ x_ix_j &\mapsto e_{i,j+n} + e_{j,i+n}, \\ \partial_i\partial_j &\mapsto -e_{i+n,i} - e_{j+n,i}. \end{aligned}$$

Here $e_{i,j}$ is the matrix whose (i, j)th entry is 1 and 0's elsewhere. Set

$$V_{even} = \text{Span}\{v_\beta : |\beta| \text{ is even}\}, \quad V_{odd} = \text{Span}\{v_\beta : |\beta| \text{ is odd}\}.$$

Then both V_{even} and V_{odd} are irreducible $\mathfrak{sp}(2n, \mathbb{R})$-modules, and

$$\mathcal{S}(\mathbb{R}^n) = V_{even} \oplus V_{odd}.$$

It follows from the condition

$$\mathcal{T} \circ x_ix_j = -\partial_i\partial_j \circ \mathcal{T} \text{ and } \mathcal{T} \circ \partial_i\partial_j = -x_ix_j \circ \mathcal{T} \ (1 \le i, j \le n)$$

that

$$\mathcal{T} \circ \left(x_i\partial_j + \frac{n}{2}\delta_i^j\right) = \mathcal{T} \circ [x_ix_j, \partial_i\partial_j] = -\left(x_i\partial_j + \frac{n}{2}\delta_i^j\right) \circ \mathcal{T}.$$

Note that $\mathcal{F}$ has the similar property. It follows that $\mathcal{T}\mathcal{F}$ commutes with action of each element in $\mathfrak{sp}(2n, \mathbb{R})$. By Schur's lemma, there exist $c_1, c_2 \in \mathbb{C}$ such that

$$\mathcal{T}\mathcal{F}|_{V_{even}} = c_1 \,\text{Id}, \quad \mathcal{T}\mathcal{F}|_{V_{odd}} = c_2 \,\text{Id}.$$

Let $a = \frac{c_1+c_2}{2}$, $b = \frac{c_1-c_2}{2}$. Note that $\mathcal{F}^2|_{V_{even}} = \text{Id}$, $\mathcal{F}^2|_{V_{odd}} = -\text{Id}$. It follows that

$$\mathcal{T}\mathcal{F} = a\mathcal{F}^2 + b\,\text{Id}.$$

Thus, $\mathcal{T} = a\mathcal{F} + b\mathcal{F}^{-1}$.

□

Remark 2.4 In the proof of (ii) of Proposition 2.3, $\mathcal{S}(\mathbb{R}^n)$ is a module (not irreducible) for the symplectic Lie algebra $\mathfrak{sp}(2n, \mathbb{R})$ generated by $\{x_ix_j, \partial_i\partial_j\}$. It can be shown that this module is actually a representation for the twofold cover of the symplectic group $Sp(2n, \mathbb{R})$ and is called the oscillator representation. One can make $\mathcal{S}(\mathbb{R}^n)$ a simple module for either the Lie superalgebra $\mathfrak{osp}(1|2n)$ generated by $\{x_j, \partial_j\}$ or the Weyl algebra $\mathbb{D}(\mathbb{R}^n)$. Then one can deduce (i) of Proposition 2.3 by the Schur's lemma.

We write I_n for the n by n identity matrix. Set

$$J_n = \begin{bmatrix} 0 & I_n \\ -I_n & 0 \end{bmatrix} \in Sp(2n, \mathbb{R}).$$

Recall that the metaplectic group $\widetilde{Sp}(2n, \mathbb{R})$ is a two-fold cover of the symplectic group $Sp(2n, \mathbb{R})$. There exists a unique central element ξ_0 of order 2 and an exact sequence of Lie groups:

$$1 \to \{I_{2n}, \xi_0\} \to \widetilde{Sp}(2n, \mathbb{R}) \to Sp(2n, \mathbb{R}) \to 1.$$

The oscillator representation mentioned above for $\mathfrak{sp}(2n, \mathbb{R})$ is actually a unitary representation for $\widetilde{Sp}(2n, \mathbb{R})$. There is an element $\omega_0' \in \widetilde{Sp}(2n, \mathbb{R})$ such that ω_0' corresponds to J_n under the covering map and $i^{\frac{n}{2}} \mathcal{F}^{-1} (\text{mod } \{I_{2n}, \xi_0\})$ is equal to the action of ω_0' in the oscillator representation (cf. [9] p. 21).

We now extend our consideration to the indefinite Fourier transform $\mathcal{F}_{p,q}$ and their corresponding elements in the metaplectic group $\widetilde{Sp}(2n, \mathbb{R})$ associated with the oscillator representation.

Let p, q be two positive integers such that $p + q = n$. Consider the indefinite bilinear form of signature (p, q) on $\mathbb{R}^{p+q}$:

$$\langle x, x' \rangle_{p,q} := \sum_{j=1}^{p} x_j x_j' - \sum_{j=p+1}^{p+q} x_j x_j', \forall x, x' \in \mathbb{R}^{p+q},$$

and define Fourier transforms $\mathcal{F}_{p,q}$ by

$$(\mathcal{F}_{p,q} f)(y) = \frac{1}{(2\pi)^{(p+q)/2}} \int_{\mathbb{R}^{p+q}} e^{-i \langle x, y \rangle_{p,q}} f(x) dx.$$

Then one has

$$i^{\frac{p-q}{2}} \mathcal{F}_{p,q}^{-1} (\text{mod } \{I_{2n}, \xi_0\}) = J_{p,q} = \begin{bmatrix} 0 & I_{p,q} \\ -I_{p,q} & 0 \end{bmatrix} \in Sp(2(p+q), \mathbb{R}),$$

where $I_{p,q} = \begin{bmatrix} I_p & 0 \\ 0 & -I_q \end{bmatrix}$ (cf [5]). One may use another indefinite bilinear form of signature (p, q) on $\mathbb{R}^{p+q}$ in different order of variables. The resulting Fourier transform is different, but conjugate to $\mathcal{F}_{p,q}$ by an element in $\widetilde{Sp}(2n, \mathbb{R})$.

The tensor product $\omega^p \otimes \omega^{q*}$ can be identified with $\mathcal{S}(\mathbb{R}^{p+q})$ (cf. [5, Chapter III §2.3]), that is

$$\mathcal{S}(\mathbb{R}^p) \widehat{\otimes} \mathcal{S}(\mathbb{R}^q) \cong \mathcal{S}(\mathbb{R}^{p+q}).$$

Then the representation $(\omega^{p,q}, \mathcal{S}(\mathbb{R}^{p+q})) \simeq (\omega^p \otimes \omega^{q*}, \mathcal{S}(\mathbb{R}^p)\widehat{\otimes}\mathcal{S}(\mathbb{R}^q))$ of $\mathfrak{sl}(2,\mathbb{R})$ is given by

$$\omega^{p,q}(e^+) = \frac{i}{2} r_{p,q}, \ \omega^{p,q}(e^-) = \frac{i}{2}\Delta_{p,q}, \ \omega^{p,q}(h) = E + \frac{n}{2}. \tag{2.4}$$

Here

$$r_{p,q} = \left(r_p^2 - r_q^2\right) = \sum_{j=1}^{p} x_j^2 - \sum_{j=1}^{q} x_{j+p}^2$$

is the standard indefinite metric of signature (p, q), and

$$\Delta_{p,q} = (\Delta_p - \Delta_q) = \sum_{j=1}^{p} \frac{\partial^2}{\partial x_j^2} - \sum_{j=1}^{q} \frac{\partial^2}{\partial x_{j+p}^2}$$

is called the indefinite Laplacian when $q > 0$. It follows that

$$\omega^{p,q}(\Bbbk) = \frac{1}{2}\left(r_{p,q}^2 - \Delta_{p,q}\right) = \frac{1}{4}\left(\sum_{j=1}^{p}\left(a_j a_j^+ + a_j^+ a_j\right) - \sum_{j=p+1}^{p+q}\left(a_j a_j^+ + a_j^+ a_j\right)\right).$$

Denote $\frac{1}{2}(r_{p,q}^2 - \Delta_{p,q})$ by $\Bbbk_{p,q}$. For $\beta \in \mathbb{N}^p$, $\gamma \in \mathbb{N}^q$, recall from section 2 that u_β, w_γ are the bases for $\mathcal{S}(\mathbb{R}^p)$ and $\mathcal{S}(\mathbb{R}^q)$ respectively. Then $\{v_{\beta,\gamma} := u_\beta w_\gamma^* | \beta \in \mathbb{N}^p, \gamma \in \mathbb{N}^q\}$ forms an orthogonal $\Bbbk_{p,q}$ eigenbasis for $\mathcal{S}(\mathbb{R}^{p+q})$. By comparing the actions of $\Bbbk_{p,q}$ and $\mathcal{F}_{p,q}$ on the basis of $\mathcal{S}(\mathbb{R}^{p+q})$, we conclude that

$$\mathcal{F}_{p,q} = i^{\frac{p-q}{2}} e^{-\frac{\pi i}{2}\Bbbk_{p,q}}.$$

Denote $e^{\frac{\pi i}{2}\Bbbk_{p,q}}$ by $\varpi_{p,q}$. Then

$$\varpi_{p,q}(\mathrm{mod}\{I_{2n}, \xi_0\}) = J_{p,q}$$

and

$$\varpi_{p,q} = \varpi_{q,p}^{-1}, \ \varpi_{p,q}\varpi_{s,t} = \varpi_{s,t}\varpi_{p,q} \text{ and } \varpi_{p,q}^2 = \varpi_{s,t}^2.$$

Let Γ be the finite abelian group generated by all $\varpi_{p,q}$. It is clear that Γ is actually generated by $\varpi_{p,q}$ with $p = 0, 1, \ldots, [\frac{p+q+1}{2}]$.

3 Generalized Fourier Transforms in $(O(n), \mathfrak{sl}(2,\mathbb{R}))$-Pair

The conditions (2.2) and (2.3) both give characterizations of the Fourier transform. For generalized Fourier transforms, we consider the following weaker condition called the

Helmholtz relations

$$\mathcal{T} \circ r^2 = -\Delta \circ \mathcal{T} \text{ and } \mathcal{T} \circ \Delta = -r^2 \circ \mathcal{T}.$$

It is readily to verify that $\{\Delta, r^2, [\Delta, r^2]\}$ forms a basis of a Lie algebra isomorphic to $\mathfrak{sl}(2,\mathbb{R})$. Both r^2 and Δ are $O(n)$-invariant. It leads us to consider the decomposition of $\mathcal{S}(\mathbb{R}^n)$ into a direct sum of irreducible $O(n) \times \widetilde{SL}(2,\mathbb{R})-$modules. We obtain some generalized Fourier transforms according to this decomposition.

Denote by $\mathcal{P}(\mathbb{R}^n)$ the space of all polynomials on $\mathbb{R}^n$ with complex coefficients, $\mathcal{P}_k(\mathbb{R}^n)$ the subspace of homogeneous polynomials in $\mathcal{P}(\mathbb{R}^n)$ of degree k, and

$$\mathcal{H}_k(\mathbb{R}^n) = \{f \in \mathcal{P}_k(\mathbb{R}^n) : \Delta f = 0\}$$

the space of homogeneous harmonic polynomials of degree k. Recall that $e^- = \frac{i}{2}\Delta$. Then $\mathcal{H}_k(\mathbb{R}^n)$ consists of $e^-{-}$null vectors.

Set $\tilde{\mathcal{H}}_k(\mathbb{R}^n) := \mathcal{H}_k(\mathbb{R}^n)e^{-\frac{r^2}{2}}$. Then it is clear that $\tilde{\mathcal{H}}_k(\mathbb{R}^n)$ is the space of $n^-{-}$null vectors with k eigenvalue $\frac{n}{2} + k$. Therefore, each function $\phi \in \tilde{\mathcal{H}}_k(\mathbb{R}^n)$ generates a $\mathfrak{sl}(2,\mathbb{R})$ module isomorphic to $V_{\frac{n}{2}+k}$, the lowest weight module with lowest weight $\frac{n}{2} + k$. This isomorphism defines an embedding

$$i_\phi : V_{\frac{n}{2}+k} \to \mathcal{S}(\mathbb{R}^n).$$

By Corollary 2.1.3 in Chapter II of [5], we have a decomposition

$$\mathcal{S}(\mathbb{R}^n) \simeq (V_{\frac{1}{2}} \oplus V_{\frac{1}{2}+1})^{\otimes n} \simeq \sum_{k \geq 0} a(n,k) V_{\frac{n}{2}+k},$$

where the multiplicity $a(n,k) = \binom{n+k-1}{k} - \binom{n+k-3}{k-2}$ is the same as the dimension of $\mathcal{H}_k(\mathbb{R}^n)$. Furthermore, the following result holds.

Proposition 3.1 *[5, Chapter III, Theorem 2.4.4] For $n \geq 2$, we have a decomposition of*

$$\mathcal{S}(\mathbb{R}^n) \simeq \sum_{k=0}^{\infty} (\tilde{\mathcal{H}}_k(\mathbb{R}^n) \otimes V_{\frac{n}{2}+k})^- \tag{3.1}$$

into irreducible $O(n) \times \widetilde{SL}(2,\mathbb{R})-$modules. Here the "–" indicates closure in $\mathcal{S}(\mathbb{R}^n)$.

(i) *The $O(n) \times \widetilde{SL}(2,\mathbb{R})-$intertwining map*

$$\begin{aligned} \tilde{\mathcal{H}}_k(\mathbb{R}^n) \otimes V_{\frac{n}{2}+k} &\to \mathcal{S}(\mathbb{R}^n) \\ \phi \otimes v &\mapsto i_\phi(v) \end{aligned}$$

defines an isomorphism of $O(n) \times \widetilde{SL}(2,\mathbb{R})-$modules.

(ii) *$\tilde{\mathcal{H}}_k(\mathbb{R}^n)$ is an irreducible $O(n)-$module. In particular, $\tilde{\mathcal{H}}_k(\mathbb{R}^n)$ are non-isomorphic $O(n)$-modules for different k.*

Remark 3.2 (i) Note that $\tilde{\mathcal{H}}_k(\mathbb{R}^n) \cong \mathcal{H}_k(\mathbb{R}^n)$ as $O(n)$-modules and we write $\mathcal{H}_k$ for $\mathcal{H}_k(\mathbb{R}^n)$. Therefore, we have the following decomposition for polynomials [5, Theorem on p. 118]:

$$\mathcal{P}(\mathbb{R}^n)|_{O(n)\times\widetilde{SL}(2,\mathbb{R})} \simeq \sum_{k=0}^{\infty} \mathcal{H}_k \otimes V_{\frac{n}{2}+k}. \tag{3.2}$$

(ii) The decomposition (3.2) describes the correspondence for the $(O(n), \mathfrak{sl}(2,\mathbb{R}))$-pair.

Note that $V_{\frac{n}{2}+k}$ is $O(n)-$invariant. Denote by $\mathfrak{J}(\mathbb{R}^n)$ the space of $O(n)$ invariant functions in $\mathcal{S}(\mathbb{R}^n)$. Thus, we have the following decomposition

$$\mathcal{S}(\mathbb{R}^n) = \sum_{k=0}^{\infty} \mathcal{H}_k(\mathbb{R}^n) \cdot \mathfrak{J}(\mathbb{R}^n).$$

This leads us to consider the map

$$\alpha_k : \mathcal{H}_k(\mathbb{R}^n) \otimes \mathcal{S}(\mathbb{R}_+) \to \mathcal{S}(\mathbb{R}^n),\ f \otimes \psi \mapsto f(x)\psi(r^2).$$

Let ρ_k be the representation of $\mathfrak{sl}(2,\mathbb{R})$ on $\mathcal{S}(\mathbb{R}_+)$ defined by

$$\alpha_k(f \otimes \rho_k(X)\psi) = X(\alpha_k(f \otimes \psi)), \forall X \in \mathfrak{sl}(2,\mathbb{R}).$$

By [5, Chapter IV, Lemma 1.2.2], ρ_k is explicitly given by:

$$h \mapsto 2r\frac{d}{dr} + k + \frac{n}{2},\ e^+ \mapsto \frac{ir}{2},\ e^- \mapsto 2i\left[\left(k + \frac{n}{2}\right)\frac{d}{dr} + r\frac{d^2}{dr^2}\right].$$

Proposition 3.3 (Bochner's Relations) [5, Chapter IV, Theorem 1.2.4] *If $g(x) = f(x)\psi(r^2)$ for some $f \in \mathcal{H}_k(\mathbb{R}^n)$ and $\psi \in \mathcal{S}(\mathbb{R}_+)$, then*

$$(\mathcal{F}g)(y) = f(y)T_k(\psi)(r^2),$$

where $T_k = \rho_k(i^{\frac{n}{2}} e^{-\frac{\pi}{2}k})$.

Remark 3.4 This proposition is a well-known fact in analysis. Here the T_k's are in fact the Hankel transforms [11, Chapter IV,Theorem 3.10].

The above proposition shows that $\mathcal{F}$ acts on each summand $\mathcal{H}_k \otimes V_{k+\frac{n}{2}}$ by $Id \otimes T_k$. In view of the decomposition (3.2), we have

$$\mathcal{F} \simeq \sum_{k=0}^{\infty} Id|_{\mathcal{H}_k} \otimes T_k. \tag{3.3}$$

Now we consider bounded operators on $\mathcal{S}(\mathbb{R}^n)$ that share some of the properties of $\mathcal{F}$.

Theorem 3.5 *Let $\mathcal{T}$ be a bounded operator on $\mathcal{S}(\mathbb{R}^n)$ satisfying the following conditions:*

(i) $\mathcal{T}$ *is* $O(n) - equivariant$, *i.e.*

$$\mathcal{T}(gf) = g(\mathcal{T}f), \forall f \in L^2(\mathbb{R}^n), g \in O(n);$$

(ii) $\mathcal{T} \circ r^2 = -\Delta \circ \mathcal{T}$ *and* $\mathcal{T} \circ \Delta = -r^2 \circ \mathcal{T}$;
(iii) $\mathcal{T}^4 = Id$.

Then in view of (3.2)

$$\mathcal{T} = \sum_{k=0}^{\infty} c_k (Id|_{\mathcal{H}_k} \otimes T_k),$$

where $c_k = \pm i$ *or* ± 1, *for* $k \in \mathbb{Z}_{\geq 0}$.

Proof Assume that c_k's are complex number such that $|c_k| = 1$. Then an operator of the form $\sum_{k=0}^{\infty} c_k(Id \otimes T_k)$ is unitary and therefore bounded. It is easy to check that such an operator satisfies (i)–(iii). We now prove any $\mathcal{T}$ under conditions (i)–(iii) can be written in this form with c_k's being fourth roots of unit. Set $\widetilde{\mathcal{T}} = \mathcal{T} \circ \mathcal{F}^{-1}$. Then $\widetilde{\mathcal{T}}$ is $O(n)$-equivariant, because both $\mathcal{T}$ and $\mathcal{F}^{-1}$ are $O(n)$-equivariant. On the other hand, $\mathcal{F}^{-1}$ satisfies the Helmholtz relations too, so

$$\widetilde{\mathcal{T}} \circ r^2 = r^2 \circ \widetilde{\mathcal{T}} \text{and } \widetilde{\mathcal{T}} \circ \Delta = \Delta \circ \widetilde{\mathcal{T}}.$$

It follows that $\widetilde{\mathcal{T}}$ commutes with $\mathfrak{sl}(2, \mathbb{R})$ generated spanned by $\{r^2, \Delta, [r^2, \Delta]\}$, and thus commutes with $O(n) \times \widetilde{SL}(2, \mathbb{R})$.

Recall from (3.2) that $\mathcal{S}(\mathbb{R}^n)$ is decomposed into irreducible representations of $O(n) \times \widetilde{SL}(2, \mathbb{R})$, where all the summands are not isomorphic to each other. By Schur's Lemma, $\widetilde{\mathcal{T}}$ preserves each summand and acts as scalar on each of them, i.e. $\widetilde{\mathcal{T}}|_{\mathcal{H}_k \otimes V_{k+\frac{n}{2}}} = c_k \in \mathbb{C}, \forall k \geq 0$.

Since $\mathcal{F}$ on $\mathcal{S}(\mathbb{R}^n)$ decomposes into $\mathcal{F} \simeq \sum_{k=0}^{\infty} Id \otimes T_k$, it follows that

$$\mathcal{T} = \widetilde{\mathcal{T}} \circ \mathcal{F} = \sum_{k=0}^{\infty} c_k (Id|_{\mathcal{H}_k} \otimes T_k).$$

As $\mathcal{F}^{-1}$ and $\mathcal{T}$ both have order 4, we have $\widetilde{\mathcal{T}}^4 = Id$ and thus $c_k = \pm i$ or ± 1 for all k. □

Although each c_k take only four different values, the sequence $(c_0, c_1, \ldots)$ is uncountable. Thus, Theorem 3.5 describes uncountably many operators that generalize the Fourier transform. Note that

$$\mathcal{F} v_\beta = (-i)^\beta v_\beta \implies \mathcal{F}|_{\mathcal{H}_k \otimes V_{k+\frac{n}{2}}} = (-i)^k.$$

Then $\mathcal{T}$ in Theorem 3.5 can be determined more explicitly if it satisfies a certain periodicity condition. Denote by $C_4 = \{\pm 1, \pm i\} = \langle i \rangle$ the cyclic group of order 4 generated by i.

Corollary 3.6 *If the map $k \mapsto c_k$ defines a group homomorphism from $\mathbb{Z}$ to C_4, then $\mathcal{T}$ is in $\{Id, \mathcal{F}, \mathcal{F}^2, \mathcal{F}^3 = \mathcal{F}^{-1}\}$.*

Proof If the map $k \mapsto c_k$ defines a group homomorphism from $\mathbb{Z}$ to C_4, then

$$c_{k+4} = c_k, c_0 = 1, c_{k+l} = c_k c_l, \forall\, k, l \in \mathbb{Z}.$$

Therefore, $\mathcal{T}$ is determined by c_1. There are four possibilities:

c_0	c_1	c_2	c_3	$\mathcal{T}$
1	1	1	1	Id
1	−i	−1	i	$\mathcal{F}$
1	−1	1	−1	$\mathcal{F}^2$
1	i	−1	−i	$\mathcal{F}^{-1}$

Thus, the corollary follows. □

Remark 3.7 The standard eigenfunction basis of the Fourier transform given by the Hermite functions in (2.1) of [2] are eigenfunctions of the generalized Fourier transform $\mathcal{T}$ in Theorem 3.5. Therefore, $\mathcal{T}$ is an integral transform as shown in [2, Proposition 2.3]:

$$(\mathcal{T} f)(y) = \frac{1}{(2\pi)^{n/2}} \int_{\mathbb{R}^n} K_n(x, y) f(x) dx,$$

where

$$K_n(x, y) = 2^{\lambda} \Gamma(\lambda) \sum_{k=0}^{\infty} c_k (k + \lambda) z^{-\lambda} J_{k+\lambda}(z) C_k^{\lambda}(w).$$

Here we adopt the notations $\lambda = (n-2)/2$, $z = |x||y|$, $w = \langle x, y \rangle / z$ and $J_{k+\lambda}$ are the Bessel functions while C_k^{λ} denote the Gegenbauser or ultraspherical polynomials in [2, Proposition 2.3].

4 Generalized Fourier Transforms in $(U(n), U(1))$-Pair

In this section, we consider another dual pair $(U(n), U(1))$ in the seesaw pairs with $O(n) \times \widetilde{SL}(2, \mathbb{R})$. Recall that (G, G') is a reductive dual pair in $Sp(2n, \mathbb{R})$ if G, G' are closed reductive subgroups of $Sp(2n, \mathbb{R})$ such that G and G' are centralizers of each other. Let K be a maximal compact subgroup of G, and let H' be the centralizer of K in $Sp(2n, \mathbb{R})$. Then (K, H') forms another dual pair in $Sp(2n, \mathbb{R})$. In this situation, (G, G') and (K, H') form seesaw pairs (cf. [3], p. 400).

Note that $U(1) = \{e^{itk} | t \in \mathbb{R}\}$ is a maximal compact subgroup of $G = SL(2, \mathbb{R})$ and $U(n)$ is the centralizer of $U(1)$ in $Sp(2n, \mathbb{R})$, so we get the following seesaw pairs (cf. [8])

$$\begin{array}{ccc} U(n) & & SL(2,\mathbb{R}) \\ & \times & \\ O(n) & & U(1). \end{array}$$

Let $\widetilde{U(1)}$ be the inverse image of $U(1)$ in $\widetilde{SL}(2,\mathbb{R})$. Each $V_{k+\frac{n}{2}}$, lowest weight module of $\mathfrak{sl}(2,\mathbb{R})$, can be decomposed into $\widetilde{U(1)}-$types of $\widetilde{SL}(2,\mathbb{R})$:

$$V_{k+\frac{n}{2}} = \sum_{l=0}^{\infty} W_{k+\frac{n}{2}+2l}.$$

By substituting the above formula into formula (3.2), we get

$$\mathcal{S}(\mathbb{R}^n) \simeq \sum_{k=0}^{\infty}\sum_{l=0}^{\infty} \mathcal{H}_k \otimes W_{k+\frac{n}{2}+2l}.$$

Set $\mathcal{L}_k = \sum_{l=0}^{[\frac{k}{2}]} \mathcal{H}_{k-2l}$. We note there is a $U(n)$-module isomorphism from $\mathcal{L}_k$ to $\mathcal{P}_k$ the space of homogeneous polynomials of degree k, since

$$\mathcal{P}_k = \sum_{l=0}^{[\frac{k}{2}]} r^{2l}\mathcal{H}_{k-2l}.$$

It follows that we have the following $(U(n), U(1))$-pair correspondence.

Proposition 4.1 *We have a decomposition of $\mathcal{S}(\mathbb{R}^n)$ into irreducible $U(n) \times \widetilde{U(1)}-$modules:*

$$\mathcal{S}(\mathbb{R}^n)|_{U(n)\times\widetilde{U(1)}} \simeq \sum_{k=0}^{\infty} \mathcal{L}_k \otimes W_{k+\frac{n}{2}}, \tag{4.1}$$

Here $W_{k+\frac{n}{2}}$ is a one-dimensional weight space of $\mathfrak{k}$ with the weight $k + \frac{n}{2}$. We denote the action of $\widetilde{U(1)}$ on it by χ_k and write $S_k := \chi_k(i^{\frac{n}{2}} e^{-\frac{\pi i}{2}\mathfrak{k}})$. It is clear that S_k is simply multiplication by $(-i)^k$. In addition, the Fourier transform $\mathcal{F}$ decomposes into

$$\mathcal{F} \simeq \sum_{k=0}^{\infty} Id|_{\mathcal{L}_k} \otimes S_k.$$

Now we consider the generalized Fourier transforms in the setting of the $(U(n), U(1))$-pair. Recall that by the end of Sect. 2 we defined Γ to be the finite abelian group generated by $\varpi_{p,q}$ corresponding to the indefinite Fourier transforms $\mathcal{F}_{p,q}$.

Theorem 4.2 *Let $\mathcal{T}$ be a bounded operator on $\mathcal{S}(\mathbb{R}^n)$ satisfying the following conditions:*

(i) $\mathcal{T}$ *is* $O(n)-$*equivariant and* $\Gamma-$*equivariant;*
(ii) $\mathcal{T}^4 = Id$.

Then $\mathcal{T}$ *is* $U(n)-$*equivariant, and in view of* (4.1)

$$\mathcal{T} = \sum_{k=0}^{\infty} c_k Id|_{\mathcal{L}_k} \otimes S_k,$$

where $c_k = \pm 1, \pm i$.

Moreover, the Helmholtz relations

$$\mathcal{T} \circ r^2 = -\Delta \circ \mathcal{T} \text{ and } \mathcal{T} \circ \Delta = -r^2 \circ \mathcal{T}$$

imply that $\mathcal{T}$ *is a multiple of* $\mathcal{F}$ *or* $\mathcal{F}^{-1}$.

Proof To show that $\mathcal{T}$ is $U(n)-$equivariant, it suffices to prove that $O(n)$ and Γ generate $U(n)$. Let H be subgroup of $U(n)$ generated by $O(n)$ and Γ. Denote by $Lie(H)$ the Lie algebra of H. The set

$$\mathcal{I} = \{iI_{s,t} : s = 0, \ldots, [(n+1)/2]\}$$

is contained in $Lie(H)$, since H contains Γ. Consider the decomposition

$$\mathfrak{su}(n) = \mathfrak{k}_0 \oplus i\mathfrak{p}_0 = \mathfrak{so}(n) \oplus i\mathfrak{p}_0,$$

where $\mathfrak{p}_0 = \{X | X \in \mathfrak{gl}(n, \mathbb{R}), X = X^t, \text{tr}(X) = 0\}$. It follows that

$$\mathfrak{u}(n) = \mathfrak{su}(n) \oplus i\mathbb{R}I_n = \mathfrak{so}(n) \oplus i\mathfrak{p}_0 \oplus i\mathbb{R}I_n.$$

Since $\mathfrak{p}_0$ is an irreducible representation of $\mathfrak{so}(n)$, for any non-zero $iX \in i\mathfrak{p}_0$, $\mathfrak{so}(n) \cup \{iX\}$ generates $\mathfrak{su}(n)$. As $iI_{s,t} (s > 0)$ and iI_n are in $Lie(H)$, we can take $iX = iI_{s,t} + i(t-s)I_n$, and thus $Lie(H) = \mathfrak{u}(n)$. Then H contains an open neighborhood of 1 in $U(n)$ and hence $H = U(n)$ (cf. [6, Cor. 2.10]).

Let $\widetilde{\mathcal{T}} = \mathcal{T} \circ \mathcal{F}^{-1}$. Since $\widetilde{\mathcal{T}}$ commutes with $\varpi_{n,0}$, we have $\widetilde{\mathcal{T}}$ is $\widetilde{U(1)}-$equivariant. So $\widetilde{\mathcal{T}}$ commutes with $U(n) \times \widetilde{U(1)}$. It follows from Schur's Lemma that

$$\widetilde{\mathcal{T}}|_{\mathcal{L}_k \otimes W_{k+\frac{n}{2}}} = c_k.$$

Thus, $\mathcal{T}$ must be in the form given above and $c_k = \pm i$ or ± 1 for all $k \geq 0$, because $\mathcal{T}$ and $\mathcal{F}$ both have order 4. In addition the Helmholtz relations imply that $\widetilde{\mathcal{T}}$ commutes with $\mathfrak{sp}(2n, \mathbb{R})$, since $\mathfrak{sp}(2n, \mathbb{R})$ is generated by $\mathfrak{u}(n)$ together with Δ and r^2. This implies that $\mathcal{T}$ is a linear combination of $\mathcal{F}$ and $\mathcal{F}^{-1}$ as we argued in the proof of Proposition 2.3. Then we deduce that $\mathcal{T}$ is a multiple $\mathcal{F}$ or $\mathcal{F}^{-1}$ by the condition that $T^4 = Id$. □

Remark 4.3 In the proof, it is enough to assume that $\mathcal{T}$ commutes with $\varpi_{n,0}$ and some $\varpi_{s,t}$ $(0 < s < n)$. Since Condition (ii) implies that $\mathcal{T}$ commutes with $\varpi_{n,0}$, '$\mathcal{T}$ is Γ-equivariant' in Condition (i) may be replaced by a weaker condition that '$\mathcal{T}$ commutes with $\varpi_{s,t}$ for a positive integer s less than n'.

Remark 4.4 By a similar consideration mentioned in Remark 3.10, the generalized Fourier transform $\mathcal{T} = \sum_{k=0}^{\infty} c_k Id|_{\mathcal{L}_k} \otimes S_k$ in Theorem 4.2 coincides with the integral transform

$$(\mathcal{T} f)(y) = \frac{1}{(2\pi)^{n/2}} \int_{\mathbb{R}^n} K_n'(x, y) f(x) dx,$$

where

$$K_n'(x, y) = 2^{\lambda}\Gamma(\lambda) \sum_{k=0}^{\infty} c_k z^{-\lambda} J_{k+\lambda}(z) \sum_{l=0}^{[\frac{k}{2}]} (k - 2l + \lambda) C_{k-2l}^{\lambda}(w).$$

Here we adopt the notations $\lambda = (n-2)/2$, $z = |x||y|$, $w = \langle x, y\rangle / z$ and $J_{k+\lambda}$ are the Bessel functions while C_k^{λ} denote the Gegenbauser or ultraspherical polynomials in [2, Prop. 2.3].

5 Generalized Fourier Transforms in $(U(p, q), U(1))$-Pair

It follows from the decomposition of ω^p (resp. ω^{q*}) by $(O(p), SL(2, \mathbb{R}))$ (resp. $(O(q), SL(2, \mathbb{R}))$) duality that

$$\begin{aligned} \omega^p \otimes \omega^{q*} &\simeq \left(\sum_{k=0}^{\infty} \mathcal{H}_k \otimes V_{k+\frac{p}{2}} \right) \otimes \left(\sum_{l=0}^{\infty} \mathcal{H}_l^* \otimes V_{l+\frac{q}{2}}^* \right) \\ &\simeq \sum_{k=0}^{\infty} \sum_{l=0}^{\infty} \mathcal{H}_k \otimes \mathcal{H}_l^* \otimes V_{k+\frac{p}{2}} \otimes V_{l+\frac{q}{2}}^*. \end{aligned}$$

There is a natural action of $O(p, q)$ on $\mathcal{S}(\mathbb{R}^{p+q})$, and it commutes with the $\mathfrak{sl}(2, \mathbb{R})$ action defined by (2.4).

Because $V_{l+\frac{q}{2}}^*$ is a highest weight discrete series with highest weight $-l - \frac{q}{2}$, the representation $V_{k+\frac{p}{2}} \otimes V_{l+\frac{q}{2}}^*$ is a tensor product of a lowest weight discrete series and a highest weight discrete series. In this case, $V_{k+\frac{p}{2}} \otimes V_{l+\frac{q}{2}}^*$ contains a copy of direct integral, when k, l large enough. Furthermore, there is a direct integral decomposition of $L^2(\mathbb{R}^{p+q})$ into irreducible $O(p, q) \times \widetilde{SL}(2, \mathbb{R})$−modules (cf. [4] for details). Our argument of using Schur's lemma cannot be applied to this direct integral decomposition. Nevertheless, we can use the decomposition (4.1) to get

$$\begin{aligned}\omega^p \otimes \omega^{q*} &\simeq \left(\sum_{k=0}^{\infty} \mathcal{L}_k \otimes W_{k+\frac{p}{2}}\right) \otimes \left(\sum_{l=0}^{\infty} \mathcal{L}_l^* \otimes W_{l+\frac{q}{2}}^*\right) \\ &\simeq \sum_{k=0}^{\infty}\sum_{l=0}^{\infty} \mathcal{L}_k \otimes \mathcal{L}_l^* \otimes W_{k+\frac{p}{2}} \otimes W_{l+\frac{q}{2}}^*.\end{aligned}$$

Note that the group $U(1) = \{e^{it\mathfrak{k}_{p,q}} | t \in \mathbb{R}\}$ gives a maximal compact subgroup of $SL(2,\mathbb{R})$, with $U(p,q)$ its centralizer. So the two dual pairs $(O(p,q), U(1))$ and $(U(p,q), U(1))$ form the seesaw pair:

$$\begin{array}{ccc} U(n) & & SL(2,\mathbb{R}) \\ & \times & \\ O(n) & & U(1). \end{array}$$

Let $\mathfrak{L}_k = \sum_{l=0}^{\infty} \mathcal{L}_{k+l} \otimes \mathcal{L}_l^*$. Since $W_{k+\frac{p-q}{2}} \simeq W_{k+l+\frac{p}{2}} \otimes W_{l+\frac{q}{2}}^*$ as $\widetilde{U(1)}$-modules, we have the following result.

Proposition 5.1 *There is a multiplicity-free decomposition of $\mathcal{S}(\mathbb{R}^{p+q})$ into irreducible $U(p,q) \times \widetilde{U(1)}$–modules:*

$$\mathcal{S}(\mathbb{R}^{p+q})|_{U(p,q)\times\widetilde{U(1)}} \simeq \sum_{k=-\infty}^{\infty} \mathfrak{L}_k \otimes W_{k+\frac{p-q}{2}}. \tag{5.1}$$

In the above decomposition the indefinite Fourier transform is decomposed as

$$\mathcal{F}_{p,q} \simeq \sum_{k=-\infty}^{\infty} Id|_{\mathfrak{L}_k} \otimes S_k.$$

Here S_k is the action of $i^{\frac{p-q}{2}} e^{-\frac{\pi i}{2}\mathfrak{k}_{p,q}}$ on the $\mathfrak{k}_{p,q}$ weight space $W_{k+\frac{p-q}{2}}$, which is readily identified with the multiplication by $(-i)^k$.

Theorem 5.2 *Let $\mathcal{T}$ be bounded operator on $\mathcal{S}(\mathbb{R}^n)$ satisfying the following conditions:*

(i) *$\mathcal{T}$ is $O(p,q)$–equivariant and Γ–equivariant;*
(ii) *$\mathcal{T}^4 = Id$.*

Then $\mathcal{T}$ is $U(p,q)$–equivariant, and in view of (5.1)

$$\mathcal{T} = \sum_{k=-\infty}^{\infty} c_k Id|_{\mathfrak{L}_k} \otimes S_k,$$

where $c_k = \pm 1, \pm i$.

Morover, the Helmholtz relations

$$\mathcal{T} \circ r_{p,q}^2 = -\Delta_{p,q} \circ \mathcal{T} \text{ and } \mathcal{T} \circ \Delta_{p,q} = -r_{p,q}^2 \circ \mathcal{T}$$

imply that $\mathcal{T}$ is a multiple of $\mathcal{F}_{p,q}$ or $\mathcal{F}_{p,q}^{-1}$.

Proof The proof is similar to that of Theorem 4.2. For the $U(p,q)-$equivariance it suffices to prove that $O(p,q)$ and Γ generate $U(p,q)$. Denote by H the subgroup of $U(p,q)$ generated by $O(p,q)$ and Γ, and by $Lie(H)$ its Lie algebra. Consider the adjoint action of $\mathfrak{su}(p,q)$ when restricted to $\mathfrak{so}(p,q)$. We conclude that $Lie(H) = \mathfrak{u}(p,q)$, since some $iI_{s,t}(0 < s \neq p)$ and $iI_{p,q}$ are in $Lie(H)$. It follows that H contains an open neighborhood of 1 in $U(p,q)$ and hence $H = U(p,q)$ (cf. [6, Cor. 2.10]).

Let $\widetilde{\mathcal{T}} = \mathcal{T} \circ \mathcal{F}^{-1}$. Since $\widetilde{\mathcal{T}}$ commutes with $\varpi_{p,q}$, we have $\widetilde{\mathcal{T}}$ is $\widetilde{U(1)}-$equivariant. So $\widetilde{\mathcal{T}}$ commutes with $U(p,q) \times \widetilde{U(1)}$. It follows from Schur's Lemma that

$$\widetilde{\mathcal{T}}|_{\mathfrak{L}_k \otimes W_{k+\frac{p-q}{2}}} = c_k.$$

Thus, $\mathcal{T}$ must be in the form given by the theorem. Since both $\mathcal{T}$ and $\mathcal{F}$ have order 4, $c_k = \pm i$ or ± 1 for all k. In addtion the Helmholtz relations imply that $\widetilde{\mathcal{T}}$ commutes with $\mathfrak{sp}(2n, \mathbb{R})$, since $\mathfrak{sp}(2n, \mathbb{R})$ is generated by $\mathfrak{u}(p,q)$ together with $\Delta_{p,q}$ and $r_{p,q}^2$. This forces $\mathcal{T}$ to be a linear cobmination of $\mathcal{F}$ and $\mathcal{F}^{-1}$ as we showed in the proof of Proposition 2.3. Then we deduce that $\mathcal{T}$ is a multiple of $\mathcal{F}_{p,q}$ or $\mathcal{F}_{p,q}^{-1}$. □

Remark 5.3 It is readily checked that Condition (ii) implies that $\mathcal{T}$ commutes with $\varpi_{p,q}$. For the proof of the theorem it is enough to assume $\mathcal{T}$ to commute with $\varpi_{s,t}$ for a non-negative integer $s \neq p$. Therefore, '$\mathcal{T}$ is $\Gamma-$equivariant' in Condition (i) may be replaced by a weaker condition that '$\mathcal{T}$ commutes with $\varpi_{s,t}$ for a non-negative integer $s < p$'.

How to relate the generalized Fourier transfroms in Theorem 5.2

$$\mathcal{T} = \sum_{k=-\infty}^{\infty} c_k Id|_{\mathfrak{L}_k} \otimes S_k$$

to integral transforms is an interesting problem, which we are not able to give an easy solution here.

References

1. Cohen, H.: A conceptual breakthrough in sphere packing. Not. AMS **64**(2), 102–115 (2017)
2. De Bie, H., Oste, R., Van der Jeugt, J.: Generalized Fourier transforms arising from the enveloping algebras of $sl(2)$ and $osp(1|2)$. Int. Math. Res. Not. **2015**, 4649–4705 (2016)
3. Goodman, R., Wallach, N.R.: Symmetry, Representations, and Invariants. Springer, New York (2006)
4. Howe, R.: On some results of Strichartz and of Rallis and Schiffman. J. Funct. Anal. **32**, 297–303 (1979)

5. Howe, R., Tan, E.C.: Non-Abelian Harmonic Analysis—Applications of $SL(2, \mathbb{R})$. Springer, New York (1992)
6. Kirillov, A.: An Introduction to Lie Groups and Lie Algebras. Cambridge University Press, New York (2008)
7. Knapp, A.W.: Lie Groups Beyond an Introduction, 1st edn. Springer, New York (1996)
8. Kudla, S.S.: Seesaw Dual Reductive Pair. Princeton University Press, Princeton (1986)
9. Kobayashi, T., Mano, G.: The Schrödinger Model for the Minimal Representation of the Indefinite Orthogonal Group $O(p, q)$. American Mathematical Society, Providence (2008)
10. Repka, J.: Tensor Products of Unitary Representations of $SL_2(\mathbb{R})$. Am. J. Math. **100**, 747–774 (1978)
11. Stein, E.M., Weiss, G.: Introduction to Fourier Analysis on Euclidean Spaces. Princeton University Press, Princeton (1971)

Publisher's Note Springer Nature remains neutral with regard to jurisdictional claims in published maps and institutional affiliations.

Journal of Fourier Analysis and Applications (2019) 25:2801–2835
https://doi.org/10.1007/s00041-019-09683-z

RESEARCH ARTICLE

Fourier Analysis for Type III Representations of the Noncommutative Torus

Francesco Fidaleo[1]

Received: 14 July 2018 / Revised: 21 January 2019 / Published online: 16 April 2019
© Springer Science+Business Media, LLC, part of Springer Nature 2019

Abstract

For the noncommutative 2-torus, we define and study Fourier transforms arising from representations of states with central supports in the bidual, exhibiting a possibly nontrivial modular structure (*i.e.* type III representations). We then prove the associated noncommutative analogous of Riemann–Lebesgue Lemma and Hausdorff–Young Theorem. In addition, the L^p-convergence result of the Cesaro means (*i.e.* the Fejer theorem), and the Abel means reproducing the Poisson kernel are also established, providing inversion formulae for the Fourier transforms in L^p spaces, $p \in [1, 2]$. Finally, in $L^2(M)$ we show how such Fourier transforms "diagonalise" appropriately some particular cases of modular Dirac operators, the latter being part of a one-parameter family of modular spectral triples naturally associated to the previously mentioned non type II_1 representations.

Keywords Noncommutative harmonic analysis · Noncommutative geometry · Noncommutative torus · Type III representations · Noncommutative measure theory · Modular spectral triples

Mathematics Subject Classification 43A99 · 46L36 · 46L51 · 46L65 · 46L87 · 58B34

1 Introduction

The present paper is devoted to extend the Fourier analysis to non-type II_1 representations of the noncommutative 2-torus, that is when the underlying "measure" on the noncommutative manifold under consideration is not the canonical trace. This would

Dedicated to Maddalena Briamonte.

Communicated by Fulvio Ricci.

✉ Francesco Fidaleo
fidaleo@mat.uniroma2.it

1 Dipartimento di Matematica, Università di Roma Tor Vergata, Via della Ricerca Scientifica 1, 00133 Rome, Italy

include also the type II_1 case when the underlying measure relative to the trace is deformed by an inner, possibly unbounded, density.

The argument of the present paper falls into the standard lines of investigation of operator algebras and noncommutative harmonic analysis, perhaps having many potential interactions with other branches of mathematics, and quantum physics. Due to the crucial role played by the necessarily nontrivial modular data, such a topic can be viewed as a generalisation of the analysis developed in [6] for the representation of the noncommutative torus associated to the tracial state.

For the noncommutative 2-torus $\mathbb{A}_\alpha$, we have recently shown in [10] that it is possible to construct explicitly type III representations, at least when α is Liouville number. In addition, if α is one such number with a faster approximation property by rationals, it is also possible to construct modular spectral triples that, following [8], are obtained by deforming the untwisted Dirac operator and the associated commutator by using the Tomita modular operator which is neither bounded nor inner.

As a quick introduction, we recall that the study of Connes' noncommutative geometry grew impetuously in the last decades in view of several potential applications to mathematics and physics. The main ingredient is the so-called spectral triple, which is the candidate to encode the most important properties of the underlying "noncommutative manifold". For an exhaustive explanation of the fundamental role played by the spectral triples in noncommutative geometry, the reader is referred to the seminal monograph [7], the expository paper [5], and the literature cited therein.

Having in mind such possible applications, the seminal paper [8] also pursued the plan to exhibit twisted spectral triples for which the Dirac operator and the associated derivation are deformed by directly using the Tomita modular operator, provided the last is nontrivial. In that definition of twisted spectral triple, also the Tomita conjugation plays a role, which can be identified as a kind of "charge conjugation" well known in quantum physics. The twisted spectral triples considered in the above mentioned paper were then called modular. It is also clear that such modular spectral triples can play a role in constructing new noncommutative geometries in a type III setting, by emphasising the need to take the modular data into account.

One of the most studied examples in noncommutative geometry is indeed the noncommutative 2-torus $\mathbb{A}_\alpha$ (see *e.g.* [3]), since it is a quite simple model, though highly nontrivial. It is related to the discrete Canonical Commutation Relations, and can be considered as a quantum deformation of the classical 2-torus $\mathbb{T}^2$, according to the angle $2\pi\alpha$ entering in the definition of the symplectic form involved in the construction. When α is irrational, the C^*-algebra $\mathbb{A}_\alpha$ is simple and admits a unique, necessarily faithful, trace τ. Such a trace represents the natural noncommutative generalisation of the Lebesgue measure on the undeformed 2-torus $\mathbb{T}^2$. Therefore, the associated GNS representation π_τ gives rise to a von Neumann factor $\pi_\tau(\mathbb{A}_\alpha)''$, isomorphic to the Murray-von Neumann hyperfinite type II_1 factor, which can be viewed as the noncommutative counterpart of the von Neumann group algebra $L^\infty(\mathbb{T}^2, m\times m)$, m being the Haar measure (*i.e.* the normalised Lebesgue measure) on the unit circle. Hence, $\mathbb{A}_\alpha$ is not a type I C^*-algebra for irrational α, and therefore it must exhibit also type II_∞ and III representations. The preliminary step to construct explicit non type II_1 representations of the noncommutative torus, and consequently examples of modular spectral triples, was carried out in [10].

In order to develop the Fourier analysis for these new models (*i.e.* when the underlying "measure" is not the canonical trace) associated to the noncommutative torus, we introduce and investigate two essentially different Fourier transforms naturally associated to the left and right embeddings, which are particular cases of the ones

$$L^\infty\big(\pi(\mathbb{A}_\alpha)''\big) \equiv \pi(\mathbb{A}_\alpha)'' \hookrightarrow \iota^{\theta}_{\infty,1}\big(\pi(\mathbb{A}_\alpha)''\big) \subset \pi(\mathbb{A}_\alpha)''_* \equiv L^1\big(\pi(\mathbb{A}_\alpha)''\big)$$

given in (3.1), due to the non-triviality of the modular structure. Here, $\theta = 0, 1$ correspond indeed to the left and right embeddings, respectively. Notice that our investigation appears new also for any representation of type II_1 (including all examples treated in [8]) and II_∞, when the modular group is non trivial, even if "inner".

Our construction can also be generalised to the noncommutative deformation $\mathbb{A}_{\boldsymbol{\alpha}}$ of the d-dimensional torus $\mathbb{T}^d$, $\boldsymbol{\alpha}$ being a real skew-symmetric $d \times d$-matrix (*e.g.* [6], Section 2.2), after constructing non type II_1 representations for such models, following the method developed in [10].

For such noncommutative examples of Fourier transforms, we prove the analogous of Riemann–Lebesgue Lemma and Hausdorff–Young Theorem. In addition, for $p \in [1, 2]$ we establish an inversion formula arising from the Cesaro mean, that is a noncommutative version of the Fejer Theorem. Similar results can be obtained for the Abel average which is associated with the Poisson kernel, and also for the other cases mentioned in [6].

The last part of the present paper is devoted to a useful application of the Fourier analysis developed here, to new modular spectral triples for the noncommutative 2-torus. Firstly, for non type II_1 cases, we extend the construction of deformed Dirac operators and corresponding modular spectral triples in [10] to a one-parameter family for $\eta \in [0, 1]$, where the case $\eta = 0$ is nothing but the one studied in the previous mentioned paper. Secondly, in $L^2(M)$ we show how such Fourier transforms "diagonalise" appropriately the particular cases of modular Dirac operators, corresponding to $\eta = 0, 1/2, 1$, part of such a one-parameter family of modular spectral triples.

To conclude the introduction, we recall other interesting questions, such as those listed in [6], which we leave open in the present analysis, postponing them for a future possible investigation. Among them, we mention the construction and the study of the noncommutative Hardy spaces relative to the situations emerging from the present paper. Other important problems are those connected to the spaces of the so-called *Fourier multipliers*. Concerning the latter problem, we note that the analysis developed in [6,20] cannot be directly carried out in the present situation, since the underlying states ω are constructed only for very particular deformation angles α, deeply depending on that angle, see [10,16]. Therefore, the deformation angle α cannot be merely considered as a parameter, and a more refined analysis should be used for the non type II_1 cases in the sequel.

2 Preliminaries

2.1 Notations

Let E be a normed space. We simply denote by $\|\cdot\|$ its norm whenever no confusion arises. In particular, $\|x\| \equiv \|x\|_{\mathcal{H}}$ will be the Hilbertian norm of $x \in \mathcal{H}$. For a continuous or measurable function f defined on the locally compact space X equipped with the Radon measure μ, $\|f\| \equiv \|f\|_\infty$ will denote the "esssup" norm (or the "sup" norm for f continuous) of the function f.

In such a situation, $C_b(X)$ will denote the C^*-algebra consisting of all bounded continuous functions defined on X, equipped with the natural algebraic operations, and norm $\|f\| \equiv \|f\|_\infty := \sup_{x\in X} |f(x)|$. The situation of a point-set X, together with the C^*-algebra $\mathcal{B}(X) = C_b(X)$ of all bounded functions defined on X, is reduced to a particular case of the previous one by considering X equipped with the discrete topology.

Let $\mathbb{T} := \{z \in \mathbb{C} \mid |z| = 1\}$ be the abelian group consisting of the unit circle. The dual topological group $\widehat{\mathbb{T}}$ is isomorphic to the discrete group $\mathbb{Z}$. The corresponding Haar measures are the normalised Lebesgue measure $\mathrm{d}m = \frac{\mathrm{d}z}{2\pi\imath z} = \frac{\mathrm{d}\theta}{2\pi}$ for $z = e^{\imath\theta}$ on the circle, and the counting measure on $\mathbb{Z}$, respectively.

The Fourier transform and anti-transform of a bounded signed Radon measure $\mu \in C(\mathbb{T})^*$ are respectively defined as

$$\widehat{\mu}(n) := \int_{\mathbb{T}} z^{-n}\mathrm{d}\mu(z)\,, \quad \check{\mu}(n) := \int_{\mathbb{T}} z^{n}\mathrm{d}\mu(z), \quad n \in \mathbb{Z}.$$

For $f \in L^1(\mathbb{T}, m)$, its Fourier transform and anti-transform are usually defined as those of the measure $f\mathrm{d}m$:

$$\widehat{f}(n) := \oint f(z)z^{-n}\frac{\mathrm{d}z}{2\pi\imath z}, \quad \check{f}(n) := \oint f(z)z^{n}\frac{\mathrm{d}z}{2\pi\imath z}, \quad n \in \mathbb{Z}.$$

The symbol D denotes the derivative w.r.t. the natural argument $z \in \mathbb{T}$, *i.e.* $D \equiv \frac{\mathrm{d}}{\mathrm{d}z}$, and if $z = e^{\imath\theta}$ then $\frac{\mathrm{d}}{\mathrm{d}\theta} = \imath z\frac{\mathrm{d}}{\mathrm{d}z}$. In particular, if $g \in C^1(\mathbb{T})$,

$$\frac{\mathrm{d}}{\mathrm{d}\theta} g(e^{\imath\theta}) = \imath z D g(z).$$

Let $(X, \mathcal{B})$ be a measurable space, together with two σ-additive positive measures μ and ν on the σ-algebra $\mathcal{B}$. If μ dominates ν in the sense of measures (*e.g.* [22], Sect. 6), we write $\nu \preceq \mu$. If $\nu \preceq \mu$ and $\mu \preceq \nu$, then μ and ν are equivalent as measures, and we write $\mu \sim \nu$.

We also write $X \sim Y$ whenever two sets X and Y, equipped with the same algebraic structure, are algebraically isomorphic. For example, if $\mathfrak{A}$ and $\mathfrak{B}$ are involutive algebras (and in particular, C^*-algebras or W^*-algebras), we write $\mathfrak{A} \sim \mathfrak{B}$ if there exists a $*$-isomorphism $\rho : \mathfrak{A} \to \mathfrak{B}$ of $\mathfrak{A}$ onto $\mathfrak{B}$.

For a function f, M_f denotes the multiplication operator acting on functions g as $M_f g := fg$. For example, if f is a measurable function on the measure space $(X, \mathcal{B}, \nu)$, bounded almost everywhere, the multiplication operator M_f is the closed operator acting on $L^2(X, \mathcal{B}, \nu)$ with domain

$$\mathcal{D}_{M_f} := \left\{ g \in L^2(X, \mathcal{B}, \nu) \mid \int_X |f(x)g(x)|^2 \mathrm{d}\nu(x) < +\infty \right\},$$

defined for $g \in \mathcal{D}_{M_f}$ as

$$(M_f g)(x) := f(x)g(x), \quad x \in X.$$

If $f : X \to X$ is an invertible map on a point-set X:

- $f^0 := \mathrm{id}_X$, and its inverse is denoted by f^{-1};
- for the n-times composition, $f^n := \underbrace{f \circ \cdots \circ f}_{n-\text{times}}$;
- for the n-times composition of the inverse, $f^{-n} := \underbrace{f^{-1} \circ \cdots \circ f^{-1}}_{n-\text{times}}$.

Therefore, $f^n : X \to X$ is meaningful for any $n \in \mathbb{Z}$ with the above convention, and provides an action of the group $\mathbb{Z}$ on the set X.

2.2 Modular Spectral Triples

Concerning the usual terminology, the main concepts and results in operator algebras theory such as the Tomita modular theory and so forth, the reader is referred to [4,24,25] and the reference cited therein.

Let $\mathfrak{A}$ be a unital C^*-algebra. Denote by $\mathcal{S}(\mathfrak{A}) \subset \mathfrak{A}^*_+$ the set of the states, that is the positive and normalised functionals on the C^*-algebra $\mathfrak{A}$.

Let $\omega \in \mathcal{S}(\mathfrak{A})$ with $s(\omega) \in Z(\mathfrak{A}^{**})$. For such states with central support in the bidual, it can be shown (*e.g.* [17]) that the cyclic vector $\xi_\omega \in \mathcal{H}_\omega$ is separating for $\pi_\omega(\mathfrak{A})''$, $(\mathcal{H}_\omega, \pi_\omega, \xi_\omega)$ being the GNS representation of ω. Therefore,

$$\mathcal{H}_\omega \ni \pi_\omega(a)\xi_\omega \mapsto \pi_\omega(a^*)\xi_\omega \in \mathcal{H}_\omega$$

is well defined on the dense subset $\{\pi_\omega(a)\xi_\omega \mid a \in \mathfrak{A}\} \subset \mathcal{H}_\omega$, and closable. The polar decomposition of its closure S_ω, named the Tomita involution, is usually written as $S_\omega = J_\omega \Delta_\omega^{1/2}$, where J_ω and Δ_ω are the Tomita conjugation and modular operator, respectively. We also report the useful relation (*cf.* [24], Section 2.12)

$$J_\omega f(\Delta_\omega) J_\omega = \bar{f}(\Delta_\omega^{-1}), \tag{2.1}$$

holding for each Borel function f on $\mathbb{R}_+$.

Here, we provide the definition of (even) modular spectral triple we will use in the sequel, which is slightly different from the analogous one in [8] and generalise Definition 2.4 in [10]. We refer the reader to [7] and the references cited therein, for general aspects and some natural applications of spectral triples.

Definition 2.1 For each $\eta \in [0, 1]$, a *modular spectral triple* associated to a unital C^*-algebra $\mathfrak{A}$ is a triplet $(\omega, \mathcal{A}, L)_\eta$, where $\omega \in \mathcal{S}(\mathfrak{A})$, $\mathcal{A} \subset \mathfrak{A}$ is a dense $*$-algebra and L is a densely defined closed operator acting on $\mathcal{H}_\omega$ satisfying the following conditions:

(i) $s(\omega) \in Z(\mathfrak{A}^{**})$;

(ii) the deformed Dirac operator

$$D_L^{(\eta)} := \begin{pmatrix} 0 & \Delta_\omega^{\eta-1} L \Delta_\omega^{-\eta} \\ \Delta_\omega^{-\eta} L^* \Delta_\omega^{\eta-1} & 0 \end{pmatrix}$$

acting on $\mathcal{H}_\omega \oplus \mathcal{H}_\omega$, uniquely defines a selfadjoint operator with compact resolvent: $D_L^{(\eta)}$ is densely defined essentially selfadjoint with $\left(1 + \left(\overline{D_L^{(\eta)}}\right)^2\right)^{-1/2}$ compact;

(iii) for each $a \in \mathcal{A}$, the deformed commutator

$$\mathcal{D}_L^{(\eta)}(\pi_\omega(a)) := \imath \begin{pmatrix} 0 & \Delta_\omega^{\eta-1}[L, \pi_\omega(a)]\Delta_\omega^{-\eta} \\ \Delta_\omega^{-\eta}[L^*, \pi_\omega(a)]\Delta_\omega^{\eta-1} & 0 \end{pmatrix}$$

uniquely defines a bounded operator: $\overline{\mathcal{D}_{\mathcal{D}_L^{(\eta)}(\pi_\omega(a))}} = \mathcal{H}_\omega \oplus \mathcal{H}_\omega$ and

$$\sup\left\{ \|\mathcal{D}_L^{(\eta)}(\pi_\omega(a))\xi\| \mid \xi \in \mathcal{D}_{\mathcal{D}_L^{(\eta)}(\pi_\omega(a))}, \ \|\xi\| \leqslant 1 \right\} < +\infty;$$

(iv) $\pi_\omega(\mathcal{A})\mathcal{D}_L \subset \mathcal{D}_L$, $\pi_\omega(\mathcal{A})\mathcal{D}_{L^*} \subset \mathcal{D}_{L^*}$.

The closure of the Dirac operator in (ii) will be also denoted as $D_L^{(\eta)}$ with an abuse of notation.

Notice that, similarly to the untwisted case in order to study the possibly associated Fredholm module (see *e.g.* [11]), condition (iv) above is included only to make the commutators $[L, \pi_\omega(a)]$, $[L^*, \pi_\omega(a)]$ appearing in (ii) directly meaningful.

2.3 The Noncommutative 2-Torus

We fix the noncommutative 2-torus based on the deformation of the classical 2-torus $\mathbb{T}^2$, corresponding to the rotation of the angle $4\pi\alpha$. The multiplicative factor 2 is introduced only for a pure matter of convenience explained in [10]. Indeed, for a fixed $\alpha \in \mathbb{R}$, the *noncommutative torus* $\mathbb{A}_{2\alpha}$ associated with the rotation by the angle $4\pi\alpha$, is the universal C^*-algebra with identity I generated by the commutation relations involving two noncommutative unitary indeterminates U, V:

$$\begin{aligned} UU^* &= U^*U = I = VV^* = V^*V, \\ UV &= e^{4\pi\imath\alpha} VU. \end{aligned} \tag{2.2}$$

From now on, without loosing generality (*cf.* [3]) we also assume that $\alpha \in (0, 1/2)$.

We express $\mathbb{A}_{2\alpha}$ in the so-called *Weyl form*. Let $\mathbf{a} := (m, n) \in \mathbb{Z}^2$ be a double sequence of integers, and define

$$W(\mathbf{a}) := e^{-2\pi\imath\alpha mn} U^m V^n, \quad \mathbf{a} \in \mathbb{Z}^2. \tag{2.3}$$

Obviously, $W(\mathbf{0}) = I = \mathbf{I}_{\mathbb{A}_{2\alpha}}$, and the commutation relations (2.2) become

$$\begin{aligned} W(\mathbf{a})W(\mathbf{A}) &= W(\mathbf{a}+\mathbf{A})e^{2\pi\imath\alpha\sigma(\mathbf{a},\mathbf{A})}, \\ W(\mathbf{a})^* &= W(-\mathbf{a}), \quad \mathbf{a},\mathbf{A} \in \mathbb{Z}^2, \end{aligned} \tag{2.4}$$

where the symplectic form σ is defined as

$$\sigma(\mathbf{a},\mathbf{A}) := (mN - Mn), \quad \mathbf{a} = (m,n),\ \mathbf{A} = (M,N) \in \mathbb{Z}^2.$$

We now fix a function $f \in \mathcal{B}(\mathbb{Z}^2)$, which we may assume to have finite support. The element $W(f) \in \mathbb{A}_{2\alpha}$ is then defined as

$$W(f) := \sum_{\mathbf{a}\in\mathbb{Z}^2} f(\mathbf{a})W(\mathbf{a}).$$

The set

$$\{W(f) \mid f \text{ is a function on } \mathbb{Z}^2 \text{ with finite support}\}$$

provides a dense $*$-algebra of $\mathbb{A}_{2\alpha}$. Indeed, the relations (2.4) are transferred on the generators $W(f)$ as follows:

$$W(f)^* = W(f^\star), \quad W(f)W(g) = W(f\star g),$$

where

$$\begin{aligned} f^\star(\mathbf{a}) &:= \overline{f(-\mathbf{a})}, \\ (f\star g)(\mathbf{a}) &= \sum_{\mathbf{A}\in\mathbb{Z}^2} f(\mathbf{A})g(\mathbf{a}-\mathbf{A})e^{-2\pi\imath\alpha\sigma(\mathbf{a},\mathbf{A})}. \end{aligned}$$

The deformed convolution above depends on the chosen number α. Since such a number is fixed in the sequel, we simply denote that with the $\star$ symbol.

For irrational numbers α, we also recall that $\mathbb{A}_{2\alpha}$ is simple and has a necessarily unique and faithful trace

$$\tau(W(f)) := f(\mathbf{0}), \quad W(f) \in \mathbb{A}_{2\alpha}.$$

In this situation, by [3], Remark 1.7, any element $A \in \mathbb{A}_{2\alpha}$ is uniquely determined by its *Fourier coefficients*

$$f(\mathbf{a}) := \tau(W(-\mathbf{a})A), \quad \mathbf{a} \in \mathbb{Z}^2. \tag{2.5}$$

We note that, for each fixed $n \in \mathbb{Z}$, $f^{(n)}(m) := f(m,n)$ defines a sequence whose Fourier anti-transforms

$$\widetilde{f^{(n)}}(z) := \sum_{m\in\mathbb{Z}} f(m,n)z^m, \quad z \in \mathbb{T},$$

provides a sequence of continuous functions $\left\{\widetilde{f^{(n)}} \mid n \in \mathbb{Z}\right\} \subset C(\mathbb{T})$.

In order to obtain explicit type III representations of the noncommutative 2-torus, we tacitly suppose that the deformation angle α is irrational, satisfying sufficiently fast approximation properties by rationals, see below.

2.4 Diffeomorphisms of the Unit Circle

The Liouville numbers, denoted here as **L**-numbers, are necessarily irrational, and satisfy by definition the fast approximation by rationals as follows.

(L) A *Liouville number* $\alpha \in (0,1)$ is a real number such that, for each $N \in \mathbb{N}$ the inequality

$$\left|\alpha - \frac{p}{q}\right| < \frac{1}{q^N}$$

has an infinite number of solutions for $p, q \in \mathbb{N}$ with $\gcd(p,q) = 1$.

We also consider numbers satisfying the following faster approximation by rationals, and denoted as **UL**-numbers.

(UL) A *Ultra-Liouville number* $\alpha \in (0,1)$ is a real number such that, for each $\lambda > 1$ and $N \in \mathbb{N}$, the inequality

$$\left|\alpha - \frac{p}{q}\right| < \frac{1}{\lambda q^N}$$

again admits an infinite number of solutions for $p, q \in \mathbb{N}$ with $\gcd(p,q) = 1$.

We refer the reader to [14,23] for properties and details on such Liouville numbers.

In order to construct states on $\mathbb{A}_{2\alpha}$ with a nontrivial modular structure, and then type III representations with associated nontrivial modular spectral triples, we consider an orientation-preserving C^∞-diffeomorphism (called simply "a diffeomorphism") $f \in C^\infty(\mathbb{T})$ with the rotation number (*e.g.* [13]) $\rho(f) = 2\alpha$. By the theorem of Denjoy, f is conjugate to the rotation $R_{2\alpha}$ of the angle $4\pi\alpha$ through an uniquely determined homeomorphism $\hbar_f$ of the unit circle satisfying $\hbar_f(1) = 1$, and

$$f = \hbar_f \circ R_{2\alpha} \circ \hbar_f^{-1}. \tag{2.6}$$

In such a situation, T is defined as a "square root" of f:

$$T := \hbar_f \circ R_\alpha \circ \hbar_f^{-1}. \tag{2.7}$$

We note that

$$\mu_f := (h_f)^* m = m \circ h_f^{-1},$$

is the unique invariant measure, which is also ergodic, for the natural action of f on $\mathbb{T}$. For a Diophantine number α, it can be shown that h_f is indeed smooth, and thus $\mu_f \sim m$, the last denoting the normalised Haar measure on the circle.

For a Liouville number α, things are quite different. In fact, there are diffeomorphisms as above for which the unique invariant measure $(h_f)^* m$ is singular w.r.t. the Haar measure m: $(h_f)^* m \perp m$.

Summarising, if α is Diophantine the construction in [10] produces representations which are equivalent to the tracial one, leading to type II_1 von Neumann factors. Conversely, if α is a **L**-number, in [16] diffeomorphisms as above were constructed with the prescription that

$$\pi_\omega(\mathbb{A}_{2\alpha})'' \sim L^\infty(\mathbb{T}, \mathrm{d}\theta/2\pi) \ltimes_\beta \mathbb{Z}$$

is a type II_∞, or III_λ von Neumann factor for each $\lambda \in [0, 1]$. Here, β is the action induced on functions g on the unit circle by the corresponding action of f on points: $\beta(g) = g \circ f^{-1}$. If in addition α is a **UL**-number, then it is also possible to explicitly construct nontrivial modular spectral triples for $\mathbb{A}_{2\alpha}$, which fulfil the prescribed properties. The reader is referred to [7,8] for the significance of the concept of spectral triples and the modular ones in noncommutative geometry, and to [10] for the construction of nontrivial (*i.e.* non type II_1) modular spectral triples.

Another natural object associated to C^1-diffeomorphisms of the unit circle which plays a crucial role in constructing modular spectral triples, is the so-called *growth sequence* $\{\Gamma_n(f) \mid n \in \mathbb{N}\}$, defined as

$$\Gamma_n(f) := \max\{\|Df^n\|_\infty, \|Df^{-n}\|_\infty\}, \quad n \in \mathbb{N}.$$

For smooth diffeomorphisms f of the circle conjugate to an irrational rotation, it can be shown that $\Gamma_n(f)$ cannot have a too wild behaviour at infinity, see [29]. If α is Diophantine, then $\Gamma_n(f)$ is bounded because h_f in (2.6) is smooth. For diffeomorphisms constructed in [16] for a **L**-number α, it is shown in [10] that $\Gamma_n(f) = o(n)$, and $\Gamma_n(f) = o(\ln n)$ if in addition α is a **UL**-number.

Consider $g \in \mathcal{B}(\mathbb{Z}^2)$ such that $W(g) \in \mathbb{A}_{2\alpha}$. For $k \in \mathbb{N}$, $l = 0, 1$, define the following sequence of seminorms

$$\rho_{k,l}\big(W(g)\big) := \sup_{n\in\mathbb{Z}} \left\{(|n|+1)^k \left\| D^l \left(\widetilde{g^{(n)}} \circ R^{-n} \circ h_f^{-1}\right)\right\|_\infty\right\},$$

provided $\widetilde{g^{(n)}} \circ R^{-n} \circ h_f^{-1} \in C^1(\mathbb{T})$, $n \in \mathbb{Z}$. Define

$$\mathbb{A}^o_{2\alpha} := \left\{W(g) \mid \rho_{k,l}\big(W(g)\big) < +\infty,\ k \in \mathbb{N},\ l = 0, 1\right\},$$

which was shown in [10], Section 11, to be a unital $*$-subalgebra of $\mathbb{A}_{2\alpha}$, stable under the entire functional calculus.

2.5 States on the Noncommutative 2-Torus

As explained before, for $\alpha \in (0, 1/2)$ being an irrational number, fix a diffeomorphism $f \in C^\infty(\mathbb{T})$ which we suppose to be always C^∞ without any specific mention, satisfying (2.6). As previously pointed out, if α is a **L**-number or a **UL**-number, it will be possible to construct among others, type III representations and in addition nontrivial modular spectral triples, respectively.

The starting point is to consider the unique homeomorphism $\hbar_f$ satisfying (2.6), together with the measure $\mu := m \circ \hbar_f$. Since

$$\mu \sim \mu \circ R^n_{2\alpha}, \quad n \in \mathbb{Z}, \tag{2.8}$$

it is shown in [10] that the state $\omega \equiv \omega_\mu$ on $\mathbb{A}_{2\alpha}$ defined as

$$\omega(W(f)) := \sum_{m\in\mathbb{Z}} \check{\mu}(m) f(m, 0)$$

has central support in the bidual $\mathbb{A}^{**}_{2\alpha}$: $s(\omega) \in Z(\mathbb{A}^{**}_{2\alpha})$. The GNS representation $(\mathcal{H}_\omega, \pi_\omega, \xi_\omega)$ associated to ω is given by

$$\begin{aligned}\mathcal{H}_\omega &= \ell^2\big(\mathbb{Z}; L^2(\mathbb{T}, m)\big),\\ (\pi_\omega(W(f))g)_n(z) &= \sum_{l\in\mathbb{Z}} \left(\widetilde{f^{(l)}} \circ \hbar^{-1}_{T^2} \circ T^{2n-l}\right)(z) g_{n-l}(z),\\ (\xi_\omega)_n(z) &= \delta_{n,0}, \quad z \in \mathbb{T},\ n \in \mathbb{Z},\end{aligned}$$

where T is the "square root" of f given in (2.7).

By (2.8), we have

$$m \sim m \circ f^n, \quad n \in \mathbb{Z}$$

(which of course holds true even for each C^1-diffeomorphism), with Radon-Nikodym derivative

$$\delta_n(z) := \frac{\mathrm{d}m \circ f^n}{\mathrm{d}m}(z) = \frac{z(Df^n)(z)}{f^n(z)}, \quad z \in \mathbb{T},\ n \in \mathbb{Z}. \tag{2.9}$$

The modular structure is then expressed as follows. With $a \in \mathbb{R}$, on

$$\mathcal{D}_{\Delta^{a/2}_\omega} := \left\{x \in \mathcal{H}_\omega \mid \sum_{n\in\mathbb{Z}} \int_{\mathbb{T}} \delta_n(z)^a |x_n(z)|^2 \mathrm{d}m(z) < +\infty\right\}$$

we have for the modular operator:

$$(\Delta_\omega^{a/2}x)_n(z) = \delta_n(z)^{a/2}x_n(z), \quad n \in \mathbb{Z},\ z \in \mathbb{T}.$$

Concerning the modular conjugation, we get

$$(J_\omega x)_n(z) = \delta_n(z)^{1/2}\overline{x_{-n}(f^n(z))}, \quad n \in \mathbb{Z},\ z \in \mathbb{T},$$

which is meaningful for each $x \in \mathcal{H}_\omega$.

2.6 Operator Spaces

We consider a normed space E equipped with a sequence of norms on $\mathbb{M}_n(E)$, the spaces of the $n \times n$ matrices with entries in E, satisfying for $x \in \mathbb{M}_n(E)$, $y \in \mathbb{M}_m(E)$, $a, b \in \mathbb{M}_n$, $n, m = 1, 2, \dots$, the following properties:

(i) $\|axb\|_{\mathbb{M}_n(E)} \leqslant \|a\|_{\mathbb{M}_n}\|x\|_{\mathbb{M}_n(E)}\|b\|_{\mathbb{M}_n}$,
(ii) $\|x \oplus y\|_{\mathbb{M}_{n+m}(E)} = \max\{\|x\|_{\mathbb{M}_n(E)}, \|y\|_{\mathbb{M}_m(E)}\}$.

Such a normed space, together with these sequence of norms, is said to be an (abstract) *operator space*. Conversely, for a subspace $E \subset \mathfrak{A}$ of a C^*-algebra $\mathfrak{A}$, the norms on $\mathbb{M}_n(E)$ inherited by the inclusions $\mathbb{M}_n(E) \subset \mathbb{M}_n(\mathfrak{A})$, automatically satisfy (i) and (ii) above, and therefore E is a (concrete) operator space in a natural way. It is proved in [21] that a normed space E, equipped with a sequence of norms on all $\mathbb{M}_n(E)$, can be faithfully represented as a subspace of a C^*-algebra if and only if these norms satisfy (i) and (ii) above.

Let E be an operator space. The topological dual E^* can be viewed also as an operator space, when $\mathbb{M}_n(E^*)$ are equipped with the norms given for $f \in \mathbb{M}_n(E^*)$ by

$$\|f\|_{\mathbb{M}_n(E^*)} := \sup\left\{\|f(a)\|_{\mathbb{M}_{nm}} \mid \|a\|_{\mathbb{M}_m(E)} \leqslant 1, m \in \mathbb{N}\right\}.$$

Here, with $a \in \mathbb{M}_m(E)$, the matrix $f(a)$ is defined as

$$f(a)_{(i,j)(k,l)} := f_{ik}(a_{jl}), \quad i, k = 1, \dots, n,\ j, l = 1, \dots, m.$$

Such an operator space E^* is called in [2] *the standard dual* of E.

Let E, F be operator spaces. A linear map $T : E \to F$ is said to be *completely bounded* if for $T \otimes \mathrm{id}_{\mathbb{M}_n} : \mathbb{M}_n(E) \to \mathbb{M}_n(F)$ we have

$$\|T\|_{cb} := \sup_n \left\|T \otimes \mathrm{id}_{\mathbb{M}_n}\right\| < +\infty.$$

Examples of linear maps which are bounded but not completely bounded, can be easily exhibited, see *e.g.* Sect. 3.

3 On Noncommutative L^p-Spaces

We fix a normal faithful state ω on the W^*-algebra M. The $*$-algebra of *analytic elements* for the action of the modular group, that is that formed by all elements for which $\mathbb{R} \ni t \mapsto \sigma_t^\omega(a) \in M$ has a analytic continuation to an entire function $\mathbb{C} \ni z \mapsto \sigma_z^\omega(a) \in M$ (see *e.g.* [24]) is denoted by M_∞.

For $\theta \in [0, 1]$, consider the noncommutative L^p-spaces $L^p_\theta(M, \omega) = L^p_\theta(M)$, as ω is kept fixed. These can be constructed by complex interpolation by considering various embeddings $M \hookrightarrow M_*$, one for each θ, see [9,15,26]. Here, $\theta = 0, 1$ correspond to the left and right embeddings

$$M \ni x \mapsto \begin{cases} \omega(\,\cdot\, x) \\ \omega(x\,\cdot\,) \end{cases} \in M_*,$$

respectively. We also mention another equivalent way of approaching L^p-spaces developed in [12].

The key-point for the construction of noncommutative L^p-spaces is the modular theory (*e.g.* [24]), and complex interpolation (*e.g.* [1]). In fact, for $x \in M$ the map

$$\mathbb{R} \ni t \mapsto \sigma_t^\omega(x)\omega = \omega\big(\,\cdot\,\sigma_t^\omega(x)\big) \in M_*$$

extends to a bounded and continuous map on the strip $\{z \in \mathbb{C} \mid -1 \leqslant \mathrm{Im}(z) \leqslant 0\}$, analytic in the interior. After putting $L^1_\theta(M) := M_*$ for each $\theta \in [0, 1]$, $L^\infty_\theta(M) := \iota^\theta_{\infty,1}(M) \sim M$ with the norm inherited by that of M, and checking that the complex interpolation functor C^θ based on such analytic functions coincides (equal norms) with the standard one C_θ (see Theorem 1.5 in [15]), it is possible to define

$$L^p_\theta(M) := C^{1/p}\big(\iota^\theta_{\infty,1}(M), M_*\big) = C_{1/p}\big(\iota^\theta_{\infty,1}(M), M_*\big), \quad 1 < p < +\infty.$$

In such a way, for $1 \leqslant p \leqslant q \leqslant +\infty$ there are contractive embeddings

$$\iota^\theta_{q,p} : L^q_\theta(M) \hookrightarrow L^p_\theta(M).$$

We also remark (*cf.* [9]) that one can equip the L^p-spaces with a natural operator space structure arising from the canonical embeddings

$$\iota^\theta_{\infty,1} \otimes \mathrm{id} : \mathbb{M}_n(M) \to \mathbb{M}_n(M_*).$$

As stated in Proposition 3 in [9], $\iota^\theta_{\infty,1}$ is indeed completely bounded. For the convenience of the reader, we provide some details about the proof of such a result omitted in the original one.

Proposition 3.1 *For each $\theta \in [0, 1]$, the embedding*

$$M \ni x \mapsto \iota^\theta_{\infty,1}(x) = \omega\big(\,\cdot\,\sigma^\omega_{-\imath\theta}(x)\big) \in M_* \tag{3.1}$$

is a complete contraction.

Proof By the Phragmén–Lindelöf Theorem (*e.g.* [22], Theorem 12.8), complete boundedness for the left and right embeddings (*i.e.* for $\theta = 0, 1$) implies complete boundedness for all $\theta \in [0, 1]$. Therefore, we reduce the matter to the left and right embeddings.

By considering the natural embedding $\mathbb{M}_m(M) \ni y \mapsto y \oplus 0 \in \mathbb{M}_n(M)$ for $n \geqslant m$, we can restrict to the cases where $m = n$. Thus, in order to check the complete boundedness, for each $x, y \in \mathbb{M}_n(M)$ and $i, j, k, l = 1, \ldots, n$ we should compute the operator norm of the numerical matrices $L, R \in \mathbb{M}_{n^2}$ whose entries are given by

$$L_{(ij),(kl)} := \omega(y_{jl}x_{ik}), \quad R_{(ij),(kl)} := \omega(x_{ik}y_{jl}).$$

Concerning the second one, we easily get

$$R = \left(\omega \otimes \mathrm{id}_{\mathbb{M}_{n^2}}\right)(x \otimes y),$$

and therefore

$$\|R\|_{\mathbb{M}_{n^2}} \leqslant \|x\|_{\mathbb{M}_n(M)}\|y\|_{\mathbb{M}_n(M)}.$$

For the first one, we can straightforwardly see that there exist a permutation (depending on n) σ_n of the set

$$(11), \ldots, (1n), \ldots, (n1), \ldots, (nn) \sim 1, \ldots, n^2,$$

and therefore a selfadjoint unitary V_n acting on $\mathbb{C}^{n^2}$ such that

$$L = V_n\left(\omega \otimes \mathrm{id}_{\mathbb{M}_{n^2}}\right)(y \otimes x)V_n.$$

For example, when $n = 2$,

$$\sigma_2 = \begin{pmatrix} 1 & 2 & 3 & 4 \\ 1 & 3 & 2 & 4 \end{pmatrix},$$

and for $n = 3$,

$$\sigma_3 = \begin{pmatrix} 1 & 2 & 3 & 4 & 5 & 6 & 7 & 8 & 9 \\ 1 & 4 & 7 & 2 & 5 & 8 & 3 & 6 & 9 \end{pmatrix}.$$

Consequently, again

$$\|L\|_{\mathbb{M}_{n^2}} \leqslant \|x\|_{\mathbb{M}_n(M)}\|y\|_{\mathbb{M}_n(M)}.$$

□

It is possible to show that such an operator space structure coincides with the Pisier OH-structure (*cf.* [19]) at level of Hilbert spaces, that is

$$C_{1/2}\big(\iota^{\theta}_{\infty,1}(M), M_*\big) \equiv L^2(M) = \mathrm{OH}(\mathcal{H}_\omega), \quad \theta \in [0, 1],$$

see [9], Theorem 5.

With such identifications, we have that the topological dual $L^p_\theta(M)'$ is completely isomorphic to $L^q_{1-\theta}(M)$ in a canonical way, see [9,15].

Such an operator space structure on the noncommutative L^p-spaces obtained by interpolation arising by the various embeddings of $M \equiv L^\infty(M)$ into $M_* \equiv L^1(M)$ as before, is called *canonical*.

To simplify the notations, we put for $p, q \in [1, +\infty]$ with $q \geqslant p$,

$$i_{q,p} := \iota^0_{q,p}, \quad j_{q,p} := \iota^1_{p,q}.$$

For $a \in M$, we denote with L_a and R_a its left and right embedding in M_*, that is

$$L_a := \iota^0_{\infty,1}(a) = i_{\infty,1}(a), \quad R_a := \iota^1_{\infty,1}(a) = j_{\infty,1}(a).$$

We note that it is also customary to equip M_* with the operator space structure arising from the various embeddings relative to $\theta \in [0, 1]$ of M into the predual of the opposite algebra $(M^\circ)_*$, see *e.g.* Section 2.1 of [6]. Denote by $\kappa^\theta_{\infty,1} : M \hookrightarrow (M^\circ)_*$ such an embedding, and consider the canonical identification $o : M_* \to (M^\circ)_*$ given by

$$o(\varphi)(x^\circ) := \varphi(x), \quad \varphi \in M_*, \; x^\circ \in M^\circ.$$

Apart from other things, this choice means to exchange the left with the right embedding as follows: $\kappa^\theta_{\infty,1} = o \circ \iota^{1-\theta}_{\infty,1}$, see *e.g.* [15], Proposition 7.3, and [9], Proposition 3. With this choice, the embeddings $\kappa^\theta_{\infty,1}$ would not be completely bounded. To this aim, we prove the following, probably known to the experts,

Proposition 3.2 *For the Hilbert space $\mathcal{H} = \ell^2(\mathbb{N})$, the map $(o^t)^{-1} : \mathcal{B}(\mathcal{H}) \to \mathcal{B}(\mathcal{H})^\circ$ is not completely bounded.*

Proof First we note that, in general, $(o^t)^{-1}(x) = x^\circ$, provides the canonical identification between M and M°. Let now τ be the normalised (*i.e.* taking value 1 on minimal projections) normal, semifinite, faithful trace on $\mathcal{B}(\mathcal{H})$. It is easily seen that $\mathcal{B}(\mathcal{H})^\circ$ is $*$-isomorphic to $\pi_\tau(\mathcal{B}(\mathcal{H}))'$, where the isomorphism is given by

$$T(x^\circ) = J_\tau \pi_\tau\big(o^t(x^\circ)\big)^* J_\tau.$$

With $j := T \circ (o^t)^{-1} : \mathcal{B}(\mathcal{H}) \to \pi_\tau(\mathcal{B}(\mathcal{H}))'$ given by $j(x) = J_\tau \pi_\tau(x)^* J_\tau$, $(o^t)^{-1}$ would be completely bounded if and only if j were so. Consider now the transpose map (*cf.* [27,28]) $\theta : \mathcal{B}(\mathcal{H}) \to \mathcal{B}(\mathcal{H})$ on numerical matrices. On $\mathcal{H}_\tau = L^2(\mathcal{H})$

corresponding to all Hilbert-Schmidt class operators acting on $\mathcal{H} = \ell^2(\mathbb{N})$, we define the unitary selfadjoint operator

$$Vx := \theta(x), \quad x \in \mathcal{H}_\tau.$$

With I_n the identity operator on $\mathbb{C}^n$, it is then straightforward to check that

$$\big(j \otimes \mathrm{id}_{\mathbb{M}_n}\big)(x) = (V \otimes I_n)\big(\pi_\tau \circ \theta \otimes \mathrm{id}_{\mathbb{M}_n}\big)(x)\big(V \otimes I_n\big),$$

and therefore by [27], Theorem 1.2,

$$\big\|\big(j \otimes \mathrm{id}_{\mathbb{M}_n}\big)(x)\big\|_{\mathbb{M}_n(\mathcal{B}(\mathcal{H}_\tau))} = \big\|\big(\theta \otimes \mathrm{id}_{\mathbb{M}_n}\big)(x)\big\|_{\mathbb{M}_n(\mathcal{B}(\mathcal{H}))} = n.$$

Hence, j cannot be completely bounded. □

4 The Fourier Transform for the Noncommutative Torus

For each fixed $\omega \in \mathcal{S}(\mathbb{A}_{2\alpha})$ put $M := \pi_\omega(\mathbb{A}_{2\alpha})''$, the von Neumann algebra generated by the GNS representation π_ω.

We are interested in states ω with central support in the bidual $\mathbb{A}_{2\alpha}^{**}$, that is the cyclic vector ξ_ω is also separating for M.

Concerning a suitable definition of the Fourier transform for the cases when the reference measure is not the trace, we have to understand what is the right replacement $w_{kl} \in \mathbb{A}_{2\alpha}$ of the "characters" $(e^{\imath\theta_1})^k(e^{\imath\theta_2})^l \in C(\mathbb{T}^2)$ in the commutative case, where for the tracial case correspond to U^kV^l. Another natural question is how to "integrate" w.r.t. the reference state. Indeed, for $x \in M$ and $i_{\infty,1}(x)$, $j_{\infty,1}(x)$ the left and right embedding respectively, for the definition of the Fourier coefficients we have at least the following alternatives corresponding to the left and the right embeddings of M into M_*:

$$\begin{aligned}\omega(w_{kl}^*x) &= \langle \pi_\omega(x)\xi_\omega, \pi_\omega(w_{kl})\xi_\omega\rangle,\\ \omega(w_{kl}\sigma_{-\imath}^\omega(x)) &= \langle \pi_\omega(w_{kl})\xi_\omega, \pi_\omega(x^*)\xi_\omega\rangle.\end{aligned}$$

In the second alternative, the choice not to consider the adjoint of the chosen characters is only a matter of convenience, see Sect. 6. Such a choice of the characters w_{kl} may depend on the selected alternative.

Notice also that for the tracial cases and even for the commutative case $C(\mathbb{T}^2)$, these alternatives collapse to only one, due to the tracial property of the underlying "measure". In the cases we are managing in the present section, one possible choice is reported as follows.

For $T^2 = f$ as in (2.7), put $f_k(m,n) := \delta_{m,0}\delta_{n,k}$, $g_l(m,n) := \widehat{(h_{T^2})^l}(m)\delta_{n,0}$, and define $\{u_{kl} \mid k,l \in \mathbb{Z}\} \subset \pi_\omega(\mathbb{A}_{2\alpha})$, where

$$u_{kl} := \pi_\omega\big(W(f_k)W(g_l)\big), \quad k,l \in \mathbb{Z}.$$

We can embed such elements in $\mathcal{H}_\omega = \bigoplus_{\mathbb{Z}} L^2(\mathbb{T}, \mathrm{d}m)$ by defining $e^{kl} := u_{kl}\xi_\omega$, obtaining

$$e_n^{kl}(z) = z^l \delta_{n,k}, \quad z \in \mathbb{T}, \ k, l \in \mathbb{Z}.$$

It is easily seen that

$$\langle e^{kl}, e^{rs} \rangle = \delta_{k,r} \oint z^{l-s} \frac{\mathrm{d}z}{2\pi \imath z} = \delta_{k,r}\delta_{l,s}, \tag{4.1}$$

that is $\{e^{kl} \mid k, l \in \mathbb{Z}\} \subset \mathcal{H}_\omega \equiv \bigoplus_{\mathbb{Z}} L^2(\mathbb{T}, m)$ forms an orthonormal system which is indeed a basis.

For $\omega \in \mathcal{S}(\mathbb{A}_{2\alpha})$, we denote with an abuse of notation, again with ω its vector state extension to all of $M = \pi_\omega(\mathbb{A}_{2\alpha})''$:

$$\omega(a) := \omega_{\xi_\omega}(a) = \langle a\xi_\omega, \xi_\omega \rangle, \quad a \in \pi_\omega(\mathbb{A}_{2\alpha})''.$$

In addition, to make uniform the notation we write $L_0^1(M) = M_*$, even if $L_\theta^1(M) = M_*$, independently on $\theta \in [0, 1]$.

Definition 4.1 For $x \in M_* = L_0^1(M)$ of the form L_a, we define the sequence $\widehat{x}(k, l)$ of the Fourier coefficients as

$$\widehat{x}(k, l) := \omega(u_{kl}^* a), \quad k, l \in \mathbb{Z}. \tag{4.2}$$

Since

$$|\omega(u_{kl}^* a)| \leqslant \|L_a\| = \|x\|, \quad k, l \in \mathbb{Z}, \tag{4.3}$$

(4.2) extends by continuity to all $L_0^1(M)$, providing a sequence $\widehat{x} \in \ell^\infty(\mathbb{Z}^2)$.

We have the following noncommutative version of the Riemann–Lebesgue Lemma.

Theorem 4.2 *We have* $\widehat{L_0^1(M)} \subset c_0(\mathbb{Z}^2)$ *with* $\|\widehat{x}\|_{c_0} \leqslant \|x\|_{L^1}$. *In addition,* $\widehat{x} = 0 \Rightarrow x = 0$.

Proof By the definition of the Fourier coefficients, if $x \in L_0^1(M)$ we see that the double sequence $\widehat{x}$ is uniformly bounded with $\|\widehat{x}\|_\infty \leqslant \|x\|$.

Suppose now that $\widehat{x} = 0$. This means that $x(y) = 0$ on the set made of elements y such that y^* is of the form

$$y^* = \sum_{|r|,|s| \leqslant n} a_{rs} u_{rs},$$

which generate a norm-dense subset of $\pi_\omega(\mathbb{A}_{2\alpha})$, and a $*$-weakly dense subset of $\pi_\omega(\mathbb{A}_{2\alpha})'' \equiv M$. Since $L_0^1(M) \equiv M_*$, by continuity we conclude that $x(y) = 0$ for all $y \in M$. Then $x = 0$.

It remains to check that

$$\lim_{|k|,|l|\to+\infty} \widehat{x}(k,l) = 0.$$

In fact, for the total set $\{L_{u_{rs}} \mid r,s \in \mathbb{Z}\}$, by (4.1),

$$\widehat{L_{u_{rs}}}(k,l) = \delta_{k,r}\delta_{l,s} \to 0,$$

as $|k|, |l| \to +\infty$.

For generic $x \in L^1_0(M)$ and $\varepsilon > 0$, we choose a finite linear combinations of the u_{rs} of the form

$$x_\varepsilon = \sum_{|r|,|s|\leqslant n} c_{rs} L_{u_{rs}},$$

such that $\|x - x_\varepsilon\| < \varepsilon$. Then

$$\begin{aligned}|\widehat{x}(k,l)| &\leqslant |\widehat{x}(k,l) - \widehat{x_\varepsilon}(k,l)| + |\widehat{x_\varepsilon}(k,l)|\\ &\leqslant \left|\omega(u^*_{kl}(x - x_\varepsilon)\right| + |\widehat{x_\varepsilon}(k,l)| \leqslant \varepsilon + |\widehat{x_\varepsilon}(k,l)| \to \varepsilon,\end{aligned}$$

because $\widehat{x_\varepsilon}(k,l) \to 0$ as $|k|, |l| \to +\infty$ as noticed before.

The proof follows as $\varepsilon > 0$ is arbitrary. □

Remark 4.3 Also for this case, it is not hard to show that $\widehat{L^1_0(M)} \subsetneq c_0(\mathbb{Z}^2)$.

For such a purpose, it is enough to consider the sequence of Dirichlet kernels $\{D_n \mid n \in \mathbb{N}\} \subset L^1(\mathbb{T}, m)$, and the associated functionals $\{X_n \mid n \in \mathbb{N}\} \subset L^1(M)$ given by

$$X_n\big(\pi_\omega(W(f))\big) := \int_{\mathbb{T}} \big(\widetilde{f^{(0)}} \circ h^{-1}_{T^2}\big)(z) D_n(z) \mathrm{d}m(z).$$

Denoting by $\widehat{D_n}$ the usual Fourier transforms of the D_n with an abuse of the notation, the Fourier coefficients and the L^1-norms of the X_n are given by

$$\widehat{X_n}(k,l) = \widehat{D_n}(l)\delta_{k,0}$$

and $\|X_n\|_{L^1} = \|D_n\|_1$, respectively. Reasoning as in [22], we have $\|\widehat{X_n}\|_\infty = 1$ for each $n \in \mathbb{N}$, but conversely $\|X_n\|_{L^1} \to +\infty$. This leads to a contradiction if we suppose that $\widehat{L^1_0(M)} = c_0(\mathbb{Z}^2)$.

Suppose now $\xi \in \mathcal{H}_\omega = L^2_0(M)$. Then its left embedding x_ξ in $M_* \equiv L^1_0(M)$ is given by

$$x_\xi(\,\cdot\,) = \langle\,\cdot\,\xi, \xi_\omega\rangle,$$

and therefore its Fourier transform is given for $k, l \in \mathbb{Z}$, by

$$\widehat{x_\xi}(k,l) = x_\xi(u_{kl}^*) = \langle u_{kl}^*\xi, \xi_\omega\rangle = \langle \xi, u_{kl}\xi_\omega\rangle = \langle \xi, e^{kl}\rangle. \tag{4.4}$$

Namely, the Fourier coefficients of x_ξ are precisely those arising from the orthogonal expansion of ξ w.r.t. the orthonormal basis $\{e^{kl} \mid k, l \in \mathbb{Z}\}$ of $\mathcal{H}_\omega$. Therefore,

$$\|\widehat{x_\xi}\|_{\ell^2} = \|\xi\|_{L^2}, \tag{4.5}$$

that is the Fourier transform provides a unitary isomorphism when restricted to $L^2_0(M)$ as expected.

For the Fourier transform of elements in $L^2_0(M)$, we simply write

$$\widehat{\xi}(k,l) := \widehat{x_\xi}(k,l) = \langle \xi, e^{kl}\rangle, \quad \xi \in \mathcal{H}_\omega, \ k, l \in \mathbb{Z}. \tag{4.6}$$

We now pass to consider the extension of the Fourier transform to L^p, $p \in (1,2)$. Indeed, by (4.3) which holds true for generic element in $L^1_0(M)$ (*cf.* Theorem 4.2), and (4.5), we can define the Fourier transform for $p \in [1,2]$ and q its conjugate exponent (*i.e.* $1/p + 1/q = 1$ with the convention $1/0 = \infty$) by complex interpolation (*cf.* [15]),

$$\mathcal{F}_{p,q} : L^p_0(M) \to \ell^q(\mathbb{Z}^2),$$

which satisfies

$$\|\mathcal{F}\|_{p,q} := \|\mathcal{F}_{p,q}\|_{\mathcal{B}(L^p,\ell^q)} \leqslant 1, \quad p \in [1,2],$$

because of

$$\|\mathcal{F}\|_{1,\infty} \leqslant 1, \quad \|\mathcal{F}\|_{2,2} = 1.$$

Concerning the involved completely bounded norms, we also have

Proposition 4.4 *The Fourier transforms $\mathcal{F}$, given in* (4.2) *for* (*a norm dense set in*) *$L^1_0(M)$ and in* (4.6) *for $L^2_0(M)$, provide complete contractions w.r.t. the canonical operator space structure for the noncommutative L^p-spaces obtained by interpolation.*

Proof The case $p = 1$ proceeds as follows. For any element $f \in \mathbb{M}_n(\ell^1(\mathbb{Z}^2))$, its operator norm is computed in the following way. As in Sect. 2.6, with $a \in \mathbb{M}_m(\ell^\infty(\mathbb{Z}^2))$ and after defining $f(a) \in \mathbb{M}_{nm}$ by

$$f(a)_{(i,j)(k,l)} := f_{ik}(a_{jl}), \quad i,k = 1,\dots,n, \ j,l = 1,\dots,m,$$

we have

$$\|f\|_{\mathbb{M}_n(\ell^1(\mathbb{Z}^2))} := \sup\left\{\|f(a)\|_{\mathbb{M}_{nm}} \mid \|a\|_{\mathbb{M}_m(\ell^\infty(\mathbb{Z}^2))} \leqslant 1, m \in \mathbb{N}\right\}.$$

Firstly, we note that such an element f uniquely defines a linear map on $M\overline{\otimes}\ell^\infty(\mathbb{Z}^2)$ given by $F := \mathrm{id}_M \otimes f$, assuming the form

$$F(a) := \sum_\alpha f_{ik}(\alpha)a_\alpha,$$

where we have denoted by $\alpha \in \mathbb{Z}^2$ a generic element of $\mathbb{Z}^2$. This provides a bounded linear map

$$F : M\overline{\otimes}\ell^\infty(\mathbb{Z}^2) \to \mathbb{M}_n(M)$$

with norm $\|F\| \leqslant \|f\|_{\mathbb{M}_n(\ell^1(\mathbb{Z}^2))}$. In order to check its boundedness, we proceed as follows.

For elements $\{\xi(i), \eta(i)\}_{i=1}^n \subset \mathcal{H}_\omega$, fix any orthonormal basis $\{e_r\}_{1\leqslant r\leqslant 2n}$ of any $2n$-dimensional space $\mathcal{K} \subset \mathcal{H}_\omega$ containing the vectors $\xi(i)$, $\eta(i)$, $i = 1, \dots, n$. Take $a \in M\overline{\otimes}\ell^\infty(\mathbb{Z}^2)$. With $a_{rs}(\alpha) := \langle a_\alpha e_r, e_s\rangle$, the matrix $\big(a_{rs}(\alpha)\big)_{\alpha\in\mathbb{Z}^2,\, 1\leqslant r,s\leqslant 2n}$ defines an element $\tilde{a} \in \mathbb{M}_{2n}(\ell^\infty(\mathbb{Z}^2))$ with norm less than that of a. As usual, we also denote with $f(a)$ the $2n^2 \times 2n^2$ matrix with entries given by

$$f(a)_{(ir),(js)} := \sum_{\alpha\in\mathbb{Z}^2} f_{ij}(\alpha)(a_{rs}(\alpha)), \quad r, s \in \mathbb{N},\ i, j = 1, \dots, n.$$

The elements $\{\xi(i), \eta(i)\}_{i=1}^n \subset \mathcal{H}_\omega$ with components $\xi_r(i) := \langle \xi(i), e_r\rangle$, $\eta_r(i) := \langle \eta(i), e_r\rangle$ w.r.t. the previous orthonormal basis, define vectors $\tilde{\xi}, \tilde{\eta} \in \mathbb{C}^n \otimes \mathbb{C}^{2n} \sim \mathbb{C}^{2n^2}$ with components

$$\tilde{\xi}_{ir} := \xi_r(i),\ \ \tilde{\eta}_{ir} := \eta_r(i), \quad 1 \leqslant r \leqslant n,\ i = 1, \dots, n.$$

We compute

$$\begin{aligned}
\left|\sum_{i,j=1}^n \langle F(a)\xi_i, \eta_j\rangle\right| &= \left|\sum_{i,j,\alpha} f_{ij}(\alpha)\langle a_\alpha \xi_i, \eta_j\rangle\right| \\
&= \left|\sum_{i,j;r,s}\left(\sum_\alpha f_{ij}(\alpha)a_{rs}(\alpha)\right)\xi_r(i)\overline{\eta_s(j)}\right| \\
&= |\langle f(a)\tilde{\xi}, \tilde{\eta}\rangle| \leqslant \|f(a)\|\|\tilde{\xi}\|\|\tilde{\eta}\| \leqslant \|f\|_{\mathbb{M}_n(\ell^1(\mathbb{Z}^2))}\|\tilde{a}\|\|\tilde{\xi}\|\|\tilde{\eta}\| \\
&\leqslant \|f\|_{\mathbb{M}_n(\ell^1(\mathbb{Z}^2))}\|a\|_{M\overline{\otimes}\ell^\infty(\mathbb{Z}^2)}\left(\sum_{i=1}^n \|\xi_i\|^2\right)^{1/2}\left(\sum_{i=1}^n \|\eta_i\|^2\right)^{1/2},
\end{aligned}$$

and the assertion follows.

Fix now $f \in \mathbb{M}_n(\ell^1(\mathbb{Z}^2))$, which can be supposed to have finite support, such that $\|f\|_{\mathbb{M}_n(\ell^1(\mathbb{Z}^2))} \leqslant 1$. After noticing that $\{u_\alpha\}_{\alpha\in\mathbb{Z}^2}$ gives rise to an element $u \in$

$M\overline{\otimes}\ell^\infty(\mathbb{Z}^2)$ with unit norm, we consider a generic element $L_x \in \mathbb{M}_m(M_*)$ for $x \in \mathbb{M}_m(M)$ and compute

$$f\big(\mathcal{F}_{1,\infty}(L_x)\big)_{(i,j)(k,l)} = \sum_\alpha f_{jl}(\alpha)\omega(u^*_\alpha x_{ik}) = \omega\left(\left(\sum_\alpha f_{jl}(\alpha)u^*_\alpha\right)x_{ik}\right)$$
$$= \omega\big(F(u^*)_{jl}x_{ik}\big) = L_{x_{ik}}\big(F(u^*)_{jl}\big).$$

Thus, by reasoning as in the proof of Proposition 3.1, we have

$$\big\| f\big(\mathcal{F}_{1,\infty}(L_x)\big)\big\|_{\mathbb{M}_{mn}} \leqslant \|L_x\|_{\mathbb{M}_m(L^1)}\,\|F\| = \|L_x\|_{\mathbb{M}_m(L^1)}\,.$$

By taking the supremum on the r.h.s. for f in the unit ball of $\mathbb{M}_n(\ell^1(\mathbb{Z}^2))$ and on n, we obtain the assertion by noticing that elements of the type L_x as above provide a set dense in norm of $\mathbb{M}_n(M_*)$.

For the case $p = 2$, pick $\xi \in \mathbb{M}_n(\mathcal{H}_\omega)$ and consider the matrices $\langle\xi,\xi\rangle$, $\langle\widehat{\xi},\widehat{\xi}\rangle \in \mathbb{M}_{n^2}$ whose entries are given by

$$\langle\xi,\xi\rangle_{(i,k),(j,l)} := \langle\xi_{ij},\xi_{kl}\rangle, \quad \langle\widehat{\xi},\widehat{\xi}\rangle_{(i,k),(j,l)} := \langle\widehat{\xi_{ij}},\widehat{\xi_{kl}}\rangle.$$

After denoting by P_v the orthogonal projection onto the one dimensional subspace spanned by the vector $v \in \mathcal{H}_\omega$, by (4.4) we compute

$$\langle\widehat{\xi_{ij}},\widehat{\xi_{kl}}\rangle = \sum_{m,n\in\mathbb{Z}} \langle\xi_{ij},e^{mn}\rangle\langle e^{mn},\xi_{kl}\rangle = \sum_{m,n\in\mathbb{Z}} \langle P_{e^{mn}}\xi_{ij},\xi_{kl}\rangle$$
$$= \left\langle\left(\sum_{m,n\in\mathbb{Z}} P_{e^{mn}}\right)\xi_{ij},\xi_{kl}\right\rangle = \langle\xi_{ij},\xi_{kl}\rangle,\quad i,j,k,l \in \mathbb{Z},$$

because the e^{mn} provide a basis for $\mathcal{H}_\omega$, and thus $\left(\sum_{m,n\in\mathbb{Z}} P_{e^{mn}}\right) \uparrow I_{\mathcal{H}_\omega}$ in the strong operator topology. Consequently, $\|\langle\widehat{\xi},\widehat{\xi}\rangle\|_{\mathbb{M}_{n^2}} = \|\langle\xi,\xi\rangle\|_{\mathbb{M}_{n^2}}$ and therefore $\|\mathcal{F}\|^{(\mathrm{cb})}_{2,2} = 1$ by [9], Proposition 1. □

Remark 4.5 When $p = 2$, we have proved that $\mathcal{F}_{2,2}$ is indeed a complete isometry between the canonical operator space structures of $L^2_0(M)$ and $\ell^2(\mathbb{Z}^2)$, both coinciding with the Pisier OH-structure (*cf.* [19]), see [9], Theorem 5.

Summarising, we have the noncommutative version of the Hausdorff–Young Theorem in our setting.

Theorem 4.6 *Let $p \in [1,2]$ with $q \in [2,+\infty]$ its conjugate exponent. The Fourier transform restricts to a complete contraction*

$$\mathcal{F}_{p,q} : L^p_0(M) \to \ell^q(\mathbb{Z}^2),$$

which is indeed a complete isometry for $p = 2$.

Proof By Proposition 4.4, the assertion follows as $C_\theta = C^\theta$ (equal norms, Theorem 1.5 in [15], and Theorem 2 in [9]) provides exact interpolation functors, see *e.g.* [1], Theorems 4.1.2 and 4.1.4. □

Notice that the last theorem can be viewed as a noncommutative version of the Riesz-Thorin Theorem.

The computation explained above allows to consider the noncommutative version of the Hausdorff–Young Theorem for the Fourier anti-transform defined as follows. In fact, fix any element $f \in \ell^1(\mathbb{Z}^2)$ and define $\check{f} \in M$ given by

$$\check{f} := \sum_{k,l\in\mathbb{Z}} f(k,l)u_{kl}. \tag{4.7}$$

The definition for the Fourier anti-transform for $p = 2$ proceeds as follows. For each $f \in \ell^1(\mathbb{Z}^2)$ we consider the maps, the latter being the left embedding of M in $\mathcal{H}_\omega$:

$$f \in \ell^1(\mathbb{Z}^2) \mapsto \check{f} \in M \mapsto \check{f}\xi_\omega = \sum_{k,l\in\mathbb{Z}} f(k,l)e^{kl} \in \mathcal{H}_\omega.$$

Therefore, $\ell^2(\mathbb{Z}^2) \ni f \mapsto \check{f} \in \mathcal{H}_\omega$ assumes the form

$$\check{f} := \sum_{k,l\in\mathbb{Z}} f(k,l)e^{kl}.$$

Concerning the L^p-spaces, $p \in [2, +\infty]$, the Fourier anti-transform is then associated to the left embedding of M inside M_*.

Denoting by T such a Fourier anti-transform, we have

Proposition 4.7 *The Fourier anti-transforms $T_{1,\infty}$ and $T_{2,2}$ are well defined as complete contractions w.r.t. the canonical operator space structures for the noncommutative (left) L^p-spaces obtained by interpolation.*

Proof For the case $p = 1$, we proceed as in Proposition 4.4. Firstly we note that, if $f \in \ell^1(\mathbb{Z}^2)$ then

$$\|\check{f}\|_{L^\infty} = \left\| \sum_{\alpha\in\mathbb{Z}^2} f(\alpha)u_a \right\| \leqslant \sum_{\alpha\in\mathbb{Z}^2} |f(\alpha)| = \|f\|_{\ell^1},$$

and therefore the Fourier anti-transform is well defined as a bounded map between $\ell^1(\mathbb{Z}^2)$ and $L^\infty(M)$.

To check complete boundedness, fix $f \in \mathbb{M}_n(\ell^1(\mathbb{Z}^2))$ and $g \in \mathbb{M}_m(M_*)$ with $\|g\|_{\mathbb{M}_m(M_*)} \leqslant 1$, With F as before, after recalling that $\|F\| \leqslant \|f\|$, we compute

$$g(\check{f})_{(i,j)(k,l)} = g_{jl}\Big(\sum_\alpha f_{ik}(\alpha)u_\alpha\Big) = g_{jl}\big(F(u)_{ik}\big).$$

Therefore, $\|g(\check{f})\|_{\mathbb{M}_{nm}} \leqslant \|f\|_{\mathbb{M}_n(\ell^1(\mathbb{Z}^2))}$ that is $T_{1,\infty}$ is completely bounded.

For $p = 2$, with $f \in \mathbb{M}_n(\ell^2(\mathbb{Z}^2))$ we compute

$$\begin{aligned}\langle \check{f}_{ik}, \check{f}_{jl}\rangle &= \left\langle \sum_{m,n\in\mathbb{Z}} f_{ik}(m,n)e^{mn}, \sum_{r,s\in\mathbb{Z}} f_{jl}(r,s)e^{rs}\right\rangle \\ &= \sum_{m,n;r,s\in\mathbb{Z}} f_{ik}(m,n)\overline{f_{jl}(r,s)}\langle e^{mn}, e^{rs}\rangle \\ &= \sum_{m,n;r,s\in\mathbb{Z}} f_{ik}(m,n)\overline{f_{jl}(r,s)}\delta_{m,r}\delta_{n,s} = \langle f_{ik}, f_{jl}\rangle.\end{aligned}$$

Therefore, $T_{2,2}$ is a complete isometry concerning the left embedding. □

Theorem 4.8 *Let $p \in [1,2]$ with $q \in [2,+\infty]$ its conjugate exponent. The Fourier anti-transform defined in* (4.7) *extends to a complete contraction*

$$T_{p,q} : \ell^p(\mathbb{Z}^2) \to L^q_0(M),$$

which is indeed a complete isometry for $p = 2$.

Proof The proof follows analogously to that of Theorem 4.6, by taking into account Proposition 4.7. □

5 The Noncommutative Version of the Fejer Theorem

In our context, we provide a noncommutative version of the Fejer Theorem based on the Cesaro mean of the Fourier coefficients. By using a suitable version of the so-called "transference method", we reduce the matter to the classical convolution with the Fejer kernel on $\mathbb{T}^2$ as for the classical (*e.g.* [18]) and the tracial (*cf.* [6]) cases. We also get the same result for the Abel mean of the Fourier coefficients by reducing the matter to the convolution with the Poisson kernel, and argue that our analysis can be used to manage all cases listed in Section 3 of [6] and many others. All such cases will provide inversion formulas for the Fourier transform in L^p-spaces, $p = [1,2]$, arising from non type II_1 representations (and also for type II_1 ones when the modular operator is not trivial including the cases treated in [8]) of the noncommutative 2-torus.

We start by noticing that the usual transference method, used in [6] for the tracial case, cannot be directly applied to the situations under consideration. This is because the 2-parameter group of $*$-automorphisms of $\mathbb{A}_{2\alpha}$ determined by

$$\mathbb{A}_{2\alpha} \ni (U,V) \mapsto (zU, wV) \in \mathbb{A}_{2\alpha}, \quad z, w \in \mathbb{T},$$

does not leave invariant the reference state ω. Yet, we can get the expected result, even for the situation studied in the present paper.

Consider the action of $\mathbb{Z}$ on $L^\infty(\mathbb{T}, m)$ generated by the powers of the $*$-automorphism β given by

$$\beta(H)(z) := H(f^{-1}(z)), \quad H \in L^\infty(\mathbb{T}, m), \ z \in \mathbb{T}.$$

The $*$-isomorphism

$$\pi_\omega(\mathbb{A}_{2\alpha})'' \sim L^\infty(\mathbb{T}, m) \ltimes_\beta \mathbb{Z}$$

is realised as follows (*cf.* [10], Section 6). Let $\pi_o : L^\infty(\mathbb{T}, m) \to \pi_\omega(\mathbb{A}_{2\alpha})''$ be the injective $*$-homomorphism given by

$$\pi_o(H)_n := M_{\beta^{-n}(H)} = M_{H \circ f^n}, \quad H \in L^\infty(\mathbb{T}, m), \ n \in \mathbb{Z},$$

where M_h is the multiplication operator for the function h.

In addition, for

$$g \in \mathcal{H}_\omega = \ell^2\big(\mathbb{Z}; L^2(\mathbb{T}, m)\big) \sim \bigoplus_{\mathbb{Z}} L^2(\mathbb{T}, m),$$

we consider the shift λ which is a unitary operator acting on $\mathcal{H}_\omega$ given by $(\lambda g)_n := g_{n-1}$, $n \in \mathbb{Z}$. We get

$$\lambda^k \pi_o(H) \lambda^{-k} = \pi_o\big(\beta^k(H)\big), \quad k \in \mathbb{Z}.$$

Therefore, $\pi_\omega(\mathbb{A}_{2\alpha})''$ is generated by the $\pi_o(H)$ and the powers of λ:

$$\pi_\omega(\mathbb{A}_{2\alpha})'' = \left\{ \sum_{|k| \leqslant l} \pi_o(H_k) \lambda^k \mid H_k \in L^\infty(\mathbb{T}, m), \ l \in \mathbb{N} \right\}'',$$

see *e.g.* [25], Section X.1, or [4], Section 2.7.1.

Now, let $x = \sum_k \pi_o(H_k)\lambda^k$ be a linear generator of $M = \pi_\omega(\mathbb{A}_{2\alpha})''$ with

$$x_n = \sum_k M_{H_k \circ f^n} \lambda^k, \quad n \in \mathbb{Z}.$$

For $\mathbf{w} = (w_1, w_2) \in \mathbb{T}^2$, we consider the following maps $\rho_{\mathbf{w}}$ defined on such generators of M by

$$\rho_{\mathbf{w}}\left(\sum_k M_{H_k \circ f^n} \lambda^k \right) := \left(\sum_k w_2^k M_{H_k \circ f^n \circ R_{w_1}} \lambda^k \right), \tag{5.1}$$

where for $w \in \mathbb{T}$, R_w denotes, with an abuse of notation, the map $R_w(z) := wz$, $z \in \mathbb{T}$.

Proposition 5.1 *The maps* (5.1) *extend to* $*$*-automorphisms of* $\pi_\omega(\mathbb{A}_{2\alpha})''$. *In addition, they uniquely define contractions*

$$\rho_{\mathbf{w}}^{(p,0)} : L_0^p(M) \to L_0^p(M), \quad 1 \leqslant p < +\infty,$$

satisfying

$$i_{q,p}\rho_{\mathbf{w}}^{(q,0)} = \rho_{\mathbf{w}}^{(p,0)} i_{q,p}, \quad 1 \leqslant p \leqslant q \leqslant +\infty. \tag{5.2}$$

Proof Notice that $\rho_{\mathbf{w}} = \rho_{1,w_1}\rho_{2,w_2} = \rho_{2,w_2}\rho_{1,w_1}$, where

$$\rho_{1,w}\Big(\sum_k M_{H_k \circ f^n}\lambda^k\Big) := \Big(\sum_k M_{H_k \circ f^n \circ R_w}\lambda^k\Big),$$
$$\rho_{2,w}\Big(\sum_k M_{H_k \circ f^n}\lambda^k\Big) := \Big(\sum_k w^k M_{H_k \circ f^n}\lambda^k\Big).$$

It is easily seen that $\rho_{2,w}$ define automorphisms of M. By Theorem X.1.7 in [25], $\rho_{1,w}$ also define automorphisms of M.

The pre-transpose map

$$_t\rho_{\mathbf{w}} := \rho_{\mathbf{w}}^t \lceil_{M_*},$$

also defines an isometry of $M_* = L_0^1(M)$ satisfying

$$_t\rho_{\mathbf{w}} \lceil_M = \rho_{\mathbf{w}},$$

because $M \sim i_{\infty,1}(M) \equiv \big\{L_x \mid x \in M\big\} = L_0^\infty(M) \subset L_0^1(M)$. Therefore, by interpolation (*cf.* [1]), $\rho_{\mathbf{w}}$ define contractions

$$\rho_{\mathbf{w}}^{(p,0)} : L_0^p(M) \to L_0^p(M)$$

satisfying (5.2). □

We note that it is possible to verify by direct inspection that the $\rho_{\mathbf{w}}^{(2,0)}$ are isometric isomorphisms. Indeed, if $x \in \bigoplus_{\mathbb{Z}} L^2(\mathbb{T}, m)$, for $\mathbf{w} = (w_1, w_2) \in \mathbb{T}^2$ we get

$$\big\|\rho_{\mathbf{w}}^{(2,0)}(x)\big\|^2 = \sum_{n\in\mathbb{Z}} \int_{\mathbb{T}} \big|w_2^n x_n(w_1 z)\big|^2 \mathrm{d}m(z) = \sum_{n\in\mathbb{Z}} \int_{\mathbb{T}} \big|x_n(z)\big|^2 \mathrm{d}m(z) = \|x\|^2.$$

The following results are crucial in the sequel.

Lemma 5.2 *Let* $x \in L_0^p(M)$, *and* $\mathbf{w} = (w_1, w_2) \in \mathbb{T}^2$, *then*

$$\widehat{\rho_{\mathbf{w}}^{(p,0)}(x)}(k,l) = w_1^{-l} w_2^{-k} \widehat{x}(k,l), \quad k, l \in \mathbb{Z}.$$

Proof If $p > 1$, one has (and it can be taken as the definition of the Fourier transform for $p > 2$)

$$\widehat{x}(k,l) = \widehat{i_{p,1}(x)}(k,l) = i_{p,1}(x)(u_{kl}^*),$$

where if $x \in M = L_0^\infty(M)$, $\widehat{x} = \widehat{L_x} \equiv \widehat{i_{\infty,1}(x)}$. Thus, we can restrict the analysis to $p = 1$.

For such a purpose, for each $x \in L_0^1(M)$ and $\varepsilon > 0$, we take $x_\varepsilon \in M$ such that $\|x - i_{\infty,1}(x_\varepsilon)\| < \varepsilon$. Since ε is arbitrary and

$$\begin{aligned}
\left|w_1^{-l}w_2^{-k}\widehat{x}(k,l) - \widehat{\rho_{\mathbf{w}}^{(1,0)}(x)}(k,l)\right| &\leqslant \left|\widehat{x}(k,l) - \widehat{i_{\infty,1}(x_\varepsilon)}(k,l)\right| \\
&\quad + \left|w_1^{-l}w_2^{-k}\widehat{i_{\infty,1}(x_\varepsilon)}(k,l) - \widehat{\rho_{\mathbf{w}}^{(1,0)}(i_{\infty,1}(x_\varepsilon))}(k,l)\right| \\
&\quad + \left|\widehat{\rho_{\mathbf{w}}^{(1,0)}(i_{\infty,1}(x_\varepsilon))}(k,l) - \widehat{\rho_{\mathbf{w}}^{(1,0)}(x)}(k,l)\right| \\
&\leqslant \|x - i_{\infty,1}(x_\varepsilon)\|_1 \\
&\quad + \left|w_1^{-l}w_2^{-k}\widehat{i_{\infty,1}(x_\varepsilon)}(k,l) - \widehat{\rho_{\mathbf{w}}^{(1,0)}(i_{\infty,1}(x_\varepsilon))}(k,l)\right| \\
&\quad + \left\|\rho_{\mathbf{w}}^{(1,0)}\big(i_{\infty,1}(x_\varepsilon) - x\big)\right\|_1 \\
&\leqslant 2\varepsilon + \left|w_1^{-l}w_2^{-k}\widehat{i_{\infty,1}(x_\varepsilon)}(k,l) - \widehat{\rho_{\mathbf{w}}^{(1,0)}(i_{\infty,1}(x_\varepsilon))}(k,l)\right|,
\end{aligned}$$

the assertion will follow if we prove it for the generators of $L^1(M) = M_*$ of the form $L_a \equiv i_{\infty,1}(a)$. We first note that

$$\rho_{\mathbf{w}}(u_{kl})\xi_\omega = w_1^l w_2^k u_{kl}\xi_\omega, \tag{5.3}$$

and then

$$\begin{aligned}
\widehat{\rho_{\mathbf{w}}^{(1,0)}\big(L_a\big)}(k,l) &= L_a\big(\rho_{\mathbf{w}}(u_{kl})^*\big) = \langle a\xi_\omega, \rho_{\mathbf{w}}(u_{kl})\xi_\omega\rangle \\
&= w_1^{-l}w_2^{-k}\langle a\xi_\omega, u_{kl}\xi_\omega\rangle = w_1^{-l}w_2^{-k}\widehat{L_a}(k,l).
\end{aligned}$$

□

Proposition 5.3 *For each $1 \leqslant p \leqslant 2$, we get*

$$L^p\text{–}\lim_{\mathbf{w}\to\mathbf{1}} \rho_{\mathbf{w}}^{(p,0)}x = x, \quad x \in L_0^p(M).$$

Proof For $x \in L_0^2(M) \sim \mathcal{H}_\omega = \oplus_{n\in\mathbb{Z}}L^2(\mathbb{T}, \mathrm{d}m)$, we get by the Lebesgue dominated convergence theorem,

$$\left\|\rho_{\mathbf{w}}^{(2,0)}x - x\right\|_2^2 = \sum_{n\in\mathbb{Z}}\int_{\mathbb{T}} \mathrm{d}m(z)\left|w_2^n x_n(w_1^1 z) - x_n(z)\right|^2 \to 0,$$

as $\mathbf{w} \to \mathbf{1} \equiv (1,1) \in \mathbb{T}^2$. Thus, the result holds true for $p = 2$. For $1 \leqslant p < 2$ and $\varepsilon > 0$, let now $x \in L_0^p(M)$. Since $i_{2,p}(L^2)$ is dense in each such L^p-spaces, we pick $x_\varepsilon \in L_0^2(M)$ such that $\|x - i_{2,p}(x_\varepsilon)\|_p < \varepsilon$. Again by Proposition 5.1, we also have $\left\|\rho_{\mathbf{w}}^{(p,0)}(x - i_{2,p}(x_\varepsilon))\right\|_p < \varepsilon$ for each $\mathbf{w} \in \mathbb{T}^2$. We then compute by the Holder inequality,

$$\begin{aligned}\left\|\rho_{\mathbf{w}}^{(p,0)}x - x\right\|_p &< 2\varepsilon + \left\|\rho_{\mathbf{w}}^{(p,0)}(i_{2,p}(x_\varepsilon)) - i_{2,p}(x_\varepsilon)\right\|_p\\ &\leqslant 2\varepsilon + \left\|i_{2,p}\big(\rho_{\mathbf{w}}^{(2,0)}(x_\varepsilon) - x_\varepsilon\big)\right\|_p\\ &\leqslant 2\varepsilon + \left\|\rho_{\mathbf{w}}^{(2,0)}x_\varepsilon - x_\varepsilon\right\|_2.\end{aligned}$$

Since $\varepsilon > 0$ is arbitrary, by the previous part $\left\|\rho_{\mathbf{w}}^{(p,0)}x - x\right\|_p \to 0$ whenever $\mathbf{w} \to \mathbf{1}$. □

Here, there is the noncommutative version of the Fejer theorem in our context. It will provide an inversion formula for the Fourier transform for $1 \leqslant p \leqslant 2$.

Theorem 5.4 *Let $1 \leqslant p \leqslant 2$ and $x \in L_0^p(M)$. Then we get,*

$$L^p\text{--}\lim_{N\to+\infty} \sum_{|k|,|l|\leqslant N} \left(1 - \frac{|k|}{N+1}\right)\left(1 - \frac{|l|}{N+1}\right)\widehat{x}(k,l)i_{\infty,p}(u_{kl}) = x.$$

Proof We first prove that, for each fixed $x \in L_0^p(M)$ and $N = 1, 2, \dots,$

$$\begin{aligned}&\sum_{|k|,|l|\leqslant N} \left(1 - \frac{|k|}{N+1}\right)\left(1 - \frac{|l|}{N+1}\right)\widehat{x}(k,l)i_{\infty,p}(u_{kl})\\ &\quad= \int_{\mathbb{T}^2} dm(w_1)dm(w_2)\Phi_N(w_1)\Phi_N(w_2)\rho_{w_1,w_2}^{(p,0)}(x),\end{aligned}$$

where Φ_N is the Fejer kernel. By a standard 3ε-argument, it is enough to check the identity for the total set (in $L_0^p(M)' \sim L_1^q(M)$, q being the conjugate exponent) of elements of the type $j_{\infty,q}(u_{rs}^*)$, $r, s \in \mathbb{Z}$. By Lemma 5.2,

$$\rho_{w_1,w_2}^{(p,0)}(x)\big(j_{\infty,q}(u_{rs}^*)\big) = \widehat{\rho_{w_1,w_2}^{(p,0)}(x)}(r,s) = w_1^{-s}w_2^{-r}\widehat{x}(r,s),$$

and thus

$$\begin{aligned}&\int_{\mathbb{T}^2} dm(w_1)dm(w_2)\Phi_N(w_1)\Phi_N(w_2)\rho_{w_1,w_2}^{(p,0)}(x)\big(j_{\infty,q}(u_{rs}^*)\big)\\ &\quad= \widehat{x}(r,s)\int_{\mathbb{T}} dm(w_1)\Phi_N(w_1)w_1^{-s}\int_{\mathbb{T}} dm(w_2)\Phi_N(w_2)w_1^{-r}\\ &\quad= \widehat{\Phi_N}(r)\widehat{\Phi_N}(s)\widehat{x}(r,s)\\ &\quad= \left(1 - \frac{|r|}{N+1}\right)\left(1 - \frac{|s|}{N+1}\right)\widehat{x}(r,s)\end{aligned}$$

$$= \sum_{|k|,|l|\leqslant N}\left(1-\frac{|k|}{N+1}\right)\left(1-\frac{|l|}{N+1}\right)\widehat{x}(k,l)\omega(u^*_{rs}u_{kl})$$
$$= \sum_{|k|,|l|\leqslant N}\left(1-\frac{|k|}{N+1}\right)\left(1-\frac{|l|}{N+1}\right)\widehat{x}(k,l)i_{\infty,p}(u_{kl})\big(j_{\infty,q}(u^*_{rs})\big).$$

The proof now follows by Proposition 5.3 as the Fejer kernel $\Phi_N(z)$ is an approximate identity (*e.g.* [18]) on $\mathbb{T}$. □

We report the following generalisation corresponding to the Abel mean associated to the Poisson kernel, whose proof is completely analogous to the previous one. The details relative to the remaining cases in [6] are left to the reader.

Theorem 5.5 *Let $1 \leqslant p \leqslant 2$ and $x \in L^p_0(M)$. Then we get,*

$$L^p\text{–}\lim_{r\uparrow 1}\sum_{k,l\in\mathbb{Z}} r^{|k|+|l|}\widehat{x}(k,l)i_{\infty,p}(u_{kl}) = x.$$

Proof We note that, as before,

$$\sum_{k,l\in\mathbb{Z}} r^{|k|+|l|}\widehat{x}(k,l)i_{\infty,p}(u_{kl})$$
$$= \int_{\mathbb{T}^2} \mathrm{d}m(w_1)\mathrm{d}m(w_2)P_r(w_1)P_r(w_2)\rho^{(p,0)}_{w_1,w_2}(x),$$

where P_r, $0 \leqslant r < 1$ is the Poisson kernel, which is again an approximate identity on $\mathbb{T}$. The assertion follows by Proposition 5.3. □

6 An Alternative Definition of the Fourier Transform

Even if it is still unclear how to construct a one-parameter group of Fourier transforms, one for each embedding in (3.1) of M in its predual M_*, we are able to provide an alternative definition of the Fourier transform, which is essentially associated to the right embedding

$$M \ni a \to j_{\infty,1}(a) = R_a \in M_*,$$

in the following way.

Definition 6.1 For $x \in M_* = L^1_1(M)$ of the form R_a, we define the sequence $\widehat{x}(k,l)$ of the Fourier coefficients as

$$\widehat{x}(k,l) := \omega(au_{kl}), \quad k,l \in \mathbb{Z}. \tag{6.1}$$

It is immediate to check that, for $a \in M$ we get

$$\widehat{L_a}(k,l) = \omega(u_{kl}a) \neq \omega(u^*_{kl}a) = \widehat{L_a}(k,l),$$

$$\widehat{R_a}(k,l)=\omega(au_{kl}^*)\neq\omega(au_{kl})=\widehat{R_a}(k,l).$$

This simply means that, for general $f\in M_*=L^1_0(M)=L^1_1(M)$ and $k,l\in\mathbb{Z}$, we get

$$\widehat{f}(k,l)=f(u_{kl}^*)\neq f(u_{kl})=\widehat{f}(k,l).$$

For $p=2$, suppose for simplicity that $\xi\in\mathcal{D}_{\Delta_\omega^{1/2}}$. Then

$$L^1_1(M)\ni x_\xi:=j_{2,1}(\xi)=\langle\,\cdot\,\Delta_\omega^{1/2}\xi,\xi_\omega\rangle.$$

Therefore, for $k,l\in\mathbb{Z}$ we obtain by $J_\omega\Delta_\omega^{1/2}=\Delta_\omega^{-1/2}J_\omega$ (*cf.* (2.1)),

$$\widehat{x_\xi}(k,l)=\langle u_{kl}\Delta_\omega^{1/2}\xi,\xi_\omega\rangle=\langle\xi,J_\omega e^{kl}\rangle,\tag{6.2}$$

which is meaningful for all $\xi\in\mathcal{H}_\omega$.

By putting

$$\epsilon^{kl}:=J_\omega e^{kl},\quad k,l\in\mathbb{Z},$$

we note that the ϵ^{kl} still provide an orthonormal basis for $\mathcal{H}_\omega$ as the conjugation J_ω is an anti-unitary involution. For the Fourier transform (6.2) of a generic vector $\xi\in\mathcal{H}_\omega$, we put

$$\widehat{\xi}(k,l):=\langle\xi,\epsilon^{kl}\rangle,\quad k,l\in\mathbb{Z}.$$

The corresponding Fourier anti-transform

$$f\in\ell^1(\mathbb{Z}^2)\to\check{f}\in M=L^\infty_1(M),$$

that is the analogous of (4.7), is defined as

$$\check{f}:=\sum_{k,l\in\mathbb{Z}}f(k,l)u_{kl}^*.\tag{6.3}$$

By taking into account

$$L^\infty_1(M)=M\ni a\hookrightarrow\Delta_\omega^{1/2}a\xi_\omega\in\mathcal{H}_\omega=L^2_1(M)$$

and $J_\omega\Delta_\omega^{1/2}=\Delta_\omega^{-1/2}J_\omega$ as before, for the Fourier anti-transform

$$f\in\ell^2(\mathbb{Z}^2)\to\check{f}\in L^2_1(M),$$

we obtain

$$\check{f} = \sum_{k,l\in\mathbb{Z}} f(k,l)\Delta_\omega^{1/2}u_{kl}^*\xi_\omega = \sum_{k,l\in\mathbb{Z}} f(k,l)\epsilon^{kl},\ k,l\in\mathbb{Z}.$$

Therefore, the Fourier anti-transform "$\check{}$" is also a (complete) unitary equivalence at level of Hilbert spaces as expected.

From the last consideration, compared with the analogous one (4.4), we can recognise why "$\widehat{}$" and "$\check{}$" are associated to the left and right embedding of M in M_*, respectively.

As noticed before, it is still unclear how we can interpolate between both definitions to obtain a one-parameter of Fourier transforms associated to the other embeddings $\iota^\theta_{\infty,1} : M \hookrightarrow M_*, \theta\in(0,1)$.

Also for this case of the Fourier transform $\widehat{}$ and the corresponding anti-transform $\check{}$, the Riemann–Lebesgue Lemma (*cf.* Theorem 4.2), the Hausdorff–Young inequality (*cf.* Theorem 4.6), and finally Theorem 4.8 hold true. We leave the details of the proofs to the reader.

The corresponding results analogous to Theorems 5.4 and 5.5 hold true as well. For this situation, we report the crucial details of the proof.

Theorem 6.2 *Let* $1\leqslant p\leqslant 2$ *and* $x\in L^p_0(M)$. *Then*

$$L^p\!-\!\lim_{N\to+\infty}\sum_{|k|,|l|\leqslant N}\left(1-\frac{|k|}{N+1}\right)\left(1-\frac{|l|}{N+1}\right)\widehat{x}(k,l)j_{\infty,p}(u^*_{kl}) = x,$$
$$L^p\!-\!\lim_{r\uparrow 1}\sum_{k,l\in\mathbb{Z}} r^{|k|+|l|}\widehat{x}(k,l)j_{\infty,p}(u^*_{kl}) = x.$$

Proof With $\rho^{(p,1)}_{\mathbf{w}} : L^p_1(M)\to L^p_1(M)$ the extension of the $*$-automorphism in (5.1) to the right L^p-spaces, by (5.3) we check for $x = j_{\infty,p}(a)$,

$$\begin{aligned}\rho^{(p,1)}_{w_1,w_2}(x)\big(i_{\infty,q}(u_{rs})\big) &= \widehat{\rho^{(p,1)}_{w_1,w_2}(x)}(r,s) = \langle a\rho_{\mathbf{w}}(u_{rs})\xi_\omega,\xi_\omega\rangle\\ &= w_1^s w_2^r\langle au_{rs}\xi_\omega,\xi_\omega\rangle = w_1^s w_2^r\widehat{x}(r,s).\end{aligned}$$

Therefore, by a standard 3ε-argument, we argue that for a generic $x\in L^p_1(M)$,

$$\begin{aligned}&\sum_{|k|,|l|\leqslant N}\left(1-\frac{|k|}{N+1}\right)\left(1-\frac{|l|}{N+1}\right)\widehat{x}(k,l)j_{\infty,p}(u^*_{kl})\\ &= \int_{\mathbb{T}^2}\mathrm{d}m(w_1)\mathrm{d}m(w_2)\Phi_N(w_1)\Phi_N(w_2)\rho^{(p,1)}_{w_1^{-1},w_2^{-1}}(x),\\ &\times\sum_{k,l\in\mathbb{Z}} r^{|k|+|l|}\widehat{x}(k,l)j_{\infty,p}(u^*_{kl})\\ &= \int_{\mathbb{T}^2}\mathrm{d}m(w_1)\mathrm{d}m(w_2)P_r(w_1)P_r(w_2)\rho^{(p,1)}_{w_1^{-1},w_2^{-1}}(x).\end{aligned}$$

Then the assertions follow analogously to those of the corresponding results in the previous section. □

We compare the definitions of the Fourier transforms in the present paper with the original ones concerning the situation associated to the canonical trace τ, and for the commutative situation $C(\mathbb{T}^2)$ corresponding to the case when the deformation angle α is zero.

Suppose that $A \in \mathbb{A}_{2\alpha}$. By taking into account our definitions of the Fourier transform and that in (2.5), for $x := L_{\pi_\tau(A)} = R_{\pi_\tau(A)} \in \pi_\tau(\mathbb{A}_{2a})''_*$ we get

$$\widehat{x}(k,l) = e^{2\pi\imath\alpha kl}\tau\big(W(-l,-k)A\big),$$
$$\check{x}(k,l) = e^{-2\pi\imath\alpha kl}\tau\big(W(l,k)A\big),$$

where the Weyl operators $W(m,n)$ are defined in (2.3).

The classical case corresponds to $\alpha = 0$, that is $\mathbb{A}_0 \sim C(\mathbb{T}^2)$, and the underlying measure corresponds to the Haar one $m \times m = \frac{\mathrm{d}\theta_1\mathrm{d}\theta_2}{4\pi^2}$. In this situation, each element in $L^1(M)$ uniquely corresponds to a (equivalence class of) function f in $L^1(\mathbb{T}^2, m\times m)$:

$$L^1(M) \ni x \equiv f_x \in L^1(\mathbb{T}^2, m\times m) \sim L^1(M).$$

Therefore, we get

$$\widehat{x}(k,l) = \int_{\mathbb{T}^2} f_x\big(e^{\imath\theta_1}, e^{\imath\theta_2}\big)e^{-\imath(l\theta_1+k\theta_2)}\frac{\mathrm{d}\theta_1\mathrm{d}\theta_2}{4\pi^2} = \widehat{f_x}(l,k),$$
$$\check{x}(k,l) = \int_{\mathbb{T}^2} f_x\big(e^{\imath\theta_1}, e^{\imath\theta_2}\big)e^{\imath(l\theta_1+k\theta_2)}\frac{\mathrm{d}\theta_1\mathrm{d}\theta_2}{4\pi^2} = \check{f_x}(l,k),$$

where $\widehat{f}$ and $\check{f}$ denote the classical Fourier transform and anti-transform of $f \in L^1(\mathbb{T}^2, m\times m)$.

We end the present section by noticing the following fact arising by algebraic similarities. Even if there is no reasonable motivation to consider $\mathbb{A}_{2\alpha}$ as a group-like quantum object with $\mathbb{Z}^2$ as its "dual object", we can consider the map (6.3) as a Fourier transform from $\ell^1(\mathbb{Z}^2)$ to M, whose corresponding anti-transform $M_* \to \ell^\infty(\mathbb{Z}^2)$ is given by (6.1). This alternative way to see the emerging situation does not cause any trouble.

7 Deformed Dirac Operators and Modular Spectral Triples

The present section is devoted to defining a one-parameter family of modular spectral triples for each $\eta \in [0,1]$, extending the ones defined in Section 9 of [10]. As an application of the previous analysis, we show how the Fourier transforms defined here, allow us to "diagonalise" the corresponding Dirac operators $\mathrm{D}^{(\eta)}$, $\eta = \{0, 1/2, 1\}$.

For a fixed diffeomorphism f of the unit circle $\mathbb{T}$ with growth sequence $\Gamma_n(f)$ as in Sect. 2.4, we start with an undeformed Dirac operator by putting

$$\mathrm{D}_n = \begin{pmatrix} 0 & \mathrm{L}_n \\ \mathrm{L}_n^* & 0 \end{pmatrix} := \begin{pmatrix} 0 & \left(\imath z \frac{\mathrm{d}}{\mathrm{d}z} - a_n I\right) \\ \left(-\imath z \frac{\mathrm{d}}{\mathrm{d}z} - a_n I\right) & 0 \end{pmatrix},$$

with $a_0 = 0$ and

$$a_n := \operatorname{sign}(n) \sum_{l=1}^{|n|} \frac{1}{\Gamma_{l - \frac{1-\operatorname{sign}(n)}{2}}(f)}, \quad n \in \mathbb{Z}\backslash\{0\}.$$

Here,

$$\operatorname{sign}(n) := \begin{cases} -1 & \text{if } n < 0, \\ 1 & \text{if } n \geqslant 0, \end{cases}$$

and $\mathrm{L} = \bigoplus_{n\in\mathbb{Z}} \left(\imath z \frac{\mathrm{d}}{\mathrm{d}z} - a_n I\right)$ is precisely the operator appearing in Definition 2.1 of the modular spectral triple. To simplify notation, we drop the subscript "L" in the foregoing analysis.

Consider the Sobolev–Hilbert space

$$H^1(\mathbb{T}) := \left\{ f \in AC(\mathbb{T}) \mid f' \in L^2(\mathbb{T}, m) \right\},$$

where $AC(\mathbb{T})$ denotes the set of all absolutely continuous complex valued functions on the unit circle.

For $\eta \in [0, 1]$ and δ_n as in (2.9), we can define deformed Dirac Operators as

$$\begin{aligned} \mathrm{D}^{(\eta)} &= \bigoplus_{n\in\mathbb{Z}} \begin{pmatrix} 0 & M_{\delta_n^{\eta-1}} \mathrm{L}_n M_{\delta_n^{-\eta}} \\ M_{\delta_n^{-\eta}} \mathrm{L}_n^* M_{\delta_n^{\eta-1}} & 0 \end{pmatrix} \\ &= \bigoplus_{n\in\mathbb{Z}} \mathrm{D}_n^{(\eta)} = \begin{pmatrix} 0 & \Delta_\omega^{\eta-1} \mathrm{L} \Delta_\omega^{-\eta} \\ \Delta_\omega^{-\eta} \mathrm{L}^* \Delta_\omega^{\eta-1} & 0 \end{pmatrix}, \end{aligned}$$

on their natural domain

$$\mathcal{D} := \left\{ \xi \in \bigoplus_{n\in\mathbb{Z}} H^1(\mathbb{T}) \oplus H^1(\mathbb{T}) \mid \sum_{n\in\mathbb{Z}} \|\mathrm{D}_n^{(\eta)} \xi_n\|^2 < +\infty \right\}.$$

As explained in [10], each $\mathrm{D}_n^{(\eta)}$ is invertible with bounded inverse, after defining for $n = 0$,

$$(\mathrm{D}_0^{(\eta)})^{-1} := (\mathrm{D}_0^{(\eta)})^{-1} P^{\perp}_{\mathrm{Ker}\left(\mathrm{D}_0^{(\eta)}\right)}.$$

In addition, each $(\mathrm{D}_n^{(\eta)})^{-1}$ is compact.

By following the lines of Theorem 9.1 of [10], we get

Proposition 7.1 *If $\Gamma_n(f) = o(\ln n)$, then for each $\eta \in [0, 1]$, $D^{(\eta)}$ has compact resolvent.*

Proof For the inverses $(\mathrm{D}_n^{(\eta)})^{-1}$, which are all compact operators, we have for each $n \in \mathbb{Z}$,

$$(\mathrm{D}_n^{(\eta)})^{-1} = \begin{pmatrix} 0 & M_{\delta_n^{1-\eta}}(\mathrm{L}_n^*)^{-1}M_{\delta_n^{\eta}} \\ M_{\delta_n^{\eta}}\mathrm{L}_n^{-1}M_{\delta_n^{1-\eta}} & 0 \end{pmatrix}.$$

Suppose now $\Gamma_n(f) = o(\ln n)$. By following the same computations in the above mentioned theorem, we get for $|n|$ sufficiently large,

$$\left\|(\mathrm{D}_n^{(\eta)})^{-1}\right\| \leqslant \Gamma_n(f)\left\|\mathrm{D}_n^{-1}\right\| \leqslant \frac{1}{\frac{1}{\ln|n|}\sum_{l=2}^{|n|}\frac{1}{\ln l}} \to 0$$

as $n \to \infty$, and the assertion follows. □

We have shown in [10] that if f is one of the diffeomorphisms constructed in [16] for α a **UL** number, then the $\mathrm{D}^{(\eta)}$ have compact resolvent. The same happens if α is Diophantine and f is any diffeomorphism of the unit circle with $\rho(f) = 2\alpha$.

In order to construct the associated modular spectral triples, we have to define the corresponding deformed commutators. For $\mathbf{t} := (t_1, t_2) \in \mathbb{R}^2$, we define

$$\Sigma_{\mathbf{t}}(A) := \begin{pmatrix} \sigma_{t_1}^{\omega}(A) & 0 \\ 0 & \sigma_{t_2}^{\omega}(A) \end{pmatrix}.$$

To simplify the analysis, we first suppose that $A \in M_\infty$, the set of the entire elements of M under the action of the modular group. For such operators and for $\eta \in [0, 1]$, we define $\mathbf{z}(\eta) := -\imath(1-\eta, \eta)$ and put

$$\mathfrak{D}^{(\eta)}(A) \equiv \imath\big[\mathrm{D}^{(\eta)}, A\big]_{(\eta)} := \imath\big(\mathrm{D}^{(\eta)}\Sigma_{\mathbf{z}(\eta)}(A) - \Sigma_{-\mathbf{z}(\eta)}(A)\mathrm{D}^{(\eta)}\big).$$

A straightforward computation yields

$$\mathfrak{D}^{(\eta)}(A) = \imath\begin{pmatrix} 0 & \Delta_\omega^{\eta-1}[\mathrm{L}, A]\Delta_\omega^{-\eta} \\ \Delta_\omega^{-\eta}[\mathrm{L}^*, A]\Delta_\omega^{\eta-1} & 0 \end{pmatrix}. \tag{7.1}$$

Notice that $\mathfrak{D}^{(\eta)}(A)$ can be meaningful even if A is not an analytic element. In the sequel, we assume (7.1) as the definition of the deformed derivation for each $\eta \in [0, 1]$.

Now we show that the deformed derivation in (7.1) provides sufficiently many bounded operators.

Proposition 7.2 *Let $a \in \mathbb{A}_{2\alpha}^{o}$. Then $\mathfrak{D}^{(\eta)}(\pi_\omega(a))$ defines a bounded operator acting on $\mathcal{H}_\omega \oplus \mathcal{H}_\omega$.*

Proof We sketch the proof, and refer the reader to Section 9 and 11 of [10] for further details. We first note that it is enough to check the assertion for the generators of

$\mathbb{A}^o_{2\alpha}$, and in particular for the one step shift λ and its inverse λ^{-1}. By considering the adjoints and all $\eta \in [0, 1]$, it is enough to check the boundedness for the following case, obtaining

$$\left\| \left(\Delta_\omega^{\eta-1}[\mathrm{L}, \lambda]\Delta_\omega^{-\eta} g\right)_n \right\| \leqslant |a_{n-1} - a_n|\Gamma_{|n|}(f)^{1-\eta}\Gamma_{|n-1|}(f)^\eta \|g\|.$$

By taking into account that $|a_{n-1} - a_n| = 1/\Gamma_{|n|}(f)$ and $\Gamma_l(f)/\Gamma_{l\pm1}(f) \leqslant \Gamma_1(f)$, the assertion follows. □

Notice that, in [10] it is shown that $\mathcal{D}^{(\eta)}(\pi_\omega(a))$ uniquely defines a bounded operator for elements a in a dense $*$-subalgebra of $\mathbb{A}_{2\alpha}$, and thus we can define (nontrivial in type III cases) modular spectral triples associated to the twisted Dirac operator $\mathrm{D}^{(\eta)}$. Indeed, the modular spectral triple satisfying (i)–(iv) in Definition 2.1 is given by $\left(\omega_\mu, \mathbb{A}^o_{2\alpha}, \mathrm{L}\right)_\eta$, provided $\Gamma_n(f) = o(\ln n)$, which happens if f is a diffeomorphism constructed according to Proposition 3.1 of [10] for a **UL**-number α, or if α is Diophantine. For the latter case, the representation π_{ω_μ} always generates the type II_1 hyperfinite factor R, whereas for the former the type is determined by the ratio set $r\left(R_{2\alpha}, [\mu]\right)$ (or equivalently by $r\left(f, [m]\right)$) if it is not of type II_1. We also note that the $\mathcal{D}^{(\eta)}$ define $*$-maps on their natural (common) domain:

$$\mathcal{D}^{(\eta)}(\pi_\omega(a))^* = \mathcal{D}^{(\eta)}(\pi_\omega(a^*)), \quad \eta \in [0, 1], \; a \in \mathbb{A}^o_{2\alpha}.$$

We now show how the previously defined Fourier transform in Sect. 4 diagonalises $\mathrm{D}^{(\eta)}$ for $\eta = 0, 1$, while the one defined in Sect. 6 diagonalises $\mathrm{D}^{(1/2)}$. Indeed, denote $\mathcal{F} : \mathcal{H}_\omega \to \ell^2(\mathbb{Z}^2)$ one of the Fourier transforms defined above. Then $(\mathcal{F}\oplus\mathcal{F})\mathrm{D}^{(\eta)}(\mathcal{F}^{-1}\oplus\mathcal{F}^{-1})$ provides hermitian matrices

$$(\mathcal{F}\oplus\mathcal{F})\mathrm{D}^{(\eta)}(\mathcal{F}^{-1}\oplus\mathcal{F}^{-1}) = \begin{pmatrix} 0 & A^{(\eta)} \\ \left(A^{(\eta)}\right)^* & 0 \end{pmatrix},$$

where the $A^{(\eta)} = \left(A^{(\eta)}_{(k,l)(r,s)}\right)_{k,l,r,s\in\mathbb{Z}}$ are (infinite) numerical matrices given, for $\eta = 0, 1/2, 1$ and $k, l, r, s \in \mathbb{Z}$, by

$$\begin{aligned} A^{(0)}_{(k,l)(r,s)} &= \langle \Delta_\omega^{-1}\mathrm{L}e^{kl}, e^{rs}\rangle \\ A^{(1)}_{(k,l)(r,s)} &= \langle \mathrm{L}\Delta_\omega^{-1}e^{kl}, e^{rs}\rangle \\ A^{(1/2)}_{(k,l)(r,s)} &= \langle \Delta_\omega^{-1/2}\mathrm{L}\Delta_\omega^{-1/2}\epsilon^{kl}, \epsilon^{rs}\rangle. \end{aligned}$$

It is then enough to compute the right upper corners $A^{(\eta)}$ of $(\mathcal{F}\oplus\mathcal{F})\mathrm{D}^{(\eta)}(\mathcal{F}^{-1}\oplus\mathcal{F}^{-1})$, obtaining

Proposition 7.3 *For $k, l, r, s \in \mathbb{Z}$, we have*

$$\left\langle \Delta_\omega^{-1} Le^{kl}, e^{rs}\right\rangle = (\imath l - a_k)\widehat{1/\delta_k}(s - l)\delta_{k,r},$$

$$\left\langle L\Delta_\omega^{-1}e^{kl}, e^{rs}\right\rangle = (\imath s - a_k)\widehat{1/\delta_k}(s-l)\delta_{k,r},$$
$$\left\langle \Delta_\omega^{-1/2}L\Delta_\omega^{-1/2}\epsilon^{kl}, \epsilon^{rs}\right\rangle = -\big(\imath l\delta_{l,s} + a_{-k}\widehat{\delta_k}(l-s)\big)\delta_{k,r}.$$

Proof For the first equality, we get

$$\begin{aligned}\left\langle \Delta_\omega^{-1}\mathrm{L}\xi_{kl}, \xi_{rs}\right\rangle =& \sum_{n\in\mathbb{Z}}(\imath l - a_n)\delta_{k,n}\delta_{r,n}\oint \frac{\mathrm{d}z}{2\pi\imath z}1/\delta_n(z)z^{-(s-l)}\\ =&(\imath l - a_k)\widehat{1/\delta_k}(s-l)\delta_{k,r}.\end{aligned}$$

The second one follows analogously.

The third equality follows after some computations. We first note that

$$\left\langle \Delta_\omega^{-1/2}\mathrm{L}\Delta_\omega^{-1/2}\epsilon^{kl}, \epsilon^{rs}\right\rangle = \langle \mathrm{L}u_{kl}^*\xi_\omega, u_{rs}^*\xi_\omega\rangle.$$

Secondly, after some computations we get

$$\left(u_{kl}^*\xi_\omega\right)_n(z) = f^n(z)^{-l}\delta_{n,-k}, \quad n\in\mathbb{Z},\ z\in\mathbb{T}.$$

We consider the elementary changes of variables

$$w := f^{-k}(z), \quad k\in\mathbb{Z}.$$

For the first addendum of L, we get

$$\begin{aligned}\left\langle\left(\bigoplus_{\mathbb{Z}}\imath z\frac{\mathrm{d}}{\mathrm{d}z}\right)u_{kl}^*\xi_\omega, u_{rs}^*\xi_\omega\right\rangle &= \imath\delta_{k,r}\oint z\frac{\mathrm{d}}{\mathrm{d}z}\big(f^{-k}(z)^{-l}\big)f^{-k}(z)^s\frac{\mathrm{d}z}{2\pi\imath z}\\ &= -\imath l\delta_{k,r}\oint w^{-(l-s)}\frac{\mathrm{d}w}{2\pi\imath w} = -\imath l\delta_{k,r}\delta_{l,s}.\end{aligned}$$

For the second addendum, after the changes of variables as above and (2.9), we get

$$\begin{aligned}&\left\langle\left(\bigoplus_{n\in\mathbb{Z}}a_nI\right)u_{kl}^*\xi_\omega, u_{rs}^*\xi_\omega\right\rangle = a_{-k}\delta_{k,r}\oint f^{-k}(z)^{-(l-s)}\frac{\mathrm{d}z}{2\pi\imath z}\\ &= a_{-k}\delta_{k,r}\oint w^{-(l-s)}\frac{w}{f^k(w)(Df^{-k})(f^k(w))}\frac{\mathrm{d}w}{2\pi\imath w}\\ &= a_{-k}\delta_{k,r}\oint w^{-(l-s)}\frac{wDf^k(w)}{f^k(w)}\frac{\mathrm{d}w}{2\pi\imath w} = a_{-k}\delta_{k,r}\widehat{\delta_k}(l-s).\end{aligned}$$

Collecting together, we obtain the assertion. □

Acknowledgements The author acknowledges the financial support of Italian INDAM-GNAMPA. The present project is part of "MIUR Excellence Department Project awarded to the Department of Mathematics, University of Rome Tor Vergata, CUP E83C18000100006". He is grateful to the referees for a very careful reading of the manuscript, and for several suggestions which contribute to improve the presentation of the paper.

References

1. Bergh, J., Löftröm, J.: Interpolation Spaces, an Introduction. Springer, New York (1976)
2. Blecher, D.: The standard dual of an operator space. Pac. J. Math. **153**, 15–30 (1992)
3. Boca, F.-P.: Rotation C^*-algebras and almost Mathieu operators. Theta Bucharest (2001)
4. Bratteli O., Robinson D. W.: Operator Algebras and Quantum Statistical Mechanics I, II. Springer, Berlin (1887 and 1997)
5. Carey A.L., Phillips J., Rennie A.: Spectral triples: examples and index theory. In: Carey (Ed.) Noncommutative Geometry and Physics: Renormalisation, Motives, Index Theory, ESI Lect. Math. Phys., pp. 175–265 (2011)
6. Chen, Z., Xu, Q., Yin, Z.: Harmonic analysis on quantum tori. Commun. Math. Phys. **322**, 755–805 (2013)
7. Connes, A.: Noncommutative Geometry. Academic Press, San Diego (1994)
8. Connes, A., Moscovici, H.: Modular curvature for noncommutative two-tori. J. Am. Math. Soc. **27**, 639–684 (2014)
9. Fidaleo, F.: Canonical operator space structures in non-commutative L^p spaces. J. Funct. Anal. **169**, 226–250 (1999)
10. Fidaleo, F., Suriano, L.: Type III representations and Modular Spectral Triples for the noncommutative torus. J. Funct. Anal. **275**, 1484–1531 (2018)
11. Forsyth, I., Mesland, B., Rennie, A.: Dense domains, symmetric operators and spectral triple. N. Y. J. Math. **20**, 1001–1020 (2014)
12. Haagerup, U.: L^p spaces associated with an arbitrary von Neumann algebra. Colloq. Int C.N.R.S. **274**, 175–184 (1979)
13. Katok, A., Hasselblatt, B.: Introduction to the Modern Theory of Dynamical Systems, Encyclopedia of Mathematics and Its Applications, vol. 54. Cambridge University Press, Cambridge (1995)
14. Khinchin, A.Ya.: Continued Fractions. The University of Chicago Press, Chicago (1964)
15. Kosaki, H.: Application of the complex interpolation method to a von Neumann algebra: noncommutative L^p spaces. J. Funct. Anal. **56**, 29–78 (1984)
16. Matsumoto, S.: Orbit equivalence types of circle diffeomorphisms with a Liouville rotation number. Nonlinearity **26**, 1401–1414 (2013)
17. Niculescu, C.P., Ströh, A., Zsidó, L.: Noncommutative estension of classical and multiple recurrence theorems. J. Oper. Theory **50**, 3–52 (2003)
18. Pinsky, M.A.: Introduction to Fourier Analysis and Wavelets. American Mathematical Society, Providence (2002)
19. Pisier, G.: The operator Hilbert space OH, complex interpolation and tensor norms. Mem. Amer. Math. Soc. **122**, (1996)
20. Ricard, È.: L^p-multipliers on quantum tori. J. Funct. Anal. **270**, 4604–4613 (2016)
21. Ruan, Z.-J.: Subspaces of C^*-algebras. J. Funct. Anal. **76**, 217–230 (1988)
22. Rudin, W.: Real and Complex Analysis. McGraw-Hill, New York (1987)
23. Sondow, J.: An irrationality measure for Liouville numbers and conditional measures for Euler's constant. arXiv:math/0307308
24. Strătilă, Ş.: Modular Theory in Operator Algebras. Abacus Press, Tunbridge Wells (1981)
25. Takesaki, M.: Theory of Operator Algebras I, II, III. Springer, New York (2003)
26. Terp, M.: Interpolation spaces between a von Numann algebra and its predual. J. Oper. Theory **9**, 327–360 (1982)
27. Tomiyama, J.: On the transpose map of matrix algebrs. Proc. Am. Math. Soc. **88**, 635–638 (1983)
28. Tomiyama, J.: Recent development of the theory of completely bounded maps between C^*-algebras. Publ. RIMS Kyoto Univ. **19**, 1283–1303 (1983)
29. Watanabe, N.: growth sequences for circle diffeomorphisms. Geom. Funct. Anal. **17**, 320–331 (2007)

Publisher's Note Springer Nature remains neutral with regard to jurisdictional claims in published maps and institutional affiliations.

Springer Link link.springer.com

SpringerLink

Knowledge Matters. Choose SpringerLink.

- The World's Leading Scholarship
- In the Most Complete Online Collection of STM Content
- Delivered on the Fastest,
  Most Intelligent Research Platform We've Ever Developed
- All from Springer – A Global Leader
  in Scientific Publishing

Visit today

Image: Alexaldo/Pesnkova/iStock [m] 014486c

Springer

springer.com

Springer eBooks

The world's largest collection of STM books

- Critical, must-have content
- Continuous access with archival rights
- No DRM – no limit on users, printing & downloads
- All on SpringerLink – anytime, anywhere, any device

#Books Change

springer.com/ebooks

015420x

CPSIA information can be obtained
at www.ICGtesting.com
Printed in the USA
BVHW040743071019
560223BV00006B/4/P